Student Solutions Manual

Beginning and Intermediate Algebra
A Combined Approach

Mark Clark
Palomar College

Cynthia Anfinson
Palomar College

Prepared by

Eve Nolan

Gina Hayes
Palomar College

Karen Mifflin
Palomar College

BROOKS/COLE
CENGAGE Learning

Australia • Brazil • Japan • Korea • Mexico • Singapore • Spain • United Kingdom • United States

BROOKS/COLE
CENGAGE Learning

ISBN-13: 978-1-133-36509-9
ISBN-10: 1-133-36509-4

Brooks/Cole
20 Davis Drive
Belmont, CA 94002-3098
USA

Cengage Learning is a leading provider of customized learning solutions with office locations around the globe, including Singapore, the United Kingdom, Australia, Mexico, Brazil, and Japan. Locate your local office at: **www.cengage.com/global**

Cengage Learning products are represented in Canada by Nelson Education, Ltd.

To learn more about Brooks/Cole, visit **www.cengage.com/brookscole**

Purchase any of our products at your local college store or at our preferred online store **www.cengagebrain.com**

Printed in the United States of America
1 2 3 4 5 6 7 16 15 14 13 12

Table of Contents

CHAPTER R REVIEW OF PREALGEBRA

Section R.1 Operations with Integers

1. The number 0 belongs to the set of whole numbers and all whole numbers also belong to the set of integers.

3. The number -20 belongs to the set of integers.

5. The number 11 belongs to the set of natural numbers. All natural numbers also belong to the set of whole numbers and to the set of integers.

7. The number -9 belongs to the set of integers.

9. The number 1 belongs to the set of natural numbers. All natural numbers also belong to the set of whole numbers and to the set of integers.

11. The value for the population of Los Angeles is a natural number. The population could be 10,000, 100,000, 1,000,000 .... Assuming the population would not be 0, the value 0 is not included as a possibility. These values are also whole numbers and integers.

13. The value for the population of Norfolk, VA is a natural number. The population could be 10,000, 100,000, 1,000,000 .... Assuming the population would not be 0, the value 0 is not included as a possibility. These values are also whole numbers and integers.

15. The value for the number of kittens in the animal shelter is a whole number. The number of kittens could be 0, 1, 2, ... Since the animal shelter could be empty of kittens on any given day (0 kittens), the value 0 is included as a possibility. These values are also integers.

17. The value for the number of hours worked weekly by a Home Depot employee is a whole number. The number of hours worked could be 0, 5, 12, ... Since the employee could have a week off (0 hours worked), the value 0 is included as a possibility. These values are also integers.

19. This is an example from the set of integers. The average daily high temperature in Missoula could range from negative numbers to positive numbers.

21. The set of all integers between –5 and 1, *including* –5 and 1 is as follows: {-5, -4, -3, -2, -1, 0, 1}

23. The set of all natural numbers between –5 and 5, **not** including –5 and 5 is as follows: $\{1, 2, 3, 4\}$. Notice that negative values and 0 are not included because this is a set of only natural numbers.

25. The set of all whole numbers between –4 and 0, including –4 and 0 is as follows: $\{0\}$. Notice that negative values are not included because this is a set of only whole numbers.

27. The tick marks on the number line are 4 units apart, so the scale of this number line is 4.

29. The tick marks on the number line are 0.1 units apart, so the scale of this number line is 0.1.

31.

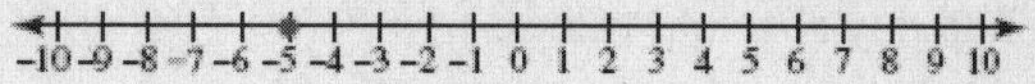

33.

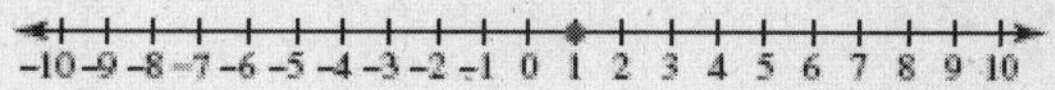

35.

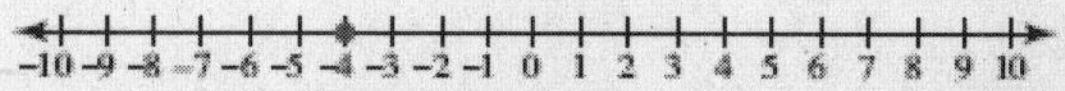

37. The number -5 is to the right of -15 on the number line, therefore -5 is greater than -15.

$-5 > -15$

39. The number -3 is at the same location as -3 on the number line, and is therefore neither less than nor greater than -3. Write an expression using inequality symbols as follows:

$-3 \le -3 \ or \ -3 \ge -3$

41. The number -4 is to the left of -3 on the number line, therefore -4 is less than -3.

$-4 < -3$

43. The number 0 is to the right of -7 on the number line, therefore 0 is greater than -7.

$0 > -7$

45. $|-3| = 3$ since the distance between -3 and 0 is 3 units.

47. $|8| = 8$ since the distance between 8 and 0 is 8 units.

49. $|16| = 16$ since the distance between 16 and 0 is 16 units.

51. $|-7| = 7$ since the distance between -7 and 0 is 7 units.

53. Order the numbers from smallest to largest (from left to right as they would appear on the number line) as follows:

$\frac{-3}{2}, \ -1, \ \frac{1}{3}, \ 0.5, 2.95, 3$

55. First, find any absolute values.

$|-5| = 5$

$-5 = -5$

$|-2| = 2$

$-2 = -2$

Then order the numbers from smallest to largest (from left to right as they would appear on the number line) writing each term in its original notation as follows:

$-5, \ -2, \ |-2|, \ |-5|$

57. First, find any absolute values.

$$|-4| = 4$$

$$-3.5 = -3.5$$

$$|-5| = 5$$

$$\frac{2}{3} = \frac{2}{3}$$

$$4.2 = 4.2$$

Then order the numbers from smallest to largest (from left to right as they would appear on the number line) writing each term in its original notation as follows:

$$-3.5,\ \frac{2}{3},\ |-4|,\ 4.2,\ |-5|$$

59. The expression translates as, "the opposite of five."

$$-(5) = -5$$

61. The expression translates as, "the opposite of the absolute value of negative 2."

$$-|-2| = -2$$

63. The expression translates as, "the opposite of negative six."

$$-(-6) = 6$$

65. The two integers have the same sign, so we add the two numbers using the rule for adding integers and keep the negative sign.

$$-3 + (-5) = -8$$

67. The two integers have different signs. Adding a negative number is the same as subtracting the number, so we will subtract 7 from 16.

$$16 + (-7) = 16 - 7 = 9$$

69. These two integers have different signs. Take the absolute value of each number and subtract the smaller absolute value from the larger. Subtract 6 from 13.

$$13 - 6 = 7$$

Attach the sign of the number that is larger in absolute value. The number that is larger in absolute value is -13, so the answer will be negative. Attach a negative sign to 7.

$$-13 + 6 = -7$$

The final answer is -7.

71. The two integers are being subtracted. Change the sign of the second number and add. Add the two numbers using the rule for adding integers, keeping the negative sign because both numbers are negative.

$$-7 + (-9) = -16$$

73. Using the rule for subtracting integers, change the sign of the second term and add the terms.

$$-4 - (-8) = -4 + 8$$

Subtracting a negative is the same as adding so we now have two integers with different signs. Take the absolute value of each number and subtract the smaller absolute value from the larger. Subtract 4 from 8.

$$8 - 4 = 4$$

Attach the sign of the number that is larger in absolute value. Since the number that is larger is positive, the answer is positive.

$$-4 - (-8) = 4$$

75. Using the rule for subtracting integers, change the sign of the second term and add the terms.

$$-12 - (-11) = -12 + 11$$

Subtracting a negative is the same as adding so we now have two integers with different signs. Take the absolute value of each number and subtract the smaller absolute value from the larger. Subtract 11 from 12.

$$12 - 11 = 1$$

Attach the sign of the number that is larger in absolute value. The number that is larger in absolute value is -12, so the answer will be negative. Attach a negative sign to 1.

$$-12 + 11 = -1$$

The final answer is -1.

77. The temperature can be found by subtracting $-23 - 3$. To subtract two integers, change the sign of the second integer and add. Therefore, we compute

$$-23 + (-3) = -26$$

The final temperature is -26° F.

79. Beginning with the depth of 30 feet below the surface, or -30 feet, and going another 20 feet deeper, or -20 feet. This can be represented by the expression

$$-30 - 20$$

To subtract two integers, change the sign of the second integer and add. Therefore, we compute

$$-30 + (-20) = -50$$

Ray and Karin are at a depth of -50 feet.

81. To find the distance between these two points we will use the formula $|b - a|$. Let $a = 7$ and $b = -2$ and substitute into the formula.

$$\begin{aligned} |b - a| &= |7 - (-2)| \\ &= |7 + 2| \\ &= |9| \\ &= 9 \end{aligned}$$

83. To find the distance between these two points we will use the formula $|b-a|$. Let $a=-1$ and $b=-9$ and substitute into the formula.

$$\begin{aligned}|b-a|&=|-1-(-9)|\\&=|-1+9|\\&=|8|\\&=8\end{aligned}$$

85. To find the distance between these two points we will use the formula $|b-a|$. Let $a=-16$ and $b=-35$ and substitute into the formula.

$$\begin{aligned}|b-a|&=|-16-(-35)|\\&=|-16+35|\\&=|19|\\&=19\end{aligned}$$

87. $3(-2)=-6$

The answer is negative since the two integers have different signs.

89. $(-4)(5)=-20$

The answer is negative since the two integers have different signs.

91. $(-3)(-6)=18$

The answer is positive since the two integers have the same sign.

93. $(-24)\div(-8)=3$

The answer is positive since the two integers have the same sign.

95. $(30)\div(-5)=-6$

The answer is negative since the two integers have different signs.

97. $9\div 0=$ *undefined*

By definition, any number divided by zero is undefined.

99. $0\div 8=0$

Zero divided by any number is always 0.

101. $\boxed{-3}+\boxed{5\cdot 6}$

This expression has 2 terms. Terms are separated by addition and subtraction symbols. The terms in the expression are outlined for problems 101-108.

103. $\boxed{-24\div 3\cdot 7}$

This expression has 1 term.

105. $\boxed{3}+\boxed{22\div 11}\boxed{-16\div 4}$

This expression has 3 terms.

107. $\boxed{92\cdot(-8)}+\boxed{26\div 13}\boxed{-1}+\boxed{7(-3)}$

This expression has 4 terms.

109. This expression has 1 term.

$\boxed{16\div(-8)\cdot 3}$ Outline the term.
$=-2\cdot 3$ Work left to right.
$=-6$

111. This expression has 2 terms.

$\boxed{-9}+\boxed{3\cdot 5}$ Outline the terms.
$=-9+15$ Do operation inside each term.
$=6$

113. This expression has 3 terms.

$\boxed{9}+\boxed{24\div 3}\boxed{-3\cdot(-2)}$ Outline the terms.
$=9+8+6$ Do operation inside each term.
$=23$ Add from left to right.

115. This expression has 3 terms.

$\boxed{-14\div 2}+\boxed{5\cdot 4\div 10}\boxed{-(-1)}$ Outline the terms.
$=-7+\boxed{20\div 10}+1$ Do operation inside each term.
$=-7+2+1$ Add/subtract from left to right.
$=-4$

117. This expression has 4 terms.

$\boxed{100\div 50\cdot 3}\boxed{-2\cdot(-7)}+\boxed{4\cdot 2}\boxed{-(-2)}$ Outline terms.
$=\boxed{2\cdot 3}+14+8+2$ Do operation inside each term.
$=6+14+8+2$ Add from left to right.
$=30$

CHAPTER R Review of Prealgebra

Section R.2 Operations with Fractions

1.

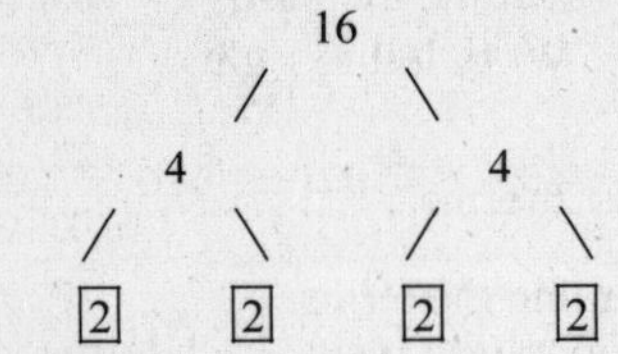

Therefore, $16 = 2\cdot 2\cdot 2\cdot 2$.

3.

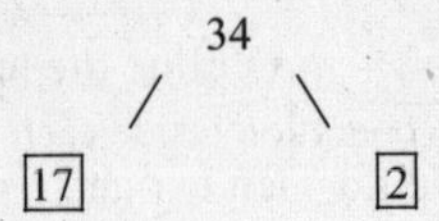

Therefore, $34 = 17\cdot 2$.

5.

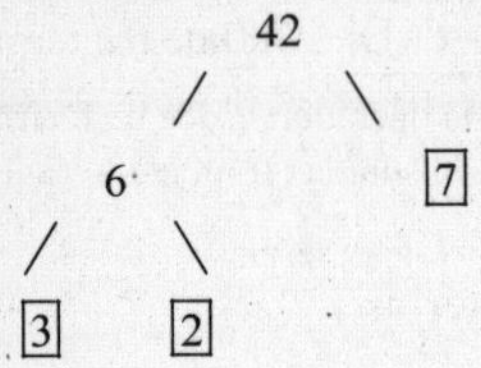

Therefore, $42 = 7\cdot 3\cdot 2$.

7.

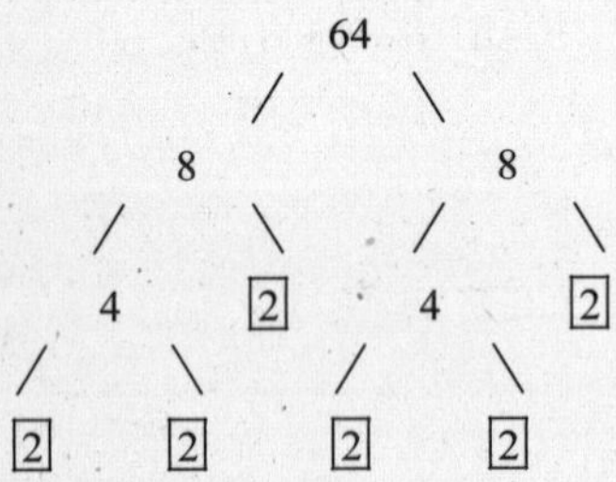

Therefore, $64 = 2\cdot 2\cdot 2\cdot 2\cdot 2\cdot 2$.

9.

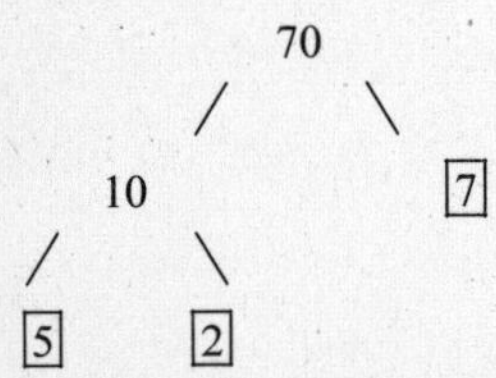

Therefore, $70 = 7\cdot 5\cdot 2$.

11. $12 = \boxed{3}\cdot 2\cdot 2$

$15 = 5\cdot\boxed{3}$

Only one factor in common, so the GCF is 3

13. $60 = \boxed{5}\cdot\boxed{3}\cdot 2\cdot 2$

$15 = \boxed{5}\cdot\boxed{3}$

The $\text{GCF} = 5\cdot 3 = 15$

15. $9 = \boxed{3}\cdot\boxed{3}$

$27 = \boxed{3}\cdot\boxed{3}\cdot 3$

$18 = \boxed{3}\cdot\boxed{3}\cdot 2$

The $\text{GCF} = 3\cdot 3 = 9$

17. $48 = \boxed{3}\cdot\boxed{2}\cdot\boxed{2}\cdot 2\cdot 2$

$36 = \boxed{3}\cdot 3\cdot\boxed{2}\cdot\boxed{2}$

$60 = 5\cdot\boxed{3}\cdot\boxed{2}\cdot\boxed{2}$

The $\text{GCF} = 3\cdot 2\cdot 2 = 12$

19. $51 = 17\cdot\boxed{3}$

$27 = 3\cdot 3\cdot\boxed{3}$

Only one factor in common, so the GCF is 3.

21. $\dfrac{6}{15} = \dfrac{\cancel{3}\cdot 2}{5\cdot\cancel{3}} = \dfrac{2}{5}$

23. $\dfrac{8}{42} = \dfrac{4\cdot\cancel{2}}{21\cdot\cancel{2}} = \dfrac{4}{21}$

25. $\dfrac{16}{48} = \dfrac{\cancel{16}\cdot 1}{\cancel{16}\cdot 3} = \dfrac{1}{3}$

27. $\dfrac{-9}{108} = -\dfrac{\cancel{9}\cdot 1}{12\cdot\cancel{9}} = -\dfrac{1}{12}$

29. $-5\dfrac{6}{21} = -5\dfrac{\cancel{3}\cdot 2}{7\cdot\cancel{3}} = -5\dfrac{2}{7}$

31. Comparing the denominators of 3 and 15, we have that $3\cdot 5 = 15$. Therefore, we multiply the numerator and denominator by 5.

$$\frac{2\cdot 5}{3\cdot 5} = \frac{10}{15}$$

33. Comparing the denominators of 7 and 56, we have that $7\cdot 8 = 56$. Therefore, we multiply the numerator and denominator by 8.

$$\frac{5\cdot 8}{7\cdot 8} = \frac{40}{56}$$

35. Comparing the denominators of 16 and 32, we have that $16\cdot 2 = 32$. Therefore, we multiply the numerator and denominator by 2.

$$\frac{-5\cdot 2}{16\cdot 2} = -\frac{10}{32}$$

37. Comparing the denominators of 9 and 45, we have that $9\cdot 5 = 45$. Therefore, we multiply the numerator and denominator by 5.

$$4\frac{2\cdot 5}{9\cdot 5} = 4\frac{10}{45}$$

39. Converting 5 to a fraction, $5 = \dfrac{5}{1}$. Now comparing the denominators of 1 and 3, we have that $1\cdot 3 = 3$. Therefore, we multiply the numerator and denominator by 3.

$$\frac{5\cdot 3}{1\cdot 3} = \frac{15}{3}$$

41.

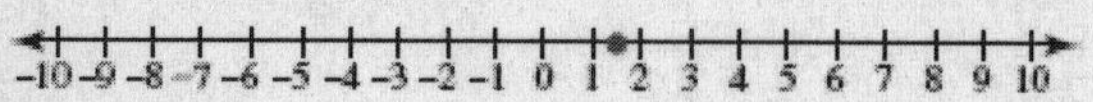

43.

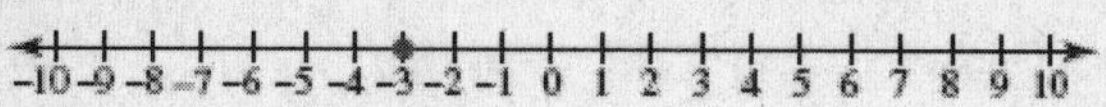

45.

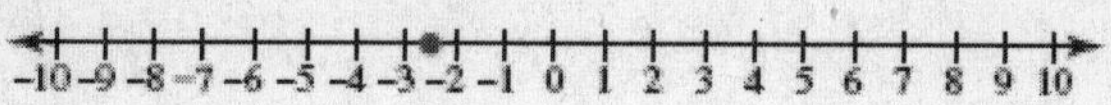

47.

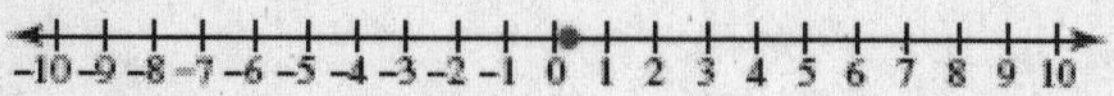

49.

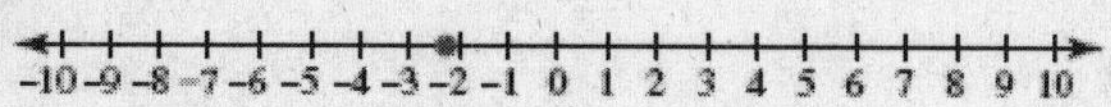

51.

$\frac{2}{3}+\frac{4}{3}$ Add the numerators.

$=\frac{6}{3}$ Reduce to lowest terms.

$=2$

53.

$\frac{7}{5}-\frac{4}{5}$ Subtract numerators.

$=\frac{3}{5}$ Answer is in lowest terms.

55.

$\frac{11}{15}-\frac{2}{15}$ Subtract numerators.

$=\frac{9}{15}$ Reduce to lowest terms.

$=\frac{3}{5}$

57.

$1\frac{1}{4}+3\frac{3}{4}$ Rewrite mixed numbers as improper fractions.

$=\frac{5}{4}+\frac{15}{4}$ Add numerators.

$=\frac{20}{4}$ Reduce to lowest terms.

$=5$

59.

$5\frac{2}{3}-2\frac{1}{3}$ Rewrite mixed numbers as improper fractions.

$=\frac{17}{3}-\frac{7}{3}$ Subtract numerators.

$=\frac{10}{3}$ Reduce to lowest terms.

$=3\frac{1}{3}$

61. The factorization of 6 is $6=3\cdot 2$ and 3 is prime. The LCD $=3\cdot 2=6$.

$\frac{2}{3}+\frac{5}{6}$

$=\frac{2\cdot 2}{3\cdot 2}+\frac{5}{6}$ Rewrite each fraction over LCD.

$=\frac{4}{6}+\frac{5}{6}$ Add numerators.

$=\frac{9}{6}$ Reduce to lowest terms.

$=\frac{3}{2}$ or $1\frac{1}{2}$

63. The factorization of 8 is $8=2\cdot 2\cdot 2$ and 3 is prime. The LCD $=3\cdot 2\cdot 2\cdot 2=24$.

$\frac{-4}{8}+\frac{1}{3}$

$=\frac{-4\cdot 3}{8\cdot 3}+\frac{1\cdot 8}{3\cdot 8}$ Rewrite each fraction over LCD.

$=\frac{-12}{24}+\frac{8}{24}$ Add/subtract numerators.

$=\frac{-4}{24}$ Reduce to lowest terms.

$=-\frac{1}{6}$

65. The factorization of 4 is $4=2\cdot 2$ and 3 is prime. The LCD $=3\cdot 2\cdot 2=12$.

$\frac{-2}{3}-\frac{3}{4}$

$=\frac{-2\cdot 4}{3\cdot 4}-\frac{3\cdot 3}{4\cdot 3}$ Rewrite each fraction over LCD.

$=\frac{-8}{12}-\frac{9}{12}$ Subtract numerators.

$=-\frac{17}{12}$ or $-1\frac{5}{12}$ Answer is in lowest terms.

67. The factorization of 6 is $6=3\cdot 2$ and 3 is prime. The LCD $=3\cdot 2=6$.

$1\frac{1}{6}+3\frac{2}{3}$ Rewrite as improper fractions

$=\frac{7}{6}+\frac{11}{3}$

$=\frac{7}{6}+\frac{11\cdot 2}{3\cdot 2}$ Rewrite each fraction over LCD.

$=\frac{7}{6}+\frac{22}{6}$ Add numerators.

$=\frac{29}{6}=4\frac{5}{6}$ Answer is in lowest terms.

69. The factorization of 4 is $4=2\cdot 2$ and 2 is prime. The LCD $=2\cdot 2=4$.

$-2\frac{1}{4}-3\frac{1}{2}$ Rewrite as improper fractions

$=-\frac{9}{4}-\frac{7}{2}$

$=-\frac{9}{4}-\frac{7\cdot 2}{2\cdot 2}$ Rewrite each fraction over LCD.

$=-\frac{9}{4}-\frac{14}{4}$ Subtract numerators.

$=-\frac{23}{4}=-5\frac{3}{4}$ Answer is in lowest terms.

71. Add the two amounts of paint.

$2\frac{1}{4}+1\frac{1}{4}$ Rewrite as improper fractions.

$=\frac{9}{4}+\frac{5}{4}$ Add numerators.

$=\frac{14}{4}=3\frac{2}{4}=3\frac{1}{2}$ Reduce to lowest terms.

Sandy needs to buy $3\frac{1}{2}$ gallons of paint.

73. Add the two amounts of salt. Looking at the denominators, the factorization of 4 is $4=2\cdot 2$ and 2 is prime. The LCD $=2\cdot 2=4$.

$\frac{1}{4}+\frac{1}{2}$

$=\frac{1}{4}+\frac{1\cdot 2}{2\cdot 2}$ Rewrite each fraction over LCD.

$=\frac{1}{4}+\frac{2}{4}$ Add numerators.

$=\frac{3}{4}$ Answer is in lowest terms.

Dominique used a total of $\frac{3}{4}$ teaspoon of salt.

75. Subtract the amount he needs from the total amount he has. Looking at the denominators, the factorization of 8 is $8=2\cdot 2\cdot 2$ and 1 is prime. The LCD $=2\cdot 2\cdot 2=8$.

$20-10\frac{3}{8}$ Rewrite as improper fractions

$=\frac{20}{1}-\frac{83}{8}$

$=\frac{20\cdot 8}{1\cdot 8}-\frac{83}{8}$ Rewrite each fraction over LCD.

$=\frac{160}{8}-\frac{83}{8}$ Subtract numerators.

$=\frac{77}{8}=9\frac{5}{8}$ Answer is in lowest terms.

Ted will have $9\frac{5}{8}$ feet of lumber left.

77. To find a reciprocal, we invert the fraction. To find the reciprocal of -3, recall that $-3=\frac{-3}{1}$. Find the reciprocal, which is $\frac{-1}{3}$.

79. Find the reciprocal of $\frac{5}{3}$, which is $\frac{3}{5}$.

81. Find the reciprocal of $\frac{1}{9}$, which is $\frac{9}{1}=9$.

83. Find the reciprocal, which is $\frac{1}{8}$.

85. The reciprocal of the number 0 does not exist. If $0=\frac{0}{1}$ and we invert $\frac{0}{1}$, we would have $\frac{1}{0}$ which is undefined.

87.

$\frac{1}{4}\cdot 8=\frac{1}{4}\cdot\frac{8}{1}$

$=\frac{1\cdot 8}{4\cdot 1}$ Multiply numerators & denominators.

$=\frac{\not{2}\cdot\not{2}\cdot 2}{\not{2}\cdot\not{2}}$ Factor. Divide out like factors.

$=\frac{2}{1}=2$

89.

$\frac{-3}{4}\cdot\frac{-5}{9}$

$=\frac{-3\cdot -5}{4\cdot 9}$ Multiply numerators & denominators.

$=\frac{5\cdot\not{3}}{\not{3}\cdot 3\cdot 2\cdot 2}$ Factor. Divide out like factors.

$=\frac{5}{12}$

91.

$\frac{2}{3}\div\frac{14}{27}$ Find reciprocal of second fraction & multiply.

$=\frac{2}{3}\cdot\frac{27}{14}$

$=\frac{2\cdot 27}{3\cdot 14}$ Factor numerator and denominator.

$=\frac{3\cdot 3\cdot\not{3}\cdot\not{2}}{7\cdot\not{3}\cdot\not{2}}$ Divide out like factors.

$=\frac{9}{7}$ or $1\frac{2}{7}$

93.

$(-100)\div(-20)$ Find reciprocal of second fraction & multiply.

$=\frac{-100}{1}\cdot\frac{-1}{20}$

$=\frac{-100\cdot -1}{1\cdot 20}$ Factor numerator and denominator.

$=\frac{\not{5}\cdot 5\cdot\not{2}\cdot\not{2}}{\not{5}\cdot\not{2}\cdot\not{2}}$ Divide out like factors.

$=\frac{5}{1}=5$

95.

$\frac{4}{15}\div\frac{-8}{45}$ Find reciprocal of second fraction & multiply.

$=\frac{4}{15}\cdot\frac{-45}{8}$

$=\frac{4\cdot -45}{15\cdot 8}$ Factor numerator and denominator.

$=\frac{-\not{5}\cdot\not{3}\cdot 3\cdot\not{2}\cdot\not{2}}{\not{5}\cdot\not{3}\cdot 2\cdot\not{2}\cdot\not{2}}$ Divide out like factors.

$=\frac{-3}{2}$ or $-1\frac{1}{2}$

97.

$\frac{-2}{9} \div (-12)$ Find reciprocal of second fraction & multiply.

$= \frac{-2}{9} \cdot \frac{-1}{12}$

$= \frac{-2 \cdot -1}{9 \cdot 12}$ Factor numerator and denominator.

$= \frac{\not{2}}{3 \cdot 3 \cdot 3 \cdot 2 \cdot \not{2}}$ Divide out like factors.

$= \frac{1}{54}$

99. This expression has 2 terms. Simplify inside each term first.

$\boxed{24 \div (-3) \cdot 2} + \boxed{\frac{3}{2}}$ Divide in the first term.

$= -8 \cdot 2 + \frac{3}{2}$ Then multiply in the first term.

$= -16 + \frac{3}{2}$ Find LCD = 2.

$= \frac{-16 \cdot 2}{1 \cdot 2} + \frac{3}{2}$ Rewrite each fraction over the LCD.

$= \frac{-32}{2} + \frac{3}{2}$ Add the two fractions.

$= \frac{-29}{2}$ or $-14\frac{1}{2}$ Fraction is in lowest terms.

101. This expression has 2 terms. Simplify inside each term first.

$\boxed{3} - \boxed{\frac{1}{3} \div \frac{5}{6}}$ Divide in the second term.

$= 3 - \frac{1}{3} \cdot \frac{6}{5}$ Find reciprocal of second fraction & multiply.

$= 3 - \frac{1 \cdot 6}{3 \cdot 5}$ Factor numerator and denominator.

$= 3 - \frac{\not{3} \cdot 2}{5 \cdot \not{3}}$ Divide out like factors.

$= 3 - \frac{2}{5}$ Find LCD = 5.

$= \frac{3 \cdot 5}{1 \cdot 5} - \frac{2}{5}$ Rewrite each fraction over the LCD.

$= \frac{15}{5} - \frac{2}{5}$ Subtract the two fractions.

$= \frac{13}{5}$ or $2\frac{3}{5}$ Fraction is in lowest terms.

103. This expression has 3 terms. Simplify inside each term first.

$\boxed{\frac{3}{8} \cdot \frac{4}{5}} + \boxed{\frac{1}{2}} \boxed{-1}$ Multiply in the first term.

$= \frac{3 \cdot 4}{8 \cdot 5} + \frac{1}{2} - 1$ Factor numerator and denominator.

$= \frac{3 \cdot \not{2} \cdot \not{2}}{\not{2} \cdot \not{2} \cdot 2 \cdot 5} + \frac{1}{2} - 1$ Divide out like factors.

$= \frac{3}{10} + \frac{1}{2} - 1$ Find LCD = 10.

$= \frac{3}{10} + \frac{1 \cdot 5}{2 \cdot 5} - \frac{1 \cdot 10}{1 \cdot 10}$ Rewrite each fraction over LCD.

$= \frac{3}{10} + \frac{5}{10} - \frac{10}{10}$ Add/Subtract from left to right.

$= \frac{-2}{10} = -\frac{1}{5}$ Reduce to lowest terms.

105. This expression has 3 terms. Simplify inside each term first.

$\boxed{\frac{5}{8} \cdot \frac{4}{15}} + \boxed{\frac{1}{2} \cdot \frac{3}{2}} + \boxed{2}$ Multiply in the first & second terms.

$= \frac{5 \cdot 4}{8 \cdot 15} + \frac{1 \cdot 3}{2 \cdot 2} + 2$ Factor numerators and denominators.

$= \frac{\not{5} \cdot \not{2} \cdot \not{2}}{\not{5} \cdot 3 \cdot \not{2} \cdot \not{2} \cdot 2} + \frac{1 \cdot 3}{2 \cdot 2} + 2$ Divide out like factors.

$= \frac{1}{6} + \frac{3}{4} + 2$ Find LCD = 12.

$= \frac{1 \cdot 2}{6 \cdot 2} + \frac{3 \cdot 3}{4 \cdot 3} + \frac{2 \cdot 12}{1 \cdot 12}$ Rewrite each fraction over LCD.

$= \frac{2}{12} + \frac{9}{12} + \frac{24}{12}$ Add/Subtract from left to right.

$= \frac{35}{12}$ or $2\frac{11}{12}$ Fraction is in lowest terms.

107. This expression has 1 term. Simplify inside each term first.

$\frac{4}{9} \div \frac{5}{18} \cdot \frac{3}{2}$ Divide first.

$= \frac{4}{9} \cdot \frac{18}{5} \cdot \frac{3}{2}$ Find reciprocal of second fraction & multiply.

$= \frac{4 \cdot 18}{9 \cdot 5} \cdot \frac{3}{2}$ Factor numerators and denominators.

$= \frac{\not{3} \cdot \not{3} \cdot 2 \cdot 2 \cdot \not{2}}{5 \cdot \not{3} \cdot \not{3}} \cdot \frac{3}{\not{2}}$ Divide out like factors.

$= \frac{12}{5}$ or $2\frac{2}{5}$ Fraction is in lowest terms.

Section R.3 Operations with Decimals and Percents

1. The digit in the tenths place is 4. $0.\boxed{4}732$

3. The digit in the ten-thousandths place is 4. $6.915\boxed{4}$

5. The digit in the ones place is 7. $\boxed{7}.25$

7. The digit in the hundredths place is 8. $6.2\boxed{8}01$

9. The digit in the thousandths place is 0. $8.95\boxed{0}4$

11. The number 2.9 is written in words as two and 9 tenths. Writing 2.9 in fraction form gives

$2.9 = 2\frac{9}{10}$ Write the decimal part in fraction form.

$= 2\frac{9}{10}$ Fraction is in lowest terms.

13. The number 0.84 is written in words as 84 hundredths. Writing 0.84 in fraction form gives

$0.84 = \frac{84}{100}$ Write the decimal part in fraction form.

$= \frac{21}{25}$ Reduce to lowest terms.

15. The number 7.71 is written in words as seven and 71 hundredths. Writing 7.71 in fraction form gives

$7.71 = 7\frac{71}{100}$ Write the decimal part in fraction form.

$= 7\frac{71}{100}$ Fraction is in lowest terms.

17. The number 1.008 is written in words as one and 8 thousandths. Writing 1.008 in fraction form gives

$1.008 = 1\frac{8}{1000}$ Write the decimal part in fraction form.

$= 1\frac{1}{125}$ Reduce to lowest terms.

19. The number 1.33 is written in words as one and 33 hundredths. Writing 1.33 in fraction form gives

$1.33 = 1\frac{33}{100}$ Write the decimal part in fraction form.

$= 1\frac{33}{100}$ Fraction is in lowest terms.

21. To convert $\frac{3}{5}$ to decimal form, use long division.

$\frac{3}{5}$ means $5\overline{)3}$. Since 3 is less than 5, add a decimal point and continue dividing.

$$\begin{array}{r} 0.6 \\ 5\overline{)3.0} \\ \underline{-30} \\ 0 \end{array}$$

Therefore, $\frac{3}{5} = 0.6$.

23. To convert $1\frac{1}{4}$ to decimal form, first convert to the improper fraction $\frac{5}{4}$ then use long division.

$\frac{5}{4}$ means $4\overline{)5}$. Add a decimal point and continue dividing.

$$\begin{array}{r} 1.25 \\ 4\overline{)5.00} \\ \underline{-4} \\ 10 \\ \underline{-8} \\ 20 \\ \underline{-20} \\ 0 \end{array}$$

Therefore, $1\frac{1}{4} = 1.25$.

25. To convert $\frac{5}{12}$ to decimal form, use long division.

$\frac{5}{12}$ means $12\overline{)5}$. Since 5 is less than 12, add a decimal point and continue dividing.

$$\begin{array}{r} 0.4166.. \\ 12\overline{)5.00000} \\ \underline{-48} \\ 20 \\ \underline{-12} \\ 80 \\ \underline{-72} \\ 80 \\ \underline{-72} \\ 80 \end{array}$$

Therefore, $\frac{5}{12} = 0.41\overline{6}$.

27. To convert $3\frac{1}{6}$ to decimal form, first convert to the improper fraction $\frac{19}{6}$ then use long division.

$\frac{19}{6}$ means $6\overline{)19}$. Add a decimal point and continue dividing.

$$\begin{array}{r} 3.166.. \\ 6\overline{)19.0000} \\ \underline{-18} \\ 10 \\ \underline{-6} \\ 40 \\ \underline{-36} \\ 40 \\ \underline{-36} \\ 4 \end{array}$$

Therefore, $\frac{19}{6} = 3.1\overline{6}$.

29. To convert $\frac{7}{8}$ to decimal form, use long division. $\frac{7}{8}$ means $8\overline{)7}$. Since 7 is less than 8, add a decimal point and continue dividing.

$$\begin{array}{r} 0.875 \\ 8\overline{)7.000} \\ \underline{-64} \\ 60 \\ \underline{-56} \\ 40 \\ \underline{-40} \\ 0 \end{array}$$

Therefore, $\frac{7}{8} = 0.875$.

31.

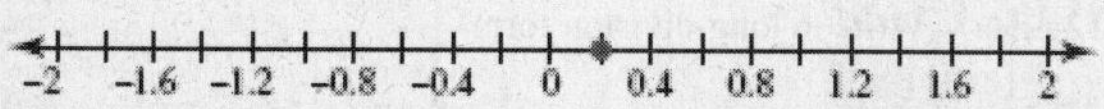

33. 1.7 lies halfway between 1.6 and 1.8.

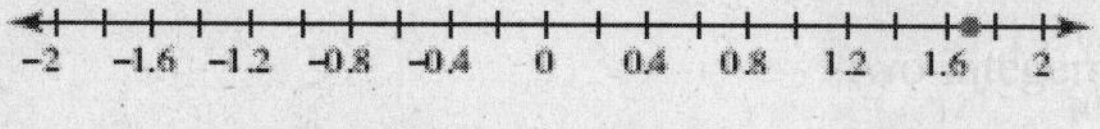

35. -0.9 lies halfway between -1.0 and -0.8.

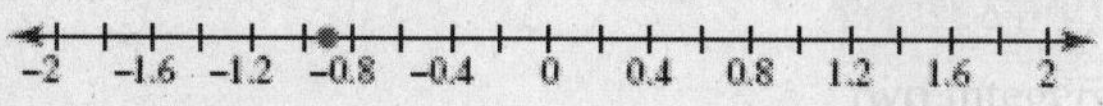

37. 1.3 lies halfway between 1.2 and 1.4.

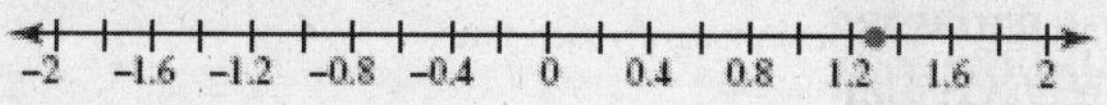

39.

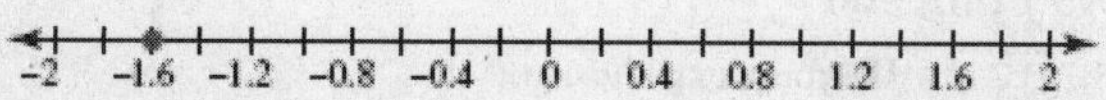

41. Round 8.25 to the tenths place.

8.$\underline{2}$5 Underline the tenths place.
The next digit to the right is 5.
8.3 Round up 1. $2+1=3$

43. Round 9.207 to the hundredths place.

9.2$\underline{0}$7 Underline the hundredths place.
The next digit to the right is 7.
9.21 Round up 1. $0+1=1$

45. Round 205.69 to the tens place.

2$\underline{0}$5.69 Underline the tens place.
The next digit to the right is 5.
210 Round up 1. $0+1=1$

47. Round 0.12999 to the tenths place.

0.$\underline{1}$2999 Underline the tenths place.
The next digit to the right is 2.
0.1 Round down.

49. Round 0.99 to the ones place.

$\underline{0}$.99 Underline the ones place.
The next digit to the right is 9.
1 Round up 1. $0+1=1$

51. Round 4.32601 to the thousandths place.

4.32$\underline{6}$01 Underline the thousandths place.
The next digit to the right is 0.
4.326 Round down.

53. Round 6.51757 to the ten-thousandths place.

6.517$\underline{5}$7 Underline the ten-thousandths place.
The next digit to the right is 7.
6.5176 Round up 1. $5+1=6$

55. Write the problem vertically, carefully matching place values, and add.

$$\begin{array}{r} 2.75 \\ \underline{+3.01} \\ 5.76 \end{array}$$

57. Write the problem vertically, carefully matching place values, and subtract.

$$\begin{array}{r} 8.\overset{5}{\cancel{6}}\,{}^{1}8 \\ \underline{-1.39} \\ 7.29 \end{array}$$

59. Write the problem vertically, carefully matching place values, and add.

$$\begin{array}{r} 96.3710 \\ \underline{+0.2275} \\ 96.5985 \end{array}$$

61. Write the problem vertically, carefully matching place values, and subtract.

$$\begin{array}{r} \overset{8}{\cancel{9}}.{}^{1}\overset{9}{\cancel{0}}\,{}^{1}0 \\ \underline{-3.67} \\ 5.33 \end{array}$$

63. Write the problem vertically, carefully matching place values, and subtract.

$$\begin{array}{r} 9.7\not{5}\,{}^{1}0 \\ -3.001 \\ \hline 6.749 \end{array}$$

65. To find the total cost of Michelle's 2 books, add the two amounts together.

$$\begin{array}{r} \$149.95 \\ +\$124.95 \\ \hline \$274.90 \end{array}$$

The two books cost Michelle \$274.90.

67. To find the amount of change Rani received, find the total cost of her purchases by adding the two amounts together then subtract the total from \$20.

$$\begin{array}{r} \$4.00 \\ +\$6.50 \\ \hline \$10.50 \end{array}$$

Now subtract the total from \$20.

$$\begin{array}{r} \$\not{2}\,\not{1}\not{0}.{}^{1}00 \\ -\ \$10.50 \\ \hline \$9.50 \end{array}$$

Rani received \$9.50 in change.

69. Multiply the numbers as if they were two integers, ignoring the decimal places until you reach the final answer.

$$\begin{array}{r} 3.01 \\ \times 1.8 \\ \hline 2408 \\ +3010 \\ \hline 5.418 \end{array}$$

The numbers 3.01 and 1.8 have a total of 3 digits after the decimal point. Move the decimal point from the right over 3 digits to get the solution 5.418.

71. Multiply the numbers as if they were two integers, ignoring the decimal places until you reach the final answer.

$$\begin{array}{r} 6.07 \\ \times 3.5 \\ \hline 3035 \\ +18210 \\ \hline 21.245 \end{array}$$

The numbers 3.5 and 6.07 have a total of 3 digits after the decimal point. Move the decimal point from the right over 3 digits to get the solution 21.245.

73. Multiply the numbers as if they were two integers, ignoring the decimal places until you reach the final answer.

$$\begin{array}{r} 0.016 \\ \times 7 \\ \hline 0.112 \end{array}$$

The numbers 7 and 0.016 have a total of 3 digits after the decimal point. Move the decimal point from the right over 3 digits to get the solution 0.112.

75. The amount you spend for downloading three songs costing \$0.99 each can be calculated as $3(0.99)$.

$$\begin{array}{r} 0.99 \\ \times 3 \\ \hline 2.97 \end{array}$$

The cost for 3 songs is \$2.97.

77. The amount Sam spent for three candy bars costing \$0.65 each can be calculated as $3(0.65)$.

$$\begin{array}{r} 0.65 \\ \times 3 \\ \hline 1.95 \end{array}$$

Sam spent \$1.95 for 3 candy bars.

79.

$8.2 \div 0.01$ Write in long-division form.

$0.01\overline{)8.2}$ Move the decimal point on 0.01 and 8.2 two places to the right.

$$\begin{array}{r} 820 \\ 1\overline{)820.} \\ \underline{-8} \\ 02 \\ \underline{-2} \\ 00 \end{array}$$

Perform long division.

81.

$0.12 \div 2$ Write in long-division form.

$2\overline{)0.12}$ The decimal does not need to be moved.

$$\begin{array}{r} 0.06 \\ 2\overline{)0.12} \\ \underline{-0} \\ 01 \\ \underline{-0} \\ 12 \\ \underline{-12} \\ 0 \end{array}$$

Perform long division.

83.

$19.32 \div 2.3$ Write in long-division form.

$2.3\overline{)19.32}$ Move the decimal point on 2.3 and 19.32 one place to the right.

$$\begin{array}{r} 8.4 \\ 23\overline{)193.2} \\ \underline{-184} \\ 92 \\ \underline{-92} \\ 0 \end{array}$$

Perform long division.

85.

$0.01073 \div 0.29$ Write in long-division form.

$0.29\overline{)0.01073}$ Move the decimal point on 0.29 and 0.01073 two places to the right.

$$\begin{array}{r} 0.037 \\ 29\overline{)1.073} \\ -87 \\ 203 \\ -203 \\ 0 \end{array}$$

Perform long division.

87. The terms are identified by being outlined. Simplify inside each term first.

$\boxed{4.2}+\boxed{1.5\cdot 6.1}$ Multiply in the 2nd term.

$=4.2+9.15$ Add the two terms.

$=13.35$

89. The terms are identified by being outlined. Simplify inside each term first.

$\boxed{4(8.95)}\boxed{-16.23}$ Multiply in the 1st term.

$=35.8-16.23$ Subtract.

$=19.57$

91. The terms are identified by being outlined. Simplify inside each term first.

$\boxed{4.6}+\boxed{8\div 2.5}\boxed{-22.1}$ Divide in the 2nd term.

$=4.6+3.2-22.1$ Add then subtract.

$=-14.3$

93. The terms are identified by being outlined. Simplify inside each term first.

$\boxed{-12.3}+\boxed{5\div 0.1}+\boxed{9.72}$ Divide in 2nd term.

$=-12.3+50+9.72$ Add/subtract from left to right.

$=47.42$

95. There is only one term in this expression.

$\boxed{24\div 1.5\cdot 8.4}$ Division is the first operation.

$=16\cdot 8.4$ Then multiply.

$=134.4$

97. **a.** To find an expression that gives the total cost of Luc's purchase, add each of the items together. Remember Luc is purchasing 2 packages of gum so we will need to multiply that cost by 2. The expression is

$1.25 + 0.99 + 2(0.59)$

b. This expression contains the operations of addition and multiplication.

c. This expression has 3 terms (the terms are separated by addition).

d. Using the order-of-operations agreement, multiplication should be done first, then addition.

e.

$1.25 + 0.99 + 2(0.59)$
$= 1.25 + 0.99 + 1.18$
$= 3.42$

The total amount of Luc's purchase is \$3.42.

99. **a.** To find an expression that gives the total cost of Van's purchase, add each of the items together. Remember Van is purchasing 2 sodas and 2 popcorns so we will need to multiply the cost of those items by 2 before adding all of the items together. The expression is

$2(4.00) + 5.25 + 2(5.00)$

b. This expression contains the operations of addition and multiplication.

c. This expression has 3 terms (the terms are separated by addition).

d. Using the order-of-operations agreement, multiplication should be done first, then addition.

e.

$2(4.00) + 5.25 + 2(5.00)$
$= 8.00 + 5.25 + 10.00$
$= 23.25$

The total amount of Van's purchase is \$23.25.

101. To convert a percentage to a decimal, move the decimal point two places to the left. $34\% = 0.34$

103. The fraction has a denominator of 100 so we can immediately write it as a percent.

$\frac{47}{100} = 47\%$

105. To convert a decimal to a percent, move the decimal point two places to the right, and add the percent sign.

$0.017 = 1.7\%$

107. This fraction is not written with a denominator of 100, so we cannot immediately write it as a percent. First, rewrite the fraction with a denominator of 100. Then write it as a percent.

$\frac{9}{25} = \frac{?}{100}$ Rewrite with a denominator of 100.

$\frac{9\cdot 4}{25\cdot 4} = \frac{36}{100}$

$\frac{36}{100} = 36\%$ Write as a percent.

109. To convert the decimal, 0.16 to a percent, move the decimal point two places to the right, and add the percent sign. $0.16 = 16\%$.

To convert 0.16 to a fraction, first figure out the place value of the last digit, 6. The place value of the last digit, 6, is the hundredths place. Rewrite 0.16 as a fraction with a denominator of 100.

$$0.16 = \frac{16}{100}$$ Write decimal as a fraction.

$$= \frac{4 \cdot \not{4}}{25 \cdot \not{4}}$$ Factor and reduce.

$$= \frac{4}{25}$$

So, the completed table is:

Percent	Decimal	Fraction
16%	0.16	$\frac{4}{25}$

111. To convert 28.5% to a decimal, move the decimal point two places to the left. $28.5\% = 0.285$

To convert 28.5% to a fraction, place 28.5 over 100 and simplify.

$$28.5\% = \frac{28.5}{100}$$ Rewrite as a fraction.

$$= \frac{28.5 \cdot 10}{100 \cdot 10}$$ Multiply to clear decimal.

$$= \frac{285}{1000}$$

$$= \frac{57 \cdot \not{5}}{200 \cdot \not{5}}$$ Factor and reduce.

$$= \frac{57}{200}$$

So, the completed table is:

Percent	Decimal	Fraction
28.5%	0.285	$\frac{57}{200}$

113. To convert 0.12% to a decimal, move the decimal point two places to the left. $0.12\% = 0.0012$

To convert 0.12% to a fraction, place 0.12 over 100 and simplify.

$$0.12\% = \frac{0.12}{100}$$ Rewrite as a fraction.

$$= \frac{0.12 \cdot 100}{100 \cdot 100}$$ Multiply to clear decimal.

$$= \frac{12}{10000}$$

$$= \frac{\not{4} \cdot 3}{2500 \cdot \not{4}}$$ Factor and reduce.

$$= \frac{3}{2500}$$

So, the completed table is:

Percent	Decimal	Fraction
0.12%	0.0012	$\frac{3}{2500}$

115. First, convert 16% to its decimal form. $16\% = 0.16$. A percent "of" means to multiply, so using the decimal form, calculate the answer.

$$16\% \cdot 64 = 0.16(64)$$
$$= 10.24$$

Therefore 16% of 64 is 10.24.

117. First, convert 0.5% to its decimal form. $0.5\% = 0.005$. A percent "of" means to multiply, so using the decimal form, calculate the answer.

$$0.5\% \cdot 112 = 0.005(112)$$
$$= 0.56$$

Therefore 0.5% of 112 is 0.56.

119. This problem asks us to find 40% of 25.

$$40\% \cdot 25 = 0.40(25)$$
$$= 10$$

There are 10 Labrador retrievers in the class.

121. This problem asks us to find 20% of 30.

$$20\% \cdot 30 = 0.20(30)$$
$$= 6$$

There are 6 diners who have a coupon.

123. This problem asks us to find 48% of 25g.

$$48\% \cdot 25 = 0.48(25)$$
$$= 12$$

One serving of yellow split peas has 12 g of dietary fiber.

125. This problem asks us to find 25% of 2400 mg.

$$25\% \cdot 2400 = 0.25(2400)$$
$$= 600$$

One serving of pasta sauce has 600 mg of sodium.

127. This problem asks us to find 6% of $25.

$$6\% \cdot 25 = 0.06(25)$$
$$= 1.5$$

Brieanna will pay $1.50 for tax.

129. This problem asks us to find 9.5% of $384.

$$9.5\% \cdot 384 = 0.095(384)$$
$$= 36.48$$

Yerica will pay $36.48 for tax.

Section R.4 The Real Number System

1.

	Whole Number	Integer	Rational Number	Irrational Number	Real Number
1.47			X		X
$\frac{16}{49}$			X		X
−7		X	X		X
$\sqrt{5}$				X	X
9	X	X	X		X

1.47 is a finite decimal, therefore it is a rational number which also belongs to the set of real numbers.

$\frac{16}{49}$ is a fraction, a ratio of integers, therefore it is a rational number which also belongs to the set of real numbers.

−7 belongs to the set of integers, therefore it also belongs to the set of rational numbers and the set of real numbers.

$\sqrt{5}$ cannot be expressed as an integer, finite decimal, or infinite repeating decimal, therefore it is an irrational number which also belongs to the set of real numbers.

9 belongs to the set of whole numbers, therefore it also belongs to the sets of integers, rational numbers, and real numbers.

3. A ratio is expressed as a fraction, therefore the ratio of boys to girls in a calculus course would be a rational number.

5. A ratio is expressed as a fraction, therefore the ratio of flour to water in a bread recipe would be a rational number.

7. Temperatures are typically expressed as whole numbers or finite decimals and can be positive or negative. Therefore, the value would belong to the set of integers and the set of rational numbers.

9. The amount owed on an electricity bill will often be a decimal number with two decimal digits, such as $73.29. Therefore, these values belong to the set of rational numbers.

11. The number of people who attend a college football game would be a whole number such as 587. Therefore, the value would belong to the set of integers and the set of rational numbers.

13. Elevations are typically measured to the nearest foot. Therefore, the elevation of Death Valley is an integer and therefore also a rational number.

15. The rational numbers in the set are as follows: −3 , 0, $\frac{5}{2}$.

The numbers π and $\sqrt{2}$ cannot be expressed as integers, finite decimals, or infinite repeating decimals therefore they are irrational.

17. The irrational numbers in the set are π and $\sqrt{2}$. These cannot be expressed as integers, finite decimals, or infinite repeating decimals therefore they are irrational.

19. The whole numbers in the set are as follows:

16, 0, 2.

Recall that the set of whole numbers is the set of natural numbers {1, 2, 3…} plus the number 0. The set of whole numbers does not include fractions, decimals, or negative numbers.

21. The integers in the set are as follows:

3, 111.

Recall that the set of integers is the set of whole numbers {0,1, 2, 3…} plus negative numbers. The set of integers does not include fractions or decimals.

23. The only natural number in the set is as follows:

56.

Recall that the set of natural numbers is the set of "counting" numbers {1, 2, 3…}. The set of natural numbers does not include 0, negative numbers, fractions, or decimals.

25. To convert $\frac{5}{8}$ to a decimal, use long division.

$\frac{5}{8}$ means $8\overline{)5}$.

$$\begin{array}{r} 0.625 \\ 8\overline{)5.000} \\ \underline{-48} \\ 20 \\ \underline{-16} \\ 40 \\ \underline{-40} \\ 0 \end{array}$$

Therefore, $\frac{5}{8} = 0.625$.

The number $\frac{5}{8}$ is rational because it is a fraction, a ratio of integers.

27. To convert 1.34 to a fraction in simplest form, place the decimal portion in the numerator of a fraction. Since the last place value of the decimal is hundredths, place 100 in the denominator.

$1.34 = 1\frac{34}{100}$

$= 1\frac{17 \cdot \cancel{2}}{50 \cdot \cancel{2}}$ Reduce to simplest form.

$= 1\frac{17}{50}$

Therefore, $1.34 = 1\frac{17}{50}$.

The number 1.34 is rational because it is a finite decimal.

29. To convert $\frac{1}{11}$ to a decimal, use long division.

$\frac{1}{11}$ means $11\overline{)1}$.

$$\begin{array}{r} 0.0909 \\ 11\overline{)1.0000} \\ -99 \\ \hline 10 \\ -0 \\ \hline 100 \\ -99 \\ \hline 1 \end{array}$$

The two-digit pattern will continue to repeat.

Therefore, $\frac{1}{11} = 0.090909... = 0.\overline{09}$.

The number $\frac{1}{11}$ is rational because it is a fraction, a ratio of integers.

31. To convert 0.0133 to a fraction in simplest form, place the decimal portion in the numerator of a fraction. Since the last place value of the decimal is ten-thousandths, place 10,000 in the denominator.

$0.0133 = \frac{133}{10000}$

Fraction is in simplest form.

The number 0.0133 is rational because it is a finite decimal.

33. Using a calculator, $\frac{3}{11} = 0.27272727....$ Recall that to write a repeating decimal, put a bar over the repeating part: $\frac{3}{11} = 0.272727... = 0.\overline{27}$. To round to the thousandths place, underline the thousandths place: $0.27\underline{2}72727...$. The digit to the right of the 2 in the thousandths place is 7, so round up. Therefore, $\frac{3}{11} \approx 0.273$.

The number $\frac{3}{11}$ is rational because it is a fraction, a ratio of integers.

35. Using a calculator, $\sqrt{11} \approx 3.3166247903....$ To round to the ten-thousandths place, underline the ten-thousandths place: $3.316\underline{6}247903....$ The digit to the right of the 6 in the ten-thousandths place is 2, so round down. Therefore, $\sqrt{11} \approx 3.3166$.

The number $\sqrt{11}$ is irrational because it cannot be expressed as a fraction, a finite decimal, or an infinite repeating decimal.

37. Using a calculator, $\pi \approx 3.141592654....$ To round to the millionths place, underline the millionths place: $3.14159\underline{2}654....$ The digit to the right of the 2 in the millionths place is 6, so round up. Therefore, $\pi \approx 3.141593$.

The number π is irrational because it cannot be expressed as a finite decimal or an infinite repeating decimal.

39. To find the unit price, divide \$9.99 by 21 ounces. Using a calculator $\frac{9.99}{21} \approx 0.4757$. Therefore the unit price is approximately \$0.48 per ounce.

41. To find the unit price, divide \$0.75 by 12 ounces. Using a calculator $\frac{0.75}{12} = 0.0625$. Therefore the unit price is approximately \$0.06 per ounce.

43. To find the unit price, divide \$2.59 by 24 ounces. Using a calculator $\frac{2.59}{24} \approx 0.1079$. Therefore the unit price is approximately \$0.11 per ounce.

45. This expression has 2 terms. The terms are outlined.

$\boxed{\frac{3}{5} \cdot \frac{1}{6}} + \boxed{\frac{1}{2}}$ Multiply within first term.

$= \frac{\cancel{3} \cdot 1}{5 \cdot \cancel{3} \cdot 2} + \frac{1}{2}$ Factor and reduce.

$= \frac{1}{10} + \frac{1}{2}$ Find LCD = 10

$= \frac{1}{10} + \frac{1 \cdot 5}{2 \cdot 5}$ Rewrite fraction over LCD.

$= \frac{1}{10} + \frac{5}{10}$ Add numerators.

$= \frac{6}{10} = \frac{3 \cdot \cancel{2}}{5 \cdot \cancel{2}}$ Factor and reduce.

$= \frac{3}{5}$ Fraction is in lowest terms.

47. This expression has 3 terms. The terms are outlined.

$\boxed{1.2\cdot 2}\boxed{-6.2\div 2}+\boxed{12.1}$ Multiply in first term.

$= 2.4-6.2\div 2+12.1$ Divide in second term.

$= 2.4-3.1+12.1$ Add/subtract from left to right.

$= 11.4$

49. This expression has 2 terms. The terms are outlined. This expression contains both fractions and decimals.

$\boxed{\frac{3}{5}}+\boxed{1.5\cdot 3}$ Multiply in second term.

$= \frac{3}{5}+4.5$ Convert fraction to decimal.

$= 0.6+4.5$ Add.

$= 5.1$

51. This expression has 2 terms. The terms are outlined. This expression contains both fractions and decimals.

$\boxed{26.1\div 6}\boxed{-\frac{17}{5}\cdot\frac{1}{20}}$ Divide in first term.

$= 4.35-\frac{17}{5}\cdot\frac{1}{20}$ Multiply in second term.

$= 4.35-\frac{17\cdot 1}{5\cdot 20}$

$= 4.35-\frac{17}{100}$ Convert fraction to decimal.

$= 4.35-0.17$ Subtract.

$= 4.18$

53. This expression has 3 terms. The terms are outlined. This expression contains both fractions and decimals.

$\boxed{5}\boxed{-6.75}+\boxed{\frac{2}{3}\div 4}$ Divide in third term.

$= 5-6.75+\frac{2}{3}\cdot\frac{1}{4}$ Invert 2nd fraction & multiply.

$= 5-6.75+\frac{2\cdot 1}{3\cdot 4}$

$= 5-6.75+\frac{\cancel{2}}{3\cdot 2\cdot\cancel{2}}$ Factor and reduce fraction.

$= 5-6.75+\frac{1}{6}$ Subtract first two terms.

$= -1.75+\frac{1}{6}$ Convert decimal to fraction.

$= -1\frac{75}{100}+\frac{1}{6}$ Simplify fraction.

$= -1\frac{3}{4}+\frac{1}{6}$ Convert to improper fraction.

$= -\frac{7}{4}+\frac{1}{6}$ Find LCD = 12.

$= -\frac{7\cdot 3}{4\cdot 3}+\frac{1\cdot 2}{6\cdot 2}$ Re write fractions over LCD.

$= -\frac{21}{12}+\frac{2}{12}$ Add/subtract.

$= -\frac{19}{12}$ or $-1\frac{1}{12}$ Fraction is in lowest terms.

55. This expression has 3 terms. The terms are outlined.

$\boxed{\frac{1}{4}}+\boxed{\frac{3}{8}\cdot\frac{16}{21}}-\boxed{4\div\frac{2}{3}}$ Multiply in second term.

$= \frac{1}{4}+\frac{3\cdot 16}{8\cdot 21}-4\div\frac{2}{3}$ Factor and reduce.

$= \frac{1}{4}+\frac{\cancel{3}\cdot\cancel{2}\cdot\cancel{2}\cdot\cancel{2}\cdot 2}{7\cdot\cancel{3}\cdot\cancel{2}\cdot\cancel{2}\cdot\cancel{2}}-4\div\frac{2}{3}$

$= \frac{1}{4}+\frac{2}{7}-4\div\frac{2}{3}$ Divide in third term.

$= \frac{1}{4}+\frac{2}{7}-\frac{4}{1}\cdot\frac{3}{2}$ Invert 2nd fraction & multiply.

$= \frac{1}{4}+\frac{2}{7}-\frac{4\cdot 3}{1\cdot 2}$

$= \frac{1}{4}+\frac{2}{7}-\frac{\cancel{2}\cdot 2\cdot 3}{\cancel{2}}$ Factor and reduce.

$= \frac{1}{4}+\frac{2}{7}-6$ Find LCD = 28.

$= \frac{1\cdot 7}{4\cdot 7}+\frac{2\cdot 4}{7\cdot 4}-\frac{6\cdot 28}{1\cdot 28}$ Rewrite fractions over LCD.

$= \frac{7}{28}+\frac{8}{28}-\frac{168}{28}$ Add/subtract from left to right.

$= -\frac{153}{28}$ or $-5\frac{13}{28}$ Fraction is in lowest terms.

CHAPTER R Review of Prealgebra

Chapter R Review

1. **a.** The number -12 belongs to the set of integers.

b. The number 0 belongs to the set of whole numbers and all whole numbers also belong to the set of integers.

c. The number 17 belongs to the set of natural numbers. All natural numbers also belong to the set of whole numbers and to the set of integers.

2. **a.** The number 9 belongs to the set of natural numbers. All natural numbers also belong to the set of whole numbers and to the set of integers.

b. The number -1 belongs to the set of integers.

c. The number 0 belongs to the set of whole numbers and all whole numbers also belong to the set of integers.

3. The value for the number of students in a chemistry lab is a whole number. The number of students could be 0, 1, 2, ... Since a chemistry lab could be empty of students at any given time (0 students), the value 0 is included as a possibility. These values are also integers.

4. The value for the number of trees in a public park is a whole number. The number of trees could be 0, 1, 2, ... Since a public park could have no trees, the value 0 is included as a possibility. These values are also integers.

5. The set of all whole numbers between -3 and 4, *including* -3 and 4 is as follows:

$\{0, 1, 2, 3, 4\}$

Notice that negative values are not included because this is a set of only whole numbers.

6. The set of all integers between -6 and -3, *including* -6 and -3 is as follows:

$\{-6, -5, -4, -3\}$

7. The tick marks on the number line are 2 units apart, therefore the scale of this number line is 2.

8. The tick marks on the number line are 15 units apart, therefore the scale of this number line is 15.

9. The numbers -3, $\frac{1}{2}$, 0, $\frac{-4}{2}$, and 4.25 are graphed as darkened circles on the number line.

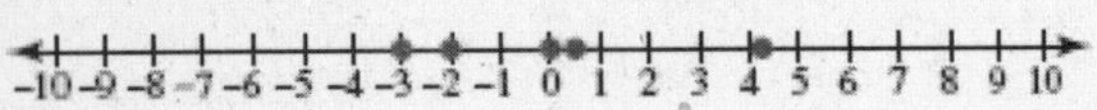

10. The numbers 2.5, $\frac{-1}{3}$, -2, $\frac{5}{2}$, and -0.8 are graphed as darkened circles on the number line.

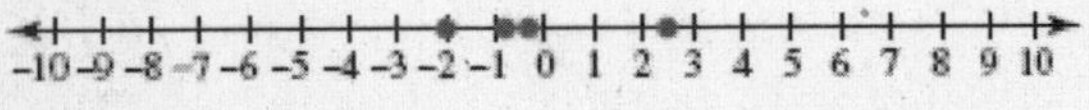

11. The number -3 is to the right of -5 on the number line, therefore -3 is greater than -5.

$-3 > -5$

12. The number -9 is to the left of -2 on the number line, therefore -9 is less than -2.

$-9 < -2$

13. $|-9| = 9$ since the distance between -9 and 0 is 9 units.

14. $|-6.7| = 6.7$ since the distance between -6.7 and 0 is 6.7 units.

15. First, find any absolute values.

$3 = 3$

$-2 = -2$

$\frac{2}{5} = \frac{2}{5}$

$|-7| = 7$

Then order the numbers from smallest to largest (from left to right as they would appear on the number line) writing each term in its original notation as follows:

$-2, \frac{2}{5}, 3, |-7|$

16. First, find any absolute values.

$-4 = -4$

$\left|\frac{-2}{3}\right| = \frac{2}{3}$

$1.8 = 1.8$

$-0.1 = -0.1$

Then order the numbers from smallest to largest (from left to right as they would appear on the number line) writing each term in its original notation as follows:

$-4, -0.1, \left|\frac{-2}{3}\right|, 1.8$

17. The expression translates as, "the opposite of negative eight."

$-(-8) = 8$

18. The expression translates as, "the opposite of negative one."

$-(-1) = 1$

19. **a.** The two integers have the same sign, so we add the two numbers using the rule for adding integers and keep the negative sign.

$-22 + (-45) = -67$

b. The two integers are being subtracted. Change the sign of the second number and add. Add the two numbers using the rule for adding integers, keeping the negative sign because both numbers are negative.

$-9+(-17)=-26$

c. These two integers have different signs. Take the absolute value of each number and subtract the smaller absolute value from the larger. Subtract 8 from 18.

$18-8=10$

Attach the sign of the number that is larger in absolute value. The number that is larger in absolute value is -18, so the answer will be negative. Attach a negative sign to 10.

$-18+8=-10$

The final answer is -10.

d. Using the rule for subtracting integers, change the sign of the second term and add the terms.

$-16-(-11)=-16+11$

Subtracting a negative is the same as adding so we now have two integers with different signs. Take the absolute value of each number and subtract the smaller absolute value from the larger. Subtract 11 from 16.

$16-11=5$

Attach the sign of the number that is larger in absolute value. Since the number that is larger is positive, the answer is positive.

$-16-(-11)=-5$

20. **a.** The two integers have the same sign, so we add the two numbers using the rule for adding integers and keep the negative sign.

$-13+(-5)=-18$

b. The two integers are being subtracted. Change the sign of the second number and add. Add the two numbers using the rule for adding integers, keeping the positive sign because both numbers are positive.

$50-(-18)=50+18=68$

c. The two integers are being subtracted. Change the sign of the second number and add.

$4+(-18)$

These two integers have different signs. Take the absolute value of each number and subtract the smaller absolute value from the larger. Subtract 4 from 18.

$18-4=14$

Attach the sign of the number that is larger in absolute value. The number that is larger in absolute value is -18, so the answer will be negative. Attach a negative sign to 14.

$4-18=-14$

The final answer is -14.

d. These two integers have different signs. Take the absolute value of each number and subtract the smaller absolute value from the larger. Subtract 12 from 19.

$19-12=7$

Attach the sign of the number that is larger in absolute value. The number that is larger in absolute value is -19, so the answer will be negative. Attach a negative sign to 7.

$-19+12=-7$

The final answer is -7.

21. **a.** $-5(-10)(-3)=-150$

Using the commutative property of multiplication, the integers can be multiplied in any order. The answer is negative because two negative numbers multiplied together will give a positive result, then multiplying that positive number by a negative number will give a negative result.

b. $-2(4)(-1)=8$

Using the commutative property of multiplication, the integers can be multiplied in any order. The answer is negative because two negative numbers multiplied together will give a positive result, then multiplying that positive number by another positive number will give a positive result.

c. $64\div(-4)=-16$

The answer is negative since the two integers have different signs.

d. $-55\div(-11)=5$

The answer is positive since the two integers have the same sign.

22. **a.** $(-3)(-5)(2)=30$

Using the commutative property of multiplication, the integers can be multiplied in any order. The answer is negative because two negative numbers multiplied together will give a positive result, then multiplying that positive number by another positive number will give a positive result.

b. $-(-3)(-6)=-18$

The answer is negative because two negative numbers multiplied together will give a positive result, then taking the opposite of that positive number will give a negative result.

c. $27\div(-9)=-3$

The answer is negative since the two integers have different signs.

d. $-24\div(-3)=8$

The answer is positive since the two integers have the same sign.

23. $\boxed{-9\cdot 4\div 6}+\boxed{18}\boxed{-3}$

This expression has 3 terms. Terms are separated by addition and subtraction symbols. The terms in the expression are outlined.

24. $\boxed{-12}+\boxed{8\div 2}\boxed{-1}$

This expression has 3 terms. Terms are separated by addition and subtraction symbols. The terms in the expression are outlined.

25. This expression has 3 terms.

$\boxed{-48\div(-12)\cdot 5}+\boxed{4\cdot 3}\boxed{-7}$ Outline terms.
$=4\cdot 5+4\cdot 3-7$ Divide in the first term.
$=20+12-7$ Multiply in the first and second terms.
$=25$ Add/subtract from left to right.

26. This expression has 3 terms.

$\boxed{9\cdot(-2)}+\boxed{17}+\boxed{32\div(-16)\cdot 3}$ Outline terms.
$=-18+17+32\div(-16)\cdot 3$ Multiply in first term.
$=-18+17-6$ Divide then multiply in the last term.
$=-7$ Add/Subtract from left to right.

27.

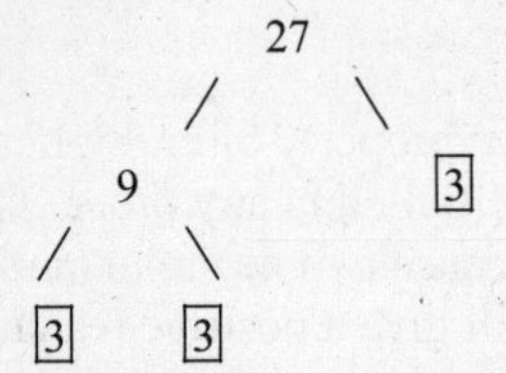

Therefore, $27=3\cdot 3\cdot 3$.

28.

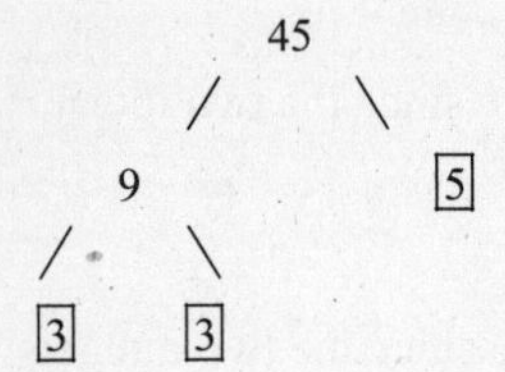

Therefore, $45=5\cdot 3\cdot 3$.

29. $8=\boxed{2}\cdot\boxed{2}\cdot 2$

$28=7\cdot\boxed{2}\cdot\boxed{2}$

The GCF $=2\cdot 2=4$

30. $15=\boxed{5}\cdot\boxed{3}$

$45=\boxed{5}\cdot\boxed{3}\cdot 3$

The GCF $=5\cdot 3=15$

31. $\frac{20}{44}=\frac{5\cdot \cancel{4}}{11\cdot \cancel{4}}=\frac{5}{11}$

32. $\frac{12}{30}=\frac{\cancel{6}\cdot 2}{\cancel{6}\cdot 5}=\frac{2}{5}$

33. $\frac{5\cdot 7}{8\cdot 7}=\frac{35}{56}$

34. $\frac{3\cdot 6}{7\cdot 6}=\frac{18}{42}$

35. **a.** $\frac{3}{5}$ is a fraction, a ratio of integers, therefore it is a rational number.

b. -5 belongs to the set of integers, therefore it also belongs to the set of rational numbers.

c. 16.1 is a finite decimal, therefore it is a rational number.

36. **a.** -0.7 is a finite decimal, therefore it is a rational number.

b. $\frac{-1}{9}$ is a fraction, a ratio of integers, therefore it is a rational number.

c. 0 belongs to the set of integers, therefore it also belongs to the set of rational numbers.

37. **a.**

$\frac{1}{6}+\frac{3}{6}$ Add the numerators.
$=\frac{4}{6}$ Reduce to lowest terms.
$=\frac{2}{3}$

b. The factorization of 6 is $6=3\cdot 2$ and 4 is $4=2\cdot 2$. The LCD $=3\cdot 2\cdot 2=12$.

$\frac{1}{6}+\frac{1}{4}$
$=\frac{1\cdot 2}{6\cdot 2}+\frac{1\cdot 3}{4\cdot 3}$ Rewrite each fraction over LCD.
$=\frac{2}{12}+\frac{3}{12}$ Add numerators.
$=\frac{5}{12}$ Reduce to lowest terms.

c. The denominators 2 and 3 are both prime. The LCD $=3\cdot 2=6$.

$\frac{1}{2}-\frac{2}{3}$
$=\frac{1\cdot 3}{2\cdot 3}-\frac{2\cdot 2}{3\cdot 2}$ Rewrite each fraction over LCD.
$=\frac{3}{6}-\frac{4}{6}$ Subtract numerators.
$=-\frac{1}{6}$ Reduce to lowest terms.

d. The denominators 2 and 3 are both prime. The LCD $=3\cdot 2=6$.

$3\frac{1}{2}-1\frac{1}{3}$ Rewrite as improper fractions
$=\frac{7}{2}-\frac{4}{3}$
$=\frac{7\cdot 3}{2\cdot 3}-\frac{4\cdot 2}{3\cdot 2}$ Rewrite each fraction over LCD.
$=\frac{21}{6}-\frac{8}{6}$ Subtract numerators.
$=\frac{13}{6}=2\frac{1}{6}$ Answer is in lowest terms.

38. **a.**

$$\frac{5}{8}-\frac{1}{8} \quad \text{Subtract the numerators.}$$
$$=\frac{4}{8} \quad \text{Reduce to lowest terms.}$$
$$=\frac{1}{2}$$

b. The factorization of 15 is $15=5\cdot3$ and 5 is prime. The LCD $=5\cdot3=15$.

$$\frac{1}{5}+\frac{1}{15}$$
$$=\frac{1\cdot3}{5\cdot3}+\frac{1}{15} \quad \text{Rewrite first fraction over LCD.}$$
$$=\frac{3}{15}+\frac{1}{15} \quad \text{Add numerators.}$$
$$=\frac{4}{15} \quad \text{Fraction is in lowest terms.}$$

c. The factorization of 4 is $4=2\cdot2$ and 5 is prime. The LCD $=5\cdot2\cdot2=20$.

$$\frac{1}{4}-\frac{2}{5}$$
$$=\frac{1\cdot5}{4\cdot5}-\frac{2\cdot4}{5\cdot4} \quad \text{Rewrite each fraction over LCD.}$$
$$=\frac{5}{20}-\frac{8}{20} \quad \text{Subtract numerators.}$$
$$=-\frac{3}{20} \quad \text{Fraction is in lowest terms.}$$

d. The denominators 5 and 2 are both prime. The LCD $=5\cdot2=10$.

$$4\frac{1}{5}-3\frac{1}{2} \quad \text{Rewrite as improper fractions}$$
$$=\frac{21}{5}-\frac{7}{2}$$
$$=\frac{21\cdot2}{5\cdot2}-\frac{7\cdot5}{2\cdot5} \quad \text{Rewrite each fraction over LCD.}$$
$$=\frac{42}{10}-\frac{35}{10} \quad \text{Subtract numerators.}$$
$$=\frac{7}{10} \quad \text{Answer is in lowest terms.}$$

39. **a.** To find a reciprocal, we invert the fraction. To find the reciprocal of 6, recall that $6=\frac{6}{1}$. Find the reciprocal, which is $\frac{1}{6}$.

b. To find a reciprocal, we invert the fraction. Find the reciprocal of $\frac{1}{4}$, which is $\frac{4}{1}=4$.

40. **a.** To find a reciprocal, we invert the fraction. To find the reciprocal of -5, recall that $-5=-\frac{5}{1}$. Find the reciprocal, which is $-\frac{1}{5}$.

b. To find a reciprocal, we invert the fraction. Find the reciprocal of $\frac{3}{2}$, which is $\frac{2}{3}$.

41. **a.**

$$\frac{2}{7}\cdot\frac{14}{22}$$
$$=\frac{2\cdot14}{7\cdot22} \quad \text{Multiply numerators \& denominators.}$$
$$=\frac{\cancel{7}\cdot\cancel{2}\cdot2}{11\cdot\cancel{7}\cdot\cancel{2}} \quad \text{Factor. Divide out like factors.}$$
$$=\frac{2}{11}$$

b.

$$\frac{1}{5}\div9=\frac{1}{5}\div\frac{9}{1} \quad \text{Invert 2nd fraction \& multiply}$$
$$=\frac{1}{5}\cdot\frac{1}{9}$$
$$=\frac{1\cdot1}{5\cdot9} \quad \text{Simplify.}$$
$$=\frac{1}{45}$$

c.

$$\frac{3}{8}\div\frac{9}{16} \quad \text{Invert 2nd fraction \& multiply.}$$
$$=\frac{3}{8}\cdot\frac{16}{9}$$
$$=\frac{3\cdot16}{8\cdot9} \quad \text{Factor numerator and denominator.}$$
$$=\frac{\cancel{3}\cdot\cancel{2}\cdot\cancel{2}\cdot\cancel{2}\cdot2}{\cancel{3}\cdot3\cdot\cancel{2}\cdot\cancel{2}\cdot\cancel{2}} \quad \text{Divide out like factors.}$$
$$=\frac{2}{3}$$

42. **a.**

$$\frac{4}{5}\cdot\frac{15}{28}$$
$$=\frac{4\cdot15}{5\cdot28} \quad \text{Multiply numerators \& denominators.}$$
$$=\frac{\cancel{5}\cdot3\cdot\cancel{2}\cdot\cancel{2}}{7\cdot\cancel{5}\cdot\cancel{2}\cdot\cancel{2}} \quad \text{Factor. Divide out like factors.}$$
$$=\frac{3}{7}$$

b.

$$\frac{3}{7}\div2=\frac{3}{7}\div\frac{2}{1} \quad \text{Invert 2nd fraction \& multiply}$$
$$=\frac{3}{7}\cdot\frac{1}{2}$$
$$=\frac{3\cdot1}{7\cdot2} \quad \text{Simplify.}$$
$$=\frac{3}{14}$$

c.

$$\frac{5}{6} \div \frac{10}{9}$$ Invert 2nd fraction & multiply.

$$= \frac{5}{6} \cdot \frac{9}{10}$$

$$= \frac{5 \cdot 9}{6 \cdot 10}$$ Factor numerator and denominator.

$$= \frac{\not{5} \cdot \not{3} \cdot 3}{\not{5} \cdot \not{3} \cdot 2 \cdot 2}$$ Divide out like factors.

$$= \frac{3}{4}$$

43. This expression has 2 terms. Simplify inside each term first.

$$\boxed{\frac{1}{3} \div \frac{2}{9}} \boxed{-7}$$ Divide in the first term.

$$= \frac{1}{3} \cdot \frac{9}{2} - 7$$ Invert 2nd fraction & multiply.

$$= \frac{1 \cdot 9}{3 \cdot 2} - 7$$ Factor and reduce.

$$= \frac{\not{3} \cdot 3}{\not{3} \cdot 2} - 7$$

$$= \frac{3}{2} - \frac{7}{1}$$ Find LCD = 2.

$$= \frac{3}{2} - \frac{7 \cdot 2}{1 \cdot 2}$$ Rewrite each fraction over the LCD.

$$= \frac{3}{2} - \frac{14}{2}$$ Add the two fractions.

$$= \frac{-11}{2} \text{ or } -5\frac{1}{2}$$ Fraction is in lowest terms.

44. This expression has 2 terms. Simplify inside each term first.

$$\boxed{3} \boxed{-\frac{4}{5} \cdot \frac{10}{3}}$$ Multiply in the second term.

$$= 3 - \frac{4 \cdot 10}{5 \cdot 3}$$

$$= 3 - \frac{\not{5} \cdot 2 \cdot 2 \cdot 2}{\not{5} \cdot 3}$$ Factor and reduce.

$$= 3 - \frac{8}{3}$$ Find LCD = 3.

$$= \frac{3 \cdot 3}{1 \cdot 3} - \frac{8}{3}$$ Rewrite each fraction over the LCD.

$$= \frac{9}{3} - \frac{8}{3}$$ Subtract.

$$= \frac{1}{3}$$ Fraction is in lowest terms.

45. The digit 9 in the number 2.497031 is in the hundredths place.

46. The digit 5 in the number 8.26571 is in the thousandths place.

47. The number 2.75 is written in words as two and 75 hundredths. Writing 2.75 in fraction form gives

$$2.75 = 2\frac{75}{100}$$ Write the decimal part in fraction form.

$$= 2\frac{3}{4}$$ Reduce to lowest terms.

48. The number 4.68 is written in words as four and 68 hundredths. Writing 4.68 in fraction form gives

$$4.68 = 4\frac{68}{100}$$ Write the decimal part in fraction form.

$$= 4\frac{17}{25}$$ Reduce to lowest terms.

49. To convert $\frac{4}{9}$ to decimal form, use long division.

$\frac{4}{9}$ means $9\overline{)4}$. Since 4 is less than 9, add a decimal point and continue dividing.

$$\begin{array}{r} 0.44.. \\ 9\overline{)4.000} \\ \underline{-36} \\ 40 \\ \underline{-36} \\ 4 \end{array}$$

Therefore, $\frac{4}{9} = 0.\overline{4}$.

50. To convert $\frac{5}{8}$ to decimal form, use long division.

$\frac{5}{8}$ means $8\overline{)5}$. Since 5 is less than 8, add a decimal point and continue dividing.

$$\begin{array}{r} 0.625 \\ 8\overline{)5.000} \\ \underline{-48} \\ 20 \\ \underline{-16} \\ 40 \\ \underline{-40} \\ 0 \end{array}$$

Therefore, $\frac{5}{8} = 0.625$.

51. **a.** Write the problem vertically, carefully matching place values, and add.

$$\begin{array}{r} \overset{1}{} \\ 3.625 \\ \underline{+7.019} \\ 10.644 \end{array}$$

b. Write the problem vertically, carefully matching place values, and subtract.

$$\begin{array}{r} 4.061 \\ -3.987 \\ \hline 0.074 \end{array}$$

Solve as follows:

$$\begin{array}{r} \overset{3}{\cancel{4}}.\overset{9}{\cancel{1}}\overset{15}{\cancel{6}}\,{}^{1}1 \\ -3.987 \\ \hline 0.074 \end{array}$$

52. **a.** Write the problem vertically, carefully matching place values, and add.

$$\begin{array}{r} \overset{1\,1\;\;1\,1}{16.998} \\ +4.073 \\ \hline 21.071 \end{array}$$

b. Write the problem vertically, carefully matching place values, and subtract.

$$\begin{array}{r} 5.072 \\ -3.826 \\ \hline 1.246 \end{array}$$

Solve as follows:

$$\begin{array}{r} \overset{4}{\cancel{5}}.{}^{1}0\overset{6}{\cancel{7}}\,{}^{1}2 \\ -3.826 \\ \hline 1.246 \end{array}$$

53. **a.** Multiply the numbers as if they were two integers, ignoring the decimal places until you reach the final answer.

$$\begin{array}{r} \overset{1}{\overset{2}{1.03}} \\ \times 4.7 \\ \hline 721 \\ +4120 \\ \hline 4.841 \end{array}$$

The numbers 4.7 and 1.03 have a total of 3 digits after the decimal point. Move the decimal point from the right over 3 digits to get the solution 4.841.

b.

$14.58 \div 5.4$ Write in long-division form.

$5.4\overline{)14.58}$ Move the decimal point on 5.4 and 14.58 one place to the right.

$$\begin{array}{r} 2.7 \\ 54\overline{)145.8} \\ -108 \\ \hline 378 \\ -378 \\ \hline 0 \end{array}$$ Perform long division.

$14.58 \div 5.4 = 2.7$

54. **a.** Multiply the numbers as if they were two integers, ignoring the decimal places until you reach the final answer.

$$\begin{array}{r} \overset{1}{9.2} \\ \times 3.8 \\ \hline \overset{1}{7}36 \\ +2760 \\ \hline 34.96 \end{array}$$

The numbers 9.2 and 3.8 have a total of 2 digits after the decimal point. Move the decimal point from the right over 2 digits to get the solution 34.96.

b.

$40.85 \div 4.3$ Write in long-division form.

$4.3\overline{)40.85}$ Move the decimal point on 4.3 and 40.85 one place to the right.

$$\begin{array}{r} 9.5 \\ 43\overline{)408.5} \\ -387 \\ \hline 215 \\ -215 \\ \hline 0 \end{array}$$ Perform long division.

$40.85 \div 4.3 = 9.5$

55. The terms are identified by being outlined. Simplify inside each term first.

$\boxed{9.6}\boxed{-11.73 \div 5.1}\boxed{-3.2}$ Divide in 2nd term.

$= 9.6 - 2.3 - 3.2$ Add/subtract from left to right.

$= 4.1$

56. The terms are identified by being outlined. Simplify inside each term first.

$\boxed{0.92 \div (-0.1)} + \boxed{6.3}\boxed{-4.75}$ Divide in 1st term.

$= -9.2 + 6.3 - 4.75$ Add/subtract from left to right.

$= -7.65$

57. Round 0.01996 to the hundredths place.

0.0<u>1</u>996 Underline the hundredths place.
The next digit to the right is 9.

0.02 Round up 1. $1+1=2$

58. Round 5.409999 to the tenths place.

5.<u>4</u>09999 Underline the tenths place.
The next digit to the right is 0.

5.4 Round down.

59. To convert a percentage to a decimal, move the decimal point two places to the left. $8.75\% = 0.0875$

60. To convert a percentage to a decimal, move the decimal point two places to the left. $5.45\% = 0.0545$

61. To convert 28% to a fraction, place 28 over 100 and simplify.

$$28\% = \frac{28}{100} \quad \text{Rewrite as a fraction.}$$
$$= \frac{7\cdot \cancel{4}}{25\cdot \cancel{4}} \quad \text{Factor and reduce.}$$
$$= \frac{7}{25}$$

62. To convert 45% to a fraction, place 45 over 100 and simplify.

$$45\% = \frac{45}{100} \quad \text{Rewrite as a fraction.}$$
$$= \frac{9\cdot \cancel{5}}{20\cdot \cancel{5}} \quad \text{Factor and reduce.}$$
$$= \frac{9}{20}$$

63. To convert the decimal, 0.0456 to a percent, move the decimal point two places to the right, and add the percent sign. $0.0456 = 4.56\%$.

64. To convert the decimal, 0.0775 to a percent, move the decimal point two places to the right, and add the percent sign. $0.0775 = 7.75\%$.

65. To convert the decimal, 1.905 to a percent, move the decimal point two places to the right, and add the percent sign. $1.905 = 190.5\%$.

66. To convert the decimal, 2.01 to a percent, move the decimal point two places to the right, and add the percent sign. $2.01 = 201\%$.

67. To convert the fraction, $\frac{4}{5}$ to a percent, rewrite the fraction with a denominator of 100. Then write it as a percent.

$$\frac{4}{5} = \frac{?}{100} \quad \text{Rewrite with a denominator of 100.}$$
$$\frac{4\cdot 20}{5\cdot 20} = \frac{80}{100}$$
$$\frac{80}{100} = 80\% \quad \text{Write as a percent.}$$

68. To convert the fraction, $\frac{3}{50}$ to a percent, rewrite the fraction with a denominator of 100. Then write it as a percent.

$$\frac{3}{50} = \frac{?}{100} \quad \text{Rewrite with a denominator of 100.}$$
$$\frac{3\cdot 2}{50\cdot 2} = \frac{6}{100}$$
$$\frac{6}{100} = 6\% \quad \text{Write as a percent.}$$

69. To convert the fraction, $1\frac{1}{5}$ to a percent, first convert to the improper fraction $\frac{6}{5}$ then rewrite the fraction with a denominator of 100. Then write it as a percent.

$$\frac{6}{5} = \frac{?}{100} \quad \text{Rewrite with a denominator of 100.}$$
$$\frac{6\cdot 20}{5\cdot 20} = \frac{120}{100}$$
$$\frac{120}{100} = 120\% \quad \text{Write as a percent.}$$

70. To convert the fraction, $3\frac{1}{3}$ to a percent, first convert to the improper fraction $\frac{10}{3}$ then rewrite the fraction with a denominator of 100. Then write it as a percent.

$$\frac{10}{3} = \frac{?}{100} \quad \text{Rewrite with a denominator of 100.}$$
$$\frac{10\cdot \frac{100}{3}}{3\cdot \frac{100}{3}} = \frac{\frac{1000}{3}}{100}$$
$$\frac{\frac{1000}{3}}{100} = \frac{1000}{3}\%$$
$$= 333\frac{1}{3}\% \quad \text{Write as a percent.}$$

71. First, convert 8.25% to its decimal form. $8.25\% = 0.0825$. A percent "of" means to multiply, so using the decimal form, calculate the answer.

$$8.25\%\cdot 45 = 0.0825(45)$$
$$= 3.7125$$

Therefore 8.25% of 45 is 3.7125.

72. First, convert 3.45% to its decimal form. $3.45\% = 0.0345$. A percent "of" means to multiply, so using the decimal form, calculate the answer.

$$3.45\%\cdot 9 = 0.0345(9)$$
$$= 0.3105$$

Therefore 3.45% of 9 is 0.3105.

73. First, convert 5.5% to its decimal form. $5.5\% = 0.055$. A percent "of" means to multiply, so using the decimal form, calculate the answer.

$$5.5\%\cdot 20 = 0.055(20)$$
$$= 1.1$$

Therefore 5.5% of 20 is 1.1.

74. First, convert 10.2% to its decimal form. $10.2\% = 0.102$. A percent "of" means to multiply, so using the decimal form, calculate the answer.

$$\begin{aligned}10.2\% \cdot 16 &= 0.102(16)\\ &= 1.632\end{aligned}$$

Therefore 10.2% of 16 is 1.632.

75. This problem asks us to find 8.5% of \$235.

$$\begin{aligned}8.5\% \cdot 235 &= 0.085(235)\\ &= 19.975\end{aligned}$$

The amount of tax on a \$235 purchase at a tax rate of 8.5% would be \$19.98.

76. The rational numbers in the set are as follows: $\frac{4}{3}, -16.9, 0$.

The number $\sqrt{5}$ cannot be expressed as an integer, finite decimal, or infinite repeating decimal therefore it is irrational.

77. The rational numbers in the set are as follows: $-7.5,\ 2,\ \frac{5}{4}$.

The number π cannot be expressed as an integer, finite decimal, or infinite repeating decimal therefore it is irrational.

78. The irrational number in the set is $\sqrt{5}$. This cannot be expressed as an integer, finite decimal, or infinite repeating decimal therefore it is irrational.

79. The irrational number in the set is π. This cannot be expressed as an integer, finite decimal, or infinite repeating decimal therefore it is irrational.

80. Using a calculator, $\frac{3}{7} \approx 0.4285714285\ldots$. To round to the hundredths place, underline the hundredths place: $0.4\underline{2}85714285\ldots$. The digit to the right of the 2 in the hundredths place is 8, so round up. Therefore, $\frac{3}{7} \approx 0.43$.

81. Using a calculator, $\sqrt{21} \approx 4.5825756949\ldots$. To round to the thousandths place, underline the thousandths place: $4.58\underline{2}5756949\ldots$. The digit to the right of the 2 in the thousandths place is 5, so round up. Therefore, $\sqrt{21} \approx 4.583$.

82. This expression has 3 terms. The terms are outlined. This expression contains both fractions and decimals.

$$\begin{aligned}&\boxed{\frac{1}{2} \div 5} + \boxed{5.2}\boxed{-7.9} && \text{Divide in first term.}\\ &= \frac{1}{2} \cdot \frac{1}{5} + 5.2 - 7.9 && \text{Invert 2nd fraction \& multiply.}\\ &= \frac{1}{10} + 5.2 - 7.9 && \text{Convert fraction to decimal.}\\ &= 0.1 + 5.2 - 7.9 && \text{Add/subtract from left to right.}\\ &= -2.6\end{aligned}$$

83. This expression has 3 terms. The terms are outlined. This expression contains both fractions and decimals.

$$\begin{aligned}&\boxed{6 \div \frac{3}{5}}\boxed{-5.69} + \boxed{8.4} && \text{Divide in first term.}\\ &= \frac{6}{1} \cdot \frac{5}{3} - 5.69 + 8.4 && \text{Invert 2nd fraction \& multiply.}\\ &= \frac{\cancel{3} \cdot 2 \cdot 5}{1 \cdot \cancel{3}} - 5.69 + 8.4 && \text{Factor and reduce fraction.}\\ &= 10 - 5.69 + 8.4 && \text{Add/subtract from left to right.}\\ &= 12.71\end{aligned}$$

84. This expression has 3 terms. The terms are outlined.

$$\begin{aligned}&\boxed{\frac{3}{5} \div \frac{9}{10}}\boxed{-\frac{3}{2}} + \boxed{4} && \text{Divide in first term.}\\ &= \frac{3}{5} \cdot \frac{10}{9} - \frac{3}{2} + 4 && \text{Invert 2nd fraction \& multiply.}\\ &= \frac{3 \cdot 10}{5 \cdot 9} - \frac{3}{2} + 4 \\ &= \frac{\cancel{3} \cdot \cancel{5} \cdot 2}{\cancel{5} \cdot \cancel{3} \cdot 3} - \frac{3}{2} + 4 && \text{Factor and reduce fraction.}\\ &= \frac{2}{3} - \frac{3}{2} + 4 && \text{Find LCD = 6.}\\ &= \frac{2 \cdot 2}{3 \cdot 2} - \frac{3 \cdot 3}{2 \cdot 3} + \frac{4 \cdot 6}{1 \cdot 6} && \text{Re write fractions over LCD.}\\ &= \frac{4}{6} - \frac{9}{6} + \frac{24}{6} && \text{Add/subtract.}\\ &= \frac{19}{6} \text{ or } 3\frac{1}{6} && \text{Fraction is in lowest terms.}\end{aligned}$$

85. This expression has 3 terms. The terms are outlined.

$$\boxed{7}\boxed{-\frac{4}{7}\div\frac{1}{14}}+\boxed{\frac{1}{3}}$$ Divide in 2nd term.

$$=7-\frac{4}{7}\cdot\frac{14}{1}+\frac{1}{3}$$ Invert 2nd fraction & multiply.

$$=7-\frac{4\cdot 14}{7\cdot 1}+\frac{1}{3}$$

$$=7-\frac{\not{7}\cdot 2\cdot 2\cdot 2}{\not{7}}+\frac{1}{3}$$ Factor and reduce fraction.

$$=7-8+\frac{1}{3}$$ Subtract first two terms.

$$=-1+\frac{1}{3}$$ Find LCD = 3.

$$=-\frac{1\cdot 3}{1\cdot 3}+\frac{1}{3}$$ Re write fractions over LCD.

$$=-\frac{3}{3}+\frac{1}{3}$$ Add/subtract.

$$=-\frac{2}{3}$$ Fraction is in lowest terms.

86. This expression has 2 terms. The terms are outlined. This expression contains both fractions and decimals.

$$\boxed{0.35}+\boxed{\frac{5}{4}\cdot\frac{3}{125}}$$ Multiply in second term.

$$=0.35+\frac{5\cdot 3}{4\cdot 125}$$ Invert 2nd fraction & multiply.

$$=0.35+\frac{\not{5}\cdot 3}{\not{5}\cdot 5\cdot 5\cdot 2\cdot 2}$$ Factor and reduce.

$$=0.35+\frac{3}{100}$$ Convert fraction to decimal.

$$=0.35+0.03$$ Add.

$$=0.38$$

87. This expression has 2 terms. The terms are outlined. This expression contains both fractions and decimals.

$$\boxed{1.27}\boxed{-\frac{2}{10}\cdot\frac{3}{20}}$$ Multiply in second term.

$$=1.27-\frac{2\cdot 3}{10\cdot 20}$$ Invert 2nd fraction & multiply.

$$=1.27-\frac{\not{2}\cdot 3}{5\cdot 5\cdot\not{2}\cdot 2\cdot 2}$$ Factor and reduce.

$$=1.27-\frac{3}{100}$$ Convert fraction to decimal.

$$=1.27-0.03$$ Subtract.

$$=1.24$$

88. This expression has 3 terms. The terms are outlined. This expression contains both fractions and decimals.

$$\boxed{44.2\div 2}+\boxed{5\cdot 7.1}\boxed{-5\div\frac{5}{2}}$$ Divide in 1st & 3rd terms.

$$=22.1+5\cdot 7.1-\frac{5}{1}\cdot\frac{2}{5}$$ Invert 2nd fraction in 3rd term & multiply.

$$=22.1+5\cdot 7.1-\frac{\not{5}\cdot 2}{1\cdot\not{5}}$$ Factor and reduce.

$$=22.1+5\cdot 7.1-2$$ Multiply in 2nd term.

$$=22.1+35.5-2$$ Add/subtract from left to right.

$$=55.6$$

89. This expression has 3 terms. The terms are outlined.

$$\boxed{25.15\div 5}+\boxed{7\div 0.2}\boxed{-6\div 12}$$ Divide in each term.

$$=5.03+35-0.5$$ Add/subtract from left to right.

$$=39.53$$

90. This expression has 1 term.

$$\boxed{\frac{4}{5}\div\frac{8}{15}\cdot\frac{4}{9}}$$ Divide first.

$$=\frac{4}{5}\cdot\frac{15}{8}\cdot\frac{4}{9}$$ Invert 2nd fraction & multiply.

$$=\frac{4\cdot 15\cdot 4}{5\cdot 8\cdot 9}$$ Multiply numerators and denominators.

$$=\frac{\not{5}\cdot\not{3}\cdot\not{2}\cdot\not{2}\cdot\not{2}\cdot 2}{\not{5}\cdot\not{3}\cdot 3\cdot\not{2}\cdot\not{2}\cdot\not{2}}$$ Factor and reduce.

$$=\frac{2}{3}$$ Fraction is in lowest terms.

91. This expression has 1 term.

$$\boxed{3\div\frac{6}{11}\cdot\frac{2}{5}}$$ Divide first.

$$=\frac{3}{1}\cdot\frac{11}{6}\cdot\frac{2}{5}$$ Invert 2nd fraction & multiply.

$$=\frac{3\cdot 11\cdot 2}{1\cdot 6\cdot 5}$$ Multiply numerators and denominators.

$$=\frac{11\cdot\not{3}\cdot\not{2}}{5\cdot\not{3}\cdot\not{2}}$$ Factor and reduce.

$$=\frac{11}{5}\text{ or }2\frac{1}{5}$$ Fraction is in lowest terms.

Chapter R Test

1.

	Whole Number	Integer	Rational Number	Irrational Number	Real Number
2.25			X		X
$\frac{2}{3}$			X		X
−6		X	X		X
$\sqrt{2}$				X	X
10	X	X	X		X

2.25 is a finite decimal, therefore it is a rational number which also belongs to the set of real numbers.

$\frac{2}{3}$ is a fraction, a ratio of integers, therefore it is a rational number which also belongs to the set of real numbers.

-6 belongs to the set of integers, therefore it also belongs to the set of rational numbers and the set of real numbers.

$\sqrt{2}$ cannot be expressed as an integer, finite decimal, or infinite repeating decimal, therefore it is an irrational number which also belongs to the set of real numbers.

10 belongs to the set of whole numbers, therefore it also belongs to the sets of integers, rational numbers, and real numbers.

2. The tick marks on the number line are 5 units apart, therefore the scale of this number line is 5.

3.

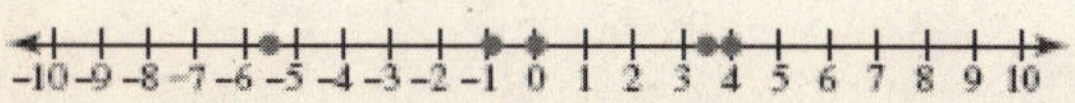

4. First, find any absolute values or simplify terms.

$-4 = -4$

$\frac{13}{5} = 2\frac{3}{5}$

$-(-6.8) = 6.8$

$|-11| = 11$

In order: $-4,\ \frac{13}{5},\ -(-6.8),\ |-11|$

5. **a.** The number -9 is to the right of -11 on the number line, therefore -9 is greater than -11.

$-9 > -11$

b. To compare a fraction and a decimal it is helpful to convert them to the same form. We will convert the decimal 0.3 to a fraction

$0.3 = \frac{3}{10}$

Now to compare the two fractions we will find the LCD and rewrite the fractions so we can compare numerators.

$\frac{3}{10} \quad \frac{1}{3}$ Find LCD = 30.

$\frac{3 \cdot 3}{10 \cdot 3} \quad \frac{1 \cdot 10}{3 \cdot 10}$ Rewrite fractions over LCD.

$\frac{9}{30} \quad \frac{10}{30}$

We now compare the numerators. The number $\frac{10}{30}$ (which is $\frac{1}{3}$) is to the right of $\frac{9}{30}$ (which is 0.3) on the number line, therefore $\frac{1}{3}$ is greater than 0.3.

$\frac{1}{3} > 0.3$

6. **a.** The expression translates as, "the opposite of the opposite of negative three."

$-(-(-3)) = -3$

b. $|-6| = 6$ since the distance between -6 and 0 is 6 units.

7.

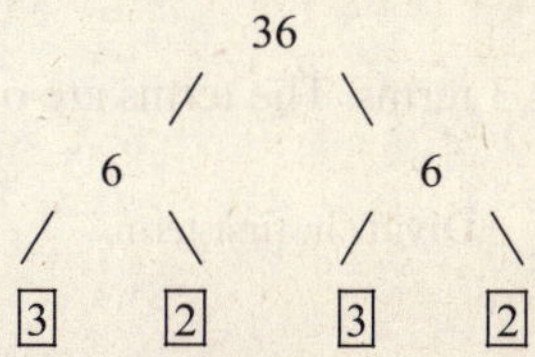

Therefore, $36 = 3 \cdot 3 \cdot 2 \cdot 2$.

8. **a.**

$-5-(-9) = -5+9$

$9-5 = 4$

b. This expression has 3 terms.

$-0.3-1.5+0.75$ Combine first two terms.
$= -0.3+(-1.5)+0.75$ Change sign and add.
$= -1.8+0.75$ Add/subtract.
$= -1.05$

c. This expression has 3 terms.

$-2+(-8)-(-6)$ Combine first two terms.
$= -10-(-6)$ Change sign and add.
$= -10+6$ Add/subtract.
$= -4$

9. **a.** $(-4)(-6)(-2)=-48$

b. $(-36)\div(-4)=9$

c. $(32.2)\div(3.5)=9.2$

10. This expression has 4 terms. The terms are outlined.

$\boxed{-6}+\boxed{3\cdot 4}\boxed{-(-5)}+\boxed{12\div(-3)}$ Multiply in 2nd term.

$=-6+12-(-5)+12\div(-3)$ Divide in last term.

$=-6+12-(-5)-4$ Change sign and add 3rd term.

$=-6+12+5-4$ Add/subtract from left to right.

$=7$

11. This expression has 3 terms. The terms are outlined.

$\boxed{24\div 2\cdot 3}+\boxed{(-2)}\boxed{-(-1)}$ Divide in 1st term.

$=12\cdot 3+(-2)-(-1)$ Now multiply in 1st term.

$=36+(-2)-(-1)$ Change sign and add 3rd term.

$=36+(-2)+1$ Add/subtract from left to right.

$=35$

12. This expression has 2 terms. The terms are outlined.

$\boxed{(-64)\div 16\cdot 5}\boxed{-9}$ Divide in 1st term.

$=-4\cdot 5-9$ Now multiply in 1st term.

$=-20-9$ Subtract.

$=-29$

13. This expression has 3 terms. The terms are outlined.

$\boxed{|-18|}\boxed{-(-5)(-7)}+\boxed{36\cdot 4}$ Evaluate absolute value.

$=18-(-5)(-7)+36\cdot 4$ Multiply in 2nd/3rd terms.

$=18-35+144$ Add/subtract from left to right.

$=127$

14. This expression has 3 terms. The terms are outlined.

$\boxed{\frac{6}{5}\div\frac{2}{15}}\boxed{-\frac{1}{3}}+\boxed{2}$ Divide in first term.

$=\frac{6}{5}\cdot\frac{15}{2}-\frac{1}{3}+2$ Invert 2nd fraction & multiply.

$=\frac{6\cdot 15}{5\cdot 2}-\frac{1}{3}+2$

$=\frac{\cancel{5}\cdot 3\cdot 3\cdot\cancel{2}}{\cancel{5}\cdot\cancel{2}}-\frac{1}{3}+2$ Factor and reduce.

$=9-\frac{1}{3}+2$ Add the whole numbers.

$=11-\frac{1}{3}$ Find LCD = 3.

$=\frac{11\cdot 3}{1\cdot 3}-\frac{1}{3}$ Re write fractions over LCD.

$=\frac{33}{3}-\frac{1}{3}$ Subtract.

$=\frac{32}{3}$ or $10\frac{2}{3}$ Fraction is in lowest terms.

15. This expression has 3 terms. The terms are outlined.

$\boxed{(1.2)(3.5)}\boxed{-6.4}+\boxed{7}$ Multiply in first term.

$=4.2-6.4+7$ Add/subtract from left to right.

$=4.8$

16.

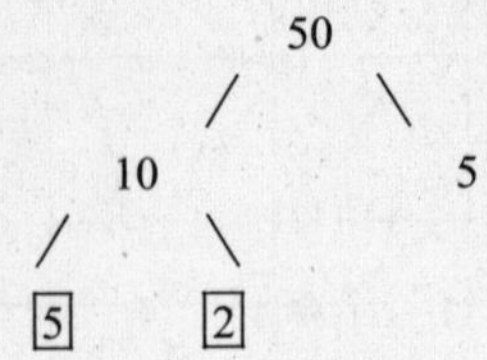

Therefore, $50=5\cdot 5\cdot 2$.

17. $18=\boxed{3}\cdot\boxed{3}\cdot\boxed{2}$

$36=\boxed{3}\cdot\boxed{3}\cdot\boxed{2}\cdot 2$

The $\text{GCF}=3\cdot 3\cdot 2=18$

18. Find the GCF for the numerator and denominator then factor out the GCF.

$\frac{48}{56}=\frac{\cancel{8}\cdot 6}{\cancel{8}\cdot 7}=\frac{6}{7}$

19. Comparing the denominators of 7 and 63, we have that $7\cdot 9=63$. Therefore, we multiply the numerator and denominator by 9.

$\frac{2\cdot 9}{7\cdot 9}=\frac{18}{63}$

20.

$\frac{3}{5}+\frac{12}{5}$ Add the numerators.

$=\frac{15}{5}$ Reduce to lowest terms.

$=3$

21. The factorization of 4 is $4=2\cdot 2$ and 5 is prime. The LCD $=5\cdot 2\cdot 2=20$.

$\frac{9}{5}-\frac{3}{4}$

$=\frac{9\cdot 4}{5\cdot 4}-\frac{3\cdot 5}{4\cdot 5}$ Rewrite each fraction over LCD.

$=\frac{36}{20}-\frac{15}{20}$ Subtract numerators.

$=\frac{21}{20}$ or $1\frac{1}{20}$ Reduce to lowest terms.

22. **a.**

$\frac{-2}{5}\cdot\frac{15}{4}$

$=\frac{-2\cdot 15}{5\cdot 4}$ Multiply numerators & denominators.

$=\frac{-\cancel{2}\cdot\cancel{5}\cdot 3}{\cancel{5}\cdot 2\cdot\cancel{2}}$ Factor. Divide out like factors.

$=-\frac{3}{2}$ or $-1\frac{1}{2}$

b.

$$\frac{3}{4} \div \frac{-9}{16} \quad \text{Invert 2nd fraction \& multiply}$$
$$= \frac{3}{4} \cdot -\frac{16}{9}$$
$$= -\frac{\not{3} \cdot \not{2} \cdot \not{2} \cdot 2 \cdot 2}{\not{3} \cdot 3 \cdot \not{2} \cdot \not{2}} \quad \text{Factor. Divide out like factors.}$$
$$= -\frac{4}{3} \text{ or } -1\frac{1}{3}$$

23. 6.02$\underline{4}$78

The digit 4 is in the thousandths place.

24. Round 5.02317 to the thousandths place.

5.02$\underline{3}$17 Underline the thousandths place.
The next digit to the right is 1.
5.023 Round down.

25. To convert 1.05% to a decimal, move the decimal point two places to the left. 1.05% = 0.0105

To convert 0.0105 to a fraction, note the place value of the last digit after the decimal, which is the ten-thousandths place.

$$0.0105 = \frac{105}{10000} \quad \text{Rewrite as a fraction.}$$
$$= \frac{21 \cdot \not{5}}{2000 \cdot \not{5}} \quad \text{Factor and reduce.}$$
$$= \frac{21}{2000}$$

So, the completed table is:

Percent	Decimal	Fraction
1.05%	0.0105	$\frac{21}{2000}$

26. First, convert 9% to its decimal form. 9% = 0.09 . A percent "of" means to multiply, so using the decimal form, calculate the answer.

$$9\% \cdot 206 = 0.09(206)$$
$$= 18.54$$

Therefore 9% of 206 is 18.54.

CHAPTER 1 BUILDING BLOCKS OF ALGEBRA

Section 1.1 Exponents, Order of Operations, and Properties of Real Numbers

1. $7\cdot7\cdot7\cdot7\cdot7=7^5$

3. $5^3=5\cdot5\cdot5$

5. $(-7)^3=(-7)\cdot(-7)\cdot(-7)$

7. $(-3)\cdot(-3)\cdot(-3)\cdot(-3)\cdot(-3)=(-3)^5$

9. $\frac{2}{3}\cdot\frac{2}{3}\cdot\frac{2}{3}=\left(\frac{2}{3}\right)^3$

11. $\left(\frac{5}{8}\right)^3=\frac{5}{8}\cdot\frac{5}{8}\cdot\frac{5}{8}$

13. $2^3=2\cdot2\cdot2=8$

15. $(-2)^4=-2\cdot-2\cdot-2\cdot-2=16$

17. In the expression -5^2 the negative is not being raised to the power of 2 so the negative will remain and the result will still be negative. $-5^2=-25$

19. Any number raised to the zero power is 1 except zero to the zero power. Therefore, $15^0=1$.

21. $3^7=2187$.

23. $(-5)^2=25$.

25. $-7^4=-2401$.

27. $(-1)^3=-1$.

29. $\left(\frac{3}{4}\right)^3=\frac{27}{64}$.

31. $(0.35)^2=0.1225$.

33. $-6^3=-216$.

35. $1.034\times10^6=1.034\cdot1{,}000{,}000=1{,}034{,}000$

1.034×10^6 has the standard form $1{,}034{,}000$.

37. $9.3\times10^7=9.3\cdot10{,}000{,}000=93{,}000{,}000$

9.3×10^7 has the standard form $93{,}000{,}000$.

39. $4.001\times10^9=4.001\cdot1{,}000{,}000{,}000$
$=4{,}001{,}000{,}000$

4.001×10^9 has the standard form $4{,}001{,}000{,}000$.

41. $-5.487\times10^4=-5.487\cdot10{,}000=54{,}870$

-5.487×10^4 has the standard form $54{,}870$.

Terms are outlined in boxes in problems 43-51.

43. This expression has 3 terms.

$\boxed{4\div2}+\boxed{3\cdot27}\boxed{-|-27|}$

45. This expression has 2 terms. $\boxed{8\cdot9}\boxed{-4\cdot16^2}$

47. This expression has 2 terms. $\boxed{\frac{3}{7}}\boxed{-\frac{1}{2}\div8}$

49. This expression has 1 term. $\boxed{(-4)^5\cdot2\div32}$

51. This expression has 1 term. $\boxed{(-64)\div4\div2}$

53. Evaluating the expression using the order of operations would be as follows:

$\boxed{16+8\div2}$	Outline the terms.
$=16+8\div2$	Division first.
$=16+4$	Then addition.
$=20$	

55. Evaluating the expression using the order of operations would be as follows:

$\boxed{\frac{4}{5}}\boxed{-\frac{2}{3}\cdot\frac{9}{5}}$	Outline the terms.
$=\frac{4}{5}-\frac{2}{3}\cdot\frac{9}{5}$	Multiplication first.
$=\frac{4}{5}-\frac{6}{5}$	Subtraction next.
$=-\frac{2}{5}$	

57. Evaluating the expression using the order of operations would be as follows:

$\boxed{3\cdot12}\boxed{-16\div4}+\boxed{1}$	Outline the terms.
$=3\cdot12-16\div4+1$	Multiplication first.
$=36-16\div4+1$	Division next.
$=36-4+1$	Add/subtract from left to right.
$=33$	

59. Evaluating the expression using the order of operations would be as follows:

$\boxed{1}+\boxed{\frac{7}{10}\cdot\frac{4}{5}}\boxed{-\frac{2}{5}}$	Outline the terms.
$=1+\frac{7}{10}\cdot\frac{4}{5}-\frac{2}{5}$	Multiplication first.
$=1+\frac{14}{25}-\frac{2}{5}$	Add/subtract from left to right after finding LCD.
$=\frac{25}{25}+\frac{14}{25}-\frac{10}{25}$	
$=\frac{29}{25}$ *or* $1\frac{4}{25}$	

61. Evaluating the expression using the order of operations would be as follows:

$\boxed{(-2)^3}+\boxed{4\cdot(-6)}$	Outline the terms.
$=(-2)^3+4\cdot(-6)$	Exponent first.
$=-8+4\cdot(-6)$	Multiplication next.
$=-8+(-24)$	Then add/subtract.
$=-32$	

63. Evaluating the expression using the order of operations would be as follows:

$\boxed{-9}+\boxed{7\cdot3}\boxed{-15\div3}\boxed{-4^2}+\boxed{(-8)}$	Outline the terms.
$=-9+7\cdot3-15\div3-4^2+(-8)$	Exponent first.
$=-9+7\cdot3-15\div3-16+(-8)$	Multiplication next.
$=-9+21-15\div3-16+(-8)$	Then division.
$=-9+21-5-16+(-8)$	Then add/subtract from left to right.
$=-17$	

65. Evaluating the expression using the order of operations would be as follows:

$\boxed{-9\cdot3}+\boxed{5}\boxed{-15\div3}\boxed{-2^3}$	Outline the terms.
$=-9\cdot3+5-15\div3-2^3$	Exponent first.
$=-9\cdot3+5-15\div3-8$	Multiplication next.
$=-27+5-15\div3-8$	Then division.
$=-27+5-5-8$	Then add/subtract from left to right.
$=-35$	

67. Evaluating the expression using the order of operations would be as follows:

$\lvert16-(3\cdot8)\rvert+9$	Evaluate innermost grouping symbol first.
$=\lvert16-24\rvert+9$	Continue evaluating inside absolute value sign.
$=\lvert-8\rvert+9$	Evaluate the absolute value.
$=8+9$	Then add.
$=17$	

69. Evaluating the expression using the order of operations would be as follows:

$\frac{4\cdot5-4}{6-4}$	Multiplication first.
$=\frac{20-4}{6-4}$	Simplify top and bottom of fraction.
$=\frac{16}{2}$	Then divide.
$=8$	

71. Evaluating the expression using the order of operations would be as follows:

$3\cdot\frac{-4+6\cdot8}{15-2^2}-3^2$	Evaluate the exponent first.
$=3\cdot\frac{-4+6\cdot8}{15-2^2}-9$	Simplify top and bottom of fraction. Start w/multiplication on top and exponent on bottom.
$=3\cdot\frac{-4+48}{15-4}-9$	Finish simplifying the fraction.
$=3\cdot4-9$	Multiply.
$=12-9$	Then subtract.
$=3$	

73. Evaluating the expression using the order of operations would be as follows:

$(18-2\cdot3)^3$	Evaluate inside the parentheses before applying the exponent. Multiply first.
$=(18-6)^3$	Continue to evaluate inside the parentheses. Subtract.
$=(12)^3$	Evaluate the exponent.
$=1728$	

75. Evaluating the expression using the order of operations would be as follows:

$4\div(-3-14\div2)-(-2)^2$	Evaluate inside parentheses first. Divide.
$=4\div(-3-7)-(-2)^2$	Continue to evaluate inside parentheses. Subtract.
$=4\div(-10)-(-2)^2$	Evaluate the exponent.
$=4\div(-10)-4$	Divide.
$=-\frac{2}{5}-4$	Then subtract.
$=-4\frac{2}{5}$ or -4.4	

77. Evaluating the expression using the order of operations would be as follows:

$(49-7\cdot2)\div5+\frac{30-(-6)\cdot2}{-4+2}$	Evaluate inside parentheses-multiply.
$=(49-14)\div5+\frac{30-(-6)\cdot2}{-4+2}$	Continue to evaluate parentheses. Subtract.
$=35\div5+\frac{30-(-6)\cdot2}{-4+2}$	Next grouping symbol to evaluate-division bar. Multiply first.
$=35\div5+\frac{30-(-12)}{-4+2}$	Simplify top and bottom of fraction.
$=35\div5+\frac{42}{-2}$	Simplify the fraction.
$=35\div5-21$	Divide.
$=7-21$	Subtract.
$=-14$	

79. Evaluating the expression using the order of operations would be as follows:

$\lvert3-5\cdot6\rvert+(-3-5)^2$	Evaluate inside grouping symbols first. Multiply inside absolute value.
$=\lvert3-30\rvert+(-3-5)^2$	Subtract inside grouping symbols.
$=\lvert-27\rvert+(-8)^2$	Evaluate absolute value & exponent.
$=27+64$	Then add.
$=91$	

81. $3+(2+5)=(3+2)+5$

Associative property of addition: The order of the numbers was not changed, but how they are grouped was changed.

83. $\frac{3}{5}\cdot24=24\cdot\frac{3}{5}$

Communicative property of multiplication: The order of the numbers was changed.

85. $-3(14+7)=-3\cdot14+(-3)\cdot7$

Distributive property: each term inside the parentheses has been multiplied by the term outside the parentheses.

87. $(-11)+19=19+(-11)$

Communicative property of Addition: the order of the terms has been changed.

89. $3(9-2)=3\cdot9-3\cdot2$

Distributive property: each term inside the parentheses has been multiplied by the term outside the parentheses.

91. $(5\cdot3)\cdot10=5(3\cdot10)$

Associative property of multiplication: The order of the numbers was not changed, but how they are grouped was changed.

93.

Operation	Commutative Property	Associative Property
Addition	√	√
Subtraction		
Multiplication	√	√
Division		

95. Distribute the 2 inside the parentheses and multiply both terms by 2. $2(5+9)=2\cdot5+2\cdot9$

97. Distribute the 16 inside the parentheses and multiply both terms by 16. $16(9-5)=16\cdot9-16\cdot5$

99. Distribute the -2 inside the parentheses and multiply both terms by -2. $-2(8+7)=-2\cdot8+(-2)\cdot7$

101. Distribute the -2 inside the parentheses and multiply both terms by -2. $-2(7-16)=-2\cdot7-(-2)\cdot16$

103. Distribute the $\frac{7}{9}$ inside the parentheses and multiply both terms by $\frac{7}{9}$. $\frac{7}{9}(11+15)=\frac{7}{9}\cdot11+\frac{7}{9}\cdot15$.

105. This expression will be evaluated using the distributive property.

$6(3+17)=6(3)+6(17)$ First, distribute 6.
$=18+102$ Add.
$=120$

107. This expression will be evaluated using the distributive property.

$-5(6+27)=-5(6)+(-5)(27)$ Distribute -5.
$=-30-135$ Subtract.
$=-165$

109. This expression will be evaluated using the order of operations agreement.

$6(7-16)+12$ Evaluate inside parentheses first.
$=6(-9)+12$ Next, multiply.
$=-54+12$ Then add/subtract.
$=-42$

111. This expression will be evaluated using the order of operations agreement.

$-(7-28)-9$ Evaluate inside parentheses first.
$=-(-21)-9$ Next, apply negative outside parentheses.
$=21-9$ Then subtract.
$=12$

113. This expression will be evaluated using the order of operations agreement.

$\frac{1}{2}(5-17)$ First, evaluate inside the parentheses.
$=\frac{1}{2}(-12)$ Then multiply.
$=-6$

Section 1.2 Algebra and Working with Variables

1. Substitute 15 for h to get $11.25(15)=168.75$.
If Sarah works 15 hours in a week, her weekly pay is \$168.75.

3. Substitute 53 for p to get $5.00(53)=265$.
If Thomas sells 53 rolls of wrapping paper, he will earn \$265.00 in revenue.

5. Substitute 1045 for s to get $0.07(1045)=73.15$.
If Marcus sells \$1045.00 worth of merchandise in a week, his commission will be \$73.15.

7. Substitute 2.5 for h to get $7.5(2.5)=18.75$.
If Vinny runs for 2.5 hours at a pace of 7.5 miles per hour, he will run 18.75 miles.

9. Substitute 5 for l to get $250(5)=1250$.
If Judy is currently listing 5 houses, she estimates spending \$1,250.00 on advertising.

11. Substitute 6.5 for h to get $75(6.5)=487.5$.
If Fusako drives for 6.5 hours at an average speed of 75 miles per hour, she will drive 487.5 miles.

13. Evaluate $-6x+9$ for $x=4$.

$-6x+9$ Substitute in $x=4$.
$=-6(4)+9$ Simplify using order-of-operations agreement.
$=-24+9$
$=-15$

15. Evaluate $\frac{2}{5}n+7$ for $n=-10$.

$\frac{2}{5}n+7$ Substitute in $n=-10$.
$=\frac{2}{5}(-10)+7$ Simplify using order-of-operations agreement.
$=-4+7$
$=3$

17. Evaluate x^2-6x+7 for $x=-1$.

x^2-6x+7 Substitute in $x=-1$ for every x in the expression.
$=(-1)^2-6(-1)+7$ Simplify using the order-of-operations agreement.
$=1+6+7$
$=14$

19. Evaluate $-y^2+7$ for $y=-4$.

$-y^2+7$ Substitute in $y=-4$.
$=-(-4)^2+7$ Simplify using the order-of-operations agreement.
$=-16+7$
$=-9$

21. Evaluate $\frac{6y-3}{2y+8}$ for $y=5$.

$\frac{6y-3}{2y+8}$ Substitute in $y=5$ for every y in the expression.
$=\frac{6(5)-3}{2(5)+8}$ Simplify using order-of-operations agreement.
$=\frac{30-3}{10+8}$
$=\frac{27}{18}=\frac{3}{2}$ or $1\frac{1}{2}$

23. Evaluate $36x^2+4y^2$ for $x=-5$ and $y=-2$.

$36x^2+4y^2$ Substitute in $x=-5$ and $y=-2$.

$=36(-5)^2+4(-2)^2$ Simplify using order-of-operations agreement.

$=36(25)+4(4)$

$=900+16$

$=916$

25. Evaluate $-22a-8b$ for $a=10$ and $b=-10$.

$-22a-8b$ Substitute in $a=10$ and $b=-10$.

$=-22(10)-8(-10)$ Simplify using the order-of-operations agreement.

$=-220+80$

$=-140$

27. Evaluate $\dfrac{x-y}{x+2y}$ for $x=3$ and $y=-4$.

$\dfrac{x-y}{x+2y}$ Substitute in $x=3$ and $y=-4$.

$=\dfrac{(3)-(-4)}{(3)+2(-4)}$ Simplify using the order-of-operations agreement.

$=\dfrac{3+4}{3-8}$

$=-\dfrac{7}{5}$ or $-1\dfrac{2}{5}$

29. Evaluate $-9x+51$ for $x=-6$.

$-9x+51$ Substitute in $x=-6$.

$=-9(-6)+51$ Simplify.

$=54+51$

$=105$

31. Evaluate $2x^2-6x+5$ for $x=3$.

$2x^2-6x+5$ Substitute in $x=3$ for every x in the expression.

$=2(3)^2-6(3)+5$ Simplify using the order-of-operations agreement.

$=18-18+5$

$=5$

33. Convert 22 feet to yards. The unity fraction to convert feet to yards is $\dfrac{1\text{ yd}}{3\text{ ft}}$. We multiply the given quantity of 22 feet by the unity fraction and simplify as follows.

$$22\text{ ft}\cdot\frac{1\text{ yd}}{3\text{ ft}}$$

$$=\frac{22\text{ ft}}{1}\cdot\frac{1\text{ yd}}{3\text{ ft}}$$

$$=\frac{22\text{ yd}}{3}\approx 7.33\text{ yd}$$

Therefore, 22 ft is equal to approximately 7.33 yd.

35. Convert 1057 feet to miles. First, form the unity fraction. We want to convert to the units of miles, so miles will go in the numerator. We want to convert from the units of feet, so feet will go in the denominator. There are 5280 ft in 1 mi, so the unity fraction is $\dfrac{1\text{ mi}}{5280\text{ ft}}$. Multiply 1057 ft by this unity fraction and simplify the expression as follows.

$$1057\text{ ft}\cdot\frac{1\text{ mi}}{5280\text{ ft}}$$

$$=\frac{1057\text{ ft}}{1}\cdot\frac{1\text{ mi}}{5280\text{ ft}}$$

$$=\frac{1057\text{ mi}}{5280}\approx 0.2\text{ mi}$$

Therefore, 1057 feet is equal to approximately 0.2 miles.

37. Convert 254 feet to meters. First, form the unity fraction. We want to convert to the units of meters, so meters will go in the numerator. We want to convert from the units of feet, so feet will go in the denominator. There are 3.28 ft in 1 m, so the unity fraction is $\dfrac{1\text{ m}}{3.28\text{ ft}}$. Multiply 254 ft by this unity fraction and simplify the expression as follows.

$$254\text{ ft}\cdot\frac{1\text{ m}}{3.28\text{ ft}}$$

$$=\frac{254\text{ ft}}{1}\cdot\frac{1\text{ m}}{3.28\text{ ft}}$$

$$=\frac{254\text{ m}}{3.28}\approx 77.44\text{ m}$$

Therefore, 254 feet is equal to approximately 77.44 m.

39. Convert 7.89 kilograms to pounds. First, form the unity fraction. We want to convert to the units of pounds, so pounds will go in the numerator. We want to convert from the units of kilograms, so kilograms will go in the denominator. There are 2.2 lb in 1 kg, so the unity fraction is $\dfrac{2.2\text{ lb}}{1\text{ kg}}$. Multiply 7.89 kg by this unity fraction and simplify the expression as follows.

$$7.89\text{ kg}\cdot\frac{2.2\text{ lb}}{1\text{ kg}}=\frac{7.89\text{ kg}}{1}\cdot\frac{2.2\text{ lb}}{1\text{ kg}}=17.358\text{ lb}$$

Therefore, 7.89 kg is equal to approximately 17.36 lb.

41. Convert 15 gallons to liters. First, form the unity fraction. We want to convert to the units of liters, so liters will go in the numerator. We want to convert from the units of gallons, so gallons will go in the denominator. There are 3.785 liters in 1 gal, so the unity fraction is $\dfrac{3.785\ l}{1\text{ gal}}$. Multiply 50 gallons by this unity fraction and simplify the expression as follows.

$15 \text{ gal} \cdot \dfrac{3.785\ l}{1 \text{ gal}}$

$= \dfrac{15 \cancel{\text{gal}}}{1} \cdot \dfrac{3.785\ l}{1 \cancel{\text{gal}}}$

$= 56.775\ l$

Therefore, 15 gal is equal to approximately 56.78 l.

43. Convert 22,000 square feet to acres. First, form the unity fraction. We want to convert to the units of acres, so acres will go in the numerator. We want to convert from the units of square feet, so square feet will go in the denominator. There are 43,560 ft^2 in 1 acre, so the unity fraction is $\dfrac{1 \text{ acre}}{43{,}560 \text{ ft}^2}$. Multiply 22,000 sq ft by this unity fraction and simplify the expression as follows.

$22{,}000 \text{ ft}^2 \cdot \dfrac{1 \text{ acre}}{43{,}560 \text{ ft}^2}$

$= \dfrac{22{,}000 \cancel{\text{ft}^2}}{1} \cdot \dfrac{1 \text{ acre}}{43{,}560 \cancel{\text{ft}^2}}$

$= \dfrac{22{,}000 \text{ acre}}{43{,}560} \approx 0.505 \text{ acre}$

Therefore, 22,000 sq ft is equal to approximately 0.51 acre.

45. Convert 601 million to thousands. First, form the unity fraction. We want to convert to the units of thousands, so thousands will go in the numerator. We want to convert from the units of millions, so millions will go in the denominator. There are 1000 thousands in 1 million, so the unity fraction is $\dfrac{1000 \text{ thou}}{1 \text{ mill}}$. Multiply 601 million by this unity fraction and simplify the expression as follows.

$601 \text{ mill} \cdot \dfrac{1000 \text{ thou}}{1 \text{ mill}}$

$= \dfrac{601 \cancel{\text{mill}}}{1} \cdot \dfrac{1000 \text{ thou}}{1 \cancel{\text{mill}}}$

$= 601{,}000 \text{ thou}$

Therefore, 601 million is equal to 601,000 thousands.

47. Convert 5397 million to billions. First, form the unity fraction. We want to convert to the units of billions, so billions will go in the numerator. We want to convert from the units of millions, so millions will go in the denominator. There are 1000 millions in 1 billion, so the unity fraction is $\dfrac{1 \text{ bill}}{1000 \text{ mill}}$. Multiply 5397 million by this unity fraction and simplify the expression as follows.

$5397 \text{ mill} \cdot \dfrac{1 \text{ bill}}{1000 \text{ mill}}$

$= \dfrac{5397 \cancel{\text{mill}}}{1} \cdot \dfrac{1 \text{ bill}}{1000 \cancel{\text{mill}}}$

$= \dfrac{5397 \text{ bill}}{1000} = 5.397 \text{ bill}$

Therefore, 5397 million is equal to 5.397 billion.

49. Convert 9 trillion to billions. First, form the unity fraction. We want to convert to the units of billions, so billions will go in the numerator. We want to convert from the units of trillions, so trillions will go in the denominator. There are 1000 billions in 1 trillion, so the unity fraction is $\dfrac{1000 \text{ bill}}{1 \text{ trill}}$. Multiply 9 trillion by this unity fraction and simplify the expression as follows.

$9 \text{ trill} \cdot \dfrac{1000 \text{ bill}}{1 \text{ trill}}$

$= \dfrac{9 \cancel{\text{trill}}}{1} \cdot \dfrac{1000 \text{ bill}}{1 \cancel{\text{trill}}}$

$= 9000 \text{ bill}$

Therefore, 9 trillion is equal to 9000 billion.

51. There are two quantities that can change. One quantity is the amount owed on a monthly electric bill and the second is the number of kilowatt hours of electricity used in the month. Let E = the amount owed in dollars on a monthly electric bill and k = the number of kilowatt hours of electricity used during that month.

53. There are two quantities that can change. One quantity is the amount you pay yearly in federal taxes and the second is your income for the year. Let T = the amount you pay yearly (in dollars) in federal taxes and d = your income (in dollars) for that year.

55. There are two quantities that can change. One quantity is the temperature outdoors and the second is the time of day. Let F = the temperature outdoors in degrees Fahrenheit and t = the time of day.

57. There are two quantities that can change. One quantity is the amount of first-class postage you owe for a package and the second is the weight of the package you are shipping. Let p = the amount of first-class postage you owe (in dollars) for the package and w = the weight of the package you are shipping (in pounds).

59. There are two quantities that can change. One quantity is the cost of attending a community college and the second is the number of credits a student enrolls in per semester. Let c = the cost in dollars of attending a community college and u = the number of credits a student enrolls in per semester.

61. Let x = a number. "Sum" translates to addition, so the sentence translates as $x + 62$ or $62 + x$.

63. Let x = a number. "More than" translates to addition and twice a number means to multiply a number by 2. The sentence translates as $2x+9$ or $9+2x$.

65. Let x = a number. Quotient means to divide, and order matters. We divide 16 by a number. Therefore, the translation is $\frac{16}{x}$.

67. Let x = a number. "Less than" translates as subtraction, but it is done in reverse order. So this translates as something minus 7. "$\frac{1}{3}$ times a number" means to multiply a number by $\frac{1}{3}$ so this translates as $\frac{1}{3}x$ and the sentence translates as $\frac{1}{3}x-7$.

69. Let x = a number. Difference means to subtract, and order matters. "8 times a number" means to multiply a number by 8 so this translates as $8x$, then we subtract 12. Therefore, the translation is $8x-12$.

71. Let x = a number. Product means to multiply so we are going to multiply -5 by the square of a number. The "square of a number" means to multiply a number by itself so this translates as x^2, then we multiply by -5. Therefore, the translation is $-5x^2$.

73. Let x = a number. "Increased by" translates to addition so we are going to add something to 16. Quotient means to divide, and order matters. The quotient of a number and 4 means to divide a number by 4, which translates as $\frac{x}{4}$, and this is what we will add to 16. Therefore the translation is $16+\frac{x}{4}$.

75. Let x = a number. Quotient means to divide, and order matters. We will divide 6 by "the difference of a number and 2." "Difference" translates to subtraction and the order matters, so the "difference of a number and 2" translates to $x-2$. This quantity will be in the denominator. Therefore, the full translation is $\frac{6}{x-2}$.

77. Let x = a number. "Half a number" means to divide a number by 2, or multiply the number by $\frac{1}{2}$. So "half a number" translates to $\frac{1}{2}x$. This is added to 6 and the full translation is $6+\frac{1}{2}x$.

79. Let x = a number. "Difference" translates to subtraction and the order matters, so we will subtract something from 7. "Twice a number" means to multiply a number by 2 which translates to $2x$. Therefore, the full translation is $7-2x$.

81. There is one quantity that can change. This variable quantity is the number of credit hours a part-time resident student is enrolled in for one semester. Let c = the number of credit hours a part-time resident student is enrolled in for one semester. The school charges \$132 per credit so the charge for taking c credits is $132c$. The school also charges \$7 per credit for the activity fee so the student will also be charged $7c$. There is an additional technology fee of \$7 per credit and the student will be charged an additional $7c$. There is an additional flat fee of \$4 per semester. The total for all of these charges is $132c+7c+7c+4$.

83. There is one quantity that can change. This quantity is the number of hours charged for labor. Let L = the number of hours charged for labor. The house painting company charges \$17 per hour for labor which translates to $17L$, plus the cost for paint which is \$175. The total cost for painting the house will be $175+17L$.

85. This problem can be written in two different ways. There is one quantity that can change, which is the number of ounces that a piece of first class mail weighs in excess of 1 ounce. One way to write an expression for the cost of mailing a letter is to let x = the number of ounces a piece of first class mail weighs *in excess of 1 ounce*. The cost is \$0.44 for the first ounce plus \$0.17 for each additional ounce. The expression representing this is $0.44+0.17x$. Another way to write the expression is to let z = the number of ounces a piece of first class mail weighs in total. The cost for the first ounce is \$0.44 plus \$0.17 for each additional ounce. Since z represents the total weight, and the additional cost of \$0.17 per ounce applies only to the number of ounces in addition to the first ounce, we would subtract 1 ounce from the total weight z, then multiply by \$0.17 and add the initial cost of \$0.44. The expression representing this is $0.44+0.17(z-1)$.

87. Let n = the number of pounds.

$$n\text{ lb}=\frac{n\ \cancel{\text{lb}}}{1}\cdot\frac{1\text{ kg}}{2.2\ \cancel{\text{lb}}}=\frac{n}{2.2}\text{ kg}$$

Notice that the units of the final output are in kilograms.

89. Let n = the number of miles.

$$n\text{ miles}=\frac{n\ \cancel{\text{miles}}}{1}\cdot\frac{5280\text{ ft}}{1\ \cancel{\text{mile}}}=5280n\text{ feet}$$

Notice that the units of the final output are in feet.

91. Let n = the number of liters.

$$n\text{ l}=\frac{n\ \cancel{\text{l}}}{1}\cdot\frac{1\text{ gal}}{3.785\ \cancel{\text{l}}}=\frac{n}{3.785}\text{ gal}$$

Notice that the units of the final output are in gallons.
Notice that the units of the final output are in gallons.

93. Let n = the number of acres.

$$n\text{ acres}=\frac{n\ \cancel{\text{acres}}}{1}\cdot\frac{43{,}560\text{ ft}^2}{1\ \cancel{\text{acre}}}=43{,}560n\text{ ft}^2$$

Notice that the units of the final output are in square feet.

95. Let n = the number of seconds.

$n \text{ sec} = \frac{n \cancel{\text{sec}}}{1} \cdot \frac{1 \text{ min}}{60 \cancel{\text{sec}}} = \frac{n}{60} \text{ min}$

Notice that the units of the final output are in minutes.

97. **a.** To find the area of a rectangle we multiply length times width. The area is
$(20 \text{ feet})(15 \text{ feet}) = 300 \text{ feet}^2$

The area of Bill's living room is 300 feet^2. The units are in square feet, or $\text{ft}^2 = \text{feet}^2$.

b. To convert units of square feet into square yards we would multiply the quantity of square feet by this unity fraction $\frac{1 \text{ yd}^2}{9 \text{ ft}^2}$. The conversion is as follows:

$\frac{300 \cancel{\text{ft}^2}}{1} \cdot \frac{1 \text{ yd}^2}{9 \cancel{\text{ft}^2}} = \frac{300}{9} \text{yd}^2 \approx 33.33 \text{ yd}^2$

Therefore, 300 ft^2 converts to 33.33 yd^2, but rounded up to the nearest full square yard it would be 34 yd^2.

c. The cost will be

$\left(\frac{34 \cancel{\text{yd}^2}}{1}\right)\left(\frac{\$19.38}{1 \cancel{\text{yd}^2}}\right) = \$658.92$

The cost for 34 square yards of the Berber carpet will be \$658.92.

d. The cost to ship the carpet will be

$\left(\frac{34 \cancel{\text{yd}^2}}{1}\right)\left(\frac{\$1.75}{1 \cancel{\text{yd}^2}}\right) = \$59.50$

The cost to ship Bill's carpet will be \$59.50.

e. To find the total cost of the carpet and shipping, we add the two amounts together.
$\$658.92 + \$59.50 = \$718.42$
The total cost of Bill's carpet and shipping will be \$718.42.

99. **a.** To find the total weight of the luggage, add the amounts of the 3 variable items, the weight of the clothing, the weight of the suitcase, and the weight of the toiletries.
$50 + 10 + 12 = 72$
Therefore, the total weight of the luggage is 72 lb.

b. To find the total weight of the luggage, add the amounts of the 3 variable items, the weight of the clothing, the weight of the suitcase, and the weight of the toiletries.
$54.6 + 10 + 9.6 = 74.2$
Therefore, the total weight of the luggage is 74.2 lb.

101.

Time to drive from San Diego to LA	Time to drive from LA to San Diego
1	$4.5 - 1 = 3.5$
2	$4.5 - 2 = 2.5$
2.5	$4.5 - 2.5 = 2$
3	$4.5 - 3 = 1.5$
t	$4.5 - t$

To find the amount of time it takes to drive the 2nd half of the round trip, subtract the amount of time it took to drive the 1st leg of the trip from 4.5 hours. The general variable expression is $4.5 - t$.

103.

Amount invested in 3.5% account	Amount invested in the 7% account
0	$2500 - 0 = 2500$
500	$2500 - 500 = 2000$
100	$2500 - 100 = 2400$
1500	$2500 - 1500 = 1000$
d	$2500 - d$

To find the amount of money Rebecca will invest in the 7% account, subtract the amount of money she invests in the 3.5% account from \$2500. The general variable expression is $2500 - d$.

105.

Pounds of Colombian coffee	Pounds of Kenyan coffee
0	$65 - 0 = 65$
10	$65 - 10 = 55$
20	$65 - 20 = 45$
50	$65 - 50 = 15$
c	$65 - c$

To find the number of pounds of Kenyan coffee needed to include in the blend, subtract the number of pounds of Columbian coffee from 65 pounds. The general variable expression is $65 - c$.

107. **a.**

p = # of people	Cost for food
25	$12 \cdot 25 = 300$
35	$12 \cdot 35 = 420$
45	$12 \cdot 45 = 540$
55	$12 \cdot 55 = 660$
p	$12p$

If we multiply \$12 times the number of people p at the party, that will give us an expression for the food cost for the party. The expression is $12p$.

b. To find the number of people Sandy can invite to the party we will use the table. Notice that the cost for food for 35 people is \$420 and the cost for 45 people is \$540. The budgeted amount of \$500 falls between these two values and is a bit closer to \$540. Extend the table to get an output value closer to \$500.

p = # of people	Cost for food
40	$12 \cdot 40 = 480$
41	$12 \cdot 41 = 492$
42	$12 \cdot 45 = 504$

The cost for 41 people is \$492 and the cost for 42 people is \$504. To stay within the budgeted amount of \$500, Sandy can invite 41 people to the party.

109. a. The catering company will charge \$18 per person for drinks and \$15 per person for appetizers.

p = # of people	Cost for drinks and appetizers
40	$18(40)+15(40)=1320$
45	$18(45)+15(45)=1485$
50	$18(50)+15(50)=1650$
55	$18(55)+15(55)=1815$
p	$18p+15p$

Therefore, the expression for the total cost of drinks and appetizers is $18p+15p$.

b. To find the number of people that can attend the conference reception and stay within Tricia's budget we will use the table. Notice that the cost for 50 people is \$1650 and the cost for 55 people is \$1815. The budgeted amount of \$1700 falls between these two values and is a bit closer to \$1650. Extend the table to get an output value closer to \$1700.

p = # of people	Cost for drinks and appetizers
51	$18(51)+15(51)=1683$
52	$18(52)+15(52)=1716$
53	$18(53)+15(53)=1749$

The cost for 51 people is \$1683 and the cost for 52 people is \$1716. To stay within the budgeted amount of \$1700, 51 people can attend the conference reception.

Section 1.3 Simplifying Expressions

1. $\boxed{2x}+8+\boxed{7x}$ There are 3 terms in this expression. The like terms are $2x$ and $7x$ because they both have x raised to the first power.

3. $\boxed{3}+\boxed{\boxed{2h}}\boxed{-5}+\boxed{\boxed{7h}}$ There are 4 terms in this expression. Both 3 and -5 are constant terms and are therefore like terms. The terms $2h$ and $7h$ are also like terms because they both have h raised to the first power.

5. $\boxed{\boxed{3x^2}}+\boxed{5x}+\boxed{\boxed{7x^2}}+\boxed{3x}$ There are 4 terms in this expression. The terms $3x^2$ and $7x^2$ are like terms because they both have x raised to the second power. The terms $5x$ and $3x$ are also like terms because they both have x raised to the first power.

7. $2x^2+5x-9$ There are three terms in this expression and there are no like terms.

9. $\boxed{7xy}+8x\boxed{-2xy}-3y$ There are 4 terms in this expression. The terms $7xy$ and $-2xy$ are like terms because they both have the same variables and each is raised to the same power.

11.

$13x+7x$ Combine like terms by adding coefficients.
$=20x$

13.

$3p-9p$ Combine like terms by subtracting coefficients.
$=-6p$

15.

$4x+5y-6x-y$ Rearrange terms using commutative property.
$=4x-6x+5y-y$ Combine like terms by adding/subtracting coefficients.
$=-2x+4y$

17.

$-18n-3+7n-5$ Rearrange terms using commutative property.
$=-18n+7n-3-5$ Combine like terms by adding/subtracting coefficients.
$=-11n-8$

19.

$3x^2+9x-1-4x^2+2x-5$ Rearrange terms using commutative property.
$=3x^2-4x^2+9x+2x-1-5$ Combine like terms by adding/subtracting coefficients.
$=-x^2+11x-6$

21. The expression is not in alphabetical order, reading left to right. Use the commutative property of addition to reorder the terms.
$-8y+3x$
$=3x-8y$

23. The expression is not in alphabetical order, reading left to right. Use the commutative property of addition to reorder the terms.
$7n-8m+12$
$=-8m+7n+12$

25. The expression is not written with descending exponents, reading left to right. Use the commutative property of addition to reorder the terms.

$5-x+7x^2$

$=7x^2-x+5$

27. The expression is not written with descending exponents, reading left to right. Use the commutative property of addition to reorder the terms.

$8+5x-7x^5-3x^2$

$=-7x^5-3x^2+5x+8$

29. This expression is not in alphabetical order and the exponents are not in descending order, reading left to right. Use the commutative property of addition to reorder the terms, first alphabetically, then in descending order.

$a-\frac{7}{5}b+3a^2-\frac{1}{2}a^6$

$=-\frac{1}{2}a^6+3a^2+a-\frac{7}{5}b$

31. **a.** It will cost $4.19 to buy each bird house kit, so multiply $4.19 times the number of children, k, to get the expression:

$4.19k$

b. It will cost $1.39 to buy each paint-with-brush set, so multiply $1.39 times k children to get the expression:

$1.39k$

c. "Total" means to add so add the expressions from part a and part b:

$4.19k+1.39k$

d. To simplify the expression from part c we combine like terms, adding coefficients:

$5.58k$

33. **a.** Julie needs supplies for k children and it will cost $8.79 to buy each pinewood derby car kit, so multiply $8.79 times the number of children, k, to get the expression:

$8.79k$

b. It will cost $1.19 to buy each child sandpaper and sealer, so multiply $1.19 times k children to get the expression:

$1.19k$

c. It will cost $1.39 to buy each child a paint-with-brush set, so multiply $1.39 times k children to get the expression:

$1.39k$

d. Add each of the expressions found in parts a, b, and c, then simplify as follows:

$8.79k+1.19k+1.39k$

$=11.37k$

35. **a.** If I represents Steve's taxable income, multiply I times 0.15 to find an expression for the amount Steve pays in federal taxes.

$0.15I$

b. As in part a, multiply I times 0.08 to find an expression for the amount Steve pays in state taxes.

$0.08I$

c. To find the total of federal and state taxes Steve pays, add the expressions from parts a and b, then simplify as follows:

$0.15I+0.08I$

$=0.23I$

37.

$3x+7+5x$ Combine like terms-add coefficients.

$=8x+7$

39.

$7t+9-3t+5$ Combine like terms-subtract coefficients. Add constant terms

$=4t+9+5$

$=4t+14$ Write in conventional form.

41.

$2x^2+3x+5x^2+6x$ Combine like terms-add coefficients.

$=7x^2+3x+6x$ Combine like terms-add coefficients.

$=7x^2+9x$ Write in conventional form.

43. $4k^2+5k-9$ There are no like terms; the expression is in simplest form.

45.

$2xy+5x-3xy+7y$ Combine like terms-add/subtract coefficients. Arrange in alphabetical order.

$=5x-xy+7y$

47.

$5a^2b^3+4ab^2+8a^2b^2+2ab^2$ Combine like terms by adding coefficients. Arrange in conventional form.

$=5a^2b^3+8a^2b^2+6ab^2$

49.

$s^5t^3+4s^3t^5+7s^2t+6s^5t^3+2s^2t$ Combine like terms-add coefficients. Arrange in conventional form.

$=7s^5t^3+4s^3t^5+9s^2t$

51. **a.** The distributive property will be used because there is more than one term in the parentheses.

b.

$2(3x+7)$ Distribute the 2 to both terms.

$=2(3x)+2(7)$ Multiply.

$=6x+14$

53. **a.** The multiplication property will be used because there is only one term in the parentheses.

b.

$5(2xy)$ Multiply.

$=10xy$

55. **a.** The distributive property will be used because there is more than one term in the parentheses.

b.

$-4(7x^2+9x-6)$ Distribute 4 to all terms.

$=-4(7x^2)+(-4)(9x)+(-4)(-6)$ Multiply.

$=-28x^2-36x+24$

57. **a.** The multiplication property will be used because there is only one term in the parentheses.

b.

$-5(4a^2bc)$ Multiply.

$=-20a^2bc$

59. **a.** The distributive property will be used because there is more than one term in the parentheses.

b.

$-2(-3xy+7y)$ Distribute -2 to both terms.

$=-2(-3xy)+(-2)(7y)$ Multiply.

$=6xy-14y$

61. **a.** The expression was simplified incorrectly. There is only one term, $3xy$, in the parentheses, therefore the term outside should be multiplied only by the coefficient 3.

b. The correct simplification is $6xy$.

63. **a.** The expression was simplified incorrectly. The term outside the parentheses, 3, should be multiplied by both terms inside the parentheses.

b. The correct simplification is as follows:

$3(2k-4)$ Distribute the 3 to both terms.

$=3(2k)+3(-4)$ Multiply.

$=6k-12$

65. **a.** The expression was simplified incorrectly. There is only one term, $3mn$, in the parentheses, therefore the term outside should be multiplied only by the coefficient 3.

b. The correct simplification is $21mn$.

67. **a.** The expression was simplified correctly using the multiplication property.

b. n/a

69. **a.** The expression was simplified incorrectly. The term outside the parentheses, $\frac{2}{3}$, should be multiplied by both terms inside the parentheses.

b. The correct simplification is as follows:

$\frac{2}{3}(6x+9)$ Distribute $\frac{2}{3}$ to both terms.

$=\frac{2}{3}(6x)+\frac{2}{3}(9)$ Multiply.

$=4x+6$

71.

$5(4x+8)+3x-10$ Evaluate parentheses first-use distributive property.

$=5(4x)+5(8)+3x-10$ Multiply.

$=20x+40+3x-10$ Combine like terms.

$=23x+30$

73.

$-3(4g+7)+2g+9$ Evaluate parentheses first-use distributive property.

$=-3(4g)+(-3)(7)+2g+9$ Multiply.

$=-12g-21+2g+9$ Combine like terms.

$=-10g-12$

75.

$(2x+5)-(5x-7)$ Evaluate parentheses first-use distributive property.

$=2x+5-5x+7$ Combine like terms.

$=-3x+12$

77.

$(4k^2+5)-3(2k^2+5k)$ Evaluate parentheses first using distributive property.

$=4k^2+5+(-3)(2k^2)+(-3)(5k)$ Multiply.

$=4k^2+5-6k^2-15k$ Combine like terms. Order in conventional form.

$=-2k^2-15k+5$

79.

$2xy+5x+2(3xy+7)$ Evaluate parentheses first-use distributive property.

$=2xy+5x+2(3xy)+2(7)$ Multiply.

$=2xy+5x+6xy+14$ Combine like terms. Order in conventional form.

$=5x+8xy+14$

81.

$3(2x^2+3x)-2(5x^2+6x)$ Evaluate parentheses first. Use distributive property.

$=3(2x^2)+3(3x)+(-2)(5x^2)+(-2)(6x)$ Multiply.

$=6x^2+9x-10x^2-12x$ Combine like terms-order in conventional form.

$=-4x^2-3x$

83.

$4(5w^2+6w-3)+3(2w^2+4w+1)$
Evaluate parentheses first using distributive property.

$=4(5w^2)+4(6w)+4(-3)+3(2w^2)+3(4w)+3(1)$
Multiply

$=20w^2+24w-12+6w^2+12w+3$
Combine like terms-order in conventional form.

$=26w^2+36w-9$

85.

$-2(4r^2+6r-9)-3(-5r^2-5r+3)$
Evaluate parentheses first using distributive property.

$=-2(4r^2)+(-2)(6r)+(-2)(-9)+(-3)(-5r^2)+(-3)(-5r)+(-3)(3)$
Multiply.

$=-8r^2-12r+18+15r^2+15r-9$
Combine like terms-order in conventional form.

$=7r^2+3r+9$

87.

$1.3(2t+6)+6.3t+8$ Evaluate parentheses first using distributive property.

$=1.3(2t)+1.3(6)+6.3t+8$ Multiply.

$=2.6t+7.8+6.3t+8$ Combine like terms, ordering in conventional form.

$=8.9t+15.8$

89.

$2.5(4h+6)-3.5(2h-8)$ Evaluate parentheses first using distributive property.

$=2.5(4h)+2.5(6)+(-3.5)(2h)+(-3.5)(-8)$ Multiply.

$=10h+15-7h+28$ Combine like terms-order in conventional form.

$=3h+43$

91.

$\frac{1}{2}(4x-8)+7x-3$ Evaluate parentheses first-use distributive property.

$=\frac{1}{2}(4x)+\frac{1}{2}(-8)+7x-3$ Multiply.

$=2x-4+7x-3$ Combine like terms-order in conventional form.

$=9x-7$

93.

$\frac{1}{3}(9a-6)+8a-14$ Evaluate parentheses first-use distributive property.

$=\frac{1}{3}(9a)+\frac{1}{3}(-6)+8a-14$ Multiply.

$=3a-2+8a-14$ Combine like terms-order in conventional form.

$=11a-16$

95.

$\frac{1}{5}(10a-5b)+\frac{1}{2}(-4a+12b)$ Evaluate parentheses first using distributive property.

$=\frac{1}{5}(10a)+\frac{1}{5}(-5b)+\frac{1}{2}(-4a)+\frac{1}{2}(12b)$ Multiply.

$=2a-b-2a+6b$ Combine like terms-order in conventional form.

$=5b$

97. To find the perimeter of a figure, add the lengths of all sides.

$l+w+l+w$ This is the sum of all 4 sides.

$=2l+2w$ Combine like terms.

99. To find the perimeter of a figure, add the lengths of all sides.

$x+2y+2x+3y$ This is the sum of all 4 sides.

$=3x+5y$ Combine like terms.

101. To find the perimeter of a figure, add the lengths of all sides. This figure is a rectangle so we know that opposite sides have equal lengths. Two sides have the length $3x-5$ and two sides have the length $8y-1$.

$2(3x-5)+2(8y-1)$ Simplify using distributive property.

$=2(3x)+2(-5)+2(8y)+2(-1)$ Multiply.

$=6x-10+16y-2$ Combine like terms.

$=6x+16y-12$

103. To find the perimeter of a figure, add the lengths of all sides. This figure is an equilateral triangle so we know that all sides have equal lengths. We can multiply the length of the given side by 3 to find the perimeter of the triangle.

$3(4x+3)$ Simplify using distributive property.

$=3(4x)+3(3)$ Multiply.

$=12x+9$

105. To find the perimeter of a figure, add the lengths of all sides. This figure is a square so we know that all 4 sides have equal lengths. We can multiply the length of the given side by 4 to find the perimeter of the square.

$4(3a)$ Simplify using multiplication property.

$=12a$

Section 1.4 Graphs and the Rectangular Coordinate System

1. **a.** The maximum percentage of violent crime reported to the police was 48.7%. This happened during the years 2004-2005.

b. The minimum percentage of property crime reported to the police was 33.4%. This happened during the years 1992-1993 and 1994-1995.

3. **a.** The fastest marathon time is 2h03:59. This happened in the year 2008.

b. Most of the top 10 runners (7 out of the 10 listed) come from Kenya.

5. **a.** The percentage of obese adults was highest during the year 2004. This point is closest in line with approximately 23 on the vertical axis, therefore approximately 23% of the adult population was obese that year.

b. The percentage of obese adults was lowest during the year 1998. This point is closest in line with approximately 18 on the vertical axis, therefore approximately 18% of the adult population was obese that year.

7. **a.** New home sales were at their lowest point during the year 2009. The line goes to just below the 400 mark on the vertical axis so there were approximately 400,000 sales made.

b. New home sales were at their highest point during the year 2005. The line goes to about half-way between the 1200 and 1400 marks on the vertical axis so there were approximately 1,300,000 sales made.

9. The input values are years 2000 through 2004, so we plot these values on the horizontal axis. The output values range from 27,000 to 31,500. Thus, we will draw the vertical axis from 0 to 35,000, counting by increments of 5000. The bar chart should have a title and the chart would be as follows:

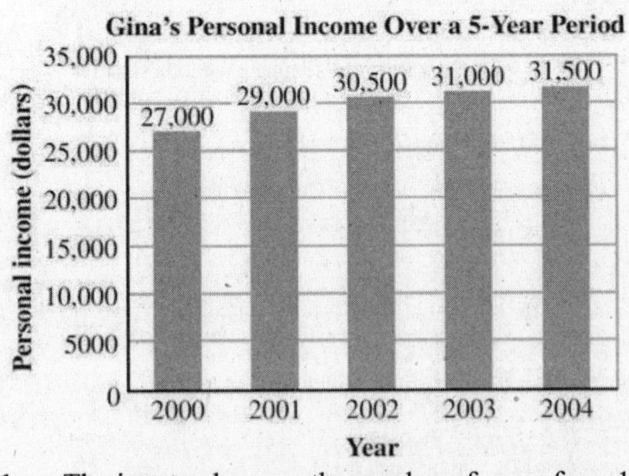

11. The input values are the number of years from 100 through 300, so we plot these values on the horizontal axis. The output values range from 1 to 3. Thus, we will draw the vertical axis from 0 to 3, counting by increments of 1. The bar chart should have a title and the chart would be as follows:

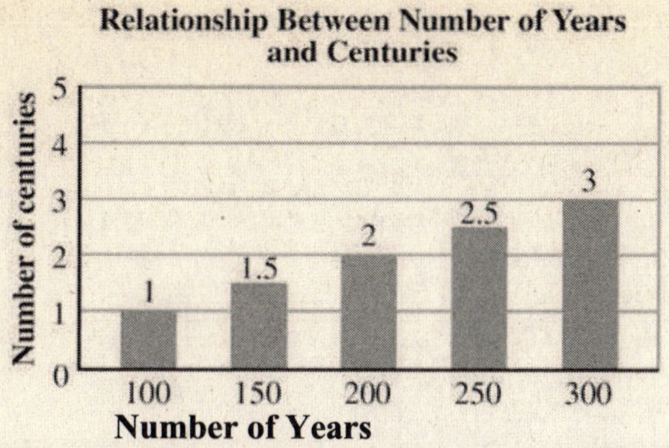

13. The output values take on both positive and negative values. When the output values take on negative values, draw the bar *upside down* under the horizontal axis. The horizontal axis will be scaled by 2, which will accommodate the values 0 through 10. The vertical axis will be scaled by 5. This will accommodate the smallest output value of -25 and the largest output value of 23. The bar chart would be as follows:

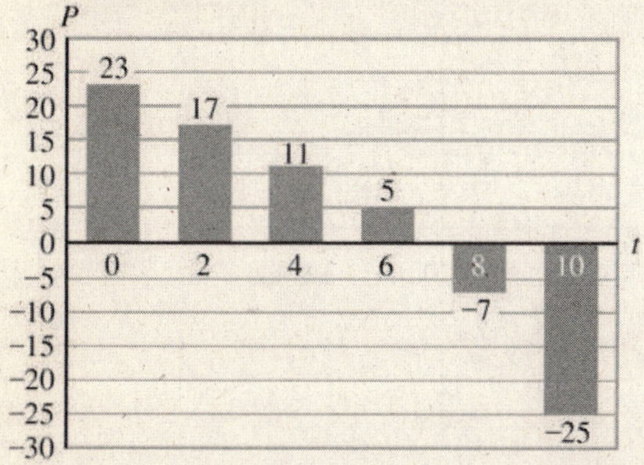

15. A reasonable vertical axis for this data would range between 0 and 35,000, with a scale of 5000, just as with the bar graph. The horizontal axis will start at 0 and will include a line break to account for the numbers between 0 and 2000 that are not part of the input values. This scatterplot should have a title. The scatterplot would be as follows:

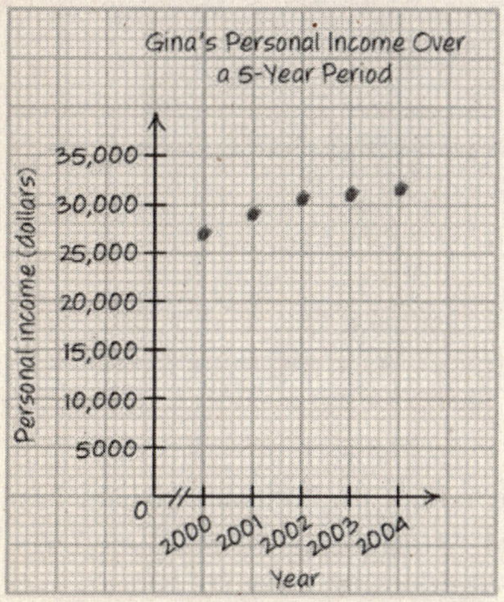

17. A reasonable horizontal axis for this data would range between -50 and 50, with a scale of 10. A reasonable vertical axis for this data would range between -25 and 25, with a scale of 5. The scatterplot would be as follows:

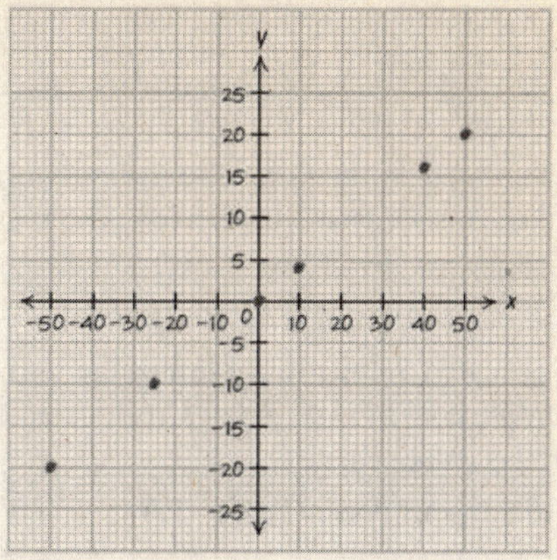

19. A reasonable vertical axis for this data would range between 0 and 80, with a scale of 10. The horizontal axis will be labeled with the months from January through June. This scatterplot should have a title:

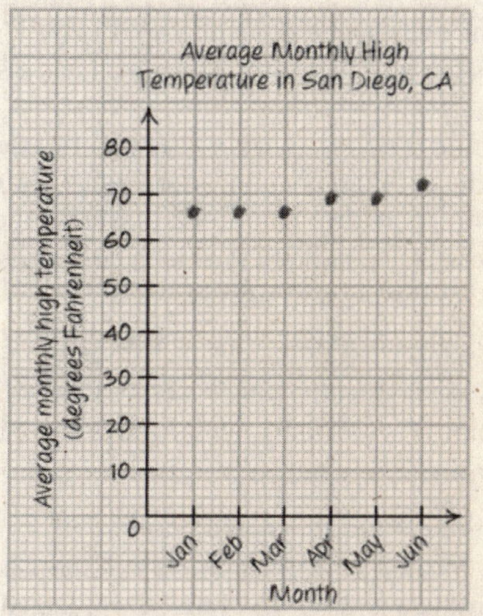

21. To determine the coordinates of the points we need to find the values of the x- and y-coordinates. Point A is located 5 units to the right of the y-axis, so the x-coordinate is $x = 5$. Point A is located 4 units above the x-axis, so the y-coordinate is $y = 4$.

The coordinates of Point A are $(5, 4)$.

Point B is located 8 units to the left of the y-axis, so the x-coordinate is $x = -8$. Point B is located 7 units above the x-axis, so the y-coordinate is $y = 7$.

The coordinates of Point B are $(-8, 7)$.

Point C is located 4 units to the left of the y-axis, so the x-coordinate is $x = -4$. Point C is located 6 units below the x-axis, so the y-coordinate is $y = -6$.

The coordinates of Point C are $(-4, -6)$.

Point D is located 2 units to the right of the y-axis, so the x-coordinate is $x = 2$. Point D is located 3 units below the x-axis, so the y-coordinate is $y = -3$.

The coordinates of Point D are $(2, -3)$.

23. To determine the coordinates of the points we need to find the values of the x- and y-coordinates. Point A is located 2 units to the right of the y-axis, so the x-coordinate is $x = 2$. Point A is located on the x-axis, so the y-coordinate is $y = 0$.

The coordinates of Point A are $(2,0)$.

Point B is located on the y-axis, so the x-coordinate is $x = 0$. Point B is located 3 units above the x-axis, so the y-coordinate is $y = 3$.

The coordinates of Point B are $(0,3)$.

Point C is located 5 units to the left of the y-axis, so the x-coordinate is $x = -5$. Point C is located on the x-axis, so the y-coordinate is $y = 0$.

The coordinates of Point C are $(-5,0)$.

Point D is located on the y-axis, so the x-coordinate is $x = 0$. Point D is located 4 units below the x-axis, so the y-coordinate is $y = -4$.

The coordinates of Point D are $(0,-4)$.

25. The x-axis has a scale of 5 because the even spacing on the x-axis is by 5's. The y-axis has a scale of 1 because the even spacing on the y-axis is by 1's.

27. The x-axis has a scale of 0.5. The y-axis has a scale of 0.1.

29. To determine the coordinates of the points we need to find the values of the x- and y-coordinates. Point A is located 24 units to the right of the y-axis, so the x-coordinate is $x = 24$. Point A is located 6 units above the x-axis, so the y-coordinate is $y = 6$.

The coordinates of Point A are $(24,6)$.

Point B is located on the y-axis, so the x-coordinate is $x = 0$. Point B is located 21 units above the x-axis, so the y-coordinate is $y = 21$.

The coordinates of Point B are $(0,21)$.

Point C is located 16 units to the left of the y-axis, so the x-coordinate is $x = -16$. Point C is located 12 units above the x-axis, so the y-coordinate is $y = 12$.

The coordinates of Point C are $(-16,12)$.

Point D is located 12 units to the right of the y-axis, so the x-coordinate is $x = 12$. Point D is located on the x-axis, so the y-coordinate is $y = 0$.

The coordinates of Point D are $(12,0)$.

Point E is located 28 units to the left of the y-axis, so the x-coordinate is $x = -28$. Point E is located 9 units below the x-axis, so the y-coordinate is $y = -9$.

The coordinates of Point E are $(-28,-9)$.

31. The input values range from -3 to 5. It is reasonable to scale the horizontal axis by 2's. Likewise, the output values range from 1 to 6 so it will also be reasonable to scale the y-axis by 2's.

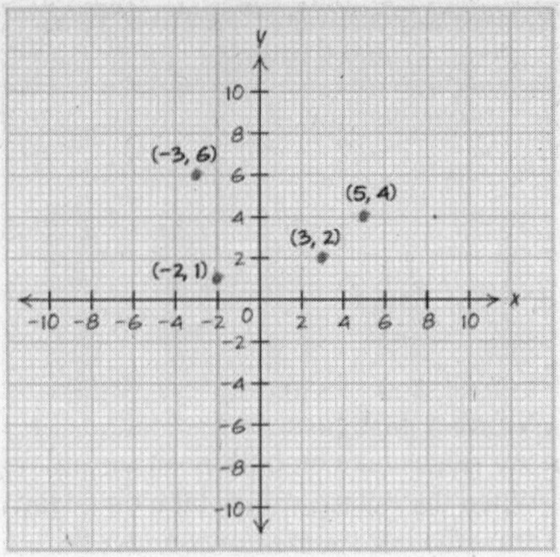

33. The input values range from 0.5 to 2.0. It is reasonable to scale the horizontal axis by 0.5. The output values range from -0.5 to 1.0 so it will be reasonable to scale the y-axis by 0.5.

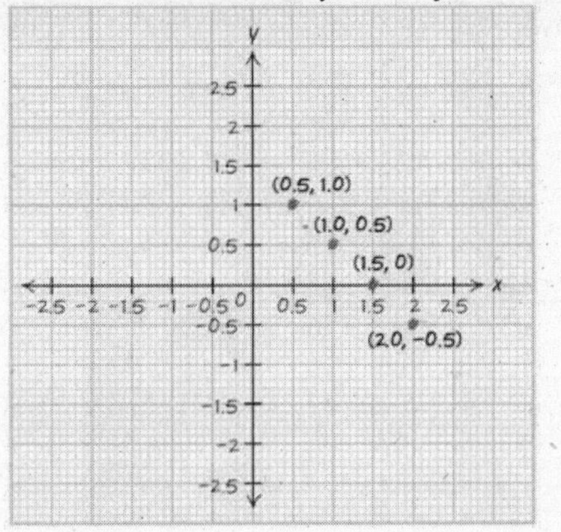

35. The input values range from 0 to 50. It is reasonable to scale the horizontal axis by 10's. The output values range from -8 to 16 so it will be reasonable to scale the y-axis by 4's.

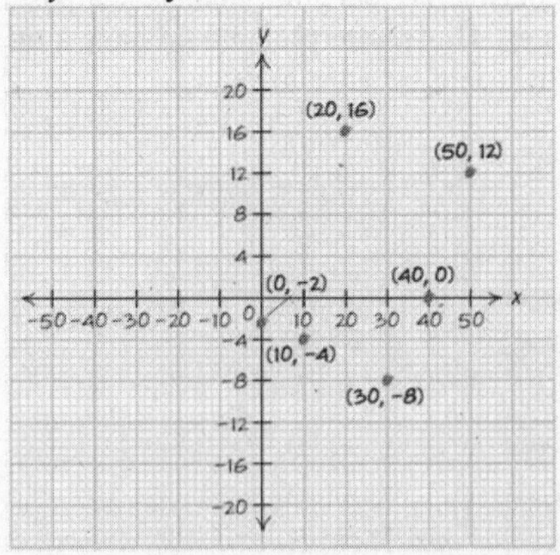

37. The input values range from -50 to 125. It is reasonable to scale the horizontal axis by 25's. The output values range from -30 to 60 so it will be reasonable to scale the y-axis by 10's.

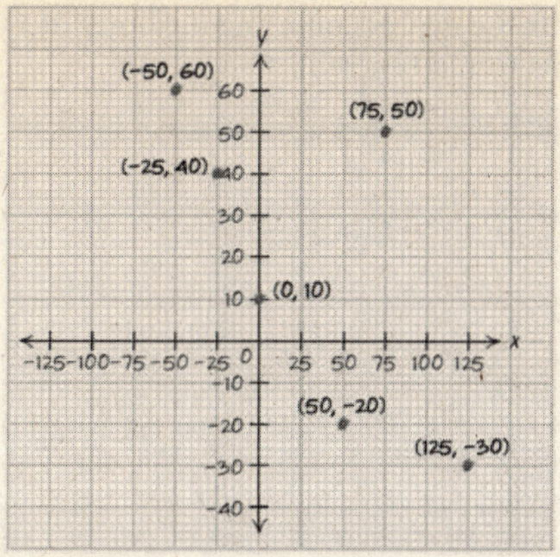

39. A reasonable vertical axis for this data would range between 0 and 220 (in billions of dollars), with a scale of 20. The horizontal axis will start at 0 and end at 2003. It will have a scale of 1 and will include a line break to account for the numbers between 0 and 1994 that are not part of the input values. This scatterplot should have a title:

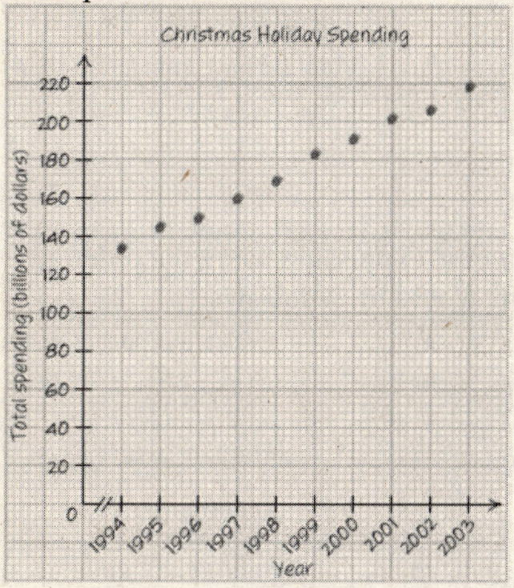

41. A reasonable vertical axis for this data would range between 0 and 4 with a scale of 0.5. The horizontal axis will start at 0 and end at 72 with a scale of 12. This scatterplot should have a title:

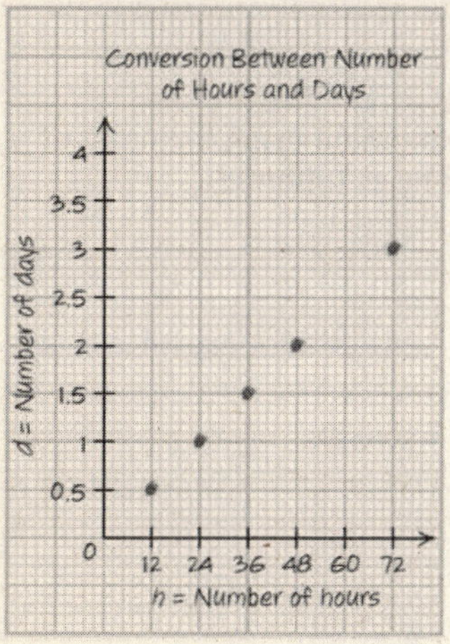

43. A reasonable vertical axis for this data would range between 0 and 9 (in trillions of dollars), with a scale of 1. The horizontal axis will start at 0 and end at 2005. It will have a scale of 2.5, with only every other value of that scale listed, and will include a line break to account for the numbers between 0 and 1990 that are not part of the input values. This scatterplot should have a title:

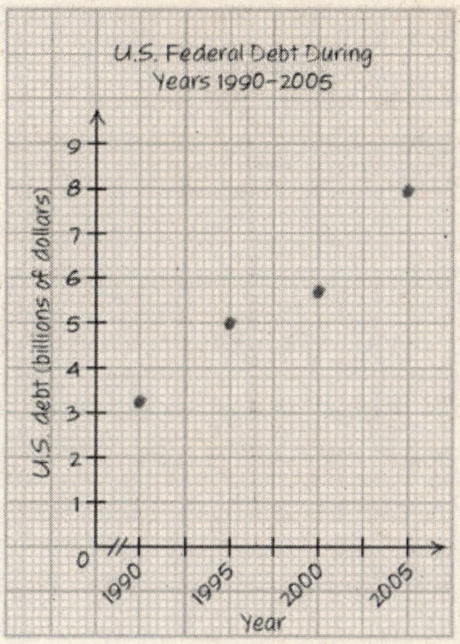

45. The student transposed the x- and y-coordinates, graphing the point $(4,3)$ instead of $(3,4)$.

47. The student is confusing the x- and y-coordinates and has transposed the coordinates for each point.

49. The student was inconsistent with the scale of the y-axis, using a scale of 10 for positive values and a scale of 1 for negative values.

CHAPTER 1 Building Blocks of Algebra

Chapter 1 Review

1. $6 \cdot 6 \cdot 6 \cdot 6 \cdot 6 = 6^5$

2. $8 \cdot 8 \cdot 8 \cdot 8 = 8^4$

3. $\left(\frac{3}{8}\right)\cdot\left(\frac{3}{8}\right)\cdot\left(\frac{3}{8}\right)=\left(\frac{3}{8}\right)^3$

4. $\left(\frac{-4}{5}\right)\cdot\left(\frac{-4}{5}\right)\cdot\left(\frac{-4}{5}\right)\cdot\left(\frac{-4}{5}\right)\cdot\left(\frac{-4}{5}\right)=\left(\frac{-4}{5}\right)^5$

5. $5^6 = 5 \cdot 5 \cdot 5 \cdot 5 \cdot 5 \cdot 5$

6. $(-2)^3 = (-2)\cdot(-2)\cdot(-2)$

7. $\left(\frac{-2}{5}\right)^4=\left(\frac{-2}{5}\right)\cdot\left(\frac{-2}{5}\right)\cdot\left(\frac{-2}{5}\right)\cdot\left(\frac{-2}{5}\right)$

8. $\left(\frac{1}{11}\right)^3=\left(\frac{1}{11}\right)\cdot\left(\frac{1}{11}\right)\cdot\left(\frac{1}{11}\right)$

9. $(-3)^3 = (-3)\cdot(-3)\cdot(-3) = -27$

10. $(-6)^2 = (-6)\cdot(-6) = 36$

11. $\left(\frac{4}{9}\right)^2=\left(\frac{4}{9}\right)\cdot\left(\frac{4}{9}\right)=\frac{16}{81}$

12. $\left(\frac{-2}{3}\right)^3=\left(\frac{-2}{3}\right)\cdot\left(\frac{-2}{3}\right)\cdot\left(\frac{-2}{3}\right)=\frac{-8}{27}$

13. $2.07\times10^5 = 2.07\cdot100{,}000 = 207{,}000$

14. $5.67\times10^6 = 5.67\cdot1{,}000{,}000 = 5{,}670{,}000$

15. Evaluating the expression using the order of operations would be as follows:

$\boxed{-3}+\boxed{4\cdot(-2)}+\boxed{24\div8}\boxed{-3^2}+\boxed{(-2)}$ Outline the terms.

$=-3+4\cdot(-2)+24\div8-3^2+(-2)$
Parentheses only for clarity. Evaluate exponent first.

$=-3+4\cdot(-2)+24\div8-9+(-2)$
Evaluate multiplication and division.

$=-3+(-8)+3-9+(-2)$ Add/Subtract from left to right.

$=-19$

16. Evaluating the expression using the order of operations would be as follows:

$\boxed{27\div3}\boxed{-5^2}+\boxed{(-10)}+\boxed{20}\boxed{-3\cdot(-5)}$ Outline the terms.

$=27\div3-5^2+(-10)+20-3\cdot(-5)$
Parentheses only for clarity. Evaluate exponent first.

$=27\div3-25+(-10)+20-3\cdot(-5)$
Evaluate multiplication and division.

$=9-25+(-10)+20+15$
Add/Subtract from left to right.

$=9$

17. Evaluating the expression using the order of operations would be as follows:

$\boxed{5\div\frac{15}{7}}+\boxed{\frac{2}{3}}\boxed{-1}$	Outline the terms.
$=5\div\frac{15}{7}+\frac{2}{3}-1$	Divide first.
$=5\cdot\frac{7}{15}+\frac{2}{3}-1$	Flip fraction & multiply.
$=\frac{7}{3}+\frac{2}{3}-1$	Add/Subtract from left to right
$=3-1$	
$=2$	

18. Evaluating the expression using the order of operations would be as follows:

$\boxed{-4}\boxed{-\frac{1}{5}}+\boxed{\frac{3}{2}\cdot\frac{4}{15}}$	Outline the terms.
$=-4-\frac{1}{5}+\frac{3}{2}\cdot\frac{4}{15}$	Multiply first.
$=-4-\frac{1}{5}+\frac{2}{5}$	Combine fractions first then Add/Subtract from left to right
$=-4+\frac{1}{5}$	
$=-3\frac{4}{5}$	

19. Evaluating the expression using the order of operations would be as follows:

$6\div(-2)\cdot5+(-3)^2$	Evaluate exponent first.
$=6\div(-2)\cdot5+9$	Then Divide.
$=-3\cdot5+9$	Multiply next.
$=-15+9$	Add/Subtract.
$=-6$	

20. Evaluating the expression using the order of operations would be as follows:

$(-2)^3-36\div9\cdot7$	Evaluate exponent first.
$=-8-36\div9\cdot7$	Then Divide.
$=-8-4\cdot7$	Multiply next.
$=-8-28$	Subtract.
$=-36$	

21. Evaluating the expression using the order of operations would be as follows:

$\frac{3}{2}\cdot\frac{4}{9}-5+\frac{2}{3}$	Multiply first.
$=\frac{2}{3}-5+\frac{2}{3}$	Add/Subtract from left to right.
$=-3\frac{2}{3}$	Subtract.

22. Evaluating the expression using the order of operations would be as follows:

$\frac{-1}{4}\cdot\frac{12}{5}+3\div\frac{5}{2}-1$	Multiply first.
$=\frac{-3}{5}+3\div\frac{5}{2}-1$	Divide next.
$=\frac{-3}{5}+3\cdot\frac{2}{5}-1$	Flip fraction and multiply.
$=\frac{-3}{5}+\frac{6}{5}-1$	Add/Subtract from left to right.
$=-\frac{2}{5}$	

23. Evaluating the expression using the order of operations would be as follows:

$-7\cdot(-4)\div2+(-1)^4+\frac{6\cdot3}{5-2}$	Simplify top & bottom of fraction first.
$=-7\cdot(-4)\div2+(-1)^4+6$	Evaluate exponent.
$=-7\cdot(-4)\div2+1+6$	Multiply next.
$=28\div2+1+6$	Divide.
$=14+1+6$	Add.
$=21$	

24. Evaluating the expression using the order of operations would be as follows:

$\frac{8-5}{4+2}+16\cdot(-3)\div12+(-3)^2$	Simplify top & bottom of fraction first.
$=\frac{1}{2}+16\cdot(-3)\div12+(-3)^2$	Evaluate exponent.
$=\frac{1}{2}+16\cdot(-3)\div12+9$	Multiply next.
$=\frac{1}{2}+(-48)\div12+9$	Divide.
$=\frac{1}{2}-4+9$	Add/Subtract from left to right.
$=5\frac{1}{2}$	

25. Evaluating the expression using the order of operations would be as follows:

$-5+7\cdot(-3)+44\div2-6^2+(-8)$	Evaluate exponent.
$=-5+7\cdot(-3)+44\div2-36+(-8)$	Then Multiply.
$=-5+(-21)+44\div2-36+(-8)$	Divide next.
$=-5+(-21)+22-36+(-8)$	Add/Subtract from. left to right.
$=-48$	

26. Evaluating the expression using the order of operations would be as follows:

$(-5)^2-24\div(-3)\cdot2-(-6-4)+9$	Simplify inside parentheses first.
$=(-5)^2-24\div(-3)\cdot2-(-10)+9$	Evaluate exponent.
$=25-24\div(-3)\cdot2-(-10)+9$	Divide next.
$=25-(-8)\cdot2-(-10)+9$	Then Multiply.
$=25+16-(-10)+9$	Add/Subtract from left to right.
$=60$	

27. Evaluating the expression using the order of operations would be as follows:

$5+\frac{22-4\cdot2}{16-3^2}-(-2)^2$	Simplify top & bottom of fraction. Multiply on top, exponent on bottom.
$=5+\frac{22-8}{16-9}-(-2)^2$	Finish simplifying the fraction.
$=5+2-(-2)^2$	Evaluate exponent.
$=5+2-4$	Add/Subtract from left to right.
$=3$	

28. Evaluating the expression using the order of operations would be as follows:

$\frac{-2+6\cdot2}{14-3^2}-(3\cdot2-7)$	Simplify inside parentheses. Multiply, then subtract.
$=\frac{-2+6\cdot2}{14-3^2}-(6-7)$	
$=\frac{-2+6\cdot2}{14-3^2}-(-1)$	Simplify top/bottom of fraction first. Multiply on top. Exponent on bottom.
$=\frac{-2+12}{14-9}-(-1)$	Finish simplifying top/bottom of fraction.
$=2-(-1)$	Subtract from left to right.
$=3$	

29. Evaluating the expression using the order of operations would be as follows:

$\lvert-2+3\cdot(-4)\rvert+(-7)^2$	Simplify inside absolute value. Multiply, then add/subtract.
$=\lvert-2+(-12)\rvert+(-7)^2$	
$=\lvert-14\rvert+(-7)^2$	Evaluate absolute value and exponent.
$=14+49$	Add.
$=63$	

30. Evaluating the expression using the order of operations would be as follows:

$\lvert-8+(-3)(-2)\rvert+3^3$	Simplify inside absolute value. Multiply, then add/subtract.
$=\lvert-8+6\rvert+3^3$	
$=\lvert-2\rvert+3^3$	Evaluate absolute value and exponent.
$=2+27$	Add.
$=29$	

31. $(-2)(-8)=(-8)(-2)$

This is an example of the commutative property of multiplication because the order of the numbers was changed.

32. $-3+(-8)=-8+(-3)$

This is an example of the commutative property of addition, because the order of the terms has been changed.

33. $-3(4+5)=-3\cdot4+(-3)\cdot5$

This is an example of the distributive property because each term inside the parentheses has been multiplied by the term outside the parentheses.

34. $2\cdot(3-7)=2\cdot3-2\cdot7$

This is an example of the distributive property because each term inside the parentheses has been multiplied by the term outside the parentheses.

35. $(-3)+(-4+9)=(-3+(-4))+9$

This is an example of the associative property of addition because the order of the numbers was not changed, but how they are grouped was changed.

36. $(5\cdot3)(-2)=5\cdot(3\cdot(-2))$

This is an example of the associative property of multiplication because the order of the numbers was not changed, but how they are grouped was changed.

37. $\frac{4}{5}+\frac{8}{9}=\frac{8}{9}+\frac{4}{5}$

This is an example of the commutative property of addition because the order of the terms has been changed.

38. $\frac{5}{7}\cdot\frac{3}{25}=\frac{3}{25}\cdot\frac{5}{7}$

This is an example of the commutative property of multiplication because the order of the numbers was changed.

For exercises 39-44, the letters chosen to represent the variables can vary.

39. There are two quantities that can change. One quantity is the amount of money you earn in a month and the second is the number of hours you worked that month. Let D = the amount of money (in dollars) you earn in a month, and h = the number of hours you work in a month.

40. There are two quantities that can change. One quantity is the amount of money a local playhouse earns for a night and the second is the number of tickets sold for that night. Let R = the amount of money (revenue in dollars) a local playhouse earns for a night, and t = the number of tickets sold for that night.

41. Let n = a number. "Six less than..." means that we will subtract 6 from something. "Three times a number" translates to $3n$ and the sentence translates as $3n-6$.

42. Let n = a number. "Eight more than..." means that we will add 8 to something. "Five times a number" translates to $5n$ and the sentence translates as $5n+8$.

43. Let n = a number. The sentence starts with "one-half of the *difference*..." Difference translates as subtraction and represents a quantity. The "difference" is "between a number and 4." Order matters and this translates as $n-4$. This is the "difference" that we will find half of, so the sentence translates as $\frac{1}{2}(n-4)$.

44. Let n = a number. Quotient means to divide, and order matters. We will divide "the sum of a number and 12" by 16. "Sum" translates to addition and "the sum of a number and 12" translates to $n+12$. This quantity will be in the numerator, with 16 in the denominator. Therefore, the full translation is $\frac{n+12}{16}$.

45. Convert 9 inches to centimeters. First, form the unity fraction. We want to convert from the units of inches, so the units of inches will go in the denominator. We want to convert to the units of centimeters, so centimeters will go in the numerator. There are 2.54 cm in 1 inch, so the unity fraction is $\frac{2.54\text{ cm}}{1\text{ in}}$.

Multiply 9 inches by this unity fraction and simplify the expression as follows.

$$9\text{ in}\cdot\frac{2.54\text{ cm}}{1\text{ in}}$$

$$=\frac{9\cancel{\text{in}}}{1}\cdot\frac{2.54\text{ cm}}{1\cancel{\text{in}}}$$

$$=22.86\text{ cm}$$

Therefore, 9 inches is equal to 22.86 cm.

46. Convert 55 kilometers to miles. First, form the unity fraction. We want to convert from the units of kilometers, so the units of kilometers will go in the denominator. We want to convert to the units of miles, so miles will go in the numerator. There are 1.609 km in 1 mile, so the unity fraction is $\frac{1\text{ mi}}{1.609\text{ km}}$. Multiply 55 km by this unity fraction and simplify the expression as follows.

$$55\text{ km}\cdot\frac{1\text{ mi}}{1.609\text{ km}}$$

$$=\frac{55\cancel{\text{km}}}{1}\cdot\frac{1\text{ mi}}{1.609\cancel{\text{km}}}$$

$$=\frac{55}{1.609}\text{ mi}\approx34.18\text{ mi}$$

Therefore, 55 km equals approximately 34.18 miles.

47. Convert 185 pounds to kilograms. First, form the unity fraction. We want to convert from the units of pounds, so the units of pounds will go in the denominator. We want to convert to the units of kilograms, so kilograms will go in the numerator. There are 2.2 lb in 1 kg, so the unity fraction is $\frac{1\text{ kg}}{2.2\text{ lb}}$. Multiply 185 pounds by this unity fraction and simplify the expression as follows.

$185 \text{ lb} \cdot \frac{1 \text{ kg}}{2.2 \text{ lb}}$

$= \frac{185 \cancel{\text{lb}}}{1} \cdot \frac{1 \text{ kg}}{2.2 \cancel{\text{lb}}}$

$= \frac{185}{2.2} \text{ kg} \approx 84.09 \text{ kg}$

Therefore, 185 lb. equals approximately 84.09 kg.

48. Convert 100 kilograms to pounds. First, form the unity fraction. We want to convert from the units of kilograms, so the units of kilograms will go in the denominator. We want to convert to the units of pounds, so pounds will go in the numerator. There are 2.2 lb in 1 kg, so the unity fraction is $\frac{2.2 \text{ lb}}{1 \text{ kg}}$.

Multiply 100 kg by this unity fraction and simplify the expression as follows.

$100 \text{ kg} \cdot \frac{2.2 \text{ lb}}{1 \text{ kg}}$

$= \frac{100 \cancel{\text{kg}}}{1} \cdot \frac{2.2 \text{ lb}}{1 \cancel{\text{kg}}}$

$= 220 \text{ lb}$

Therefore, 100 kg. equals 220 lb.

49. Convert 55 grams to ounces. First, form the unity fraction. We want to convert from the units of grams, so the units of grams will go in the denominator. We want to convert to the units of ounces, so ounces will go in the numerator. There are 28.35 g in 1 oz, so the unity fraction is $\frac{1 \text{ oz}}{28.35 \text{ g}}$. Multiply 55 grams by this unity fraction and simplify the expression as follows.

$55 \text{ g} \cdot \frac{1 \text{ oz}}{28.35 \text{ g}}$

$= \frac{55 \cancel{\text{g}}}{1} \cdot \frac{1 \text{ oz}}{28.35 \cancel{\text{g}}}$

$= \frac{55}{28.35} \text{ oz} \approx 1.94 \text{ oz}$

Therefore, 55 g equals approximately 1.94 oz.

50. Convert 5 ounces to grams. First, form the unity fraction. We want to convert from the units of ounces, so the units of ounces will go in the denominator. We want to convert to the units of grams, so grams will go in the numerator. There are 28.35 g in 1 oz, so the unity fraction is $\frac{28.35 \text{ g}}{1 \text{ oz}}$.

Multiply 5 ounces by this unity fraction and simplify the expression as follows.

$5 \text{ oz} \cdot \frac{28.35 \text{ g}}{1 \text{ oz}}$

$= \frac{5 \cancel{\text{oz}}}{1} \cdot \frac{28.35 \text{ g}}{1 \cancel{\text{oz}}}$

$= 141.75 \text{ g}$

Therefore, 5 ounces equals 141.75 g.

51. Evaluate $\frac{-5h}{18} + 9$ for $h = -3$.

$\frac{-5h}{18} + 9$	Substitute in $h = -3$.
$= \frac{-5(-3)}{18} + 9$	Simplify using the order-of-operations agreement.
$= \frac{15}{18} + 9$	
$= \frac{5}{6} + 9$	
$= 9\frac{5}{6}$	

52. Evaluate $-4r + \frac{r}{2}$ for $r = 8$.

$-4r + \frac{r}{2}$	Substitute in $r = 8$.
$= -4(8) + \frac{8}{2}$	Simplify using the order-of-operations agreement.
$= -32 + 4$	
$= -28$	

53. Evaluate $5x - 2y$ for $x = -3$, $y = -5$.

$5x - 2y$	Substitute in $x = -3$ and $y = -5$.
$= 5(-3) - 2(-5)$	Simplify using the order-of-operations agreement.
$= -15 + 10$	
$= -5$	

54. Evaluate $5y - 6x$ for $x = -2$, $y = -7$.

$5y - 6x$	Substitute in $x = -2$ and $y = -7$.
$= 5(-7) - 6(-2)$	Simplify using the order-of-operations agreement.
$= -35 + 12$	
$= -23$	

55. Evaluate $3xy + 2x - y$ for $x = -2$, $y = 3$.

$3xy + 2x - y$	Substitute in $x = -2$ and $y = 3$.
$= 3(-2)(3) + 2(-2) - (3)$	Simplify using the order-of-operations agreement.
$= -18 - 4 - 3$	
$= -25$	

56. Evaluate $l + k - 10$ for $l = 4$, $k = -3$.

$l + k - 10$	Substitute in $l = 4$ and $k = -3$.
$= 4 + (-3) - 10$	Simplify using the order-of-operations agreement.
$= -9$	

57. Evaluate $0.25a+1.75b$ for $a=0.8,\ b=4.6$.

$0.25a+1.75b$ Substitute in $a=0.8$ and $b=4.6$.

$=0.25(0.8)+1.75(4.6)$ Simplify using order-of-operations agreement.

$=0.2+8.05$

$=8.25$

58. Evaluate $0.5h-2.5k$ for $h=8,\ k=-10$.

$0.5h-2.5k$ Substitute in $h=8$ and $k=-10$.

$=0.5(8)-2.5(-10)$ Simplify using order-of-operations agreement.

$=4+25$

$=29$

59. Evaluate $15ab+c$ for $a=2,\ b=-3,\ c=4$.

$15ab+c$ Substitute in $a=2,\ b=-3,\ c=4$.

$=15(2)(-3)+4$ Simplify using the order-of-operations agreement.

$=-90+4$

$=-86$

60. Evaluate $a+b+c$ for $a=3,\ b=4,\ c=5$.

$a+b+c$ Substitute in $a=3,\ b=4$ and $c=5$.

$=3+4+5$ Simplify using the order-of-operations agreement.

$=12$

61.

Amount invested in first account	Amount invested in the second account
0	$4500-0=4500$
1000	$4500-1000=3500$
2000	$4500-2000=2500$
3500	$4500-3500=1000$
d	$4500-d$

To find the amount of money Wang Li will invest in the second account, subtract the amount of money he invests in the first account from $4500. The general variable expression is $4500-d$.

62.

First piece	Second piece
0	$20-0=20$
4	$20-4=16$
8	$20-8=12$
16	$20-16=4$
d	$20-d$

To find the length (in feet) of the 2nd piece of lumber, subtract the length (in feet) of the 1st piece of lumber from 20 feet. The general variable expression is $20-d$.

63. Convert 800 million to thousands. First, form the unity fraction. We want to convert to the units of thousands, so thousands will go in the numerator. We want to convert from the units of millions, so millions will go in the denominator. There are 1000 thousands in 1 million, so the unity fraction is $\dfrac{1000 \text{ thou}}{1 \text{ mill}}$. Multiply 800 million by this unity fraction and simplify the expression as follows.

$$800 \text{ mill}\cdot\frac{1000 \text{ thou}}{1 \text{ mill}}$$

$$=\frac{800 \cancel{\text{mill}}}{1}\cdot\frac{1000 \text{ thou}}{1 \cancel{\text{mill}}}$$

$$=800{,}000 \text{ thou}$$

Therefore, 800 million is equal to 800,000 thousands.

64. Convert 700 billion to millions. First, form the unity fraction. We want to convert to the units of millions, so millions will go in the numerator. We want to convert from the units of billions, so billions will go in the denominator. There are 1000 millions in 1 billion, so the unity fraction is $\dfrac{1000 \text{ mill}}{1 \text{ bill}}$. Multiply 700 billion by this unity fraction and simplify the expression as follows.

$$700 \text{ bill}\cdot\frac{1000 \text{ mill}}{1 \text{ bill}}$$

$$=\frac{700 \cancel{\text{bill}}}{1}\cdot\frac{1000 \text{ mill}}{1 \cancel{\text{bill}}}$$

$$=700{,}000 \text{ mill}$$

Therefore, 700 billion is equal to 700,000 million.

65. The input values are years 1960 through 2008, so we plot these values on the horizontal axis. The output values range from 47,699 to 62,635. Thus, we will draw the vertical axis from 0 to 64,000 with a break from 0 to 30,000, counting by increments of 2000. The bar chart should have a title and the chart would be as follows:

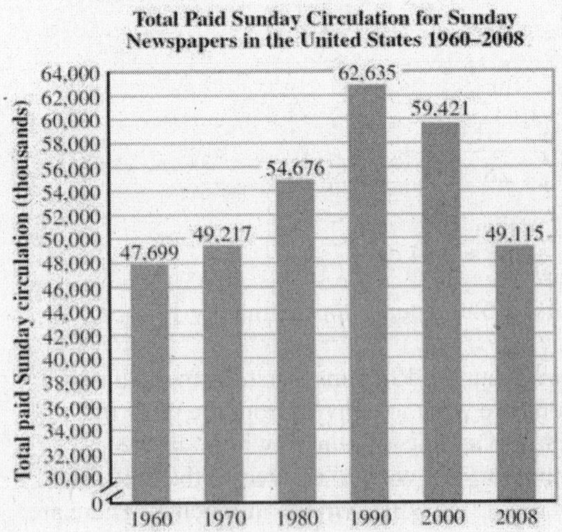

66. A reasonable horizontal axis for this data would range between -25 and 25, with a scale of 5. A reasonable vertical axis for this data would range between −5 and 5, with a scale of 1. The scatterplot would be as follows:

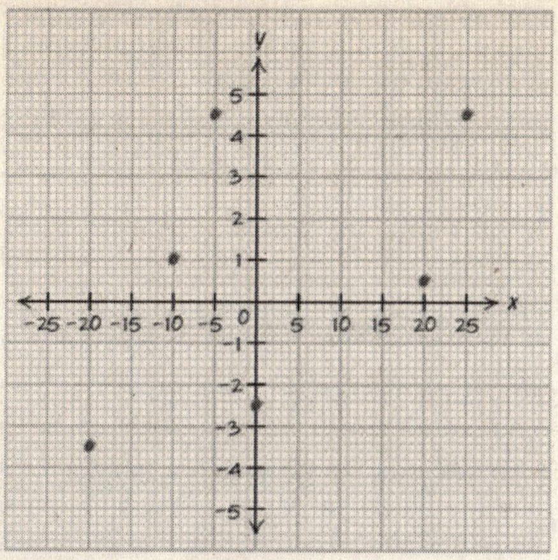

67. $\boxed{-6p}+\boxed{4}+\boxed{7p}\boxed{-9}$ There are 4 terms in this expression. The terms $-6p$ and $7p$ are like terms because they both have p raised to the first power. The terms 4 and −9 are constants and are therefore like terms.

68. $\boxed{5x}+\boxed{16}\boxed{-\frac{2}{3}x}+\boxed{12}$ There are 4 terms in this expression. The terms $5x$ and $-\frac{2}{3}x$ are like terms because they both have x raised to the first power. The terms 16 and 12 are constants and are therefore like terms.

69. $\boxed{-3x^2}+2x\boxed{-4x^2}+1$ There are 4 terms in this expression. The terms $-3x^2$ and $-4x^2$ are like terms because they both have x raised to the second power.

70. $\boxed{-\frac{2}{5}y}+y^2\boxed{-6y}+12$ There are 4 terms in this expression. The terms $-\frac{2}{5}y$ and $-6y$ are like terms because they both have y raised to the first power.

71. **a.** There are s students in the class and it will cost Ms. Welch \$3.25 to buy each book. We multiply \$3.25 times the number of children, s, to get the expression:

$3.25s$

b. It will cost \$1.25 to buy each candy bar, so multiply \$1.25 times s children to get the expression:

$1.25s$

c. "Total" means to add so we will add the expressions from parts a and b to get the total cost to buy each student a book and educational toy:

$3.25s+1.25s$

d. To simplify the expression from part c we combine like terms:

$4.5s$

72. **a.** Yong Park makes a monthly salary of d dollars and his monthly rent is subtracted from that. His monthly rent is \$500, therefore an expression to represent how much he has left after paying rent is:

$d-500$

b. He pays \$85 per week for groceries. Assuming there are 4 weeks in a month, an expression representing the amount he pays each month for groceries is:

$4(85)$

c. To find the total he has left after paying for rent and groceries for the month, subtract the expressions from parts a and b from his total salary:

$d-500-4(85)$

d. Simplify the expression from part c as follows:

$d-500-4(85)$ Multiply first.
$=d-500-340$ Combine like terms.
$=d-840$

73. Combine like terms by adding or subtracting coefficients. Arrange in alphabetical order.

$y^2-4y^3+3y^2-8y^3$
$=-12y^3+4y^2$

74. Combine like terms by adding or subtracting coefficients. Arrange in alphabetical order.

$-8a^2b+3ab-16ab+12a^2b$
$=4a^2b-13ab$

75. Combine like terms by adding or subtracting coefficients. Arrange in alphabetical order.

$-\frac{3}{4}x+7-\frac{1}{4}x-12$
$=-x-5$

76. Combine like terms by adding or subtracting coefficients. Arrange in alphabetical order.

$-5+\frac{2}{3}x-7+\frac{1}{3}x$
$=x-12$

77. Combine like terms by adding or subtracting coefficients. Arrange in alphabetical order.

$1.25a-6b-7.5a+8.1b$
$=-6.25a+2.1b$

78. Combine like terms by adding or subtracting coefficients. Arrange in alphabetical order.

$4.7x+3.2y-5x-1.3y$
$=-0.3x+1.9y$

79. Combine like terms by adding or subtracting coefficients. Arrange in alphabetical order.

$5x^2-6x+4-3x^2+7x+8$
$=2x^2+x+12$

80. Combine like terms by adding or subtracting coefficients. Arrange in alphabetical order.

$12-y+3y^2-6y-2y^2+7$
$=y^2-7y+19$

81. **a.** The multiplication property will be used because there is only one term in the parentheses.

b. $16(-3ab^2)$ Multiply.
$=-48ab^2$

82. **a.** The multiplication property will be used because there is only one term in the parentheses.

b. $-4(3xy)$ Multiply.
$=-12xy$

83. **a.** The distributive property will be used because there is more than one term in the parentheses.

b.
$-5(3x-6y)$ Distribute -5 to both terms.
$=-5(3x)+(-5)(-6y)$ Multiply.
$=-15x+30y$

84. **a.** The distributive property will be used because there is more than one term in the parentheses.

b.
$2(3a-8b)$ Distribute 2 to both terms.
$=2(3a)+2(-8b)$ Multiply.
$=6a-16b$

85. **a.** The multiplication property will be used because there is only one term in the parentheses.

b.
$-(-13x)$ Multiply.
$=13x$

86. **a.** The distributive property will be used because there is more than one term in the parentheses.

b.
$-(5y-7)$ Distribute 2 to both terms.
$=-1(5y)+(-1)(-7)$ Multiply.
$=-5y+7$

87.
$9-3(2x-6)$ Evaluate parentheses using distributive property.
$=9-3(2x)+(-3)(-6)$ Multiply.
$=9-6x+18$ Combine like terms.
$=-6x+27$

88.
$-2(3x-4)+7x-1$ Evaluate parentheses using distributive property.
$=-2(3x)+(-2)(-4)+7x-1$ Multiply.
$=-6x+8+7x-1$ Combine like terms.
$=x+7$

89.
$-(-2x+3)$ Evaluate parentheses using distributive property.
$=-(-2x)+(-1)(3)$ Multiply.
$=2x-3$

90.
$-(9y-5)$ Evaluate parentheses using distributive property.
$=-(9y)+(-1)(-5)$ Multiply.
$=-9y+5$

91.
$(-2a+3b)-(4a-5b)$ Evaluate parentheses using distributive property.
$=-2a+3b-(4a)+(-1)(-5b)$ Multiply.
$=-2a+3b-4a+5b$ Combine like terms.
$=-6a+8b$

92.
$3x+2y-4(-x+5y)$ Evaluate parentheses using distributive property.
$=3x+2y-4(-x)+(-4)(5y)$ Multiply.
$=3x+2y+4x-20y$ Combine like terms.
$=7x-18y$

93.
$\frac{1}{2}(4x-12)+9x-3$ Evaluate parentheses-use distributive property.
$=\frac{1}{2}(4x)+\frac{1}{2}(-12)+9x-3$ Multiply.
$=2x-6+9x-3$ Combine like terms.
$=11x-9$

94.
$3h-2k+\frac{1}{4}(4h-16k)$ Evaluate parentheses using distributive property.
$=3h-2k+\frac{1}{4}(4h)+\frac{1}{4}(-16k)$ Multiply.
$=3h-2k+h-4k$ Combine like terms.
$=4h-6k$

95.
$-0.2(10x-5)+2x-1$ Evaluate parentheses using distributive property.
$=-0.2(10x)+(-0.2)(-5)+2x-1$ Multiply.
$=-2x+1+2x-1$ Combine like terms.
$=0$

96.
$0.25(7x-6y)-1.25x+9.8y$ Evaluate parentheses-use distributive property.
$=0.25(7x)+0.25(-6y)-1.25x+9.8y$ Multiply.
$=1.75x-1.5y-1.25x+9.8y$ Combine like terms.
$=0.5x+8.3y$

97. To find the perimeter of a figure, add the lengths of all sides.
$2x+(3y+1)+(4x+5)+5y$ Sum of all sides.
$=2x+3y+1+4x+5+5y$ Combine like terms.
$=6x+8y+6$

98. To find the perimeter of a figure, add the lengths of all sides.
$w+4m+(w-3)+(2m+3)+3+(2m-3)$ Sum of all sides.
$=w+4m+w-3+2m+3+3+2m-3$ Combine like terms. Constant terms amount to 0.
$=8m+2w$

99. To determine the coordinates of the points we need to find the values of the x- and y-coordinates. The first point is located 6 units to the right of the y-axis, so the x-coordinate is $x=6$. The point is located 1 unit above the x-axis, so the y-coordinate is $y=1$.

The coordinates of this point are $(6,1)$.

The next point is located 4 units to the right of the y-axis, so the x-coordinate is $x=4$. The point is located 5 units above the x-axis, so the y-coordinate is $y=5$.

The coordinates of this point are $(4,5)$.

The next point is located 1 unit to the left of the y-axis, so the x-coordinate is $x=-1$. The point is located 6 units above the x-axis, so the y-coordinate is $y=6$.

The coordinates of this point are $(-1,6)$.

The next point is located 2 units to the left of the y-axis, so the x-coordinate is $x=-2$. The point is located on the x-axis, so the y-coordinate is $y=0$.

The coordinates of this point are $(-2,0)$.

The next point is located 5 units to the left of the y-axis, so the x-coordinate is $x=-5$. The point is located 2 units below the x-axis, so the y-coordinate is $y=-2$.

The coordinates of this point are $(-5,-2)$.

The next point is located 3 units to the right of the y-axis, so the x-coordinate is $x=3$. The point is located 4 units below the x-axis, so the y-coordinate is $y=-4$.

The coordinates of this point are $(3,-4)$.

100. To determine the coordinates of the points we need to find the values of the x- and y-coordinates. The first point is located 5 units to the right of the y-axis, so the x-coordinate is $x=5$. The point is located on the x-axis, so the y-coordinate is $y=0$.

The coordinates of this point are $(5,0)$.

The next point is located 6 units to the right of the y-axis, so the x-coordinate is $x=6$. The point is located 3 units above the x-axis, so the y-coordinate is $y=3$.

The coordinates of this point are $(6,3)$.

The next point is located on the y-axis, so the x-coordinate is $x=0$. The point is located 6 units above the x-axis, so the y-coordinate is $y=6$.

The coordinates of this point are $(0,6)$.

The next point is located 4 units to the left of the y-axis, so the x-coordinate is $x=-4$. The point is located 3 units above the x-axis, so the y-coordinate is $y=3$.

The coordinates of this point are $(-4,3)$.

The next point is located 3 units to the left of the y-axis, so the x-coordinate is $x=-3$. The point is located 5 units below the x-axis, so the y-coordinate is $y=-5$.

The coordinates of this point are $(-3,-5)$.

The next point is located 2 units to the right of the y-axis, so the x-coordinate is $x=2$. The point is located 1 unit below the x-axis, so the y-coordinate is $y=-1$.

The coordinates of this point are $(2,-1)$.

101. The x-axis has a scale of 30 because the even spacing on the x-axis is by 30's. The y-axis has a scale of 7 because the even spacing on the y-axis is by 7's.

102. The x-axis has a scale of 10 because the even spacing on the x-axis is by 10's. The y-axis has a scale of 200 because the even spacing on the y-axis is by 200's.

103. The input values range from -5 to 4. It is reasonable to scale the horizontal axis by 1's. The output values range from -7 to 9 so it will be reasonable to scale the y-axis by 2's. The points are plotted on the graph below.

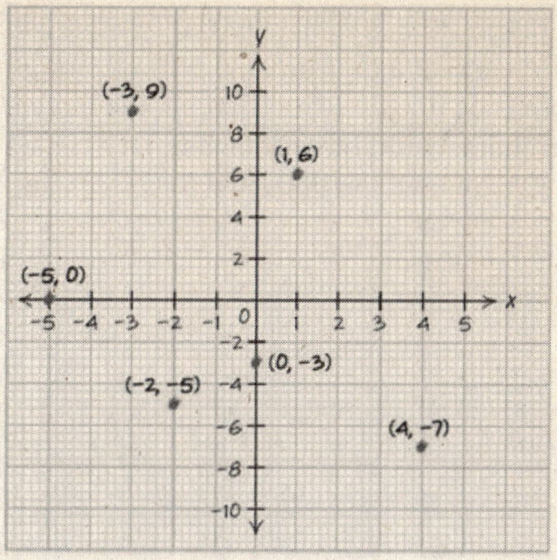

104. The input values range from -9 to 4. It is reasonable to scale the horizontal axis by 2's. The output values range from -5 to 5 so it will be reasonable to scale the y-axis by 1's. The points are plotted on the graph below.

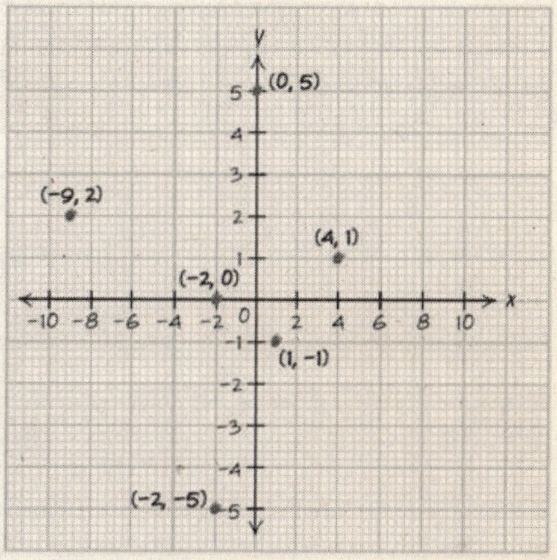

105. The input values are various states within the U.S., so we plot these values on the horizontal axis. The output values range from 8.3% to 14.6%. Thus, we will draw the vertical axis from 0 to 15, counting by increments of 1. The bar chart should have a title and the chart would be as follows:

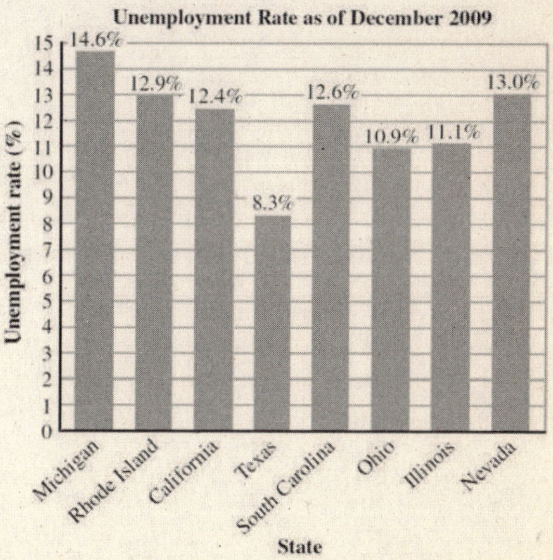

106. The input values are years 200 through 2007, so we plot these values on the horizontal axis. The output values range from 3.99% to 6.0%. Thus, we will draw the vertical axis from 0 to 6.5, counting by increments of 0.5. The bar chart should have a title and the chart would be as follows:

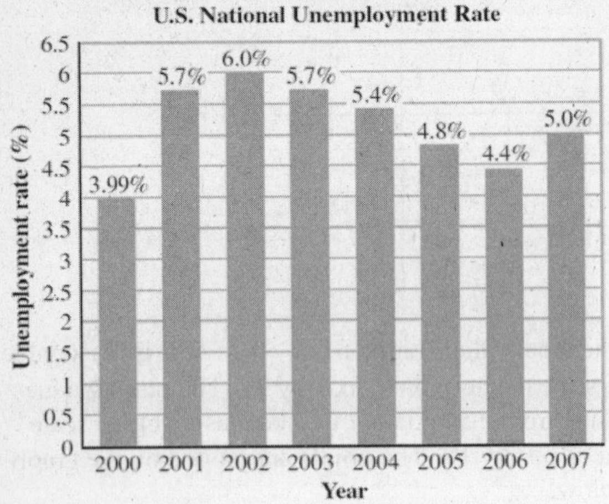

Chapter 1 Test

1. $3\cdot3\cdot3\cdot3\cdot3\cdot3=3^6$

2. $\left(\frac{-5}{8}\right)\cdot\left(\frac{-5}{8}\right)\cdot\left(\frac{-5}{8}\right)=\left(\frac{-5}{8}\right)^3$

3. $(-2)^3=(-2)\cdot(-2)\cdot(-2)=-8$

4. $4.299\times10^7=4.299\cdot10{,}000{,}000=42{,}990{,}000$

5. $\boxed{-6}+\boxed{3\cdot4}\boxed{-(-5)}+\boxed{12\div(-3)}$ There are 4 terms in this expression. Simplify as follows:

$-6+3\cdot4-(-5)+12\div(-3)$ Multiply first.

$=-6+12-(-5)+12\div(-3)$ Then Divide.

$=-6+12-(-5)+(-4)$ Add/Subtract from left to right.

$=7$

6. $\boxed{24\div2\cdot3}+\boxed{(-2)^2}\boxed{-(-1)}$ There are 3 terms in this expression. Simplify as follows:

$24\div2\cdot3+(-2)^2-(-1)$ Evaluate exponent first.

$=24\div2\cdot3+4-(-1)$ Then Divide.

$=12\cdot3+4-(-1)$ Multiply next.

$=36+4-(-1)$ Add/Subtract from left to right.

$=41$

7. $\boxed{\frac{12-3\cdot2}{4-2}}+\boxed{8\div4}$ There are 2 terms in this expression. Simplify as follows:

$\frac{12-3\cdot2}{4-2}+8\div4$ Simplify top/bottom of fraction first. Multiply.

$=\frac{12-6}{4-2}+8\div4$ Finish simplifying the fraction.

$=3+8\div4$ Divide.

$=3+2$ Add.

$=5$

8. $\boxed{-3}+\boxed{|7\cdot(-2)|}\boxed{-(-5)}+\boxed{3\cdot6}$ There are 4 terms in this expression. Simplify as follows:

$-3+|7\cdot(-2)|-(-5)+3\cdot6$ Simplify inside absolute value. Multiply.

$=-3+|-14|-(-5)+3\cdot6$ Evaluate absolute value.

$=-3+14-(-5)+3\cdot6$ Multiply.

$=-3+14-(-5)+18$ Add/Subtract from left to right.

$=34$

9. $(-3+9)+(-6)=-3+(9+(-6))$

This is an example of the associative property of addition, because the order of the numbers was not changed, but how they are grouped was changed.

10. $4(59-11)=4\cdot59-4\cdot11$

This is an example of the distributive property because each term inside the parentheses has been multiplied by the term outside the parentheses.

11. $72(-89) = (-89) \cdot 72$

This is an example of the commutative property of multiplication because the order of the numbers was changed.

12. If Lucas works 24 hours in a week, then we can substitute 24 for h and we get $13.75(24) = 330$.

If Lucas works 24 hours in a week, his weekly pay is \$330.

13. There is one quantity that can change. This quantity is the number of hours charged for labor. Let L = the number of hours charged for labor. The plumber charges \$65 per hour for labor which translates to $65L$, plus a charge of \$20. The total cost for the plumber's services will be $20 + 65L$.

14. Convert 100 yards to meters. First, form the unity fraction. We want to convert from the units of yards, so the units of yards will go in the denominator. We want to convert to the units of meters, so meters will go in the numerator. There are 1.09 yd in 1 m, so the unity fraction is $\frac{1\text{ m}}{1.09\text{ yd}}$. Multiply 100 yards by this unity fraction and simplify the expression as follows.

$$100\text{ yd} \cdot \frac{1\text{ m}}{1.09\text{ yd}}$$

$$= \frac{100\ \cancel{\text{yd}}}{1} \cdot \frac{1\text{ m}}{1.09\ \cancel{\text{yd}}}$$

$$= \frac{100}{1.09}\text{ m} \approx 91.74\text{ m}$$

Therefore, 100 yd. equals approximately 91.74 m.

15. Evaluate $x^2 + 2x - 1$ for $x = -3$.

$x^2 + 2x - 1$ — Substitute in $x = -3$.

$= (-3)^2 + 2(-3) - 1$ — Simplify using the order-of-operations agreement.

$= 9 - 6 - 1$

$= 2$

16.

$-3(-2x^2 + 5x - 1)$ — Use distributive property, multiply each term by -3.

$= -3(-2x^2) + (-3)(5x) + (-3)(-1)$ — Multiply.

$= 6x^2 - 15x + 3$

17. Combine like terms.

$-5hk + 3k^2 - 2h + 7 + 2hk + 9h + 15$
$= 7h - 3hk + 3k^2 + 22$

18.

$-2(x + 3y) + 4(2y) + 5x$ — Evaluate parentheses. Use distributive prop. and multiplication prop.

$= -2(x) + (-2)(3y) + 8y + 5x$ — Multiply.

$= -2x - 6y + 8y + 5x$ — Combine like terms.

$= 3x + 2y$

19. **a.** The line goes to its highest point between 2003 and 2004 on the horizontal axis, slightly past the mid-way point, therefore the index was at its highest during the third quarter of 2003.

b. The line goes to its lowest point between 2006 and 2007 on the horizontal axis, slightly past the mid-way point, therefore the index was at its lowest during the third quarter of 2006.

c. The vertical axis is spaced evenly by increments of 5's therefore the scale is 5.

d. The horizontal axis is spaced evenly by increments of 1's therefore the scale is 1.

20. The x-axis has a scale of 12 because the even spacing on the x-axis is by 12's. The y-axis has a scale of 5 because the even spacing on the y-axis is by 5's.

21. The input values range from -50 to 40. It is reasonable to scale the horizontal axis by 10's. The output values range from -3.0 to 4.5 so it will be reasonable to scale the y-axis by 1's. The points are plotted on the graph below.

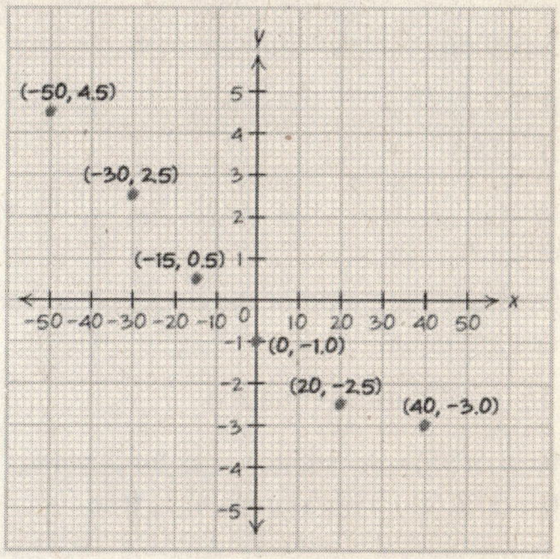

22. To determine the coordinates of the points we need to find the values of the x- and y-coordinates. The first point is located 4 units to the right of the y-axis, so the x-coordinate is $x = 4$. The point is located 50 units above the x-axis, so the y-coordinate is $y = 50$.

The coordinates of this point are $(4, 50)$.

The next point is located 4 units to the left of the y-axis, so the x-coordinate is $x = -4$. The point is located 50 units above the x-axis, so the y-coordinate is $y = 50$.

The coordinates of this point are $(-4, 50)$.

The next point is located 10 units to the left of the y-axis, so the x-coordinate is $x = -10$. The point is located on the x-axis, so the y-coordinate is $y = 0$.

The coordinates of this point are $(-10, 0)$.

The next point is located on the y-axis, so the x-coordinate is $x = 0$. The point is located 50 units below the x-axis, so the y-coordinate is $y = -50$.

The coordinates of this point are $(0, -50)$.

The next point is located 4 units to the right of the y-axis, so the x-coordinate is $x = 4$. The point is located 10 units below the x-axis, so the y-coordinate is $y = -10$.

The coordinates of this point are $(4, -10)$.

CHAPTER 2 LINEAR EQUATIONS AND INEQUALITIES WITH ONE VARIABLE

Section 2.1 Addition and Subtraction Properties of Equality

1. This is an expression. There is no equal sign.
3. This is an equation. There is an equal sign.
5. This is an expression. There is no equal sign.
7. This is an equation. There is an equal sign.
9. This is an equation. There is an equal sign.
11. This is an expression. There is no equal sign.

13. **a.** Using the given equation $S = 9.5h$ replace h with 20 and solve for S.

$S = 9.5h$

$S = 9.5 \cdot 20$

$S = 190$

If Joe works 20 hours in a week, his salary will be \$190.00.

b. Evaluate each side of the equation and see if the two sides are equal.

$S = 9.5h$ Original equation.

$332.50 \stackrel{?}{=} 9.5(35)$ Substitute values for S and h.

$332.50 = 332.50$ This statement is true.

Because the final statement is true, $h = 35$ and $S = 332.50$ is a solution to the equation $S = 9.5h$. If Joe works 35 hours, then he will earn a salary of \$332.50 for the week.

c. Evaluate each side of the equation and see if the two sides are equal.

$S = 9.5h$ Original equation.

$240 \stackrel{?}{=} 9.5(25)$ Substitute values for S and h.

$240 \neq 237.50$ The statement is not true.

Because the final results are not equal, $h = 25$ and $S = 240$ do not give a solution to the equation $S = 9.5h$. These values state that if Joe works 25 hours in a week, then his salary would be \$240. These values do not make the equation true.

d. Evaluate each side of the equation and see if the two sides are equal.

$S = 9.5h$ Original equation.

$380 \stackrel{?}{=} 9.5(40)$ Substitute values for S and h.

$380 = 380$ This statement is true.

Because the final statement is true, $h = 40$ and $S = 380$ is a solution to the equation $S = 9.5h$. If Joe works 40 hours, then he will earn a salary of \$380 for the week.

15. **a.** Evaluate each side of the equation and see if the two sides are equal.

$D = 60t$ Original equation.

$240 \stackrel{?}{=} 60(4)$ Substitute values for D and t.

$240 = 240$ This statement is true.

Because the final statement is true, $t = 4$ and $D = 240$ is a solution to the equation $D = 60t$.

b. These values represent that 240 miles is the distance traveled when driving for 4 hours.

17. **a.** Evaluate each side of the equation and see if the two sides are equal.

$C = 2\pi r$ Original equation.

$37.7 \stackrel{?}{=} 2(3.14)(6)$ Substitute values for D & t.

Substitue 3.14 for π and round answer to one decimal place.

$37.7 = 37.7$ This statement is true.

Because the final statement is true, $r = 6$ and $C = 37.7$ is a solution to the equation $C = 2\pi r$.

b. These values represent that a circle with a radius of 6 has a circumference of 37.7 (rounded to one decimal place).

19. **a.** Evaluate each side of the equation and see if the two sides are equal.

$F = 15L$ Original equation.

$4500 \stackrel{?}{=} 15(300)$ Substitute values for F and L.

$4500 = 4500$ This statement is true.

Because the final statement is true, $L = 300$ and $F = 4500$ is a solution to the equation $F = 15L$.

b. These values represent that a loan of \$300,000.00 would have an origination fee of \$4,500.00.

21. **a.** Evaluate each side of the equation and see if the two sides are equal.

$F = 10L$ Original equation.

$2000 \stackrel{?}{=} 10(225)$ Substitute values for F and L.

$2000 \stackrel{?}{=} 2250$ This statement is not true.

$2000 \neq 2250$

Because the final results are not equal, $L = 225$ and $F = 2000$ do not give a solution to the equation $F = 10L$.

b. These values state that a loan of \$225,000.00 would have an origination fee of \$2,000.00. These values do not make the equation true, so they do not make sense in this situation.

23. **a.** Evaluate each side of the equation and see if the two sides are equal.

$S = 77.8f$ Original equation.

$155.6 \stackrel{?}{=} 77.8(2)$ Substitute values for S and f.

$155.6 = 155.6$ This statement is true.

Because the final statement is true, $f = 2$ and $S = 155.6$ is a solution to the equation $S = 77.8f$.

b. These values represent that it would require 155.6 cubic yards of sand to fill a volleyball court 2 feet deep.

25. Evaluate each side of the equation and see if the two sides are equal.

$2x+8=30$	Original equation.
$2(11)+8\stackrel{?}{=}30$	Substitute $x=11$.
$22+8\stackrel{?}{=}30$	Simplify both sides.
$30=30$	This statement is true.

Because the final statement is true, $x=11$ is a solution to the equation $2x+8=30$.

27. Evaluate each side of the equation and see if the two sides are equal.

$7m+15=10m$	Original equation.
$7(8)+15\stackrel{?}{=}10(8)$	Substitute $m=8$.
$56+15\stackrel{?}{=}80$	Simplify both sides.
$71\stackrel{?}{=}80$	This statement is not true.
$71\neq 80$	

Because the final results are not equal, $m=8$ is not a solution to the equation $7m+15=10m$.

29. Evaluate each side of the equation and see if the two sides are equal.

$\frac{3}{4}x-13=8$	Original equation.
$\frac{3}{4}(28)-13\stackrel{?}{=}8$	Substitute $x=28$.
$21-13\stackrel{?}{=}8$	Simplify both sides.
$8=8$	This statement is true.

Because the final statement is true, $x=28$ is a solution to the equation $\frac{3}{4}x-13=8$.

31. Evaluate each side of the equation and see if the two sides are equal.

$\frac{1}{2}x-10=\frac{-15}{2}$	Original equation.
$\frac{1}{2}(5)-10\stackrel{?}{=}\frac{-15}{2}$	Substitute $x=5$.
$\frac{5}{2}-10\stackrel{?}{=}\frac{-15}{2}$	Simplify both sides.
$\frac{5}{2}-\frac{20}{2}\stackrel{?}{=}\frac{-15}{2}$	
$\frac{-15}{2}=\frac{-15}{2}$	This statement is true.

Because the final statement is true, $x=5$ is a solution to the equation $\frac{1}{2}x-10=\frac{-15}{2}$.

33. Evaluate each side of the equation and see if the two sides are equal.

$0.15x+1.2=1.35$	Original equation.
$0.15(1)+1.2\stackrel{?}{=}1.35$	Substitute $x=1$.
$0.15+1.2\stackrel{?}{=}1.35$	Simplify both sides.
$1.35=1.35$	This statement is true.

Because the final statement is true, $x=1$ is a solution to the equation $0.15x+1.2=1.35$.

35. Evaluate each side of the equation and see if the two sides are equal.

$0.5t+3.2=1.7$	Original equation.
$0.5(3)+3.2\stackrel{?}{=}1.7$	Substitute $t=3$.
$1.5+3.2\stackrel{?}{=}1.7$	Simplify both sides.
$4.7\stackrel{?}{=}1.7$	This statement is not true.
$4.7\neq 1.7$	

Because the final results are not equal, $t=3$ is not a solution to the equation $0.5t+3.2=1.7$.

37. Evaluate each side of the equation and see if the two sides are equal.

$3x+5y=20$	Original equation.
$3(5)+5(1)\stackrel{?}{=}20$	Substitute $x=5$ and $y=1$.
$15+5\stackrel{?}{=}20$	Simplify both sides.
$20=20$	This statement is true.

Because the final statement is true, $x=5$ and $y=1$ are a solution to the equation $3x+5y=20$.

39. Evaluate each side of the equation and see if the two sides are equal.

$-x+2y=4$	Original equation.
$-(-4)+2(0)\stackrel{?}{=}4$	Substitute $x=-4$ and $y=0$.
$4+0\stackrel{?}{=}4$	Simplify both sides.
$4=4$	This statement is true.

Because the final statement is true, $x=-4$ and $y=0$ are a solution to the equation $-x+2y=4$.

41. Evaluate each side of the equation and see if the two sides are equal.

$25m+15=10n+40$	Original equation.
$25(3)+15=10(5)+40$	Substitute $m=3$ & $n=5$.
$75+15\stackrel{?}{=}50+40$	Simplify both sides.
$90=90$	This statement is true.

Because the final statement is true, $m=3$ and $n=5$ are a solution to the equation $25m+15=10n+40$.

43. Evaluate each side of the equation and see if the two sides are equal.

$-5x+3y=-15$	Original equation.
$-5(0)+3(5)\stackrel{?}{=}-15$	Substitute $x=0$ & $y=5$.
$0+15\stackrel{?}{=}-15$	Simplify both sides.

$15 \stackrel{?}{=} -15$ This statement is not true.

$15 \neq -15$

Because the final results are not equal, $x = 0$ and $y = 5$ are not a solution to the equation $-5x + 3y = -15$.

45. **a.** Using the equation $s + l + c = 220$ substitute $s = 75$ for the weight of the cargo storage system and $l = 100$ for the weight of the luggage, then solve for c to find the number of pounds of other cargo you can store on the roof of the car.

$s + l + c = 220$ Given equation.
$(75) + (100) + c = 220$ Substitute $s = 75$ & $l = 100$.
$175 + c = 220$ Simplify on left side.
$-175 \quad -175$ Subtract 175 from both sides.
$c = 45$

The solution is $c = 45$ which represents that an additional 45 lb. of other cargo could be stored on the roof of the car.

b. Using the equation $s + l + c = 220$ substitute $s = 75$ for the weight of the cargo storage system and $l = 60$ for the weight of the luggage, then solve for c to find the number of pounds of other cargo you can store on the roof of the car.

$s + l + c = 220$ Given equation.
$(75) + (60) + c = 220$ Substitute $s = 75$ & $l = 60$.
$135 + c = 220$ Simplify on left side.
$-135 \quad -135$ Subtract 135 from both sides.
$c = 85$

The solution is $c = 85$ which represents that an additional 85 lb. of other cargo could be stored on the roof of the car.

47. Using the equation $P = R - C$ substitute $R = 47000$ for the revenue, $C = 39000$ for costs, then solve for P to find the company's profit.

$P = R - C$ Given equation.
$P = (47000) - (39000)$ Substitute value of variables.
$P = 8000$ Simplify on right side.

The solution is $P = 8000$ which represents that the company has monthly profit of \$8,000.

49. Using the equation $P = R - C$ substitute $R = 38000$ for the revenue, $C = 41000$ for costs, then solve for P to find the company's profit.

$P = R - C$ Given equation.
$P = (38000) - (41000)$ Substitute value of variables.
$P = -3000$ Simplify on right side.

The solution is $P = -3000$ which represents that the company has monthly profit of $-\$3{,}000$, representing that the company is operating at a loss.

51. Using the equation $P = R - C$ substitute $P = 4000$ for the revenue, $C = 25000$ for costs, then solve for R to find the company's revenue.

$P = R - C$ Given equation.
$(4000) = R - (25000)$ Substitute value of variables.
$+25000 \quad +25000$ Add 25000 to both sides.
$29000 = R$

The solution is $R = 29000$ which represents that the company would need to generate \$29,000 in revenue to earn the desired profit.

53.

$\boxed{x} + 5 = 30$ Identify the variable term.
$-5 \quad -5$ Subtract 5 from both sides.
$x = 25$

$(25) + 5 \stackrel{?}{=} 30$ Check the answer.
$30 = 30$ The answer works.

55.

$\boxed{x} + 20.5 = 45$ Identify the variable term.
$-20.5 \quad -20.5$ Subtract 20.5 from both sides.
$x = 24.5$

$(24.5) + 20.5 \stackrel{?}{=} 45$ Check the answer.
$45 = 45$ The answer works.

57.

$\boxed{m} - 14 = 5$ Identify the variable term.
$+14 \quad +14$ Add 14 to both sides.
$m = 19$

$(19) - 14 \stackrel{?}{=} 5$ Check the answer.
$5 = 5$ The answer works.

59.

$-4 + \boxed{p} = 23$ Identify the variable term.
$+4 \quad +4$ Add 4 to both sides.
$p = 27$

$-4 + (27) \stackrel{?}{=} 23$ Check the answer.
$23 = 23$ The answer works.

61.

$-1 + \boxed{k} = -16$ Identify the variable term.
$+1 \quad +1$ Add 1 to both sides.
$k = -15$

$-1 + (-15) \stackrel{?}{=} -16$ Check the answer.
$-16 = -16$ The answer works.

63.

$5=-3+\boxed{x}$ Identify the variable term.

$+3\ \ +3$ Add 3 to both sides.

$8=x$

$5\overset{?}{=}-3+(8)$ Check the answer.

$5=5$ The answer works.

65.

$\boxed{a}+0.25=2.75$ Identify the variable term.

$-0.25\ \ -0.25$ Subtract 0.25 from both sides.

$a=2.5$

$(2.5)+0.25\overset{?}{=}2.75$ Check the answer.

$2.75=2.75$ The answer works.

67.

$\frac{3}{2}=\frac{1}{2}+\boxed{x}$ Identify the variable term.

$-\frac{1}{2}\ \ -\frac{1}{2}$ Subtract $\frac{1}{2}$ from both sides.

$\frac{2}{2}=x$

$1=x$

$\frac{3}{2}\overset{?}{=}\frac{1}{2}+1$ Check the answer.

$\frac{3}{2}\overset{?}{=}\frac{1}{2}+\frac{2}{2}$

$\frac{3}{2}=\frac{3}{2}$ The answer works.

69.

$\boxed{m}+\frac{2}{3}=\frac{7}{3}$ Identify the variable term.

$-\frac{2}{3}\ \ -\frac{2}{3}$ Subtract $\frac{2}{3}$ from both sides.

$m=\frac{5}{3}$

$\left(\frac{5}{3}\right)+\frac{2}{3}\overset{?}{=}\frac{7}{3}$ Check the answer.

$\frac{7}{3}=\frac{7}{3}$ The answer works.

71.

$\frac{4}{9}+\boxed{m}=\frac{5}{9}$ Identify the variable term.

$-\frac{4}{9}\ \ -\frac{4}{9}$ Subtract $\frac{4}{9}$ from both sides.

$m=\frac{1}{9}$

$\frac{4}{9}+\left(\frac{1}{9}\right)\overset{?}{=}\frac{5}{9}$ Check the answer.

$\frac{5}{9}=\frac{5}{9}$ The answer works.

73.

$5+\boxed{x}=14$ Identify the variable term.

$-5\ \ -5$ Subtract 5 from both sides.

$x=9$

$5+(9)\overset{?}{=}14$ Check the answer.

$14=14$ The answer works.

75.

$\boxed{d}-1.75=6.5$ Identify the variable term.

$+1.75\ \ +1.75$ Add 1.75 to both sides.

$d=8.25$

$(8.25)-1.75\overset{?}{=}6.5$ Check the answer.

$6.5=6.5$ The answer works.

77.

$-1.3+\boxed{x}=-4.7$ Identify the variable term.

$+1.3\ \ +1.3$ Add 1.3 to both sides.

$x=-3.4$

$-1.3+(-3.4)\overset{?}{=}-4.7$ Check the answer.

$-4.7=-4.7$ The answer works.

79.

$\frac{2}{5}+\boxed{x}=\frac{3}{7}$ Identify the variable term.

$-\frac{2}{5}\ \ -\frac{2}{5}$ Subtract $\frac{2}{5}$ from both sides.

$x=\frac{3}{7}-\frac{2}{5}$ Find LCD = 35.

$x=\frac{15}{35}-\frac{14}{35}$ Rewrite over LCD.

$x=\frac{1}{35}$

$\frac{2}{5}+\left(\frac{1}{35}\right)\overset{?}{=}\frac{3}{7}$ Check the answer.

$\frac{14}{35}+\frac{1}{35}\overset{?}{=}\frac{3}{7}$ Find LCD = 35.

$\frac{15}{35}\overset{?}{=}\frac{3}{7}$ Reduce on left side.

$\frac{3}{7}=\frac{3}{7}$ The answer works.

81.

$\frac{4}{5}+\boxed{y}=\frac{3}{10}$ Identify the variable term.

$-\frac{4}{5} \quad -\frac{4}{5}$ Subtract $\frac{4}{5}$ from both sides.

$y=\frac{3}{10}-\frac{4}{5}$ Find LCD = 10.

$x=\frac{3}{10}-\frac{8}{10}$ Rewrite over LCD.

$x=-\frac{5}{10}$ Reduce.

$x=-\frac{1}{2}$

$\frac{4}{5}+\left(-\frac{1}{2}\right)\stackrel{?}{=}\frac{3}{10}$ Check the answer.

$\frac{8}{10}-\frac{5}{10}\stackrel{?}{=}\frac{3}{10}$ Find LCD = 10.

$\frac{3}{10}=\frac{3}{10}$ The answer works.

83.

$P=\boxed{R}-C$ Identify the variable term to isolate.

$+C \quad +C$ Add C to both sides.

$C+P=R$ or $R=C+P$

85.

$s+\boxed{l}+c=220$ Identify the variable term to isolate.

$-s \quad -s$ Subtract s from both sides.

$l+c=220-s$

$-c \quad -c$ Subtract c from both sides.

$l=220-c-s$

87.

$\boxed{A}+B+90^\circ=180^\circ$ Identify the variable term to isolate.

$-B \quad -B$ Subtract B from both sides.

$A+90^\circ=180^\circ-B$

$-90^\circ \quad -90^\circ$ Subtract 90 from both sides.

$A=90^\circ-B$

89.

$T=45+\boxed{M}$ Identify the variable term to isolate.

$-45 \quad -45$ Subtract 45 from both sides.

$T-45=M$ or $M=T-45$

91.

$x+\boxed{y}=-4$ Identify the variable term to isolate.

$-x \quad -x$ Subtract x from both sides.

$y=-x-4$

93.

$D+10=3F+\boxed{G}-20$ Identify variable term to isolate.

$-3F \quad -3F$ Subtract $3F$ from both sides.

$D-3F+10=\boxed{G}-20$

$+20 \quad +20$ Add 20 to both sides.

$D-3F+30=G$ or $G=D-3F+30$

95.

$t=-3u+\boxed{w}$ Identify variable term to isolate.

$+3u \quad +3u$ Add $3u$ to both sides.

$t+3u=w$ or $w=t+3u$

97.

$P=a+b+c$ Identify variable term to isolate.

$-a-c \quad -a \quad -c$ Subtract a and c from both sides.

$P-a-c=b$ or $b=P-a-c$

99.

$3y=\boxed{x}-5$ Identify variable term to isolate.

$+5 \quad +5$ Add 5 to both sides.

$3y+5=x$ or $x=3y+5$

101. The statement has an equal sign therefore it is an equation.

$23=\boxed{x}+9$ Identify variable term to isolate.

$-9 \quad -9$ Subtract 9 from both sides.

$14=x$

$x=14$

103. The statement does not have an equal sign, therefore it is an expression. Simplify:

$\boxed{5x}+7\boxed{-3x}$ Combine like terms

$2x+7$

105. The statement does not have an equal sign, therefore it is an expression. Simplify:

$\boxed{20a^2}+6a\boxed{-3a^2}+10$ Combine like terms

$17a^2+6a+10$

107. The statement has an equal sign therefore it is an equation.

$-14+\boxed{c}=-10$ Identify variable term to isolate.

$+14 \quad +14$ Add 14 to both sides.

$c=4$

109. The statement has an equal sign therefore it is an equation.

$\boxed{p}-14=20$ Identify variable term to isolate.

$+14 \quad +14$ Add 14 to both sides.

$p=34$

Section 2.2 Multiplication and Division Properties of Equality

1. Divide both sides of the equation by 3.

$3x=-81$ Variable term is isolated.

$\frac{3x}{3}=\frac{-81}{3}$ Divide both sides by 3.

$\frac{\cancel{3}x}{\cancel{3}}=-27$

$x=-27$

$3(-27)\overset{?}{=}-81$ Check the answer.

$-81=-81$ The answer works.

3. Divide both sides of the equation by -10.

$-10y=-55$ Variable term is isolated.

$\frac{-10y}{-10}=\frac{-55}{-10}$ Divide both sides by -10.

$\frac{\cancel{-10}y}{\cancel{-10}}=\frac{11}{2}$

$y=5\frac{1}{2}$

$-10\left(5\frac{1}{2}\right)\overset{?}{=}-55$ Check the answer.

$-10\left(\frac{11}{2}\right)\overset{?}{=}-55$

$-55=-55$ The answer works.

5. Divide both sides of the equation by 7.

$7g=91$ Variable term is isolated.

$\frac{7g}{7}=\frac{91}{7}$ Divide both sides by 7.

$\frac{\cancel{7}g}{\cancel{7}}=13$

$g=13$

$7(13)\overset{?}{=}91$ Check the answer.

$91=91$ The answer works.

7. Divide both sides of the equation by -1.

$-x=-9$ Variable term is isolated.

$\frac{-1x}{-1}=\frac{-9}{-1}$ Divide both sides by -1.

$x=9$

Check the answer as in problems 1-5

9. Divide both sides of the equation by 2.8.

$42=2.8x$ Variable term is isolated.

$\frac{42}{2.8}=\frac{2.8x}{2.8}$ Divide both sides by 2.8.

$15=\frac{\cancel{2.8}x}{\cancel{2.8}}$

$15=x$

Check the answer as in problems 1-5

11. Divide both sides of the equation by 6.1.

$15.25=6.1n$ Variable term is isolated.

$\frac{15.25}{6.1}=\frac{6.1n}{6.1}$ Divide both sides by 6.1.

$2.5=\frac{\cancel{6.1}n}{\cancel{6.1}}$

$2.5=n$

Check the answer as in problems 1-5

13. Divide both sides of the equation by -6.

$-6c=42$ Variable term is isolated.

$\frac{-6c}{-6}=\frac{42}{-6}$ Divide both sides by -6.

$\frac{\cancel{-6}c}{\cancel{-6}}=-7$

$c=-7$

Check the answer as in problems 1-5

15. Multiply both sides of the equation by 5.

$\frac{t}{5}=14$ Variable term is isolated.

$5\left(\frac{t}{5}\right)=5(14)$ Multiply both sides by 5.

$\cancel{5}\left(\frac{t}{\cancel{5}}\right)=70$

$t=70$

Check the answer as in problems 1-5

17. Multiply both sides by the reciprocal.

$\frac{2}{3}x=-10$ Variable term is isolated.

$\frac{3}{2}\cdot\frac{2}{3}x=-10\cdot\frac{3}{2}$ Multiply both sides by reciprocal of $\frac{2}{3}$.

$\frac{\cancel{3}}{\cancel{2}}\cdot\frac{\cancel{2}}{\cancel{3}}x=\frac{-5\cdot\cancel{2}\cdot 3}{\cancel{2}}$ Divide out like factors.

$x=-15$

Check the answer as in problems 1-5

19. Multiply both sides by the reciprocal.

$\frac{5n}{7}=\frac{-2}{5}$ Variable term is isolated.

$\frac{7}{5}\cdot\frac{5n}{7}=\frac{-2}{5}\cdot\frac{7}{5}$ Multiply both sides by reciprocal of $\frac{5}{7}$.

$\frac{\cancel{7}}{\cancel{5}}\cdot\frac{\cancel{5}}{\cancel{7}}n=\frac{-14}{25}$ Divide out like factors.

$x=\frac{-14}{25}$

Check the answer as in problems 1-5

21. Multiply both sides of the equation by -3.5.

$$\frac{k}{-3.5} = -4 \quad \text{Variable term is isolated.}$$

$$-3.5\left(\frac{k}{-3.5}\right) = -3.5(-4) \quad \text{Multiply both sides by } -3.5.$$

Ch

$$\not{-3.5}\left(\frac{k}{\not{-3.5}}\right) = 14$$

$$k = 14$$

eck the answer as in problems 1-5

23. **a.**

$S = 17h$ Given equation.
$S = 17(80)$ Substitute $h = 80$.
$S = 1360$ Multiply & solve for S.

If Victoria works 80 hours, her salary will be \$1360.

b.

$S = 17h$ Given equation.
$2040 = 17h$ Substitute $S = 2040$ & solve for h.
$\frac{2040}{17} = \frac{17h}{17}$ Divide both sides by 17.
$120 = \frac{\not{17}h}{\not{17}}$
$120 = h$

Victoria has to work 120 hours to make \$2040 a month.

c.

$S = 17h$ Given equation.
$700 = 17h$ Substitute $S = 700$ & solve for h.
$\frac{700}{17} = \frac{17h}{17}$ Divide both sides by 17.
$41.18 \approx \frac{\not{17}h}{\not{17}}$
$41.18 \approx h$

Victoria has to work approximately 41.18 hours to earn \$700.

25. **a.**

$F = 0.015L$ Given equation.
$F = 0.015(180000)$ Substitute $L = 180000$.
$F = 2700$ Multiply & solve for F.

The loan fees will be \$2700 for a loan of \$180,000.

b.

$F = 0.015L$ Given equation.
$2000 = 0.015L$ Substitute $F = 2000$.
$\frac{2000}{0.015} = \frac{0.015L}{0.015}$ Divide both sides by 0.015.
$133333.33 \approx \frac{\not{0.015}L}{\not{0.015}}$
$133333.33 = L$

Pablo can suggest a maximum loan amount of approximately \$133,333.33 to keep the fees down to only \$2000.

27. **a.**

$D = 18b$ Given equation.
$D = 18(500)$ Substitute $b = 500$.
$D = 9000$ Multiply & solve for D.

There will be 9000 bricks that do not meet the quality standards.

b.

$D = 18b$ Given equation.
$D = 18(40.8)$ Substitute $b = 40.8$.
$D = 734.4$ Multiply & solve for D.

Rounding to the nearest whole number, approximately 734 bricks will not meet the quality standards of the 40.8 million bricks produced in a day.

c.

$D = 18b$ Given equation.
$21600 = 18b$ Substitute $D = 21600$.
$\frac{21600}{18} = \frac{18b}{18}$ Divide both sides by 18.
$1200 = \frac{\not{18}b}{\not{18}}$
$1200 = b$

LEGO produced 1200 million (or 1.2 billion) bricks if there are 21,600 bricks that do not meet the company's quality standards.

29. **a.**

$B = \frac{P}{60}$ Given equation.
$B = \frac{300}{60}$ Substitute $P = 300$. Solve for B.
$B = 5$

A total of 5 buses would be needed to transport 300 people.

b.

$B = \frac{P}{60}$ Given equation.
$350 = \frac{P}{60}$ Substitute $B = 350$.
$60 \cdot 350 = \frac{P}{60} \cdot 60$ Multiply both sides by 60.
$21000 = \frac{P}{\not{60}} \cdot \not{60}$ Solve for P.
$21000 = P$

Public transportation can handle 21,000 people at a time with 350 city buses.

c.

$B = \frac{P}{60}$ Given equation.

$600 = \frac{P}{60}$ Substitute $B = 600$.

$60 \cdot 600 = \frac{P}{60} \cdot 60$ Multiply both sides by 60.

$36000 = \frac{P}{\cancel{60}} \cdot \cancel{60}$ Solve for P.

$36000 = P$

The city can transport 36,000 people at a time with 600 school and city buses.

31. **a.**

$M = -237t + 3796$ Given equation.
$M = -237(6) + 3796$ Substitute $t = 6$.
$M = -1422 + 3796$ Multiply & solve for M.
$M = 2374$

The total amount of money taken out is $2374 billion.

b.

$M = -237t + 3796$ Given equation.
$1900 = -237t + 3796$ Substitute $M = 1900$.
$-3796 \quad -3796$ Isolate variable term.
$1896 = -237t$
$\frac{-1896}{-237} = \frac{\cancel{-237}}{\cancel{-237}}t$ Divide both sides by – 237.
$8 = t$

In the year 2008 the total amount of money in mortgages for single and multi-family homes in the United States was only $1900 billion.

33. **a.**

$f = 0.01p + 20$ Given equation.
$f = 0.01(15250) + 20$ Substitute $p = 15250$.
$f = 152.5 + 20$ Multiply & solve for f.
$f = 172.5$

The fee will be $172.50.

b.

$f = 0.01p + 20$ Given equation.
$124 = 0.01p + 20$ Substitute $f = 124$.
$-20 \quad -20$ Isolate variable term.
$104 = 0.01p$
$\frac{104}{0.01} = \frac{\cancel{0.01}p}{\cancel{0.01}}$ Divide both sides by 0.01.
$10400 = p$

The final price of the log splitter was $10,400.

c.

$f = 0.01p + 20$ Given equation.
$270 = 0.01p + 20$ Substitute $f = 270$.
$-20 \quad -20$ Isolate variable term.
$250 = 0.01p$
$\frac{250}{0.01} = \frac{\cancel{0.01}p}{\cancel{0.01}}$ Divide both sides by 0.01.
$25000 = p$

The final price of the tractor was $25,000.

35. **a.**

$F = 0.25 + 0.025p$ Given equation.
$F = 0.25 + 0.025(100)$ Substitute $p = 100$.
$F = 0.25 + 2.5$ Multiply & solve for f.
$F = 2.75$

The fee for a $100 transaction is $2.75.

b.

$F = 0.25 + 0.025p$ Given equation.
$9.25 = 0.25 + 0.025p$ Substitute $F = 9.25$.
$-0.25 \quad -0.25$ Isolate variable term.
$9 = 0.025p$
$\frac{9}{0.025} = \frac{\cancel{0.025}p}{\cancel{0.025}}$ Divide both sides by 0.025.
$360 = p$

The total transaction was $360 if the fee charged was $9.25.

c.

$F = 0.25 + 0.025p$ Given equation.
$37.75 = 0.25 + 0.025p$ Substitute $F = 37.75$.
$-0.25 \quad -0.25$ Isolate variable term.
$37.5 = 0.025p$
$\frac{37.5}{0.025} = \frac{\cancel{0.025}p}{\cancel{0.025}}$ Divide both sides by 0.025.
$1500 = p$

The total transaction was $1500.

37.

$\boxed{5x} + 20 = 60$ Identify variable term to isolate.
$-20 \quad -20$ Subtract 20 from both sides.
$5x = 40$
$\frac{\cancel{5}x}{\cancel{5}} = \frac{40}{5}$ Divide both sides by 5.
$x = 8$

$5(8) + 20 \stackrel{?}{=} 60$ Check the answer.
$40 + 20 \stackrel{?}{=} 60$
$60 = 60$ The answer works.

39.

$\boxed{-3p} + 50 = -85$ Identify variable term to isolate.
$-50 \quad -50$ Subtract 50 from both sides.
$-3p = -135$
$\frac{\cancel{-3}p}{\cancel{-3}} = \frac{-135}{-3}$ Divide both sides by – 3.
$p = 45$

$-3(45)+50 \stackrel{?}{=} -85$ Check the answer.

$-135+50 \stackrel{?}{=} -85$

$-85=-85$ The answer works.

41.

$\boxed{-2.5c}+15=-5$ Identify variable term to isolate.

$\underline{\quad -15 \quad -15}$ Subtract 15 from both sides.

$-2.5c=-20$

$\frac{\cancel{-2.5}c}{\cancel{-2.5}}=\frac{-20}{-2.5}$ Divide both sides by -2.5.

$c=8$

Check the answer as in problems 37-40.

43.

$\boxed{-2.3t}+20=26.9$ Identify variable term to isolate.

$\underline{\quad -20 \quad -20}$ Subtract 20 from both sides.

$-2.3t=6.9$

$\frac{\cancel{-2.3}t}{\cancel{-2.3}}=\frac{6.9}{-2.3}$ Divide both sides by -2.3.

$t=-3$

Check the answer as in problems 37-40.

45.

$48=\boxed{12m}-30$ Identify variable term to isolate.

$\underline{+30 \qquad +30}$ Add 30 to both sides.

$78=12m$

$\frac{78}{12}=\frac{\cancel{12}m}{\cancel{12}}$ Divide both sides by 12.

$6.5=m$

Check the answer as in problems 37-40.

47.

$-251.8=\boxed{-8.6h}+14.8$ Identify variable term to isolate.

$\underline{-14.8 \qquad -14.8}$ Subtract 14.8 from both sides.

$-266.6=-8.6h$

$\frac{-266.6}{-8.6}=\frac{\cancel{-8.6}h}{\cancel{-8.6}}$ Divide both sides by -8.6.

$31=h$

Check the answer as in problems 37-40.

49.

$-134.0=-4.7x-16.5$ Identify variable term to isolate.

$\underline{+16.5 \qquad +16.5}$ Add 16.5 to both sides.

$-117.5=-4.7x$

$\frac{-117.5}{-4.7}=\frac{\cancel{-4.7}x}{\cancel{-4.7}}$ Divide both sides by -4.7.

$25=x$

Check the answer as in problems 37-40.

51.

$\boxed{-2n}+3.8=4.85$ Identify variable term to isolate.

$\underline{\quad -3.8 \quad -3.8}$ Subtract 3.8 from both sides.

$-2n=1.05$

$\frac{\cancel{-2}n}{\cancel{-2}}=\frac{1.05}{-2}$ Divide both sides by -2.

$n=-0.525$ Round answer to 2 decimal places.

$n\approx-0.53$

Check the answer as in problems 37-40.

53.

$\boxed{7m}+8=31.8$ Identify variable term to isolate.

$\underline{\quad -8 \quad -8}$ Subtract 8 from both sides.

$7m=23.8$

$\frac{\cancel{7}m}{\cancel{7}}=\frac{23.8}{7}$ Divide both sides by 7.

$m=3.4$

Check the answer as in problems 37-40.

55.

$\boxed{-5z}+16=32$ Identify variable term to isolate.

$\underline{\quad -16 \quad -16}$ Subtract 16 from both sides.

$-5z=16$

$\frac{\cancel{-5}z}{\cancel{-5}}=\frac{16}{-5}$ Divide both sides by -5.

$z=-3.2$

Check the answer as in problems 37-40.

57.

$\boxed{\frac{x}{4}}+6=9$ Identify variable term to isolate.

$\underline{\quad -6 \quad -6}$ Subtract 6 from both sides.

$\frac{x}{4}=3$

$\cancel{4}\cdot\frac{x}{\cancel{4}}=3\cdot4$ Multiply both sides by 4.

$x=12$

Check the answer as in problems 37-40.

59.

$20.1=\boxed{\frac{w}{8}}+17$ Identify variable term to isolate.

$\underline{-17 \qquad -17}$ Subtract 17 from both sides.

$3.1=\frac{w}{8}$

$8\cdot3.1=\frac{w}{\cancel{8}}\cdot\cancel{8}$ Multiply both sides by 8.

$24.8=w$

Check the answer as in problems 37-40.

61.

$\boxed{\frac{x}{-16}}+5=2.5$ Identify variable term to isolate.

$\underline{\quad -5 \quad -5}$ Subtract 5 from both sides.

$\frac{x}{-16}=-2.5$

$\cancel{-16}\cdot\frac{x}{\cancel{-16}}=-2.5\cdot-16$ Multiply both sides by -16.

$x=40$

Check the answer as in problems 37-40.

63.

$$\boxed{\frac{2x}{3}}+\frac{1}{5}=\frac{4}{5} \quad \text{Identify variable term to isolate.}$$

$$-\frac{1}{5} \quad -\frac{1}{5} \quad \text{Subtract } \frac{1}{5} \text{ from both sides.}$$

$$\frac{2x}{3}=\frac{3}{5} \quad \text{Multiply by reciprocal of coefficient.}$$

$$\frac{\cancel{3}}{\cancel{2}}\cdot\frac{\cancel{2}x}{\cancel{3}}=\frac{3}{5}\cdot\frac{3}{2} \quad \text{Multiply both sides by } \frac{3}{2}.$$

$$x=\frac{9}{10}$$

Check the answer as in problems 37-40.

65. **a.** Let n = the number of items to be shipped.

Let C = the total cost in dollars for shipping.

$$C=6.99+1.99n$$

b. Substitute $n=5$ into the equation from part a, then solve for C.

$$C=6.99+1.99n \quad \text{Given equation.}$$
$$C=6.99+1.99(5) \quad \text{Substitute } n=5.$$
$$C=6.99+9.95 \quad \text{Simplify.}$$
$$C=16.94$$

It costs $16.94 to ship 5 items UPS Ground.

c. Substitute $C=12.96$ into the equation from part a, then solve for n.

$$C=6.99+1.99n \quad \text{Substitute } C=12.96.$$
$$12.96=6.99+1.99n \quad \text{Isolate variable term.}$$
$$-6.99 \quad -6.99 \quad \text{Subtract 6.99 from both sides.}$$
$$5.97=1.99n \quad \text{Divide both sides by 1.99.}$$
$$\frac{5.97}{1.99}=\frac{\cancel{1.99}n}{\cancel{1.99}}$$
$$3=n$$

A total of 3 items can be shipped UPS Ground for $12.96.

67. **a.** Let c = the number of cranks to be produced.

Let T = the total time in minutes it will take the machine shop to produce the cranks.

$$T=360+5c$$

b. Substitute $c=50$ into the equation from part a, then solve for T.

$$T=360+5c \quad \text{Given equation.}$$
$$T=360+5(50) \quad \text{Substitute } n=50.$$
$$T=360+250 \quad \text{Simplify.}$$
$$T=610$$

It will take 610 minutes, or 10 hours and 10 minutes, to produce 50 cranks.

c. Substitute $T=1175$ into the equation from part a, then solve for c.

$$T=360+5c \quad \text{Substitute } T=1175.$$
$$1175=360+5c \quad \text{Isolate variable term.}$$
$$-360 \quad -360 \quad \text{Subtract 360 from both sides.}$$
$$815=5c \quad \text{Divide both sides by 5.}$$
$$\frac{815}{5}=\frac{\cancel{5}c}{\cancel{5}}$$
$$163=c$$

The machine shop made 163 cranks.

69. Let x = "a number." "Times" means to multiply so we have $3x$, and "is" translates to "equals" so the equation is

$$3x=45$$

Solve the equation as follows

$$3x=45 \quad \text{Divide both sides by 3.}$$
$$\frac{\cancel{3}x}{\cancel{3}}=\frac{45}{3}$$
$$x=15$$
$$3(15)\overset{?}{=}45 \quad \text{Check the answer.}$$
$$45=45 \quad \text{The answer works.}$$

71. Recall that the perimeter of a rectangle can be calculated using the formula $P=2l+2w$ where P is the perimeter of the rectangle with length, l, and width, w. Substitute $P=56$ and $w=8$. Solve the equation for l.

$$56=2l+2(8) \quad \text{Substitute in known values.}$$
$$56=\boxed{2l}+16 \quad \text{Simplify \& isolate variable term.}$$
$$-16 \quad -16 \quad \text{Subtract 16 from both sides.}$$
$$40=2l \quad \text{Divide both sides by 2.}$$
$$\frac{40}{2}=\frac{\cancel{2}l}{\cancel{2}}$$
$$20=l$$
$$56\overset{?}{=}2(20)+16 \quad \text{Check the answer.}$$
$$56\overset{?}{=}40+16$$
$$56=56 \quad \text{The answer works.}$$

The length of the garden is 20 feet.

73. Recall that the area of a triangle can be calculated using the formula $A=\frac{1}{2}bh$ where A is the area of the triangle with base, b, and height, h. Substitute $A=40$ and $b=8$. Solve the equation for h.

$$40=\frac{1}{2}(8)h \quad \text{Substitute in known values.}$$
$$40=\boxed{4h} \quad \text{Simplify \& isolate variable term.}$$
$$\frac{40}{4}=\frac{\cancel{4}h}{\cancel{4}} \quad \text{Divide both sides by 4.}$$
$$10=h$$

Check the answer as in problems 69-72.

The height of the triangle is 10 inches.

75. Let x = "a number." “Quotient” represents division and “is” represents the equals sign. Remember that the order of division is important. Read from left to right. Write the quotient in the same order as it is read. The equation translates as

$$\frac{x}{7}=13$$

Solve the equation as follows

$\frac{x}{7}=13$ Multiply both sides by 7.

$$\cancel{7}\cdot\frac{x}{\cancel{7}}=13\cdot 7$$

$$x=91$$

Check the answer as in problems 69-72.

77. Let $x=$ "a number." "Plus" means to add and "product" tells us to multiply. We are adding to 7 times a number. "Is equal to" translates to the equal sign so the equation is

$$6+7x=62$$

Solve the equation as follows

$6+\boxed{7x}=62$ Isolate variable term.
$-6 \quad -6$ Subtract 6 from both sides.
$7x=56$ Divide both sides by 7.

$$\frac{\cancel{7}x}{\cancel{7}}=\frac{56}{7}$$

$$x=8$$

Check the answer as in problems 69-72.

79. Let $x=$ "a number." "Of" means to multiply so we will multiply $\frac{1}{3}$ times x, subtract 8 then set it equal to 28.

$$\frac{1}{3}x-8=28$$

Solve the equation as follows

$\boxed{\frac{1}{3}x}-8=28$ Isolate variable term.
$+8 \quad +8$ Add 8 to both sides.
$\frac{1}{3}x=36$ Multiply both sides by reciprocal.

$$\frac{\cancel{3}}{1}\cdot\frac{1}{\cancel{3}}x=36\cdot\frac{3}{1}$$

$$x=108$$

Check the answer as in problems 69-72

81. Let $x=$ "a number." "Difference" represents subtraction, "twice a number" means to multiply x by 2, and "is" represents the equal sign. The equation translates as

$$2x-15=40$$

Solve the equation as follows

$\boxed{2x}-15=40$ Isolate variable term.
$+15 \quad +15$ Add 15 to both sides
$2x=55$

$\frac{\cancel{2}x}{\cancel{2}}=\frac{55}{2}$ Divide both sides by 2.

$$x=27.5$$

Check the answer as in problems 69-72.

83. Let $x=$ " an unknown number." We will subtract the unknown number from 8 then set it equal to 19.

$$8-x=19$$

Solve the equation as follows

$8\boxed{-x}=19$ Isolate variable term.
$-8 \quad -8$ Subtract 8 from both sides
$-x=11$
$-1\cdot -x=11\cdot -1$ Multiply both sides by -1.
$x=-11$

Check the answer as in problems 69-72.

85. Let $x=$ "a number." "Less than" translates to subtraction so we will subtract 7 from twice (2 times) a number then set it equal to 11.

$$2x-7=11$$

Solve the equation as follows

$\boxed{2x}-7=11$ Isolate variable term.
$+7 \quad +7$ Add 7 to both sides
$2x=18$

$\frac{\cancel{2}x}{\cancel{2}}=\frac{18}{2}$ Divide both sides by 2.

$$x=9$$

Check the answer as in problems 69-72.

87.

$P=\boxed{2l}+2w$ Identify variable term to isolate.
$-2w \quad -2w$ Subtract $2w$ from both sides.
$P-2w=2\boxed{l}$ Identify variable to isolate

$\frac{P-2w}{2}=\frac{\cancel{2}l}{\cancel{2}}$ Divide both sides by 2.

$$\frac{P-2w}{2}=l$$

OR

$\frac{P}{2}-\frac{\cancel{2}w}{\cancel{2}}=l$ Simplify on left side.

$$\frac{P}{2}-w=l$$

89.

$V = \pi r^2 \boxed{h}$ Identify variable to isolate.

$\frac{V}{\pi r^2} = \frac{\not{\pi} \not{r^2} h}{\not{\pi} \not{r^2}}$ Divide both sides by πr^2.

$\frac{V}{\pi r^2} = h$

91.

$A = \frac{1}{2}\boxed{b}h$ Identify variable to isolate.

$2 \cdot A = \frac{1}{\not{2}} bh \cdot \not{2}$ Multiply both sides by 2.

$2A = \boxed{b}h$

$\frac{2A}{h} = \frac{b\not{h}}{\not{h}}$ Divide both sides by h.

$\frac{2A}{h} = b$

93.

$W = \boxed{3t} - 20$ Identify variable term to isolate.
$+20 \quad +20$ Add 20 to both sides.

$W + 20 = 3\boxed{t}$ Identify variable to isolate.

$\frac{W+20}{3} = \frac{\not{3}t}{\not{3}}$ Divide both sides by 3.

$\frac{W+20}{3} = t$

95.

$P = 2b + \boxed{2B}$ Identify variable term to isolate.
$-2b \quad -2b$ Subtract $2b$ from both sides.

$P - 2b = 2\boxed{B}$ Identify variable to isolate

$\frac{P-2b}{2} = \frac{\not{2}B}{\not{2}}$ Divide both sides by 2.

$\frac{P-2b}{2} = B$

OR

$\frac{P}{2} - \frac{\not{2}b}{\not{2}} = B$ Simplify on left side.

$\frac{P}{2} - b = B$

97.

$5x\boxed{-3y} = -15$ Identify variable term to isolate.
$-5x \quad -5x$ Subtract $5x$ from both sides.

$-3\boxed{y} = -5x - 15$ Identify variable to isolate

$\frac{\not{-3}y}{\not{-3}} = \frac{-5x-15}{-3}$ Divide both sides by -3.

$y = \frac{-5x}{-3} - \frac{15}{-3}$ Simplify on right side.

$y = \frac{5}{3}x + 5$

99.

$x\boxed{-5y} = 9$ Identify variable term to isolate.
$-x \quad -x$ Subtract x from both sides.

$-5\boxed{y} = -x + 9$ Identify variable to isolate

$\frac{\not{-5}y}{\not{-5}} = \frac{-x+9}{-5}$ Divide both sides by -5.

$y = \frac{x-9}{5}$ or $y = \frac{x}{5} - \frac{9}{5}$

101.

$y - 2 = 3(\boxed{x} + 7)$ Identify variable to isolate.

$y - 2 = \boxed{3x} + 21$ Distribute on right side.
$-21 \quad -21$ Subtract 21 from both sides.

$y - 23 = 3\boxed{x}$ Identify variable to isolate

$\frac{y-23}{3} = \frac{\not{3}x}{\not{3}}$ Divide both sides by 3.

$\frac{y}{3} - \frac{23}{3} = x$ or $x = \frac{y-23}{3}$

103.

$y + 1 = \frac{3}{4}(\boxed{12x} + 8)$ Identify variable term to isolate.

$y + 1 = \frac{3}{\not{4}}(\not{12}^{3}x + \not{8}^{2})$ Distribute on right side.

$y + 1 = 9x + 6$ Distribute on right side.
$-6 \quad -6$ Subtract 6 from both sides.

$y - 5 = 9\boxed{x}$ Identify variable to isolate

$\frac{y-5}{9} = \frac{\not{9}x}{\not{9}}$ Divide both sides by 9.

$\frac{y}{9} - \frac{5}{9} = x$ or $x = \frac{y-5}{9}$

105.

$I = P\boxed{r}t$ Identify variable to isolate.

$\frac{I}{Pt} = \frac{\not{P}r\not{t}}{\not{P}\not{t}}$ Divide both sides by Pt.

$\frac{I}{Pt} = r$

107. The statement does not have an equal sign, therefore it is an expression.

$5a^2 + \boxed{6a}\boxed{\boxed{-8}} + \boxed{12a} + \boxed{\boxed{3}}$ Identify like terms.

$5a^2 + 18a - 5$ Combine like terms.

109. The statement has an equal sign, therefore it is an equation.

$\boxed{2x} + 7 = 21$ Isolate variable term.
$-7 \quad -7$ Subtract 7 from both sides.
$2x = 14$ Divide both sides by 2.

$\frac{\not{2}x}{\not{2}} = \frac{14}{2}$

$x = 7$

111. The statement has an equal sign, therefore it is an equation.

$\boxed{\frac{1}{3}r}+8\boxed{-\frac{5}{6}r}=-2$ Combine like terms on left side.

$\frac{2}{6}r-\frac{5}{6}r+8=-2$ Rewrite fraction with LCD.

$-\frac{3}{6}r+8=-2$ Reduce fraction.

$\boxed{-\frac{1}{2}r}+8=-2$ Now isolate variable term.

$\quad -8 \quad -8$ Subtract 8 from both sides.

$-\frac{1}{2}r=-10$ Multiply both sides by reciprocal.

$-\frac{\cancel{2}}{1}\cdot-\frac{1}{\cancel{2}}r=-10\cdot-\frac{2}{1}$

$r=20$

113. The statement does not have an equal sign, therefore it is an expression.

$\boxed{5n}+\boxed{7}\boxed{-8n}+20m+\boxed{4}$ Identify like terms.

$20m-3n+11$ Combine like terms.

Arrange in conventional form.

115. The statement has an equal sign, therefore it is an equation.

$\boxed{3.4y}+4\boxed{-2.1y}=11.28$ Combine like terms on left side.

$\boxed{1.3y}+4=11.28$ Isolate variable term.

$\quad -4 \quad -4$ Subtract 4 from both sides.

$1.3y=7.28$ Divide both sides by 1.3.

$\frac{1.3y}{1.3}=\frac{7.28}{1.3}$

$y=5.6$

Section 2.3 Solving Equations with Variables on Both Sides

1. $4x=x+33$

$\boxed{4x}=\boxed{x}+33$ Identify variable terms.

$-x \quad -x$ Subtract x from both sides

$3x=33$

$\frac{\cancel{3}x}{\cancel{3}}=\frac{33}{3}$ Divide both sides by 3.

$x=11$

$4(11)\stackrel{?}{=}(11)+33$ Check the answer.

$44=44$ The answer works.

3. $5x+8=3x-4$

$\boxed{5x}+\boxed{8}=\boxed{3x}\boxed{-4}$ Identify like terms.

$-3x \quad -3x$ Subtract $3x$ from both sides.

$2x+8=-4$

$\quad -8 \quad -8$ Subtract 8 from both sides.

$2x=-12$ Divide both sides by 2.

$\frac{\cancel{2}x}{\cancel{2}}=\frac{-12}{2}$

$x=-6$

$5(-6)+8\stackrel{?}{=}3(-6)-4$ Check the answer.

$-30+8\stackrel{?}{=}-18-4$

$-22=-22$ The answer works.

5. $2x=5(x+6)$

$2x=5(x+6)$ Distribute on the right side.

$\boxed{2x}=\boxed{5x}+30$ Identify like terms.

$-5x \quad -5x$ Subtract $5x$ from both sides.

$-3x=30$

$\frac{\cancel{-3}x}{\cancel{-3}}=\frac{30}{-3}$ Divide both sides by 3.

$x=-10$

$2(-10)\stackrel{?}{=}5([-10]+6)$ Check the answer.

$-20\stackrel{?}{=}5(-4)$

$-20=-20$ The answer works.

7. $20-3x=4x-22$

$\boxed{20}\boxed{-3x}=\boxed{4x}\boxed{-22}$ Identify like terms.

$-4x \quad -4x$ Subtract $4x$ from both sides.

$20-7x=-22$

$-20 \quad -20$ Subtract 20 from both sides.

$-7x=-42$ Divide both sides by -7.

$\frac{\cancel{-7}x}{\cancel{-7}}=\frac{-42}{-7}$

$x=6$

Check the answer as in problems 1-5.

9. $\frac{1}{2}x = x + 5$

$(\cancel{2})\frac{1}{\cancel{2}}x = (2)x + 5(2)$

$\boxed{x} = \boxed{2x} + 10$ Identify like terms.

$-2x \quad -2x$ Subtract $2x$ from both sides.

$-x = 10$

$-1 \cdot -x = 10 \cdot -1$ Multiply both sides by -1.

$x = -10$

Check the answer as in problems 1-5.

11. $5 + \frac{x}{2} = x - 7$

Multiply each term by 2 to clear fraction.

$(2)5 + (\cancel{2})\frac{x}{\cancel{2}} = (2)x - 7(2)$

$\boxed{\boxed{10}} + \boxed{x} = \boxed{2x}\boxed{\boxed{-14}}$ Identify like terms.

$-2x \quad -2x$ Subtract $2x$ from both sides.

$10 - x = -14$

$-10 \quad -10$ Subtract 10 from both sides.

$-x = -24$

$-1 \cdot -x = -24 \cdot -1$ Multiply both sides by -1.

$x = 24$

Check the answer as in problems 1-5.

13. $x - 8 = 5x + 4$

$\boxed{x}\boxed{\boxed{-8}} = \boxed{5x} + \boxed{\boxed{4}}$ Identify like terms.

$-5x \quad -5x$ Subtract $5x$ from both sides.

$-4x - 8 = 4$

$+8 \quad +8$ Add 8 to both sides.

$-4x = 12$ Divide both sides by -7.

$\frac{\cancel{-4}x}{\cancel{-4}} = \frac{12}{-4}$

$x = -3$

Check the answer as in problems 1-5.

15. $\frac{1}{4}x - 8 = 2x - 29$

Multiply each term by 4 to clear fraction.

$(\cancel{4})\frac{1}{\cancel{4}}x - (4)8 = (4)2x - 29(4)$

$\boxed{x}\boxed{\boxed{-32}} = \boxed{8x}\boxed{\boxed{-116}}$ Identify like terms.

$-8x \quad -8x$ Subtract $8x$ from both sides.

$-7x - 32 = -116$

$+32 \quad +32$ Add 32 to both sides.

$-7x = -84$

$\frac{\cancel{-7}x}{\cancel{-7}} = \frac{-84}{-7}$ Divide both sides by -7.

$x = 12$

Check the answer as in problems 1-5.

17.

$\boxed{3x} + 12 = \boxed{7x} - 28$ Combine like terms across = sign.

$-7x \quad -7x$ Subtract $7x$ from both sides.

$-4x + 12 = -28$

$-12 \quad -12$ Subtract 12 from both sides.

$-4x = -40$ Divide both sides by -4.

$\frac{\cancel{-4}x}{\cancel{-4}} = \frac{-40}{-4}$

$x = 10$

$3(10) + 12 \stackrel{?}{=} 7(10) - 28$ Check the answer.

$30 + 12 \stackrel{?}{=} 70 - 28$

$42 = 42$ The answer works.

19.

$\boxed{12t} - 50 = \boxed{4t} + 14$ Combine like terms across = sign.

$-4t \quad -4t$ Subtract $4t$ from both sides.

$8t - 50 = 14$

$+50 \quad +50$ Add 50 to both sides.

$8t = 64$ Divide both sides by 8.

$\frac{\cancel{8}t}{\cancel{8}} = \frac{64}{8}$

$t = 8$

$12(8) - 50 \stackrel{?}{=} 4(8) + 14$ Check the answer.

$96 - 50 \stackrel{?}{=} 32 + 14$

$46 = 46$ The answer works.

21.

$\boxed{4b} + 36 = \boxed{9b} + 66$ Combine like terms across = sign.

$-9b \quad -9b$ Subtract $9b$ from both sides.

$-5b + 36 = 66$

$-36 \quad -36$ Subtract 36 from both sides.

$-5b = 30$ Divide both sides by -5.

$\frac{\cancel{-5}b}{\cancel{-5}} = \frac{30}{-5}$

$b = -6$

Check the answer as in problems 17-20.

23.

$\boxed{2c} + 5 + \boxed{c} = 3c - 4$ Combine like terms on left side.

$3c + 5 = 3c - 4$ Combine like terms across = sign.

$-3c \quad -3c$ Subtract $3c$ from both sides.

$5 \stackrel{?}{=} -4$ This is a false statement.

This equation has no solution.

25.

$4t+5=\boxed{3t}+5+\boxed{t}$ Combine like terms on right side.

$\boxed{4t}+5=\boxed{4t}+5$ Combine like terms across = sign.

$\underline{-4t \qquad -4t}$ Subtract $4t$ from both sides.

$5=5$ This is a true statement.

The equation is an identity therefore the solution is all real numbers or $\mathbb{R}$.

Substitute two random numbers to check.

$4(-4)+5\overset{?}{=}3(-4)+5+(-4)$ Check the answer using -4.

$-16+5\overset{?}{=}-12+5-4$

$-11=-11$ The answer works.

$4(11)+5\overset{?}{=}3(11)+5+(11)$ Check the answer using 11.

$44+5\overset{?}{=}33+5+11$

$49=49$ The answer works.

27.

$\boxed{3.5h}=\boxed{2h}-12$ Combine like terms across = sign.

$\underline{-2h \quad -2h}$ Subtract $2h$ from both sides.

$1.5h=-12$ Divide both sides by 1.5.

$\frac{\cancel{1.5}h}{\cancel{1.5}}=\frac{-12}{1.5}$

$h=-8$

Check the answer as in problems 17-20..

29.

$15\boxed{-x}=8$ Isolate variable.

$\underline{-15 \quad -15}$ Subtract 15 from both sides.

$-x=-7$ Multiply both sides by -1.

$-1\cdot -x=-7\cdot -1$

$x=7$

Check the answer as in problems 17-20.

31.

$\boxed{3h}+8\boxed{-4h}=13$ Combine like terms on left side.

$\boxed{-h}+8=13$ Isolate variable.

$\underline{-8 \quad -8}$ Subtract 8 from both sides.

$-h=5$

$-1\cdot -h=5\cdot -1$ Multiply both sides by -1.

$h=-5$

Check the answer as in problems 17-20.

33.

$2(3x+5)=4x+22$ Simplify on left - distribute.

$\boxed{6x}+10=\boxed{4x}+22$ Combine like terms across = sign.

$\underline{-4x \qquad -4x}$ Subtract $4x$ from both sides.

$2x+10=22$

$\underline{-10 \quad -10}$ Subtract 10 from both sides.

$2x=12$ Divide both sides by 2.

$\frac{\cancel{2}x}{\cancel{2}}=\frac{12}{2}$

$x=6$

Check the answer as in problems 17-20.

35.

$5(2c+3)-8=3(4c+2)-2c$ Distribute on both sides.

$10c+\boxed{\boxed{15}}\boxed{\boxed{-8}}=\boxed{12c}+6\boxed{-2c}$ Simplify on both sides.

$\boxed{10c}+7=\boxed{10c}+6$ Combine like terms across = sign.

$\underline{-10c \qquad -10c}$ Subtract $10c$ from both sides.

$7\overset{?}{=}6$ This is a false statement.

This equation has no solution.

37.

$\boxed{7d}+20\boxed{-3d}=9d+50$ Combine like terms on left side.

$\boxed{4d}+20=\boxed{9d}+50$ Combine like terms across = sign.

$\underline{-9d \qquad -9d}$ Subtract $9d$ from both sides.

$-5d+20=50$

$\underline{-20 \quad -20}$ Subtract 20 from both sides.

$-5d=30$ Divide both sides by -5.

$\frac{\cancel{-5}d}{\cancel{-5}}=\frac{30}{-5}$

$d=-6$

Check the answer as in problems 17-20.

39.

$\frac{1}{2}x+5=3x-45$ Multiply by 2 to remove fraction.

$(\cancel{2})\frac{1}{\cancel{2}}x+(2)5=(2)3x-45(2)$

$\boxed{x}+10=\boxed{6x}-90$ Combine like terms across = sign.

$\underline{-6x \qquad -6x}$ Subtract $6x$ from both sides.

$-5x+10=-90$

$\underline{-10 \quad -10}$ Subtract 10 from both sides.

$-5x=-100$ Divide both sides by -5.

$\frac{\cancel{-5}x}{\cancel{-5}}=\frac{-100}{-5}$

$x=20$

Check the answer as in problems 17-20.

41.

$\frac{1}{3}x+20=2x+20$ Multiply by 3 to remove fraction.

$(\cancel{3})\frac{1}{\cancel{3}}x+(3)20=2x(3)+20(3)$

$\boxed{x}+60=\boxed{6x}+60$ Combine like terms across = sign.

$\underline{-x \qquad -x}$ Subtract x from both sides.

$60=5x+60$

$\underline{-60 \qquad -60}$ Subtract 60 from both sides.

$0=5x$ Divide both sides by 5.

$\frac{0}{5}=\frac{\cancel{5}x}{\cancel{5}}$

$0=x$

Check the answer as in problems 17-20.

43.

$\boxed{9y}-2.1=\boxed{5y}+3.5$ Combine like terms across = sign.

$\underline{-5y \quad\quad -5y}$ Subtract $5y$ from both sides.

$4y-2.1=3.5$

$\underline{+2.1 \quad +2.1}$ Add 2.1 to both sides.

$4y=5.6$ Divide both sides by 4.

$\frac{\cancel{4}y}{\cancel{4}}=\frac{5.6}{4}$

$y=1.4$

Check the answer as in problems 17-20.

45.

$\boxed{35t}+10.2=\boxed{5t}+9.9$ Combine like terms across = sign.

$\underline{-5t \quad\quad -5t}$ Subtract $5t$ from both sides.

$30t+10.2=9.9$

$\underline{-10.2 \quad -10.2}$ Subtract 10.2 from both sides.

$30y=-0.3$ Divide both sides by 4.

$\frac{\cancel{30}y}{\cancel{30}}=\frac{-0.3}{30}$

$y=-0.01$

Check the answer as in problems 17-20.

47. Add the measures of the given angles and set them equal to $180°$ then solve for x.

$x+2x+5x=180$ Combine like terms.

$8x=180$ Divide both sides by 8.

$\frac{\cancel{8}x}{\cancel{8}}=\frac{180}{8}$

$x=22.5$

Substitute $x=22.5$ into the expressions that represent the. The measures of the 3 angles are $22.5°$, $45°$, and $112.5°$.

To check the solution, the measures of the angles should add up to 180.

$22.5°+45°+112.5°\overset{?}{=}180°$ Check the answer.

$180°=180°$ The answer works.

49. Substitute $P=66$, $l=3x+1$, and $w=5x$ into the formula for perimeter and solve for x.

$P=2l+2w$ Formula for perimeter.

$66=2(3x+1)+2(5x)$ Substitute in values.

$66=6x+2+10x$ Distribute/mulitply on right side.

$66=\boxed{16x}+2$ Simplify & isolate variable term.

$\underline{-2 \quad\quad -2}$ Subtract 2 from both sides.

$64=16x$ Divide both sides by 16.

$\frac{64}{16}=\frac{\cancel{16}x}{\cancel{16}}$

$4=x$

Now substitute $x=4$ into the expressions for length and width.

$l=3(4)+1 \quad w=5(4)$ Substitute $x=4$.

$l=12+1 \quad w=20$ Simplify each equation.

$l=13$

The length and width of the yard measure 13 ft and 20 ft.

To check the solution, substitute these values into the formula for perimeter and set it equal to 66.

$P=2l+2w$

$66\overset{?}{=}2(13)+2(20)$ Check the answer.

$66\overset{?}{=}26+40$

$66=66$ The answer works.

51. Substitute $P=83$, and $s=3x-2$ into the formula for perimeter and solve for x.

$P=4s$ Formula for perimeter.

$83=4(3x-2)$ Substitute in value for s.

$83=\boxed{12x}-8$ Distribute/mulitply on right side.

$\underline{+8 \quad\quad +8}$ Add 8 to both sides.

$91=12x$ Divide both sides by 12.

$\frac{91}{12}=\frac{\cancel{12}x}{\cancel{12}}$

$7\frac{7}{12}=x$ or $x=\frac{91}{12}$

Now substitute $x=\frac{91}{12}$ into the expression for the length of each side.

$s=3(\frac{91}{12})-2$ Substitute $x=\frac{91}{12}$.

$s=22\frac{3}{4}-2$ Simplify.

$s=20\frac{3}{4}$ or 20.75

To check the solution, substitute this value into the formula for perimeter and set it equal to 83.

$P=4s$

$83\overset{?}{=}4(20.75)$ Check the answer.

$83=83$ The answer works.

53. Substitute $P=100$, $l=x-5$, and $w=4x$ into the formula for perimeter and solve for x.

$P=2l+2w$ Formula for perimeter.

$100=2(x-5)+2(4x)$ Substitute in values.

$100=2x-10+8x$ Distribute/mulitply on right side.

$100=\boxed{10x}-10$ Simplify & isolate variable term.

$\underline{+10 \quad\quad +10}$ Add 10 to both sides.

$110=10x$ Divide both sides by 10.

$\frac{110}{10}=\frac{\cancel{10}x}{\cancel{10}}$

$11=x$

Now substitute $x = 11$ into the expressions for length and width.

$l = (11) - 5 \quad w = 4(11)$ Substitute $x = 11$.

$l = 6 \quad w = 44$ Simplify each equation.

The length and width of the lawn measure 6 ft and 44 ft.

To check the solution, substitute these values into the formula for perimeter and set it equal to 100.

$P = 2l + 2w$

$100 \stackrel{?}{=} 2(6) + 2(44)$ Check the answer.

$100 \stackrel{?}{=} 12 + 88$

$100 = 100$ The answer works.

55.

$x + 2x + 2x = 180$ Combine like terms.

$5x = 180$ Divide both sides by 5.

$\frac{\cancel{5}x}{\cancel{5}} = \frac{180}{5}$

$x = 36$

Now substitute $x = 36$ into the expressions that represent the angles to find the measure of each angle. The measures of the 3 angles are $36°$, $72°$, and $72°$.

To check the solution, the measures of the angles should add up to 180.

$36° + 72° + 72° \stackrel{?}{=} 180°$ Check the answer.

$180° = 180°$ The answer works.

57.

$3(x - 0.5) + 1 = 3x - 0.5$ Simplify on left - distribute.

$3x - 1.5 + 1 = 3x - 0.5$ Combine like terms on left side.

$\boxed{3x} - 0.5 = \boxed{3x} - 0.5$ Combine like terms across = sign.

$-3x \qquad -3x$ Subtract $3x$ from both sides.

$-0.5 = -0.5$ This is a true statement.

The equation is an identity therefore the solution is all real numbers or $\mathbb{R}$.

To check this we will randomly choose two real numbers -2 and 2, and substitute these into the equation.

$3([-2] - 0.5) + 1 \stackrel{?}{=} 3(-2) - 0.5$ Check the answer using -2.

$3(-2.5) + 1 \stackrel{?}{=} -6 - 0.5$

$-7.5 + 1 \stackrel{?}{=} -6.5$

$-6.5 = -6.5$ The answer works.

$3([2] - 0.5) + 1 \stackrel{?}{=} 3(2) - 0.5$ Check the answer using 2.

$3(1.5) + 1 \stackrel{?}{=} 6 - 0.5$

$4.5 + 1 \stackrel{?}{=} 5.5$

$5.5 = 5.5$ The answer works.

59.

$0.3x + 0.3 = \boxed{0.5x} + 0.1 \boxed{-0.2x}$ Simplify on right side.

$\boxed{0.3x} + 0.3 = \boxed{0.3x} + 0.1$ Combine like terms across = sign.

$-0.3x \qquad -0.3x$ Subtract $0.3x$ from both sides.

$0.3 \stackrel{?}{=} 0.1$ This is a false statement.

This equation has no solution.

61.

$\frac{1}{2}r + 5 = \frac{7}{5}r + 5$ Multiply by 10 $(5 \cdot 2)$ to clear fractions.

$(10)\frac{1}{2}r + (10)5 = (10)\frac{7}{5}r + 5(10)$

$(^{5}\cancel{10})\frac{1}{\cancel{2}}r + 50 = (^{2}\cancel{10})\frac{7}{\cancel{5}}r + 50$

$\boxed{5r} + 50 = \boxed{14r} + 50$ Combine like terms across = sign.

$-14r \qquad -14r$ Subtract $14r$ from both sides.

$-9r + 50 = 50$

$-50 \quad -50$ Subtract 50 from both sides.

$-9r = 0$ Divide both sides by -9.

$\frac{\cancel{-9}r}{\cancel{-9}} = \frac{0}{-9}$

$r = 0$

$\frac{1}{2}(0) + 5 \stackrel{?}{=} \frac{7}{5}(0) + 5$ Check the answer.

$0 + 5 \stackrel{?}{=} 0 + 5$

$5 = 5$ The answer works.

63.

$\frac{3}{5}d - 1 = \frac{1}{10}(6d - 10)$ Multiply by 10 to clear fractions.

$(^{2}\cancel{10})\frac{3}{\cancel{5}}d - (10)1 = (\cancel{10})\frac{1}{\cancel{10}}(6d - 10)$

$\boxed{6d} - 10 = \boxed{6d} - 10$ Combine like terms across = sign.

$-6d \qquad -6d$ Subtract $6d$ from both sides.

$-10 = -10$ This is a true statement.

The equation is an identity therefore the solution is all real numbers or $\mathbb{R}$.

To check this we will randomly choose two real numbers -20 and 10, and substitute these into the equation.

$\frac{3}{5}(-20) - 1 \stackrel{?}{=} \frac{1}{10}(6[-20] - 10)$ Check the answer using -20.

$-12 - 1 \stackrel{?}{=} \frac{1}{10}(-120 - 10)$

$-13 \stackrel{?}{=} \frac{1}{10}(-130)$

$-13 = -13$ The answer works.

$\frac{3}{5}(10) - 1 \stackrel{?}{=} \frac{1}{10}(6[10] - 10)$ Check the answer using 10.

$6 - 1 \stackrel{?}{=} \frac{1}{10}(60 - 10)$

$5 \stackrel{?}{=} \frac{1}{10}(50)$

$5 = 5$ The answer works.

65.

$$\frac{1}{5}g+7=2+\frac{2}{5}g \quad \text{Multiply by 5 to clear fractions.}$$

$$(\cancel{5})\frac{1}{\cancel{5}}g+(5)7=(5)2+\frac{2}{\cancel{5}}g(\cancel{5})$$

$$\boxed{g}+35=10+\boxed{2g} \quad \text{Combine like terms across = sign.}$$

$$\underline{-2g \qquad -2g} \quad \text{Subtract } 2g \text{ from both sides.}$$

$$-g+35=10$$

$$\underline{-35 \; -35} \quad \text{Subtract 35 from both sides.}$$

$$-g=-25 \quad \text{Multiply both sides by } -1.$$

$$-1\cdot -g=-25\cdot -1$$

$$g=25$$

$$\frac{1}{5}(25)+7\overset{?}{=}2+\frac{2}{5}(25) \quad \text{Check the answer.}$$

$$5+7\overset{?}{=}2+10$$

$$12=12 \quad \text{The answer works.}$$

67.

$$\frac{1}{2}x+7=\frac{2}{5}x+9 \quad \text{Multiply by 10 } (5\cdot 2) \text{ to clear fractions.}$$

$$(\,^{5}\cancel{10})\frac{1}{\cancel{2}}x+(10)7=(\,^{2}\cancel{10})\frac{2}{\cancel{5}}x+9(10)$$

$$\boxed{5x}+70=\boxed{4x}+90 \quad \text{Combine like terms across = sign.}$$

$$\underline{-4x \qquad -4x} \quad \text{Subtract } 4x \text{ from both sides.}$$

$$x+70=90$$

$$\underline{-70 \; -70} \quad \text{Subtract 70 from both sides.}$$

$$x=20$$

$$\frac{1}{2}(20)+7\overset{?}{=}\frac{2}{5}(20)+9 \quad \text{Check the answer.}$$

$$10+7\overset{?}{=}8+9$$

$$17=17 \quad \text{The answer works.}$$

69.

$$\frac{1}{2}x+5=\frac{1}{2}x+3 \quad \text{Multiply by 2 to clear fractions.}$$

$$(\cancel{2})\frac{1}{\cancel{2}}x+(2)5=(\cancel{2})\frac{1}{\cancel{2}}x+3(2)$$

$$\boxed{x}+10=\boxed{x}+6 \quad \text{Combine like terms across = sign.}$$

$$\underline{-x \qquad -x} \quad \text{Subtract } x \text{ from both sides.}$$

$$10\overset{?}{=}6 \quad \text{This is a false statement.}$$

This equation has no solution.

71.

$$75+x+(x-5)=180 \quad \text{Combine like terms.}$$

$$2x+70=180$$

$$\underline{-70 \; -70} \quad \text{Subtract 70 from both sides.}$$

$$2x=110$$

$$\frac{\cancel{2}x}{\cancel{2}}=\frac{110}{2} \quad \text{Divide both sides by 2.}$$

$$x=55$$

Now substitute $x=55$ into the expressions that represent the angles. The measures of the 3 angles are 75°, 55°, and 50°.

To check the solution, the measures of the angles should add up to 180.

$$75°+55°+50°\overset{?}{=}180° \quad \text{Check the answer.}$$

$$180°=180° \quad \text{The answer works.}$$

73.

$$x+(x+5)+(x-14)=180 \quad \text{Combine like terms.}$$

$$3x-9=180$$

$$\underline{+9 \; +9} \quad \text{Add 9 to both sides.}$$

$$3x=189$$

$$\frac{\cancel{3}x}{\cancel{3}}=\frac{189}{3} \quad \text{Divide both sides by 3.}$$

$$x=63$$

Now substitute $x=63$ into the expressions that represent the angles. The measures of the 3 angles are 63°, 68°, and 49°.

To check the solution, the measures of the angles should add up to 180.

$$63°+68°+49°\overset{?}{=}180° \quad \text{Check the answer.}$$

$$180°=180° \quad \text{The answer works.}$$

75. The sum of the measures of the angles in a quadrilateral must equal 360°. Add the measures of the given angles and set them equal to 360° then solve for x.

$$x+(x-25)+65+(x+20)=360$$

$$3x+60=360 \quad \text{Combine like terms.}$$

$$\underline{-60 \; -60} \quad \text{Subtract 60 from both sides.}$$

$$3x=300$$

$$\frac{\cancel{3}x}{\cancel{3}}=\frac{300}{3} \quad \text{Divide both sides by 3.}$$

$$x=100$$

Now substitute $x=100$ into the expressions that represent the angles. The measures of the 4 angles are 100°, 75°, 65°, and 120°.

To check the solution, the measures of the angles should add up to 360.

$$100°+75°+65°+120°\overset{?}{=}360° \quad \text{Check the answer.}$$

$$360°=360° \quad \text{The answer works.}$$

77. Add the given expressions for side lengths, set them equal to 342, and solve for p.

$$(p-3)+(p+20)+p+(p+5)=342$$

$$4p+22=342 \quad \text{Combine like terms on left side.}$$

$$\underline{-22 \; -22} \quad \text{Subtract 22 from both sides.}$$

$$4p=320 \quad \text{Divide both sides by 4.}$$

$$\frac{\cancel{4}p}{\cancel{4}}=\frac{320}{4}$$

$$p=80$$

Now substitute $p=80$ into the expressions that represent the lengths of each side. The lengths of the sides are 77 ft., 100 ft., 80 ft., and 85 ft.

To check the solution, the sum of the lengths of the sides should equal 342.

$77+100+80+85 \overset{?}{=} 342$ Check the answer.
$342=342$ The answer works.

79. Add the given expressions for side lengths, set them equal to 345, and solve for s.

$s+2s+(s-5)+(2s-10)=345$
$6s-15=345$ Combine like terms on left side.
$+15 \quad +15$ Add 15 to both sides.
$6s=360$ Divide both sides by 6.
$\frac{6s}{6}=\frac{360}{6}$
$s=60$

Now substitute $s=60$ into the expressions that represent the lengths of each side. The lengths of the sides are 60 ft., 120 ft., 55 ft., and 110 ft.

To check the solution, the sum of the lengths of the sides should equal 345.

$60+120+55+110 \overset{?}{=} 345$ Check the answer.
$345=345$ The answer works.

81. Let $s=$ the number of students going on the bike tour.

- $145(s)$ dollars for the bikes
- $2(2s)$ dollars for the water bottles
- $4.50(3s)$ dollars for the food sacks

We write out the following equation equal to Carlotta's budget of $2275.00

$$145s+2(2s)+4.50(3s)=2275$$

Now solve the equation for s.

$145s+2(2s)+4.50(3s)=2275$ Multiply on left side.
$145s+4s+13.5s=2275$ Combine like terms on left side.
$162.5s=2275$
$\frac{162.5s}{162.5}=\frac{2275}{162.5}$ Divide both sides by 162.5.
$s=14$

Carlotta can supply 14 students for the tour.

83. **a.** First consider how many of each item is needed for k kids.

- $k+8$ pine 2X2, 1foot for each kid plus 8 extra feet.
- $1.5k+10$ pine 1X3, 1.5 feet for each kid plus 10 extra feet.
- $k+12$ sets of nails, 1 set for each kid plus 12 extra sets.

Now we multiply each of these expressions by how much they cost.

- $0.45(k+8)$ dollars for pine 2X2
- $0.55(1.5k+10)$ dollars for pine 1X3
- $0.05(k+12)$ dollars for nails

Let $S=$ the total amount in dollars needed for supplies.

$S=0.45(k+8)+0.55(1.5k+10)+0.05(k+12)$
$S=0.45k+3.6+0.825k+5.5+0.05k+0.6$ Distribute.
$S=1.325k+9.7$ Combine like terms.

b. Substitute $k=75$ into the equation and solve for S.

$S=1.325(75)+9.7$ Substitute $k=75$.
$S=99.375+9.7$ Multiply then add.
$S=109.075$

The cost would be $109.08 (rounded to the nearest cent).

c. Substitute $S=310$ into the equation and solve for k.

$310=1.325k+9.7$ Substitute $S=310$.
$-9.7 \quad -9.7$ Subtract 9.7 from both sides.
$300.3=1.325k$ Divide both sides by 1.325.
$\frac{300.3}{1.325}=\frac{1.325k}{1.325}$
$226.64 \approx k$ Rounded to hundredths

The manager can buy supplies for 226 projects.

85. The statement does not have an equal sign, therefore it is an expression.

$\boxed{5ab}+6a^2-4a+\boxed{12ab}+3$ Identify like terms.
$6a^2-4a+17ab+3$ Arrange in conventional form.

87. The statement has an equal sign, therefore it is an equation.

$\boxed{0.5x}\boxed{-4}=\boxed{2x}+\boxed{7}$ Identify like terms.
$-2x \quad -2x$ Subtract $2x$ from both sides.
$-1.5x-4=7$
$+4 \quad +4$ Add 4 to both sides.
$-1.5x=11$ Divide both sides by -1.5.
$\frac{-1.5x}{-1.5}=\frac{11}{-1.5}$
$x=-7.\overline{3}$

89. The statement has an equal sign, therefore it is an equation.

Multiply each term by 24 to clear fractions.

$\frac{1}{8}n+12=\frac{3}{24}n-2$
$(24)\frac{1}{8}n+(24)12=(24)\frac{3}{24}n-2(24)$
$\boxed{3n}+\boxed{288}=\boxed{3n}\boxed{-48}$ Identify like terms.
$-3n \quad -3n$ Subtract $3n$ from both sides.
$288 \overset{?}{=} -48$ This is a false statement.

This equation has no solution.

91. The statement does not have an equal sign, therefore it is an expression.

$\boxed{2x}+\boxed{\boxed{11}}\boxed{-4x}+\boxed{\boxed{8}}$ Identify like terms.

$-2x+19$ Arrange in conventional form.

93. The statement has an equal sign, therefore it is an equation.

Multiply each term by 6 to clear fractions.

$$\frac{1}{2}(3x+4)=\frac{1}{3}x+20$$

$$(\overset{3}{\cancel{6}})\frac{1}{\cancel{2}}(3x+4)=(\overset{2}{\cancel{6}})\frac{1}{\cancel{3}}x+20(6)$$

$3(3x+4)=2x+120$ Distribute on left side.

$\boxed{9x}+\boxed{\boxed{12}}=\boxed{2x}+\boxed{\boxed{120}}$ Identify like terms.

$-2x \quad -2x$ Subtract $2x$ from both sides.

$7x+12=120$

$-12 \quad -12$ Subtract 12 from both sides.

$7x=108$ Divide both sides by 7.

$$\frac{\cancel{7}x}{\cancel{7}}=\frac{108}{7}$$

$$x=15\frac{3}{7}$$

Section 2.4 Solving and Graphing Linear Inequalities on a Number Line

1.

$3x\geq 21$ Divide both sides by 3.

$$\frac{\cancel{3}x}{\cancel{3}}\geq\frac{21}{3}$$

$x\geq 7$

$3(7)\overset{?}{=}21$ Check the answer.
$21=21$ The answer works.

Check a number greater than 7 ($x=8$) to test the direction of the inequality symbol.

$3(8)\overset{?}{\geq}21$ Check the direction of the inequality.
$24\geq 21$ The direction of the inequality works.

3.

$2d<15$ Divide both sides by 2.

$$\frac{\cancel{2}d}{\cancel{2}}<\frac{15}{2}$$

$d<7.5$

$2(7.5)\overset{?}{=}15$ Check the answer.
$15=15$ The answer works.

Check a number less than 7.5 ($d=7$) to test the direction of the inequality symbol.

$2(7)\overset{?}{<}15$ Check the direction of the inequality.
$14<15$ The direction of the inequality works.

5.

$-4x\geq 48$ Divide both sides by -14.

$$\frac{\cancel{-4}x}{\cancel{-4}}\leq\frac{48}{-4}$$ Divide by neg., reverse inequality.

$x\leq -12$

$-4(-12)\overset{?}{=}48$ Check the answer.
$48=48$ The answer works.

Check a number less than -12 ($x=-13$) to test the direction of the inequality symbol.

$-4(-13)\overset{?}{\geq}48$ Check the direction of the inequality.
$52\geq 48$ The direction of the inequality works.

7.

$-a<-61$ Divide both sides by -1.

$$\frac{-a}{-1}>\frac{-61}{-1}$$ Divide by neg., reverse inequality.

$a>61$

Check using problems 1-5 as a guide.

9.

$$2g+5\le 35$$
$$\underline{-5\quad -5}\qquad \text{Subtract 5 from both sides.}$$
$$2g\le 30\qquad \text{Divide both sides by 2.}$$
$$\frac{2g}{2}\le\frac{30}{2}$$
$$g\le 15$$

Check using problems 1-5 as a guide.

11.

$$5m-13>33$$
$$\underline{+13\ +13}\qquad \text{Add 13 to both sides.}$$
$$5m>46\qquad \text{Divide both sides by 5.}$$
$$\frac{5m}{5}>\frac{46}{5}$$
$$m>9.2 \text{ or } m>9\frac{1}{5}$$

Check using problems 1-5 as a guide.

13.

$$-2x+9\le 25$$
$$\underline{-9\ -9}\qquad \text{Subtract 9 from both sides.}$$
$$-2x\le 16\qquad \text{Divide both sides by}-2.$$
$$\frac{-2x}{-2}\ge\frac{16}{-2}\qquad \text{Divide by neg., reverse inequality}$$
$$x\ge -8$$

Check using problems 1-5 as a guide.

15.

$$4x+7>6x+3$$
$$\underline{-6x\qquad -6x}\qquad \text{Subtract } 6x \text{ from both sides.}$$
$$-2x+7>3$$
$$\underline{-7\ -7}\qquad \text{Subtract 7 from both sides.}$$
$$-2x>-4\qquad \text{Divide both sides by}-2.$$
$$\frac{-2x}{-2}<\frac{-4}{-2}\qquad \text{Divide by neg., reverse inequality}$$
$$x<2$$

Check using problems 1-5 as a guide.

17.

$$-4x+7\ge 5x+25$$
$$\underline{-5x\qquad -5x}\qquad \text{Subtract } 5x \text{ from both sides.}$$
$$-9x+7\ge 25$$
$$\underline{-7\ -7}\qquad \text{Subtract 7 from both sides.}$$
$$-9x\ge 18\qquad \text{Divide both sides by}-9.$$
$$\frac{-9x}{-9}\le\frac{18}{-9}\qquad \text{Divide by neg., reverse inequality}$$
$$x\le -2$$

Check using problems 1-5 as a guide.

19.

$$2(k+5)-3<k+18\qquad \text{Distribute on left side.}$$
$$2k+10-3<k+18$$
$$2k+7<k+18$$
$$\underline{-k\qquad -k}\qquad \text{Subtract } k \text{ from both sides.}$$
$$k+7<18$$
$$\underline{-7\ -7}\qquad \text{Subtract 7 from both sides.}$$
$$k<11$$

Check using problems 1-5 as a guide.

21.

$$7(x-2)+1>8x-6\qquad \text{Distribute on left side.}$$
$$7x-14+1>8x-6$$
$$7x-13>8x-6$$
$$\underline{-8x\qquad -8x}\qquad \text{Subtract } 8x \text{ from both sides.}$$
$$-x-13>-6$$
$$\underline{+13\ +13}\qquad \text{Add 13 to both sides.}$$
$$-x>7\qquad \text{Divide both sides by}-1.$$
$$\frac{-x}{-1}<\frac{7}{-1}\qquad \text{Divide by neg., reverse inequality}$$
$$x<-7$$

Check using problems 1-5 as a guide.

23. **a.** Write the inequality as follows:

$$11h\ge 275$$

b.

$$11h\ge 275\qquad \text{Divide both sides by 11.}$$
$$\frac{11h}{11}\ge\frac{275}{11}$$
$$h\ge 25$$

Kati needs to work at least 25 hours this week.

c. Rewrite the inequality and solve as follows:

$$11h\ge 400\qquad \text{Divide both sides by 11.}$$
$$\frac{11h}{11}\ge\frac{400}{11}$$
$$h\ge 36.\overline{36}$$

Kati must work at least 36.36 hours.

25. **a.** Kevin's monthly cell phone charges are given by the expression $30+0.50m$. The inequality that represents Kevin keeping his monthly cell phone bill to at most $45 is:

$$30+0.5m\le 45$$

b.

$$30+0.5m\le 45\qquad \text{Isolate variable term.}$$
$$\underline{-30\qquad -30}\qquad \text{Subtract 30 from both sides.}$$
$$0.5m\le 15$$
$$\frac{0.5m}{0.5}\le\frac{15}{0.5}\qquad \text{Divide both sides by 0.5.}$$
$$m\le 30$$

Kevin can use no more than 30 minutes over his allowed 200 minutes.

c.

$30 + 0.5m \le 40$ Isolate variable term.

$-30 \quad -30$ Subtract 30 from both sides.

$0.5m \le 10$

$\frac{0.5m}{0.5} \le \frac{10}{0.5}$ Divide both sides by 0.5.

$m \le 20$

He can use no more than 20 minutes over his allowed 200 minutes.

27. **a.** The inequality that represents Warren keeping his monthly texting bill to at most \$18 is:

$9 + 0.1t \le 18$

b.

$9 + 0.1t \le 18$ Isolate variable term.

$-9 \quad -9$ Subtract 9 from both sides.

$0.1t \le 9$

$\frac{0.1t}{0.1} \le \frac{9}{0.1}$ Divide both sides by 0.1.

$t \le 90$

To stay within his budget, Warren can send no more than 90 texts over his 200 texts allowed.

c.

$9 + 0.1t \le 15$ Isolate variable term.

$-9 \quad -9$ Subtract 9 from both sides.

$0.1t \le 6$

$\frac{0.1t}{0.1} \le \frac{6}{0.1}$ Divide both sides by 0.1.

$t \le 60$

He can send no more than 60 texts over his 200 texts allowed.

29. **a.** Add the costs in an inequality less than or equal to 115.

$75 + 3m \le 115$

b.

$75 + 3m \le 115$ Isolate variable term.

$-75 \quad -75$ Subtract 75 from both sides.

$3m \le 40$

$\frac{3m}{3} \le \frac{40}{3}$ Divide both sides by 3.

$m \le 13\frac{1}{3}$

Alicia can have her car towed up to 13 miles, over the initial 10 miles.

31.

$0.02t + 2.67 \ge 3$ Isolate variable term.

$-2.67 \quad -2.67$ Subtract 2.67 from both sides.

$0.02t \ge 0.33$

$\frac{0.02t}{0.02} \ge \frac{0.33}{0.02}$ Divide both sides by 0.02.

$t \ge 16.5$

The population of Arkansas is projected to be 3 million or more after the year 2016.

33.

$0.35t + 16.01 \ge 18$ Isolate variable term.

$-16.01 \quad -16.01$ Subtract 16.01 from both sides.

$0.35t \ge 1.99$

$\frac{0.35t}{0.35} \ge \frac{1.99}{0.35}$ Divide both sides by 0.35.

$t \ge 5.69$ Rounded to hundredths.

The population of Florida is projected to be 18 million or more after the year 2006.

35. If $4 > x$, then x must be less than 4. This means we can rewrite the inequality as

$x < 4$

37. If $-9 < n$, then n must be greater than -9. This means we can rewrite the inequality as

$n > -9$

39. If $0 \ge t$, then t must be less than or equal to 0. This means we can rewrite the inequality as

$t \le 0$

41. If $6 > 2x$, then $2x$ must be less than 6. This means we can rewrite the inequality as

$2x < 6$

To solve, divide both sides by 2.

$\frac{2x}{2} < \frac{6}{2}$

$x < 3$

43. If $5 \le 2x + 1$, then $2x + 1$ must be greater than or equal to 5. This means we can rewrite the inequality as

$2x + 1 \ge 5$

To solve, isolate the variable as follows:

$2x + 1 \ge 5$ Isolate variable term.

$-1 \quad -1$ Subtract 1 from both sides.

$2x \ge 4$

$\frac{2x}{2} \ge \frac{4}{2}$ Divide both sides by 2.

$x \ge 2$

45.

$4x > 20$ Divide both sides by 4.

$\frac{4x}{4} > \frac{20}{4}$

$x > 5$

$4(5) \stackrel{?}{=} 20$ Check the answer.
$20 = 20$ The answer works.

Check a number greater than 5 ($x = 6$) to test the direction of the inequality symbol.

$4(6) \stackrel{?}{>} 20$ Check the direction of the inequality.
$24 > 20$ The direction of the inequality works.

Using interval notation, the answer is written as $(5, \infty)$.

Using a number line:

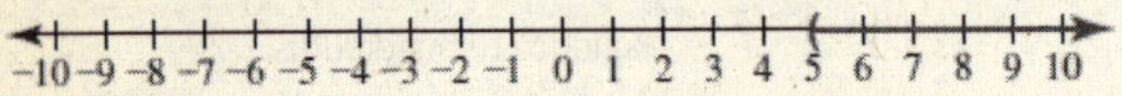

47. $\frac{m}{3} < -2$ Multiply both sides by 3.

$3 \cdot \frac{m}{3} < -2 \cdot 3$

$m < -6$

$\frac{(-6)}{3} \stackrel{?}{=} -2$ Check the answer.

$-2 = -2$ The answer works.

Check a number less than -6 ($m = -9$) to test the direction of the inequality symbol.

$\frac{(-9)}{3} \stackrel{?}{<} -2$ Check the direction of the inequality.

$-3 < -2$ The direction of the inequality works.

Using interval notation, the answer is written as $(-\infty, -6)$.

Using a number line:

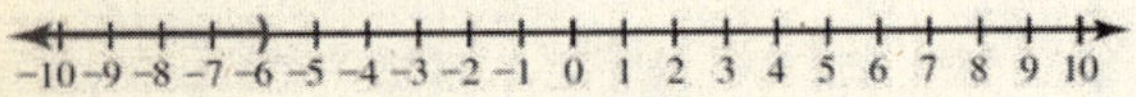

49.

$-3x \geq 12$ Divide both sides by -3.

$\frac{-3x}{-3} \leq \frac{12}{-3}$ Divide by neg., reverse inequality.

$x \leq -4$

Check using problems 45-47 as a guide.

Using interval notation, the answer is written as $(-\infty, -4]$.

Using a number line:

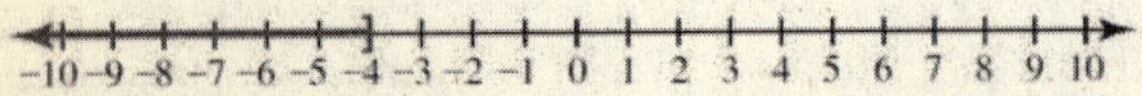

51.

$-\frac{2}{5}x < -\frac{1}{10}$ Multiply both sides by $-\frac{5}{2}$.

$-\frac{5}{2} \cdot -\frac{2}{5}x > -\frac{1}{10_2} \cdot -\frac{5}{2}$ Multiply by neg., reverse inequality.

$x > \frac{1}{4}$

Check using problems 45-47 as a guide.

Using interval notation, the answer is written as $\left(\frac{1}{4}, \infty\right)$.

Using a number line:

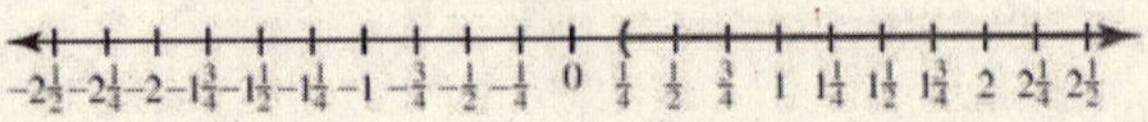

53.

$\frac{-a}{3} \leq -4$ Multiply both sides by -3.

$-3 \cdot \frac{-a}{3} \geq -4 \cdot -3$ Multiply by neg., reverse inequality.

$a \geq 12$

Check using problems 45-47 as a guide.

Using interval notation, the answer is written as $[12, \infty)$.

Using a number line:

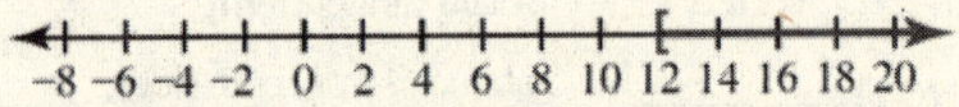

55.

$2(k+5) \leq k+13$ Distribute on left side.
$2k + 10 \leq k + 13$
$\underline{-k \quad -k}$ Subtract k from both sides.
$k + 10 \leq 13$
$\underline{-10 \quad -10}$ Subtract 10 from both sides.
$k \leq 3$

Check using problems 45-47 as a guide.

Using interval notation, the answer is written as $(-\infty, 3]$.

Using a number line:

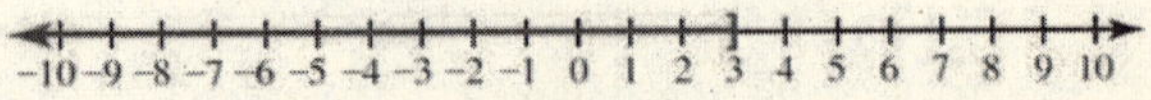

57.

$2.5z + 8 \geq 3.5z - 4$
$\underline{-3.5z \quad -3.5z}$ Subtract 3.5z from both sides.
$-z + 8 \geq -4$
$\underline{-8 \quad -8}$ Subtract 8 from both sides.
$-z \geq -12$ Divide both sides by -1.
$\frac{-z}{-1} \leq \frac{-12}{-1}$ Divide by neg., reverse inequality.
$z \leq 12$

Check using problems 45-47 as a guide.

Using interval notation, the answer is written as $(-\infty, 12]$.

Using a number line:

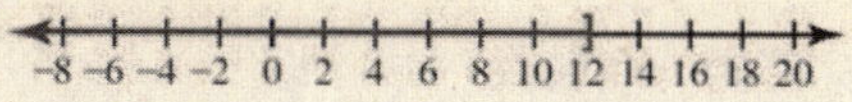

59.

$$\begin{aligned} 2c+7 &\le 6c+17 \\ -6c \quad & \quad -6c \\ -4c+7 &\le 17 \\ -7 \quad & \quad -7 \\ -4c &\le 10 \\ \frac{-4c}{-4} &\ge \frac{10}{-4} \\ c &\ge -2.5 \end{aligned}$$

Subtract $6c$ from both sides.
Subtract 7 from both sides.
Divide both sides by – 4.
Divide by neg., reverse inequality

Check using problems 45-47 as a guide.

Using interval notation, the answer is written as $[-2.5,\infty)$.

Using a number line:

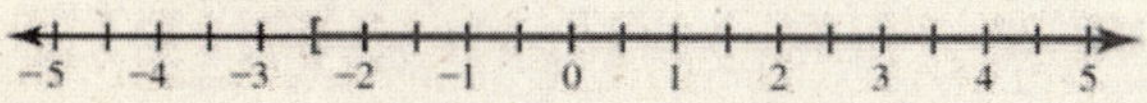

61.

$$\begin{aligned} 5w-12 &\ge 2(5w-10)-30 \\ 5w-12 &\ge 10w-20-30 \\ 5w-12 &\ge 10w-50 \\ -10w \quad & \quad -10w \\ -5w-12 &\ge -50 \\ +12 \quad & \quad +12 \\ -5w &\ge -38 \\ \frac{-5w}{-5} &\le \frac{-38}{-5} \\ w &\le 7.6 \end{aligned}$$

Distribute on right side.
Simplify on right side
Subtract $10w$ from both sides.
Add 12 to both sides.
Divide both sides by – 14.
Divide by neg., reverse inequality

Check using problems 45-47 as a guide.

Using interval notation, the answer is written as $(-\infty,7.6]$.

Using a number line:

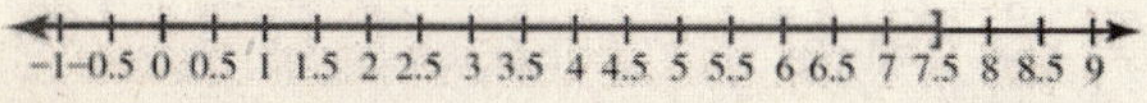

63.

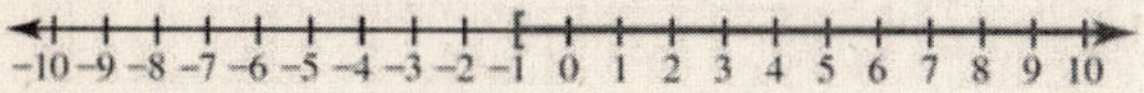

This is equivalent to the inequality $x \ge -1$.

65.

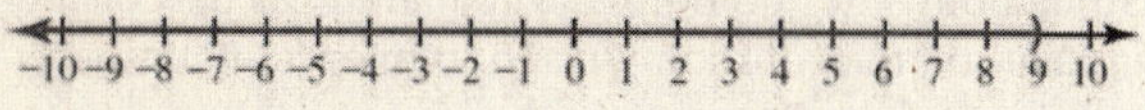

This is equivalent to the inequality $x < 9$.

67.

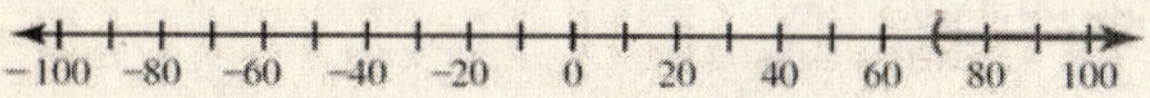

This is equivalent to the inequality $x > 69$.

69.

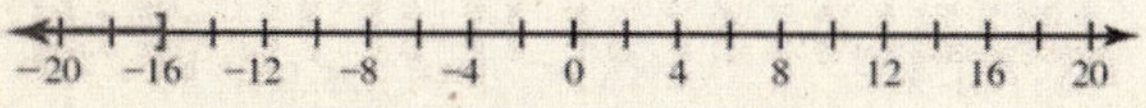

This is equivalent to the inequality $x \le -16$.

71. Let $s =$ the number of students enrolled in the beginning algebra class.

$$16 \le s \le 38$$

73. Let $s =$ the number of students that can ride on school bus #12.

$$0 \le s \le 71$$

75. Let $p =$ the number of people the camp can accommodate.

$$200 \le p \le 450$$

77. Let $h =$ the height requirements in inches for children to ride on some attractions at LEGOLAND with an adult.

$$42 \le h \le 52$$

79. Let $r =$ the average adult's resting heart rate in beats per minute.

$$66 \le r \le 100$$

81.

$$\begin{aligned} &6 < 2x < 8 \\ &\frac{6}{2} < \frac{2x}{2} < \frac{8}{2} \\ &3 < x < 4 \end{aligned}$$

Isolate x by dividing all sides by 2.

Since both sides do not have "equal to" as part of the inequality, the solution written in interval notation will have parentheses on both sides. Always write intervals from lowest to highest numbers, so we get the interval $(3,4)$.

Using a number line:

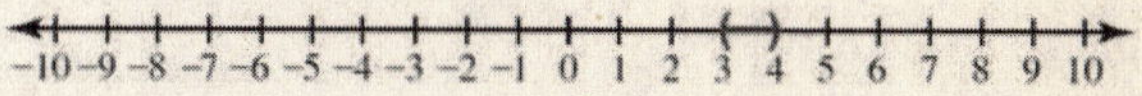

83.

$$\begin{aligned} -7 &< y-1 < -3 \\ +1 \quad & \quad +1 \quad +1 \\ -6 &< y < -2 \end{aligned}$$

Isolate y.
Add 1 to all sides.

Since both sides do not have "equal to" as part of the inequality, the solution written in interval notation will have parentheses on both sides. Always write intervals from lowest to highest numbers, so we get the interval $(-6,-2)$.

Using a number line:

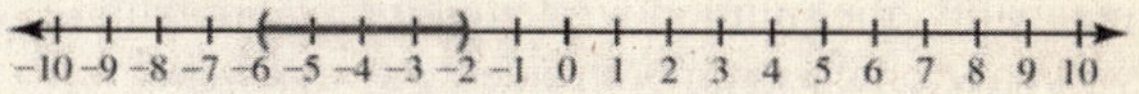

85.

$$\begin{aligned} &12 < 3x+3 < 21 \\ &-3 \qquad -3 \quad -3 \\ &9 < 3x < 18 \\ &\frac{9}{3} < \frac{3x}{3} < \frac{18}{3} \\ &3 < x < 6 \end{aligned}$$

Isolate x.
Subtract 3 from all sides.
Divide all sides by 3.

Since both sides do not have "equal to" as part of the inequality, the solution written in interval notation will have parentheses on both sides. Always write intervals from lowest to highest numbers, so we get the interval $(3,6)$.

Using a number line:

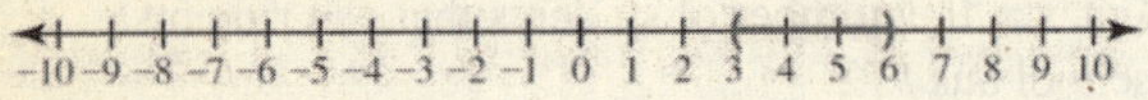

87.

$3<\frac{k}{2}<5$ Isolate k by multiplying all sides by 2.

$(2)3<(2)\frac{k}{2}<(2)5$

$6<k<10$

Since both sides do not have "equal to" as part of the inequality, the solution written in interval notation will have parentheses on both sides. Always write intervals from lowest to highest numbers, so we get the interval $(6,10)$.

Using a number line:

89.

$40\le 6x+4\le 64$ Isolate x.

$\underline{-4 \quad -4 \quad -4}$ Subtract 4 from all sides.

$36\le 6x\le 60$ Divide all sides by 6.

$\frac{36}{6}\le\frac{6x}{6}\le\frac{60}{6}$

$6\le x\le 10$

Since both sides have "equal to" as part of the inequality, the solution written in interval notation will have brackets on both sides. Always write intervals from lowest to highest numbers, so we get the interval $[6,10]$.

Using a number line:

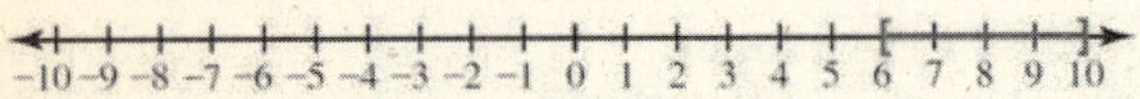

91.

$-1.5\le\frac{m}{6}\le -0.5$ Isolate m.

$-1.5(6)\le(6)\frac{m}{6}\le -0.5(6)$ Multiply all sides by 6.

$-9\le m\le -3$

Since both sides have "equal to" as part of the inequality, the solution written in interval notation will have brackets on both sides. Always write intervals from lowest to highest numbers, so we get the interval $[-9,-3]$.

Using a number line:

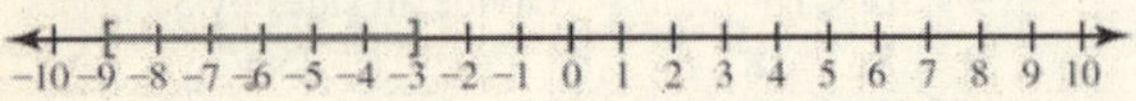

93.

$-12\le\frac{1}{2}x-2\le 0$ Isolate x.

$\underline{+2 \quad +2 \quad +2}$ Add 2 to all sides.

$-10\le\frac{1}{2}x\le 2$ Multiply all sides by 2.

$-10(2)\le(2)\frac{1}{2}x\le(2)2$

$-20\le x\le 4$

Since both sides have "equal to" as part of the inequality, the solution written in interval notation will have brackets on both sides. Always write intervals from lowest to highest numbers, so we get the interval $[-20,4]$.

Using a number line:

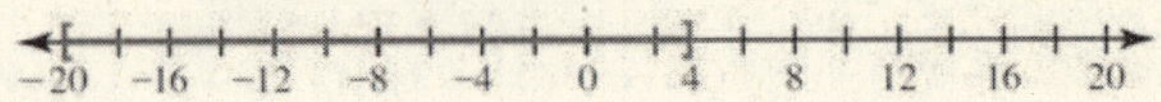

95.

$4<\frac{x}{2}<30$ Isolate x by multiplying all sides by 2.

$(2)4<(2)\frac{x}{2}<(2)30$

$8<x<60$

Since both sides do not have "equal to" as part of the inequality, the solution written in interval notation will have parentheses on both sides. Always write intervals from lowest to highest numbers, so we get the interval $(8,60)$.

Using a number line:

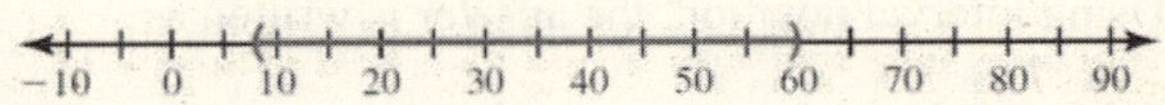

97.

$-7<\frac{3x}{4}<-1$ Isolate x.

$\left(\frac{4}{3}\right)(-7)<\left(\frac{4}{3}\right)\frac{3x}{4}<\left(\frac{4}{3}\right)(-1)$ Multiply by $\frac{4}{3}$

$-\frac{28}{3}<x<-\frac{4}{3}$

Since both sides do not have "equal to" as part of the inequality, the solution written in interval notation will have parentheses on both sides. Always write intervals from lowest to highest numbers, so we get the interval $\left(-\frac{28}{3},-\frac{4}{3}\right)$.

Using a number line:

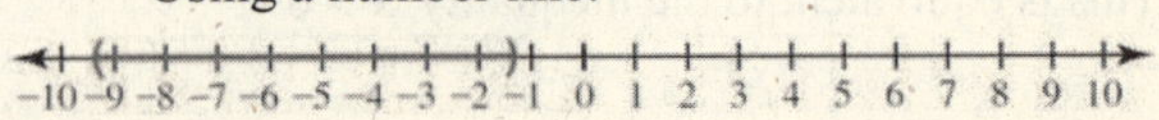

99.

$$3.5 \le 2x - 2 \le 8.5 \quad \text{Isolate } x.$$
$$\underline{+2 \qquad +2 \; +2} \quad \text{Add 2 to all sides.}$$
$$5.5 \le 2x \le 10.5 \quad \text{Divide all sides by 2.}$$
$$\frac{5.5}{2} \le \frac{2x}{2} \le \frac{10.5}{2}$$
$$2.75 \le x \le 5.25$$

Since both sides have "equal to" as part of the inequality, the solution written in interval notation will have brackets on both sides. Always write intervals from lowest to highest numbers, so we get the interval $[2.75, 5.25]$.

Using a number line:

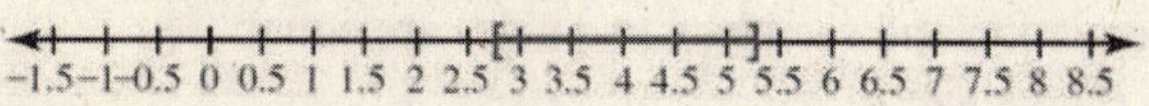

101.

$$-9 \le 3v + 6 < 3 \quad \text{Isolate } x.$$
$$\underline{-6 \qquad -6 \; -6} \quad \text{Subtract 6 from all sides.}$$
$$-15 \le 3v < -3 \quad \text{Divide all sides by 3.}$$
$$\frac{-15}{3} \le \frac{3v}{3} < \frac{-3}{3}$$
$$-5 \le v < -1$$

In interval notation, the solution set will have a bracket on the left side, since it is less than or equal to, and a parenthesis on the right side, since it is less than. This gives us the interval $[-5, -1)$.

Using a number line:

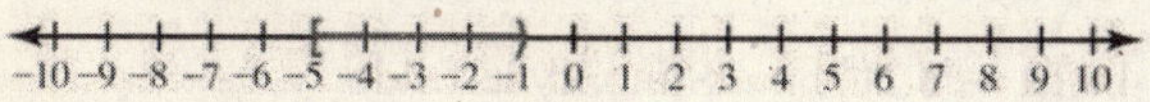

103.

$$-7 < \frac{5a}{3} - 8 \le 3 \quad \text{Isolate } a.$$
$$\underline{+8 \qquad +8 \; +8} \quad \text{Add 8 to all sides.}$$
$$1 < \frac{5a}{3} \le 11 \quad \text{Multiply all sides by } \frac{3}{5}.$$
$$\left(\frac{3}{5}\right)1 < \left(\frac{3}{5}\right)\frac{5a}{3} \le \left(\frac{3}{5}\right)11$$
$$\frac{3}{5} < a \le \frac{33}{5}$$

In interval notation, the solution set will have a parenthesis on the left side, since it is less than, and a bracket on the right side, since it is less than or equal to. This gives us the interval $\left(\frac{3}{5}, \frac{33}{5}\right]$.

Using a number line:

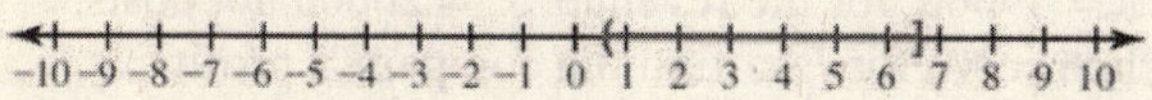

105. The statement has an inequality symbol, therefore it is an inequality.

$$2x + 7 > 41 \quad \text{Isolate variable term.}$$
$$\underline{-7 \quad -7} \quad \text{Subtract 7 from both sides.}$$
$$2x > 34$$
$$\frac{\cancel{2}x}{\cancel{2}} > \frac{34}{2} \quad \text{Divide both sides by 2.}$$
$$x > 17$$

Using interval notation, the answer is written as $(17, \infty)$. Use a parenthesis next to the 17 because the inequality is not also "equal to" 17. Always use a parenthesis next to the infinity symbol.

107. The statement does not have an equal sign or an inequality symbol, therefore it is an expression.

$$\boxed{4x} + 7 \boxed{-12x} \quad \text{Combine like terms.}$$
$$-8x + 7$$

109. The statement has an equal sign, therefore it is an equation.

$$5h + 12 = 3h - 16 \quad \text{Isolate variable term.}$$
$$\underline{-3h \quad -3h} \quad \text{Subtract } 3h \text{ from both sides.}$$
$$2h + 12 = -16$$
$$\underline{-12 \quad -12} \quad \text{Subtract 12 from both sides.}$$
$$2h = -28$$
$$\frac{\cancel{2}h}{\cancel{2}} = \frac{-28}{2} \quad \text{Divide both sides by 2.}$$
$$h = -14$$

111. The statement has an inequality symbol, therefore it is an inequality.

$$3a + 4 \ge 7a - 20 \quad \text{Isolate variable term.}$$
$$\underline{-7a \quad -7a} \quad \text{Subtract } 7a \text{ from both sides.}$$
$$-4a + 4 \ge -20$$
$$\underline{-4 \quad -4} \quad \text{Subtract 4 from both sides.}$$
$$-4a \ge -24 \quad \text{Divide both sides by } -4.$$
$$\frac{\cancel{-4}a}{\cancel{-4}} \le \frac{-24}{-4} \quad \text{Divide by neg. reverse inequality.}$$
$$a \le 6$$

Using interval notation, the answer is written as $(-\infty, 6]$. Use a bracket next to the 6 because the inequality is also "equal to" 6. Always use a parenthesis next to the infinity symbol.

113. The statement has an equal sign, therefore it is an equation. Multiply each term by 3 first to clear the denominators.

$$\frac{2}{3}x + \frac{4}{3} = \frac{1}{3}x + 7 \quad \text{Multiply by 3.}$$
$$(\cancel{3})\frac{2}{\cancel{3}}x + (\cancel{3})\frac{4}{\cancel{3}} = (\cancel{3})\frac{1}{\cancel{3}}x + 7(3)$$

$2x+4=x+21$ Isolate variable term.

$-x \quad -x$ Subtract x from both sides.

$x+4=21$

$-4 \quad -4$ Subtract 4 from both sides.

$x=17$

115. The statement has an inequality symbol, therefore it is an inequality.

$-20 < 4a + 2 < 12$ Isolate a.

$-2 \quad -2 \quad -2$ Subtract 2 from all sides.

$-22 < 4a < 10$ Divide all sides by 4.

$\frac{-22}{4} < \frac{4a}{4} < \frac{10}{4}$

$-5.5 < a < 2.5$

The answer in interval notation is $(-5.5, 2.5)$.

117. The statement does not have an equal sign or an inequality symbol, therefore it is an expression.

$\boxed{\frac{2}{3}b} + \boxed{\boxed{5}} + \boxed{\frac{4}{9}b} \boxed{\boxed{-2}}$ Combine like terms.

$\frac{6}{9}b+\frac{4}{9}b+3$ Rewrite fractions with LCD = 9.

$\frac{10}{9}b+3$ Write in conventional form.

119. The statement has an inequality symbol, therefore it is an inequality.

$4<-\frac{g}{2}\le 7$ Isolate g.

$(-2)4>(-2)-\frac{g}{2}\ge 7(-2)$ Multiply all sides by -2.

$-8>g\ge -14$ Multiply by neg., reverse inequalities.

$-14\le g<-8$ Reorder compound inequality from smallest to greatest.

The answer, in interval notation, is $[-14,-8)$.

Chapter 2 Review

1. **a.** To check if the given values of the variables are a solution to the equation, evaluate each side of the equation and see if the two sides are equal.

$D = 75t$ Original equation.

$262.5 \stackrel{?}{=} 75(3.5)$ Substitute values for D and t.

$262.5 = 262.5$ This statement is true.

Because the final statement is true, $t = 3.5$ and $D = 262.5$ is a solution to the equation $D = 75t$.

b. These values represent that 262.5 miles is the distance traveled when driving for 3.5 hours.

2. **a.** To check if the given values of the variables are a solution to the equation, evaluate each side of the equation and see if the two sides are equal.

$F = 11.5L$ Original equation.

$6900 \stackrel{?}{=} 11.5(600)$ Substitute values for F and L.

$6900 = 6900$ This statement is true.

Because the final statement is true, $L = 600$ and $F = 6900$ is a solution to the equation $F = 11.5L$.

b. These values represent that a loan of \$600,000.00 would have an origination fee of \$6,900.00.

3. To check if the given values of the variables are a solution to the equation, evaluate each side of the equation and see if the two sides are equal.

$-2x+4y=-8$ Original equation.

$-2(-2)+4(3) \stackrel{?}{=} -8$ Substitute $x=-2$ and $y=3$.

$4+12 \stackrel{?}{=} -8$ Simplify left side.

$16 \stackrel{?}{=} -8$ This statement is not true.

$16 \neq -8$

Because the final results are not equal, $x=-2$ and $y=3$ is not a solution to the equation $-2x+4y=-8$.

4. To check if the given values of the variables are a solution to the equation, evaluate each side of the equation and see if the two sides are equal.

$y=-3x+5$ Original equation.

$(14) \stackrel{?}{=} -3(-3)+5$ Substitute $x=-3$ and $y=14$.

$14 \stackrel{?}{=} 9+5$ Simplify both sides.

$14=14$ This statement is true.

Because the final statement is true, $x=-3$ and $y=14$ is a solution to the equation $y=-3x+5$.

5. Using the equation $P=R-C$ substitute $R=49000$ for the revenue, $C=26000$ for costs, then solve for P to find the company's profit.

$P=R-C$ Given equation.

$P=(49000)-(26000)$ Substitute value of variables.

$P=23000$ Simplify on right side.

The solution is $P = 23000$ which represents that the company has monthly profit of $23,000.

6. Using the equation $P = R - C$ substitute $R = 225000$ for the revenue, $C = 186000$ for costs, then solve for P to find the company's profit.

$P = R - C$ Given equation.
$P = (225000) - (186000)$ Substitute value of variables.
$P = 39000$ Simplify on right side.

The solution is $P = 39000$ which represents that the company has monthly profit of $39,000.

7.

$$\begin{aligned} x - 3 &= -9 \\ +3 \quad &\ \ +3 \\ x &= -6 \end{aligned}$$ Add 3 to both sides.

$(-6) - 3 \stackrel{?}{=} -9$ Check the answer.
$-9 = -9$ The answer works.

8.

$$\begin{aligned} p - 16 &= -9 \\ +16 \quad &\ \ +16 \\ p &= 7 \end{aligned}$$ Add 16 to both sides.

$(7) - 16 \stackrel{?}{=} -9$ Check the answer.
$-9 = -9$ The answer works.

9.

$$\begin{aligned} -8 &= a + 9 \\ -9 \quad &\ \ -9 \\ -17 &= a \end{aligned}$$ Subtract 9 from both sides.

$-8 \stackrel{?}{=} (-17) + 9$ Check the answer.
$-8 = -8$ The answer works.

10.

$$\begin{aligned} -16 &= b + 7 \\ -7 \quad &\ \ -7 \\ -23 &= b \end{aligned}$$ Subtract 7 from both sides.

$-16 \stackrel{?}{=} (-23) + 7$ Check the answer.
$-16 = -16$ The answer works.

11.

$$\begin{aligned} k - \frac{1}{2} &= \frac{3}{2} \\ +\frac{1}{2} \quad &\ \ +\frac{1}{2} \\ k &= \frac{4}{2} \\ k &= 2 \end{aligned}$$ Add $\frac{1}{2}$ to both sides.

$(2) - \frac{1}{2} \stackrel{?}{=} \frac{3}{2}$ Check the answer.
$\frac{4}{2} - \frac{1}{2} \stackrel{?}{=} \frac{3}{2}$
$\frac{3}{2} = \frac{3}{2}$ The answer works.

12.

$$\begin{aligned} \frac{-2}{5} + x &= \frac{4}{5} \\ +\frac{2}{5} \quad &\ \ +\frac{2}{5} \\ x &= \frac{6}{5} \end{aligned}$$ Add $\frac{2}{5}$ to both sides.

$\frac{-2}{5} + \left(\frac{6}{5}\right) \stackrel{?}{=} \frac{4}{5}$ Check the answer.
$\frac{4}{5} = \frac{4}{5}$ The answer works.

13.

$$\begin{aligned} y - 9.8 &= 4.6 \\ +9.8 \quad &\ \ +9.8 \\ y &= 14.4 \end{aligned}$$ Add 9.8 to both sides.

$(14.4) - 9.8 \stackrel{?}{=} 4.6$ Check the answer.
$4.6 = 4.6$ The answer works.

14.

$$\begin{aligned} 0.02 + h &= 1.04 \\ -0.02 \quad &\ \ -0.02 \\ h &= 1.02 \end{aligned}$$ Subtract 0.02 from both sides.

$0.02 + (1.02) \stackrel{?}{=} 1.04$ Check the answer.
$1.04 = 1.04$ The answer works.

15. To solve for q means to isolate q on one side of the equal sign.

$$\begin{aligned} 2p + \boxed{q} &= 9 \\ -2p \quad &\ \ -2p \\ q &= -2p + 9 \end{aligned}$$ Identify the variable term to isolate. Subtract $2p$ from both sides.

16. To solve for x means to isolate x on one side of the equal sign.

$$\begin{aligned} \boxed{x} - y &= 2 \\ +y \quad &\ \ +y \\ x &= y + 2 \end{aligned}$$ Identify the variable term to isolate. Add y to both sides.

17. To solve for y means to isolate y on one side of the equal sign.

$$\begin{aligned} 8x \boxed{-24y} &= 48 \\ -8x \quad &\ \ -8x \\ -24y &= -8x + 48 \end{aligned}$$ Identify the variable term to isolate. Subtract $8x$ from both sides.

$$\frac{-24y}{-24}=\frac{-8x+48}{-24}$$ Divided both sides by – 24.

$$y=\frac{-8x}{-24}+\frac{48}{-24}$$

$$y=\frac{1}{3}x-2$$

18. To solve for p means to isolate p on one side of the equal sign.

$$\boxed{3p}-5q=-30$$ Identify the variable term to isolate.

$$+5q \quad +5q$$ Add $5q$ to both sides.

$$3p=5q-30$$

$$\frac{3p}{3}=\frac{5q-30}{3}$$ Divided both sides by 3.

$$p=\frac{5q}{3}-\frac{30}{3}$$

$$p=\frac{5}{3}q-10$$

19.

$$7y=84$$

$$\frac{7y}{7}=\frac{84}{7}$$ Divide both sides by 7.

$$y=12$$

$$7(12)\overset{?}{=}84$$ Check the answer.

$$84=84$$ The answer works.

20.

$$4x=-64$$

$$\frac{4x}{4}=\frac{-64}{4}$$ Divide both sides by 4.

$$x=-16$$

$$4(-16)\overset{?}{=}-64$$ Check the answer.

$$-64=-64$$ The answer works.

21.

$$-a=7$$

$$-1\cdot -a=7\cdot -1$$ Multiply both sides by – 1.

$$a=-7$$

$$-(-7)\overset{?}{=}7$$ Check the answer.

$$7=7$$ The answer works.

22.

$$-8=-x$$

$$-1\cdot -8=-x\cdot -1$$ Multiply both sides by – 1.

$$8=x$$

$$-8\overset{?}{=}-(8)$$ Check the answer.

$$-8=-8$$ The answer works.

23.

$$-2b=-22$$

$$\frac{-2b}{-2}=\frac{-22}{-2}$$ Divide both sides by – 2.

$$b=11$$

Check the answer as in problems 19-22.

24.

$$-5y=35$$

$$\frac{-5y}{-5}=\frac{35}{-5}$$ Divide both sides by – 5.

$$y=-7$$

Check the answer as in problems 19-22.

25.

$$\frac{2}{3}x=-16$$

$$\frac{3}{2}\cdot\frac{2}{3}x=-16\cdot\frac{3}{2}$$ Multiply both sides by $\frac{3}{2}$.

$$x=-24$$

Check the answer as in problems 19-22.

26.

$$\frac{-3}{4}n=-9$$

$$-\frac{4}{3}\cdot\frac{-3}{4}n=-9\cdot -\frac{4}{3}$$ Multiply both sides by $-\frac{4}{3}$.

$$n=12$$

Check the answer as in problems 19-22.

27.

$$6.2t=-8.4$$

$$\frac{6.2t}{6.2}=\frac{-8.4}{6.2}$$ Divide both sides by 6.2.

$$t\approx -1.35$$ Answer rounded to hundredths.

Check the answer as in problems 19-22.

28.

$$-0.5k=-16.5$$

$$\frac{-0.5k}{-0.5}=\frac{-16.5}{-0.5}$$ Divide both sides by – 0.5.

$$k=33$$

Check the answer as in problems 19-22.

29. **a.** To find the number of airplanes that are needed to transport 520 people, substitute $P=520$ into the equation $J=\frac{P}{126}$ and solve for J.

$$J=\frac{P}{126}$$ Substitute $P=520$.

$$J=\frac{520}{126}$$

$$J\approx 4.13$$ Answer rounded to hundredths.

Since you cannot have 4.13 airplanes, we round the final answer up to the nearest whole number. To transport 520 people, 5 airplanes would be needed.

b. To find the number of people that the carrier can transport at a time with 15 planes, substitute $J = 15$ into the equation $J = \frac{P}{126}$ and solve for P.

$$J = \frac{P}{126} \quad \text{Substitute } J = 15.$$
$$15 = \frac{P}{126}$$
$$126 \cdot 15 = \frac{P}{126} \cdot 126 \quad \text{Multiply both sides by 126.}$$
$$1890 = P$$

If the airline carrier has 15 planes, they can transport 1890 people at a time.

30. **a.** To find Devora's monthly salary if she works 140 hours, substitute $h = 140$ into the equation $S = 14h$ and solve for S.

$$S = 14h \quad \text{Substitute } h = 140.$$
$$S = 14(140)$$
$$S = 1960$$

Devora's monthly salary will be \$1960 if she works 140 hours.

b. To find the number of hours Devora needs to work a month to make \$2100 a month, substitute $S = 2100$ into the equation $S = 14h$ and solve for h.

$$S = 14h \quad \text{Substitute } S = 2100.$$
$$2100 = 14h$$
$$\frac{2100}{14} = \frac{14h}{14} \quad \text{Divide both sides by 14}$$
$$150 = h$$

Devora has to work 150 hours a month to make \$2100 a month.

c. To find the number of hours Devora needs to work a month to earn \$1330 a month, substitute $S = 1330$ into the equation $S = 14h$ and solve for h.

$$S = 14h \quad \text{Substitute } S = 1330.$$
$$1330 = 14h$$
$$\frac{1330}{14} = \frac{14h}{14} \quad \text{Divide both sides by 14}$$
$$95 = h$$

Devora has to work 95 hours a month to earn \$1330 a month.

31.

$$\begin{array}{r} 6x - 9 = -63 \\ \underline{+9 \quad +9} \\ 6x = -54 \end{array} \quad \text{Add 9 to both sides.}$$
$$\frac{6x}{6} = \frac{-54}{6} \quad \text{Divide both sides by 6.}$$
$$x = -9$$

$$6(-9) - 9 \stackrel{?}{=} -63 \quad \text{Check the answer.}$$
$$-54 - 9 \stackrel{?}{=} -63$$
$$-63 = -63 \quad \text{The answer works.}$$

32.

$$\begin{array}{r} 5y - 17 = 33 \\ \underline{+17 \quad +17} \\ 5y = 50 \end{array} \quad \text{Add 17 to both sides.}$$
$$\frac{5y}{5} = \frac{50}{5} \quad \text{Divide both sides by 5.}$$
$$y = 10$$

$$5(10) - 17 \stackrel{?}{=} 33 \quad \text{Check the answer.}$$
$$50 - 17 \stackrel{?}{=} 33$$
$$33 = 33 \quad \text{The answer works.}$$

33.

$$\begin{array}{r} -2a + 9 = -5 \\ \underline{-9 \quad -9} \\ -2a = -14 \end{array} \quad \text{Subtract 9 from both sides.}$$
$$\frac{-2a}{-2} = \frac{-14}{-2} \quad \text{Divide both sides by } -2.$$
$$a = 7$$

Check the answer as in problems 31-32.

34.

$$\begin{array}{r} -8k - 19 = 5 \\ \underline{+19 \quad +19} \\ -8k = 24 \end{array} \quad \text{Add 19 to both sides.}$$
$$\frac{-8k}{-8} = \frac{24}{-8} \quad \text{Divide both sides by } -8.$$
$$k = -3$$

Check the answer as in problems 31-32.

35.

$$\begin{array}{r} \frac{x}{2} - 1 = -11 \\ \underline{+1 \quad +1} \\ \frac{x}{2} = -10 \end{array} \quad \text{Add 1 to both sides.}$$
$$2 \cdot \frac{x}{2} = -10 \cdot 2 \quad \text{Multiply both sides by 2.}$$
$$x = -20$$

Check the answer as in problems 31-32.

36.

$$\frac{-x}{3}+2=-4$$
$$-2 \quad -2 \qquad \text{Subtract 2 from both sides.}$$
$$\frac{-x}{3}=-6$$
$$-3\cdot\frac{-x}{3}=-6\cdot-3 \qquad \text{Multiply both sides by } -3.$$
$$x=18$$

Check the answer as in problems 31-32.

37.

$$-8.6-1.2x=10.6$$
$$+8.6 \qquad +8.6 \qquad \text{Add 8.6 to both sides.}$$
$$-1.2x=19.2$$
$$\frac{-1.2x}{-1.2}=\frac{19.2}{-1.2} \qquad \text{Divide both sides by } -1.2.$$
$$x=-16$$

Check the answer as in problems 31-32.

38.

$$0.25y-1=-7$$
$$+1 \quad +1 \qquad \text{Add 1 to both sides.}$$
$$0.25y=-6$$
$$\frac{0.25y}{0.25}=\frac{-6}{0.25} \qquad \text{Divide both sides by 0.25.}$$
$$y=-24$$

Check the answer as in problems 31-32.

39. Let x = "a number." The sentence translates as follows

$$9x=76$$

Solve:

$$\frac{9x}{9}=\frac{76}{9} \qquad \text{Divide both sides by 9.}$$
$$x=8\frac{4}{9}$$

$$9\left(8\frac{4}{9}\right)\stackrel{?}{=}76 \qquad \text{Check the answer.}$$
$$9\left(\frac{76}{9}\right)\stackrel{?}{=}76$$
$$76=76 \qquad \text{The answer works.}$$

40. Let x = "a number." The sentence translates as follows

$$12=-2x$$

Solve:

$$\frac{12}{-2}=\frac{-2x}{-2} \qquad \text{Divide both sides by } -2.$$
$$-6=x$$

$$12\stackrel{?}{=}-2(-6) \qquad \text{Check the answer.}$$
$$12=12 \qquad \text{The answer works.}$$

41. Let x = "a number." The sentence translates as follows

$$2x+16=-44$$

Solve:

$$2x+16=-44$$
$$-16 \quad -16 \qquad \text{Subtract 16 from both sides.}$$
$$2x=-60$$
$$\frac{2x}{2}=\frac{-60}{2} \qquad \text{Divide both sides by 2.}$$
$$x=-30$$

Check the answer as in problems 39-40.

42. Let x = "a number." The sentence translates as follows

$$\frac{1}{2}x+4=-18$$

Solve:

$$\frac{1}{2}x+4=-18$$
$$-4 \quad -4 \qquad \text{Subtract 4 from both sides.}$$
$$\frac{1}{2}x=-22$$
$$2\cdot\frac{1}{2}x=-22\cdot 2 \qquad \text{Multiply both sides by 2.}$$
$$x=-44$$

Check the answer as in problems 39-40.

43. Let x = "a number." The sentence translates as follows

$$\frac{x}{4}=0.25$$

Solve:

$$4\cdot\frac{x}{4}=0.25\cdot 4 \qquad \text{Multiply both sides by 4.}$$
$$x=1$$

Check the answer as in problems 39-40.

44. Let x = "a number." The sentence translates as follows

$$-\frac{x}{3}=0.75$$

Solve:

$$-3\cdot-\frac{x}{3}=0.75\cdot-3 \qquad \text{Multiply both sides by } -3.$$
$$x=-2.25$$

Check the answer as in problems 39-40.

45. Let $x =$ "a number." The sentence translates as follows

$8-3x=17$

Solve:

$8-3x=17$

$\underline{-8 \quad\quad -8}$ Subtract 8 from both sides.

$-3x=9$

$\frac{-3x}{-3}=\frac{9}{-3}$ Divide both sides by – 3.

$x=-3$

Check the answer as in problems 39-40.

46. Let $x =$ "a number." The sentence translates as follows $2x-9=7$.

Solve:

$2x-9=7$

$\underline{+9 \quad +9}$ Add 9 to both sides.

$2x=16$

$\frac{2x}{2}=\frac{16}{2}$ Divide both sides by 2.

$x=8$

Check the answer as in problems 39-40.

47.

$D=c\boxed{m}a$ Identify the variable to isolate.

$\frac{D}{ca}=\frac{\cancel{c}m\cancel{a}}{\cancel{c}\cancel{a}}$ Divide both sides by *ca*.

$\frac{D}{ca}=m$

48.

$V=lw\boxed{h}$ Identify the variable to isolate.

$\frac{V}{lw}=\frac{\cancel{l}\cancel{w}h}{\cancel{l}\cancel{w}}$ Divide both sides by *lw*.

$\frac{V}{lw}=h$

49.

$C=\pi\cdot\boxed{d}$ Identify the variable to isolate.

$\frac{C}{\pi}=\frac{\cancel{\pi}d}{\cancel{\pi}}$ Divide both sides by π.

$\frac{C}{\pi}=d$

50.

$A=\frac{1}{2}\boxed{b}\cdot h$ Identify the variable to isolate.

$(2)A=\frac{1}{\cancel{2}}b\cdot h(\cancel{2})$ Multiply both sides by 2.

$\frac{2A}{h}=\frac{b\cdot\cancel{h}}{\cancel{h}}$ Divide both sides by *h*.

$\frac{2A}{h}=b$

51.

$9x-3\boxed{y}=-12$ Identify the variable to isolate.

$\underline{-9x \quad\quad -9x}$ Subtract 9*x* from both sides.

$-3y=-9x-12$ Divide both sides by – 3.

$\frac{-3y}{-3}=\frac{-9x-12}{-3}$

$y=\frac{-9x}{-3}-\frac{12}{-3}$

$y=3x+4$

52.

$-8x+4\boxed{y}=-24$ Identify the variable to isolate.

$\underline{+8x \quad\quad +8x}$ Add 8*x* to both sides.

$4y=8x-24$ Divide both sides by – 3.

$\frac{4y}{4}=\frac{8x-24}{4}$

$y=\frac{8x}{4}-\frac{24}{4}$

$y=2x-6$

53. Let $x =$ "a number." The sentence translates as follows

$2x+1=3x-4$

Solve:

$2x+1=3x-4$

$\underline{-3x \quad\quad -3x}$ Subtract 3*x* from both sides.

$-x+1=-4$

$\underline{\quad -1 \quad -1}$ Subtract 1 from both sides.

$-x=-5$

$-1\cdot -x=-5\cdot -1$ Multiply both sides by – 1.

$x=5$

$2(5)+1\stackrel{?}{=}3(5)-4$ Check the answer.

$10+1\stackrel{?}{=}15-4$

$11=11$ The answer works.

54. Let $x =$ "a number." The sentence translates as follows

$\frac{1}{2}x+10=x-4$

Solve:

$(2)\frac{1}{2}x+(2)10=(2)x-4(2)$ Multiply each term by 2.

$x+20=2x-8$

$\underline{-2x \quad -2x}$ Subtract $2x$ from both sides.

$-x+20=-8$

$\underline{-20 \quad -20}$ Subtract 20 from both sides.

$-x=-28$

$-1\cdot -x=-28\cdot -1$ Multiply both sides by -1.

$x=28$

$\frac{1}{2}(28)+10\overset{?}{=}(28)-4$ Check the answer.

$14+10\overset{?}{=}28-4$

$24=24$ The answer works.

55.

$4x-8=-3x+20$

$\underline{+3x \quad +3x}$ Add $3x$ to both sides.

$7x-8=20$

$\underline{+8 \quad +8}$ Add 8 to both sides.

$7x=28$

$\frac{7x}{7}=\frac{28}{7}$ Divide both sides by 7.

$x=4$

$4(4)-8\overset{?}{=}-3(4)+20$ Check the answer.

$16-8\overset{?}{=}-12+20$

$8=8$ The answer works.

56.

$9y+12=7y+6$

$\underline{-7y \quad -7y}$ Subtract $7y$ from both sides.

$2y+12=6$

$\underline{-12 \quad -12}$ Subtract 12 from both sides.

$2y=-6$

$\frac{2y}{2}=\frac{-6}{2}$ Divide both sides by 2.

$y=-3$

$9(-3)+12\overset{?}{=}7(-3)+6$ Check the answer.

$-27+12\overset{?}{=}-21+6$

$-15=-15$ The answer works.

57.

$5(a+5)=12a-13$ Distribute on left side.

$5a+25=12a-13$

$\underline{-12a \quad -12a}$ Subtract $12a$ from both sides.

$-7a+25=-13$

$\underline{-25 \quad -25}$ Subtract 25 from both sides.

$-7a=-38$

$\frac{-7a}{-7}=\frac{-38}{-7}$ Divide both sides by -7.

$a=\frac{38}{7}$ or $a=5\frac{3}{7}$

$5(\left[\frac{38}{7}\right]+5)=12\left(\frac{38}{7}\right)-13$ Check the answer.

$5(\left[\frac{38}{7}\right]+\frac{35}{7})=12\left(\frac{38}{7}\right)-\frac{91}{7}$

$5\left(\frac{73}{7}\right)\overset{?}{=}\frac{456}{7}-\frac{91}{7}$

$\frac{365}{7}=\frac{365}{7}$ The answer works.

58.

$11+7x+7=5x-2$ Simplify on left side.

$7x+18=5x-2$

$\underline{-5x \quad -5x}$ Subtract $5x$ from both sides.

$2x+18=-2$

$\underline{-18 \quad -18}$ Subtract 18 from both sides.

$2x=-20$

$\frac{2x}{2}=\frac{-20}{2}$ Divide both sides by 2.

$x=-10$

$11+7(-10)+7\overset{?}{=}5(-10)-2$ Check the answer.

$18-70\overset{?}{=}-50-2$

$-52=-52$ The answer works.

59.

$4k+1=-5k+3(3k-1)$ Distribute on right side.

$4k+1=-5k+9k-3$ Simplify on right side.

$4k+1=4k-3$

$\underline{-4k \quad -4k}$ Subtract $4k$ from both sides.

$1\overset{?}{=}-3$ This is a false statement.

This equation has no solution.

60.

$9h-7(h+1)=2h+9$ Distribute on left side.

$9h-7h-7=2h+9$ Simplify on left side.

$2h-7=2h+9$

$\underline{-2h \quad -2h}$ Subtract $2h$ from both sides.

$-7\stackrel{?}{=}9$ This is a false statement.

This equation has no solution.

61.

$-5x+3=13-5(x+2)$ Distribute on right side.

$-5x+3=13-5x-10$ Simplify on right side.

$-5x+3=-5x+3$

$\underline{+5x \quad +5x}$ Add $5x$ to both sides.

$3=3$ This is a true statement.

The equation is an identity therefore the solution is all real numbers or $\mathbb{R}$.

To check this we will randomly choose two real numbers -3 and 1, and substitute these into the equation.

$-5(-3)+3\stackrel{?}{=}13-5([-3]+2)$ Check the answer using -3.

$15+3\stackrel{?}{=}13-5(-1)$

$18=18$ The answer works.

$-5(0)+3\stackrel{?}{=}13-5([0]+2)$ Check the answer using 0.

$0+3\stackrel{?}{=}13-5(2)$

$3=3$ The answer works.

62.

$7y-10+2(-3y+2)=y-6$ Distribute-left side.

$7y-10-6y+4=y-6$ Simplify on left side.

$y-6=y-6$

$\underline{+y \quad +y}$ Add y to both sides.

$-6=-6$ This is a true statement.

The equation is an identity therefore the solution is all real numbers or $\mathbb{R}$.

To check this we will randomly choose two real numbers -1 and 5, and substitute these into the equation.

Check the answer using -1.

$7(-1)-10+2(-3[-1]+2)\stackrel{?}{=}(-1)-6$

$-7-10+2(3+2)\stackrel{?}{=}-7$

$-17+2(5)\stackrel{?}{=}-7$

$-7=-7$ The answer works.

Check the answer using 5.

$7(5)-10+2(-3[5]+2)\stackrel{?}{=}(5)-6$

$35-10+2(-15+2)\stackrel{?}{=}-1$

$25+2(-13)\stackrel{?}{=}-1$

$-1=-1$ The answer works.

63. The sum of the measures of the angles in a triangle must equal $180°$. Add the measures of the given angles and set them equal to $180°$ then solve for x.

$x+(x+6)+(x-12)=180$ Combine like terms.

$3x-6=180$

$\underline{+6 \quad +6}$ Add 6 to both sides.

$3x=186$

$\frac{3x}{3}=\frac{186}{3}$ Divide both sides by 3.

$x=62$

Now substitute $x=62$ into the expressions that represent the angles to find the measure of each angle. The measures of the 3 angles are $62°,\ 68°,$ and $50°$.

To check the solution, the measures of the angles should add up to 180.

$62°+68°+50°\stackrel{?}{=}180°$ Check the answer.

$180°=180°$ The answer works.

64. The sum of the measures of the angles in a triangle must equal $180°$. Add the measures of the given angles and set them equal to $180°$ then solve for x.

$3x+(x+20)+(x-10)=180$ Combine like terms.

$5x+10=180$

$\underline{-10 \quad -10}$ Subtract 10 from both sides.

$5x=170$

$\frac{5x}{5}=\frac{170}{5}$ Divide both sides by 5.

$x=34$

Now substitute $x=34$ into the expressions that represent the angles to find the measure of each angle. The measures of the 3 angles are $102°,\ 54°,$ and $24°$.

To check the solution, the measures of the angles should add up to 180.

$102°+54°+24°\stackrel{?}{=}180°$ Check the answer.

$180°=180°$ The answer works.

65.

$4x+1>49$

$\underline{-1 \quad -1}$ Subtract 1 from both sides.

$4x>48$ Divide both sides by 4.

$$\frac{4x}{4} > \frac{48}{4}$$
$$x > 12$$

$4(12)+1\stackrel{?}{=}49$ Check the answer.

$48+1\stackrel{?}{=}49$

$49 = 49$ The answer works.

Check a number greater than 12 ($x = 13$) to test the direction of the inequality symbol.

$4(13)+1\stackrel{?}{>}49$ Check the direction of the inequality.

$52+1\stackrel{?}{>}49$

$53 > 49$ The direction of the inequality works.

Using interval notation, the answer is written as $(12,\infty)$. Use a parenthesis next to the 12 because x is not also "equal to" 12. Always use a parenthesis next to the infinity symbol.

Using a number line:

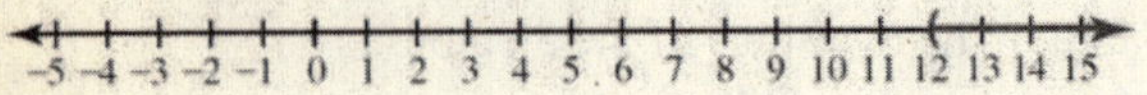

66.

$3x-2 \le 19$

$\quad +2 \quad +2$ Add 2 to both sides.

$3x \le 21$ Divide both sides by 3.

$$\frac{3x}{3} \le \frac{21}{3}$$
$$x \le 7$$

$3(7)-2\stackrel{?}{=}19$ Check the answer.

$21-2\stackrel{?}{=}19$

$19 = 19$ The answer works.

Check a number less than 7 ($x = 6$) to test the direction of the inequality symbol.

$3(6)-2\stackrel{?}{\le}19$ Check the direction of the inequality.

$18-2\stackrel{?}{\le}19$

$16 \le 19$ The direction of the inequality works.

Using interval notation, the answer is written as $(-\infty,7]$. Use a bracket next to the 7 because x is also "equal to" 7. This means that 7 is part of the solution set. Always use a parenthesis next to the infinity symbol.

Using a number line:

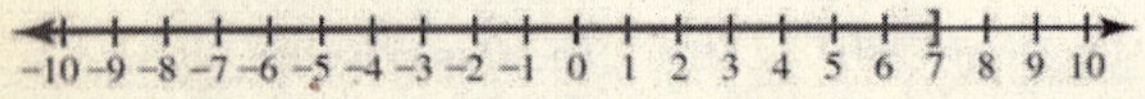

67.

$-x \le 7$ Multiply both sides by -1.

$-1\cdot -x \ge 7\cdot -1$ Multiply by neg., reverse inequality.

$x \ge -7$

Check the answer as in problems 65-66.

Using interval notation, the answer is written as $[-7,\infty)$. Use a bracket next to the -7 because x is also "equal to" -7. This means that -7 is part of the solution set. Always use a parenthesis next to the infinity symbol.

Using a number line:

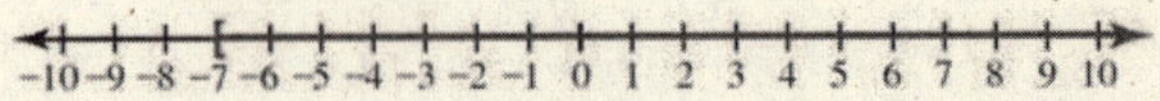

68.

$-y \ge 3$ Multiply both sides by -1.

$-1\cdot -y \le 3\cdot -1$ Multiply by neg., reverse inequality.

$y \le -3$

$-(-3)\stackrel{?}{=}3$ Check the answer.

$3 = 3$ The answer works.

Check the answer as in problems 65-66.

Using interval notation, the answer is written as $(-\infty,-3]$. Use a bracket next to the -3 because y is also "equal to" -3. This means that -3 is part of the solution set. Always use a parenthesis next to the infinity symbol.

Using a number line:

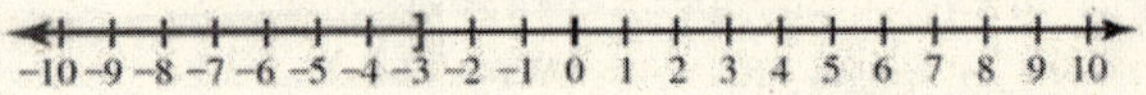

69.

$-5y+1 \ge -16$

$\quad -1 \quad -1$ Subtract 1 from both sides.

$-5y \ge -17$ Divide both sides by -5.

$\frac{-5y}{-5} \le \frac{-17}{-5}$ Divide by neg., reverse inequality.

$$y \le \frac{17}{5}$$

Check the answer as in problems 65-66.

Using interval notation, the answer is written as $\left(-\infty,\frac{17}{5}\right]$. Use a bracket next to the $\frac{17}{5}$ because y is also "equal to" $\frac{17}{5}$. This means that $\frac{17}{5}$ is part of the solution set. Always use a parenthesis next to the infinity symbol.

Using a number line:

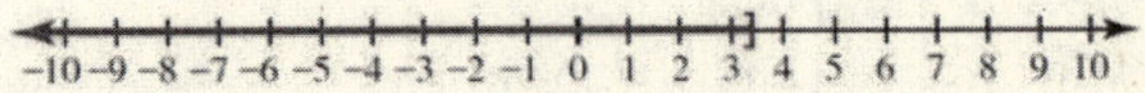

70.

$-6x-7 < -3$

$\quad +7 \quad +7$ Add 7 to both sides.

$-6x < 4$ Divide both sides by -6.

$$\frac{-6x}{-6} > \frac{4}{-6} \quad \text{Divide by neg., reverse inequality.}$$

$$x > -\frac{2}{3}$$

Using interval notation, the answer is written as $\left(-\frac{2}{3},\infty\right)$. Use a parenthesis next to the $-\frac{2}{3}$ because x is not also "equal to" $-\frac{2}{3}$. Always use a parenthesis next to the infinity symbol.

Using a number line:

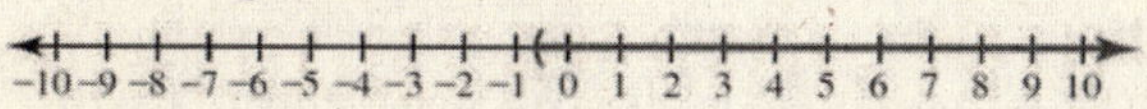

71.

$$\frac{-1}{4}y+3 \le -2$$

$$\underline{\quad -3 \quad -3} \quad \text{Subtract 3 from both sides.}$$

$$\frac{-1}{4}y \le -5 \quad \text{Multiply both sides by } -4.$$

$$-4\cdot\frac{-1}{4}y \ge -5\cdot -4 \quad \text{Multiply by neg., reverse inequality.}$$

$$y \ge 20$$

Check the answer as in problems 65-66.

Using interval notation, the answer is written as $[20,\infty)$. Use a bracket next to the 20 because y is also "equal to" 20. This means that 20 is part of the solution set. Always use a parenthesis next to the infinity symbol.

Using a number line:

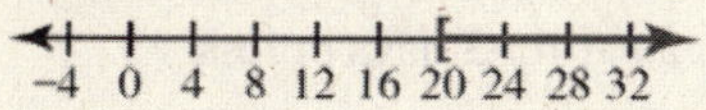

72.

$$4-\frac{2}{5}t > -8$$

$$\underline{-4 \qquad -4} \quad \text{Subtract 4 from both sides.}$$

$$-\frac{2}{5}t > -12 \quad \text{Multiply both sides by } -\frac{5}{2}.$$

$$-\frac{5}{2}\cdot-\frac{2}{5}t < -12\cdot-\frac{5}{2} \quad \text{Multiply by neg., reverse inequality.}$$

$$t < 30$$

Check the answer as in problems 65-66.

Using interval notation, the answer is written as $(-\infty,30)$. Use a parenthesis next to the 30 because t is not also "equal to" 30. Always use a parenthesis next to the infinity symbol.

Using a number line:

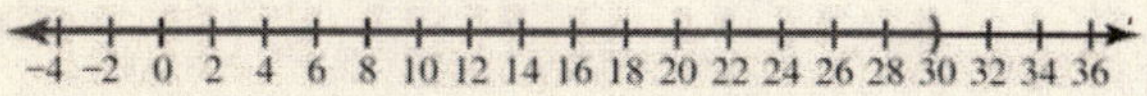

73.

$$-3x-3 < -2x+9$$

$$\underline{+2x \qquad +2x} \quad \text{Add } 2x \text{ to both sides.}$$

$$-x-3 < 9$$

$$\underline{\quad +3 \ +3} \quad \text{Add 3 to both sides.}$$

$$-x < 12 \quad \text{Multiply both sides by } -1.$$

$$-1\cdot -x > 12\cdot -1 \quad \text{Multiply by neg., reverse inequality.}$$

$$x > -12$$

Check the answer as in problems 65-66.

Using interval notation, the answer is written as $(-12,\infty)$. Use a parenthesis next to the -12 because x is not also "equal to" -12. Always use a parenthesis next to the infinity symbol.

Using a number line:

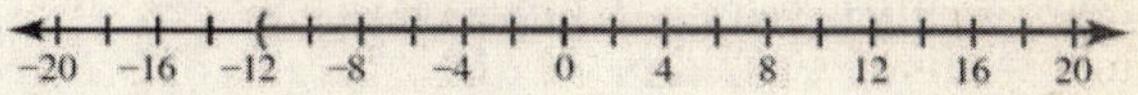

74.

$$-8y-1 > -3y+9$$

$$\underline{+3y \qquad +3y} \quad \text{Add } 3y \text{ to both sides.}$$

$$-5y-1 > 9$$

$$\underline{\quad +1 \ +1} \quad \text{Add 1 to both sides.}$$

$$-5y > 10 \quad \text{Divide both sides by } -5.$$

$$\frac{-5y}{-5} < \frac{10}{-5} \quad \text{Divide by neg., reverse inequality.}$$

$$y < -2$$

Using interval notation, the answer is written as $(-\infty,-2)$.

Using a number line:

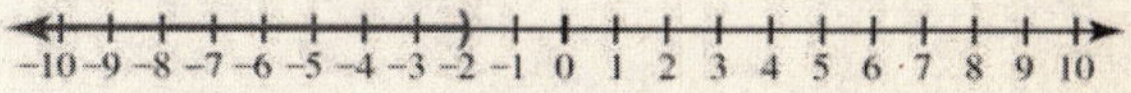

75.

$$-3 < 2x+3 < 11 \quad \text{Isolate } x.$$

$$\underline{-3 \qquad -3 \ -3} \quad \text{Subtract 3 from all sides.}$$

$$-6 < 2x < 8 \quad \text{Divide all sides by 2.}$$

$$\frac{-6}{2} < \frac{2x}{2} < \frac{8}{2}$$

$$-3 < x < 4$$

Check the answer as in problems 65-66.

The solution in interval notation is $(-3,4)$.

Using a number line:

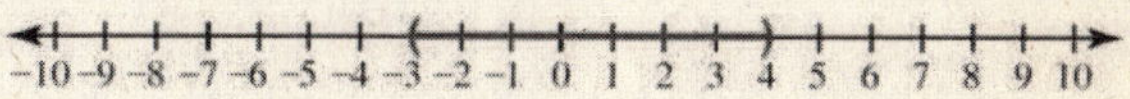

76.

$$-1 < 5x-6 < 24 \quad \text{Isolate } x.$$

$$\underline{+6 \qquad +6 \ +6} \quad \text{Add 6 to all sides.}$$

$$5 < 5x < 30 \quad \text{Divide all sides by 5.}$$

$$\frac{5}{5}<\frac{5x}{5}<\frac{30}{5}$$

$$1<x<6$$

Check the answer as in problems 65-66.

The solution in interval notation is $(1,6)$.

Using a number line:

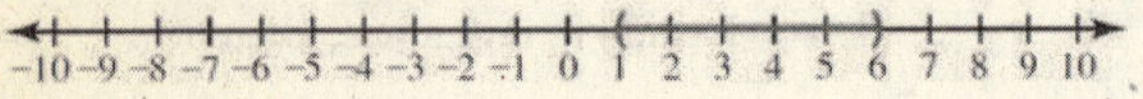

77.

$-2<\frac{x}{3}<9$ Isolate x.

$(3)-2<(3)\frac{x}{3}<9(3)$ Multiply each side by 3.

$-6<x<27$

$-2\overset{?}{<}\frac{(3)}{3}\overset{?}{<}9$ Check direction using a number between – 6 and 27.

$-2<1<9$ The direction of the inequality symbols works.

The solution in interval notation is $(-6,27)$.

Using a number line:

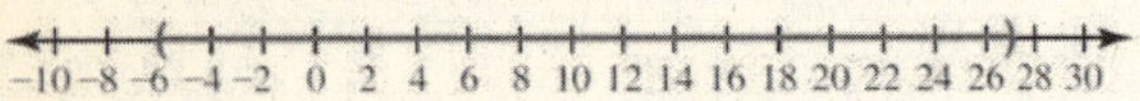

78.

$-8<\frac{y}{2}+11<-5$ Isolate x.

$\underline{-11 \quad -11 \; -11}$ Subtract 11 from all sides.

$-19<\frac{y}{2}<-16$

$(2)-19<(2)\frac{y}{2}<-16(2)$ Multiply each side by 2.

$-38<y<-32$

The solution in interval notation is $(-38,-32)$.

Using a number line:

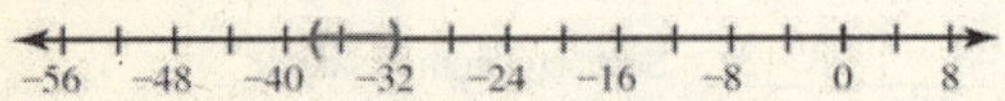

79. The graph of the interval $[-3,\infty)$ is below:

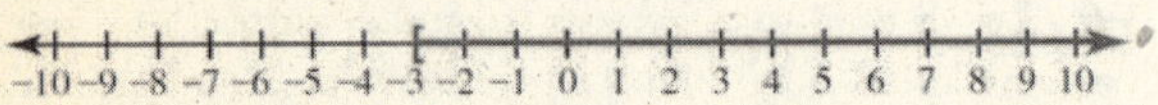

The values that are shown start at -3, include -3, and all numbers greater than -3.

This is equivalent to the inequality $x\geq-3$.

80. The graph of the interval $(5,\infty)$ is below:

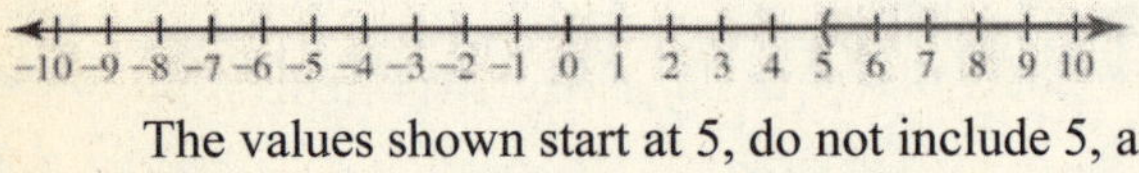

The values shown start at 5, do not include 5, and include all numbers greater than 5.

This is equivalent to the inequality $x>5$.

81. The graph of the interval $(-\infty,6)$ is below:

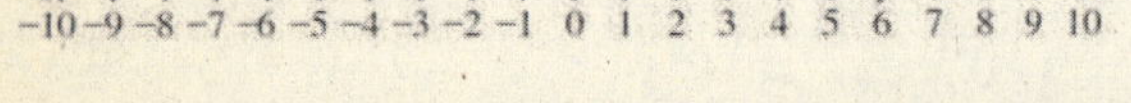

The values that are shown start at 6, do not include 6, and include all numbers less than 6.

This is equivalent to the inequality $x<6$.

82. The graph of the interval $(-\infty,-12]$ is below:

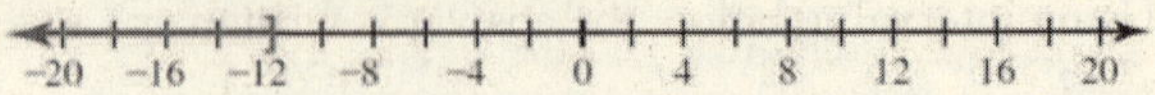

The values that are shown start at -12, include -12, and all numbers less than -12.

This is equivalent to the inequality $x\leq-12$.

83. If $-6>x$, then x must be less than -6. This means we can rewrite the inequality as $x<-6$

84. If $0\leq y$, then y must be greater than or equal to 0. This means we can rewrite the inequality as $y\geq0$

85. $80n-2400>0$ Isolate variable term.

$\underline{\quad +2400 \; +2400}$ Add 2400 to both sides.

$80n>2400$

$\frac{80n}{80}>\frac{2400}{80}$ Divide both sides by 80.

$n>30$

Kiano must sell more than 30 stained-glass windows in order to make a profit.

86. $32.49d+21.65\leq119.12$ Isolate variable term.

$\underline{\quad -21.65 \; -21.65}$ Subtract 21.65 from both sides.

$32.49d\leq97.47$

$\frac{32.49d}{32.49}\leq\frac{97.47}{32.49}$ Divide both sides by 32.49.

$d\leq3$

Jessie can rent a car for 3 days or less.

87. **a.** The inequality that represents Karyn keeping her monthly texting bill to at most \$10 is: $7.5+0.1t\leq10$

b.

$7.5+0.1t\leq10$ Isolate variable term.

$\underline{-7.5 \qquad -7.5}$ Subtract 7.5 from both sides.

$0.1t\leq2.5$

$\frac{0.1t}{0.1}\leq\frac{2.5}{0.1}$ Divide both sides by 0.1.

$t\leq25$

Karyn can send no more than 25 texts over her 200 texts allowed.

88. **a.** Add the costs in an inequality less than or equal to 65.

$50+1.5m\leq65$

b. $50+1.5m \le 65$ Isolate variable term.

$-50 \quad -50$ Subtract 50 from both sides.

$1.5m \le 15$

$\frac{1.5m}{1.5} \le \frac{15}{1.5}$ Divide both sides by 1.5.

$m \le 10$

Trey can have the TV delivered and stay within his budget for the delivery charge if he lives 10 miles or less from the store.

Chapter 2 Test

1. To check if the given value of the variable is a solution to the equation, evaluate each side of the equation and see if the two sides are equal.

$5b+8=2$ Original equation.

$5(-2)+8\overset{?}{=}2$ Substitute $b=-2$.

$-10+8\overset{?}{=}2$ Simplify both sides.

$-2\overset{?}{=}2$ This statement is not true.

$-2 \ne 2$

Because the final results are not equal, $b=-2$ is not a solution to the equation $5b+8=2$.

2. To check if the given values of the variables are a solution to the equation, evaluate each side of the equation and see if the two sides are equal.

$D=70t$ Original equation.

$294\overset{?}{=}70(4.2)$ Substitute values for D and t.

$294=294$ This statement is true.

Because the final statement is true, $t=4.2$ and $D=294$ is a solution to the equation $D=70t$.

3.

$-5+\boxed{x}=-17$ Identify the variable term.

$+5 \quad +5$ Add 5 to both sides.

$x=-12$

$-5+(-12)\overset{?}{=}-17$ Check the answer.

$-17=-17$ The answer works.

4. To solve for c means to isolate c on one side of the equal sign.

$P=a+b+\boxed{c}$ Identify the variable term to isolate.

$-a-b \quad -a-b$ Subtract a and b from both sides.

$P-a-b=c$ or $c=P-a-b$

5. Using the equation $P=R-C$ substitute $R=250000$ for the revenue, $C=88000$ for costs, then solve for P to find the company's profit.

$P=R-C$ Given equation.

$P=(250000)-(88000)$ Substitute value of variables.

$P=162000$ Simplify on right side.

The solution is $P=162000$ which represents that the company has monthly profit of $162,000.

6.

$-5w=450$ Variable term is isolated.

$\frac{-5w}{-5}=\frac{450}{-5}$ Divide both sides by -5.

$w=-90$

$-5(-90)\overset{?}{=}450$ Check the answer.

$450=450$ The answer works.

7.

$\boxed{-4p}+25=-119$ Identify variable term to isolate.

$\underline{\quad -25 \quad -25}$ Subtract 25 from both sides.

$-4p=-144$

$\frac{-4p}{-4}=\frac{-14}{-4}$ Divide both sides by -4.

$p=36$

$-4(36)+25\overset{?}{=}-119$ Check the answer.

$-144+25\overset{?}{=}-119$

$-119=-119$ The answer works.

8.

$\boxed{\frac{x}{3}}+7=-5$ Identify variable term to isolate.

$\underline{\quad -7 \quad -7}$ Subtract 7 from both sides.

$\frac{x}{3}=-12$

$3\cdot\frac{x}{3}=-12\cdot 3$ Multiply both sides by 3.

$x=-36$

$\frac{(-36)}{3}+7\overset{?}{=}-5$ Check the answer.

$-12+7\overset{?}{=}-5$

$-5=-5$ The answer works.

9.

$4(x+4)-5=4x+11$ Distribute-left side.

$4x+16-5=4x+11$ Simplify on left side.

$4x+11=4x+11$

$\underline{-4x \qquad -4x}$ Subtract $4x$ from both sides.

$11=11$ This is a true statement.

The equation is an identity therefore the solution is all real numbers or $\mathbb{R}$.

To check this we will randomly choose two real numbers -6 and 10, and substitute these into the equation.

Check the answer using -6.

$4([-6]+4)-5\overset{?}{=}4(-6)+11$

$4(-2)-5\overset{?}{=}-24+11$

$-8-5\overset{?}{=}-13$

$-13=-13$ The answer works.

Check the answer using 10.

$4([10]+4)-5\overset{?}{=}4(10)+11$

$4(14)-5\overset{?}{=}40+11$

$56-5\overset{?}{=}51$

$51=51$ The answer works.

10.

$3x+20=20+3(x-5)$ Distribute on right side.

$3x+20=20+3x-15$ Simplify on right side.

$3x+20=3x+5$

$\underline{-3x \qquad -3x}$ Subtract $3x$ from both sides.

$20\overset{?}{=}5$ This is a false statement.

This equation has no solution.

11. Recall that the perimeter of a rectangle can be calculated using the formula $P=2l+2w$ where P is the perimeter of the rectangle with length, l, and width, w. We are given the perimeter and width of the rectangular lot, so substitute $P=330$ and $w=40$. After substituting in these values, solve the equation for l.

$330=2l+2(40)$ Substitute in known values.

$330=\boxed{2l}+80$ Simplify & isolate variable term.

$\underline{-80 \qquad -80}$ Subtract 80 from both sides.

$250=2l$ Divide both sides by 2.

$\frac{250}{2}=\frac{2l}{2}$

$125=l$

$330\overset{?}{=}2(125)+80$ Check the answer.

$330\overset{?}{=}250+80$

$330=330$ The answer works.

The length of the building lot is 125 feet.

12. To solve for x means to isolate x on one side of the equal sign.

$y=m\boxed{x}+b$ Identify the variable to isolate.

$\underline{-b \qquad -b}$ Subtract b from both sides.

$y-b=mx$ Divide both sides by m.

$\frac{y-b}{m}=\frac{\not{m}x}{\not{m}}$

$\frac{y-b}{m}=x$

13. Let $x=$ "a number." The sentence translates as follows

$4x+7=-17$

Solve:

$4x+7=-17$

$\underline{\quad -7 \quad -7}$ Subtract 7 from both sides.

$4x=-24$

$\frac{4x}{4}=\frac{-24}{4}$ Divide both sides by 4.

$x=-6$

$4(-6)+7\stackrel{?}{=}-17$ Check the answer.

$-24+7\stackrel{?}{=}-17$

$-17=-17$ The answer works.

14.

$10x-15=7x-9$

$\underline{-7x \quad -7x}$ Subtract $7x$ from both sides.

$3x-15=-9$

$\underline{+15 \quad +15}$ Add 15 to both sides.

$3x=6$

$\frac{3x}{3}=\frac{6}{3}$ Divide both sides by 3.

$x=2$

$10(2)-15\stackrel{?}{=}7(2)-9$ Check the answer.

$20-15\stackrel{?}{=}14-9$

$5=5$ The answer works.

15.

$4x+3=4(x-1)+7$ Distribute-right side.

$4x+3=4x-4+7$ Simplify on right side.

$4x+3=4x+3$

$\underline{-4x \quad -4x}$ Subtract $4x$ from both sides.

$3=3$ This is a true statement.

The equation is an identity therefore the solution is all real numbers or $\mathbb{R}$.

To check this we will randomly choose two real numbers -2 and 7, and substitute these into the equation.

Check the answer using -2.

$4(-2)+3\stackrel{?}{=}4([-2]-1)+7$

$-8+3\stackrel{?}{=}4(-3)+7$

$-5\stackrel{?}{=}-12+7$

$-5=-5$ The answer works.

Check the answer using 7.

$4(7)+3\stackrel{?}{=}4([7]-1)+7$

$28+3\stackrel{?}{=}4(6)+7$

$31\stackrel{?}{=}24+7$

$31=31$ The answer works.

16. The sum of the measures of the angles in a triangle must equal $180°$. Add the measures of the given angles and set them equal to $180°$ then solve for x.

$x+11x+6x=180$ Combine like terms.

$18x=180$

$\frac{18x}{18}=\frac{180}{18}$ Divide both sides by 18.

$x=10$

Now substitute $x=10$ into the expressions that represent the angles to find the measure of each angle. The measures of the 3 angles are $10°$, $110°$, and $60°$.

To check the solution, the measures of the angles should add up to 180.

$10°+110°+60°\stackrel{?}{=}180°$ Check the answer.

$180°=180°$ The answer works.

17. If $-16 \le x$, then x must be greater than or equal to -16. This means we can rewrite the inequality as

$x \ge -16$

18.

$2x+1>4x-3$

$\underline{-4x \quad -4x}$ Subtract $4x$ from both sides.

$-2x+1>-3$

$\underline{-1 \quad -1}$ Subtract 1 from both sides.

$-2x>-4$ Divide both sides by -2.

$\frac{-2x}{-2}<\frac{-4}{-2}$ Multiply by neg., reverse inequality.

$x<2$

Graphing the solution on a number line:

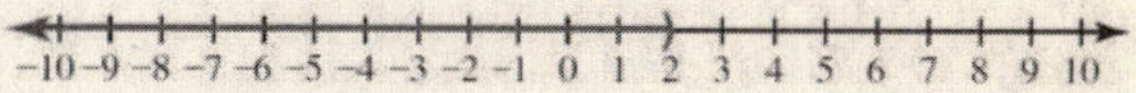

19.

$5 \le 2x-7 \le 15$ Isolate x.

$\underline{+7 \quad +7 \quad +7}$ Add 7 to all sides.

$12 \le 2x \le 22$ Divide all sides by 2.

$\frac{12}{2} \le \frac{2x}{2} \le \frac{22}{2}$

$6 \le x \le 11$

20. **a.** Aiden's hourly wage, \$10.25, times the number of hours h he works in a week must be greater than \$287 in order to meet his expenses. Write the inequality as follows:

$10.25h>287$

b.

$10.25h>287$ Divide both sides by 10.25.

$\frac{10.25h}{10.25}>\frac{287}{10.25}$

$h>28$

Aiden needs to work more than 28 hours a week to meet his expenses.

Cumulative Review Chapters 1-2

1. The base is $\frac{9}{4}$ and we multiply $\frac{9}{4}$ by itself 6 times. So the exponent is 6. The exponential form is $\left(\frac{9}{4}\right)^6$.

$$\frac{9}{4}\cdot\frac{9}{4}\cdot\frac{9}{4}\cdot\frac{9}{4}\cdot\frac{9}{4}\cdot\frac{9}{4}=\left(\frac{9}{4}\right)^6$$

2. The base is -4, and the exponent of 3 tells us we are multiplying -4 by itself 3 times. The expanded form is $(-4)\cdot(-4)\cdot(-4)$.

$$(-4)^3=(-4)\cdot(-4)\cdot(-4)$$

3. The base is -1 and we multiply -1 by itself 5 times.

$$-1\cdot-1\cdot-1\cdot-1\cdot-1=-1$$

4. The exponent on 10 is 8. Move the decimal point 8 places to the right.

$$3.045\times10^8=3.045\cdot100{,}000{,}000=304{,}500{,}000$$

Therefore, 3.045×10^8 has the standard form 304,500,000.

5. Evaluating the expression using the order of operations would be as follows:

$$\boxed{\frac{84}{10-6}\cdot3}\,\boxed{-2^2}\,\boxed{-(-5)}+\boxed{20\div4\cdot(-3)} \quad \text{Outline the terms.}$$
$$=\frac{84}{10-6}\cdot3-2^2-(-5)+20\div4\cdot(-3) \quad \text{Simplify fraction first.}$$
$$=\frac{84}{4}\cdot3-2^2-(-5)+20\div4\cdot(-3)$$
$$=21\cdot3-2^2-(-5)+20\div4\cdot(-3) \quad \text{Evaluate exponent.}$$
$$=21\cdot3-4-(-5)+20\div4\cdot(-3) \quad \text{Simplify within each term.}$$
$$=63-4-(-5)+5\cdot(-3) \quad \text{Multiply in 1st term, divide in last.}$$
$$=63-4+5-15 \quad \text{Multiply in last term.}$$
$$=49 \quad \text{Add/subtract from left to right.}$$

6. Evaluating the expression using the order of operations would be as follows:

$$\boxed{18\div(-2)}+\boxed{|6-7\cdot(-2)|}+\boxed{(-3)^2} \quad \text{Outline the terms.}$$
$$=18\div(-2)+|6+14|+(-3)^2 \quad \text{Evaluate absolute value.}$$
$$=18\div(-2)+|20|+(-3)^2$$
$$=18\div(-2)+20+9 \quad \text{Evaluate exponent.}$$
$$=-9+20+9 \quad \text{Divide.}$$
$$=20 \quad \text{Add/subtract from left to right.}$$

7. $(-16)+(3-7)=(3-7)+(-16)$

This is an example of the commutative property of addition, because the order of the terms has been changed.

8. In this problem, there are 2 quantities that can change. One quantity is the amount of money (revenue) the soccer club makes, and the second is the number of candy bars sold. Let R = the amount of revenue in dollars the soccer club makes and c = the number of candy bars sold.

9. Let x = a number. "Less than" translates as subtraction, but it is done in reverse order. So this translates as something minus 7. "8 times a number" means to multiply a number by 8 so this translates as $8x$ and the sentence translates as $8x-7$.

10. Convert 95 feet to meters. First, form the unity fraction. We want to convert to the units of meters, so meters will go in the numerator. We want to convert from the units of feet, so feet will go in the denominator. There are 3.28 ft in 1 m, so the unity fraction is $\frac{1\text{ m}}{3.28\text{ ft}}$. Multiply 95 ft by this unity fraction and simplify the expression as follows.

$$95\text{ ft}\cdot\frac{1\text{ m}}{3.28\text{ ft}}$$
$$=\frac{95\,\cancel{\text{ft}}}{1}\cdot\frac{1\text{ m}}{3.28\,\cancel{\text{ft}}}$$
$$=\frac{95\text{ m}}{3.28}\approx28.96\text{ m}$$

Therefore, 95 feet is equal to approximately 28.96 m.

11. Evaluate $\frac{-x}{3}+7x$ for $x=-6$.

$$\frac{-x}{3}+7x \quad \text{Substitute in } x=-6.$$
$$=\frac{-(-6)}{3}+7(-6) \quad \text{Simplify.}$$
$$=\frac{6}{3}-42$$
$$=2-42$$
$$=-40$$

12.

Number of Biscuits for Scout	Number of Biscuits for Shadow
0	50 - 0 = 50
5	50 - 5 = 45
10	50 - 10 = 40
15	50 - 15 = 35
n	50 - n

To find number of dog biscuits Lisa will give to Shadow, subtract the number of biscuits given to Scout from 50 biscuits. The general variable expression is $50-n$.

13. Identify like terms first. Then combine like terms and arrange in conventional form.

$$\boxed{\boxed{7xy}}\,\boxed{-3x^2}+\boxed{\boxed{6x}}\,\boxed{-9x^2}+\boxed{\boxed{\boxed{8xy}}}\,\boxed{-x}$$
$$=-12x^2+5x+15xy$$

14. Evaluate parentheses first. Then identify like terms, combine like terms, and arrange in conventional form.

$$6x-(3x-5)+8 \quad \text{Evaluate parentheses.}$$
$$=\boxed{6x}\,\boxed{-3x}+\boxed{\boxed{5}}+\boxed{\boxed{8}} \quad \text{Identify like terms.}$$
$$=3x+13 \quad \text{Combine like terms.}$$

15. Evaluate parentheses first. Then identify like terms, combine like terms, and arrange in conventional form.

$(3b-c)-(-b+7c)$ Evaluate parentheses.
$=3b-c+b-7c$ Identify like terms.
$=4b-8c$ Combine like terms.

16. To determine the coordinates of the points we need to find the values of the x- and y-coordinates. Beginning in the top right quadrant with the point on the x-axis, then following counterclockwise for the other points, the first point is located 4 units to the right of the y-axis, so the x-coordinate is $x=4$. The point is located on the x-axis, so the y-coordinate is $y=0$.

The coordinates of this point are $(4,0)$.

The next point is located 1 unit to the right of the y-axis, so the x-coordinate is $x=1$. The point is located 2 units above the x-axis, so the y-coordinate is $y=2$.

The coordinates of this point are $(1,2)$.

The next point is located 3 units to the left of the y-axis, so the x-coordinate is $x=-3$. The point is located 4 units above the x-axis, so the y-coordinate is $y=4$.

The coordinates of this point are $(-3,4)$.

The next point is located 3 units to the left of the y-axis, so the x-coordinate is $x=-3$. The point is located 5 units below the x-axis, so the y-coordinate is $y=-5$.

The coordinates of this point are $(-3,-5)$.

The next point is located on the y-axis, so the x-coordinate is $x=0$. The point is located 3 units below the x-axis, so the y-coordinate is $y=-3$.

The coordinates of this point are $(0,-3)$.

The next point is located 6 units to the right of the y-axis, so the x-coordinate is $x=6$. The point is located 2 units below the x-axis, so the y-coordinate is $y=-2$.

The coordinates of this point are $(6,-2)$.

17. The input values range from -3 to 7. It is reasonable to scale the horizontal axis by 2's. The output values range from -3 to 6 so it will be reasonable to scale the y-axis by 2's. The points are plotted on the graph below.

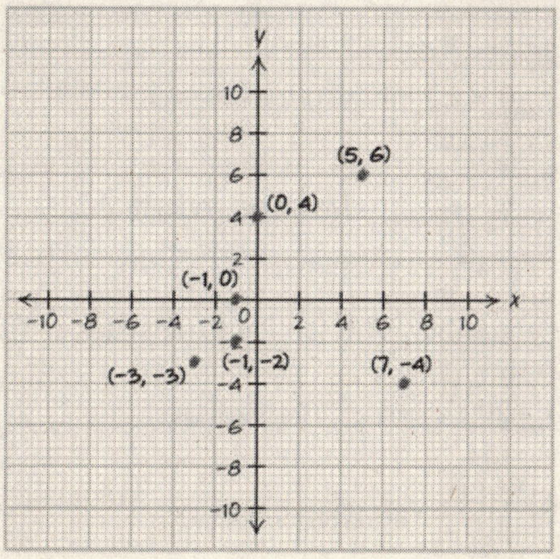

18. A reasonable vertical axis for this data would range between 0 and 9.2 with a scale of 0.1 and a break from 0 to 8. The horizontal axis will start at 0 and end at 1990 with a scale of 3 and a break from 0 to 1960. For a scatterplot, instead of using a bar to represent the height, use a large dot. This scatterplot should have a title and would be as follows:

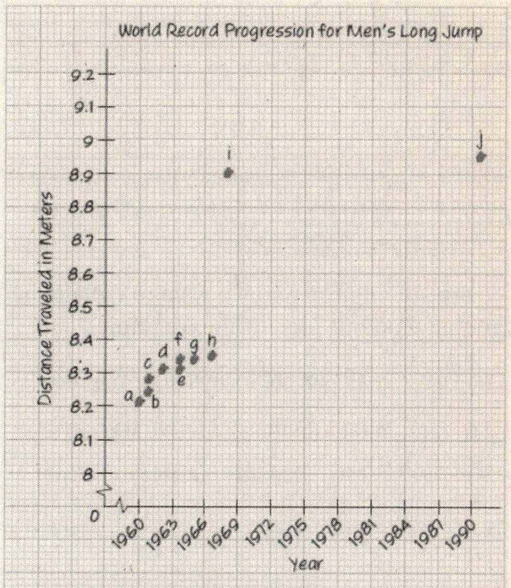

19. To check if the given values of the variables are a solution to the equation, evaluate each side of the equation and see if the two sides are equal.

$4x-7y=-6$ Original equation.
$4(-1.5)-7(0)\stackrel{?}{=}-6$ Substitute $x=-1.5$ and $y=0$.
$-6-0\stackrel{?}{=}-6$ Simplify both sides.
$-6=-6$ This statement is true.

Because the final statement is true, $x=-1.5$ and $y=0$ are a solution to the equation $4x-7y=-6$.

20.

$-2+\boxed{y}=-7$ Identify the variable term.
$+2 \quad\quad +2$ Add 2 to both sides.
$y=-5$

21.

$\boxed{x}-\frac{1}{3}=\frac{5}{6}$ Identify the variable term.

$+\frac{1}{3}\quad+\frac{1}{3}$ Add $\frac{1}{3}$ to both sides.

$x=\frac{5}{6}+\frac{2}{6}$ Find LCD = 6.

$x=\frac{7}{6}$ or $x=1\frac{1}{6}$

22.

$-0.051+\boxed{n}=4.006$ Identify the variable term.

$+0.051\qquad +0.051$ Add 0.051 to both sides.

$n=4.057$

23. We are solving for y so we want to isolate the variable y on one side of the equation.

$3x\boxed{-5y}=-30$ Identify variable term to isolate.

$-3x\qquad -3x$ Subtract $3x$ from both sides.

$-5\boxed{y}=-3x-30$ Identify variable to isolate

$\frac{-5y}{-5}=\frac{-3x-30}{-5}$ Divide both sides by -5.

$y=\frac{-3x}{-5}-\frac{30}{-5}$ Simplify on right side.

$y=\frac{3}{5}x+6$

24.

$-7x=-21$

$\frac{-7x}{-7}=\frac{-21}{-7}$ Divide both sides by -7.

$x=3$

25.

$\frac{2}{3}n=-8$

$\frac{3}{2}\cdot\frac{2}{3}n=-8\cdot\frac{3}{2}$ Multiply both sides by $\frac{3}{2}$.

$n=-12$

26.

$-5z=7.95$

$\frac{-5z}{-5}=\frac{7.95}{-5}$ Divide both sides by -5.

$z=-1.59$

27. **a.** Substitute $h=120$ into the equation $S=12h$ and solve for S.

$S=12h$ Given equation.

$S=12(120)$ Substitute $h=120$.

$S=1440$ Multiply & solve for S.

If Corinthia works 120 hours, her salary will be \$1440.

b. Substitute $S=1680$ into the equation $S=12h$ and solve for h.

$S=12h$ Given equation.

$1680=12h$ Substitute $S=1680$ & solve for h.

$\frac{1680}{12}=\frac{12h}{12}$ Divide both sides by 12.

$140=h$

Corinthia needs to work 140 hours in a month to make \$1680 a month.

c. Substitute $S=1200$ into the equation $S=12h$ and solve for h.

$S=12h$ Given equation.

$1200=12h$ Substitute $S=1200$ & solve for h.

$\frac{1200}{12}=\frac{12h}{12}$ Divide both sides by 12.

$100=h$

Corinthia needs to work 100 hours in a month if she only wants to earn \$1200 a month.

28.

$8y-11=21$

$+11\ +11$ Add 11 to both sides.

$8y=32$

$\frac{8y}{8}=\frac{32}{8}$ Divide both sides by 8.

$y=4$

29.

$\frac{x}{5}-4=-9$

$+4\ +4$ Add 4 to both sides.

$\frac{x}{5}=-5$

$5\cdot\frac{x}{5}=-5\cdot5$ Multiply both sides by 5.

$x=-25$

30.

$-1.2+3.7x=15.45$

$+1.2\qquad +1.2$ Add 1.2 to both sides.

$3.7x=16.65$

$\frac{3.7x}{3.7}=\frac{16.65}{3.7}$ Divide both sides by 3.7.

$x=4.5$

31. Let x = a number. The sentence translates as follows:

$2x+20=6$

$-20\ -20$ Subtract 20 from both sides.

$2x=-14$

$\frac{2x}{2}=\frac{-14}{2}$ Divide both sides by 8.

$x=-7$

32. Isolate the variable m.0

$P = rg\boxed{m}$ Identify variable to isolate.

$\frac{P}{rg} = \frac{\not{r}\not{g}m}{\not{r}\not{g}}$ Divide both sides by lh.

$\frac{P}{rg} = m$

33. Isolate the variable y.

$6x\boxed{-3y} = -30$ Identify variable term to isolate.

$\underline{-6x \qquad -6x}$ Subtract $6x$ from both sides.

$-3\boxed{y} = -6x - 30$ Identify variable to isolate

$\frac{-3y}{-3} = \frac{-6x-30}{-3}$ Divide both sides by -3.

$y = \frac{-6x}{-3} - \frac{30}{-3}$ Simplify on right side.

$y = 2x + 10$

34. Let x = a number. The sentence translates as follows:

$2x+8=4x-18$

$\underline{-4x \qquad -4x}$ Subtract $4x$ from both sides.

$-2x+8=-18$

$\underline{-8 \quad -8}$ Subtract 8 from both sides.

$-2x=-26$

$\frac{-2x}{-2}=\frac{-26}{-2}$ Divide both sides by -2.

$x=13$

35.

$8y+4=12y+20$

$\underline{-12y \quad -12y}$ Subtract $12y$ from both sides.

$-4y+4=20$

$\underline{-4 \quad -4}$ Subtract 4 from both sides.

$-4y=16$

$\frac{-4y}{-4}=\frac{16}{-4}$ Divide both sides by -4.

$y=-4$

36.

$7n+8=-3n+2(4n-12)$ Distribute on right side.

$7n+8=-3n+8n-24$ Simplify on right side.

$7n+8=5n-24$

$\underline{-5n \quad -5n}$ Subtract $5n$ from both sides.

$2n+8=-24$

$\underline{-8 \quad -8}$ Subtract 8 from both sides.

$2n=-32$

$\frac{2n}{2}=\frac{-32}{2}$ Divide both sides by 2.

$n=-16$

37.

$5w-6+3(-4w+9)=w+5$ Distribute on left side.

$5w-6-12w+27=w+5$ Simplify on left side.

$-7w+21=w+5$

$\underline{-w \quad -w}$ Subtract w from both sides.

$-8w+21=5$

$\underline{-21 \quad -21}$ Subtract 21 from both sides.

$-8w=-16$

$\frac{-8w}{-8}=\frac{-16}{-8}$ Divide both sides by -8.

$w=2$

38.

$x+(x+18)+(x-3)=180$ Combine like terms.

$3x+15=180$

$\underline{-15 \quad -15}$ Subtract 15 from both sides.

$\frac{3x}{3}=\frac{165}{3}$ Divide both sides by 3.

$x=55$

Now substitute $x=55$ into the expressions that represent the angles to find the measure of each angle. The measures of the 3 angles are $55°$, $73°$, and $52°$.

To check the solution, the measures of the angles should add up to 180.

$55°+73°+52° \stackrel{?}{=} 180°$ Check the answer.

$180°=180°$ The answer works.

39.

$6x+1>49$

$\underline{-1 \quad -1}$ Subtract 1 from both sides.

$6x>48$ Divide both sides by 6.

$\frac{6x}{6}>\frac{48}{6}$

$x>8$

40.

$\frac{-1}{7}r+5\le 9$

$\underline{-5 \quad -5}$ Subtract 5 from both sides.

$\frac{-1}{7}r\le 4$ Multiply both sides by -7.

$-7\cdot\frac{-1}{7}r\ge 4\cdot -7$ Multiply by neg., reverse inequality.

$r\ge -28$

41.

$$\begin{array}{ll} 14 < 3x+5 < 26 & \text{Isolate } x. \\ \underline{-5 \quad\; -5 \;\; -5} & \text{Subtract 5 from all sides.} \\ 9 < 3x < 21 & \text{Divide all sides by 3.} \\ \frac{9}{3} < \frac{3x}{3} < \frac{21}{3} & \\ 3 < x < 7 & \end{array}$$

42.

$$\begin{array}{ll} 10-\frac{3}{8}t > -4 & \\ \underline{-10 \qquad -10} & \text{Subtract 10 from both sides.} \\ -\frac{3}{8}t > -14 & \text{Multiply both sides by } -\frac{8}{3}. \\ -\frac{8}{3}\cdot -\frac{3}{8}t < -14\cdot -\frac{8}{3} & \text{Multiply by neg., reverse inequality.} \\ t < \frac{112}{3} \text{ or } t < 37\frac{1}{3} & \end{array}$$

43. The English class can have between 15 students and 25 students enrolled. The maximum enrollment is 25, so we include 25 in the interval. The class must have at least 15 students, so we include 15 in the interval. Let $s =$ the number of students enrolled in the class.

$$15 \le s \le 25$$

44. The upper-division courses are numbered between 200 and 400. The problem states that 200 is included in the interval, but 400 is not included. Let $n =$ the course number.

$$200 \le n < 400$$

CHAPTER 3 LINEAR EQUATIONS WITH TWO VARIABLES

Section 3.1 Graphing Equations with Two Variables

1. We are given some sales values to start with, so use the equation to calculate the monthly salary for the given sales amounts. The last two rows could be filled with any sales amounts; these are just two possibilities.

s = Sales (dollars)	P = Monthly pay (dollars)	Using the equation to find s
0	300	$P = 0.05(0) + 300$ $P = 300$
1000	350	$P = 0.05(1000) + 300$ $P = 350$
8000	700	$P = 0.05(8000) + 300$ $P = 700$
30,000	1800	$P = 0.05(30000) + 300$ $P = 1800$
50,000	2800	$P = 0.05(50000) + 300$ $P = 2800$

3. We are given some input values to start with, so use the equation to calculate the total number of customers for the given values. The last two rows could be filled with any reasonable values; these are just two possibilities.

h = Hours since opening	C = # of customers for the day	Use equation to find C
0	2	$C = 8(0) + 2$ $C = 2$
1	10	$C = 8(1) + 2$ $C = 10$
3	26	$C = 8(3) + 2$ $C = 26$
6	50	$C = 8(6) + 2$ $C = 50$
10	82	$C = 8(10) + 2$ $C = 82$

5. We are given some input values to start with, so use the equation to calculate the cost for framing a wall for the given values. The last two rows could be filled with any reasonable values; these are just two possibilities.

n = Number of studs purchased	C = Cost (dollars)	Use equation to find C
4	14	$C = 3.5(4)$ $C = 14$
6	21	$C = 3.5(6)$ $C = 21$
8	28	$C = 3.5(8)$ $C = 28$
11	38.50	$C = 3.5(11)$ $C = 38.5$
18	63	$C = 3.5(18)$ $C = 63$

7. We are given some input values to start with, so use the equation to calculate the number of calories burned for the given values. The last two rows could be filled with any reasonable values; these are just two possibilities.

h = Hours of bowling	C = Calories burned	Using equation to find C
1	259	$C = 259(1)$ $C = 259$
2	518	$C = 259(2)$ $C = 518$
3	777	$C = 259(3)$ $C = 777$
5	1295	$C = 259(5)$ $C = 1295$
6	1554	$C = 259(6)$ $C = 1554$

9.

s = Sales (dollars)	P = Monthly pay (dollars)	(s, P)
0	300	$(0, 300)$
1000	350	$(1000, 350)$
8000	700	$(8000, 700)$
30,000	1800	$(30000, 1800)$
50,000	2800	$(50000, 2800)$

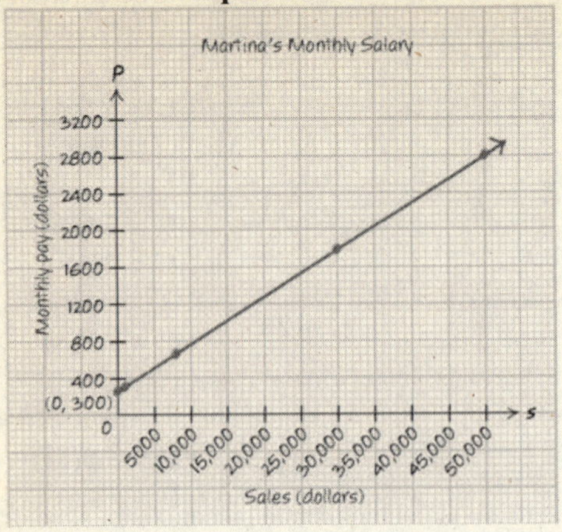

11.

h = Hours since opening	C = # of customers for the day	(h, C)
0	2	$(0,2)$
1	10	$(1,10)$
3	26	$(3,26)$
6	50	$(6,50)$
10	82	$(10,82)$

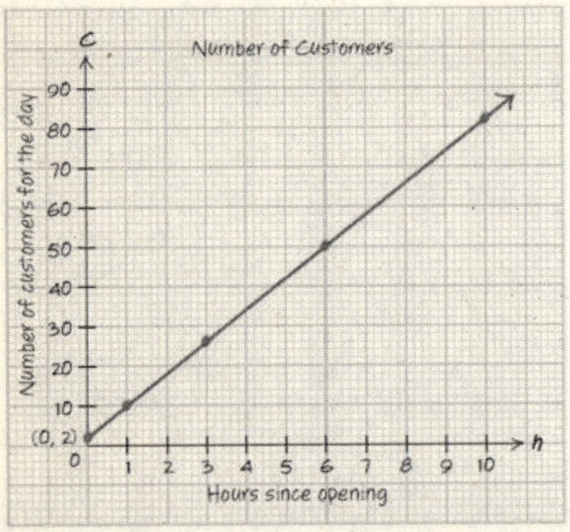

13.

n = Number of studs purchased	C = Cost (dollars)	(n, C)
4	14	$(4,14)$
6	21	$(6,21)$
8	28	$(8,28)$
11	38.50	$(11,38.5)$
18	63	$(18,63)$

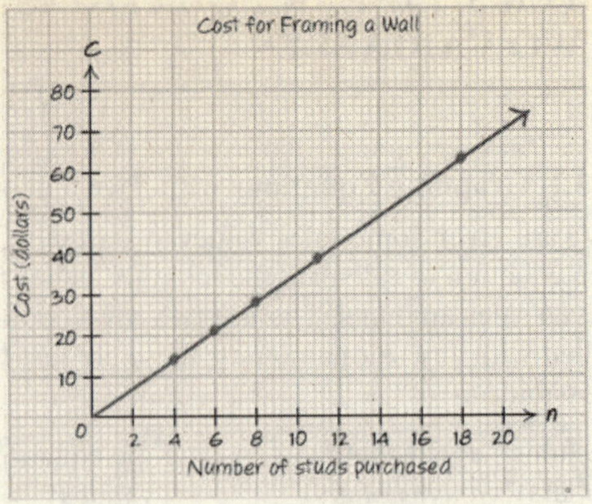

15.

h = Hours of bowling	C = Calories burned	(h, C)
1	259	$(1,259)$
2	518	$(2,518)$
3	777	$(3,777)$
5	1295	$(5,1295)$
6	1554	$(6,1554)$

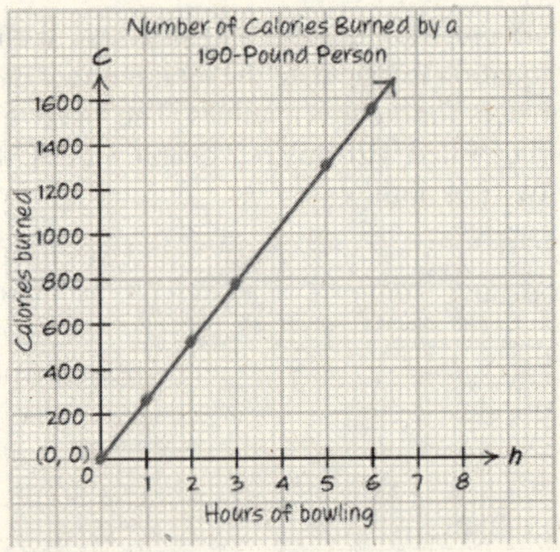

17. Fill in the table using the given values for t and calculate the value for P. Graph the points and connect them with a smooth curve. All of the points lie on the line, therefore the graph is linear.

t	$P=-8t+17$	(t,P)
−4	49	$(-4,49)$
−2	33	$(-2,33)$
0	17	$(0,17)$
2	1	$(2,1)$
4	−15	$(4,-15)$

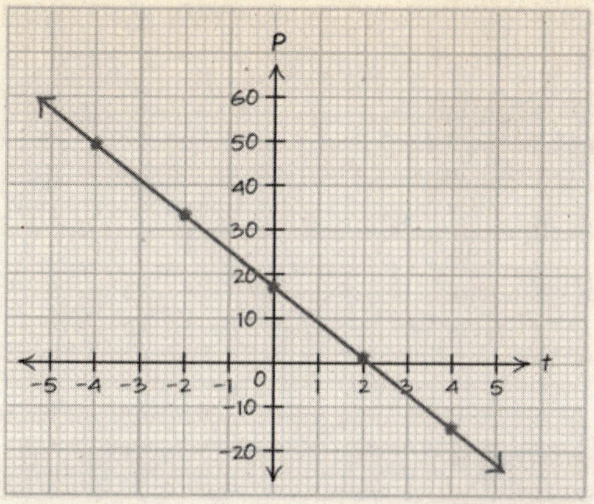

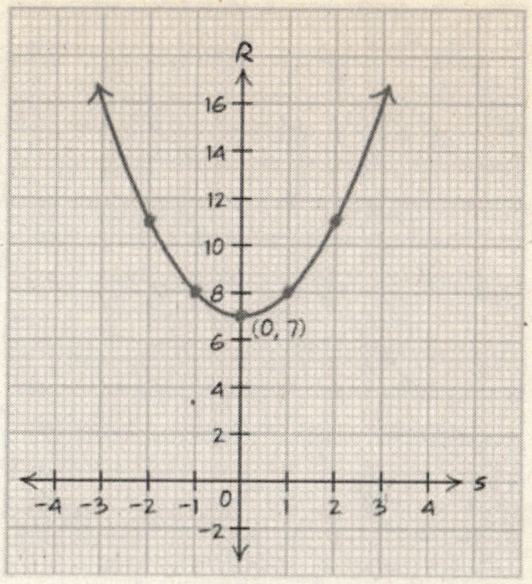

19. Fill in the table using the given values for x and calculate the value for y. Graph the points and connect them with a smooth curve. The points do not lie on a straight line, therefore the graph is nonlinear.

x	$y=\|x-2\|$	(x,y)
−6	8	(−6,8)
−3	5	(−3,5)
0	2	(0,2)
2	0	(2,0)
3	1	(3,1)
6	4	(6,4)

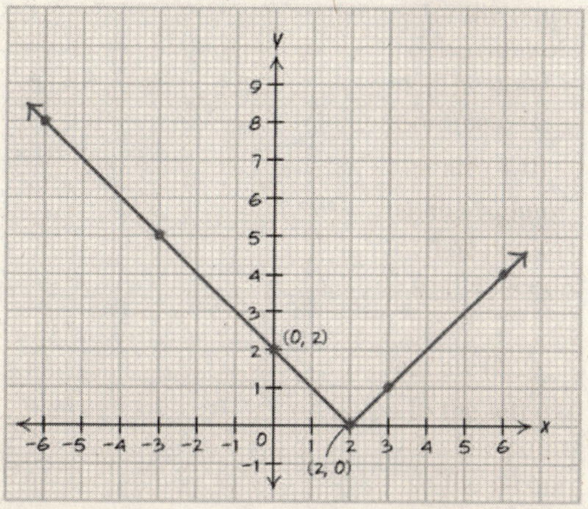

21. Fill in the table using the given values for s and calculate the value for R. Graph the points and connect them with a smooth curve. The points do not lie on a straight line, therefore the graph is nonlinear.

s	$R=s^2+7$	(s,R)
−2	11	(−2,11)
−1	8	(−1,8)
0	7	(0,7)
1	8	(1,8)
2	11	(2,11)

23. Fill in the table using the given values for x and calculate the value for y. Graph the points and connect them with a smooth curve. The points do not lie on a straight line, therefore the graph is nonlinear.

x	$y=-2x^2+3$	(x,y)
−2	−5	(−2,−5)
−1	1	(−1,1)
0	3	(0,3)
1	1	(1,1)
2	−5	(2,−5)

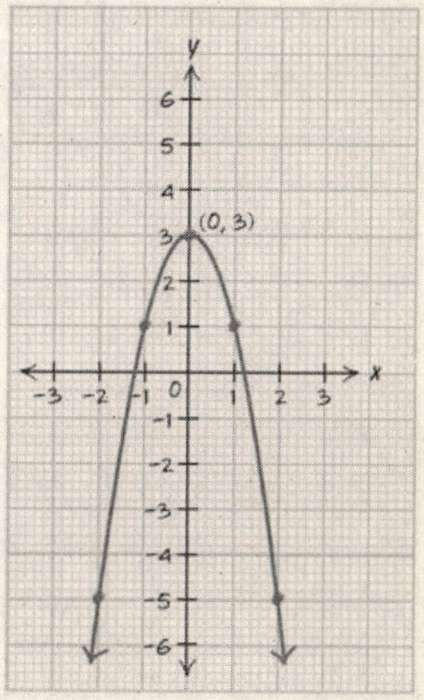

25. To fill in the table, select values for x that are easy to work with and calculate the value for y. Graph the points and connect them with a smooth curve. All of the points lie on the line, therefore the graph is linear.

x	$y=2x-8$	(x,y)
−1	−10	(−1,−10)
0	−8	(0,−8)
2	−4	(2,−4)
4	0	(4,0)
7	6	(7,6)

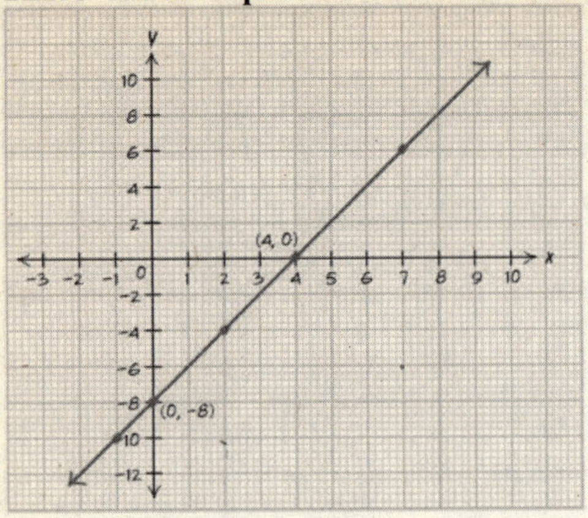

27. To fill in the table, select values for x that are easy to work with and calculate the value for y. Graph the points and connect them with a smooth curve. All of the points lie on the line, therefore the graph is linear.

x	$y=0.5x+3$	(x,y)
−8	−1	(−8,−1)
−4	1	(−4,1)
0	3	(0,3)
4	5	(4,5)
8	7	(8,7)

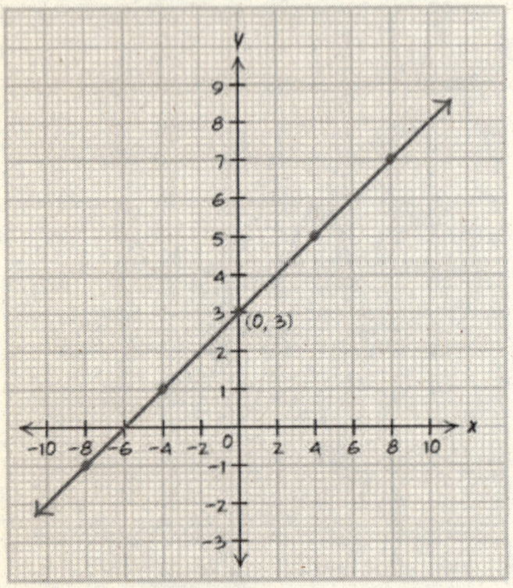

29. To fill in the table, select values for x that are easy to work with and calculate the value for y. Graph the points and connect them with a smooth curve. All of the points lie on the line, therefore the graph is linear.

x	$y=-2x+10$	(x,y)
−2	14	(−2,14)
0	10	(0,10)
2	6	(2,6)
4	2	(4,2)
7	−4	(7,−4)

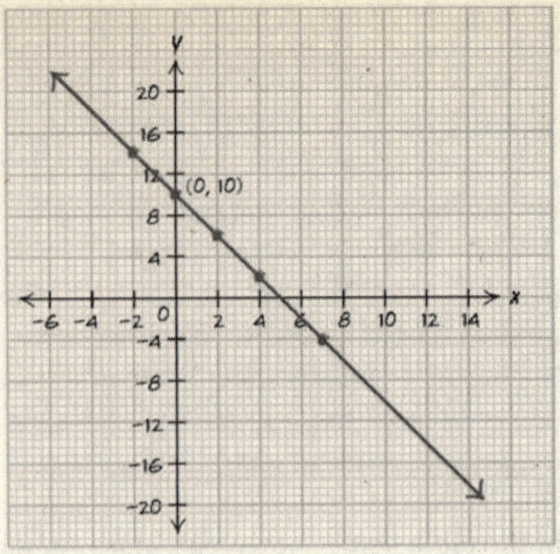

31. To fill in the table, select values for x that are easy to work with and calculate the value for y. Graph the points and connect them with a smooth curve. All of the points lie on the line, therefore the graph is linear.

x	$y=-0.4x+2$	(x,y)
−10	6	(−10,6)
−5	4	(−5,4)
0	2	(0,2)
5	0	(5,0)
10	−2	(10,−2)

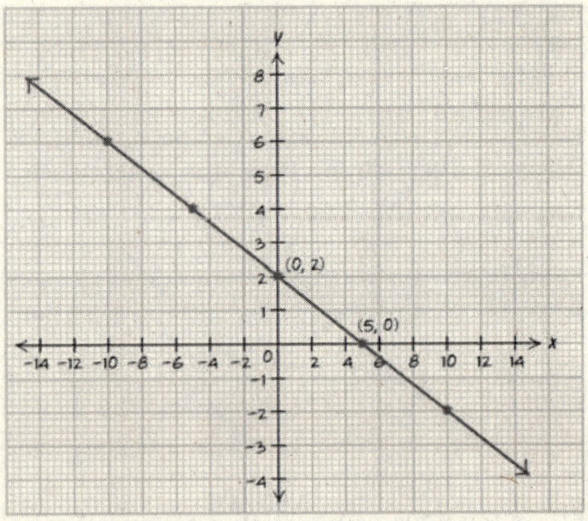

33. To fill in the table, select values for x that are easy to work with and calculate the value for y. Graph the points and connect them with a smooth curve. All of the points lie on the line, therefore the graph is linear.

x	$y=3x$	(x,y)
−3	−9	(−3,−9)
−1	−3	(−1,−3)
0	0	(0,0)
1	3	(1,3)
3	9	(3,9)

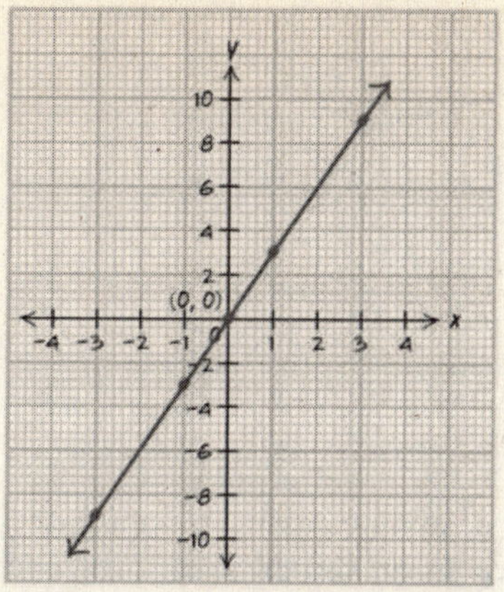

35. To fill in the table, select values for x that are easy to work with and calculate the value for y. Graph the points and connect them with a smooth curve. All of the points lie on the line, therefore the graph is linear.

x	$y=-3x$	(x,y)
−3	9	(−3,9)
−1	3	(−1,3)
0	0	(0,0)
1	−3	(1,−3)
3	−9	(3,−9)

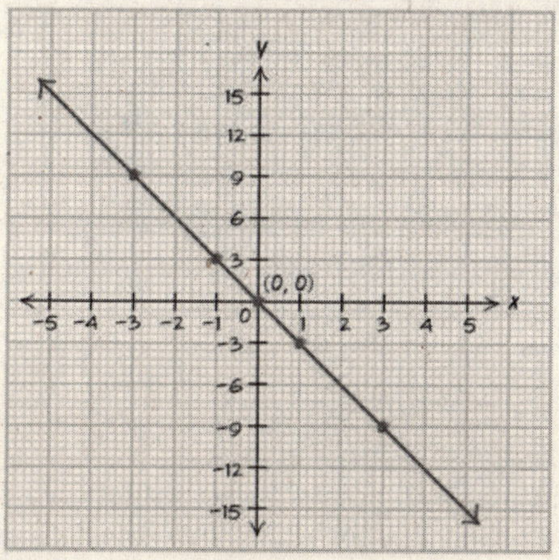

37. To fill in the table, select values for x that are easy to work with and calculate the value for y. Graph the points and connect them with a smooth curve. All of the points lie on the line, therefore the graph is linear.

x	$y=\frac{1}{2}x+3$	(x,y)
−10	−2	(−10,−2)
−6	0	(−6,0)
0	3	(0,3)
6	6	(6,6)
10	8	(10,8)

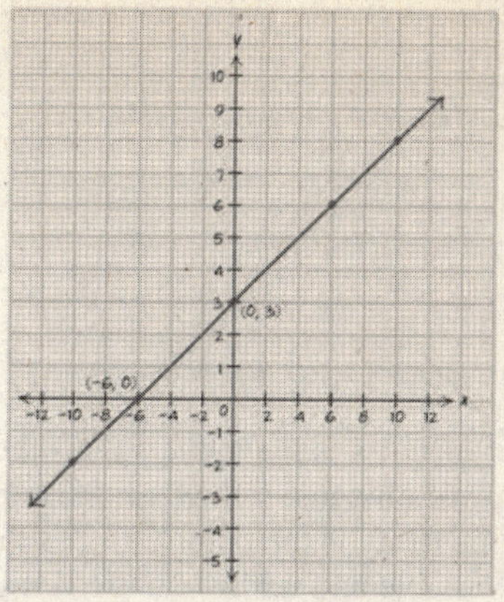

39. To fill in the table, select values for x that are easy to work with and calculate the value for y. Graph the points and connect them with a smooth curve. All of the points lie on the line, therefore the graph is linear.

x	$y=\frac{2}{3}x-5$	(x,y)
−6	−9	(−6,−9)
0	−5	(0,−5)
3	−3	(3,−3)
6	−1	(6,−1)
9	1	(9,1)

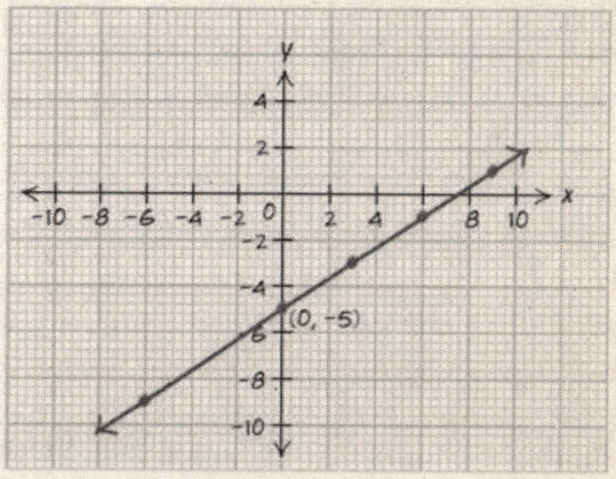

41. To fill in the table, select values for x that are easy to work with and calculate the value for y. Graph the points and connect them with a smooth curve. All of the points lie on the line, therefore the graph is linear.

x	$y=-\frac{1}{5}x-2$	(x,y)
−10	0	(−10,−9)
−5	−1	(−6,−7)
0	−2	(0,−4)
5	−3	(6,−1)
10	−4	(10,1)

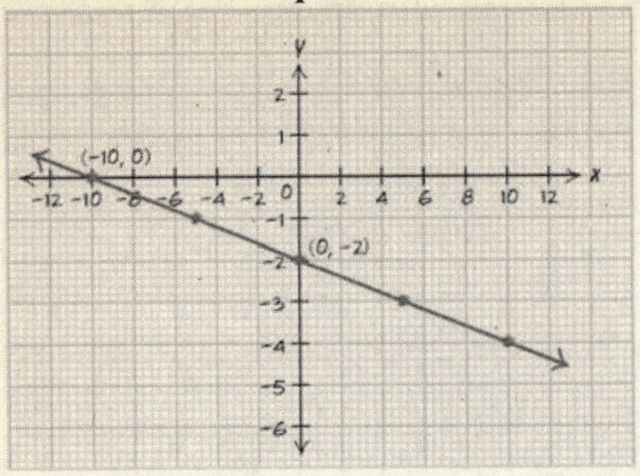

43. The number of miles is the input, so d is the independent variable and will go on the horizontal axis. The time delay is the output, so T is the dependent variable and will go on the vertical axis. Select different numbers of miles for the input and calculate the different time delays based on the distance. A negative number of miles or a negative amount of time does not make sense, so we will only graph positive values on both axes.

d	$T=5d$
0	0
1	5
2	10
5	25
7	35

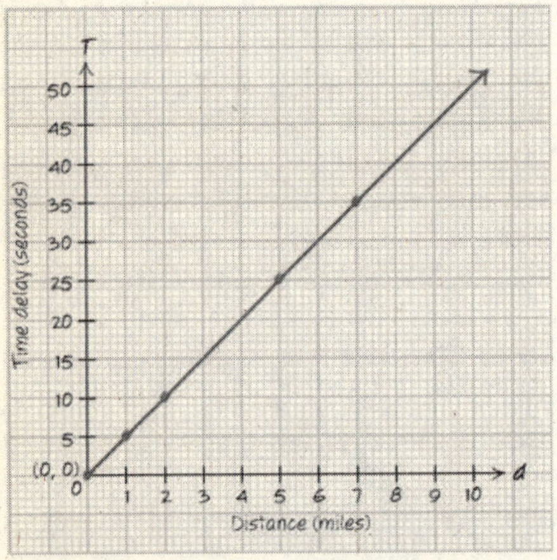

45. The number of hours after 12 noon is the input, so h is the independent variable and will go on the horizontal axis. The temperature outside is the output, so T is the dependent variable and will go on the vertical axis. Select different numbers of hours for the input and calculate the different temperatures based on the number of hours. A negative number of hours does not make sense and a negative temperature is not possible in the given situation, so we will only graph positive values on both axes.

h	$T=75-3h$
0	75
1	72
2	69
4	63
6	57
8	51

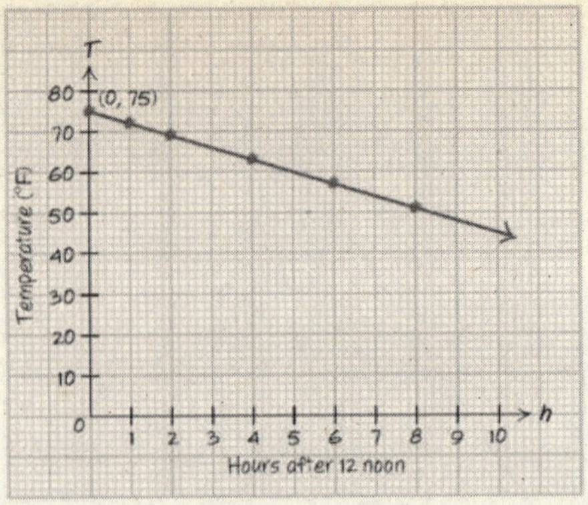

47. The number of minutes spent mowing is the input, so m is the independent variable and will go on the horizontal axis. The number of square feet of grass still to be mowed is the output, so g is the dependent variable and will go on the vertical axis. Select different numbers of minutes for the input and calculate the different numbers of square fee still to be mowed based on the time. A negative number of minutes or a negative number of square feet does not make sense, so we will only graph positive values on both axes.

m	$g=24{,}000-100m$
0	24,000
50	19,000
100	14,000
175	6500
240	0

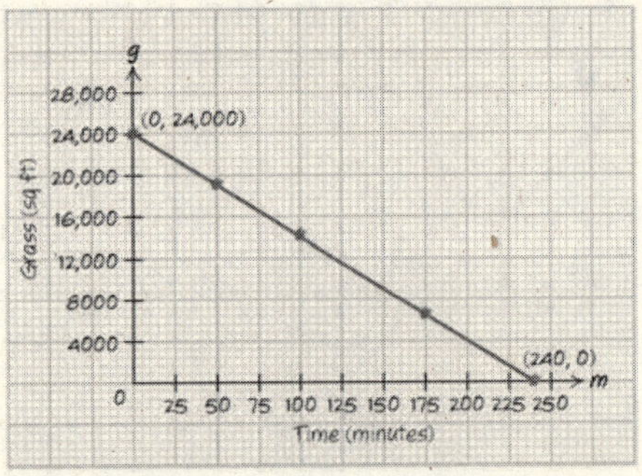

49. Prepare a table of points. Select values for x and calculate the value for y. For the values of x, select values that are easy to work with, using positive values, negative values, and 0. Use the coordinates to graph the points and connect them with a smooth curve.

x	$y=x^2+1$
-5	26
-3	10
-2	5
-1	2
0	1
1	2
2	5
3	10
5	26

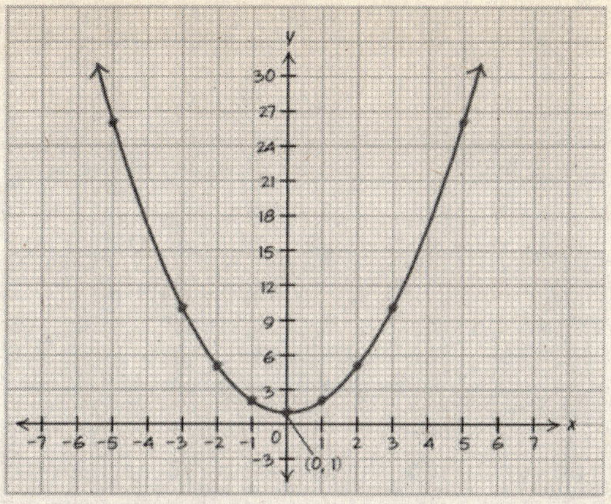

The equation of the line is nonlinear, the points do not lie on a straight line, and therefore the graph is nonlinear.

51. Prepare a table of points. Select values for x and calculate the value for y. For the values of x, select values that are easy to work with, using positive values, negative values, and 0. Use the coordinates to graph the points and connect them with a smooth curve.

x	$y = x^2 - 4$
-5	21
-3	5
-2	0
-1	-3
0	-4
1	-3
2	0
3	5
5	21

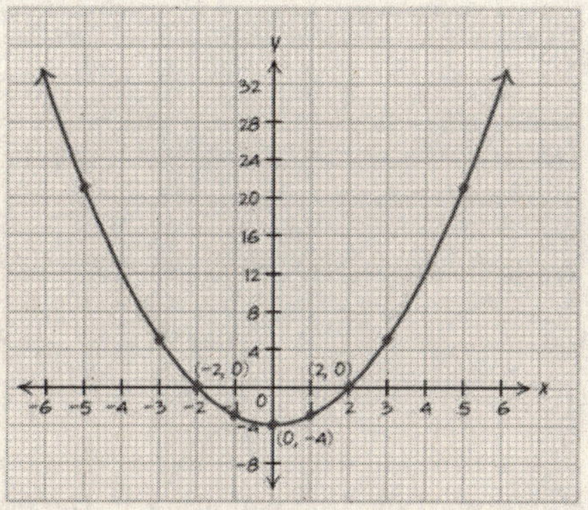

The equation of the line is nonlinear, the points do not lie on a straight line, and therefore the graph is nonlinear.

53. Prepare a table of points. Select values for x and calculate the value for y. For the values of x, select values that are easy to work with, using positive values, negative values, and 0. Use the coordinates to graph the points and connect them with a smooth curve.

x	$y = (x+3)^2$
-7	16
-6	9
-5	4
-4	1
-3	0
-2	1
-1	4
0	9
1	16

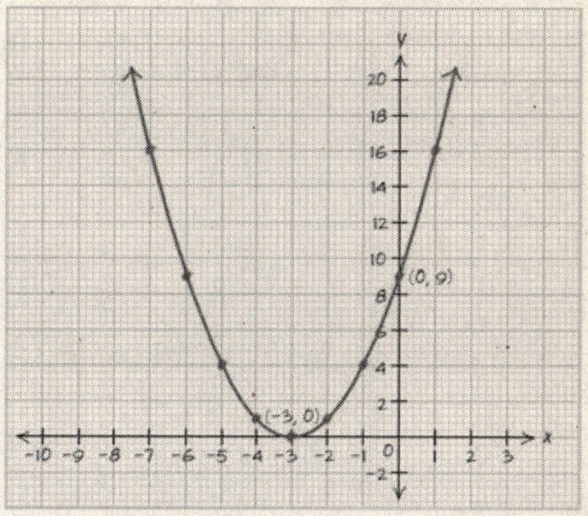

The equation of the line is nonlinear, the points do not lie on a straight line, and therefore the graph is nonlinear.

55. Prepare a table of points. Select values for x and calculate the value for y. For the values of x, select values that are easy to work with, using positive values, negative values, and 0. Use the coordinates to graph the points and connect them with a smooth curve.

x	$y = (x-2)^2$
-2	16
-1	9
0	4
1	1
2	0
3	1
4	4
5	9
6	16

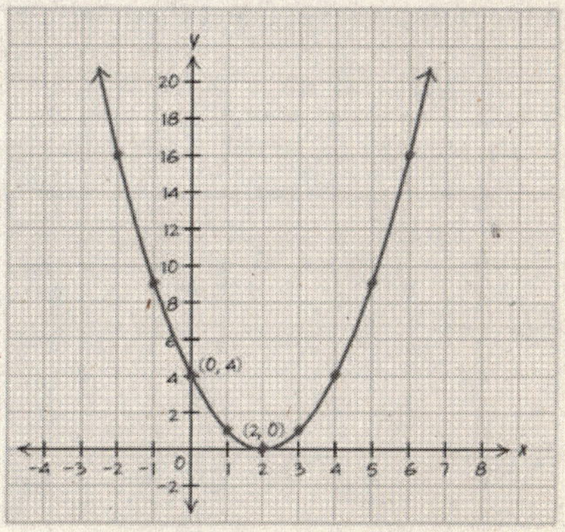

The equation of the line is nonlinear, the points do not lie on a straight line, and therefore the graph is nonlinear.

57. Prepare a table of points. Select values for x and calculate the value for y. For the values of x, select values that are easy to work with, using positive values, negative values, and 0. Use the coordinates to graph the points and connect them with a smooth curve.

x	$y=\lvert x\rvert+2$
-8	10
-6	8
-4	6
-2	4
0	2
2	4
4	6
6	8
8	10

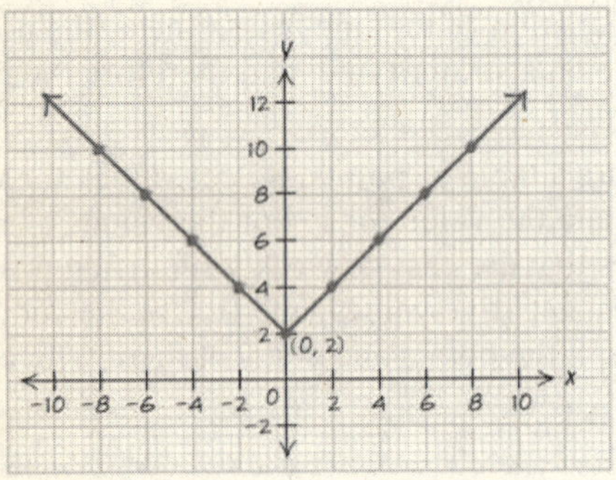

The equation of the line is nonlinear, the points do not lie on a straight line, and therefore the graph is nonlinear.

59. Prepare a table of points. Select values for x and calculate the value for y. For the values of x, select values that are easy to work with, using positive values, negative values, and 0. Use the coordinates to graph the points and connect them with a smooth curve.

x	$y=\lvert x+2\rvert$
-9	7
-7	5
-5	3
-3	1
-2	0
-1	1
1	3
3	5
5	7

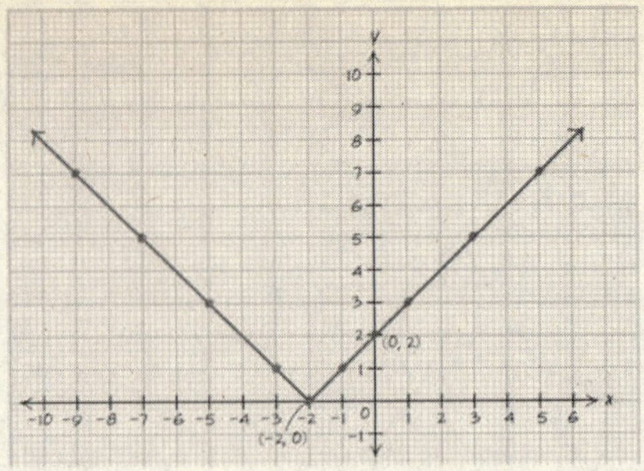

The equation of the line is nonlinear, the points do not lie on a straight line, and therefore the graph is nonlinear.

61. Prepare a table of points. Select values for x and calculate the value for y. For the values of x, select values that are easy to work with, using positive values, negative values, and 0. Use the coordinates to graph the points and connect them with a smooth curve.

x	$y=\lvert x-3\rvert$
-3	6
-1	4
0	3
2	1
3	0
4	1
6	3
7	4
9	6

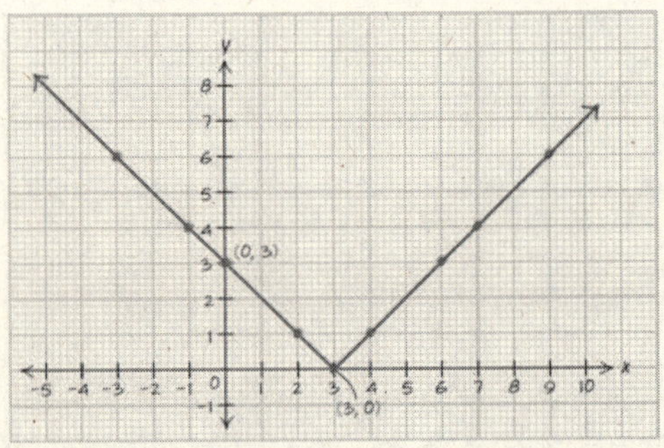

The equation of the line is nonlinear, the points do not lie on a straight line, and therefore the graph is nonlinear.

63. Prepare a table of points. Select values for x and calculate the value for y. For the values of x, select values that are easy to work with, using positive values, negative values, and 0 if possible. Use the coordinates to graph the points and connect them with a smooth curve.

x	$y = 6$
-9	6
-5	6
0	6
5	6
9	6

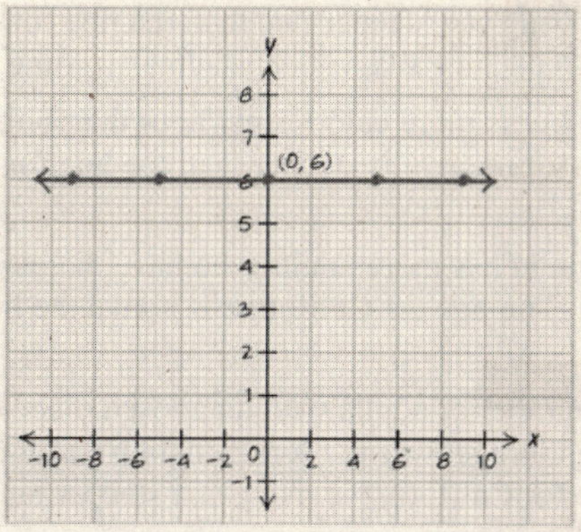

The equation of the line is linear, all of the points lie on the line, and therefore the graph is linear.

65. Prepare a table of points. Select values for x and calculate the value for y. For the values of x, select values that are easy to work with, using positive values, negative values, and 0 if possible. Use the coordinates to graph the points and connect them with a smooth curve.

$x = 7$	y
7	-9
7	-5
7	0
7	5
7	9

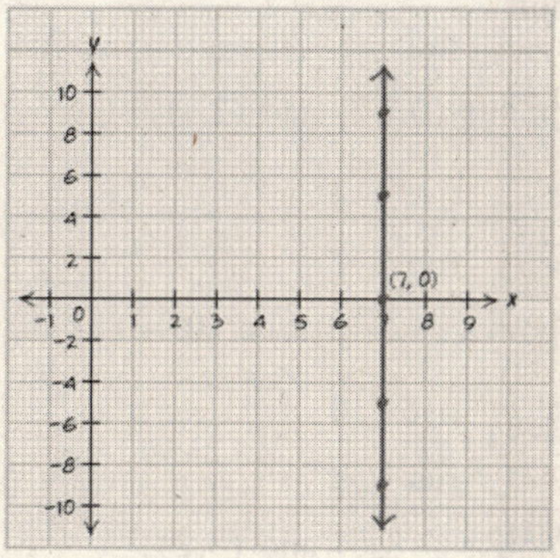

The equation of the line is linear, all of the points lie on the line, and therefore the graph is linear.

67. Prepare a table of points. Select values for x and calculate the value for y. For the values of x, select values that are easy to work with, using positive values, negative values, and 0 if possible. Use the coordinates to graph the points and connect them with a smooth curve.

x	$y = -3.5$
-9	-3.5
-5	-3.5
0	-3.5
5	-3.5
9	-3.5

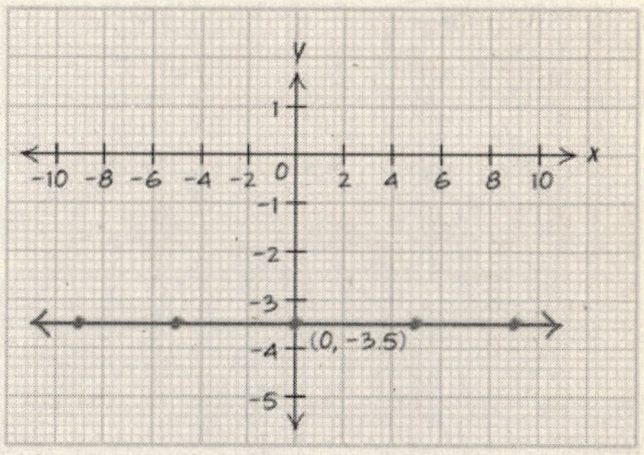

The equation of the line is linear, all of the points lie on the line, and therefore the graph is linear.

69. Prepare a table of points. Select values for x and calculate the value for y. For the values of x, select values that are easy to work with, using positive values, negative values, and 0 if possible. Use the coordinates to graph the points and connect them with a smooth curve.

$x = -1.5$	y
-1.5	-9
-1.5	-5
-1.5	0
-1.5	5
-1.5	9

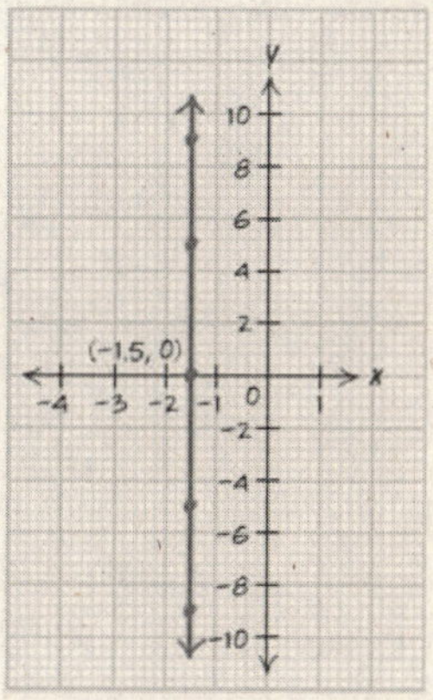

The equation of the line is linear, all of the points lie on the line, and therefore the graph is linear.

71. Prepare a table of points. Select values for x and calculate the value for y. For the values of x, select values that are easy to work with, using positive values, negative values, and 0 if possible. Use the coordinates to graph the points and connect them with a smooth curve.

$x=0$	y
0	-9
0	-5
0	0
0	5
0	9

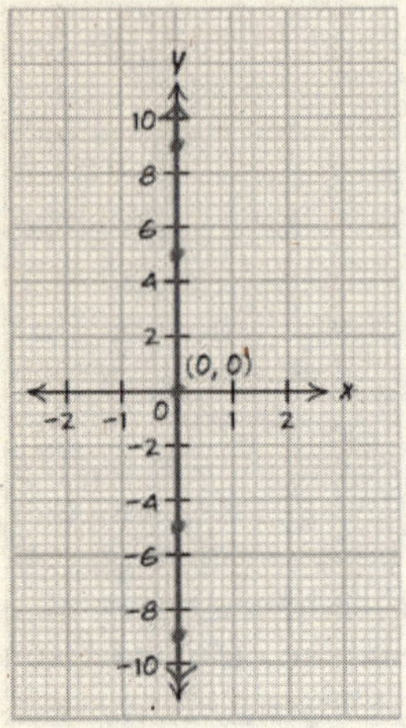

The equation of the line is linear, all of the points lie on the line, and therefore the graph is linear.

Section 3.2 Finding and Interpreting Slope

1. **a.** Buffalo-Niagara Falls, NY has the larger population than El Paso, TX during these years. The line for Buffalo-Niagara Falls is above the line for El Paso.

b. El Paso, TX is growing faster than Buffalo-Niagara Falls, NY during these years. The line for El Paso is rising at a steeper rate than the line for Buffalo-Niagara Falls, which is actually going down.

3. Archery burns more calories per hour than billiards. The line for archery is steeper than the line for billiards.

5. **a.** California has more adults who have diagnosed diabetes. The line for California is above the line for Florida.

b. The number of adults with diagnosed diabetes is growing more slowly in Florida than in California. The line for California is rising at a steeper rate than the line for Florida.

7. **a.** Rock on Live Tour had more tickets available at the start of sales. The line for Rock on Live Tour at the start of sales is above the line for Classics Rule Tour.

b. The Rock on Live Tour sold out first. The line reaches zero on the graph before the line for the Classics Rule Tour.

c. The available tickets for the Rock on Live Tour decreased faster than the available tickets for the Classics Rule Tour. The line for Rock on Live Tour is decreasing at a steeper rate than the line for Classics Rule Tour.

9. Using the points $(-1,-1)$ and $(0,1)$.

$$\text{slope} = \frac{\text{rise}}{\text{run}} = \frac{2}{1} = 2$$

11. Using the points $(0,-2)$ and $(2,1)$.

$$\text{slope} = \frac{\text{rise}}{\text{run}} = \frac{3}{2}$$

13. Using the points $(-2,-3)$ and $(0,-3)$.

$$\text{slope} = \frac{\text{rise}}{\text{run}} = \frac{0}{2} = 0$$

15. Using the points $(-8,0)$ and $(0,2)$.

$$\text{slope} = \frac{\text{rise}}{\text{run}} = \frac{2}{8} = \frac{1}{4}$$

17. Using the points $(1,0)$ and $(1,2)$.

$$\text{slope} = \frac{\text{rise}}{\text{run}} = \frac{2}{0} = \text{undefined}$$

A zero in the denominator of a fraction is undefined, therefore the slope is undefined.

19. Using the points $(0,3)$ and $(6,0)$.

$$\text{slope} = \frac{\text{rise}}{\text{run}} = \frac{-3}{6} = -\frac{1}{2}$$

21. Using the points $(0,2)$ and $(2,-4)$.

$$\text{slope} = \frac{\text{rise}}{\text{run}} = \frac{-6}{2} = -3$$

23. Substitute (2, 5) and (6, 13) into the slope formula:

$$\text{slope} = \frac{y_2 - y_1}{x_2 - x_1} = \frac{13-(5)}{6-(2)} = \frac{8}{4} = 2$$

25. Substitute (0, 2) and (4, 3) into the slope formula:

$$\text{slope} = \frac{y_2 - y_1}{x_2 - x_1} = \frac{3-(2)}{4-(0)} = \frac{1}{4}$$

27. Substitute (–4,–7) and (2, 11) into the slope formula:

$$\text{slope} = \frac{y_2 - y_1}{x_2 - x_1} = \frac{11-(-7)}{2-(-4)} = \frac{18}{6} = 3$$

29. Substitute $(4,8)$ and $(1,8)$ into the slope formula:

$$\text{slope} = \frac{y_2 - y_1}{x_2 - x_1} = \frac{8-(8)}{1-(4)} = \frac{0}{-3} = 0$$

31. Substitute $(-3,-7.5)$ and $(1,-5.5)$ into the slope formula:

$$\text{slope} = \frac{y_2 - y_1}{x_2 - x_1} = \frac{-5.5-(-7.5)}{1-(-3)} = \frac{2}{4} = \frac{1}{2}$$

33. Substitute $(2,5)$ and $(2,6)$ into the slope formula:

$$\text{slope} = \frac{y_2 - y_1}{x_2 - x_1} = \frac{6-(5)}{2-(2)}$$
$$= \frac{1}{0} = \text{undefined}$$

A fraction with a zero in the denominator is undefined, therefore the slope of the line that passes through the given points is undefined.

35. Substitute $(-3,18)$ and $(3,-6)$ into the slope formula:

$$\text{slope} = \frac{y_2 - y_1}{x_2 - x_1} = \frac{-6-(18)}{3-(-3)} = \frac{-24}{6} = -4$$

37. The red line is steeper than the blue line and the blue line is closer to being horizontal, therefore the blue line has a slope closer to zero.

39. The blue line is steeper than the red line and the blue line is closer to being vertical, therefore the blue line has a slope closer to being undefined.

41. Let $(x_1, y_1) = (1,-17)$ and $(x_2, y_2) = (2,-14)$. Substituting these points into the slope formula yields:

$$\text{slope} = \frac{y_2 - y_1}{x_2 - x_1} = \frac{-14-(-17)}{2-(1)} = \frac{3}{1} = 3$$

43. Let $(x_1, y_1) = (2,-2)$ and $(x_2, y_2) = (10,10)$. Substituting these points into the slope formula yields:

$$\text{slope} = \frac{y_2 - y_1}{x_2 - x_1} = \frac{10-(-2)}{10-(2)} = \frac{12}{8} = \frac{3}{2}$$

45. Let $(x_1, y_1) = (-4,2)$ and $(x_2, y_2) = (-4,-3)$. Substituting these points into the slope formula yields:

$$\text{slope} = \frac{y_2 - y_1}{x_2 - x_1} = \frac{-3-(2)}{-4-(-4)} = \frac{-5}{0} = \text{undefined}$$

A fraction with a zero in the denominator is undefined, therefore the slope of the line that passes through the given points is undefined.

47. Let $(x_1, y_1) = (3,8)$ and $(x_2, y_2) = (15,12)$. Substituting these points into the slope formula yields:

$$\text{slope} = \frac{y_2 - y_1}{x_2 - x_1} = \frac{12-(8)}{15-(3)} = \frac{4}{12} = \frac{1}{3}$$

49. Let $(x_1, y_1) = (1,2)$ and $(x_2, y_2) = (3,-6)$. Substituting these points into the slope formula yields:

$$\text{slope} = \frac{y_2 - y_1}{x_2 - x_1} = \frac{-6-(2)}{3-(1)} = \frac{-8}{2} = -4$$

51. Let $(x_1, y_1) = (1,10)$ and $(x_2, y_2) = (7,10)$. Substituting these points into the slope formula yields:

$$\text{slope} = \frac{y_2 - y_1}{x_2 - x_1} = \frac{10-(10)}{7-(1)} = \frac{0}{6} = 0$$

53. Let $(x_1, y_1) = (-1,8.5)$ and $(x_2, y_2) = (5,5.5)$. Substituting these points into the slope formula yields:

$$\text{slope} = \frac{y_2 - y_1}{x_2 - x_1} = \frac{5.5-(8.5)}{5-(-1)} = \frac{-3}{6} = -\frac{1}{2}$$

55. **a.** Let $(x_1, y_1) = (0,1169)$ and $(x_2, y_2) = (6,1138)$.

$$\text{slope} = \frac{y_2 - y_1}{x_2 - x_1} = \frac{1138-(1169)}{6-(0)} = \frac{-31}{6} = -5\frac{1}{6}$$

Therefore the slope for the Buffalo-Niagara, NY line is $-5\frac{1}{6}$.

b. The population of Buffalo-Niagara Falls, NY is declining at a rate of approximately $5\frac{1}{6}$ thousand, or 5167 people per year.

57. **a.** Let $(x_1, y_1) = (1,1107)$ and $(x_2, y_2) = (5,1154)$.

$$\text{slope} = \frac{y_2 - y_1}{x_2 - x_1} = \frac{1154-(1107)}{5-(1)} = \frac{47}{4} = 11\frac{3}{4}$$

Therefore the slope for the Oklahoma City, OK line is $11\frac{3}{4}$.

b. The population of Oklahoma City, OK is increasing at a rate of approximately $11\frac{3}{4}$ thousand, or 11,750 people per year.

59. **a.** Estimate the points $(0,0)$ and $(5,1000)$ from the graph to find the slope of the archery line. Let $(x_1, y_1) = (0,0)$ and $(x_2, y_2) = (5,1000)$.

$$\text{slope} = \frac{y_2 - y_1}{x_2 - x_1} = \frac{1000-(0)}{5-(0)} = \frac{1000}{5} = 200$$

Therefore the slope for the archery line is 200.

b. A 130-lb person burns approximately 200 calories per hour during archery exercise.

61. **a.** Estimate the points $(0,0)$ and $(5,2000)$ from the graph to find the slope of the competitive badminton line. Let $(x_1, y_1) = (0,0)$ and $(x_2, y_2) = (5,2000)$.

$$\text{slope} = \frac{y_2 - y_1}{x_2 - x_1} = \frac{2000-(0)}{5-(0)} = \frac{2000}{5} = 400$$

Therefore the slope for the competitive badminton line is 400.

b. A 130-lb person burns approximately 400 calories per hour while playing competitive badminton.

63. **a.** Estimate the points $(0,700)$ and $(4,1000)$ to find the slope of the line for California. Let $(x_1, y_1) = (0,700)$ and $(x_2, y_2) = (4,1000)$.

$$\text{slope} = \frac{y_2 - y_1}{x_2 - x_1} = \frac{1000-(700)}{4-(0)} = \frac{300}{4} = 75$$

Therefore the slope for the California line is approximately 75.

b. The number of adults with diagnosed diabetes in California is increasing by approximately 75,000 adults per year.

65. **a.** Estimate the points $(4,100)$ and $(10,200)$ to find the slope of the line for South Carolina. Let $(x_1, y_1) = (4,100)$ and $(x_2, y_2) = (10,200)$.

$$\text{slope} = \frac{y_2 - y_1}{x_2 - x_1} = \frac{200-(100)}{10-(4)} = \frac{100}{6} = 16\frac{2}{3}$$

Therefore the slope for the South Carolina line is approximately $16\frac{2}{3}$.

b. The number of adults with diagnosed diabetes in South Carolina is increasing by approximately 16,667 adults per year.

67. **a.** We will use the points $(0,2500)$ and $(14,0)$ to find the slope of the Rock-On Live Tour line. Let $(x_1, y_1) = (0,2500)$ and $(x_2, y_2) = (14,0)$.

$$\text{slope} = \frac{y_2 - y_1}{x_2 - x_1} = \frac{0-(2500)}{14-(0)} = \frac{-2500}{14} = -178\frac{4}{7}$$

Therefore the slope for the Rock-On Live Tour line is approximately $-178\frac{4}{7}$.

b. The number of available tickets for the Rock On Live Tour declined by approximately 179 tickets per hour.

69. **a.** We will use the points $(0,4500)$ and $(49,0)$ to find the slope of the line for Deb. Let $(x_1, y_1) = (0,4500)$ and $(x_2, y_2) = (49,0)$.

$$\text{slope} = \frac{y_2 - y_1}{x_2 - x_1} = \frac{0-(4500)}{49-(0)} = \frac{-4500}{49} = -91\frac{41}{49}$$

Therefore the slope of the line for Deb's fundraising is approximately $-91\frac{41}{49}$.

b. The amount Deb still needed to raise for breast cancer declined by approximately $92 each day after she began fundraising.

71. The blue line has the greater slope. The blue line is steeper than the red line.

73. The red line has the greater slope. The red line is steeper than the blue line.

75. The line is rising from left to right therefore it is increasing.

77. The line is falling from left to right, therefore it is decreasing.

79. The line is horizontal, therefore it is neither increasing nor decreasing.

81. Create a table with a few points for the line $y = -3x + 19$ by selecting values for x and calculating the values for y. For example:

x	y
−2	25
0	19
1	16
3	10

When these points are graphed, the line is falling from left to right, therefore the line is decreasing and the slope is negative.

83. The equation has no value for x, only a value for y. For any value of x, the y value would remain the same, therefore the line is horizontal and it is neither increasing nor decreasing. We can create a table with a few points for the line $y = \frac{-5}{4}$ by selecting values for x; the values for y will remain the same regardless of the value for x. For example:

x	y
−2	$\frac{-5}{4}$
0	$\frac{-5}{4}$
2	$\frac{-5}{4}$
4	$\frac{-5}{4}$

When these points are graphed, the line is horizontal, therefore the line is neither increasing nor decreasing.

85. Create a table with a few points for the line $y = \frac{6}{5}x - 7$ by selecting values for x and calculating the values for y. For example:

x	y
−5	−13
0	−7
5	−1
10	5

When these points are graphed, the line is rising from left to right, therefore the line is increasing and the slope is positive.

87. Create a table with a few points for the equation $P = 12.5h$ to determine whether the output quantity is increasing, decreasing or neither. Select increasing values for h and calculate the values for P. For example:

h	P
0	0
1	12.5
2	25
4	50

For each higher input value, the output value becomes higher, therefore the output value is increasing.

89. Create a table with a few points for the equation $T = 89 - 3h$ to determine whether the output quantity is increasing, decreasing or neither. Select increasing values for h and calculate the values for P. For example:

h	T
0	89
1	86
2	83
4	77

For each higher input value, the output value goes down, therefore the output value is decreasing.

91. Create a table with a few points for the equation $W = 189 - 2t$ to determine whether the output quantity is increasing, decreasing or neither. Select increasing values for t and calculate the values for W. For example:

t	W
0	189
1	187
2	185
4	181

For each higher input value, the output value goes down, therefore the output value is decreasing.

93. Create a table with a few points for the equation $D = 66h$ to determine whether the output quantity is increasing, decreasing or neither. Select increasing values for h and calculate the values for D. For example

h	D
0	0
1	66
2	132
4	264

For each higher input value, the output value becomes higher, therefore the output value is increasing.

95. **a.** The red line is steeper than the blue line and appears to be decreasing more rapidly.

b. The points identified on the red line are $(0,3)$ and $(1,2)$. Let $(x_1, y_1) = (0,3)$ and $(x_2, y_2) = (1,2)$.

$$\text{slope} = \frac{y_2 - y_1}{x_2 - x_1} = \frac{2-(3)}{1-(0)} = \frac{-1}{1} = -1$$

Therefore the slope of the red line is -1.

The points identified on the blue line are $(-2,-1)$ and $(0,-2)$. Let $(x_1, y_1) = (-2,-1)$ and $(x_2, y_2) = (0,-2)$.

$$\text{slope} = \frac{y_2 - y_1}{x_2 - x_1} = \frac{-2-(-1)}{0-(-2)} = \frac{-1}{2} = -\frac{1}{2}$$

Therefore the slope of the blue line is $-\frac{1}{2}$.

c. The line that is decreasing more rapidly has a slope that is less (has a greater negative value) than the other line. The red line is decreasing more rapidly and has a slope that is less than the blue line.

Section 3.3 Slope-Intercept Form of Lines

1. The line intersects the x-axis at 2 therefore the x-intercept is the point $(2,0)$. The line intersects the y-axis at -1, therefore the y-intercept is the point $(0,-1)$.

3. The x-intercept is the point $(3,0)$. The y-intercept is the point $(0,2)$.

5. The x-intercept is the point $(2,0)$. The y-intercept is the point $(0,10)$.

7. The x-intercepts are the points $(-6,0)$ and $(0,0)$. The y-intercept is the point $(0,0)$.

9. The x-intercepts are the points $(-4,0)$, $(-2,0)$ and $(4,0)$. The y-intercept is the point $(0,-50)$.

11. **a.** The vertical intercept is the point (0,1169), meaning that the population of Buffalo-Niagara Falls, NY was about 1,169,000 people in the year 2000.

b. The vertical intercept is the point (0,682), meaning that the population of El Paso, TX was about 682,000 people in the year 2000.

13. The intercept for these lines is the point (0,0), meaning that zero calories are burned in zero hours of exercise.

15. **a.** The vertical intercept is the point (0,741), meaning that there were about 741,000 adults with diagnosed diabetes in California in the year 1990.

b. The vertical intercept is the point (0,262), meaning that there were about 262,000 adults with diagnosed diabetes in Florida in the year 1990.

17. **a.** The vertical intercept for the Rock On Live Tour line is the point (0,2500), meaning that there were 2500 tickets available at the start of ticket sales. The horizontal intercept is the point (14,0), meaning that 14 hours after going on sale, the tickets were sold out.

b. The vertical intercept for the Classics Rule Tour line is the point (0,2000), meaning that there were 2000 tickets available at the start of ticket sales. The horizontal intercept is the point (18,0), meaning that 18 hours after going on sale, the tickets were sold out.

19. Substitute $x = 0$ and find y.

$y = 5(0) + 20$ Substitute in $x = 0$.
$y = 20$ Solve for y.

The y-intercept is the point (0, 20).

Substitute $y = 0$ and find x.

$(0) = 5x + 20$ Substitute in $y = 0$.
$-20 = 5x$ Solve for x.

$$\frac{-20}{5} = \frac{5x}{5}$$

$$x = -4$$

The x-intercept is the point $(-4,0)$.

21. Substitute $x=0$ and find y.

$y=4(0)-36$ Substitute in $x=0$.
$y=-36$ Solve for y.

The y-intercept is the point $(0,-36)$.

Substitute $y=0$ and find x.

$(0)=4x-36$ Substitute in $y=0$.
$36=4x$ Solve for x.
$\frac{36}{4}=\frac{4x}{4}$
$x=9$

The x-intercept is the point $(9,0)$.

23. Substitute $x=0$ and find y.

$y=\frac{1}{2}(0)+3$ Substitute in $x=0$.
$y=3$ Solve for y.

The y-intercept is the point $(0,3)$.

Substitute $y=0$ and find x.

$(0)=\frac{1}{2}x+3$ Substitute in $y=0$.
$-3=\frac{1}{2}x$ Solve for x.
$\frac{2}{1}\cdot -3=\frac{1}{2}x\cdot\frac{2}{1}$
$x=-6$

The x-intercept is the point $(-6,0)$.

25. Substitute $x=0$ and find y.

$y=\frac{2}{3}(0)+6$ Substitute in $x=0$.
$y=6$ Solve for y.

The y-intercept is the point $(0,6)$.

Substitute $y=0$ and find x.

$(0)=\frac{2}{3}x+6$ Substitute in $y=0$.
$-6=\frac{2}{3}x$ Solve for x.
$\frac{3}{2}\cdot -6=\frac{2}{3}x\cdot\frac{3}{2}$
$x=-9$

The x-intercept is the point $(-9,0)$.

27. Substitute $x=0$ and find y.

$y=0.25(0)+3.75$ Substitute in $x=0$.
$y=3.75$ Solve for y.

The y-intercept is the point $(0,3.75)$.

Substitute $y=0$ and find x.

$(0)=0.25x+3.75$ Substitute in $y=0$.
$\frac{-3.75}{0.25}=\frac{0.25x}{0.25}$ Solve for x.
$x=-15$

The x-intercept is the point $(-15,0)$.

29. Substitute $x=0$ and find y.

$y=(0)$ Substitute in $x=0$.

The y-intercept is the point $(0,0)$.

Substitute $y=0$ and find x.

$(0)=x$ Substitute in $y=0$.
$x=0$

The x-intercept is the point $(0,0)$.

31. Substitute $x=0$ and find y.

$(0)+5y=15$ Substitute in $x=0$.
$5y=15$ Solve for y.
$\frac{5y}{5}=\frac{15}{5}$
$y=3$

The y-intercept is the point $(0,3)$.

Substitute $y=0$ and find x.

$x+5(0)=15$ Substitute in $y=0$.
$x=15$ Solve for x.

The x-intercept is the point $(15,0)$.

33. Substitute $x=0$ and find y.

$4(0)-2y=10$ Substitute in $x=0$.
$-2y=10$ Solve for y.
$\frac{-2y}{-2}=\frac{10}{-2}$
$y=-5$

The y-intercept is the point $(0,-5)$.

Substitute $y=0$ and find x.

$4x-2(0)=10$ Substitute in $y=0$.
$4x=10$ Solve for x.
$\frac{4x}{4}=\frac{10}{4}$
$x=2\frac{1}{2}$

The x-intercept is the point $\left(2\frac{1}{2},0\right)$.

35. Substitute $x=0$ and find y.

$4(0)+3y=30$ Substitute in $x=0$.
$3y=30$ Solve for y.
$\frac{3y}{3}=\frac{30}{3}$
$y=10$

The y-intercept is the point $(0,10)$.

Substitute $y=0$ and find x.

$4x+3(0)=30$ Substitute in $y=0$.
$4x=30$ Solve for x.
$\frac{4x}{4}=\frac{30}{4}$
$x=7\frac{1}{2}$

The x-intercept is the point $\left(7\frac{1}{2},0\right)$.

37. In the equation $y=6$, the y-intercept is given. There is no value for x to solve for, therefore there is no x-intercept. The y-intercept is $(0,6)$.

39. In the equation $x=-5$, the x-intercept is given. There is no value for y to solve for, therefore there is no y-intercept. The x-intercept is $(-5,0)$.

41. **a.** To find the P-intercept, we will substitute zero for t and find P.

$P=38(0)+2674$ Substitute in $t=0$.
$P=2674$ Simplify.

Therefore, the P-intercept is the point $(0,2674)$ and means that in the year 2000, there were about 2,674,000 American Indians and Alaska Natives in the United States.

b. To find the t-intercept we will substitute zero for P and solve for t.

$(0)=38t+2674$ Substitute in $P=0$.
$-2674=38t$ Solve for t.
$\frac{-2674}{38}=\frac{38t}{38}$
$t\approx-70.37$.

Therefore, the t-intercept is the point $(-70.37,0)$ and means that in the year 1929 there were zero American Indians and Alaska Natives in the United States.

c. The t-intercept does not make sense because there must have been more than zero American Indians and Alaska Natives in the United States in the year 1929. The P-intercept makes sense in this situation, as it is reasonable that there were about 2,674,000 American Indians and Alaska Natives in the United States in the year 2000.

43. **a.** To find the P-intercept, we will substitute zero for t and find P.

$P=0.05(0)+0.65$ Substitute in $t=0$.
$P=0.65$ Simplify.

Therefore, the P-intercept is the point $(0,0.65)$ and means that the price of navel oranges was \$0.65 per pound in the year 2000.

b. To find the t-intercept we will substitute zero for P and solve for t.

$(0)=0.05t+0.65$ Substitute in $P=0$.
$-0.65=0.05t$ Solve for t.
$\frac{-0.65}{0.05}=\frac{0.05t}{0.05}$
$t=-13$

Therefore, the t-intercept is the point $(-13,0)$ and means that there was no charge for navel oranges in the year 1987.

c. The t-intercept does not make sense because navel oranges would not have been offered free of charge in 1987. The P-intercept makes sense in this situation, as it is reasonable that navel oranges cost \$0.65 per pound in the year 2000.

45. **a.** To find the C-intercept, we will substitute zero for w and find C.

$C=-2(0)+25$ Substitute in $w=0$.
$C=25$ Simplify.

Therefore, the C-intercept is the point $(0,25)$ and means that the there were 25 cars sold during week 0 of the year.

b. To find the w-intercept we will substitute zero for C and solve for w.

$(0)=-2w+25$ Substitute in $C=0$.
$-25=-2w$ Solve for w.
$\frac{-25}{-2}=\frac{-2w}{-2}$
$w=12\frac{1}{2}$

Therefore, the w-intercept is the point $\left(12\frac{1}{2},0\right)$ and means that there were no cars sold during week 12/13, of the year.

c. The C-intercept does not make sense because there is no week 0 of the year; it must be between weeks 1 through 52.

The w-intercept makes sense because it is possible that the lot might not sell any cars during a given week.

47. **a.** To find the D-intercept, we will substitute zero for m and find D.

$D=-250(0)+12500$ Substitute in $m=0$.
$D=12500$ Simplify.

Therefore, the D-intercept is the point $(0,12500)$ and means that Jose had \$12,500 in credit card debt before he began to pay them off.

b. To find the *m*-intercept we will substitute zero for *D* and solve for *m*.

$$\begin{aligned} (0) &= -250m + 12500 && \text{Substitute in } D = 0. \\ -12500 &= -250m && \text{Solve for } m. \\ \frac{-12500}{-250} &= \frac{-250m}{-250} \\ m &= 50 \end{aligned}$$

Therefore, the *m*-intercept is the point $(50,0)$ and means that Jose had no credit card debt 50 months after starting to pay them off.

c. The *D*-intercept makes sense because it is reasonable to have $12,500 in credit card debt. The *m*-intercept makes sense in this situation, as it is reasonable that it would take Jose 50 months to pay off the credit cards and have no debt.

49. This equation is in slope-intercept form, so the slope is the coefficient of the *x* term and the *y*-intercept is the point (0, b).

Therefore, the slope is -3 and the *y*-intercept is $(0,27)$.

Substitute $y = 0$ and find *x*.

$$\begin{aligned} 0 &= -3x + 27 && \text{Substitute in } y = 0. \\ -27 & \quad\;\; -27 && \text{Solve for } x. \\ -27 &= -3x \\ \frac{-27}{-3} &= \frac{-3x}{-3} \\ x &= 9 \end{aligned}$$

The *x*-intercept is the point $(9,0)$.

51. This equation is in slope-intercept form, so the slope is the coefficient of the *x* term and the *y*-intercept is the point (0, b).

Therefore, the slope is 8 and the *y*-intercept is $(0,-16)$.

Substitute $y = 0$ and find *x*.

$$\begin{aligned} 0 &= 8x - 16 && \text{Substitute in } y = 0. \\ +16 & \quad\;\; +16 && \text{Solve for } x. \\ 16 &= 8x \\ \frac{16}{8} &= \frac{8x}{8} \\ x &= 2 \end{aligned}$$

The *x*-intercept is the point $(2,0)$.

53. This equation is in slope-intercept form, so the slope is the coefficient of the *x* term and the *y*-intercept is the point (0, b).

Therefore, the slope is $\frac{1}{2}$ and the *y*-intercept is $(0,6)$.

Substitute $y = 0$ and find *x*.

$$\begin{aligned} 0 &= \frac{1}{2}x + 6 && \text{Substitute in } y = 0. \\ -6 & \quad\;\; -6 && \text{Solve for } x. \\ -6 &= \frac{1}{2}x \\ 2 \cdot -6 &= \frac{1}{2}x \cdot 2 \\ x &= -12 \end{aligned}$$

The *x*-intercept is the point $(-12,0)$.

55. This equation is in slope-intercept form, so the slope is the coefficient of the *x* term and the *y*-intercept is the point (0, b).

Therefore, the slope is $-\frac{1}{3}$ and the *y*-intercept is $(0,5)$.

Substitute $y = 0$ and find *x*.

$$\begin{aligned} 0 &= -\frac{1}{3}x + 5 && \text{Substitute in } y = 0. \\ -5 & \quad\;\; -5 && \text{Solve for } x. \\ -5 &= -\frac{1}{3}x \\ -3 \cdot -5 &= -\frac{1}{3}x \cdot -3 \\ x &= 15 \end{aligned}$$

The *x*-intercept is the point $(15,0)$.

57. This equation is in slope-intercept form, so the slope is the coefficient of the *x* term and the *y*-intercept is the point (0, b).

Therefore, the slope is $\frac{2}{3}$ and the *y*-intercept is $(0,-4)$.

Substitute $y = 0$ and find *x*.

$$\begin{aligned} 0 &= \frac{2}{3}x - 4 && \text{Substitute in } y = 0. \\ +4 & \quad\;\; +4 && \text{Solve for } x. \\ 4 &= \frac{2}{3}x \\ \frac{3}{2} \cdot 4 &= \frac{2}{3}x \cdot \frac{3}{2} \\ x &= 6 \end{aligned}$$

The *x*-intercept is the point $(6,0)$.

59. This equation is in slope-intercept form, so the slope is the coefficient of the *x* term and the *y*-intercept is the point (0, b).

Therefore, the slope is 1.5 and the *y*-intercept is $(0,-3)$.

Substitute $y=0$ and find x.

$$\begin{aligned} 0 &= 1.5x-3 && \text{Substitute in } y=0.\\ +3 &\quad\;\; +3 && \text{Solve for } x.\\ 3 &= 1.5x \\ \frac{3}{1.5} &= \frac{1.5x}{1.5} \\ x &= 2 \end{aligned}$$

The x-intercept is the point $(2,0)$.

61. This equation is in slope-intercept form, so the slope would be the coefficient of the x term and the y-intercept is the point (0, b). There is no x term which means the slope (m) is 0. The line does not intersect the x-axis so there is no x-intercept and the y-intercept is $(0,-5)$.

63. This equation cannot be written in slope-intercept form because there is no value for y. This means the line does not intersect the y-axis. The line is vertical, there is no y-intercept and the slope is undefined. The x-intercept is $(6,0)$.

65. The equation is written in slope-intercept form with the slope being the coefficient of the x term, or input variable, which in this case is t. The slope is 38, which represents that the number of American Indians and Alaska Natives in the United States is rising by about 38,000 people per year.

67. The equation is written in slope-intercept form with the slope being the coefficient of the x term, or input variable, which in this case is t. The slope is 0.05, which represents that the price for navel oranges is increasing by $0.05 per year.

69. The equation is written in slope-intercept form with the slope being the coefficient of the x term, or input variable, which in this case is w. The slope is -2, which represents that during each successive week of the year, Big Jim's Car Lot sells 2 fewer cars than the week before.

71. This equation is written in slope-intercept form but using different variables than x and y, so the slope would be the coefficient of the t term and the R-intercept, or vertical intercept, is the point (0, b).

a. The R-intercept is the point (0.29.4), which represents that the total revenue earned by Cisco Systems, Inc. during the year 2005 was $29.4 billion.

b. The slope is 5, which represents that the company's revenue is increasing by $5 billion each year since 2005.

73. First, rearrange the equation into slope-intercept form.

$$\begin{aligned} 3x+y &= 18 && \text{Solve for } y.\\ -3x \quad &\quad -3x \\ y &= -3x+18 \end{aligned}$$

Now that the equation is in slope-intercept form, we can see that the slope is -3 and the y-intercept is $(0,18)$.

75. Put the equation into slope-intercept form.

$$\begin{aligned} 3x+2y &= 6 && \text{Solve for } y.\\ -3x \quad &\quad -3x \\ 2y &= -3x+6 \\ \frac{2y}{2} &= \frac{-3x+6}{2} \\ y &= \frac{-3x}{2}+\frac{6}{2} \\ y &= -\frac{3}{2}x+3 \end{aligned}$$

The slope is $-\frac{3}{2}$ and the y-intercept is $(0,3)$.

77. Put the equation into slope-intercept form.

$$\begin{aligned} 2x-7y &= 21 && \text{Solve for } y.\\ -2x \quad &\quad -2x \\ -7y &= -2x+21 \\ \frac{-7y}{-7} &= \frac{-2x+21}{-7} \\ y &= \frac{2}{7}x-3 \end{aligned}$$

The slope is $\frac{2}{7}$ and the y-intercept is $(0,-3)$.

79. Put the equation into slope-intercept form.

$$\begin{aligned} x-2y &= -7 && \text{Solve for } y.\\ -x \quad &\quad -x \\ -2y &= -x-7 \\ \frac{-2y}{-2} &= \frac{-x-7}{-2} \\ y &= \frac{1}{2}x+\frac{7}{2} \end{aligned}$$

The slope is $\frac{1}{2}$ and the y-intercept is $\left(0,\frac{7}{2}\right)$.

81. Put the equation into slope-intercept form.

$$\begin{aligned} y+3 &= 2(x-5) && \text{Distribute on the right.}\\ y+3 &= 2x-10 && \text{Solve for } y.\\ -3 &\quad\;\; -3 \\ y &= 2x-13 \end{aligned}$$

The slope is 2 and the y-intercept is $(0,-13)$.

83. Put the equation into slope-intercept form.

$$\begin{aligned} y-2 &= \frac{1}{2}(x+20) && \text{Distribute on the right.}\\ y-2 &= \frac{1}{2}x+10 && \text{Solve for } y.\\ +2 &\quad\;\; +2 \\ y &= \frac{1}{2}x+12 \end{aligned}$$

The slope is $\frac{1}{2}$ and the y-intercept is $(0,12)$.

85. Put the equation into slope-intercept form.

$$\begin{aligned} y-7&=0 \quad \text{Solve for } y.\\ +7\ &+7\\ y&=7 \end{aligned}$$

The slope is 0 and the y-intercept is $(0,7)$.

Section 3.4 Linear Equations and Their Graphs

1. For $y=2x+3$, the slope is 2 and the y-intercept is $(0,3)$. Only Graph A and Graph B could have the slope of 2 (Graph C has a negative slope), and of those two, only Graph A could have a y-intercept of $(0,3)$

Therefore $y=2x+3$ matches Graph A.

For $y=-2x+3$, the slope is -2 and the y-intercept is $(0,3)$. Only Graph C could have the slope of -2 ,and Graph C could have the y-intercept of $(0,3)$.

Therefore $y=-2x+3$ matches Graph C.

For $y=2x-3$, the slope is 2 and the y-intercept is $(0,-3)$. Only Graph A and Graph B could have the slope of 2, and of those two, only Graph B could have a y-intercept of $(0,-3)$ (Graph A has a positive y-intercept).

Therefore $y=2x-3$ matches Graph B.

3. For $y=\frac{1}{2}x+2$, the slope is $\frac{1}{2}$ and the y-intercept is $(0,2)$. Any of the 3 graphs could have the slope of $\frac{1}{2}$ so we will look at the y-intercepts. Graph B or Graph C could have a y-intercept of $(0,2)$. So we will compare the equations to see which one would have a line closer to the origin. The only other equation listed that has a positive y-intercept is $y=\frac{1}{2}x+5$ which has a y-intercept of $(0,5)$. So the first equation has a y-intercept of $(0,2)$ and the second equation has a y-intercept of $(0,5)$. The first equation has a y-intercept that is closer to the origin, so that will match the graph with the line closer to the origin, which is Graph C.

Therefore $y=\frac{1}{2}x+2$ matches Graph C.

For $y=\frac{1}{2}x-3$, the slope is $\frac{1}{2}$ and the y-intercept is $(0,-3)$. Only Graph A could have a y-intercept of $(0,-3)$ (the other 2 graphs have positive y-intercepts).

Therefore $y=\frac{1}{2}x-3$ matches Graph A.

For $y=\frac{1}{2}x+5$, we already looked at this in comparison with the first equation and we know that this equation matches Graph B because the y-intercept of this graph is farther from the origin than Graph C, and the y-intercept of this equation is farther from the origin than the y-intercept of the first equation.

Therefore $y=\frac{1}{2}x+5$ matches Graph B.

5. From the equation $y=\frac{1}{2}x-3$, we see that the slope is $m=\frac{1}{2}=\frac{rise}{run}$ and the y-intercept is the point $(0,-3)$. Start by plotting the y-intercept, then use the slope to rise 1 and run 2 from the y-intercept.

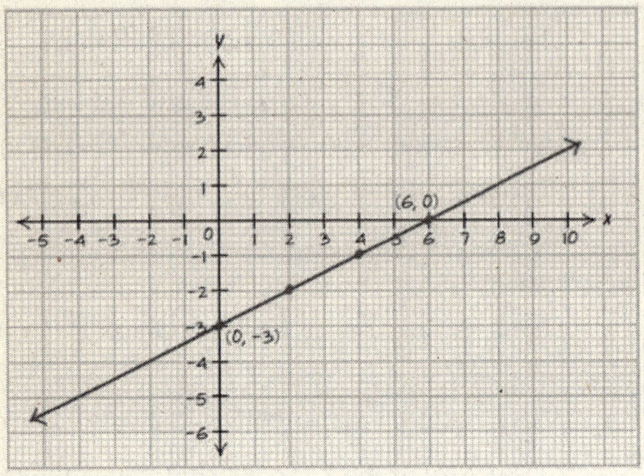

7. From the equation $y=\frac{1}{3}x+2$, we see that the slope is $m=\frac{1}{3}=\frac{rise}{run}$ and the y-intercept is the point $(0,2)$. Start by plotting the y-intercept, then use the slope to rise 1 and run 3 from the y-intercept.

9. From the equation $y=-\frac{3}{5}x+6$, we see that the slope is $m=\frac{-3}{5}=\frac{rise}{run}$ and the y-intercept is the point $(0,6)$. Start by plotting the y-intercept, then use the slope to rise -3 and run 5 from the y-intercept. Since the rise is negative, we go down 3 instead of up.

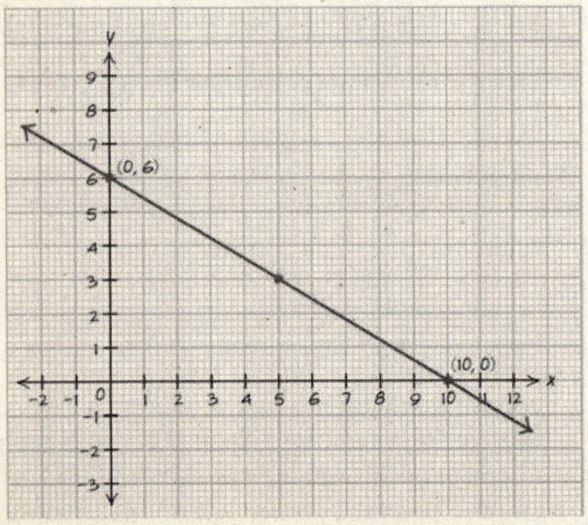

11. From the equation $y=3x-8$, we see that the slope is $m=\frac{3}{1}=\frac{rise}{run}$ and the y-intercept is the point $(0,-8)$. Start by plotting the y-intercept, then use the slope to rise 3 and run 1 from the y-intercept.

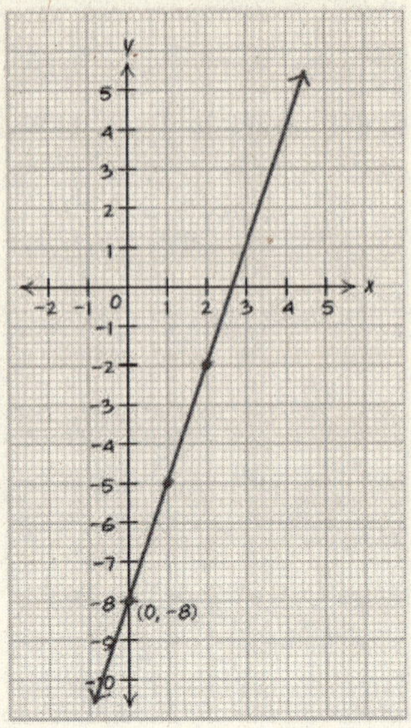

13. From the equation $y=-2x+4$, we see that the slope is $m=\frac{-2}{1}=\frac{rise}{run}$ and the y-intercept is the point $(0,4)$. Start by plotting the y-intercept, then use the slope to rise -2 and run 1 from the y-intercept. Since the rise is negative, we go down 2 instead of up.

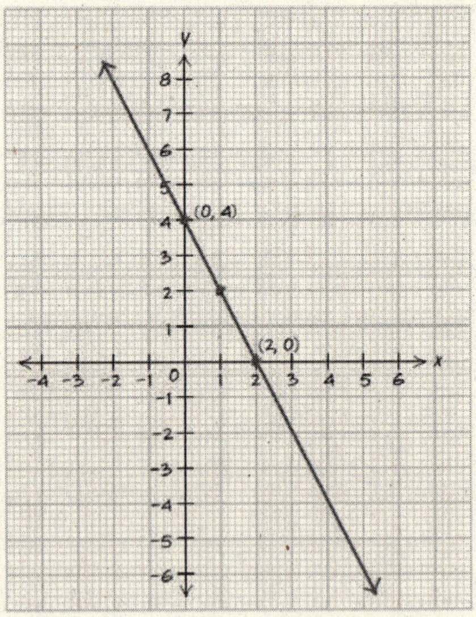

15. Put the equation into slope-intercept form.

$$y+5=\frac{1}{2}(x+26) \quad \text{Distribute on the right side.}$$

$$y+5=\frac{1}{2}x+13 \quad \text{Solve for } y.$$

$$\underline{-5 \qquad -5}$$

$$y=\frac{1}{2}x+8$$

Now we can see that the slope is $m = \frac{1}{2} = \frac{rise}{run}$ and the y-intercept is the point $(0,8)$. Start by plotting the y-intercept, then use the slope to rise 1 and run 2 from the y-intercept.

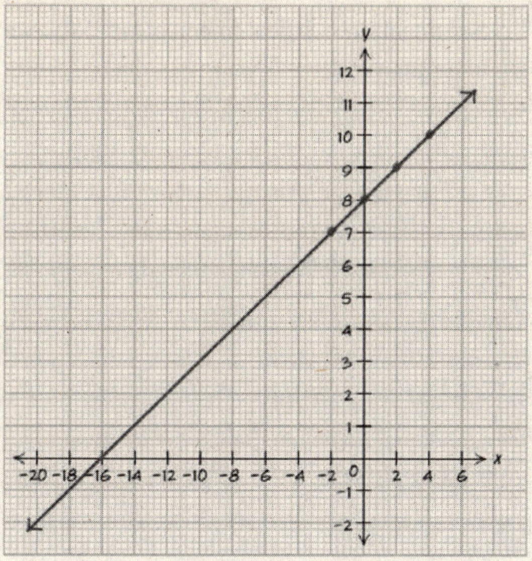

17. Put the equation into slope-intercept form.

$y+6=-2(x+1)$ Distribute on the right side.

$y+6=-2x-2$ Solve for y.

$\underline{-6 \qquad -6}$

$y=-2x-8$

Now we can see that the slope is $m = \frac{-2}{1} = \frac{rise}{run}$ and the y-intercept is the point $(0,-8)$. Start by plotting the y-intercept, then use the slope to rise -2 and run 1 from the y-intercept. Since the rise is negative, we go down 2 instead of up.

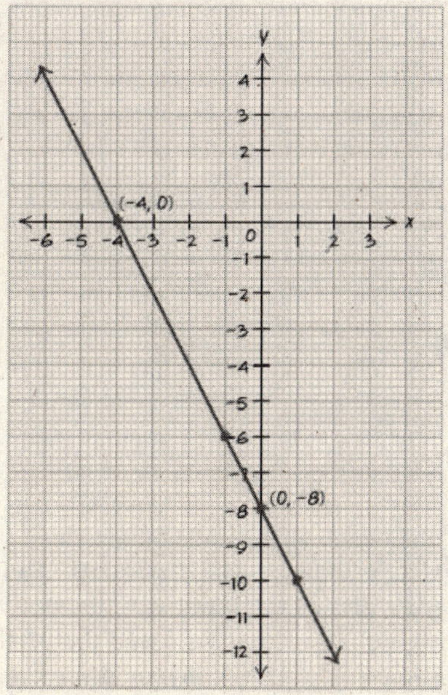

19. Put the equation into slope-intercept form.

$5y=4x-30$ Solve for y.

$$\frac{5y}{5}=\frac{4x-30}{5}$$

$$y=\frac{4x}{5}-\frac{30}{5}$$

$$y=\frac{4}{5}x-6$$

Now we can see that the slope is $m = \frac{4}{5} = \frac{rise}{run}$ and the y-intercept is the point $(0,-6)$. Start by plotting the y-intercept, then use the slope to rise 4 and run 5 from the y-intercept.

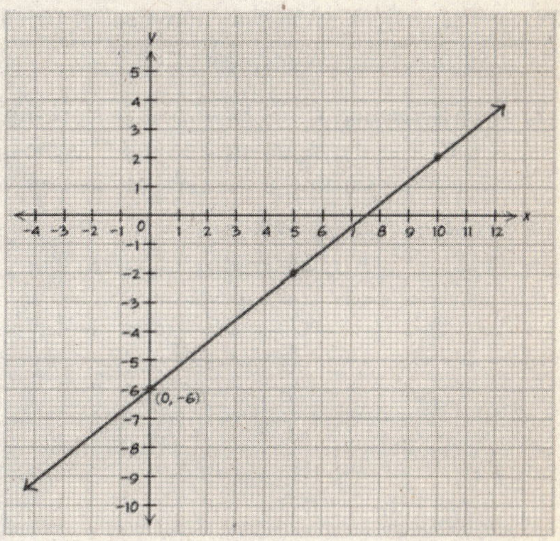

21. Put the equation into slope-intercept form.

$-4y=3x-12$ Solve for y.

$$\frac{-4y}{-4}=\frac{3x-12}{-4}$$

$$y=-\frac{3}{4}x+3$$

Now we can see that the slope is $m = \frac{-3}{4} = \frac{rise}{run}$ and the y-intercept is the point $(0,3)$. Start by plotting the y-intercept, then use the slope to rise -3 and run 4 from the y-intercept. Since the rise is negative, we go down 3 instead of up.

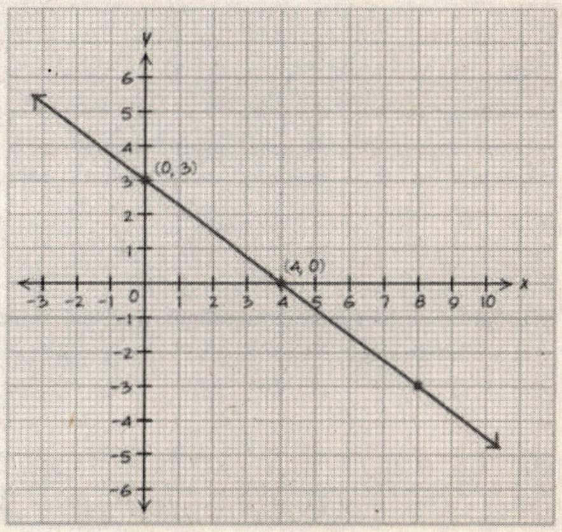

23. The student graphed a slope of $-\frac{1}{2}$ instead of the slope of $\frac{1}{2}$. When using the slope to "rise" and "run," "rise" is always the first step and the rise is always up (unless the slope is negative). "Run" is always second and the run is always to the right. The student should have gone up 1 then right 2.

25. The student graphed a slope of 2 instead of the slope of -2.

27. The student graphed a slope of $\frac{1}{2}$ instead of the slope of 2.

29. The student graphed a y-intercept of $(0,-1)$ instead of the correct y-intercept of $(0,2)$.

31. The equation $C=-2w+25$ is in slope-intercept form with w as the input variable and C as the output variable. Starting at the vertical intercept, $(0,25)$, we rise -2 and run over 1.

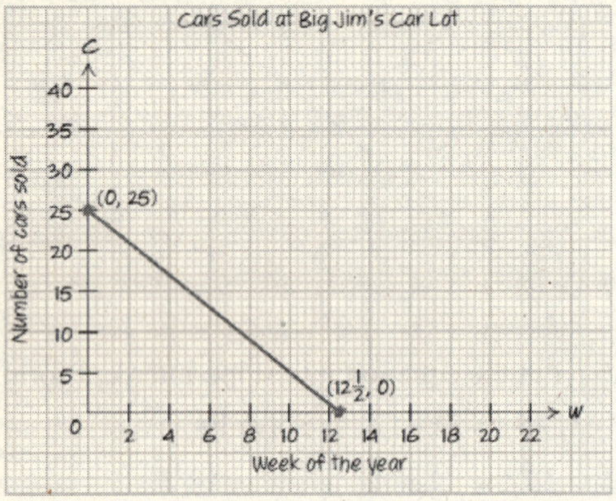

33. The equation $D=-250m+12500$ is in slope-intercept form with m as the input variable and D as the output variable. Starting at the vertical intercept, $(0,12500)$, we rise -250 and run over 1.

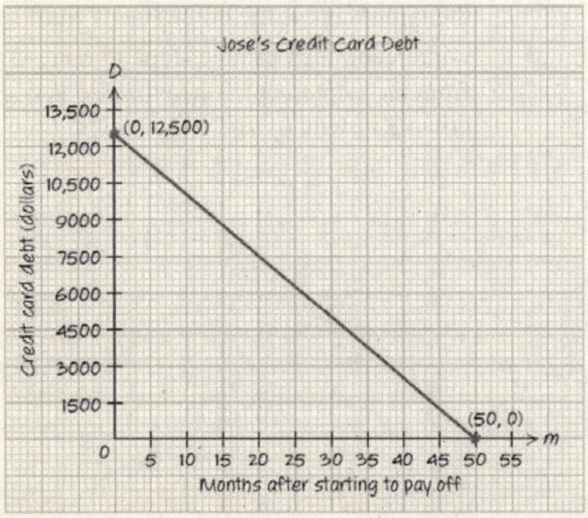

35. The equation $P=1.5t+5.0$ is in slope-intercept form with t as the input variable and P as the output variable. The slope is 1.5 and the vertical intercept is the point $(0,5.0)$. Convert the slope to a fraction to determine the rise and run.

$$m=1.5=\frac{15}{10}=\frac{3}{2}=\frac{rise}{run}$$

Starting at the vertical intercept, $(0,5.0)$, we rise 3 and run over 2.

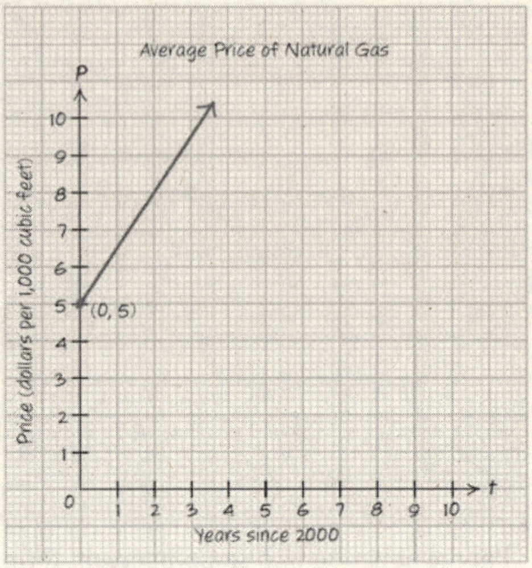

37. The equation $R=1.6t+13$ is in slope-intercept form with t as the input variable and R as the output variable. The slope is 1.6 and the vertical intercept is the point $(0,13)$. Convert the slope to a fraction to determine the rise and run.

$$m=1.6=\frac{16}{10}=\frac{8}{5}=\frac{rise}{run}$$

Starting at the vertical intercept, $(0,13)$, we rise 8 and run over 5.

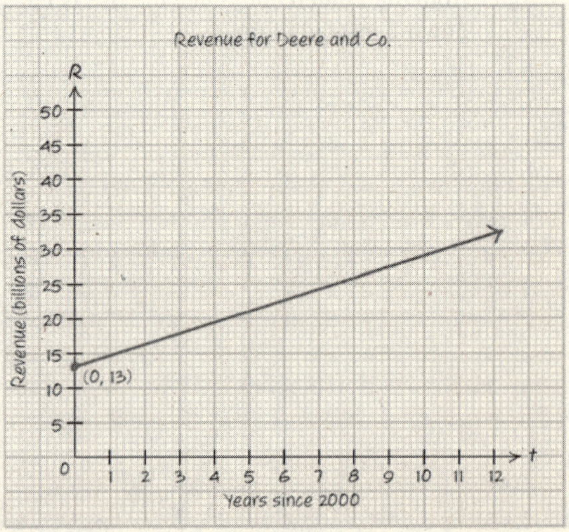

39. The equation $P=25t+238$ is in slope-intercept form with t as the input variable and P as the output variable. The slope is 25 and the vertical intercept is the point $(0,238)$. Starting at the vertical intercept, $(0,238)$, we rise 25 and run over 1.

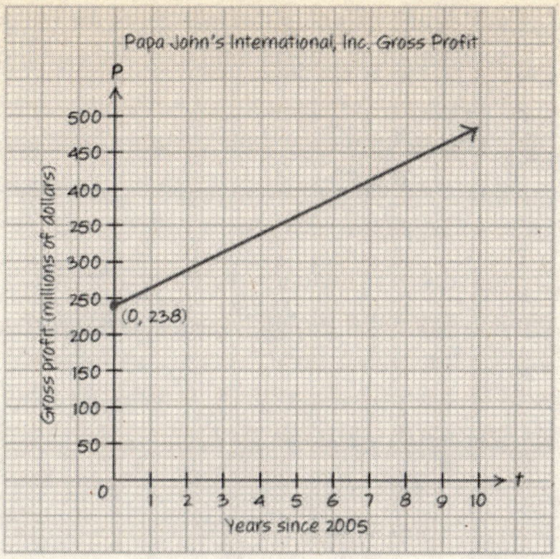

41. To find the y-intercept, substitute in $x=0$ and solve for y.

$$2(0)+5y=20$$
$$5y=20$$
$$y=4$$

Therefore, the point $(0,4)$ is the y-intercept. To find the x-intercept, substitute in $y=0$ and solve for x.

$$2x+5(0)=20$$
$$2x=20$$
$$x=10$$

Therefore, the point $(10,0)$ is the x-intercept. Use these two points to graph the line.

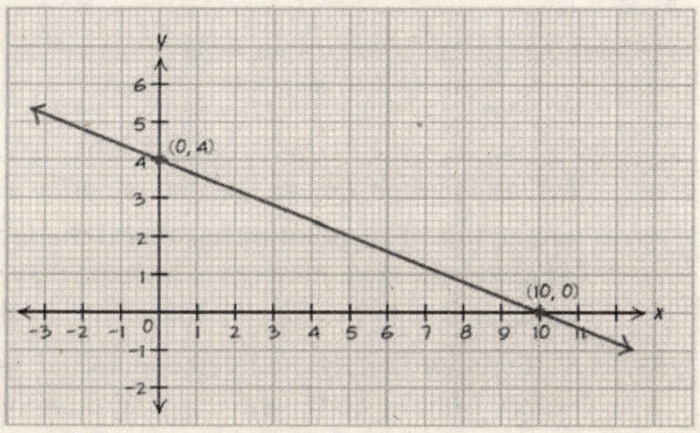

43. To find the y-intercept, substitute in $x=0$ and solve for y.

$$(0)+2y=6$$
$$2y=6$$
$$y=3$$

Therefore, the point $(0,3)$ is the y-intercept. To find the x-intercept, substitute in $y=0$ and solve for x.

$$x+2(0)=6$$
$$x=6$$

Therefore, the point $(6,0)$ is the x-intercept. Use these two points to graph the line.

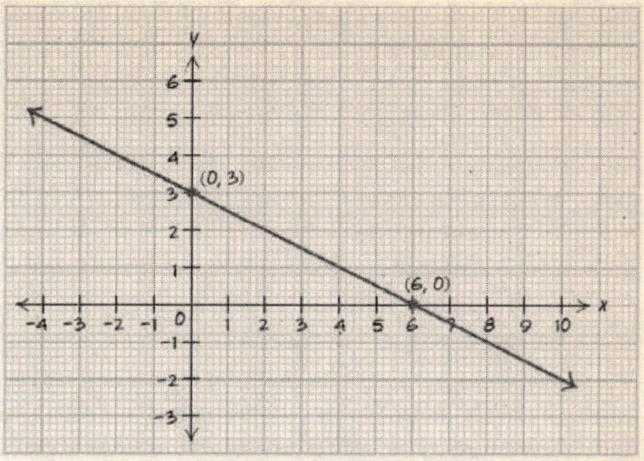

45. To find the y-intercept, substitute in $x=0$ and solve for y.

$$4(0)-2y=10$$
$$-2y=10$$
$$y=-5$$

Therefore, the point $(0,-5)$ is the y-intercept. To find the x-intercept, substitute in $y=0$ and solve for x.

$$4x-2(0)=10$$
$$4x=10$$
$$x=2\frac{1}{2}$$

Therefore, the point $\left(2\frac{1}{2},0\right)$ is the x-intercept. Use these two points to graph the line.

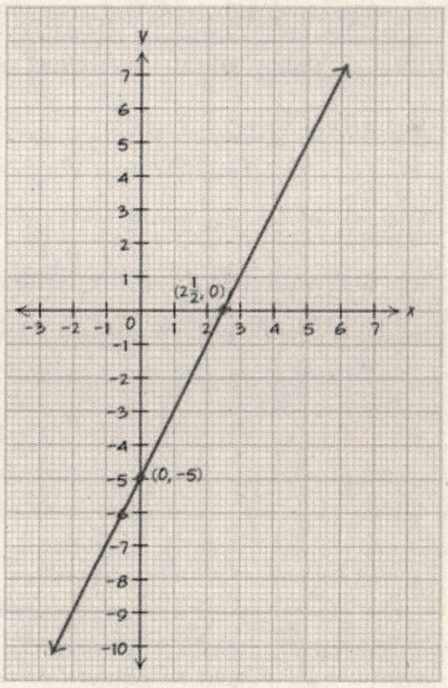

47. To find the y-intercept, substitute in $x=0$ and solve for y.

$$4(0)-5y=15$$
$$-5y=15$$
$$y=-3$$

Therefore, the point $(0,-3)$ is the y-intercept. To find the x-intercept, substitute in $y=0$ and solve for x.

$$4x - 5(0) = 15$$
$$4x = 15$$
$$x = 3\frac{3}{4}$$

Therefore, the point $\left(3\frac{3}{4}, 0\right)$ is the x-intercept. Use these two points to graph the line.

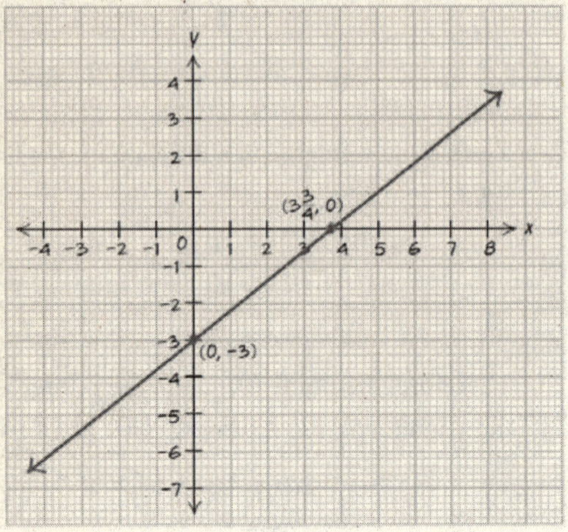

49. To find the y-intercept, substitute in $x = 0$ and solve for y.

$$10(0) + 35y = 210$$
$$35y = 210$$
$$y = 6$$

Therefore, the point $(0, 6)$ is the y-intercept. To find the x-intercept, substitute in $y = 0$ and solve for x.

$$10x + 35(0) = 210$$
$$10x = 210$$
$$x = 21$$

Therefore, the point $(21, 0)$ is the x-intercept. Use these two points to graph the line.

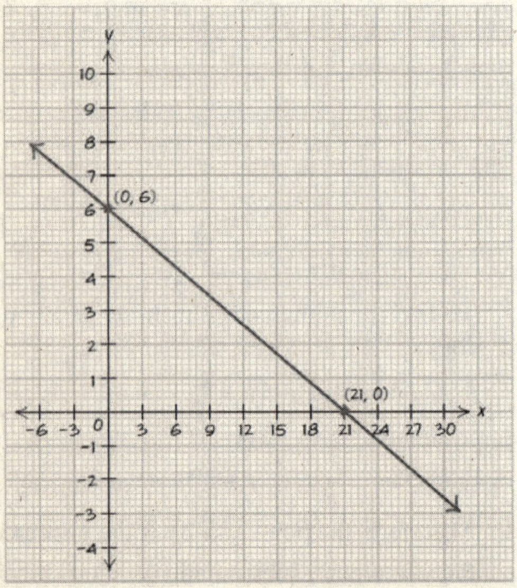

51. **a.** Begin by finding the intercepts for the graph. We select m on the horizontal axis and t on the vertical axis.

$$2(0) + 3t = 80$$
$$3t = 80$$
$$t = 26\frac{2}{3}$$
$$\left(0, 26\frac{2}{3}\right) \quad \text{vertical intercept}$$

$$2m + 3(0) = 80$$
$$2m = 80$$
$$m = 40$$
$$(40, 0) \quad \text{horizontal intercept}$$

Using these intercepts, we get the following graph.

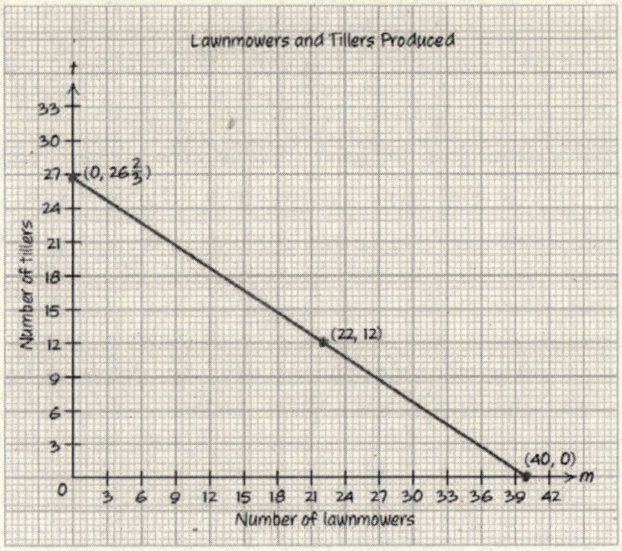

b. On the graph we can see that the point $(22, 12)$ is on the line. This point means that if the factory produces 22 lawn mowers in a week, they will have enough time to produce 12 tillers that week.

53. **a.** Begin by finding the intercepts for the graph. We select A on the horizontal axis and B on the vertical axis.

$$0.16(0) + 0.24B = 800$$
$$0.24B = 800$$
$$B = 3333.33$$
$$(0, 3333.33) \quad \text{vertical intercept}$$

$$0.16A + 0.24(0) = 800$$
$$0.16A = 800$$
$$A = 5000$$
$$(5000, 0) \quad \text{horizontal intercept}$$

Using these intercepts, we get the following graph.

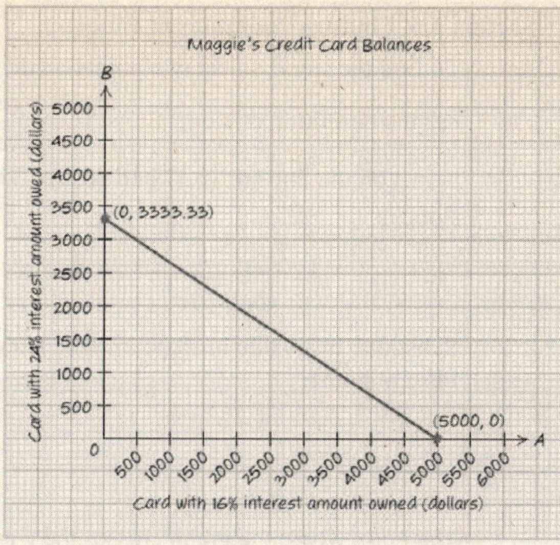

b. The A-intercept represents one possible distribution of credit card balances, that if the card with 16% interest had a balance of \$5000, the card with 24% interest had a balance of \$0.

55. This equation can be simplified to slope-intercept form or general form, so it is a linear equation. Change to slope-intercept form.

$$\begin{aligned} 2y+7&=3x+21\\ -7\quad&\quad\;\, -7\\ 2y&=3x+14\\ \frac{2y}{2}&=\frac{3x+14}{2}\\ y&=\frac{3}{2}x+7 \end{aligned}$$

57. This equation has the variables x and y *multiplied* together, so this is not a linear equation.

59. This equation has the variable x inside an absolute value, so this is not a linear equation.

61. This equation can be simplified to slope-intercept form or general form, so it is a linear equation. Change to slope-intercept form.

63. This equation has the variable x raised to the second power, so this is not a linear equation.

65. This equation has the variable x in a denominator, so this is not a linear equation.

67. This equation can be simplified to slope-intercept form or general form, so it is a linear equation. Change to slope-intercept form.

$$\begin{aligned} 3x+8&=4y-9\\ +9\quad&\quad\;\, +9\\ 3x+17&=4y\\ \frac{3x+17}{4}&=\frac{4y}{4}\\ y&=\frac{3}{4}x+4\frac{1}{4} \end{aligned}$$

69. This equation has the variable x raised to the third power, so this is not a linear equation.

71. This equation is already written in slope-intercept form so it would make sense to graph using the slope and y-intercept.

73. This equation is written in general form so it may be simplest to graph using the x- and y-intercepts.

75. This equation has x raised to the second power so it would be reasonable to graph using a table of points.

77. This equation is not in general form or slope-intercept form, so you could transform the equation to slope-intercept form then solve using the slope and y-intercept, or transform to general form and graph using x- and y-intercepts.

79. This equation is in slope-intercept form, but the slope is in decimal form, so it may be simplest to graph using x- and y-intercepts.

81. This equation has x inside an absolute value so it would be reasonable to graph using a table of points.

83. These 2 equations both have the same slope of -3, therefore these lines are parallel.

85. These equations have slopes of -7 and $\frac{1}{7}$ which are negative reciprocals of one another, therefore the lines are perpendicular.

87. These equations have slopes of $\frac{2}{5}$ and $\frac{5}{2}$, which are reciprocals of one another but not *negative* reciprocals, therefore these lines are neither parallel nor perpendicular.

89. The second equation $6x+3y=27$, is in general form. Change to slope-intercept form.

$$\begin{aligned} 6x+3y&=27\\ -6x\quad&\quad -6x\\ 3y&=-6x+27\\ \frac{3y}{3}&=\frac{-6x+27}{3}\\ y&=-2x+9 \end{aligned}$$

We can now compare the slopes of the two equations. The equations have slopes of -2 and $\frac{1}{2}$ which are negative reciprocals of one another, therefore the lines are perpendicular.

91. The first equation $2x-12y=12$, is in general form. Change to slope-intercept form.

$$\begin{aligned} 2x-12y&=12\\ -2x\quad&\quad -2x\\ -12y&=-2x+12\\ \frac{-12y}{-12}&=\frac{-2x+12}{-12}\\ y&=\frac{1}{6}x-1 \end{aligned}$$

We can now compare the slopes of the two equations. The equations both have the same slope of $\frac{1}{6}$, therefore these lines are parallel.

93. The first equation $x-6y=12$, is in general form. Change to slope-intercept form.

$$\begin{aligned} x-6y&=12 \\ \underline{-x\quad} & \underline{\quad -x} \\ -6y&=-x+12 \\ \frac{-6y}{-6}&=\frac{-x+12}{-6} \\ y&=\frac{1}{6}x-2 \end{aligned}$$

We can now compare the slopes of the two equations. The equations have slopes of $\frac{1}{6}$ and 6, which are reciprocals of one another but not *negative* reciprocals, therefore these lines are neither parallel nor perpendicular.

Section 3.5 Finding Equations of Lines

1. To find the slope, select two points on the line and use the definition of slope. Slope is $m=\frac{\text{rise}}{\text{run}}$. Two points on the line are $(0,1)$ and $(6,2)$. Therefore the slope is

$$m=\frac{rise}{run}=\frac{1}{6}$$

Substitute this value of slope into the slope-intercept form of a line.

$$y=mx+b$$
$$y=\frac{1}{6}x+b$$

The point where the graph crosses the *y*-axis is the *y*-intercept (0, *b*). We know that point is $(0,1)$ so $b=1$. Substituting this value into our equation yields:

$$y=\frac{1}{6}x+1$$

Therefore, the equation of the line is $y=\frac{1}{6}x+1$.

3. To find the slope, select two points on the line and use the definition of slope. Slope is $m=\frac{\text{rise}}{\text{run}}$. Two points on the line are $(-8,20)$ and $(4,-10)$. Therefore the slope is

$$m=\frac{rise}{run}=\frac{-30}{12}=-\frac{5}{2}$$

Substitute this value of slope into the slope-intercept form of a line.

$$y=mx+b$$
$$y=-\frac{5}{2}x+b$$

To find the *y*-intercept, (0, *b*), substitute either point into the equation. Selecting $(4,-10)$ and substituting it into the equation yields

$$\begin{aligned} y&=-\frac{5}{2}x+b & \\ -10&=-\frac{5}{2}(4)+b & \text{Substitute } x=4,\ y=-10 \\ -10&=-10+b & \text{Solve for } b. \\ \underline{+10}&\ \underline{\quad +10} & \\ 0&=b & \end{aligned}$$

Now that we have found the slope $m=-\frac{5}{2}$ and $b=0$, write the equation of the line using these values. Since $b=0$ we do not need to include $+0$ as part of the equation.

$$y=-\frac{5}{2}x$$

5. To find the slope, select two points on the line and use the definition of slope. Slope is $m = \frac{\text{rise}}{\text{run}}$. Two points on the line are $(-2,4)$ and $(4,4)$. Therefore the slope is

$$m = \frac{rise}{run} = \frac{0}{6} = 0$$

Substitute this value of slope into the slope-intercept form of a line.

$$y = mx + b$$
$$y = 0x + b$$
$$y = b$$

To find the y-intercept, (0, b), substitute either point into the equation. For the purposes of this example we will use the equation $y = 0x + b$ before it was simplified to $y = b$. Selecting $(4,4)$ and substituting it into the equation yields

$$y = 0x + b$$
$$4 = 0(4) + b \quad \text{Substitute } x = 4, y = 4$$
$$4 = 0 + b \quad \text{Solve for } b.$$
$$4 = b$$

Now that we have found the slope $m = 0$ and $b = 4$, write the equation of the line using these values.

$$y = 0x + 4$$
$$y = 4$$

7. To find the slope, select two points on the line and use the definition of slope. Slope is $m = \frac{\text{rise}}{\text{run}}$. Two points on the line are $(1,-50)$ and $(4,60)$. Therefore the slope is

$$m = \frac{rise}{run} = \frac{110}{3}$$

Substitute this value of slope into the slope-intercept form of a line.

$$y = mx + b$$
$$y = \frac{110}{3}x + b$$

To find the y-intercept, (0, b), substitute either point into the equation. Selecting $(1,-50)$ and substituting it into the equation yields:

$$y = \frac{110}{3}x + b$$
$$-50 = \frac{110}{3}(1) + b \quad \text{Substitute } x = 1, y = -50$$
$$-50 = \frac{110}{3} + b \quad \text{Solve for } b.$$
$$\underline{-\frac{110}{3} \quad -\frac{110}{3}}$$
$$-50 - \frac{110}{3} = b \quad \text{Find like denominators.}$$
$$-\frac{150}{3} - \frac{110}{3} = b$$
$$-\frac{260}{3} = b$$
$$b = -86\frac{2}{3}$$

Now that we have found the slope $m = \frac{110}{3}$ and $b = -86\frac{2}{3}$, write the equation of the line using these values.

$$y = \frac{110}{3}x + \left(-86\frac{2}{3}\right)$$
$$y = \frac{110}{3}x - 86\frac{2}{3}$$

9. To find the slope, select two points on the line and use the definition of slope. Slope is $m = \frac{\text{rise}}{\text{run}}$. Two points on the line are $(3,4.5)$ and $(21,2)$. Therefore the slope is

$$m = \frac{rise}{run} = \frac{-2.5}{18} = -\frac{5}{36}$$

Substitute this value of slope into the slope-intercept form of a line.

$$y = mx + b$$
$$y = -\frac{5}{36}x + b$$

To find the y-intercept, (0, b), substitute either point into the equation. Selecting $(21,2)$ and substituting it into the equation yields:

$$y = -\frac{5}{36}x + b \quad \text{Substitute } x = 21, y = 2$$
$$2 = -\frac{5}{36}(21) + b$$
$$2 = -\frac{35}{12} + b \quad \text{Solve for } b.$$
$$\underline{+\frac{35}{12} \quad +\frac{35}{12}}$$
$$2 + \frac{35}{12} = b \quad \text{Find like denominators.}$$
$$\frac{24}{12} + \frac{35}{12} = b$$
$$\frac{59}{12} = b$$
$$b = 4\frac{11}{12}$$

Now that we have found the slope $m=-\frac{5}{36}$ and $b=4\frac{11}{12}$, write the equation of the line using these values.

$$y=-\frac{5}{36}x+4\frac{11}{12}$$

11. Use the points $(0,0)$ and $(100,1)$ to compute the slope of the line.

$$m=\frac{y_2-y_1}{x_2-x_1}=\frac{c_2-c_1}{y_2-y_1}$$
$$=\frac{1-0}{100-0}=\frac{1}{100}$$

Substitute this value of slope into the slope-intercept form of a line:

$y=mx+b$ Slope-intercept form of a line.

$c=my+b$ Use the variables from the problem.

$c=\frac{1}{100}y+b$

To find the y-intercept, $(0, b)$, substitute either point into the equation. Selecting $(0,0)$ and substituting it into the equation yields:

$c=\frac{1}{100}y+b$

$(0)=\frac{1}{100}(0)+b$ Substitute $y=0$, $c=0$.

$0=0+b$ Solve for b.

$0=b$

Using the values we found for the slope and b yields the equation of the line

$$c=\frac{1}{100}y+0$$
$$c=\frac{1}{100}y$$

13. Use the points $(1,75)$ and $(3,71)$ to compute the slope of the line.

$$m=\frac{y_2-y_1}{x_2-x_1}=\frac{F_2-F_1}{t_2-t_1}$$
$$=\frac{75-71}{1-3}=\frac{4}{-2}=-2$$

Substitute this value of slope into the slope-intercept form of a line:

$y=mx+b$ Slope-intercept form of a line.

$F=mt+b$ Use the variables from the problem.

$F=-2t+b$

To find the y-intercept, $(0, b)$, substitute either point into the equation. Selecting $(1,75)$ and substituting it into the equation yields:

$F=-2t+b$

$(75)=-2(1)+b$ Substitute $t=1$, $F=75$.

$75=-2+b$ Solve for b.

$+2\quad +2$

$77=b$

Using the values we found for the slope and b yields the equation of the line

$$F=-2t+b$$
$$F=-2t+77$$

15. Use the points $(0,165)$ and $(4,157)$ to compute the slope of the line.

$$m=\frac{y_2-y_1}{x_2-x_1}=\frac{W_2-W_1}{t_2-t_1}$$
$$=\frac{157-165}{4-0}=\frac{-8}{4}=-2$$

Substitute this value of slope into the slope-intercept form of a line:

$y=mx+b$ Slope-intercept form of a line.

$W=mt+b$ Use the variables from the problem.

$W=-2t+b$

To find the y-intercept, $(0, b)$, substitute either point into the equation. Selecting $(0,165)$ and substituting it into the equation yields:

$W=-2t+b$

$(165)=-2(0)+b$ Substitute $t=0$, $W=165$.

$165=0+b$ Solve for b.

$165=b$

Using the values we found for the slope and b yields the equation of the line

$$W=-2t+b$$
$$W=-2t+165$$

17. The slope is $m=\frac{1}{100}$. Interpret the slope:

$$m=\frac{1}{100}=\frac{\text{change in } c}{\text{change in } y}=\frac{1\text{ century}}{100\text{ years}}$$

Therefore the slope represents that there is 1 century per 100 years.

19. The slope is $m=-2$. Interpret the slope:

$$m=\frac{-2}{1}=\frac{\text{change in } F}{\text{change in } t}=\frac{-2°\text{ Fahrenheit}}{1\text{ hour}}$$

Therefore the slope represents that the temperature changes by -2 degrees Fahrenheit per hour.

21. The slope is $m=-2$. Interpret the slope:

$$m=\frac{-2}{1}=\frac{\text{change in } W}{\text{change in } t}=\frac{-2\text{ pounds}}{1\text{ week}}$$

Therefore the slope represents Irina's weight changes by -2 pounds per week.

23. One point given is $(c, y) = (0,0)$. This point is both the c-intercept because $y = 0$ and the y-intercept because $c = 0$. This means that for 0 centuries, there are 0 years.

25. We need to find the intercepts for the equation $F = -2t + 77$. To find the t-intercept, substitute in $F = 0$ and solve for t.

$$\begin{aligned}(0) &= -2t + 77\\ -77 &\quad -77\\ \hline -77 &= -2t\\ t &= 38.5\end{aligned}$$

Therefore, the point $(38.5, 0)$ is the t-intercept.

The interpretation of this point is that 38.5 hours after noon the temperature would be 0 degrees Fahrenheit. This value does not make sense in the given situation.

To find the F-intercept, substitute in $t = 0$ and solve for F.

$$\begin{aligned}F &= -2(0) + 77\\ F &= 0 + 77\\ F &= 77\end{aligned}$$

Therefore, the point $(0, 77)$ is the F-intercept. The interpretation of this point is that at hour 0, which would be noon, the temperature would be 77 degrees Fahrenheit.

27. We need to find the intercepts for the equation $W = -2t + 165$. To find the t-intercept, substitute in $W = 0$ and solve for t.

$$\begin{aligned}(0) &= -2t + 165\\ -165 &\quad -165\\ \hline -165 &= -2t\\ \frac{-165}{-2} &= \frac{-2t}{-2}\\ t &= 82.5\end{aligned}$$

Therefore, the point $(82.5, 0)$ is the t-intercept.

The interpretation of this point is that 82.5 weeks after starting her diet, Irina will weigh 0 pounds. This value does not make sense in the given situation.

To find the W-intercept, substitute in $t = 0$ and solve for W.

$$\begin{aligned}W &= -2(0) + 165\\ W &= 0 + 165\\ W &= 165\end{aligned}$$

Therefore, the point $(0, 165)$ is the W-intercept. The interpretation of this point is that at the start of her diet (0 weeks into her diet), Irina weighs 165 pounds.

29. **a.** We will choose two of the given points $(2, 90)$ and $(4, 180)$, and use these points to find the slope of the line.

$$\begin{aligned}m &= \frac{y_2 - y_1}{x_2 - x_1} = \frac{d_2 - d_1}{g_2 - g_1}\\ &= \frac{180 - 90}{4 - 2} = \frac{90}{2} = 45\end{aligned}$$

Substitute this value of slope into the slope-intercept form of a line:

$y = mx + b$ Slope-intercept form of a line.

$d = mg + b$ Use the variables from the problem.

$d = 45g + b$

To find the d-intercept, (0, b), substitute either point into the equation. Selecting $(2, 90)$ and substituting it into the equation yields:

$d = 45g + b$

$(90) = 45(2) + b$ Substitute $g = 2$, $d = 90$.

$90 = 90 + b$ Solve for b.

$-90 \quad -90$

$0 = b$

Using the values we found for the slope and b yields the equation of the line

$$\begin{aligned}d &= 45g + b\\ d &= 45g + 0\\ d &= 45g\end{aligned}$$

b. The slope is $m = 45$. To interpret the slope, attach the proper units to the numerator and denominator:

$$m = \frac{45}{1} = \frac{\text{change in } d}{\text{change in } g} = \frac{45 \text{ miles}}{1 \text{ gallon}}$$

Therefore, the slope of 45 represents that for 45 highway miles driven, the Prius uses 1 gallon of gas, or the Prius gets 45 mpg.

c. We need to find the intercepts for the equation $d = 45g$. To find the d-intercept, substitute in $g = 0$ and solve for d.

$$\begin{aligned}d &= 45(0)\\ d &= 0\end{aligned}$$

Therefore, the point $(0, 0)$ is both the d-intercept because $g = 0$ and the g-intercept because $d = 0$. The interpretation of this point is that for 0 miles driven, there are 0 gallons of gas used.

31. **a.** The membership fee is \$45 per month which can be represented by $45m$ plus a one-time sign-up fee of \$125. Therefore, the total cost c can be represented by the equation of the line $c = 45m + 125$.

b. The equation $c = 45m + 125$ is in slope-intercept form, therefore we can see that the slope is 45. The slope represents that the cost increases by \$45 for every month of membership.

c. To find the m-intercept, substitute in $c = 0$ and solve for m.

$$(0) = 45m + 125$$
$$\underline{-125 \quad -125}$$
$$-125 = 45m$$
$$\frac{-125}{45} = \frac{45m}{45}$$
$$m = -2\frac{7}{9}$$

Therefore, the point $\left(-2\frac{7}{9}, 0\right)$ is the m-intercept.

The interpretation of this point is that $2\frac{7}{9}$ weeks before beginning a membership, the cost would be \$0. This value does not make sense in the problem situation.

To find the c-intercept, substitute in $m = 0$ and solve for c.

$$c = 45(0) + 125$$
$$c = 0 + 125$$
$$c = 125$$

Therefore, the point $(0,125)$ is the c-intercept. The interpretation of this point is the initial sign-up fee of \$125.

33. Let $(x_1, y_1) = (0,-9)$ and $(x_2, y_2) = (4,-5)$. Substituting these points into the slope formula yields:

$$\text{slope} = \frac{y_2 - y_1}{x_2 - x_1} = \frac{-5-(-9)}{4-(0)} = \frac{4}{4} = 1$$

Therefore the slope of the line is 1.

Find the y-intercept using the value for slope and one of the given points $(0,-9)$.

$y = mx + b$

$(-9) = (1)(0) + b$ Substitute $m = 1,\ x = 0,\ y = -9.$

$-9 = 0 + b$ Solve for b.

$b = -9$

Using the values we found for the slope and b yields the equation of the line

$$y = x - 9$$

35. Let $(x_1, y_1) = (-3,-1)$ and $(x_2, y_2) = (0,1)$. Substituting these points into the slope formula yields:

$$\text{slope} = \frac{y_2 - y_1}{x_2 - x_1} = \frac{1-(-1)}{0-(-3)} = \frac{2}{3}$$

Therefore the slope of the line is $\frac{2}{3}$.

Find the y-intercept using the value for slope and one of the given points $(0,1)$.

$y = mx + b$

$(1) = (\frac{2}{3})(0) + b$ Substitute $m = \frac{2}{3},\ x = 0,\ y = 1.$

$1 = 0 + b$ Solve for b.

$1 = b$

Using the values we found for the slope and b yields the equation of the line

$$y = \frac{2}{3}x + 1$$

37. Let $(x_1, y_1) = (1,-6)$ and $(x_2, y_2) = (-1,2)$. Substituting these points into the slope formula yields:

$$\text{slope} = \frac{y_2 - y_1}{x_2 - x_1} = \frac{2-(-6)}{-1-(1)} = \frac{8}{-2} = -4$$

Therefore the slope of the line is -4.

Find the y-intercept using the value for slope and one of the given points $(-1,2)$.

$y = mx + b$ Substitute $m = -4,\ x = -1,\ y = 2.$

$(2) = (-4)(-1) + b$

$2 = 4 + b$ Solve for b.

$\underline{-4 \quad -4}$

$-2 = b$

Using the values we found for the slope and b yields the equation of the line

$$y = -4x - 2$$

39. Let $(x_1, y_1) = (3,5)$ and $(x_2, y_2) = (-4,5)$. Substituting these points into the slope formula yields:

$$\text{slope} = \frac{y_2 - y_1}{x_2 - x_1} = \frac{5-(5)}{-4-(3)} = \frac{0}{-7} = 0$$

Therefore the slope of the line is 0.

Find the y-intercept using the value for slope and one of the given points $(-4,5)$.

$y = mx + b$ Substitute $m = 0,\ x = -4,\ y = 5.$

$(5) = (0)(-4) + b$

$5 = 0 + b$ Solve for b.

$5 = b$

Using the values we found for the slope and b yields the equation of the line

$$y = 0x + 5$$
$$y = 5$$

41. Let $(x_1, y_1) = (6,3)$ and $(x_2, y_2) = (7,6)$. Substituting these points into the slope formula yields:

$$\text{slope} = \frac{y_2 - y_1}{x_2 - x_1} = \frac{6-(3)}{7-(6)} = \frac{3}{1} = 3$$

Therefore the slope of the line is 3.

Find the y-intercept using the value for slope and one of the given points $(7,6)$.

$y = mx + b$ Substitute $m = 3$, $x = 7$, $y = 6$.
$(6) = (3)(7) + b$
$6 = 21 + b$ Solve for b.
$-21 \ -21$
$-15 = b$

Using the values we found for the slope and b yields the equation of the line

$$y = 3x - 15$$

43. Substituting $m = -2$ and $(x_1, y_1) = (0,-7)$ into the point-slope form of a line yields

$y - y_1 = m(x - x_1)$ Substitute in $m = -2$ and
$y - (-7) = -2(x - 0)$ $(x_1, y_1) = (0,-7)$
$y + 7 = -2x - (-2)(0)$ Distribute and simplify.
$y + 7 = -2x$ Solve for y.
$-7 \ -7$
$y = -2x - 7$

Therefore, the equation of the line is $y = -2x - 7$.

45. Substituting $m = 6$ and $(x_1, y_1) = (2,3)$ into the point-slope form of a line yields

$y - y_1 = m(x - x_1)$ Substitute in $m = 6$ and
$y - (3) = 6(x - 2)$ $(x_1, y_1) = (2,3)$
$y - 3 = 6x - (6)(2)$ Distribute and simplify.
$y - 3 = 6x - 12$ Solve for y.
$+3 \ +3$
$y = 6x - 9$

Therefore, the equation of the line is $y = 6x - 9$.

47. Substituting $m = \frac{2}{3}$ and $(x_1, y_1) = (6,4)$ into the point-slope form of a line yields

$y - y_1 = m(x - x_1)$ Substitute in $m = \frac{2}{3}$ and
$y - (4) = \frac{2}{3}(x - 6)$ $(x_1, y_1) = (6,4)$
$y - 4 = \frac{2}{3}x - (\frac{2}{3})(6)$ Distribute and simplify.
$y - 4 = \frac{2}{3}x - 4$ Solve for y.
$+4 \ +4$
$y = \frac{2}{3}x$

Therefore, the equation of the line is $y = \frac{2}{3}x$.

49. Substituting $m = 0$ and $(x_1, y_1) = (-3,5)$ into the point-slope form of a line yields

$y - y_1 = m(x - x_1)$ Substitute in $m = 0$ and
$y - (5) = 0(x - (-3))$ $(x_1, y_1) = (-3,5)$
$y - 5 = 0x - (0)(-3)$ Distribute and simplify.
$y - 5 = 0$ Solve for y.
$+5 \ +5$
$y = 5$

Therefore, the equation of the line is $y = 5$.

51. We must first determine the slope of the line.

$$m = \frac{y_2 - y_1}{x_2 - x_1} = \frac{-1-(5)}{-1-1} = \frac{-6}{-2} = 3$$

Substituting slope and the point $(1,5)$ into the point-slope form of a line yields:

$y - y_1 = m(x - x_1)$ Substitute in $m = 3$ and
$y - (5) = 3(x - 1)$ $(x_1, y_1) = (1,5)$
$y - 5 = 3x - (3)(1)$ Distribute and simplify.
$y - 5 = 3x - 3$ Solve for y.
$+5 \ +5$
$y = 3x + 2$

Therefore, the equation of the line is $y = 3x + 2$.

53. We must first determine the slope of the line.

$$m = \frac{y_2 - y_1}{x_2 - x_1} = \frac{-5-(7)}{3-(-1)} = \frac{-12}{4} = -3$$

Substituting slope and the point $(3,-5)$, into the point-slope form of a line yields:

$y - y_1 = m(x - x_1)$ Substitute in $m = -3$ and
$y - (-5) = -3(x - 3)$ $(x_1, y_1) = (3,-5)$
$y + 5 = -3x - (-3)(3)$ Distribute and simplify.
$y + 5 = -3x + 9$ Solve for y.
$-5 \ -5$
$y = -3x + 4$

Therefore, the equation of the line is $y = -3x + 4$.

55. We must first determine the slope of the line.

$$m = \frac{y_2 - y_1}{x_2 - x_1} = \frac{0-(2)}{-2-(2)} = \frac{-2}{-4} = \frac{1}{2}$$

Substituting slope and the point $(2,2)$, into the point-slope form of a line yields:

$y - y_1 = m(x - x_1)$ Substitute in $m = \frac{1}{2}$ and
$y - (2) = \frac{1}{2}(x - 2)$ $(x_1, y_1) = (2,2)$
$y - 2 = \frac{1}{2}x - (\frac{1}{2})(2)$ Distribute and simplify.
$y - 2 = \frac{1}{2}x - 1$ Solve for y.
$+2 \ +2$
$y = \frac{1}{2}x + 1$

Therefore, the equation of the line is $y = \frac{1}{2}x + 1$.

57. We must first determine the slope of the line.

$$m=\frac{y_2-y_1}{x_2-x_1}=\frac{-6-(-6)}{-2-(3)}=\frac{0}{-5}=0$$

Substituting slope and the point $(-2,-6)$, into the point-slope form of a line yields:

$y-y_1=m(x-x_1)$ Substitute in $m=0$ and
$y-(-6)=0(x-(-2))$ $(x_1,y_1)=(-2,-6)$
$y+6=0x-(0)(-2)$ Distribute and simplify.
$y+6=0$ Solve for y.
$-6\quad -6$
$y=-6$

Therefore, the equation of the line is $y=-6$.

59. Rewrite $y=\frac{-2}{3}x+7$ in general form.

$y=\frac{-2}{3}x+7$ Multiply each term by 3 to
$3\cdot y=\frac{-2}{3}(3)x+7\cdot 3$ clear denominator.
$3y=-2x+21$
$+2x\quad +2x$ Group x and y on left side.
$2x+3y=21$

61. Rewrite $-3x+4y=-24$ in slope-intercept form.

$-3x+4y=-24$ Isolate y term on left side.
$+3x\quad +3x$
$4y=3x-24$
$\frac{4y}{4}=\frac{3x-24}{4}$
$y=\frac{3}{4}x-6$

63. Rewrite $y-3=-6(x+11)$ in slope-intercept form.

$y-3=-6(x+11)$ Evaluate parentheses first.
$y-3=-6x-66$ Isolate y term on left side.
$+3\quad +3$
$y=-6x-63$

65. Rewrite $y-2=7(x+1)$ in general form.

$y-2=7(x+1)$ Evaluate parentheses first.
$y-2=7x+7$ Group x and y on left side.
$-7x\quad -7x$
$-7x+y-2=7$
$-7x+y-2=7$ Isolate constants on right side.
$+2\quad +2$
$-7x+y=9$

67. We must first determine the slope of the line.

$$m=\frac{y_2-y_1}{x_2-x_1}=\frac{3-(15)}{5-(5)}=\frac{-12}{0}=\textit{undefined}$$

We know that a line with an undefined slope is a vertical line that intersects only the x-axis in one point. Looking at the given points $(5,15)$ and $(5,3)$, we notice that $x=5$ for both points. If we were to graph these two points and connect them with a line, it would be a vertical line intersecting the x-axis at the point $(5,0)$, therefore the equation of the line is $x=5$.

69. We must first determine the slope of the line.

$$m=\frac{y_2-y_1}{x_2-x_1}=\frac{-10-(-6)}{0-(0)}=\frac{-4}{0}=\textit{undefined}$$

We know that a line with an undefined slope is a vertical line that intersects only the x-axis in one point. Looking at the given points $(0,-6)$ and $(0,-10)$, we notice that $x=0$ for both points. If we were to graph these two points and connect them with a line, it would be a vertical line intersecting the x-axis at the point $(0,0)$, therefore the equation of the line is $x=0$.

71. Parallel lines have the same slope. Therefore, since the slope of the given line is $m=6$, the slope of the line that we are trying to find is the same. Substituting this into slope-intercept form yields

$y=mx+b$ Substitute in $m=6$.
$y=6x+b$

To find the value of b, substitute in the given point $(-7,0)$ and solve for b.

$y=6x+b$ Substitute in the given point $(-7,0)$.
$0=6(-7)+b$ Simplify the right side.
$0=-42+b$ Solve for b.
$+42\ +42$
$42=b$

$y=6x+42$ The equation of the line.

73. Change the equation into slope-intercept form.

$x-5y=9$ Isolate y term on left side.
$-x\quad -x$
$-5y=-x+9$
$\frac{-5y}{-5}=\frac{-x+9}{-5}$
$y=\frac{1}{5}x-\frac{9}{5}$

Parallel lines have the same slope. Therefore, since the slope of the given line is $m=\frac{1}{5}$, the slope of the line that we are trying to find is the same. Substituting this into slope-intercept form yields

$y=mx+b$ Substitute in $m=\frac{1}{5}$.
$y=\frac{1}{5}x+b$

To find the value of b, substitute in the given point $(-10,2)$ and solve for b.

$y=\frac{1}{5}x+b$ Substitute in the given point $(-10,2)$.

$2=\frac{1}{5}(-10)+b$ Simplify the right side.

$2=-2+b$ Solve for b.

$+2\ \ +2$

$4=b$

$y=\frac{1}{5}x+4$ The equation of the line.

75. Perpendicular lines have slopes that are negative reciprocals. The slope of the given line is $m=\frac{3}{8}$. To find the negative reciprocal of the given slope, we invert the fraction and find its opposite:

$-\left(\frac{8}{3}\right)=-\frac{8}{3}$

Substituting this into slope-intercept form yields:

$y=mx+b$ Substitute in $m=-\frac{8}{3}$.

$y=-\frac{8}{3}x+b$

To find the value of b, substitute in the given point $(-12,4)$ and solve for b.

$y=-\frac{8}{3}x+b$ Substitute in the given point $(-12,4)$.

$4=-\frac{8}{3}(-12)+b$ Simplify the right side.

$4=32+b$ Solve for b.

$-32-32$

$-28=b$

$y=-\frac{8}{3}x-28$ The equation of the line.

77. We will first rewrite the given equation into slope-intercept form so we can easily find the slope.

$-5x+4y=-9$ Isolate y term on left side.

$+5x\qquad +5x$

$4y=5x-9$

$\frac{4y}{4}=\frac{5x-9}{4}$

$y=\frac{5}{4}x-\frac{9}{4}$

Perpendicular lines have slopes that are negative reciprocals. The slope of the given line is $m=\frac{5}{4}$. To find the negative reciprocal of the given slope, we invert the fraction and find its opposite:

$-\left(\frac{4}{5}\right)=-\frac{4}{5}$

Substituting this into slope-intercept form yields

$y=mx+b$ Substitute in $m=-\frac{4}{5}$.

$y=-\frac{4}{5}x+b$

To find the value of b, substitute in the given point $(-5,1)$ and solve for b.

$y=-\frac{4}{5}x+b$ Substitute in the given point $(-5,1)$.

$1=-\frac{4}{5}(-5)+b$ Simplify the right side.

$1=4+b$ Solve for b.

$-4\ \ -4$

$-3=b$

$y=-\frac{4}{5}x-3$ The equation of the line.

79. Parallel lines have the same slope and we want to find the equation of a line that is parallel to the line $y=4$. This line has a slope of 0 (the equation of the line before being simplified would be $y=0x+4$). We will follow the same steps as with any other line to fine the equation of the parallel line. Since the slope of the given line is $m=0$, the slope of the line that we are trying to find is the same. Substituting $m=0$ into slope-intercept form yields

$y=mx+b$ Substitute in $m=0$.

$y=0x+b$

To find the value of b, substitute in the given point $(-3,6)$ and solve for b.

$y=0x+b$ Substitute in the given point $(-3,6)$.

$6=0(-3)+b$ Simplify the right side.

$6=0+b$ Solve for b.

$6=b$

$y=0x+6$

$y=6$ The equation of the line.

81. The given line is $x=3$ which is a vertical line with an undefined slope. A line that is perpendicular to a vertical line is a horizontal line. We know that the slope of a horizontal line is 0, therefore we are looking for the equation of a line whose slope is 0. Substituting this value for slope into slope-intercept form yields

$y=mx+b$ Substitute in $m=0$.

$y=0x+b$

To find the value of b, substitute in the given point $(2,5)$ and solve for b.

$y = 0x + b$ Substitute in the given point $(2,5)$.
$5 = 0(2) + b$ Simplify the right side.
$5 = 0 + b$ Solve for b.
$5 = b$

$y = 0x + 5$
$y = 5$ The equation of the line

83. Parallel lines have the same slope and we want to find the equation of a line that is parallel to the line $x = -5$. This is a vertical line which has an undefined slope so the line we are looking for is also vertical with an undefined slope. We know that a vertical line intersects only the x-axis in one point. The equation of the line we are looking for passes through the point $(1,-3)$. We notice for the given point that $x = 1$. If we were to graph a vertical line that passes through the point $(1,-3)$, it would intersect the x-axis at the point $(1,0)$, therefore the equation of the line is $x = 1$.

85. The given line is $y = -3$ which is a horizontal line with slope of 0. A line that is perpendicular to a horizontal line is a vertical line. We know that the slope of a vertical line is undefined. We are looking for a vertical line with an undefined slope that passes through the point $(5,2)$. We notice for the given point that $x = 5$. If we were to graph a vertical line that passes through the point $(5,2)$, it would intersect the x-axis at the point $(5,0)$, therefore the equation of the line is $x = 5$.

Chapter 3 Review

1. Complete the table; insert any reasonable values for the last two rows.

h = Hours of walking	C = Calories Burned
1	87
2	174
3	261
4	348
6	522

2. Complete the table; insert any reasonable values for the last two rows.

n = Number of plants purchased	C = Cost in dollars
2	10.50
4	21.00
8	42.00
14	73.50
20	105.00

3. Graph the points in the table.

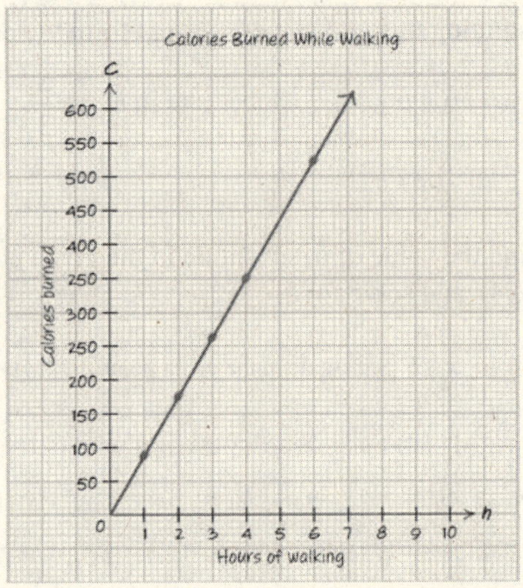

4. Graph the points in the table.

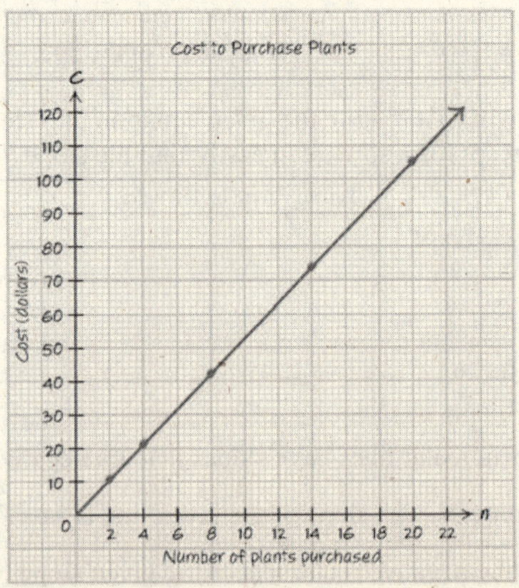

5. Fill in the table selecting values for x that are easy to work with and calculating the value for y. Graph the points.

x	$y=-3x+8$	(x,y)
–1	11	(–1,11)
0	8	(0,8)
1	5	(1,5)
3	–1	(3,–1)
5	–7	(5,–7)

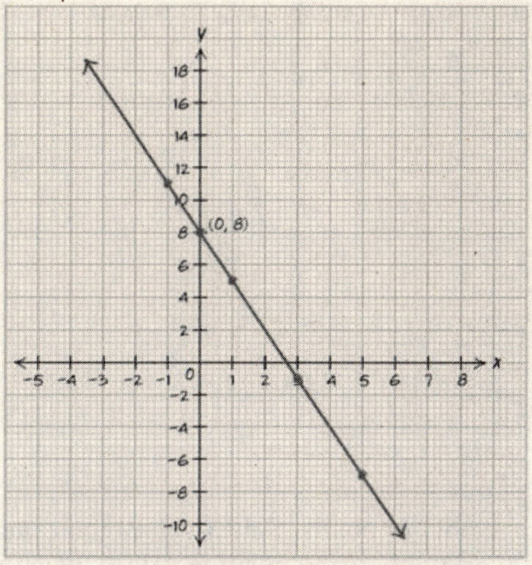

6. Complete the table as in exercise 5:

x	$y=-5x-3$	(x,y)
–2	7	(–2,7)
–1	2	(–1,2)
0	–3	(0,–3)
1	–8	(1,–8)
2	–13	(2,–13)

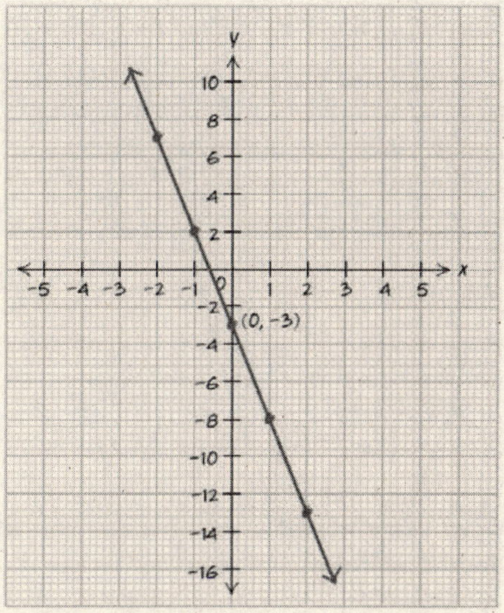

7. Complete the table as in exercise 5:

x	$y=\frac{3}{4}x+1$	(x,y)
–8	–5	(–8,–5)
–4	–2	(–4,–2)
0	1	(0,1)
4	4	(4,4)
8	7	(8,7)

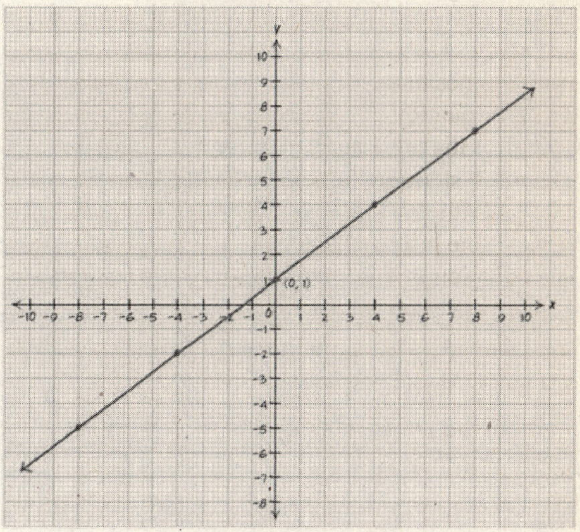

8. Complete the table as in exercise 5:

x	$y=\frac{-1}{5}x-6$	(x,y)
–30	0	(–30,0)
–10	–4	(–10,–4)
0	–6	(0,–6)
10	–8	(10,–8)
20	–10	(20,–10)

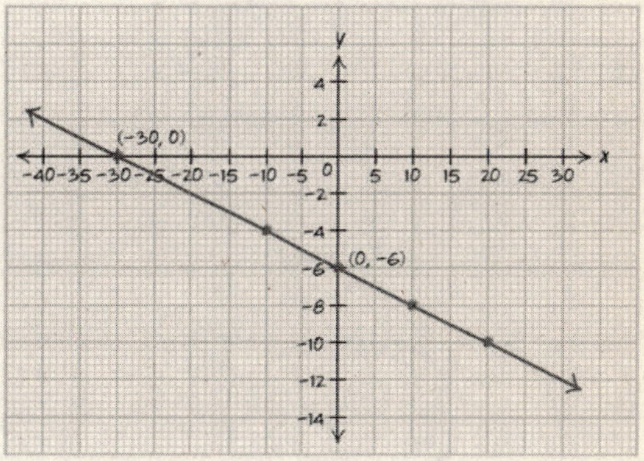

9. Complete the table as in exercise 5:

x	$y=5$	(x,y)
–9	5	(–9,5)
–5	5	(–5,5)
0	5	(0,5)
5	5	(5,5)
9	5	(9,5)

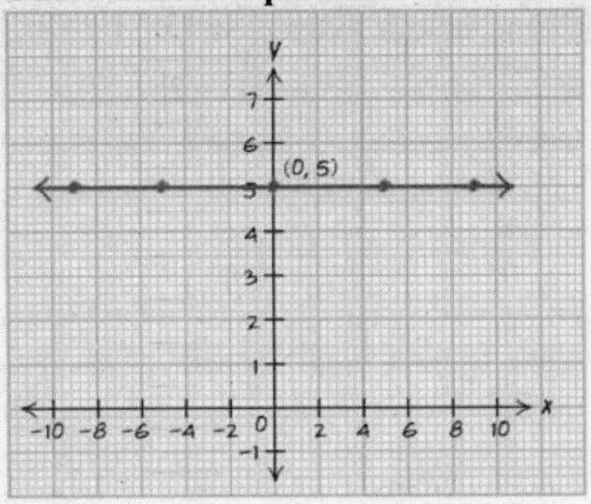

10. Complete the table as in exercise 5:

x	$y=-4$	(x,y)
−9	−4	(−9,−4)
−5	−4	(−5,−4)
0	−4	(0,−4)
5	−4	(5,−4)
9	−4	(9,−4)

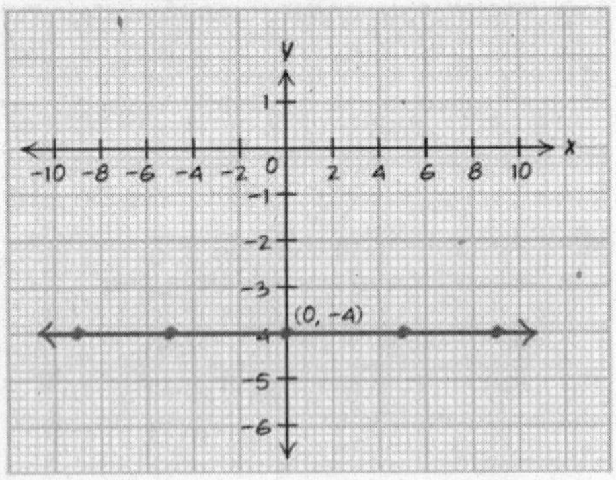

11. Complete the table as in exercise 5:

$x=-3$	y	(x,y)
−3	−8	(−3,−8)
−3	−5	(−3,−5)
−3	0	(−3,0)
−3	5	(−3,5)
−3	8	(−3,8)

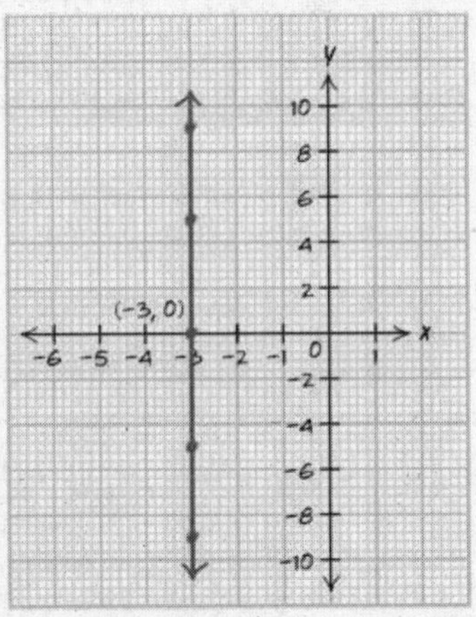

12. Complete the table as in exercise 5:

$x=5$	y	(x,y)
5	−8	(5,−8)
5	−5	(5,−5)
5	0	(5,0)
5	5	(5,5)
5	8	(5,8)

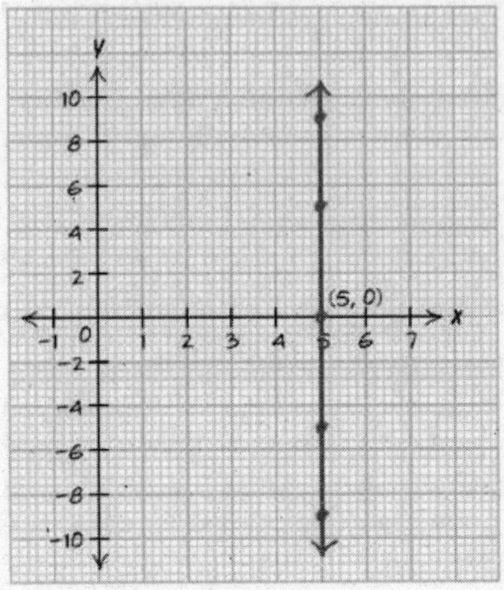

13. Prepare a table of points. Select values for x that are easy to work with, using positive values, negative values, and 0, then calculate the value for y. Graph the points.

x	$y=x^2+2$
−4	18
−3	11
−2	6
−1	3
0	2
1	3
2	6
3	11
4	18

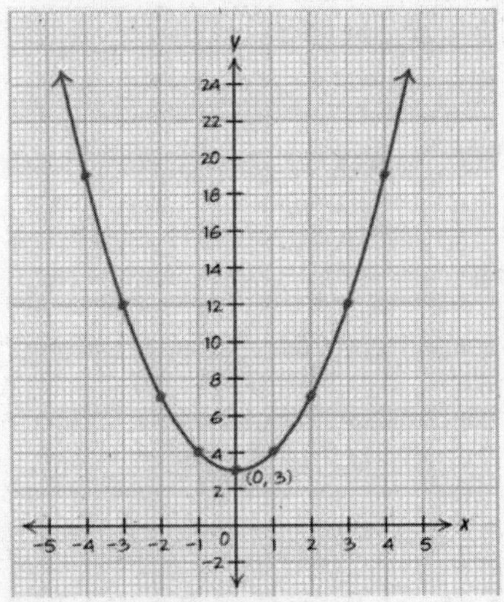

14. Complete the table as in exercise 13:

x	$y=-x^2+5$
−4	−11
−3	−4
−2	1
−1	4
0	5
1	4
2	1
3	−4
4	−11

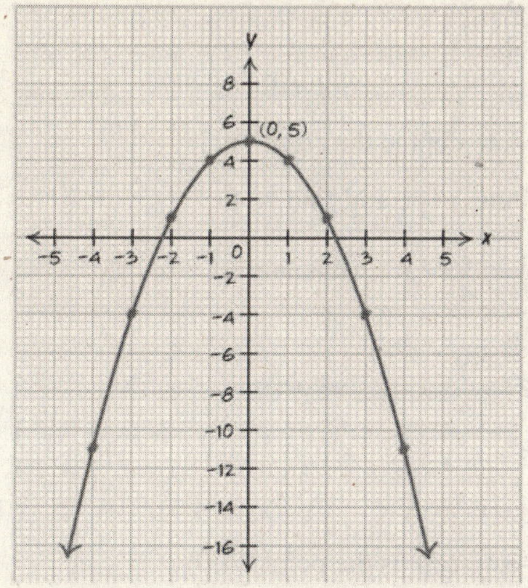

15. Complete the table as in exercise 13:

x	$y=\|9-3x\|$
−1	12
0	9
1	6
2	3
3	0
4	3
5	6
6	9
7	12

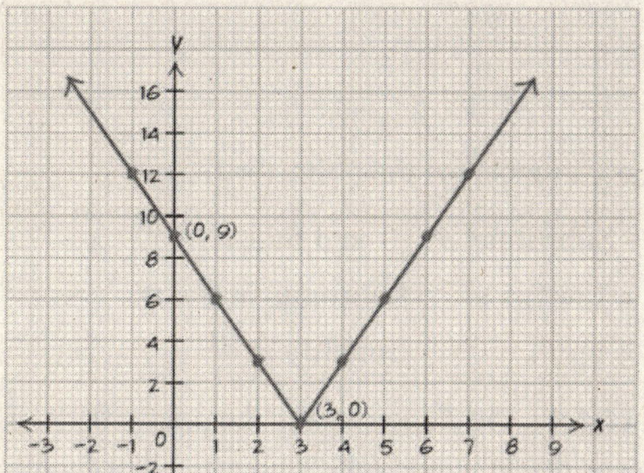

16. Complete the table as in exercise 13:

x	$y=\|x+2\|$
−5	3
−4	2
−3	1
−2	0
−1	1
0	2
1	3
2	4
3	5

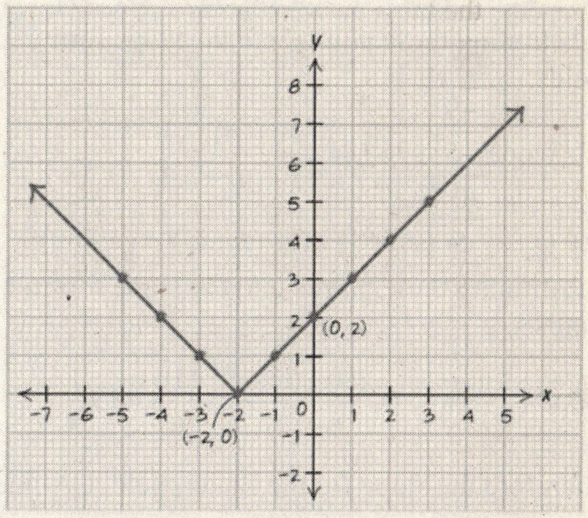

17. **a.** Bakersfield, CA has the larger population than Wichita, KS during these years. The line for Bakersfield is above the line for Wichita.

b. Bakersfield, CA is growing faster than Wichita, KS during these years. The line for Bakersfield is rising at a steeper rate than the line for Wichita.

18. **a.** St. Louis, MO-IL has the larger population than San Antonio, TX during these years. The line for St. Louis is above the line for San Antonio.

b. San Antonio, TX is growing faster than St. Louis, MO-IL during these years. The line for San Antonio is rising at a steeper rate than the line for St. Louis.

19. Using the points $(0,-3)$ and $(6,0)$.

$$\text{slope}=\frac{\text{rise}}{\text{run}}=\frac{3}{6}=\frac{1}{2}$$

20. Using the points $(0,2)$ and $(8,0)$.

$$\text{slope}=\frac{\text{rise}}{\text{run}}=\frac{-2}{8}=-\frac{1}{4}$$

21. Let $(x_1,y_1)=(-3,4)$ and $(x_2,y_2)=(5,-2)$. Substituting these points into the slope formula yields:

$$\text{slope}=\frac{y_2-y_1}{x_2-x_1}=\frac{-2-(4)}{5-(-3)}=\frac{-6}{8}=-\frac{3}{4}$$

22. Let $(x_1, y_1) = (-8, 2)$ and $(x_2, y_2) = (0, 5)$. Substituting these points into the slope formula yields:

$$\text{slope} = \frac{y_2 - y_1}{x_2 - x_1} = \frac{5-(2)}{0-(-8)} = \frac{3}{8}$$

23. Let $(x_1, y_1) = (0, -7)$ and $(x_2, y_2) = (5, -5)$. Substituting these points into the slope formula yields:

$$\text{slope} = \frac{y_2 - y_1}{x_2 - x_1} = \frac{-5-(-7)}{5-(0)} = \frac{2}{5}$$

24. Let $(x_1, y_1) = (0, -4)$ and $(x_2, y_2) = (3, -3)$. Substituting these points into the slope formula yields:

$$\text{slope} = \frac{y_2 - y_1}{x_2 - x_1} = \frac{-3-(-4)}{3-(0)} = \frac{1}{3}$$

25. Let $(x_1, y_1) = (0, 664)$ and $(x_2, y_2) = (6, 780)$. Substituting these points into the slope formula yields:

$$\text{slope} = \frac{y_2 - y_1}{x_2 - x_1} = \frac{780-(664)}{6-(0)} = \frac{116}{6} = 19\frac{1}{3}$$

Therefore the slope for the Bakersfield, CA line is $19\frac{1}{3}$. This means that the population of Bakersfield is increasing at a rate of approximately 19,333 people per year.

26. Let $(x_1, y_1) = (0, 1711)$ and $(x_2, y_2) = (6, 1942)$. Substituting these points into the slope formula yields:

$$\text{slope} = \frac{y_2 - y_1}{x_2 - x_1} = \frac{1942-(1711)}{6-(0)} = \frac{231}{6} = 38\frac{1}{2}$$

Therefore the slope for the San Antonio, TX line is $38\frac{1}{2}$. This means that the population of San Antonio is increasing at a rate of approximately 38,500 people per year.

27. The line intersects the x-axis at -3, therefore the x-intercept is the point $(-3, 0)$. The line intersects the y-axis at -1, therefore the y-intercept is the point $(0, -1)$.

28. The curve intersects the x-axis at -2, 0, 3, and 4. Therefore the x-intercepts are the points $(-2, 0)$, $(0, 0)$, $(3, 0)$, and $(4, 0)$. The line intersects the y-axis at 0, therefore the y-intercept is the point $(0, 0)$.

29. **a.** The vertical intercept for the Bakersfield line is $(0, 664)$. This means that the population of Bakersfield was approximately 664,000 people in the year 2000.

b. The vertical intercept for the Wichita line is $(0, 572)$. This means that the population of Wichita was approximately 572,000 people in the year 2000.

30. **a.** The vertical intercept is $(0, 2000)$. This means that the balance of the credit card debt is \$2000 at the beginning of the year.

b. The horizontal intercept is $(16, 0)$. This means that the balance of the credit card debt was paid down to zero 16 weeks after the beginning of the year.

31. Substitute $x = 0$ and find y.

$y = -\frac{3}{4}(0) + 3$ Substitute in $x = 0$.

$y = 3$ Solve for y.

The y-intercept is the point $(0, 3)$.

Substitute $y = 0$ and find x.

$(0) = -\frac{3}{4}x + 3$ Substitute in $y = 0$.

$$\begin{aligned} -3 \quad & \quad -3 \\ -3 &= -\frac{3}{4}x \quad \text{Solve for } x. \\ -\frac{4}{3} \cdot -3 &= -\frac{3}{4}x \cdot -\frac{4}{3} \\ 4 &= x \end{aligned}$$

The x-intercept is the point $(4, 0)$.

32. Substitute $x = 0$ and find y.

$-6(0) + 3y = -18$ Substitute in $x = 0$.

$3y = -18$ Solve for y.

$$\frac{3y}{3} = \frac{-18}{3}$$

$$y = -6$$

The y-intercept is the point $(0, -6)$.

Substitute $y = 0$ and find x.

$-6x + 3(0) = -18$ Substitute in $x = 0$.

$-6x = -18$ Solve for y.

$$\frac{-6x}{-6} = \frac{-18}{-6}$$

$$x = 3$$

The x-intercept is the point $(3, 0)$.

33. **a.** Substitute $d = 0$ and find T.

$T = 1.5(0) + 1$ Substitute in $d = 0$.

$T = 1$ Simplify.

Therefore, the T-intercept is the point $(0, 1)$. This means that it would take 1 hour to paint 0 dolls. In this problem situation, it means that in addition to the time it takes to paint each doll, there is an additional hour of work to do.

b. Substitute $T = 0$ and solve for d.

$(0) = 1.5d + 1$	Substitute in $T = 0$.
$-1 = 1.5d$	Solve for d.
$\frac{-1}{1.5} = \frac{1.5d}{1.5}$	
$d = -\frac{2}{3}$	

Therefore, the d-intercept is the point $\left(-\frac{2}{3}, 0\right)$. This means that it would take 0 hours to paint $-\frac{2}{3}$ dolls. This does not make sense in terms of the problem situation.

34. **a.** Substitute $d = 0$ and find C.

$C = 25(0) + 10$	Substitute in $d = 0$.
$C = 10$	Simplify.

Therefore, the C-intercept is the point $(0,10)$. This means that it would cost \$10 to produce 0 dolls. This represents that there is a base cost of \$10 to produce the dolls regardless of how many dolls are painted.

b. Substitute $C = 0$ and solve for d.

$(0) = 25d + 10$	Substitute in $C = 0$.
$-10 = 25d$	Solve for d.
$\frac{-10}{25} = \frac{25d}{25}$	
$d = -\frac{2}{5}$	

Therefore, the d-intercept is the point $\left(-\frac{2}{5}, 0\right)$. This means that it would cost nothing to paint $-\frac{2}{5}$ dolls. This does not make sense in terms of the problem situation.

35. Substitute $y = 0$ and find x.

$0 = -3x + 9$	Substitute in $y = 0$.
$\underline{-9 \quad\; -9}$	Solve for x.
$-9 = -3x$	
$\frac{-9}{-3} = \frac{-3x}{-3}$	
$x = 3$	

The x-intercept is the point $(3,0)$.

To find the y-intercept and slope, look at the equation in slope-intercept form. The y-intercept is the point (0, b) and the slope is the coefficient of the x term.

The y-intercept is $(0,9)$ and the slope is -3.

36. Substitute $y = 0$ and find x.

$0 = 2x + 12$	Substitute in $y = 0$.
$\underline{-12 \quad -12}$	Solve for x.
$-12 = 2x$	
$\frac{-12}{2} = \frac{2x}{2}$	
$x = -6$	

The x-intercept is the point $(-6,0)$.

To find the y-intercept and slope, look at the equation in slope-intercept form. The y-intercept is the point (0, b) and the slope is the coefficient of the x term.

The y-intercept is $(0,12)$ and the slope is 2.

37. Substitute $y = 0$ and find x.

$0 = 1.005x - 7.112$	Substitute in $y = 0$.
$\underline{+7.112 \qquad +7.112}$	Solve for x.
$7.112 = 1.005x$	
$\frac{7.112}{1.005} = \frac{1.005x}{1.005}$	
$7.077 \approx x$	

The x-intercept is the point $(7.077,0)$.

To find the y-intercept and slope, look at the equation in slope-intercept form. The y-intercept is the point (0, b) and the slope is the coefficient of the x term.

The y-intercept is $(0,-7.112)$ and the slope is 1.005.

38. Substitute $y = 0$ and find x.

$0 = -0.25x + 9.75$	Substitute in $y = 0$.
$\underline{-9.75 \qquad -9.75}$	Solve for x.
$-9.75 = -0.25x$	
$\frac{-9.75}{-0.25} = \frac{-0.25x}{-0.25}$	
$39 = x$	

The x-intercept is the point $(39,0)$.

To find the y-intercept and slope, look at the equation in slope-intercept form. The y-intercept is the point (0, b) and the slope is the coefficient of the x term.

The y-intercept is $(0,9.75)$ and the slope is -0.25.

39. Substitute $y = 0$ and find x.

$4x + 8(0) = -16$	Substitute in $y = 0$.
$4x = -16$	Solve for x.
$\frac{4x}{4} = \frac{-16}{4}$	
$x = -4$	

The x-intercept is the point $(-4,0)$.

Change the equation to slope-intercept form.

$$\begin{aligned} 4x+8y&=-16 \quad \text{Isolate } y \text{ term on left side.}\\ -4x\quad &\quad -4x \\ 8y&=-4x-16\\ \frac{8y}{8}&=\frac{-4x-16}{8}\\ y&=\frac{-4x}{8}-\frac{16}{8}\\ y&=-\frac{1}{2}x-2 \end{aligned}$$

Now identify the y-intercept which is the point (0, b) and the slope which is the coefficient of the x term.

The y-intercept is $(0,-2)$ and the slope is $-\frac{1}{2}$.

40. Substitute $y=0$ and find x.

$$\begin{aligned} -x-3(0)&=12 \quad \text{Substitute in } y=0.\\ -x&=12 \quad \text{Solve for } x.\\ -1\cdot -x&=12\cdot -1\\ x&=-12 \end{aligned}$$

The x-intercept is the point $(-12,0)$.

Change the equation to slope-intercept form.

$$\begin{aligned} -x-3y&=12 \quad \text{Isolate } y \text{ term on left side.}\\ +x\quad &\quad +x\\ -3y&=x+12\\ \frac{-3y}{-3}&=\frac{x+12}{-3}\\ y&=\frac{x}{-3}+\frac{12}{-3}\\ y&=-\frac{1}{3}x-4 \end{aligned}$$

Now identify the y-intercept which is the point (0, b) and the slope which is the coefficient of the x term.

The y-intercept is $(0,-4)$ and the slope is $-\frac{1}{3}$.

41. Substitute $y=0$ and find x.

$$\begin{aligned} 4x-5(0)&=-20 \quad \text{Substitute in } y=0.\\ 4x&=-20 \quad \text{Solve for } x.\\ \frac{4x}{4}&=\frac{-20}{4}\\ x&=-5 \end{aligned}$$

The x-intercept is the point $(-5,0)$.

Change the equation to slope-intercept form.

$$\begin{aligned} 4x-5y&=-20 \quad \text{Isolate } y \text{ term on left side.}\\ -4x\quad &\quad -4x\\ -5y&=-4x-20\\ \frac{-5y}{-5}&=\frac{-4x-20}{-5}\\ y&=\frac{-4x}{-5}-\frac{20}{-5}\\ y&=\frac{4}{5}x+4 \end{aligned}$$

Now identify the y-intercept which is the point (0, b) and the slope which is the coefficient of the x term.

The y-intercept is $(0,4)$ and the slope is $\frac{4}{5}$.

42. Substitute $y=0$ and find x.

$$\begin{aligned} 2x-10(0)&=-20 \quad \text{Substitute in } y=0.\\ 2x&=-20 \quad \text{Solve for } x.\\ \frac{2x}{2}&=\frac{-20}{2}\\ x&=-10 \end{aligned}$$

The x-intercept is the point $(-10,0)$.

Change the equation to slope-intercept form.

$$\begin{aligned} 2x-10y&=-20 \quad \text{Isolate } y \text{ term on left side.}\\ -2x\quad &\quad -2x\\ -10y&=-2x-20\\ \frac{-10y}{-10}&=\frac{-2x-20}{-10}\\ y&=\frac{1}{5}x+2 \end{aligned}$$

Now identify the y-intercept which is the point (0, b) and the slope which is the coefficient of the x term.

The y-intercept is $(0,2)$ and the slope is $\frac{1}{5}$.

43. Using slope-intercept form, the slope is the coefficient of the t term and the P-intercept, or vertical intercept, is the point (0, b).

a. The slope is 7.4 thousand, or 7,400 people per year. This means that the population of Alaska is increasing at a rate of approximately 7,400 people per year.

b. The P-intercept is $(0,626)$. This means that the population of Alaska was approximately 626 thousand, or 626,000 people in the year 2000.

44. Using slope-intercept form, the slope is the coefficient of the s term and the C-intercept, or vertical intercept, is the point (0, b).

a. The slope is 2.5. This means that the cost increases by $2.50 for every set of clamps produced.

b. The C-intercept is $(0,200)$. This means that there is a base cost of $200 regardless of how many sets of clamps are produced.

45. Slope is $m=\frac{1}{4}=\frac{rise}{run}$ and the y-intercept is the point $(0,-5)$. Start by plotting the y-intercept, then use the slope to rise 1 and run 4 from the y-intercept.

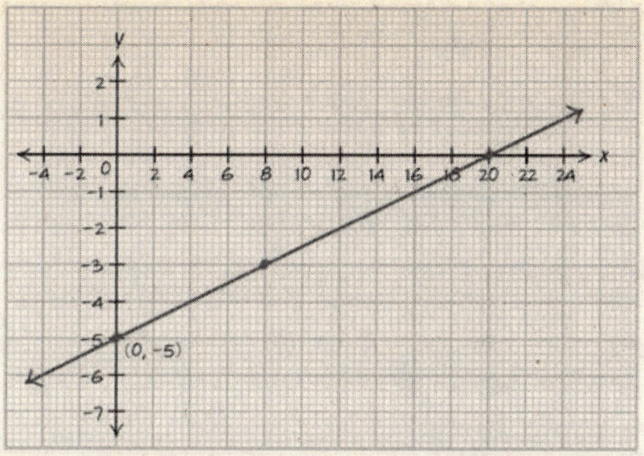

46. Graph using the slope and y-intercept as in exercise 45.

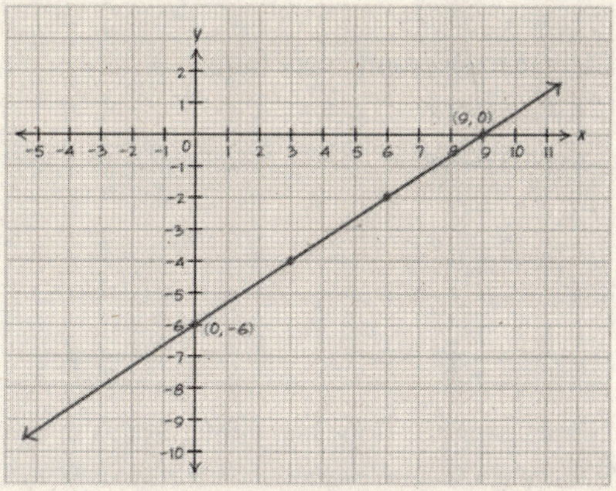

47. Graph using the slope and y-intercept as in exercise 45.

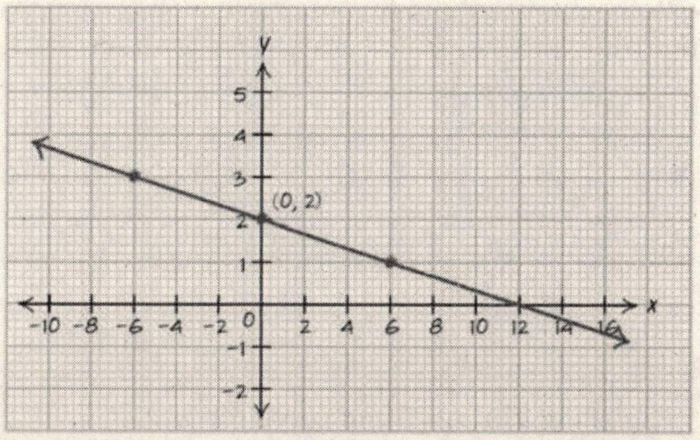

48. Graph using the slope and y-intercept as in exercise 45.

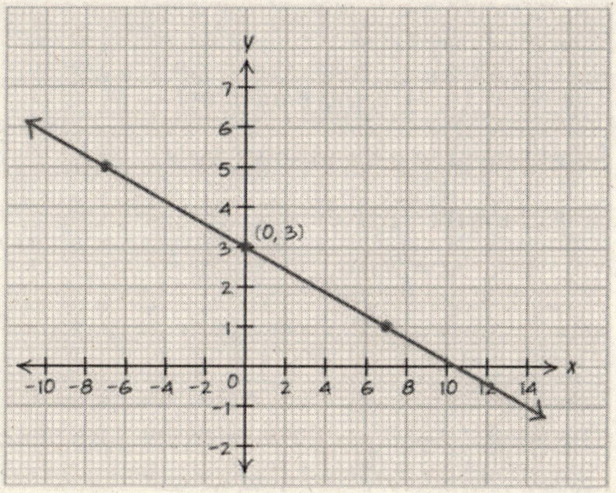

49. Graph using the slope and y-intercept as in exercise 45.

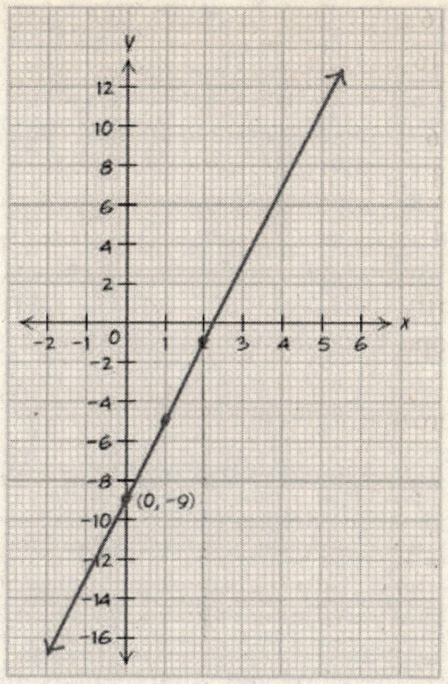

50. Graph using the slope and y-intercept as in exercise 45.

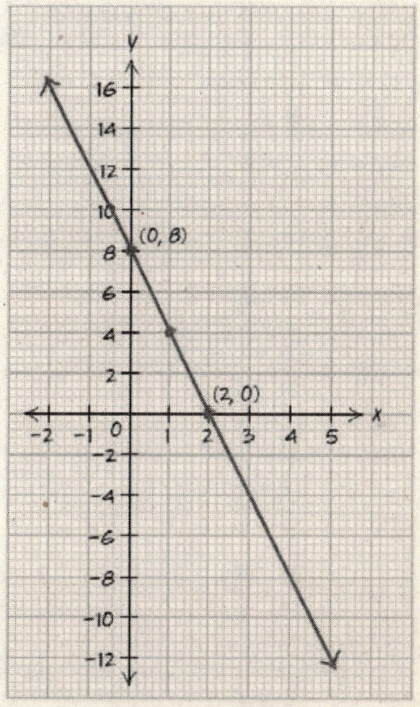

51. Change to slope-intercept form.

$y+3=\frac{1}{2}(x+12)$ Distribute on the right side.

$y+3=\frac{1}{2}x+6$ Solve for y.

$\underline{-3 \qquad -3}$

$y=\frac{1}{2}x+3$

Graph using the slope and y-intercept as in exercise 45.

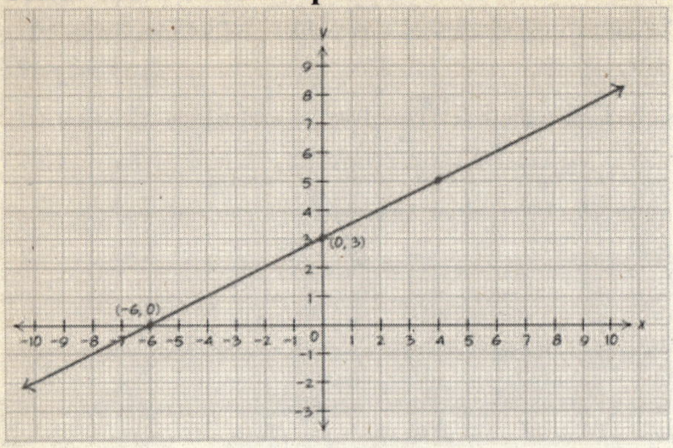

52. Change to slope-intercept form.

$y-5=-3(x+4)$ Distribute on the right side.

$y-5=-3x-12$ Solve for y.

$\underline{+5 \qquad\quad +5}$

$y=-3x-7$

Graph using the slope and y-intercept as in exercise 45.

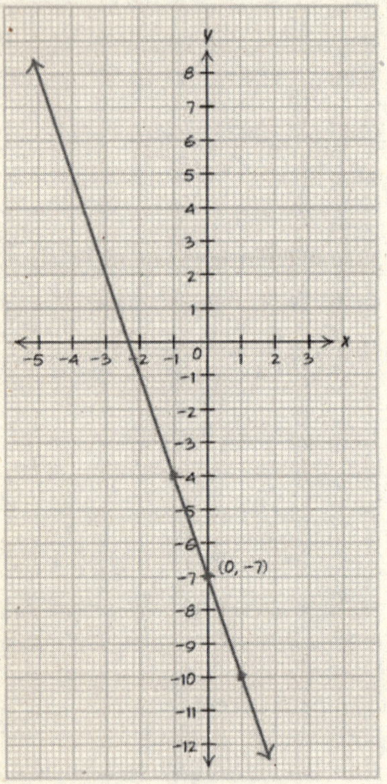

53. Change to slope-intercept form.

$3y=2x-30$ Solve for y.

$$\frac{3y}{3}=\frac{2x-30}{3}$$

$$y=\frac{2x}{3}-\frac{30}{3}$$

$$y=\frac{2}{3}x-10$$

Graph using the slope and y-intercept as in exercise 45.

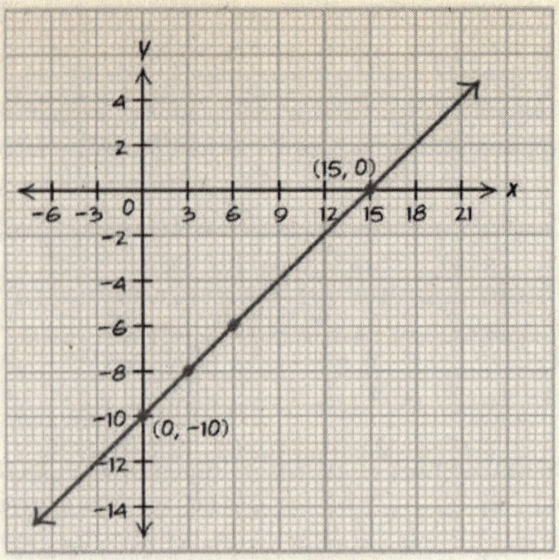

54. Change to slope-intercept form.

$-6y=3x-18$ Solve for y.

$$\frac{-6y}{-6}=\frac{3x-18}{-6}$$

$$y=\frac{3x}{-6}-\frac{18}{-6}$$

$$y=-\frac{1}{2}x+3$$

Graph using the slope and y-intercept as in exercise 45.

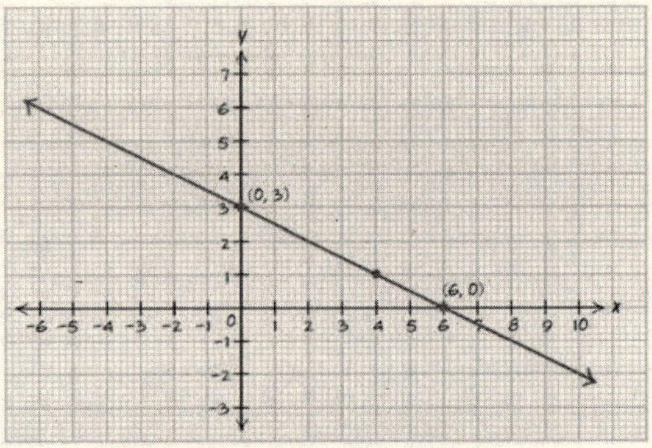

55. Substitute in $x=0$ and solve for y.

$$3(0)+4y=28$$
$$4y=28$$
$$y=7$$

The point $(0,7)$ is the y-intercept.

Substitute in $y=0$ and solve for x.

$$3x+4(0)=28$$
$$3x=28$$
$$x=9\frac{1}{3}$$

The point $\left(9\frac{1}{3},0\right)$ is the x-intercept. Use these two points to graph the line.

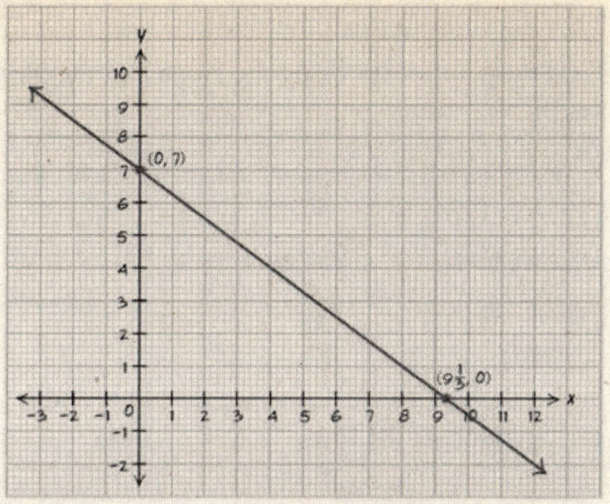

The point $(9,0)$ is the x-intercept. Use these two points to graph the line.

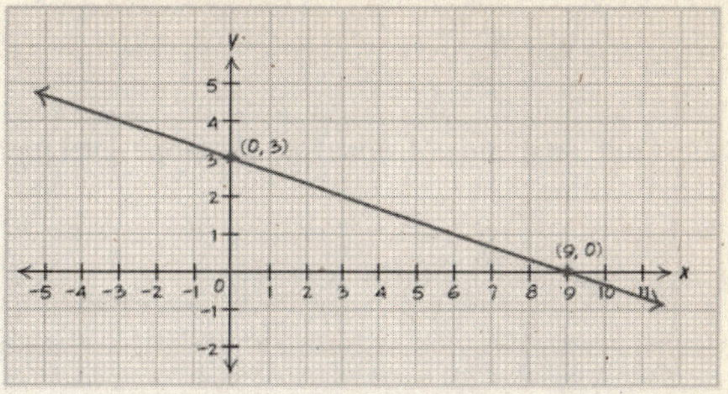

56. Substitute in $x=0$ and solve for y.

$$\begin{aligned}2(0)+5y&=30\\5y&=30\\y&=6\end{aligned}$$

The point $(0,6)$ is the y-intercept.

Substitute in $y=0$ and solve for x.

$$\begin{aligned}2x+5(0)&=30\\2x&=30\\x&=15\end{aligned}$$

The point $(15,0)$ is the x-intercept. Use these two points to graph the line.

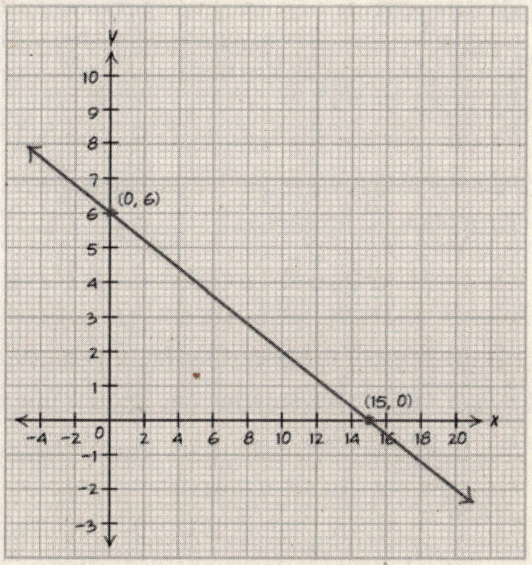

57. Substitute in $x=0$ and solve for y.

$$\begin{aligned}(0)+3y&=9\\3y&=9\\y&=3\end{aligned}$$

The point $(0,3)$ is the y-intercept.

Substitute in $y=0$ and solve for x.

$$\begin{aligned}x+3(0)&=9\\x&=9\end{aligned}$$

58. Substitute in $x=0$ and solve for y.

$$\begin{aligned}-(0)+6y&=24\\6y&=24\\y&=4\end{aligned}$$

The point $(0,4)$ is the y-intercept.

Substitute in $y=0$ and solve for x.

$$\begin{aligned}-x+6(0)&=24\\-x&=24\\-1\cdot -x&=24\cdot -1\\x&=-24\end{aligned}$$

The point $(-24,0)$ is the x-intercept. Use these two points to graph the line.

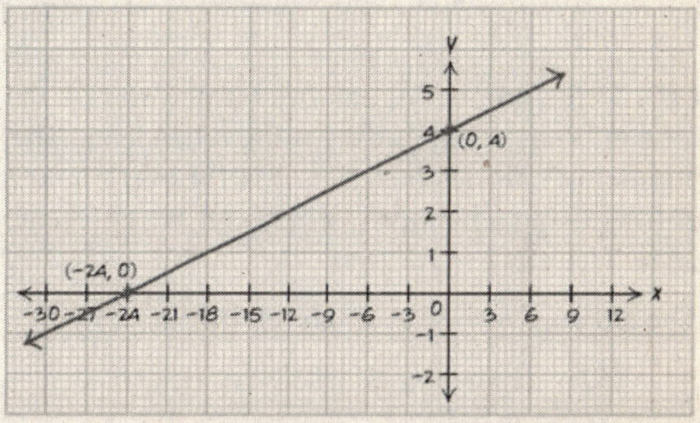

59. To find the slope, use the points $(-3,15)$ and $(0,3)$.

$$m=\frac{rise}{run}=\frac{-12}{3}=-4$$

Substitute this value of slope into the slope-intercept form of a line.

$$\begin{aligned}y&=mx+b\\y&=-4x+b\end{aligned}$$

The point where the graph crosses the y-axis is the y-intercept $(0, b)$. We know that point is $(0,3)$ so $b=3$. Substituting this value into our equation yields:

$$y=-4x+b \qquad \text{Substitute } b=3.$$
$$y=-4x+3$$

Therefore, the equation of the line is $y=-4x+3$.

60. To find the slope, use the points $(-3,-8)$ and $(6,-2)$.

$$m = \frac{rise}{run} = \frac{6}{9} = \frac{2}{3}$$

Substitute this value of slope into the slope-intercept form of a line.

$$y = mx + b \quad \text{Substitute } m = \frac{2}{3}.$$

$$y = \frac{2}{3}x + b$$

The point where the graph crosses the y-axis is the y-intercept (0, b). Looking at the graph, that point is $(0,-6)$ so $b = -6$. Substituting this value into our equation yields:

$$y = \frac{2}{3}x + b \quad \text{Substitute } b = -6.$$

$$y = \frac{2}{3}x - 6$$

Therefore, the equation of the line is $y = \frac{2}{3}x - 6$.

61. Let C = the number of calories burned and t = the time in hours spent running. The number of calories burned is 852 for every hour t of running at a pace of 8 mph, which can be represented by $852t$. Therefore, $C = 852t$.

62. Let C = the number of calories burned and t = the time in hours spent golfing using a handcart. The number of calories burned is 264 for every hour t of golfing using a handcart, which can be represented by $264t$. Therefore, $C = 264t$

63. The slope is $m = 852$. Interpret the slope:

$$m = \frac{852}{1} = \frac{\text{change in } C}{\text{change in } t} = \frac{852 \text{ calories}}{1 \text{ hour}}$$

Therefore, the slope of 852 means that a 160-170 lb person will burn 852 calories for every hour spent running (at a pace of 8 mph).

64. The slope is $m = 264$. Interpret the slope:

$$m = \frac{264}{1} = \frac{\text{change in } C}{\text{change in } t} = \frac{264 \text{ calories}}{1 \text{ hour}}$$

Therefore, the slope of 264 means that a 160-170 lb person will burn 264 calories for every hour spent golfing using a handcart.

65. Find the slope of the line using $(x_1, y_1) = (6,2)$ and $(x_2, y_2) = (11,-3)$. Substituting these points into the slope formula yields:

$$\text{slope} = \frac{y_2 - y_1}{x_2 - x_1} = \frac{-3-(2)}{11-(6)} = \frac{-5}{5} = -1$$

Therefore the slope of the line is -1.

Find the y-intercept using the slope and the point $(6,2)$.

$$y = mx + b \quad \text{Substitute } m = -1,\ x = 6,\ y = 2.$$

$$(2) = (-1)(6) + b$$

$$2 = -6 + b \quad \text{Solve for } b.$$

$$\underline{+6 \quad +6}$$

$$8 = b$$

Using the values we found for the slope and b yields the equation of the line

$$y = -x + 8$$

66. Find the slope of the line using $(x_1, y_1) = (6,1)$ and $(x_2, y_2) = (9,-1)$. Substituting these points into the slope formula yields:

$$\text{slope} = \frac{y_2 - y_1}{x_2 - x_1} = \frac{-1-(1)}{9-(6)} = \frac{-2}{3} = -\frac{2}{3}.$$

Therefore the slope of the line is $-\frac{2}{3}$.

Find the y-intercept using the value for slope and the point $(6,1)$.

$$y = mx + b \quad \text{Substitute } m = -\frac{2}{3},\ x = 6,\ y = 1.$$

$$(1) = (-\frac{2}{3})(6) + b$$

$$1 = -4 + b \quad \text{Solve for } b.$$

$$\underline{+4 \quad +4}$$

$$5 = b$$

Using the values we found for the slope and b yields the equation of the line

$$y = -\frac{2}{3}x + 5$$

67. Find the slope of the line using $(x_1, y_1) = (-10,4)$ and $(x_2, y_2) = (0,5)$. Substituting these points into the slope formula yields:

$$\text{slope} = \frac{y_2 - y_1}{x_2 - x_1} = \frac{5-(4)}{0-(-10)} = \frac{1}{10}$$

Therefore the slope of the line is $\frac{1}{10}$.

Find the y-intercept using the value for slope and the point $(0,5)$.

$$y = mx + b \quad \text{Substitute } m = \frac{1}{10},\ x = 0,\ y = 5.$$

$$(5) = (\frac{1}{10})(0) + b$$

$$5 = 0 + b \quad \text{Solve for } b.$$

$$5 = b$$

Using the values we found for the slope and b yields the equation of the line

$$y=\frac{1}{10}x+5$$

68. Find the slope of the line using $(x_1,y_1)=(-2,-8)$ and $(x_2,y_2)=(2,-6)$. Substituting these points into the slope formula yields:

$$\text{slope}=\frac{y_2-y_1}{x_2-x_1}=\frac{-6-(-8)}{2-(-2)}=\frac{2}{4}=\frac{1}{2}$$

Therefore the slope of the line is $\frac{1}{2}$.

Find the y-intercept using the value for slope and the point $(2,-6)$.

$y=mx+b$ Substitute $m=\frac{1}{2}$, $x=2, y=-6$.

$(-6)=(\frac{1}{2})(2)+b$

$-6=1+b$ Solve for b.

$\underline{-1\ \ -1}$

$-7=b$

Using the values we found for the slope and b yields the equation of the line

$$y=\frac{1}{2}x-7$$

69. Find the slope of the line using $(x_1,y_1)=(3,-11)$ and $(x_2,y_2)=(-2,-11)$. Substituting these points into the slope formula yields:

$$\text{slope}=\frac{y_2-y_1}{x_2-x_1}=\frac{-11-(-11)}{-2-(3)}=\frac{0}{-5}=0$$

Therefore the slope of the line is 0.

Find the y-intercept using the value for slope and the point $(3,-11)$.

$y=mx+b$ Substitute $m=0$, $x=3, y=-11$.

$(-11)=(0)(3)+b$

$-11=0+b$ Solve for b.

$-11=b$

Using the values we found for the slope and b yields the equation of the line

$$y=0x-11$$
$$y=-11$$

70. Find the slope of the line using $(x_1,y_1)=(-10,12)$ and $(x_2,y_2)=(16,12)$. Substituting these points into the slope formula yields:

$$\text{slope}=\frac{y_2-y_1}{x_2-x_1}=\frac{12-(12)}{16-(-10)}=\frac{0}{26}=0$$

Therefore the slope of the line is 0.

Find the y-intercept using the value for slope and the point $(-10,12)$.

$y=mx+b$ Substitute $m=0$, $x=-10, y=12$.

$(12)=(0)(-10)+b$

$12=0+b$ Solve for b.

$12=b$

Using the values we found for the slope and b yields the equation of the line

$$y=0x+12$$
$$y=12$$

71. Find the slope of the line using $(x_1,y_1)=(1,-8)$ and $(x_2,y_2)=(1,7)$. Substituting these points into the slope formula yields:

$$m=\frac{y_2-y_1}{x_2-x_1}=\frac{7-(-8)}{1-(1)}=\frac{15}{0}=\textit{undefined}$$

We know that a line with an undefined slope is a vertical line that intersects only the x-axis in one point. Looking at the given points $(1,-8)$ and $(1,7)$, we notice that $x=1$ for both points. If we were to graph these two points and connect them with a line, it would be a vertical line intersecting the x-axis at the point $(1,0)$, therefore the equation of the line is $x=1$.

72. Find the slope of the line $(x_1,y_1)=(-4,-3)$ and $(x_2,y_2)=(-4,17)$. Substituting these points into the slope formula yields:

$$m=\frac{y_2-y_1}{x_2-x_1}=\frac{17-(-3)}{-4-(-4)}=\frac{20}{0}=\textit{undefined}$$

We know that a line with an undefined slope is a vertical line that intersects only the x-axis in one point. Looking at the given points $(-4,-3)$ and $(-4,17)$, we notice that $x=-4$ for both points. If we were to graph these two points and connect them with a line, it would be a vertical line intersecting the x-axis at the point $(-4,0)$, therefore the equation of the line is $x=-4$.

73. We have that $m=\frac{1}{2}$ and $(x_1,y_1)=(-6,3)$.

Substituting these values into the point-slope form of a line yields

$y-y_1=m(x-x_1)$ Substitute in $m=\frac{1}{2}$ and

$y-(3)=\frac{1}{2}(x-(-6))$ $(x_1,y_1)=(-6,3)$

$y-3=\frac{1}{2}x-(\frac{1}{2})(-6)$ Distribute and simplify.

$y-3=\frac{1}{2}x+3$ Solve for y.

$+3 \quad +3$

$y=\frac{1}{2}x+6$

Therefore, the equation of the line is $y=\frac{1}{2}x+6$.

74. We have that $m=\frac{-1}{3}$ and $(x_1,y_1)=(9,-1)$.

Substituting these values into the point-slope form of a line yields

$y-y_1=m(x-x_1)$ Substitute in $m=\frac{-1}{3}$ and

$y-(-1)=\frac{-1}{3}(x-9)$ $(x_1,y_1)=(9,-1)$

$y+1=-\frac{1}{3}x-(\frac{-1}{3})(9)$ Distribute and simplify.

$y+1=-\frac{1}{3}x+3$ Solve for y.

$-1 \quad -1$

$y=-\frac{1}{3}x+2$

Therefore, the equation of the line is $y=-\frac{1}{3}x+2$.

75. Determine the slope of the line.

$$m=\frac{y_2-y_1}{x_2-x_1}=\frac{-1-(5)}{4-(-4)}=\frac{-6}{8}=-\frac{3}{4}$$

Substituting the slope and the point $(4,-1)$ into the point-slope form of a line yields:

$y-y_1=m(x-x_1)$ Substitute in $m=-\frac{3}{4}$ and

$y-(-1)=-\frac{3}{4}(x-4)$ $(x_1,y_1)=(4,-1)$

$y+1=-\frac{3}{4}x-(-\frac{3}{4})(4)$ Distribute and simplify.

$y+1=-\frac{3}{4}x+3$ Solve for y.

$-1 \quad -1$

$y=-\frac{3}{4}x+2$

Therefore, the equation of the line is $y=-\frac{3}{4}x+2$.

76. Determine the slope of the line.

$$m=\frac{y_2-y_1}{x_2-x_1}=\frac{2-(4)}{6-(-6)}=\frac{-2}{12}=-\frac{1}{6}$$

Substituting the slope and the point $(6,2)$ into the point-slope form of a line yields:

$y-y_1=m(x-x_1)$ Substitute in $m=-\frac{1}{6}$ and

$y-(2)=-\frac{1}{6}(x-6)$ $(x_1,y_1)=(6,2)$

$y-2=-\frac{1}{6}x-(-\frac{1}{6})(6)$ Distribute and simplify.

$y-2=-\frac{1}{6}x+1$ Solve for y.

$+2 \quad +2$

$y=-\frac{1}{6}x+3$

Therefore, the equation of the line is $y=-\frac{1}{6}x+3$.

77. To determine whether the two lines are parallel or perpendicular, we will compare the slopes. The first equation $y=-3x+7$ is in slope-intercept form so we can see that the slope $m=-3$. The second equation is in general form so we will first change it to slope-intercept form:

$6x+2y=11$ Isolate y term on left side.

$-6x \quad -6x$

$2y=-6x+11$

$\frac{2y}{2}=\frac{-6x+11}{2}$

$y=-3x+5\frac{1}{2}$

With the second equation in slope-intercept form, $y=-3x+5\frac{1}{2}$, we can now see the slope $m=-3$.

Both lines have the same slope of -3 therefore the lines are parallel.

78. To determine whether the two lines are parallel or perpendicular, we will compare the slopes. The first equation $y=\frac{1}{3}x-13$ is in slope-intercept form so we can see that the slope $m=\frac{1}{3}$. The second equation is in general form so we will first change it to slope-intercept form:

$5x-15y=19$ Isolate y term on left side.

$-5x \quad -5x$

$-15y=-5x+19$

$\frac{-15y}{-15}=\frac{-5x+19}{-15}$

$y=\frac{1}{3}x-1\frac{4}{15}$

With the second equation in slope-intercept form, $y=\frac{1}{3}x-1\frac{4}{15}$, we can now see the slope $m=\frac{1}{3}$.

Both lines have the same slope of $\frac{1}{3}$ therefore the lines are parallel.

79. To determine whether the two lines are parallel or perpendicular, we will compare the slopes. The first equation is in general form so we will first change it to slope-intercept form:

$$5x+2y=6 \quad \text{Isolate } y \text{ term on left side.}$$
$$\underline{-5x \quad -5x}$$
$$2y=-5x+6$$
$$\frac{2y}{2}=\frac{-5x+6}{2}$$
$$y=-\frac{5}{2}x+3$$

With the second equation in slope-intercept form, $y=-\frac{5}{2}x+3$, we can now see that the slope $m=-\frac{5}{2}$. The second equation $y=\frac{2}{5}x-1$ is already in slope-intercept form so we can see the slope $m=\frac{2}{5}$. The two slopes $-\frac{5}{2}$ and $\frac{2}{5}$ are negative reciprocals. Therefore the lines are perpendicular.

80. To determine whether the two lines are parallel or perpendicular, we will compare the slopes. The first equation $y=-3x+9$ is in slope-intercept form so we can see that the slope $m=-3$. The second equation is in general form so we will first change it to slope-intercept form:

$$x-3y=-15 \quad \text{Isolate } y \text{ term on left side.}$$
$$\underline{-x \quad -x}$$
$$-3y=-x-15$$
$$\frac{-3y}{-3}=\frac{-x-15}{-3}$$
$$y=\frac{1}{3}x+5$$

With the second equation in slope-intercept form, $y=\frac{1}{3}x+5$, we can now see the slope $m=\frac{1}{3}$. The two slopes -3 and $\frac{1}{3}$ are negative reciprocals. Therefore the lines are perpendicular.

81. Parallel lines have the same slope. Change the equation to slope-intercept form.

$$-x+8y=1 \quad \text{Isolate } y \text{ term on left side.}$$
$$\underline{+x \quad +x}$$
$$8y=x+1$$
$$\frac{8y}{8}=\frac{x+1}{8}$$
$$y=\frac{1}{8}x+\frac{1}{8}$$

The slope $m=\frac{1}{8}$. The slope of the line that we are trying to find is the same. Substituting $m=\frac{1}{8}$ into slope-intercept form yields

$$y=mx+b \quad \text{Substitute in } m=\frac{1}{8}.$$
$$y=\frac{1}{8}x+b$$

To find the value of b, substitute in the given point $(-8,3)$ and solve for b.

$$y=\frac{1}{8}x+b \quad \text{Substitute in the given point } (-8,3).$$
$$3=\frac{1}{8}(-8)+b \quad \text{Simplify the right side.}$$
$$3=-1+b \quad \text{Solve for } b.$$
$$4=b$$

$$y=\frac{1}{8}x+4 \quad \text{The equation of the line.}$$

82. Parallel lines have the same slope. Change the equation to slope-intercept form.

$$5x-2y=7 \quad \text{Isolate } y \text{ term on left side.}$$
$$\underline{-5x \quad -5x}$$
$$-2y=-5x+7$$
$$\frac{-2y}{-2}=\frac{-5x+7}{-2}$$
$$y=\frac{5}{2}x-3\frac{1}{2}$$

The slope $m=\frac{5}{2}$. The slope of the line that we are trying to find is the same. Substituting $m=\frac{5}{2}$ into slope-intercept form yields

$$y=mx+b \quad \text{Substitute in } m=\frac{5}{2}.$$
$$y=\frac{5}{2}x+b$$

To find the value of b, substitute in the given point $(10,-1)$ and solve for b.

$$y=\frac{5}{2}x+b \quad \text{Substitute in the given point } (10,-1).$$
$$-1=\frac{5}{2}(10)+b \quad \text{Simplify the right side.}$$
$$-1=25+b \quad \text{Solve for } b.$$
$$-26=b$$

$$y=\frac{5}{2}x-26 \quad \text{The equation of the line.}$$

83. Perpendicular lines have slopes that are negative reciprocals. The slope of the given line is $m=\frac{-4}{3}$.

The negative reciprocal of $\frac{-4}{3}$ is $\frac{3}{4}$. Substituting $m=\frac{3}{4}$ into slope-intercept form yields

$y=mx+b$ Substitute in $m=\frac{3}{4}$.

$y=\frac{3}{4}x+b$

To find the value of b, substitute in the given point $(12,-5)$ and solve for b.

$y=\frac{3}{4}x+b$ Substitute in the given point (12, –5).
$-5=\frac{3}{4}(12)+b$ Simplify the right side.
$-5=9+b$ Solve for b.
$-14=b$

$y=\frac{3}{4}x-14$ The equation of the line.

84. Perpendicular lines have slopes that are negative reciprocals. The slope of the given line is $m=-\frac{1}{2}$.

The negative reciprocal of $-\frac{1}{2}$ is 2. Substituting $m=2$ into slope-intercept form yields

$y=mx+b$ Substitute in $m=2$.
$y=2x+b$

To find the value of b, substitute in the given point $(-3,-7)$ and solve for b.

$y=2x+b$ Substitute in the given point (–3, –7).
$-7=2(-3)+b$ Simplify the right side.
$-7=-6+b$ Solve for b.
$-1=b$

$y=2x-1$ The equation of the line.

85. Change to general form.

$y=\frac{-5}{6}x+7$ Multiply each term by 6 to
$6\cdot y=\frac{-5}{6}(6)x+7\cdot 6$ clear denominator.
$6y=-5x+42$
$+5x\quad +5x$ Group x and y on left side.
$5x+6y=42$

86. Change to general form.

$y=\frac{-4}{5}x-9$

$y=\frac{-4}{5}x-9$ Multiply each term by 5 to
$5\cdot y=\frac{-4}{5}(5)x-9\cdot 5$ clear denominator.
$5y=-4x-45$
$+4x\quad +4x$ Group x and y on left side.
$4x+5y=-45$

87. Change to slope-intercept form.

$-6x+5y=-30$ Isolate y term on left side.
$+6x\qquad +6x$
$5y=6x-30$
$\frac{5y}{5}=\frac{6x-30}{5}$
$y=\frac{6}{5}x-6$

88. Change to slope-intercept form.

$5x-10y=-100$ Isolate y term on left side.
$-5x\qquad -5x$
$-10y=-5x-100$
$\frac{-10y}{-10}=\frac{-5x-100}{-10}$
$y=\frac{1}{2}x+10$

89. Change to general form.

$y-1=-2(x+4)$ Evaluate parentheses first.
$y-1=-2x-8$ Group x and y on left side.
$+2x\quad +2x$
$2x+y-1=-8$ Isolate constants on right side.
$+1\quad +1$
$2x+y=-7$

90. Change to general form.

$y+3=-7(x-5)$ Evaluate parentheses first.
$y+3=-7x+35$ Group x and y on left side.
$+7x\quad +7x$
$7x+y+3=35$ Isolate constants on right side.
$-3\quad -3$
$7x+y=32$

91. Change to slope-intercept form.

$y+4=\frac{2}{3}(x-6)$ Evaluate parentheses first.
$y+4=\frac{2}{3}x-4$ Isolate y term on left side.
$-4\qquad -4$
$y=\frac{2}{3}x-8$

92. Change to slope-intercept form.

$y-7=\frac{-5}{7}(x+14)$ Evaluate parentheses first.
$y-7=\frac{-5}{7}x-10$ Isolate y term on left side.
$+7\qquad +7$
$y=-\frac{5}{7}x-3$

Chapter 3 Test

1. Use the given equation to create and fill in a table, select values for x and calculate the value for y.

x	$y=-5x+2$	(x,y)
–2	12	(–2,12)
–1	7	(–1,7)
0	2	(0,2)
1	–3	(1,–3)
2	–8	(2,–8)

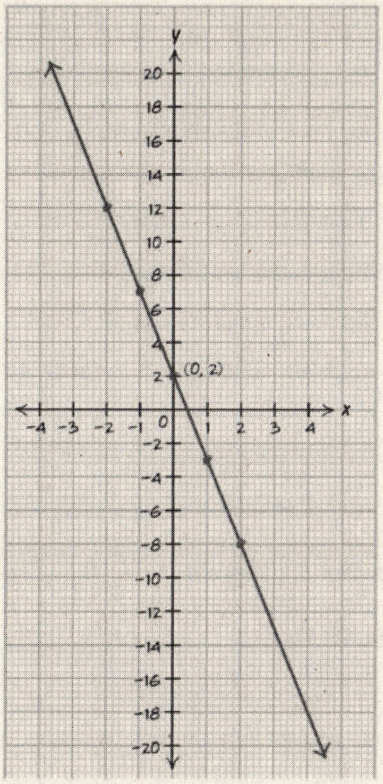

2. Prepare a table of points. Select values for x and calculate the value for y. For the values of x, select values that are easy to work with, using positive values, negative values, and 0. Use the coordinates to graph the points and connect them with a smooth curve.

x	$y=x^2-1$
–4	15
–3	8
–2	3
–1	0
0	–1
1	0
2	3
3	8
4	15

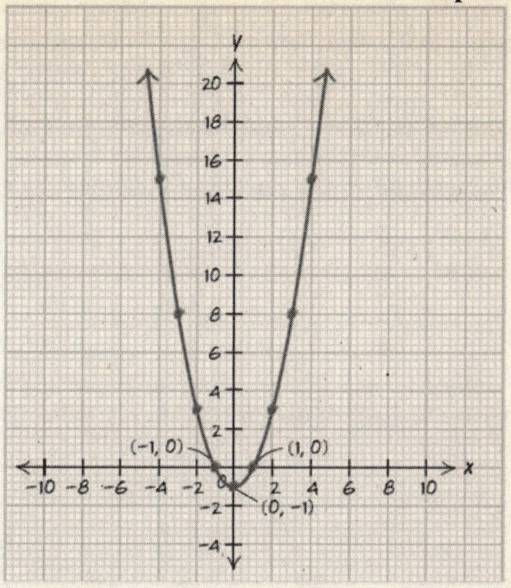

3. Let $(x_1,y_1)=(3,7)$ and $(x_2,y_2)=(-4,0)$. Substituting these points into the slope formula yields:

$$\text{slope}=\frac{y_2-y_1}{x_2-x_1}=\frac{0-(7)}{-4-(3)}=\frac{-7}{-7}=1$$

4. Let $(x_1,y_1)=(1999,66043)$ and $(x_2,y_2)=(2003,141022)$. Substituting these points into the slope formula yields:

$$\text{slope}=\frac{y_2-y_1}{x_2-x_1}=\frac{141022-(66043)}{2003-(1999)}$$
$$=\frac{74979}{4}=18744\frac{3}{4}$$

This means that the number of children in the United States diagnosed with autism is growing by about 18,745 children per year.

5. To find the y-intercept we will substitute zero for x and find y.

$$-2(0)+3y=-6 \quad \text{Substitute in } x=0.$$
$$3y=-6 \quad \text{Solve for } y.$$
$$y=-2$$

Therefore, the y-intercept is the point $(0,-2)$.

To find the x-intercept, substitute zero for y and find x.

$$-2x+3(0)=-6 \quad \text{Substitute in } y=0.$$
$$-2x=-6 \quad \text{Solve for } x.$$
$$x=3$$

Therefore, the x-intercept is the point $(3,0)$.

6. **a.** The vertical intercept is $(0,5.6)$ which represents that 5.6 billion kilowatt hours of electricity were generated by wind in 2000.

b. The vertical intercept is $(0,14)$ which represents that 14 billion kilowatt hours of electricity were generated by geothermal sources in the year 2000.

7. Looking at the equation $y=\frac{3}{5}x-\frac{2}{5}$ we see that the y-intercept is $\left(0,-\frac{2}{5}\right)$ and the slope is $\frac{3}{5}$. To find the x-intercept, substitute zero for y and solve for x.

$$0=\frac{3}{5}x-\frac{2}{5} \quad \text{Substitute in } y=0.$$

$$+\frac{2}{5} \qquad +\frac{2}{5}$$

$$\frac{2}{5}=\frac{3}{5}x \quad \text{Solve for } x.$$

$$\frac{5}{3}\cdot\frac{2}{5}=\frac{3}{5}x\cdot\frac{5}{3}$$

$$x=\frac{2}{3}$$

The x-intercept is the point $\left(\frac{2}{3},0\right)$.

8. The slope is $m=-\frac{1}{2}=\frac{rise}{run}$ and the y-intercept is the point $(0,7)$. Start by plotting the y-intercept, then use the slope to rise -1 and run 2 from the y-intercept. Since the rise is negative, we go down 1 instead of up.

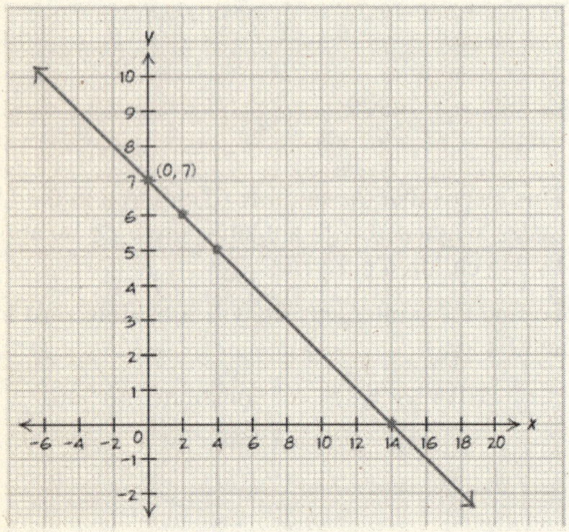

9. The slope is 12and the W-intercept is $(0,0)$. Start by plotting the W-intercept, then use the slope to rise 12 and run 1 from the W-intercept.

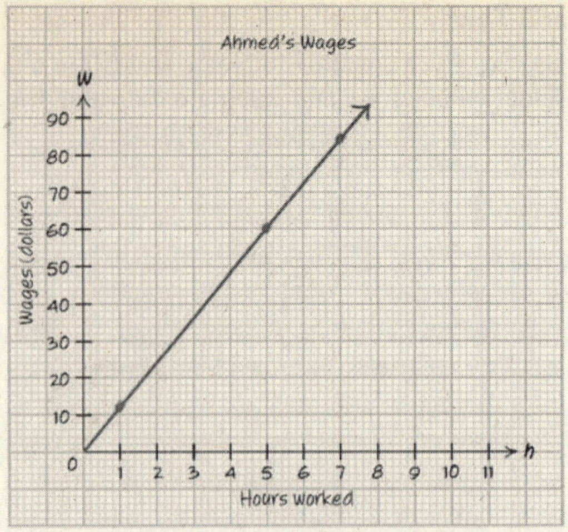

10. Substitute in $x=0$ and solve for y.

$$4(0)+y=-7$$

$$y=-7$$

The y-intercept is the point $(0,-7)$.

Substitute in $y=0$ and solve for x.

$$4x+(0)=-7$$

$$4x=-7$$

$$x=-1\frac{3}{4}$$

The x-intercept is the point $\left(-1\frac{3}{4},0\right)$. Use these two points to graph the line.

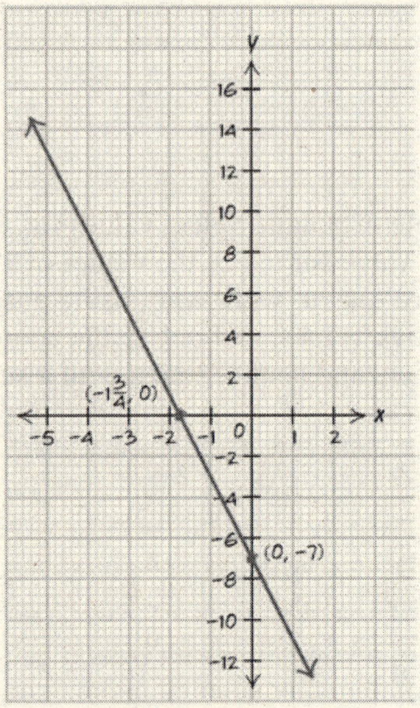

11. **a.** This equation has the variable x raised to the second power, so this is not a linear equation.

b. This equation can be simplified to slope-intercept form or general form, so it is a linear equation. Change to slope-intercept form.

$$y-2=3(x+4)$$
$$y-2=3x+12$$
$$\underline{\quad +2 \qquad +2}$$
$$y=3x+14$$

12. Find the slope using $(x_1, y_1)=(1,-2)$ and $(x_2, y_2)=(-2,-17)$. Substituting these points into the slope formula yields:

$$\text{slope}=\frac{y_2-y_1}{x_2-x_1}=\frac{-17-(-2)}{-2-(1)}=\frac{-15}{-3}=5$$

Therefore the slope of the line is 5.

Find the y-intercept using the value for slope and the point $(1,-2)$.

$y=mx+b$ Substitute $m=5,\ x=1,\ y=-2$.

$(-2)=(5)(1)+b$

$-2=5+b$ Solve for b.

$\underline{-5 \quad -5}$

$-7=b$

Using the values we found for the slope and b yields the equation of the line

$$y=5x-7$$

13. Find the slope using the points $(-6,1)$ and $(2,3)$.

$$m=\frac{rise}{run}=\frac{2}{8}=\frac{1}{4}$$

Substitute this value of slope into the slope-intercept form of a line.

$$y=mx+b$$
$$y=\frac{1}{4}x+b$$

The point where the graph crosses the y-axis is the y-intercept (0, b). Now we will substitute the slope and the point $(2,3)$ into the equation to find the y-intercept.

$y=\frac{1}{4}x+b$ Substitute $x=2,\ y=3$.

$(3)=\frac{1}{4}(2)+b$

$3=\frac{1}{2}+b$

$2\frac{1}{2}=b$

Now substituting the slope and $b=2\frac{1}{2}$ into our equation yields:

$y=\frac{1}{4}x+b$ Substitute $b=2\frac{1}{2}$.

$y=\frac{1}{4}x+2\frac{1}{2}$

Therefore, the equation of the line is $y=\frac{1}{4}x+2\frac{1}{2}$.

14. Use the points $(0,235)$ and $(5,210)$ to find the slope of line.

$$m=\frac{y_2-y_1}{x_2-x_1}=\frac{W_2-W_1}{t_2-t_1}=\frac{210-235}{5-0}=\frac{-25}{5}=-5$$

Substitute this value of slope into the slope-intercept form of a line:

$y=mx+b$ Slope-intercept form of a line.

$W=mt+b$ Use the variables from the problem.

$W=-5t+b$

To find the y-intercept, (0, b), substitute either point into the equation. Selecting $(0,235)$ and substituting it into the equation yields:

$W=-5t+b$

$(235)=-5(0)+b$ Substitute $t=0,\ W=235$.

$235=0+b$ Solve for b.

$235=b$

Using the values we found for the slope and b yields the equation of the line

$$W=-5t+b$$
$$W=-5t+235$$

15. The slope is $m=-5$. Interpret the slope:

$$m=\frac{-5}{1}=\frac{\text{change in } W}{\text{change in } t}=\frac{-5 \text{ pounds}}{1 \text{ week}}$$

Therefore, the slope of -5 means that Paul is losing weight at a rate of 5 pounds per week.

16. Determine the slope of the line.

$$m=\frac{y_2-y_1}{x_2-x_1}=\frac{-11-(-2)}{-8-(4)}=\frac{-9}{-12}=\frac{3}{4}$$

Substituting the slope and the point $(4,-2)$, into the point-slope form of a line yields:

$y-y_1=m(x-x_1)$ Substitute in $m=\frac{3}{4}$ and

$y-(-2)=\frac{3}{4}(x-4)$ $(x_1,y_1)=(4,-2)$

$y+2=\frac{3}{4}x-(\frac{3}{4})(4)$ Distribute and simplify.

$y+2=\frac{3}{4}x-3$ Solve for y.

$\underline{-2 \qquad -2}$

$y=\frac{3}{4}x-5$

Therefore, the equation of the line is $y=\frac{3}{4}x-5$.

17. To determine whether the two lines are parallel or perpendicular, we will compare the slopes. The first equation $y=-7x+11$ is in slope-intercept form so we can see that the slope $m=-7$. The second equation is in general form so we will first change it to slope-intercept form:

$2x-14y=3$ Isolate y term on left side.

$\underline{-2x \quad -2x}$

$-14y=-2x+3$

$\dfrac{-14y}{-14}=\dfrac{-2x+3}{-14}$

$y=\dfrac{1}{7}x-\dfrac{3}{14}$

The slope is $m=\dfrac{1}{7}$. The two slopes -7 and $\dfrac{1}{7}$ are negative reciprocals. Therefore the lines are perpendicular.

18. Parallel lines have the same slope. The slope of the given line is $m=\dfrac{2}{5}$. The slope of the line that we are trying to find is the same. Substituting $m=\dfrac{2}{5}$ into slope-intercept form yields

$y=mx+b$ Substitute in $m=\dfrac{2}{5}$.

$y=\dfrac{2}{5}x+b$

To find the value of *b*, substitute in the given point $(-5,-3)$ and solve for *b*.

$y=\dfrac{2}{5}x+b$ Substitute in the given point $(-5,-3)$.

$-3=\dfrac{2}{5}(-5)+b$ Simplify the right side.

$-3=-2+b$ Solve for *b*.

$-1=b$

$y=\dfrac{2}{5}x-1$ The equation of the line.

CHAPTER 4 SYSTEMS OF LINEAR EQUATIONS

Section 4.1 Identifying Systems of Linear Equations

1. There is one input and output pair that is the same, $(-5,6)$. Therefore, this point is the solution to the system.

3. There is one input and output pair that is the same, $(2,6)$. This point is the solution.

5. There is one input and output pair that is the same, $(0,3)$. This point is the solution.

7. There is one input and output pair that is the same, $(9,1)$. This point is the solution.

9. **a.** Substitute $x=-3$ and $C=2$ into both equations to see if both are true.

First equation:

$(2)\overset{?}{=}3(-3)-11$ Substitute values for x and C.

$2\overset{?}{=}-9-11$ Simplify.

$2\neq-20$ This is a false statement.

Second equation:

$(2)\overset{?}{=}-2(-3)+14$ Substitute values for x and C.

$2\overset{?}{=}6+14$ Simplify.

$2\neq 20$ This is a false statement.

The point is not on either line. It is *not* a solution of the system.

b. Substitute $x=5$ and $C=4$ into both equations to see if both are true.

First equation:

$(4)\overset{?}{=}3(5)-11$ Substitute values for x and C.

$4\overset{?}{=}15-11$ Simplify.

$4=4$ This is a true statement.

Second equation:

$(4)\overset{?}{=}-2(5)+14$ Substitute values for x and C.

$4\overset{?}{=}-10+14$ Simplify.

$4=4$ This is a true statement.

Since both statements are true, $(5,4)$ is a solution of the system.

11. **a.** Substitute $a=0$ and $b=3$ into both equations to see if both are true.

First equation:

$5(0)+3(3)\overset{?}{=}-15$ Substitute values for a and b.

$0+9\overset{?}{=}-15$ Simplify.

$9\neq-15$ This is a false statement.

Second equation:

$(0)+3(3)\overset{?}{=}3$ Substitute values for a and b.

$0+9\overset{?}{=}3$ Simplify.

$9\neq 3$ This is a false statement.

The point is not on either line. It is *not* a solution of the system.

b. Substitute $a=3$ and $b=0$ into both equations to see if both are true.

First equation:

$5(3)+3(0)\overset{?}{=}-15$ Substitute values for a and b.

$15+3\overset{?}{=}-15$ Simplify.

$18\neq-15$ This is a false statement.

Second equation:

$(3)+3(0)\overset{?}{=}3$ Substitute values for a and b.

$3+0\overset{?}{=}3$ Simplify.

$3=3$ This is a true statement.

The point is on the second line but not on the first. It is *not* a solution of the system.

13. **a.** Substitute $x=0$ and $y=-8$ into both equations to see if both are true.

First equation:

$-12(0)+3(-8)\overset{?}{=}-24$ Substitute values for x and y.

$0-24\overset{?}{=}-24$ Simplify.

$-24=-24$ This is a true statement.

Second equation:

$4(0)-(-8)\overset{?}{=}8$ Substitute values for x and y.

$0+8\overset{?}{=}8$ Simplify.

$8=8$ This is a true statement.

Since both statements are true, $(0,-8)$ is a solution of the system.

b. Substitute $x=1$ and $y=-4$ into both equations to see if both are true.

First equation:

$-12(1)+3(-4)\overset{?}{=}-24$ Substitute values for x and y.

$-12-12\overset{?}{=}-24$ Simplify.

$-24=-24$ This is a true statement.

Second equation:

$4(1)-(-4)\overset{?}{=}8$ Substitute values for x and y.

$4+4\overset{?}{=}8$ Simplify.

$8=8$ This is a true statement.

Since both statements are true, $(1,-4)$ is a solution of the system.

15. **a.** Substitute $x=6$ and $y=0$ into both equations to see if both are true.

First equation:

$-2(6)+(0)\overset{?}{=}-12$ Substitute values for x and y.

$-12+0\overset{?}{=}-12$

$-12=-12$ This is a true statement.

Second equation:

$(0)\stackrel{?}{=}2(6)+16$ Substitute values for x and y.

$0\stackrel{?}{=}12+16$ Simplify.

$0\neq 28$ This is a false statement.

The point is on the first line but not on the second. It is *not* a solution of the system.

b. Substitute $x=-8$ and $y=0$ into both equations to see if both are true.

First equation:

$-2(-8)+(0)\stackrel{?}{=}-12$ Substitute values for x and y.

$16+0\stackrel{?}{=}-12$

$16\neq -12$ This is a false statement.

Second equation:

$(0)\stackrel{?}{=}2(-8)+16$ Substitute values for x and y.

$0\stackrel{?}{=}-16+16$ Simplify.

$0=0$ This is a true statement.

The point is on the second line but not on the first. It is *not* a solution of the system.

17. **a.** Substitute $x=3$ and $y=1$ into both equations to see if both are true.

First equation:

$(3)\stackrel{?}{=}3$ Substitute value for x.

$3=3$ This is a true statement.

Second equation:

$(1)\stackrel{?}{=}-1$ Substitute value for y.

$1\neq -1$ This is a false statement.

The point is on the first line but not on the second. It is *not* a solution of the system.

b. Substitute $x=3$ and $y=-1$ into both equations to see if both are true.

First equation:

$(3)\stackrel{?}{=}3$ Substitute value for x.

$3=3$ This is a true statement.

Second equation:

$(-1)\stackrel{?}{=}-1$ Substitute value for y.

$-1=-1$ This is a true statement.

Since both statements are true, $(3,-1)$ is a solution of the system.

19. **a.** Substitute $x=1$ and $y=1$ into both equations to see if both are true.

First equation:

$(1)\stackrel{?}{=}-5(1)+6$ Substitute values for x and y.

$1\stackrel{?}{=}-5+6$ Simplify.

$1=1$ This is a true statement.

Second equation:

$5(1)+(1)\stackrel{?}{=}6$ Substitute values for x and y.

$5+1\stackrel{?}{=}6$ Simplify.

$6=6$ This is a true statement.

Since both statements are true, $(1,1)$ is a solution of the system.

b. Substitute $x=2$ and $y=-4$ into both equations to see if both are true.

First equation:

$(-4)\stackrel{?}{=}-5(2)+6$ Substitute values for x and y.

$-4\stackrel{?}{=}-10+6$ Simplify.

$-4=-4$ This is a true statement.

Second equation:

$5(2)+(-4)\stackrel{?}{=}6$ Substitute values for x and y.

$10-4\stackrel{?}{=}6$ Simplify.

$6=6$ This is a true statement.

Since both statements are true, $(2,-4)$ is a solution of the system.

21. Using the graph, we see that the CHD mortality rate for Serbia and Japan are equal at the serum cholesterol level of 125 mg/dL.

23. **a.** Let $t=$ the number of tracks, and let $C=$ the monthly cost. Find the equation for the cost of downloading music from Music OnLine.

$C=0.80t+10$

Find the equation of the cost of downloading music from GotTunes.

$C=1t+8$ or $C=t+8$

b. The completed tables are as follows:

# of tracks downloaded	Monthly Cost for Music OnLine
0	$10.00
2	$11.60
4	$13.20
6	$14.80
8	$16.40
10	$18.00
12	$19.60

# of tracks downloaded	Monthly Cost for GotTunes
0	$8.00
2	$10.00
4	$12.00
6	$14.00
8	$16.00
10	$18.00
12	$20.00

c. Looking at the tables, the number of tracks downloaded that make the costs of the two services equal is 10 tracks. The solution is 10 tracks and cost of \$18.00.

d. If you download an average of 8 tracks per month, the cost for Music OnLine is \$16.40 and the cost for GotTunes is \$16.00. It will be cheaper for you to use GotTunes.

25. **a.** Let $m =$ the number of months, and let $C =$ the total cost. First find the equation for the cost of joining Fitness Plus. $C = 15m + 100$

Find the equation for the cost of joining Rock's Gym. $C = 20m + 80$

b. The completed tables are as follows:

# of months	Cost of Fitness Plus
0	\$100
1	\$115
2	\$130
3	\$145
4	\$160
5	\$175
6	\$190

# of months	Cost of Rock's Gym
0	\$80
1	\$100
2	\$120
3	\$140
4	\$160
5	\$180
6	\$200

c. Looking at the tables, the costs of joining the two fitness clubs are equal at 4 months. The solution is 4 months and cost of \$160.00.

d. If you plan to join for 6 months, the cost for Fitness Plus is \$190.00 and the cost for Rock's Gym is \$200.00. It will be cheaper for you to use Fitness Plus.

27. The point of intersection is at $x = 1$ and $y = 1$, or as an ordered pair, $(1,1)$.

29. The point of intersection is at $x = 100$ and $y = -4$, or as an ordered pair, $(100,-4)$.

31. The point of intersection is at $x = -3$ and $y = 2$, or as an ordered pair, $(-3,2)$.

33. The two lines do not intersect and appear to be parallel. Therefore there is no solution to the system and the system in inconsistent.

35. **a.** Substitute $m = 10$ and $C = 27.85$ into both equations to see if both are true.

First equation:

$(27.85) \stackrel{?}{=} 19.95 + 0.79(10)$ Substitute values for m and C.

$27.85 \stackrel{?}{=} 19.95 + 7.9$

$27.85 = 27.85$ This is a true statement.

Second equation:

$(27.85) \stackrel{?}{=} 21.95 + 0.59(10)$ Substitute values for m and C.

$27.85 \stackrel{?}{=} 21.95 + 5.9$

$27.85 = 27.85$ This is a true statement.

Since both statements are true, $(10, 27.85)$ is a solution of the system.

b. Substitute $m = 15$ and $C = 31.80$ into both equations to see if both are true.

First equation:

$(31.80) \stackrel{?}{=} 19.95 + 0.79(15)$ Substitute values for m and C.

$31.80 \stackrel{?}{=} 19.95 + 11.85$

$31.80 = 31.8$ This is a true statement.

Second equation:

$(31.80) \stackrel{?}{=} 21.95 + 0.59(15)$ Substitute values for m and C.

$31.80 \stackrel{?}{=} 21.95 + 8.85$

$31.80 \neq 30.75$ This is a false statement.

The point is on the first line but not on the second. It is *not* a solution of the system.

37. **a.** Substitute $x = 45$ and $S = 8.45$ into both equations to see if both are true.

First equation:

$(8.45) \stackrel{?}{=} 7.00 + 0.03(45)$ Substitute values for x and S.

$8.45 \stackrel{?}{=} 7.00 + 1.35$

$8.45 \neq 8.35$ This is a false statement.

Second equation:

$(8.45) \stackrel{?}{=} 8.00 + 0.01(45)$ Substitute values for x and S.

$8.45 \stackrel{?}{=} 8.00 + 0.45$

$8.45 = 8.45$ This is a true statement.

The point is on the second line but not on the first. It is *not* a solution of the system.

b. Substitute $x = 50$ and $S = 8.50$ into both equations to see if both are true.

First equation:

$(8.50) \stackrel{?}{=} 7.00 + 0.03(50)$ Substitute values for x and S.

$8.50 \stackrel{?}{=} 7.00 + 1.5$

$8.50 = 8.5$ This is a true statement.

Second equation:

$(8.50)\overset{?}{=}8.00+0.01(50)$ Substitute values for x and S.

$8.50\overset{?}{=}8.00+0.5$

$8.50=8.5$ This is a true statement.

Since both statements are true, $(50,8.50)$ is a solution of the system.

39. Graph both lines.

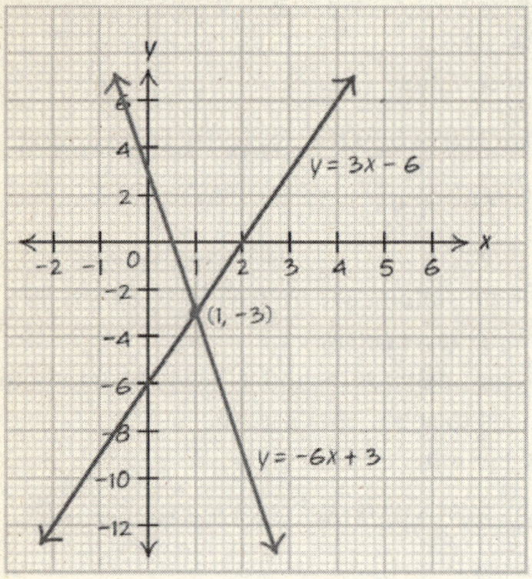

The two lines intersect at the point $(1,-3)$.

Now check the point by substituting it into both equations.

First equation:

$(-3)\overset{?}{=}3(1)-6$ Substitute values for x and y.

$(-3)\overset{?}{=}3-6$

$-3=-3$ This is a true statement.

Second equation:

$(-3)\overset{?}{=}-6(1)+3$ Substitute values for x and y.

$(-3)\overset{?}{=}-6+3$

$-3=-3$ This is a true statement.

Since both statements are true, $(1,-3)$ is a solution of the system. The system is consistent.

41. Graph both lines.

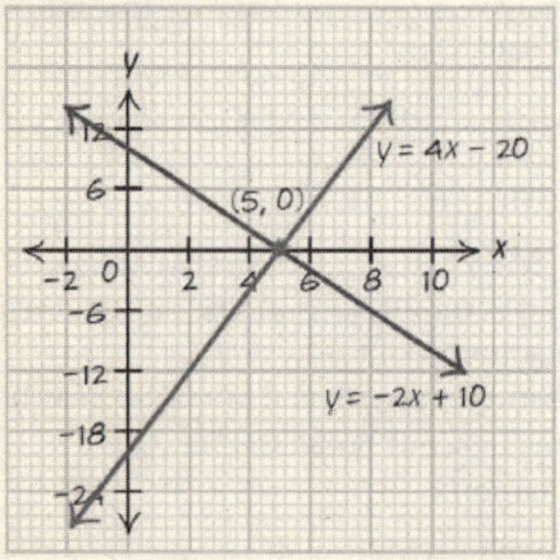

The two lines intersect at the point $(5,0)$. Check the point as in exercise 39. The system is consistent.

43. Graph both lines.

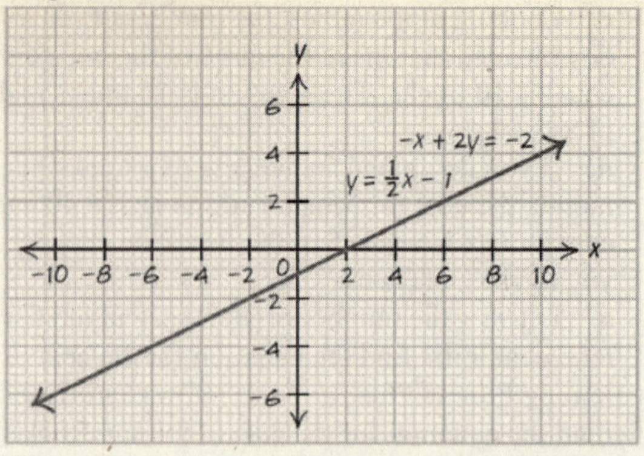

The two linear equations graph as the same line. Therefore, the two equations intersect everywhere and there are infinitely many solutions. This system is consistent.

45. Graph both lines.

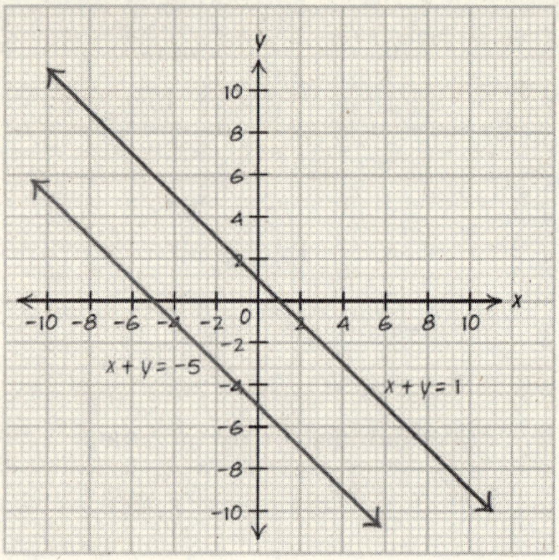

The two lines appear to be parallel on the graph. Solve for slope-intercept form and compare slopes. The two lines have the same slope of -1, therefore they are parallel.

There is no solution to this system of equations. The system is inconsistent.

47. Graph both lines.

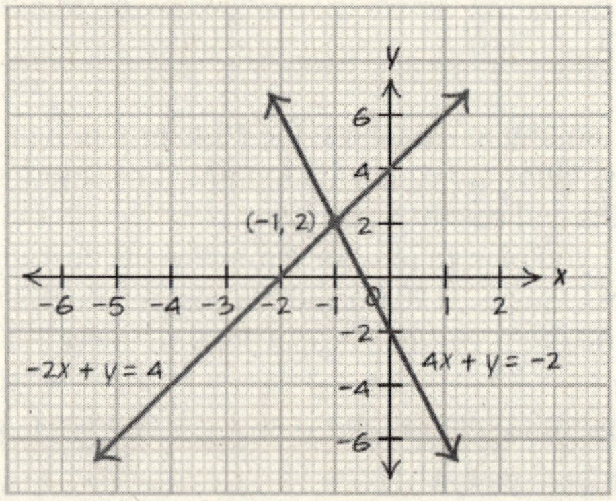

The two lines intersect at the point $(-1,2)$. Check the point as in exercise 39. The system is consistent.

49. Graph both lines.

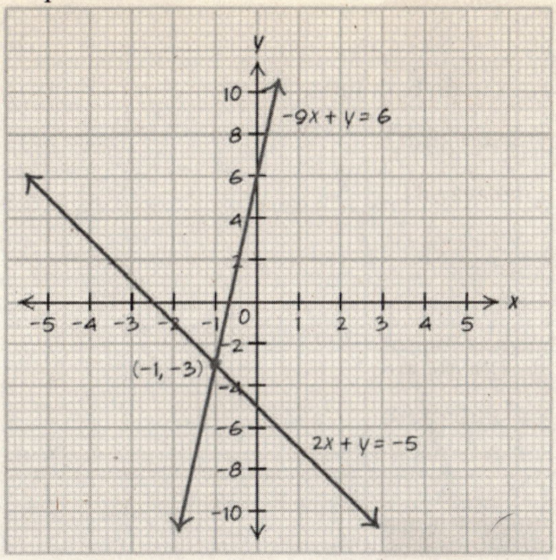

The two lines intersect at the point $(-1,-3)$. Check the point as in exercise 39. The system is consistent.

51. Graph both lines.

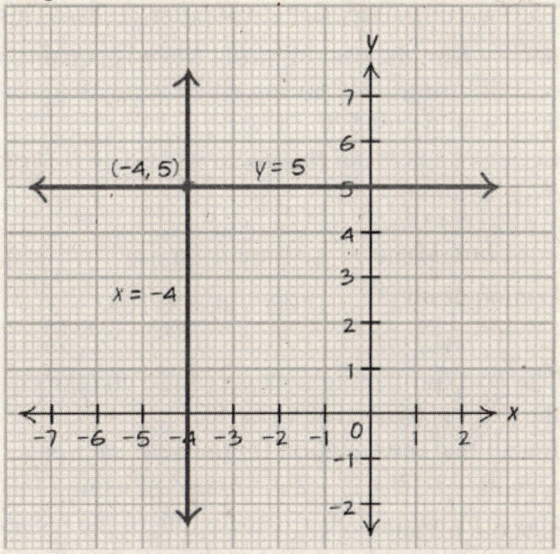

The two lines intersect at the point $(-4,5)$. Check the point as in exercise 39. The system is consistent.

53. Graph both lines.

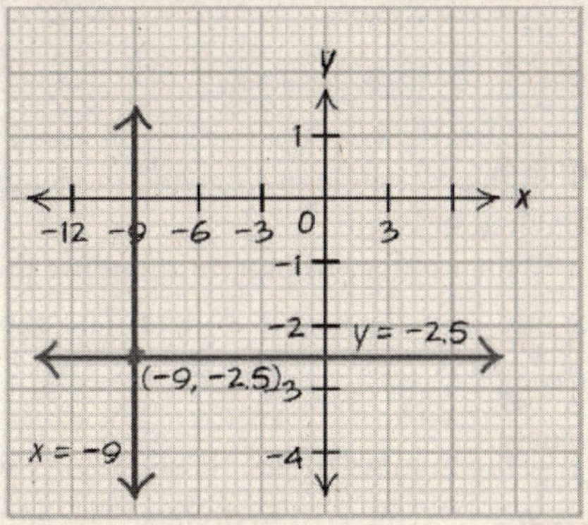

The two lines intersect at the point $(-9,-2.5)$. Check the point as in exercise 39. The system is

55. Graph both lines.

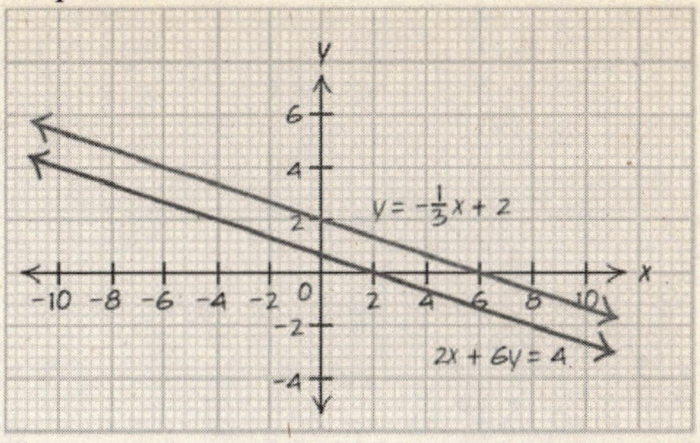

The two lines appear to be parallel on the graph. Solve for slope-intercept form and compare slopes. The two lines have the same slope of $-\frac{1}{3}$, therefore they are parallel.
There is no solution to this system of equations. The system is inconsistent.

57. Graph both lines.

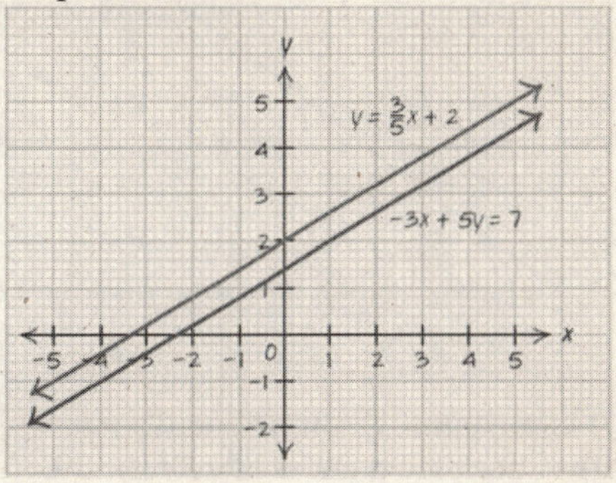

The two lines appear to be parallel on the graph. Solve for slope-intercept form and compare slopes. The two lines have the same slope of $\frac{3}{5}$, therefore they are parallel.
There is no solution to this system of equations. The system is inconsistent.

59. Graph both lines.

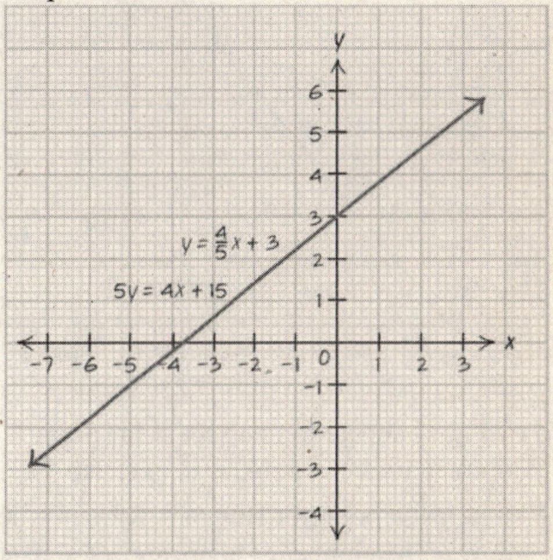

The two linear equations graph as the same line. Therefore, the two equations intersect everywhere and there are infinitely many solutions. This system is consistent.

61. Graph both lines.

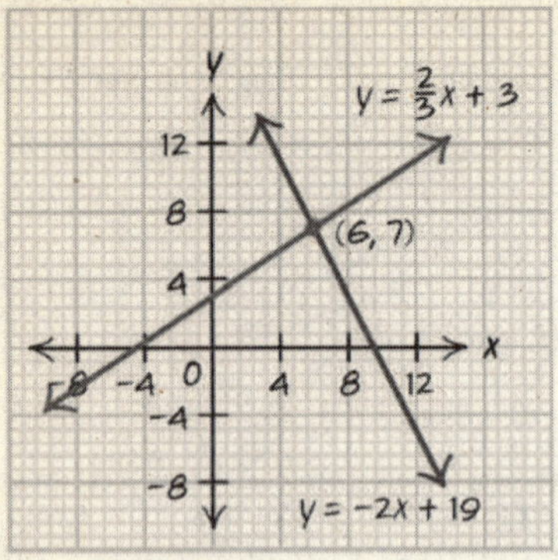

The two lines intersect at the point $(6,7)$. Check the point as in exercise 39. The system is consistent.

63. Graph both lines.

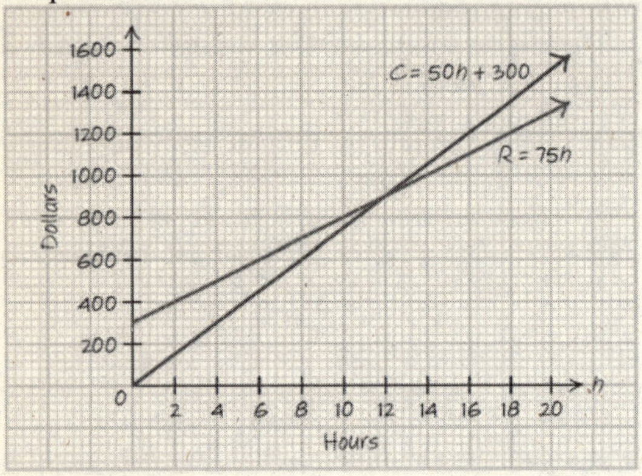

The two lines appear to intersect at the point $(12,900)$. Check the point as in exercise 39.

The break-even point for Jason's mobile auto repair is when he does 12 hours of repair work a week and earns \$900.

65. Graph both lines.

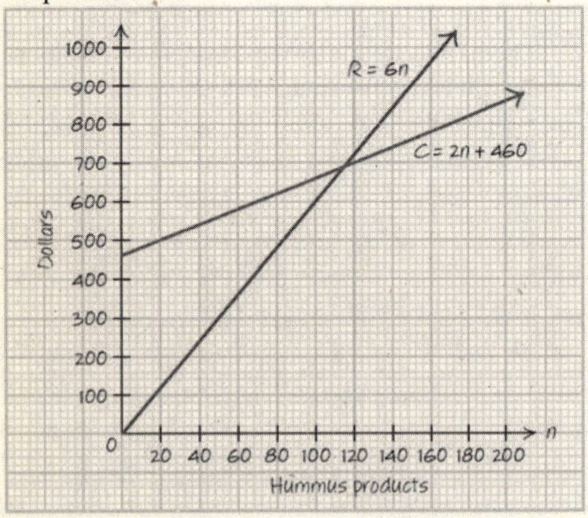

The two lines appear to intersect at the point $(115,690)$. Check the point as in exercise 39.

The break-even point for Hummus R Us is when they

67. **a.** Let $h =$ the number of hours and $C =$ the cost in dollars. The equation for the cost of hiring Mike for one day is

$$C = 60h + 10$$

b. The equation for the cost of hiring Josh for one day is

$$C = 55h + 15$$

c. Graph both lines.

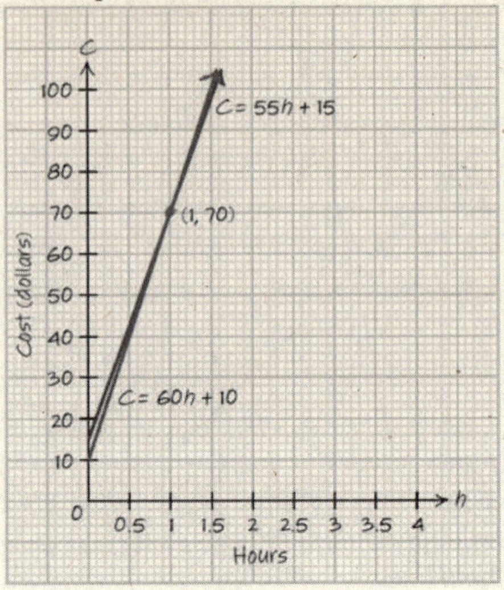

The two lines intersect at the point $(1,70)$. This is the point where the cost of hiring Mike is equal to the cost of hiring Josh. Hiring either handyman for 1 hour would cost \$70.

69. Let $x =$ the cost of the microwave. The cost of the oven would be $2x$.

$x + 2x = 960$

$3x = 960$ Simplify on left side.

$\frac{3x}{3} = \frac{960}{3}$ Divide both sides by 3.

$x = 320$

The cost for the microwave is \$320 and the cost for the oven is twice that, or \$640.

71. Recall that the equation for the perimeter of a rectangle is $P = 2l + 2w$. Instead of using l for length and w for width, let $x =$ the width of the rectangle. The length is 3 times that so the length is $3x$. Substitute these values, along with the given value for perimeter $P = 24$, and solve for x.

$24 = 2(3x) + 2(x)$

$24 = 6x + 2x$ Simplify on right side.

$24 = 8x$

$\frac{24}{8} = \frac{8x}{8}$ Divide both sides by 8.

$3 = x$

The width of the rectangle is 3 cm and the length is three times that, or 9 cm.

73. **a.** Let $p =$ the number of people attending the fundraiser and $C =$ the cost in dollars. The equation for the total cost is

$$C = 5.75p + 255$$

b. Let $p =$ the number of people attending the fundraiser and $R =$ the cost in dollars. The equation for the total revenue is

$$R = 10p$$

c. Graph both lines.

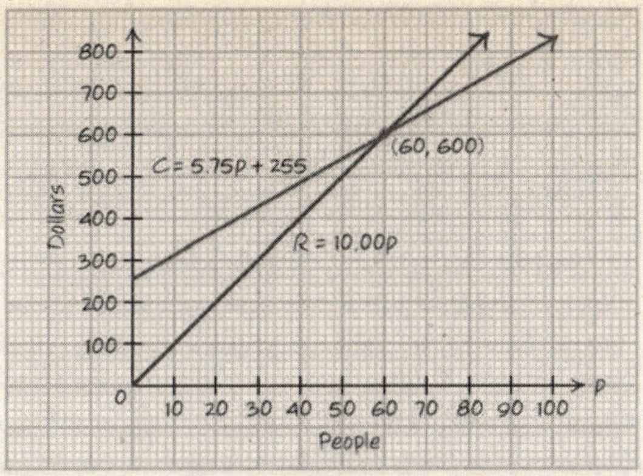

The two lines intersect at the point $(60,600)$. This is when the values for cost and revenue are equal, so this would be the break-even point.

d. The track team would have to sell 60 tickets to break even.

75. These 2 equations both have the same slope of -3, therefore these lines are parallel.

77. These equations have slopes of -7 and $-\frac{1}{7}$ which are reciprocals of one another but not *negative* reciprocals, therefore these lines are neither parallel nor perpendicular.

79. These equations have slopes of $\frac{-5}{6}$ and $\frac{6}{5}$ which are negative reciprocals of one another, therefore the lines are perpendicular.

81. Convert the first equation to slope-intercept form so we can easily see the value for slope: $y=-x+3$. These 2 equations both have the same slope of -1, therefore these lines are parallel.

83. Convert both equations to slope-intercept form so we can easily see the values for slope and compare them.

$$y=-\frac{2}{5}x+1$$

$$y=-\frac{2}{5}x-\frac{1}{5}$$

We can now see that these 2 equations both have the same slope of $-\frac{2}{5}$, therefore these lines are parallel.

85. Two lines can intersect at exactly one point therefore this system will have one solution.

87. Two lines that are parallel will never intersect. This type of system has no solutions.

89. Two linear equations can be the same line. They intersect everywhere on that line. This type of system has infinitely many solutions.

Section 4.2 Solving Systems Using the Substitution Method

1. Solve for y: isolate y on one side of equation.

$3x+\boxed{y}=-4$ Identify the variable term to isolate.

$\underline{-3x \qquad -3x}$ Subtract $3x$ from both sides.

$y=-3x-4$

3. Solve for y: isolate y on one side of equation.

$\boxed{x}+4y=-3$ Identify the variable term to isolate.

$\underline{-4y \quad -4y}$ Subtract $4y$ from both sides.

$x=-4y-3$

5. Solve for y: isolate y on one side of equation.

$5x\boxed{-10y}=30$ Identify the variable term to isolate.

$\underline{-5x \qquad -5x}$ Subtract $5x$ from both sides.

$-10y=-5x+30$

$\frac{-10y}{-10}=\frac{-5x+30}{-10}$ Divide both sides by -10.

$y=\frac{-5x}{-10}+\frac{30}{-10}$

$y=\frac{1}{2}x-3$

7. Solve the second equation for x.

$\boxed{x}+3y=7$ Identify the variable term to isolate.

$\underline{-3y \quad -3y}$ Subtract $3y$ from both sides.

$x=-3y+7$

9. Solve the second equation for a.

$\boxed{-a}+9b=6$ Identify the variable term to isolate.

$\underline{-9b \quad -9b}$ Subtract $9b$ from both sides.

$-a=-9b+6$

$\frac{-a}{-1}=\frac{-9b+6}{-1}$ Divide both sides by -1.

$a=9b-6$

11. The second equation already has y isolated on one side.

$y=-2x+1$

13. Substitute $x=3y-1$ into the second equation.

$2x+y=12$ Substitute in $x=3y-1$ for x.

$2(3y-1)+y=12$ Distribute.

$6y-2+y=12$ Simplify.

$7y-2=12$

$\underline{+2 \quad +2}$

$7y=14$ Solve for y.

$y=2$

$x=3y-1$

$x=3(2)-1$ Substitute in $y=2$ to find x.

$x=5$

$(x,y)=(5,2)$ This is the solution.

Check $(x,y)=(5,2)$ in both equations.

Equation 1	Equation 2
$x=3y-1$	$2x+y=12$
$(5)\overset{?}{=}3(2)-1$	$2(5)+(2)\overset{?}{=}12$
$5\overset{?}{=}6-1$	$10+2\overset{?}{=}12$
$5=5$	$12=12$

Since both equations are true, the solution checks.

15. Substitute $x=-5y+1$ into the second equation.

$-3x+2y=31$ Substitute in $x=-5y+1$ for x.
$-3(-5y+1)+2y=31$ Distribute.
$15y-3+2y=31$ Simplify.
$17y-3=31$
$\underline{\quad +3 \quad +3}$
$17y=34$ Solve for y.
$y=2$
$x=-5y+1$
$x=-5(2)+1$ Substitute in $y=2$ to find x.
$x=-9$
$(x,y)=(-9,2)$ This is the solution.

Check $(x,y)=(-9,2)$ in both equations.

Equation 1	Equation 2
$x=-5y+1$	$-3x+2y=31$
$(-9)\overset{?}{=}-5(2)+1$	$-3(-9)+2(2)\overset{?}{=}31$
$-9\overset{?}{=}-10+1$	$27+4\overset{?}{=}31$
$-9=-9$	$31=31$

Since both equations are true, the solution checks.

17. Solve the first equation for y.

$2x+\boxed{y}=10$ Identify the variable term to isolate.
$\underline{-2x \qquad -2x}$ Subtract $2x$ from both sides.
$y=-2x+10$

Substitute $y=-2x+10$ into second equation.

$-6x+y=-5$ Substitute in $y=-2x+10$ for y.
$-6x+(-2x+10)=-5$ Simplify.
$-8x+10=-5$
$\underline{\quad -10 \quad -10}$
$-8x=-15$ Solve for x.
$x=\frac{15}{8}$

$y=-2x+10$ Substitute in $x=\frac{15}{8}$ to find y.
$y=-2\left(\frac{15}{8}\right)+10$
$y=-\frac{15}{4}+\frac{40}{4}$
$y=\frac{25}{4}$
$(x,y)=\left(\frac{15}{8},\frac{25}{4}\right)$ This is the solution.

Check the solution as in exercises 13-15.

19. Substitute $y=-6x+5$ into second equation.

$y=4x-15$ Substitute in $y=-6x+5$ for y.
$(-6x+5)=4x-15$ Simplify.
$\underline{-4x \qquad -4x}$ Subtract $4x$ from both sides.
$-10x+5=-15$
$\underline{\quad -5 \quad -5}$
$-10x=-20$ Solve for x.
$x=2$
$y=-6x+5$ Substitute in $x=2$ to find y.
$y=-6(2)+5$
$y=-12+5$
$y=-7$
$(x,y)=(2,-7)$ This is the solution.

Check the solution as in exercises 13-15.

21. Substitute $x=y$ into second equation.

$2x+y=-6$ Substitute in $x=y$ for x.
$2(y)+y=-6$ Simplify.
$3y=-6$ Solve for x.
$y=-2$
$x=y$ Substitute in $y=-2$ to find x.
$x=(-2)$
$x=-2$
$(x,y)=(-2,-2)$ This is the solution.

Check the solution as in exercises 13-15.

23. **a.** Let $x=$ the first number and let $y=$ the second number.

b. The equations translate as follows:
$x+y=65$
$x=12y$

c. Substitute $x=12y$ into first equation.
$x+y=65$ Substitute in $x=12y$ for x.
$(12y)+y=65$ Simplify.
$13y=65$ Solve for y.
$y=5$
$x=12y$ Substitute in $y=5$ to find x.
$x=12(5)$
$x=60$
$(x,y)=(60,5)$ This is the solution.

25. Let $x=$ the first number and let $y=$ the second number. The equations translate as follows:
$x+y=25$
$x=4y$

Substitute $x=4y$ into first equation.
$x+y=25$ Substitute in $x=4y$ for x.
$(4y)+y=25$ Simplify.
$5y=25$ Solve for y.
$y=5$

$x=4y$ Substitute in $y=5$ to find x.
$x=4(5)$
$x=20$
The two numbers are 5 and 20.

27. Let $x =$ the first number and let $y =$ the second number. The equations translate as follows:

$$x + y = 12$$
$$x = 5y$$

Substitute $x = 5y$ into the first equation.

$x + y = 12$ Substitute in $x = 5y$ for x.
$(5y) + y = 12$ Simplify.
$6y = 12$ Solve for y.
$y = 2$

$x = 5y$ Substitute in $y = 2$ to find x.
$x = 5(2)$
$x = 10$

The two numbers are 2 and 10.

29. Let $x =$ the first number and let $y =$ the second number. The equations translate as follows:

$$x - y = -7$$
$$1 + 4x = y$$

Substitute $y = 1 + 4x$ into first equation.

$x - y = -7$ Substitute in $y = 1 + 4x$ for y.
$x - (1 + 4x) = -7$ Simplify.
$x - 1 - 4x = -7$
$-3x - 1 = -7$
$\underline{\quad +1 \quad +1}$ Add 1 to both sides.
$-3x = -6$ Solve for x.
$x = 2$

$y = 1 + 4x$ Substitute in $x = 2$ to find y.
$y = 1 + 4(2)$
$y = 9$

The two numbers are 2 and 9.

31. Let $x =$ the first number and let $y =$ the second number. The equations translate as follows:

$$x + y = 12$$
$$x = \frac{y}{2}$$

Substitute $x = \frac{y}{2}$ into first equation.

$x + y = 12$ Substitute in $x = \frac{y}{2}$ for x.
$(\frac{y}{2}) + y = 12$ Multiply by 2 to clear denominator.
$(2)\frac{y}{2} + (2)y = 12(2)$
$y + 2y = 24$ Simplify.
$3y = 24$ Solve for y.
$y = 8$

$x = \frac{y}{2}$ Substitute in $y = 8$ to find x.
$x = \frac{(8)}{2}$
$x = 4$

The two numbers are 4 and 8.

33. **a.** Let $w =$ the amount of weekly sales in dollars and let $P =$ the weekly pay in dollars.

b. The equations translate as follows:

$$P = 0.05w + 100$$
$$P = 0.04w + 120$$

c. Substitute $P = 0.04w + 120$ into first equation.

$P = 0.05w + 100$ Substitute in $P = 0.04w + 120$ for P.
$(0.04w + 120) = 0.05w + 100$
$\underline{-0.05w \qquad -0.05w}$ Subtract $0.05w$ from both sides.
$-0.01w + 120 = 100$
$\underline{\qquad -120 \quad -120}$ Subtract 120 from both sides.
$-0.01w = -20$ Solve for w.
$w = 2000$

$P = 0.04w + 120$ Substitute in $w = 2000$ to find P.
$P = 0.04(2000) + 120$
$P = 80 + 120$
$P = 200$

The solution to the system is $w = 2000$ and $P = 200$. Ethan's weekly pay would be $200 and would be the same at either job if he had $2000 in weekly sales at either job.

35. **a.** Let $f =$ the number of hours per week Marcus works at his first job and let $s =$ the number of hours per week Marcus works at his second job.

b. For the first equation, the total number of hours worked at both jobs is equal to 20. For the second equation, the hourly pay of each job times the number of hours worked at each job should equal his desired weekly pay of $176. The system of equations defined by the problem is as follows:

$$f + s = 20$$
$$8f + 9s = 176$$

c. Transform the first equation so that s is isolated on one side. Substitute $s = 20 - f$ into the second equation.

$8f + 9s = 176$ Substitute in $s = 20 - f$ for s.
$8f + 9(20 - f) = 176$ Distribute.
$8f + 180 - 9f = 176$ Simplify.
$-f + 180 = 176$
$\underline{\qquad -180 \quad -180}$ Subtract 180 from both sides.
$-f = -4$ Solve for f.
$f = 4$

$s = 20 - f$ Substitute in $f = 4$ to find s.
$s = 20 - (4)$
$s = 16$

The solution to the system is $s = 16$ and $f = 4$. If Marcus wants to earn $176 working 20 hours in a week, he should work 4 hours at the first job and 16 hours at the second job.

37. Substitute $y = 2x - 1$ into second equation.

$y = -4x + 5$ Substitute in $y = 2x - 1$ for y.

$(2x - 1) = -4x + 5$ Simplify.

$+4x \quad +4x$ Add $4x$ to both sides.

$6x - 1 = 5$

$+1 \quad +1$

$6x = 6$ Solve for x.

$x = 1$

$y = 2x - 1$ Substitute in $x = 1$ to find y.

$y = 2(1) - 1$

$y = 2 - 1$

$y = 1$

$(x, y) = (1,1)$ This is the solution.

Check the solution as in exercises 13-15. The solution checks and the system is consistent.

39. Substitute $y = 4x + 5$ into second equation.

$y = \frac{-1}{2}x - 4$ Substitute in $y = 4x + 5$ for y.

$(4x + 5) = \frac{-1}{2}x - 4$ Multiply by 2 to clear denominator.

$(2)4x + (2)5 = (2)\frac{-1}{2}x - 4(2)$

$8x + 10 = -x - 8$

$+x \quad +x$ Add x to both sides.

$9x + 10 = -8$

$-10 \quad -10$

$9x = -18$ Solve for x.

$x = -2$

$y = 4x + 5$ Substitute in $x = -2$ to find y.

$y = 4(-2) + 5$

$y = -8 + 5$

$y = -3$

$(x, y) = (-2, -3)$ This is the solution.

Check the solution as in exercises 13-15. The solution checks and the system is consistent.

41. Solve the second equation for y: $y = -2x + 5$.

$-x + 3y = 1$ Substitute in $y = -2x + 5$ for y.

$-x + 3(-2x + 5) = 1$ Distribute.

$-x - 6x + 15 = 1$ Simplify.

$-7x + 15 = 1$

$-15 \quad -15$

$-7x = -14$ Solve for x.

$x = 2$

$y = -2x + 5$ Substitute in $x = 2$ to find y.

$y = -2(2) + 5$

$y = -4 + 5$

$y = 1$

$(x, y) = (2,1)$ This is the solution.

Check the solution as in exercises 13-15. The solution checks and the system is consistent.

43. Solve the first equation for x: $x = -2y + 3$.

$2x - y = 4$ Substitute in $x = -2y + 3$ for x.

$2(-2y + 3) - y = 4$ Distribute.

$-4y + 6 - y = 4$ Simplify.

$-5y + 6 = 4$

$-6 \quad -6$ Subtract 6 from both sides.

$-5y = -2$ Solve for y.

$y = \frac{2}{5}$

$x = -2y + 3$ Substitute in $y = \frac{2}{5}$ to find x.

$x = -2\left(\frac{2}{5}\right) + 3$

$x = -\frac{4}{5} + \frac{15}{5}$

$x = \frac{11}{5}$

$(x, y) = \left(\frac{11}{5}, \frac{2}{5}\right)$ This is the solution.

Check the solution as in exercises 13-15. The solution checks and the system is consistent.

45. Substitute $y = 2.5x - 4$ into first equation.

$3x + 2y = 6$ Substitute in $y = 2.5x - 4$ for y.

$3x + 2(2.5x - 4) = 6$ Distribute.

$3x + 5x - 8 = 6$ Simplify.

$8x - 8 = 6$

$+8 \quad +8$

$8x = 14$ Solve for x.

$x = 1.75$

$y = 2.5x - 4$ Substitute in $x = 1.75$ to find y.

$y = 2.5(1.75) - 4$

$y = 4.375 - 4$

$y = 0.375$

$(x, y) = (1.75, 0.375)$ This is the solution.

Check the solution as in exercises 13-15. The solution checks and the system is consistent.

47. Substitute $y = 1$ into first equation.

$x + 5y = 6$ Substitute in $y = 1$ for y.

$x + 5(1) = 6$ Simplify.

$x + 5 = 6$

$-5 \quad -5$

$x = 1$

We are already given the value $y = 1$, therefore $(x, y) = (1,1)$ is the solution.

Check the solution as in exercises 13-15. The solution checks and the system is consistent.

49. Solve the second equation for y: $y = x + 5$.

$5x + 2y = -4$ Substitute in $y = x + 5$ for y.

$5x + 2(x + 5) = -4$ Distribute.

$5x + 2x + 10 = -4$ Simplify.

$$\begin{aligned} 7x+10&=-4 \\ -10\ \ -10& \\ \hline 7x&=-14 \end{aligned}$$ Solve for x.

$x=-2$

$y=x+5$ Substitute in $x=-2$ to find y.

$y=(-2)+5$

$y=3$

$(x,y)=(-2,3)$ This is the solution.

Check the solution as in exercises 13-15. The solution checks and the system is consistent.

51. Solve the second equation for y: $y=\frac{3}{4}x-1$.

$-3x+4y=12$ Substitute in $y=\frac{3}{4}x-1$ for y.

$-3x+4(\frac{3}{4}x-1)=12$ Distribute.

$-3x+3x-4=12$ Simplify.

$-4\overset{?}{=}12$ This is a false statement.

Since this is a false statement ($-4\overset{?}{=}12$), there are no solutions to the system. The lines are parallel. To prove that these two lines are parallel, show their slopes are equal. The second line has slope $m=\frac{3}{4}$. To find the slope of the first line, rewrite the equation in slope-intercept form: $y=\frac{3}{4}x+3$. The slope of this line is also $m=\frac{3}{4}$. Therefore, the lines are parallel. The system is inconsistent and has no solution.

53. Substitute $x=2y+3$ into first equation.

$5x-10y=15$ Substitute in $x=2y+3$ for x.

$5(2y+3)-10y=15$ Distribute.

$10y+15-10y=15$ Simplify.

$15=15$ This is a true statement.

Because this equation is always true ($15=15$), there are infinitely many solutions. These two equations graph as the same line. This is an example of the dependent case. To prove that these really are the same lines, put both lines in slope-intercept form. Rewriting the first equation in slope-intercept form yields: $y=\frac{1}{2}x-\frac{3}{2}$. Rewriting the second line in slope-intercept form yields: $y=\frac{1}{2}x-\frac{3}{2}$. These two lines are the same line, therefore there are infinitely many solutions to the system.

55. Solve the first equation for x: $x=-3y+18$.

$2x+6y=-12$ Substitute in $x=-3y+18$ for x.

$2(-3y+18)+6y=-12$ Distribute.

$-6y+36+6y=-12$ Simplify.

$36\overset{?}{=}-12$ This is a false statement.

Since this is a false statement ($36\overset{?}{=}-12$), there are no solutions to the system. The lines are parallel. To prove that these two lines are parallel, show their slopes are equal. Rewrite both equations in slope-intercept forms so we can compare the slopes. The first line has slope $m=-\frac{1}{3}$. The second line also has slope $m=-\frac{1}{3}$. Therefore, the lines are parallel. The system is inconsistent and has no solution.

57. Solve the second equation for y: $y=-3x-5$.

$-x+3y=6$ Substitute in $y=-3x-5$ for y.

$-x+3(-3x-5)=6$ Distribute.

$-x-9x-15=6$ Simplify.

$$\begin{aligned} -10x-15&=6 \\ +15\ \ +15& \\ \hline -10x&=21 \end{aligned}$$ Solve for x.

$x=-2.1$

$y=-3x-5$ Substitute in $x=-2.1$ to find y.

$y=-3(-2.1)-5$

$y=6.3-5$

$y=1.3$

$(x,y)=(-2.1,1.3)$ This is the solution.

Check the solution as in exercises 13-15. The solution checks and the system is consistent.

59. Solve the first equation for x: $x=-10$. Solve the second equation for y: $y=0$.

We have the values for both x and y and the solution is $(x,y)=(-10,0)$.

Check the solution as in exercises 13-15. The solution checks and the system is consistent.

61. Substitute $y=7x+1$ into second equation..

$14x-2y=-2$ Substitute in $y=7x+1$ for y.

$14x-2(7x+1)=-2$ Distribute.

$14x-14x-2=-2$ Simplify.

$-2=-2$ This is a true statement.

Because this equation is always true ($-2=-2$), there are infinitely many solutions. These two equations graph as the same line. This is an example of the dependent case. To prove that these really are the same lines, put both lines in slope-intercept form. The first equation is already in slope-intercept form: $y=7x+1$. Rewriting the second equation in slope-intercept form yields: $y=7x+1$. These two lines are the same line, therefore there are infinitely many solutions to the system.

63. Solve the first equation for x: $x=\frac{1}{4}y+9$.

$2x+y=1$ Substitute in $x=\frac{1}{4}y+9$ for x.

$2\left(\frac{1}{4}y+9\right)+y=1$ Distribute.

$\frac{1}{2}y+18+y=1$ Simplify.

$\frac{3}{2}y+18=1$

$\underline{-18 \quad -18}$

$\frac{3}{2}y=-17$ Solve for y.

$\frac{2}{3}\cdot\frac{3}{2}y=-17\cdot\frac{2}{3}$

$y=-\frac{34}{3}$

$x=\frac{1}{4}y+9$ Substitute in $y=-\frac{34}{3}$ to find x.

$x=\frac{1}{4}\left(-\frac{34}{3}\right)+9$

$x=-\frac{34}{12}+\frac{108}{12}$

$x=\frac{74}{12}=\frac{37}{6}$

$(x,y)=\left(\frac{37}{6},-\frac{34}{3}\right)$ This is the solution.

Check the solution as in exercises 13-15. The solution checks and the system is consistent.

65. With a headwind, the wind is slowing the plane down. The rate of the plane with the headwind is $120-15=105$, or 105 mph.

67. With a tailwind, the wind is increasing the speed of the plane. The rate of the plane with the tailwind is $100+5=105$, or 105 mph.

69. When the boat is traveling downstream, the current is pushing the boat, so its rate is faster. The rate of the boat traveling downstream is $7+1.5=8.5$, or 8.5 knots.

71. When the boat is traveling upstream, the current is pushing opposite of the direction of the boat, so its rate is slower. The rate of the boat traveling upstream is $9-2=7$, or 7 knots.

73. **a.** The value of 5.5 lb. of tea is
$5.5\cdot\$7.95\approx\$43.73$ Rounded to nearest cent.
b. The value of t lb. of tea is $t\cdot\$7.95=\$7.95t$.

75. **a.** The value of 26 lb. of coffee is
$26\cdot\$11.95=\$310.70$
b. The value of c lb. of coffee is
$c\cdot\$11.95=\$11.95c$.

77. The unit price for 50 lb. of coffee priced at \$547.50 can be found by dividing the total cost by the number of units (lb.).
$\$547.50\div50=\$10.95$.
The unit price for the coffee is \$10.95 per lb.

79. **a.** Let $s=$ the rate of the boat (in knots) in still water and $c=$ the rate of the current in knots.
b. When the boat is travelling downstream, the current speeds the boat up, so the boat's rate downstream is $s+c$. Using the formula $d=r\cdot t$ with the given distance $d=50$ and time $t=2.5$ yields: $2.5(s+c)=50$. When the boat is travelling upstream, the current slows the boat down, so the boat's rate upstream is $s-c$. Using the formula $d=r\cdot t$ with the given distance $d=50$ and time $t=5$ yields: $5(s-c)=50$.
c. To solve the system using the substitution method, isolate a variable of one of the equations.
System equations:
$2.5(s+c)=50$
$5(s-c)=50$
Solve $5(s-c)=50$ for s.
$5(s-c)=50$ Distribute.
$5s-5c=50$ Add $5c$ to both sides.
$\underline{\quad +5c \quad +5c}$
$5s=5c+50$ Solve for s.
$\frac{5s}{5}=\frac{5c+50}{5}$
$s=c+10$
Now substitute $s=c+10$ into the first equation.
$2.5(s+c)=50$ Distribute.
$2.5s+2.5c=50$ Substitute in $s=c+10$ for s.
$2.5(c+10)+2.5c=50$ Distribute.
$2.5c+25+2.5c=50$ Simplify.
$5c+25=50$
$\underline{\quad -25 \quad -25}$
$5c=25$ Solve for c.
$c=5$
$s=c+10$ Substitute in $c=5$ to find s.
$s=(5)+10$
$s=15$
Since $s=15$ and $c=5$, the rate of the boat in still water is 15 knots and the rate of the current is 5 knots.

81. **a.** Let $p=$ the rate of the plane (in mph) in still air and $w=$ the rate of the wind in mph.
b. When the plane is travelling with a tailwind, the wind speeds the plane up, so the plane's rate with a tailwind is $p+w$. Using the formula $d=r\cdot t$ with the given distance $d=1275$ and time $t=3$ yields: $3(p+w)=1275$. When the plane is travelling with a headwind, the wind slows the plane down, so the plane's rate with a headwind is $p-w$. Using the formula $d=r\cdot t$ with the given distance $d=1275$ and time $t=3.4$ yields: $3.4(p-w)=1275$.
c. To solve the system using the substitution method, isolate a variable of one of the equations.
System equations:
$3(p+w)=1275$
$3.4(p-w)=1275$

Solve $3.4(p-w)=1275$ for p.

$3.4(p-w)=1275$ Distribute.

$3.4p-3.4w=1275$ Add $3.4w$ to both sides.

$\underline{\quad +3.4w \quad +3.4w}$

$3.4p=3.4w+1275$ Solve for p.

$\frac{3.4p}{3.4}=\frac{3.4w+1275}{3.4}$

$p=w+375$

Now substitute $p=w+375$ into the first equation.

$3(p+w)=1275$ Distribute.

$3p+3w=1275$ Substitute in $p=w+375$ for p.

$3(w+375)+3w=1275$ Distribute.

$3w+1125+3w=1275$ Simplify.

$6w+1125=1275$

$\underline{\quad -1125 \quad -1125}$

$6w=150$ Solve for w.

$w=25$

$p=w+375$ Substitute in $w=25$ to find p.

$p=(25)+375$

$p=400$

Since $p=400$ and $w=25$, the rate of the plane in still air is 400 mph and the rate of the wind is 25 mph.

83. **a.** Let $c=$ the number of pounds of Ceylon tea and $b=$ the number of pounds of black currant tea.

b. The total number of pounds in the mix is 50 lb. so our first equation is: $c+b=50$. The price for the mix is \$7.25 lb. with a total of 50 lb. for a total cost of \$362.50, Ceylon is \$5.75 lb. and black currant is \$8.75 lb., so our second equation is: $5.75c+8.75b=362.50$.

c. To solve the system using the substitution method, isolate a variable of one of the equations.
System equations:

$c+b=50$

$5.75c+8.75b=362.50$

Solve $c+b=50$ for c.

$c+b=50$

$\underline{\quad -b \quad -b}$

$c=-b+50$

Now substitute $c=-b+50$ into the second equation.

$5.75c+8.75b=362.50$ Substitute in $c=-b+50$ for c.

$5.75(-b+50)+8.75b=362.50$ Distribute.

$-5.75b+287.5+8.75b=362.50$ Simplify.

$3b+287.5=362.50$

$\underline{\quad -287.5 \quad -287.5}$

$3b=75$ Solve for b.

$b=25$

$c=-b+50$ Substitute in $b=25$ to find c.

$c=-(25)+50$

$c=25$

Since $b=25$ and $c=25$, Julie needs to use 25 lb. of each type of tea for the blend.

85. **a.** Let $p=$ the number of pounds of dried papaya and $g=$ the number of pounds of crystallized ginger.

b. The total number of pounds in the mix is 25 lb. so our first equation is: $p+g=25$. The price for the mix is \$7.49 lb. with a total of 25 lb. for a total cost of \$187.25, dried papaya is \$9.99 lb. and crystallized ginger is \$3.75 lb., so our second equation is: $9.99p+3.75g=187.25$.

c. To solve the system using the substitution method, isolate a variable of one of the equations.
System equations:

$p+g=25$

$9.99p+3.75g=187.25$

Solve $p+g=25$ for p.

$p+g=25$

$\underline{\quad -g \quad -g}$

$p=-g+25$

Now substitute $p=-g+25$ into the second equation.

$9.99p+3.75g=187.25$ Substitute in $p=-g+25$ for p.

$9.99(-g+25)+3.75g=187.25$ Distribute.

$-9.99g+249.75+3.75g=187.25$ Simplify.

$-6.24g+249.75=187.25$

$\underline{\quad -249.75 \quad -249.75}$

$-6.24g=-62.5$ Solve for g.

$g\approx 10.02$

$p=-g+25$ Substitute in $g=10.02$ to find p.

$p\approx -(10.02)+25$

$p\approx 14.98$

Since $p\approx 14.98$ and $g\approx 10.02$, Graciela needs to use 14.98 lb. of dried papaya and 10.02 lb. of crystallized ginger for the mix.

87. The break-even point is when revenue is equal to costs, or $R=C$. In the given equations, if $R=C$, then $150n=100n+1400$. This represents when revenue is equal to costs. Solve the equation to find the break-even point.

$150n=100n+1400$

$\underline{-100n \quad -100n}$

$50n=1400$ Solve for n.

$n=28$

The break-even point for Rita's auto detail is when she details 28 cars in a week and earns \$4200.

89. The break-even point is when revenue is equal to costs, or $R=C$. In the given equations, if $R=C$, then $24.50n=8.50n+4300$. This represents when revenue is equal to costs. Solve the equation to find the break-even point.

$24.50n = 8.50n + 4300$

$\underline{-8.50n \quad -8.50n}$

$16n = 4300$ Solve for n.

$n = 268.75$

The break-even point for Mobile Style is when the company sells about 269 accessories in a month (rounded up to the nearest whole number) and has an average cost/revenue of \$6588.50 (the average of the cost and the revenue when $n = 269$).

91. **a.** Substitute $y = \frac{3}{5}x + 3$ into first equation.

$-3x + 5y = 15$ Substitute in $y = \frac{3}{5}x + 3$.

$-3x + 5(\frac{3}{5}x + 3) = 15$ Distribute.

$-3x + 3x + 15 = 15$ Simplify.

$15 = 15$ This is a true statement.

Because this equation is always true ($15 = 15$), there are infinitely many solutions.

b. Graph as follows:

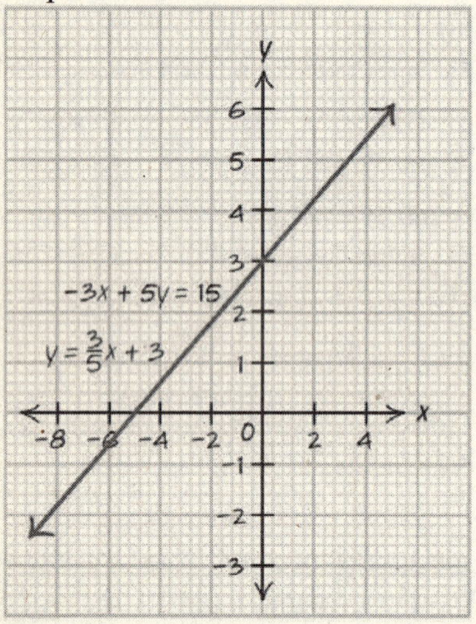

c. Since the two points are on the line, they are solutions to this system.

d. Since the two points are not on the line, they are not solutions to this system.

e. The solutions are only the points on the line. The line is infinitely long, so there are infinitely many solutions as there are infinitely many points on the line.

93. **a.** Substitute $y = -6x + 8$ into second equation.

$18x + 3y = -21$ Substitute in $y = -6x + 8$.

$18x + 3(-6x + 8) = -21$ Distribute.

$18x - 18x + 24 = -21$ Simplify.

$24 \stackrel{?}{=} -21$ This is a false statement.

Since this is a false statement ($24 \stackrel{?}{=} -21$), there are no solutions to the system.

b. Graph as follows:

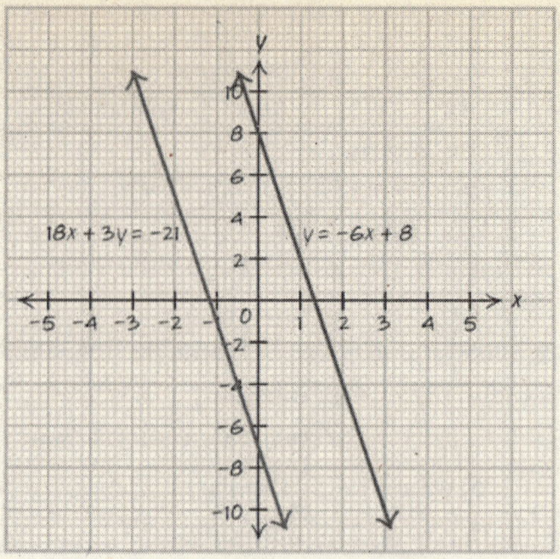

c. No, if two points are on the first line, they can not also be on the second line since the lines do not intersect and the points will not be solutions to the system.

d. No, if two points are on the second line, they can not also be on the first line since the lines do not intersect and the points will not be solutions to the system.

e. There are no solutions to this system because the lines are parallel and never intersect. The system is inconsistent.

95. In this step:

$2x - (4x - 8) = 6$

$2x - 4x - 8 = 6$

The student did not apply the negative to the 8 in the parentheses. The equation after this step should be:

$2x - 4x + 8 = 6$

97. The student did not completely isolate the x when preparing the first equation to be substituted into the second equation:

$$\begin{aligned} 3x &= -4y + 11 \\ -5x + 2y &= -1 \end{aligned} \quad \text{should be} \quad \begin{aligned} x &= -\frac{4}{3}y + \frac{11}{3} \\ -5x + 2y &= -1 \end{aligned}$$

Now the first equation is ready to be used for substitution into the second equation as follows:

$$-5\left(-\frac{4}{3}y + \frac{11}{3}\right) + 2y = -1$$

99. If the conclusion of solving a system is a false statement such as $0 = 4$, the system is inconsistent and there is no solution.

101. If two lines are parallel, the system is inconsistent and there is no solution because the lines never intersect.

103. If the graph of a system results in two intersecting lines, the system is consistent and has exactly one solution, which is the point of intersection.

Section 4.3 Solving Systems Using the Elimination Method

1. The coefficients on y are -1 and 1: that is, the coefficients are opposites.

$2x - y = 16$ Add the two equations vertically.
$3x + y = 9$
$5x = 25$
$\frac{5x}{5} = \frac{25}{5}$
$x = 5$

Substitute $x = 5$ into second equation:
$3(5) + y = 9$ Substitute in $x = 5$
$15 + y = 9$
$-15 \quad -15$
$y = -6$

Check the solution as in Section 4.2, exercises 13-15. Both equations check, so the solution is $(5, -6)$.

3. The coefficients on x are -1 and 1: that is, the coefficients are opposites.

$-x + 6y = -3$ Add the two equations vertically.
$x - 2y = -5$
$4y = -8$ Solve for y.
$y = -2$

Substitute $y = -2$ into second equation yields:
$x - 2(-2) = -5$ Substitute in $y = -2$
$x + 4 = -5$
$-4 \quad -4$
$x = -9$

Check the solution as in Section 4.2, exercises 13-15. Both equations check, so the solution is $(-9, -2)$.

5. The coefficients on y are -2 and 2: that is, the coefficients are opposites.

$-4x + 2y = 12$ Add the two equations vertically.
$8x - 2y = -40$
$4x = -28$ Solve for x.
$x = -7$

Substitute $x = -7$ into first equation:
$-4(-7) + 2y = 12$ Substitute in $x = -7$
$28 + 2y = 12$
$-28 \quad -28$
$2y = -16$
$y = -8$

Check the solution as in Section 4.2, exercises 13-15. Both equations check, so the solution is $(-7, -8)$.

7. Neither of the coefficients on x or y is an opposite. To eliminate the variable x, multiply both sides of the second equation by -1 so that the coefficients on x are 4 and -4.

$4x + 2y = 8$ and $-(4x - y) = -(4)$ is now $4x + 2y = 8$ and $-4x + y = 4$

$4x + 2y = 8$ Add the two equations vertically.
$-4x + y = 4$
$3y = 12$ Solve for y.
$y = 4$

Substitute $y = 4$ into first equation:
$4x + 2(4) = 8$ Substitute in $y = 4$
$4x + 8 = 8$
$-8 \quad -8$
$4x = 0$ Solve for x.
$x = 0$

Check the solution as in Section 4.2, exercises 13-15. Both equations check, so the solution is $(0, 4)$.

9. Neither of the coefficients on x or y is an opposite. To eliminate the variable y, multiply both sides of the second equation by 3 so that the coefficients on y are 3 and -3.

$4x + 3y = -5$ and $3(2x - y) = 3(5)$ is now $4x + 3y = -5$ and $6x - 3y = 15$

$4x + 3y = -5$ Add the two equations vertically.
$6x - 3y = 15$
$10x = 10$ Solve for x.
$x = 1$

Substitute $x = 1$ into first equation:
$4(1) + 3y = -5$ Substitute in $x = 1$
$4 + 3y = -5$
$-4 \quad -4$
$3y = -9$ Solve for y.
$y = -3$

Check the solution as in Section 4.2, exercises 13-15. Both equations check, so the solution is $(1, -3)$.

11. Neither of the coefficients on x or y is an opposite. To eliminate the variable y, multiply both sides of the second equation by 2 so that the coefficients on y are 2 and -2.

$2(3x - y) = 2(-1)$ and $-6x + 2y = 2$ is now $6x - 2y = -2$ and $-6x + 2y = 2$

$6x - 2y = -2$ Add the two equations vertically.
$-6x + 2y = 2$
$0 = 0$

Because this equation is always true ($0 = 0$), there are infinitely many solutions. These two equations graph as the same line. This is an example of the dependent case. To prove that these really are the same lines, put both lines in slope-intercept form. Rewriting the first equation in slope-intercept form yields: $y = 3x + 1$. Rewriting the second equation in slope-intercept form yields: $y = 3x + 1$. These two lines are the same line, therefore there are infinitely many solutions to the system.

13. Neither of the coefficients on x or y is an opposite. Considering the coefficients on y, the least common multiple of 2 and 3 is 6. To get coefficients of 6 and -6 on the variable y, multiply the first equation by 3 and the second equation by 2.

$$\begin{array}{ccc} 3(3x-2y)=3(-14) & \text{is now} & 9x-6y=-42 \\ 2(2x+3y)=2(8) & & 4x+6y=16 \end{array}$$

$$\begin{array}{rl} 9x-6y=-42 & \text{Add the equations vertically.} \\ \underline{4x+6y=16} & \\ 13x=-26 & \text{Solve for } x. \\ x=-2 & \end{array}$$

Substitute $x=-2$ into first equation:

$$\begin{array}{rl} 3(-2)-2y=-14 & \text{Substitute in } x=-2 \\ -6-2y=-14 & \\ \underline{+6 \qquad +6} & \\ -2y=-8 & \text{Solve for } y. \\ y=4 & \end{array}$$

Check the solution as in Section 4.2, exercises 13-15. Both equations check, so the solution is $(-2,4)$.

15. Neither of the coefficients on x or y is an opposite. Considering the coefficients on y, the least common multiple of 2 and 3 is 6. To get coefficients of 6 and -6 on the variable y, multiply the first equation by 3 and the second equation by 2.

$$\begin{array}{ccc} 3(13x+2y)=3(-40) & \text{is now} & 39x+6y=-120 \\ 2(10x-3y)=2(1) & & 20x-6y=2 \end{array}$$

$$\begin{array}{rl} 39x+6y=-120 & \text{Add the equations vertically.} \\ \underline{20x-6y=2} & \\ 59x=-118 & \text{Solve for } x. \\ x=-2 & \end{array}$$

Substitute $x=-2$ into first equation:

$$\begin{array}{rl} 13(-2)+2y=-40 & \text{Substitute in } x=-2 \\ -26+2y=-40 & \\ \underline{+26 \qquad +26} & \\ 2y=-14 & \text{Solve for } y. \\ y=-7 & \end{array}$$

Check the solution as in Section 4.2, exercises 13-15. Both equations check, so the solution is $(-2,-7)$.

17. Neither of the coefficients on x or y is an opposite. Considering the coefficients on y, the least common multiple of 3 and 2 is 6. To get coefficients of 6 and -6 on the variable y, multiply the first equation by 2 and the second equation by -3.

$$\begin{array}{ccc} 2(10x+3y)=2(22) & \text{is now} & 20x+6y=44 \\ 3(3x-2y)=3(24) & & 9x-6y=72 \end{array}$$

$$\begin{array}{rl} 20x+6y=44 & \text{Add the equations vertically.} \\ \underline{9x-6y=72} & \\ 29x=116 & \text{Solve for } x. \\ x=4 & \end{array}$$

Substitute $x=4$ into first equation:

$$\begin{array}{rl} 10(4)+3y=22 & \text{Substitute in } x=4 \\ 40+3y=22 & \\ \underline{-40 \qquad -40} & \\ 3y=-18 & \text{Solve for } y. \\ y=-6 & \end{array}$$

Check the solution as in Section 4.2, exercises 13-15. Both equations check, so the solution is $(4,-6)$.

19. Neither of the coefficients on x or y is an opposite. Considering the coefficients on x, the least common multiple of 6 and 3 is 6. To get coefficients of 6 and -6 on the variable x, multiply the second equation by -2 and leave the first equation as it is.

$$\begin{array}{ccc} 6x+8y=-8 & \text{is now} & 6x+8y=-8 \\ -2(3x+4y)=-2(36) & & -6x-8y=-72 \end{array}$$

$$\begin{array}{rl} 6x+8y=-8 & \text{Add the equations vertically.} \\ \underline{-6x-8y=-72} & \\ 0\overset{?}{=}-80 & \end{array}$$

Since this is a false statement ($0\overset{?}{=}-80$), there are no solutions to the system. The lines are parallel. To prove that these two lines are parallel, show their slopes are equal. Rewrite each equation in slope-intercept form so we can compare slopes. The first line written in slope-intercept form is: $y=-\frac{3}{4}x-1$, so we can see that the slope is $m=-\frac{3}{4}$. The second line written in slope-intercept form is: $y=-\frac{3}{4}x+9$, and we can see the line has slope $m=-\frac{3}{4}$. Therefore, the lines are parallel. The system is inconsistent and has no solution.

21. Neither of the coefficients on x or y is an opposite. Considering the coefficients on y, the least common multiple of 2 and 6 is 6. To get coefficients of 6 and -6 on the variable y, multiply the first equation by 3 and leave the second equation as it is.

$$\begin{array}{ccc} 3(7x+2y)=3(-2) & \text{is now} & 21x+6y=-6 \\ -21x-6y=6 & & -21x-6y=6 \end{array}$$

$$\begin{array}{rl} 21x+6y=-6 & \text{Add the two equations vertically.} \\ \underline{-21x-6y=6} & \\ 0=0 & \end{array}$$

Because this equation is always true ($0=0$), there are infinitely many solutions. These two equations graph as the same line. This is an example of the dependent case. To prove that these really are the same lines, put both lines in slope-intercept form. Rewriting the first equation in slope-intercept form yields: $y=-\frac{7}{2}x-1$. Rewriting the second equation in slope-intercept form yields: $y=-\frac{7}{2}x-1$. These two lines are the same line, therefore there are infinitely many solutions to the system.

23. Neither of the coefficients on x or y is an opposite. Considering the coefficients on x, the least common multiple of 6 and 3 is 6. To get coefficients of 6 and -6 on the variable x, multiply the second equation by 2 and leave the first equation as it is.

$$\begin{matrix} -6x+2y=-16 \\ 2(3x-y)=2(-2) \end{matrix} \text{ is now } \begin{matrix} -6x+2y=-16 \\ 6x-2y=-4 \end{matrix}$$

$$\begin{aligned} -6x+2y&=-16 && \text{Add the equations vertically.} \\ 6x-2y&=-4 \\ \hline 0&\overset{?}{=}-20 \end{aligned}$$

Since this is a false statement ($0\overset{?}{=}-20$), there are no solutions to the system. The lines are parallel. To prove that these two lines are parallel, show their slopes are equal. Rewrite each equation in slope-intercept form so we can compare slopes. The first line written in slope-intercept form is: $y=3x-8$, so we can see that the slope is $m=3$. The second line written in slope-intercept form is: $y=3x+2$, and we can see the line has slope $m=3$.Therefore, the lines are parallel. The system is inconsistent and has no solution.

25. **a.** Let $f=$ the first integer and let $s=$ the second integer.

b. The equations translate as follows:

$$f+s=95$$
$$4f-s=15$$

c. The coefficients on s are -1 and 1: that is, the coefficients are opposites.

$$\begin{aligned} f+s&=95 && \text{Add the two equations vertically.} \\ 4f-s&=15 \\ \hline 5f&=110 && \text{Solve for } f. \\ \frac{5f}{5}&=\frac{110}{5} && \text{Divide both sides by 5.} \\ f&=22 \\ f+s&=95 && \text{Substitute in } f=22 \text{ to find } s. \\ (22)+s&=95 \\ -22\quad &\ \ -22 \\ \hline s&=73 \end{aligned}$$

Check the solution as in Section 4.2, exercises 13-15. Both equations check, so the solution is $(22,73)$.

27. **a.** Let $t=$ the cost in dollars for each t-shirt and let $w=$ the cost in dollars for each water bottle.

b. The equations translate as follows:

$$8t+6w=186$$
$$10t+8w=237$$

c. Neither of the coefficients on t or w is an opposite. Considering the coefficients on w, the least common mulpiple of 6 and 8 is 24. To get coefficients of 24 and -24 on the variable w, multiply the first equation by 4 and multiply the second equation by -3.

$$\begin{matrix} 4(8t+6w)=4(186) \\ -3(10t+8w)=-3(237) \end{matrix} \text{ is now } \begin{matrix} 32t+24w=744 \\ -30t-24w=-711 \end{matrix}$$

$$\begin{aligned} 32t+24w&=744 && \text{Add the equations vertically.} \\ -30t-24w&=-711 \\ \hline 2t&=33 && \text{Solve for } t. \\ \frac{2t}{2}&=\frac{33}{2} && \text{Divide both sides by 2.} \\ t&=16.5 \\ 8t+6w&=186 && \text{Substitute in } t=16.5 \text{ to find } w. \\ 8(16.5)+6w&=186 \\ 132+6w&=186 \\ -132\quad &\ \ -132 \\ \hline 6w&=54 \\ \frac{6w}{6}&=\frac{54}{6} \\ w&=9 \end{aligned}$$

Each individual t-shirt costs \$16.50 and each individual water bottle costs \$9.00.
Check the solution by substituting the values into both of the original equations.

29. **a.** Let $m=$ the cost in dollars for each gallon of milk and let $b=$ the cost in dollars for each loaf of French bread.

b. The equations translate as follows:

$$3m+4b=23.4$$
$$2m+3b=16.35$$

c. Neither of the coefficients on m or b is an opposite. Considering the coefficients on m, the least common multiple of 3 and 2 is 6. To get coefficients of 6 and -6 on the variable m, multiply the first equation by -2 and multiply the second equation by 3.

$$\begin{matrix} -2(3m+4b)=-2(23.4) \\ 3(2m+3b)=3(16.35) \end{matrix} \text{ is now } \begin{matrix} -6m-8b=-46.8 \\ 6m+9b=49.05 \end{matrix}$$

$$\begin{aligned} -6m-8b&=-46.8 && \text{Add the equations vertically.} \\ 6m+9b&=49.05 \\ \hline b&=2.25 \\ 2m+3b&=16.35 && \text{Substitute in } b=2.25 \text{ to find } m. \\ 2m+3(2.25)&=16.35 \\ 2m+6.75&=16.35 \\ -6.75\quad &\ \ -6.75 \\ \hline 2m&=9.6 \\ \frac{2m}{2}&=\frac{9.6}{2} \\ m&=4.8 \end{aligned}$$

Each gallon of milk costs \$4.80 and each loaf of French bread costs \$2.25.

Check the solution by substituting the values into both of the original equations.

31. The first step is to put the second equation in general form. Once this has been done, the general form of the system is:

$$-2x+5y=10$$
$$2x-5y=25$$

$$\begin{array}{r} -2x+5y=10 \\ 2x-5y=25 \\ \hline 0\overset{?}{=}35 \end{array}$$ Add the equations vertically.

Since this is a false statement ($0\overset{?}{=}35$), there are no solutions to the system. The lines are parallel. To prove that these two lines are parallel, show their slopes are equal. Rewrite each equation in slope-intercept form so we can compare slopes. The first line written in slope-intercept form is: $y=\frac{2}{5}x+2$, so we can see that the slope is $m=\frac{2}{5}$. The second line written in slope-intercept form is: $y=\frac{2}{5}x-5$, and we can see the line has slope $m=\frac{2}{5}$. Therefore, the lines are parallel. The system is inconsistent and has no solution.

33. To eliminate the variable y, multiply both sides of the second equation by -1 so that the coefficients on y are 2 and -2.

$$\begin{array}{lcl} x+2y=18 & \text{is now} & x+2y=18 \\ -(-3x+2y)=-(2) & & 3x-2y=-2 \end{array}$$

$$\begin{array}{r} x+2y=18 \\ 3x-2y=-2 \\ \hline 4x=16 \end{array}$$ Add the two equations vertically. Solve for x.

$$\frac{4x}{4}=\frac{16}{4}$$ Divide both sides by 4.

$$x=4$$

$x+2y=18$ Substitute in $x=4$ to find y.

$$\begin{array}{r} (4)+2y=18 \\ -4 \quad\quad -4 \\ \hline 2y=14 \end{array}$$

$$\frac{2y}{2}=\frac{14}{2}$$

$$y=7$$

The solution is $(4,7)$.

Check the solution by substituting the values into both of the original equations.

35. To eliminate the variable y, multiply both sides of the second equation by 3 so that the coefficients on y are 3 and -3.

$$\begin{array}{lcl} 4x-3y=-24 & \text{is now} & 4x-3y=-24 \\ 3(x+y)=3(1) & & 3x+3y=3 \end{array}$$

$$\begin{array}{r} 4x-3y=-24 \\ 3x+3y=3 \\ \hline 7x=-21 \end{array}$$ Add the two equations vertically. Solve for x.

$$\frac{7x}{7}=\frac{-21}{7}$$ Divide both sides by 7.

$$x=-3$$

$x+y=1$ Substitute in $x=-3$ to find y.

$$\begin{array}{r} (-3)+y=1 \\ +3 \quad\quad +3 \\ \hline y=4 \end{array}$$

The solution is $(-3,4)$.

Check the solution by substituting the values into both of the original equations.

37. To eliminate the variable y, multiply both sides of the second equation by 5 so that the coefficients on y are 5 and -5.

$$\begin{array}{lcl} 4x-5y=-15 & \text{is now} & 4x-5y=-15 \\ 5(2x+y)=5(-11) & & 10x+5y=-55 \end{array}$$

$$\begin{array}{r} 4x-5y=-15 \\ 10x+5y=-55 \\ \hline 14x=-70 \end{array}$$ Add the two equations vertically. Solve for x.

$$\frac{14x}{14}=\frac{-70}{14}$$ Divide both sides by 14.

$$x=-5$$

$2x+y=-11$ Substitute in $x=-5$ to find y.

$$\begin{array}{r} 2(-5)+y=-11 \\ -10+y=-11 \\ +10 \quad\quad +10 \\ \hline y=-1 \end{array}$$

The solution is $(-5,-1)$.

Check the solution by substituting the values into both of the original equations.

39. To eliminate the variable x, multiply both sides of the second equation by 7 so that the coefficients on x are 14 and -14.

$$\begin{array}{lcl} -14x+35y=21 & \text{is now} & -14x+35y=21 \\ 7(2x-5y)=7(-3) & & 14x-35y=-21 \end{array}$$

$$\begin{array}{r} -14x+35y=21 \\ 14x-35y=-21 \\ \hline 0=0 \end{array}$$ Add the two equations vertically.

Because this equation is always true ($0=0$), there are infinitely many solutions. These two equations graph as the same line. To prove that these really are the same lines, put both equations in slope-intercept form. Rewriting the first equation in slope-intercept form yields: $y=\frac{2}{5}x+\frac{3}{5}$. Rewriting the second equation in slope-intercept form yields: $y=\frac{2}{5}x+\frac{3}{5}$.

These two equations are the same line, therefore there are infinitely many solutions to the system.

41. The first step is to put the second equation in general form:

$$3x+5y=30$$
$$\frac{3}{5}x+y=6$$

Neither of the coefficients on x or y is an opposite. To eliminate the variable y, multiply both sides of the second equation by -5 so that the coefficients on y are 5 and -5.

$$\begin{matrix} 3x+5y=30 \\ -5(\frac{3}{5}x+y)=-5(6) \end{matrix} \text{ is now } \begin{matrix} 3x+5y=30 \\ -3x-5y=-30 \end{matrix}$$

$3x+5y=30$ Add the two equations vertically.
$-3x-5y=-30$
$0=0$

Because this equation is always true ($0=0$), there are infinitely many solutions. These two equations graph as the same line. To prove that these really are the same lines, put both equations in slope-intercept form. Rewriting the first equation in slope-intercept form yields: $y=-\frac{3}{5}x+6$. The second equation was given in slope-intercept form: $y=\frac{-3}{5}x+6$. These two equations are the same line, therefore there are infinitely many solutions to the system.

43. To eliminate the variable y, multiply both sides of the second equation by -2 so that the coefficients on y are 2 and -2.

$$\begin{matrix} 2x-2y=-14 \\ -2(4x-y)=-2(-1) \end{matrix} \text{ is now } \begin{matrix} 2x-2y=-14 \\ -8x+2y=2 \end{matrix}$$

$2x-2y=-14$ Add the two equations vertically.
$-8x+2y=2$
$-6x=-12$ Solve for x.
$\frac{-6x}{-6}=\frac{-12}{-6}$ Divide both sides by -6.
$x=2$
$2x-2y=-14$ Substitute in $x=2$ to find y.
$2(2)-2y=-14$
$4-2y=-14$
$-4 \quad -4$
$-2y=-18$
$\frac{-2y}{-2}=\frac{-18}{-2}$
$y=9$

The solution is $(2,9)$.

Check the solution by substituting the values into both of the original equations.

45. The first step is to put the first equation in general form:

$$-\frac{5}{2}x+y=-3$$
$$5x-2y=3$$

To eliminate the variable y, multiply both sides of the first equation by 2 so that the coefficients on y are 2 and -2.

$$2\left(-\frac{5}{2}x+y\right)=2(-3) \text{ is now } \begin{matrix} -5x+2y=-6 \\ 5x-2y=3 \end{matrix}$$
$$5x-2y=3$$

$-5x+2y=-6$ Add the equations vertically.
$5x-2y=3$
$0\overset{?}{=}-3$

Since this is a false statement ($0\overset{?}{=}-3$), there are no solutions to the system. The lines are parallel. To prove that these two lines are parallel, show their slopes are equal. Rewrite each equation in slope-intercept form so we can compare slopes. The first line is given in slope-intercept form: $y=\frac{5}{2}x-3$, so we can see that the slope is $m=\frac{5}{2}$. The second line written in slope-intercept form is: $y=\frac{5}{2}x-\frac{3}{2}$, and we can see the line has slope $m=\frac{5}{2}$. Therefore, the lines are parallel. The system is inconsistent and has no solution.

47. To eliminate the variable y, multiply both sides of the first equation by 3 so that the coefficients on y are 3 and -3.

$$\begin{matrix} 3(3x-y)=3(7) \\ -9x+3y=-21 \end{matrix} \text{ is now } \begin{matrix} 9x-3y=21 \\ -9x+3y=-21 \end{matrix}$$

$9x-3y=21$ Add the two equations vertically.
$-9x+3y=-21$
$0=0$

Because this equation is always true ($0=0$), there are infinitely many solutions. These two equations graph as the same line. To prove that these really are the same lines, put both equations in slope-intercept form. Rewriting the first equation in slope-intercept form yields: $y=3x-7$. Rewriting the second equation in slope-intercept form yields: $y=3x-7$. These two equations are the same line, therefore there are infinitely many solutions to the system.

49. To eliminate the variable y, multiply both sides of the second equation by 2 so that the coefficients on y are 2 and -2.

$$\begin{matrix} 2x-2y=10 \\ 2(-x+y)=2(8) \end{matrix} \text{ is now } \begin{matrix} 2x-2y=10 \\ -2x+2y=16 \end{matrix}$$

$2x-2y=10$ Add the equations vertically.
$-2x+2y=16$

$0\overset{?}{=}26$

Since this is a false statement ($0\overset{?}{=}26$), there are no solutions to the system. The lines are parallel. To prove that these two lines are parallel, show their slopes are equal. Rewrite each equation in slope-intercept form so we can compare slopes. The first line written in slope-intercept form is: $y=x-5$, so we can see that the slope is $m=1$. The second line written in slope-intercept form is: $y=x+8$, and we can see the line has slope $m=1$. Therefore, the lines are parallel. The system is inconsistent and has no solution.

51. The coefficients of the y terms are opposites, therefore it would be most efficient to solve using the elimination method.

53. The second equation has y isolated on one side, therefore it would be most efficient to solve using the substitution method.

55. Both equations have y isolated on one side, therefore it would be most efficient to solve using the substitution method.

57. The coefficients of the y terms are opposites, therefore it would be most efficient to solve using the elimination method.

59. The principal P is the original investment, so $P=500$. The interest rate r of 4.25% has to be converted to decimal form: $r=0.0425$. The time t is 4 years, so $t=4$. Substituting these values into the formula yields:

$I=P\cdot r\cdot t$
$I=(500)(0.0425)(4)$
$I=85$

This account earns $85.00 in interest in 4 years.

61. The principal P is the original investment, so $P=8720$. The interest rate r of 4.25% has to be converted to decimal form: $r=0.0425$. The time t is 3 years, so $t=3$. Substituting these values into the formula yields:

$I=P\cdot r\cdot t$
$I=(8720)(0.0425)(3)$
$I=1111.8$

This account earns $1111.80 in interest in 3 years.

63. To find the amount of a given substance in a solution, use the formula:

given substance = % • amt. of solution

The amount of saline solution is 10 liters. The percentage of sodium is $8.5\%=0.085$.

amount of sodium $=0.085\cdot 10=0.85$

There is 0.85 liter of sodium in the solution.

65. To find the amount of a given substance in a solution, use the formula:

given substance = % • amt. of solution

The amount of solution is 8 liters. The percentage of sucrose is $8\%=0.08$.

amount of sucrose $=0.08\cdot 8=0.64$

There is 0.64 liter of sucrose in the solution.

67. The sum of two complementary angles is 90° .
$90^\circ-25^\circ=65^\circ$ The measure of the other angle is 65° .

69. The sum of two supplementary angles is 180° .
$180^\circ-110^\circ=70^\circ$ The measure of the other angle is 70° .

71. If Teresa has $10,000 to invest and she puts n dollars in the stock account, the expression that represents the remainder that she puts in the bond account is $10{,}000-n$.

73. If Alexis has $3,800 to invest and he puts n dollars in the CD, the expression that represents the remainder that he puts in the money market is $3800-n$.

75. **a.** Let $p=$ the rate of the plane (in mph) in still air and $w=$ the rate of the wind in mph.

b. For explanation determining the equations see Section 4.2, exercises 80 and 81. The equations are:

$3(p+w)=375$
$5(p-w)=375$

c. Use the substitution method to solve the system, isolate a variable of one of the equations.
Solve $5(p-w)=375$ for p.

$5(p-w)=375$ Divide both sides by 5.
$\frac{\not{5}(p-w)}{\not{5}}=\frac{375}{5}$
$p-w=75$
$+w \quad +w$ Add w to both sides.
$p=w+75$

Now substitute $p=w+75$ into the first equation.

$3(p+w)=375$ Distribute.
$3p+3w=375$ Substitute in $p=w+75$.
$3(w+75)+3w=375$ Distribute.
$3w+225+3w=375$ Simplify.
$6w+225=375$
$-225 \quad -225$
$6w=150$ Solve for w.
$\frac{6w}{6}=\frac{150}{6}$
$w=25$

$p=w+75$ Substitute in $w=25$ to find p.
$p=(25)+75$
$p=100$

Since $p=100$ and $w=25$, the rate of the plane in still air is 100 mph and the rate of the wind is 25 mph.
Check the solution by substituting the values into both of the original equations.

77. **a.** Let $b =$ the rate of the boat (in knots) in still water and $c =$ the rate of the current in knots.
b. For explanation determining the equations see Section 4.2, exercises 79 and 82. The equations are:
$$3(b+c)=60$$
$$6(b-c)=60$$
c. Rewrite the equations in general form then use the elimination method to solve.
System equations:
$$3b+3c=60$$
$$6b-6c=60$$
Multiply the first equation by 2 so the coefficients of the c terms are opposites.
$$\begin{aligned}2(3b+3c)&=2(60)\\6b-6c&=60\end{aligned}\quad\text{is now}\quad\begin{aligned}6b+6c)&=120\\6b-6c&=60\end{aligned}$$
Now that the coefficients on c are opposites, add the two equations vertically to eliminate c.
$$6b+6c)=120 \quad \text{Add the two equations vertically.}$$
$$\underline{6b-6c=60}$$
$$12b=180 \quad \text{Solve for } b.$$
$$\frac{12b}{12}=\frac{180}{12} \quad \text{Divide both sides by 12.}$$
$$b=15$$
$$3b+3c=60 \quad \text{Substitute in } b=15 \text{ to find } c.$$
$$3(15)+3c=60$$
$$45+3c=60$$
$$\underline{-45 \quad -45}$$
$$3c=15$$
$$\frac{3c}{3}=\frac{15}{3}$$
$$c=5$$
Since $b=15$ and $c=5$, the rate of the boat in still water is 15 knots and the rate of the current is 5 knots.
Check the solution by substituting the values into both of the original equations.

79. **a.** Let $w =$ the number of pounds of whole-wheat flour and let $f =$ the number of pounds of white flour.
b. For explanation determining the equations see Chapter 4 Example 8. The equations are:
$$w+f=50$$
$$4.25w+3.75f=200$$
c. Use the substitution method to solve the system, isolate a variable of one of the equations.
Solve $w+f=50$ for w.
$$w+f=50$$
$$\underline{-f \quad -f} \quad \text{Subtract } f \text{ from both sides.}$$
$$w=-f+50$$
Now substitute $w=-f+50$ into the second equation.
$$4.25w+3.75f=200 \quad \text{Substitute in } w=-f+50.$$
$$4.25(-f+50)+3.75f=200 \quad \text{Distribute.}$$
$$-4.25f+212.5+3.75f=200 \quad \text{Simplify.}$$
$$-0.5f+212.5=200$$
$$\underline{-212.5 \quad -212.5}$$
$$-0.5f=-12.5 \quad \text{Solve for } f.$$
$$\frac{-0.5f}{-0.5}=\frac{-12.5}{-0.5}$$
$$f=25$$
$$w=-f+50 \quad \text{Substitute in } f=25 \text{ to find } w.$$
$$w=-(25)+50$$
$$w=25$$
Since $w=25$ and $f=25$, the baker should use 25 lb. of whole-wheat flour and 25 lb. of white flour in the mix.
Check the solution by substituting the values into both of the original equations.

81. **a.** Let $f =$ the number of hours per week Renata works at her first job and let $s =$ the number of hours per week Renata works at her second job.
b. For explanation determining the equations see Section 4.2 exercises 35 and 36. The equations are:
$$f+s=25$$
$$8.5f+9s=225$$
c. Transform the first equation so that s is isolated on one side. Substitute $s=-f+25$ into the second equation.
$$8.5f+9s=225 \quad \text{Substitute in } s=-f+25 \text{ for } s.$$
$$8.5f+9(-f+25)=225 \quad \text{Distribute.}$$
$$8.5f-9f+225=225 \quad \text{Simplify.}$$
$$-0.5f+225=225$$
$$\underline{-225 \quad -225} \quad \text{Subtract 225 from both sides.}$$
$$-0.5f=0 \quad \text{Solve for } f.$$
$$f=0$$
$$s=-f+25 \quad \text{Substitute in } f=0 \text{ to find } s.$$
$$s=-(0)+25$$
$$s=25$$
The solution to the system is $s=25$ and $f=0$. If Renata wants to earn \$225 working 25 hours in a week, she should work 25 hours at the second job and no hours at the first job.

83. Let $A =$ the amount (in dollars) invested in a CD earning 4% interest and let $B =$ the amount (in dollars) invested in a bond earning 5% interest.
For explanation determining the equations see Chapter 4 Example 6. The equations are:
$$A+B=4000$$
$$0.04A+0.05B=185$$
Transform the first equation so that A is isolated on one side. Substitute $A=-B+4000$ into the second equation.

$0.04A+0.05B=185$ Substitute in $A=-B+4000$.

$0.04(-B+4000)+0.05B=185$ Distribute.

$-0.04B+160+0.05B=185$ Simplify.

$0.01B+160=185$

$-160 \quad -160$ Subtract 160 from both sides.

$0.01B=25$ Solve for B.

$\frac{0.01B}{0.01}=\frac{25}{0.01}$

$B=2500$

$A=-B+4000$ Substitute in $B = 2500$ to find A.

$A=-(2500)+4000$

$A=1500$

The solution to the system is $A=1500$ and $B=2500$. If Cate wants to earn \$185 in interest every year, she should invest \$1500 in the 4% interest-earning CD and \$2500 in the 5% interest-earning bond fund.

85. Let $A=$ the amount (in dollars) Michael owes on the 18% APR MasterCard and let $B=$ the amount (in dollars) Michael owes on the 22% APR store credit card.

For explanation determining the equations see Chapter 4 Example 6. The equations are:

$A+B=3500$

$0.18A+0.22B=730$

Transform the first equation so that A is isolated on one side. Substitute $A=-B+3500$ into the second equation.

$0.18A+0.22B=730$ Substitute in $A=-B+3500$.

$0.18(-B+3500)+0.22B=730$ Distribute.

$-0.18B+630+0.22B=730$ Simplify.

$0.04B+630=730$

$-630 \quad -630$ Subtract 630 from both sides.

$0.04B=100$ Solve for B.

$\frac{0.04B}{0.04}=\frac{100}{0.04}$

$B=2500$

$A=-B+3500$ Substitute in $B = 2500$ to find A.

$A=-(2500)+3500$

$A=1000$

The solution to the system is $A=1000$ and $B=2500$. Michael owes \$1000 on the MasterCard and \$2500 on the store credit card.

87. Let $A=$ the amount (in liters) Greg needs of the 15% hydrochloric-acid and let $B=$ the amount (in liters) Greg needs of the 20% hydrochloric-acid.

For explanation determining the equations see Chapter 4 Example 8.

The equations are:

$A+B=20$

$0.15A+0.20B=3.75$

Transform the first equation so that A is isolated on one side. Substitute $A=-B+20$ into the second equation.

$0.15A+0.20B=3.75$ Substitute in $A=-B+20$.

$0.15(-B+20)+0.20B=3.75$ Distribute.

$-0.15B+3+0.20B=3.75$ Simplify.

$0.05B+3=3.75$

$-3 \quad -3$ Subtract 3 from both sides.

$0.05B=0.75$ Solve for B.

$\frac{0.05B}{0.05}=\frac{0.75}{0.05}$

$B=15$

$A=-B+20$ Substitute in $B = 15$ to find A.

$A=-(15)+20$

$A=5$

The solution to the system is $A=5$ and $B=15$. Greg needs to use 5 liters of 15% hydrochloric-acid and 15 liters of 20% hydrochloric-acid solution to make an 18.75% hydrochloric-acid solution.

89. Let $A=$ the amount (in liters) Christine needs of the 30% antiseptic alcohol solution and let $B=$ the amount (in liters) she needs of the 50% antiseptic alcohol solution.

For explanation determining the equations see Chapter 4 Example 8. The equations are:

$A+B=100$

$0.3A+0.5B=42$

Transform the first equation so that A is isolated on one side. Substitute $A=-B+100$ into the second equation.

$0.3A+0.5B=42$ Substitute in $A=-B+100$.

$0.3(-B+100)+0.5B=42$ Distribute.

$-0.3B+30+0.5B=42$ Simplify.

$0.2B+30=42$

$-30 \quad -30$ Subtract 30 from both sides.

$0.2B=12$ Solve for B.

$\frac{0.2B}{0.2}=\frac{12}{0.2}$

$B=60$

$A=-B+100$ Substitute in $B = 60$ to find A.

$A=-(60)+100$

$A=40$

The solution to the system is $A=40$ and $B=60$. Christine needs to use 40 liters of 30% antiseptic alcohol solution and 60 liters of 50% antiseptic alcohol solution to make a 42% antiseptic alcohol solution.

91. Complementary angles have a sum of 90°. Let $x=$ the degree measure of the first angle and let $y=$ the degree measure of the second angle.

For explanation determining the equations see Chapter 4 Example 9. The equations are:

$x+y=90^\circ$

$y=2x+3$

Substitute $y=2x+3$ into the first equation.

$x+y=90^\circ$ Substitute in $y=2x+3$.

$x+(2x+3)=90$ Simplify.

$3x+3=90$

$-3\ \ -3$ Subtract 3 from both sides.

$3x=87$ Solve for x.

$\frac{3x}{3}=\frac{87}{3}$

$x=29$

$y=2x+3$ Substitute in $x=29$ to find y.

$y=2(29)+3$

$y=58+3$

$y=61$

The measures of the angles are 29° and 61°.

93. Supplementary angles have a sum of 180°. Let $x=$ the degree measure of the first angle and let $y=$ the degree measure of the second angle.
For explanation determining the equations see Chapter 4 Example 9. The equations are:

$x+y=180^\circ$

$y=4x-5$

Substitute $y=4x-5$ into the first equation.

$x+y=180^\circ$ Substitute in $y=4x-5$.

$x+(4x-5)=180$ Simplify.

$5x-5=180$

$+5\ \ +5$ Add 5 to both sides.

$5x=185$ Solve for x.

$\frac{5x}{5}=\frac{185}{5}$

$x=37$

$y=4x-5$ Substitute in $x=37$ to find y.

$y=4(37)-5$

$y=148-5$

$y=143$

The measures of the angles are 37° and 143°.

95. The student did not multiply one of the equations by -1 to change the coefficients of the x terms to be opposites before adding the equations vertically.

97. When multiplying the second equation by -2, the student did not multiply on the right side, making it a 4 not -2. The equation after that step should be $-2x-10y=4$ (not $-2x-10y=-2$).

Section 4.4 Solving Linear Inequalities in Two Variables Graphically

1. **a.** $y<2x-3$ Is $(x,y)=(5,0)$ in the solution set?

$0\overset{?}{<}2(5)-3$ Substitute values for x and y.

$0\overset{?}{<}10-3$

$0<7$ This statement is true.

Yes, the point is in the solution set.

b. $y<2x-3$ Is $(x,y)=(6,0)$ in the solution set?

$0\overset{?}{<}2(6)-3$ Substitute values for x and y.

$0\overset{?}{<}12-3$

$0<9$ This statement is true.

Yes, the point is in the solution set.

3. **a.** $x\le-5$ Is $(x,y)=(-2,1)$ in the solution set?

$-2\overset{?}{\le}-5$ Substitute values for x and y.

$-2\overset{?}{\le}-5$ This statement is false.

No, the point is not in the solution set.

b. $x\le-5$ Is $(x,y)=(6,6)$ in the solution set?

$6\overset{?}{\le}-5$ Substitute values for x and y.

$6\overset{?}{\le}-5$ This statement is false.

No, the point is not in the solution set.

5. **a.** $y>33$ Is $(x,y)=(0,0)$ in the solution set?

$0\overset{?}{>}33$ Substitute values for x and y.

$0\overset{?}{>}33$ This statement is false.

No, the point is not in the solution set.

b. $y>33$ Is $(x,y)=(-3,58)$ in the solution set?

$58\overset{?}{>}33$ Substitute values for x and y.

$58>33$ This statement is true.

Yes, the point is in the solution set.

7. **a.** $-4x+y\le4$ Is $(x,y)=(0,7)$ in the solution set?

$-4(0)+(7)\overset{?}{\le}4$ Substitute values for x and y.

$0+7\overset{?}{\le}4$

$7\overset{?}{\le}4$ This statement is false.

No, the point is not in the solution set.

b. $-4x+y\le4$ Is $(x,y)=(1,-2)$ in the solution set?

$-4(1)+(2)\overset{?}{\le}4$ Substitute values for x and y.

$-4+2\overset{?}{\le}4$

$-2\le4$ This statement is true.

Yes, the point is in the solution set.

9. **a.** $x > -6.5$ Is $(x, y) = (1,1)$ in the solution set?

$1 \stackrel{?}{>} -6.5$ Substitute values for x and y.

$1 > -6.5$ This statement is true.

Yes, the point is in the solution set.

b. $x > -6.5$ Is $(x, y) = (-8,10)$ in the solution set?

$-8 \stackrel{?}{>} -6.5$ Substitute values for x and y.

$-8 \stackrel{?}{>} -6.5$ This statement is false.

No, the point is not in the solution set.

11. **a.** The inequality is of the form "less than" (<) but not equal to, therefore the graph of the solution will use a dashed line.

b. $y < \frac{3}{5}x + 8$ Is $(x, y) = (0,8)$ in the solution set?

$(8) \stackrel{?}{<} \frac{3}{5}(0) + 8$ Substitute values for x and y.

$8 \stackrel{?}{<} 0 + 8$

$8 \stackrel{?}{<} 8$ This statement is false.

No, the point is not in the solution set.

13. **a.** The inequality is of the form "less than or equal to" ($\le$), therefore the graph of the solution will use a solid line.

b. $x \le -5$ Is $(x, y) = (-5, 2)$ in the solution set?

$(-5) \stackrel{?}{\le} -5$ Substitute values for x and y.

$-5 \le -5$ This statement is true.

Yes, the point is in the solution set.

15. **a.** The inequality is of the form "less than or equal to" ($\le$), therefore the graph of the solution will use a solid line.

b. $-4x + y \le 4$ Is $(x, y) = (1,8)$ in the solution set?

$-4(1) + (8) \stackrel{?}{\le} 4$ Substitute values for x and y.

$-4 + 8 \stackrel{?}{\le} 4$

$4 \le 4$ This statement is true.

Yes, the point is in the solution set.

17. **a.** The inequality is of the form "greater than" (>) but not equal to, therefore the graph of the solution will use a dashed line.

b. $x - 6y > 9$ Is $(x, y) = (9,0)$ in the solution set?

$(9) - 6(0) \stackrel{?}{>} 9$ Substitute values for x and y.

$9 - 0 \stackrel{?}{>} 9$

$9 \stackrel{?}{>} 9$ This statement is false.

No, the point is not in the solution set.

19. **a.** Select the test point $(0,0)$.

b. $y < 3x - 3$ Is $(x, y) = (0,0)$ in the solution set?

$(0) \stackrel{?}{<} 3(0) - 3$ Substitute values for x and y.

$0 \stackrel{?}{<} 0 - 3$

$0 \stackrel{?}{<} -3$ This statement is false.

Because this is not true, we do not shade in the region containing the point $(0,0)$.

c. Based on the result of the point tested in part b., shade in the region below the line.

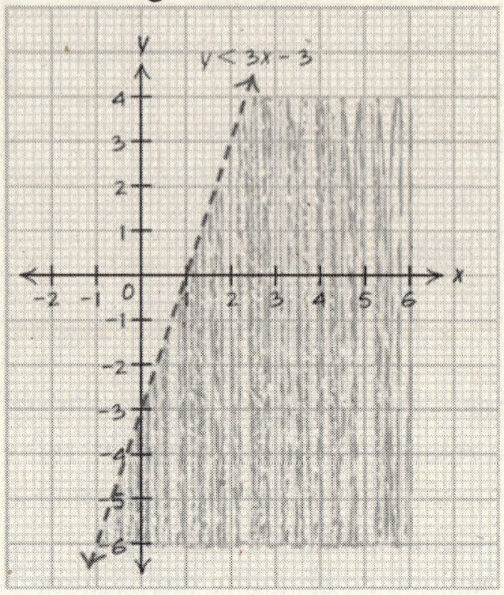

21. **a.** Select the test point $(0,0)$.

b. $4x - 3y \ge 12$ Is $(x, y) = (0,0)$ in the solution set?

$4(0) - 3(0) \stackrel{?}{\ge} 12$ Substitute values for x and y.

$0 - 0 \stackrel{?}{\ge} 12$

$0 \stackrel{?}{\ge} 12$ This statement is false.

Because this is not true, we do not shade in the region containing the point $(0,0)$.

c. Based on the result of the point tested in part b., shade in the region below the line.

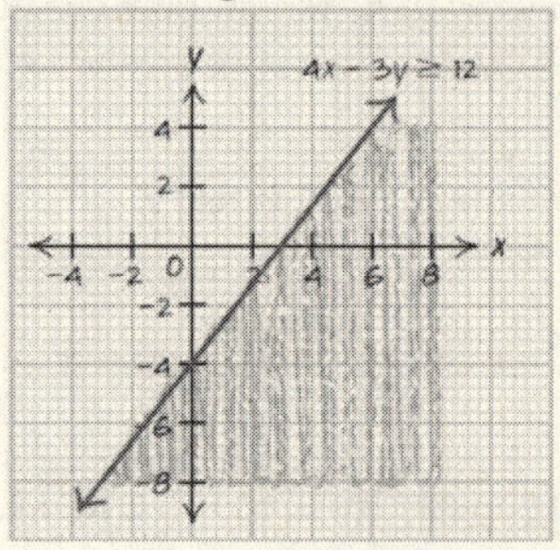

23. **a.** Select the test point $(1,1)$.

b. $-2y < 3x$ Is $(x, y) = (1,1)$ in the solution set?

$-2(1) \stackrel{?}{<} 3(1)$ Substitute values for x and y.

$-2 \stackrel{?}{<} 3$

$-2 < 3$ This statement is true.

Because this is true, we will shade in the region containing the point $(1,1)$.

c. Based on the result of the point tested in part b., shade in the region above the line.

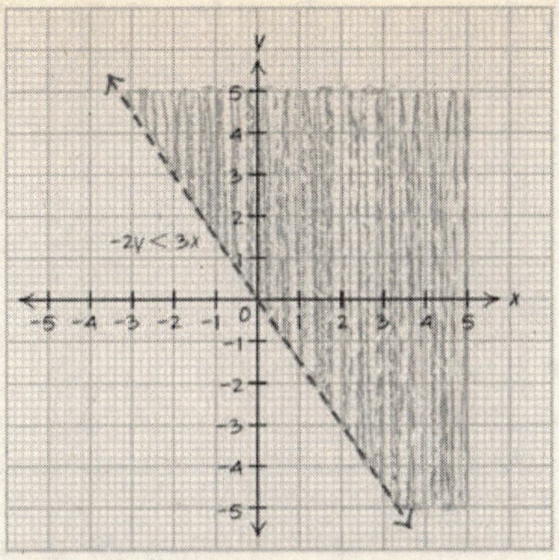

25. **a.** Select the test point $(0,0)$.

b. $x > 2$ Is $(x, y) = (0,0)$ in the solution set?

$(0) \overset{?}{>} 2$ Substitute values for x and y.

$0 \overset{?}{>} 2$ This statement is false.

Because this is not true, we do not shade in the region containing the point $(0,0)$.

c. Based on the result of the point tested in part b., shade in the region above/to the right of the line.

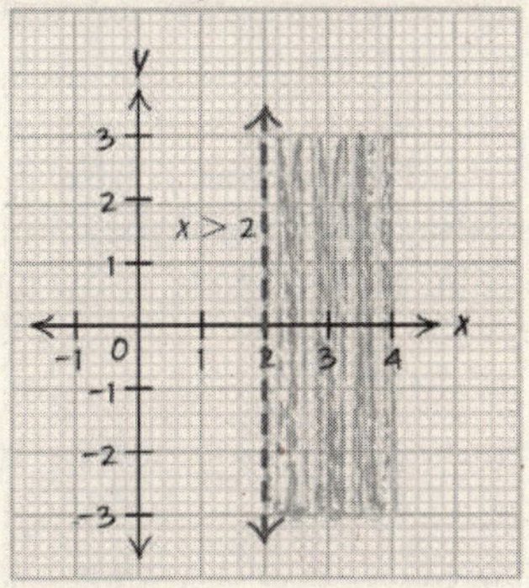

27. **a.** Select the test point $(0,0)$.

b. $y \geq -2$ Is $(x, y) = (0,0)$ in the solution set?

$(0) \overset{?}{\geq} -2$ Substitute values for x and y.

$0 \geq -2$ This statement is true.

Because this is true, we will shade in the region containing the point $(0,0)$.

c. Based on the result of the point tested in part b., shade in the region above the line.

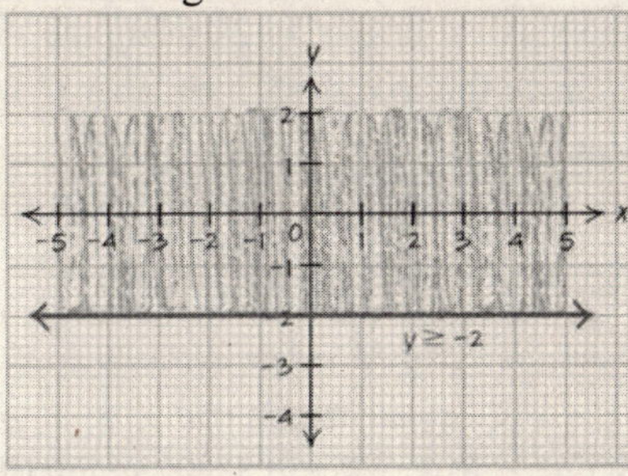

29. Graph the line $y < -2x + 6$, using the y-intercept $(0,6)$ and the slope $m = \frac{-2}{1} = \frac{rise}{run}$. Since this is a strict inequality (<), graph a dashed line.

Select the test point $(0,0)$. Is $(x, y) = (0,0)$ in the solution set?

$(0) \overset{?}{<} -2(0) + 6$ Substitute values for x and y.

$0 \overset{?}{<} 0 + 6$

$0 < 6$ This statement is true.

Because this is true, we will shade in the region containing the point $(0,0)$. Shade below the dashed line.

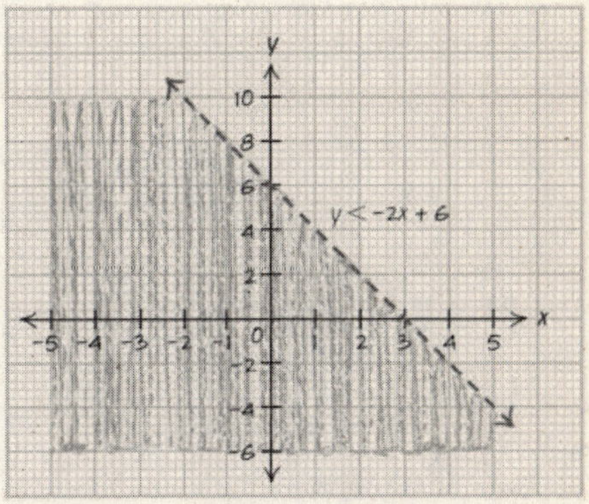

31. Graph the line $y \geq 3x - 7$, using the y-intercept $(0,-7)$ and the slope $m = \frac{3}{1} = \frac{rise}{run}$. Since this is not a strict inequality ($\geq$), graph a solid line.

Select the test point $(0,0)$. Is $(x, y) = (0,0)$ in the solution set?

$(0) \overset{?}{\geq} 3(0) - 7$ Substitute values for x and y.

$0 \overset{?}{>} 0 - 7$

$0 \geq -7$ This statement is true.

Because this is true, we will shade in the region containing the point $(0,0)$. Shade above the solid line.

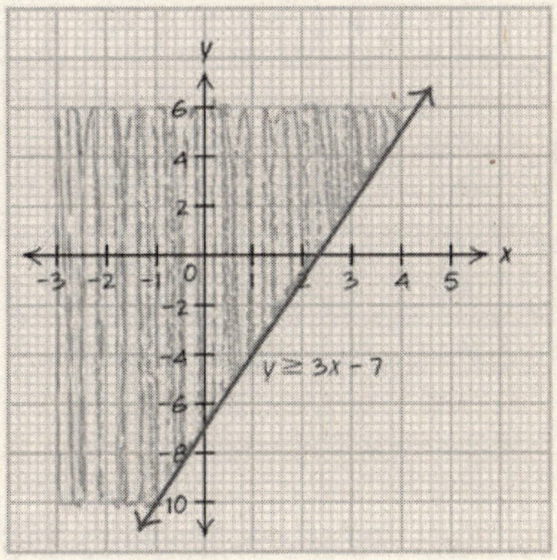

33. Graph the line $y \leq \frac{2}{3}x - 4$, using the y-intercept $(0,-4)$ and the slope $m = \frac{2}{3} = \frac{rise}{run}$. Since this is not a strict inequality ($\leq$), graph a solid line.

Select the test point $(0,0)$. Is $(x, y) = (0,0)$ in the solution set?

$(0)\overset{?}{\le}\frac{2}{3}(0)-4$ Substitute values for x and y.

$0\overset{?}{\le}0-4$

$0\le -4$ This statement is false.

Because this is false, we do not shade in the region containing the point $(0,0)$. Shade below the solid line.

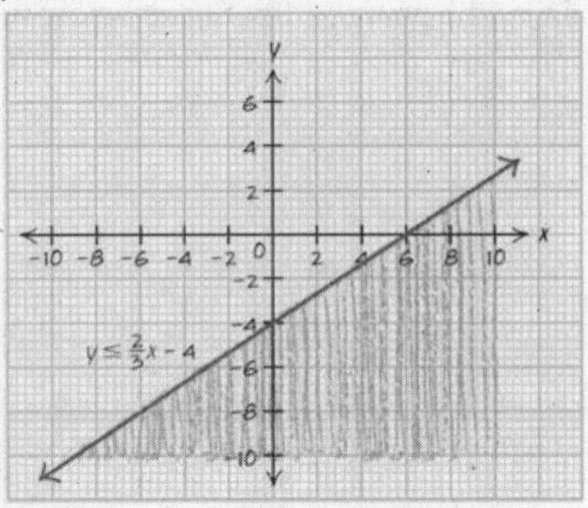

35. Graph the line $y<\frac{-1}{4}x+3$, using the y-intercept $(0,3)$ and the slope $m=\frac{-1}{4}=\frac{rise}{run}$. Since this is a strict inequality (<), graph a dashed line.

Select the test point $(0,0)$. Is $(x,y)=(0,0)$ in the solution set?

$(0)\overset{?}{<}\frac{-1}{4}(0)+3$ Substitute values for x and y.

$0\overset{?}{<}0+3$

$0<3$ This statement is true.

Because this is true, we will shade in the region containing the point $(0,0)$. Shade below the dashed line.

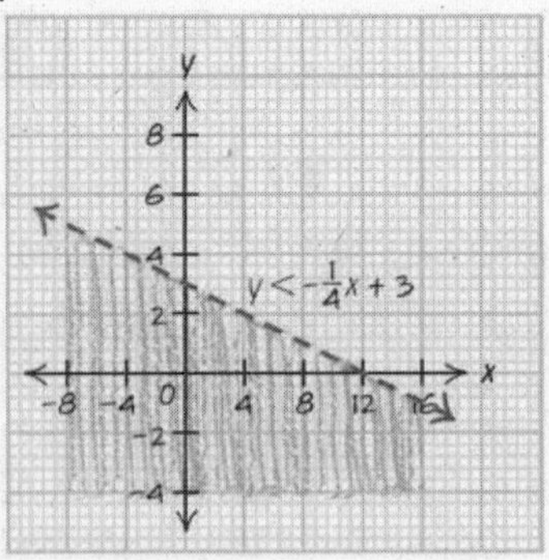

37. Graph the line $4x+2y\le 8$, using the x- and y-intercepts $(2,0)$ and $(0,4)$. Since this is not a strict inequality ($\le$), graph a solid line.

Select the test point $(0,0)$. Is $(x,y)=(0,0)$ in the solution set?

$4(0)+2(0)\overset{?}{\le}8$ Substitute values for x and y.

$0+0\overset{?}{\le}8$

$0\le 8$ This statement is true.

Because this is true, we will shade in the region containing the point $(0,0)$. Shade below the solid line.

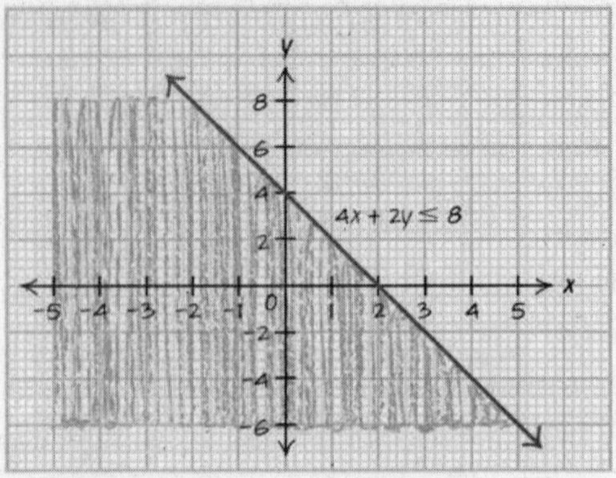

39. Graph the line $4x+3y<-12$, using the x- and y-intercepts $(-3,0)$ and $(0,-4)$. Since this is a strict inequality (<), graph a dashed line.

Select the test point $(0,0)$. Is $(x,y)=(0,0)$ in the solution set?

$4(0)+3(0)\overset{?}{<}-12$ Substitute values for x and y.

$0+0\overset{?}{<}-12$

$0\overset{?}{<}-12$ This statement is false.

Because this is false, we do not shade in the region containing the point $(0,0)$. Shade below the dashed line.

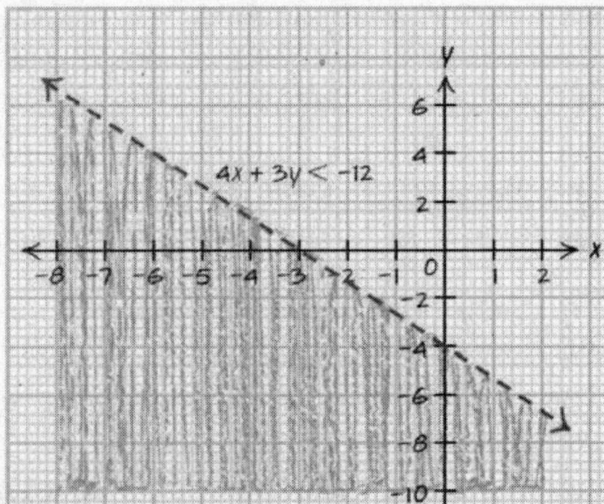

41. Graph the vertical line $x>-2$, using the x- intercept $(-2,0)$. Since this is a strict inequality (>), graph a dashed line.

Select the test point $(0,0)$. Is $(x,y)=(0,0)$ in the solution set?

$(0)\overset{?}{>}-2$ Substitute values for x and y.

$0>-2$ This statement is true.

Because this is true, we will shade in the region containing the point $(0,0)$. Shade above/to the right of the dashed line.

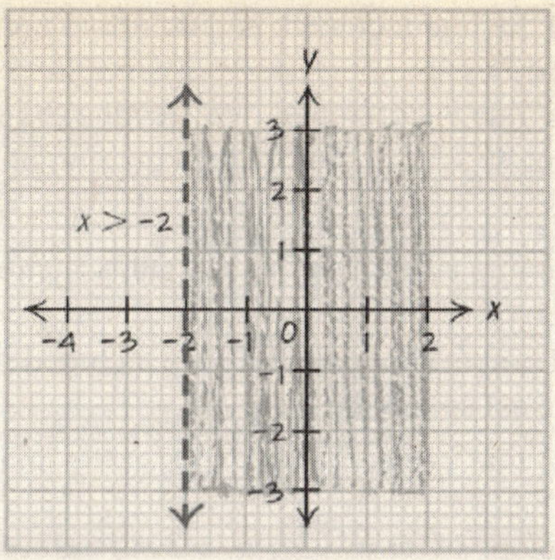

43. Graph the horizontal line $y \geq 3.5$, using the y-intercept $(0, 3.5)$. Since this is not a strict inequality ($\geq$), graph a solid line.
Select the test point $(0, 0)$. Is $(x, y) = (0, 0)$ in the solution set?

$(0) \overset{?}{\geq} 3.5$ Substitute values for x and y.

$0 \overset{?}{\geq} 3.5$ This statement is false.

Because this is false, we do not shade in the region containing the point $(0, 0)$. Shade above the solid line.

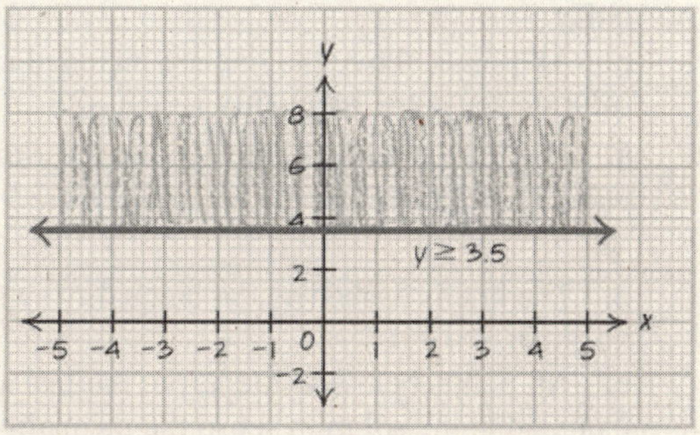

45. Graph the vertical line $x \geq 0$, using the x-intercept $(0, 0)$. Since this is not a strict inequality ($\geq$), graph a solid line.
Select the test point $(1, 1)$. Is $(x, y) = (1, 1)$ in the solution set?

$(1) \overset{?}{\geq} 0$ Substitute values for x and y.

$1 \geq 0$ This statement is true.

Because this is true, we will shade in the region containing the point $(1, 1)$. Shade above/to the right of the solid line.

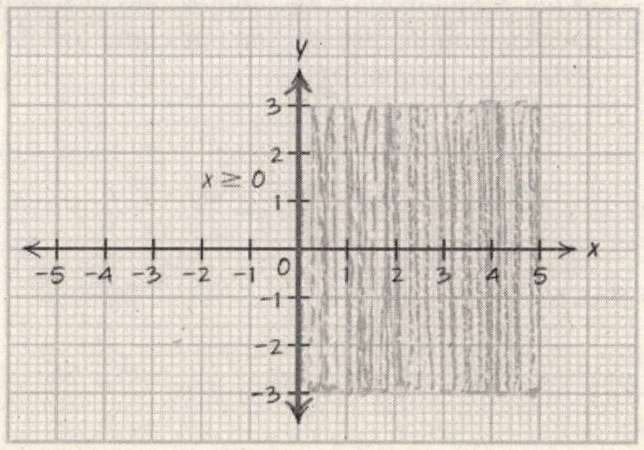

47. **a.** Let m = the number of boxes of oatmeal cookies, and let c = the number of boxes of chocolate chip cookies.

b. $4m + 5.5c \leq 25$

c. Graph the solution set using the m- and c-intercepts $(6.25, 0)$ and $(0, 4.\overline{54})$ and shade below the solid line (negative values do not make sense in this situation so we will not graph them).

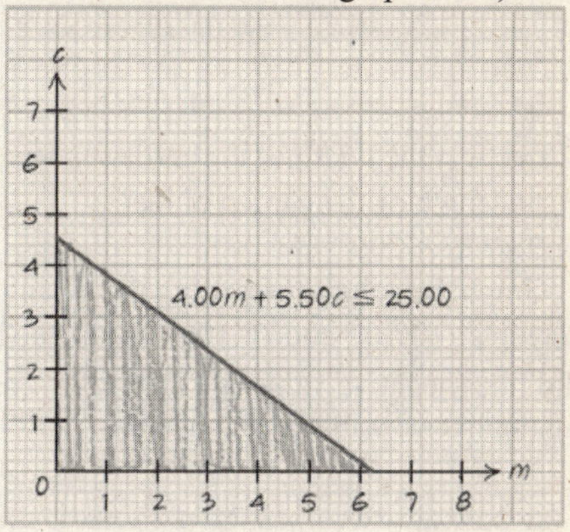

49. **a.** Let f = the number of hours Shahab works at his first job, and let s = the number of hours Shahab works at his second job.

b. $9f + 8s > 300$

c. Graph the solution set using the f- and s-intercepts $(33.\overline{3}, 0)$ and $(0, 37.5)$ and shade above the dashed line (negative values do not make sense in this situation so we will not graph them).

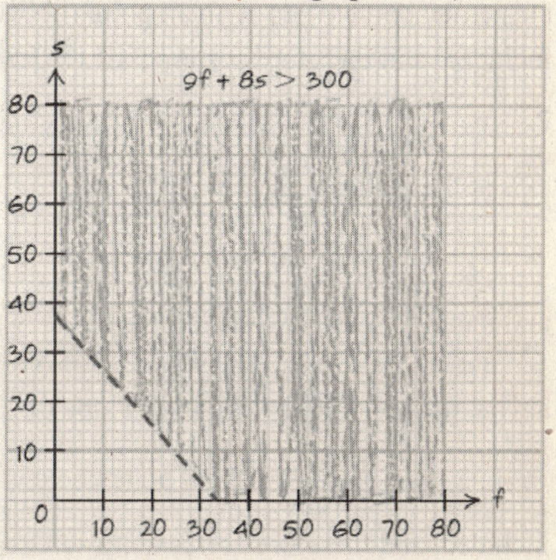

51. **a.** Let f = the number of dollars Paul invests in the first account, and let s = the number of dollars Paul invests in the second account.

b. $0.05f + 0.06s > 500$

c. Graph the solution set using the f- and s-intercepts $(10000, 0)$ and $(0, 8333.\overline{3})$ and shade above the dashed line (negative values do not make sense in this situation so we will not graph them).

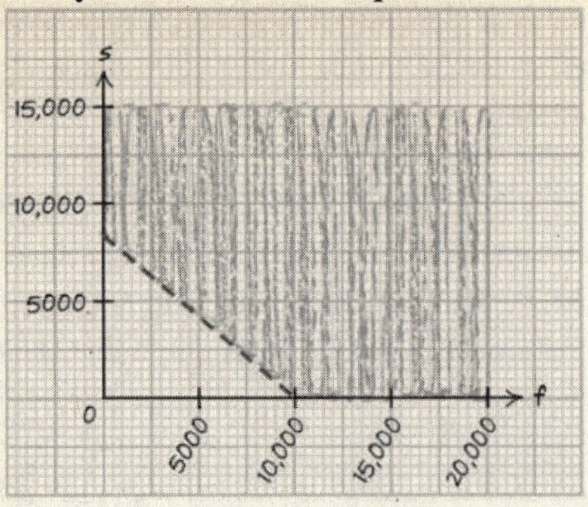

53. **a.** Let $w=$ the width of the garden in feet, and let $l=$ the length of the garden in feet.
b. $2w+2l \le 200$
c. Graph the solution set using the w- and l-intercepts $(100,0)$ and $(0,100)$ and shade below the solid line (negative values do not make sense in this situation so we will not graph them).

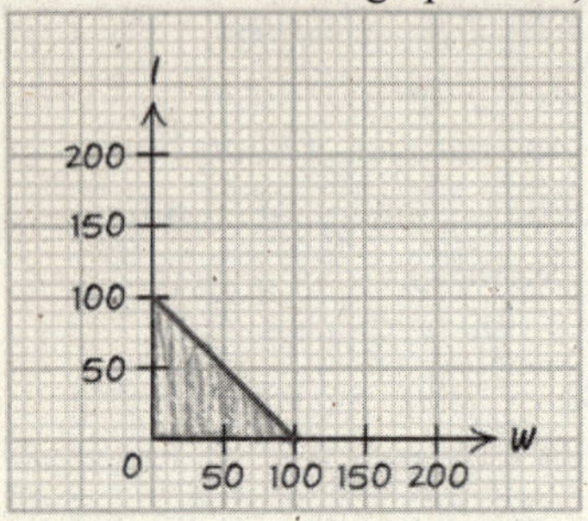

55. **1.** $y \ge 2$ is a solid horizontal line intersecting the y-axis at the point $(0,2)$ with all points above the line included, and therefore corresponds to Graph B.
2. $y > \frac{1}{2}x+2$ is a dashed line intersecting the y-axis at the point $(0,2)$, having a slope of $m=\frac{1}{2}=\frac{rise}{run}$ with all points above the line included, and therefore corresponds to Graph A.

57. **1.** $y < -2x+2$ is a dashed line intersecting the y-axis at the point $(0,2)$, having a slope of $m=-2=-\frac{2}{1}=\frac{rise}{run}$ with all points below the line included, and therefore corresponds to Graph A.
2. $y \le -2x+2$ is a solid line intersecting the y-axis at the point $(0,2)$, having a slope of $m=-2=-\frac{2}{1}=\frac{rise}{run}$ with all points below the line included, and therefore corresponds to Graph B.

59. **1.** $y \ge 4$ is a solid horizontal line intersecting the y-axis at the point $(0,4)$ with all points above the line included, and therefore corresponds to Graph A.
2. $x \ge 4$ is a solid vertical line intersecting the x-axis at the point $(4,0)$ with all points above/to the right of the line included, and therefore corresponds to Graph B.

61. **a.** The line is horizontal, intersecting the y-axis at the point $(0,-1)$, therefore the equation of the line is $y=-1$.
b. The line is solid so the inequality symbol will be either $\le$ or $\ge$.
c. All points above the line are included in the solution set so the inequality is $y \ge -1$.

63. **a.** The line intersects the y-axis at the point $(0,0)$ with a slope of $m=\frac{rise}{run}=\frac{3}{1}=3$, therefore the equation of the line is $y=3x$.
b. The line is dashed so the inequality symbol will be either $<$ or $>$.
c. All points above the line are included in the solution set so the inequality is $y > 3x$.

65. **a.** The line intersects the y-axis at the point $(0,0)$ with a slope of $m=\frac{rise}{run}=-\frac{2}{1}=-2$, therefore the equation of the line is $y=-2x$.
b. The line is solid so the inequality symbol will be either $\le$ or $\ge$.
c. All points above the line are included in the solution set so the inequality is $y \ge -2x$.

67. **a.** The line is vertical, intersecting the x-axis at the point $(2,0)$, therefore the equation of the line is $x=2$.
b. The line is dashed so the inequality symbol will be either $<$ or $>$.
c. All points below/to the left of the line are included in the solution set so the inequality is $x<2$.

69. The amount should be *at least* \$150.00, so anything over \$150.00 would also be acceptable and this would correspond to "greater than or equal to." Therefore, this is an inequality.

71. The amount should be exactly 20 hours, not any more or less, therefore this is an equality.

73. The amount should be *at least* 15 hours, so anything over 15 hours would also be acceptable and this would correspond to "greater than or equal to." Therefore, this is an inequality.

75. The amount should be *at least* 95%, so anything over 95% would also be acceptable and this would correspond to "greater than or equal to." Therefore, this is an inequality.

Chapter 4 Review

1. The solution to this system is the intersection of the two lines. Looking at the graph, we see the point of intersection is at $x=2$ and $y=-4$, or as an ordered pair, $(2,-4)$.

2. The solution to this system is the intersection of the two lines. Looking at the graph, we see the point of intersection is at $x=2$ and $y=3$, or as an ordered pair, $(2,3)$.

3. **a.** Substitute the points into both equations to see if both are true.
First equation:

$(-8)\stackrel{?}{=}\frac{-4}{3}(6)$ Sustitute values for x and y.

$-8=-8$ This is a true statement.

Second equation:

$(-8)\stackrel{?}{=}(6)-7$ Sustitute values for x and y.

$-8\stackrel{?}{=}-1$ This is a false statement.

The point is a solution for the first equation but not the second. Since $(6,-8)$ does not make both equations true, it is *not* a solution of the system.

b. Substitute the points into both equations to see if both are true.
First equation:

$(-4)\stackrel{?}{=}\frac{-4}{3}(3)$ Sustitute values for x and y.

$-4=-4$ This is a true statement.

Second equation:

$(-4)\stackrel{?}{=}(3)-7$ Sustitute values for x and y.

$-4=-4$ This is a true statement.

The point is a solution for both equations, therefore it is a solution of the system.

4. **a.** Substitute the points into both equations to see if both are true.
First equation:

$(4)\stackrel{?}{=}(-2)+6$ Sustitute values for x and y.

$4=4$ This is a true statement.

Second equation:

$(4)\stackrel{?}{=}-2(-2)$ Sustitute values for x and y.

$4=4$ This is a true statement.

The point is a solution for both equations, therefore it is a solution of the system.

b. Substitute the points into both equations to see if both are true.
First equation:

$(-4)\stackrel{?}{=}(2)+6$ Sustitute values for x and y.

$-4\stackrel{?}{=}8$ This is a false statement.

Second equation:

$(-4)\stackrel{?}{=}-2(2)$ Sustitute values for x and y.

$-4=-4$ This is a true statement.

The point is a solution for the second equation but not the first. Therefore, it is *not* a solution of the system.

5. Graph both lines using slope-intercept form.

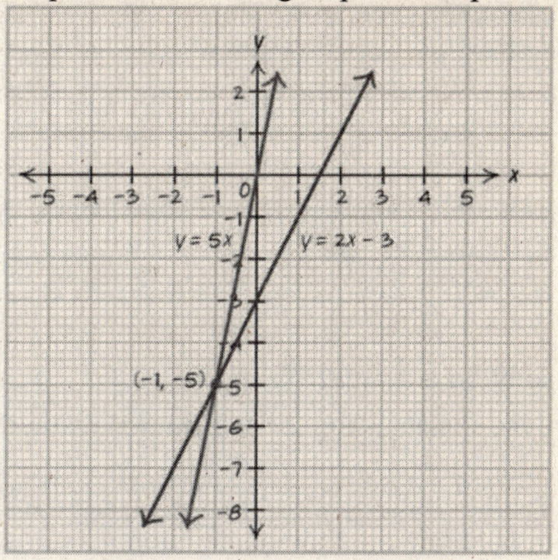

The solution is the point $(-1,-5)$.

6. Graph both lines using slope-intercept form.

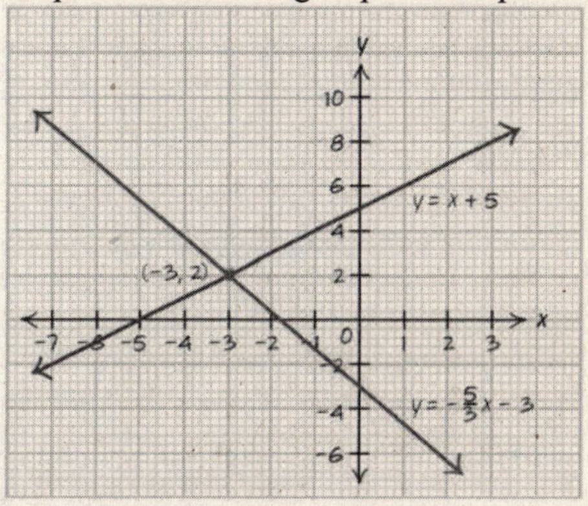

The solution is the point $(-3,2)$.

7. Solve the system using the substitution method. Solve the first equation for y, changing it to $y=3x+1$.

$y=3x+1$ Substitute $y=3x+1$ into second equation.
$x+y=13$

$x+(3x+1)=13$ Simplify.

$4x+1=13$
$\underline{\;\;-1\;\;-1}$
$4x=12$ Solve for x.

$\frac{4x}{4}=\frac{12}{4}$ Divide both sides by 4.

$x=3$

$y=3x+1$ Substitute in $x=3$ to find y.
$y=3(3)+1$
$y=9+1$
$y=10$

The solution is $(3,10)$.

8. Solve the system using the substitution method. Solve the first equation for y, changing it to $y=-2x-4$.

$y=-2x-4$ Substitute $y=-2x-4$ into second equation.
$3x+2y=7$

$3x+2(-2x-4)=7$ Distribute.

$3x-4x-8=7$ Simplify.

$-x-8=7$

$+8\ \ +8$

$-x=15$ Solve for x.

$\frac{-x}{-1}=\frac{15}{-1}$ Divide both sides by -1.

$x=-15$

$y=-2x-4$ Substitute in $x=-15$ to find y.

$y=-2(-15)-4$

$y=30-4$

$y=26$

The solution is $(-15,26)$.

9. Solve the system using the substitution method.

$y=\frac{-2}{3}x+1$ Substitute $y=\frac{-2}{3}x+1$ into second equation.

$4x-6y=-6$

$4x-6\left(\frac{-2}{3}x+1\right)=-6$ Distribute.

$4x+4x-6=-6$ Simplify.

$8x-6=-6$

$+6\ \ +6$

$8x=0$ Solve for x.

$\frac{8x}{8}=\frac{0}{8}$ Divide both sides by 8.

$x=0$

$y=\frac{-2}{3}x+1$ Substitute in $x=0$ to find y.

$y=\frac{-2}{3}(0)+1$

$y=0+1$

$y=1$

The solution is $(0,1)$.

10. Solve the system using the substitution method.

$x=4y+12$ Substitute $x=4y+12$ into second equation.

$-2x+8y=-24$

$-2(4y+12)+8y=-24$ Distribute.

$-8y-24+8y=-24$ Simplify.

$-24=-24$ This is a true statement.

Because this equation is always true ($-24=-24$), there are infinitely many solutions. These two equations graph as the same line.

11. Solve the system using the substitution method. Solve the second equation for y, changing it to $y=-3x+1$.

$6x+2y=14$ Substitute $y=-3x+1$ into first equation.

$y=-3x+1$

$6x+2(-3x+1)=14$ Distribute.

$6x-6x+2=14$ Simplify.

$2\overset{?}{=}14$ This is a false statement.

Since this is a false statement ($2\overset{?}{=}14$), there are no solutions to the system. The lines are parallel.

12. Solve the system using the substitution method.

$5y=2x+5$ Substitute $y=\frac{2}{5}x-5$ into first equation.

$y=\frac{2}{5}x-5$

$5\left(\frac{2}{5}x-5\right)=2x+5$ Distribute.

$2x-25=2x+5$ Simplify.

$-2x\ \ \ -2x$

$-25\overset{?}{=}5$ This is a false statement.

Since this is a false statement ($-25\overset{?}{=}5$), there are no solutions to the system. The lines are parallel.

13. Let $c=$ the rate of the cyclist (in mph) in still air and $w=$ the rate of the wind in mph.

The system of equations is:

$2(c+w)=48$

$3(c-w)=48$

Solve the system using the substitution method. Solve the first equation for w.

$2(c+w)=48$ Distribute.

$2c+2w=48$

$-2c\ \ \ -2c$

$2w=-2c+48$

$\frac{2w}{2}=\frac{-2c+48}{2}$

$w=-c+24$

Now substitute $w=-c+24$ into the second equation.

$3(c-w)=48$ Distribute.

$3c-3w=48$ Substitute in $w=-c+24$ for w.

$3c-3(-c+24)=48$ Distribute.

$3c+3c-72=48$ Simplify.

$6c-72=48$

$+72\ \ +72$

$6c=120$ Solve for c.

$\frac{6c}{6}=\frac{120}{6}$

$c=20$

$w=-c+24$ Substitute in $c=20$ to find w.

$w=-(20)+24$

$w=4$

The rate of the cyclist is 20 mph and the rate of the wind is 4 mph.

14. Let $c=$ the rate of the cyclist (in mph) in still air and $w=$ the rate of the wind in mph.

The system of equations is:

$3(c+w)=45$

$5(c-w)=45$

Solve the system using the substitution method. Solve the second equation for c.

$5(c-w)=45$ Distribute.

$5c-5w=45$
$\underline{+5w\ +5w}$
$5c=5w+45$

$\frac{5c}{5}=\frac{5w+45}{5}$

$c=w+9$

Now substitute $c=w+9$ into the first equation.

$3(c+w)=45$ Distribute.

$3c+3w=45$ Substitute in $c=w+9$ for c.

$3(w+9)+3w=45$ Distribute.

$3w+27+3w=45$ Simplify.

$6w+27=45$
$\underline{-27\ -27}$
$6w=18$ Solve for w.

$\frac{6w}{6}=\frac{18}{6}$

$w=3$

$c=(3)+9$ Substitute in $w=3$ to find c.

$c=12$

The rate of the cyclist is 12 mph and the rate of the wind is 3 mph.

15. Let $p=$ the rate of the plane (in mph) in still air and $w=$ the rate of the wind in mph.
The system of equations is:

$2(p+w)=240$

$3(p-w)=240$

Solve the system using the substitution method.
Solve the second equation for p.

$3(p-w)=240$ Distribute.

$3p-3w=240$
$\underline{+3w\ +3w}$
$3p=3w+240$

$\frac{3p}{3}=\frac{3w+240}{3}$

$p=w+80$

Now substitute $p=w+80$ into the first equation.

$2(p+w)=240$ Distribute.

$2p+2w=240$ Substitute in $p=w+80$ for p.

$2(w+80)+2w=240$ Distribute.

$2w+160+2w=240$ Simplify.

$4w+160=240$
$\underline{-160\ -160}$
$4w=80$ Solve for w.

$\frac{4w}{4}=\frac{80}{4}$

$w=20$

$p=(20)+80$ Substitute in $w=20$ to find p.

$p=100$

The rate of the plane is 100 mph and the rate of the wind is 20 mph.

16. Let $b=$ the rate of the boat (in knots) in still water and $c=$ the rate of the current in knots.
The system of equations is:

$2(b+c)=28$

$4(b-c)=28$

Solve the system using the substitution method.
Solve the second equation for b.

$4(b-c)=28$ Distribute.

$4b-4c=28$
$\underline{+4c\ +4c}$
$4b=4c+28$

$\frac{4b}{4}=\frac{4c+28}{4}$

$b=c+7$

Now substitute $b=c+7$ into the first equation.

$2(b+c)=28$ Distribute.

$2b+2c=28$ Substitute in $b=c+7$ for b.

$2(c+7)+2c=28$ Distribute.

$2c+14+2c=28$ Simplify.

$4c+14=28$
$\underline{-14\ -14}$
$4c=14$ Solve for c.

$\frac{4c}{4}=\frac{14}{4}$

$c=3.5$

$b=(3.5)+7$ Substitute in $c=3.5$ to find b.

$b=10.5$

The rate of the boat is 10.5 knots and the rate of the current is 3.5 knots.

17. Let $x=$ the first number and $y=$ the second number. The system of equations is:

$x-y=8$

$2x+y=37$

Solve the system using the substitution method.
Solve the first equation for x.

$x-y=8$
$\underline{+y\ +y}$
$x=y+8$

Now substitute $x=y+8$ into the second equation.

$2x+y=37$ Substitute in $x=y+8$ for x.

$2(y+8)+y=37$ Distribute.

$2y+16+y=37$

$3y+16=37$ Simplify.
$\underline{-16\ -16}$
$3y=21$ Solve for y.

$\frac{3y}{3}=\frac{21}{3}$

$y=7$

$x=(7)+8$ Substitute in $y=7$ to find x.

$x=15$

The numbers are 15 and 7.

18. Let $x =$ the first number and $y =$ the second number. The system of equations is:

$x - y = 20$

$3x + y = 16$

Solve the system using the substitution method. Solve the first equation for x.

$$\begin{aligned} x - y &= 20 \\ +y\ &\ +y \\ \hline x &= y + 20 \end{aligned}$$

Now substitute $x = y + 20$ into the second equation.

$3x + y = 16$ Substitute in $x = y + 20$ for x.

$3(y + 20) + y = 16$ Distribute.

$3y + 60 + y = 16$

$4y + 60 = 16$ Simplify.

$-60\ \ -60$

$4y = -44$ Solve for y.

$\frac{4y}{4} = \frac{-44}{4}$

$y = -11$

$x = (-11) + 20$ Substitute in $y = -11$ to find x.

$x = 9$

The numbers are 9 and -11.

19. Let $x =$ the first number and $y =$ the second number. The system of equations is:

$x + y = 52$

$x = 3y$

Solve the system using the substitution method.

$x + y = 52$ Substitute in $x = 3y$ for x.

$(3y) + y = 52$ Simplify.

$4y = 52$ Solve for y.

$\frac{4y}{4} = \frac{52}{4}$

$y = 13$

$x = 3(13)$ Substitute in $y = 13$ to find x.

$x = 39$

The numbers are 39 and 13.

20. Let $x =$ the first number and $y =$ the second number. The system of equations is:

$x + y = 28$

$x = 6y$

Solve the system using the substitution method.

$x + y = 28$ Substitute in $x = 6y$ for x.

$(6y) + y = 28$ Simplify.

$7y = 28$ Solve for y.

$\frac{7y}{7} = \frac{28}{7}$

$y = 4$

$x = 6(4)$ Substitute in $y = 4$ to find x.

$x = 24$

The numbers are 24 and 4.

21. Solve the system using the elimination method.

$x - 2y = 7$ Add the two equations vertically.

$3x + 2y = -3$

$4x = 4$

$\frac{4x}{4} = \frac{4}{4}$

$x = 1$

$3(1) + 2y = -3$ Substitute in $x = 1$ to find y.

$3 + 2y = -3$

$-3\ \ \ -3$

$2y = -6$

$\frac{2y}{2} = \frac{-6}{2}$

$y = -3$

The solution is $(1, -3)$.

22. Solve the system using the elimination method.

$5x + 7y = 27$ Add the two equations vertically.

$-5x - y = 9$

$6y = 36$

$\frac{6y}{6} = \frac{36}{6}$

$y = 6$

$5x + 7(6) = 27$ Substitute in $y = 6$ to find x.

$5x + 42 = 27$

$-42\ \ -42$

$5x = -15$

$\frac{5x}{5} = \frac{-15}{5}$

$x = -3$

The solution is $(-3, 6)$.

23. Solve the system using the elimination method. Multiply the first equation by 2.

$2(3x + 2y) = 2(16)$, $10x - 4y = 0$ is now $6x + 4y = 32$, $10x - 4y = 0$

$6x + 4y = 32$ Add the two equations vertically.

$10x - 4y = 0$

$16x = 32$

$\frac{16x}{16} = \frac{32}{16}$

$x = 2$

$3(2) + 2y = 16$ Substitute in $x = 2$ to find y.

$6 + 2y = 16$

$-6\ \ \ -6$

$2y = 10$

$\frac{2y}{2} = \frac{10}{2}$

$y = 5$

The solution is $(2,5)$.

24. Solve the system using the elimination method. Multiply the second equation by 2.

$$\begin{aligned}2x+y&=-7\\2(-x+3y)&=2(0)\end{aligned}\quad\text{is now}\quad\begin{aligned}2x+y&=-7\\-2x+6y&=0\end{aligned}$$

$$\begin{aligned}2x+y&=-7 \quad \text{Add the two equations vertically.}\\-2x+6y&=0\\\hline 7y&=-7\\\frac{7y}{7}&=\frac{-7}{7}\\y&=-1\end{aligned}$$

$$\begin{aligned}2x+(-1)&=-7 \quad \text{Substitute in } y=-1 \text{ to find } x.\\+1\quad&+1\\\hline 2x&=-6\\\frac{2x}{2}&=\frac{-6}{2}\\x&=-3\end{aligned}$$

The solution is $(-3,-1)$.

25. Solve the system using the elimination method. Multiply the first equation by -1.

$$\begin{aligned}-1(3x+y)&=-1(5)\\3x+y&=-8\end{aligned}\quad\text{is now}\quad\begin{aligned}-3x-y&=-5\\3x+y&=-8\end{aligned}$$

$$\begin{aligned}-3x-y&=-5 \quad \text{Add the two equations vertically.}\\3x+y&=-8\\\hline 0&\stackrel{?}{=}-13 \quad \text{This is a false statement.}\end{aligned}$$

Since this is a false statement ($0\stackrel{?}{=}-13$), there are no solutions to the system. The lines are parallel.

26. Solve the system using the elimination method.

$$\begin{aligned}6x+y&=-5 \quad \text{Add the two equations vertically.}\\-6x-y&=-1\\\hline 0&\stackrel{?}{=}-6 \quad \text{This is a false statement.}\end{aligned}$$

Since this is a false statement ($0\stackrel{?}{=}-6$), there are no solutions to the system. The lines are parallel.

27. Solve the system using the elimination method.

Multiply the first equation by 2. $\begin{aligned}2(x+2y)&=2(3)\\-2x-4y&=-6\end{aligned}$ is now $\begin{aligned}2x+4y&=6\\-2x-4y&=-6\end{aligned}$

$$\begin{aligned}2x+4y&=6 \quad \text{Add the two equations vertically.}\\-2x-4y&=-6\\\hline 0&=0\end{aligned}$$

Because this equation is always true ($0=0$), there are infinitely many solutions. These two equations graph as the same line.

28. Solve the system using the elimination method. Multiply the first equation by 2.

$$\begin{aligned}2(-2x+3y)&=2(18)\\4x-6y&=-36\end{aligned}\quad\text{is now}\quad\begin{aligned}-4x+6y&=36\\4x-6y&=-36\end{aligned}$$

$$\begin{aligned}-4x+6y&=36 \quad \text{Add the two equations vertically.}\\4x-6y&=-36\\\hline 0&=0\end{aligned}$$

Because this equation is always true ($0=0$), there are infinitely many solutions. These two equations graph as the same line.

29. The coefficients of the x terms are opposites, therefore it would be most efficient to solve using the elimination method.

30. Isolating the x term in the first equation would be more efficient than finding the least common multiples of either of the coefficients, therefore it would be most efficient to solve using the substitution method.

31. Let $J=$ the cost in dollars for each box of jerky treats and let $L=$ the cost in dollars for each leash. The system of equations is as follows:

$$\begin{aligned}4J+2L&=65\\3J+3L&=73.5\end{aligned}$$

Solve the system using the substitution method. Solve for L in the first equation.

$$\begin{aligned}4J+2L&=65\\-4J\quad&-4J\\\hline 2L&=-4J+65\\\frac{2L}{2}&=\frac{-4J+65}{2}\\L&=-2J+32.5\end{aligned}$$

Now substitute $L=-2J+32.5$ into the second equation.

$$\begin{aligned}3J+3L&=73.5 \quad \text{Substitute in } L=-2J+32.5 \text{ for } J.\\3J+3(-2J+32.5)&=73.5 \quad \text{Distribute.}\\3J-6J+97.5&=73.5 \quad \text{Simplify.}\\-3J+97.5&=73.5\\-97.5\quad&-97.5\\\hline -3J&=-24 \quad \text{Solve for } J.\\\frac{-3J}{-3}&=\frac{-24}{-3}\\J&=8\end{aligned}$$

$$\begin{aligned}L&=-2(8)+32.5 \quad \text{Substitute in } J=8 \text{ to find } L.\\L&=-16+32.5\\L&=-16.5\end{aligned}$$

The cost of each box of jerky treats is \$8.00 and the cost of each leash is \$16.50.

Check the solution by substituting the values into both of the original equations. For steps on how to check the solution see exercise 25 in this section.

32. Let $c=$ the cost in dollars for each bag of candy and let $m=$ the cost in dollars for each metronome. The system of equations is as follows:

$$\begin{aligned}3c+m&=67.75\\2c+2m&=78.5\end{aligned}$$

Solve the system using the elimination method. Multiply the first equation by -2.

$-2(3c+m)=-2(67.75)$ is now $-6c-2m=-135.5$

$2c+2m=78.5$ $\quad$ $2c+2m=78.5$

$-6c-2m=-135.5$ Add the two equations vertically.

$\underline{2c+2m=78.5}$

$-4c=-57$

$\frac{-4c}{-4}=\frac{-57}{-4}$

$c=14.25$

$3(14.25)+m=67.75$ Substitute in $c=14.25$ to find m.

$42.75+m=67.75$

$\underline{-42.75 \quad -42.75}$

$m=25$

The cost of each bag of candy is \$14.25 and the cost of each metronome is \$16.50.

33. Let $x=$ the amount (in dollars) invested in the bond fund earning 5% interest and let $y=$ the amount (in dollars) invested in the stock fund earning 6% interest.

The system of equations is:

$x+y=2000$

$0.05x+0.06y=105$

Solve the system using the substitution method. Solve for y in the first equation.

$x+y=2000$

$\underline{-x \quad\quad -x}$

$y=-x+2000$

Now substitute $y=-x+2000$ into the second equation.

$0.05x+0.06y=105$ Substitute in $y=-x+2000$.

$0.05x+0.06(-x+2000)=105$ Distribute.

$0.05x-0.06x+120=105$ Simplify.

$-0.01x+120=105$

$\underline{-120 \quad -120}$ Subtract 120 from both sides.

$-0.01x=-15$ Solve for x.

$\frac{-0.01x}{-0.01}=\frac{-15}{-0.01}$

$x=1500$

$y=-(1500)+2000$ Substitute in $x=1500$ to find y.

$y=500$

If Lance wants to earn \$105 in simple interest every year, he should invest \$1500 in the 5% interest-earning bond fund and \$500 in the 6% interest-earning stock fund.

34. Let $x=$ the amount (in dollars) invested in the bond fund earning 6% interest and let $y=$ the amount (in dollars) invested in the stock fund earning 9% interest.

The system of equations is:

$x+y=5000$

$0.06x+0.09y=350$

Solve the system using the substitution method. Solve for y in the first equation.

$x+y=5000$

$\underline{-x \quad\quad -x}$

$y=-x+5000$

Now substitute $y=-x+5000$ into the second equation.

$0.06x+0.09y=350$ Substitute in $y=-x+5000$.

$0.06x+0.09(-x+5000)=350$ Distribute.

$0.06x-0.09x+450=350$ Simplify.

$-0.03x+450=350$

$\underline{-450 \quad -450}$ Subtract 450 from both sides.

$-0.03x=-100$ Solve for x.

$\frac{-0.03x}{-0.03}=\frac{-100}{-0.03}$

$x\approx 3333.33$ Rounded to nearest cent.

$y\approx -(3333.33)+5000$ Substitute in $x\approx 3333.33$.

$y\approx 1666.67$

If Ming-Yue wants to earn \$350 in simple interest every year, she should invest \$3333.33 in the 6% interest-earning bond fund and \$1666.67 in the 9% interest-earning stock fund.

35. Let $x=$ the degree measure of the first angle and let $y=$ the degree measure of the second angle. The system of equations is:

$x+y=180^\circ$

$y=5x$

Substitute $y=5x$ into the first equation.

$x+y=180^\circ$ Substitute in $y=5x$.

$x+(5x)=180$ Simplify.

$6x=180$ Solve for x.

$\frac{6x}{6}=\frac{180}{6}$

$x=30$

$y=5x$ Substitute in $x=30$ to find y.

$y=5(30)$

$y=150$

The measures of the angles are 30° and 150°.

36. Let $x=$ the degree measure of the first angle and let $y=$ the degree measure of the second angle. The system of equations is:

$x+y=180^\circ$

$y=4x+5$

Substitute $y=4x+5$ into the first equation.

$x+y=180^\circ$ Substitute in $y=4x+5$.

$x+(4x+5)=180$ Simplify.

$5x+5=180$

$-5 \quad -5$ Subtract 5 from both sides.

$5x=175$ Solve for x.

$\frac{5x}{5}=\frac{175}{5}$

$x=35$

$y=4x+5$ Substitute in $x=35$ to find y.

$y=4(35)+5$

$y=140+5$

$y=145$

The measures of the angles are 35° and 145°.

37. Let $x=$ the degree measure of the first angle and let $y=$ the degree measure of the second angle. The system of equations is:

$x+y=90^\circ$

$y=9x$

Substitute $y=9x$ into the first equation.

$x+y=90^\circ$ Substitute in $y=9x$.

$x+(9x)=90$ Simplify.

$10x=90$ Solve for x.

$\frac{10x}{10}=\frac{90}{10}$

$x=9$

$y=9x$ Substitute in $x=9$ to find y.

$y=9(9)$

$y=81$

The measures of the angles are 9° and 81°.

38. Let $x=$ the degree measure of the first angle and let $y=$ the degree measure of the second angle. The system of equations is:

$x+y=90^\circ$

$y=2x$

Substitute $y=2x$ into the first equation.

$x+y=90^\circ$ Substitute in $y=2x$.

$x+(2x)=90$ Simplify.

$3x=90$ Solve for x.

$\frac{3x}{3}=\frac{90}{3}$

$x=30$

$y=2x$ Substitute in $x=30$ to find y.

$y=2(30)$

$y=60$

The measures of the angles are 30° and 60°.

39. Let $x=$ the number of ounces of 5% saline solution and let $y=$ the number of ounces of 30% saline solution. The system of equations is:

$x+y=20$

$0.05x+0.3y=5$

Solve the system using the substitution method. Solve for y in the first equation.

$x+y=20$

$-x \quad -x$

$y=-x+20$

Now substitute $y=-x+20$ into the second equation.

$0.05x+0.3y=5$ Substitute in $y=-x+20$.

$0.05x+0.3(-x+20)=5$ Distribute.

$0.05x-0.3x+6=5$ Simplify.

$-0.25x+6=5$

$-6 \quad -6$ Subtract 6 from both sides.

$-0.25x=-1$ Solve for x.

$\frac{-0.25x}{-0.25}=\frac{-1}{-0.25}$

$x=4$

$y=-x+20$ Substitute in $x=4$ to find y.

$y=-(4)+20$

$y=16$

Tarin needs to use 4 oz. of 5% saline solution and 16 oz. of 30% saline solution to make 20 oz. of a 25% saline solution.

40. Let $x=$ the number of gallons of 6% fruit juice solution and let $y=$ the number of gallons of 12% fruit juice solution. The system of equations is:

$x+y=20$

$0.06x+0.12y=2$

Solve the system using the substitution method. Solve for y in the first equation.

$x+y=20$

$-x \quad -x$

$y=-x+20$

Now substitute $y=-x+20$ into the second equation.

$0.06x+0.12y=2$ Substitute in $y=-x+20$.

$0.06x+0.12(-x+20)=2$ Distribute.

$0.06x-0.12x+2.4=2$ Simplify.

$-0.06x+2.4=2$

$-2.4 \quad -2.4$ Subtract 6 from both sides.

$-0.06x=-0.4$ Solve for x.

$\frac{-0.06x}{-0.06}=\frac{-0.4}{-0.06}$

$x\approx 6.67$ Rounded to hundredths.

$y=-x+20$ Substitute in $x\approx 6.67$ to find y.

$y\approx -(6.67)+20$

$y\approx 13.33$

Shelby needs to use 6.67 gallons of 6% fruit juice solution and 13.33 gallons of 12% fruit juice solution to make 20 gallons of a 10% fruit juice solution.

41. Let $r =$ the number of pounds of raisins and let $p =$ the number of pounds of peanuts. The system of equations is:

$r+p=10$

$3.6r+2p=26.4$

Use the substitution method to solve the system. Solve for r in the first equation.

$$\begin{array}{rl} r+p=10 & \\ \underline{-p \quad -p} & \text{Subtract } p \text{ from both sides.} \\ r=-p+10 & \end{array}$$

Now substitute $r=-p+10$ into the second equation.

$$\begin{array}{ll} 3.6r+2p=26.4 & \text{Substitute in } r=-p+10. \\ 3.6(-p+10)+2p=26.4 & \text{Distribute.} \\ -3.6p+36+2p=26.4 & \text{Simplify.} \\ -1.6p+36=26.4 & \\ \underline{-36 \quad -36} & \\ -1.6p=-9.6 & \text{Solve for } p. \\ \frac{-1.6p}{-1.6}=\frac{-9.6}{-1.6} & \\ p=6 & \end{array}$$

$$\begin{array}{ll} r=-p+10 & \text{Substitute in } p=6 \text{ to find } r. \\ r=-(6)+10 & \\ r=4 & \end{array}$$

The trail mix should include 4 lb. of raisins and 6 lb. of peanuts.

42. Let $f =$ the number of pounds of French Roast coffee and let $g =$ the number of pounds of Guatemalan coffee. The system of equations is:

$f+g=20$

$9.95f+12.95g=219$

Use the substitution method to solve the system. Solve for f in the first equation.

$$\begin{array}{rl} f+g=20 & \\ \underline{-g \quad -g} & \text{Subtract } g \text{ from both sides.} \\ f=-g+20 & \end{array}$$

Now substitute $f=-g+20$ into the second equation.

$$\begin{array}{ll} 9.95f+12.95g=219 & \text{Substitute in } f=-g+20. \\ 9.95(-g+20)+12.95g=219 & \text{Distribute.} \\ -9.95g+199+12.95g=219 & \text{Simplify.} \\ 3g+199=219 & \\ \underline{-199 \quad -199} & \\ 3g=20 & \text{Solve for } p. \\ \frac{3g}{3}=\frac{20}{3} & \\ g=6\frac{2}{3} & \end{array}$$

$$\begin{array}{ll} f=-g+20 & \text{Substitute in } g=6\frac{2}{3} \text{ to find } f. \\ f=-(6\frac{2}{3})+20 & \\ f=-\frac{20}{3}+\frac{60}{3} & \\ f=\frac{40}{3}=13\frac{1}{3} & \end{array}$$

The blend should include $13\frac{1}{3}$ lb. of French Roast coffee and $6\frac{2}{3}$ lb. of Guatemalan coffee.

43. **a.** $y<-3x+5$ Is $(x,y)=(0,5)$ in the solution set?

$$\begin{array}{ll} (5)\overset{?}{<}-3(0)+5 & \text{Substitute values for } x \text{ and } y. \\ 5\overset{?}{<}0+5 & \\ 5\overset{?}{<}5 & \text{This statement is false.} \end{array}$$

No, the point is not in the solution set.

b. $y<-3x+5$ Is $(x,y)=(-2,-1)$ in the solution set?

$$\begin{array}{ll} (-1)\overset{?}{<}-3(-2)+5 & \text{Substitute values for } x \text{ and } y. \\ -1\overset{?}{<}6+5 & \\ -1<11 & \text{This statement is true.} \end{array}$$

Yes, the point is in the solution set.

44. **a.** $y\geq\frac{2}{3}x-1$ Is $(x,y)=(0,-1)$ in the solution set?

$$\begin{array}{ll} (-1)\overset{?}{\geq}\frac{2}{3}(0)-1 & \text{Substitute values for } x \text{ and } y. \\ -1\overset{?}{\geq}0-1 & \\ -1\geq-1 & \text{This statement is true.} \end{array}$$

Yes, the point is in the solution set.

b. $y \geq \frac{2}{3}x - 1$ Is $(x, y) = (-3, 3)$ in the solution set?

$(3) \overset{?}{\geq} \frac{2}{3}(-3) - 1$ Substitute values for x and y.

$3 \overset{?}{\geq} -2 - 1$

$3 \geq -3$ This statement is true.

Yes, the point is in the solution set.

45. Select the test point $(0,0)$. Is $(x, y) = (0,0)$ in the solution set for $y > -\frac{2}{3}x + 4$?

$(0) \overset{?}{>} -\frac{2}{3}(0) + 4$ Substitute values for x and y.

$0 \overset{?}{>} 0 + 4$

$0 \overset{?}{>} 4$ This statement is false.

Because this is not true, we do not shade in the region containing the point $(0,0)$. Shade the region above the line.

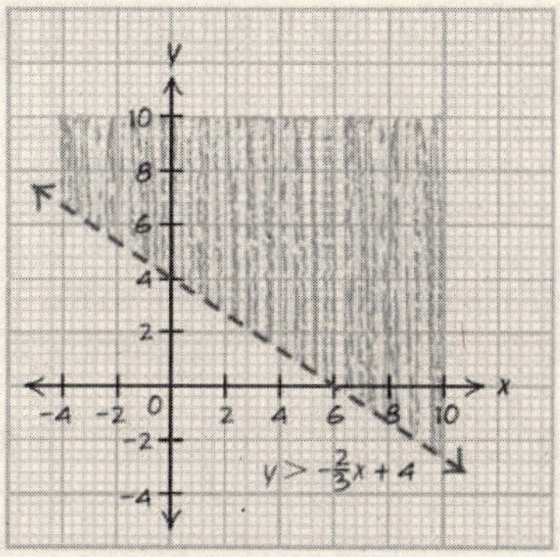

46. Select the test point $(0,0)$. Is $(x, y) = (0,0)$ in the solution set for $y < \frac{1}{4}x - 3$?

$(0) \overset{?}{<} \frac{1}{4}(0) - 3$ Substitute values for x and y.

$0 \overset{?}{<} 0 - 3$

$0 \overset{?}{<} -3$ This statement is false.

Because this is not true, we do not shade in the region containing the point $(0,0)$. Shade the region below the line.

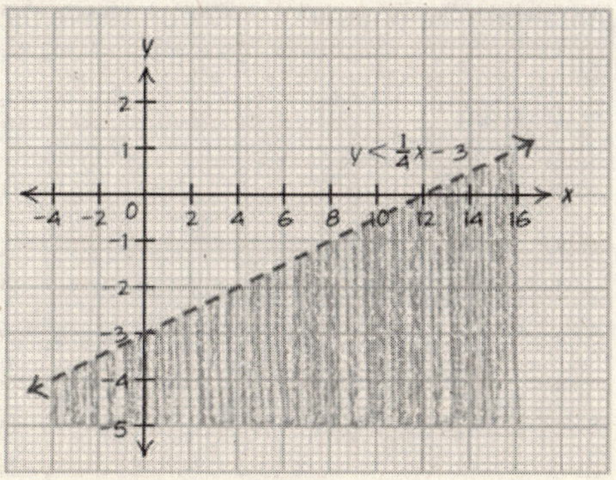

47. Select the test point $(0,0)$. Is $(x, y) = (0,0)$ in the solution set?

$(0) \overset{?}{<} -3(0) + 7$ Substitute values for x and y.

$0 \overset{?}{<} 0 + 7$

$0 < 7$ This statement is true.

Because this is true, we will shade in the region containing the point $(0,0)$. Shade below the dashed line.

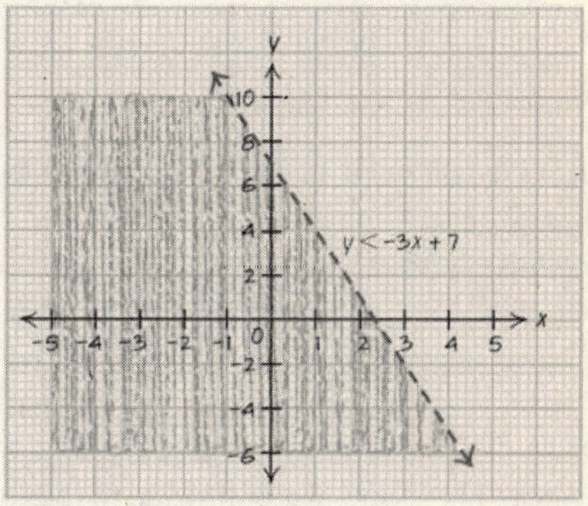

48. Select the test point $(0,0)$. Is $(x, y) = (0,0)$ in the solution set?

$(0) \overset{?}{>} \frac{1}{2}(0) - 1$ Substitute values for x and y.

$0 \overset{?}{>} 0 - 1$

$0 > -1$ This statement is true.

Because this is true, we will shade in the region containing the point $(0,0)$. Shade above the dashed line.

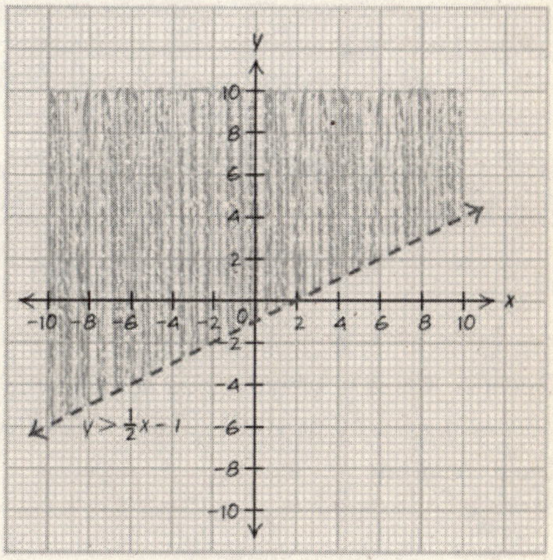

49. Select the test point $(0,0)$. Is $(x, y) = (0,0)$ in the solution set?

$(0) \overset{?}{\geq} 2(0) - 1$ Substitute values for x and y.

$0 \overset{?}{\geq} 0 - 1$

$0 \geq -1$ This statement is true.

Because this is true, we will shade in the region containing the point $(0,0)$. Shade above the solid line.

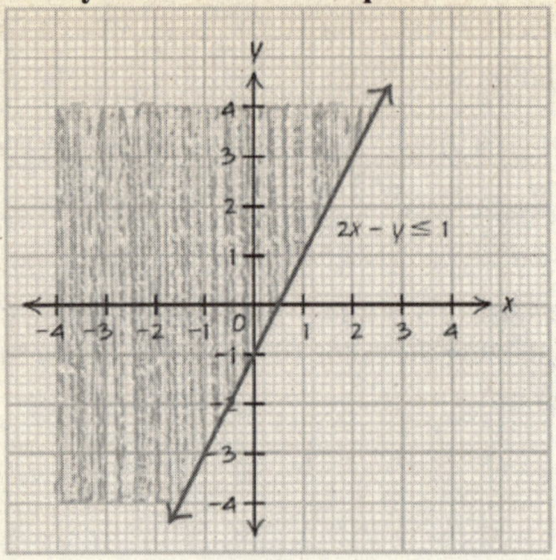

50. Select the test point $(0,0)$. Is $(x,y)=(0,0)$ in the solution set?

$(0)\overset{?}{\geq}-\frac{1}{2}(0)-2$ Substitute values for x and y.

$0\overset{?}{\geq}0-2$

$0\geq -2$ This statement is true.

Because this is true, we will shade in the region containing the point $(0,0)$. Shade above the solid line.

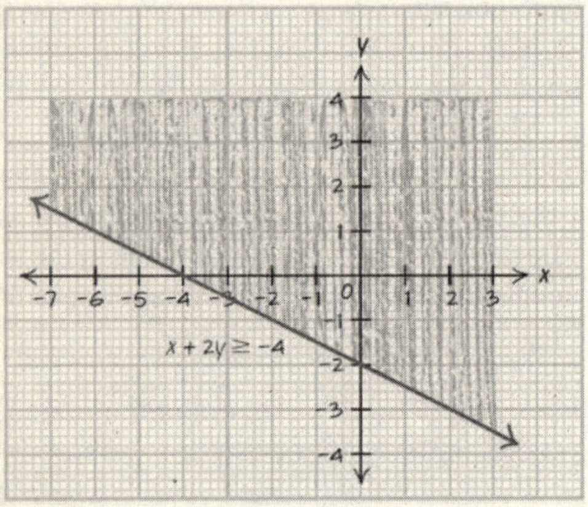

51. Select the test point $(0,0)$. Is $(x,y)=(0,0)$ in the solution set?

$(0)\overset{?}{<}-3$ Substitute values for x and y.

$0\overset{?}{<}-3$ This statement is false.

Because this is not true, we do not shade in the region containing the point $(0,0)$. Shade below/to the left of the dashed line.

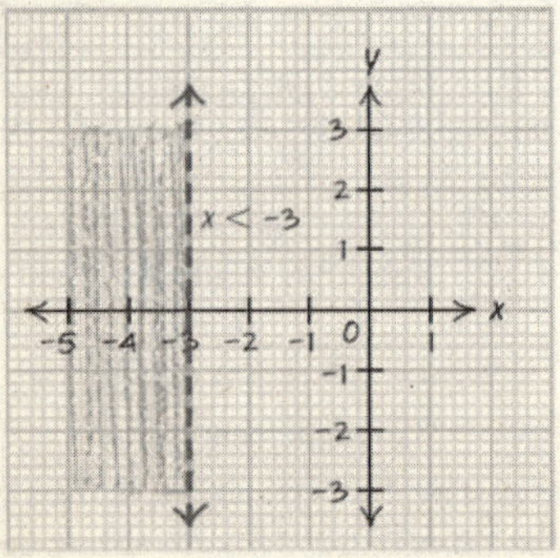

52. Select the test point $(0,0)$. Is $(x,y)=(0,0)$ in the solution set?

$(0)\overset{?}{\geq}2.5$ Substitute values for x and y.

$0\overset{?}{\geq}2.5$ This statement is false.

Because this is not true, we do not shade in the region containing the point $(0,0)$. Shade above/to the right of the solid line.

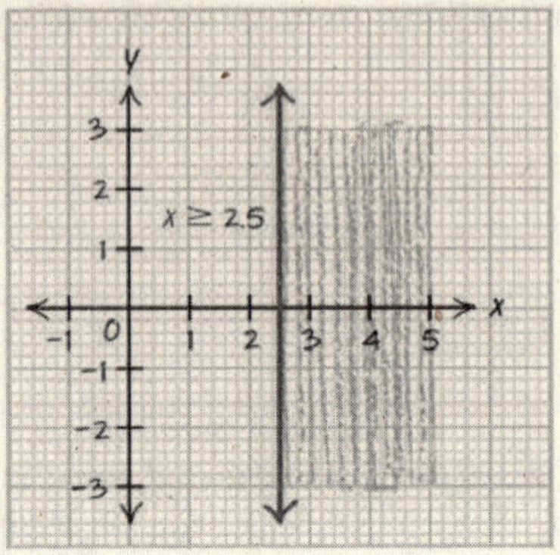

53. Select the test point $(0,0)$. Is $(x,y)=(0,0)$ in the solution set?

$(0)\overset{?}{\leq}-2$ Substitute values for x and y.

$0\overset{?}{\leq}-2$ This statement is false.

Because this is not true, we do not shade in the region containing the point $(0,0)$. Shade below the solid line.

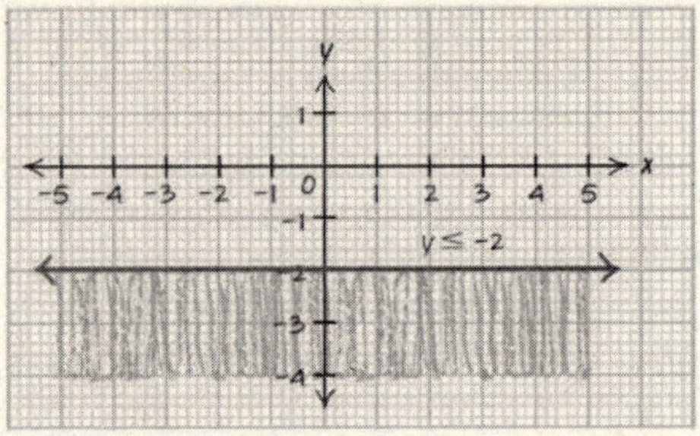

54. Select the test point $(1,1)$. Is $(x,y)=(1,1)$ in the solution set?

$(1)\overset{?}{>}0$ Substitute values for x and y.

$1>0$ This statement is true.

Because this is true, we will shade in the region containing the point $(1,1)$. Shade above the dashed line.

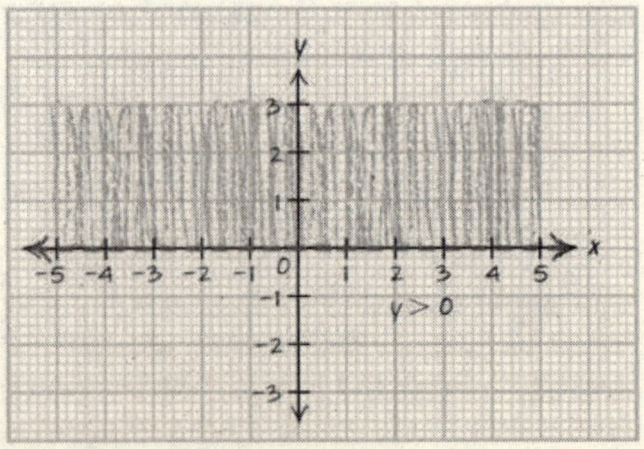

55. **a.** Let f = the number of hours Zita works at her first job, and let s = the number of hours Zita works at her second job.

b. $8f+9s>400$

c. Graph the solution set using the *f*- and *s*-intercepts rounded to hundredths $(50,0)$ and $(0,44.44)$ and shade above the dashed line (negative values do not make sense in this situation so we will not graph them).

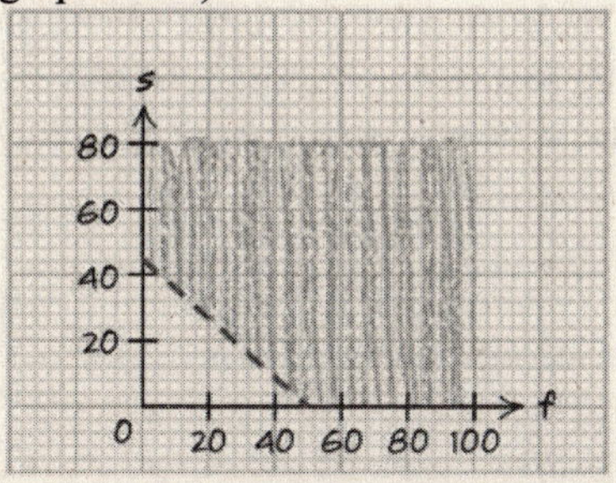

56. **a.** Let f = the number of hours Aaron works at his first job, and let s = the number of hours Aaron works at his second job.

b. $8.25f+9s>300$

c. Graph the solution set using the *f*- and *s*-intercepts rounded to hundredths $(36.36,0)$ and $(0,33.33)$ and shade above the dashed line (negative values do not make sense in this situation so we will not graph them).

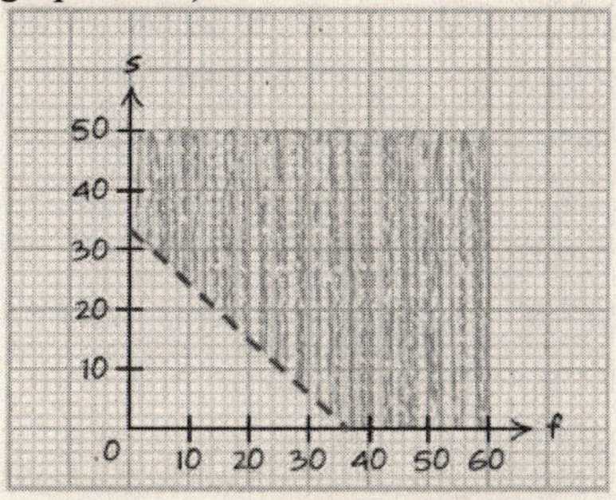

57. **1.** $y\le -\frac{2}{3}x+4$ is a solid line intersecting the *y*-axis at the point $(0,4)$, having a slope of $m=-\frac{2}{3}=\frac{rise}{run}$ with all points below the line included, and therefore corresponds to Graph B.

2. $y<-\frac{1}{3}x+6$ is a dashed line intersecting the *y*-axis at the point $(0,6)$, having a slope of $m=-\frac{1}{3}=\frac{rise}{run}$ with all points below the line included, and therefore corresponds to Graph A.

58. **1.** $2x-4y>12$ corresponds to $y<\frac{1}{2}x-3$ in slope-intercept form. The graph is a dashed line intersecting the *y*-axis at the point $(0,-3)$, having a slope of $m=\frac{1}{2}=\frac{rise}{run}$ with all points below the line included, and therefore corresponds to Graph A.

2. $6x-3y<15$ corresponds to $y>2x-5$ in slope-intercept form. The graph is a dashed line intersecting the *y*-axis at the point $(0,-5)$, having a slope of $m=\frac{2}{1}=\frac{rise}{run}$ with all points above the line included, and therefore corresponds to Graph B.

CHAPTER 4 Systems of Linear Equations

Chapter 4 Test

1.

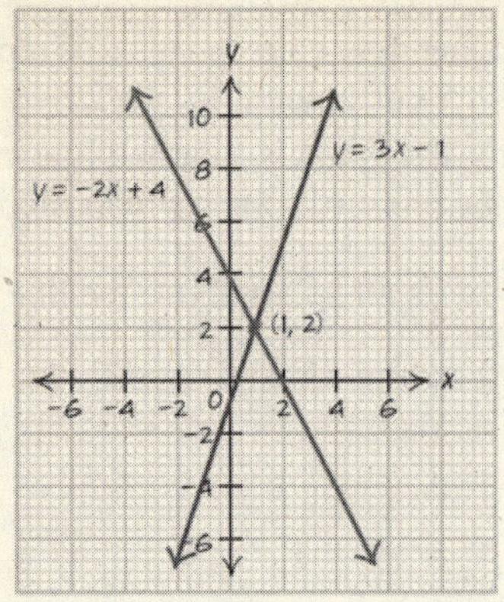

The two lines intersect at the point $(1,2)$.

See Section 4.1, exercises 39-42 for the steps to check the solution.

The system is consistent.

2. Solve $7x-3y=-21$ for y.

$7x\boxed{-3y}=-21$ Identify variable term to isolate.

$-7x \quad -7x$ Subtract $7x$ from both sides.

$-3y=-7x-21$ Solve for y.

$\frac{-3y}{-3}=\frac{-7x-21}{-3}$ Divide by -3.

$y=\frac{7}{3}x+7$

3. Solve the system using the substitution method. Solve the first equation for y, changing it to $y=2x+1$.

$y=2x+1$ Substitute $y=2x+1$ into second equation.

$2x-3y=5$

$2x-3(2x+1)=5$ Distribute.

$2x-6x-3=5$ Simplify.

$-4x-3=5$

$+3 \quad +3$

$-4x=8$ Solve for x.

$\frac{-4x}{-4}=\frac{8}{-4}$ Divide both sides by -4.

$x=-2$

$y=2x+1$ Substitute in $x=-2$ to find y.

$y=2(-2)+1$

$y=-4+1$

$y=-3$

The solution is $(-2,-3)$.

4. Solve the system using the elimination method.

$4x-5y=10$ Add the two equations vertically.

$-2x+5y=0$

$2x=10$

$\frac{2x}{2}=\frac{10}{2}$

$x=5$

$-2(5)+5y=0$ Substitute in $x=5$ to find y.

$-10+5y=0$

$+10 \quad +10$

$5y=10$

$\frac{5y}{5}=\frac{10}{5}$

$y=2$

The solution is $(5,2)$.

5. Let $m=$ the cost in dollars for heart rate monitor and let $b=$ the cost in dollars for each BPA-free water bottle. The system of equations is as follows:

$3m+5b=159$

$2m+4b=114$

Solve the system using the substitution method. Solve the second equation for m.

$2m+4b=114$

$-4b \quad -4b$

$2m=-4b+114$

$\frac{2m}{2}=\frac{-4b+114}{2}$

$m=-2b+57$

$3m+5b=159$ Substitute in $m=-2b+57$.

$3(-2b+57)+5b=159$ Distribute.

$-6b+171+5b=159$ Simplify.

$-b+171=159$

$-171 \quad -171$ Subtract 171 from both sides.

$-b=-12$ Solve for b.

$\frac{-b}{-1}=\frac{-12}{-1}$

$b=12$

$m=-2b+57$ Substitute in $b=12$ to find m.

$m=-2(12)+57$

$m=-24+57$

$m=33$

The cost of each heart rate monitor is \$33 and the cost of each BPA-free water bottle is \$12.

6. Let $b=$ the rate of the boat (in knots) in still water and $c=$ the rate of the current in knots.

The system of equations is:

$2(b+c)=24$

$3(b-c)=24$

Solve the system using the substitution method. Solve the second equation for b.

$3(b-c)=24$ Distribute.

$3b-3c=24$

$+3c \quad +3c$

$3b=3c+24$

$\frac{3b}{3}=\frac{3c+24}{3}$

$b=c+8$

$2(b+c)=24$ Distribute.
$2b+2c=24$ Substitute in $b=c+8$ for b.
$2(c+8)+2c=24$ Distribute.
$2c+16+2c=24$ Simplify.
$4c+16=24$
$\underline{-16\ -16}$
$4c=8$ Solve for c.
$\frac{4c}{4}=\frac{8}{4}$
$c=2$
$b=(2)+8$ Substitute in $c=2$ to find b.
$b=10$
The rate of the boat is 10 knots and the rate of the current is 2 knots.

7. Let $x=$ the amount (in dollars) invested in the bond fund earning 5% interest and let $y=$ the amount (in dollars) invested in the stock fund earning 7% interest.
The system of equations is:
$x+y=2000$
$0.05x+0.07y=124$
Solve the system using the substitution method.
Solve for y in the first equation.
$x+y=2000$
$\underline{-x\quad -x}$
$y=-x+2000$
Now substitute $y=-x+2000$ into the second equation.
$0.05x+0.07y=124$ Substitute in $y=-x+2000$.
$0.05x+0.07(-x+2000)=124$ Distribute.
$0.05x-0.07x+140=124$ Simplify.
$-0.02x+140=124$
$\underline{-140\ -140}$ Subtract 140 from both sides.
$-0.02x=-16$ Solve for x.
$\frac{-0.02x}{-0.02}=\frac{-16}{-0.02}$
$x=800$
$y=-(800)+2000$ Substitute in $x=800$ to find y.
$y=1200$
If Laurent wants to earn \$124 in simple interest every year, he should invest \$800 in the 5% interest-earning bond fund and \$1200 in the 7% interest-earning stock fund.

8. Solve the system using the substitution method.
Solve the first equation for x.
$-x+5y=15$
$\underline{-5y\ -5y}$
$-x=-5y+15$
$\frac{-x}{-1}=\frac{-5y+15}{-1}$
$x=5y-15$
$2x-10y=10$ Substitute in $x=5y-15$ for x.
$2(5y-15)-10y=10$ Distribute.
$10y-30-10y=10$ Simplify.
$-30\overset{?}{=}10$ This statement is false.
Since this is a false statement ($-30\overset{?}{=}10$), there are no solutions to the system. The lines are parallel.

9. Solve the system using the elimination method.
Multiply the second equation by 2. $\begin{aligned}-2x-8y&=-8\\2(x+4y)&=2(4)\end{aligned}$
is now $\begin{aligned}-2x-8y&=-8\\2x+8y&=8\end{aligned}$
$-2x-8y=-8$ Add the two equations vertically.
$\underline{2x+8y=8}$
$0=0$
Because this equation is always true ($0=0$), there are infinitely many solutions. These two equations graph as the same line.

10. Graph the line. Select the test point $(0,0)$. Is $(x,y)=(0,0)$ in the solution set?
$(0)\overset{?}{<}-4(0)+7$ Substitute values for x and y.
$0\overset{?}{<}0+7$
$0<7$ This statement is true.
Because this is true, we will shade in the region containing the point $(0,0)$. Shade below the dashed line.

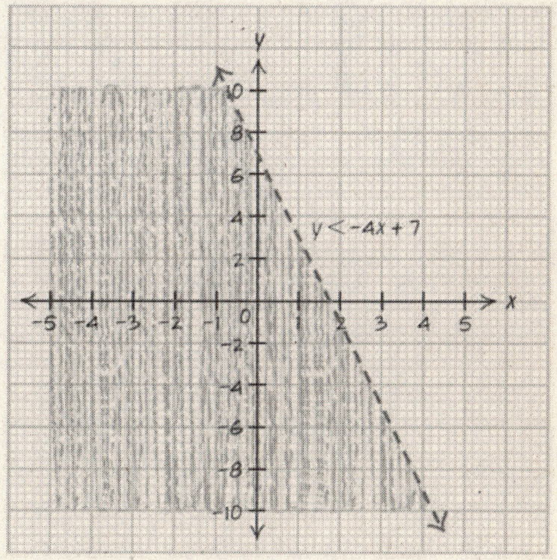

11. Graph the line. Select the test point $(0,0)$. Is $(x,y)=(0,0)$ in the solution set?
$(0)\overset{?}{>}5$ Substitute values for x and y.
$0\overset{?}{>}5$ This statement is false.
Because this is not true, we do not shade in the region containing the point $(0,0)$. Shade above/to the right of the dashed line.

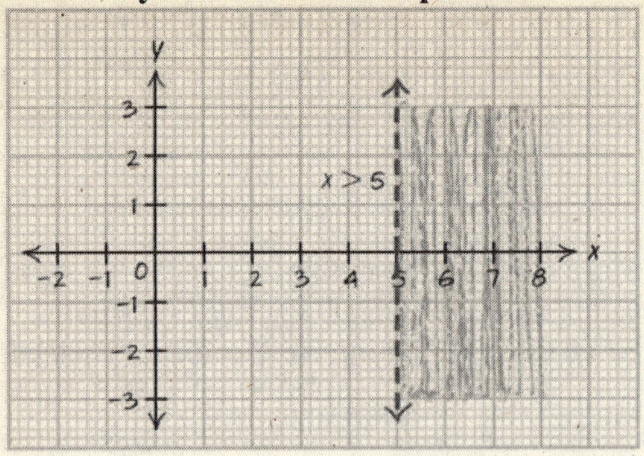

Cumulative Review Chapters 1-4

1. $\left(\frac{-1}{5}\right)\cdot\left(\frac{-1}{5}\right)\cdot\left(\frac{-1}{5}\right)\cdot\left(\frac{-1}{5}\right)=\left(\frac{-1}{5}\right)^4$

2. $(-2)^4 = -2\cdot -2\cdot -2\cdot -2 = 16$

3. $6.103\times 10^6 = 6.103\cdot 1{,}000{,}000 = 6{,}103{,}000$

Therefore, 6.103×10^6 has the standard form 6,103,000.

4. Evaluating the expression using the order of operations would be as follows:

$\boxed{\left|\frac{-12}{11-7}\right|}+\boxed{3^3}\boxed{-(-1-8)}+\boxed{18\div 3\cdot(9)}$ Outline the terms.

$=\left|\frac{-12}{4}\right|+3^3-(-1-8)+18\div 3\cdot(9)$ Simplify fraction first.

$=|-3|+3^3-(-9)+18\div 3\cdot(9)$ Evaluate grouping symbols.

$=3+3^3+9+18\div 3\cdot(9)$ Evaluate exponent.

$=3+27+9+18\div 3\cdot(9)$ Divide in last term.

$=3+27+9+6\cdot(9)$ Multiply in last term.

$=3+27+9+54$ Add from left to right.

$=93$

5. Let c = the amount of commission in dollars the sales person earns and let n = the number of sales made in dollars.

6. Let x = a number. The sentence translates as $3(9+x)$.

7. Convert 12 meters to feet. First, form the unity fraction. We want to convert to the units of feet, so feet will go in the numerator. We want to convert from the units of meters, so meters will go in the denominator. There is 1 ft in 0.3048 m, so the unity fraction is $\frac{1\text{ ft}}{0.3048\text{ m}}$. Multiply 12 m. by this unity fraction and simplify the expression as follows.

$$12\text{ m}\cdot\frac{1\text{ ft}}{0.3048\text{ m}}$$

$$=\frac{12\not{\text{m}}}{1}\cdot\frac{1\text{ ft}}{0.3048\not{\text{m}}}$$

$$=\frac{12\text{ ft}}{0.3048}\approx 39.37\text{ ft}$$

Therefore, 12 m. is approximately equal to 39.37 ft.

8. Evaluate $-x^2+3x-1$ for $x=-2$.

$-x^2+3x-1$ Substitute in $x=-2$.

$=-(-2)^2+3(-2)-1$ Simplify.

$=-4+3(-2)-1$

$=-4-6-1$

$=-11$

9. Simplify.

$3(5x-1)-4(3x-7)$ Distribute.

$=\boxed{15x}\boxed{-3}\boxed{-12x}+\boxed{28}$ Identify like terms.

$=3x+25$ Combine like terms.

10. Simplify.

$-(-(-5x))=-5x$

11.

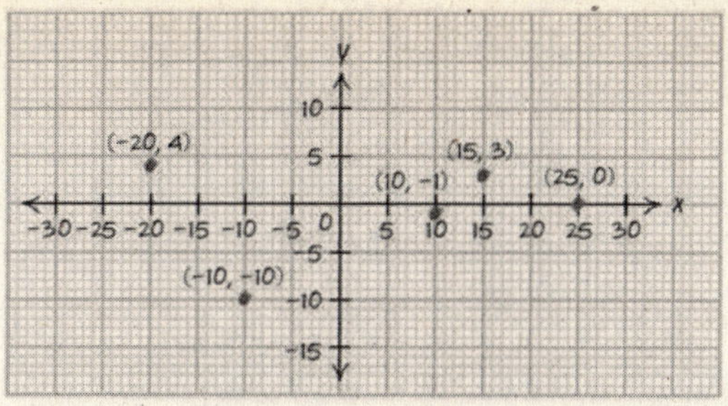

12. To check if the given values of the variables are a solution to the equation, evaluate each side of the equation and see if the two sides are equal.

$4x+3y=-12$ Original equation.

$4(0)+3(-4)\stackrel{?}{=}-12$ Substitute $x=0$ and $y=-4$.

$0-12\stackrel{?}{=}-12$ Simplify both sides.

$-12=-12$ This statement is true.

Because the final statement is true, $x=0$ and $y=-4$ is a solution to the equation $4x+3y=-12$.

13. Solve:

$\frac{-3}{4}x=18$ Multiply both sides by $-\frac{4}{3}$.

$-\frac{4}{3}\cdot\frac{-3}{4}x=18\cdot-\frac{4}{3}$

$x=-24$

14. Solve for y.

$-9x\boxed{-3y}=27$ Identify variable term to isolate.

$\underline{+9x \quad\quad +9x}$ Add $9x$ to both sides.

$-3\boxed{y}=9x+27$ Identify variable to isolate

$\frac{-3y}{-3}=\frac{9x+27}{-3}$ Divide both sides by -3.

$y=-3x-9$

15. Solve:

$3(x-1)-2(x+2)=-3+\frac{1}{3}x$ Distribute on left side.

$3x-3-2x-4=-3+\frac{1}{3}x$ Simplify on left side.

$3(x-7)=3(-3+\frac{1}{3}x)$ Multiply by 3 to clear denom.

$3x-21=-9+x$

$\underline{-x \quad\quad -x}$ Subtract x from both sides.

$2x-21=-9$

$\underline{+21 \quad +21}$ Add 21 to both sides.

$2x=12$

$\frac{2x}{2}=\frac{12}{2}$ Divide both sides by 2.

$x=6$

16. **a.** To find Daniel's monthly salary if he works 110 hours, substitute $h=110$ into the equation $S=13h$ and solve for S.

$S=13h$ Given equation.

$S=13(110)$ Substitute $h=110$.

$S=1430$ Multiply & solve for S.

If Daniel works 110 hours, his salary will be $1430.

b. To find the number of hours Daniel needs to work to make $1820, substitute $S=1820$ into the equation $S=13h$ and solve for h.

$S=13h$ Given equation.

$1820=13h$ Substitute $S=1820$ & solve for h.

$\frac{1820}{13}=\frac{13h}{13}$ Divide both sides by 13.

$140=h$

Daniel needs to work 140 hours in a month to make $1820 a month.

17. Solve.

$12x+9=5x+26.5$

$\underline{-5x \quad\quad -5x}$ Subtract $5x$ from both sides.

$7x+9=26.5$

$\underline{\quad -9 \quad -9}$ Subtract 9 from both sides.

$7x=17.5$

$\frac{7x}{7}=\frac{17.5}{7}$ Divide both sides by 7.

$x=2.5$

18. Let x = a number. The sentence translates as $8+3x=2x-(-6)$.

$8+3x=2x-(-6)$ Simplify on right side.

$8+3x=2x+6$

$\underline{-2x \quad -2x}$ Subtract $2x$ from both sides.

$8+x=6$

$\underline{-8 \quad\quad -8}$ Subtract 8 from both sides.

$x=-2$

19. Solve:

$9y-8=6y+2+3(y+1)-10$ Distribute on left side.

$9y-8=6y+2+3y+3-10$ Simplify on right side.

$9y-8=9y-5$

$\underline{-9y \quad\quad -9y}$ Subtract $9y$ from both sides.

$-8\stackrel{?}{=}-5$

This equation has no solution.

20. Solve:

$-5x+2>37$

$\underline{\quad -2 \quad -2}$ Subtract 2 from both sides.

$-5x>35$ Divide both sides by -5.

$\frac{-5x}{-5}<\frac{35}{-5}$ Multiply by neg., reverse ineq.

$x<-7$

21. **a.**

x	$y = \frac{-3}{5}x - 7$
−5	−4
0	−7
5	−10
15	−16
20	−19

b. Plot the points and connect them with a smooth curve.

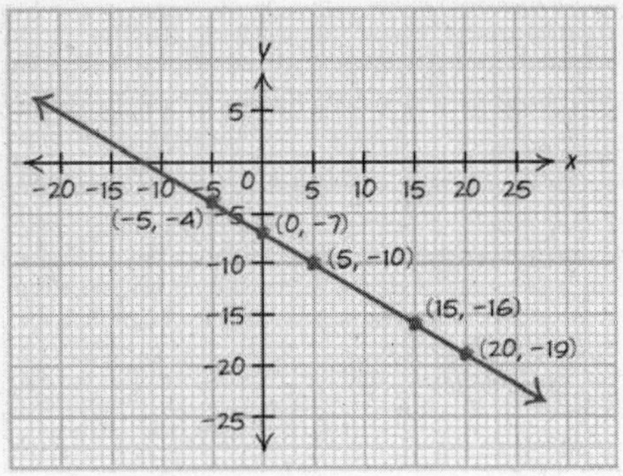

c. The graph is linear.

22.

x	$y = \frac{-2}{3}x + 4$
−3	6
0	4
3	2
6	0
9	−2

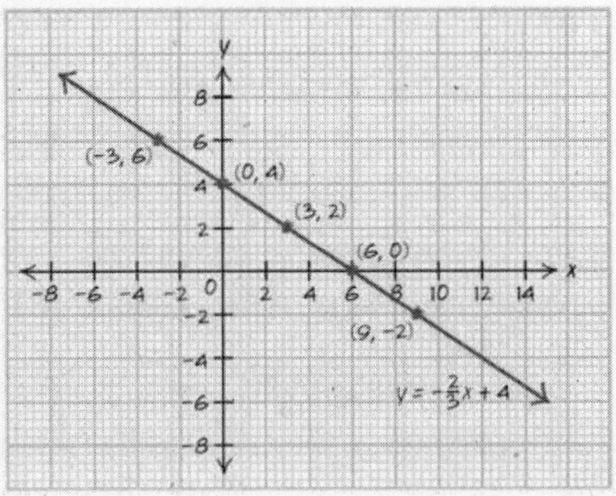

The graph is linear.

23.

d	$C = 4000d$
0	0
1	4000
2	8000
3	12,000
4	16,000

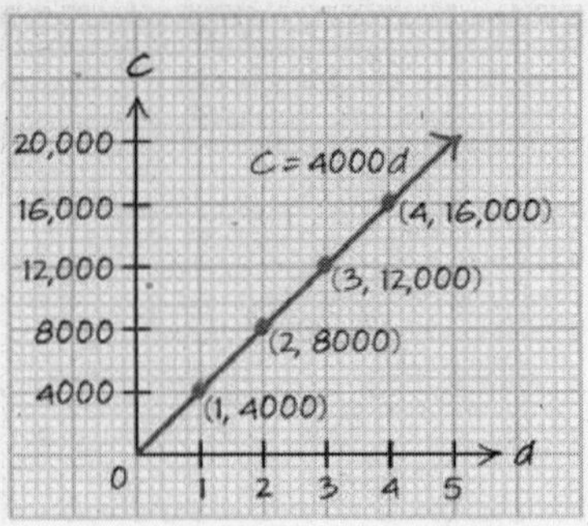

24.

x	$y = x^2 - 2$
−4	14
−3	7
−2	2
−1	−1
0	−2
1	−1
2	2
3	7
4	14

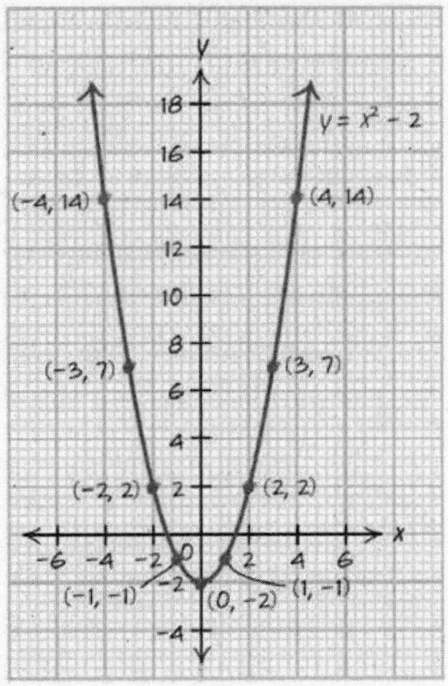

The graph is non-linear.

25.

x	$y = -5$
−2	−5
−1	−5
0	−5
1	−5
2	−5
3	−5

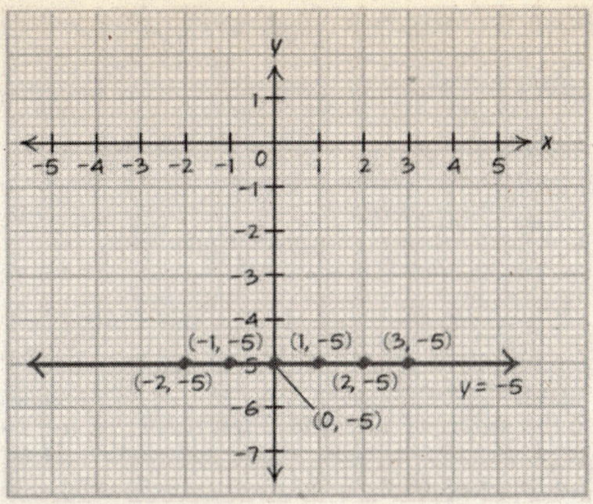

The graph is linear.

26.

$x=6$	y
6	0
6	1
6	2
6	3
6	4

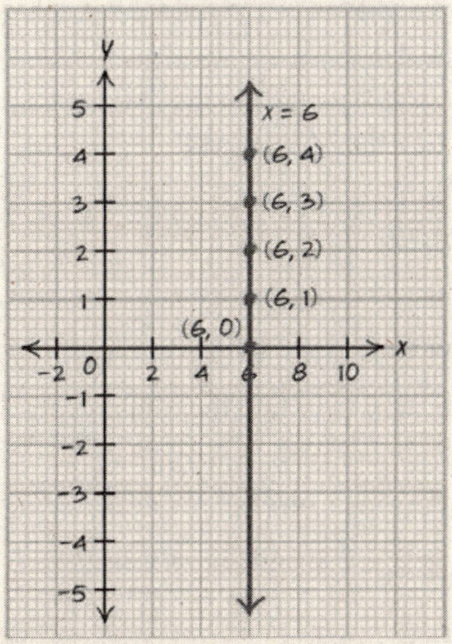

The graph is linear.

27. The slope is $m=\dfrac{rise}{run}=\dfrac{2}{3}$.

28. The slope of the line that goes through the points $(-2,5)$ and $(3,-7)$. Substituting these points into the slope formula yields:

$$\text{slope}=\frac{y_2-y_1}{x_2-x_1}=\frac{-7-(5)}{3-(-2)}=\frac{-12}{5}=-\frac{12}{5}$$

The slope of the line that passes through the given points is $-\dfrac{12}{5}$.

29. Using the points $(0,9)$ and $(5,7)$ and substituting these points into the slope formula yields:

$$\text{slope}=\frac{y_2-y_1}{x_2-x_1}=\frac{7-(9)}{5-(0)}=\frac{-2}{5}=-\frac{2}{5}$$

The slope of the line that passes through the given points is $-\dfrac{2}{5}$.

30. Using the points $(5,-7)$ and $(7,-7)$ and substituting these points into the slope formula yields:

$$\text{slope}=\frac{y_2-y_1}{x_2-x_1}=\frac{-7-(-7)}{7-(5)}=\frac{0}{2}=0$$

The slope of the line that passes through the given points is 0.

31. The red line is steeper than the blue line, therefore, the red line has the greater slope.

32. The line is going down from left to right, therefore, the line is decreasing.

33. The equation of the line is in slope-intercept form so we can see that the slope $m=3=\dfrac{3}{1}$. This is a positive slope, therefore the line is increasing.

34. Looking at the graph, we see that the x-intercept is the point $(-3,0)$ and the y-intercept is the point $(0,-2)$.

35. **a.** The vertical intercept for the Minnesota line is the point $(0,4.94)$ which means that the population of Minnesota was 4.94 million in the year 2000.

b. The vertical intercept for the Washington line is the point $(0,5.89)$ which means that the population of Washington was 5.89 million in the year 2000.

c. Using the given points for the Minnesota line $(0,4.94)$ and $(8,5.22)$ and substituting these points into the slope formula yields:

$$m=\frac{5.22-(4.94)}{8-(0)}=\frac{0.28}{8}=0.035$$

The slope of the Minnesota line is 0.035 which means that the population of Minnesota is growing by 0.035 million, or 35,000, people per year.

d. Using the given points for the Washington line $(0,5.89)$ and $(10,6.69)$ and substituting these points into the slope formula yields:

$$m=\frac{6.69-(5.89)}{10-(0)}=\frac{0.8}{10}=0.08$$

The slope of the Washington line is 0.08 which means that the population of Washington is growing by 0.08 million, or 80,000, people per year.

36. To find the y-intercept, substitute $x=0$ and find y.

$y=-\dfrac{3}{5}(0)+15$ Substitute in $x=0$.

$y=0+15$ Solve for y.

$y=15$

The y-intercept is the point $(0,15)$.

To find the x-intercept, substitute $y=0$ and find x.

$$(0)=-\frac{3}{5}x+15 \qquad \text{Substitute in } y=0.$$

$$\underline{-15 \qquad -15} \qquad \text{Subtract 15 from both sides.}$$

$$-15=-\frac{3}{5}x \qquad \text{Multiply both sides by } -\frac{5}{3}.$$

$$-\frac{5}{\cancel{3}}\cdot -\overset{5}{\cancel{15}}=-\frac{\cancel{3}}{\cancel{5}}x\cdot -\frac{\cancel{5}}{\cancel{3}} \qquad \text{Solve for } x.$$

$$25=x$$

The x-intercept is the point $(25,0)$.

37. Change the given equation to slope-intercept form.

$$-4x+8y=-12$$

$$\underline{+4x \qquad +4x} \qquad \text{Add } 4x \text{ to both sides.}$$

$$8y=4x-12$$

$$\frac{8y}{8}=\frac{4x-12}{8} \qquad \text{Divide both sides by 8.}$$

$$y=\frac{1}{2}x-\frac{3}{2}$$

Using the slope-intercept form of the equation, we see the slope is $\frac{1}{2}$ and the y-intercept is $\left(0,-\frac{3}{2}\right)$.

To find the x-intercept, substitute $y=0$ and solve for x.

$$-4x+8y=-12 \qquad \text{Substitute } y=0.$$

$$-4x+8(0)=-12$$

$$-4x+0=-12$$

$$-4x=-12$$

$$\frac{-4x}{-4}=\frac{-12}{-4} \qquad \text{Divide both sides by } -4.$$

$$x=3$$

The x-intercept is the point $(3,0)$.

38. Graph $y=\frac{-3}{2}x+5$ using slope $m=-\frac{3}{2}$ and y-intercept is $(0,5)$.

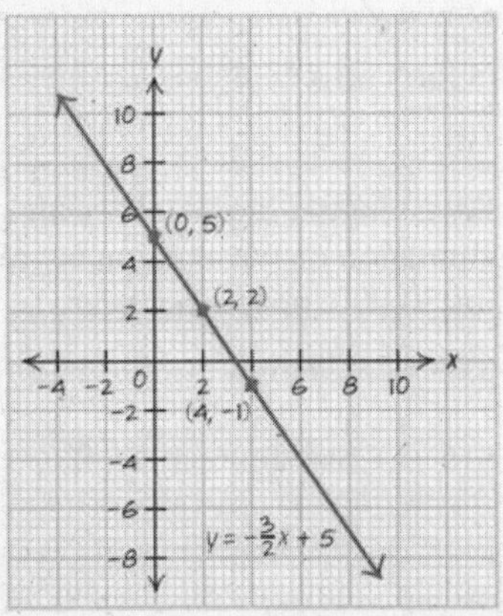

39. Graph $2x-4y=3$. Change to slope-intercept form.

$$2x-4y=3$$

$$\underline{-2x \qquad -2x} \qquad \text{Subtract } 2x \text{ from both sides.}$$

$$-4y=-2x+3$$

$$\frac{-4y}{-4}=\frac{-2x+3}{-4} \qquad \text{Divide both sides by } -4.$$

$$y=\frac{1}{2}x-\frac{3}{4}$$

Use the slope $m=\frac{1}{2}$ and y-intercept $\left(0,-\frac{3}{4}\right)$ to graph the line.

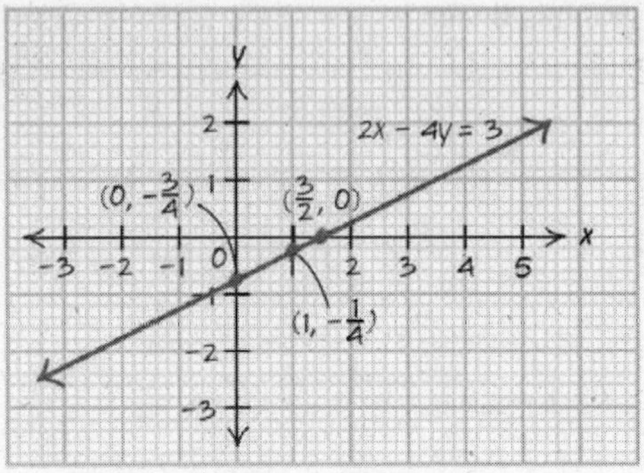

40. Graph $M=220-a$ using the intercepts. Substitute $a=0$ to find the M-intercept:

$$M=220-a \qquad \text{Substitute } a=0.$$

$$M=220-(0)$$

$$M=220$$

The M-intercept is the point $(0,220)$.

Substitute $M=0$ to find the a-intercept:

$$M=220-a \qquad \text{Substitute } M=0.$$

$$(0)=220-a$$

$$\underline{+a \qquad +a} \qquad \text{Add } a \text{ to both sides.}$$

$$a=220$$

The a-intercept is the point $(220,0)$.

Use these intercepts to produce the following graph:

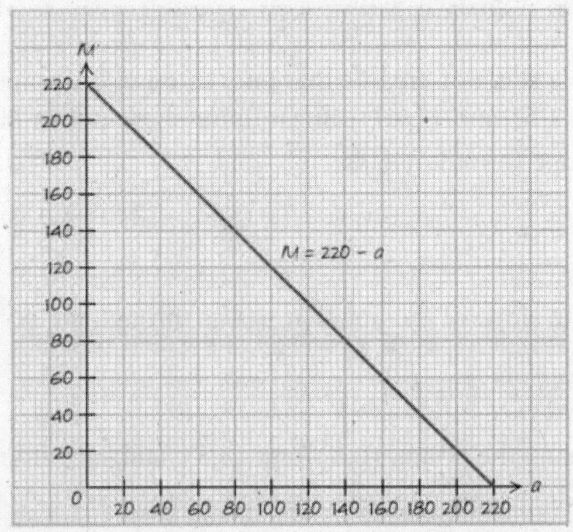

41. Graph $3x-2y=6$ using the x- and y-intercepts.
Substitute $x=0$ to find the y-intercept:

$$3x-2y=6 \quad \text{Substitute } x=0.$$
$$3(0)-2y=6$$
$$-2y=6$$
$$\frac{-2y}{-2}=\frac{6}{-2}$$
$$y=-3$$

The y-intercept is the point $(0,-3)$.

Substitute $y=0$ to find the x-intercept:

$$3x-2y=6 \quad \text{Substitute } y=0.$$
$$3x-2(0)=6$$
$$3x=6$$
$$\frac{3x}{3}=\frac{6}{3}$$
$$x=2$$

The x-intercept is the point $(2,0)$.

Use these intercepts to produce the following graph:

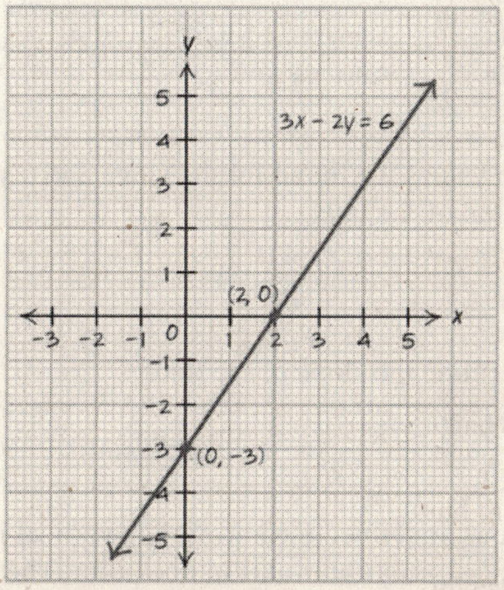

42. **a.** This equation is in general form, so it is a linear equation.

b. This equation has the variable x raised to the second power, so this is not a linear equation.

c. This equation has the variable x inside an absolute value, so this is not a linear equation.

d. This equation is in slope-intercept form, so it is a linear equation.

43. The first equation $y=-5x+12$ has a slope of -5. To find the slope of the second equation $10x+2y=6$, change to slope-intercept form.

$$10x+2y=6$$
$$-10x \qquad -10x$$
$$2y=-10x+6$$
$$\frac{2y}{2}=\frac{-10x+6}{2}$$
$$y=-5x+6$$

This equation has a slope of -5. Comparing the slopes of the two equations, both have the same slope of -5, therefore these lines are parallel.

44. To find the slope of the first equation $5x+2y=6$, change to slope-intercept form.

$$5x+2y=6$$
$$-5x \qquad -5x$$
$$2y=-5x+6$$
$$\frac{2y}{2}=\frac{-5x+6}{2}$$
$$y=-\frac{5}{2}x+3$$

This equation has a slope of $-\frac{5}{2}$. The second equation $y=\frac{2}{5}x+1$ has a slope of $\frac{2}{5}$. Comparing the slopes, the equations have slopes of $-\frac{5}{2}$ and $\frac{2}{5}$ which are negative reciprocals of one another, therefore the lines are perpendicular.

45. Find the slope using two points on the line $(0,-2)$ and $(2,5)$. "Rise" from $(0,-2)$ and "run" to $(2,5)$. Therefore the slope is

$$m=\frac{rise}{run}=\frac{7}{2}$$

Substitute this value of slope into the slope-intercept form of a line.

$$y=mx+b$$
$$y=\frac{7}{2}x+b$$

The y-intercept is the point $(0,-2)$ so $b=-2$. Substitute this value into our equation and we have the equation of the line:

$$y=\frac{7}{2}x-2$$

46. We are given two points on the line $(0,155)$ and $(6,130)$. Find the slope.

$$m=\frac{y_2-y_1}{x_2-x_1}=\frac{W_2-W_1}{t_2-t_1}$$
$$=\frac{155-130}{0-6}=\frac{25}{-6}=-\frac{25}{6}$$

Substitute this value of slope into the slope-intercept form of a line:

$y=mx+b$ Slope-intercept form of a line.

$W=mt+b$ Use the variables from the problem.

$$W=-\frac{25}{6}t+b$$

To find the y-intercept, $(0, b)$, substitute one of the points $(6,130)$ into the equation.

$$W=-\frac{25}{6}t+b$$

$(130)=-\frac{25}{6}(6)+b$ Substitute $t=6$, $W=130$.

$130=-25+b$ Solve for b.

$+25 \quad +25$ Add 25 to both sides.

$155=b$

Using the values we found for the slope and b yields the equation of the line

$$W=-\frac{25}{6}t+b$$
$$W=-\frac{25}{6}t+155$$

47. To determine the equation of the line, we will first find the slope of the line using the given points $(-5,11)$ and $(5,7)$.

$$\text{slope}=\frac{7-(11)}{5-(-5)}=\frac{-4}{10}=-\frac{2}{5}$$

Therefore the slope of the line is $-\frac{2}{5}$.

Find the y-intercept using the value for slope and one of the given points $(5,7)$.

$y=mx+b$

$(7)=(-\frac{2}{5})(5)+b$ Substitute $m=-\frac{2}{5}$, $x=5, y=7$.

$7=-2+b$ Solve for b.

$+2 \quad +2$

$9=b$

Using the values we found for the slope and b yields the equation of the line

$$y=-\frac{2}{5}x+9$$

48. We have that $m=\frac{2}{3}$ and $(x_1,y_1)=(-6,0)$. Substitute these values into the point-slope form of a line and simplify, yielding the equation of the line.

$y-y_1=m(x-x_1)$ Substitute in $m=\frac{2}{3}$ and

$y-(0)=\frac{2}{3}\left[x-(-6)\right]$ $(x_1,y_1)=(-6,0)$

$y=\frac{2}{3}(x+6)$ Distribute.

$y=\frac{2}{3}x+4$

49. To determine the equation of the line, we will first find the slope of the line using the given points $(2,-5)$ and $(-3,-5)$.

$$\text{slope}=\frac{-5-(-5)}{-3-(2)}=\frac{0}{-5}=0$$

Therefore the slope of the line is 0.

Find the y-intercept using the value for slope and one of the given points $(2,-5)$.

$y=mx+b$

$(-5)=(0)(2)+b$ Substitute $m=0$, $x=2, y=-5$.

$-5=0+b$ Solve for b.

$-5=b$

Using the values we found for the slope and b yields the equation of the line

$$y=0x-5$$
$$y=-5$$

50. First determine the slope of the line.

$$m=\frac{y_2-y_1}{x_2-x_1}=\frac{2-(-5)}{3-(3)}=\frac{7}{0}=\textit{undefined}$$

The slope is undefined meaning it is a vertical line that intersects only the x-axis in one point. The given points $(3,-5)$ and $(3,2)$ have $x=3$ for both points. If we were to graph these two points and connect them with a line, it would be a vertical line intersecting the x-axis at the point $(3,0)$, therefore the equation of the line is $x=3$.

51. To find the equation of a line, we need both the slope and a point. Parallel lines have the same slope. Therefore, since the slope of the given line $y=-5x+7$ is $m=-5$, the slope of the line that we are trying to find is the same. Substituting this into slope-intercept form yields

$y=mx+b$ Substitute in $m=-5$.

$y=-5x+b$

To find the value of b, substitute in the given point $(-10,3)$ and solve for b.

$y=-5x+b$ Substitute in the given point $(-10,3)$.

$3=-5(-10)+b$ Simplify the right side.

$3=50+b$ Solve for b.

$\underline{-50 \quad -50}$

$-47=b$

$y=-5x-47$ The equation of the line.

52. To find the equation of a line, we need both the slope and a point. Perpendicular lines have slopes that are negative reciprocals. The slope of the given line $y=3x-11$ is $m=3$. The negative reciprocal of 3 is $-\frac{1}{3}$.

The slope of the line we are trying to find is $m=-\frac{1}{3}$.

Substituting this into slope-intercept form yields

$y=mx+b$ Substitute in $m=-\frac{1}{3}$.

$y=-\frac{1}{3}x+b$

To find the value of b, substitute in the given point $(6,-4)$ and solve for b.

$y=-\frac{1}{3}x+b$ Substitute in the given point $(6,-4)$.

$-4=-\frac{1}{3}(6)+b$ Simplify the right side.

$-4=-2+b$ Solve for b.

$\underline{+2 \quad +2}$

$-2=b$

$y=-\frac{1}{3}x-2$ The equation of the line.

53. From the two tables, we see that there is one input and output pair that is the same, $(-1,6)$. Therefore, this point is the solution to the system.

54. **a.** To determine if $(2,-1)$ is a solution of the system, substitute the values into both equations and find if both are true.

First equation:

$(-1)\overset{?}{=}-2(2)+3$ Substitute values for x and y.

$-1\overset{?}{=}-4+3$ Simplify.

$-1=-1$ This is a true statement.

Second equation:

$(-1)\overset{?}{=}4(2)-3$ Substitute values for x and y.

$-1\overset{?}{=}8-3$ Simplify.

$-1\overset{?}{=}5$ This is a false statement.

The point is on the first line but not on the second line. Since the point does not make both equations true statements, it is *not* a solution of the system.

b. To determine if $(1,1)$ is a solution of the system, substitute the values into both equations and find if both are true.

First equation:

$(1)\overset{?}{=}-2(1)+3$ Substitute values for x and y.

$1\overset{?}{=}-2+3$ Simplify.

$1=1$ This is a true statement.

Second equation:

$(1)\overset{?}{=}4(1)-3$ Substitute values for x and y.

$1\overset{?}{=}4-3$ Simplify.

$1=1$ This is a true statement.

Since both statements are true, $(1,1)$ is a solution of the system.

55. The solution to this system is the intersection of the two lines. Looking at the graph, we see the point of intersection is at $x=0$ and $y=1$, or as an ordered pair, $(0,1)$.

56. **a.** To determine if $(100,11.25)$ is a solution of the system, substitute the values into both equations and find if both are true.

First equation:

$(11.25)\overset{?}{=}9.25+0.02(100)$ Substitute values for x and S.

$11.25\overset{?}{=}9.25+2$ Simplify.

$11.25=11.25$ This is a true statement.

Second equation:

$(11.25)\overset{?}{=}10.50+0.01(100)$ Substitute values for x and S.

$11.25\overset{?}{=}10.5+1$ Simplify.

$11.25\overset{?}{=}11.5$ This is a false statement.

The point is on the first line but not on the second line. Since the point does not make both equations true statements, it is *not* a solution of the system.

b. To determine if $(125,11.75)$ is a solution of the system, substitute the values into both equations and find if both are true.

First equation:

$(11.75)\overset{?}{=}9.25+0.02(125)$ Substitute values for x and S.

$11.75\overset{?}{=}9.25+2.5$ Simplify.

$11.75=11.75$ This is a true statement.

Second equation:

$(11.75)\overset{?}{=}10.50+0.01(125)$ Substitute values for x and S.

$11.75\overset{?}{=}10.5+1.25$ Simplify.

$11.75=11.75$ This is a true statement.

Since both statements are true, $(125,11.75)$ is a solution of the system.

57. Graph the equations to determine the solution of the system.

$y=-2x+5$

$y=2x-3$

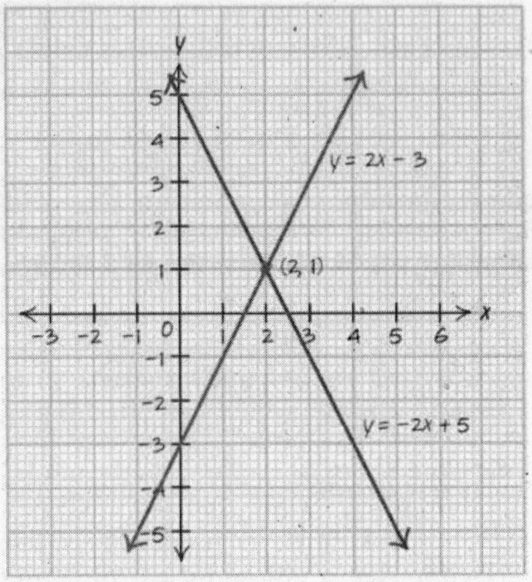

The two lines intersect at the point $(2,1)$.

58. Graph the equations to determine the solution of the system.

$y=3x-4$

$y=2$

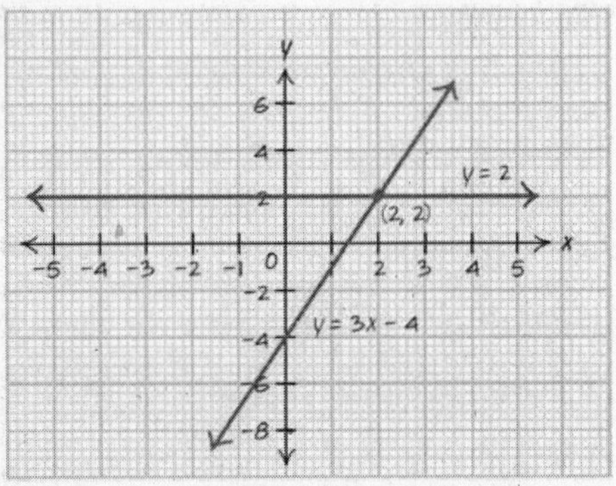

The two lines intersect at the point $(2,2)$.

59. Solve the system using the substitution method. Solve the second equation for y, changing it to $y=-2x-2$.

$5x+4y=4$ Substitute $y=-2x-2$ into first equation.

$y=-2x-2$

$5x+4(-2x-2)=4$ Distribute.

$5x-8x-8=4$ Simplify.

$$\begin{aligned}-3x-8&=4\\ +8\ \ &+8\\ \hline -3x&=12\end{aligned}$$ Solve for x.

$\frac{-3x}{-3}=\frac{12}{-3}$ Divide both sides by -3.

$x=-4$

$y=-2x-2$ Substitute in $x=-4$ to find y.

$y=-2(-4)-2$

$y=8-2$

$y=6$

The solution is $(-4,6)$.

60. Solve the system using the substitution method. Solve the first equation for y, changing it to $y=4x+7$.

$y=4x+7$ Substitute $y=4x+7$ into second equation.

$x-y=8$

$x-(4x+7)=8$ Distribute.

$x-4x-7=8$ Simplify.

$$\begin{aligned}-3x-7&=8\\ +7\ \ &+7\\ \hline -3x&=15\end{aligned}$$ Solve for x.

$\frac{-3x}{-3}=\frac{15}{-3}$ Divide both sides by -3.

$x=-5$

$y=4x+7$ Substitute in $x=-5$ to find y.

$y=4(-5)+7$

$y=-20+7$

$y=-13$

The solution is $(-5,-13)$.

61. Solve the system using the substitution method.

$-2x+3y=6$ Substitute $y=\frac{2}{3}x-4$ into first equation.

$y=\frac{2}{3}x-4$

$-2x+3\left(\frac{2}{3}x-4\right)=6$ Distribute.

$-2x+2x-12=6$ Simplify.

$-12\overset{?}{=}6$ This is a false statement.

Since this is a false statement ($-12\overset{?}{=}6$), there are no solutions to the system. The lines are parallel.

62. Let $b=$ the cost in dollars for each BPA-free water bottle and let $e=$ the cost in dollars for each package of energy bars. The system of equations is as follows:

$15b+7e=265$

$9b+5e=179$

Solve the system using the elimination method. To make the coefficients on e opposites of one another, multiply the first equation by 5 and the second equation by -7.

$$\begin{aligned} 5(15b+7e) &= 5(265) \\ -7(9b+5e) &= -7(179) \end{aligned} \text{ is now } \begin{aligned} 75b+35e &= 1325 \\ -63b-35e &= -1253 \end{aligned}$$

Now that the coefficients on e are opposites, add the two equations vertically to eliminate e.

$75b+35e=1325$ Add the equations vertically.
$\underline{-63b-35e=-1253}$
$12b=72$ Solve for b.
$b=6$

Now that we know $b=6$, substitute this value into either of the original equations to find the value of e. Substituting into the second equation yields:

$9(6)+5e=179$ Substitute in $b=6$
$54+5e=179$
$\underline{-54 \quad -54}$
$5e=125$ Solve for e.
$e=25$

The cost of each BPA-free water bottle is \$6 and the cost of each package of energy bars is \$25.

63. Let $x=$ the first number and let $y=$ the second number. The system of equations is:

$x+y=32$
$x=y-6$

Solve the system using the substitution method. Substitute $x=y-6$ into the first equation.

$x+y=32$ Substitute in $x=y-6$ for x.
$(y-6)+y=32$ Simplify.
$2y-6=32$
$\underline{+6 \quad +6}$
$2y=38$
$\frac{2y}{2}=\frac{38}{2}$ Solve for y.
$y=19$

$x=y-6$ Substitute in $y=19$ to find x.
$x=(19)-6$
$x=13$

The numbers are 13 and 19.

64. Let $p=$ the rate of the plane (in mph) in still air and $w=$ the rate of the wind in mph.

The system of equations is:

$3(p+w)=450$
$4.5(p-w)=450$

Solve the system using the substitution method. Solve the second equation for p.

$4.5(p-w)=450$ Distribute.
$4.5p-4.5w=450$
$\underline{+4.5w \quad +4.5w}$
$4.5p=4.5w+450$
$\frac{4.5p}{4.5}=\frac{4.5w+450}{4.5}$
$p=w+100$

Now substitute $p=w+100$ into the first equation.

$3(p+w)=450$ Distribute.
$3p+3w=450$ Substitute in $p=w+100$.
$3(w+100)+3w=450$ Distribute.
$3w+300+3w=450$ Simplify.
$6w+300=450$
$\underline{-300 \quad -300}$
$6w=150$ Solve for w.
$\frac{6w}{6}=\frac{150}{6}$
$w=25$

$p=w+100$ Substitute in $w=20$ to find p.
$p=(25)+100$
$p=125$

The rate of the plane is 125 mph and the rate of the wind is 25 mph.

65. Let $t=$ the number of pounds of turmeric and let $c=$ the number of pounds of ground coriander. The system of equations is:

$t+c=5$
$8.9t+11.2c=49.1$

Use the substitution method to solve the system. Solve for t in the first equation.

$t+c=5$
$\underline{-c \quad -c}$ Subtract c from both sides.
$t=-c+5$

Now substitute $t=-c+5$ into the second equation.

$8.9t+11.2c=49.1$ Substitute in $t=-c+5$.
$8.9(-c+5)+11.2c=49.1$ Distribute.
$-8.9c+44.5+11.2c=49.1$ Simplify.
$2.3c+44.5=49.1$
$\underline{-44.5 \quad -44.5}$
$2.3c=4.6$ Solve for c.
$\frac{2.3c}{2.3}=\frac{4.6}{2.3}$
$c=2$

$t=-c+5$ Substitute in $c=2$ to find t.
$t=-(2)+5$
$t=3$

The spice mix should include 3 lb. of turmeric and 2 lb. of ground coriander.

66. Solve the system using the elimination method. Multiply the second equation by -1. After this step, the coefficients on y will be opposites so the y variable should be eliminated after adding the equations vertically.

$$\begin{aligned} 3x+y &= 7 \\ -1(2x+y) &= -1(3) \end{aligned} \text{ is now } \begin{aligned} 3x+y &= 7 \\ -2x-y &= -3 \end{aligned}$$

$3x+y=7$ Add the two equations vertically.
$\underline{-2x-y=-3}$
$x=4$

$2(4)+y=3$ Substitute in $x=4$ to find y.

$8+y=3$

$\underline{-8 \quad -8}$

$y=-5$

The solution is $(4,-5)$.

67. Solve the system using the elimination method. Multiply the first equation by 3 and multiply the second equation by 2. After this step, the coefficients on y will be opposites so the y variable should be eliminated after adding the equations vertically.

$3(5x-2y)=3(3)$ is now $15x-6y=9$

$2(7x+3y)=2(-19)$ is now $14x+6y=-38$

$15x-6y=9$ Add the two equations vertically.

$\underline{14x+6y=-38}$

$29x=-29$

$\frac{29x}{29}=\frac{-29}{29}$

$x=-1$

$7(-1)+3y=-19$ Substitute in $x=-1$ to find y.

$-7+3y=-19$

$\underline{+7 \quad +7}$

$3y=-12$

$\frac{3y}{3}=\frac{-12}{3}$

$y=-4$

The solution is $(-1,-4)$.

68. Solve the system using the elimination method. Multiply the first equation by 2.

$2(2x+3y)=2(18)$ is now $4x+6y=36$

$-4x-6y=-36$ is now $-4x-6y=-36$

Now that the coefficients on y are opposites, add the two equations vertically to eliminate y.

$4x+6y=36$ Add the two equations vertically.

$\underline{-4x-6y=-36}$

$0=0$

Because this equation is always true ($0=0$), there are infinitely many solutions. These two equations graph as the same line.

69. Let $b=$ the cost in dollars for each scrapbook and let $s=$ the cost in dollars for each sticker set. The system of equations is as follows:

$10b+8s=280$

$4b+6s=147$

Use the substitution method to solve the system. Solve for b in the first equation.

$10b+8s=280$

$\underline{-8s \quad -8s}$ Subtract $8s$ from both sides.

$10b=-8s+280$

$\frac{10b}{10}=\frac{-8s+280}{10}$

$b=-0.8s+28$

Now substitute $b=-0.8s+28$ into the second equation.

$4b+6s=147$ Substitute in $b=-0.8s+28$.

$4(-0.8s+28)+6s=147$ Distribute.

$-3.2s+112+6s=147$ Simplify.

$2.8s+112=147$

$\underline{-112 \quad -112}$

$2.8s=35$ Solve for s.

$\frac{2.8s}{2.8}=\frac{35}{2.8}$

$s=12.5$

$b=-0.8s+28$ Substitute in $s=12.5$ to find b.

$b=-0.8(12.5)+28$

$b=-10+28$

$b=18$

The cost of each scrapbook is \$18 and the cost of each sticker set is \$12.50.

70. Supplementary angles have a sum of 180°. Let $x=$ the degree measure of the first angle and let $y=$ the degree measure of the second angle. The system of equations is:

$x+y=180^\circ$

$y=5x$

Substitute $y=5x$ into the first equation.

$x+y=180^\circ$ Substitute in $y=4x+5$.

$x+(5x)=180$ Simplify.

$6x=180$ Solve for x.

$\frac{6x}{6}=\frac{180}{6}$

$x=30$

$y=5x$ Substitute in $x=30$ to find y.

$y=5(30)$

$y=150$

The measures of the angles are 30° and 150°.

71. Let $x=$ the amount (in dollars) invested in the CD earning 3% interest and let $y=$ the amount (in dollars) invested in the bond fund earning 5% interest.

The system of equations is:

$x+y=12000$

$0.03x+0.05y=440$

Solve the system using the substitution method. Solve for y in the first equation.

$x+y=12000$

$\underline{-x \quad\quad -x}$

$y=-x+12000$

Now substitute $y=-x+12000$ into the second equation.

$0.03x+0.05y=440$ Substitute in $y=-x+12000$.

$0.03x+0.05(-x+12000)=440$ Distribute.

$0.03x-0.05x+600=440$ Simplify.

$-0.02x+600=440$

$-600\ -600$ Subtract 600 from both sides.

$-0.02x=-160$ Solve for x.

$\frac{-0.02x}{-0.02}=\frac{-160}{-0.02}$

$x=8000$

$y=-x+12000$ Substitute in $x=8000$ to find y.

$y=-(8000)+12000$

$y=4000$

If Eric wants to earn \$440 in interest every year, he should invest \$8000 in the 3% interest-earning CD and \$4000 in the 5% interest-earning bond fund.

72. $3x-y<7$ Is $(x,y)=(3,10)$ in the solution set?

$3(3)-(10)\overset{?}{<}7$ Substitute values for x and y.

$9-10\overset{?}{<}7$

$-1<7$ This statement is true.

Yes, the point is in the solution set.

73. Graph the line $-2x+3y>6$ by converting to slope-intercept form $y>\frac{2}{3}x+2$. Use the y-intercept $(0,2)$ and the slope $m=\frac{2}{3}=\frac{rise}{run}$. Since this is a strict inequality (>), graph a dashed line.

Select the test point $(0,0)$. Is $(x,y)=(0,0)$ in the solution set?

$(0)\overset{?}{>}\frac{2}{3}(0)+2$ Substitute values for x and y.

$0\overset{?}{>}0+2$

$0\overset{?}{>}2$ This statement is false.

Because this is not true, we do not shade in the region containing the point $(0,0)$. Shade above the dashed line.

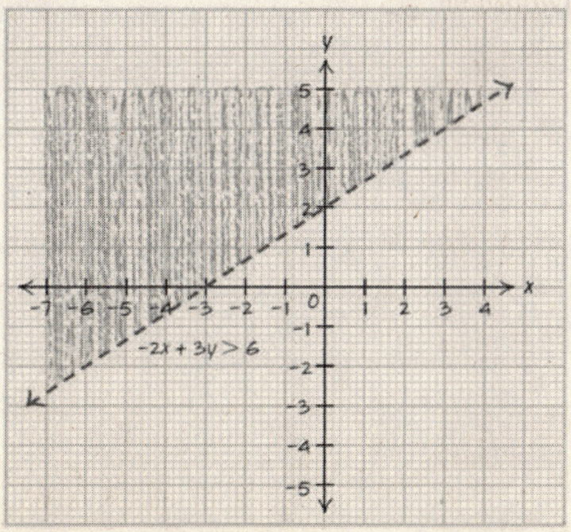

74. Graph the line $3x-10y\le 20$ by converting to slope-intercept form $y\ge\frac{3}{10}x-2$. Use the y-intercept $(0,-2)$ and the slope $m=\frac{3}{10}=\frac{rise}{run}$. Since this is not a strict inequality ($\ge$), graph a solid line.

Select the test point $(0,0)$. Is $(x,y)=(0,0)$ in the solution set?

$(0)\overset{?}{\ge}\frac{3}{10}(0)-2$ Substitute values for x and y.

$0\overset{?}{\ge}0-2$

$0\ge -2$ This statement is true.

Because this is true, we will shade in the region containing the point $(0,0)$. Shade above the solid line.

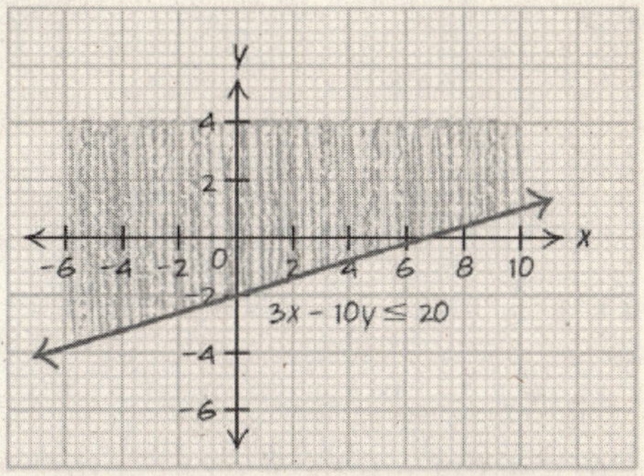

75. **a.** Let $r=$ the number of spring rolls Mai purchases, and let $g=$ the number of chicken gyoza Mai purchases.

b. $3.99r+4.25g\le 20$

c. Graph the solution set using the r- and g-intercepts rounded to hundredths $(5.01,0)$ and $(0,4.71)$ and shade above the solid line (negative values do not make sense in this situation so we will not graph them).

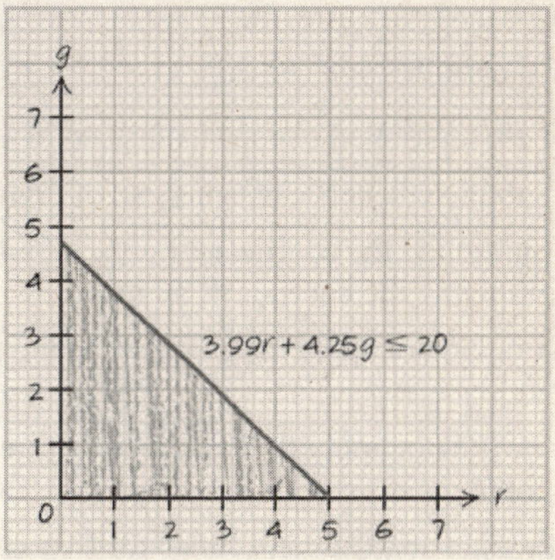

CHAPTER 5 EXPONENTS AND POLYNOMIALS

Section 5.1 Rules for Exponents

1. When multiplying exponential expressions that have the same base, you add exponents.

3. When raising an exponential expression to another power, you multiply the exponents.

5. Applying exponents does NOT work over the operations addition and subtraction.

7. $(-2)^2 = (-2)(-2) = 4$

9. $(-2)^3 = (-2)(-2)(-2) = -8$

11. $-5^2 = -(5)(5) = -25$

13. $-1^3 = -(1)(1)(1) = -1$

15. $x^7 x^2 = x^{7+2} = x^9$

17. $7^3 \cdot 7^5 = 7^8 = 5{,}764{,}801$

19. $5w^4 \cdot 3w^8 = 5 \cdot 3 \cdot w^4 \cdot w^8 = 15w^{12}$

21. $(5xy^2)(8x^5y^3) = 5 \cdot 8 \cdot x \cdot x^5 \cdot y^2 \cdot y^3 = 40x^6y^5$

23. $(-3m^3n^4)(5m^7n^3) = -3 \cdot 5 \cdot m^3 \cdot m^7 \cdot n^4 \cdot n^3 = -15m^{10}n^7$

25. $(-6x^2y)(3x^4y^3) = -6 \cdot 3 \cdot x^2 \cdot x^4 \cdot y \cdot y^3 = -18x^6y^4$

27. $(6x)^5(6x)^3 = (6x)^8 = (6)^8(x)^8 = 1{,}679{,}616x^8$

29. $(7b)^2 \cdot (7b)^4 = (7b)^6 = (7)^6(b)^6 = 117{,}649b^6$

31. $(-4m)^3(-4m)^2 = (-4m)^5 = (-4)^5(m)^5 = -1024m^5$

33. $\dfrac{x^{12}}{x^5} = x^{12-5} = x^7$

35. $\dfrac{20k^{17}}{4k^3} = 5k^{17-3} = 5k^{14}$

37. $(x^2)^5 = x^{2 \cdot 5} = x^{10}$

39. $(a^2b^6)^5 = a^{2 \cdot 5}b^{6 \cdot 5} = a^{10}b^{30}$

41. Elizabeth multiplied exponents instead of adding.

43. Elba subtracted coefficients instead of dividing.

45. Luis multiplied the coefficient by the exponent instead of multiplying the coefficient by itself.

47. Yomaira did not apply the exponent to the negative sign.

49. Andrew did not apply the exponents before dividing, or he ignored the exponents altogether.

51. The Product and Quotient Rules for Exponents

$$\frac{x^2x^5y^5}{x^4y^2} = \frac{x^7y^5}{x^4y^2} = x^{7-4}y^{5-2} = x^3y^3$$

53. The Product and Quotient Rules for Exponents

$$\frac{m^2m^3n^2}{mn} = \frac{m^5n^2}{mn} = m^{5-1}n^{2-1} = m^4n$$

55. The Quotient Rule for Exponents

$$\frac{5a^5b^3}{10a^5b^2} = \frac{a^{5-5}b^{3-2}}{2} = \frac{b}{2}$$

57. The Quotient Rule for Exponents

$$\frac{30x^5y^2}{14xy^2} = \frac{15x^{5-1}y^{2-2}}{7} = \frac{15x^4}{7}$$

59. The Quotient Rule for Exponents

$$\frac{x^4}{25} \cdot \frac{5}{x^2} = \frac{x^{4-2}}{5} = \frac{x^2}{5}$$

61. The Product and Quotient Rules for Exponents

$$\frac{a^3b}{a^2b} \cdot \frac{b^3}{a} = \frac{a^3b^4}{a^3b} = a^{3-3}b^{4-1} = b^3$$

63. The Product and Quotient Rules for Exponents

$$\frac{16a^2b^6}{ab^3} \cdot \frac{a^3b^2}{48ab^2} = \frac{a^5b^8}{3a^2b^5} = \frac{a^{5-2}b^{8-5}}{3} = \frac{a^3b^3}{3}$$

65. The Power Rule for Exponents and Powers of Products Rules

$$(3c^2d^5)^2 = (3)^2(c^2)^2(d^5)^2 = 9c^4d^{10}$$

67. The Power Rule for Exponents and Powers of Products Rules

$$(-5a^7b^5)^2 = (-5)^2(a^7)^2(b^5)^2 = 25a^{14}b^{10}$$

69. The definition of Zero as an Exponent

$$(15p^{10}q^5)^0 = 1$$

71. The Power Rule for Exponents and Powers of Products and Quotients Rules

$$\left(\frac{12x^5y^4}{2x^2y}\right)^2 = \left(6x^{5-2}y^{4-1}\right)^2 = \left(6x^3y^3\right)^2$$

$$= (6)^2\left(x^3\right)^2\left(y^3\right)^2 = 36x^6y^6$$

73. The Power Rule for Exponents and Powers of Products and Quotients Rules

$$\left(\frac{-26x^3y^5}{6xy}\right)^2 = \left(\frac{-13x^{3-1}y^{5-1}}{3}\right)^2 = \left(\frac{-13x^2y^4}{3}\right)^2$$

$$= \left(\frac{-13}{3}\right)^2\left(x^2\right)^2\left(y^4\right)^2 = \frac{169x^4y^8}{9}$$

75. The definition of Zero as an Exponent

$$\left(\frac{9x^{-3}}{3y^5}\right)^0 = 1$$

77. The Power Rule for Exponents and Powers of Products Rules

$\left(3xy^2\right)^2\left(5x^3y^7\right)=(3)^2(x)^2\left(y^2\right)^2\left(5x^3y^7\right)$
$=9x^2y^4\cdot 5x^3y^7=45x^{2+3}y^{4+7}=45x^5y^{11}$

79. The Power Rule for Exponents and Powers of Products rules are used to simplify.

$\left(4m^4n^3\right)^2\left(3mn^5\right)^3=(4)^2\left(m^4\right)^2\left(n^3\right)^2(3)^3(m)^3\left(n^5\right)^3$
$=16m^8n^6\cdot 27m^3n^{15}=432m^{8+3}n^{6+15}=432m^{11}n^{21}$

81. The Power Rule for Exponents and Powers of Products rules are used to simplify.

$(-3a^2)^4(2ab^3)^3=(-3)^4(a^2)^4(2)^3(a)^3(b^3)^3=81a^8\cdot 8a^3b^9$
$=648a^{8+3}b^9=648a^{11}b^9$

83. The Power Rule for Exponents and Powers of Products and Quotients rules are used to simplify.

$$\frac{\left(2x^3y\right)^4}{\left(5xy^2\right)^2}=\frac{(2)^4\left(x^3\right)^4(y)^4}{(5)^2(x)^2\left(y^2\right)^2}=\frac{16x^{12}y^4}{25x^2y^4}=\frac{16x^{12-2}y^{4-4}}{25}$$
$$=\frac{16x^{10}}{25}$$

85. The Power Rule for Exponents and Powers of Products and Quotients rules are used to simplify.

$$\frac{\left(-4x^5y^3\right)^2}{\left(2x^2y\right)^3}=\frac{(-4)^2\left(x^5\right)^2\left(y^3\right)^2}{(2)^3\left(x^2\right)^3(y)^3}=\frac{16x^{10}y^6}{8x^6y^3}$$
$=2x^{10-6}y^{6-3}=2x^4y^3$

87. The student mistakenly applied the power to the negative sign.

89. The student multiplied the bases before applying the exponent, and included the exponent incorrectly.

91. The student applied the exponent over addition, but exponents can only be applied over operations of multiplication or division.

93. The student divided the bases before applying the exponent, and included the exponent incorrectly.

95. **a.** Yes, the expression $(x+3)^2(x+3)^5$ can be simplified to $(x+3)^7$ using the Product Rule for Exponents. In this case, $(x+3)$ is the base and the exponents are added.

b. No, the exponent 7 in the expression $(x+3)^7$ can not be applied to x and 3 to get x^7+3^7. This would be applying the Power Rule to a sum, but the Power Rule only applies to products and quotients.

97. No, x^2+x^3 can not be simplified. These are not like terms because they do not have the same exponent and therefore can not be combined using addition.

99. $A=s^2$ Area = side^2

$A=(7x)^2$ Substitute $s=7x$.
$A=(7)^2(x)^2$
$A=49x^2$

101. $V=s^3$ Volume = side^3

$V=(5x)^3$ Substitute $s=5x$.
$V=(5)^3(x)^3$
$V=125x^3$

103. $A=\pi r^2$

$A=\pi(2x)^2$ Substitute $r=2x$.
$A=\pi(2)^2(x)^2$
$A=4\pi x^2$

105. $A=\frac{1}{2}bh$

$A=\frac{1}{2}(3x)(4x)$ Substitute $b=3x$ and $h=4x$.
$A=6x^2$

Section 5.2 Negative Exponents and Scientific Notation

1. $x^{-3} = \frac{1}{x^3}$

3. $3^{-2} = \frac{1}{3^2} = \frac{1}{9}$

5. $2^{-5} = \frac{1}{2^5} = \frac{1}{32}$

7. $4d^{-6} = 4 \cdot \frac{1}{d^6} = \frac{4}{d^6}$

9. $-5a^{-4} = -5 \cdot \frac{1}{a^4} = \frac{-5}{a^4}$

11. $\frac{5x^2}{2y^{-3}} = \frac{5x^2y^3}{2}$

13. $\frac{1}{5x^{-4}} = \frac{x^4}{5}$

15. $\frac{a^{-3}}{2} = \frac{1}{2a^3}$

17. $\frac{3^{-2}}{x} = \frac{1}{3^2x} = \frac{1}{9x}$

19. $\frac{x^{-2}}{y^{-3}} = \frac{y^3}{x^2}$

21. The student treated the negative sign as though it were a negative exponent. The expression is already in simplest form.

$$\frac{-2x}{y}$$

23. The student applied the negative exponent to the coefficient but it only applies to the variable.

$$42a^{-1} = 42 \cdot \frac{1}{a} = \frac{42}{a}$$

25. The student applied the negative exponent only to the variable but it applies to the whole term.

$$(5x)^{-2} = \frac{1}{(5x)^2} = \frac{1}{(5)^2(x)^2} = \frac{1}{25x^2}$$

27. The student applied a positive exponent to the coefficient and applied the negative exponent only to the variable, but the negative exponent applies to the whole term.

$$(-3x)^{-2} = \frac{1}{(-3x)^2} = \frac{1}{(-3)^2(x)^2} = \frac{1}{9x^2}$$

29. $\frac{x^3}{x^7} = x^{3-7} = x^{-4} = \frac{1}{x^4}$

31. $\frac{m^2n}{m^5n^2} = m^{2-5}n^{1-2} = m^{-3}n^{-1} = \frac{1}{m^3n}$

33. $\frac{10x^5y^7}{6x^{11}y^2} = \frac{5x^{5-11}y^{7-2}}{3}$

$= \frac{5x^{-6}y^5}{3}$

$= \frac{5y^5}{3x^6}$

35. $\frac{4x^3y^{-2}}{x^5y^6} = 4x^{3-5}y^{-2-6}$

$= 4x^{-2}y^{-8}$

$= \frac{4}{x^2y^8}$

37. $\frac{24m^{-3}n^5}{10m^2n^{-3}} = \frac{12m^{-3-2}n^{5-(-3)}}{5}$

$= \frac{12m^{-5}n^8}{5}$

$= \frac{12n^8}{5m^5}$

39. $\frac{-15x^2y^{-4}}{6xy^3} = \frac{-5x^{2-1}y^{-4-3}}{2}$

$= \frac{-5x^1y^{-7}}{2}$

$= \frac{-5x}{2y^7}$

41. $(2x^{-2})^3 = (2)^3(x^{-2})^3$

$= 8x^{-6}$

$= \frac{8}{x^6}$

43. $(-5x^{-3})^2 = (-5)^2(x^{-3})^2$

$= 25x^{-6}$

$= \frac{25}{x^6}$

45. $(-3a^{-4})^2(-6a^3b^{-3}) = (-3)^2(a^{-4})^2(-6a^3b^{-3})$

$= 9a^{-8} \cdot -6a^3b^{-3}$

$= -54a^{-8+3}b^{-3}$

$= -54a^{-5}b^{-3}$

$= \frac{-54}{a^5b^3}$

47. $\left(\frac{-4x^5y^3}{6x^7y^{-1}}\right)^2 = \left(\frac{-2x^{5-7}y^{3-(-1)}}{3}\right)^2$

$= \left(\frac{-2x^{-2}y^4}{3}\right)^2$

$$=\left(\frac{-2}{3}\right)^2\left(x^{-2}\right)^2\left(y^4\right)^2$$
$$=\frac{4x^{-4}y^8}{9}$$
$$=\frac{4y^8}{9x^4}$$

49. $\left(\frac{a}{b^3}\right)^{-2}=\left(\frac{b^3}{a}\right)^2$
$$=\frac{\left(b^3\right)^2}{(a)^2}$$
$$=\frac{b^6}{a^2}$$

51. The student subtracted 5 when subtracting the exponent in the denominator rather than subtracting -5.
$$\frac{x^3}{x^{-5}}=x^{3-(-5)}=x^8$$

53. The student did not retain the negative sign for the coefficient and incorrectly moved the coefficient in the denominator to the numerator.
$$\frac{-2x^{-4}}{16x^3}=\frac{-x^{-4-3}}{8}=\frac{-x^{-7}}{8}=\frac{-1}{8x^7}$$

55. The student did not apply the Power of Products rule correctly, missing the first variable.
$$(x^3y)^{-2}=(x^3)^{-2}(y)^{-2}=x^{-6}y^{-2}=\frac{1}{x^6y^2}$$

57. $6.23\times10^8=6.23\cdot100{,}000{,}000=623{,}000{,}000$

59. 5.98×10^{24}
$=5.98\cdot1{,}000{,}000{,}000{,}000{,}000{,}000{,}000{,}000$
$=5{,}980{,}000{,}000{,}000{,}000{,}000{,}000{,}000$

The value for the mass of the Earth has the standard form 5,980,000,000,000,000,000,000,000 kg.

61. $3.745\times10^{-9}=0.000000003745$

63. $1.67\times10^{-27}=0.00000000000000000000000000167$

The value for the mass of an electron has the standard form 0.00000000000000000000000000167 kg.

65. $62{,}000{,}000{,}000=6.2\times10^{10}$

67. $0.0000043=4.3\times10^{-6}$

69. $5{,}700{,}000{,}000=5.7\times10^9$

The 2008 fourth-quarter revenue for Google Inc.: $\$5.7\times10^9$.

71. $14{,}300{,}000{,}000{,}000=1.43\times10^{13}$

The 2008 United States GDP: $\$1.43\times10^{13}$.

73. $0.00007=7.0\times10^{-5}$

The average width of a human hair: 7.0×10^{-5} meter.

75. $0.0000005=5.0\times10^{-7}$

The width of pits where data on compact discs is stored: 5.0×10^{-7} meter.

77. $2(4.2\times10^{11})$
$=2(4.2)\times10^{11}$
$=8.4\times10^{11}$

79. $5(1.5\times10^{-12})$
$=5(1.5)\times10^{-12}$
$=7.5\times10^{-12}$

81. $4(6.3\times10^{13})$
$=4(6.3)\times10^{13}$
$=25.2\times10^{13}$ Decimal moves left 1.
$=2.52\times10^{14}$ Exponent increases 1.

83. $12(3.0\times10^{-7})$
$=12(3.0)\times10^{-7}$
$=36.0\times10^{-7}$ Decimal moves left 1.
$=3.6\times10^{-6}$ Exponent increases 1.

85. $\frac{2.5\times10^8}{5}$
$=\frac{2.5}{5}\times10^8$
$=0.5\times10^8$ Decimal moves right 1.
$=5.0\times10^7$ Exponent decreases 1.

87. $\frac{1.4\times10^{15}}{7}$
$=\frac{1.4}{7}\times10^{15}$
$=0.2\times10^{15}$ Decimal moves right 1.
$=2.0\times10^{14}$ Exponent decreases 1.

89. $25(6.0\times10^{23})$
$=25(6.0)\times10^{23}$
$=150.0\times10^{23}$ Decimal moves left 2.
$=1.5\times10^{25}$ Exponent increases 2.

91. $50(7.2\times10^{-24})$
$=50(7.2)\times10^{-24}$
$=360.0\times10^{-24}$ Decimal moves left 2.
$=3.6\times10^{-22}$ Exponent increases 2.

93. $\dfrac{7.2\times10^{30}}{1200}$

$=\dfrac{7.2}{1200}\times10^{30}$

$=0.006\times10^{30}$ Decimal moves right 3.

$=6.0\times10^{27}$ Exponent decreases 3.

95. $\dfrac{7.2\times10^{-14}}{3000}$

$=\dfrac{7.2}{3000}\times10^{-14}$

$=0.0024\times10^{-14}$ Decimal moves right 3.

$=2.4\times10^{-17}$ Exponent decreases 3.

97. $-3(4.5\times10^{11})$

$=-3(4.5)\times10^{11}$

$=-13.5\times10^{11}$ Decimal moves left 1.

$=-1.35\times10^{12}$ Exponent increases 1.

99. $5(-3.6\times10^{23})$

$=5(-3.6)\times10^{23}$

$=-18.0\times10^{23}$ Decimal moves left 1.

$=-1.8\times10^{24}$ Exponent increases 1.

101. $95(5.98\times10^{24}\text{ kg})$

$=95(5.98)\times10^{24}\text{ kg}$

$=568.1\times10^{24}\text{ kg}$ Decimal moves left 2.

$=5.681\times10^{26}\text{ kg}$ Exponent increases 2.

The mass of Saturn is approximately 5.681×10^{26} kg.

103. $0.055(5.98\times10^{24}\text{ kg})$

$=0.055(5.98)\times10^{24}\text{ kg}$

$=0.3289\times10^{24}\text{ kg}$ Decimal moves right 1.

$=3.289\times10^{23}\text{ kg}$ Exponent decreases 1.

The mass of Mercury is approximately 3.289×10^{23} kg.

105. One AU $=1.496\times10^{8}$ km. Find 39.5 AU.

$39.5(1.496\times10^{8}\text{ km})$

$=39.5(1.496)\times10^{8}\text{ km}$

$=59.092\times10^{8}\text{ km}$ Decimal moves left 1.

$=5.9092\times10^{9}\text{ km}$ Exponent increases 1.

The average distance from Pluto to the Sun is approximately 5.9092×10^{9} km.

107. One AU $=1.496\times10^{8}$ km. Find 1.5

$1.5(1.496\times10^{8}\text{ km})$

$\text{AU.}=1.5(1.496)\times10^{8}\text{ km}$

$=2.244\times10^{8}\text{ km}$

The average distance from Mars to the Sun is approximately 2.244×10^{8} km.

109. $\dfrac{1.28\times10^{4}\text{ km}}{2}$

$=\dfrac{1.28}{2}\times10^{4}\text{ km}$

$=0.64\times10^{4}\text{ km}$ Decimal moves right 1.

$=6.4\times10^{3}\text{ km}$ Exponent decreases 1.

The diameter of Mars is approximately 6.4×10^{3} km.

111. $200(1.602\times10^{-19}\text{ coulombs})$

$=200(1.602)\times10^{-19}$

$=320.4\times10^{-19}$ Decimal moves left 2.

$=3.204\times10^{-17}$ Exponent increases 2.

The electric charge of 200 protons is 3.204×10^{-17} coulombs.

113. $500(1.67\times10^{-27}\text{ kg})$

$=500(1.67)\times10^{-27}\text{ kg}$

$=835.0\times10^{-27}\text{ kg}$ Decimal moves left 2.

$=8.35\times10^{-25}\text{ kg}$ Exponent increases 2.

The mass of 500 electrons is 8.35×10^{-25} kg.

115. $3,000,000(1.67\times10^{-27}\text{ kg})$

$=3,000,000(1.67)\times10^{-27}\text{ kg}$

$=5,010,000.0\times10^{-27}\text{ kg}$ Decimal moves left 6.

$=5.01\times10^{-21}\text{ kg}$ Exponent increases 6.

The mass of 3 million electrons is 5.01×10^{-21} kg.

117. $(2.0\times10^{13})(4.2\times10^{11})$

$=(2.0)(4.2)\times10^{13+11}$

$=8.4\times10^{24}$

119. $(2.5\times10^{17})(1.5\times10^{-12})$

$=(2.5)(1.5)\times10^{17+(-12)}$

$=3.75\times10^{5}$

121. $(6.0\times10^{20})(6.3\times10^{13})$

$=(6.0)(6.3)\times10^{20+13}$

$=37.8\times10^{33}$ Decimal moves left 1.

$=3.78\times10^{34}$ Exponent increases 1.

123. $(7.1\times10^{-8})(3.0\times10^{-7})$

$=(7.1)(3.0)\times10^{-8+(-7)}$

$=21.3\times10^{-15}$ Decimal moves left 1.

$=2.13\times10^{-14}$ Exponent increases 1.

125. $\dfrac{2.5\times10^{8}}{1.25\times10^{4}}$

$=\dfrac{2.5}{1.25}\times10^{8-4}$

$=2.0\times10^{4}$

127. $\dfrac{2.7\times10^{15}}{3.0\times10^{28}}$

$=\dfrac{2.7}{3.0}\times10^{15-28}$

$=0.9\times10^{-13}$ Decimal moves right 1.

$=9.0\times10^{-14}$ Exponent decreases 1.

129. $\dfrac{7.2\times10^{30}}{1.5\times10^{11}}$

$=\dfrac{7.2}{1.5}\times10^{30-11}$

$=4.8\times10^{19}$

131. $\dfrac{4.87\times10^{24}\text{ kg}}{5.98\times10^{24}\text{ kg}}$

$=\dfrac{4.87}{5.98}\times10^{24-24}$

≈0.814

The mass of Venus is approximately 0.81 times as large as the mass of Earth.

133. $\dfrac{1.428\times10^{5}\text{ km}}{1.28\times10^{4}\text{ km}}$

$=\dfrac{1.428}{1.28}\times10^{5-4}$

$=1.115625\times10^{1}$ Convert to standard notation.

$=11.15625$

The diameter of Jupiter is approximately 11.156 times larger than the diameter of Earth.

135. $\dfrac{1.06\times10^{13}}{3.06\times10^{8}}$

$=\dfrac{1.06}{3.06}\times10^{13-8}$

$\approx0.346405\times10^{5}$ Convert to standard notation.

$\approx34{,}641$

Each person's share of the U.S. national debt in January 2009 was approximately \$34,641.

Section 5.3 Adding and Subtracting Polynomials

1. The terms are: x^2, $2x$, and 5.

This is a trinomial.

3. The terms are: $4x$ and 8.

This is a binomial.

5. The term is: $8x^3y$.

This is a monomial.

7. The terms are: $7m^2$ and -9.

This is a binomial.

9. The terms are: $14x^3y$, $7x^2y^2$, and $-9xy$.

This is a trinomial.

11. The terms are: $2m^7n^3$ and $-4m^2n^5$.

This is a binomial.

13. The terms are: $-45mn^2$ and 56.

This is a binomial.

15. The polynomial has 2 terms.

Written in descending order: $\frac{3}{2}x-12$

The highest exponent is 1 so the degree is 1.

17. The polynomial has 3 terms.

Written in descending order: $-6x^2+3x-9$

The highest exponent is 2 so the degree is 2.

19. The polynomial has 1 term.

Written in descending order: -45

This expression has no exponents so the highest exponent is 0 and the degree is 0.

21. The polynomial has 4 terms.

Written in descending order: $-y^3+9y^2+4y+7$

The highest exponent is 3 so the degree is 3.

23. The polynomial has 3 terms.

Written in descending order: $-3b^8+2b^7+9b$

The highest exponent is 8 so the degree is 8.

25. The degree of the term $-6xy^2$ is 3.

The single term has a degree of 3, so the degree of the polynomial is 3.

27. The degree of the term $3a$ is 1.

The degree of the term $2ab$ is 2.

The degree of the term $-6b^2$ is 2.

The highest degree term has a degree of 2, so the degree of the polynomial is 2.

29. The degree of the term 1 is 0.

The degree of the term $-x^3y^3$ is 6.

The highest degree term has a degree of 6, so the degree of the polynomial is 6.

31. The degree of the term $-24x^2yz^3$ is 6.

The degree of the term $6z$ is 1.

The degree of the term $-7y$ is 1.

The degree of the term 5 is 0.

The highest degree term has a degree of 6, so the degree of the polynomial is 6.

33. The degree of the term -29 is 0.

The single term has a degree of 0, so the degree of the polynomial is 0.

35. x^2+2x+5 is an expression that has a second degree term, so it is not linear. There is no equal sign so it is an expression, not an equation.

37. $4x+8$ is a polynomial in 1 variable with degree 1, so it is a linear expression in 1 variable. There is no equal sign so it is an expression, not an equation.

39. $y=4x+3$ is an equation in 2 variables that has degree 1, so it is a linear equation in 2 variables. It is an equation because there is an equal sign.

41. $5x-(2x+8)=4x$ is a linear equation in 1 variable that has degree 1. It is an equation because there is an equal sign.

43. $4xy+7x-9xy$ is a polynomial in 2 variables that has degree 2 terms. It is not linear because is it not degree 1. There is no equal sign so it is an expression, not an equation.

45. $y=2x^2+3x-20$ is an equation that has a second degree term, so it is not linear. It is an equation because there is an equal sign.

47. $2x+7y=68$ is an equation in 2 variables that has degree 1, so it is a linear equation in 2 variables. It is an equation because there is an equal sign.

49. $(3x+5)+(4x+9)$

$=3x+4x+5+9$

$=7x+14$

51. $(t^2+5t-9)+(4t^2+3t+4)$

$=t^2+4t^2+5t+3t-9+4$

$=5t^2+8t-5$

53. $(-5m^2+6m-3)+(-4m^2+7m-6)$

$=-5m^2-4m^2+6m+7m-3-6$

$=-9m^2+13m-9$

55. $(6x+9)-(4x+3)$

$=6x+9-4x-3$

$=6x-4x+9-3$

$=2x+6$

57. $(n^2+8n-6)-(n^2-5n+3)$

$=n^2+8n-6-n^2+5n-3$

$=n^2-n^2+8n+5n-6-3$

$=13n-9$

59. $(5h^2-7h+12)-(3h^2-5h+8)$

$=5h^2-7h+12-3h^2+5h-8$

$=5h^2-3h^2-7h+5h+12-8$

$=2h^2-2h+4$

61. The student treated this as a product, adding exponents, rather than a sum and should have added coefficients.

$2x+x=3x$

63. The student did not subtract the entire quantity of $(4x+5)$, only subtracting the first term and not the second term.

$(3x-2)-(4x+5)$

$=3x-2-4x-5$

$=3x-4x-2-5$

$=-x-7$

65. The student did not subtract the entire quantity of $(2x^2-5x)$, only subtracting the first term and not the second term (subtracting a negative changes it to adding the term).

$(x^2+4x-8)-(2x^2-5x)$

$=x^2+4x-8-2x^2+5x$

$=x^2-2x^2+4x+5x-8$

$=-x^2+9x-8$

67. **a.** The total cost for a class to see the life of a pumpkin demo and visit the petting zoo is as follows

$(1.5k+5)+(2k+10)$

$=1.5k+2k+5+10$

$=3.5k+15$

b. The cost for 33 students to see the life of a pumpkin demo and visit the petting zoo is \$130.50, calculated as follows

$3.5k+15$ Substitute $k=33$.

$=3.5(33)+15$

$=115.5+15$

$=130.5$

69. **a.** The total cost for a class to see the life of a pumpkin demo, take a tractor ride, and take home a pumpkin is as follows

$(1.5k+5)+(1.25k+5)+(4.5k)$
$=1.5k+1.25k+4.5k+5+5$
$=7.25k+10$

b. The cost for 60 students to see the life of a pumpkin demo, take a tractor ride, and take home a pumpkin is \$445, calculated as follows

$7.25k+10$ Substitute $k=60$.
$=7.25(60)+10$
$=435+10$
$=445$

71. **a.** The total cost for the Grizzlies to purchase all the gear is as follows

$(7.5m+55)+(8m+15)+(10.5m+25)$
$=7.5m+8m+10.5m+55+15+25$
$=26m+95$

b. The cost for 15 members to purchase all the gear is \$485, calculated as follows

$26m+95$ Substitute $m=15$.
$=26(15)+95$
$=390+95$
$=485$

73. **a.** The total cost for the Grizzlies to purchase T-shirts and go on the Dos Picos campout is as follows

$(7.5m+55)+(30m+100)$
$=7.5m+30m+55+100$
$=37.5m+155$

b. The cost for 12 members to purchase T-shirts and go on the Dos Picos campout is \$605, calculated as follows

$37.5m+155$ Substitute $m=12$.
$=37.5(12)+155$
$=450+155$
$=605$

75. **a.** The total cost for a 5-day trip is as follows

$(275m+500)-(125m+150)$
$=275m+500-125m-150$
$=275m-125m+500-150$
$=150m+350$

b. The cost for 30 ski club member to go on the 5-day trip is \$4850, calculated as follows

$150m+350$ Substitute $m=30$.
$=150(30)+350$
$=4500+350$
$=4850$

77. $(3x^2+25)+(2x+5)$
$=3x^2+2x+25+5$
$=3x^2+2x+30$

79. $P=2l+2w$ Substitue values for l and w.
$P=2(5x+10)+2(x+3)$ Distribute.
$P=10x+20+2x+6$ Combine like terms.
$P=12x+26$

81. $(4x^2+21x)+(3x^2+5x)$
$=4x^2+3x^2+21x+5x$
$=7x^2+26x$

83. $(a^2+5ab-8b)+(4a^2-9ab+6b)$
$=a^2+4a^2+5ab-9ab-8b+6b$
$=5a^2-4ab-2b$

85. $(-2t^2+6t-14)+(-8t^2-6)$
$=-2t^2-8t^2+6t-14-6$
$=-10t^2+6t-20$

87. $(6a+11b)-(18a+5b)$
$=6a+11b-18a-5b$
$=6a-18a+11b-5b$
$=-12a+6b$

89. $(m^2+2mn-n)-(7m^2-11mn+3n)$
$=m^2+2mn-n-7m^2+11mn-3n$
$=m^2-7m^2+2mn+11mn-n-3n$
$=-6m^2+13mn-4n$

91. $(45a^2+17a)-(8a+24)$
$=45a^2+17a-8a-24$
$=45a^2+9a-24$

93. $(7x^2+8x-9)+(4x+10)-(8x^2-3x+4)$
$=7x^2+8x-9+4x+10-8x^2+3x-4$
$=7x^2-8x^2+8x+4x+3x-9+10-4$
$=-x^2+15x-3$

95. $(-5m^3+3m^2+8m)+(7m^3+9m-4)-(6m^2-3m+2)$
$=-5m^3+3m^2+8m+7m^3+9m-4-6m^2+3m-2$
$=-5m^3+7m^3+3m^2-6m^2+8m+9m+3m-4-2$
$=2m^3-3m^2+20m-6$

Section 5.4 Multiplying Polynomials

1. $5(2x+7)$
$=5(2x)+5(7)$
$=10x+35$

3. $2x(7x+6)$
$=2x(7x)+2x(6)$
$=14x^2+12x$

5. $-3(4x+2)$
$=-3(4x)-3(2)$
$=-12x-6$

7. $6(2x^2+3x-5)$
$=6(2x^2)+6(3x)+6(-5)$
$=12x^2+18x-30$

9. $-7(x^2+4x-3)$
$=-7(x^2)-7(4x)-7(-3)$
$=-7x^2-28x+21$

11. $3x(5x^2-3x+4)$
$=3x(5x^2)+3x(-3x)+3x(4)$
$=15x^3-9x^2+12x$

13. The new expression to represent the number of screws needed to hold down a lid on a container is found as follows $2(3P)=6P$.

15. $(x+5)(x^2+2x-3)$
$=x(x^2)+x(2x)+x(-3)+5(x^2)+5(2x)+5(-3)$
$=x^3+2x^2-3x+5x^2+10x-15$
$=x^3+7x^2+7x-15$

17. $(x-2)(5x^2-3x-8)$
$=x(5x^2)+x(-3x)+x(-8)-2(5x^2)-2(-3x)-2(-8)$
$=5x^3-3x^2-8x-10x^2+6x+16$
$=5x^3-13x^2-2x+16$

19. $(2x+7)(4x^2-6x+3)$
$=2x(4x^2)+2x(-6x)+2x(3)+7(4x^2)+7(-6x)+7(3)$
$=8x^3-12x^2+6x+28x^2-42x+21$
$=8x^3+16x^2-36x+21$

21. $(4x-3)(4x^2-6x-3)$
$=4x(4x^2)+4x(-6x)+4x(-3)-3(4x^2)-3(-6x)-3(-3)$
$=16x^3-24x^2-12x-12x^2+18x+9$
$=16x^3-36x^2+6x+9$

23. $(x^2+3)(2x^2+3x-4)$
$=x^2(2x^2)+x^2(3x)+x^2(-4)+3(2x^2)+3(3x)+3(-4)$
$=2x^4+3x^3-4x^2+6x^2+9x-12$
$=2x^4+3x^3+2x^2+9x-12$

25. $(x+3)(x+5)$
$=x(x)+x(5)+3(x)+3(5)$
$=x^2+5x+3x+15$
$=x^2+8x+15$

27. $(x+6)(x-2)$
$=x(x)+x(-2)+6(x)+6(-2)$
$=x^2-2x+6x-12$
$=x^2+4x-12$

29. $(a+2)(a+5)$
$=a(a)+a(5)+2(a)+2(5)$
$=a^2+5a+2a+10$
$=a^2+7a+10$

31. $(m+n)(2m+n)$
$=m(2m)+m(n)+n(2m)+n(n)$
$=2m^2+mn+2mn+n^2$
$=2m^2+3mn+n^2$

33. $(x^2+3)(5x-4)$
$=x^2(5x)+x^2(-4)+3(5x)+3(-4)$
$=5x^3-4x^2+15x-12$

35. **a.** The total amount of high fructose corn syrup consumed by people in the U.S. is represented by the expression $-1.925t^2-25.25t+17,653.2$ million pounds, found as follows

$(-0.7t+62.6)(2.75t+282)$
$=-0.7t(2.75t)-0.7t(282)+62.6(2.75t)+62.6(282)$
$=-1.925t^2-197.4t+172.15t+17,653.2$
$=-1.925t^2-25.25t+17,653.2$

b. The total amount of high fructose corn syrup consumed by Americans in 2010 is approximately 17,208.2 million pounds, or 17,208,200,000 pounds, calculated as follows

$-1.925t^2-25.25t+17,653.2$ Substitute $t=10$.
$=-1.925(10)^2-25.25(10)+17,653.2$
$=-1.925(100)-25.25(10)+17,653.2$
$=-192.5-252.5+17,653.2$
$=17,208.2$ million pounds

37. **a.** The average income of a family in the U.S. in 2010 is estimated at \$65,000, calculated as follows

$1.7t+48$ Substitute $t=10$.

$=1.7(10)+48$

$=17+48$

$=65$ (in thousands of dollars)

b. The number of families in the U.S. in 2010 is estimated at 86,000,000, calculated as follows

$2.5t+61$ Substitute $t=10$.

$=2.5(10)+61$

$=25+61$

$=86$ (in millions)

c. The total income in billions of dollars of all families in the U.S. can be represented by the expression $4.25t^2+223.7t+2928$, found as follows:

$(1.7t+48)(2.5t+61)$

$=1.7t(2.5t)+1.7t(61)+48(2.5t)+48(61)$

$=4.25t^2+103.7t+120t+2928$

$=4.25t^2+223.7t+2928$

d. The total income of all families in the U.S. in 2012 is estimated at \$6,224,400,000,000, calculated as follows

$4.25t^2+223.7t+2928$ Substitute $t=12$.

$=4.25(12)^2+223.7(12)+2928$

$=4.25(144)+223.7(12)+2928$

$=612+2684.4+2928$

$=6224.4$ billion dollars

39. $A=lw$ Substitue values for l and w.

$A=(5x-8)(3x+4)$

$A=5x(3x)+5x(4)-8(3x)-8(4)$

$A=15x^2+20x-24x-32$

$A=15x^2-4x-32$

41. Use the formula $(a+b)^2=a^2+2ab+b^2$.

$(x+6)^2$

$=x^2+2(x)(6)+6^2$

$=x^2+12x+36$

43. Use the formula $(a-b)^2=a^2-2ab+b^2$.

$(x-3)^2$

$=x^2-2(x)(3)+3^2$

$=x^2-6x+9$

45. Use the formula $(a+b)^2=a^2+2ab+b^2$.

$(3x+5)^2$

$=(3x)^2+2(3x)(5)+5^2$

$=9x^2+30x+25$

47. Use the formula $(a-b)^2=a^2-2ab+b^2$.

$(5x-3)^2$

$=(5x)^2-2(5x)(3)+3^2$

$=25x^2-30x+9$

49. Use the formula $(a+b)^2=a^2+2ab+b^2$.

$(2a+7b)^2$

$=(2a)^2+2(2a)(7b)+(7b)^2$

$=4a^2+28ab+49b^2$

51. Use the formula $(a+b)^2=a^2+2ab+b^2$.

$(3x^2+5)^2$

$=(3x^2)^2+2(3x^2)(5)+5^2$

$=9x^4+30x^2+25$

53. Use the formula $(a-b)^2=a^2-2ab+b^2$.

$(4d^3-5c)^2$

$=(4d^3)^2-2(4d^3)(5c)+(5c)^2$

$=16d^6-40cd^3+25c^2$

55. Using $(a+b)(a-b)=a^2-b^2$

$(x+5)(x-5)$

$=(x)^2-5^2$

$=x^2-25$

57. Using $(a+b)(a-b)=a^2-b^2$

$(2a+5)(2a-5)$

$=(2a)^2-5^2$

$=4a^2-25$

59. Using $(a+b)(a-b)=a^2-b^2$

$\left(\frac{x}{2}+1\right)\left(\frac{x}{2}-1\right)$

$=\left(\frac{x}{2}\right)^2-1^2$

$=\frac{x^2}{4}-1$

61. Use the formula $(a-b)^2=a^2-2ab+b^2$.

$\left(\frac{4}{5}x-3\right)^2$

$=\left(\frac{4}{5}x\right)^2-2\left(\frac{4}{5}x\right)(3)+3^2$

$=\frac{16}{25}x^2-\frac{24}{5}x+9$

63. Using $(a+b)(a-b)=a^2-b^2$

$(3a+2b)(3a-2b)$
$=(3a)^2-(2b)^2$
$=9a^2-4b^2$

65. Using $(a+b)(a-b)=a^2-b^2$

$(x^2+5)(x^2-5)$
$=(x^2)^2-5^2$
$=x^4-25$

67. Using $(a+b)(a-b)=a^2-b^2$

$(2a^5+3b)(2a^5-3b)$
$=(2a^5)^2-(3b)^2$
$=4a^{10}-9b^2$

69. The expression in the parentheses is a product.

$(-2x)^2=(-2)^2(x)^2=4x^2$

71. The expression in the parentheses is a sum.

Use the formula $(a+b)^2=a^2+2ab+b^2$.

$(y+6)^2$
$=y^2+2(y)(6)+6^2$
$=y^2+12y+36$

73. The expression in the parentheses is a sum.

Use the formula $(a+b)^2=a^2+2ab+b^2$.

$(x+y)^2$
$=x^2+2(x)(y)+y^2$
$=x^2+2xy+y^2$

75. The expression in the parentheses is a product.

$\left(\frac{1}{2}n\right)^2=\left(\frac{1}{2}\right)^2(n)^2=\frac{1}{4}n^2$

77. The expression in the parentheses is a product.

$\left(\frac{1}{5}a\right)^2=\left(\frac{1}{5}\right)^2(a)^2=\frac{1}{25}a^2$

79. The student left out the exponent when multiplying the first terms, which should be x^2.

$(x+4)(x-3)$
$=x(x)+x(-3)+4(x)+4(-3)$
$=x^2-3x+4x-12$
$=x^2+x-12$

81. The student left out the middle step when squaring a binomial, adding twice the product of the 2 terms.

Use the formula $(a+b)^2=a^2+2ab+b^2$.

$(x+4)^2$
$=x^2+2(x)(4)+4^2$
$=x^2+8x+16$

83. The student did not square the coefficient of the first term, and left out the middle step when squaring a binomial, adding twice the product of the 2 terms.

Use the formula $(a-b)^2=a^2-2ab+b^2$.

$\left(\frac{2}{3}a-1\right)^2$
$=\left(\frac{2}{3}a\right)^2-2\left(\frac{2}{3}a\right)(1)+1^2$
$=\frac{4}{9}a^2-\frac{4}{3}a+1$

85. $5x(x+7)(x+3)$
$=[5x(x)+5x(7)](x+3)$
$=(5x^2+35x)(x+3)$
$=5x^2(x)+5x^2(3)+35x(x)+35x(3)$
$=5x^3+15x^2+35x^2+105x$
$=5x^3+50x^2+105x$

87. $2x(3x+4)(4x-7)$
$=[2x(3x)+2x(4)](4x-7)$
$=(6x^2+8x)(4x-7)$
$=6x^2(4x)+6x^2(-7)+8x(4x)+8x(-7)$
$=24x^3-42x^2+32x^2-56x$
$=24x^3-10x^2-56x$

89. $7a(a+2)(a-2)$
$=[7a(a)+7a(2)](a-2)$
$=(7a^2+14a)(a-2)$
$=7a^2(a)+7a^2(-2)+14a(a)+14a(-2)$
$=7a^3-14a^2+14a^2-28a$
$=7a^3-28a$

91. $7h(h+5)^2$
$=7h\left[h^2+2(h)(5)+5^2\right]$
$=7h(h^2+10h+25)$
$=7h(h^2)+7h(10h)+7h(25)$
$=7h^3+70h^2+175h$

93. $8x^2(5x-4)^2$

$=8x^2\left[(5x)^2-2(5x)(4)+4^2\right]$

$=8x^2(25x^2-40x+16)$

$=8x^2(25x^2)+8x^2(-40x)+8x^2(16)$

$=200x^4-320x^3+128x^2$

95. $2x+5+(x+3)(x-7)$ Evaluate parentheses first.

$=2x+5+x(x)+x(-7)+3(x)+3(-7)$

$=2x+5+x^2-7x+3x-21$

$=x^2-2x-16$

97. $5x^2+10x-(2x+4)(x+3)$ Evaluate parentheses first.

$=5x^2+10x+(-2x-4)(x+3)$

$=5x^2+10x-2x(x)-2x(3)-4(x)-4(3)$

$=5x^2+10x-2x^2-6x-4x-12$

$=3x^2-12$

99. $-(3x-7)(x+1)+4x^2-7x$ Evaluate parentheses first.

$=(-3x+7)(x+1)+4x^2-7x$

$=-3x(x)-3x(1)+7(x)+7(1)+4x^2-7x$

$=-3x^2-3x+7x+7+4x^2-7x$

$=x^2-3x+7$

101. $(x+4)^2-(x-3)^2$

$=x^2+2(x)(4)+4^2-\left[x^2-2(x)(3)+3^2\right]$

$=x^2+8x+16-(x^2-6x+9)$

$=x^2+8x+16-x^2+6x-9$

$=14x+7$

103. $(2x+5)(2x-5)+(3x^2+6x-8)$

$=(2x)^2-5^2+3x^2+6x-8$

$=4x^2-25+3x^2+6x-8$

$=7x^2+6x-33$

105. $(2t+7)(3t-5)-4(7t^2+3t-5)$

$=(2t+7)(3t-5)-4(7t^2)-4(3t)-4(-5)$

$=(2t+7)(3t-5)-28t^2-12t+20$

$=2t(3t)+2t(-5)+7(3t)+7(-5)-28t^2-12t+20$

$=6t^2-10t+21t-35-28t^2-12t+20$

$=-22t^2-t-15$

Section 5.5 Dividing Polynomials

1. The divisor is a monomial.

3. The divisor is a monomial.

5. The divisor is a polynomial of more than 1 term.

7. The divisor is a polynomial of more than 1 term.

9. The divisor is a polynomial of more than 1 term.

11.

$\dfrac{12x^3+16x^2-8x}{2x}$

$=\dfrac{12x^{\cancel{3}2}}{2\cancel{x}}+\dfrac{16x^{\cancel{2}1}}{2\cancel{x}}-\dfrac{8\cancel{x}}{2\cancel{x}}$

$=6x^2+8x-4$

13.

$\dfrac{30m^4-24m^3+72m^2}{6m^2}$

$=\dfrac{30m^{\cancel{4}2}}{6\cancel{m^2}}-\dfrac{24m^{\cancel{3}1}}{6\cancel{m^2}}+\dfrac{72\cancel{m^2}}{6\cancel{m^2}}$

$=5m^2-4m+12$

15.

$\dfrac{2x^5+7x^3-15x^2+6x}{4x}$

$=\dfrac{2x^{\cancel{5}4}}{4\cancel{x}}+\dfrac{7x^{\cancel{3}2}}{4\cancel{x}}-\dfrac{15x^{\cancel{2}1}}{4\cancel{x}}+\dfrac{6\cancel{x}}{4\cancel{x}}$

$=\dfrac{1}{2}x^4+\dfrac{7}{4}x^2-\dfrac{15}{4}x+\dfrac{3}{2}$

17.

$\dfrac{21h^5+12h^3-2h^2}{7h^2}$

$=\dfrac{21h^{\cancel{5}3}}{7\cancel{h^2}}+\dfrac{12h^{\cancel{3}1}}{7\cancel{h^2}}-\dfrac{2\cancel{h^2}}{7\cancel{h^2}}$

$=3h^3+\dfrac{12}{7}h-\dfrac{2}{7}$

19.

$\dfrac{25x^3+30x^2-5x}{5x^2}$

$=\dfrac{25x^{\cancel{3}1}}{5\cancel{x^2}}+\dfrac{30\cancel{x^2}}{5\cancel{x^2}}-\dfrac{5\cancel{x}}{5x^{\cancel{2}1}}$

$=5x+6-\dfrac{1}{x}$

21.

$$\frac{24d^5-21d^3+14d}{14d^2}$$

$$=\frac{24d^{5\,3}}{14d^2}-\frac{21d^{3\,1}}{14d^2}+\frac{14d}{14d^{2\,1}}$$

$$=\frac{12}{7}d^3-\frac{3}{2}d+\frac{1}{d}$$

23.

$$\frac{20x^2-10xy+15y^2}{5xy}$$

$$=\frac{20x^{2\,1}}{5xy}-\frac{10xy}{5xy}+\frac{15y^{2\,1}}{5xy}$$

$$=\frac{4x}{y}-2+\frac{3y}{x}$$

25.

$$\frac{18m^4n^3-20m^2n^2+15mn}{6mn^2}$$

$$=\frac{18m^{4\,3}n^{3\,1}}{6mn^2}-\frac{20m^{2\,1}n^2}{6mn^2}+\frac{15mn}{6mn^{2\,1}}$$

$$=3m^3n-\frac{10}{3}m+\frac{5}{2n}$$

27. The result is correct.

$3(x-2)=3x-6$

29. The result is correct.

$(x+3)(x-2)$

$=x^2-2x+3x-6$

$=x^2+x-6$

31. The result is correct.

$(x-1)(x^2+x+1)$

$=x^3+x^2+x-x^2-x-1$

$=x^3-1$

33.

$$\begin{array}{r} x+9 \\ x+2\overline{)x^2+11x+18} \\ \underline{-x^2-2x} \\ 9x+18 \\ \underline{-9x-18} \\ 0 \end{array}$$

$(x^2+11x+18)\div(x+2)=x+9$

Check the result:

$(x+9)(x+2)$

$=x^2+2x+9x+18$

$=x^2+11x+18$

35.

$$\begin{array}{r} x-7 \\ x+4\overline{)x^2-3x-28} \\ \underline{-x^2-4x} \\ -7x-28 \\ \underline{7x+28} \\ 0 \end{array}$$

$(x^2-3x-28)\div(x+4)=x-7$

37.

$$\begin{array}{r} m-4 \\ m-3\overline{)m^2-7m+12} \\ \underline{-m^2+3m} \\ -4m+12 \\ \underline{4m-12} \\ 0 \end{array}$$

$(m^2-7m+12)\div(m-3)=m-4$

39.

$$\begin{array}{r} 2x+5 \\ 7x+3\overline{)14x^2+41x+15} \\ \underline{-14x^2-6x} \\ 35x+15 \\ \underline{-35x-15} \\ 0 \end{array}$$

$(14x^2+41x+15)\div(7x+3)=2x+5$

41.

$$\begin{array}{r} 5d-7 \\ 6d-4\overline{)30d^2-62d+28} \\ \underline{-30d^2+20d} \\ -42d+28 \\ \underline{42d-28} \\ 0 \end{array}$$

$(30d^2-62d+28)\div(6d-4)=5d-7$

43.

$A=lw$ Substitute values for A and w.

$2x^2+9x-35=l(2x-5)$

$$\frac{2x^2+9x-35}{2x-5}=l$$

$$\begin{array}{r} x+7 \\ 2x-5\overline{)2x^2+9x-35} \\ \underline{-2x^2+5x} \\ 14x-35 \\ \underline{-14x+35} \\ 0 \end{array}$$

$l=x+7$

45.

$A = lw$ Substitute values for A and w.

$14a^2 + 5a - 24 = l(2a+3)$

$\frac{14a^2 + 5a - 24}{2a+3} = l$

$$\begin{array}{r} 7a - 8 \\ 2a+3 \overline{)\,14a^2 + 5a - 24} \\ \underline{-14a^2 - 21a} \\ -16a - 24 \\ \underline{16a + 24} \\ 0 \end{array}$$

$l = 7a - 8$

47.

$$\begin{array}{r} x + 9 \\ x-4 \overline{)\,x^2 + 5x - 29} \\ \underline{-x^2 + 4x} \\ 9x - 29 \\ \underline{-9x + 36} \\ 7 \end{array}$$

$(x^2 + 5x - 29) \div (x-4) = x + 9 + \frac{7}{x-4}$

49.

$$\begin{array}{r} b - 7 \\ b-3 \overline{)\,b^2 - 10b + 13} \\ \underline{-b^2 + 3b} \\ -7b + 13 \\ \underline{7b - 21} \\ -8 \end{array}$$

$(b^2 - 10b + 13) \div (b-3) = b - 7 - \frac{8}{b-3}$

51.

$$\begin{array}{r} x + 4 \\ 2x+5 \overline{)\,2x^2 + 13x + 9} \\ \underline{-2x^2 - 5x} \\ 8x + 9 \\ \underline{-8x - 20} \\ -11 \end{array}$$

$(2x^2 + 13x + 9) \div (2x+5) = x + 4 - \frac{11}{2x+5}$

53.

$$\begin{array}{r} 3a - 2 \\ 4a+7 \overline{)\,12a^2 + 13a - 9} \\ \underline{-12a^2 - 21a} \\ -8a - 9 \\ \underline{8a + 14} \\ 5 \end{array}$$

$(12a^2 + 13a - 9) \div (4a+7) = 3a - 2 + \frac{5}{4a+7}$

55.

$$\begin{array}{r} 3x - 7 \\ 5x-2 \overline{)\,15x^2 - 41x + 8} \\ \underline{-15x^2 + 6x} \\ -35x + 8 \\ \underline{35x - 14} \\ -6 \end{array}$$

$(15x^2 - 41x + 8) \div (5x-2) = 3x - 7 - \frac{6}{5x-2}$

57.

$$\begin{array}{r} x^2 + 5x + 7 \\ 2x-3 \overline{)\,2x^3 + 7x^2 - x - 21} \\ \underline{-2x^3 + 3x^2} \\ 10x^2 - x \\ \underline{-10x^2 + 15x} \\ 14x - 21 \\ \underline{-14x + 21} \\ 0 \end{array}$$

$(2x^3 + 7x^2 - x - 21) \div (2x-3) = x^2 + 5x + 7$

59.

$$\begin{array}{r} 2a^2 + 4a - 5 \\ 5a+3 \overline{)\,10a^3 + 26a^2 - 13a - 15} \\ \underline{-10a^3 - 6a^2} \\ 20a^2 - 13a \\ \underline{-20a^2 - 12a} \\ -25a - 15 \\ \underline{25a + 15} \\ 0 \end{array}$$

$(10a^3 + 26a^2 - 13a - 15) \div (5a+3) = 2a^2 + 4a - 5$

61.

$$\begin{array}{r} 5x^2 + 15x + 47 \\ x-3 \overline{)\,5x^3 + 0x^2 + 2x + 12} \\ \underline{-5x^3 + 15x^2} \\ 15x^2 + 2x \\ \underline{-15x^2 + 45x} \\ 47x + 12 \\ \underline{-47x + 141} \\ 153 \end{array}$$

$(5x^3 + 2x + 12) \div (x-3) = 5x^2 + 15x + 47 + \frac{153}{x-3}$

63.

$$\begin{array}{r} a^2-2a+4 \\ a+2\overline{\smash{)}a^3+0a^2+0a+8} \\ \underline{-a^3-2a^2} \\ -2a^2+0a \\ \underline{2a^2+4a} \\ 4a+8 \\ \underline{-4a-8} \\ 0 \end{array}$$

$(a^3+8)\div(a+2)=a^2-2a+4$

65.

$$\begin{array}{r} 4x^2+6x+9 \\ 2x-3\overline{\smash{)}8x^3+0x^2+0x-27} \\ \underline{-8x^3+12x^2} \\ 12x^2+0x \\ \underline{-12x^2+18x} \\ 18x-27 \\ \underline{-18x+27} \\ 0 \end{array}$$

$(8x^3-27)\div(2x-3)=4x^2+6x+9$

67.

$$\begin{array}{r|rrr} 7 & 1 & -3 & -28 \\ & & 7 & 28 \\ \hline & 1 & 4 & 0 \end{array}$$

$(x^2-3x-28)\div(x-7)=x+4$

69.

$$\begin{array}{r|rrr} 6 & 2 & -5 & -42 \\ & & 12 & 42 \\ \hline & 2 & 7 & 0 \end{array}$$

$(2m^2-5m-42)\div(m-6)=2m+7$

71.

$$\begin{array}{r|rrr} -3 & 4 & 17 & 15 \\ & & -12 & -15 \\ \hline & 4 & 5 & 0 \end{array}$$

$(4x^2+17x+15)\div(x+3)=4x+5$

73.

$$\begin{array}{r|rrr} -4 & 3 & 17 & 25 \\ & & -12 & -20 \\ \hline & 3 & 5 & 5 \end{array}$$

$(3x^2+17x+25)\div(x+4)=3x+5+\dfrac{5}{x+4}$

75.

$$\begin{array}{r|rrrr} -4 & 2 & 11 & 17 & 20 \\ & & -8 & -12 & -20 \\ \hline & 2 & 3 & 5 & 0 \end{array}$$

$(2x^3+11x^2+17x+20)\div(x+4)=2x^2+3x+5$

77.

$$\begin{array}{r|rrrr} -3 & 1 & 3 & -12 & 16 \\ & & -3 & 0 & 36 \\ \hline & 1 & 0 & -12 & 52 \end{array}$$

$(t^3+3t^2-12t+16)\div(t+3)=t^2-12+\dfrac{52}{t+3}$

79.

$$\begin{array}{r|rrrr} 2 & 1 & 4 & 0 & -24 \\ & & 2 & 12 & 24 \\ \hline & 1 & 6 & 12 & 0 \end{array}$$

$(t^3+4t^2-24)\div(t-2)=t^2+6t+12$

81.

$$\begin{array}{r|rrrr} -2 & 4 & 0 & 2 & 40 \\ & & -8 & 16 & -36 \\ \hline & 4 & -8 & 18 & 4 \end{array}$$

$(4a^3+2a+40)\div(a+2)=4a^2-8a+18+\dfrac{4}{a+2}$

Chapter 5 Review

1. $4x^5 \cdot 7x^3 = 4 \cdot 7x^{5+3} = 28x^8$

2. $-3y^4 \cdot 8y = -3 \cdot 8 \cdot y^{4+1} = -24y^5$

3. $(3x^4y^3)(8x^5y^2) = 24x^{4+5}y^{3+2} = 24x^9y^5$

4. $(-4x^2y^7)(7xy^3) = -28x^{2+1}y^{7+3} = -28x^3y^{10}$

5. $\dfrac{-4x^5y^3}{16x^4y} = \dfrac{-x^{5-4}y^{3-1}}{4} = \dfrac{-xy^2}{4}$

6. $\dfrac{7x^7y^3}{-14xy^3} = \dfrac{-x^{7-1}y^{3-3}}{2} = \dfrac{-x^6}{2}$

7. $\dfrac{c^5c^9d^5d^7}{c^2cd^8} = \dfrac{c^{14}d^{12}}{c^3d^8} = c^{14-3}d^{12-8} = c^{11}d^4$

8. $\dfrac{25a^3ab^6}{50a^5b^8} = \dfrac{a^4b^6}{2a^5b^8} = \dfrac{a^{4-5}b^{6-8}}{2} = \dfrac{a^{-1}b^{-2}}{2} = \dfrac{1}{2ab^2}$

9. $(5a^2b^5)^3 = (5)^3(a^2)^3(b^5)^3 = 125a^6b^{15}$

10. $(-2x^2y^3)^2 = (-2)^2(x^2)^2(y^3)^2 = 4x^4y^6$

11. $\left(\dfrac{-6a^3b^4}{12a^2b}\right)^3 = \left(\dfrac{-a^{3-2}b^{4-1}}{2}\right)^3$

$= \left(\dfrac{-ab^3}{2}\right)^3$

$= \left(-\dfrac{1}{2}\right)^3(a)^3(b^3)^3$

$= \dfrac{-a^3b^9}{8}$

12. $\left(\dfrac{20m^5n^3}{14m^2n}\right)^4 = \left(\dfrac{10m^{5-2}n^{3-1}}{7}\right)^4$

$= \left(\dfrac{10m^3n^2}{7}\right)^4$

$= \left(\dfrac{10}{7}\right)^4(m^3)^4(n^2)^4$

$= \dfrac{10,000m^{12}n^8}{2401}$

13. $(-7x^2y^5)^3(3x^3y)^2$

$= (-7)^3(x^2)^3(y^5)^3(3)^2(x^3)^2(y)^2$

$= -343x^6y^{15} \cdot 9x^6y^2$

$= -3087x^{6+6}y^{15+2}$

$= -3087x^{12}y^{17}$

14. $(6x^3y^2)^2(-5x^3y)^4$

$= (6)^2(x^3)^2(y^2)^2(-5)^4(x^3)^4(y)^4$

$= 36x^6y^4 \cdot 625x^{12}y^4$

$= 22,500x^{6+12}y^{4+4}$

$= 22,500x^{18}y^8$

15. $(-5x^3y^8)^0 = 1$

16. $\left(\dfrac{-6a^5b}{3ab^3}\right)^0 = 1$

17.

$\dfrac{-15x^3y^9}{6x^8y^5} = \dfrac{-5x^{3-8}y^{9-5}}{2}$

$= \dfrac{-5x^{-5}y^4}{2}$

$= \dfrac{-5y^4}{2x^5}$

18.

$\dfrac{10a^{10}b^3}{30a^7b^5} = \dfrac{a^{10-7}b^{3-5}}{3}$

$= \dfrac{a^3b^{-2}}{3}$

$= \dfrac{a^3}{3b^2}$

19.

$\dfrac{-9x^8y^{-4}}{x^{12}y^9} = -9x^{8-12}y^{-4-9}$

$= -9x^{-4}y^{-13}$

$= \dfrac{-9}{x^4y^{13}}$

20.

$\dfrac{-12m^5n^{-7}}{mn^3} = -12m^{5-1}n^{-7-3}$

$= -12m^4n^{-10}$

$= \dfrac{-12m^4}{n^{10}}$

21.

$\dfrac{c^4d^{-5}}{c^{-2}d^3} = c^{4-(-2)}d^{-5-3}$

$= c^6d^{-8}$

$= \dfrac{c^6}{d^8}$

22.

$$\frac{y^{-3}z^{8}}{y^{-7}z^{-5}} = y^{-3-(-7)}z^{8-(-5)}$$
$$= y^{4}z^{13}$$

23. Move the decimal 10 places to the left; the exponent of 10 will be positive.

$52{,}600{,}000{,}000 = 5.26\times10^{10}$

24. Move the decimal 7 places to the left; the exponent of 10 will be positive.

$65{,}120{,}000 = 6.512\times10^{7}$

25. Move the decimal 9 places to the left; the exponent of 10 will be positive.

$4{,}550{,}000{,}000 = 4.55\times10^{9}$

The average distance from Neptune to the Sun is about 4.55×10^{9} km.

26. Move the decimal 9 places to the right; the exponent of 10 will be negative.

$0.000000002 = 2.0\times10^{-9}$

The width of one helix of the DNA molecule is 2.0×10^{-9} meter.

27. The exponent on 10 is -6. Move the decimal point 6 places to the *left* since the exponent is negative.
$6.024\times10^{-6} = 0.000006024$

Therefore, 6.024×10^{-6} has the standard form 0.000006024.

28. The exponent on 10 is 7. Move the decimal point 7 places to the right.
$4.3\times10^{7} = 4.3\cdot10{,}000{,}000 = 43{,}000{,}000$

Therefore, 4.3×10^{7} has the standard form 43,000,000.

29. The exponent on 10 is -7. Move the decimal point 7 places to the *left* since the exponent is negative.
$3.0\times10^{-7} = 0.0000003$

Therefore, the value for the width of a flu virus has the standard form 0.0000003 meter.

30. The exponent on 10 is -9. Move the decimal point 9 places to the *left* since the exponent is negative.
$7.5\times10^{-9} = 0.0000000075$

Therefore, the value for the approximate width of a carbon nanotube shaft of one of the first nanomotors created has the standard form 0.0000000075 meter.

31. $40(3.5\times10^{18})$

$= 40(3.5)\times10^{18}$
$= 140\times10^{18}$ Decimal moves left 2.
$= 1.4\times10^{20}$ Exponent increases 2.

32. $-30(5.7\times10^{-8})$

$= -30(5.7)\times10^{-8}$
$= -171\times10^{-8}$ Decimal moves left 2.
$= -1.71\times10^{-6}$ Exponent increases 2.

33. $\dfrac{1.5\times10^{-12}}{3.0\times10^{8}}$

$= \dfrac{1.5}{3.0}\times10^{-12-8}$
$= 0.5\times10^{-20}$ Decimal moves right 1.
$= 5.0\times10^{-21}$ Exponent decreases 1.

34. $\dfrac{6.2\times10^{20}}{2.0\times10^{4}}$

$= \dfrac{6.2}{2.0}\times10^{20-4}$
$= 3.1\times10^{16}$

35. $(2.5\times10^{10})(6.0\times10^{8})$

$= (2.5)(6.0)\times10^{10+8}$
$= 15\times10^{18}$ Decimal moves left 1.
$= 1.5\times10^{19}$ Exponent increases 1.

36. $(4.0\times10^{-12})(3.6\times10^{-5})$

$= (4.0)(3.6)\times10^{-12+(-5)}$
$= 14.4\times10^{-17}$ Decimal moves left 1.
$= 1.44\times10^{-16}$ Exponent increases 1.

37. The polynomial has 3 terms.
The highest exponent is 2 so the degree is 2.

38. The polynomial has 2 terms.
The highest exponent is 1 so the degree is 1.

39. Degree of the term $-5a^3b$ is 4

Degree of the term $6a^5$ is 5

Degree of the term $-7b^2$ is 2

Degree of the term 8 is 0

The highest degree term has a degree of 5, so the degree of the polynomial is 5.

40. Degree of the term $20a^2b^3c$ is 6

Degree of the term $18ab^2c^2$ is 5

Degree of the term $-9bc^8$ is 9

The highest degree term has a degree of 9, so the degree of the polynomial is 9.

41. $7x-16y=-23$ is linear because it has degree 1.

42. $y=3x-8$ is linear because it has degree 1.

43. $y = 3x^2 + 10$ is not linear because it has a second degree term.

44. $y = 25 - x^2$ is not linear because it has a second degree term.

45. $P = (x+8)+(3x-4)+(7x+15)$

$P = x+3x+7x+8-4+15$

$P = 11x+19$

46. $P = 2l+2w$ Substitue values for l and w.

$P = 2(2x+3)+2(x+1)$ Distribute.

$P = 4x+6+2x+2$ Combine like terms.

$P = 6x+8$

47. $(42-0.5w)-(25-0.4w)$

$= 42-0.5w-25+0.4w$

$= 17-0.1w$

There are $17-0.1w$ more students that attend the math class than the English class.

48. **a.** $350+5(100)$

$= 350+500$

$= 850$

The cost for one student to go on the trip for 5 days is \$850.

b. The cost for n students to go on the trip for 5 days is $\$850n$.

49. $(5a^2+6a-12)+(3a^2-10a+5)$

$= 5a^2+3a^2+6a-10a-12+5$

$= 8a^2-4a-7$

50. $(3x^2-8x+14)+(7x^2+3x-2)$

$= 3x^2+7x^2-8x+3x+14-2$

$= 10x^2-5x+12$

51. $(5x^2-8x+2)-(2x^2-8)$

$= 5x^2-8x+2-2x^2+8$

$= 3x^2-8x+10$

52. $(8n^2-7n-5)-(15n^2-4n)$

$= 8n^2-7n-5-15n^2+4n$

$= -7n^2-3n-5$

53. $A = lw$ Substitue values for l and w.

$A = (9x+5)(4x-7)$

$A = 36x^2-63x+20x-35$

$A = 36x^2-43x-35$

54. $(2.15+0.04w)(150+5w)$

$= (2.15+0.04w)(150+5w)$

$= 322.5+10.75w+6w+0.2w^2$

$= 322.5+16.75w+0.2w^2$

The cost for gasoline used by Vista Unified School District during week w of the year can be represented by the expression $0.2w^2+16.75w+322.5$.

55. $5x(7x+8)$

$= 5x(7x)+5x(8)$

$= 35x^2+40x$

56. $-8x(7x^2+2x-4)$

$= -8x(7x^2)-8x(2x)-8x(-4)$

$= -56x^3-16x^2+32x$

57. $(2x+5)(3x-6)$

$= 6x^2-12x+15x-30$

$= 6x^2+3x-30$

58. $(4x-7)(2x-3)$

$= 8x^2-12x-14x+21$

$= 8x^2-26x+21$

59. Using $(a+b)(a-b) = a^2-b^2$

$(3x-7)(3x+7)$

$= (3x)^2-7^2$

$= 9x^2-49$

60. Using $(a+b)(a-b) = a^2-b^2$

$(4a+5b)(4a-5b)$

$= (4a)^2-(5b)^2$

$= 16a^2-25b^2$

61. $(x+2)(x^2-3x+7)$

$= x^3-3x^2+7x+2x^2-6x+14$

$= x^3-x^2+x+14$

62. $(x-4)(x^2+5x-3)$

$= x^3+5x^2-3x-4x^2-20x+12$

$= x^3+x^2-23x+12$

63. Use the formula $(a+b)^2 = a^2+2ab+b^2$.

$(3x+7)^2$

$= (3x)^2+2(3x)(7)+7^2$

$= 9x^2+42x+49$

64. Use the formula $(a-b)^2 = a^2 - 2ab + b^2$.

$$(6x-9)^2$$
$$= (6x)^2 - 2(6x)(9) + 9^2$$
$$= 36x^2 - 108x + 81$$

65. $5m(m+4)(3m-8)$

$$= (5m^2 + 20m)(3m-8)$$
$$= 15m^3 - 40m^2 + 60m^2 - 160m$$
$$= 15m^3 + 20m^2 - 160m$$

66. $-2a(3a+5)(5a-8)$

$$= (-6a^2 - 10a)(5a-8)$$
$$= -30a^3 + 48a^2 - 50a^2 + 80a$$
$$= -30a^3 - 2a^2 + 80a$$

67. $A = lw$ Substitute values for A and w.

$$6x^2 - 11x - 35 = l(3x+5)$$
$$\frac{6x^2 - 11x - 35}{3x+5} = l$$

$$\begin{array}{r} 2x-7 \\ 3x+5\overline{)6x^2-11x-35} \\ \underline{-6x^2-10x} \\ -21x-35 \\ \underline{21x+35} \\ 0 \end{array}$$

$$l = 2x - 7$$

68. $A = lw$ Substitute values for A and w.

$$7x^2 - 31x + 12 = l(x-4)$$
$$\frac{7x^2 - 31x + 12}{x-4} = l$$

$$\begin{array}{r} 7x-3 \\ x-4\overline{)7x^2-31x+12} \\ \underline{-7x^2+28x} \\ -3x+12 \\ \underline{3x-12} \\ 0 \end{array}$$

$$l = 7x - 3$$

69.

$$\frac{20x^2 + 6x - 8}{2x}$$
$$= \frac{20x^{\cancel{2}1}}{2\cancel{x}} + \frac{6\cancel{x}}{2\cancel{x}} - \frac{8}{2x}$$
$$= 10x + 3 - \frac{4}{x}$$

70.

$$\frac{15a^2b^3 - 18ab^2 + 9b}{3ab}$$
$$= \frac{15a^{\cancel{2}1}b^{\cancel{3}2}}{3\cancel{a}\cancel{b}} - \frac{18\cancel{a}b^{\cancel{2}1}}{3\cancel{a}\cancel{b}} + \frac{9\cancel{b}}{3a\cancel{b}}$$
$$= 5ab^2 - 6b + \frac{3}{a}$$

71.

$$\begin{array}{r} x+7 \\ x-3\overline{)x^2+4x-21} \\ \underline{-x^2+3x} \\ 7x-21 \\ \underline{-7x+21} \\ 0 \end{array}$$

$$(x^2 + 4x - 21) \div (x-3) = x + 7$$

72.

$$\begin{array}{r} 3a+7 \\ 2a+4\overline{)6a^2+26a+28} \\ \underline{-6a^2-12a} \\ 14a+28 \\ \underline{-14a-28} \\ 0 \end{array}$$

$$(6a^2 + 26a + 28) \div (2a+4) = 3a + 7$$

73.

$$\begin{array}{r} m-5 \\ m-3\overline{)m^2-8m+20} \\ \underline{-m^2+3m} \\ -5m+20 \\ \underline{5m-15} \\ 5 \end{array}$$

$$(m^2 - 8m + 20) \div (m-3) = m - 5 + \frac{5}{m-3}$$

74.

$$\begin{array}{r} 4p^2-10p+25 \\ 2p+5\overline{)8p^3+0p^2+0p+125} \\ \underline{-8p^3-20p^2} \\ -20p^2+0p \\ \underline{20p^2+50p} \\ 50p+125 \\ \underline{-50p-125} \\ 0 \end{array}$$

$$(8p^3 + 125) \div (2p+5) = 4p^2 - 10p + 25$$

75.

$$\begin{array}{r|rrr} 3 & 2 & -1 & -15 \\ & & 6 & 15 \\ \hline & 2 & 5 & 0 \end{array}$$

$$\left(2t^2-t-15\right)\div\left(t-3\right)=2t+5$$

76.

$$\begin{array}{r|rrr} -4 & 5 & 17 & -10 \\ & & -20 & 12 \\ \hline & 5 & -3 & 2 \end{array}$$

$$\left(5r^2+17r-10\right)\div\left(r+4\right)=5r-3+\frac{2}{r+4}$$

77.

$$\begin{array}{r|rrrr} -2 & 4 & 15 & 12 & -4 \\ & & -8 & -14 & 4 \\ \hline & 4 & 7 & -2 & 0 \end{array}$$

$$\left(4x^3+15x^2+12x-4\right)\div\left(x+2\right)=4x^2+7x-2$$

78.

$$\begin{array}{r|rrrr} 3 & 5 & 0 & -20 & -75 \\ & & 15 & 45 & 75 \\ \hline & 5 & 15 & 25 & 0 \end{array}$$

$$\left(5y^3-20y-75\right)\div\left(y-3\right)=5y^2+15y+25$$

Chapter 5 Test

1. $(5x^3y^5)(4x^7y^2)=5\cdot 4\cdot x^{3+7}y^{5+2}=20x^{10}y^7$

2. $\dfrac{m^3m^4n^2n^4}{m^5n^5}=\dfrac{m^7n^6}{m^5n^5}=m^{7-5}n^{6-5}=m^2n$

3. $(2a^{-4}b^3)^3=(2)^3(a^{-4})^3(b^3)^3$

$=8a^{-12}b^9$

$=\dfrac{8b^9}{a^{12}}$

4. $\dfrac{20x^{-3}y^4}{14x^5y^{-3}}=\dfrac{10x^{-3-5}y^{4-(-3)}}{7}$

$=\dfrac{10x^{-8}y^7}{7}$

$=\dfrac{10y^7}{7x^8}$

5. $\left(12a^4b^{10}\right)^0=1$

6. $\left(\dfrac{-8xy^2}{64x^2y}\right)^0=1$

7. Move the decimal 6 places to the right; the exponent of 10 will be negative.

$0.000007=7.0\times 10^{-6}$

The diameter of an average red blood cell is 7.0×10^{-6} meter.

8. The exponent on 10 is 9. Move the decimal point 9 places to the right.

$5.9\times 10^9=5.9\cdot 1{,}000{,}000{,}000=5{,}900{,}000{,}000$

The average distance from the Sun to Pluto is roughly 5,900,000,000 km.

9. $5(7.3\times 10^9)$

$=5(7.3)\times 10^9$

$=36.5\times 10^9$ Decimal moves left 1.

$=3.65\times 10^{10}$ Exponent increases 1.

10. $(3.0\times 10^7)(4.5\times 10^4)$

$=(3.0)(4.5)\times 10^{7+4}$

$=13.5\times 10^{11}$ Decimal moves left 1.

$=1.35\times 10^{12}$ Exponent increases 1.

11. $\dfrac{6.5\times 10^{-8}}{5.0\times 10^6}$

$=\dfrac{6.5}{5.0}\times 10^{-8-6}$

$=1.3\times 10^{-14}$

12. $(5x^2+2x+8)+(3x-12)$

$=5x^2+2x+3x+8-12$

$=5x^2+5x-4$

13. $(7a^2-8a+10)-(4a^2+5a-3)$

$=7a^2-8a+10-4a^2-5a+3$

$=7a^2-4a^2-8a-5a+10+3$

$=3a^2-13a+13$

14. $7x(2x^2+3x-7)$

$=7x(2x^2)+7x(3x)+7x(-7)$

$=14x^3+21x^2-49x$

15. Using $(a+b)(a-b)=a^2-b^2$

$(2x+5)(2x-5)$

$=(2x)^2-5^2$

$=4x^2-25$

16. $(3x-4)(4x^2+2x+5)$

$=12x^3+6x^2+15x-16x^2-8x-20$

$=12x^3-10x^2+7x-20$

17. Use the formula $(a-b)^2=a^2-2ab+b^2$.

$(6x-7)^2$

$=(6x)^2-2(6x)(7)+7^2$

$=36x^2-84x+49$

18.

$\dfrac{4m^2+6m-8}{2m}$

$=\dfrac{4m^{2\,1}}{2m}+\dfrac{6m}{2m}-\dfrac{8}{2m}$

$=2m+3-\dfrac{4}{m}$

19.

$$\begin{array}{r} 2d-3 \\ d+5\overline{)2d^2+7d-15} \\ \underline{-2d^2-10d} \\ -3d-15 \\ \underline{3d+15} \\ 0 \end{array}$$

$(2d^2+7d-15)\div(x+2)=2d-3$

20.

$$\begin{array}{r} 3x^2+x-2 \\ x+2\overline{)3x^3+7x^2+0x-18} \\ \underline{-3x^3-6x^2} \\ x^2+0x \\ \underline{-x^2-2x} \\ -2x-18 \\ \underline{2x+4} \\ -14 \end{array}$$

$(3x^3+7x^2-18)\div(x+2)=3x^2+x-2-\dfrac{14}{x+2}$

21. $A=lw$ Substitue values for l and w.

$A=(8x+3)(5x-4)$

$A=40x^2-32x+15x-12$

$A=40x^2-17x-12$

22. $P=2l+2w$ Substitue values for l and w.

$P=2(8x+3)+2(5x-4)$ Distribute.

$P=16x+6+10x-8$ Combine like terms.

$P=26x-2$

23.

$$\begin{array}{r|rrrrr} -5 & 1 & 5 & 3 & 19 & 20 \\ & & -5 & 0 & -15 & -20 \\ \hline & 1 & 0 & 3 & 4 & 0 \end{array}$$

$(x^4+5x^3+3x^2+19x+20)\div(x+5)=x^3+3x+4$

CHAPTER 6 FACTORING AND QUADRATIC EQUATIONS

Section 6.1 What It Means to Factor

1. $3 = \boxed{3}$

$12 = \boxed{3} \cdot 2 \cdot 2$

$24 = \boxed{3} \cdot 2 \cdot 2 \cdot 2$

$\text{GCF} = 3$

3. $16 = 2 \cdot \boxed{2 \cdot 2 \cdot 2}$

$24 = 3 \cdot \boxed{2 \cdot 2 \cdot 2}$

$48 = 3 \cdot 2 \cdot \boxed{2 \cdot 2 \cdot 2}$

$\text{GCF} = 8$

5. $15a = 5 \cdot \boxed{3 \cdot a}$

$30a^2 = 5 \cdot \boxed{3} \cdot 2 \cdot \boxed{a} \cdot a$

$9a = 3 \cdot \boxed{3 \cdot a}$

$\text{GCF} = 3a$

7. $27y^2 = \boxed{3 \cdot 3 \cdot 3 \cdot y \cdot y}$

$54y^4 = \boxed{3 \cdot 3 \cdot 3} \cdot 2 \cdot y \cdot y \cdot \boxed{y \cdot y}$

$\text{GCF} = 27y^2$

9. $15xy^3 = 5 \cdot 3 \cdot \boxed{x \cdot y} \cdot y \cdot y$

$45x^2y^2 = 5 \cdot 3 \cdot 3 \cdot x \cdot \boxed{x \cdot y} \cdot y$

$16xy = 2 \cdot 2 \cdot 2 \cdot 2 \cdot \boxed{x \cdot y}$

$\text{GCF} = xy$

11. $9x = \boxed{3} \cdot 3 \cdot x$

$21 = 7 \cdot \boxed{3}$

$\text{GCF} = 3$

13. $5ab = \boxed{5 \cdot a} \cdot b$

$-35a^2 = -1 \cdot 7 \cdot \boxed{5 \cdot a} \cdot a$

$\text{GCF} = 5a$

15. $-3x = \boxed{-1 \cdot 3 \cdot x}$

$-9x^2 = \boxed{-1 \cdot 3} \cdot 3 \cdot x \cdot \boxed{x}$

$\text{GCF} = -3x$

17. $-2ab^5 = -1 \cdot \boxed{2 \cdot a \cdot b} \cdot b \cdot b \cdot b \cdot b$

$16ab^2 = 2 \cdot 2 \cdot 2 \cdot \boxed{2 \cdot a \cdot b} \cdot b$

$-28ab = -1 \cdot 7 \cdot 2 \cdot \boxed{2 \cdot a \cdot b}$

$\text{GCF} = 2ab$

19. $30x^2y = 5 \cdot \boxed{3} \cdot 2 \cdot x \cdot \boxed{x} \cdot y$

$9xyz = 3 \cdot \boxed{3 \cdot x} \cdot y \cdot z$

$-15xz = -1 \cdot 5 \cdot \boxed{3 \cdot x} \cdot z$

$\text{GCF} = 3x$

21. $\text{GCF} = 5$

$$\begin{aligned} 10x + 55 &= 5 \cdot 2x + 11 \cdot 5 \\ &= 5(2x + 11) \end{aligned}$$

23. $\text{GCF} = 2$

$$\begin{aligned} 6x + 2 &= 2 \cdot 3x + 2 \cdot 1 \\ &= 2(3x + 1) \end{aligned}$$

25. $\text{GCF} = -5y$

$$\begin{aligned} -15y - 20y^2 &= -5y \cdot 3 + (-5y) \cdot 4y \\ &= -5y(3 + 4y) \end{aligned}$$

27. $\text{GCF} = 16a^2$

$$\begin{aligned} 48a^4 - 16a^2 &= 16a^2 \cdot 3a^2 - 1 \cdot 16a^2 \\ &= 16a^2(3a^2 - 1) \end{aligned}$$

29. $\text{GCF} = -6$

$$\begin{aligned} -6x - 6y &= -6 \cdot x + (-6) \cdot y \\ &= -6(x + y) \end{aligned}$$

31. $\text{GCF} = 3p$

$$\begin{aligned} 12p^6 - 6p^3 + 3p &= 3p \cdot 4p^5 - 2p^2 \cdot 3p + 1 \cdot 3p \\ &= 3p(4p^5 - 2p^2 + 1) \end{aligned}$$

33. $\text{GCF} = 16a^2b^2$

$$\begin{aligned} 48a^4b^2 - 16a^2b^4 &= 16a^2b^2 \cdot 3a^2 - b^2 \cdot 16a^2b^2 \\ &= 16a^2b^2(3a^2 - b^2) \end{aligned}$$

35. $\text{GCF} = x + 4$

$2(x + 4) + y(x + 4) = (x + 4)(2 + y)$

37. $\text{GCF} = x - 2$

$9x(x - 2) + 5(x - 2) = (x - 2)(9x + 5)$

39. $\text{GCF} = x + 8$

$$\begin{aligned} 2x(x + 8) - (x + 8) &= 2x(x + 8) - 1(x + 8) \\ &= (x + 8)(2x - 1) \end{aligned}$$

41. $\text{GCF} = y + 5$

$3(y + 5) - 2y(y + 5) = (y + 5)(3 - 2y)$

43. $\text{GCF} = 5y + 7$

$-3x(5y + 7) + 2(5y + 7) = (5y + 7)(-3x + 2)$

45. Commutative property of multiplication: The order of the factors has changed

47. Commutative property of addition: The order of the terms has changed

49. Distributive property: Seth factored out -1 while Sonia did not. The more conventional answer is $-6(x + 2)$.

51. Factor out -9.

$$-9x+27=-9\cdot x-(-9)\cdot 3$$
$$=-9(x-3)$$

53. Factor out -1.

$$-y-2=-1\cdot y+(-1)\cdot 2$$
$$=-(y+2)$$

55. Factor out -1.

$$5-x=-x+5$$
$$=-1\cdot x-(-1)\cdot 5$$
$$=-(x-5)$$

57. Factor out -1.

$$-x^2+7=-1\cdot x^2-(-1)\cdot 7$$
$$=-(x^2-7)$$

59. $x^2+5x+4x+20$

$$=(x^2+5x)+(4x+20)$$
$$=x(x+5)+4(x+5)$$
$$=(x+4)(x+5)$$

61. $x^2+7x+11x+77$

$$=(x^2+7x)+(11x+77)$$
$$=x(x+7)+11(x+7)$$
$$=(x+7)(x+11)$$

63. $x^2-2x+6x-12$

$$=(x^2-2x)+(6x-12)$$
$$=x(x-2)+6(x-2)$$
$$=(x-2)(x+6)$$

65. $x^2+8x-3x-24$

$$=(x^2+8x)-(3x+24)$$
$$=x(x+8)-3(x+8)$$
$$=(x+8)(x-3)$$

67. $x^2y+8x-3xy-24$

$$=(x^2y+8x)-(3xy+24)$$
$$=x(xy+8)-3(xy+8)$$
$$=(xy+8)(x-3)$$

69. $2mn-2m-5n+5$

$$=(2mn-2m)-(5n-5)$$
$$=2m(n-1)-5(n-1)$$
$$=(n-1)(2m-5)$$

71. $3ab-2a-9b+6$

$$=(3ab-2a)-(9b-6)$$
$$=a(3b-2)-3(3b-2)$$
$$=(3b-2)(a-3)$$

73. $2xy-y+6x-3$

$$=(2xy-y)+(6x-3)$$
$$=y(2x-1)+3(2x-1)$$
$$=(2x-1)(y+3)$$

75. $2a^2-6ab+ab-3b^2$

$$=(2a^2-6ab)+(ab-3b^2)$$
$$=2a(a-3b)+b(a-3b)$$
$$=(a-3b)(2a+b)$$

77. Joey factored the expression completely. Javier left a common factor of 2 within the parentheses.

79. Abi factored the expression completely. Noah left a common factor of 2.

81. Kyla factored the expression completely. Lucia did not factor out -1 so that the leading coefficient would be positive.

83. $\text{GCF}=6b^2$

$$6ab^3-12a^2b^2+18b^2$$
$$=6b^2\cdot ab-2a^2\cdot 6b^2+6b^2\cdot 3$$
$$=6b^2(ab-2a^2+3)$$

85. $\text{GCF}=-3x$

$$-6x^2+3x$$
$$=-3x\cdot 2x-1(-3x)$$
$$=-3x(2x-1)$$

87. $\text{GCF}=-17$

$$-17p+34q$$
$$=-17\cdot p-(-17)\cdot 2q$$
$$=-17(p-2q)$$

89. $\text{GCF}=27x$

$$81x^3-27x$$
$$=27x\cdot 3x^2-1(27x)$$
$$=27x(3x^2-1)$$

91. $\text{GCF}=2$

$$2p^2+6p+4p+12$$
$$=2(p^2+3p+2p+6)$$
$$=2\left[(p^2+3p)+(2p+6)\right]$$
$$=2\left[p(p+3)+2(p+3)\right]$$
$$=2(p+2)(p+3)$$

93. $\text{GCF}=6$

$$6n^2-12n-6n+12$$
$$=6(n^2-2n-n+2)$$
$$=6\left[(n^2-2n)-(n-2)\right]$$

$= 6[n(n-2)-1(n-2)]$

$= 6(n-1)(n-2)$

95. The student did not factor, but rather multiplied and simplified. It should be:

$x(x+2)+3(x+2)$ Factor.

$= (x+2)(x+3)$

97. The student lost the last term while factoring, the expression should be factored as follows:

$6x^2 - 8xy + 2x$ Factor.

$= 2x \cdot 3x - 4y \cdot 2x + 2x \cdot 1$

$= 2x(3x - 4y + 1)$

99. The student did not multiply, as the directions had stated, but rather factored. It should be as follows:

$5(n-4)+m(n-4)$ Multiply.

$= 5n - 20 + mn - 4m$

$= -4m + mn + 5n - 20$

Section 6.2 Factoring Trinomials

1. A binomial has <u>2</u> terms.

3. A monomial has <u>1</u> term.

5. In the equation $3 \cdot 7 = 21$, 3 and 7 are called <u>factors</u> of 21.

7. In the equation $3 + 7 = 10$, 10 is known as the <u>sum</u> of 3 and 7.

9. To factor a polynomial means to rewrite the expression as a product.

11. The order of the terms has changed; this is justified by the commutative property of addition.

13. $a = 1,\ b = -6,\ c = 5$

15. $a = 1,\ b = -6,\ c = -1$

17. $a = 1,\ b = 0,\ c = -7$

19. $a = 1,\ b = -2,\ c = 0$

21. $a = 2,\ b = 6,\ c = -8$

23. $a = -1,\ b = -6,\ c = 7$

25. Rearrange the terms: $-6m^2 - 3m + 5$.

$a = -6,\ b = -3,\ c = 5$

27. Rearrange the terms: $x^2 - 2x + 16$.

$a = 1,\ b = -2,\ c = 16$

29. Factors of 5 that add to 6 are 5 and 1.

$x^2 + 6x + 5 = (x+5)(x+1)$

31. Factors of -18 that add to 3 are 6 and -3.

$x^2 + 3x - 18 = (x+6)(x-3)$

33. Factors of -40 that add to -3 are -8 and 5.

$x^2 - 3x - 40 = (x-8)(x+5)$

35. Factors of 56 that add to -15 are -8 and -7.

$x^2 - 15x + 56 = (x-8)(x-7)$

37. There are no factors of 2 that add to -6, therefore $x^2 - 6x + 2$ is prime.

39. $a \cdot c = 2 \cdot 6 = 12$ Factors of 12 that add to 7 are 4 and 3. Factor by grouping.

$$\begin{aligned} 2x^2 + 7x + 6 &= 2x^2 + 4x + 3x + 6 \\ &= (2x^2 + 4x) + (3x + 6) \\ &= 2x(x+2) + 3(x+2) \\ &= (2x+3)(x+2) \end{aligned}$$

41. $a \cdot c = 2 \cdot 1 = 2$ There are no factors of 2 that add to 9, therefore $2x^2 + 9x + 1$ is prime.

43. $a \cdot c = 7 \cdot -6 = -42$ Factors of -42 that add to 41 are 42 and -1. Factor by grouping.

$$\begin{aligned}7x^2+41x-6 &= 7x^2+42x-x-6\\ &= (7x^2+42x)-(x+6)\\ &= 7x(x+6)-1(x+6)\\ &= (7x-1)(x+6)\end{aligned}$$

45. $a \cdot c = 3 \cdot 6 = 18$ Factors of 18 that add to -19 are -18 and -1. Factor by grouping.

$$\begin{aligned}3x^2-19x+6 &= 3x^2-18x-x+6\\ &= (3x^2-18x)-(x-6)\\ &= 3x(x-6)-1(x-6)\\ &= (3x-1)(x-6)\end{aligned}$$

47. $a \cdot c = 4 \cdot 15 = 60$ Factors of 60 that add to -17 are -12 and -5. Factor by grouping.

$$\begin{aligned}4x^2-17x+15 &= 4x^2-12x-5x+15\\ &= (4x^2-12x)-(5x-15)\\ &= 4x(x-3)-5(x-3)\\ &= (4x-5)(x-3)\end{aligned}$$

49. Factor out -1. $-x^2+x+6 = -(x^2-x-6)$

Factors of -6 that add to -1 are -3 and 2.

$-(x^2-x-6) = -(x-3)(x+2)$

51. Factor out -1. $-2x^2+4x-3 = -(2x^2-4x+3)$

$a \cdot c = 2 \cdot 3 = 6$ There are no factors of 6 that add to -4, therefore $-2x^2+4x-3$ is prime.

53. Factor out GCF $4y$.

$$\begin{aligned}&4xy-8y^2\\ &= 4y \cdot x - 2y \cdot 4y\\ &= 4y(x-2y)\end{aligned}$$

55. Factor out $2b$.

$6a^2b-12ab+2b = 2b(3a^2-6a+1)$

Factoring $(3a^2-6a+1)$ $a \cdot c = 3 \cdot 1 = 3$ There are no factors of 3 that add to -6, therefore the complete factorization is

$6a^2b-12ab+2b = 2b(3a^2-6a+1)$

57. Factor out 3.

$3x^2-6x-72 = 3(x^2-2x-24)$

Factoring $(x^2-2x-24)$ Factors of -24 that add to -2 are -6 and 4. Factor by grouping.

$$\begin{aligned}3(x^2-2x-24) &= 3(x^2-6x+4x-24)\\ &= 3\left[(x^2-6x)+(4x-24)\right]\\ &= 3\left[x(x-6)+4(x-6)\right]\\ &= 3(x+4)(x-6)\end{aligned}$$

59. Factor out 5.

$$\begin{aligned}&15a^2-15ab+5ab-5b^2\\ &= 5(3a^2-3ab+ab-b^2)\end{aligned}$$

Factor by grouping.

$$\begin{aligned}&5(3a^2-3ab+ab-b^2)\\ &= 5\left[(3a^2-3ab)+(ab-b^2)\right]\\ &= 5\left[3a(a-b)+b(a-b)\right]\\ &= 5(3a+b)(a-b)\end{aligned}$$

61. Factor out 3.

$12x^2+21x-6 = 3(4x^2+7x-2)$

Factoring $(4x^2+7x-2)$ $a \cdot c = 4 \cdot -2 = -8$ Factors of -8 that add to 7 are 8 and -1. Factor by grouping.

$$\begin{aligned}3(4x^2+7x-2) &= 3(4x^2+8x-x-2)\\ &= 3\left[(4x^2+8x)-(x+2)\right]\\ &= 3\left[4x(x+2)-1(x+2)\right]\\ &= 3(4x-1)(x+2)\end{aligned}$$

63. Factor out *t*.

$t^3-t^2-30t = t(t^2-t-30)$

Factoring (t^2-t-30) Factors of -30 that add to -1 are -6 and 5. Factor by grouping.

$$\begin{aligned}t(t^2-t-30) &= t(t^2-6t+5t-30)\\ &= t\left[(t^2-6t)+(5t-30)\right]\\ &= t\left[t(t-6)+5(t-6)\right]\\ &= t(t+5)(t-6)\end{aligned}$$

65. Factor out $2a$.

$2a^3-18a^2+40a = 2a(a^2-9a+20)$

Factoring $(a^2-9a+20)$ Factors of 20 that add to -9 are -5 and -4.

$2a(a^2-9a+20) = 2a(a-5)(a-4)$

67. Factor out $4c$.

$24c^3-4c^2-160c = 4c(6c^2-c-40)$

Factoring $(6c^2-c-40)$ $a \cdot c = 6 \cdot -40 = -240$ Factors of -240 that add to -1 are -16 and 15. Factor by grouping.

$$4c(6c^2-c-40)=4c(6c^2-16c+15c-40)$$
$$=4c\left[(6c^2-16c)+(15c-40)\right]$$
$$=4c[2c(3c-8)+5(3c-8)]$$
$$=4c(2c+5)(3c-8)$$

69. Factor out y.

$x^2y-16xy+63y=y(x^2-16x+63)$

Factoring $(x^2-16x+63)$ Factors of 63 that add to -16 are -9 and -7.

$y(x^2-16x+63)=y(y-9)(y-7)$

71. $a\cdot c=10\cdot-27=-270$ Factors of -270 that add to 3 are -18 and 15. Factor by grouping.

$10m^2+3mn-27n^2$
$=10m^2-18mn+15mn-27n^2$
$=(10m^2-18mn)+(15mn-27n^2)$
$=2m(5m-9n)+3n(5m-9n)$
$=(2m+3n)(5m-9n)$

Section 6.3 Factoring Special Forms

1. Two terms: $-12x^2$ and 2

3. One term: $-6ab^2$

5. Four terms: y, $-2x$, $8xy$ and $-3y$

7. Difference of squares: Both terms are perfect squares and the terms are being subtracted.

9. NOT a difference of squares: Both terms are perfect squares but the terms are being added

11. Difference of squares: Both terms are perfect squares and the terms are being subtracted

13. Using $a^2-b^2=(a-b)(a+b)$.

$x^2-144=(x-12)(x+12)$

15. Using $a^2-b^2=(a-b)(a+b)$.

$1-y^2=(1-y)(1+y)$

17. $9+p^2$ is prime. The expression is a sum not a difference.

19. Using $a^2-b^2=(a-b)(a+b)$.

$100x^2-49=(10x-7)(10x+7)$

21. x^2+1 is prime. The expression is a sum not a difference.

23. Using $a^2-b^2=(a-b)(a+b)$.

$81-a^2=(9-a)(9+a)$

25. Using $a^2-b^2=(a-b)(a+b)$.

$16x^2-49=(4x-7)(4x+7)$

27. Using $a^2-b^2=(a-b)(a+b)$.

$9t^2-4=(3t-2)(3t+2)$

29. $a^2+12a+36$ is a perfect square trinomial.

$a^2+12a+36$	Original trinomial.
$a^2+2ab+b^2$	General form.
$a=a$ & $b=6$	Compare first & last terms.
	(finding b: $6^2=36$)
$12a\overset{?}{=}2ab$	Check the middle term.
$12a\overset{?}{=}2(a)(6)$	Substitute in $a=a$ and $b=6$.
$12a=12a$	It checks.

31. $25x^2-10x+1$ is a perfect square trinomial.

$25x^2-10x+1$	Original trinomial.
$a^2-2ab+b^2$	General form.

$a = 5x$ & $b = 1$ Compare first & last terms.

(finding a & b: $(5x)^2 = 25x^2$ $1^2 = 1$)

$10x \stackrel{?}{=} 2ab$ Check the middle term.

$10x \stackrel{?}{=} 2(5x)(1)$ Substitute in $a = 5x$ and $b = 1$.

$10x = 10x$ It checks.

33. $x^2 - 50x + 25$ is *not* a perfect square trinomial.

$x^2 - 50x + 25$ Original trinomial.

$a^2 - 2ab + b^2$ General form.

$a = x$ & $b = 5$ Compare first & last terms.

(finding b: $5^2 = 25$)

$50x \stackrel{?}{=} 2ab$ Check the middle term.

$50x \stackrel{?}{=} 2(x)(5)$ Substitute in $a = x$ and $b = 5$.

$50x \neq 10x$ It does not check.

35. $a^2 - 8ab + 16b^2$ is a perfect square trinomial.

$a^2 - 8ab + 16b^2$ Original trinomial.

$a^2 - 2ab + b^2$ General form.

$a = a$ & $b = 4b$ Compare first & last terms.

(finding b: $(4b)^2 = 16b^2$)

$8ab \stackrel{?}{=} 2ab$ Check the middle term.

$8ab \stackrel{?}{=} 2(a)(4b)$ Substitute in $a = a$ and $b = 4b$.

$8ab = 8ab$ It checks.

37. $x^2 + 2x + 1$ is a perfect square trinomial. Factor using $a^2 + 2ab + b^2 = (a+b)^2$.

$x^2 + 2x + 1 = (x+1)^2$

39. $y^2 - 8y + 16$ is a perfect square trinomial. Factor using $a^2 - 2ab + b^2 = (a-b)^2$.

$y^2 - 8y + 16 = (y-4)^2$

41. $9p^2 - 6p + 1$ is a perfect square trinomial. Factor using $a^2 - 2ab + b^2 = (a-b)^2$.

$9p^2 - 6p + 1 = (3p-1)^2$

43. Factors of 6 that add to 5 are 3 and 2.

$t^2 + 5t + 6 = (t+3)(t+2)$

45. $a \cdot c = 6 \cdot -20 = -120$ Factors of -120 that add to 7 are 15 and -8. Factor by grouping.

$$\begin{aligned} 6x^2 + 7x - 20 &= 6x^2 + 15x - 8x - 20 \\ &= (6x^2 + 15x) - (8x + 20) \\ &= 3x(2x+5) - 4(2x+5) \\ &= (3x-4)(2x+5) \end{aligned}$$

47. $4x^2 + 20xy + 25y^2$ is a perfect square trinomial. Factor using $a^2 + 2ab + b^2 = (a+b)^2$.

$4x^2 + 20xy + 25y^2 = (2x+5y)^2$

49. $9x^2 + 24xy + 16y^2$ is a perfect square trinomial. Factor using $a^2 + 2ab + b^2 = (a+b)^2$.

$9x^2 + 24xy + 16y^2 = (3x+4y)^2$

51. Factor out -1. $-t^2 + 6t - 9 = -(t^2 - 6t + 9)$

$t^2 - 6t + 9$ is a perfect square trinomial. Factor using $a^2 - 2ab + b^2 = (a-b)^2$.

$$\begin{aligned} -t^2 + 6t - 9 &= -(t^2 - 6t + 9) \\ &= -(t-3)^2 \end{aligned}$$

53. $y^2 + 9$ is prime.

55.

$$\begin{aligned} m^3 - 64 &= m^3 - 4^3 \\ &= (m-4)(m^2 + 4 \cdot m + 4^2) \\ &= (m-4)(m^2 + 4m + 16) \end{aligned}$$

57.

$$\begin{aligned} x^3 + 125 &= x^3 + 5^3 \\ &= (x+5)(x^2 - 5 \cdot x + 5^2) \\ &= (x+5)(x^2 - 5x + 25) \end{aligned}$$

59.

$$\begin{aligned} 3g^3 - 24 &= 3(g^3 - 8) \\ &= 3(g^3 - 2^3) \\ &= 3(g-2)(g^2 + 2 \cdot g + 2^2) \\ &= 3(g-2)(g^2 + 2g + 4) \end{aligned}$$

61. Prime. The sum of two squares will not factor with real numbers.

63. Prime. The polynomial is not a perfect square trinomial, because the middle term should be $8m$.

65. Factor out 3. $3x^2 - 75 = 3(x^2 - 25)$

Now using $a^2 - b^2 = (a-b)(a+b)$

$3(x^2 - 25) = 3(x-5)(x+5)$

67. Factor out GCF $17x$.

$17xy - 34x^2 = 17x(y - 2x)$

69. Factor out 5. $5x^2 - 125 = 5(x^2 - 25)$

Now using $a^2 - b^2 = (a-b)(a+b)$

$5(x^2 - 25) = 5(x-5)(x+5)$

71. Factor out 2.

$4pr+12p-2r-6=2(2pr+6p-r-3)$

Factor by grouping.

$$\begin{aligned}2(2pr+6p-r-3)&=2[(2pr+6p)-(r+3)]\\&=2[2p(r+3)-1(r+3)]\\&=2(2p-1)(r+3)\end{aligned}$$

73. Factor out 3.

$6x^2-27x-15=3(2x^2-9x-5)$

Factoring $(2x^2-9x-5)$ $a\cdot c=2\cdot -5=-10$ Factors of -10 that add to -9 are -10 and 1. Factor by grouping.

$$\begin{aligned}3(2x^2-9x-5)&=3(2x^2-10x+x-5)\\&=3\left[(2x^2-10x)+(x-5)\right]\\&=3[2x(x-5)+1(x-5)]\\&=3(2x+1)(x-5)\end{aligned}$$

75. Factor out -2.

$-18a^2-12a-2=-2(9a^2+6a+1)$

$9a^2+6a+1$ is a perfect square trinomial. Factor using $a^2+2ab+b^2=(a+b)^2$.

$-2(9a^2+6a+1)=-2(3a+1)^2$

77. Factor out 5.

$5x^2+20=5(x^2+4)$

79. Factor out 3. $3t^2+6t+3=3(t^2+2t+1)$

(t^2+2t+1) is a perfect square trinomial. Factor using $a^2+2ab+b^2=(a+b)^2$.

$$\begin{aligned}3t^2+6t+3&=3(t^2+2t+1)\\&=3(t+1)^2\end{aligned}$$

81. Factor out $5x$.

$5xy^2+40xy+80x=5x(y^2+8y+16)$

$(y^2+8y+16)$ is a perfect square trinomial. Factor using $a^2+2ab+b^2=(a+b)^2$.

$$\begin{aligned}5xy^2+40xy+80x&=5x(y^2+8y+16)\\&=5x(y+4)^2\end{aligned}$$

83. Factor out 5.

$5x^2-20x-105=5(x^2-4x-21)$

Factoring $(x^2-4x-21)$ $a\cdot c=1\cdot -21=-21$ Factors of -21 that add to -4 are -7 and 3.

$5(x^2-4x-21)=5(x+3)(x-7)$

85. Factor out -2.

$-20x^2+46x-24=-2(10x^2-23x+12)$

Factoring $(10x^2-23x+12)$ $a\cdot c=12\cdot 10=120$
Factors of 120 that add to -23 are -15 and -8.
Factor by grouping.

$$\begin{aligned}&-2(10x^2-23x+12)\\&=-2(10x^2-15x-8x+12)\\&=-2\left[(10x^2-15x)-(8x-12)\right]\\&=-2[5x(2x-3)-4(2x-3)]\\&=-2(5x-4)(2x-3)\end{aligned}$$

87. Factor out a.

$3a^3+7a^2-20a=a(3a^2+7a-20)$

Factoring $(3a^2+7a-20)$ $a\cdot c=3\cdot -20=-60$
Factors of -60 that add to 7 are 12 and -5. Factor by grouping.

$$\begin{aligned}a(3a^2+7a-20)&=a(3a^2+12a-5a-20)\\&=a\left[(3a^2+12a)-(5a+20)\right]\\&=a[3a(a+4)-5(a+4)]\\&=a(3a-5)(a+4)\end{aligned}$$

89.

$$\begin{aligned}r^6-64&=(r^3)^2-8^2\\&=(r^3+8)(r^3-8)\\&=(r+2)(r^2-2r+4)\cdot(r-2)(r^2+2r+4)\\&=(r+2)(r-2)(r^2-2r+4)(r^2+2r+4)\end{aligned}$$

91.

$$\begin{aligned}16b^4-625c^4&=(4b^2)^2-(25c^2)^2\\&=(4b^2+25c^2)(4b^2-25c^2)\\&=(4b^2+25c^2)((2b)^2-(5c)^2)\\&=(4b^2+25c^2)(2b+5c)(2b-5c)\end{aligned}$$

93. Let $u=h^2$.

$$\begin{aligned}h^4+3h^2-10&=(h^2)^2+3h^2-10\\&=u^2+3u-10\\&=(u+5)(u-2)\end{aligned}$$

Replace u with h^2.

$h^4+3h^2-10=(h^2+5)(h^2-2)$

95. Let $u = g^4$.

$$g^8 + 6g^4 + 9 = u^2 + 6u + 9$$
$$= (u+3)(u+3)$$

Replace u with g^4

$$g^8 + 6g^4 + 9 = (g^4 + 3)^2$$

97. Let $u = t^6$.

$$2t^{12} + 13t^6 + 15 = 2u^2 + 13u + 15$$
$$= (2u^2 + 3u) + (10u + 15)$$
$$= u(2u+3) + 5(2u+3)$$
$$= (2u+3)(u+5)$$

Replace u with t^6.

$$2t^{12} + 13t^6 + 15 = (2t^6 + 3)(t^6 + 5)$$

99. $8x^8 + 12x^4 + 18$

$2(4x^8 + 6x^4 + 9)$

101.

$$7g^4 - 567h^4 = 7(g^4 - 81h^4)$$
$$= 7(g^2 + 9h^2)(g^2 - 9h^2)$$
$$= 7(g^2 + 9h^2)(g + 3h)(g - 3h)$$

103. This is repeatedly a difference of two squares.

$x^{16} - 1$

$(x^8 + 1)(x^8 - 1)$

$(x^8 + 1)(x^4 + 1)(x^4 - 1)$

$(x^8 + 1)(x^4 + 1)(x^2 + 1)(x^2 - 1)$

$(x^8 + 1)(x^4 + 1)(x^2 + 1)(x + 1)(x - 1)$

Section 6.4 Solving Quadratic Equations by Factoring

1. This is an equation of degree 1, so it is linear.

3. This is an equation of degree 3, so it is neither linear nor quadratic.

5. This is an equation of degree 2, so it is quadratic.

7. This is an equation of degree 2, so it is quadratic.

9. This is an equation of degree 1, so it is linear.

11. This is an equation of degree 1, so it is linear.

13. This is an equation of degree 2, so it is quadratic.

15. $5(x+3) = 0$

$5 \neq 0$

$$x + 3 = 0$$
$$\underline{-3 \quad -3}$$
$$x = -3$$

17. $-4(2n-5) = 0$

$-4 \neq 0$

$$2n - 5 = 0$$
$$\underline{+5 \quad +5}$$
$$2n = 5$$
$$\frac{2n}{2} = \frac{5}{2}$$
$$n = \frac{5}{2}$$

19. $a(a-8) = 0$

$a = 0$

$$a - 8 = 0$$
$$\underline{+8 \quad +8}$$
$$a = 8$$

$a = 0$ and $a = 8$

21. $y(2y+7) = 0$

$y = 0$

$$2y + 7 = 0$$
$$\underline{-7 \quad -7}$$
$$2y = -7$$
$$\frac{2y}{2} = \frac{-7}{2}$$
$$y = -\frac{7}{2}$$

$y = 0$ and $y = -\frac{7}{2}$

23. $3t(2t+5) = 0$

$$3t = 0$$
$$\frac{3t}{3} = \frac{0}{3}$$
$$t = 0$$

$$2t + 5 = 0$$
$$\underline{-5 \quad -5}$$
$$2t = -5$$
$$\frac{2t}{2} = \frac{-5}{2}$$
$$t = -\frac{5}{2}$$

$t = 0$ and $t = -\frac{5}{2}$

25. $(x+5)(x-6)=0$

$$\begin{array}{ll} x+5=0 & x-6=0 \\ \underline{-5\ \ -5} & \underline{+6\ \ +6} \\ x=-5 & x=6 \end{array}$$

$x=-5$ and $x=6$

27. $4p(2p+3)(p-9)=0$

$$\begin{array}{lll} 4p=0 & 2p+3=0 & p-9=0 \\ \frac{4p}{4}=\frac{0}{4} & \underline{-3\ \ -3} & \underline{+9\ \ +9} \\ p=0 & 2p=-3 & p=9 \\ & \frac{2p}{2}=\frac{-3}{2} & \\ & p=-\frac{3}{2} & \end{array}$$

$p=0$, $p=-\frac{3}{2}$ and $p=9$

29. Factors of 6 that add to 5 are 3 and 2.

$x^2+5x+6=0$

$(x+3)(x+2)=0$

$$\begin{array}{ll} x+3=0 & x+2=0 \\ \underline{-3\ \ -3} & \underline{-2\ \ -2} \\ x=-3 & x=-2 \end{array}$$

$x=-3$ and $x=-2$

31. Factors of 12 that add to 7 are 4 and 3.

$a^2+7a+12=0$

$(a+4)(a+3)=0$

$$\begin{array}{ll} a+4=0 & a+3=0 \\ \underline{-4\ \ -4} & \underline{-3\ \ -3} \\ a=-4 & a=-3 \end{array}$$

33. Multiply both sides by -1 so leading term is not negative.

$-1(-x^2-2x+3)=-1(0)$

$x^2+2x-3=0$

Factors of -3 that add to 2 are 3 and -1.

$x^2+2x-3=0$

$(x+3)(x-1)=0$

$$\begin{array}{ll} x+3=0 & x-1=0 \\ \underline{-3\ \ -3} & \underline{+1\ \ +1} \\ x=-3 & x=1 \end{array}$$

$x=-3$ and $x=1$

35. Factors of 12 that add to -7 are -4 and -3.

$t^2-7t+12=0$

$(t-4)(t-3)=0$

$$\begin{array}{ll} t-4=0 & t-3=0 \\ \underline{+4\ \ +4} & \underline{+3\ \ +3} \\ t=4 & t=3 \end{array}$$

37. Factor out 2.

$2x^2+6x+4=0$

$2(x^2+3x+2)=0$

Factoring (x^2+3x+2) Factors of 2 that add to 3 are 2 and 1.

$2(x^2+3x+2)=0$

$2(x+2)(x+1)=0$

$$\begin{array}{lll} & x+2=0 & x+1=0 \\ 2\neq 0 & \underline{-2\ \ -2} & \underline{-1\ \ -1} \\ & x=-2 & x=-1 \end{array}$$

$x=-2$ and $x=-1$

39. $a\cdot c=2\cdot -5=-10$ Factors of -10 that add to 9 are 10 and -1. Factor by grouping.

$2x^2+9x-5=0$

$2x^2+10x-x-5=0$

$(2x^2+10x)-(x+5)=0$

$2x(x+5)-1(x+5)=0$

$(2x-1)(x+5)=0$

$$\begin{array}{ll} 2x-1=0 & x+5=0 \\ \underline{+1\ \ +1} & \underline{-5\ \ -5} \\ 2x=1 & x=-5 \\ \frac{2x}{2}=\frac{1}{2} & \\ x=\frac{1}{2} & \end{array}$$

$x=\frac{1}{2}$ and $x=-5$

41. $a\cdot c=3\cdot -8=-24$ Factors of -24 that add to -10 are -12 and 2. Factor by grouping.

$3x^2-10x-8=0$

$3x^2-12x+2x-8=0$

$(3x^2-12x)+(2x-8)=0$

$3x(x-4)+2(x-4)=0$

$(3x+2)(x-4)=0$

$$\begin{array}{ll} 3x+2=0 & x-4=0 \\ \underline{-2\ \ -2} & \underline{+4\ \ +4} \\ 3x=-2 & x=4 \\ \frac{3x}{3}=\frac{-2}{3} & \\ x=-\frac{2}{3} & \end{array}$$

$x=-\frac{2}{3}$ and $x=4$

43. $a \cdot c = 6 \cdot 20 = 120$ Factors of 120 that add to 23 are 15 and 8. Factor by grouping.

$$6x^2 + 23x + 20 = 0$$
$$6x^2 + 15x + 8x + 20 = 0$$
$$(6x^2 + 15x) + (8x + 20) = 0$$
$$3x(2x + 5) + 4(2x + 5) = 0$$
$$(3x + 4)(2x + 5) = 0$$

$$\begin{array}{ll} 3x + 4 = 0 & 2x + 5 = 0 \\ \underline{\;-4 \;\; -4} & \underline{\;-5 \;\; -5} \\ 3x = -4 & 2x = -5 \\ \frac{3x}{3} = \frac{-4}{3} & \frac{2x}{2} = \frac{-5}{2} \\ x = -\frac{4}{3} & x = -\frac{5}{2} \end{array}$$

$x = -\frac{4}{3}$ and $x = -\frac{5}{2}$

45. $a \cdot c = 24 \cdot 6 = 144$ Factors of 144 that add to 25 are 16 and 9. Factor by grouping.

$$24a^2 + 25a + 6 = 0$$
$$24a^2 + 16a + 9a + 6 = 0$$
$$(24a^2 + 16a) + (9a + 6) = 0$$
$$8a(3a + 2) + 3(3a + 2) = 0$$
$$(8a + 3)(3a + 2) = 0$$

$$\begin{array}{ll} 8a + 3 = 0 & 3a + 2 = 0 \\ \underline{\;-3 \;\; -3} & \underline{\;-2 \;\; -2} \\ 8a = -3 & 3a = -2 \\ \frac{8a}{8} = \frac{-3}{8} & \frac{3a}{3} = \frac{-2}{3} \\ a = -\frac{3}{8} & a = -\frac{2}{3} \end{array}$$

$a = -\frac{3}{8}$ and $a = -\frac{2}{3}$

47. **a.** $P = -4a^2 + 80a$

$P = -4(4)^2 + 80(4)$ Substitute $a = 4$.

$P = -4(16) + 320$ Solve for P.

$P = -64 + 320$

$P = 256$

The manufacturer's average profit is \$256 per appliance when they sell 4 appliances in a single order.

b. $300 = -4a^2 + 80a$ Substitute $P = 300$.

Factor and use the zero-product property to solve for a.

$$\begin{array}{l} 300 = -4a^2 + 80a \\ \underline{-300 \quad -300} \\ 0 = -4a^2 + 80a - 300 \end{array}$$

$$-4a^2 + 80a - 300 = 0$$
$$-4(a^2 - 20a + 75) = 0$$

Factors of 75 that add to -20 are -15 and -5.

$$-4(a^2 - 20a + 75) = 0$$
$$-4(a - 15)(a - 5) = 0$$

$$\begin{array}{lll} & a - 15 = 0 & a - 5 = 0 \\ -4 \neq 0 & \underline{\;+15 \;\; +15} & \underline{\;+5 \;\; +5} \\ & a = 15 & a = 5 \end{array}$$

The company needs to sell either 5 or 15 appliances in a single order to have an average profit of \$300 per appliance.

49. **a.** $P = -n^2 + 16n$

$P = -(3)^2 + 16(3)$ Substitute $n = 3$.

$P = -9 + 48$ Solve for P.

$P = 39$

The manufacturer's average profit per plane is \$39 million when an airline orders 3 planes.

b. $60 = -n^2 + 16n$ Substitute $P = 60$.

Factor and use the zero-product property to solve for n.

$$\begin{array}{l} 60 = -n^2 + 16n \\ \underline{-60 \quad -60} \\ 0 = -n^2 + 16n - 60 \end{array}$$

$$-n^2 + 16n - 60 = 0$$
$$-(n^2 - 16n + 60) = 0$$

Factors of 60 that add to -16 are -10 and -6.

$$-(n^2 - 16n + 60) = 0$$
$$-1(n - 10)(n - 6) = 0$$

$$\begin{array}{lll} & n - 10 = 0 & n - 6 = 0 \\ -1 \neq 0 & \underline{\;+10 \;\; +10} & \underline{\;+6 \;\; +6} \\ & n = 10 & n = 6 \end{array}$$

The manufacturer needs to sell either 6 or 10 planes to an airline for the average profit per plane to be \$60 million.

51. Let $w =$ the width of the rug in feet, therefore the length would be $w + 10$. Substitute these values into the formula for area.

$$A = lw$$
$$24 = (w + 10)(w)$$
$$24 = w^2 + 10w$$

Factor and use the zero-product property to solve for w.

$$\begin{array}{l} 24 = w^2 + 10w \\ \underline{-24 \quad -24} \\ 0 = w^2 + 10w - 24 \end{array}$$

$$w^2 + 10w - 24 = 0$$

Factors of -24 that add to 10 are 12 and -2.

$$w^2+10w-24=0$$
$$(w+12)(w-2)=0$$

$$\begin{array}{rr} w+12=0 & w-2=0 \\ \underline{-12\ -12} & \underline{+2\ +2} \\ w=-12 & w=2 \end{array}$$

Width cannot be a negative value, therefore the width of the rug is 2 feet. The length is $w+10$, giving us a length of 12 feet.

53. Let $w=$ the width of the rectangle in cm, therefore the length would be $w+7$. Substitute these values into the formula for area.

$$A=lw$$
$$144=(w+7)(w)$$
$$144=w^2+7w$$

Factor and use the zero-product property to solve for w.

$$\begin{array}{r} 144=w^2+7w \\ \underline{-144\quad -144} \\ 0=w^2+7w-144 \end{array}$$
$$w^2+7w-144=0$$

Factors of -144 that add to 7 are 16 and -9.

$$w^2+7w-144=0$$
$$(w+16)(w-9)=0$$

$$\begin{array}{rr} w+16=0 & w-9=0 \\ \underline{-16\ -16} & \underline{+9\ +9} \\ w=-16 & w=9 \end{array}$$

Width cannot be a negative value, therefore the width of the rectangle is 9 cm. The length is $w+7$, giving us a length of 16 cm.

55. Let $b=$ the length in inches of the base of the triangle, therefore the height would be $b-4$. Substitute these values into the formula for area of a triangle.

$$A=\frac{1}{2}bh$$
$$30=\frac{1}{2}(b)(b-4)$$
$$30=\frac{1}{2}(b^2-4b) \qquad \text{Multiply both sides by 2.}$$
$$2\cdot 30=\frac{1}{\not 2}(b^2-4b)\cdot \not 2$$
$$60=b^2-4b$$

Factor and use the zero-product property to solve for b.

$$\begin{array}{r} 60=b^2-4b \\ \underline{-60\quad -60} \\ 0=b^2-4b-60 \end{array}$$
$$b^2-4b-60=0$$

Factors of -60 that add to -4 are -10 and 6.

$$b^2-4b-60=0$$
$$(b+6)(b-10)=0$$

$$\begin{array}{rr} b+6=0 & b-10=0 \\ \underline{-6\ -6} & \underline{+10\ +10} \\ b=-6 & b=10 \end{array}$$

A length cannot be a negative value, therefore the base of the triangle is 10 in. The height is $b-4$, giving us a height of 6 in.

57. $h=-16t^2+32t+48$ Substitute $h=0$.

$$0=-16t^2+32t+48$$
$$-16t^2+32t+48=0$$
$$-16(t^2-2t-3)=0$$

Factors of -3 that add to -2 are -3 and 1.

$$-16(t^2-2t-3)=0$$
$$-16(t+1)(t-3)=0$$

$$\begin{array}{rrr} & t+1=0 & t-3=0 \\ -16\neq 0 & \underline{-1\ -1} & \underline{+3\ +3} \\ & t=-1 & t=3 \end{array}$$

A value for time cannot be a negative value, therefore the ball will hit the ground 3 seconds after being thrown from the roof.

59. $h=-16t^2+64t+80$ Substitute $h=128$.

$$\begin{array}{r} 128=-16t^2+64t+80 \\ \underline{-128\qquad\qquad\quad -128} \end{array}$$
$$0=-16t^2+64t-48$$
$$-16t^2+64t-48=0$$
$$-16(t^2-4t+3)=0$$

Factors of 3 that add to -4 are -3 and -1.

$$-16(t^2-4t+3)=0$$
$$-16(t-1)(t-3)=0$$

$$\begin{array}{rrr} & t-1=0 & t-3=0 \\ -16\neq 0 & \underline{+1\ +1} & \underline{+3\ +3} \\ & t=1 & t=3 \end{array}$$

The ball will reach a height of 128 ft at 1 second after being thrown and 3 seconds after being thrown.

61. $x^2+4x=-3$

$$\begin{array}{r} x^2+4x=-3 \\ \underline{+3\quad +3} \\ x^2+4x+3=0 \end{array}$$

Factors of 4 that add to 4 are 3 and 1.

$$x^2+4x+3=0$$
$$(x+3)(x+1)=0$$

$$\begin{array}{rr} x+3=0 & x+1=0 \\ \underline{-3\ -3} & \underline{-1\ -1} \\ x=-3 & x=-1 \end{array}$$

$x=-3$ and $x=-1$

63. $2a^2+9=9a$

$$\begin{array}{r} 2a^2+9=9a \\ \underline{-9a\ -9a} \\ 2a^2-9a+9=0 \end{array}$$

$a\cdot c=2\cdot 9=18$ Factors of 18 that add to -9 are -6 and -3. Factor by grouping.

$$2a^2-9a+9=0$$
$$2a^2-6a-3a+9=0$$
$$(2a^2-6a)-(3a-9)=0$$
$$2a(a-3)-3(a-3)=0$$
$$(2a-3)(a-3)=0$$

$$\begin{array}{rr} 2a-3=0 & \\ \underline{+3\ +3} & a-3=0 \\ 2a=3 & \underline{+3\ +3} \\ \frac{2a}{2}=\frac{3}{2} & a=3 \\ a=\frac{3}{2} & \end{array}$$

$a=\frac{3}{2}$ and $a=3$

65. $x(x-5)=14$

$$\begin{array}{r} x^2-5x=14 \\ \underline{-14\ -14} \\ x^2-5x-14=0 \end{array}$$

Factors of -14 that add to -5 are -7 and 2.

$$x^2-5x-14=0$$
$$(x-7)(x+2)=0$$

$$\begin{array}{rr} x-7=0 & x+2=0 \\ \underline{+7\ +7} & \underline{-2\ -2} \\ x=7 & x=-2 \end{array}$$

$x=7$ and $x=-2$

67. $n(n+9)=-20$

$$\begin{array}{r} n^2+9n=-20 \\ \underline{+20\ \ +20} \\ n^2+9n+20=0 \end{array}$$

Factors of 20 that add to 9 are 5 and 4.

$$n^2+9n+20=0$$
$$(n+5)(n+4)=0$$

$$\begin{array}{rr} n+5=0 & n+4=0 \\ \underline{-5\ -5} & \underline{-4\ -4} \\ n=-5 & n=-4 \end{array}$$

$n=-5$ and $n=-4$

69. $(x+3)(x-2)=6$

$$x^2-2x+3x-6=6$$
$$\begin{array}{r} x^2+x-6=6 \\ \underline{-6\ -6} \\ x^2+x-12=0 \end{array}$$

Factors of -12 that add to 1 are 4 and -3.

$$x^2+x-12=0$$
$$(x+4)(x-3)=0$$

$$\begin{array}{rr} x+4=0 & x-3=0 \\ \underline{-4\ -4} & \underline{+3\ +3} \\ x=-4 & x=3 \end{array}$$

$x=-4$ and $x=3$

71. $(y+4)(y+2)=24$

$$y^2+2y+4y+8=24$$
$$\begin{array}{r} y^2+6y+8=24 \\ \underline{-24\ -24} \\ y^2+6y-16=0 \end{array}$$

Factors of -16 that add to 6 are 8 and -2.

$$y^2+6y-16=0$$
$$(y+8)(y-2)=0$$

$$\begin{array}{rr} y+8=0 & y-2=0 \\ \underline{-8\ -8} & \underline{+2\ +2} \\ y=-8 & y=2 \end{array}$$

$y=-8$ and $y=2$

73. Using $a^2-b^2=(a-b)(a+b)$.

$$t^2-25=0$$
$$(t+5)(t-5)=0$$

$$\begin{array}{rr} t+5=0 & t-5=0 \\ \underline{-5\ -5} & \underline{+5\ +5} \\ t=-5 & t=5 \end{array}$$

$t=-5$ and $t=5$

75. $g^2=100$

$$\begin{array}{r} g^2=100 \\ \underline{-100\ -100} \\ g^2-100=0 \end{array}$$

Using $a^2-b^2=(a-b)(a+b)$.

$$g^2-100=0$$
$$(g+10)(g-10)=0$$

$$\begin{array}{cc} g+10=0 & g-10=0 \\ \underline{-10\ -10} & \underline{+10\ +10} \\ g=-10 & g=10 \end{array}$$

$g=-10$ and $g=10$

77. Factor out 2.

$$6x^2-22x-120=0$$

$$2(3x^2-11x-60)=0$$

$a\cdot c=3\cdot-60=-180$ Factors of -180 that add to -11 are -20 and 9. Factor by grouping.

$$2(3x^2-11x-60)=0$$
$$2(3x^2-20x+9x-60)=0$$
$$2\left[(3x^2-20x)+(9x-60)\right]=0$$
$$2\left[x(3x-20)+3(3x-20)\right]=0$$
$$2(x+3)(3x-20)=0$$

$$\begin{array}{ccc} & & 3x-20=0 \\ & x+3=0 & \underline{+20\ +20} \\ 2\neq 0 & \underline{-3\ -3} & 3x=20 \\ & x=-3 & \frac{3x}{3}=\frac{20}{3} \\ & & x=\frac{20}{3} \end{array}$$

$x=-3$ and $x=\frac{20}{3}$

79. Factor out t.

$$t^3-9t=0$$

$$t(t^2-9)=0$$

Using $a^2-b^2=(a-b)(a+b)$.

$$t(t^2-9)=0$$
$$t(t+3)(t-3)=0$$

$$\begin{array}{ccc} & t+3=0 & t-3=0 \\ t=0 & \underline{-3\ -3} & \underline{+3\ +3} \\ & t=-3 & t=3 \end{array}$$

$t=0$, $t=-3$ and $t=3$

81. Factor out $2x$.

$$2x^3+14x^2+24x=0$$

$$2x(x^2+7x+12)=0$$

Factors of 12 that add to 7 are 4 and 3.

$$2x(x^2+7x+12)=0$$
$$2x(x+4)(x+3)=0$$

$$\begin{array}{ccc} 2x=0 & & \\ & x+4=0 & x+3=0 \\ \frac{2x}{2}=\frac{0}{2} & \underline{-4\ -4} & \underline{-3\ -3} \\ & x=-4 & x=-3 \\ x=0 & & \end{array}$$

$x=0$, $x=-4$ and $x=-3$

83. $8t^3-24t^2=40t^2-120t$

$$\begin{array}{c} \underline{-40t^2\quad -40t^2} \\ 8t^3-64t^2=-120t \\ \underline{+120t\quad +120t} \\ 8t^3-64t^2+120t=0 \end{array}$$

Factor out $8t$.

$$8t^3-64t^2+120t=0$$
$$8t(t^2-8t+15)=0$$

Factors of 15 that add to -8 are -5 and -3.

$$8t(t^2-8t+15)=0$$
$$8t(t-5)(t-3)=0$$

$$\begin{array}{ccc} 8t=0 & & \\ & t-5=0 & t-3=0 \\ \frac{8t}{8}=\frac{0}{8} & \underline{+5\ +5} & \underline{+3\ +3} \\ & t=5 & t=3 \\ t=0 & & \end{array}$$

$t=0$, $t=5$ and $t=3$

85. **a.** The statement does not have an equal sign, therefore it is an expression.

b. $x^2+16x+64$ is a perfect square trinomial. Factor using $a^2+2ab+b^2=(a+b)^2$.

$$x^2+16x+64=(x+8)^2$$

87. **a.** The statement has an equal sign, therefore it is an equation.

b. $4n^2-25=0$

$$(2n+5)(2n-5)=0$$

$$\begin{array}{cc} 2n+5=0 & 2n-5=0 \\ \underline{-5\ -5} & \underline{+5\ +5} \\ 2n=-5 & 2n=5 \\ \frac{2n}{2}=\frac{-5}{2} & \frac{2n}{2}=\frac{5}{2} \\ n=-\frac{5}{2} & n=\frac{5}{2} \end{array}$$

$n=-\frac{5}{2}$ and $n=\frac{5}{2}$

89. **a.** The statement does not have an equal sign, therefore it is an expression.

b. Factors of 7 that add to 8 are 7 and 1.

$$a^2+8a+7=(a+7)(a+1)$$

91. **a.** The statement has an equal sign, therefore it is an equation.

b. Factor out $4x$.

$$4x^3 + 4x^2 - 24x = 0$$
$$4x(x^2 + x - 6) = 0$$

Factors of -6 that add to 1 are 3 and -2.

$$4x(x^2 + x - 6) = 0$$
$$4x(x+3)(x-2) = 0$$

$$4x = 0 \quad \frac{4x}{4} = \frac{0}{4} \quad x = 0$$
$$x + 3 = 0 \quad \underline{-3 \ -3} \quad x = -3$$
$$x - 2 = 0 \quad \underline{+2 \ +2} \quad x = 2$$

$x = 0$, $x = -3$ and $x = 2$

93. **a.** The statement has an equal sign, therefore it is an equation.

b. $4m^2 - 4m = 3$

$$\underline{\quad -3 \ -3}$$
$$4m^2 - 4m - 3 = 0$$

$a \cdot c = 4 \cdot -3 = -12$ Factors of -12 that add to -4 are -6 and 2. Factor by grouping.

$$4m^2 - 4m - 3 = 0$$
$$4m^2 - 6m + 2m - 3 = 0$$
$$(4m^2 - 6m) + (2m - 3) = 0$$
$$2m(2m-3) + 1(2m-3) = 0$$
$$(2m+1)(2m-3) = 0$$

$$2m + 1 = 0 \quad \underline{-1 \ -1} \quad 2m = -1 \quad \frac{2m}{2} = \frac{-1}{2} \quad m = -\frac{1}{2}$$
$$2m - 3 = 0 \quad \underline{+3 \ +3} \quad 2m = 3 \quad \frac{2m}{2} = \frac{3}{2} \quad m = \frac{3}{2}$$

$m = -\frac{1}{2}$ and $m = \frac{3}{2}$

95. **a.** The statement does not have an equal sign, therefore it is an expression.

b. $a \cdot c = 40 \cdot -15 = -600$ Factors of -600 that add to 38 are 50 and -12. Factor by grouping.

$$\begin{aligned} 40b^2 + 38b - 15 &= 40b^2 + 50b - 12b - 15 \\ &= (40b^2 + 50b) - (12b + 15) \\ &= 10b(4b+5) - 3(4b+5) \\ &= (10b-3)(4b+5) \end{aligned}$$

97. **a.** The statement has an equal sign, therefore it is an equation.

b. Factor out 2.

$$2m^2 - 50 = 0$$
$$2(m^2 - 25) = 0$$
$$2(m+5)(m-5) = 0$$

$$2 \neq 0$$
$$m + 5 = 0 \quad \underline{-5 \ -5} \quad m = -5$$
$$m - 5 = 0 \quad \underline{+5 \ +5} \quad m = 5$$

$m = -5$ and $m = 5$

99. $h^2 + 100$ is prime.

Chapter 6 Review

1. GCF $= 2x^2$

$12x^5 + 10x^3 - 6x^2$
$= 2x^2 \cdot 6x^3 + (2x^2) \cdot 5x - (2x^2) \cdot 3$
$= 2x^2(6x^3 + 5x - 3)$

2. GCF $= 3b^2$

$3a^3b^2 + 6ab^2 - 9b^2$
$= 3b^2 \cdot a^3 + (3b^2) \cdot 2a - (3b^2) \cdot 3$
$= 3b^2(a^3 + 2a - 3)$

3. GCF $= 3a + 5$

$2a(3a+5) - 7(3a+5)$
$= (3a+5)(2a-7)$

4. GCF $= x + 3$

$2y(x+3) - 5(x+3)$
$= (x+3)(2y-5)$

5. $4x^2 + 8x + 3x + 6$
$= (4x^2 + 8x) + (3x + 6)$
$= 4x(x+2) + 3(x+2)$
$= (x+2)(4x+3)$

6. $2t^2 - 12t + 7t - 42$
$= (2t^2 - 12t) + (7t - 42)$
$= 2t(t-6) + 7(t-6)$
$= (t-6)(2t+7)$

7. $5m^2 + 10mn - 7m - 14n$
$= (5m^2 + 10mn) - (7m + 14n)$
$= 5m(m+2n) - 7(m+2n)$
$= (m+2n)(5m-7)$

8. $6xy + 10x - 3y - 5$
$= (6xy + 10x) - (3y + 5)$
$= 2x(3y+5) - 1(3y+5)$
$= (3y+5)(2x-1)$

9. Factors of 24 that add to 10 are 6 and 4.

$a^2 + 10a + 24 = (a+6)(a+4)$

10. $m^2 + 16m + 64$ is a perfect square trinomial. Factor using $a^2 + 2ab + b^2 = (a+b)^2$.

$m^2 + 16m + 64 = (m+8)^2$

11. Factors of 20 that add to -9 are -5 and -4.

$x^2 - 9x + 20 = (x-5)(x-4)$

12. Factors of 36 that add to -13 are -9 and -4.

$h^2 - 13h + 36 = (h-9)(h-4)$

13. Factors of -21 that add to -4 are -7 and 3.

$x^2 - 4x - 21 = (x-7)(x+3)$

14. Factors of -18 that add to -7 are -9 and 2.

$y^2 - 7y - 18 = (y-9)(y+2)$

15. $a \cdot c = 2 \cdot 15 = 30$ Factors of 30 that add to 13 are 10 and 3. Factor by grouping.

$2x^2 + 13x + 15 = 2x^2 + 10x + 3x + 15$
$= (2x^2 + 10x) + (3x + 15)$
$= 2x(x+5) + 3(x+5)$
$= (x+5)(2x+3)$

16. $a \cdot c = 3 \cdot 6 = 18$ Factors of 18 that add to 19 are 18 and 1. Factor by grouping.

$3a^2 + 19a + 6 = 3a^2 + 18a + a + 6$
$= (3a^2 + 18a) + (a + 6)$
$= 3a(a+6) + 1(a+6)$
$= (a+6)(3a+1)$

17. $a \cdot c = 5 \cdot -6 = -30$ Factors of -30 that add to 13 are 15 and -2. Factor by grouping.

$5x^2 + 13x - 6 = 5x^2 + 15x - 2x - 6$
$= (5x^2 + 15x) - (2x + 6)$
$= 5x(x+3) - 2(x+3)$
$= (x+3)(5x-2)$

18. $a \cdot c = 3 \cdot -28 = -84$ Factors of -84 that add to 17 are 21 and -4. Factor by grouping.

$3b^2 + 17b - 28 = 3b^2 + 21b - 4b - 28$
$= (3b^2 + 21b) - (4b + 28)$
$= 3b(b+7) - 4(b+7)$
$= (b+7)(3b-4)$

19. $a \cdot c = 4 \cdot -40 = -160$ Factors of -160 that add to -27 are -32 and 5. Factor by grouping.

$4x^2 - 27x - 40 = 4x^2 - 32x + 5x - 40$
$= (4x^2 - 32x) + (5x - 40)$
$= 4x(x-8) + 5(x-8)$
$= (x-8)(4x+5)$

20. $9y^2 - 24y + 16$ is a perfect square trinomial. Factor using $a^2 - 2ab + b^2 = (a-b)^2$.

$9y^2 - 24y + 16 = (3y-4)^2$

21. Factor out -1.

$-10b^2 + 19b - 6 = -(10b^2 - 19b + 6)$ $a \cdot c = 10 \cdot 6 = 60$
Factors of 60 that add to -19 are -15 and -4. Factor by grouping.

$$-(10b^2-19b+6)=-(10b^2-15b-4b+6)$$
$$=-\left[(10b^2-15b)-(4b-6)\right]$$
$$=-\left[5b(2b-3)-2(2b-3)\right]$$
$$=-(2b-3)(5b-2)$$

22. Factor out -1.

$-2t^2+15t-25=-(2t^2-15t+25)$ $a\cdot c=2\cdot 25=50$

Factors of 50 that add to -15 are -10 and -5. Factor by grouping.

$$-(2t^2-15t+25)=-(2t^2-10t-5t+25)$$
$$=-\left[(2t^2-10t)-(5t-25)\right]$$
$$=-\left[2t(t-5)-5(t-5)\right]$$
$$=-(t-5)(2t-5)$$

23. Factor out g.

$g^3-8g^2+15g=g(g^2-8g+15)$

Factors of 15 that add to -8 are -5 and -3.

$g(g^2-8g+15)=g(g-5)(g-3)$

24. $a\cdot c=6\cdot -20=-120$ Factors of -120 that add to 7 are 15 and -8. Factor by grouping.

$$6x^2+7xy-20y^2=6x^2+15xy-8xy-20y^2$$
$$=(6x^2+15xy)-(8xy+20y^2)$$
$$=3x(2x+5y)-4y(2x+5y)$$
$$=(2x+5y)(3x-4y)$$

25. n^2-81 is a difference of squares.

Using $a^2-b^2=(a-b)(a+b)$.

$n^2-81=(n-9)(n+9)$

26. t^2-64 is a difference of squares.

Using $a^2-b^2=(a-b)(a+b)$.

$t^2-64=(t-8)(t+8)$

27. $9a^2-100$ is a difference of squares.

Using $a^2-b^2=(a-b)(a+b)$.

$9a^2-100=(3a-10)(3a+10)$

28. $4x^2-49$ is a difference of squares.

Using $a^2-b^2=(a-b)(a+b)$.

$4x^2-49=(2x-7)(2x+7)$

29. b^2+36 is prime.

30. x^2+16 is prime.

31. $3x^2-27x-7x+63$
$$=(3x^2-27x)-(7x-63)$$
$$=3x(x-9)-7(x-9)$$
$$=(x-9)(3x-7)$$

32. $4x^2y+8y-5x^2-10$
$$=(4x^2y+8y)-(5x^2+10)$$
$$=4y(x^2+2)-5(x^2+2)$$
$$=(x^2+2)(4y-5)$$

33. $6abc-24ac+3bc-12c=3c(2ab-8a+b-4)$
$$=3c\left[(2ab-8a)+(b-4)\right]$$
$$=3c\left[2a(b-4)+1(b-4)\right]$$
$$=3c(b-4)(2a+1)$$

34. $4xyz+2xz-24yz-12z=2z(2xy+x-12y-6)$
$$=2z\left[(2xy+x)-(12y+6)\right]$$
$$=2z\left[x(2y+1)-6(2y+1)\right]$$
$$=2z(2y+1)(x-6)$$

35.

$$27x^3-1=(3x)^3-1^3$$
$$=(3x-1)\left((3x)^2+1\cdot 3x+1^2\right)$$
$$=(3x-1)\left(9x^2+3x+1\right)$$

36.

$$125w^3+8x^3=(5w)^3+(2x)^3$$
$$=(5w+2x)\left((5w)^2-5w\cdot 2x+(2x)^2\right)$$
$$=(5w+2x)\left(25w^2-10wx+4x^2\right)$$

37.

$$t^6-64=\left(t^3\right)^2-8^2$$
$$=\left(t^3+8\right)\left(t^3-8\right)$$
$$=(t+2)\left(t^2-2t+4\right)\cdot(t-2)\left(t^2+2t+4\right)$$
$$=(t+2)(t-2)\left(t^2-2t+4\right)\left(t^2+2t+4\right)$$

38. Let $u=m^3$.

$$10m^6-29m^3+21=10\left(m^3\right)^2-29m^3+21$$
$$=10u^2-29u+21$$
$$=(5u-7)(2u-3)$$

Replace u with m^3

$$=\left(5m^3-7\right)\left(2m^3-3\right)$$

39.

$$\begin{aligned}80h^2-125&=5\left(16h^2-25\right)\\&=5\left((4h)^2-5^2\right)\\&=5(4h+5)(4h-5)\end{aligned}$$

40.

$$\begin{aligned}18a^2b+84ab+98b&=2b\left(9a^2+42a+49\right)\\&=2b\left((3a)^2+42a+7^2\right)\end{aligned}$$

Since the middle term is $42a=2(7\cdot 3a)$

$$=2b(3a+7)^2$$

41.

$$\begin{aligned}3r^4-48&=3\left(r^4-16\right)\\&=3\left(\left(r^2\right)^2-4^2\right)\\&=3\left(r^2+4\right)\left(r^2-4\right)\\&=3\left(r^2+4\right)\left(r^2-2^2\right)\\&=3\left(r^2+4\right)(r+2)(r-2)\end{aligned}$$

42. $(3x-5)(x+7)=0$

$$\begin{aligned}3x-5&=0\\+5\ \ +5&\\3x&=5\\\frac{3x}{3}&=\frac{5}{3}\\x&=\frac{5}{3}\end{aligned}\qquad\begin{aligned}x+7&=0\\-7\ \ -7&\\x&=-7\end{aligned}$$

$x=\frac{5}{3}$ and $x=-7$

43. $(2x-9)(x-3)=0$

$$\begin{aligned}2x-9&=0\\+9\ \ +9&\\2x&=9\\\frac{2x}{2}&=\frac{9}{2}\\x&=\frac{9}{2}\end{aligned}\qquad\begin{aligned}x-3&=0\\+3\ \ +3&\\x&=3\end{aligned}$$

$x=\frac{9}{2}$ and $x=3$

44. $y(4y-6)=0$

$$y=0\qquad\begin{aligned}4y-6&=0\\+6\ \ +6&\\4y&=6\\\frac{4y}{4}&=\frac{6}{4}\\y&=\frac{3}{2}\end{aligned}$$

$y=0$ and $y=\frac{3}{2}$

45. $x(3x-1)=0$

$$x=0\qquad\begin{aligned}3x-1&=0\\+1\ \ +1&\\3x&=1\\\frac{3x}{3}&=\frac{1}{3}\\x&=\frac{1}{3}\end{aligned}$$

$x=0$ and $x=\frac{1}{3}$

46. Factors of -28 that add to 3 are 7 and -4.

$x^2+3x-28=0$

$(x+7)(x-4)=0$

$$\begin{aligned}x+7&=0\\-7\ \ -7&\\x&=-7\end{aligned}\qquad\begin{aligned}x-4&=0\\+4\ \ +4&\\x&=4\end{aligned}$$

$x=-7$ and $x=4$

47. Factors of -54 that add to -3 are -9 and 6.

$n^2-3n-54=0$

$(n-9)(n+6)=0$

$$\begin{aligned}n-9&=0\\+9\ \ +9&\\n&=9\end{aligned}\qquad\begin{aligned}n+6&=0\\-6\ \ -6&\\n&=-6\end{aligned}$$

$n=9$ and $n=-6$

48. $a\cdot c=2\cdot -35=-70$ Factors of -70 that add to 9 are 14 and -5. Factor by grouping.

$2x^2+9x-35=0$

$2x^2+14x-5x-35=0$

$(2x^2+14x)-(5x+35)=0$

$2x(x+7)-5(x+7)=0$

$(x+7)(2x-5)=0$

$$x+7=0 \Rightarrow x=-7 \qquad 2x-5=0 \Rightarrow 2x=5 \Rightarrow \frac{2x}{2}=\frac{5}{2} \Rightarrow x=\frac{5}{2}$$

$x=-7$ and $x=\frac{5}{2}$

49. $a \cdot c = 3 \cdot -4 = -12$ Factors of -12 that add to 11 are 12 and -1. Factor by grouping.

$$3y^2+11y-4=0$$
$$3y^2+12y-y-4=0$$
$$(3y^2+12y)-(y+4)=0$$
$$3y(y+4)-1(y+4)=0$$
$$(y+4)(3y-1)=0$$

$$y+4=0 \Rightarrow y=-4 \qquad 3y-1=0 \Rightarrow 3y=1 \Rightarrow \frac{3y}{3}=\frac{1}{3} \Rightarrow y=\frac{1}{3}$$

$y=-4$ and $y=\frac{1}{3}$

50. $m^2+4m=32$

$$m^2+4m-32=0$$

Factors of -32 that add to 4 are 8 and -4.

$$m^2+4m-32=0$$
$$(m+8)(m-4)=0$$

$$m+8=0 \Rightarrow m=-8 \qquad m-4=0 \Rightarrow m=4$$

$m=-8$ and $m=4$

51. $n^2+2n=35$

$$n^2+2n-35=0$$

Factors of -35 that add to 2 are 7 and -5.

$$n^2+2n-35=0$$
$$(n+7)(n-5)=0$$

$$n+7=0 \Rightarrow n=-7 \qquad n-5=0 \Rightarrow n=5$$

$n=-7$ and $n=5$

52. Using $a^2-b^2=(a-b)(a+b)$.

$$a^2-81=0$$
$$(a+9)(a-9)=0$$

$$a+9=0 \Rightarrow a=-9 \qquad a-9=0 \Rightarrow a=9$$

$a=-9$ and $a=9$

53. Using $a^2-b^2=(a-b)(a+b)$.

$$x^2-49=0$$
$$(x+7)(x-7)=0$$

$$x+7=0 \Rightarrow x=-7 \qquad x-7=0 \Rightarrow x=7$$

$x=-7$ and $x=7$

54. Factor out 3.

$$3y^2-12=0$$
$$3(y^2-4)=0$$

Using $a^2-b^2=(a-b)(a+b)$.

$$3(y^2-4)=0$$
$$3(y+2)(y-2)=0$$

$$3 \neq 0 \qquad y+2=0 \Rightarrow y=-2 \qquad y-2=0 \Rightarrow y=2$$

$y=-2$ and $y=2$

55. Factor out 4.

$$4a^2-36=0$$
$$4(a^2-9)=0$$

Using $a^2-b^2=(a-b)(a+b)$.

$$4(a^2-9)=0$$
$$4(a+3)(a-3)=0$$

$$4 \neq 0 \qquad a+3=0 \Rightarrow a=-3 \qquad a-3=0 \Rightarrow a=3$$

$a=-3$ and $a=3$

56. Factor out $2t$.

$$2t^3+2t^2-60t=0$$
$$2t(t^2+t-30)=0$$

Factors of -30 that add to 1 are 6 and -5.

$$2t(t^2+t-30)=0$$
$$2t(t+6)(t-5)=0$$

$$\begin{array}{lll} 2t=0 & t+6=0 & t-5=0 \\ & \underline{-6\ \ -6} & \underline{+5\ \ +5} \\ t=0 & t=-6 & t=5 \end{array}$$

$t=0$, $t=-6$ and $t=5$

57. Factor out x.

$$2x^3-7x^2-15x=0$$
$$x(2x^2-7x-15)=0$$

$a\cdot c=2\cdot -15=-30$ Factors of -30 that add to -7 are -10 and 3. Factor by grouping.

$$x(2x^2-7x-15)=0$$
$$x(2x^2-10x+3x-15)=0$$
$$x\left[(2x^2-10x)+(3x-15)\right]=0$$
$$x\left[2x(x-5)+3(x-5)\right]=0$$
$$x(x-5)(2x+3)=0$$

$$\begin{array}{lll} & & 2x+3=0 \\ & x-5=0 & \underline{-3\ \ -3} \\ x=0 & \underline{+5\ \ +5} & 2x=-3 \\ & x=5 & \frac{2x}{2}=\frac{-3}{2} \\ & & x=-\frac{3}{2} \end{array}$$

$x=0$, $x=5$ and $x=-\frac{3}{2}$

58. $x(x+5)=36$

$$\begin{array}{l} x^2+5x=36 \\ \underline{\ \ -36\ \ -36} \\ x^2+5x-36=0 \end{array}$$

Factors of -36 that add to 5 are 9 and -4.

$$x^2+5x-36=0$$
$$(x+9)(x-4)=0$$

$$\begin{array}{ll} x+9=0 & x-4=0 \\ \underline{-9\ \ -9} & \underline{+4\ \ +4} \\ x=-9 & x=4 \end{array}$$

$x=-9$ and $x=4$

59. $x(3x+2)=1$

$$\begin{array}{l} 3x^2+2x=1 \\ \underline{\ \ -1\ \ -1} \\ 3x^2+2x-1=0 \end{array}$$

$a\cdot c=3\cdot -1=-3$

Factors of -3 that add to 2 are 3 and -1. Factor by grouping.

$$3x^2+2x-1=0$$
$$3x^2+3x-x-1=0$$
$$(3x^2+3x)-(x+1)=0$$
$$3x(x+1)-1(x+1)=0$$
$$(x+1)(3x-1)=0$$

$$\begin{array}{ll} & 3x-1=0 \\ x+1=0 & \underline{+1\ \ +1} \\ \underline{-1\ \ -1} & 3x=1 \\ x=-1 & \frac{3x}{3}=\frac{1}{3} \\ & x=\frac{1}{3} \end{array}$$

$x=-1$ and $x=\frac{1}{3}$

60. Let $w=$ the width of the rectangle in cm, therefore the length would be $2w+1$. Substitute these values into the formula for area.

$$A=lw$$
$$36=(2w+1)(w)$$
$$36=2w^2+w$$

Factor and use the zero-product property to solve for w.

$$\begin{array}{l} 36=2w^2+w \\ \underline{-36\ \ -36} \\ 0=2w^2+w-36 \end{array}$$
$$2w^2+w-36=0$$

$a\cdot c=2\cdot -36=-72$ Factors of -72 that add to 1 are 9 and -8. Factor by grouping.

$$2w^2+w-36=0$$
$$2w^2+9w-8w-36=0$$
$$(2w^2+9w)-(8w+36)=0$$
$$w(2w+9)-4(2w+9)=0$$
$$(2w+9)(w-4)=0$$

$$\begin{array}{ll} 2w+9=0 & \\ \underline{-9\ \ -9} & w-4=0 \\ 2w=-9 & \underline{+4\ \ +4} \\ \frac{2w}{2}=\frac{-9}{2} & w=4 \\ w=-\frac{9}{2} & \end{array}$$

Width cannot be a negative value, therefore, the width of the rectangle is 4 cm. The length is $2w+1$, giving us a length of 9 cm.

61. **a.** $P = -2t^2 + 34t$ where t represents the number of toys in thousands so 8000 toys is represented by $t = 8$.

$P = -2(8)^2 + 34(8)$ Substitute $t = 8$.

$P = -2(64) + 272$ Solve for P.

$P = -128 + 272$

$P = 144$

The profit is $144,000 for 8000 toys sold.

b. $120 = -2t^2 + 34t$ Substitute $P = 120$.

Factor and use the zero-product property to solve for t.

$$\begin{aligned} 120 &= -2t^2 + 34t \\ \underline{-120} & \quad \underline{-120} \\ 0 &= -2t^2 + 34t - 120 \\ -2t^2 + 34t - 120 &= 0 \\ -2(t^2 - 17t + 60) &= 0 \end{aligned}$$

Factors of 60 that add to -17 are -12 and -5.

$$-2(t^2 - 17t + 60) = 0$$
$$-2(t-12)(t-5) = 0$$

$$-2 \neq 0 \qquad \begin{aligned} t - 12 &= 0 \\ \underline{+12 \ \ +12} \\ t &= 12 \end{aligned} \qquad \begin{aligned} t - 5 &= 0 \\ \underline{+5 \ \ +5} \\ t &= 5 \end{aligned}$$

RC Toys, Inc. needs to sell either 5000 or 12,000 toys to have a profit of $120,000.

Chapter 6 Test

1. Factors of -24 that add to 2 are 6 and -4.

$x^2 + 2x - 24 = (x+6)(x-4)$

2. Factors of 48 that add to -14 are -8 and -6.

$a^2 - 14a + 48 = (a-8)(a-6)$

3. Using $a^2 - b^2 = (a-b)(a+b)$.

$x^2 - 100 = (x-10)(x+10)$

4. Using $a^2 - b^2 = (a-b)(a+b)$.

$25m^2 - 1 = (5m-1)(5m+1)$

5. $a \cdot c = 6 \cdot -3 = -18$ Factors of -18 that add to 7 are 9 and -2. Factor by grouping.

$$\begin{aligned} 6x^2 + 7x - 3 &= 6x^2 + 9x - 2x - 3 \\ &= (6x^2 + 9x) - (2x + 3) \\ &= 3x(2x+3) - 1(2x+3) \\ &= (2x+3)(3x-1) \end{aligned}$$

6. $4y^2 + 20y + 25$ is a perfect square trinomial. Factor using $a^2 + 2ab + b^2 = (a+b)^2$.

$4y^2 + 20y + 25 = (2y+5)^2$

7. Factor out t.

$4t^3 + 5t^2 - 21t = t(4t^2 + 5t - 21)$

$a \cdot c = 4 \cdot -21 = -84$ Factors of -84 that add to 5 are 12 and -7. Factor by grouping.

$$\begin{aligned} t(4t^2 + 5t - 21) &= t(4t^2 + 12t - 7t - 21) \\ &= t\left[(4t^2 + 12t) - (7t + 21)\right] \\ &= t\left[4t(t+3) - 7(t+3)\right] \\ &= t(t+3)(4t-7) \end{aligned}$$

8. Factor out $2x$.

$2x^3 - 6x^2 - 8x = 2x(x^2 - 3x - 4)$

Factors of -4 that add to -3 are -4 and 1.

$2x(x^2 - 3x - 4) = 2x(x-4)(x+1)$

9. Factor by grouping.

$$\begin{aligned} 5ab - 10a + 2b - 4 &= (5ab - 10a) + (2b - 4) \\ &= 5a(b-2) + 2(b-2) \\ &= (b-2)(5a+2) \end{aligned}$$

10. $2m(n+8) - 3(n+8) = (n+8)(2m-3)$

11. $t^2-20t+100=t^2-20t+10^2$

Since the middle term $20t=2(10\cdot t)$ this is a perfect square trinomial.

$$=(t-10)^2$$

12.

$$7x^3-28x=7x(x^2-4)$$
$$=7x(x^2-2^2)$$
$$=7x(x+2)(x-2)$$

13.

$$8m^3+125n^3=(2m)^3+(5n)^3$$
$$=(2m+5n)\left((2m)^2-2m\cdot 5n+(5n)^2\right)$$
$$=(2m+5n)(4m^2-10mn+25n^2)$$

14.

$3x^2+x+10$

$ac=3\times 10=30$

Since there are no factors of 30 that sum to 1, the quadratic is prime.

15. Let $u=m^4$.

$$6m^8-29m^4+28=6(m^4)^2-29m^4+28$$
$$=6u^2-29u+28$$
$$=(3u-4)(2u-7)$$

Replace u with m^4.

$$=(3m^4-4)(2m^4-7)$$

16.

$$a^3-125b^3=a^3-(5b)^3$$
$$=(a-5b)\left(a^2+a\cdot 5b+(5b)^2\right)$$
$$=(a-5b)(a^2+5ab+25b^2)$$

17. Factors of 30 that add to 13 are 10 and 3.

$t^2+13t+30=0$

$(t+10)(t+3)=0$

$$\begin{array}{ll} t+10=0 & t+3=0 \\ \underline{-10\ \ -10} & \underline{-3\ \ -3} \\ t=-10 & t=-3 \end{array}$$

$t=-10$ and $t=-3$

18. Factor out 9.

$9w^2-36=0$

$9(w^2-4)=0$

Using $a^2-b^2=(a-b)(a+b)$.

$9(w^2-4)=0$

$9(w+2)(w-2)=0$

$$\begin{array}{lll} & w+2=0 & w-2=0 \\ 9\neq 0 & \underline{-2\ \ -2} & \underline{+2\ \ +2} \\ & w=-2 & w=2 \end{array}$$

$w=-2$ and $w=2$

19. $x^3+2x^2=48x$

$\underline{\quad -48x\ \ -48x}$

$x^3+2x^2-48x=0$

Factor out x.

$x^3+2x^2-48x=0$

$x(x^2+2x-48)=0$

Factors of -48 that add to 2 are 8 and -6.

$x(x^2+2x-48)=0$

$x(x+8)(x-6)=0$

$$\begin{array}{lll} & x+8=0 & x-6=0 \\ x=0 & \underline{-8\ \ -8} & \underline{+6\ \ +6} \\ & x=-8 & x=6 \end{array}$$

$x=0$, $x=-8$ and $x=6$

Cumulative Review Chapters 1-6

1. $(2x)\cdot(2x)\cdot(2x)\cdot(2x)\cdot(2x)=(2x)^5$

2. $-\frac{1}{2}\cdot-\frac{1}{2}\cdot-\frac{1}{2}=-\frac{1}{8}$

3. $\boxed{24\div 8\cdot(-5)}+\boxed{(-3)^3}\boxed{-6\div(-2)}+\boxed{|-1-4(5)|}$

$=24\div 8\cdot(-5)+(-3)^3-6\div(-2)+|-1-20|$
$=24\div 8\cdot(-5)-27-6\div(-2)+|-21|$
$=3\cdot(-5)-27-6\div(-2)+21$
$=3\cdot(-5)-27+3+21$
$=-15-27+3+21$
$=-18$

4. Evaluate x^3+2x^2-3x+7 for $x=-1$.

x^3+2x^2-3x+7 Substitute in $x=-1$.
$=(-1)^3+2(-1)^2-3(-1)+7$ Simplify.
$=-1+2+3+7$
$=11$

5. $-3(4a+2b)-(3a-7b)$
$=-12a-6b-3a+7b$
$=-15a+b$

6. $-(5x-6x)+3(x+2)$
$=-5x+6x+3x+6$
$=4x+6$

7. $6x\boxed{-5y}=35$ Identify variable term to isolate.

$\underline{-6x\quad -6x}$ Subtract $6x$ from both sides.

$-5\boxed{y}=-6x+35$ Identify variable to isolate.

$\frac{-5y}{-5}=\frac{-6x+35}{-5}$ Divide both sides by -5.

$y=\frac{6}{5}x-7$

8. $-16=\frac{4}{5}y$

$\frac{5}{4}\cdot-16=\frac{\not{4}}{\not{5}}y\cdot\frac{\not{5}}{\not{4}}$

$-20=y$

9. $P=a+\boxed{b}+c$

$\underline{-a-c\quad -a\quad -c}$

$P-a-c=b$

10. $5(n+1)-6=2n+1$ Distribute on left side.

$5n+5-6=2n+1$ Simplify on left side.

$5n-1=2n+1$

$\underline{-2n\quad -2n}$ Subtract $2n$ from both sides.

$3n-1=1$

$\underline{+1\quad +1}$ Add 1 to both sides.

$3n=2$

$\frac{3n}{3}=\frac{2}{3}$ Divide both sides by 3.

$n=\frac{2}{3}$

11. **a.** Substitute $t=6.5$ into the equation $D=65t$ and solve for D.

$D=65t$ Given equation.
$D=65(6.5)$ Substitute $t=6.5$.
$D=422.5$ Multiply & solve for D.

Helena will travel 422.5 miles if she drives for 6.5 hours.

b. Substitute $D=455$ into the equation $D=65t$ and solve for t.

$D=65t$ Given equation.
$455=65t$ Substitute $D=455$ & solve for t.
$\frac{455}{65}=\frac{65t}{65}$ Divide both sides by 65.
$7=t$

Helena must drive for 7 hours to travel 455 miles.

12. $\frac{1}{2}x+3=\frac{13}{4}x+11$ Multiply by 4 to clear denom.

$4\left(\frac{1}{2}x+3\right)=4\left(\frac{13}{4}x+11\right)$

$2x+12=13x+44$

$\underline{-13x\quad -13x}$ Subtract $13x$ from both sides.

$-11x+12=44$

$\underline{-12\quad -12}$ Subtract 12 from both sides.

$-11x=32$

$\frac{-11x}{-11}=\frac{32}{-11}$ Divide both sides by -11.

$x=-\frac{32}{11}$

13. $-2(x-3)+6<-8$ Distribute on the left.

$-2x+6+6<-8$ Simplify.

$-2x+12<-8$

$\underline{-12\quad -12}$ Subtract 12 from both sides.

$-2x<-20$ Divide both sides by -2.

$\frac{-2x}{-2}>\frac{-20}{-2}$ Multiply by neg., reverse ineq.

$x>10$

14. **a.**

x	$y=\frac{2}{3}x-4$
–6	–8
–3	–6
0	–4
3	–2
6	0

b.

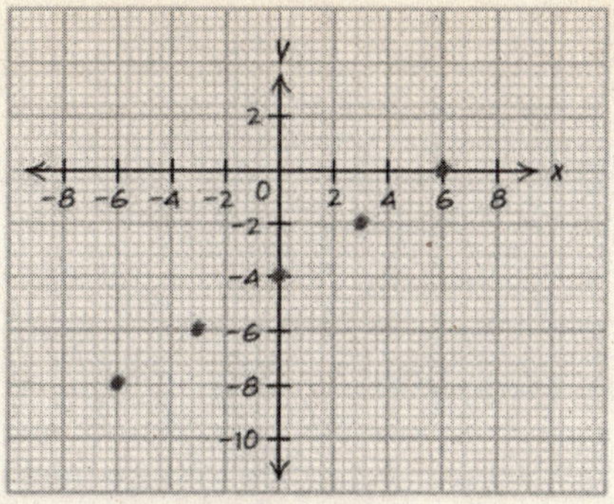

c. The points are linear.

15.

m	$C = 124m$
0	0
1	124
2	248
3	372
4	496

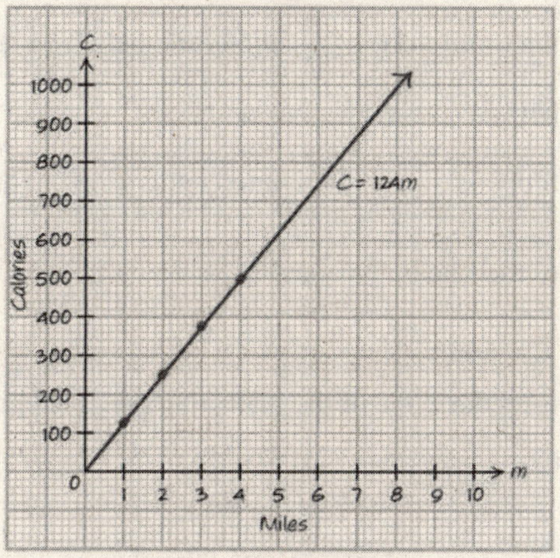

16.

x	$y = x^2 + 3$
−4	19
−3	12
−2	7
−1	4
0	3
1	4
2	7
3	12
4	19

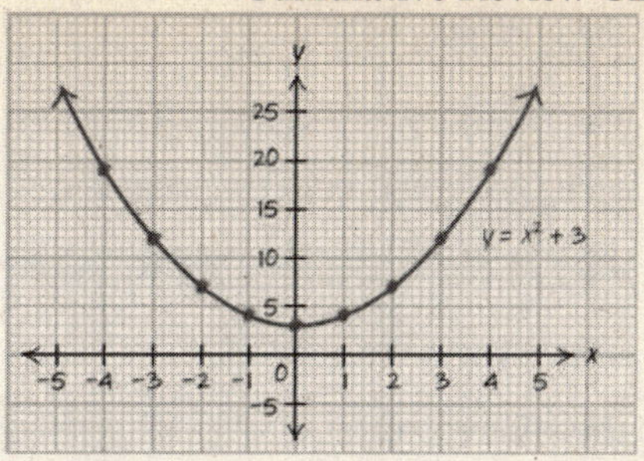

The graph is non-linear.

17. Substituting the points $(4,-3)$ and $(-2,9)$ into the slope formula yields:

$$\text{slope} = \frac{y_2 - y_1}{x_2 - x_1} = \frac{9-(-3)}{-2-(4)} = \frac{12}{-6} = -2$$

The slope of the line that passes through the given points is $m = -2$.

18. Find the y-intercept:

$$-3(0) + 8y = -24 \qquad \text{Substitute in } x = 0.$$
$$8y = -24$$
$$\frac{8y}{8} = \frac{-24}{8} \qquad \text{Solve for } y.$$
$$y = -3$$

The y-intercept is the point $(0,-3)$.

Find the x-intercept:

$$-3x + 8(0) = -24 \qquad \text{Substitute in } y = 0.$$
$$-3x = -24$$
$$\frac{-3x}{-3} = \frac{-24}{-3} \qquad \text{Solve for } x.$$
$$x = 8$$

The x-intercept is the point $(8,0)$.

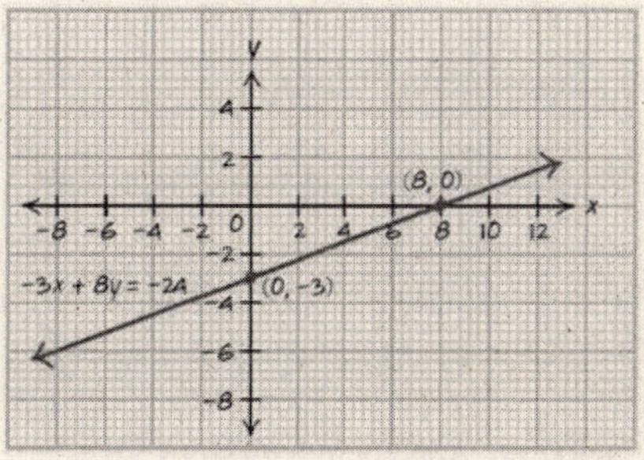

19. Find the slope (change to slope-intercept form):

$$2x + 7y = 14$$
$$\underline{-2x \qquad -2x} \qquad \text{Subtract } 2x \text{ from both sides.}$$
$$7y = -2x + 14$$
$$\frac{7y}{7} = \frac{-2x+14}{7} \qquad \text{Divide both sides by 7.}$$
$$y = -\frac{2}{7}x + 2$$

The slope is $m = -\frac{2}{7}$.

Find the y-intercept:

$$y = -\frac{2}{7}(0) + 2 \qquad \text{Substitute in } x = 0.$$
$$y = 0 + 2$$
$$y = 2$$

The y-intercept is the point $(0,2)$.

Find the x-intercept:

$$2x + 7(0) = 14 \qquad \text{Substitute in } y = 0.$$
$$2x = 14$$
$$\frac{2x}{2} = \frac{14}{2} \qquad \text{Solve for } x.$$
$$x = 7$$

The x-intercept is the point $(7,0)$.

20. Slope is $\frac{-3}{5}$; y-intercept is $(0,-2)$.

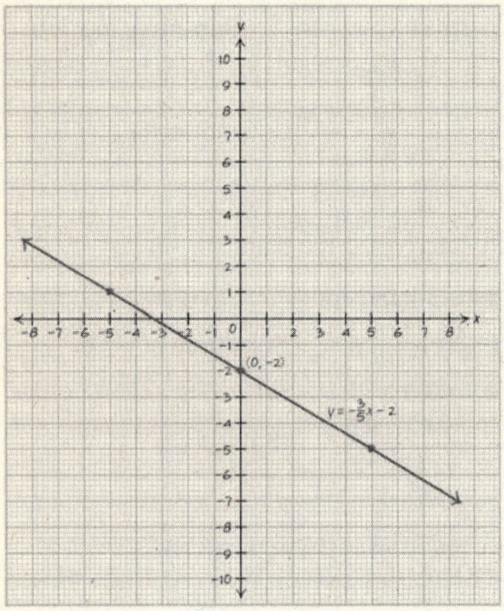

21. Change to slope-intercept form.

$$\frac{-2}{3}x + \frac{1}{6}y = -2 \qquad \text{Multiply by 6 to clear denom.}$$
$$6\left(\frac{-2}{3}x + \frac{1}{6}y\right) = 6(-2)$$
$$-4x + y = -12$$
$$+4x \qquad +4x \qquad \text{Add } 4x \text{ to both sides.}$$
$$y = 4x - 12$$

Graph using slope $m = 4$ and y-intercept $(0,-12)$.

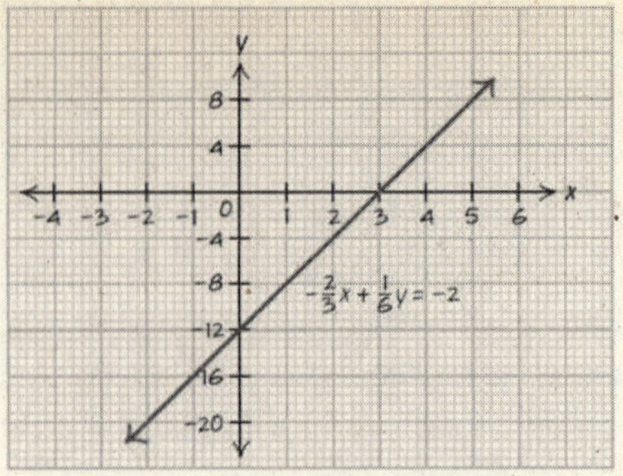

22. Change the second equation to slope-intercept form.

$$4x - 5y = -1$$
$$-4x \qquad -4x \qquad \text{Subtract } 4x \text{ from both sides.}$$
$$-5y = -4x - 1$$
$$\frac{-5y}{-5} = \frac{-4x-1}{-5}$$
$$y = \frac{4}{5}x + \frac{1}{5}$$

Comparing the slopes of the two equations, both have the same slope of $\frac{4}{5}$, therefore these lines are parallel.

23. Slope is $\frac{2}{3}$ and the y-intercept is $(0,3)$. The equation of the line is $y = \frac{2}{3}x + 3$.

24. $C = 600h$

25. Find the slope of the line using the given points.

$$\text{slope} = \frac{-4-(21)}{4-(-6)} = \frac{-25}{10} = -\frac{5}{2}$$

Find the y-intercept using the value for slope and one of the given points $(4,-4)$.

$$y = mx + b \qquad \text{Substitute } m = -\frac{5}{2},\ x = 4,\ y = -4$$
$$(-4) = \left(-\frac{5}{2}\right)(4) + b$$
$$-4 = -10 + b \qquad \text{Solve for } b.$$
$$+10 \quad +10$$
$$6 = b$$

Using the values we found for the slope and b yields the equation of the line

$$y = -\frac{5}{2}x + 6$$

26. Using point-slope form:

$y-y_1=m(x-x_1)$ Substitute in $m=0$ and

$y-(4)=0[x-(-3)]$ $(x_1,y_1)=(-3,4)$

$y-4=0(x+3)$

$y-4=0$

$+4\ \ +4$ Add 4 to both sides.

$y=4$

27. Find the slope of the line using the given points.

$$\text{slope}=\frac{17-(-4)}{-4-(2)}=\frac{21}{-6}=-\frac{7}{2}$$

Find the y-intercept using the value for slope and one of the given points $(2,-4)$.

$y=mx+b$ Substitute $m=-\frac{7}{2}$, $x=2, y=-4$

$(-4)=\left(-\frac{7}{2}\right)(2)+b$

$-4=-7+b$ Solve for b.

$+7\ \ +7$

$3=b$

Using the values we found for the slope and b yields the equation of the line

$$y=-\frac{7}{2}x+3$$

28. Parallel lines have the same slope. The slope of the line that we are trying to find is $\frac{1}{2}$. Using point-slope form:

$y-y_1=m(x-x_1)$ Substitute in $m=\frac{1}{2}$ and

$y-(-5)=\frac{1}{2}[x-(-4)]$ $(x_1,y_1)=(-4,-5)$

$y+5=\frac{1}{2}(x+4)$

$y+5=\frac{1}{2}x+2$

$-5\ \ \ \ -5$ Subtract 5 from both sides.

$y=\frac{1}{2}x-3$

29. Perpendicular lines have slopes that are negative reciprocals. The slope of the given line is $m=\frac{3}{5}$. The negative reciprocal of $\frac{3}{5}$ is $-\frac{5}{3}$. Using point-slope form:

$y-y_1=m(x-x_1)$ Substitute in $m=-\frac{5}{3}$ and

$y-(7)=-\frac{5}{3}[x-(-3)]$ $(x_1,y_1)=(-3,7)$

$y-7=-\frac{5}{3}(x+3)$

$y-7=-\frac{5}{3}x-5$

$+7\ \ \ \ +7$ Add 7 to both sides.

$y=-\frac{5}{3}x+2$

30. Looking at the tables, there is one input and output pair that is the same, $(-5,3)$. This point is the solution to the system.

31. **a.** First equation:

$(8)\overset{?}{=}\frac{-5}{7}(-7)+3$ Substitute values for x and y.

$8\overset{?}{=}5+3$ Simplify.

$8=8$ This is a true statement.

Second equation:

$(8)\overset{?}{=}\frac{4}{7}(-7)-6$ Substitute values for x and y.

$8\overset{?}{=}-4-6$ Simplify.

$8\overset{?}{=}-10$ This is a false statement.

Since only one of the statements is true, $(-7,8)$ is not a solution of the system.

b. First equation:

$(-2)\overset{?}{=}\frac{-5}{7}(7)+3$ Substitute values for x and y.

$-2\overset{?}{=}-5+3$ Simplify.

$-2=-2$ This is a true statement.

Second equation:

$(-2)\overset{?}{=}\frac{4}{7}(7)-6$ Substitute values for x and y.

$-2\overset{?}{=}4-6$ Simplify.

$-2=-2$ This is a true statement.

Since both statements are true, $(7,-2)$ is a solution of the system.

32.

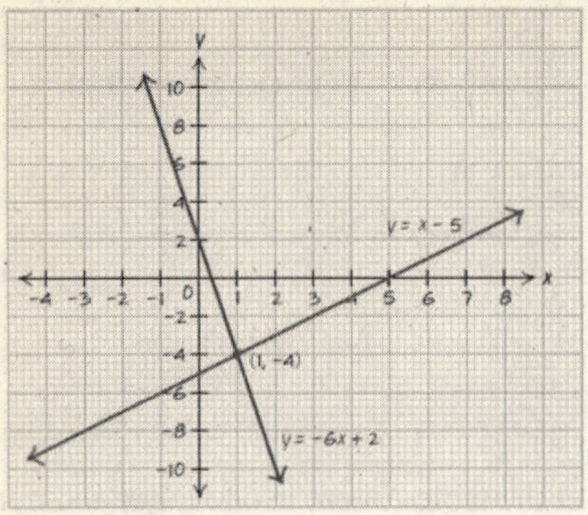

The solution is the point $(1,-4)$.

33. Substitute $y=\frac{-3}{5}x+2$ into the second equation.

$$3x+5\left(\frac{-3}{5}x+2\right)=2$$

$$3x-3x+10=2$$

$$10\overset{?}{=}2 \quad \text{This is a false statement.}$$

The system is inconsistent; it has no solution. The lines are parallel.

34. Change the first equation and substitute $y=-2x+2$ into the second equation.

$$7x+2(-2x+2)=-17$$

$$7x-4x+4=-17$$

$$3x+4=-17$$

$$\underline{\quad -4 \quad -4} \quad \text{Subtract 4 from both sides.}$$

$$3x=-21$$

$$\frac{3x}{3}=\frac{-21}{3} \quad \text{Divide both sides by 3.}$$

$$x=-7$$

Substitute $x=-7$ into the equation to find y.

$$y=-2x+2$$

$$y=-2(-7)+2$$

$$y=14+2$$

$$y=16$$

The solution to the system is $(-7,16)$.

35. Let $t=$ the cost in dollars for each t-shirt and let $b=$ the cost in dollars for each basketball. The system of equations is as follows:

$$12t+7b=220$$

$$10t+8b=218$$

$$\begin{array}{l} 8(12t+7b)=8(220) \\ -7(10t+8b)=-7(218) \end{array} \text{ is now } \begin{array}{l} 96t+56b=1760 \\ -70t-56b=-1526 \end{array}$$

Add the equations vertically to eliminate b.

$$96t+56b=1760 \quad \text{Add the equations vertically.}$$

$$\underline{-70t-56b=-1526}$$

$$26t=234 \quad \text{Solve for } t.$$

$$t=9$$

Substitute $t=9$ into either original equation to find the value of b.

$$10(9)+8b=218 \quad \text{Substitute in } t=9$$

$$90+8b=218$$

$$\underline{-90 \qquad -90}$$

$$8b=128 \quad \text{Solve for } b.$$

$$b=16$$

The cost of each t-shirt is \$9 and the cost of each basketball is \$16.

36. Let $b=$ the rate of the boat (in knots) in still water and $c=$ the rate of the current in knots.

The system of equations is:

$$4(b+c)=40$$

$$5(b-c)=40$$

Solve the second equation for b.

$$5(b-c)=40 \quad \text{Distribute.}$$

$$5b-5c=40$$

$$\underline{\quad +5c \quad +5c}$$

$$5b=5c+40$$

$$\frac{5b}{5}=\frac{5c+40}{5}$$

$$b=c+8$$

Now substitute $b=c+8$ into the first equation.

$$4(b+c)=40 \quad \text{Distribute.}$$

$$4b+4c=40 \quad \text{Substitute in } b=c+8.$$

$$4(c+8)+4c=40 \quad \text{Distribute.}$$

$$4c+32+4c=40 \quad \text{Simplify.}$$

$$8c+32=40$$

$$\underline{\quad -32 \; -32}$$

$$8c=8 \quad \text{Solve for } c.$$

$$\frac{8c}{8}=\frac{8}{8}$$

$$c=1$$

$$b=c+8 \quad \text{Substitute in } c=1 \text{ to find } b.$$

$$b=(1)+8$$

$$b=9$$

The rate of the boat is 9 knots and the rate of the current is 1 knot.

37. Let $y=$ the number of pounds of dried papaya and let $n=$ the number of pounds of dried pineapple. The system of equations is:

$$y+n=40$$

$$7.99y+11.99n=379.6$$

Solve for y in the first equation.

$$y+n=40$$

$$\underline{\; -n \quad -n} \quad \text{Subtract } n \text{ from both sides.}$$

$$y=-n+40$$

Now substitute $y=-n+40$ into the second equation.

$7.99y+11.99n=379.6$ Substitute in $y=-n+40$.

$7.99(-n+40)+11.99n=379.6$ Distribute.

$-7.99n+319.6+11.99n=379.6$ Simplify.

$$4n+319.6=379.6$$
$$\underline{-319.6\ \ -319.6}$$
$$4n=60 \quad \text{Solve for } n.$$
$$\frac{4n}{4}=\frac{60}{4}$$
$$n=15$$

$y=-n+40$ Substitute in $n=15$ to find y.

$y=-(15)+40$

$y=25$

Emma should put 25 lb. of papaya and 15 lb. of pineapple in the mix.

38. Multiply the second equation by -2:

$3x+2y=8$ and $-2(-2x+y)=-2(-17)$ is now $3x+2y=8$ and $4x-2y=34$

Add the equations vertically to eliminate y.

$3x+2y=8$ Add the equations vertically.

$\underline{4x-2y=34}$

$7x=42$ Solve for x.

$$\frac{7x}{7}=\frac{42}{7}$$
$$x=6$$

Substitute $x=6$ into either original equation to find the value of y.

$3x+2y=8$ Substitute in $x=6$

$3(6)+2y=8$

$18+2y=8$

$\underline{-18\ \ \ \ -18}$

$2y=-10$ Solve for y.

$y=-5$

The solution to the system is $(6,-5)$.

39. Multiply the first equation by 2:

$2(3x+4y)=2(7)$ and $-6x-8y=4$ is now $6x+8y=14$ and $-6x-8y=4$

Add the equations vertically.

$6x+8y=14$ Add the equations vertically.

$\underline{-6x-8y=4}$

$0\overset{?}{=}18$ This is a false statement.

The system is inconsistent; it has no solution. The lines are parallel.

40. Let $x=$ the degree measure of the first angle and let $y=$ the degree measure of the second angle. The system of equations is:

$x+y=90^\circ$

$y=x+20$

Substitute $y=x+20$ into the first equation.

$x+y=90^\circ$ Substitute in $y=x+20$.

$x+(x+20)=90$ Simplify.

$2x+20=90$ Solve for x.

$\underline{-20\ \ -20}$ Subtract 20 from both sides.

$$\frac{2x}{2}=\frac{70}{2}$$
$$x=35$$

$y=x+20$ Substitute in $x=35$ to find y.

$y=(35)+20$

$y=55$

The measures of the angles are 35° and 55°.

41. Let $x=$ the amount (in dollars) invested in the CD earning 3% interest and let $y=$ the amount (in dollars) invested in the bond fund earning 3.5% interest.

The system of equations is:

$x+y=10000$

$0.03x+0.035y=340$

Solve for y in the first equation.

$x+y=10000$

$\underline{-x\ \ \ \ \ \ -x}$

$y=-x+10000$

Now substitute $y=-x+10000$ into the second equation.

$0.03x+0.035y=340$ Substitute in $y=-x+10000$.

$0.03x+0.035(-x+10000)=340$ Distribute.

$0.03x-0.035x+350=340$ Simplify.

$-0.005x+350=340$

$\underline{-350\ -350}$ Subtract 350 from both sides.

$-0.005x=-10$ Solve for x.

$$\frac{-0.005x}{-0.005}=\frac{-10}{-0.005}$$
$$x=2000$$

$y=-x+10000$ Substitute in $x=2000$ to find y.

$y=-(2000)+10000$

$y=8000$

Seamus should invest \$2000 in the 3% interest-earning CD and \$8000 in the 3.5% interest-earning bond fund.

42. $6x-9y\geq 10$ Is $(x,y)=(2,4)$ in the solution set?

$6(2)-9(4)\overset{?}{\geq}10$ Substitute values for x and y.

$12-36\overset{?}{\geq}10$

$-24\overset{?}{\geq}10$ This statement is false.

No, the point is not in the solution set.

43. Graph using the slope and y-intercept. This is a strict inequality (<) therefore the line is dashed and we shade below the line.

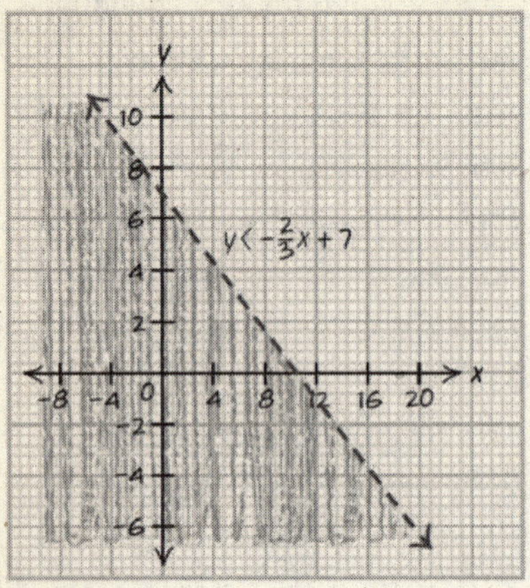

44. $(-3)^2 = (-3)\cdot(-3) = 9$

45. $-3^2 = -(3)\cdot(3) = -9$

46. **a.** $x^3 \cdot x^5 = x^{3+5} = x^8$

b. $3a^4 \cdot 4a = 12\cdot a^{4+1} = 12a^5$

c. $(-3ab^3)(-5a^4b) = (-3)(-5)\cdot a^{1+4}\cdot b^{3+1} = 15a^5b^4$

47. **a.** $n^5 \cdot n = n^{5+1} = n^6$

b. $5t^3 \cdot 7t^2 = 35\cdot t^{3+2} = 35t^5$

c. $(9x^2y)(-3xy^6) = (9)(-3)\cdot x^{2+1}\cdot y^{1+6} = -27x^3y^7$

48. **a.** $(3x^4)^3 = (3)^3\cdot(x^4)^3$

$= 27x^{4\cdot 3}$

$= 27x^{12}$

b. $(-2x^2)^3 = (-2)^3\cdot(x^2)^3$

$= -8x^{2\cdot 3}$

$= -8x^6$

c. $(2x^3y^4)^2 = (2)^2\cdot(x^3)^2\cdot(y^4)^2$

$= 4x^{3\cdot 2}y^{4\cdot 2}$

$= 4x^6y^8$

49. **a.** $(5y^4)^2 = (5)^2\cdot(y^4)^2$

$= 25y^{4\cdot 2}$

$= 25y^8$

b. $(-3y^2)^5 = (-3)^5\cdot(y^2)^5$

$= -243y^{2\cdot 5}$

$= -243y^{10}$

c. $(4a^5b^2)^3 = (4)^3\cdot(a^5)^3\cdot(b^2)^3$

$= 64a^{5\cdot 3}b^{2\cdot 3}$

$= 64a^{15}b^6$

50. **a.** $(-7x)(3x^2)^2 = (-7x)\cdot(3)^2\cdot(x^2)^2$

$= (-7x)\cdot 9x^4$

$= -63x^{1+4}$

$= -63x^5$

b. $(-3xy)^2(-2y^3)^3 = (-3)^2(x)^2(y)^2(-2)^3(y^3)^3$

$= 9x^2y^2\cdot -8y^9$

$= -72x^2y^{2+9}$

$= -72x^2y^{11}$

51. **a.** $(8n)(-2n^6)^3 = (8n)(-2)^3(n^6)^3$

$= 8n\cdot -8n^{18}$

$= -64n^{1+18}$

$= -64n^{19}$

b. $(-xy^4)^5(-5x^2y)^2 = (-x)^5(y^4)^5(-5)^2(x^2)^2(y)^2$

$= -x^5y^{20}\cdot 25x^4y^2$

$= -25x^{5+4}y^{20+2}$

$= -25x^9y^{22}$

52. **a.** $\dfrac{x^9}{x^3} = x^{9-3} = x^6$

b. $\dfrac{-8a^5}{16a} = -\dfrac{1}{2}a^{5-1} = -\dfrac{1}{2}a^4$

c. $\dfrac{x^3xy^5}{x^2y^3} = \dfrac{x^4y^5}{x^2y^3} = x^{4-2}y^{5-3} = x^2y^2$

53. **a.** $\dfrac{y^3}{y} = y^{3-1} = y^2$

b. $\dfrac{-25y^4}{5y^3} = -5y^{4-3} = -5y$

c. $\dfrac{a^4bb^6}{a^2b^3} = \dfrac{a^4b^7}{a^2b^3} = a^{4-2}b^{7-3} = a^2b^4$

54. $\left(\dfrac{-15x^4y^2}{50x^3y}\right)^2 = \left(\dfrac{-3x^{4-3}y^{2-1}}{10}\right)^2$

$= \left(\dfrac{-3xy}{10}\right)^2$

$= \left(\dfrac{-3}{10}\right)^2(x)^2(y)^2$

$= \dfrac{9x^2y^2}{100}$

55. $\left(\dfrac{49n^4m^3}{-7nm^2}\right)^3 = \left(-7n^{4-1}m^{3-2}\right)^3$

$= \left(-7n^3m\right)^3$

$= (-7)^3\left(n^3\right)^3(m)^3$

$= -343n^9m^3$

56. Substitute $s = 4x$ into the formula:

$A = s^2$

$A = (4x)^2$

$A = (4)^2(x)^2$

$A = 16x^2$

57. Substitute $r = 3n$ into the formula:

$A = \pi r^2$

$A = \pi(3n)^2$

$A = \pi(3)^2(n)^2$

$A = 9\pi n^2$

58. **a.** $5t^{-2} = 5\cdot\dfrac{1}{t^2} = \dfrac{5}{t^2}$

b. $\dfrac{-2}{x^{-3}} = -2x^3$

59. **a.** $6n^{-5} = 6\cdot\dfrac{1}{n^5} = \dfrac{6}{n^5}$

b. $\dfrac{-7}{y^{-4}} = -7y^4$

60. **a.** $\dfrac{x^4y}{x^2y^2} = x^{4-2}y^{1-2} = x^2y^{-1} = \dfrac{x^2}{y}$

b. $\dfrac{-3x^{-3}y^3}{9xy^2} = \dfrac{-x^{-3-1}y^{3-2}}{3} = \dfrac{-x^{-4}y^1}{3} = \dfrac{-y}{3x^4}$

c. $(-3n^{-3})^2 = (-3)^2(n^{-3})^2 = 9n^{-6} = \dfrac{9}{n^6}$

d. $\left(\dfrac{x^2}{y}\right)^{-2} = \left(\dfrac{y}{x^2}\right)^2 = \dfrac{(y)^2}{\left(x^2\right)^2} = \dfrac{y^2}{x^4}$

61. **a.** $\dfrac{r^5t^2}{r^6t^{-2}} = r^{5-6}t^{2-(-2)} = r^{-1}t^4 = \dfrac{t^4}{r}$

b. $\dfrac{-16n^5n}{24n^7m^{-2}} = \dfrac{-2n^6}{3n^7m^{-2}}$

$= \dfrac{-2n^{6-7}m^2}{3}$

$= \dfrac{-2n^{-1}m^2}{3}$

$= \dfrac{-2m^2}{3n}$

c. $(-4m^{-2})^3 = (-4)^3(m^{-2})^3 = -64m^{-6} = \dfrac{-64}{m^6}$

d. $\left(\dfrac{x}{y^4}\right)^{-3} = \left(\dfrac{y^4}{x}\right)^3 = \dfrac{\left(y^4\right)^3}{(x)^3} = \dfrac{y^{12}}{x^3}$

62. **a.** $895{,}100{,}000{,}000 = 8.951\times10^{11}$

b. $0.000413 = 4.13\times10^{-4}$

63. **a.** $648{,}700{,}000 = 6.487\times10^8$

b. $0.00000692 = 6.92\times10^{-6}$

64. **a.** $6.75\times10^7 = 6.75\cdot10{,}000{,}000 = 67{,}500{,}000$

b. $3.58\times10^{-5} = 0.0000358$

65. **a.** $4.01\times10^8 = 4.01\cdot100{,}000{,}000 = 401{,}000{,}000$

b. $8.43\times10^{-4} = 0.000843$

66. $1{,}000{,}000{,}000 = 1.0\times10^9$

67. $2.5\times10^{-9}\text{ m} = 0.0000000025\text{ m}$

68. **a.** $9(3.2\times10^8)$

$= 9(3.2)\times10^8$

$= 28.8\times10^8$ Decimal moves left 1.

$= 2.88\times10^9$ Exponent increases 1.

b. $\dfrac{6.8\times10^{12}}{4}$

$= \dfrac{6.8}{4}\times10^{12}$

$= 1.7\times10^{12}$

c. $34(2.5\times10^{-9})$

$= 34(2.5)\times10^{-9}$

$= 85\times10^{-9}$ Decimal moves left 1.

$= 8.5\times10^{-8}$ Exponent increases 1.

d. $(6.2\times10^{13})(4.3\times10^8)$

$= (6.2)(4.3)\times10^{13+8}$

$= 26.66\times10^{21}$ Decimal moves left 1.

$= 2.666\times10^{22}$ Exponent increases 1.

e. $\dfrac{3.4\times10^{12}}{1.7\times10^4}$

$= \dfrac{3.4}{1.7}\times10^{12-4}$

$= 2.0\times10^8$

69. The polynomial has 4 terms. The highest exponent is 3 so the degree is 3.

70. The polynomial has 2 terms. The highest exponent is 2 so the degree is 2.

71. The degree of the term $4xy^2$ is 3.

The degree of the term $-xy$ is 2.

The degree of the term $2y^4$ is 4.

The degree of the term $9x$ is 1.

The highest degree term has a degree of 4, so the degree of the polynomial is 4.

72. The degree of the term $8a^5b$ is 6.

The degree of the term $3a^3$ is 3.

The degree of the term $-6b^5$ is 5.

The degree of the term 5 is 0.

The highest degree term has a degree of 6, so the degree of the polynomial is 6.

73. $(x^2-x-4)+(-2x^2-3x+1)$

$=x^2-2x^2-x-3x-4+1$

$=-x^2-4x-3$

74. $(9y-7)+(2y^2-3y-5)$

$=2y^2+9y-3y-7-5$

$=2y^2+6y-12$

75. $(-6x^2+3x-7)-(-2x+9)$

$=-6x^2+3x-7+2x-9$

$=-6x^2+3x+2x-7-9$

$=-6x^2+5x-16$

76. $(5a^2-2a+7)-3(4a^2-11)$

$=5a^2-2a+7-3(4a^2)-3(-11)$

$=5a^2-2a+7-12a^2+33$

$=5a^2-12a^2-2a+7+33$

$=-7a^2-2a+40$

77. $-2(5n-9)$

$=-2(5n)-2(-9)$

$=-10n+18$

78. $-3(-2n+1)$

$=-3(-2n)-3(1)$

$=6n-3$

79. **a.** $(x-4)(x+7)$

$=x^2+7x-4x-28$

$=x^2+3x-28$

b. $(3x+1)(2x-5)$

$=6x^2-15x+2x-5$

$=6x^2-13x-5$

c. $(x+2)(x^2-3x+4)$

$=x^3-3x^2+4x+2x^2-6x+8$

$=x^3-x^2-2x+8$

d. $(n+4)^2=n^2+8n+16$

e. $(2x+1)(2x-1)=4x^2-1$

80. **a.** $(y+9)(y-3)$

$=y^2-3y+9y-27$

$=y^2+6y-27$

b. $(4x+7)(-2x+1)$

$=-8x^2+4x-14x+7$

$=-8x^2-10x+7$

c. $(x-1)(x^2+4x-2)$

$=x^3+4x^2-2x-x^2-4x+2$

$=x^3+3x^2-6x+2$

d. $(n-3)^2=n^2-6n+9$

e. $(3x+5)(3x-5)=9x^2-25$

81. $5x-4-2(6x+8)$

$=5x-4-12x-16$

$=-7x-20$

82. $-4y+7-3(2y-5)$

$=-4y+7-6y+15$

$=-10y+22$

83. $\dfrac{-6n^2+12n-15}{3n}$

$=\dfrac{-6n^{\not 2\,1}}{3\not n}+\dfrac{12\not n}{3\not n}-\dfrac{15}{3n}$

$=-2n+4-\dfrac{5}{n}$

84. $\dfrac{-4y^2+8y-16}{4y}$

$=\dfrac{-4y^{\not 2\,1}}{4\not y}+\dfrac{8\not y}{4\not y}-\dfrac{16}{4y}$

$=-y+2-\dfrac{4}{y}$

85.

$$\begin{array}{r} 2x-1 \\ x+4\overline{)2x^2+7x-4} \\ \underline{-2x^2-8x} \\ -x-4 \\ \underline{+x+4} \\ 0 \end{array}$$

$(2x^2+7x-4)\div(x+4)=2x-1$

86.

$$\begin{array}{r} 3x+1 \\ x-3\overline{)3x^2-8x-3} \\ \underline{-3x^2+9x} \\ x-3 \\ \underline{-x+3} \\ 0 \end{array}$$

$(3x^2-8x-3)\div(x-3)=3x+1$

87.

$$\begin{array}{r|rrrr} -4 & 3 & 28 & 52 & -48 \\ & & -12 & -64 & 48 \\ \hline & 3 & 16 & -12 & 0 \end{array}$$

$\left(3x^3+28x^2+52x-48\right)\div\left(x+4\right)=3x^2+16-12$

88.

$$\begin{array}{r|rrrr} 3 & 4 & 15 & -46 & -86 \\ & & 12 & 81 & 105 \\ \hline & 4 & 27 & 35 & 19 \end{array}$$

$\left(4x^3+15x^2-46x-86\right)\div\left(x-3\right)$

$=4x^2+27x+35+\dfrac{19}{x-3}$

89. **a.** $\text{GCF}=-3ab$

$-12a^2b+3ab^2-30ab=-3ab(4a-b+10)$

b. $\text{GCF}=(n+3)$

$2n(n+3)-1(n+3)=(n+3)(2n-1)$

90. **a.** $\text{GCF}=16x^2y^2$

$16x^3y^3+48x^2y^2=16x^2y^2(xy+3)$

b. $\text{GCF}=(m-4)$

$-3m(m-4)+2(m-4)=(m-4)(-3m+2)$

91. $3ab-12a+b-4$

$=(3ab-12a)+(b-4)$

$=3a(b-4)+1(b-4)$

$=(b-4)(3a+1)$

92. $2xy+12x-5y-30$

$=(2xy+12x)-(5y+30)$

$=2x(y+6)-5(y+6)$

$=(y+6)(2x-5)$

93. $3x^2-15x+3x-15$

$=3x^2-12x-15$

$=3(x^2-4x-5)$

Factors of -5 that add to -4 are -5 and 1.

$3(x^2-4x-5)=3(x-5)(x+1)$

94. $2y^2-4y-14y+28$

$=2y^2-18y+28$

$=2(y^2-9y+14)$

Factors of 14 that add to -9 are -7 and -2.

$2(y^2-9y+14)=2(y-7)(y-2)$

95. **a.** Factors of -30 that add to 1 are 6 and -5.

$x^2+x-30=(x+6)(x-5)$

b. $-x^2-x+6=-(x^2+x-6)$

Factors of -6 that add to 1 are 3 and -2.

$-(x^2+x-6)=-(x+3)(x-2)$

c. $a\cdot c=4\cdot -3=-12$ Factors of -12 that add to 4 are 6 and -2. Factor by grouping.

$$\begin{aligned} 4x^2+4x-3 &= 4x^2+6x-2x-3 \\ &= (4x^2+6x)-(2x+3) \\ &= 2x(2x+3)-1(2x+3) \\ &= (2x+3)(2x-1) \end{aligned}$$

96. **a.** Factors of -36 that add to -5 are -9 and 4.

$x^2-5x-36=(x-9)(x+4)$

b. $-x^2+x+20=-(x^2-x-20)$

Factors of -20 that add to -1 are -5 and 4.

$-(x^2-x-20)=-(x-5)(x+4)$

c. $a\cdot c=3\cdot -3=-9$ There are no factors of -9 that add to -10. $3x^2-10x-3$ is prime.

97. **a.** $2n^3-6n^2-20n=2n(n^2-3n-10)$

Factors of -10 that add to -3 are -5 and 2.

$2n(n^2-3n-10)=2n(n-5)(n+2)$

b. $10x^2+5xy+30xy+15y^2$

$=5(2x^2+xy+6xy+3y^2)$

$=5\left[(2x^2+xy)+(6xy+3y^2)\right]$

$=5[x(2x+y)+3y(2x+y)]$

$=5(2x+y)(x+3y)$

98. **a.** $3t^3-45t^2+162t=3t(t^2-15t+54)$

Factors of 54 that add to -15 are -9 and -6.

$3t(t^2-15t+54)=3t(t-9)(t-6)$

b. $6x^2+3xy-42xy-21y^2$

$=3(2x^2+xy-14xy-7y^2)$

$=3\left[(2x^2+xy)-(14xy+7y^2)\right]$

$=3[x(2x+y)-7y(2x+y)]$

$=3(2x+y)(x-7y)$

99. **a.** Using $a^2-b^2=(a-b)(a+b)$.

$t^2-49=(t-7)(t+7)$

b. Using $a^2-b^2=(a-b)(a+b)$.

$9-4y^2=(3-2y)(3+2y)$

c. Factor using $a^2+2ab+b^2=(a+b)^2$.

$4x^2+20x+25=(2x+5)^2$

d. a^2+36 is prime.

e. $5x^2-45=5(x^2-9)$

Using $a^2-b^2=(a-b)(a+b)$.

$5(x^2-9)=5(x-3)(x+3)$

100. **a.** Using $a^2-b^2=(a-b)(a+b)$.

$y^2-36=(y-6)(y+6)$

b. Using $a^2-b^2=(a-b)(a+b)$.

$1-25n^2=(1-5n)(1+5n)$

c. Factor using $a^2-2ab+b^2=(a-b)^2$.

$9y^2-12y+4=(3y-2)^2$

d. b^2+16 is prime.

e. $3y^2-48=3(y^2-16)$

Using $a^2-b^2=(a-b)(a+b)$.

$3(y^2-16)=3(y-4)(y+4)$

101. **a.**

$125x^3-64=\left((5x)^3-4^3\right)$

$=(5x-4)\left((5x)^2+4\cdot 5x+4^2\right)$

$=(5x-4)\left(25x^2+20x+16\right)$

b.

$8m^3+125=\left((2m)^3+5^3\right)$

$=(2m+5)\left((2m)^2-5\cdot 2m+5^2\right)$

$=(2m+5)\left(4m^2-10m+25\right)$

102. **a.** Let $u=x^2$.

$20x^4+23x^2-21=20u^2+23u-21$

$=20u^2+35u-12u-21$

$=5u(4u+7)-3(4u+7)$

$=(4u+7)(5u-3)$

Replace u with x^2

$20x^4+23x^2-21=\left(4x^2+7\right)\left(5x^2-3\right)$

b. $a^6-729b^6=\left(a^3\right)^2-\left(27b^3\right)^2$

$=\left(a^3+27b^3\right)\left(a^3-27b^3\right)$

$=\left(a^3+(3b)^3\right)\left(a^3-27b^3\right)$

$=(a+3b)\left(a^2-3b\cdot a+(3b)^2\right)\left(a^3-27b^3\right)$

$=(a+3b)\left(a^2-3ab+9b^2\right)\left(a^3-27b^3\right)$

$=(a+3b)\left(a^2-3ab+9b^2\right)\left(a^3-(3b)^3\right)$

$=(a+3b)\left(a^2-3ab+9b^2\right)(a-3b)\left(a^2+3b\cdot a+(3b)^2\right)$

$=(a+3b)\left(a^2-3ab+9b^2\right)(a-3b)\left(a^2+3ab+9b^2\right)$

103. **a.** This is an equation of degree 2 (the variable has an exponent of 2), therefore it is quadratic.

b. This is an equation of degree 3 (it has a variable with an exponent of 3), therefore it is neither linear nor quadratic.

c. This is an equation of degree 1, therefore it is linear.

104. **a.** This is an equation of degree 1, therefore it is linear.

b. This is an equation of degree 2 (the variable has an exponent of 2), therefore it is quadratic.

c. This is an equation of degree 4 (it has a variable with an exponent of 4), therefore it is neither linear nor quadratic.

105. **a.** Factors of 32 that add to 12 are 8 and 4.

$$x^2+12x+32=0$$
$$(x+8)(x+4)=0$$

$$\begin{array}{ll} x+8=0 & x+4=0 \\ \underline{-8\ \ -8} & \underline{-4\ \ -4} \\ x=-8 & x=-4 \end{array}$$

$x=-8$ and $x=-4$

b. $a\cdot c=2\cdot -15=-30$ Factors of -30 that add to 7 are 10 and -3. Factor by grouping.

$$2x^2+7x-15=0$$
$$2x^2+10x-3x-15=0$$
$$(2x^2+10x)-(3x+15)=0$$
$$2x(x+5)-3(x+5)=0$$
$$(x+5)(2x-3)=0$$

$$\begin{array}{ll} x+5=0 & 2x-3=0 \\ \underline{-5\ \ -5} & \underline{+3\ \ +3} \\ x=-5 & 2x=3 \\ & \frac{2x}{2}=\frac{3}{2} \\ & x=\frac{3}{2} \end{array}$$

$x=-5$ and $x=\frac{3}{2}$

c. $x(x-2)=35$

$$x^2-2x=35$$
$$\underline{-35\ \ -35}$$
$$x^2-2x-35=0$$

Factors of -35 that add to -2 are -7 and 5.

$$x^2-2x-35=0$$
$$(x-7)(x+5)=0$$

$$\begin{array}{ll} x-7=0 & x+5=0 \\ \underline{+7\ \ +7} & \underline{-5\ \ -5} \\ x=7 & x=-5 \end{array}$$

$x=7$ and $x=-5$

d. Using $a^2-b^2=(a-b)(a+b)$.

$$y^2-49=0$$
$$(y-7)(y+7)=0$$

$$\begin{array}{ll} y-7=0 & y+7=0 \\ \underline{+7\ \ +7} & \underline{-7\ \ -7} \\ y=7 & y=-7 \end{array}$$

$y=7$ and $y=-7$

106. **a.** Factors of 21 that add to 10 are 7 and 3.

$$x^2+10x+21=0$$
$$(x+7)(x+3)=0$$

$$\begin{array}{ll} x+7=0 & x+3=0 \\ \underline{-7\ \ -7} & \underline{-3\ \ -3} \\ x=-7 & x=-3 \end{array}$$

$x=-7$ and $x=-3$

b. $a\cdot c=4\cdot -6=-24$ Factors of -24 that add to -23 are -24 and 1. Factor by grouping.

$$4x^2-23x-6=0$$
$$4x^2-24x+x-6=0$$
$$(4x^2-24x)+(x-6)=0$$
$$4x(x-6)+1(x-6)=0$$
$$(x-6)(4x+1)=0$$

$$\begin{array}{ll} x-6=0 & 4x+1=0 \\ \underline{+6\ \ +6} & \underline{-1\ \ -1} \\ x=6 & 4x=-1 \\ & \frac{4x}{4}=\frac{-1}{4} \\ & x=-\frac{1}{4} \end{array}$$

$x=6$ and $x=-\frac{1}{4}$

c. $x(x-1)=6$

$$x^2-x=6$$
$$\underline{-6\ \ -6}$$
$$x^2-x-6=0$$

Factors of -6 that add to -1 are -3 and 2.

$$x^2-x-6=0$$
$$(x-3)(x+2)=0$$

$$\begin{array}{ll} x-3=0 & x+2=0 \\ \underline{+3\ \ +3} & \underline{-2\ \ -2} \\ x=3 & x=-2 \end{array}$$

$x=3$ and $x=-2$

d. Using $a^2-b^2=(a-b)(a+b)$.

$$n^2-25=0$$
$$(n-5)(n+5)=0$$

$$\begin{array}{ll} n-5=0 & n+5=0 \\ \underline{+5\ \ +5} & \underline{-5\ \ -5} \\ n=5 & n=-5 \end{array}$$

$n=5$ and $n=-5$

CHAPTER 7 RATIONAL EXPRESSIONS & EQUATIONS

Section 7.1 The Basics of Rational Expressions and Equations

1. $\dfrac{(12)+6}{(12)-3}$ Substitute in $x=12$.

$=\dfrac{18}{9}=2$ Simplify.

3. $\dfrac{2(10)+5}{3(10)+4}$ Substitute in $a=10$.

$=\dfrac{20+5}{30+4}$ Simplify.

$=\dfrac{25}{34}$

5. $\dfrac{(5)-1}{(5)^2+2(5)+5}$ Substitute in $x=5$.

$=\dfrac{4}{25+10+5}$ Simplify.

$=\dfrac{4}{40}=\dfrac{1}{10}$

7. $\dfrac{(3)^2+6}{(3)^2-5(3)+6}$ Substitute in $r=3$.

$=\dfrac{9+6}{9-15+6}$ Simplify.

$=\dfrac{15}{0}$ This expression is undefined.

9. $\dfrac{(-1)^2-6(-1)-7}{3(-1)+6}$ Substitute in $b=-1$.

$=\dfrac{1+6-7}{-3+6}$ Simplify.

$=\dfrac{0}{3}=0$

11. **a.** $c=\dfrac{250}{(10)}$ Substitute in $n=10$.

$c=25$

The cost per person is \$25 if 10 people agree to help pay for the gift.

b. $c=\dfrac{250}{(50)}$ Substitute in $n=50$.

$c=5$

The lowest possible cost per person is \$5 if there are 50 employees in the company.

13. **a.** $c=\dfrac{400+25(8)}{(8)}$ Substitute in $p=8$.

$c=\dfrac{400+200}{8}$

$c=\dfrac{600}{8}=75$

The cost per person is \$75 if 8 people go on the trip.

b. $c=\dfrac{400+25(25)}{(25)}$ Substitute in $p=25$.

$c=\dfrac{400+625}{25}$

$c=\dfrac{1025}{25}=41$

The lowest possible cost per person is \$41 if the trip is limited to 25 people.

15. **a.** $c=\dfrac{800+100(8)}{(8)}$ Substitute in $b=8$.

$c=\dfrac{800+800}{8}$

$c=\dfrac{1600}{8}=200$

The cost of each surfboard is \$200 if 8 surfboards are made in a week.

b. $c=\dfrac{800+100(10)}{(10)}$ Substitute in $b=10$.

$c=\dfrac{800+1000}{10}$

$c=\dfrac{1800}{10}=180$

The lowest cost per surfboard for a week is \$180 if 10 surfboards can be made in a week.

17. $s=\dfrac{2557(10)+50957}{0.37(10)+20.92}$ Substitute in $t=10$.

$s=\dfrac{25570+50957}{3.7+20.92}$

$s=\dfrac{76527}{24.62}\approx 3108.33$

Spending per person in 2010 by the state of Texas is estimated at \$3108.33.

19. Set denominator equal to 0 and solve:

$\dfrac{5x+6}{x}$

$x=0$ This is the excluded value.

21. Set denominator equal to 0 and solve:

$\frac{x+6}{x-3}$

$x-3=0$

$x=3$ This is the excluded value.

23. Set denominator equal to 0 and solve:

$\frac{2a+5}{3a+4}$

$3a+4=0$

$3a=-4$

$\frac{3a}{3}=\frac{-4}{3}$

$a=-\frac{4}{3}$ This is the excluded value.

25. Set denominator equal to 0 and solve:

$\frac{x^2+4}{7}$

$7\stackrel{?}{=}0$ There are no excluded values.

27. Set denominator equal to 0 and solve:

$\frac{x+5}{(x-3)(x+2)}$

$x-3=0$ $\quad$ $x+2=0$

$x=3$ $\quad$ $x=-2$

$x=3,-2$ These are the excluded values.

29. Set denominator equal to 0 and solve:

$\frac{25}{(h+4)(h-8)}$

$h+4=0$ $\quad$ $h-8=0$

$h=-4$ $\quad$ $h=8$

$h=-4, 8$ These are the excluded values.

31. Set denominator equal to 0 and solve:

$\frac{r^2+6}{r^2-5r+6}$

$r^2-5r+6=0$

$(r-3)(r-2)=0$

$r-3=0$ $\quad$ $r-2=0$

$r=3$ $\quad$ $r=2$

$r=3, 2$ These are the excluded values.

33. Set denominator equal to 0 and solve:

$\frac{n-5}{n^2+4n+4}$

$n^2+4n+4=0$

$(n+2)^2=0$

$n+2=0$

$n=-2$

$n=-2$ This is the excluded value.

35. Set denominator equal to 0 and solve:

$\frac{b^2+5b-7}{3b+6}$

$3b+6=0$

$3(b+2)=0$

$3\stackrel{?}{=}0$ $\quad$ $b+2=0$

$b=-2$

$b=-2$ This is the excluded value.

37. $\frac{4\cancel{(x-3)}}{x\cancel{(x-3)}}$ Divide out common factor.

$=\frac{4}{x}$

39. $\frac{7n\cancel{(n+6)}}{\cancel{(n+6)}}$ Divide out common factor.

$=7n$

41. $\frac{\cancel{(a+8)}}{(a+5)\cancel{(a+8)}}$ Divide out common factor.

$=\frac{1}{a+5}$

43. $\frac{4x\cancel{(x+2)}}{(x-3)\cancel{(x+2)}}$ Divide out common factor.

$=\frac{4x}{x-3}$

45. $\frac{(h+6)\cancel{(h-8)}}{(h+4)\cancel{(h-8)}}$ Divide out common factor.

$=\frac{h+6}{h+4}$

47. $\frac{r-3}{r^2-9}$

$=\frac{\cancel{r-3}}{\cancel{(r-3)}(r+3)}$ Divide out common factor.

$=\frac{1}{r+3}$

49. **a.** $T = \dfrac{4(6)}{4+(6)}$ Substitute in $x = 6$.

$T = \dfrac{24}{10} = 2.4$

The two landscapers working together can complete the job in 2.4 hours.

b. $T = \dfrac{4(5.5)}{4+(5.5)}$ Substitute in $x = 5.5$.

$T = \dfrac{22}{9.5} \approx 2.32$

The two landscapers working together can complete the job in approximately 2.32 hours.

51. **a.** $T = \dfrac{5(8)}{5+(8)}$ Substitute in $x = 8$.

$T = \dfrac{40}{13} \approx 3.08$

The two staff members working together can complete the job in approximately 3.08 hours.

b. $T = \dfrac{5(6.5)}{5+(6.5)}$ Substitute in $x = 6.5$.

$T = \dfrac{32.5}{11.5} \approx 2.83$

The two staff members working together can complete the job in 2.83 hours.

53. **a.** $y = \dfrac{30(70)}{(70)-30}$ Substitute in $x = 70$.

$y = \dfrac{2100}{40} = 52.5$

Jack's speed on his return trip is 52.5 mph if he drives at 70 mph to the city.

b. $y = \dfrac{30(55)}{(55)-30}$ Substitute in $x = 55$.

$y = \dfrac{1650}{25} = 66$

Jack's speed on his return trip is 66 mph if he drives at 55 mph to the city.

55. **a.** $y = \dfrac{225(500)}{(500)-225}$ Substitute in $x = 500$.

$y = \dfrac{112{,}500}{275} \approx 409.09$

The pilot's speed on the return trip is approximately 409.09 mph if she flies at 500 mph to the city.

b. $y = \dfrac{225(425)}{(425)-225}$ Substitute in $x = 425$.

$y = \dfrac{95{,}625}{200} = 478.125$

The pilot's speed on her return trip is 478.125 mph if she flies at 425 mph to the city.

57. The student divided out common terms, but only common *factors* can be divided.

Correctly simplified: $\dfrac{x+2}{x+4}$

59. The student divided out common terms, but only common *factors* can be divided.

Correctly simplified: $\dfrac{x}{x+7}$

61. The student divided out common terms, but only common *factors* can be divided.

Correctly simplified: $\dfrac{x^2-4}{x-2} = \dfrac{\cancel{(x-2)}(x+2)}{\cancel{x-2}} = x+2$

63. $\dfrac{n+2}{n^2+4n+4}$

$= \dfrac{\cancel{n+2}}{(n+2)\cancel{(n+2)}}$ Divide out common factor.

$= \dfrac{1}{n+2}$

65. $\dfrac{2b+8}{b^2+9b+20}$

$= \dfrac{2\cancel{(b+4)}}{(b+5)\cancel{(b+4)}}$ Divide out common factor.

$= \dfrac{2}{b+5}$

67. $\dfrac{8+(x+7)}{(x+2)(x+7)}$

$= \dfrac{x+15}{(x+2)(x+7)}$ No common factors.

69. $\dfrac{5+(a-3)}{(a+2)(a-3)}$

$= \dfrac{\cancel{a+2}}{\cancel{(a+2)}(a-3)}$ Divide out common factor.

$= \dfrac{1}{a-3}$

71. $\frac{(r+2)(r-9)}{r^2-5r-14}$

$=\frac{(r+2)(r-9)}{(r-7)(r+2)}$ Divide out common factor.

$=\frac{r-9}{r-7}$

73. $\frac{y^2+11y-12}{(y-1)(y-11)}$

$=\frac{(y+12)(y-1)}{(y-1)(y-11)}$ Divide out common factor.

$=\frac{y+12}{y-11}$

75. $1-x=-1(-1+x)$

$=-(x-1)$

77. $4-y=-1(-4+y)$

$=-(y-4)$

79. $5-2x-3x^2=-1(-5+2x+3x^2)$

$=-(3x^2+2x-5)$

81. $3x-8-2x^2=-1(-3x+8+2x^2)$

$=-(2x^2-3x+8)$

83. $4-8x^2=-1(-4+8x^2)$

$=-(8x^2-4)$

85. $\frac{15-3x}{x-5}=\frac{-3x+15}{x-5}$

$=\frac{-3(x-5)}{x-5}=-3$

87. $\frac{y-1}{1-y}=\frac{y-1}{-y+1}$

$=\frac{y-1}{-(y-1)}=\frac{1}{-1}=-1$

89. $\frac{4t+20}{3(t+5)}=\frac{4(t+5)}{3(t+5)}=\frac{4}{3}$

91. $\frac{w^2-49}{(w+3)(7-w)}=\frac{(w-7)(w+7)}{(-w+7)(w+3)}$

$=\frac{(w-7)(w+7)}{-(w-7)(w+3)}$

$=-\frac{w+7}{w+3}$

93. $\frac{a+5}{(a+5)(a+4)}=\frac{1}{a+4}$

95. $\frac{5-x}{x^2-10x+25}=\frac{-x+5}{(x-5)(x-5)}$

$=\frac{-(x-5)}{(x-5)(x-5)}$

$=\frac{-1}{x-5}$

97. $\frac{m^2+5m-6}{3m+18}=\frac{(m+6)(m-1)}{3(m+6)}=\frac{m-1}{3}$

99. $\frac{12-3t}{t^2-10t+24}=\frac{-3t+12}{(t-6)(t-4)}$

$=\frac{-3(t-4)}{(t-6)(t-4)}$

$=\frac{-3}{t-6}$

Section 7.2 Multiplication and Division of Rational Expressions

1. $\frac{12}{25}\cdot\frac{40}{9}=\frac{2\cdot2\cdot\cancel{3}}{5\cdot\cancel{5}}\cdot\frac{2\cdot2\cdot2\cdot\cancel{5}}{\cancel{3}\cdot3}=\frac{32}{15}$

3. $\frac{-120}{230}\cdot\frac{45}{150}=-\frac{\cancel{2}\cdot\cancel{2}\cdot2\cdot\cancel{3}\cdot\cancel{5}}{\cancel{2}\cdot\cancel{5}\cdot23}\cdot\frac{3\cdot3\cdot\cancel{5}}{\cancel{2}\cdot\cancel{3}\cdot\cancel{5}\cdot5}=-\frac{18}{115}$

5. $\frac{3x}{4y^2}\cdot\frac{6y}{15x}=\frac{\cancel{3}\cdot\cancel{x}}{\cancel{2}\cdot2\cdot y\cdot\cancel{y}}\cdot\frac{\cancel{2}\cdot3\cdot\cancel{y}}{\cancel{3}\cdot5\cdot\cancel{x}}=\frac{3}{10y}$

7. $\frac{-8mn^2}{25m^2}\cdot\frac{5mn}{4n^2}$

$=-\frac{\cancel{2}\cdot\cancel{2}\cdot2\cdot\cancel{m}\cdot\cancel{n}\cdot\cancel{n}}{\cancel{5}\cdot5\cdot\cancel{m}\cdot\cancel{m}}\cdot\frac{\cancel{5}\cdot\cancel{m}\cdot n}{\cancel{2}\cdot\cancel{2}\cdot\cancel{n}\cdot\cancel{n}}=-\frac{2n}{5}$

9. $\frac{(x+5)}{\cancel{(x-8)}}\cdot\frac{\cancel{(x-8)}}{(x+2)}=\frac{x+5}{x+2}$

11. $\frac{(a+7)}{(a+2)}\cdot\frac{(a+3)}{(a-5)}=\frac{(a+7)(a+3)}{(a+2)(a-5)}$

13. $\frac{(x+2)\cancel{(x-3)}}{\cancel{(x-3)}\cancel{(x+5)}}\cdot\frac{\cancel{(x+5)}}{(x-4)}=\frac{x+2}{x-4}$

15. $\frac{\cancel{(m+2)}(m-5)}{\cancel{(m+3)}\cancel{(m+2)}}\cdot\frac{\cancel{(m+3)}}{(m+9)}=\frac{m-5}{m+9}$

17. $\frac{\cancel{(h+5)}\cancel{(h+7)}}{\cancel{(h+4)}\cancel{(h+5)}}\cdot\frac{\cancel{(h+4)}(h+3)}{(h+2)\cancel{(h+7)}}=\frac{h+3}{h+2}$

19. $\frac{(b+6)}{(b+3)(b+4)}\cdot\frac{(b+4)}{b^2-36}$

$=\frac{\cancel{(b+6)}}{(b+3)\cancel{(b+4)}}\cdot\frac{\cancel{(b+4)}}{(b-6)\cancel{(b+6)}}=\frac{1}{(b+3)(b-6)}$

21. $\frac{(k+2)(k-5)}{(k+3)}\cdot\frac{(k+4)}{k^2+6k+8}$

$=\frac{\cancel{(k+2)}(k-5)}{(k+3)}\cdot\frac{\cancel{(k+4)}}{\cancel{(k+4)}\cancel{(k+2)}}=\frac{k-5}{k+3}$

23. $\frac{(h+10)}{(h-3)}\cdot\frac{(h+4)(h-1)}{h^2+9h-10}$

$=\frac{\cancel{(h+10)}}{(h-3)}\cdot\frac{(h+4)\cancel{(h-1)}}{\cancel{(h+10)}\cancel{(h-1)}}=\frac{h+4}{h-3}$

25. $\frac{x^2+7x+10}{x^2+8x+15}\cdot\frac{x^2+9x+18}{x^2+3x+2}$

$=\frac{\cancel{(x+5)}\cancel{(x+2)}}{\cancel{(x+5)}\cancel{(x+3)}}\cdot\frac{(x+6)\cancel{(x+3)}}{\cancel{(x+2)}(x+1)}=\frac{x+6}{x+1}$

27. $\frac{t^2-t-12}{t^2-11t+30}\cdot\frac{t^2-3t-10}{t^2+5t+6}$

$=\frac{(t-4)\cancel{(t+3)}}{(t-6)\cancel{(t-5)}}\cdot\frac{\cancel{(t-5)}\cancel{(t+2)}}{\cancel{(t+3)}\cancel{(t+2)}}=\frac{t-4}{t-6}$

29. $\frac{(x-2)}{20(x-3)}\cdot\frac{5x-15}{(x^2-4)}$

$=\frac{\cancel{(x-2)}}{2\cdot2\cdot\cancel{5}\cdot\cancel{(x-3)}}\cdot\frac{\cancel{5}\cancel{(x-3)}}{\cancel{(x-2)}(x+2)}=\frac{1}{4(x+2)}$

31. $\frac{2-x}{x^2-1}\cdot\frac{4x+4}{x^2-4}=\frac{-x+2}{(x-1)(x+1)}\cdot\frac{4(x+1)}{(x-2)(x+2)}$

$=\frac{-\cancel{(x-2)}}{(x-1)\cancel{(x+1)}}\cdot\frac{4\cancel{(x+1)}}{\cancel{(x-2)}(x+2)}=\frac{-4}{(x-1)(x+2)}$

33. $\frac{x^2-49}{x^2+2x-8}\cdot\frac{3x-6}{x^2+7x}$

$=\frac{(x-7)\cancel{(x+7)}}{(x+4)\cancel{(x-2)}}\cdot\frac{3\cancel{(x-2)}}{x\cancel{(x+7)}}=\frac{3(x-7)}{x(x+4)}$

35. $\frac{30\text{ mg}}{1\ \cancel{\text{kg}}}\cdot\frac{1\ \cancel{\text{kg}}}{2.2\text{ lb}}=\frac{30\text{ mg}}{2.2\text{ lb}}\approx 13.6\text{ mg/lb}$

The child's drug dosage is 13.6 mg/lb.

37. $\frac{45\text{ mg}}{1\ \cancel{\text{kg}}}\cdot\frac{1\ \cancel{\text{kg}}}{2.2\text{ lb}}=\frac{45\text{ mg}}{2.2\text{ lb}}\approx 20.5\text{ mg/lb}$

The adult's drug dosage is 20.5 mg/lb.

39. $\frac{\$1.60}{1\ \cancel{\text{lb}}}\cdot\frac{1\ \cancel{\text{lb}}}{16\text{ oz}}=\frac{\$1.60}{16\text{ oz}}=\$0.10\text{ per oz.}$

The price for the 5-pound bag of rice is $0.10 per ounce.

41. $\frac{\$3.85}{1\ \cancel{\text{lb}}}\cdot\frac{1\ \cancel{\text{lb}}}{16\text{ oz}}=\frac{\$3.85}{16\text{ oz}}\approx\$0.24\text{ per oz.}$

The price for the 1-pound jar of peanut butter is $0.24 per ounce.

43. $\frac{60\ \cancel{\text{mile}}}{1\text{ hr}}\cdot\frac{1.609\text{ km}}{1\ \cancel{\text{mile}}}=\frac{96.54\text{ km}}{1\text{ hr}}\approx 96.5\text{ km/hr}$

The car's speed is 96.5 kilometers per hour.

45. $\frac{50\ \cancel{\text{mile}}}{1\text{ hr}}\cdot\frac{5280\text{ ft}}{1\ \cancel{\text{mile}}}=\frac{264{,}000\text{ ft}}{1\text{ hr}}$

$\frac{264{,}000\text{ ft}}{1\ \cancel{\text{hr}}}\cdot\frac{1\ \cancel{\text{hr}}}{60\text{ min}}=\frac{264{,}000\text{ ft}}{60\text{ min}}=4400\text{ ft/min}$

The car's speed is 4400 feet per minute.

47. $\dfrac{4400\ \cancel{ft}}{1\ \text{min}}\cdot\dfrac{1\ \text{m}}{3.28\ \cancel{ft}}=\dfrac{4400\ \text{m}}{3.28\ \text{min}}$

$\dfrac{4400\ \text{m}}{3.28\ \cancel{\text{min}}}\cdot\dfrac{1\ \cancel{\text{min}}}{60\ \text{sec}}=\dfrac{4400\ \text{m}}{196.8\ \text{sec}}\approx 22.4\ \text{m/sec}$

The car's speed is 22.4 meters per second.

49. **a.** $\dfrac{150\ \mu g}{1\ \cancel{\text{min}}}\cdot\dfrac{60\ \cancel{\text{min}}}{1\ \text{hr}}=\dfrac{9000\ \mu g}{1\ \text{hr}}$

$\dfrac{9000\ \cancel{\mu g}}{1\ \text{hr}}\cdot\dfrac{1\ \text{mg}}{1000\ \cancel{\mu g}}=\dfrac{9000\ \text{mg}}{1000\ \text{hr}}=9\ \text{mg/hr}$

The dosage is 9 mg per hour.

b. $\dfrac{9\ \text{mg}}{1\ \cancel{\text{hr}}}\cdot 24\ \cancel{\text{hr}}=216\ \text{mg}$

The patient will receive 216 mg of dopamine in a 24 hour period.

51. **a.** $\dfrac{40\ \cancel{\text{km}}}{1\ \text{hr}}\cdot\dfrac{1\ \text{mile}}{1.609\ \cancel{\text{km}}}=\dfrac{40\ \text{mile}}{1.609\ \text{hr}}\approx 24.86\ \text{mi/hr}$

The cyclist's rate is 24.86 miles per hour.

b. $\dfrac{40\ \text{mile}}{1.609\ \cancel{\text{hr}}}\cdot 6\ \cancel{\text{hr}}\approx 149.16\ \text{miles}$

The cyclist can ride 149.16 miles during a 6-hour ride.

53. $\dfrac{8\ \text{kWh}}{100\ \cancel{\text{km}}}\cdot\dfrac{1.609\ \cancel{\text{km}}}{1\ \text{mile}}$

$=\dfrac{12.872\ \text{kWh}}{100\ \text{mile}}\approx 0.13\ \text{kWh/mile}$

$\dfrac{12.872\ \text{kWh}}{\cancel{100}\ \cancel{\text{mile}}}\cdot\overset{5}{\cancel{500}}\ \cancel{\text{mile}}=64.36\ \text{kWh}$

The Solartaxi will use about 64.36 kWh when traveling 500 miles.

55. $\dfrac{24}{15}\div\dfrac{20}{9}=\dfrac{24}{15}\cdot\dfrac{9}{20}=\dfrac{\cancel{2}\cdot\cancel{2}\cdot 2\cdot\cancel{3}}{\cancel{3}\cdot 5}\cdot\dfrac{3\cdot 3}{\cancel{2}\cdot\cancel{2}\cdot 5}=\dfrac{18}{25}$

57. $\dfrac{85}{22}\div\dfrac{30}{55}=\dfrac{85}{22}\cdot\dfrac{55}{30}=\dfrac{5\cdot 17}{2\cdot\cancel{11}}\cdot\dfrac{\cancel{5}\cdot\cancel{11}}{2\cdot 3\cdot\cancel{5}}=\dfrac{85}{12}$

59. $\dfrac{5x}{4y}\div\dfrac{55x^3}{16y}=\dfrac{5x}{4y}\cdot\dfrac{16y}{55x^3}$

$=\dfrac{\cancel{5}\cdot\cancel{x}}{\cancel{2}\cdot\cancel{2}\cdot\cancel{y}}\cdot\dfrac{\cancel{2}\cdot\cancel{2}\cdot 2\cdot 2\cdot\cancel{y}}{\cancel{5}\cdot 11\cdot\cancel{x}\cdot x\cdot x}=\dfrac{4}{11x^2}$

61. $\dfrac{-4m^3}{5n}\div\dfrac{12m}{25n^2}=\dfrac{-4m^3}{5n}\cdot\dfrac{25n^2}{12m}$

$=-\dfrac{\cancel{2}\cdot\cancel{2}\cdot\cancel{m}\cdot m\cdot m}{\cancel{5}\cdot\cancel{n}}\cdot\dfrac{\cancel{5}\cdot 5\cdot\cancel{n}\cdot n}{\cancel{2}\cdot\cancel{2}\cdot 3\cdot\cancel{m}}=-\dfrac{5m^2n}{3}$

63. $\dfrac{(x+5)}{(x-3)}\div\dfrac{(x+5)}{(x+4)}=\dfrac{\cancel{(x+5)}}{(x-3)}\cdot\dfrac{(x+4)}{\cancel{(x+5)}}=\dfrac{x+4}{x-3}$

65. $\dfrac{(a+3)}{(a+4)}\div\dfrac{(a+1)}{(a+3)}=\dfrac{(a+3)}{(a+4)}\cdot\dfrac{(a+3)}{(a+1)}=\dfrac{(a+3)^2}{(a+4)(a+1)}$

67. $\dfrac{(x+2)(x-3)}{(x-3)(x+5)}\div\dfrac{(x+2)}{(x-7)}$

$=\dfrac{\cancel{(x+2)}\cancel{(x-3)}}{\cancel{(x-3)}(x+5)}\cdot\dfrac{(x-7)}{\cancel{(x+2)}}=\dfrac{x-7}{x+5}$

69. $\dfrac{(n+2)(n-5)}{(n+8)(n+2)}\div\dfrac{(n+3)}{(n+8)}$

$=\dfrac{\cancel{(n+2)}(n-5)}{\cancel{(n+8)}\cancel{(n+2)}}\cdot\dfrac{\cancel{(n+8)}}{(n+3)}=\dfrac{n-5}{n+3}$

71. $\dfrac{(h+8)(h+9)}{(h+4)(h+7)}\div\dfrac{(h+2)(h+8)}{(h+4)(h+9)}$

$=\dfrac{\cancel{(h+8)}(h+9)}{\cancel{(h+4)}(h+7)}\cdot\dfrac{\cancel{(h+4)}(h+9)}{(h+2)\cancel{(h+8)}}=\dfrac{(h+9)^2}{(h+7)(h+2)}$

73. $\dfrac{(b+7)}{(b-7)(b+5)}\div\dfrac{(b+4)}{b^2-49}$

$=\dfrac{(b+7)}{\cancel{(b-7)}(b+5)}\cdot\dfrac{\cancel{(b-7)}(b+7)}{(b+4)}=\dfrac{(b+7)^2}{(b+5)(b+4)}$

75. $\dfrac{k^2+8k-9}{k^2-3k-28}\div\dfrac{k^2+5k-6}{k^2+6k+8}$

$=\dfrac{(k+9)\cancel{(k-1)}}{(k-7)\cancel{(k+4)}}\cdot\dfrac{\cancel{(k+4)}(k+2)}{(k+6)\cancel{(k-1)}}=\dfrac{(k+9)(k+2)}{(k-7)(k+6)}$

77. $\dfrac{\frac{4x}{9y^2}}{\frac{-6x^2}{27y}}=\dfrac{4x}{9y^2}\div\dfrac{-6x^2}{27y}=\dfrac{4x}{9y^2}\cdot\dfrac{27y}{-6x^2}$

$=\dfrac{\cancel{2}\cdot 2\cdot\cancel{x}}{\cancel{3}\cdot\cancel{3}\cdot\cancel{y}\cdot y}\cdot-\dfrac{\cancel{3}\cdot\cancel{3}\cdot\cancel{3}\cdot\cancel{y}}{\cancel{2}\cdot\cancel{3}\cdot\cancel{x}\cdot x}=-\dfrac{2}{xy}$

79. $\dfrac{\frac{5x-15}{6x}}{\frac{2x-6}{12x}}=\dfrac{5x-15}{6x}\div\dfrac{2x-6}{12x}=\dfrac{5x-15}{6x}\cdot\dfrac{12x}{2x-6}$

$=\dfrac{5\cancel{(x-3)}}{\cancel{2}\cdot\cancel{3}\cdot\cancel{x}}\cdot\dfrac{\cancel{2}\cdot\cancel{2}\cdot\cancel{3}\cdot\cancel{x}}{\cancel{2}\cancel{(x-3)}}=5$

81. $\frac{\frac{x+5}{x^2-9}}{\frac{5x+25}{x+3}} = \frac{x+5}{x^2-9} \div \frac{5x+25}{x+3} = \frac{x+5}{x^2-9} \cdot \frac{x+3}{5x+25}$

$= \frac{\cancel{x+5}}{(x-3)\cancel{(x+3)}} \cdot \frac{\cancel{x+3}}{5\cancel{(x+5)}} = \frac{1}{5(x-3)}$

83. $\frac{-9n^3}{5m} \cdot \frac{25m^3}{6n} = -\frac{3\cdot\cancel{3}\cdot n^{\cancel{3}2}}{\cancel{5}\cancel{m}} \cdot \frac{\cancel{5}\cdot 5\cdot m^{\cancel{3}2}}{2\cdot\cancel{3}\cdot\cancel{n}} = -\frac{15m^2n^2}{2}$

85. $\frac{(x+4)}{(x-8)} \cdot \frac{(x+7)}{2x+8} = \frac{\cancel{(x+4)}}{(x-8)} \cdot \frac{(x+7)}{2\cancel{(x+4)}} = \frac{x+7}{2(x-8)}$

87. $\frac{x^2-9}{12} \cdot \frac{28}{3-x} = \frac{\cancel{(x-3)}(x+3)}{\cancel{2}\cdot\cancel{2}\cdot 3} \cdot \frac{\cancel{2}\cdot\cancel{2}\cdot 7}{-\cancel{(x-3)}} = -\frac{7(x+3)}{3}$

89. $\frac{-9a}{5b} \div \frac{21a^2}{20b^2} = \frac{-9a}{5b} \cdot \frac{20b^2}{21a^2}$

$= -\frac{3\cdot\cancel{3}\cdot\cancel{a}}{\cancel{5}\cancel{b}} \cdot \frac{2\cdot 2\cdot\cancel{5}\cdot b^{\cancel{2}1}}{\cancel{3}\cdot 7\cdot a^{\cancel{2}1}} = -\frac{12b}{7a}$

91. $\frac{(z+2)}{(z-6)} \div \frac{(z+5)}{(z-6)} = \frac{(z+2)}{\cancel{(z-6)}} \cdot \frac{\cancel{(z-6)}}{(z+5)} = \frac{z+2}{z+5}$

93. $\frac{5x+10}{(x-3)} \div \frac{(x+2)}{6x-18} = \frac{5\cancel{(x+2)}}{\cancel{(x-3)}} \cdot \frac{6\cancel{(x-3)}}{\cancel{(x+2)}} = 30$

95. $\frac{a^2-9}{(a-3)} \div \frac{(a+3)}{(a+2)} = \frac{\cancel{(a-3)}\cancel{(a+3)}}{\cancel{(a-3)}} \cdot \frac{(a+2)}{\cancel{(a+3)}} = a+2$

97. $\frac{a^2+6a+9}{a^2-9} \cdot \frac{a^2-5a-24}{a^2-64}$

$= \frac{(a+3)\cancel{(a+3)}}{(a-3)\cancel{(a+3)}} \cdot \frac{\cancel{(a-8)}(a+3)}{\cancel{(a-8)}(a+8)} = \frac{(a+3)^2}{(a-3)(a+8)}$

99. $\frac{b^2+10b+25}{b^2+8b+16} \div \frac{b^2-25}{b^2-16} = \frac{b^2+10b+25}{b^2+8b+16} \cdot \frac{b^2-16}{b^2-25}$

$= \frac{(b+5)\cancel{(b+5)}}{(b+4)\cancel{(b+4)}} \cdot \frac{(b-4)\cancel{(b+4)}}{(b-5)\cancel{(b+5)}} = \frac{(b+5)(b-4)}{(b+4)(b-5)}$

Section 7.3 Addition and Subtraction of Rational Expressions

1. $\frac{4}{7c^2} + \frac{6}{7c^2} = \frac{10}{7c^2}$

3. $\frac{16}{5a^2} - \frac{1}{5a^2} = \frac{15}{5a^2} = \frac{3\cdot\cancel{5}}{\cancel{5}a^2} = \frac{3}{a^2}$

5. $\frac{2x}{x+5} + \frac{3}{x+5} = \frac{2x+3}{x+5}$

7. $\frac{4h}{h-9} - \frac{3}{h-9} = \frac{4h-3}{h-9}$

9. $\frac{10}{x+2} + \frac{5x}{x+2} = \frac{5x+10}{x+2} = \frac{5\cancel{(x+2)}}{\cancel{x+2}} = 5$

11. $\frac{4a^2}{2a-1} - \frac{2a}{2a-1} = \frac{4a^2-2a}{2a-1} = \frac{2a\cancel{(2a-1)}}{\cancel{2a-1}} = 2a$

13. $\frac{h+3}{(h+4)(h+7)} + \frac{4}{(h+4)(h+7)}$

$= \frac{h+3+4}{(h+4)(h+7)} = \frac{\cancel{h+7}}{(h+4)\cancel{(h+7)}} = \frac{1}{h+4}$

15. $\frac{z+7}{(z+2)(z-3)} - \frac{5}{(z+2)(z-3)}$

$= \frac{z+7-5}{(z+2)(z-3)} = \frac{\cancel{z+2}}{\cancel{(z+2)}(z-3)} = \frac{1}{z-3}$

17. $\frac{3x+8}{(x+2)(2x+5)} - \frac{x+3}{(x+2)(2x+5)}$

$= \frac{3x+8-x-3}{(x+2)(2x+5)} = \frac{\cancel{2x+5}}{(x+2)\cancel{(2x+5)}} = \frac{1}{x+2}$

19. $\frac{7a-4}{(2a+3)(a-4)} - \frac{5a-8}{(2a+3)(a-4)}$

$= \frac{7a-4-5a+8}{(2a+3)(a-4)} = \frac{2a+4}{(2a+3)(a-4)}$

21. $\frac{3}{5^2\cdot 2}$ $\quad \frac{8}{3\cdot 5}$ $\quad$ LCD $= 5^2\cdot 2\cdot 3 = 150$

23. $\frac{2}{7xy^3}$ $\quad \frac{5}{yz^2}$ $\quad$ LCD $= 7\cdot x\cdot y^3\cdot z^2 = 7xy^3z^2$

25. $\frac{4h}{h-9}$ $\quad \frac{3}{h-6}$ $\quad$ LCD $= (h-9)(h-6)$

27. $\frac{5}{2(a-3)}$ $\quad \frac{2}{a-3}$ $\quad$ LCD $= 2(a-3)$

29. $\dfrac{5}{(h+4)(h+7)}$ $\quad \dfrac{4}{(h+4)(h+2)}$

LCD $=(h+4)(h+7)(h+2)$

31. $\dfrac{z+7}{(z+2)(z-3)}$ $\quad \dfrac{z+4}{(z-5)(z-3)}$

LCD $=(z+2)(z-3)(z-5)$

33. $\dfrac{5}{(m+10)(m-2)}$ $\quad \dfrac{3}{(m+5)(m-2)}$

LCD $=(m+10)(m-2)(m+5)$

35. $\dfrac{3}{2^3\cdot 7}$ $\quad \dfrac{4}{3\cdot 7}$ $\quad$ LCD $=2^3\cdot 3\cdot 7$

$$\frac{3}{2^3\cdot 7}\cdot\frac{3}{3} \qquad \frac{4}{3\cdot 7}\cdot\frac{2^3}{2^3}$$
$$\frac{9}{168} \qquad \frac{32}{168}$$

37. $\dfrac{7n}{2^2\cdot 5\cdot m^3}$ $\quad \dfrac{5}{2\cdot 3^2\cdot mn^2}$

LCD $=2^2\cdot 3^2\cdot 5\cdot m^3n^2$

$$\frac{7n}{2^2\cdot 5\cdot m^3}\cdot\frac{3^2\cdot n^2}{3^2\cdot n^2} \qquad \frac{5}{2\cdot 3^2\cdot mn^2}\cdot\frac{2\cdot 5\cdot m^2}{2\cdot 5\cdot m^2}$$
$$\frac{63n^3}{180m^3n^2} \qquad \frac{50m^2}{180m^3n^2}$$

39. $\dfrac{10}{x+4}$ $\quad \dfrac{5x}{x+2}$ $\quad$ LCD $=(x+4)(x+2)$

$$\frac{10}{x+4}\cdot\frac{x+2}{x+2} \qquad \frac{5x}{x+2}\cdot\frac{x+4}{x+4}$$
$$\frac{10x+20}{(x+4)(x+2)} \qquad \frac{5x^2+20x}{(x+4)(x+2)}$$

41. $\dfrac{5}{7(a-2)}$ $\quad \dfrac{9}{a-2}$ $\quad$ LCD $=7(a-2)$

$$\frac{5}{7(a-2)} \qquad \frac{9}{a-2}\cdot\frac{7}{7}$$
$$\frac{5}{7(a-2)} \qquad \frac{63}{7(a-2)}$$

43. $\dfrac{z+4}{(z+5)(z-2)}$ $\quad \dfrac{z+1}{(z-8)(z-2)}$

LCD $=(z+5)(z-2)(z-8)$

$$\frac{z+4}{(z+5)(z-2)}\cdot\frac{z-8}{z-8} \qquad \frac{z+1}{(z-8)(z-2)}\cdot\frac{z+5}{z+5}$$
$$\frac{z^2-4z-32}{(z+5)(z-2)(z-8)} \qquad \frac{z^2+6z+5}{(z+5)(z-2)(z-8)}$$

45. $\dfrac{2x}{(x-3)(x+3)}$ $\quad \dfrac{x}{(x+3)(x+2)}$

LCD $=(x-3)(x+3)(x+2)$

$$\frac{2x}{(x-3)(x+3)}\cdot\frac{x+2}{x+2} \qquad \frac{x}{(x+3)(x+2)}\cdot\frac{x-3}{x-3}$$
$$\frac{2x^2+4x}{(x-3)(x+3)(x+2)} \qquad \frac{x^2-3x}{(x-3)(x+3)(x+2)}$$

47. $\dfrac{15}{(m+8)(m-2)}$ $\quad \dfrac{3}{(m+8)(m-1)}$

LCD $=(m+8)(m-2)(m-1)$

$$\frac{15}{(m+8)(m-2)}\cdot\frac{m-1}{m-1} \qquad \frac{3}{(m+8)(m-1)}\cdot\frac{m-2}{m-2}$$
$$\frac{15m-15}{(m+8)(m-2)(m-1)} \qquad \frac{3m-6}{(m+8)(m-2)(m-1)}$$

49. $\dfrac{8}{3\cdot 3\cdot 5}+\dfrac{6}{3\cdot 5}$ $\quad$ LCD $=3^2\cdot 5=45$

$$=\frac{8}{3\cdot 3\cdot 5}+\frac{6}{3\cdot 5}\cdot\frac{3}{3}$$
$$=\frac{8}{45}+\frac{18}{45}=\frac{8+18}{45}=\frac{26}{45}$$

51. $\dfrac{5}{2\cdot 3\cdot xy}+\dfrac{8}{yz^2}$ $\quad$ LCD $=2\cdot 3\cdot x\cdot y\cdot z^2=6xyz^2$

$$=\frac{5}{2\cdot 3\cdot xy}\cdot\frac{z^2}{z^2}+\frac{8}{yz^2}\cdot\frac{6x}{6x}$$
$$=\frac{5z^2}{6xyz^2}+\frac{48x}{6xyz^2}=\frac{5z^2+48x}{6xyz^2}=\frac{48x+5z^2}{6xyz^2}$$

53. $\dfrac{4n}{3\cdot 3\cdot m^3}-\dfrac{\not{5}}{3\cdot\not{5}\cdot mn^2}$

$$=\frac{4n}{3\cdot 3\cdot m^3}-\frac{1}{3\cdot mn^2} \qquad \text{LCD}=3^2\cdot m^3\cdot n^2=9m^3n^2$$
$$=\frac{4n}{3\cdot 3\cdot m^3}\cdot\frac{n^2}{n^2}-\frac{1}{3\cdot mn^2}\cdot\frac{3m^2}{3m^2}$$
$$=\frac{4n^3}{9m^3n^2}-\frac{3m^2}{9m^3n^2}=\frac{4n^3-3m^2}{9m^3n^2}=\frac{-3m^2+4n^3}{9m^3n^2}$$

55. $\dfrac{4h}{h-9}+\dfrac{3}{h-6}$ $\quad$ LCD $=(h-9)(h-6)$

$$=\frac{4h}{h-9}\cdot\frac{h-6}{h-6}+\frac{3}{h-6}\cdot\frac{h-9}{h-9}$$
$$=\frac{4h^2-24h}{(h-9)(h-6)}+\frac{3h-27}{(h-9)(h-6)}$$
$$=\frac{4h^2-24h+3h-27}{(h-9)(h-6)}=\frac{4h^2-21h-27}{(h-9)(h-6)}$$

57. $\dfrac{4a-14}{a-5}+\dfrac{6}{-(a-5)}=\dfrac{4a-14}{a-5}+\dfrac{-6}{a-5}$

$=\dfrac{4a-14-6}{a-5}=\dfrac{4a-20}{a-5}=\dfrac{4\cancel{(a-5)}}{\cancel{a-5}}=4$

59. $\dfrac{3b-1}{b-7}+\dfrac{6}{-(b-7)}=\dfrac{3b-1}{b-7}+\dfrac{-6}{b-7}$

$=\dfrac{3b-1-6}{b-7}=\dfrac{3b-7}{b-7}$

61. $\dfrac{5}{4(a-2)}+\dfrac{3}{a-2}$ LCD $=4(a-2)$

$=\dfrac{5}{4(a-2)}+\dfrac{3}{a-2}\cdot\dfrac{4}{4}=\dfrac{5}{4(a-2)}+\dfrac{12}{4(a-2)}$

$=\dfrac{5+12}{4(a-2)}=\dfrac{17}{4(a-2)}$

63. $\dfrac{x}{x+4}-\dfrac{x-3}{(x+5)(x+4)}$ LCD $=(x+5)(x+4)$

$=\dfrac{x}{x+4}\cdot\dfrac{x+5}{x+5}-\dfrac{x-3}{(x+5)(x+4)}$

$=\dfrac{x^2+5x}{(x+5)(x+4)}-\dfrac{x-3}{(x+5)(x+4)}$

$=\dfrac{x^2+5x-x+3}{(x+5)(x+4)}=\dfrac{x^2+4x+3}{(x+5)(x+4)}$

65. $\dfrac{m}{m+2}-\dfrac{2m-15}{(m+10)(m+2)}$ LCD $=(m+10)(m+2)$

$=\dfrac{m}{m+2}\cdot\dfrac{m+10}{m+10}-\dfrac{2m-15}{(m+10)(m+2)}$

$=\dfrac{m^2+10m}{(m+10)(m+2)}-\dfrac{2m-15}{(m+10)(m+2)}$

$=\dfrac{m^2+10m-2m+15}{(m+10)(m+2)}=\dfrac{m^2+8m+15}{(m+10)(m+2)}$

67. $\dfrac{2x}{(x-3)(x+3)}+\dfrac{x}{(x+3)(x+2)}$

LCD $=(x-3)(x+3)(x+2)$

$=\dfrac{2x}{(x-3)(x+3)}\cdot\dfrac{x+2}{x+2}+\dfrac{x}{(x+3)(x+2)}\cdot\dfrac{x-3}{x-3}$

$=\dfrac{2x^2+4x}{(x-3)(x+3)(x+2)}+\dfrac{x^2-3x}{(x-3)(x+3)(x+2)}$

$=\dfrac{2x^2+4x+x^2-3x}{(x-3)(x+3)(x+2)}=\dfrac{3x^2+x}{(x-3)(x+3)(x+2)}$

69. $\dfrac{5}{(m+10)(m-2)}+\dfrac{3}{(m+5)(m-2)}$

LCD $=(m+10)(m+5)(m-2)$

$=\dfrac{5}{(m+10)(m-2)}\cdot\dfrac{m+5}{m+5}+\dfrac{3}{(m+5)(m-2)}\cdot\dfrac{m+10}{m+10}$

$=\dfrac{5m+25}{(m+10)(m+5)(m-2)}+\dfrac{3m+30}{(m+10)(m+5)(m-2)}$

$=\dfrac{5m+25+3m+30}{(m+10)(m+5)(m-2)}=\dfrac{8m+55}{(m+10)(m+5)(m-2)}$

71. **a.** $\dfrac{1}{a}+\dfrac{1}{b}$ LCD $=a\cdot b=ab$

$=\dfrac{1}{a}\cdot\dfrac{b}{b}+\dfrac{1}{b}\cdot\dfrac{a}{a}=\dfrac{b}{ab}+\dfrac{a}{ab}=\dfrac{a+b}{ab}$

b. $\dfrac{(3)+(2)}{(3)(2)}$ Substitute $a=3$ and $b=2$.

$=\dfrac{5}{6}$

The two brothers working together can mow $\dfrac{5}{6}$ of the lawn in one hour.

73. **a.** $\dfrac{1}{m}+\dfrac{1}{n}$ LCD $=m\cdot n=mn$

$=\dfrac{1}{m}\cdot\dfrac{n}{n}+\dfrac{1}{n}\cdot\dfrac{m}{m}=\dfrac{n}{mn}+\dfrac{m}{mn}=\dfrac{m+n}{mn}$

b. $\dfrac{(4)+(5)}{(4)(5)}$ Substitute $m=4$ and $n=5$.

$=\dfrac{9}{20}$

The two employees working together can file $\dfrac{9}{20}$ of the fall applications in one day.

75. $\dfrac{6}{x+3}+\dfrac{5}{x+1}$ LCD $=(x+3)(x+1)$

$=\dfrac{6}{x+3}\cdot\dfrac{x+1}{x+1}+\dfrac{5}{x+1}\cdot\dfrac{x+3}{x+3}$

$=\dfrac{6x+6}{(x+3)(x+1)}+\dfrac{5x+15}{(x+3)(x+1)}$

$=\dfrac{6x+6+5x+15}{(x+3)(x+1)}=\dfrac{11x+21}{(x+3)(x+1)}$

77. $\dfrac{4x}{x+3}\cdot\dfrac{8}{x+2}=\dfrac{4x\cdot 8}{(x+3)(x+2)}=\dfrac{32x}{(x+3)(x+2)}$

79. $\dfrac{n^2-9}{n^2+6n+9}\div\dfrac{2n-6}{n+5}=\dfrac{n^2-9}{n^2+6n+9}\cdot\dfrac{n+5}{2n-6}$

$=\dfrac{\cancel{(n-3)}\,\cancel{(n+3)}}{\cancel{(n+3)}(n+3)}\cdot\dfrac{n+5}{2\cancel{(n-3)}}=\dfrac{n+5}{2(n+3)}$

81. $\dfrac{8h}{h-7}+\dfrac{6}{h-4}$ LCD $=(h-7)(h-4)$

$=\dfrac{8h}{h-7}\cdot\dfrac{h-4}{h-4}+\dfrac{6}{h-4}\cdot\dfrac{h-7}{h-7}$

$$=\frac{8h^2-32h}{(h-7)(h-4)}+\frac{6h-42}{(h-7)(h-4)}$$
$$=\frac{8h^2-32h+6h-42}{(h-7)(h-4)}=\frac{8h^2-26h-42}{(h-7)(h-4)}$$

83. $\frac{7n-15}{n-3}+\frac{6}{-(n-3)}=\frac{7n-15}{n-3}+\frac{-6}{n-3}$

$$=\frac{7n-15-6}{n-3}=\frac{7n-21}{n-3}=\frac{7(n-3)}{n-3}=7$$

85. $\frac{c^2+3c+2}{c-1}\cdot\frac{5c+3}{c+2}=\frac{(c+2)(c+1)}{c-1}\cdot\frac{5c+3}{c+2}$

$$=\frac{(c+1)(5c+3)}{c-1}=\frac{5c^2+3c+5c+3}{c-1}=\frac{5c^2+8c+3}{c-1}$$

87. $\frac{\frac{2x+1}{x-5}}{\frac{4x+2}{3x-15}}=\frac{2x+1}{x-5}\div\frac{4x+2}{3x-15}=\frac{2x+1}{x-5}\cdot\frac{3(x-5)}{2(2x+1)}=\frac{3}{2}$

89. $\frac{\frac{a+2}{3b}}{\frac{a^2-4}{7b^2}}=\frac{a+2}{3b}\div\frac{a^2-4}{7b^2}$

$$=\frac{a+2}{3b}\cdot\frac{7b^2}{(a-2)(a+2)}=\frac{7b}{3(a-2)}$$

91. $\frac{\frac{3}{4}-\frac{1}{6}}{\frac{1}{12}+\frac{1}{3}}\cdot\frac{\frac{12}{1}}{\frac{12}{1}}$ Multiply by LCD = 12.

$$=\frac{\frac{3}{4}\cdot\frac{12}{1}-\frac{1}{6}\cdot\frac{12}{1}}{\frac{1}{12}\cdot\frac{12}{1}+\frac{1}{3}\cdot\frac{12}{1}}=\frac{9-2}{1+4}=\frac{7}{5}$$

93. $\frac{\frac{3}{a}-4}{\frac{5}{a}+7}\cdot\frac{\frac{a}{1}}{\frac{a}{1}}$ Multiply by LCD = a.

$$=\frac{\frac{3}{a}\cdot\frac{a}{1}-4\cdot\frac{a}{1}}{\frac{5}{a}\cdot\frac{a}{1}+7\cdot\frac{a}{1}}=\frac{3-4a}{5+7a}=\frac{-4a+3}{7a+5}$$

95. $\frac{\frac{4}{x}-16}{\frac{8}{x^2}-12}\cdot\frac{\frac{x^2}{1}}{\frac{x^2}{1}}$ Multiply by LCD = x^2.

$$=\frac{\frac{4}{x}\cdot\frac{x^2}{1}-16\cdot\frac{x^2}{1}}{\frac{8}{x^2}\cdot\frac{x^2}{1}-12\cdot\frac{x^2}{1}}=\frac{4x-16x^2}{8-12x^2}$$
$$=\frac{4x(4x-1)}{-4(3x^2-2)}=\frac{x(4x-1)}{3x^2-2}=\frac{4x^2-x}{3x^2-2}$$

97. $\frac{8p+\frac{3}{p}}{8p+\frac{3}{p}}\cdot\frac{\frac{p}{1}}{\frac{p}{1}}$ Multiply by LCD = p.

$$=\frac{8p\cdot\frac{p}{1}+\frac{3}{p}\cdot\frac{p}{1}}{8p\cdot\frac{p}{1}+\frac{3}{p}\cdot\frac{p}{1}}=\frac{8p^2+3}{8p^2+3}=1$$

99. $\frac{\frac{4}{x}+\frac{5}{y}}{\frac{6}{x}-\frac{7}{y}}\cdot\frac{\frac{xy}{1}}{\frac{xy}{1}}$ Multiply by LCD = xy.

$$=\frac{\frac{4}{x}\cdot\frac{xy}{1}+\frac{5}{y}\cdot\frac{xy}{1}}{\frac{6}{x}\cdot\frac{xy}{1}-\frac{7}{y}\cdot\frac{xy}{1}}=\frac{4y+5x}{6y-7x}=\frac{5x+4y}{-7x+6y}$$

101. a. $\frac{50(20)+400}{(20)+2}$ Substitute $n = 20$.

$$=\frac{1000+400}{22}=\frac{1400}{22}\approx 63.64$$

The average cost for food for 20 players (and 2 coaches) is \$63.64.

b. $\frac{75(20)+500}{(20)+2}$ Substitute $n = 20$.

$$=\frac{1500+500}{22}=\frac{2000}{22}\approx 90.91$$

The average cost for transportation and lodging is \$90.91 if 20 players go on the trip.

c. $\frac{50n+400}{n+2}+\frac{75n+500}{n+2}$

$$=\frac{50n+400+75n+500}{n+2}=\frac{125n+900}{n+2}$$

d. $\frac{125(25)+900}{(25)+2}$ Substitute $n = 25$.

$$=\frac{3125+900}{27}=\frac{4025}{27}\approx 149.07$$

The average cost for all expenses is \$149.07 if 25 players go on the trip.

Section 7.4 Solving Rational Equations

1. $$\frac{x}{1}\left(\frac{10}{x}+3\right)=\left(7-\frac{2}{x}\right)\frac{x}{1}$$

$$\frac{x}{1}\cdot\frac{10}{x}+\frac{x}{1}\cdot 3=\frac{x}{1}\cdot 7-\frac{x}{1}\cdot\frac{2}{x}$$

$$\begin{aligned}10+3x&=7x-2\\ \underline{-10}\quad&\quad\underline{-10}\\ 3x&=7x-12\\ \underline{-7x}\ &\ \underline{-7x}\\ -4x&=-12\\ \frac{-4x}{-4}&=\frac{-12}{-4}\\ x&=3\end{aligned}$$

3. $$\frac{n}{1}\left(4+\frac{16}{n}\right)=\left(20-\frac{112}{n}\right)\frac{n}{1}$$

$$\frac{n}{1}\cdot 4+\frac{n}{1}\cdot\frac{16}{n}=\frac{n}{1}\cdot 20-\frac{n}{1}\cdot\frac{112}{n}$$

$$\begin{aligned}4n+16&=20n-112\\ \underline{-16}\quad&\quad\underline{-16}\\ 4n&=20n-128\\ \underline{-20n}\ &\ \underline{-20n}\\ -16n&=-128\\ \frac{-16n}{-16}&=\frac{-128}{-16}\\ n&=8\end{aligned}$$

5. $$\frac{x}{1}\left(\frac{3}{x}+3\right)=\left(5-\frac{1}{x}\right)\frac{x}{1}$$

$$\frac{x}{1}\cdot\frac{3}{x}+\frac{x}{1}\cdot 3=\frac{x}{1}\cdot 5-\frac{x}{1}\cdot\frac{1}{x}$$

$$\begin{aligned}3+3x&=5x-1\\ \underline{-3}\quad&\quad\underline{-3}\\ 3x&=5x-4\\ \underline{-5x}\ &\ \underline{-5x}\\ -2x&=-4\\ \frac{-2x}{-2}&=\frac{-4}{-2}\\ x&=2\end{aligned}$$

7. $$\frac{3x}{1}\left(2-\frac{7}{3x}\right)=\left(1-\frac{4}{3x}\right)\frac{3x}{1}$$

$$\frac{3x}{1}\cdot 2-\frac{3x}{1}\cdot\frac{7}{3x}=\frac{3x}{1}\cdot 1-\frac{3x}{1}\cdot\frac{4}{3x}$$

$$\begin{aligned}6x-7&=3x-4\\ \underline{-3x}\quad&\quad\underline{-3x}\\ 3x-7&=-4\\ \underline{+7}\ &\ \underline{+7}\\ 3x&=3\\ \frac{3x}{3}&=\frac{3}{3}\\ x&=1\end{aligned}$$

9. $$\frac{5x}{1}\left(-1+\frac{8}{5x}\right)=\left(-\frac{7}{5x}+2\right)\frac{5x}{1}$$

$$\frac{5x}{1}\cdot -1+\frac{5x}{1}\cdot\frac{8}{5x}=\frac{5x}{1}\cdot -\frac{7}{5x}+\frac{5x}{1}\cdot 2$$

$$\begin{aligned}-5x+8&=-7+10x\\ \underline{-10x}\quad&\quad\underline{-10x}\\ -15x+8&=-7\\ \underline{-8}\ &\ \underline{-8}\\ -15x&=-15\\ \frac{-15x}{-15}&=\frac{-15}{-15}\\ x&=1\end{aligned}$$

11. $$\frac{x+3}{1}\left(\frac{4}{x+3}\right)=\left(2+\frac{10}{x+3}\right)\frac{x+3}{1}$$

$$\frac{x+3}{1}\cdot\frac{4}{x+3}=\frac{x+3}{1}\cdot 2+\frac{x+3}{1}\cdot\frac{10}{x+3}$$

$$\begin{aligned}4&=2(x+3)+10\\ 4&=2x+6+10\\ 4&=2x+16\\ \underline{-16}\ &\ \underline{-16}\\ -12&=2x\\ \frac{-12}{2}&=\frac{2x}{2}\\ -6&=x\end{aligned}$$

13. $$\frac{r+2}{1}\left(\frac{10}{r+2}\right)=\left(9-\frac{8}{r+2}\right)\frac{r+2}{1}$$

$$\frac{r+2}{1}\cdot\frac{10}{r+2}=\frac{r+2}{1}\cdot 9-\frac{r+2}{1}\cdot\frac{8}{r+2}$$

$$\begin{aligned}10&=9(r+2)-8\\ 10&=9r+18-8\\ 10&=9r+10\\ \underline{-10}\ &\ \underline{-10}\\ 0&=9r\\ \frac{0}{9}&=\frac{9r}{9}\\ 0&=r\end{aligned}$$

15. $$\frac{n+8}{1}\left(\frac{11}{n+8}+4\right)=(n+2)\frac{n+8}{1}$$

$$\frac{n+8}{1}\cdot\frac{11}{n+8}+\frac{n+8}{1}\cdot 4=\frac{n+8}{1}\cdot n+\frac{n+8}{1}\cdot 2$$

$$\begin{aligned}11+4(n+8)&=n(n+8)+2(n+8)\\ 11+4n+32&=n^2+8n+2n+16\\ 4n+43&=n^2+10n+16\\ \underline{-4n-43}\ &\ \underline{-4n-43}\\ 0&=n^2+6n-27\\ 0&=(n+9)(n-3)\end{aligned}$$

$$n+9=0 \qquad n-3=0$$
$$n=-9 \qquad n=3$$

17. $\frac{k+7}{1}\left(\frac{-2}{k+7}-8\right)=(k-4)\frac{k+7}{1}$

$$\frac{k+7}{1}\cdot\frac{-2}{k+7}-\frac{k+7}{1}\cdot 8=\frac{k+7}{1}\cdot k-\frac{k+7}{1}\cdot 4$$
$$-2-8(k+7)=k(k+7)-4(k+7)$$
$$-2-8k-56=k^2+7k-4k-28$$
$$-8k-58=k^2+3k-28$$
$$\underline{+8k+58 \quad +8k+58}$$
$$0=k^2+11k+30$$
$$0=(k+6)(k+5)$$

$$k+6=0 \qquad k+5=0$$
$$k=-6 \qquad k=-5$$

19. **a.** $C=\frac{2500}{(100)}$ Substitute $p=100$.

$C=25$

The cost per person is \$25 if 100 people attend the banquet.

b. $(6.25)=\frac{2500}{p}$ Substitute $C=6.25$.

$$\frac{p}{1}\cdot(6.25)=\frac{2500}{p}\cdot\frac{p}{1}$$
$$6.25p=2500$$
$$\frac{6.25p}{6.25}=\frac{2500}{6.25}$$
$$p=400$$

There must be 400 people attending the banquet for the cost per person to be \$6.25.

21. **a.** $c=\frac{150}{(10)}$ Substitute $n=10$.

$c=15$

The cost per person is \$15 if 10 people agree to help pay for the gift.

b. $(5)=\frac{150}{n}$ Substitute $c=5$.

$$\frac{n}{1}\cdot(5)=\frac{150}{n}\cdot\frac{n}{1}$$
$$5n=150$$
$$\frac{5n}{5}=\frac{150}{5}$$
$$n=30$$

There must be 30 people who agree to help pay for the gift for the cost per person to be \$5.

23. **a.** $c=\frac{500+25(8)}{(8)}$ Substitute $p=8$.

$$c=\frac{500+200}{8}=\frac{700}{8}=87.5$$

The cost per person is \$87.50 if 8 people go on the trip.

b. $(75)=\frac{500+25p}{p}$ Substitute $c=75$.

$$\frac{p}{1}\cdot(75)=\frac{500+25p}{p}\cdot\frac{p}{1}$$
$$75p=500+25p$$
$$\underline{-25p \qquad -25p}$$
$$50p=500$$
$$\frac{50p}{50}=\frac{500}{50}$$
$$p=10$$

If they want the cost per person to be \$75, they should recruit 10 people to go on the trip.

25. **a.** $c=\frac{1200+250(4)}{(4)}$ Substitute $n=4$.

$$c=\frac{1200+1000}{4}=\frac{2200}{4}=550$$

The cost for each large shutter is \$550 if 4 large shutters are made in a week.

b. $(400)=\frac{1200+250n}{n}$ Substitute $c=400$.

$$\frac{n}{1}\cdot(400)=\frac{1200+250n}{n}\cdot\frac{n}{1}$$
$$400n=1200+250n$$
$$\underline{-250n \qquad -250n}$$
$$150n=1200$$
$$\frac{150n}{150}=\frac{1200}{150}$$
$$n=8$$

For the cost per shutter to be \$400, there must be 8 large shutters made in a week.

27. $\frac{6r}{1}\left(\frac{7}{3}+\frac{8}{6r}\right)=(2)\frac{6r}{1}$

$$\frac{6r}{1}\cdot\frac{7}{3}+\frac{6r}{1}\cdot\frac{8}{6r}=\frac{6r}{1}\cdot 2$$
$$14r+8=12r$$
$$\underline{-14r \qquad -14r}$$
$$8=-2r$$
$$\frac{8}{-2}=\frac{-2r}{-2}$$
$$-4=r$$

29. $\frac{30g}{1}\left(\frac{2}{6g}\right)=\left(3+\frac{3}{10}\right)\frac{30g}{1}$

$$\frac{30g}{1}\cdot\frac{2}{6g}=\frac{30g}{1}\cdot 3+\frac{30g}{1}\cdot\frac{3}{10}$$
$$10=90g+9g$$
$$10=99g$$
$$\frac{10}{99}=\frac{99g}{99}$$
$$\frac{10}{99}=g$$

31.
$$\frac{x}{1}\left(6-\frac{3}{x}\right)=\left(7+\frac{9}{x}\right)\frac{x}{1}$$
$$\begin{aligned}
\frac{x}{1}\cdot 6-\frac{x}{1}\cdot\frac{3}{x}&=\frac{x}{1}\cdot 7+\frac{x}{1}\cdot\frac{9}{x}\\
6x-3&=7x+9\\
-7x\quad&\quad-7x\\
-x-3&=9\\
+3\quad&\quad+3\\
-x&=12\\
\frac{-x}{-1}&=\frac{12}{-1}\\
x&=-12
\end{aligned}$$

33.
$$\frac{n}{1}\left(\frac{3}{n}-3\right)=\left(2+\frac{3}{n}\right)\frac{n}{1}$$
$$\begin{aligned}
\frac{n}{1}\cdot\frac{3}{n}-\frac{n}{1}\cdot 3&=\frac{n}{1}\cdot 2+\frac{n}{1}\cdot\frac{3}{n}\\
3-3n&=2n+3\\
-2n\quad&\quad-2n\\
3-5n&=3\\
-3\quad&\quad-3\\
-5n&=0\\
\frac{-5n}{-5}&=\frac{0}{-5}\\
n&=0 \text{ This is an excluded value}
\end{aligned}$$

No solution

35.
$$\frac{\cancel{(x+5)}(x-4)}{1}\cdot\frac{10}{\cancel{x+5}}=\frac{4}{\cancel{x-4}}\cdot\frac{(x+5)\cancel{(x-4)}}{1}$$
$$\begin{aligned}
10(x-4)&=4(x+5)\\
10x-40&=4x+20\\
+40\quad&\quad+40\\
10x&=4x+60\\
-4x\quad&\quad-4x\\
6x&=60\\
\frac{6x}{6}&=\frac{60}{6}\\
x&=10
\end{aligned}$$

37.
$$\frac{\cancel{(p+7)}(p-5)}{1}\cdot\frac{3}{\cancel{p+7}}=\frac{-6}{\cancel{p-5}}\cdot\frac{(p+7)\cancel{(p-5)}}{1}$$
$$\begin{aligned}
3(p-5)&=-6(p+7)\\
3p-15&=-6p-42\\
+15\quad&\quad+15\\
3p&=-6p-27\\
+6p\quad&\quad+6p\\
9p&=-27\\
\frac{9p}{9}&=\frac{-27}{9}\\
p&=-3
\end{aligned}$$

39.
$$\frac{\cancel{(t-3)}(t+2)}{1}\cdot\frac{8}{\cancel{t-3}}=\frac{9}{\cancel{t+2}}\cdot\frac{(t-3)\cancel{(t+2)}}{1}$$
$$\begin{aligned}
8(t+2)&=9(t-3)\\
8t+16&=9t-27\\
-16\quad&\quad-16\\
8t&=9t-43\\
-9t\quad&\quad-9t\\
-t&=-43\\
\frac{-t}{-1}&=\frac{-43}{-1}\\
t&=43
\end{aligned}$$

41.
$$\frac{y+5}{1}\cdot\left(\frac{-3}{y+5}\right)=\left(6-\frac{3}{y+5}\right)\cdot\frac{y+5}{1}$$
$$\begin{aligned}
\frac{\cancel{y+5}}{1}\cdot\frac{-3}{\cancel{y+5}}&=\frac{y+5}{1}\cdot 6-\frac{\cancel{y+5}}{1}\cdot\frac{3}{\cancel{y+5}}\\
-3&=6(y+5)-3\\
-3&=6y+30-3\\
-3&=6y+27\\
-27\quad&\quad-27\\
-30&=6y\\
\frac{-30}{6}&=\frac{6y}{6}\\
-5&=y \text{ This is an excluded value}
\end{aligned}$$

No solution

43.
$$\frac{m-2}{1}\cdot\left(\frac{9}{m-2}\right)=\left(\frac{9}{m-2}-4\right)\cdot\frac{m-2}{1}$$
$$\begin{aligned}
\frac{\cancel{m-2}}{1}\cdot\frac{9}{\cancel{m-2}}&=\frac{\cancel{m-2}}{1}\cdot\frac{9}{\cancel{m-2}}-\frac{m-2}{1}\cdot 4\\
9&=9-4(m-2)\\
9&=9-4m+8\\
9&=-4m+17\\
-17\quad&\quad-17\\
8&=-4m\\
\frac{8}{-4}&=\frac{-4m}{-4}\\
-2&=m \text{ This is an excluded value}
\end{aligned}$$

No solution

45.
$$\frac{\cancel{(n+8)}(n-3)}{1}\cdot\frac{4}{\cancel{n+8}}=\frac{n+3}{\cancel{(n+8)}\cancel{(n-3)}}\cdot\frac{\cancel{(n+8)}\cancel{(n-3)}}{1}$$
$$\begin{aligned}
4(n-3)&=n+3\\
4n-12&=n+3\\
-n\quad&\quad-n\\
3n-12&=3\\
+12\quad&\quad+12\\
3n&=15\\
\frac{3n}{3}&=\frac{15}{3}\\
n&=5
\end{aligned}$$

47. $$\frac{\cancel{(k+7)}(k+4)}{1}\cdot\frac{-8}{\cancel{k+7}}=\frac{k-5}{\cancel{(k+7)}\cancel{(k+4)}}\cdot\frac{\cancel{(k+7)}\cancel{(k+4)}}{1}$$

$$-8(k+4)=k-5$$
$$-8k-32=k-5$$
$$\underline{-k \quad -k}$$
$$-9k-32=-5$$
$$\underline{+32 \quad +32}$$
$$-9k=27$$
$$\frac{-9k}{-9}=\frac{27}{-9}$$
$$k=-3$$

49. $$\frac{\cancel{(x+2)}(x-4)}{1}\cdot\frac{3}{\cancel{x+2}}=\frac{x^2-4x}{\cancel{(x+2)}\cancel{(x-4)}}\cdot\frac{\cancel{(x+2)}\cancel{(x-4)}}{1}$$

$$3(x-4)=x^2-4x$$
$$3x-12=x^2-4x$$
$$\underline{-x^2+4x \quad -x^2+4x}$$
$$-x^2+7x-12=0$$
$$-(x^2-7x+12)=0$$
$$-(x-4)(x-3)=0$$
$$x-4=0 \qquad x-3=0$$
$$x=4 \qquad x=3$$

$x=4$ is an excluded value, therefore $x=3$

51. $$\frac{(m-5)(m+12)}{1}\cdot\left(\frac{1}{m-5}+\frac{7}{m+12}\right)=\frac{m+5}{(m-5)(m+12)}\cdot\frac{(m-5)(m+12)}{1}$$

$$m+12+7(m-5)=m+5$$
$$m+12+7m-35=m+5$$
$$8m-23=m+5$$
$$\underline{+23 \quad +23}$$
$$8m=m+28$$
$$8m=m+28$$
$$\underline{-m \quad -m}$$
$$7m=28$$
$$\frac{7m}{7}=\frac{28}{7}$$
$$m=4$$

53. $$\frac{(a-5)(a+3)}{1}\left(\frac{6}{a-5}+\frac{8}{a+3}\right)=\frac{a^2+2}{(a-5)(a+3)}\cdot\frac{(a-5)(a+3)}{1}$$

$$6(a+3)+8(a-5)=a^2+2$$
$$6a+18+8a-40=a^2+2$$
$$14a-22=a^2+2$$
$$\underline{-a^2-2 \quad -a^2-2}$$
$$-a^2+14a-24=0$$
$$-(a^2-14a+24)=0$$
$$-(a-12)(a-2)=0$$
$$a-12=0 \qquad a-2=0$$
$$a=12 \qquad a=2$$

55. $$\frac{(b+1)(b+4)}{1}\left(\frac{2}{b+1}+\frac{3}{b+4}\right)=\frac{b^2+6b-1}{(b+1)(b+4)}\cdot\frac{(b+1)(b+4)}{1}$$

$$2(b+4)+3(b+1)=b^2+6b-1$$
$$2b+8+3b+3=b^2+6b-1$$
$$5b+11=b^2+6b-1$$
$$\underline{-5b-11 \quad -5b-11}$$
$$0=b^2+b-12$$
$$0=(b+4)(b-3)$$
$$b+4=0 \qquad b-3=0$$
$$b=-4 \qquad b=3$$

$b=-4$ is an excluded value, therefore $b=3$

57. $$\frac{4}{x+6}=\frac{28}{x^2+8x+12}$$

$$\frac{(x+6)(x+2)}{1}\cdot\frac{4}{x+6}=\frac{28}{(x+6)(x+2)}\cdot\frac{(x+6)(x+2)}{1}$$
$$4(x+2)=28$$
$$4x+8=28$$
$$\underline{-8 \quad -8}$$
$$4x=20$$
$$\frac{4x}{4}=\frac{20}{4}$$
$$x=5$$

59. $$\frac{n}{n+2}+\frac{5}{n-3}=\frac{9}{n^2-n-6}$$

$$\frac{(n+2)(n-3)}{1}\cdot\left(\frac{n}{n+2}+\frac{5}{n-3}\right)=\frac{9}{(n+2)(n-3)}\cdot\frac{(n+2)(n-3)}{1}$$
$$n(n-3)+5(n+2)=9$$
$$n^2-3n+5n+10=9$$
$$n^2+2n+10=9$$
$$\underline{-9 \quad -9}$$
$$n^2+2n+1=0$$
$$(n+1)^2=0$$
$$n+1=0$$
$$n=-1$$

61. $$\frac{8}{h-4}+\frac{h}{h-6}=\frac{12}{h^2-10h+24}$$

$$\frac{(h-6)(h-4)}{1}\left(\frac{8}{h-4}+\frac{h}{h-6}\right)=\frac{12}{(h-6)(h-4)}\cdot\frac{(h-6)(h-4)}{1}$$
$$8(h-6)+h(h-4)=12$$
$$8h-48+h^2-4h=12$$
$$h^2+4h-48=12$$
$$\underline{-12 \quad -12}$$
$$h^2+4h-60=0$$
$$(h+10)(h-6)=0$$
$$h+10=0 \qquad h-6=0$$
$$h=-10 \qquad h=6$$

$h=6$ is an excluded value, therefore $h=-10$

63. Time $= 30$ min $= 0.5$ hr

Rate $= \frac{1}{0.5} \cdot \frac{2}{2} = \frac{2}{1} = 2$

Brittany's work rate is 2 per hour.

65. Time $= 12$ hr

Rate $= \frac{1}{12}$

The pump's work rate is $\frac{1}{12}$ per hour.

67. Time $= 2$ hr

Rate $= \frac{1}{2}$

Kasim's work rate is $\frac{1}{2}$ per hour.

69. Sherri's work time $= 100$ min $= \frac{5}{3}$ hr

Sherri's rate $= \frac{1}{\frac{5}{3}} = 1 \cdot \frac{3}{5} = \frac{3}{5}$

Jen's work time $= 90$ min $= \frac{3}{2}$ hr

Jen's rate $= \frac{1}{\frac{3}{2}} = 1 \cdot \frac{2}{3} = \frac{2}{3}$

Work time for both $= t$ hr . Rate $= \frac{1}{t}$

$$\begin{aligned}
\frac{3}{5} + \frac{2}{3} &= \frac{1}{t} \\
\frac{15t}{1}\left(\frac{3}{5} + \frac{2}{3}\right) &= \left(\frac{1}{t}\right)\frac{15t}{1} \\
\frac{15t}{1} \cdot \frac{3}{5} + \frac{15t}{1} \cdot \frac{2}{3} &= \frac{15t}{1} \cdot \frac{1}{t} \\
9t + 10t &= 15 \\
19t &= 15 \\
\frac{19t}{19} &= \frac{15}{19} \\
t &= \frac{15}{19}
\end{aligned}$$

Sherri and Jen can scan the file box of medical records in $\frac{15}{19}$ of an hour (about 47 min) if they work together.

71. Martin's work time $= 4$ hr . Rate $= \frac{1}{4}$

Bill's work time $= 16$ hr . Rate $= \frac{1}{16}$

Work time for both $= t$ hr . Rate $= \frac{1}{t}$

$$\begin{aligned}
\frac{1}{4} + \frac{1}{16} &= \frac{1}{t} \\
\frac{16t}{1}\left(\frac{1}{4} + \frac{1}{16}\right) &= \left(\frac{1}{t}\right)\frac{16t}{1} \\
\frac{16t}{1} \cdot \frac{1}{4} + \frac{16t}{1} \cdot \frac{1}{16} &= \frac{16t}{1} \cdot \frac{1}{t} \\
4t + t &= 16 \\
5t &= 16 \\
\frac{5t}{5} &= \frac{16}{5} \\
t &= \frac{16}{5} = 3.2
\end{aligned}$$

Martin and Bill can mow the lawn in 3.2 hours (3 hours 12 min.)if they work together.

73. New copier's work time $= 8$ hr . Rate $= \frac{1}{8}$

Old copier's work time $= 10$ hr . Rate $= \frac{1}{10}$

Work time for both $= t$ hr . Rate $= \frac{1}{t}$

$$\begin{aligned}
\frac{1}{8} + \frac{1}{10} &= \frac{1}{t} \\
\frac{40t}{1}\left(\frac{1}{8} + \frac{1}{10}\right) &= \left(\frac{1}{t}\right)\frac{40t}{1} \\
\frac{40t}{1} \cdot \frac{1}{8} + \frac{40t}{1} \cdot \frac{1}{10} &= \frac{40t}{1} \cdot \frac{1}{t} \\
5t + 4t &= 40 \\
9t &= 40 \\
\frac{9t}{9} &= \frac{40}{9} \\
t &= \frac{40}{9} = 4\frac{4}{9}
\end{aligned}$$

Using both copy machines, the bulletins can be copied in $4\frac{4}{9}$ hours (about 4 hours 27 min.).

75. **a.** The statement is an expression because it does not have an equal sign.

b. $\frac{\cancel{x+5}}{\cancel{x-6}} \cdot \frac{(x+3)\cancel{(x-6)}}{\cancel{(x+5)}(x+1)} = \frac{x+3}{x+1}$

77. **a.** The statement is an equation because it has an equal sign.

b. $\dfrac{5x}{x+3}+\dfrac{2x+3}{x+3}=\dfrac{5}{x+3}$

$$\frac{5x+2x+3}{x+3}=\frac{5}{x+3}$$

$$\frac{\cancel{x+3}}{1}\cdot\frac{7x+3}{\cancel{x+3}}=\frac{5}{\cancel{x+3}}\cdot\frac{\cancel{x+3}}{1}$$

$$\begin{aligned}7x+3&=5\\ -3\ \ &\ \ -3\\ 7x&=2\\ \frac{7x}{7}&=\frac{2}{7}\\ x&=\frac{2}{7}\end{aligned}$$

79. **a.** The statement is an expression because it does not have an equal sign.

b. $\dfrac{4}{x+8}-\dfrac{5}{x+1}$ LCD $=(x+8)(x+1)$

$$=\frac{4}{x+8}\cdot\frac{x+1}{x+1}-\frac{5}{x+1}\cdot\frac{x+8}{x+8}$$

$$=\frac{4x+4}{(x+8)(x+1)}-\frac{5x+40}{(x+8)(x+1)}$$

$$=\frac{4x+4-5x-40}{(x+8)(x+1)}=\frac{-x-36}{(x+8)(x+1)}$$

81. **a.** The statement is an expression because it does not have an equal sign.

b. $\dfrac{3x+12}{x+4}\div\dfrac{3}{5x}=\dfrac{3x+12}{x+4}\cdot\dfrac{5x}{3}$

$$=\frac{\cancel{3}\,\cancel{(x+4)}}{\cancel{x+4}}\cdot\frac{5x}{\cancel{3}}=5x$$

83. **a.** The statement is an equation because it has an equal sign.

b. $\dfrac{6}{x+1}+\dfrac{4}{x+2}=\dfrac{36}{x^2+3x+2}$

$$\frac{(x+1)(x+2)}{1}\left(\frac{6}{x+1}+\frac{4}{x+2}\right)=\frac{36}{(x+1)(x+2)}\cdot\frac{(x+1)(x+2)}{1}$$

$$\begin{aligned}6(x+2)+4(x+1)&=36\\ 6x+12+4x+4&=36\\ 10x+16&=36\\ -16\ \ &\ \ -16\\ 10x&=20\\ \frac{10x}{10}&=\frac{20}{10}\\ x&=2\end{aligned}$$

85. **a.** The statement is an expression because it does not have an equal sign.

b. $\dfrac{(x+2)(x+4)}{(x+5)(x-2)}\div\dfrac{(x+4)(x-2)}{(x+5)(x+1)}$

$$=\frac{(x+2)\cancel{(x+4)}}{\cancel{(x+5)}(x-2)}\cdot\frac{\cancel{(x+5)}(x+1)}{\cancel{(x+4)}(x-2)}$$

$$=\frac{(x+2)(x+1)}{(x-2)^2}$$

Section 7.5 Proportions, Similar Triangles, and Variation

1. The ratio of 4 to 15 $= \frac{4}{15}$

3. The ratio of 6 to 20 $= \frac{6}{20} = \frac{3}{10}$

5. $\frac{4\text{yd}}{1} \cdot \frac{3\text{ft}}{1\text{yd}} = 12\text{ft}$

5 feet to 4 yards = 5 feet to 12 feet

5 feet to 4 yards $= \frac{5}{12}$

7. $\frac{2\text{hr}}{1} \cdot \frac{60\text{min}}{1\text{hr}} = 120\text{min}$

45 minutes to 2 hours = 45 min. to 120 min.

45 minutes to 2 hours $= \frac{45}{120} = \frac{3}{8}$

9. $\frac{2\text{km}}{1} \cdot \frac{1000\text{m}}{1\text{km}} = 2000\text{m}$

150 m to 2 km = 150 m to 2000 m.

150 meters to 2 kilometers $= \frac{150}{2000} = \frac{3}{40}$

11. \$35 for 9 ounces $= \frac{\$35}{9\text{oz.}} \approx \$3.89\text{/oz.}$

13. \$1800 for 2 ounces $= \frac{\$1800}{2\text{oz.}} = \$900\text{/oz.}$

15. 15 miles in 2 hours $= \frac{15\text{mi}}{2\text{hr}} = 7.5$ mph

17. 40 ft in 30 min. $= \frac{40\text{ft}}{30\text{min}} \approx 1.33$ ft/min.

19. 24 Duracell AA batteries for \$14.49

$= \frac{\$14.49}{24} \approx \$0.60$ for each battery.

16 Duracell AA batteries for \$10.99

$= \frac{\$10.99}{16} \approx \$0.69$ for each battery.

24 Duracell AA batteries for \$14.49 are the better buy.

21. An 11 oz bag of Lay's Classic potato chips for \$3.29

$= \frac{\$3.29}{11\text{oz}} \approx \$0.30$ per ounce.

A 14 oz bag of Lay's Classic potato chips for \$4.49

$= \frac{\$4.49}{14\text{oz}} \approx \$0.32$ per ounce.

An 11 oz bag of Lay's Classic potato chips for \$3.29 is the better buy.

23. A 16 oz bottle of Herdez salsa casera for \$2.50

$= \frac{\$2.50}{16\text{oz}} \approx \$0.16$ per ounce.

A 24 oz bottle of Herdez salsa casera for \$4.19

$= \frac{\$4.19}{24\text{oz}} \approx \$0.17$ per ounce.

A 16 oz bottle of Herdez salsa casera for \$2.50 is the better buy.

25. $\frac{1}{12} = \frac{x}{40}$

$12x = 40$

$\frac{12x}{12} = \frac{40}{12}$

$x = \frac{40}{12} = \frac{10}{3} = 3\frac{1}{3}$

The toy couch will be $3\frac{1}{3}$ inches tall.

27. $\frac{1}{4} = \frac{x}{12}$

$4x = 12$

$\frac{4x}{4} = \frac{12}{4}$

$x = 3$

The length of a 12-ft wall will be 3 inches on the blueprint.

29. $\frac{1}{8} = \frac{2.5}{x}$

$x = 8 \cdot 2.5 = 20$

The length of the window in the stadium will be 20 feet.

31. $\frac{8}{152} = \frac{304}{x}$

$8x = 152 \cdot 304$

$\frac{8x}{8} = \frac{46,208}{8}$

$x = 5776$

The cost of 304 ft of vinyl fencing is \$5776.

33. $\frac{78}{31} = \frac{3900}{x}$

$78x = 31 \cdot 3900$

$\frac{78x}{78} = \frac{120,900}{78}$

$x = 1550$

The cost of 3900 square feet of shingles is \$1550.

35. **a.** $\frac{11}{224}=\frac{x}{370}$

$224x=11\cdot 370$

$\frac{224x}{224}=\frac{4070}{224}$

$x\approx 18.17$

The total fat in a medium order of french fries is about 18 g.

b. $\frac{1}{224}=\frac{x}{370}$

$224x=1\cdot 370$

$\frac{224x}{224}=\frac{370}{224}$

$x\approx 1.65$

The amount of saturated fat in a medium order of french fries is about 2 g.

37. $\frac{3}{5}=\frac{9}{x}$

$3x=5\cdot 9$

$\frac{3x}{3}=\frac{45}{3}$

$x=15$

39. $\frac{-2}{3}=\frac{-8}{b}$

$-2b=3\cdot -8$

$\frac{-2b}{-2}=\frac{-24}{-2}$

$b=12$

41. $\frac{y+6}{3}=\frac{5}{3}$

$3(y+6)=3\cdot 5$

$3y+18=15$

$\underline{\quad -18\ \ -18}$

$3y=-3$

$\frac{3y}{3}=\frac{-3}{3}$

$y=-1$

43. $\frac{-4}{3}=\frac{6x-8}{6}$

$3(6x-8)=-4\cdot 6$

$18x-24=-24$

$\underline{\quad +24\ \ +24}$

$18x=0$

$\frac{18x}{18}=\frac{0}{18}$

$x=0$

45. $\frac{y+2}{6}=\frac{y-2}{3}$

$6(y-2)=3(y+2)$

$6y-12=3y+6$

$\underline{\quad +12\quad +12}$

$6y=3y+18$

$\underline{-3y\ -3y}$

$3y=18$

$\frac{3y}{3}=\frac{18}{3}$

$y=6$

47. $\frac{x-3}{x+4}=\frac{8}{4}$

$8(x+4)=4(x-3)$

$8x+32=4x-12$

$\underline{\quad -32\qquad -32}$

$8x=4x-44$

$\underline{-4x\ -4x}$

$4x=-44$

$\frac{4x}{4}=\frac{-44}{4}$

$x=-11$

49. $\frac{25\text{ m}}{x\text{ km}}=\frac{1\text{ m}}{1.609\text{ km}}$

$x=25\cdot 1.609$

$x\approx 40.23$

25 miles is approximately equal to 40.23 km.

51. $\frac{165\text{ lb}}{x\text{ kg}}=\frac{2.2\text{ lb}}{1\text{ kg}}$

$2.2x=165$

$\frac{2.2x}{2.2}=\frac{165}{2.2}$

$x=75$

165 pounds is equal to 75 kg.

53. $\frac{45\text{ in}}{x\text{ cm}}=\frac{1\text{ in}}{2.54\text{ cm}}$

$x=45\cdot 2.54$

$x=114.3$

45 inches is equal to 114.3 cm.

55. $\frac{60\text{ acre}}{x\text{ ft}^2}=\frac{1\text{ acre}}{43,560\text{ ft}^2}$

$x=60\cdot 43,560$

$x=2,613,600$

60 acres is equal to 2,613,600 square feet.

57. $\dfrac{8}{10} = \dfrac{10}{x}$

$8x = 10 \cdot 10$

$8x = 100$

$\dfrac{8x}{8} = \dfrac{100}{8}$

$x = 12.5$

The width of the enlarged photo will be 12.5 in.

59. $\dfrac{x}{3} = \dfrac{4.5}{1.8}$

$1.8x = 3 \cdot 4.5$

$1.8x = 13.5$

$\dfrac{1.8x}{1.8} = \dfrac{13.5}{1.8}$

$x = 7.5$

61. $\dfrac{5}{3} = \dfrac{x}{4}$

$3x = 5 \cdot 4$

$3x = 20$

$\dfrac{3x}{3} = \dfrac{20}{3}$

$x = \dfrac{20}{3} = 6\dfrac{2}{3}$

63. $\dfrac{x}{1.6} = \dfrac{709}{8}$

$8x = 1.6 \cdot 709$

$8x = 1134.4$

$\dfrac{8x}{8} = \dfrac{1134.4}{8}$

$x = 141.8$

The Victory Obelisk is 141.8 meters tall.

65. $\dfrac{\frac{1}{4}}{5} = \dfrac{x}{16}$

$5x = \dfrac{1}{4} \cdot 16$

$5x = 4$

$\dfrac{5x}{5} = \dfrac{4}{5}$

$x = \dfrac{4}{5}$

The amount of unsalted butter needed to make 16 servings is $\dfrac{4}{5}$ cup.

67. $\dfrac{1\frac{1}{2}}{5} = \dfrac{x}{12}$

$5x = 1\dfrac{1}{2} \cdot 12$

$5x = \dfrac{3}{2} \cdot 12$

$5x = 18$

$\dfrac{5x}{5} = \dfrac{18}{5}$

$x = 3\dfrac{3}{5}$

The amount of chicken broth needed to make 12 servings is $3\dfrac{3}{5}$ cup.

69. $\dfrac{4}{4} = \dfrac{x}{16}$

$4x = 4 \cdot 16$

$4x = 64$

$\dfrac{4x}{4} = \dfrac{64}{4}$

$x = 16$

The amount of milk needed to make 16 servings is 16 cups.

71. $\dfrac{\frac{3}{4}}{4} = \dfrac{x}{12}$

$4x = \dfrac{3}{4} \cdot 12$

$4x = 9$

$\dfrac{4x}{4} = \dfrac{9}{4}$

$x = 2\dfrac{1}{4}$

The amount of maple syrup needed to make 12 servings is $2\dfrac{1}{4}$ cups.

73. $y = kx$ Substitute given values for x and y.

$(30) = k(9)$

$\frac{30}{9} = \frac{9k}{9}$

$\frac{10}{3} = k$

$y = \frac{10}{3}x$ Substitute values for k and x.

$y = \frac{10}{3}(15)$

$y = 50$

75. $d = kt$ Substitute given values for d and t.

$(21) = k(7)$

$\frac{21}{7} = \frac{7k}{7}$

$3 = k$

$d = 3t$ Substitute values for k and x.

$d = 3(10)$

$d = 30$

77. $y = kx$ Substitute given values for x and y.

$(5) = k(8)$

$\frac{5}{8} = \frac{8k}{8}$

$0.625 = k$

$y = 0.625x$ Substitute values for k and x.

$y = 0.625(17)$

$y = 10.625$

79. **a.** Let $h =$ the number of hours spent biking.

$C = kh$ Substitute given values for C and h.

$(1152) = k(2)$

$\frac{1152}{2} = \frac{2k}{2}$

$576 = k$

The equation that expresses the relationship is $C = 576h$

b. $C = 576h$ Substitute values for k and h.

$C = 576(3.5)$

$C = 2016$

A 150-lb person burns 2016 calories during 3.5 hours of mountain biking.

81. **a.** Let $D =$ the distance travelled in miles and let $t =$ the time (in hours) spent driving.

$D = kt$ Substitute given values for D and t.

$(130) = k(2)$

$\frac{130}{2} = \frac{2k}{2}$

$65 = k$

The equation that expresses the relationship is $D = 65t$

b. $D = 65t$ Substitute values for k and t.

$D = 65(5.5)$

$D = 357.5$

357.5 miles are traveled in 5.5 hours.

83. **a.** $T = kc$ Substitute given values for T and c.

$(256) = k(8)$

$\frac{256}{8} = \frac{8k}{8}$

$32 = k$

The equation that expresses the relationship is $T = 32c$

b. $T = 32c$ Substitute values for k and c.

$T = 32(15)$

$T = 480$

A student who enrolls in 15 credits will owe \$480 for tuition.

85. **a.** $F = kx$ Substitute given values for F and x.

$(6) = k(5)$

$\frac{6}{5} = \frac{5k}{5}$

$\frac{6}{5} = k$

The equation that expresses the relationship is $F = \frac{6}{5}x$

b. $F = \frac{6}{5}x$ Substitute values for k and F.

$(10) = \frac{6}{5}x$

$\frac{5}{6} \cdot 10 = \frac{6}{5}x \cdot \frac{5}{6}$

$\frac{25}{3} = x$

$8\frac{1}{3} = x$

A force of 10 lb. will stretch the spring $8\frac{1}{3}$ inches.

87. **a.** $W = kt$ Substitute given values for W and t.

$(290) = k(20)$

$\frac{290}{20} = \frac{20k}{20}$

$14.5 = k$

The equation that expresses the relationship is

$W = 14.5t$

b. $W = 14.5t$ Substitute values for k and t.

$W = 14.5(25)$

$W = 362.5$

The amount of the paycheck would be \$362.50 for 25 hours of work.

89. $y = \frac{k}{x}$ Substitute given values for x and y.

$(6) = \frac{k}{(2.5)}$ Cross multiply.

$k = 6 \cdot 2.5$

$k = 15$

$y = \frac{15}{x}$ Substitute values for k and x.

$y = \frac{15}{(5.5)}$

$y \approx 2.73$

91. $P = \frac{k}{n}$ Substitute given values for n and P.

$(200) = \frac{k}{(5)}$ Cross multiply.

$k = 200 \cdot 5$

$k = 1000$

$P = \frac{1000}{n}$ Substitute values for k and n.

$P = \frac{1000}{(50)}$

$P = 20$

93. $C = \frac{k}{h}$ Substitute given values for h and C.

$(30) = \frac{k}{(6)}$ Cross multiply.

$k = 30 \cdot 6$

$k = 180$

$C = \frac{180}{h}$ Substitute values for k and h.

$C = \frac{180}{(3)}$

$C = 60$

95. **a.** $P = \frac{k}{V}$ Substitute given values for V and P.

$(15) = \frac{k}{(2)}$ Cross multiply.

$k = 2 \cdot 15$

$k = 30$

b. $P = \frac{30}{V}$ Substitute value for V.

$P = \frac{30}{(5)}$

$P = 6$

The inside air pressure of the balloon is 6 psi when the volume of the balloon is 5 in^3.

97. **a.** $s = \frac{k}{t}$ Substitute given values for t and s.

$(55) = \frac{k}{(5)}$ Cross multiply.

$k = 55 \cdot 5$

$k = 275$

$s = \frac{275}{t}$

b. $s = \frac{275}{t}$ Substitute value for t.

$s = \frac{275}{(3.5)}$

$s \approx 78.57$

The speed is approximately 78.57 mph for 3.5 hours of driving.

99. **a.** $I = \frac{k}{d^2}$ Substitute given values for d and I.

$(36) = \frac{k}{(5)^2}$ Cross multiply.

$k = 36 \cdot 25$

$k = 900$

$I = \frac{900}{d^2}$

b. $I = \frac{900}{d^2}$ Substitute value for d.

$I = \frac{900}{(10)^2}$

$I = \frac{900}{100}$

$I = 9$

The illumination is 9 foot-candles at a distance of 10 feet from the light.

Chapter 7 Review

1. $\dfrac{(2)+7}{(2)^2+3(2)-8}$ Substitute $m = 2$.

$\dfrac{9}{4+6-8}=\dfrac{9}{2}$

2. $\dfrac{(1)-5}{(1)^2-6(1)+5}$ Substitute $d = 1$.

$\dfrac{-4}{1-6+5}=\dfrac{-4}{0}$ Undefined.

3. $c=\dfrac{1100+50(50)}{(50)}$ Substitute $m = 50$.

$c=\dfrac{1100+2500}{50}=\dfrac{3600}{50}=72$

The cost per person is $72 when 50 members go to the competition.

4. $s=\dfrac{500(10)+23100}{0.06(10)+5.91}$ Substitute $t = 10$.

$s=\dfrac{5000+23100}{0.6+5.91}=\dfrac{28100}{6.51}\approx 4316.44$

The average spending per person by the state of Washington in 2010 is $4316.44.

5. Set denominator equal to 0 and solve:

$\dfrac{84}{x+7}$

$x+7=0$

$x=-7$ This is the excluded value.

6. Set denominator equal to 0 and solve:

$\dfrac{65}{(n+2)(n-6)}$

$(n+2)(n-6)=0$

$n+2=0$ $\quad$ $n-6=0$

$n=-2$ $\quad$ $n=6$

$n=-2, 6$ These are the excluded values.

7. Set denominator equal to 0 and solve:

$\dfrac{b+2}{b^2+4b+3}$

$b^2+4b+3=0$

$(b+3)(b+1)=0$

$b+3=0$ $\quad$ $b+1=0$

$b=-3$ $\quad$ $b=-1$

$b=-3, -1$ These are the excluded values.

8. Set denominator equal to 0 and solve:

$\dfrac{t^2+2t-3}{t^2-25}$

$t^2-25=0$

$(t-5)(t+5)=0$

$t-5=0$ $\quad$ $t+5=0$

$t=5$ $\quad$ $t=-5$

$t=\pm 5$ These are the excluded values.

9. $\dfrac{-8x^3y}{64x^4y}=\dfrac{-1}{8}x^{3-4}y^{1-1}=\dfrac{-1}{8}x^{-1}=-\dfrac{1}{8x}$

10. $\dfrac{1-x}{x^2-1}=\dfrac{-\cancel{(x-1)}}{\cancel{(x-1)}(x+1)}=\dfrac{-1}{x+1}$

11. $\dfrac{\cancel{(y+2)}(y-8)}{(y+3)\cancel{(y+2)}}=\dfrac{y-8}{y+3}$

12. $\dfrac{g-8}{g^2-64}=\dfrac{\cancel{g-8}}{\cancel{(g-8)}(g+8)}=\dfrac{1}{g+8}$

13. $\dfrac{5-h}{(h-5)(h+3)}=\dfrac{-\cancel{(h-5)}}{\cancel{(h-5)}(h+3)}=\dfrac{-1}{h+3}$

14. $\dfrac{4n-24}{n^2-8n+12}=\dfrac{4\cancel{(n-6)}}{\cancel{(n-6)}(n-2)}=\dfrac{4}{n-2}$

15. $\dfrac{708\text{ cal}}{1\ \cancel{\text{hr}}}\cdot\dfrac{1\ \cancel{\text{hr}}}{60\text{ min}}=\dfrac{708\text{ cal}}{60\text{ min}}=11.8\text{ cal/min}$

$11.8\text{ cal/}\cancel{\text{min}}\cdot 6\ \cancel{\text{min}}=70.8\text{ cal}$

A 130-lb rower will burn approximately 70.8 calories during a 6-minute event.

16. $\dfrac{10\ \cancel{\text{mile}}}{1\text{ hr}}\cdot\dfrac{5280\text{ ft}}{1\ \cancel{\text{mile}}}=\dfrac{52{,}800\text{ ft}}{1\text{ hr}}$

$\dfrac{52{,}800\text{ ft}}{1\ \cancel{\text{hr}}}\cdot\dfrac{1\ \cancel{\text{hr}}}{60\text{ min}}=\dfrac{52{,}800\text{ ft}}{60\text{ min}}=880\text{ ft/min}$

The runner's speed is 880 feet per minute.

17. $\dfrac{3h^2}{25hk^4}\cdot\dfrac{-5k^3}{6h}=\dfrac{\cancel{3}\cancel{h^2}}{\cancel{5}\cdot 5\cdot\cancel{h}k^{\cancel{4}1}}\cdot\dfrac{-\cancel{5}\cancel{k^3}}{2\cdot\cancel{3}\cdot\cancel{h}}=\dfrac{-1}{10k}$

18. $\dfrac{-7xy^3}{6y^2}\cdot\dfrac{4xy}{14y^2}=\dfrac{-\cancel{7}x\cancel{y^3}}{\cancel{2}\cdot 3\cdot\cancel{y^2}}\cdot\dfrac{\cancel{2}\cdot\cancel{2}\cdot x\cancel{y}}{\cancel{2}\cdot\cancel{7}\cdot\cancel{y^2}}=\dfrac{-x^2}{3}$

19. $\dfrac{-10n^2}{9m}\div\dfrac{35n}{3m^2}=\dfrac{-10n^2}{9m}\cdot\dfrac{3m^2}{35n}$

$=-\dfrac{2\cdot\cancel{5}\cdot n^{\cancel{2}1}}{\cancel{3}\cdot 3\cdot\cancel{m}}\cdot\dfrac{\cancel{3}m^{\cancel{2}1}}{\cancel{5}\cdot 7\cdot\cancel{n}}=-\dfrac{2mn}{21}$

20. $\dfrac{2p^5}{-7q}\div\dfrac{16p^7}{63q^2}=\dfrac{2p^5}{-7q}\cdot\dfrac{63q^2}{16p^7}$

$=-\dfrac{\cancel{2}\,\cancel{p^5}}{\cancel{7}\,\cancel{q}}\cdot\dfrac{3\cdot3\cdot\cancel{7}\cdot q^{\cancel{2}1}}{\cancel{2}\cdot2\cdot2\cdot2\cdot p^{\cancel{7}2}}=-\dfrac{9q}{8p^2}$

21. $\dfrac{\cancel{(x-4)}}{(x+3)}\cdot\dfrac{(x+2)}{\cancel{(x-4)}}=\dfrac{x+2}{x+3}$

22. $\dfrac{(d-7)}{(d+6)}\cdot\dfrac{(d+9)}{d^2+2d-63}$

$=\dfrac{\cancel{(d-7)}}{(d+6)}\cdot\dfrac{\cancel{(d+9)}}{\cancel{(d+9)}\cancel{(d-7)}}=\dfrac{1}{d+6}$

23. $\dfrac{(k+5)}{(k+2)}\div\dfrac{(k+3)(k+5)}{(k-2)(k+2)}$

$=\dfrac{\cancel{(k+5)}}{\cancel{(k+2)}}\cdot\dfrac{(k-2)\cancel{(k+2)}}{(k+3)\cancel{(k+5)}}=\dfrac{k-2}{k+3}$

24. $\dfrac{(p+8)}{(p+3)}\div\dfrac{p^2+10p+16}{p^2-9}$

$=\dfrac{\cancel{(p+8)}}{\cancel{(p+3)}}\cdot\dfrac{(p-3)\cancel{(p+3)}}{\cancel{(p+8)}(p+2)}=\dfrac{p-3}{p+2}$

25. $\dfrac{z^2+4z-5}{z^2-3z+2}\div\dfrac{z^2-8z+15}{z^2-4}$

$=\dfrac{(z+5)\cancel{(z-1)}}{\cancel{(z-2)}\cancel{(z-1)}}\cdot\dfrac{\cancel{(z-2)}(z+2)}{(z-5)(z-3)}$

$=\dfrac{(z+5)(z+2)}{(z-5)(z-3)}=\dfrac{z^2+7z+10}{(z-5)(z-3)}$

26. $\dfrac{g^2-7g+10}{g^2+3g-10}\div\dfrac{g^2-25}{g^2+10g+25}$

$=\dfrac{\cancel{(g-5)}\cancel{(g-2)}}{\cancel{(g+5)}\cancel{(g-2)}}\cdot\dfrac{\cancel{(g+5)}\cancel{(g+5)}}{\cancel{(g-5)}\cancel{(g+5)}}=1$

27. $\dfrac{5x}{x+2}+\dfrac{2x+7}{x+2}=\dfrac{5x+2x+7}{x+2}=\dfrac{7x+7}{x+2}$

28. $\dfrac{6n}{5n+1}+\dfrac{3+n}{5n+1}=\dfrac{6n+3+n}{5n+1}=\dfrac{7n+3}{5n+1}$

29. $\dfrac{2x+5}{(x-1)(x+4)}+\dfrac{x-8}{(x-1)(x+4)}=\dfrac{2x+5+x-8}{(x-1)(x+4)}$

$=\dfrac{3x-3}{(x-1)(x+4)}=\dfrac{3\cancel{(x-1)}}{\cancel{(x-1)}(x+4)}=\dfrac{3}{x+4}$

30. $\dfrac{4h-9}{(h-5)(h-4)}+\dfrac{h-11}{(h-5)(h-4)}=\dfrac{4h-9+h-11}{(h-5)(h-4)}$

$=\dfrac{5h-20}{(h-5)(h-4)}=\dfrac{5\cancel{(h-4)}}{(h-5)\cancel{(h-4)}}=\dfrac{5}{h-5}$

31. $\dfrac{8x-9}{x+6}-\dfrac{2x-4}{x+6}=\dfrac{8x-9-2x+4}{x+6}=\dfrac{6x-5}{x+6}$

32. $\dfrac{5y+1}{y-3}-\dfrac{3y-4}{y-3}=\dfrac{5y+1-3y+4}{y-3}=\dfrac{2y+5}{y-3}$

33. $\dfrac{8n+10}{(n+3)(n-4)}-\dfrac{2n-8}{(n+3)(n-4)}=\dfrac{8n+10-2n+8}{(n+3)(n-4)}$

$=\dfrac{6n+18}{(n+3)(n-4)}=\dfrac{6\cancel{(n+3)}}{\cancel{(n+3)}(n-4)}=\dfrac{6}{n-4}$

34. $\dfrac{3x-2}{(x+3)(x+5)}-\dfrac{x-7}{(x+3)(x+5)}=\dfrac{3x-2-x+7}{(x+3)(x+5)}$

$=\dfrac{2x+5}{(x+3)(x+5)}$

35. $\dfrac{4}{3a^2b}\qquad\dfrac{11a}{2\cdot3^2\cdot ab^3c}$

$\text{LCD}=2\cdot3^2\cdot a^2b^3c=18a^2b^3c$

$\dfrac{4}{3a^2b}\cdot\dfrac{2\cdot3\cdot b^2c}{2\cdot3\cdot b^2c}\qquad\dfrac{11a}{2\cdot3^2\cdot ab^3c}\cdot\dfrac{a}{a}$

$\dfrac{24b^2c}{18a^2b^3c}\qquad\dfrac{11a^2}{18a^2b^3c}$

36. $\dfrac{-9}{8xy^3}\qquad\dfrac{7}{4x^3y^2}$

$\text{LCD}=2^3\cdot x^3y^3=8x^3y^3$

$\dfrac{-9}{8xy^3}\cdot\dfrac{x^2}{x^2}\qquad\dfrac{7}{4x^3y^2}\cdot\dfrac{2y}{2y}$

$\dfrac{-9x^2}{8x^3y^3}\qquad\dfrac{14y}{8x^3y^3}$

37. $\dfrac{7}{2(v+4)}\qquad\dfrac{11}{v+4}\qquad\text{LCD}=2(v+4)$

$\dfrac{7}{2(v+4)}\qquad\dfrac{11}{v+4}\cdot\dfrac{2}{2}$

$\dfrac{7}{2(v+4)}\qquad\dfrac{22}{2(v+4)}$

38. $\dfrac{-\cancel{3}}{\cancel{3}(y-3)}\qquad\dfrac{11}{y-3}\qquad\text{LCD}=y-3$

$\dfrac{-1}{y-3}\qquad\dfrac{11}{y-3}$

39. $\dfrac{5}{(p+2)(p-3)}$ $\qquad$ $\dfrac{3p}{(p+2)(p-8)}$

LCD $=(p+2)(p-3)(p-8)$

$\dfrac{5}{(p+2)(p-3)}\cdot\dfrac{p-8}{p-8}$ $\qquad$ $\dfrac{3p}{(p+2)(p-8)}\cdot\dfrac{p-3}{p-3}$

$\dfrac{5p-40}{(p+2)(p-3)(p-8)}$ $\qquad$ $\dfrac{3p^2-9p}{(p+2)(p-3)(p-8)}$

40. $\dfrac{-1}{(3x-1)(x+4)}$ $\qquad$ $\dfrac{3x}{(3x-1)(x-2)}$

LCD $=(3x-1)(x+4)(x-2)$

$\dfrac{-1}{(3x-1)(x+4)}\cdot\dfrac{x-2}{x-2}$ $\qquad$ $\dfrac{3x}{(3x-1)(x-2)}\cdot\dfrac{x+4}{x+4}$

$\dfrac{-x+2}{(3x-1)(x+4)(x-2)}$ $\qquad$ $\dfrac{3x^2+12x}{(3x-1)(x+4)(x-2)}$

41. $\dfrac{z-5}{(z+7)(z+2)}$ $\qquad$ $\dfrac{4}{(z-7)(z+7)}$

LCD $=(z+7)(z+2)(z-7)$

$\dfrac{z-5}{(z+7)(z+2)}\cdot\dfrac{z-7}{z-7}$ $\qquad$ $\dfrac{4}{(z-7)(z+7)}\cdot\dfrac{z+2}{z+2}$

$\dfrac{z^2-12z+35}{(z+7)(z+2)(z-7)}$ $\qquad$ $\dfrac{4z+8}{(z+7)(z+2)(z-7)}$

42. $\dfrac{x-4}{(x+9)(x+5)}$ $\qquad$ $\dfrac{-3}{(x-9)(x+9)}$

LCD $=(x+9)(x+5)(x-9)$

$\dfrac{x-4}{(x+9)(x+5)}\cdot\dfrac{x-9}{x-9}$ $\qquad$ $\dfrac{-3}{(x-9)(x+9)}\cdot\dfrac{x+5}{x+5}$

$\dfrac{x^2-13x+36}{(x+9)(x+5)(x-9)}$ $\qquad$ $\dfrac{-3x-15}{(x+9)(x+5)(x-9)}$

43. $\dfrac{7c}{2\cdot 5\cdot a^3b}-\dfrac{4}{5\cdot 5\cdot ab^2}$

LCD $=2\cdot 5^2\cdot a^3b^2=50a^3b^2$

$=\dfrac{7c}{2\cdot 5\cdot a^3b}\cdot\dfrac{5b}{5b}-\dfrac{4}{5\cdot 5\cdot ab^2}\cdot\dfrac{2a^2}{2a^2}$

$=\dfrac{35bc}{50a^3b^2}-\dfrac{8a^2}{50a^3b^2}=\dfrac{35bc-8a^2}{50a^3b^2}=\dfrac{-8a^2+35bc}{50a^3b^2}$

44. $\dfrac{\cancel{5}\cancel{x}}{3\cdot\cancel{5}\cdot\cancel{x}y}+\dfrac{2\cdot\cancel{5}}{\cancel{5}xy^2}$ $\qquad$ LCD $=3xy^2$

$=\dfrac{1}{3y}\cdot\dfrac{xy}{xy}+\dfrac{2}{xy^2}\cdot\dfrac{3}{3}=\dfrac{xy}{3xy^2}+\dfrac{6}{3xy^2}=\dfrac{xy+6}{3xy^2}$

45. $\dfrac{5}{y+2}+\dfrac{8}{y-7}$ $\qquad$ LCD $=(y+2)(y-7)$

$=\dfrac{5}{y+2}\cdot\dfrac{y-7}{y-7}+\dfrac{8}{y-7}\cdot\dfrac{y+2}{y+2}$

$=\dfrac{5y-35}{(y+2)(y-7)}+\dfrac{8y+16}{(y+2)(y-7)}$

$=\dfrac{5y-35+8y+16}{(y+2)(y-7)}=\dfrac{13y-19}{(y+2)(y-7)}$

46. $\dfrac{4}{t-3}-\dfrac{5}{t+6}$ $\qquad$ LCD $=(t-3)(t+6)$

$=\dfrac{4}{t-3}\cdot\dfrac{t+6}{t+6}-\dfrac{5}{t+6}\cdot\dfrac{t-3}{t-3}$

$=\dfrac{4t+24}{(t-3)(t+6)}-\dfrac{5t-15}{(t-3)(t+6)}$

$=\dfrac{4t+24-5t+15}{(t-3)(t+6)}=\dfrac{-t+39}{(t-3)(t+6)}$

47. $\dfrac{k}{k+1}-\dfrac{1}{(k+1)(k+2)}$ $\qquad$ LCD $=(k+1)(k+2)$

$=\dfrac{k}{k+1}\cdot\dfrac{k+2}{k+2}-\dfrac{1}{(k+1)(k+2)}$

$=\dfrac{k^2+2k}{(k+1)(k+2)}-\dfrac{1}{(k+1)(k+2)}=\dfrac{k^2+2k-1}{(k+1)(k+2)}$

48. $\dfrac{w+3}{w+7}+\dfrac{7w+5}{(w+7)(w-4)}$ $\qquad$ LCD $=(w+7)(w-4)$

$=\dfrac{w+3}{w+7}\cdot\dfrac{w-4}{w-4}+\dfrac{7w+5}{(w+7)(w-4)}$

$=\dfrac{w^2-w-12}{(w+7)(w-4)}+\dfrac{7w+5}{(w+7)(w-4)}$

$=\dfrac{w^2-w-12+7w+5}{(w+7)(w-4)}=\dfrac{w^2+6w-7}{(w+7)(w-4)}$

$=\dfrac{\cancel{(w+7)}(w-1)}{\cancel{(w+7)}(w-4)}=\dfrac{w-1}{w-4}$

49. $\dfrac{x+1}{(x-5)(x+2)}+\dfrac{x-3}{x-5}$ $\qquad$ LCD $=(x-5)(x+2)$

$=\dfrac{x+1}{(x-5)(x+2)}+\dfrac{x-3}{x-5}\cdot\dfrac{x+2}{x+2}$

$=\dfrac{x+1}{(x-5)(x+2)}+\dfrac{x^2-x-6}{(x-5)(x+2)}$

$=\dfrac{x+1+x^2-x-6}{(x-5)(x+2)}=\dfrac{x^2-5}{(x-5)(x+2)}$

50. $\frac{y-1}{y-3}-\frac{10y-12}{(y+6)(y-3)}$ LCD $=(y+6)(y-3)$

$=\frac{y-1}{y-3}\cdot\frac{y+6}{y+6}-\frac{10y-12}{(y+6)(y-3)}$

$=\frac{y^2+5y-6}{(y+6)(y-3)}-\frac{10y-12}{(y+6)(y-3)}$

$=\frac{y^2+5y-6-10y+12}{(y+6)(y-3)}=\frac{y^2-5y+6}{(y+6)(y-3)}$

$=\frac{(y-3)(y-2)}{(y+6)(y-3)}=\frac{y-2}{y+6}$

51. $(100)=\frac{1100+50m}{m}$ Substitute $c=100$.

$100m=1100+50m$ Cross multiply.

$-50m \quad -50m$

$50m=1100$

$\frac{50m}{50}=\frac{1100}{50}$

$m=22$

There must be 22 members of the band who travel to the competition in order for the cost per person to be \$100.

52. $(4241)=\frac{500t+23100}{0.06t+5.91}$ Substitute $s=4241$.

$500t+23100=4241(0.06t+5.91)$ Cross multiply.

$500t+23100=254.46t+25064.31$

$-23100 \quad -23100$

$500t=254.46t+1964.31$

$-254.46t \quad -254.46t$

$245.54t=1964.31$

$\frac{245.54t}{245.54}=\frac{1964.31}{245.54}$

$t\approx 8$

In the year 2008 the state of Washington spent \$4241 per resident.

53. $\frac{h}{1}\left(5+\frac{14}{h}\right)=\left(\frac{-6}{h}\right)\frac{h}{1}$

$\frac{h}{1}\cdot 5+\frac{h}{1}\cdot\frac{14}{h}=\frac{h}{1}\cdot\frac{-6}{h}$

$5h+14=-6$

$-14 \quad -14$

$5h=-20$

$\frac{5h}{5}=\frac{-20}{5}$

$h=-4$

54. $\frac{f+2}{1}\left(\frac{4}{f+2}\right)=\left(5-\frac{1}{f+2}\right)\frac{f+2}{1}$

$\frac{f+2}{1}\cdot\frac{4}{f+2}=\frac{f+2}{1}\cdot 5-\frac{f+2}{1}\cdot\frac{1}{f+2}$

$4=5f+10-1$

$4=5f+9$

$-9 \quad -9$

$-5=5f$

$\frac{-5}{5}=\frac{5f}{5}$

$-1=f$

55. $\frac{15r}{1}\left(\frac{2}{15}+4\right)=\left(\frac{5}{3r}\right)\frac{15r}{1}$

$\frac{15r}{1}\cdot\frac{2}{15}+\frac{15r}{1}\cdot 4=\frac{15r}{1}\cdot\frac{5}{3r}$

$2r+60r=25$

$62r=25$

$\frac{62r}{62}=\frac{25}{62}$

$r=\frac{25}{62}$

56. $\frac{6x}{1}\left(\frac{2}{3x}-6\right)=\left(\frac{1}{2}\right)\frac{6x}{1}$

$\frac{6x}{1}\cdot\frac{2}{3x}-\frac{6x}{1}\cdot 6=\frac{6x}{1}\cdot\frac{1}{2}$

$4-36x=3x$

$+36x \quad +36x$

$4=39x$

$\frac{4}{39}=\frac{39x}{39}$

$\frac{4}{39}=x$

57. $\frac{7}{w-3}=\frac{2}{w+1}$ Cross multiply.

$2(w-3)=7(w+1)$

$2w-6=7w+7$

$+6 \quad +6$

$2w=7w+13$

$-7w \quad -7w$

$-5w=13$

$\frac{-5w}{-5}=\frac{13}{-5}$

$w=-\frac{13}{5}$

58. $\dfrac{-5}{b-4}=\dfrac{5}{b+3}$ Cross multiply.

$$5(b-4)=-5(b+3)$$
$$5b-20=-5b-15$$
$$\underline{+20\qquad +20}$$
$$5b=-5b+5$$
$$\underline{+5b\quad +5b}$$
$$10b=5$$
$$\frac{10b}{10}=\frac{5}{10}$$
$$b=\frac{1}{2}$$

59. $\dfrac{2}{n+5}=\dfrac{10}{(n-3)(n+5)}$ Cross multiply.

$$10(n+5)=2(n-3)(n+5)$$
$$10n+50=2(n^2+2n-15)$$
$$10n+50=2n^2+4n-30$$
$$\underline{-50\qquad\qquad -50}$$
$$10n=2n^2+4n-80$$
$$\underline{-10n\qquad -10n}$$
$$0=2n^2-6n-80$$
$$0=2(n^2-3n-40)$$
$$0=2(n-8)(n+5)$$
$$n-8=0 \qquad n+5=0$$
$$n=8 \qquad n=-5$$

$n=-5$ is an excluded value, therefore $n=8$

60. $\dfrac{4}{x+1}=\dfrac{8}{(x+1)(x-2)}$ Cross multiply.

$$8(x+1)=4(x+1)(x-2)$$
$$8x+8=4(x^2-x-2)$$
$$8x+8=4x^2-4x-8$$
$$\underline{-8\qquad\qquad -8}$$
$$8x=4x^2-4x-16$$
$$\underline{-8x\qquad -8x}$$
$$0=4x^2-12x-16$$
$$0=4(x^2-3x-4)$$
$$0=4(x-4)(x+1)$$
$$x-4=0 \qquad x+1=0$$
$$x=4 \qquad x=-1$$

$x=-1$ is an excluded value, therefore $x=4$

61. $\dfrac{4}{a-2}+\dfrac{5}{a+1}=\dfrac{a^2+4a}{(a-2)(a+1)}$

$$\frac{(a-2)(a+1)}{1}\cdot\frac{4}{a-2}+\frac{5}{a+1}=\frac{a^2+4a}{(a-2)(a+1)}\cdot\frac{(a-2)(a+1)}{1}$$
$$4(a+1)+5(a-2)=a^2+4a$$
$$4a+4+5a-10=a^2+4a$$
$$9a-6=a^2+4a$$
$$\underline{-9a+6\quad -9a+6}$$
$$0=a^2-5a+6$$
$$0=(a-3)(a-2)$$
$$a-3=0 \qquad a-2=0$$
$$a=3 \qquad a=2$$

$a=2$ is an excluded value, therefore $a=3$

62. $\dfrac{d}{d-3}+\dfrac{2}{d+4}=\dfrac{21}{(d-3)(d+4)}$

$$\frac{(d-3)(d+4)}{1}\cdot\frac{d}{d-3}+\frac{2}{d+4}=\frac{21}{(d-3)(d+4)}\cdot\frac{(d-3)(d+4)}{1}$$
$$d(d+4)+2(d-3)=21$$
$$d^2+4d+2d-6=21$$
$$d^2+6d-6=21$$
$$\underline{-21\quad -21}$$
$$d^2+6d-27=0$$
$$(d+9)(d-3)=0$$
$$d+9=0 \qquad d-3=0$$
$$d=-9 \qquad d=3$$

$d=3$ is an excluded value, therefore $d=-9$

63. $\dfrac{6+\frac{2}{x}}{8-\frac{3}{x}}\cdot\dfrac{\frac{x}{1}}{\frac{x}{1}}$ Multiply by LCD = x.

$$=\frac{6\cdot\frac{x}{1}+\frac{2}{x}\cdot\frac{x}{1}}{8\cdot\frac{x}{1}-\frac{3}{x}\cdot\frac{x}{1}}=\frac{6x+2}{8x-3}$$

64. $\dfrac{\frac{5}{x}+4}{2-\frac{3}{x}}\cdot\dfrac{\frac{x}{1}}{\frac{x}{1}}$ Multiply by LCD = x.

$$=\frac{\frac{5}{x}\cdot\frac{x}{1}+4\cdot\frac{x}{1}}{2\cdot\frac{x}{1}-\frac{3}{x}\cdot\frac{x}{1}}=\frac{5+4x}{2x-3}=\frac{4x+5}{2x-3}$$

65. $\dfrac{9-\frac{3}{n}}{9-\frac{1}{n^2}}\cdot\dfrac{\frac{n^2}{1}}{\frac{n^2}{1}}$ Multiply by LCD $= n^2$.

$$=\frac{9\cdot\frac{n^2}{1}-\frac{3}{n}\cdot\frac{n^2}{1}}{9\cdot\frac{n^2}{1}-\frac{1}{n^2}\cdot\frac{n^2}{1}}=\frac{9n^2-3n}{9n^2-1}=\frac{3n\cancel{(3n-1)}}{\cancel{(3n-1)}(3n+1)}=\frac{3n}{3n+1}$$

66. $\dfrac{1-\frac{16}{n^2}}{1+\frac{4}{n}}\cdot\dfrac{\frac{n^2}{1}}{\frac{n^2}{1}}$ Multiply by LCD $= n^2$.

$$=\frac{1\cdot\frac{n^2}{1}-\frac{16}{n^2}\cdot\frac{n^2}{1}}{1\cdot\frac{n^2}{1}+\frac{4}{n}\cdot\frac{n^2}{1}}=\frac{n^2-16}{n^2+4n}=\frac{(n-4)\cancel{(n+4)}}{n\cancel{(n+4)}}=\frac{n-4}{n}$$

67. Amin's work time $=12$ hr. Rate $=\dfrac{1}{12}$

Danya's work time $=10$ hr. Rate $=\dfrac{1}{10}$

Work time for both $=t$ hr. Rate $=\dfrac{1}{t}$

$$\frac{1}{12}+\frac{1}{10}=\frac{1}{t}$$
$$\frac{60t}{1}\left(\frac{1}{12}+\frac{1}{10}\right)=\left(\frac{1}{t}\right)\frac{60t}{1}$$
$$\frac{60t}{1}\cdot\frac{1}{12}+\frac{60t}{1}\cdot\frac{1}{10}=\frac{60t}{1}\cdot\frac{1}{t}$$
$$5t+6t=60$$
$$11t=60$$
$$\frac{11t}{11}=\frac{60}{11}$$
$$t=\frac{60}{11}=5\frac{5}{11}$$

Working together, Amin and Danya can typeset a chapter in $5\frac{5}{11}$ hours, or about 5 hours 27 minutes.

68. Mike's work time $=18$ hr. Rate $=\dfrac{1}{18}$

Assistant's work time $=27$ hr. Rate $=\dfrac{1}{27}$

Work time for both $=t$ hr. Rate $=\dfrac{1}{t}$

$$\frac{1}{18}+\frac{1}{27}=\frac{1}{t}$$
$$\frac{54t}{1}\left(\frac{1}{18}+\frac{1}{27}\right)=\left(\frac{1}{t}\right)\frac{54t}{1}$$
$$\frac{54t}{1}\cdot\frac{1}{18}+\frac{54t}{1}\cdot\frac{1}{27}=\frac{54t}{1}\cdot\frac{1}{t}$$
$$3t+2t=54$$
$$5t=54$$
$$\frac{5t}{5}=\frac{54}{5}$$
$$t=10.8$$

Working together, Mike and his assistant can paint the house in 10.8 hours, or 10 hours 48 minutes.

69. $\dfrac{1\text{ft}}{1}\cdot\dfrac{12\text{in}}{1\text{ft}}=12\text{in}$

7 inches to 1 ft $=$ 7 inches to 12 inches

7 inches to 1 ft $=\dfrac{7}{12}$

70. The ratio of 3 to 8 $=\dfrac{3}{8}$

71. 12lb for \$9 $=\dfrac{\$9}{12\text{lb.}}=\$0.75/\text{lb.}$

72. 42 miles in 2 hours $=\dfrac{42\text{mi}}{2\text{hr}}=21$ mph

73. 946 ml of Concrobium Mold Control for \$9.49 $=\dfrac{\$9.49}{946\text{ml}}\approx\$0.010$ per ml.

3784 ml of Concrobium Mold Control for \$35.49 $=\dfrac{\$35.49}{3784\text{ml}}\approx\$0.009$ per ml.

3784 ml of Concrobium Mold Control for \$35.49 is the better buy.

74. 12 extra large grade AA eggs for \$2.59 $=\dfrac{\$2.59}{12}\approx\$0.22$ per egg.

18 extra large grade AA eggs for \$4.49 $=\dfrac{\$4.49}{18}\approx\$0.25$ per egg.

The package of 12 extra large grade AA eggs for \$2.59 is the better buy.

75. $\frac{35}{150}=\frac{x}{400}$

$150x=35\cdot 400$

$\frac{150x}{150}=\frac{14,000}{150}$

$x=93\frac{1}{3}$

There are approximately 93 g of sodium in a 32 fl. oz. serving of Pepsi.

76. $\frac{41}{150}=\frac{x}{400}$

$150x=41\cdot 400$

$\frac{150x}{150}=\frac{16,400}{150}$

$x=109\frac{1}{3}$

There are approximately 109 g of sugars in a 32 fl. oz. serving of Pepsi.

77. $\frac{x-1}{4}=\frac{x-2}{6}$ Cross multiply.

$4(x-2)=6(x-1)$

$4x-8=6x-6$

$\underline{\quad +8 \qquad +8}$

$4x=6x+2$

$\underline{-6x\ -6x}$

$-2x=2$

$\frac{-2x}{-2}=\frac{2}{-2}$

$x=-1$

78. $\frac{x+3}{8}=\frac{x-5}{4}$ Cross multiply.

$8(x-5)=4(x+3)$

$8x-40=4x+12$

$\underline{\quad +40 \qquad +40}$

$8x=4x+52$

$\underline{-4x\ -4x}$

$4x=52$

$\frac{4x}{4}=\frac{52}{4}$

$x=13$

79. $\frac{x+1}{2x-14}=\frac{9}{2}$ Cross multiply.

$9(2x-14)=2(x+1)$

$18x-126=2x+2$

$\underline{\quad +126 \qquad +126}$

$18x=2x+128$

$\underline{-2x\ -2x}$

$16x=128$

$\frac{16x}{16}=\frac{128}{16}$

$x=8$

80. $\frac{2x+3}{3x+18}=\frac{-7}{3}$ Cross multiply.

$-7(3x+18)=3(2x+3)$

$-21x-126=6x+9$

$\underline{\quad +126 \qquad +126}$

$-21x=6x+135$

$\underline{-6x\ -6x}$

$-27x=135$

$\frac{-27x}{-27}=\frac{135}{-27}$

$x=-5$

81. $\frac{x}{12}=\frac{20}{8}$ Cross multiply.

$8x=12\cdot 20$

$8x=240$

$\frac{8x}{8}=\frac{240}{8}$

$x=30$

82. $\frac{h}{6.35}=\frac{115.28}{5}$ Cross multiply.

$5h=6.35\cdot 115.28$

$5h=7320.28$

$\frac{5h}{5}=\frac{7320.28}{5}$

$h\approx 146.41$

The original height of the Great Pyramid of Giza is approximately 146.41 m.

83. $y = kx$ Substitute given values for x and y.

$$(12) = k(36)$$
$$\frac{12}{36} = \frac{36k}{36}$$
$$\frac{1}{3} = k$$

$y = \frac{1}{3}x$ Substitute values for k and x.

$$y = \frac{1}{3}(10)$$
$$y = \frac{10}{3}$$

84. $y = kx$ Substitute given values for x and y.

$$(9) = k\left(\frac{1}{2}\right)$$
$$2 \cdot 9 = \frac{1}{2}k \cdot 2$$
$$18 = k$$

$y = 18x$ Substitute values for k and x.

$$y = 18(6)$$
$$y = 108$$

85. $y = \frac{k}{x}$ Substitute given values for x and y.

$\left(-\frac{1}{3}\right) = \frac{k}{(12)}$ Cross multiply.

$$3k = -1 \cdot 12$$
$$3k = -12$$
$$\frac{3k}{3} = \frac{-12}{3}$$
$$k = -4$$

$y = \frac{-4}{x}$ Substitute values for k and x.

$$y = \frac{-4}{(-2)}$$
$$y = 2$$

86. $y = \frac{k}{x}$ Substitute given values for x and y.

$\left(-\frac{1}{2}\right) = \frac{k}{(14)}$ Cross multiply.

$$2k = -1 \cdot 14$$
$$2k = -14$$
$$\frac{2k}{2} = \frac{-14}{2}$$
$$k = -7$$

$y = \frac{-7}{x}$ Substitute values for k and x.

$$y = \frac{-7}{(21)}$$
$$y = -\frac{1}{3}$$

Chapter 7 Test

1. $\dfrac{(b+2)\cancel{(b-9)}}{(b-4)\cancel{(b-9)}}=\dfrac{b+2}{b-4}$

2. $\dfrac{(c-7)\cancel{(c+7)}}{\cancel{(c+7)}(c+2)}=\dfrac{c-7}{c+2}$

3. Set denominator equal to 0 and solve:

$\dfrac{x+4}{x+7}$

$x+7=0$

$x=-7$ This is the excluded value.

4. Set denominator equal to 0 and solve:

$\dfrac{x+4}{x^2-5x-24}$

$x^2-5x-24=0$

$(x-8)(x+3)=0$

$x-8=0 \qquad x+3=0$

$x=8 \qquad x=-3$

$x=8,-3$ These are the excluded values.

5. $\dfrac{150\ \cancel{\text{mile}}}{1\text{ hr}}\cdot\dfrac{5280\text{ ft}}{1\ \cancel{\text{mile}}}=\dfrac{792{,}000\text{ ft}}{1\text{ hr}}$

$\dfrac{792{,}000\text{ ft}}{1\ \cancel{\text{hr}}}\cdot\dfrac{1\ \cancel{\text{hr}}}{3600\text{ sec}}=\dfrac{792{,}000\text{ ft}}{3600\text{ sec}}=220\text{ ft/sec}$

The rocket is traveling at 220 feet per second.

6. $\dfrac{\cancel{(d-3)}}{\cancel{(d+8)}}\cdot\dfrac{(d+2)\cancel{(d+8)}}{\cancel{(d-3)}(d+7)}=\dfrac{d+2}{d+7}$

7. $\dfrac{\cancel{(p+1)}}{(p-6)}\cdot\dfrac{(p+3)}{(p+4)\cancel{(p+1)}}=\dfrac{p+3}{(p-6)(p+4)}$

8. $\dfrac{(k-11)}{(k+8)}\div\dfrac{2k-22}{k^2+5k-24}=\dfrac{(k-11)}{(k+8)}\cdot\dfrac{k^2+5k-24}{2k-22}$

$=\dfrac{\cancel{(k-11)}}{\cancel{(k+8)}}\cdot\dfrac{\cancel{(k+8)}(k-3)}{2\cancel{(k-11)}}=\dfrac{k-3}{2}$

9. $\dfrac{2}{x+7} \quad \dfrac{4x}{(x+3)(x+7)}$ LCD $=(x+3)(x+7)$

10. $\dfrac{5x}{x+3}+\dfrac{7}{x+3}=\dfrac{5x+7}{x+3}$

11. $\dfrac{2}{n-7}+\dfrac{4}{n-8}$ LCD $=(n-7)(n-8)$

$=\dfrac{2}{n-7}\cdot\dfrac{n-8}{n-8}+\dfrac{4}{n-8}\cdot\dfrac{n-7}{n-7}$

$=\dfrac{2n-16}{(n-7)(n-8)}+\dfrac{4n-28}{(n-7)(n-8)}$

$=\dfrac{2n-16+4n-28}{(n-7)(n-8)}=\dfrac{6n-44}{(n-7)(n-8)}$

12. $\dfrac{a}{a-2}+\dfrac{4}{(a-4)(a-2)}$

LCD $=(a-4)(a-2)$

$=\dfrac{a}{a-2}\cdot\dfrac{a-4}{a-4}+\dfrac{4}{(a-4)(a-2)}$

$=\dfrac{a^2-4a}{(a-4)(a-2)}+\dfrac{4}{(a-4)(a-2)}$

$=\dfrac{a^2-4a+4}{(a-4)(a-2)}=\dfrac{(a-2)\cancel{(a-2)}}{(a-4)\cancel{(a-2)}}=\dfrac{a-2}{a-4}$

13. $\dfrac{4}{g-3}-\dfrac{3}{g+2}$

LCD $=(g-3)(g+2)$

$=\dfrac{4}{g-3}\cdot\dfrac{g+2}{g+2}-\dfrac{3}{g+2}\cdot\dfrac{g-3}{g-3}$

$=\dfrac{4g+8}{(g-3)(g+2)}-\dfrac{3g-9}{(g-3)(g+2)}$

$=\dfrac{4g+8-3g+9}{(g-3)(g+2)}=\dfrac{g+17}{(g-3)(g+2)}$

14. $\dfrac{4}{(t-3)(t+1)}-\dfrac{2}{(t-3)}$

LCD $=(t-3)(t+1)$

$=\dfrac{4}{(t-3)(t+1)}-\dfrac{2}{(t-3)}\cdot\dfrac{t+1}{t+1}$

$=\dfrac{4}{(t-3)(t+1)}-\dfrac{2t+2}{(t-3)(t+1)}$

$=\dfrac{4-2t-2}{(t-3)(t+1)}=\dfrac{-2t+2}{(t-3)(t+1)}$

15. $\dfrac{\dfrac{-2}{x+6}}{\dfrac{4}{x^2}+1}\cdot\dfrac{\dfrac{x^2(x+6)}{1}}{\dfrac{x^2(x+6)}{1}}$ Multiply by LCD $=x^2(x+6)$.

$=\dfrac{\dfrac{-2}{\cancel{x+6}}\cdot\dfrac{x^2\cancel{(x+6)}}{1}}{\dfrac{4}{\cancel{x^2}}\cdot\dfrac{\cancel{x^2}(x+6)}{1}+1\cdot\dfrac{x^2(x+6)}{1}}$

$=\dfrac{-2x^2}{4(x+6)+x^2(x+6)}=\dfrac{-2x^2}{(x+6)(x^2+4)}$

16. **a.** $c=\dfrac{150+20(10)}{(10)}$ Substitute $m=10$.

$$c=\frac{150+200}{10}=\frac{350}{10}=35$$

The average cost per foot is \$35 when 10 ft of mow curb is installed.

b. $(25)=\dfrac{150+20m}{m}$ Substitute $c=25$.

$$25m=150+20m \quad \text{Cross multiply.}$$
$$\underline{-20m \qquad -20m}$$
$$5m=150$$
$$\frac{5m}{5}=\frac{150}{5}$$
$$m=30$$

There must be 30 feet of mow curb installed for the average cost to be \$25 per foot.

17. $$\frac{5}{m}+3=\frac{17}{m}$$
$$\frac{m}{1}\left(\frac{5}{m}+3\right)=\left(\frac{17}{m}\right)\frac{m}{1}$$
$$\frac{m}{1}\cdot\frac{5}{m}+\frac{m}{1}\cdot 3=\frac{m}{1}\cdot\frac{17}{m}$$
$$5+3m=17$$
$$\underline{-5 \qquad -5}$$
$$3m=12$$
$$\frac{3m}{3}=\frac{12}{3}$$
$$m=4$$

18. $$\frac{3}{r+5}=2+\frac{1}{r+5}$$
$$\frac{r+5}{1}\left(\frac{3}{r+5}\right)=\left(2+\frac{1}{r+5}\right)\frac{r+5}{1}$$
$$\frac{r+5}{1}\cdot\frac{3}{r+5}=\frac{r+5}{1}\cdot 2+\frac{r+5}{1}\cdot\frac{1}{r+5}$$
$$3=2r+10+1$$
$$3=2r+11$$
$$\underline{-11 \qquad -11}$$
$$-8=2r$$
$$\frac{-8}{2}=\frac{2r}{2}$$
$$-4=r$$

19. $\dfrac{4}{z+2}=\dfrac{3}{z-8}$ Cross multiply.

$$3(z+2)=4(z-8)$$
$$3z+6=4z-32$$
$$\underline{-6 \qquad -6}$$
$$3z=4z-38$$
$$\underline{-4z \quad -4z}$$
$$-z=-38$$
$$\frac{-z}{-1}=\frac{-38}{-1}$$
$$z=38$$

20. $$\frac{h}{h+3}+\frac{2}{h+5}=\frac{-4}{(h+5)(h+3)}$$
$$\frac{(h+5)(h+3)}{1}\cdot\frac{h}{h+3}+\frac{2}{h+5}=\frac{-4}{(h+5)(h+3)}\cdot\frac{(h+5)(h+3)}{1}$$
$$h(h+5)+2(h+3)=-4$$
$$h^2+5h+2h+6=-4$$
$$h^2+7h+6=-4$$
$$\underline{+4 \quad +4}$$
$$h^2+7h+10=0$$
$$(h+5)(h+2)=0$$
$$h+5=0 \qquad h+2=0$$
$$h=-5 \qquad h=-2$$

$h=-5$ is an excluded value, therefore $h=-2$

21. This work rate problem will be solved in minutes rather than hours.

First pump's fill time $=25$ min. Rate $=\dfrac{1}{25}$

Larger pump's fill time $=10$ min. Rate $=\dfrac{1}{10}$

Fill time for both $=t$ min. Rate $=\dfrac{1}{t}$

$$\frac{1}{25}+\frac{1}{10}=\frac{1}{t}$$
$$\frac{50t}{1}\left(\frac{1}{25}+\frac{1}{10}\right)=\left(\frac{1}{t}\right)\frac{50t}{1}$$
$$\frac{50t}{1}\cdot\frac{1}{25}+\frac{50t}{1}\cdot\frac{1}{10}=\frac{50t}{1}\cdot\frac{1}{t}$$
$$2t+5t=50$$
$$7t=50$$
$$\frac{7t}{7}=\frac{50}{7}$$
$$t=7\frac{1}{7}$$

It would take $7\frac{1}{7}$ minutes, or about 7 minutes 9 seconds, to fill the truck if both pumps were used.

22. **a.** 5ft of lights for \$1.35 $= \dfrac{\$1.35}{5\text{ft.}} = \$0.27\text{ /ft.}$

b. 8ft of lights for \$2.24 $= \dfrac{\$2.24}{8\text{ft.}} = \$0.28\text{ /ft.}$

5 ft of lights for \$1.35 is cheaper.

23. $\dfrac{12}{10} = \dfrac{x}{12.5}$ Cross multiply.

$10x = 12 \cdot 12.5$

$10x = 150$

$\dfrac{10x}{10} = \dfrac{150}{10}$

$x = 15$

24. $\dfrac{40}{4} = \dfrac{x}{5.2}$ Cross multiply.

$4x = 40 \cdot 5.2$

$4x = 208$

$\dfrac{4x}{4} = \dfrac{208}{4}$

$x = 52$

It takes 52 seconds to download a 5.2 megabyte song.

25. $y = \dfrac{k}{x}$ Substitute given values for x and y.

$(2) = \dfrac{k}{(5)}$ Cross multiply.

$k = 2 \cdot 5$

$k = 10$

$y = \dfrac{10}{x}$ Substitute values for k and x.

$y = \dfrac{10}{(40)}$

$y = \dfrac{1}{4}$

Cumulative Review Chapters 1-7

1. $(-3a)\cdot(-3a)\cdot(-3a)\cdot(-3a) = (-3a)^4$

2. $-6^2 = -6 \cdot 6 = -36$

3. $(8 - 5 \cdot 6 \div 10)^2 - (-1)^3 + 17$

$= (8 - 30 \div 10)^2 - (-1)^3 + 17$

$= (8 - 3)^2 - (-1)^3 + 17$

$= (5)^2 - (-1)^3 + 17$

$= 25 - (-1) + 17$

$= 25 + 1 + 17$

$= 43$

4. Convert 68 meters to feet. First, form the unity fraction. We want to convert to the units of feet, so feet will go in the numerator. We want to convert from the units of meters, so meters will go in the denominator. There are 3.28 ft in 1 m, so the unity fraction is $\dfrac{3.28\text{ ft}}{1\text{ m}}$. Multiply 68 m by this unity fraction and simplify the expression as follows.

$68\text{ m} \cdot \dfrac{3.28\text{ ft}}{1\text{ m}}$

$= \dfrac{68\ \cancel{\text{m}}}{1} \cdot \dfrac{3.28\text{ ft}}{1\ \cancel{\text{m}}}$

$= 223.04\text{ ft}$

Therefore, 68 meters is equal to 223.04 feet.

5. Evaluate $\dfrac{3a+12}{a+4}$ for $a = 4$.

$\dfrac{3a+12}{a+4}$ Substitute in $a = 4$.

$= \dfrac{3(4)+12}{(4)+4}$ Simplify using order-of-operations agreement.

$= \dfrac{12+12}{4+4}$

$= \dfrac{24}{8} = 3$

6. Let n = a number. The sentence translates as $2n + 16$.

7. $-(3t-1) - 2t + \dfrac{1}{2}(6-4t)$

$= -(3t-1) - 2t + \dfrac{1}{2}(6-4t)$

$= -3t + 1 - 2t + \dfrac{1}{2}(6-4t)$

$= -3t + 1 - 2t + 3 - 2t$

$= -7t + 4$

8. $2x\boxed{-5y} = -10$ Identify variable term to isolate.

$\underline{-2x \qquad -2x}$ Subtract $2x$ from both sides.

$-5\boxed{y} = -2x - 10$ Identify variable to isolate.

$\dfrac{-5y}{-5} = \dfrac{-2x-10}{-5}$ Divide both sides by -5.

$y = \dfrac{2}{5}x + 2$

9. $\frac{-6}{7}n = \frac{2}{3}$

$\frac{7}{-6} \cdot \frac{-6}{7}n = \frac{2}{3} \cdot \frac{7}{-6}$

$n = -\frac{14}{18} = -\frac{7}{9}$

10. The theater will take in $17t$ dollars.

11. $P = R\boxed{-C}$ Identify variable term to isolate.

$\underline{-R \quad -R}$ Subtract R from both sides.

$P - R = -C$

$\frac{P-R}{-1} = \frac{-C}{-1}$ Divide both sides by -1.

$-P + R = C$

$C = R - P$

12. $-5t + 16.4 = 9.8$

$\underline{-16.4 \quad -16.4}$ Subtract 16.4 from both sides.

$-5t = -6.6$

$\frac{-5t}{-5} = \frac{-6.6}{-5}$ Divide both sides by -5.

$t = 1.32$

13. $P = 2l + 2w$

$102 = 2l + 2(22)$

$102 = 2l + 44$

$\underline{-44 \qquad -44}$

$58 = 2l$

$\frac{58}{2} = \frac{2l}{2}$

$29 = l$

The length of the garden is 29 feet.

14. $-6a + 10 = 6(3a - 9) + 16$

$-6a + 10 = 18a - 54 + 16$

$-6a + 10 = 18a - 38$

$\underline{-18a \qquad -18a}$ Subtract $18a$ from both sides.

$-24a + 10 = -38$

$\underline{-10 \quad -10}$ Subtract 10 from both sides.

$-24a = -48$

$\frac{-24a}{-24} = \frac{-48}{-24}$ Divide both sides by -24.

$a = 2$

15. $2x - 5 - 2(5x - 6) > 8$

$2x - 5 - 10x + 12 > 8$ Simplify.

$-8x + 7 > 8$

$\underline{-7 \quad -7}$ Subtract 7 from both sides.

$-8x > 1$ Divide both sides by -8.

$\frac{-8x}{-8} < \frac{1}{-8}$ Multiply by neg., reverse ineq.

$x < -\frac{1}{8}$

16. **a.**

x	$y = \lvert x + 5\rvert$
-8	3
-5	0
-2	3
0	5
2	7

b. The graph of $y = \lvert x + 5\rvert$ follows:

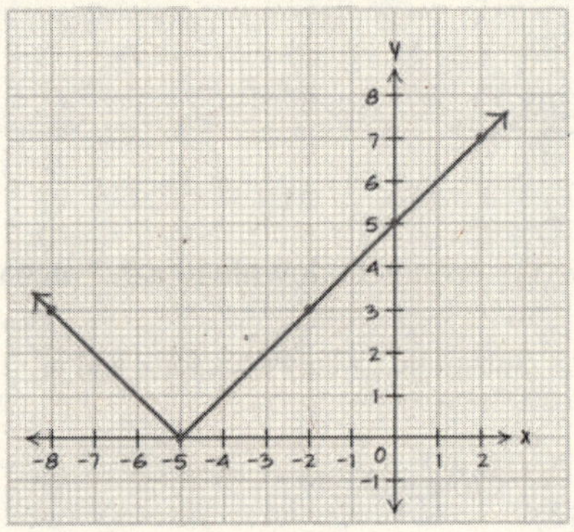

c. The graph is non-linear.

17.

h	$C = 508h$
0	0
0.5	254
1	508
1.5	762
2	1016

The graph of $C = 508h$ follows:

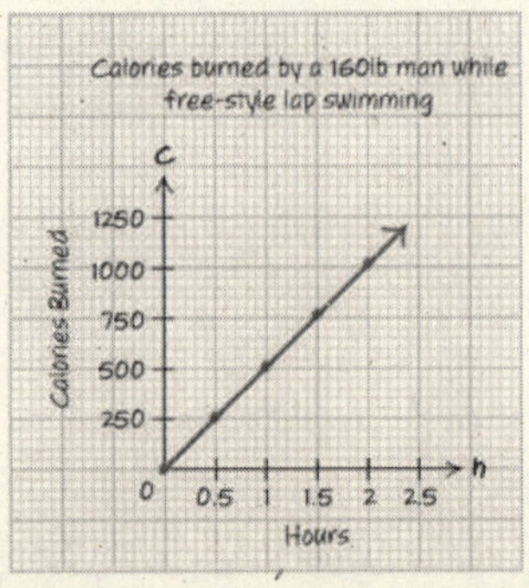

18. **a.** Let T = the total charges paid and let c = the number of credits. The equations translates as $T = 20 + 36c$.

b. $\text{slope} = \dfrac{\$36}{1 \text{ credit}} = 36$

c. The rate of increase is \$36.00 per credit hour.

19. $\text{slope} = \dfrac{y_2 - y_1}{x_2 - x_1} = \dfrac{-5-(3)}{8-(-6)} = \dfrac{-8}{14} = -\dfrac{4}{7}$

The slope of the line that passes through the given points is $m = -\dfrac{4}{7}$.

20. Find the y-intercept:

$4(0) - 6y = 18$ Substitute in $x = 0$.

$-6y = 18$

$\dfrac{-6y}{-6} = \dfrac{18}{-6}$ Solve for y.

$y = -3$

The y-intercept is the point $(0,-3)$.

Find the x-intercept:

$4x - 6(0) = 18$ Substitute in $y = 0$.

$4x = 18$

$\dfrac{4x}{4} = \dfrac{18}{4}$ Solve for x.

$x = \dfrac{9}{2}$

The x-intercept is the point $\left(\dfrac{9}{2}, 0\right)$.

The graph of $4x - 6y = 18$ follows:

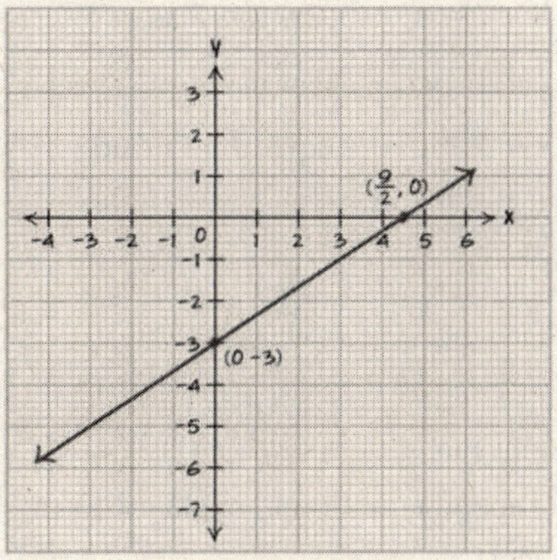

21. Slope $m = \dfrac{7}{5}$ and y-intercept $(0,-6)$.

The graph of $y = \dfrac{7}{5}x - 6$ follows:

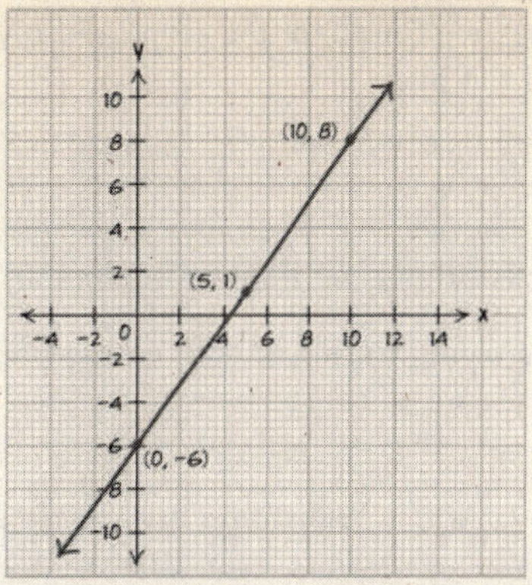

22. Change the first equation to slope-intercept form.

$2x + 2y = -9$

$-2x \quad -2x$ Subtract $2x$ from both sides.

$2y = -2x - 9$

$\dfrac{2y}{2} = \dfrac{-2x-9}{2}$

$y = -x - \dfrac{9}{2}$

Comparing the slopes of the two equations, the first equation has slope $m = -1$ and the second equation has slope $m = 1$. These are negative reciprocals, therefore these lines are perpendicular.

23. Find the slope of the line using the given points.

$\text{slope} = \dfrac{7-(4)}{0-(-8)} = \dfrac{3}{8}$

Find the y-intercept using the value for slope and one of the given points $(0,7)$.

$y = mx + b$ Substitute $m = \dfrac{3}{8}$, $x = 0, y = 7$

$(7) = \left(\dfrac{3}{8}\right)(0) + b$

$7 = 0 + b$ Solve for b.

$7 = b$

Using the values we found for the slope and b yields the equation of the line

$$y = \frac{3}{8}x + 7$$

24. Find the slope of the line using the given points.

$\text{slope} = \dfrac{5-(5)}{-3-(2)} = \dfrac{0}{-5} = 0$

Find the y-intercept using the value for slope and one of the given points $(2,5)$.

$y = mx + b$ Substitute $m = 0$, $x = 2, y = 5$

$(5) = (0)(2) + b$

$5 = 0 + b$ Solve for b.

$5 = b$

Using the values we found for the slope and b yields the equation of the line

$$y = 0x + 5$$
$$y = 5$$

25. Parallel lines have the same slope. The slope of the line that we are trying to find is $-\frac{3}{2}$. Using point-slope form:

$y - y_1 = m(x - x_1)$ Substitute in $m = -\frac{3}{2}$ and

$y - (7) = -\frac{3}{2}[x - (-2)]$ $(x_1, y_1) = (-2, 7)$

$y - 7 = -\frac{3}{2}(x + 2)$

$y - 7 = -\frac{3}{2}x - 3$

$+7 \quad +7$ Add 7 to both sides.

$y = -\frac{3}{2}x + 4$

26. Perpendicular lines have slopes that are negative reciprocals. The slope of the given line is $\frac{1}{5}$, therefore the slope of the line that we are trying to find is -5. Using point-slope form:

$y - y_1 = m(x - x_1)$ Substitute in $m = -5$ and

$y - (5) = -5[x - (-3)]$ $(x_1, y_1) = (-3, 5)$

$y - 5 = -5(x + 3)$

$y - 5 = -5x - 15$

$+5 \quad +5$ Add 5 to both sides.

$y = -5x - 10$

27. Looking at the tables, there is one input and output pairing that is the same, $(-2, 7)$. This point is the solution to the system.

28. **a.** First equation:

$(-1) \stackrel{?}{=} -2(4) + 7$ Substitute values for x and y.

$-1 \stackrel{?}{=} -8 + 7$ Simplify.

$-1 = -1$ This is a true statement.

Second equation:

$(-1) \stackrel{?}{=} -(4) + 2$ Substitute values for x and y.

$-1 \stackrel{?}{=} -4 + 2$ Simplify.

$-1 \stackrel{?}{=} -2$ This is a false statement.

Since only one of the statements is true, $(4, -1)$ is not a solution of the system.

b. First equation:

$(-3) \stackrel{?}{=} -2(5) + 7$ Substitute values for x and y.

$-3 \stackrel{?}{=} -10 + 7$ Simplify.

$-3 = -3$ This is a true statement.

Second equation:

$(-3) \stackrel{?}{=} -(5) + 2$ Substitute values for x and y.

$-3 \stackrel{?}{=} -5 + 2$ Simplify.

$-3 = -3$ This is a true statement.

Since both statements are true, $(5, -3)$ is a solution of the system.

29. The graph of $y = -x + 6$ and $y = \frac{5}{4}x - 3$ follows:

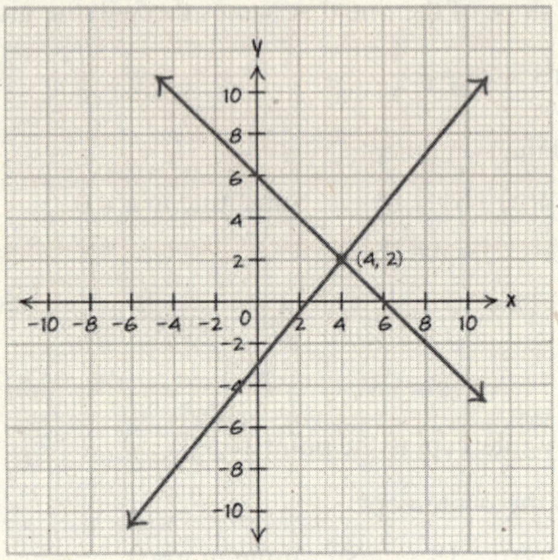

The solution is the point $(4, 2)$.

30. Change the second equation and substitute $y = 2x - 3$ into the first equation.

$x + 2(2x - 3) = 14$

$x + 4x - 6 = 14$

$5x - 6 = 14$

$+6 \quad +6$ Add 6 to both sides.

$5x = 20$

$\frac{5x}{5} = \frac{20}{5}$ Divide both sides by 5.

$x = 4$

Substitute $x = 4$ into the equation to find y.

$$y = 2x - 3$$
$$y = 2(4) - 3$$
$$y = 8 - 3$$
$$y = 5$$

The solution to the system is $(4, 5)$.

31. Change the second equation to standard form:

$y = \frac{3}{4}x - 2$

$-\frac{3}{4}x \quad -\frac{3}{4}x$ Subtract $\frac{3}{4}x$ from both sides.

$-\frac{3}{4}x + y = -2$ Multiply both sides by -4 to clear denom.

$-4\left(-\frac{3}{4}x + y\right) = -2 \cdot -4$

$3x - 4y = 8$

Add the equations vertically.

$$\begin{array}{rl} -3x+4y=4 & \text{Add the equations vertically.} \\ 3x-4y=8 & \\ \hline 0\stackrel{?}{=}12 & \text{This is a false statement.} \end{array}$$

The system is inconsistent; it has no solution. The lines are parallel.

32. Let $x =$ the cost (in dollars) of each package of glue sticks and let $y =$ the cost (in dollars) of each package of whiteboard markers.

The system of equations is:

$$6x+12y=126.00$$
$$4x+7y=76.50$$

Solve for y in the first equation.

$$\begin{array}{rl} 6x+12y=126.00 \\ -6x \quad\quad -6x \\ \hline 12y=-6x+126.00 \end{array}$$

$$\frac{12y}{12}=\frac{-6x+126.00}{12}$$

$$y=-0.5x+10.5$$

Now substitute $y=-0.5x+10.5$ into the second equation.

$$\begin{array}{rl} 4x+7y=76.50 & \text{Substitute in } y=-0.5x+10.5 \\ 4x+7(-0.5x+10.5)=76.50 & \text{Distribute.} \\ 4x-3.5x+73.5=76.50 & \text{Simplify.} \\ 0.5x+73.5=76.50 & \\ -73.5 \quad -73.5 & \text{Subtract 73.5 from both sides.} \\ \hline 0.5x=3 & \text{Solve for } x. \end{array}$$

$$\frac{0.5x}{0.5}=\frac{3}{0.5}$$

$$x=6$$

$$\begin{array}{rl} y=-0.5x+10.5 & \text{Substitute in } x=6 \text{ to find } y. \\ y=-0.5(6)+10.5 & \\ y=-3+10.5 & \\ y=7.5 & \end{array}$$

Each package of glue sticks costs \$6.00 and each package of whiteboard markers costs \$7.50.

33. Let $p =$ the rate of the plane (in mph) in still air and $w =$ the rate of the wind in mph.

The system of equations is:

$$2.2(p+w)=330$$
$$3(p-w)=330$$

Solve the second equation for p.

$$\begin{array}{rl} 3(p-w)=330 & \text{Distribute.} \\ 3p-3w=330 & \\ +3w \quad +3w & \\ \hline 3p=3w+330 & \end{array}$$

$$\frac{3p}{3}=\frac{3w+330}{3}$$

$$p=w+110$$

Now substitute $p=w+110$ into the first equation.

$$\begin{array}{rl} 2.2(p+w)=330 & \text{Distribute.} \\ 2.2p+2.2w=330 & \text{Substitute in } p=w+110. \\ 2.2(w+110)+2.2w=330 & \text{Distribute.} \\ 2.2w+242+2.2w=330 & \text{Simplify.} \\ 4.4w+242=330 & \\ -242 \quad -242 & \\ \hline 4.4w=88 & \text{Solve for } w. \end{array}$$

$$\frac{4.4w}{4.4}=\frac{88}{4.4}$$

$$w=20$$

$$\begin{array}{rl} p=w+110 & \text{Substitute in } w=20 \text{ to find } p. \\ p=(20)+110 & \\ p=130 & \end{array}$$

The rate of the plane is 130 mph and the rate of the wind is 20 mph.

34. Let $t =$ the number of pounds of white tea and let $b =$ the number of pounds of blueberry herbal tea.

The system of equations is:

$$t+b=10$$
$$24t+16b=196$$

Solve for t in the first equation.

$$\begin{array}{rl} t+b=10 & \\ -b \quad -b & \text{Subtract } b \text{ from both sides.} \\ \hline t=-b+10 & \end{array}$$

Now substitute $t=-b+10$ into the second equation.

$$\begin{array}{rl} 24t+16b=196 & \text{Substitute in } t=-b+10. \\ 24(-b+10)+16b=196 & \text{Distribute.} \\ -24b+240+16b=196 & \text{Simplify.} \\ -8b+240=196 & \\ -240 \quad -240 & \\ \hline -8b=-44 & \text{Solve for } b. \end{array}$$

$$\frac{-8b}{-8}=\frac{-44}{-8}$$

$$b=5.5$$

$$\begin{array}{rl} t=-b+10 & \text{Substitute in } b=5.5 \text{ to find } t. \\ t=-(5.5)+10 & \\ t=4.5 & \end{array}$$

Chi-Lan needs to put 4.5 lb. of white tea and 5.5 lb. of blueberry herbal tea in the mix.

35. Multiply the first equation by -4:

$$\begin{array}{rcl} -4(3x+y)=-4(-18) & \text{is now} & -12x-4y=72 \\ -5x+4y=-4 & & -5x+4y=-4 \end{array}$$

Add the equations vertically to eliminate y.

$$-12x-4y=72 \quad \text{Add the equations vertically.}$$
$$\underline{-5x+4y=-4}$$
$$-17x=68 \quad \text{Solve for } x.$$
$$\frac{-17x}{-17}=\frac{68}{-17}$$
$$x=-4$$

Substitute $x=4$ into either original equation to find the value of y.

$$3x+y=-18 \quad \text{Substitute in } x=-4$$
$$3(-4)+y=-18$$
$$-12+y=-18$$
$$\underline{+12 \qquad +12}$$
$$y=-6$$

The solution to the system is $(-4,-6)$.

36. Multiply the first equation by 2:

$$\begin{matrix} 2(-x-5y)=2(4) \\ 2x+10y=-8 \end{matrix} \text{ is now } \begin{matrix} -2x-10y=8 \\ 2x+10y=-8 \end{matrix}$$

Add the equations vertically to eliminate y.

$$-2x-10y=8 \quad \text{Add the equations vertically.}$$
$$\underline{2x+10y=-8}$$
$$0=0$$

Because this equation is always true ($0=0$), there are infinitely many solutions. These two equations graph as the same line.

37. Supplementary angles have a sum of 180°. Let $x=$ the degree measure of the first angle, therefore the second angle will measure $3x$.

$$x+3x=180^\circ$$
$$4x=180$$
$$\frac{4x}{4}=\frac{180}{4}$$
$$x=45$$

The measure of the first angle is 45° and the measure of the second angle is 135°.

38. Let $x=$ the amount (in dollars) invested in the CD earning 1.5% interest and let $y=$ the amount (in dollars) invested in the money market fund earning 2.0% interest.

The system of equations is:

$$x+y=5000$$
$$0.015x+0.02y=82.5$$

Solve for y in the first equation.

$$x+y=5000$$
$$\underline{-x \qquad -x}$$
$$y=-x+5000$$

Now substitute $y=-x+5000$ into the second equation.

$$0.015x+0.02y=82.5 \quad \text{Substitute in } y=-x+5000.$$
$$0.015x+0.02(-x+5000)=82.5 \quad \text{Distribute.}$$
$$0.015x-0.02x+100=82.5 \quad \text{Simplify.}$$
$$-0.005x+100=82.5$$
$$\underline{-100 \quad -100} \quad \text{Subtract 100 from both sides.}$$
$$-0.005x=-17.5 \quad \text{Solve for } x.$$
$$\frac{-0.005x}{-0.005}=\frac{-17.5}{-0.005}$$
$$x=3500$$

$$y=-x+5000 \quad \text{Substitute in } x=3500 \text{ to find } y.$$
$$y=-(3500)+5000$$
$$y=1500$$

Conrad should invest \$3500 in the CD and \$1500 in the money market fund.

39. $-3x+4y>7$ Is $(x,y)=(0,0)$ in the solution set?

$$-3(0)+4(0)\overset{?}{>}7 \quad \text{Substitute values for } x \text{ and } y.$$
$$0+0\overset{?}{>}7$$
$$0\overset{?}{>}7 \quad \text{This statement is false.}$$

No, the point is not in the solution set.

40. The graph of $3x-2y\le 6$ follows:

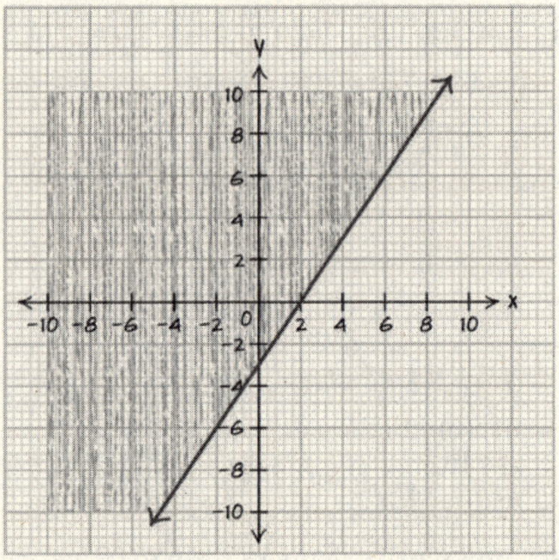

41. The graph of $y > \frac{3}{5}x - 4$ follows:

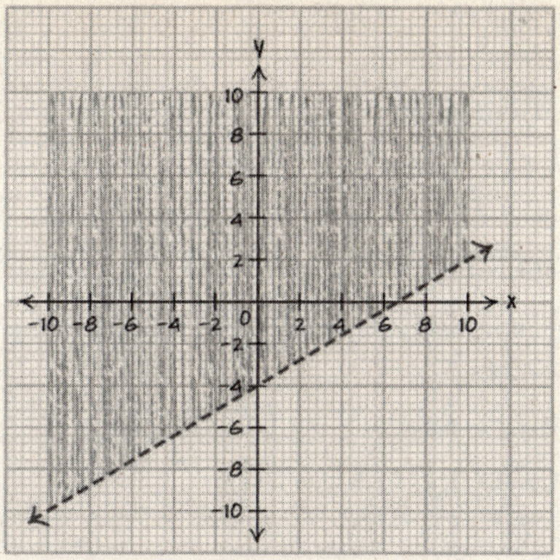

42. $(-1)^4 = (-1)\cdot(-1)\cdot(-1)\cdot(-1) = 1$

43. $-1^4 = -(1)\cdot(1)\cdot(1)\cdot(1) = -1$

44. **a.** $a^4 \cdot a \cdot a^4 = a^{4+1+4} = a^9$

b. $(-3b^3)(5b^6) = (-3)(5)\cdot b^{3+6} = -15b^9$

c. $(-4xy^2)(-x^3y) = (-4)(-1)\cdot x^{1+3}\cdot y^{2+1} = 4x^4y^3$

45. **a.** $-t^4(-9t) = (-1)(-9)t^{4+1} = 9t^5$

b. $(16n)(-3n^3) = (16)(-3)\cdot n^{1+3} = -48n^4$

c. $(5ab^3)(-2a^4b^2) = (5)(-2)\cdot a^{1+4}\cdot b^{3+2} = -10a^5b^5$

46. **a.** $(4t^2)^3 = (4)^3\cdot(t^2)^3 = 64t^{2\cdot 3} = 64t^6$

b. $(-5n^5)^2 = (-5)^2\cdot(n^5)^2 = 25n^{5\cdot 2} = 25n^{10}$

c. $(6ab^7)^2 = (6)^2\cdot(a)^2\cdot(b^7)^2 = 36a^2b^{7\cdot 2} = 36a^2b^{14}$

47. **a.** $(3a^7)^3 = (3)^3\cdot(a^7)^3 = 27a^{7\cdot 3} = 27a^{21}$

b. $(-9b^6)^2 = (-9)^2\cdot(b^6)^2 = 81b^{6\cdot 2} = 81b^{12}$

c. $(-2x^3y^4)^3 = (-2)^3\cdot(x^3)^3\cdot(y^4)^3 = -8x^{3\cdot 3}y^{4\cdot 3}$

$= -8x^9y^{12}$

48. **a.** $(-5x^3)^2(3xy^4) = (-5)^2(x^3)^2(3xy^4) = 25x^6\cdot 3xy^4$

$= 75x^{6+1}y^4 = 75x^7y^4$

b. $(7a^2b^3)^2(2ab^4)^3 = (7)^2(a^2)^2(b^3)^2(2)^3(a)^3(b^4)^3$

$= 49a^4b^6\cdot 8a^3b^{12} = 392a^{4+3}b^{6+12} = 392a^7b^{18}$

49. **a.** $(-12p^4)(-2pq^3)^4 = (-12p^4)(-2)^4(p)^4(q^3)^4$

$= -12p^4\cdot 16p^4q^{12} = -192p^{4+4}q^{12} = -192p^8q^{12}$

b. $(-3g^2h)^2(-5gh^5)^2 = (-3)^2(g^2)^2(h)^2(-5)^2(g)^2(h^5)^2$

$= 9g^4h^2\cdot 25g^2h^{10} = 225g^{4+2}h^{2+10} = 225g^6h^{12}$

50. **a.** $\frac{a^5}{a^2} = a^{5-2} = a^3$

b. $\frac{-9b^4}{36b^3} = -\frac{b^{4-3}}{4} = -\frac{b}{4}$

c. $\frac{3a^3\cdot 2a^3}{4a\cdot 8a^2} = \frac{6a^6}{32a^3} = \frac{3a^{6-3}}{16} = \frac{3a^3}{16}$

51. **a.** $\frac{n^8}{n^5} = n^{8-5} = n^3$

b. $\frac{7a^9}{-21a^8} = -\frac{a^{9-8}}{3} = -\frac{a}{3}$

c. $\frac{r^3r^2t}{r^4t} = \frac{r^5t}{r^4t} = r^{5-4}t^{1-1} = r$

52. $\left(\frac{-8a^5b^8}{24a^3b^5}\right)^2 = \left(-\frac{a^{5-3}b^{8-5}}{3}\right)^2$

$= \left(-\frac{a^2b^3}{3}\right)^2 = \left(-\frac{1}{3}\right)^2(a^2)^2(b^3)^2 = \frac{a^4b^6}{9}$

53. $\left(\frac{-24x^5y^3}{6x^2y}\right)^2 = \left(-4x^{5-2}y^{3-1}\right)^2$

$= \left(-4x^3y^2\right)^2 = (-4)^2(x^3)^2(y^2)^2 = 16x^6y^4$

54. $A = \frac{1}{2}bh$

$A = \frac{1}{2}(4x)(6x)$

$A = (2x)(6x)$

$A = 12x^2$

55. $A = \pi r^2$

$A = \pi(6x)^2$

$A = 36\pi x^2$

56. **a.** $-6y^{-3} = -6\cdot\frac{1}{y^3} = -\frac{6}{y^3}$

b. $\frac{-1}{a^{-2}} = -a^2$

57. **a.** $-7t^{-1} = -7\cdot\frac{1}{t} = -\frac{7}{t}$

b. $\frac{-5}{r^{-3}} = -5r^3$

58. **a.** $\frac{ab^5}{a^4b^2} = a^{1-4}b^{5-2} = a^{-3}b^3 = \frac{b^3}{a^3}$

b. $\frac{-6r^{-2}t^9}{15rt^{-3}} = \frac{-2r^{-2-1}t^{9-(-3)}}{5} = \frac{-2r^{-3}t^{12}}{5} = \frac{-2t^{12}}{5r^3}$

c. $(-2x^{-2})^4 = (-2)^4(x^{-2})^4 = 16x^{-8} = \dfrac{16}{x^8}$

d. $\left(\dfrac{a^2}{b}\right)^{-2} = \left(\dfrac{b}{a^2}\right)^2 = \dfrac{(b)^2}{(a^2)^2} = \dfrac{b^2}{a^4}$

59. a. $\dfrac{x^5y^2}{x^7y} = x^{5-7}y^{2-1} = x^{-2}y = \dfrac{y}{x^2}$

b. $\dfrac{-30p^{-2}q^8}{45p^6q^3} = \dfrac{-2p^{-2}q^8}{3p^6q^3} = \dfrac{-2p^{-2-6}q^{8-3}}{3}$

$= \dfrac{-2p^{-8}q^5}{3} = \dfrac{-2q^5}{3p^8}$

c. $(-3x^{-5})^2 = (-3)^2(x^{-5})^2 = 9x^{-10} = \dfrac{9}{x^{10}}$

d. $\left(\dfrac{x^3}{y^4}\right)^{-3} = \left(\dfrac{y^4}{x^3}\right)^3 = \dfrac{(y^4)^3}{(x^3)^3} = \dfrac{y^{12}}{x^9}$

60. a. $6{,}213{,}000{,}000 = 6.213\times10^9$

b. $0.0000468 = 4.68\times10^{-5}$

61. a. $72{,}910{,}000{,}000 = 7.291\times10^{10}$

b. $0.00273 = 2.73\times10^{-3}$

62. a. $1.06\times10^5 = 1.06\cdot100{,}000 = 106{,}000$

b. $5.071\times10^{-4} = 0.0005071$

63. a. $3.157\times10^7 = 3.157\cdot10{,}000{,}000 = 31{,}570{,}000$

b. $5.163\times10^{-5} = 0.00005163$

64. 0.0001 oz. $= 1.0\times10^{-4}$ oz.

65. $5{,}913{,}000{,}000$ km. $= 5.913\times10^9$ km.

66. The polynomial has 4 terms. The highest exponent is 5 so the degree is 5.

67. The polynomial has 1 term. The highest exponent is 9 so the degree is 9.

68. The degree of the term $6ab^3$ is 4.

The degree of the term $-12a^5b^2$ is 7.

The degree of the term $16b^6$ is 6.

The highest degree term has a degree of 7, so the degree of the polynomial is 7.

69. The degree of the term $-17x^4y^2$ is 6.

The degree of the term $2xy^3$ is 4.

The degree of the term -8 is 0.

The highest degree term has a degree of 6, so the degree of the polynomial is 6.

70. $(-x^2+2x+7)+(3x^2-x-9)$

$= -x^2+3x^2+2x-x+7-9$

$= 2x^2+x-2$

71. $(-4y+3)-(-8y-1)$

$= -4y+8y+3+1$

$= 4y+4$

72. $-4(3x-8)$

$= -4(3x)-4(-8)$

$= -12x+32$

73. $-2x(x^2-6x+2)$

$= -2x(x^2)-2x(-6x)-2x(2)$

$= -2x^3+12x^2-4x$

74. a. $(5a-1)(a+2)$

$= 5a^2+10a-a-2$

$= 5a^2+9a-2$

b. $(y-3)(y^2+2y-4)$

$= y^3+2y^2-4y-3y^2-6y+12$

$= y^3-y^2-10y+12$

c. $(3x-1)^2 = 9x^2-6x+1$

d. $(5x+2)(5x-2) = 25x^2-4$

75. a. $(x+7)(3x-1)$

$= 3x^2-x+21x-7$

$= 3x^2+20x-7$

b. $(t+2)(3t^2-t+8)$

$= 3t^3-t^2+8t+6t^2-2t+16$

$= 3t^3+5t^2+6t+16$

c. $(5y+2)^2 = 25y^2+20y+4$

d. $(n-6)(n+6) = n^2-36$

76. $2a+7-3(5a-6)$

$= 2a+7-15a+18$

$= -13a+25$

77. $-8t+1-4(3t-9)$

$= -8t+1-12t+36$

$= -20t+37$

78. **a.** $\dfrac{-4x^2-6x+2}{2x}$

$=\dfrac{-4x^{2\,1}}{2x}-\dfrac{6x}{2x}+\dfrac{2}{2x}$

$=-2x-3+\dfrac{1}{x}$

b. $\dfrac{5n^3-10n^2+15}{5n}$

$=\dfrac{5n^{3\,2}}{5n}-\dfrac{10n^{2\,1}}{5n}+\dfrac{15}{5n}$

$=n^2-2n+\dfrac{3}{n}$

79. **a.**

$$\begin{array}{r} 4x+1 \\ 3x-1\overline{)12x^2-x-1} \\ \underline{-12x^2+4x} \\ 3x-1 \\ \underline{-3x+1} \\ 0 \end{array}$$

$(12x^2-x-1)\div(3x-1)=4x+1$

b.

$$\begin{array}{r} x-4 \\ 2x+3\overline{)2x^2-5x-12} \\ \underline{-2x^2-3x} \\ -8x-12 \\ \underline{8x+12} \\ 0 \end{array}$$

$(2x^2-5x-12)\div(2x+3)=x-4$

80. **a.**

$$\begin{array}{r|rrr} 3 & 3 & -11 & 6 \\ & & 9 & -6 \\ \hline & 3 & -2 & 0 \end{array}$$

$(3x^2-11x+6)\div(x-3)=3x-2$

b.

$$\begin{array}{r|rrr} 7 & 5 & -33 & -14 \\ & & 35 & 14 \\ \hline & 5 & 2 & 0 \end{array}$$

$(5x^2-33x-14)\div(x-7)=5x+2$

81. **a.** GCF $=5x^2$

$-10x^5+20x^3+5x^2=5x^2(-2x^3+4x+1)$

b. GCF $=(b-5)$

$3a(b-5)-2(b-5)=(b-5)(3a-2)$

82. **a.** GCF $=25ab^3$

$-25a^3b^4+75ab^3=25ab^3(-a^2b+3)$

b. GCF $=(h+2)$

$-7g(h+2)+3(h+2)=(h+2)(-7g+3)$

83. $xy+4x-2y-8$

$=(xy+4x)-(2y+8)$

$=x(y+4)-2(y+4)$

$=(y+4)(x-2)$

84. $ab-5b+2a-10$

$=(ab-5b)+(2a-10)$

$=b(a-5)+2(a-5)$

$=(a-5)(b+2)$

85. $2ab-10b+12a-60$

$=2(ab-5b+6a-30)$

$=2\left[(ab-5b)+(6a-30)\right]$

$=2\left[b(a-5)+6(a-5)\right]$

$=2(a-5)(b+6)$

86. $3xy-6x+12y-24$

$=3(xy-2x+4y-8)$

$=3\left[(xy-2x)+(4y-8)\right]$

$=3\left[x(y-2)+4(y-2)\right]$

$=3(y-2)(x+4)$

87. **a.** Factors of -28 that add to -3 are 4 and -7.

$a^2-3a-28=(a+4)(a-7)$

b. $-y^2+9y-18=-(y^2-9y+18)$

Factors of 18 that add to -9 are -6 and -3.

$-(y^2-9y+18)=-(y-6)(y-3)$

c. $a\cdot c=6\cdot-6=-36$ Factors of -36 that add to -5 are 4 and -9. Factor by grouping.

$6x^2-5x-6=6x^2+4x-9x-6$

$=(6x^2+4x)-(9x+6)$

$=2x(3x+2)-3(3x+2)$

$=(3x+2)(2x-3)$

88. **a.** Factors of 21 that add to -10 are -7 and -3.

$a^2-10a+21=(a-7)(a-3)$

b. $-b^2+13b-42=-(b^2-13b+42)$

Factors of 42 that add to -13 are -7 and -6.

$-(b^2-13b+42)=-(b-7)(b-6)$

c. $a\cdot c=10\cdot -7=-70$ Factors of -70 that add to 33 are 35 and -2. Factor by grouping.

$$\begin{aligned}10t^2+33t-7&=10t^2+35t-2t-7\\&=(10t^2+35t)-(2t+7)\\&=5t(2t+7)-(2t+7)\\&=(2t+7)(5t-1)\end{aligned}$$

89. **a.** $10t^3+28t^2-6t=2t(5t^2+14t-3)$

$a\cdot c=5\cdot -3=-15$ Factors of -15 that add to 14 are 15 and -1. Factor by grouping.

$$\begin{aligned}2t(5t^2+14t-3)&=2t(5t^2+15t-t-3)\\&=2t\left[(5t^2+15t)-(t+3)\right]\\&=2t\left[5t(t+3)-(t+3)\right]\\&=2t(t+3)(5t-1)\end{aligned}$$

b. $6a^2-12ab+2ab-4b^2$

$$\begin{aligned}&=2(3a^2-6ab+ab-2b^2)\\&=2\left[(3a^2-6ab)+(ab-2b^2)\right]\\&=2\left[3a(a-2b)+b(a-2b)\right]\\&=2(a-2b)(3a+b)\end{aligned}$$

90. **a.** $6n^3-27n^2-105n=3n(2n^2-9n-35)$

$a\cdot c=2\cdot -35=-70$ Factors of -70 that add to -9 are -14 and 5. Factor by grouping.

$$\begin{aligned}3n(2n^2-9n-35)&=3n(2n^2-14n+5n-35)\\&=3n\left[(2n^2-14n)+(5n-35)\right]\\&=3n\left[2n(n-7)+5(n-7)\right]\\&=3n(n-7)(2n+5)\end{aligned}$$

b. $6rt-3t+24r-12$

$$\begin{aligned}&=3(2rt-t+8r-4)\\&=3\left[(2rt-t)+(8r-4)\right]\\&=3\left[t(2r-1)+4(2r-1)\right]\\&=3(2r-1)(t+4)\end{aligned}$$

91. **a.** Using $a^2-b^2=(a-b)(a+b)$.

$x^2-64=(x-8)(x+8)$

b. Using $a^2-b^2=(a-b)(a+b)$.

$16-25t^2=(4-5t)(4+5t)$

c. Factor using $a^2-2ab+b^2=(a-b)^2$.

$9x^2-6x+1=(3x-1)^2$

d. x^2+64 is prime.

e. $3t^2-27=3(t^2-9)$

Using $a^2-b^2=(a-b)(a+b)$.

$3(t^2-9)=3(t-3)(t+3)$

92. **a.** Using $a^2-b^2=(a-b)(a+b)$.

$y^2-81=(y-9)(y+9)$

b. Using $a^2-b^2=(a-b)(a+b)$.

$1-4a^2=(1-2a)(1+2a)$

c. Factor using $a^2+2ab+b^2=(a+b)^2$.

$4x^2+12x+9=(2x+3)^2$

d. y^2+81 is prime.

e. $2n^2-50=2(n^2-25)$

Using $a^2-b^2=(a-b)(a+b)$.

$2(n^2-25)=2(n-5)(n+5)$

93. **a.** This is an equation of degree 1, therefore it is linear.

b. This is an equation of degree 2 (the variable has an exponent of 2), therefore it is quadratic.

c. This is an equation of degree 1, therefore it is linear.

94. **a.** Factors of -35 that add to 2 are 7 and -5.

$a^2+2a-35=0$

$(a+7)(a-5)=0$

$$\begin{array}{rcl}a+7=0 & \qquad & a-5=0\\ \underline{-7\ \ -7} & & \underline{+5\ \ +5}\\ a=-7 & & a=5\end{array}$$

$a=-7$ and $a=5$

b. $a \cdot c = 5 \cdot -6 = -30$ Factors of -30 that add to -29 are -30 and 1. Factor by grouping.

$$5x^2 - 29x - 6 = 0$$
$$5x^2 - 30x + x - 6 = 0$$
$$(5x^2 - 30x) + (x - 6) = 0$$
$$5x(x-6) + (x-6) = 0$$
$$(x-6)(5x+1) = 0$$

$$x - 6 = 0 \qquad 5x + 1 = 0$$
$$\underline{+6 \ +6} \qquad \underline{-1 \ -1}$$
$$x = 6 \qquad 5x = -1$$
$$\frac{5x}{5} = \frac{-1}{5}$$
$$x = -\frac{1}{5}$$

$x = 6$ and $x = -\frac{1}{5}$

c. $x(x-11) = -24$

$$x^2 - 11x = -24$$
$$\underline{+24 \ \ +24}$$
$$x^2 - 11x + 24 = 0$$

Factors of 24 that add to -11 are -8 and -3.

$$x^2 - 11x + 24 = 0$$
$$(x-8)(x-3) = 0$$

$$x - 8 = 0 \qquad x - 3 = 0$$
$$\underline{+8 \ +8} \qquad \underline{+3 \ +3}$$
$$x = 8 \qquad x = 3$$

$x = 8$ and $x = 3$

d. Using $a^2 - b^2 = (a-b)(a+b)$.

$$b^2 - 1 = 0$$
$$(b-1)(b+1) = 0$$

$$b - 1 = 0 \qquad b + 1 = 0$$
$$\underline{+1 \ +1} \qquad \underline{-1 \ -1}$$
$$b = 1 \qquad b = -1$$

$b = 1$ and $b = -1$

95. a. Factors of 48 that add to 14 are 8 and 6.

$$b^2 + 14b + 48 = 0$$
$$(b+8)(b+6) = 0$$

$$b + 8 = 0 \qquad b + 6 = 0$$
$$\underline{-8 \ -8} \qquad \underline{-6 \ -6}$$
$$b = -8 \qquad b = -6$$

$b = -8$ and $b = -6$

b. $a \cdot c = 5 \cdot -8 = -40$ Factors of -40 that add to 6 are 10 and -4. Factor by grouping.

$$5x^2 + 6x - 8 = 0$$
$$5x^2 + 10x - 4x - 8 = 0$$
$$(5x^2 + 10x) - (4x + 8) = 0$$
$$5x(x+2) - 4(x+2) = 0$$
$$(x+2)(5x-4) = 0$$

$$x + 2 = 0 \qquad 5x - 4 = 0$$
$$\underline{-2 \ -2} \qquad \underline{+4 \ +4}$$
$$x = -2 \qquad 5x = 4$$
$$\frac{5x}{5} = \frac{4}{5}$$
$$x = \frac{4}{5}$$

$x = -2$ and $x = \frac{4}{5}$

c. $x(x-15) = -56$

$$x^2 - 15x = -56$$
$$\underline{+56 \ \ +56}$$
$$x^2 - 15x + 56 = 0$$

Factors of 56 that add to -15 are -8 and -7.

$$x^2 - 15x + 56 = 0$$
$$(x-8)(x-7) = 0$$

$$x - 8 = 0 \qquad x - 7 = 0$$
$$\underline{+8 \ +8} \qquad \underline{+7 \ +7}$$
$$x = 8 \qquad x = 7$$

$x = 8$ and $x = 7$

d. Using $a^2 - b^2 = (a-b)(a+b)$.

$$r^2 - 100 = 0$$
$$(r-10)(r+10) = 0$$

$$r - 10 = 0 \qquad r + 10 = 0$$
$$\underline{+10 \ +10} \qquad \underline{-10 \ -10}$$
$$r = 10 \qquad r = -10$$

$r = 10$ and $r = -10$

96. $\frac{(4)-3}{(4)^2+6}$ Substitute in $x = 4$.

$= \frac{1}{16+6}$ Simplify.

$= \frac{1}{22}$

97. **a.** Set denominator equal to 0 and solve:

$$\frac{3x-16}{5x+7}$$

$$5x+7=0$$
$$\underline{\quad -7 \quad -7}$$
$$5x=-7$$
$$\frac{5x}{5}=\frac{-7}{5}$$
$$x=-\frac{7}{5}$$ This is the excluded value.

b. Set denominator equal to 0 and solve:

$$\frac{4x}{x^2+3x-18}$$

$$x^2+3x-18=0$$
$$(x+6)(x-3)=0$$
$$x+6=0 \qquad x-3=0$$
$$x=-6 \qquad x=3$$
$$x=-6, 3$$ These are the excluded values.

98. **a.** Set denominator equal to 0 and solve:

$$\frac{-2y+5}{6y-12}$$

$$6y-12=0$$
$$\underline{\quad +12 \quad +12}$$
$$6y=12$$
$$\frac{6y}{6}=\frac{12}{6}$$
$$y=2$$ This is the excluded value.

b. Set denominator equal to 0 and solve:

$$\frac{4x+3}{2x^2+9x-5}$$

$$2x^2+9x-5=0$$
$$(2x-1)(x+5)=0$$
$$2x-1=0$$
$$\underline{\quad +1 \quad +1}$$
$$2x=1$$
$$\frac{2x}{2}=\frac{1}{2}$$
$$x=\frac{1}{2}$$

$$x+5=0$$
$$\underline{\quad -5 \quad -5}$$
$$x=-5$$

$$x=\frac{1}{2}, -5$$ These are the excluded values.

99. **a.** $\frac{t-5}{t^2-4t-5}$

$$=\frac{\cancel{t-5}}{(t+1)\cancel{(t-5)}}$$ Divide out common factor.

$$=\frac{1}{t+1}$$

b. $\frac{x+2}{x^2-4}$

$$=\frac{\cancel{x+2}}{(x-2)\cancel{(x+2)}}$$ Divide out common factor.

$$=\frac{1}{x-2}$$

c. $\frac{x-2}{2-x}$

$$=\frac{\cancel{x-2}}{-\cancel{(x-2)}}$$ Divide out common factor.

$$=-1$$

100. **a.** $\frac{5n-1}{5n^2+9n-2}$

$$=\frac{\cancel{5n-1}}{\cancel{(5n-1)}(n+2)}$$ Divide out common factor.

$$=\frac{1}{n+2}$$

b. $\frac{x+1}{4x^2-4}$

$$=\frac{x+1}{4(x^2-1)}$$

$$=\frac{\cancel{x+1}}{4(x-1)\cancel{(x+1)}}$$ Divide out common factor.

$$=\frac{1}{4(x-1)}$$

c. $\frac{5-n}{n-5}$

$$=\frac{-\cancel{(n-5)}}{\cancel{n-5}}$$ Divide out common factor.

$$=-1$$

101. **a.** $T=\frac{3(5)}{3+(5)}$ Substitute in $x = 5$.

$$T=\frac{15}{8}=1.875$$

The two carpet cleaners working together can complete the job in 1.875 hours.

b. $T=\frac{3(4)}{3+(4)}$ Substitute in $x = 4$.

$$T=\frac{12}{7}\approx 1.71$$

The two carpet cleaners working together can complete the job in approximately 1.71 hours.

102. **a.** $\frac{n+5}{(n-1)(n+3)}\cdot\frac{n-1}{n^2-25}$

$=\frac{\cancel{n+5}}{\cancel{(n-1)}(n+3)}\cdot\frac{\cancel{n-1}}{(n-5)\cancel{(n+5)}}$

$=\frac{1}{(n+3)(n-5)}$

b. $\frac{x+2}{2x^2-5x-3}\cdot\frac{2x+1}{x^2-4}$

$=\frac{\cancel{x+2}}{\cancel{(2x+1)}(x-3)}\cdot\frac{\cancel{2x+1}}{(x-2)\cancel{(x+2)}}$

$=\frac{1}{(x-3)(x-2)}$

103. **a.** $\frac{2x-9}{(x+5)(x-8)}\cdot\frac{x-8}{4x^2-81}$

$=\frac{\cancel{2x-9}}{(x+5)\cancel{(x-8)}}\cdot\frac{\cancel{x-8}}{\cancel{(2x-9)}(2x+9)}$

$=\frac{1}{(x+5)(2x+9)}$

b. $\frac{2t+10}{t^2-12t+27}\cdot\frac{t-9}{t^2-25}$

$=\frac{2\cancel{(t+5)}}{\cancel{(t-9)}(t-3)}\cdot\frac{\cancel{t-9}}{(t-5)\cancel{(t+5)}}$

$=\frac{2}{(t-3)(t-5)}$

104. $\frac{15\ \cancel{\text{mile}}}{1\ \text{hr}}\cdot\frac{5280\ \text{ft}}{1\ \cancel{\text{mile}}}=\frac{79{,}200\ \text{ft}}{1\ \text{hr}}$

$\frac{79{,}200\ \text{ft}}{1\ \cancel{\text{hr}}}\cdot\frac{1\ \cancel{\text{hr}}}{60\ \text{min}}=\frac{79{,}200\ \text{ft}}{60\ \text{min}}$

$\frac{79{,}200\ \text{ft}}{60\ \cancel{\text{min}}}\cdot\frac{1\ \cancel{\text{min}}}{60\ \text{sec}}=\frac{79{,}200\ \text{ft}}{3600\ \text{sec}}=22\ \text{ft/sec.}$

The runner's speed is 22 feet per second.

105. $\frac{42\ \text{mg}}{1\ \cancel{\text{kg}}}\cdot\frac{1\ \cancel{\text{kg}}}{2.2\ \text{lb}}=\frac{42\ \text{mg}}{2.2\ \text{lb}}\approx 19.09\ \text{mg/lb}$

The drug dosage is approximately 19.09 mg/lb.

106. **a.** $\frac{a-4}{(a-1)(a+5)}\div\frac{3a-12}{a-1}$

$=\frac{\cancel{a-4}}{\cancel{(a-1)}(a+5)}\cdot\frac{\cancel{a-1}}{3\cancel{(a-4)}}$

$=\frac{1}{3(a+5)}$

b. $\frac{z-5}{(z-6)(z+3)}\div\frac{z^2-25}{z^2+5z+6}$

$=\frac{z-5}{(z-6)(z+3)}\cdot\frac{z^2+5z+6}{z^2-25}$

$=\frac{\cancel{z-5}}{(z-6)\cancel{(z+3)}}\cdot\frac{\cancel{(z+3)}(z+2)}{\cancel{(z-5)}(z+5)}$

$=\frac{z+2}{(z-6)(z+5)}$

107. **a.** $\frac{x+4}{(x+7)(x-1)}\div\frac{5x+20}{x+7}$

$=\frac{\cancel{x+4}}{\cancel{(x+7)}(x-1)}\cdot\frac{\cancel{x+7}}{5\cancel{(x+4)}}$

$=\frac{1}{5(x-1)}$

b. $\frac{n-6}{(n+4)(n-7)}\div\frac{5n-30}{n^2-8n+7}$

$=\frac{n-6}{(n+4)(n-7)}\cdot\frac{n^2-8n+7}{5n-30}$

$=\frac{\cancel{n-6}}{(n+4)\cancel{(n-7)}}\cdot\frac{\cancel{(n-7)}(n-1)}{5\cancel{(n-6)}}$

$=\frac{n-1}{5(n+4)}$

108. **a.** $\frac{\frac{-15h}{4k^2}}{\frac{3h^3}{16k}}=\frac{-15h}{4k^2}\div\frac{3h^3}{16k}=\frac{-15h}{4k^2}\cdot\frac{16k}{3h^3}$

$=-\frac{\cancel{3}\cdot 5\cdot\cancel{h}}{\cancel{2}\cdot\cancel{2}\cdot\cancel{k}\cdot k}\cdot\frac{\cancel{2}\cdot\cancel{2}\cdot 2\cdot 2\cdot\cancel{k}}{\cancel{3}\cdot\cancel{h}\cdot h\cdot h}=-\frac{20}{h^2k}$

b. $\frac{\frac{10x+15}{3y^2}}{\frac{2x+3}{9y}}=\frac{10x+15}{3y^2}\div\frac{2x+3}{9y}=\frac{10x+15}{3y^2}\cdot\frac{9y}{2x+3}$

$=\frac{5\cancel{(2x+3)}}{\cancel{3}\cdot\cancel{y}\cdot y}\cdot\frac{\cancel{3}\cdot 3\cdot\cancel{y}}{\cancel{2x+3}}=\frac{15}{y}$

109. **a.** $\frac{\frac{49x^2}{5y^3}}{\frac{7x}{50y}}=\frac{49x^2}{5y^3}\div\frac{7x}{50y}=\frac{49x^2}{5y^3}\cdot\frac{50y}{7x}$

$=\frac{\cancel{7}\cdot 7\cdot\cancel{x}\cdot x}{\cancel{5}\cdot\cancel{y}\cdot y\cdot y}\cdot\frac{2\cdot\cancel{5}\cdot 5\cdot\cancel{y}}{\cancel{7}\cdot\cancel{x}}=\frac{70x}{y^2}$

b.

$$\frac{\frac{2y-1}{6y}}{\frac{8y-4}{18y^2}} = \frac{2y-1}{6y} \div \frac{8y-4}{18y^2} = \frac{2y-1}{6y} \cdot \frac{18y^2}{8y-4}$$

$$= \frac{\cancel{2y-1}}{2 \cdot \cancel{3} \cdot \cancel{y}} \cdot \frac{\cancel{2} \cdot \cancel{3} \cdot 3 \cdot \cancel{y} \cdot y}{\cancel{2} \cdot 2(\cancel{2y-1})} = \frac{3y}{4}$$

110. a. $\frac{4x+7}{x+3} + \frac{5}{x+3}$

$$= \frac{4x+7+5}{x+3} = \frac{4x+12}{x+3} = \frac{4\cancel{(x+3)}}{\cancel{x+3}} = 4$$

b. $\frac{3x+4}{(x-1)(x+2)} - \frac{7}{(x-1)(x+2)}$

$$= \frac{3x+4-7}{(x-1)(x+2)} = \frac{3x-3}{(x-1)(x+2)}$$

$$= \frac{3\cancel{(x-1)}}{\cancel{(x-1)}(x+2)} = \frac{3}{x+2}$$

111. a. $\frac{3a-10}{a-5} - \frac{5}{a-5}$

$$= \frac{3a-10-5}{a-5} = \frac{3a-15}{a-5} = \frac{3\cancel{(a-5)}}{\cancel{a-5}} = 3$$

b. $\frac{4b-6}{3b-5} + \frac{2b-4}{3b-5}$

$$= \frac{4b-6+2b-4}{3b-5} = \frac{6b-10}{3b-5} = \frac{2\cancel{(3b-5)}}{\cancel{3b-5}} = 2$$

112. a. $\frac{4}{x-2} \qquad \frac{3}{x+1}$

$\text{LCD} = (x-2)(x+1)$

$\frac{4}{x-2} \cdot \frac{x+1}{x+1} \qquad \frac{3}{x+1} \cdot \frac{x-2}{x-2}$

$\frac{4x+4}{(x-2)(x+1)} \qquad \frac{3x-6}{(x-2)(x+1)}$

b. $\frac{6t}{t+7} \qquad \frac{-5}{3(t+7)}$

$\text{LCD} = 3(t+7)$

$\frac{6t}{t+7} \cdot \frac{3}{3} \qquad \frac{-5}{3(t+7)}$

$\frac{18t}{3(t+7)} \qquad \frac{-5}{3(t+7)}$

c. $\frac{-2}{(x-3)(x+1)} \qquad \frac{6}{(x+5)(x-3)}$

$\text{LCD} = (x-3)(x+1)(x+5)$

$\frac{-2}{(x-3)(x+1)} \cdot \frac{x+5}{x+5} \qquad \frac{6}{(x+5)(x-3)} \cdot \frac{x+1}{x+1}$

$\frac{-2x-10}{(x-3)(x+1)(x+5)} \qquad \frac{6x+6}{(x-3)(x+1)(x+5)}$

113. a. $\frac{-9}{n-6} \qquad \frac{12}{n+4}$

$\text{LCD} = (n-6)(n+4)$

$\frac{-9}{n-6} \cdot \frac{n+4}{n+4} \qquad \frac{12}{n+4} \cdot \frac{n-6}{n-6}$

$\frac{-9n-36}{(n-6)(n+4)} \qquad \frac{12n-72}{(n-6)(n+4)}$

b. $\frac{7y}{y-3} \qquad \frac{-3y}{6(y-3)}$

$\text{LCD} = 6(y-3)$

$\frac{7y}{y-3} \cdot \frac{6}{6} \qquad \frac{-3y}{6(y-3)}$

$\frac{42y}{6(y-3)} \qquad \frac{-3y}{6(y-3)}$

c. $\frac{5}{(x-2)(x+1)} \qquad \frac{-7}{(x-4)(x-2)}$

$\text{LCD} = (x-2)(x+1)(x-4)$

$\frac{5}{(x-2)(x+1)} \cdot \frac{x-4}{x-4} \qquad \frac{-7}{(x-4)(x-2)} \cdot \frac{x+1}{x+1}$

$\frac{5x-20}{(x-2)(x+1)(x-4)} \qquad \frac{-7x-7}{(x-2)(x+1)(x-4)}$

114. a. $\frac{6}{b-5} + \frac{a}{3(b-5)} \qquad \text{LCD} = 3(b-5)$

$$= \frac{6}{b-5} \cdot \frac{3}{3} + \frac{a}{3(b-5)}$$

$$= \frac{18}{3(b-5)} + \frac{a}{3(b-5)} = \frac{a+18}{3(b-5)}$$

b. $\frac{2x-1}{x-2} + \frac{4x}{-(x-2)} \qquad \text{LCD} = -(x-2)$

$$= \frac{2x-1}{x-2} \cdot \frac{-1}{-1} + \frac{4x}{-(x-2)}$$

$$= \frac{-2x+1}{-(x-2)} + \frac{4x}{-(x-2)}$$

$$= \frac{-2x+1+4x}{-(x-2)} = \frac{2x+1}{-(x-2)} = \frac{-2x-1}{x-2}$$

c. $\dfrac{5}{m+1}+\dfrac{m-6}{(m+1)(m-4)}$ LCD $=(m+1)(m-4)$

$=\dfrac{5}{m+1}\cdot\dfrac{m-4}{m-4}+\dfrac{m-6}{(m+1)(m-4)}$

$=\dfrac{5m-20}{(m+1)(m-4)}+\dfrac{m-6}{(m+1)(m-4)}$

$=\dfrac{5m-20+m-6}{(m+1)(m-4)}=\dfrac{6m-26}{(m+1)(m-4)}$

115. a. $\dfrac{-3}{2(5x-3)}+\dfrac{4}{5x-3}$ LCD $=2(5x-3)$

$=\dfrac{-3}{2(5x-3)}+\dfrac{4}{5x-3}\cdot\dfrac{2}{2}$

$=\dfrac{-3}{2(5x-3)}+\dfrac{8}{2(5x-3)}=\dfrac{-3+8}{2(5x-3)}=\dfrac{5}{2(5x-3)}$

b. $\dfrac{3y+2}{-(y-7)}+\dfrac{5y-1}{y-7}$ LCD $=-(y-7)$

$=\dfrac{3y+2}{-(y-7)}+\dfrac{5y-1}{y-7}\cdot\dfrac{-1}{-1}$

$=\dfrac{3y+2}{-(y-7)}+\dfrac{-5y+1}{-(y-7)}$

$=\dfrac{3y+2-5y+1}{-(y-7)}=\dfrac{-2y+3}{-(y-7)}=\dfrac{2y-3}{y-7}$

c. $\dfrac{3x-1}{(x+4)(x-6)}+\dfrac{7}{x+4}$ LCD $=(x+4)(x-6)$

$=\dfrac{3x-1}{(x+4)(x-6)}+\dfrac{7}{x+4}\cdot\dfrac{x-6}{x-6}$

$=\dfrac{3x-1}{(x+4)(x-6)}+\dfrac{7x-42}{(x+4)(x-6)}$

$=\dfrac{3x-1+7x-42}{(x+4)(x-6)}=\dfrac{10x-43}{(x+4)(x-6)}$

116. $\dfrac{\frac{4}{x}-2}{\frac{3}{x}+1}\cdot\dfrac{\frac{x}{1}}{\frac{x}{1}}$ Multiply by LCD $=x$.

$=\dfrac{\frac{4}{x}\cdot\frac{x}{1}-2\cdot\frac{x}{1}}{\frac{3}{x}\cdot\frac{x}{1}+1\cdot\frac{x}{1}}=\dfrac{4-2x}{3+x}=\dfrac{-2x+4}{x+3}$

117. $\dfrac{\frac{5}{t+6}}{\frac{3}{t}-2\cdot\frac{t}{t}}$ LCD for denom. $=t$

$=\dfrac{\frac{5}{t+6}}{\frac{3}{t}-\frac{2t}{t}}=\dfrac{\frac{5}{t+6}}{\frac{3-2t}{t}}=\dfrac{5}{t+6}\div\dfrac{3-2t}{t}$

$=\dfrac{5}{t+6}\cdot\dfrac{t}{3-2t}=\dfrac{5t}{(t+6)(3-2t)}$

118. a. $C(12)=\dfrac{75+10(12)}{12}$ Substitute $n=12$.

$C(12)=\dfrac{75+120}{12}=\dfrac{195}{12}=16.25$

The average cost per person is $16.25 if 12 people attend the retirement lunch.

b. $15=\dfrac{75+10n}{n}$ Substitute $C(n)=15$.

$\dfrac{n}{1}\cdot 15=\dfrac{75+10n}{n}\cdot\dfrac{n}{1}$

$15n=75+10n$

$-10n \quad -10n$

$5n=75$

$\dfrac{5n}{5}=\dfrac{75}{5}$

$n=15$

There must be 15 people who attend the retirement lunch for the average cost per person to be $15.00.

119. a. $\dfrac{x}{1}\left(\dfrac{12}{x}+1\right)=\left(\dfrac{2}{x}-4\right)\dfrac{x}{1}$

$\dfrac{x}{1}\cdot\dfrac{12}{x}+1\cdot\dfrac{x}{1}=\dfrac{x}{1}\cdot\dfrac{2}{x}-4\cdot\dfrac{x}{1}$

$12+x=2-4x$

$+4x \quad +4x$

$12+5x=2$

$-12 \quad -12$

$5x=-10$

$\dfrac{5x}{5}=\dfrac{-10}{5}$

$x=-2$

b. $\frac{x-2}{1}\left(\frac{-3}{x-2}\right)=\left(5-\frac{3}{x-2}\right)\frac{x-2}{1}$

$$\frac{x-2}{1}\cdot\frac{-3}{x-2}=\frac{x-2}{1}\cdot 5-\frac{3}{x-2}\cdot\frac{x-2}{1}$$

$$-3=5x-10-3$$
$$-3=5x-13$$
$$\underline{+13\qquad +13}$$
$$10=5x$$
$$\frac{10}{5}=\frac{5x}{5}$$
$$2=x$$ This is an excluded value.

No solution

c. $\frac{t-1}{1}\left(\frac{4}{t-1}+2\right)=(t-2)\frac{t-1}{1}$

$$\frac{t-1}{1}\cdot\frac{4}{t-1}+2\cdot\frac{t-1}{1}=(t-2)(t-1)$$
$$4+2t-2=t^2-t-2t+2$$
$$2t+2=t^2-3t+2$$
$$\underline{-2t\qquad -2t}$$
$$2=t^2-5t+2$$
$$\underline{-2\qquad -2}$$
$$0=t^2-5t$$
$$t(t-5)=0$$

$t=0$ $\qquad$ $t-5=0$, $\underline{+5\ +5}$, $t=5$

$t=0$ and $t=5$

d. $\frac{(x+5)(x-3)}{1}\left(\frac{x}{x-3}+\frac{2}{x+5}\right)=\left(\frac{2x+18}{(x+5)(x-3)}\right)\frac{(x+5)(x-3)}{1}$

$$\frac{(x+5)(x-3)}{1}\cdot\frac{x}{x-3}+\frac{2}{x+5}\cdot\frac{(x+5)(x-3)}{1}=\left(\frac{2x+18}{(x+5)(x-3)}\right)\frac{(x+5)(x-3)}{1}$$
$$x^2+5x+2x-6=2x+18$$
$$x^2+7x-6=2x+18$$
$$\underline{-2x\qquad -2x}$$
$$x^2+5x-6=18$$
$$\underline{-18\quad -18}$$
$$x^2+5x-24=0$$
$$(x+8)(x-3)=0$$

$x+8=0$, $\underline{-8\ -8}$, $x=-8$ $\qquad$ $x-3=0$, $\underline{+3\ +3}$, $x=3$

$x=3$ is an excluded value, therefore $x=-8$

120. a. $\frac{n}{1}\left(\frac{-16}{n}+3\right)=\left(\frac{6}{n}-1\right)\frac{n}{1}$

$$\frac{n}{1}\cdot\frac{-16}{n}+3\cdot\frac{n}{1}=\frac{n}{1}\cdot\frac{6}{n}-1\cdot\frac{n}{1}$$
$$-16+3n=6-n$$
$$\underline{+n\qquad +n}$$
$$-16+4n=6$$
$$\underline{+16\quad +16}$$
$$4n=22$$
$$\frac{4n}{4}=\frac{22}{4}$$
$$n=\frac{11}{2}=5.5$$

b. $\frac{a+6}{1}\left(7+\frac{5}{a+6}\right)=\left(\frac{5}{a+6}\right)\frac{a+6}{1}$

$$\frac{a+6}{1}\cdot 7+\frac{5}{a+6}\cdot\frac{a+6}{1}=\left(\frac{5}{a+6}\right)\frac{a+6}{1}$$
$$7a+42+5=5$$
$$7a+47=5$$
$$\underline{-47\ -47}$$
$$7a=-42$$
$$\frac{7a}{7}=\frac{-42}{7}$$
$$a=-6$$

This is an excluded value.

No solution

c. $\frac{x+1}{1}\left(\frac{5}{x+1}-2\right)=(x+3)\frac{x+1}{1}$

$$\frac{x+1}{1}\cdot\frac{5}{x+1}-2\cdot\frac{x+1}{1}=(x+3)(x+1)$$
$$5-2x-2=x^2+x+3x+3$$
$$-2x+3=x^2+4x+3$$
$$\underline{+2x\qquad +2x}$$
$$3=x^2+6x+3$$
$$\underline{-3\qquad -3}$$
$$0=x^2+6x$$
$$x(x+6)=0$$

$x=0$ $\qquad$ $x+6=0$, $\underline{-6\ -6}$, $x=-6$

$x=0$ and $x=-6$

d. $\frac{(x+6)(x-4)}{1}\left(\frac{x}{x-4}+\frac{3}{x+6}\right)=\left(\frac{5x+20}{(x+6)(x-4)}\right)\frac{(x+6)(x-4)}{1}$

$$\frac{(x+6)(x-4)}{1}\cdot\frac{x}{x-4}+\frac{3}{x+6}\cdot\frac{(x+6)(x-4)}{1}=\left(\frac{5x+20}{(x+6)(x-4)}\right)\frac{(x+6)(x-4)}{1}$$

$$x^2+6x+3x-12=5x+20$$
$$x^2+9x-12=5x+20$$
$$\underline{-5x\quad -5x}$$
$$x^2+4x-12=20$$
$$\underline{-20\ -20}$$
$$x^2+4x-32=0$$
$$(x+8)(x-4)=0$$

$x+8=0$, $\underline{-8\ -8}$, $x=-8$ $\qquad$ $x-4=0$, $\underline{+4\ +4}$, $x=4$

$x=4$ is an excluded value, therefore $x=-8$

121. a. $w=\frac{8,500,000}{(15)}$ $\quad$ Substitute $p=15$.

$w\approx 566,666.67$

The winnings per person will be about $566,666.67 if 15 people pool their ticket money.

b. Substitute $w=425,000$.

$$(425,000)=\frac{8,500,000}{p}$$
$$\frac{p}{1}\cdot(425,000)=\frac{8,500,000}{p}\cdot\frac{p}{1}$$
$$425,000p=8,500,000$$
$$\frac{425,000p}{425,000}=\frac{8,500,000}{425,000}$$
$$p=20$$

If the group wants the minimum winnings per person to be $425,000, there must be no more than 20 people who pitch in to the pool.

122. Tonya's work time $=15$ min . Rate $=\frac{1}{15}$

Ron's work time $=20$ min . Rate $=\frac{1}{20}$

Work time for both $=t$ min . Rate $=\frac{1}{t}$

$$\frac{1}{15}+\frac{1}{20}=\frac{1}{t}$$
$$\frac{60t}{1}\left(\frac{1}{15}+\frac{1}{20}\right)=\left(\frac{1}{t}\right)\frac{60t}{1}$$
$$\frac{60t}{1}\cdot\frac{1}{15}+\frac{60t}{1}\cdot\frac{1}{20}=\frac{60t}{1}\cdot\frac{1}{t}$$
$$4t+3t=60$$
$$7t=60$$
$$\frac{7t}{7}=\frac{60}{7}$$
$$t=\frac{60}{7}\approx 8.57$$

Tonya and Ron can shred the box of documents in about 8.57 minutes if they work together.

123. $\frac{4\text{yd}}{1}\cdot\frac{3\text{ft}}{1\text{yd}}=12\text{ft}$

9 feet to 4 yards $=9$ feet to 12 feet

9 feet to 4 yards $=\frac{9}{12}=\frac{3}{4}$

124. $4.59 for one gallon of milk $=\frac{\$4.59}{1\text{ gal}}$

125. $\frac{1}{33.95}=\frac{115}{x}$

$x=33.95\cdot 115$

$x=3904.25$

The cost of 115 square yards of carpet is $3904.25.

126. a. $\frac{3}{a+2}=\frac{5}{a}$

$$3\cdot a=5(a+2)$$
$$3a=5a+10$$
$$\underline{-5a\ -5a}$$
$$-2a=10$$
$$\frac{-2a}{-2}=\frac{10}{-2}$$
$$a=-5$$

b. $\frac{a+5}{3}=-\frac{(a+11)}{6}$

$$-6(a+5)=3(a+11)$$
$$-6a-30=3a+33$$
$$\underline{\;+30\qquad +30\;}$$
$$-6a=3a+63$$
$$\underline{-3a\;-3a}$$
$$-9a=63$$
$$\frac{-9a}{-9}=\frac{63}{-9}$$
$$a=-7$$

127. a. $\frac{b+6}{4}=-\frac{1}{2}$

$$2(b+6)=4\cdot -1$$
$$2b+12=-4$$
$$\underline{-12\;-12}$$
$$2b=-16$$
$$\frac{2b}{2}=\frac{-16}{2}$$
$$b=-8$$

b. $\frac{n+5}{2}=\frac{-2n+5}{6}$

$$6(n+5)=2(-2n+5)$$
$$6n+30=-4n+10$$
$$\underline{-30\qquad -30}$$
$$6n=-4n-20$$
$$\underline{+4n\;+4n}$$
$$10n=-20$$
$$\frac{10n}{10}=\frac{-20}{10}$$
$$n=-2$$

128. $\frac{4}{x}=\frac{3}{6}$

$$3x=4\cdot 6$$
$$3x=24$$
$$\frac{3x}{3}=\frac{24}{3}$$
$$x=8$$

The value of the unknown side is 8.

129. $\frac{5}{x}=\frac{8}{6}$

$$8x=5\cdot 6$$
$$8x=30$$
$$\frac{8x}{8}=\frac{30}{8}$$
$$x=3.75$$

The value of the unknown side is 3.75.

130. $y=kx$ Substitute given values for x and y.

$$(-12)=k(2)$$
$$-12=2k$$
$$\frac{-12}{2}=\frac{2k}{2}$$
$$-6=k$$

$y=-6x$ Substitute values for k and x.

$$y=-6(-4)$$
$$y=24$$

131. $y=\frac{k}{x}$ Substitute given values for x and y.

$(4)=\frac{k}{(3)}$ Cross multiply.

$$k=12$$

$y=\frac{12}{x}$ Substitute values for k and x.

$$y=\frac{12}{(6)}$$
$$y=2$$

CHAPTER 8 MODELING DATA AND FUNCTIONS

Section 8.1 Solving Linear Applications

1. **a.**

$C = 10h + 20$

$C = 10(1) + 20$

$C = 30$

After 1 hour of training, a new employee can make 30 candies an hour.

b.

$C = 10h = 20$

$C = 10(4) + 20$

$C = 40 + 20 = 60$

After 4 hour of training, a new employee can make 60 candies an hour.

c.

Let $C = 150$

$150 = 10h + 20$

$150 - 20 = 10h + 20 - 20$

$130 = 10h$

$\frac{130}{10} = \frac{10h}{10}$

$13 = h$

A new employee can make 150 candies an hour after 13 hours of training.

3. **a.**

$N = -315.9t + 4809.8$

$N = -315.9(2) + 4809.8$

$N = -631.8 + 4809.8$

$N = 4178$

In 1992, there were 4178 homicides of 15-19 year olds in the United States.

b.

$N = -315.9t + 4809.8$

$N = -315.9(12) + 4809.8$

$N = -3790.8 + 4809.8$

$N = 1019$

In 2002, there were 1019 homicides of 15-19 year olds in the United States.

c.

Let $N = 7337$

$7337 - 4809.8 = -315.9t + 4809.8 - 4809.8$

$2527.2 = -315.9t$

$\frac{2527.2}{-315.9} = \frac{-315.9t}{-315.9}$

$-8 = t$

In 1982, there were 7337 homicides of 15-19 year olds in the United States.

5. **a.**

$P = 1.5t - 300$

$P = 1.5(100) - 300$

$P = 150 - 300$

$P = -150$

There is a loss of \$150 for selling only 100 T-shirts.

b.

$P = 1.5t - 300$

$P = 1.5(400) - 300$

$P = 600 - 300$

$P = 300$

There is a profit of \$300 for selling 400 T-shirts.

c.

Let $P = 1000$

$1000 = 1.5t - 300$

$1000 + 300 = 1.5t - 300 + 300$

$1300 = 1.5t$

$\frac{1300}{1.5} = \frac{1.5t}{1.5}$

$866.67 \approx t$

To make \$1000 profit, you must sell 867 T-shirts.

7. **a.**

$C = 2.50 + 2.0m$

$C = 2.50 + 2.0(25)$

$C = 2.50 + 50.0$

$C = 52.50$

It costs \$52.50 to take a 25-mile taxi ride in NYC.

b.

$100 = 2.50 + 2.0m$

$100 - 2.50 = 2.50 - 2.50 + 2.0m$

$97.50 = 2.0m$

$\frac{97.50}{2.0} = \frac{2.0m}{2.0}$

$48.75 = m$

For \$100, you can take about a 48-mile taxi ride in NYC.

9. **a.** $P = 3.5$. This is too few people. This would mean that only 3500 people live in Kentucky.

b. $P = 4200$. This answer is most reasonable. This would mean that 4,200,000 people live in Kentucky.

c. $P = -210$. This is not possible. This would mean that $-210{,}000$ people live in Kentucky.

11. **a.** $T = -50$. This answer is most reasonable.

b. $T = 75$. This temperature is too warm for South Pole temperatures.

c. $T = 82$. This temperature is too warm for South Pole temperatures.

13. **a.**

$P = 0.08(s - 1000)$

$P = 0.08(2000 - 1000)$

$P = 0.08(1000) = 80$

On \$2000 in sales, you will make \$80 in commissions.

b.

$P = 0.08(s - 1000)$

$P = 0.08(50{,}000 - 1000)$

$P = 0.08(49{,}000)$

$P = 3920$

On sales of \$50,000, you will make \$3920 in commissions.

c.

$P = 0.08(s-1000)$
$500 = 0.08s - 80$
$500 + 80 = 0.08s - 80 + 80$
$580 = 0.08s$
$\frac{580}{0.08} = \frac{0.08s}{0.08}$
$7250 = s$

To make \$500 per week, you will need \$7250 in sales each week.

15. **a.** $B = 29.95 + 0.55m$

b.

$B = 29.95 + 0.55(75)$
$B = 29.95 + 41.25$
$B = 71.20$

If you drive the 10-foot truck 75 miles, it will cost you \$71.20.

c.

$B = 29.95 + 0.55m$
$100 = 29.95 + 0.55m$
$100 - 29.95 = 29.95 - 29.95 + 0.55m$
$70.05 = 0.55m$
$\frac{70.05}{0.55} = \frac{0.55m}{0.55}$
$127.36 \approx m$

\$37 will buy 135 minutes.
For a total of \$100, you can rent the 10-foot truck from Budget and drive it 127 miles.

17. **a.** $P = 250 + 0.07s$

b. Let $s = 2000$

$P = 250 + 0.07(2000)$
$P = 250 + 140 = 390$

If you have sales of \$2000 in a week, your pay will be \$390.

c. Let $P = 650$

$650 = 250 + 0.07s$
$650\text{-}250 = 250 - 250 + 0.07s$
$400 = 0.07s$
$\frac{400}{0.07} = \frac{0.07s}{0.07}$
$5714.29 = s$

To earn \$650 per week, you must have \$5714.29 in sales each week.

19. Let C be the total cost of a trip to Las Vegas for d days.

a. $C = 125 + 100d$

b.

$C = 125 + 100(3)$
$C = 125 + 300 = 425$

A three day trip to Las Vegas will cost \$425.

c.

$\frac{\$700}{2} = \$350$
$C = 125 + 100d$
$350 = 125 + 100d$
$350 - 125 = 125 - 125 + 100d$
$225 = 100d$
$\frac{225}{100} = \frac{100d}{100}$
$2.25 = d$

If you have \$700 and gamble half of it, you can stay in Las Vegas for only two days.

21. Let C be the total cost (in dollars) of shooting a wedding, and p be the number of proofs the photographer edits and prints.

a. $C = 5.29p + 400$

b.

$C = 5.29(100) + 400$
$C = 529 + 400$
$C = 929$

If the photographer edits and prints 100 proofs the cost will be \$929.

c.

Let $C = 1250$
$1250 = 5.29p + 400$
$1250 - 400 = 5.29p + 400 - 400$
$850 = 5.29p$
$\frac{850}{5.29} = \frac{5.29p}{5.29}$
$160.68 \approx p$

The photographer can edit and print 160 proofs with a budget of \$1250.

23. **a.** Let C be the total cost (in dollars) for selling s snow cones for a month.
Fixed costs are: $5500 + 1500 = 7000$.
$C = 7000 + 0.45s$

b.

$C = 7000 + 0.45(3000)$
$C = 8350$

The monthly cost for selling 3000 snow cones is \$8350.

c.

$10,600 = 7000 + 0.45s$
$10,600 - 7000 = 7000 - 7000 + 0.45s$
$3600 = 0.45s$
$\frac{3600}{0.45} = \frac{0.45s}{0.45}$
$8000 = s$

For a \$10,600 budget, the vendor can sell up to 8000 snow cones.

25. **a.** Let C be the total cost (in dollars) for the Squeaky Clean Window Company to clean windows for a day when w windows are cleaned.
$C = 1.50w + 230$.

b.

$$C = 1.50(60) + 230$$
$$C = 90 + 230$$
$$C = 320$$

If the Squeaky Clean Window Company cleans 60 windows in a day, it will cost the company \$320.

c.

$$450 = 1.50w + 230$$
$$450 - 230 = 1.50w + 230 - 230$$
$$220 = 1.50w$$
$$\frac{220}{1.50} = \frac{1.50w}{1.50}$$
$$146.7 \approx w$$

To stay within a budget of \$450, the Squeaky Clean Window Company can clean up to 146 windows.

27. Maria's work is correct. Javier needs a decimal to correctly represent 55 cents per bottle in terms of dollars per bottle.

29. **a.** Let C be the total cost (in dollars) for pest management from Enviro-Safe Pest Management when m monthly treatments are done.
$C = 150 + 38m$.

b. There are 18 months in 1.5 years.

$$C = 150 + 38(18)$$
$$C = 150 + 684$$
$$C = 834$$

If the house has an initial treatment, and then is treated for an additional 1.5 years, it will cost \$834.

31. **a.** Let C be the total monthly cost (in dollars) for a manufacturer to produce g sets of golf clubs.
$C = 23{,}250 + 145g$.

b.

$$C = 23{,}250 + 145(100)$$
$$C = 23{,}250 + 14{,}500$$
$$C = 37{,}750$$

It costs the manufacturer \$37,750 to produce 100 sets of golf clubs.

c.

$$20{,}000 - 23{,}250 = 23{,}250 - 23{,}250 + 145g$$
$$-3250 = 145g$$
$$\frac{-3250}{145} = \frac{145g}{145}$$
$$-22.41 \approx g$$

This does not make sense. Their costs can never be lower than their fixed costs of \$23,250.

d.

$$\frac{\$37{,}750}{100 \text{ sets}} = \$377.50 \text{ per set}$$

To break even selling 100 sets of golf clubs per month, the manufacturer must sell each set for \$377.50.

33. **a.** $C = 1500 + 1.50n$ for $n \le 500$.

b.

$$C = 1500 + 1.50(250)$$
$$C = 1500 + 375$$
$$C = 1875$$

It costs Rockon \$1875 to make 250 CDs.

c.

$$2000 = 1500 + 1.50n$$
$$2000 - 1500 = 1500 - 1500 + 1.50n$$
$$500 = 1.50n$$
$$\frac{500}{1.50} = \frac{1.50n}{1.50}$$
$$333.3 \approx n$$

With a budget of \$2000, Rockon can order 333 CDs.

d.

$$3000 = 1500 + 1.50n$$
$$3000 - 1500 = 1500 - 1500 + 1.50n$$
$$1500 = 1.50n$$
$$\frac{1500}{1.50} = \frac{1.50n}{1.50}$$
$$1000 = n$$

With a budget of \$3000, Rockon can order 1000 CDs. This does not make sense. They can only order up to 500 CDs.

35.

$$5x + 60 = 2x + 90$$
$$5x + 60 - 2x = 2x + 90 - 2x$$
$$3x + 60 = 90$$
$$3x + 60 - 60 = 90 - 60$$
$$3x = 30$$
$$\frac{3x}{3} = \frac{30}{3}$$
$$x = 10$$

37.

$$\frac{2}{5}d + 6 = 14$$
$$\frac{2}{5}d + 6 - 6 = 14 - 6$$
$$\frac{2}{5}d = 8$$
$$\frac{5}{2}\left(\frac{2}{5}d\right) = \frac{5}{2}(8)$$
$$d = 20$$

39.

$$\frac{1}{3}m + \frac{4}{3} = 4$$
$$3\left(\frac{1}{3}m + \frac{4}{3}\right) = 3(4)$$
$$m + 4 = 12$$
$$m + 4 - 4 = 12 - 4$$
$$m = 8$$

41.

$$\begin{aligned}
-3x-6&=14+8x\\
-3x-6-8x&=14+8x-8x\\
-11x-6&=14\\
-11x-6+6&=14+6\\
-11x&=20\\
\frac{-11x}{-11}&=\frac{20}{-11}\\
x&=-\frac{20}{11}
\end{aligned}$$

43.

$$\begin{aligned}
\frac{5}{7}d-\frac{3}{10}&=\frac{4}{7}d+4\\
70\left(\frac{5}{7}d-\frac{3}{10}\right)&=70\left(\frac{4}{7}d+4\right)\\
50d-21&=40d+280\\
50d-21-40d&=40d+280-40d\\
10d-21&=280\\
10d-21+21&=280+21\\
10d&=301\\
\frac{10d}{10}&=\frac{301}{10}\\
d&=\frac{301}{10}
\end{aligned}$$

45.

$$\begin{aligned}
1.25d-3.4&=-2.3(5d+4)\\
1.25d-3.4&=-11.5d-9.2\\
1.25d-3.4+11.5d&=-11.5d-9.2+11.5d\\
12.75d-3.4&=-9.2\\
12.75d-3.4+3.4&=-9.2+3.4\\
12.75d&=-5.8\\
\frac{12.75d}{12.75}&=\frac{-5.8}{12.75}\\
d&\approx-0.45
\end{aligned}$$

47.

$$\begin{aligned}
3(c+5)-21&=107\\
3c+15-21&=107\\
3c-6&=107\\
3c-6+6&=107+6\\
3c&=113\\
\frac{3c}{3}&=\frac{113}{3}\\
c&=\frac{113}{3}
\end{aligned}$$

49.

$$\begin{aligned}
1.7d+5.7&=29.7+5d\\
1.7d+5.7-5d&=29.7+5d-5d\\
-3.3d+5.7&=29.7\\
-3.3d+5.7-5.7&=29.7-5.7\\
-3.3d&=24\\
\frac{-3.3d}{-3.3}&=\frac{24}{-3.3}\\
d&\approx-7.27
\end{aligned}$$

51.

$$\begin{aligned}
\frac{3}{7}(2z-5)&=\frac{4}{7}(-3z+9)\\
\frac{6}{7}z-\frac{15}{7}&=-\frac{12}{7}z+\frac{36}{7}\\
7\left(\frac{6}{7}z-\frac{15}{7}\right)&=7\left(-\frac{12}{7}z+\frac{36}{7}\right)\\
6z-15&=-12z+36\\
6z-15+12z&=-12z+36+12z\\
18z-15&=36\\
18z-15+15&=36+15\\
18z&=51\\
\frac{18z}{18}&=\frac{51}{18}\\
z&=\frac{51}{18}\\
z&=\frac{17}{6}
\end{aligned}$$

53.

$$\begin{aligned}
-3(2v+9)-3(3v-7)&=4v+6(2v-8)\\
-6v-27-9v+21&=4v+12v-48\\
-15v-6&=16v-48\\
-15v-6-16v&=16v-48-16v\\
-31v-6&=-48\\
-31v-6+6&=-48+6\\
-31v&=-42\\
\frac{-31v}{-31}&=\frac{-42}{-31}\\
v&=\frac{42}{31}
\end{aligned}$$

55.

$$\begin{aligned}
-\frac{8}{9}(3t+5)&=\frac{2}{3}t-12\\
-\frac{24}{9}t-\frac{40}{9}&=\frac{2}{3}t-12\\
9\left(-\frac{24}{9}t-\frac{40}{9}\right)&=9\left(\frac{2}{3}t-12\right)\\
-24t-40&=6t-108\\
-24t-40-6t&=6t-108-6t\\
-30t-40&=-108\\
-30t-40+40&=-108+40\\
-30t&=-68\\
\frac{-30t}{-30}&=\frac{-68}{-30}\\
t&=\frac{68}{30}\\
t&=\frac{34}{15}
\end{aligned}$$

57. Rounding the outside temperature to 73°F is appropriate because a difference of 0.4°F would not be felt.

59. A result of \$236.5725 would be rounded to \$236.57 because our monetary units extend to the nearest hundredth.

61. The company would need to wash 313 cars (312.25 rounded up) to make a profit of \$400. Anything less would result in a profit of less than \$400.

63. $P = 75.8t + 5906.2$

a. Slope $= 75.8$. The population of Washington state increases 75.8 thousand each year.

b. The vertical intercept is (0, 5906.2). In 2000, the population of Washington state was 5906.2 thousand.

c. Horizontal intercept:

$$0 = 75.8t + 5906.2$$
$$-5906.2 = 75.8t$$
$$\frac{-5906.2}{75.8} = \frac{75.8t}{75.8}$$
$$-77.9 \approx t$$

The horizontal intercept is $(-77.9, 0)$. In 1922, the population of Washington State was zero. This does not make sense.

65. $M = 4.02t + 25.8$

a. Slope $= 4.02$. The annual amount of public expenditure on medical research in the United States increases \$4.02 billion each year.

b. The vertical intercept is (0, 25.8). The annual amount of public expenditure on medical research in the United States in 2000 was \$25.8 billion.

c. Horizontal intercept:

$$0 = 4.02t + 25.8$$
$$-25.8 = 4.02t$$
$$\frac{-25.8}{4.02} = \frac{4.02t}{4.02}$$
$$-6.42 \approx t$$

The horizontal intercept is $(-6.42, 0)$. The annual amount of public expenditure on medical research in the United States was \$0 in 1994. This does not make sense.

67. $P = 500c - 6000$

a. Slope $= 500$. The monthly profit for a small used car lot increases by \$500 for every car that is sold.

b. The vertical intercept is $(0, -6000)$. If the small used car lot sells 0 cars, it will lose \$6000.

c. Horizontal intercept:

$$0 = 500c - 6000$$
$$6000 = 500c$$
$$\frac{6000}{500} = \frac{500c}{500}$$
$$12 = c$$

The horizontal intercept is (12, 0). The small used car lot will need to sell 12 cars in order to break even in their monthly profit.

69. **a.** Let P the total profit in dollars Dan earns teaching surfing lessons, and let s be the number of one-hour surf lessons Dan gives. This can be represented by the equation $P = 30s - 700$.

b. Slope $= 30$. Dan's earnings increase by \$30 for every one-hour surf lesson he gives.

c. The vertical intercept is $(0, -700)$. If Dan doesn't give any surf lessons, he will lose \$700 in profit.

d. Horizontal intercept:

$$0 = 30s - 700$$
$$700 = 30s$$
$$\frac{700}{30} = \frac{30s}{30}$$
$$23.33 \approx c$$

The horizontal intercept is (23.33, 0). Dan will need to give 24 one-hour surf lessons in order to break even.

71. **a.** Let T be the total amount of money raised by the PTA for a new track at Mission Meadows Elementary School in dollars, and let a be the total pounds of aluminum cans the PTA recycles. $T = 1.24a + 2000$.

b. Slope $= 1.24$. The funds raised by the PTA at Mission Meadows Elementary School increases by \$1.24 for every pound of aluminum cans they recycle.

c. The vertical intercept is (0, 2000). The PTA at Mission Meadows Elementary School will have \$2000 to put toward a new track if they recycle 0 pounds of aluminum cans.

d. Horizontal intercept:

$$0 = 1.24a + 2000$$
$$-2000 = 1.24a$$
$$\frac{-2000}{1.24} = \frac{1.24a}{1.24}$$
$$-1612.9 \approx a$$

The horizontal intercept is $(-1612.9, 0)$. The PTA at Mission Meadows Elementary School will need to recycle -1612.9 pounds of aluminum cans in order to earn \$0 for a new track. This does not make sense.

73. **a.** Let P be the total percentage of Americans who have been diagnosed with diabetes, and let t be years since 2000.

Use $(2, 4.8)$ and $(4, 5.1)$:

$$\text{Slope} = \frac{5.1 - 4.8}{4 - 2} = \frac{0.3}{2} = 0.15$$

Use the point slope form of the equation:

$$P - 4.8 = 0.15(t - 2)$$
$$P - 4.8 = 0.15t - 0.3$$
$$P - 4.8 + 4.8 = 0.15t - 0.3 + 4.8$$
$$P = 0.15t + 4.5$$

b.

$$P = 0.15(10) + 4.5$$
$$P = 6$$

This implies 6% of Americans will be diagnosed with diabetes in 2010.

c. Slope = 0.15. The percentage of Americans diagnosed with diabetes increases 0.15 percentage points each year.

75. **a.** Let C be the total cost of shirts in dollars, and let n be the total number of shirts purchased.

$(10,110)$ & $(30,280)$

Slope: $m=\dfrac{280-110}{30-10}=\dfrac{170}{20}=8.5$

Use the point slope form of the equation:

$C-110=8.5(n-10)$
$C-110=8.5n-85$
$C-110+110=8.5n-85+110$
$C=8.5n+25$

b.

$C=8.5(50)+25$
$C=425+25$
$C=450$

It will cost \$450 for 50 shirts.

c. Slope = 8.5. The cost per shirt increases by \$8.50 per shift.

d. The vertical intercept is (0,25). It costs \$25 if 0 shirts are purchased. This does not make sense.

e.

$0=8.5n+25$
$-25=8.5n$
$\dfrac{-25}{8.5}=\dfrac{8.5n}{8.5}$
$-2.94\approx n$

The horizontal intercept is $(-2.94,0)$. To pay \$0, you would have to purchase -3 shirts. This does not make sense.

77. **a.** Let P be the total population of Washington State in thousands, and let t be the years since 2000.

$(15,6951)$ & $(25,7996)$

Slope: $m=\dfrac{7996-6951}{25-15}=\dfrac{1045}{10}=104.5$

Use the point slope form of the equation:

$P-6951=104.5(t-15)$
$P-6951=104.5t-1567.5$
$P-6951+6951=104.5t-1567.5+6951$
$P=104.5t+5383.5$

b.

$P=104.5(30)+5383.5$
$P=3135+5383.5$
$P=8518.5$

In 2030, the population of Washington state will be about 8518.5 thousand.

c. Slope = 104.5. The population of Washington state increases by 104.5 thousand each year.

d. The vertical intercept is (0, 5383.5). In 2000, the population of Washington state was 5383.5 thousand.

e.

$0=104.5t+5383.5$
$-5383.5=104.5t$
$\dfrac{-5383.5}{104.5}=\dfrac{104.5t}{104.5}$
$-51.52\approx t$

The horizontal intercept is $(-51.52,0)$. In 1949, the population of Washington State was zero. This does not make sense.

79. **a.** Let W be a woman's optimal weight in pounds, and let h be a woman's height in inches above 5 feet.

$(0,100)$ & $(6,130)$

Slope: $m=\dfrac{130-100}{6-0}=\dfrac{30}{6}=5$

The vertical intercept is $(0,100)$

$W=5h+100$

b.

6 feet is 12 inches over 5 feet.

Use $h=12$:

$W=5(12)+100$
$W=60+100$
$W=160$

The optimal weight of a woman who is 6 feet tall is 160 pounds.

c. Slope = 5. For every inch taller than 5 feet, a woman's optimal weight increases 5 pounds.

d. The vertical intercept is (0, 100). A woman's optimal weight at 5 feet tall is 100 pounds.

e.

$0=5h+100$
$-100=5h$
$\dfrac{-100}{5}=\dfrac{5h}{5}$
$-20=h$

The horizontal intercept is $(-20,0)$. At 3 feet 4 inches, a woman's optimal weight is 0 pounds. This does not make sense.

81. **a.** Let G be the total number of teenagers who underwent gastric bypass surgery, and let t be years since 2000.

$(3,771)$ & $(1,408)$

Slope: $m=\dfrac{771-408}{3-1}=\dfrac{363}{2}=181.5$

Use the point slope form of the equation:

$G-408=181.5(t-1)$
$G-408=181.5t-181.5$
$G-408+408=181.5t-181.5+408$
$G=181.5t+226.5$

b.

$$G = 181.5(5) + 226.5$$
$$G = 907.5 + 226.5$$
$$G = 1134$$

The number of teenagers who had gastric bypass surgery in 2005 was 1134.

c. Slope = 181.5. The number of teenagers who undergo gastric bypass surgery increases by about 182 each year.

d. The vertical intercept is (0, 226.5). In 2000, 227 teenagers underwent gastric bypass surgery.

e.

$$0 = 181.5t + 226.5$$
$$-181.5t = 226.5$$
$$\frac{-181.5t}{-181.5} = \frac{226.5}{-181.5}$$
$$t \approx -1.25$$

The horizontal intercept is $(-1.25, 0)$. In 1998, 0 teenagers underwent gastric bypass surgery. This does not make sense.

Section 8.2 Using Data to Create Scatterplots

1. **a.** The total prize money at Wimbledon in 2002 was 8.5 million British pounds.
b. In 2003, the total prize money at Wimbledon was 9 million British pounds.
c. $(0,7)$. In the year 2000, the winnings were 7 million British pounds
d. Domain: $[2,11]$, range: $[8.5,14]$.

3. (A) and (F)

5. (C) and (E)

7. 50

9. 0.01

11. 50

13. 10,000

15. **a.** $(0,2)$. The point that crosses the y-axis.
b. $(-3,0)$. The point that crosses the x-axis.
c. $x = 3$ from the point $(3,4)$.
d. $y = 2.6$ from the point $(1,2.6)$.

17. **a.** $(0,5)$. The point that crosses the y-axis.
b. $(4,0)$. The point that crosses the x-axis.
c. The input value is -10 when the output value is 18. That is, $x = -10$ when $y = 18$.
d. The input value is 15 when the output value is -15. That is, $x = 15$ when $y = -15$.
e. The output value is -8 when the input value is 10. That is, $y = -8$ when $x = 10$.

19. **a.** $(0,-8)$. The point that crosses the y-axis.
b. $(30,0)$. The point that crosses the x-axis.
c. $x = 11$ from the point $(11,-5)$.
d. $x = -9$ from the point $(-9,-10)$.
e. $y = -3$ from the point $(20,-3)$.

21. **a.** False, vertical intercept $= (0,-2)$
b. True.
c. False, $x = -4.5$
d. False, $x = -1.5$
e. True.

23. **a.** $(-15,0)$. The point that crosses the x-axis.
b. $(0,-10)$ The point that crosses the y-axis.
c. $x = 15$ from the point $(15,-20)$.
d. $x = 7.5$ from the point $(7.5,-15)$.
e. $y = -3$ from the point $(-10,-3)$.

25. **a.** $(0,1)$ The point that crosses the y-axis.
b. $(-2,0)$ The point that crosses the x-axis.
c. The input value is 1 when the output value is 1.5. That is, $x = 1$ when $y = 1.5$.
d. The input value is -4 when the output value is -1. That is, $x = -4$ when $y = -1$.
e. The output value is 2.5 when the input value is 3. That is, $y = 2.5$ when $x = 3$.

27. **a.** Dependent variable: C = Cost in dollars for producing chocolate dipped key lime pie bars.
Independent variable: b = Number of bars produced.

b.

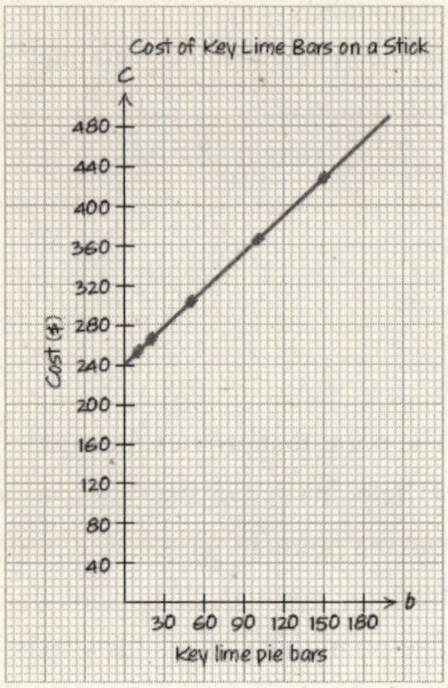

c. When $b = 200$, then $C = \$485$. The cost of producing 200 key lime bars on a stick is about \$485.

d. Domain: $0 \le b \le 200$, range: $240 \le C \le 485$

29. **a.** Dependent variable: G = Gross Profit for Quicksilver, Inc. in millions of dollars.
Independent variable: t = years since 2000.

b.

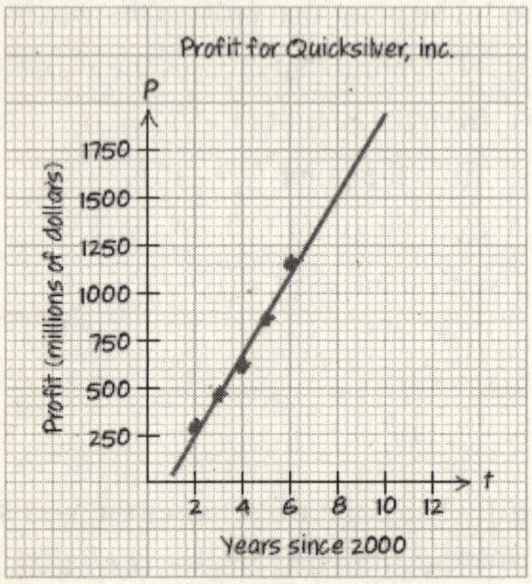

c. When $t = 10$, then $G = 1850$. The gross profit for Quicksilver, Inc. in 2010 will be approximately \$1850 million.

d. Domain: $1 \le t \le 10$, range: $50 \le G \le 1850$.

31. Maria's model fits the data better. There is a smaller variance between the data points and the line of best fit.

33. $0 \le t \le 10$ is not a reasonable domain because the model predicts a negative value of cases per 100,000 population for t-values of greater than 8.2. This does not make sense. Domain: $0 \le t \le 7$, range: $0.6 \le A \le 4.1$.

35. **a.** Dependent variable: P = Population of the United States in millions
Independent variable: t = Years since 1995

b.

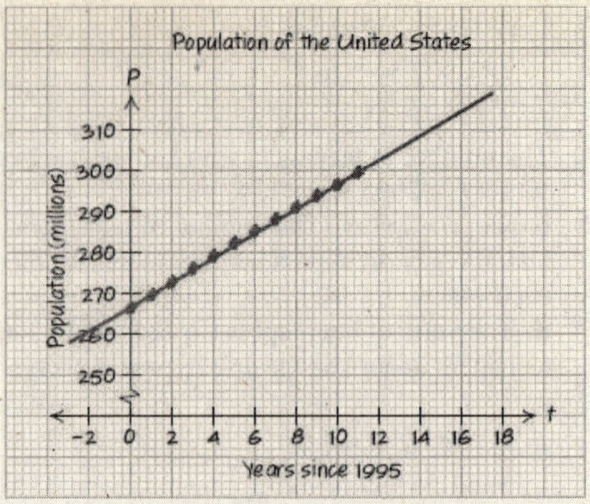

c. When $t = 14$ then $P \approx 308$ The population of the United States in 2009 will be around 308 million people.

d. Domain: $[0,14]$, range: $[266,308]$

e. $(0,266)$

f. In 1995, there were approximately 266 million people in the United States.

37. **a.** Let W be the number of deaths of women induced by illegal drugs in the United States t years since 2000.

t	W
0	6583
1	7452
2	9306
3	10,297
4	11,349

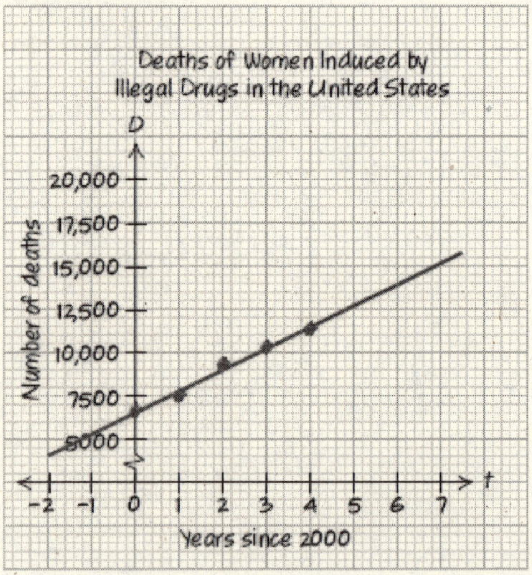

b. When $t = 7$, then $W \approx 15{,}000$. The number of drug-induced deaths of females in the United States in 2007 was around 15,000.

c. The number of drug-induced deaths of females in the United States reached 4000 in 1998.

d. Domain: $-1 \le t \le 7$, range: $5200 \le W \le 15{,}000$

39. **a.** Let T be the number of years someone is expected live if they are a years in age.

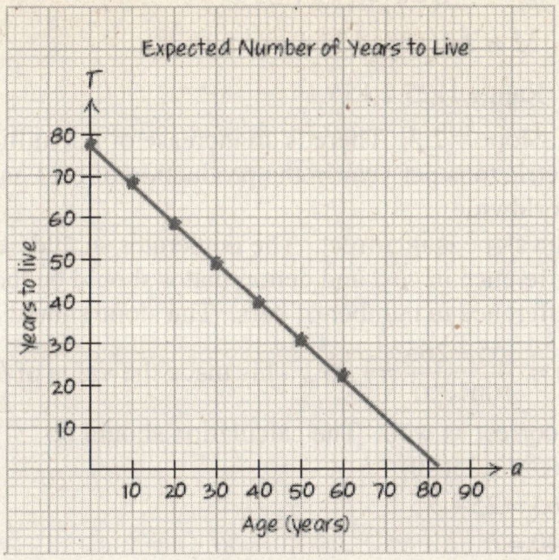

b. When $a = 45$, then $T = 35$. According to the model, a 45-year-old person will live approximately 35 years.

c. Domain: $[0,80]$, range: $[3,77]$

d. The graphical model predicts a 90-year-old to live negative years. This does not make sense.

e. The vertical intercept is $(0,77)$.

f. At birth, a person will live about 77 years.

Section 8.3 Finding Linear Models

1. b

3. c

5. c

7. a

9. x-scl = 1 , y-scl = 5

11. x-scl = 1 , y-scl = 0.1

13. The student switched the min and max for the x-values.

```
WINDOW
 Xmin=-45
 Xmax=-5
 Xscl=1
 Ymin=-20
 Ymax=35
 Yscl=1
↓Xres=■
```

15. **a.** Let R be the revenue in thousands of dollars for Quick Tire Repair, Inc. and t be the years since 2000.

t	R
6	608
7	611
8	616
9	620
10	625

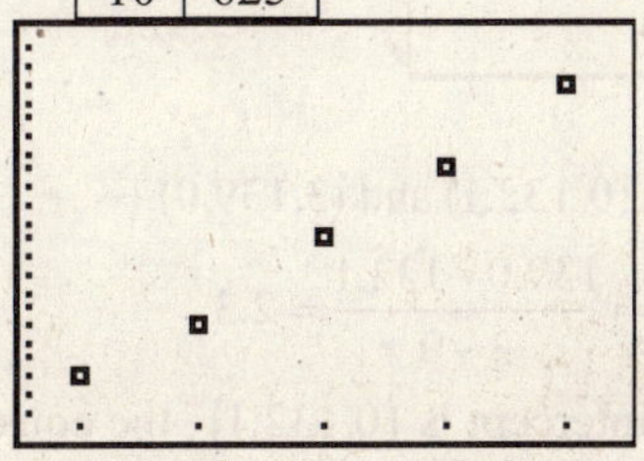

$R = mt + b$

Using two points: (6,608) and (9,620)

Find the slope: $m = \dfrac{620-608}{9-6} = \dfrac{12}{3} = 4$

$R = 4t + b$

$b = 608 - 4(6)$

$b = 584$

$R = 4t + 584$

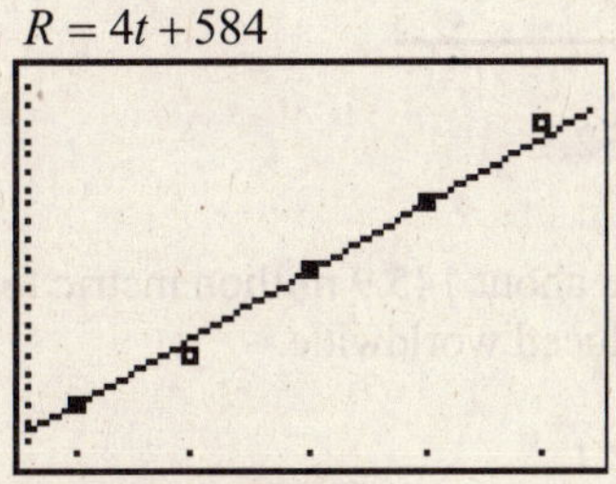

b. Domain [2,14]

$R = 4(2) + 584 = 592$

$R = 4(14) + 584 = 640$

Range [592,640]

c.

$700 = 4t + 584$

$700 - 584 = 4t + 584 - 584$

$116 = 4t$

$\dfrac{116}{4} = \dfrac{4t}{4}$

$29 = t$

In 2029, the revenue for Quick Tire Repair, Inc. will be $700 thousand.

d. S = 4. Quick Tire Repair, Inc. revenue is increasing about $4 thousand per year.

17. $y = 2.5x + 8$; The graph must shift up so b needs to be increased.

19. $y = -0.75x + 20$; The slope needs to be less negative so m needs to be increased.

21. **a.** Let M be millions of metric tons of beef and pork, and let t be years since 2000.

t	M
0	132.1
1	133.2
2	137.7
3	139.0
4	141.2
5	144.5

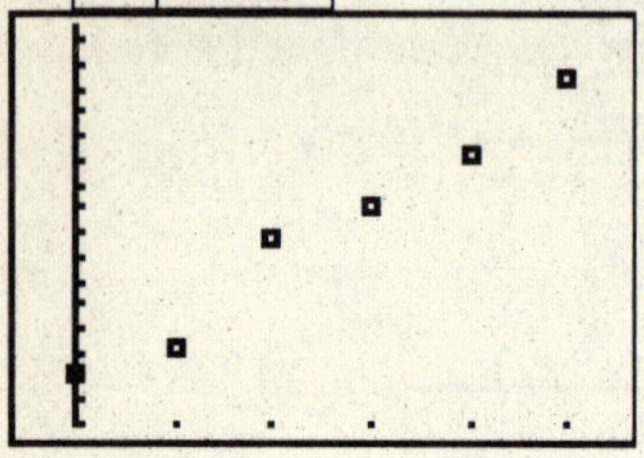

$M = mt + b$

Using two points: (0,132.1) and (3,139.0)

Find the slope: $m = \dfrac{139.0 - 132.1}{3 - 0} = 2.3$

Since the vertical intercept is $(0,132.1)$, the equation becomes

$M = 2.3t + 132.1$

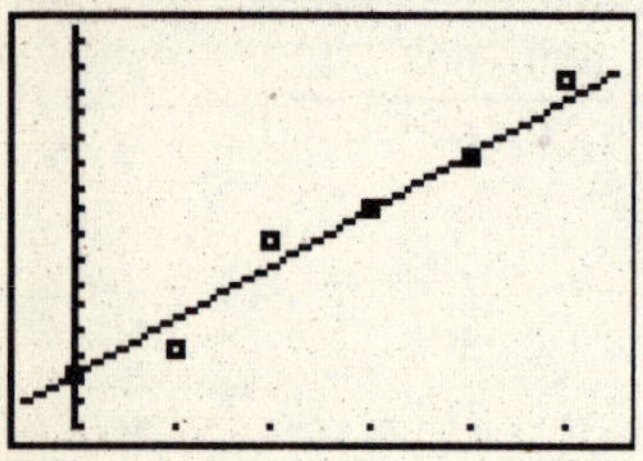

b.

$$M = 2.3(6) + 132.1$$
$$M = 145.9$$

In 2006, there were about 145.9 million metric tons of beef and pork produced worldwide.

c.

$$175 = 2.3t + 132.1$$
$$175 - 132.1 = 2.3t + 132.1 - 132.1$$
$$42.9 = 2.3t$$
$$\frac{42.9}{2.3} = \frac{2.3t}{2.3}$$
$$18.65 \approx t$$

In 2018, the worldwide production of beef and pork will be about 175 million metric tons.

d. Domain $[-3,10]$

$$M = 2.3(-3) + 132.1 = 125.2$$
$$M = 2.3(10) + 132.1 = 155.1$$

Range $[125.2,155.1]$

e. Slope = 2.3. There is an increase of 2.3 million metric tons per year in the production of beef and pork worldwide.

23. The slope is incorrect. The numerator should be the difference in cost and denominator should be the difference in number of bars. The slope should be $m = \dfrac{365 - 265}{100 - 20} = 1.25$. This made the value of b incorrect as well. The calculation should be

$$265 = 1.25(20) + b$$
$$265 = 25 + b$$
$$240 = b$$
$$C = 1.25b + 240$$

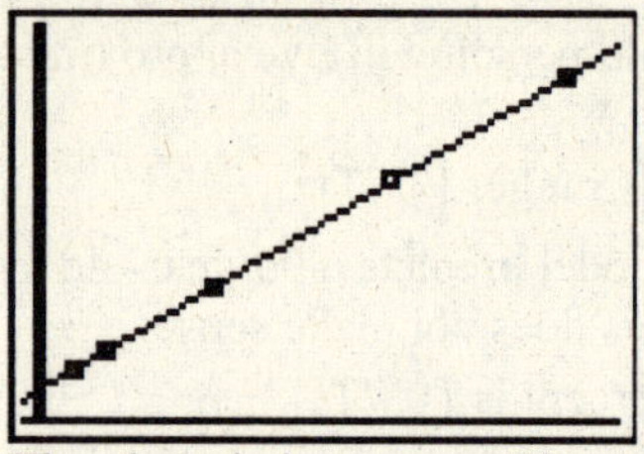

25. The slope is incorrect. The order of subtraction is reversed in the numerator. The slope should be $m = \dfrac{1105 - 90}{3000 - 100} = 0.35$. This made the value of b incorrect as well. The calculation should be:

$$90 = 0.35(100) + b$$
$$90 = 35 + b$$
$$55 = b$$
$$C = 0.35p + 55$$

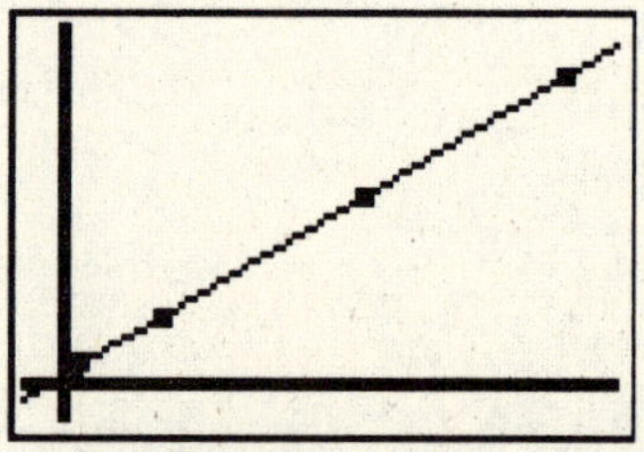

27. **a.** Let C be the number of reported Chlamydia cases in thousands, and t be the number of years since 2000.

t	C
0	702
1	783
2	835
3	877
4	929

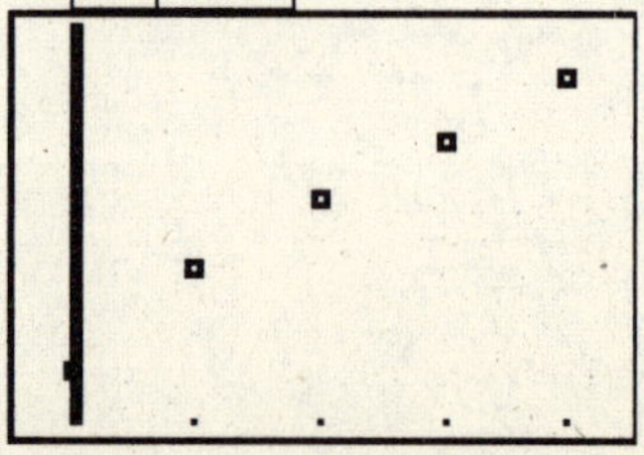

$C = mt + b$

Using two points: (0,702) and (3,877)

Find the slope: $m = \dfrac{877-702}{3-0} = \dfrac{175}{3} \approx 58.3$

Since the vertical intercept is $(0,702)$, the equation becomes:

$C = 58.3t + 702$

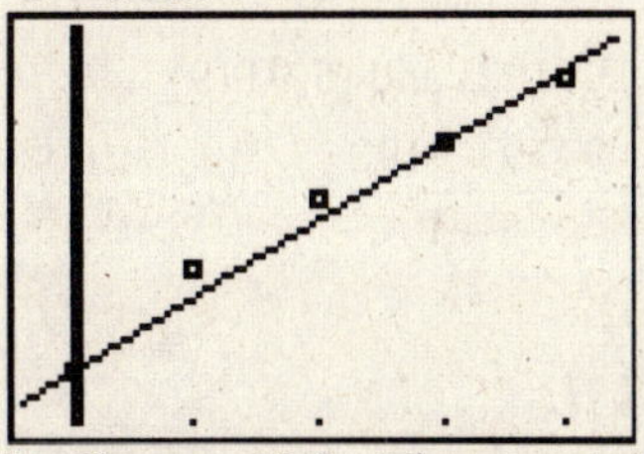

b. Slope = 58.3. The reported number of Chlamydia cases increases by 58.3 thousand cases per year.

c.

$C = 58.3(6) + 702$

$C = 349.8 + 702$

$C = 1051.8$

There are 1,051,800 estimated Chlamydia cases for 2006.

29. **a.** Let E be the total egg production in the United States (in billions), and let t be years since 2000.

t	E
0	84.7
1	86.1
2	87.3
3	87.5
4	89.1
5	90.0

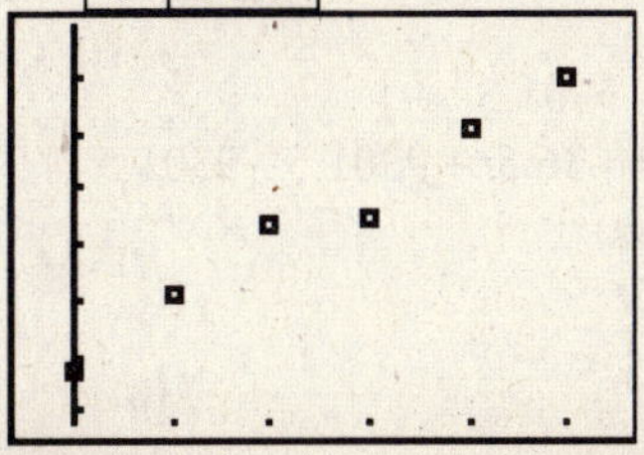

$E = mt + b$

Using two points: (0,84.7) and (4,89.1)

Find the slope: $m = \dfrac{89.1-84.7}{4-0} = \dfrac{4.4}{4} = 1.1$

Since the vertical intercept is $(0,84.7)$, the equation becomes

$E = 1.1t + 84.7$

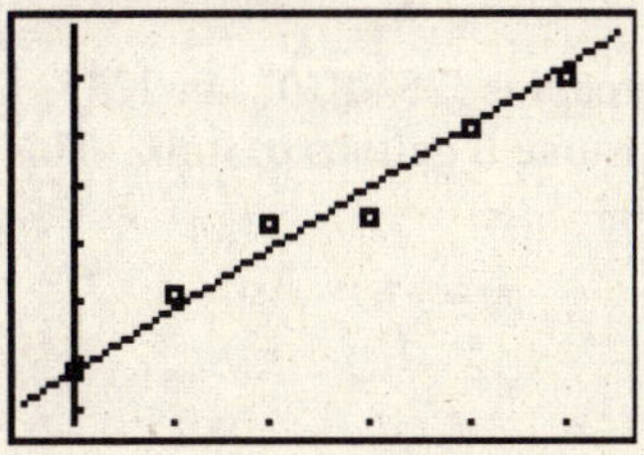

b.

$E = 1.1(8) + 84.7$

$E = 8.8 + 84.7$

$E = 93.5$

In 2008, the total egg production in the United States was 93.5 billion.

c. Domain $[-4,10]$

$E = 1.1(-4) + 84.7 = 80.3$

$E = 1.1(10) + 84.7 = 95.7$

Range $[80.3, 95.7]$

d. Slope = 1.1. The egg production is increased by 1.1 billion eggs per year in the United States.

31. Let U, S, and F be the number of Internet users in the United Kingdom, Spain, and Finland in millions, respectively, and let t be years since 1990.

t	U	S	F
8	6.7	0.6	1.4
9	12.5	2.8	1.7
10	18.0	5.4	1.9
11	24.0	7.7	2.2
14	38.9	14.3	2.9

United Kingdom : $U = mt + b$

Using two points: (9,12.5) and (10,18.0)

$m = \dfrac{18.0-12.5}{10-9} = 5.5$

$b = 18.0 - 5.5(10) = 37$

$\boxed{U = 5.5t + 37}$

Spain : $S = mt + b$

Using two points: (8,0.6) and (11,7.4)

$m = \dfrac{7.4-0.6}{11-8} = \dfrac{6.8}{3} \approx 2.27$

$b = 7.4 - 2.27(11) = -17.57 \approx -17.6$

$\boxed{S = 2.27t - 17.6}$

Finland : $F = mt + b$

Using two points: (8,1.4) and (14,2.9)

$m = \dfrac{2.9-1.4}{14-8} = \dfrac{1.5}{6} = 0.25$

$b = 1.4 - 0.25(8) = -0.6$

$\boxed{F = 0.25t - 0.6}$

33. By comparing the slopes of the models we can see that the United Kingdom has the greatest growth. United Kingdom (5.5) > Spain (2.27) > Finland (0.25)

35. **a.** Let R be total revenue for FedEx in millions of dollars, and let t be the number of years since 2000

t	R
0	18,257
1	19,629
2	20,607
3	22,487
4	24,710
5	29,363
6	32,294
7	35,214

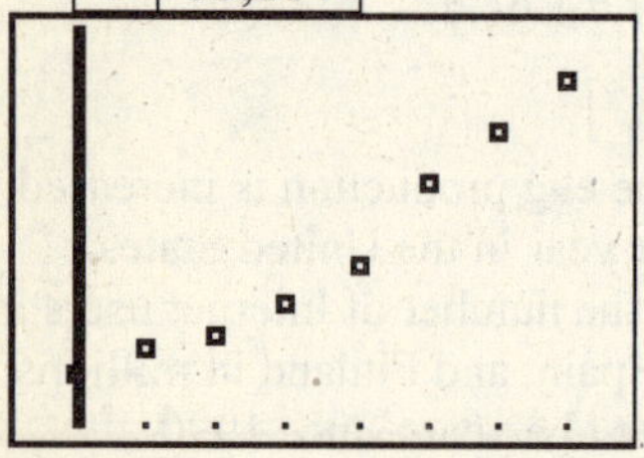

$R = mt + b$

Using two points: (1,19629) and (5,29363)

Find the slope: $m = \dfrac{29{,}363 - 19{,}629}{5-1} = 2433.5$

$R = 2433.5t + b$

$b = 19{,}629 - 2433.5(1)$

$b = 17{,}195.5$

$R = 2433.5t + 17{,}195.5$

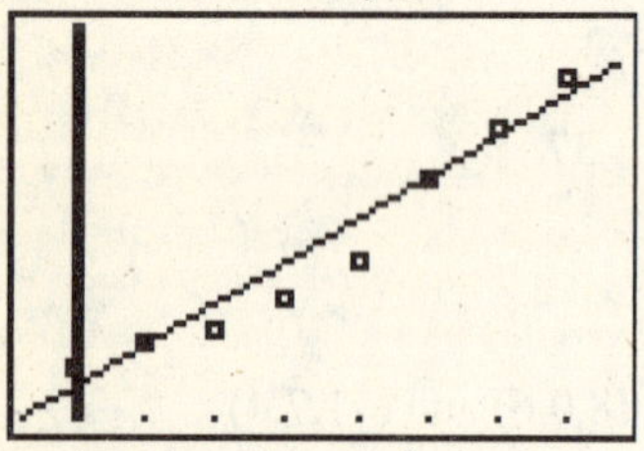

b. Domain $[-3,10]$

$R = 2433.5(-3) + 17{,}195.5 = 9895$

$R = 2433.5(10) + 17{,}195.5 = 41{,}531$

Range $[9895, 41531]$

c. The slope is 2433.5. The total revenue for FedEx increases by $2433.5 million each year.

d.

$R = 2433.5(10) + 17{,}195.5$

$R = 41{,}530.5$

In 2010, the total revenue for FedEx will be $41530.5 million.

37. **a.** Let M be the total consumption of milk in the United States in millions of gallons, and let t be years since 2000.

t	M
0	9297
2	9244
3	9192
4	9141
5	9119

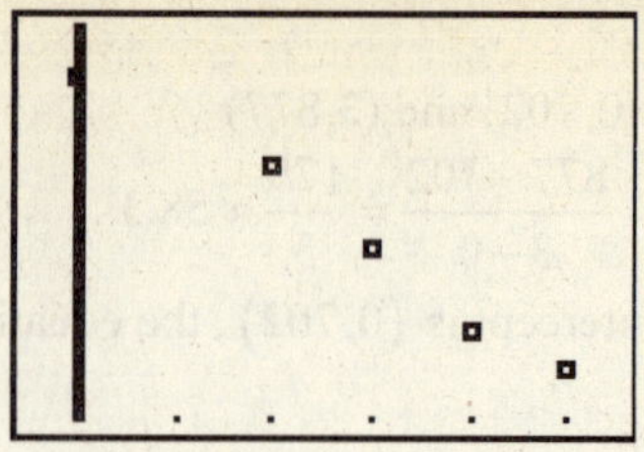

$M = mt + b$

Using two points: (3,9192) and (5,9119)

Find the slope: $m = \dfrac{9119 - 9192}{5-3} = -36.5$

$M = -36.5t + b$

$b = 9192 - (-36.5)(3)$

$b = 9301.5$

$M = -36.5t + 9301.5$

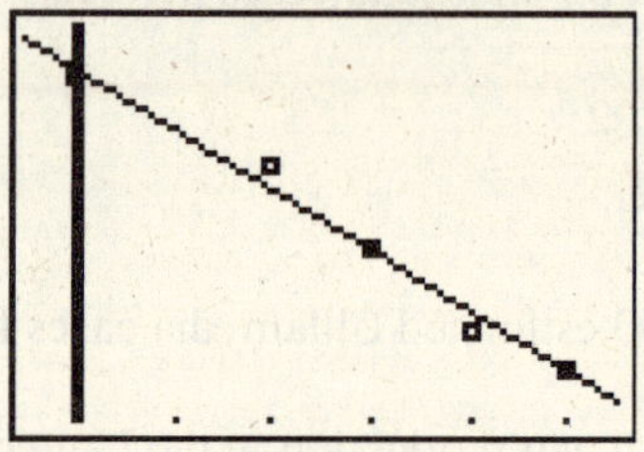

b. Domain $[-3,10]$

$M = -36.5(-3) + 9301.5 = 9411$

$M = -36.5(10) + 9301.5 = 8936.5$

Range $[8936.5, 9411]$

c. The vertical intercept is (0, 9301.5). In 2000, the total consumption of milk in the United States was 9301.5 million gallons of milk.

d.

$9000 = -36.5t + 9301.5$

$9000 - 9301.5 = -36.5t + 9301.5 - 9301.5$

$-301.5 = -36.5t$

$\dfrac{-301.5}{-36.5} = \dfrac{-36.5t}{-36.5}$

$8.3 \approx t$

In 2008, Americans consumed approximately 9000 million gallons of milk.

e.

$0 = -36.5t + 9301.5$

$36.5t = 9301.5$

$\dfrac{36.5t}{36.5} = \dfrac{9301.5}{36.5}$

$t \approx 254.8$

The horizontal intercept is (254.8, 0). In 2255, Americans will consume 0 gallons of milk. This does not make sense.

39. **a.** Let H be the total amount spent by individuals on health care expenses in the United States in billions of dollars, and let t be the years since 2000.

t	H
6	251
7	266
8	281
9	299
10	317
11	335

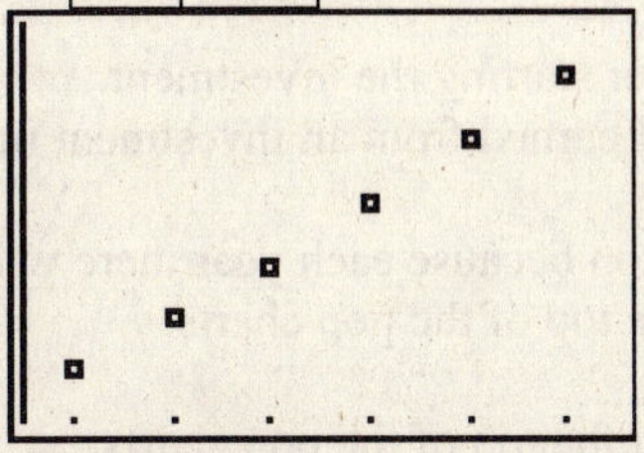

$H = mt + b$

Using two points: (7,266) and (10,317)

Find the slope: $m = \dfrac{317-266}{10-7} = 17$

$H = 17t + b$

$b = 266 - 17(7)$

$b = 147$

$H = 17t + 147$

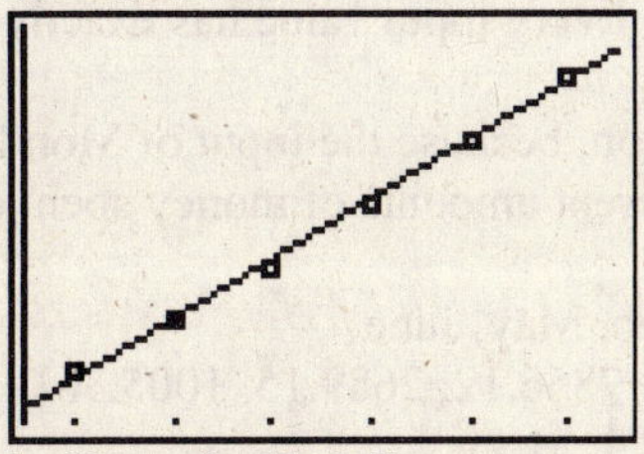

b. Domain [3, 14]

$H = 17(3) + 147 = 198$

$H = 17(14) + 147 = 385$

Range [198, 385]

c. The slope is 17. The amount of money Americans spend on health care expenses increases \$17 billion each year.

d.

$H = 17(15) + 147$

$H = 402$

In 2015, Americans will spend \$402 billion on health care expenses.

e.

$500 = 17t + 147$

$500 - 147 = 17t + 147 - 147$

$353 = 17t$

$\dfrac{353}{17} = \dfrac{17t}{17}$

$20.8 \approx t$

The amount that Americans will spend on health care expenses will reach \$500 billion in 2021.

41.

Find the slope: $m = \dfrac{40-4}{10-4} = 6$

$4 = 6(4) + b$

$4 = 24 + b$

$b = -20$

Since $m = 6$ and $b = -20$

$y = 6x - 20$

43.

Using two points: $(1,5)$ and $(5,-7)$

Find the slope: $m = \dfrac{5-(-7)}{1-5} = -3$

$5 = -3(1) + b$

$5 = -3 + b$

$b = 8$

Since $m = -3$ and $b = 8$

$y = -3x + 8$

45.

Using two points: $(3,-4)$ and $(12,-1)$

Find the slope: $m = \dfrac{-1-(-4)}{12-3} = \dfrac{1}{3}$

$-4 = \dfrac{1}{3}(3) + b$

$-4 = 1 + b$

$b = -5$

Since $m = \dfrac{1}{3}$ and $b = -5$

$y = \dfrac{1}{3}x - 5$

47.

Using two points: $(-8,4.167)$ and $(7,1.333)$

Find the slope: $m = \dfrac{1.333-4.167}{7-(-8)} = -\dfrac{2.834}{11} \approx -0.189$

$4.167 = -0.189(-8) + b$

$4.167 = 1.512 + b$

$b \approx 2.66$

Since $m \approx -0.19$ and $b = 2.66$

$y = -0.19x + 2.66$

49.

Using two points: $(10,492)$ and $(19,627)$

Find the slope: $m = \dfrac{627-492}{19-10} = 15$

$492 = 15(10) + b$

$492 = 150 + b$

$b = 342$

Since $m = 15$ and $b = 342$

$y = 15x + 342$

51.

Using two points: $(7,-3755)$ and $(22,-4370)$

Find the slope: $m=\dfrac{-4370-(-3755)}{22-7}=-41$

$-3755=-41(7)+b$

$-3755=-287+b$

$b=-3468$

Since $m=-41$ and $b=-3468$

$y=-41x-3468$

Section 8.4 Functions and Function Notation

1. This is not a function because two children the same age may not be in the same grade.
Input: a = years old.
Output: G = grade level of student.

3. This is not a function because two children the same age may not be the same height.
Input: a = years old.
Output: H = heights in inches of children attending Mission Meadows Elementary School.

5. This is a function.
Input: t = years after starting the investment.
Output: I = interest earned from an investment in dollars.

7. This is not a function because each year there will be several songs at the top of the pop charts.
Input: y = year.
Output: S = song at the top of the pop charts.

9. This is a function.
Input: t = year.
Output: T = amount of taxes you paid in dollars.

11. This is a function. Every input value has exactly one output value.

13. This is not a function, because the input of 1 hour of poker play is related to two different amounts of winnings.

15. This is a function. Every input value has exactly one output value.

17. This is not a function, because the input of Monday is related to two different amounts of money spent on lunch.

19. Domain = {Jan, Feb, May, June}
Range = {5689.35, 7856.12, 2689.15, 1005.36}

21. Domain = {1, 2, 3, 4, 5}
Range = $\{-650,-100,-150,60,125,200,300\}$

23. The death rate from HIV for 20-year-olds is 0.5 per 100,000.

25. In 2002, 14,304 small business loans were made to minority-owned small businesses.

27. The amount spent on lunch on Tuesday was \$5.95 and \$6.33.

29. Yes, this is a function. The graph passes the Vertical line test.

31. No, this is not a function. The graph fails the Vertical line test.

33. Yes, this is a function. The graph passes the Vertical line test.

35. Yes, this is a function. The graph passes the Vertical line test.

37. Yes, this is a function. The graph passes the Vertical line test.

39. No, this is not a function. The graph fails the Vertical line test.

41. **a.** $W(0)=86.5$. At the start of a diet, a person's weight will be 86.5 kg.
b. $W(10)=82$. Ten days after starting a diet, a person's weight will be 82 kg.

c. $W = 75$ when $d = 30$. Thirty days after starting a diet, a person's weight will be 75 kg.

d. $W(100) = 88$. One hundred days after starting a diet, a person's weight will be 88 kg.

43. **a.** $C(5) = 8$. Eight ounces of chocolate were consumed in the Clark household on the fifth day of the month.

b. $C = 20$ when $d = 15$. Twenty ounces of chocolate were consumed in the Clark household on the 15th day of the month.

c. $C(30) = 28$. Twenty-eight ounces of chocolate were consumed in the Clark household on the 30th day of the month.

45. **a.** $P(\text{Ohio}) = 11.48$

b. $P(\text{Texas}) = 23.507783$

c. $P(\text{Wyoming}) = 0.515004$

47. Let $B(t)$ be the average monthly Social Security benefit in dollars t years since 2000.

a.

t	$B(t)$
2	834
3	862
4	894
5	938
6	978

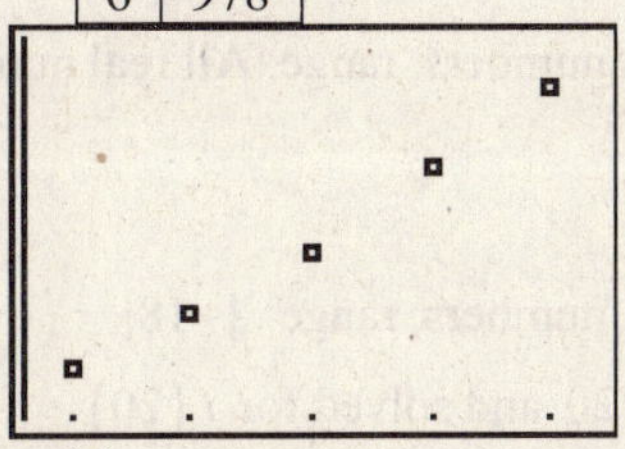

$B(t) = mt + b$

Using two points: (3,862) and (5,938)

Find the slope: $m = \dfrac{938 - 862}{5 - 3} = 38$

$B(t) = 38t + b$

$b = 862 - 38(3)$

$b = 748$

$B(t) = 38t + 748$

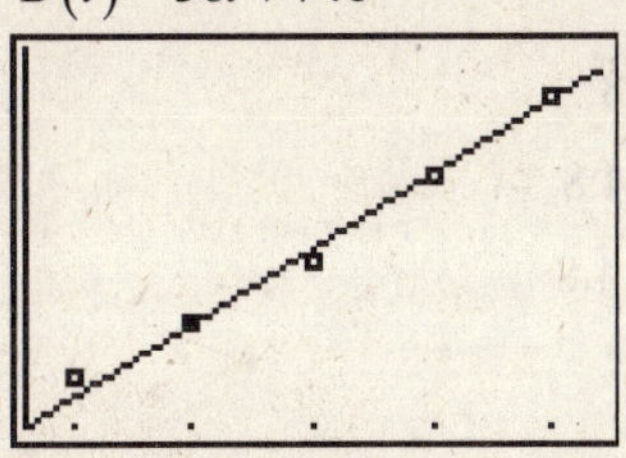

b. $B(9) = 38(9) + 748 = 1090$ In 2009, the average social security benefit for retired workers was \$1090.

c.

$$2000 = 38t + 748$$
$$2000 - 748 = 38t + 748 - 748$$
$$1252 = 38t$$
$$\frac{1252}{38} = \frac{38t}{38}$$
$$33 \approx t$$

The average monthly Social Security benefit will be \$2000 in 2033.

d. Domain: $[-1,10]$

$B(-1) = 38(-1) + 748 = 710$

$B(10) = 38(10) + 748 = 1128$

Range: [710, 1128]

e. $(0,748)$. In 2000, the average monthly Social Security benefit was \$748.

49. Let $M(t)$ be the number of Medicare enrollees t years since 2000.

a.

T	$M(t)$
2	40.5
3	41.2
4	41.9
5	42.5

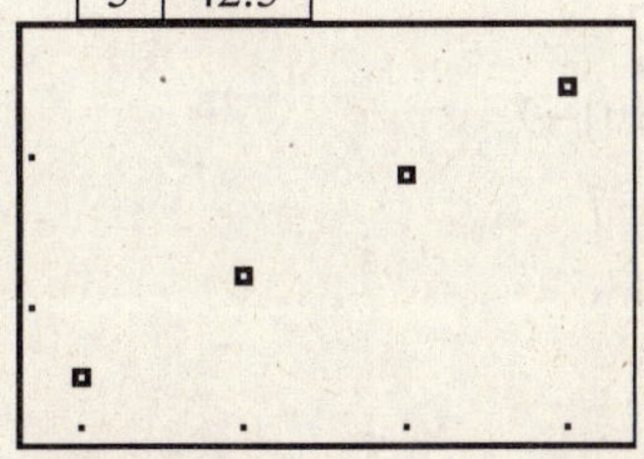

$M(t) = mt + b$

Using two points: (2,40.5) and (5,42.5)

Find the slope: $m = \dfrac{42.5 - 40.5}{5 - 2} \approx 0.67$

$M(t) = 0.67t + b$

$b = 40.5 - 0.67(2)$

$b = 39.16$

$M(t) = 0.67t + 39.16$

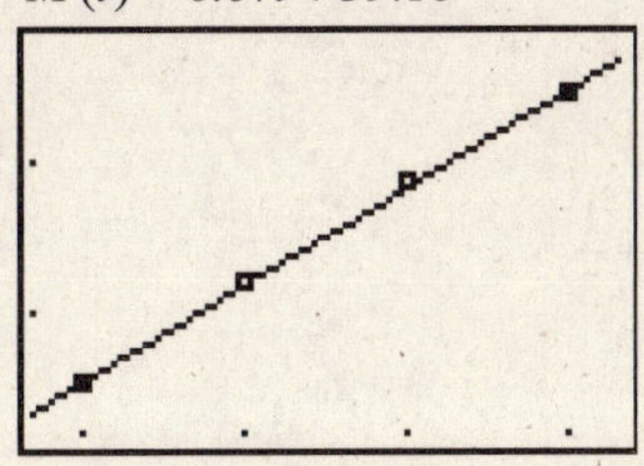

b. Domain: $[-3,10]$

$M(-3) = 0.67(-3) + 39.16 = 37.15$

$M(10) = 0.67(10) + 39.16 = 45.86$

Range: $[37.15, 45.86]$

c. $(0,39.16)$ In 2000, there were 39.16 million people enrolled in Medicare.

d.

$$M(10) = 0.67(10) + 39.16$$
$$M(10) = 45.86$$

In 2010, there were 45.86 million people enrolled in Medicare.

e.

$$50 = 0.67t + 39.16$$
$$50 - 39.16 = 0.67t + 39.16 - 39.16$$
$$10.84 = 0.67t$$
$$\frac{10.84}{0.67} = \frac{0.67t}{0.67}$$
$$16.18 \approx t$$

In 2016, there will be approximately 50 million people enrolled in Medicare.

51. **a.**

$$f(x) = 2x - 7$$
$$f(\ ) = 2(\ ) - 7$$
$$f(5) = 2(5) - 7$$
$$f(5) = 10 - 7$$
$$f(5) = 3$$

b.

$$f(x) = 2x - 7$$
$$f(\ ) = 2(\ ) - 7$$
$$f(-10) = 2(-10) - 7$$
$$f(-10) = -20 - 7$$
$$f(-10) = -27$$

c.

$$f(x) = 2x - 7$$
$$-1 = 2x - 7$$
$$-1 + 7 = 2x - 7 + 7$$
$$6 = 2x$$
$$x = 3$$

d. Domain: All real numbers, range: All real numbers

53. **a.**

$$h(x) = \frac{2}{3}x + \frac{1}{3}$$
$$h(\ ) = \frac{2}{3}(\ ) + \frac{1}{3}$$
$$h(15) = \frac{2}{3}(15) + \frac{1}{3}$$
$$h(15) = \frac{30}{3} + \frac{1}{3}$$
$$h(15) = \frac{31}{3}$$

b.

$$h(x) = \frac{2}{3}x + \frac{1}{3}$$
$$h(\ ) = \frac{2}{3}(\ ) + \frac{1}{3}$$
$$h(-9) = \frac{2}{3}(-9) + \frac{1}{3}$$
$$h(-9) = \frac{-18}{3} + \frac{1}{3}$$
$$h(-9) = \frac{-17}{3}$$

c.

$$h(x) = \frac{2}{3}x + \frac{1}{3}$$
$$4 = \frac{2}{3}x + \frac{1}{3}$$
$$3(4) = 3\left(\frac{2}{3}x + \frac{1}{3}\right)$$
$$12 = 2x + 1$$
$$12 - 1 = 2x + 1 - 1$$
$$11 = 2x$$
$$\frac{11}{2} = \frac{2x}{2}$$
$$x = \frac{11}{2}$$

d. Domain: All real numbers, range: All real numbers

55. **a.** $g(2) = -18$

b. $g(-11) = -18$

c. Domain: All real numbers, range: $\{-18\}$

57. The student set $x = 20$ and solved for $f(20)$.

The correct solution is:

$$f(x) = 4x + 8$$
$$20 = 4x + 8$$
$$20 - 8 = 4x + 8 - 8$$
$$12 = 4x$$
$$x = 3$$

59. The student restricted the domain and range.
The correct solution is: Domain: All real numbers, range: All real numbers

61. **a.**

$$f(x) = 3.2x - 4.8$$
$$f(\ ) = 3.2(\ ) - 4.8$$
$$f(2) = 3.2(2) - 4.8$$
$$f(2) = 6.4 - 4.8$$
$$f(2) = 1.6$$

b.

$f(x)=3.2x-4.8$
$f(\ )=3.2(\ )-4.8$
$f(-14)=3.2(-14)-4.8$
$f(-14)=-44.8-4.8$
$f(-14)=-49.6$

c.

$f(x)=3.2x-4.8$
$-10=3.2x-4.8$
$-10+4.8=3.2x-4.8+4.8$
$-5.2=3.2x$
$\frac{-5.2}{3.2}=\frac{3.2x}{3.2}$
$x=-1.625$

d. Domain: All real numbers, range: All real numbers

63. **a.**

$h(x)=14x+500$
$h(\ )=14(\ )+500$
$h(105)=14(105)+500$
$h(105)=1470+500$
$h(105)=1970$

b.

$h(x)=14x+500$
$-140=14x+500$
$-140-500=14x+500-500$
$-640=14x$
$\frac{-640}{14}=\frac{14x}{14}$
$x=-\frac{320}{7}$

c. Domain: All real numbers, range: All real numbers

65. **a.** $f(3)=2$
b. $f(-2)=-8$
c. $x=-1$
d. Domain: All real numbers, range: All real numbers
e. Vertical intercept: $(0,-4)$. This is the point the graph crosses the y-axis.
Horizontal intercept: $(2,0)$. This is the point the graph crosses the x-axis.

67. **a.** $f(25)=50$
b. $f(100)=275$
c. $x=75$
d. Domain: All real numbers, range: All real numbers
e. Vertical intercept $(0,-25)$. This is the point the graph crosses the y-axis.
Horizontal intercept $(8,0)$. This is the point the graph crosses the x-axis.

69. **a.** $h(2)=3$
b. $h(5)=-4$
c. $x=-0.2$ and $x=4.2$
d. Domain: All real numbers, range: $(-\infty,3]$
e. Vertical intercept $(0,0)$. This is the point the graph crosses the y-axis.
Horizontal intercepts $(0,0)$ and $(4,0)$. These are the points the graph crosses the x-axis.

CHAPTER 8 Modeling Data and Functions

Section 8.5 Combining Functions

1. **a.** $T(0)=-4.5(0)+712.3=712.3$. In the year 2000, the average American ate 712.3 pounds of fruits and vegetables.

b. $F(0)=-3.5(0)+289.6=289.6$. In the year 2000, the average American ate 289.6 pounds of fruit.

c. $T(0)-F(0)=712.3-289.6=422.7$. In the year 2000, the average American ate 422.7 pounds of vegetables.

d. Let $V(y)$ be the average number of pounds of vegetables each American eats per year y years since 2000.

$$\begin{aligned}V(y)&=T(y)-F(y)\\&=(-4.5y+712.3)-(-3.5y+289.6)\\&=-4.5y+712.3+3.5y-289.6\\V(y)&=-y+422.7\end{aligned}$$

e. $V(0)=422.7$. In the year 2000, the average American ate 422.7 pounds of vegetables.

f.

$$V(10)=-10+422.7=412.7$$
$$V(15)=-15+422.7=407.7$$
$$V(20)=-20+422.7=402.7$$

The average American will eat approximately 412.7 pounds of vegetables in 2010, 407.7 pounds of vegetables in 2015, and 402.7 pounds of vegetables in 2020.

3. **a.** Let $O(t)$ be the per capita consumption of milk products other than whole milk in the U.S. (in gallons per person) t years since 2000.

$$\begin{aligned}O(t)&=M(t)-W(t)\\&=(-0.29t+22.42)-(-0.21t+8.09)\\&=-0.29t+22.42+0.21t-8.09\\O(t)&=-0.08t+14.33\end{aligned}$$

b. $W(5)=-0.21(5)+8.09=7.04$. In 2005, 7.04 gallons of whole milk were consumed per person in the United States.

c.

$$O(5)=-0.08(5)+14.33=13.93$$
$$O(10)=-0.08(10)+14.33=13.53$$

In the United States, the per capita consumption of milk products other than whole milk was 13.93 gallons in 2005 and 13.53 gallons in 2010.

d. Slope $=-0.21$. The amount of whole milk consumed per person in the United States is decreasing by approximately 0.21 gallons per year.

e. M -intercept: (0, 22.42). In the year 2000, the per capita consumption of milk products in the United States was 22.42 gallons.

5. $U(t)-M(t)$ is the number of women in the United States in millions t years since 1900.

7. The total distance can be found by multiplying $\frac{\text{miles}}{\text{hour}}\times\frac{\text{hour}}{1}=\text{miles}$. Let $S(d)T(d)=ST(d)$ be the number of miles driven on day d of a cross country trip.

9. The average amount of national debt per person is given by $\frac{D(t)}{U(t)}=\frac{D}{U}(t)$ dollars per person, t years since 1900.

11. To get the personal dept use $(\#\text{persons})\left(\frac{\text{dollars}}{\text{person}}\right)$. Let $U(t)\cdot D(t)$ be the total amount of personal debt in dollars in the United States t years since 1900.

13. **a.** $M(t)+F(t)=100$, because the percent must add up to 100 if it is to represent all of the jail inmates.

b.

$$\begin{aligned}M(t)+F(t)&=100\\(-0.24t+88.64)+F(t)&=100\\F(t)&=100-(-0.24t+88.64)\\F(t)&=100+0.24t-88.64\\F(t)&=0.24t+11.36\end{aligned}$$

c. $F(5)=0.24(5)+11.36=12.56$. In 2005, approximately 12.56% of jail inmates in the U.S. federal and state prisons were female.

d. Slope $=0.24$. The percentage of jail inmates who are female in U.S. federal and state prisons is increasing by approximately0.24 percentage points per year.

e. M intercept = (0, 88.64). The percentage of jail inmates who are male in U.S federal and state prisons was 88.64% in 2000.

15. **a.**

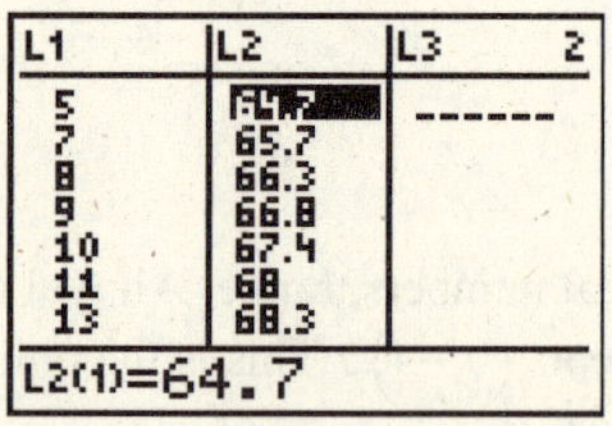

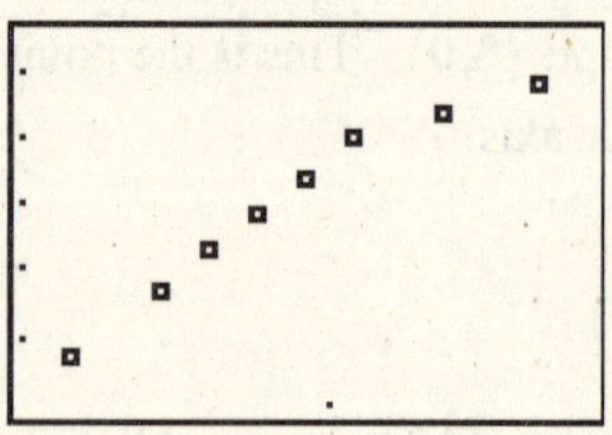

Using two points: $(7,65.7)$ and $(13,68.3)$

Find the slope: $m=\dfrac{68.3-65.7}{13-7}\approx 0.433$

$$O(t)=0.433t+b$$

$$65.7=0.433(7)+b$$
$$65.7=3.031+b$$
$$65.7-3.031=3.031-3.031+b$$
$$62.669=b$$
$$O(t)=0.433t+62.67$$

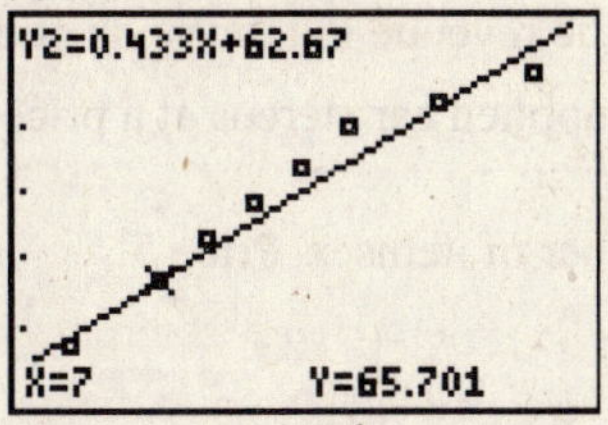

b.

$$R(t)=100-O(t)$$
$$R(t)=100-(0.433t+62.67)$$
$$R(t)=100-0.433t-62.67$$
$$R(t)=-0.433t+37.33$$

c. $R(20)=-0.433(20)+37.33=28.67$. In 2010, approximately 28.67% of occupied housing units should be renter occupied.

d. Slope $=-0.433$. The number of housing units that are renter occupied is decreasing by 0.433 percentage point each year.

17. **a.** Let $N(w)$ be the total area in square inches of a Norman window whose rectangular part has a height of 75 inches and a width of w inches.

Area of window = Area of half circle + Area of rectangle

$$N(w)=\frac{1}{2}\cdot\pi\left(\frac{1}{2}w\right)^2+75w$$

$$N(w)=\frac{\pi}{8}w^2+75w$$

b.

$$N(36)=\frac{\pi}{8}(36)^2+75(36)$$

$$N(36)=3208.94\ in^2$$

A Norman window whose rectangular part has a height of 75 inches and a width of 36 inches will have a total area of 3208.94 square inches.

19. $f(x)=3x+8$ and $g(x)=-6x+9$

a.

$$f(x)+g(x)=(3x+8)+(-6x+9)$$
$$=3x+8+(-6x)+9$$
$$=-3x+17$$

b.

$$f(x)-g(x)=(3x+8)-(-6x+9)$$
$$=3x+8+6x-9$$
$$=9x-1$$

c.

$$f(x)g(x)=(3x+8)(-6x+9)$$
$$=-18x^2+27x-48x+72$$
$$=-18x^2-21x+72$$

d.

$$\frac{f(x)}{g(x)}=\frac{3x+8}{-6x+9}$$

21. $f(x)=\frac{1}{2}x+\frac{1}{5}$ and $g(x)=2x-\frac{3}{5}$

a.

$$f(x)+g(x)=\left(\frac{1}{2}x+\frac{1}{5}\right)+\left(2x-\frac{3}{5}\right)$$
$$=\frac{1}{2}x+\frac{1}{5}+2x-\frac{3}{5}$$
$$=\frac{5}{2}x-\frac{2}{5}$$

b.

$$g(x)-f(x)=\left(2x-\frac{3}{5}\right)-\left(\frac{1}{2}x+\frac{1}{5}\right)$$
$$=2x-\frac{3}{5}-\frac{1}{2}x-\frac{1}{5}$$
$$=\frac{3}{2}x-\frac{4}{5}$$

c.

$$f(x)g(x)=\left(\frac{1}{2}x+\frac{1}{5}\right)\left(2x-\frac{3}{5}\right)$$
$$=x^2-\frac{3}{10}x+\frac{2}{5}x-\frac{3}{25}$$
$$=x^2+\frac{1}{10}x-\frac{3}{25}$$

d

$$\frac{f(x)}{g(x)}=\frac{\frac{1}{2}x+\frac{1}{5}}{2x-\frac{3}{5}}$$

23. $f(x)=3x+5$ and $g(x)=x^2+4x+10$

a.

$$f(x)+g(x)=(3x+5)+(x^2+4x+10)$$
$$=3x+5+x^2+4x+10$$
$$=x^2+7x+15$$

b.

$$g(x)-f(x)=(x^2+4x+10)-(3x+5)$$
$$=x^2+4x+10-3x-5$$
$$=x^2+x+5$$

c.

$$f(x)g(x) = (3x+5)(x^2+4x+10)$$
$$= 3x^3+12x^2+30x+5x^2+20x+50$$
$$= 3x^3+17x^2+50x+50$$

d.

$$\frac{f(x)}{g(x)} = \frac{3x+5}{x^2+4x+10}$$

25. Jose set the expression equal to zero and solved for x.

$$f(x)+g(x) = (3x-10)+(7x+8)$$
$$= 3x-10+7x+8$$
$$= 10x-2$$

27. a.

$$R(10) = 3.1(10)^2 - 33.7(10) + 180.9$$
$$= 310 - 337 + 180.9$$
$$= 153.9$$

In 2010, IBM had revenue of $153.9 billion for goods sold.

$$C(10) = 2.5(10)^2 - 28.9(10) + 136.1$$
$$= 250 - 289 + 136.1$$
$$= 97.1$$

In 2010, IBM had a cost of $97.1 billion for selling goods.

b. Let $P(t)$ be the profit of IBM in billions of dollars t years since 2000.

$$P(t) = R(t) - C(t)$$
$$= (3.1t^2 - 33.7t + 180.9) - (2.5t^2 - 28.9t + 136.1)$$
$$= 3.1t^2 - 33.7t + 180.9 - 2.5t^2 + 28.9t - 136.1$$
$$P(t) = 0.6t^2 - 4.8t + 44.8$$

c.

$$P(10) = 0.6(10)^2 - 4.8(10) + 44.8$$
$$= 60 - 48 + 44.8$$
$$= 56.8$$

According to the model, IBM's profit in 2010 was approximately $56.8 billion.

29. a.

$$R(11) = 164.1(11)^2 - 1353.9(11) + 6380.0$$
$$= 11{,}343.2$$
$$P(11) = 87.9(11)^2 - 708.9(11) + 3322.8$$
$$= 6160.8$$

R(11) = 11343.2. The revenue earned by Pearson Publishing Co. in 2011, is approximately 11,343.2 million British pounds. P(11) = 6160.8. The profit earned by Pearson Publishing Co. in 2011, is approximately 6160.8 million British pounds.

b. Let $C(t)$ be the publishing costs at Pearson Publishing in millions of British pounds t years since 2000.

$$C(t) = (164.1t^2 - 1353.9t + 6380.0) - (87.9t^2 - 708.9t + 3322.8)$$
$$= 164.1t^2 - 1353.9t + 6380.0 - 87.9t^2 + 708.9t - 3322.8$$
$$C(t) = 76.2t^2 - 645t + 3057.2$$

c. $C(11) = 76.2(11)^2 - 645(11) + 3057.2 = 5182.4.$

The costs at Pearson Publishing Co. for 2011 are approximately 5182.4 million British pounds.

31. a. $S(100) = 0.009(100)^2 - 0.5(100) + 20 = 60.$ Car stereo producers are willing to supply 60 thousand car stereos if the price is set at $100.

b. Let $R(p)$ be the revenue in thousands of dollars from selling the supplied car stereos at a price of p dollars.

Revenue = Number of items × Price

$$R(p) = S(p) \times p$$
$$R(p) = 0.009p^3 - 0.5p^2 + 20p$$

c. $R(90) = 0.009(90)^3 - 0.5(90)^2 + 20(90) = 4311.$

If car stereos are sold for $90 each, the projected revenue will be about $4,311 thousand.

33. a. $S(25) = 0.11(25)^2 - 3.2(25) + 45 = 33.75.$ Car manufacturers are willing to supply about 33,750 minivans if they can be sold for $25,000 each.

b. Let $R(p)$ be the revenue in millions of dollars from selling minivans at a price of p thousand dollars.

Revenue = Number of items × Price

$$R(p) = S(p) \times p$$
$$R(p) = 0.11p^3 - 3.2p^2 + 45p$$

c. $R(31) = 0.11(31)^3 - 3.2(31)^2 + 45(31) = 1596.81.$

If minivans sell for $31,000 each, the projected revenue will be about $1.6 billion.

35. a.

$$I(5) = -17{,}854(5)^2 + 212{,}449(5) - 501{,}784$$
$$I(5) = 114{,}111$$

There were about 114,111 immigrants admitted to the United States as permanent residents under refugee acts in 1995.

b.

$$E(5) = -6432(5)^2 + 74{,}006(5) - 161{,}816$$
$$E(5) = 47{,}414$$

There were about 47,414 European immigrants admitted to the United States as permanent residents under refugee acts in 1995.

c. Let $N(t)$ be the number of non-European immigrants admitted to the United States as permanent residents under refugee acts t years since 1990.

$$N(t)=I(t)-E(t)$$
$$=-17{,}854t^2+212{,}449t-501{,}784$$
$$\underline{-\left(-6432t^2+74{,}006t-161{,}816\right)}$$

Distribute the negative to each of the terms in $E(t)$.

$$N(t)=I(t)-E(t)$$
$$=-17{,}854t^2+212{,}449t-501{,}784$$
$$\underline{+6432t^2\ -74{,}006t\ +161{,}816}$$
$$N(t)=-11{,}442t^2+138{,}443t-339{,}968$$

d.

$$N(8)=-11{,}422(8)^2+138{,}443(8)-339{,}968$$
$$N(8)=36{,}568$$

There were about 36,568 non-European immigrants admitted to the United States as permanent residents under refugee acts in 1998.

37. **a.** Let $U(t)$ be the number of United States residents in millions who are Caucasian and under 18 years old, t years since 1990.

$$U(t)=C(t)-O(t)$$
$$=\quad 1.693t+209.107$$
$$\underline{-\left(0.014t^2+1.169t+157.528\right)}$$

Distribute the negative to each of the terms in $O(t)$.

$$U(t)=C(t)-O(t)$$
$$=\quad 1.693t+209.107$$
$$\underline{-0.014t^2-1.169t-157.528}$$
$$U(t)=-0.014t^2+0.524t+51.579$$

b.

$$U(12)=-0.014(12)^2+0.524(12)+51.579$$
$$U(12)=55.851$$

In 2002, there were approximately 56 million Caucasian United States residents who were under 18 years old.

39. $f(x)=3x+8$ and $g(x)=-6x+9$

a.

$$f(5)+g(5)=[3(5)+8]+[-6(5)+9]$$
$$=(23)+(-21)$$
$$=2$$

b.

$$f(5)-g(5)=[3(5)+8]-[-6(5)+9]$$
$$=(23)-(-21)$$
$$=44$$

c.

$$f(5)g(5)=[3(5)+8][-6(5)+9]$$
$$=(23)(-21)$$
$$=-483$$

d.

$$\frac{f(5)}{g(5)}=\frac{3(5)+8}{-6(5)+9}$$
$$=-\frac{23}{21}$$

41. $f(x)=\frac{1}{2}x+\frac{1}{5}$ and $g(x)=2x-\frac{3}{5}$

a.

$$f(2)+g(2)=\left[\frac{1}{2}(2)+\frac{1}{5}\right]+\left[2(2)-\frac{3}{5}\right]$$
$$=\left(1+\frac{1}{5}\right)+\left(4-\frac{3}{5}\right)$$

Use a common denominator of 5.

$$=\frac{5}{5}+\frac{1}{5}+\frac{20}{5}-\frac{3}{5}$$
$$=\frac{23}{5}=4.6$$

b.

$$g(2)-f(2)=\left[2(2)-\frac{3}{5}\right]-\left[\frac{1}{2}(2)+\frac{1}{5}\right]$$
$$=\left(4-\frac{3}{5}\right)-\left(1+\frac{1}{5}\right)$$

Use a common denominator of 5.

$$=\frac{20}{5}-\frac{3}{5}-\frac{5}{5}-\frac{1}{5}$$
$$=\frac{11}{5}=2.2$$

c.

$$f(2)+g(2)=\left[\frac{1}{2}(2)+\frac{1}{5}\right]\left[2(2)-\frac{3}{5}\right]$$
$$=\left(1+\frac{1}{5}\right)\left(4-\frac{3}{5}\right)$$
$$=4-\frac{3}{5}+\frac{4}{5}-\frac{3}{25}$$

Use a common denominator of 25.

$$=4\left(\frac{25}{25}\right)-\frac{3}{5}\left(\frac{5}{5}\right)+\frac{4}{5}\left(\frac{5}{5}\right)-\frac{3}{25}$$
$$=\frac{100}{25}-\frac{15}{25}+\frac{20}{25}-\frac{3}{25}$$
$$=\frac{102}{25}=4.08$$

d.

$$\frac{f(2)}{g(2)}=\frac{\frac{1}{2}(2)+\frac{1}{5}}{2(2)-\frac{3}{5}}$$
$$=\frac{1+\frac{1}{5}}{4-\frac{3}{5}}$$
$$=\frac{1.2}{3.4}$$
$$\approx 0.35$$

43. $f(x)=3x+5$ and $g(x)=x^2+4x+10$

a.

$$\begin{aligned} f(5)+g(5)&=[3(5)+5]+[(5)^2+4(5)+10] \\ &=(20)+(55) \\ &=75 \end{aligned}$$

b.

$$\begin{aligned} g(5)-f(5)&=[(5)^2+4(5)+10]-[3(5)+5] \\ &=(55)-(20) \\ &=35 \end{aligned}$$

c.

$$\begin{aligned} f(5)g(5)&=[3(5)+5][(5)^2+4(5)+10] \\ &=(20)(55) \\ &=1100 \end{aligned}$$

d.

$$\begin{aligned} \frac{f(5)}{g(5)}&=\frac{3(5)+5}{(5)^2+4(5)+10} \\ &=\frac{20}{55} \\ &=\frac{4}{11} \end{aligned}$$

45. c

47. a

49. $f(x)=5x+7$ and $g(x)=4x+1$

a.

$$\begin{aligned} g(x)+f(x)&=(4x+1)+(5x+7) \\ &=4x+1+5x+7 \\ &=9x+8 \end{aligned}$$

b.

$$\begin{aligned} g(2)+f(2)&=[4(2)+1]+[5(2)+7] \\ &=(9)+(17) \\ &=26 \end{aligned}$$

c.

$$\begin{aligned} f(x)g(x)&=(5x+7)(4x+1) \\ &=20x^2+5x+28x+7 \\ &=20x^2+33x+7 \end{aligned}$$

d.

$$\begin{aligned} f(8)g(8)&=[5(8)+7][4(8)+1] \\ &=(47)(33) \\ &=1551 \end{aligned}$$

51. $f(x)=x+2$ and $g(x)=4x^2+x+1$

a.

$$\begin{aligned} g(x)-f(x)&=(4x^2+x+1)-(x+2) \\ &=4x^2+x+1-x-2 \\ &=4x^2-1 \end{aligned}$$

b.

$$\begin{aligned} g(2)-f(2)&=[4(2)^2+(2)+1]-[(2)+2] \\ &=(19)-(4) \\ &=15 \end{aligned}$$

c.

$$\begin{aligned} f(x)g(x)&=(x+2)(4x^2+x+1) \\ &=4x^3+x^2+x+8x^2+2x+2 \\ &=4x^3+9x^2+3x+2 \end{aligned}$$

d.

$$\begin{aligned} f(2)g(2)&=[(2)+2][4(2)^2+(2)+1] \\ &=(4)(19) \\ &=76 \end{aligned}$$

53. **a.** The domain of

$f(x)+g(x)=(3x+1)+(-2x+4)=2x+5$

is the set of all real numbers.

b. The domain of

$f(x)-g(x)=(3x+1)-(-2x+4)=5x-3$

is the set of all real numbers.

c. The domain of

$f(x)g(x)=(3x+1)(-2x+4)=-6x^2+10x+4$

is the set of all real numbers.

55. **a.** The domain of

$f(x)+g(x)=(x^2)+(3x+1)=x^2+3x+1$

is the set of all real numbers.

b. The domain of

$f(x)-g(x)=(x^2)-(3x+1)=x^2-3x-1$

is the set of all real numbers.

c. The domain of

$f(x)g(x)=(x^2)(3x+1)=3x^3+x^2$

is the set of all real numbers.

57. **a.** The domain of

$f(x)+g(x)=(x^2)+(x^2-1)=2x^2-1$

is the set of all real numbers.

b. The domain of

$f(x)-g(x)=(x^2)-(x^2-1)=1$

is the set of all real numbers.

c. The domain of

$f(x)g(x)=(x^2)(x^2-1)=x^4-x^2$

is the set of all real numbers.

59.

$$g(x)=\frac{25}{x}$$

The denominator cannot equal zero.

Set the linear factors not equal to zero and solve:

$x\neq 0$

Domain: x is all real numbers except $x\neq 0$.

61.

$$f(x)=\frac{x+5}{x-3}$$

The denominator cannot equal zero.
Set the linear factors not equal to zero and solve:
$x-3\neq 0$
$x\neq 3$
Domain: x is all real numbers except $x\neq 3$.

63.

$m(b)=\dfrac{b+9}{b+11}$
The denominator cannot equal zero.
Set the linear factors not equal to zero and solve:
$b+11\neq 0$
$b\neq -11$
Domain: b is all real numbers except $b\neq -11$.

65.

$p(t)=\dfrac{3t-7}{4t+5}$
The denominator cannot equal zero.
Set the linear factors not equal to zero and solve:
$4t+5\neq 0$
$t\neq -\dfrac{5}{4}$
Domain: t is all real numbers except $t\neq -\dfrac{5}{4}$.

67.

$h(a)=\dfrac{3a-1}{(2a+7)(a-3)}$
The denominator cannot equal zero.
Set the linear factors not equal to zero and solve:
$2a+7\neq 0,\ a-3\neq 0$
$a\neq -\dfrac{7}{2},\ a\neq 3$
Domain: a is all real numbers except $a\neq -\dfrac{7}{2},\ a\neq 3$.

69.

$C(n)=\dfrac{n+10}{(n+4)(n+10)}$
The denominator cannot equal zero.
Set the linear factors not equal to zero and solve:
$n+4\neq 0,\ n+10\neq 0$
$n\neq -4,\ n\neq -10$
Domain: n is all real numbers except $n\neq -4,\ n\neq -10$.

71.

$h(m)=\dfrac{4m^2+2m-9}{m^2+7m+12}=\dfrac{4m^2+2m-9}{(m+3)(m+4)}$
The denominator cannot equal zero.
Set the linear factors not equal to zero and solve:
$m+3\neq 0$ and $m+4\neq 0$
$m\neq -3$ and $m\neq -4$
Domain: m is all real numbers except $m\neq -3$ and $m\neq -4$.

73.

$f(x)=\dfrac{2x+1}{x^2+3x+19}$
The denominator cannot equal zero.
Set the linear factors not equal to zero and solve:
$x^2+3x+19\neq 0$
$x\neq \dfrac{-3\pm\sqrt{3^2-4(1)(19)}}{2(1)}$
$x\neq \dfrac{-3\pm\sqrt{-67}}{2}$
This is an imaginary number, meaning there is no real number such that the denominator will equal zero.
Domain: x is all real numbers.

75.

$h(t)=\dfrac{t+7}{t^2-49}=\dfrac{t+7}{(t+7)(t-7)}$
The denominator cannot equal zero.
Set the linear factors not equal to zero and solve:
$t+7\neq 0$ and $t-7\neq 0$
$t\neq -7$ and $t\neq 7$
Domain: t is all real numbers except $t\neq -7$ and $t\neq 7$.

77.

$b(r)=\dfrac{r+3}{r^2+5r+6}=\dfrac{r+3}{(r+3)(r+2)}$
The denominator cannot equal zero.
Set the linear factors not equal to zero and solve:
$r+3\neq 0$ and $r+2\neq 0$
$r\neq -3$ and $r\neq -2$
Domain: r is all real numbers except $r\neq -3$ and $r\neq -2$.

79.

$h(t)=\dfrac{t^2-3t+5}{t^2-7t+10}=\dfrac{t^2-3t+5}{(t-5)(t-2)}$
The denominator cannot equal zero.
Set the linear factors not equal to zero and solve:
$t-5\neq 0$ and $t-2\neq 0$
$t\neq 5$ and $t\neq 2$
Domain: t is all real numbers except $t\neq 5$ and $t\neq 2$.

81.

$P(a)=\dfrac{a^2+2a+1}{a^2+9}$
The denominator cannot equal zero.
Set the linear factors not equal to zero and solve:
$a^2+9\neq 0$
$a\neq \dfrac{0\pm\sqrt{0^2-4(1)(9)}}{2(1)}$
$a\neq \dfrac{\pm\sqrt{-36}}{2}$
This is an imaginary number, meaning there is no real number such that the denominator will equal zero.
Domain: a is all real numbers.

83. **a.** Domain: x is all real numbers but $x \neq -5$ and $x \neq 7$.

b. $f(0) = 0$

c. $f(x) = 5$ when $x \approx -4.5$ and $x \approx 7.5$.

85. **a.** Domain: x is all real numbers but $x \neq -3$ and $x \neq 4$.

b. $h(1) \approx -1$

c. $h(-4) \approx 1.9$

d. $h(x) = -2$ when $x \approx -1.9$ and $x \approx 2.8$.

87. Domain: x is all real numbers but $x \neq -3$.
The value represents a vertical asymptote because the factor $(x + 3)$ does not simplify with the numerator.

89. Domain: x is all real numbers but $x \neq -3$ and $x \neq -1$.
The values represent vertical asymptotes because the factors $(x + 3)$ and $(x + 1)$ do not simplify with the numerator.

91. Domain: x is all real numbers but $x \neq -5$ and $x \neq -2$.
The value $x = -5$ represents a vertical asymptote because the factor $(x + 5)$ does not simplify with the numerator.
The value $x = -2$ represents a hole because the factor $(x + 2)$ simplifies with the numerator.

93.

$$g(x) = \frac{3x+9}{6x^2+7x-20} = \frac{3(x+3)}{(3x-4)(2x+5)}$$

Domain: x is all real numbers but $x \neq \frac{4}{3}$ and $x \neq -\frac{5}{2}$.
The values represent vertical asymptotes because the factors $(3x - 4)$ and $(2x + 5)$ do not simplify with the numerator.

95.

$$f(x) = \frac{x+2}{x^2+3x+2} = \frac{x+2}{(x+1)(x+2)}$$

Domain: x is all real numbers but $x \neq -1$ and $x \neq -2$.
The value $x = -1$ represents a vertical asymptote because the factor $(x + 1)$ does not simplify with the numerator.
The value $x = -2$ represents a hole because the factor $(x + 2)$ simplifies with the numerator.

97.

$$h(x) = \frac{5x}{x^3-2x^2-8x} = \frac{5x}{x(x+2)(x-4)}$$

Domain: x is all real numbers but $x \neq 0, x \neq -2$ and $x \neq 4$.
The values $x = -2$ and $x = 4$ represent vertical asymptotes because the factors $(x + 2)$ and $(x - 4)$ do not simplify with the numerator. The value $x = 0$ represents a hole because the factor (x) simplifies with the numerator.

99.

$$g(x) = \frac{3x+12}{x^2-16} = \frac{3(x+4)}{(x-4)(x+4)}$$

Domain: x is all real numbers but $x \neq 4$ and $x \neq -4$.
The value $x = 4$ represents a vertical asymptote because the factor $(x - 4)$ does not simplify with the numerator.
The value $x = -4$ represents a hole because the factor $(x + 4)$ simplifies with the numerator.

101.

$$g(x) = \frac{x+5}{x^2+25}$$

The denominator cannot equal zero.
Set the linear factors not equal to zero and solve:

$$x^2 + 25 \neq 0$$

$$x \neq \frac{0 \pm \sqrt{0^2 - 4(1)(25)}}{2(1)}$$

$$x \neq \frac{\pm\sqrt{-100}}{2}$$

This is an imaginary number, meaning there is no real number such that the denominator will equal zero.

Domain: x is all real numbers.

Section 8.6 Composing Functions

1. $f(x)=2x+4$ and $g(x)=6x+9$

a.

$$\begin{aligned}f(g(x))&=2(6x+9)+4\\&=12x+18+4\\&=12x+22\end{aligned}$$

b.

$$\begin{aligned}g(f(x))&=6(2x+4)+9\\&=12x+24+9\\&=12x+33\end{aligned}$$

3. $f(x)=3x+8$ and $g(x)=-6x+9$

a.

$$\begin{aligned}(f\circ g)(x)&=3(-6x+9)+8\\&=-18x+27+8\\&=-18x+35\end{aligned}$$

b.

$$\begin{aligned}(g\circ f)(x)&=-6(3x+8)+9\\&=-18x-48+9\\&=-18x-39\end{aligned}$$

5. $f(x)=4x+5$ and $g(x)=2x+6$

a.

$$\begin{aligned}f(g(x))&=f(2x+6)\\&=4(2x+6)+5\\&=8x+24+5\\&=8x+29\end{aligned}$$

b.

$$\begin{aligned}g(f(x))&=g(4x+5)\\&=2(4x+5)+6\\&=8x+10+6\\&=8x+16\end{aligned}$$

7. $f(x)=7x+11$ and $g(x)=-8x+15$

a.

$$\begin{aligned}(f\circ g)(x)&=f(g(x))\\&=f(-8x+15)\\&=7(-8x+15)+11\\&=-56x+105+11\\&=-56x+116\end{aligned}$$

b.

$$\begin{aligned}(g\circ f)(x)&=g(f(x))\\&=g(7x+11)\\&=-8(7x+11)+15\\&=-56x-88+15\\&=-56x-73\end{aligned}$$

9. **a.** Let $K(w)$ be the total weekly cost in dollars for week w of production.

$$\begin{aligned}K(w)&=C(T(w))\\&=1.75(500w+3000)+5000\\&=875w+5250+5000\\&=875w+10,250\end{aligned}$$

b. $T(5)=500(5)+3000=5500$. There were 5500 toys produced during week five.

c. $C(5000)=1.75(5000)+5000=13,750$. The total weekly cost from the production of 5000 toys per week, is \$13,750.

d. $K(7)=875(7)+10,250=16,375$. The total weekly cost of production for week 7 was \$16,375.

e.

$$\begin{aligned}18,500&=875w+10,250\\-10,250&\quad\quad\;\;-10,250\\8250&=875w\\\frac{8250}{875}&=\frac{875w}{875}\\9.4&\approx w\end{aligned}$$

The weekly cost will reach, and surpass, \$18,500 in week 10 of production.

11. **a.** $v(10)=0.25(10)+1.5=4$. A West Tech employee will get 4 weeks of vacation per year after working with the company for ten years.

b. $C(4)=1500(4)+575=6575$. West Tech's cost for a ten-year employee's vacation is \$6575.

c. Let $K(y)$ be the cost for vacation taken by an employee who has been with the company y years.

$$\begin{aligned}K(y)&=C(v(y))\\&=1500(0.25y+1.5)+575\\&=375y+2250+575\\&=375y+2825\end{aligned}$$

d. $K(20)=375(20)+2825=10,325$. West Tech's cost for a 20-year employee's vacation is \$10,325.

e. $K(30)=375(30)+2825=14,075$. West Tech's cost for a 30-year employee's vacation is \$14,075.

13. **a.** $A(4)=-400(4)^2+3500(4)-1000=6600$. On the 4th day of the Renaissance fair, there were 6,600 people in attendance.

b. $P(6600)=2(6600)-2500=10,700$. On the 4th day of the Renaissance fair, with 6600 in attendance, a profit of \$10700 was made.

c. Let $T(d)$ be the profit made at the Renaissance fair d days after the fair opens.

$$\begin{aligned}T(d)&=P(A(d))\\&=2(-400d^2+3500d-1000)-2500\\&=-800d^2+7000d-2000-2500\\&=-800d^2+7000d-4500\end{aligned}$$

d. $T(3) = -800(3)^2 + 7000(3) - 4500 = 9300$. The Renaissance fair made $9300 profit on the 3rd day of the fair.

e.

d	Profit
1	1700
2	6300
3	9300
4	10700
5	10500
6	8700
7	5300

$$T(1)+T(2)+T(3)+T(4) +T(5)+T(6)+T(7) = 52{,}500$$

If the Renaissance fair lasts a total of 7 days, the total profit would be $52,500.

15. $P(B(t))$

$P =$ profit in thousands of $

$b =$ number of bikes, $B =$ number of bikes } same

$t =$ year

Use $B(t)$ in place of b in $P(b)$ to get $P(B(t))$.

Input units: the year
Output units: profit in thousands of dollars

17. $H(T(d))$

$H =$ number of homeless who seek space

$t =$ low night-time temperature in $^\circ F$, $T =$ low night-time temperature in $^\circ F$ } same

$d =$ day of the week

Use $T(d)$ in place of t in $H(t)$ to get $H(T(d))$.

Input units: day of the week
Output units: number of homeless who seek space

19. $f(x) = \frac{1}{2}x + \frac{1}{5}$ and $g(x) = 2x - \frac{3}{5}$

a.

$$\begin{aligned} f(g(x)) &= f\left(2x - \frac{3}{5}\right) \\ &= \frac{1}{2}\left(2x - \frac{3}{5}\right) + \frac{1}{5} \\ &= x - \frac{3}{10} + \frac{1}{5} \\ &= x - \frac{1}{10} \end{aligned}$$

b.

$$\begin{aligned} g(f(x)) &= g\left(\frac{1}{2}x + \frac{1}{5}\right) \\ &= 2\left(\frac{1}{2}x + \frac{1}{5}\right) - \frac{3}{5} \\ &= x + \frac{2}{5} - \frac{3}{5} \\ &= x - \frac{1}{5} \end{aligned}$$

21. $f(x) = 0.68x + 2.36$ and $g(x) = 3.57x + 6.49$

a.

$$\begin{aligned} f(g(x)) &= f(3.57x + 6.49) \\ &= 0.68(3.57x + 6.49) + 2.36 \\ &= 2.4276x + 4.4132 + 2.36 \\ &= 2.4276x + 6.7732 \end{aligned}$$

b.

$$\begin{aligned} g(f(x)) &= g(0.68x + 2.36) \\ &= 3.57(0.68x + 2.36) + 6.49 \\ &= 2.4276x + 8.4252 + 6.49 \\ &= 2.4276x + 14.9152 \end{aligned}$$

23. $f(x) = x + 1$ and $g(x) = x^2 + 5$

a.

$$\begin{aligned} f(g(x)) &= f(x^2 + 5) \\ &= (x^2 + 5) + 1 \\ &= x^2 + 6 \end{aligned}$$

b.

$$\begin{aligned} g(f(x)) &= g(x + 1) \\ &= (x + 1)^2 + 5 \\ &= x^2 + 2x + 1 + 5 \\ &= x^2 + 2x + 6 \end{aligned}$$

25. $f(x) = x - 1$ and $g(x) = x^2 + 6$

a.

$$\begin{aligned} (f \circ g)(x) &= f(g(x)) \\ &= f(x^2 + 6) \\ &= (x^2 + 6) - 1 \\ &= x^2 + 5 \end{aligned}$$

b.

$$\begin{aligned} (g \circ f)(x) &= g(f(x)) \\ &= g(x - 1) \\ &= (x - 1)^2 + 6 \\ &= x^2 - 2x + 1 + 6 \\ &= x^2 - 2x + 7 \end{aligned}$$

27. $f(x)=3x+5$ and $g(x)=x^2+4x+10$

a.

$$\begin{aligned}f(g(x))&=f(x^2+4x+10)\\&=3(x^2+4x+10)+5\\&=3x^2+12x+30+5\\&=3x^2+12x+35\end{aligned}$$

b.

$$\begin{aligned}g(f(x))&=g(3x+5)\\&=(3x+5)^2+4(3x+5)+10\\&=(3x+5)(3x+5)+12x+20+10\\&=9x^2+15x+15x+25+12x+30\\&=9x^2+42x+55\end{aligned}$$

29. $f(x)=x+2$ and $g(x)=4x^2+x+1$

a.

$$\begin{aligned}(f\circ g)(x)&=f(g(x))\\&=f(4x^2+x+1)\\&=(4x^2+x+1)+2\\&=4x^2+x+1+2\\&=4x^2+x+3\end{aligned}$$

b.

$$\begin{aligned}(g\circ f)(x)&=g(f(x))\\&=g(x+2)\\&=4(x+2)^2+(x+2)+1\\&=4(x+2)(x+2)+x+2+1\\&=4(x^2+2x+2x+4)+x+3\\&=4(x^2+4x+4)+x+3\\&=4x^2+16x+16+x+3\\&=4x^2+17x+19\end{aligned}$$

31. Lilybell multiplied $f(x)$ times $g(x)$.

$f(x)=3x-10$ and $g(x)=7x+8$

$$\begin{aligned}f(g(x))&=f(7x+8)\\&=3(7x+8)-10\\&=21x+24-10\\&=21x+14\end{aligned}$$

33. Colter reversed the composition and found $g(f(x))$.

$f(x)=3x-10$ and $g(x)=7x+8$

$$\begin{aligned}f(g(x))&=f(7x+8)\\&=3(7x+8)-10\\&=21x+24-10\\&=21x+14\end{aligned}$$

35. $f(x)=3x+8$ and $g(x)=-6x+9$

a.

$$\begin{aligned}f(g(5))&=f(-21)\quad\text{*Since } g(5)=-6(5)+9=-21\\&=3(-21)+8\\&=-63+8\\&=-55\end{aligned}$$

b.

$$\begin{aligned}g(f(5))&=g(23)\quad\text{*Since } f(5)=3(5)+8=23\\&=-6(23)+9\\&=-138+9\\&=-129\end{aligned}$$

37. $f(x)=\frac{1}{2}x+\frac{1}{5}$ and $g(x)=2x-\frac{3}{5}$

a.

$$\begin{aligned}f(g(2))&=f\left(\frac{17}{5}\right)\quad\text{*Since } g(2)=2(2)-\frac{3}{5}=\frac{17}{5}\\&=\frac{1}{2}\left(\frac{17}{5}\right)+\frac{1}{5}\\&=\frac{17}{10}+\frac{1}{5}\\&=\frac{19}{10}=1.9\end{aligned}$$

b.

$$\begin{aligned}g(f(2))&=g\left(\frac{6}{5}\right)\quad\text{*Since } f(2)=\frac{1}{2}(2)+\frac{1}{5}=\frac{6}{5}\\&=2\left(\frac{6}{5}\right)-\frac{3}{5}\\&=\frac{12}{5}-\frac{3}{5}\\&=\frac{9}{5}=1.8\end{aligned}$$

39. $f(x)=0.68x+2.36$ and $g(x)=3.57x+6.49$

a.

$$\begin{aligned}(f\circ g)(-3)&=f(g(-3))\\&=f(-4.22)\quad\text{*Since } g(-3)=3.57(-3)+6.49=-4.22\\&=0.68(-4.22)+2.36\\&=-2.8696+2.36\\&=-0.5096\end{aligned}$$

b.

$$\begin{aligned}(g\circ f)(-3)&=g(f(-3))\\&=g(0.32)\quad\text{*Since } f(-3)=0.68(-3)+2.36=0.32\\&=3.57(0.32)+6.49\\&=1.1424+6.49\\&=7.6324\end{aligned}$$

41. $f(x)=x+1$ and $g(x)=x^2+5$

a.

$$\begin{aligned} f(g(5)) &= f(30) \quad \text{*Since } g(5)=(5)^2+5=30 \\ &= (30)+1 \\ &= 31 \end{aligned}$$

b.

$$\begin{aligned} g(f(5)) &= g(6) \quad \text{*Since } f(5)=(5)+1=6 \\ &= (6)^2+5 \\ &= 36+5 \\ &= 41 \end{aligned}$$

43. $f(x)=x+3$ and $g(x)=x^2+7$

a.

$$\begin{aligned} (f\circ g)(3) &= f(g(3)) \\ &= f(15) \quad \text{*Since } g(3)=(3)^2+6=15 \\ &= (15)-1 \\ &= 14 \end{aligned}$$

b.

$$\begin{aligned} (g\circ f)(3) &= g(f(3)) \\ &= g(2) \quad \text{*Since } f(3)=(3)-1=2 \\ &= (2)^2+6 \\ &= 4+6 \\ &= 10 \end{aligned}$$

45. $f(x)=3x+5$ and $g(x)=x^2+4x+10$

a.

$$\begin{aligned} f(g(5)) &= f(55) \quad \text{*Since } g(5)=(5)^2+4(5)+10=55 \\ &= 3(55)+5 \\ &= 165+5 \\ &= 170 \end{aligned}$$

b.

$$\begin{aligned} g(f(5)) &= g(20) \quad \text{*Since } f(5)=3(5)+5=20 \\ &= (20)^2+4(20)+10 \\ &= 400+80+10 \\ &= 490 \end{aligned}$$

47. $f(x)=x+2$ and $g(x)=4x^2+x+1$

a.

$$\begin{aligned} f(g(2)) &= f(19) \quad \text{*Since } g(2)=4(2)^2+(2)+1=19 \\ &= (19)+2 \\ &= 21 \end{aligned}$$

b.

$$\begin{aligned} g(f(2)) &= g(4) \quad \text{*Since } f(2)=(2)+2=4 \\ &= 4(4)^2+(4)+1 \\ &= 64+4+1 \\ &= 69 \end{aligned}$$

49. $T(t)-K(t)$

Input units: the year

Output units: number of people over 12-years-old who go on the tour

51. $A(T(t)-K(t))$

Input units: the year

Output units: cost for all people over 12-years-old who go on the tour

53. $C(K(t))-A(T(t)-K(t))$

Input units: the year

Output units: difference between the cost for all children under 12-years-old and the cost for all people over 12-years-old who traveled on the tour

55. $f(x)=(9x+2)$ and $g(x)=(3x+7)$

a.

$$\begin{aligned} f(x)g(x) &= (9x+2)(3x+7) \\ &= 27x^2+63x+6x+14 \\ &= 27x^2+69x+14 \end{aligned}$$

b.

$$\begin{aligned} f(x)+g(x) &= (9x+2)+(3x+7) \\ &= 12x+9 \end{aligned}$$

c.

$$\begin{aligned} f(g(x)) &= 9(3x+7)+2 \\ &= 27x+63+2 \\ &= 27x+65 \end{aligned}$$

57. $f(x)=-9x+1$ and $g(x)=x+8$

a.

$$\begin{aligned} f(3)+g(3) &= (-9(3)+1)+((3)+8) \\ &= -27+1+3+8 \\ &= -15 \end{aligned}$$

b.

$$\begin{aligned} f(4)g(4) &= (-9(4)+1)(4+8) \\ &= (-35)(12) \\ &= -420 \end{aligned}$$

c.

$$g(7)=7+8=15$$

$$f(g(7))=f(15)=-9(15)+1$$

$$f(g(7))=-134$$

59. $f(x)=\frac{2}{3}x+\frac{1}{3}$ and $g(x)=\frac{1}{4}x+\frac{3}{4}$

a.

$$\begin{aligned} f(g(x)) &= \frac{2}{3}\left(\frac{1}{4}x+\frac{3}{4}\right)+\frac{1}{3} \\ &= \frac{2}{12}x+\frac{6}{12}+\frac{1}{3} \\ &= \frac{1}{6}x+\frac{1}{2}+\frac{1}{3} \\ &= \frac{1}{6}x+\frac{3}{6}+\frac{2}{6} \\ &= \frac{1}{6}x+\frac{5}{6} \end{aligned}$$

b.

$$\begin{aligned} f(42)g(42) &= \left(\frac{2}{3}(42)+\frac{1}{3}\right)\left(\frac{1}{4}(42)+\frac{3}{4}\right) \\ &= \left(\frac{85}{3}\right)\left(\frac{45}{4}\right) \\ &= 318.75 \end{aligned}$$

c.

$$\begin{aligned} f(x)+g(x) &= \left(\frac{2}{3}x+\frac{1}{3}\right)+\left(\frac{1}{4}x+\frac{3}{4}\right) \\ &= \frac{8}{12}x+\frac{4}{12}+\frac{3}{12}x+\frac{9}{12} \\ &= \frac{11}{12}x+\frac{13}{12} \end{aligned}$$

61. $f(x)=1.5x+4.5$ and $g(x)=3.5x+9$

a.

$$\begin{aligned} f(g(x)) &= 1.5(3.5x+9)+4.5 \\ &= 5.25x+13.5+4.5 \\ &= 5.25x+18 \end{aligned}$$

b.

$$f(6)=1.5(6)+4.5=13.5$$
$$g(f(6))=g(13.5)=3.5(13.5)+9$$
$$g(f(6))=56.25$$

c.

$$\begin{aligned} g(x)-f(x) &= (3.5x+9)-(1.5x+4.5) \\ &= 3.5x+9-1.5x-4.5 \\ &= 2x+4.5 \end{aligned}$$

Chapter 8 Review Exercises

1. $h=0.75d+4.5$

a.

$$h=0.75(7)+4.5$$
$$h=9.75$$

The height after 1 week (7 days) is 9.75 inches.

b.

$$12=0.75d+4.5$$
$$12-4.5=0.75d+4.5-4.5$$
$$7.5=0.75d$$
$$\frac{7.5}{0.75}=\frac{0.75d}{0.75}$$
$$10=d$$

The course grass attendants should cut the grass in the rough 10 days before the tournament.

2. a. Let C be the cost of the satellite phone service in dollars, and m be the number of minutes used.
$C=10m+10$

b.

$$C=10(3)+10$$
$$C=40$$

A 3-minute call cost $40.

c.

$$300=10m+10$$
$$300-10=10m+10-10$$
$$290=10m$$
$$\frac{290}{10}=\frac{10m}{10}$$
$$29=m$$

The satellite call can last 29 minutes for $300.

3. Let C be the cost in dollars of renting a Bobcat for h hours.

a. $C=40h+15$

b.

$$C=40(2)+15$$
$$C=95$$

Renting a Bobcat for 2 hours will cost $95.

c.

$$(3\text{ days})\left(\frac{8\text{ hours}}{\text{day}}\right)=24\text{ hours}$$
$$C=40(24)+15$$
$$C=975$$

A three-day rental of a Bobcat will cost $975.

4. a. Let C be the cost in dollars to produce n putters.

n	C
5	600.00
10	725.00
20	950.00
25	1100.00
50	1700.00

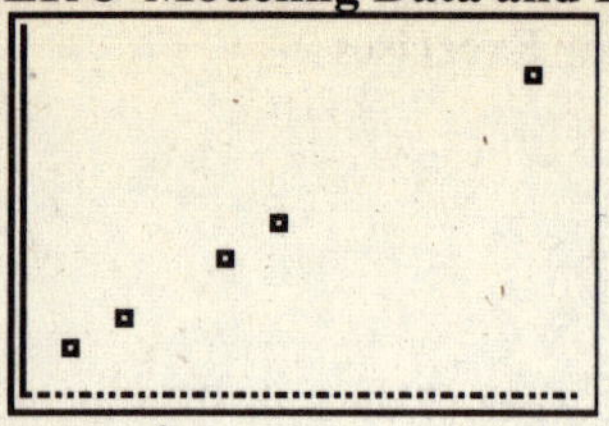

$C = mt + b$

Using two points: $(5,600)$ and $(50,1700)$

Find the slope: $m = \dfrac{1700-600}{50-5} \approx 24.44$

$C = 24.44t + b$

$b = 600 - 24.44(5)$

$b = 477.80$

$C = 24.44n + 477.80$

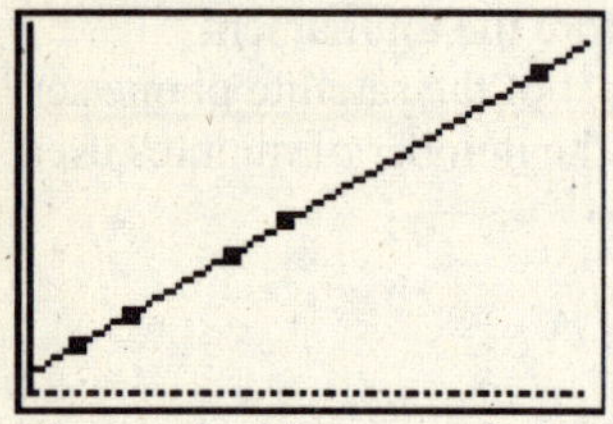

b. The vertical intercept is (0,477.80). It cost $477.80 to produce 0 putters.

c.

$C = 24.44(100) + 477.80$

$C = 2921.80$

The cost to produce 100 putters is $2921.80.

d. Domain: [0,100],

$C = 24.44(100) + 477.80 = 2921.80$

Range: [477.80, 2921.80]

e. The slope is 24.44. For every putter produced the cost increases by $24.44.

5. **a.** Let P be the profit for Costco Wholesale Corporation in millions of dollars, t years since 2000.

t	P
1	4199
2	4779
3	5310
4	6018
5	6605
6	7406
7	7950

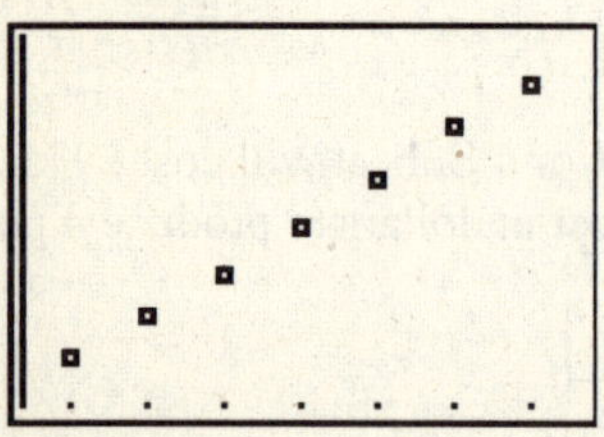

$P = mt + b$

Using two points: (2,4779) and (7,7950)

Find the slope: $m = \dfrac{7950-4779}{7-2} = 634.2$

$P = 634.2t + b$

$b = 4779 - 634.2(2)$

$b = 3510.6$

$P = 634.2t + 3510.6$

b.

$9000 = 634.2t + 3510.6$

$9000 - 3510.6 = 634.2t + 3510.6 - 3510.6$

$5489.4 = 634.2t$

$\dfrac{5489.4}{634.2} = \dfrac{634.2t}{634.2}$

$8.66 \approx t$

The gross profit for Costco Wholesale will be $9 billion in 2008.

c.

$P = 634.2(9) + 3510.6$

$P = 9218.4$

The gross profit for Costco Wholesale in 2009 will be $9.2184 billion.

d. Domain [0,9]

$P = 634.2(9) + 3510.6 = 9218.4$

Range [3510.6, 9218.4]

e. The slope is 634.2. The gross profit for Costco Wholesale increases by $634.2 million per year.

6. **a.** Let P be the percent of full-time workers in private industry who filed an injury case t years since 2000.

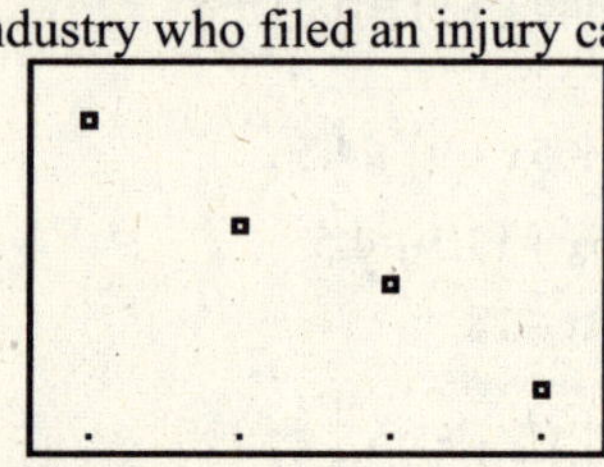

$P(t) = mt + b$

Using two points: (3,4.7) and (6,4.2)

Find the slope: $m = \dfrac{4.2-4.7}{6-3} \approx -0.167$

$P(t) = -0.167t + b$

$b = 4.7 - (-0.167)(3)$

$b \approx 5.2$

$P(t) = -0.167t + 5.2$

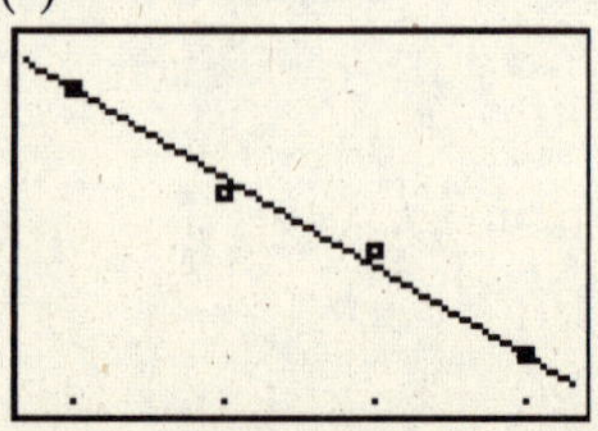

b. The slope is -0.167. The percent of full-time workers in private industry who filed an injury case is decreasing by about 0.167 percentage points per year.

c.

$$P(3) = -0.167(3) + 5.2$$
$$P(3) = 4.699$$

In 2003, about 4.7 percent of full-time workers in private industry filed an injury case.

d. Vertical intercept is $(0, 5.2)$. In 2000, about 5.2% of full-time workers in private industry filed an injury case.

e.

$$0 = -0.167t + 5.2$$
$$0.167t = 5.2$$
$$\frac{0.167t}{0.167} = \frac{5.2}{0.167}$$
$$t \approx 31.14$$

The horizontal intercept is $(31.14, 0)$. In 2031, there will be no injury cases filed by full-time workers in private industry. This is probably model breakdown since there will most likely always be some injury cases.

7. **a.** Let S be the number of kindergarten through twelfth grade students, in millions, in Texas public schools t years since 2000.

t	S
0	4.00
1	4.07
2	4.16
3	4.26
4	4.33
5	4.40
6	4.52

$S(t) = mt + b$

Using two points: $(0, 4.00)$ and $(4, 4.33)$

Find the slope: $m = \frac{4.33 - 4.00}{4 - 0} \approx 0.08$

Since the vertical intercept is $(0, 4.00)$, the equation becomes

$S(t) = 0.08t + 4$

b. The slope is 0.08. The number of kindergarten through twelfth grade students in millions in Texas public schools is increasing by about 0.08 million (80 thousand) per year.

c.

$$S(11) = 0.08(11) + 4$$
$$S(11) = 4.88$$

In 2011, Texas can expect to have about 4.88 million kindergarten through twelfth grade students in public schools.

8. $C(t) = 0.56t + 4.3$

a.

$$2015 \rightarrow t = 5$$
$$C(5) = 0.56(5) + 4.3$$
$$C(5) = 7.1$$

7.1 million candles will be produced in 2015.

b.

$$C(3) = 0.56(3) + 4.3$$
$$C(3) = 5.98$$

5.98 million candles will be produced in 2013

c.

$$10 = 0.56t + 4.3$$
$$10 - 4.3 = 0.56t + 4.3 - 4.3$$
$$5.7 = 0.56t$$
$$\frac{5.7}{0.56} = \frac{0.56t}{0.56}$$
$$10.18 \approx t$$

10 million candles will be produced in 2020.

9. **a.** $(0, 4)$ is the point where the graph crosses the y-axis.

b. $(1.3, 0)$ is the point where the graph crosses the x-axis.

c.

Use two points from the graph $(0, 4)$ and $(4, -8)$.

$$\text{slope} = \frac{-8 - 4}{4 - 0} = \frac{-12}{4} = -3$$

d. $f(4) = -8$ Since when $x = 4$ $y = -8$.

e. $x = -4$ when $f(x) = 16$

f.

Since the y-intercept is $(0, 4)$, then $b = 4$.

Since the slope is -3, then $m = -3$.

Use $y = mx + b$ where $m = -3$ and $b = 4$.

The equation of the line is $f(x) = -3x + 4$.

10. **a.** $(0, -5)$ is the point where the graph crosses the y-axis.

b. $(3.3, 0)$ is the point where the graph crosses the x-axis.

c.

Use two points from the graph $(-2, -8)$ and $(2, -2)$.

$$\text{slope} = \frac{-2 - (-8)}{2 - (-2)} = \frac{6}{4} = \frac{3}{2}$$

d. $h(-2) = -8$ since when $x = -2$, $y = -8$.

e. $x = 2$ when $h(x) = -2$

f.

Since the y-intercept is $(0, -5)$, then $b = -5$.

Since the slope is $\frac{3}{2}$, then $m = \frac{3}{2}$.

Use $y = mx + b$ where $m = \frac{3}{2}$ and $b = -5$.

The equation of the line is $h(x) = \frac{3}{2}x - 5$.

11. The electricity cost for using holiday lights 6 hours a day is \$13.

12. $P(t) = 30t + 9979$

a.

$$P(10) = 30(10) + 9979$$
$$P(10) = 10{,}279$$

In 2010, the population of Michigan will be about 10,279 thousand.

b.

$$11{,}000 = 30t + 9979$$
$$11{,}000 - 9979 = 30t + 9979 - 9979$$
$$1021 = 30t$$
$$\frac{1021}{30} = \frac{30t}{30}$$
$$34.03 \approx t$$

The population of Michigan will reach 11,000 thousand in about 2034. This may be model breakdown if the current population growth does not continue.

13. $f(x) = 2x - 8$

a.

$$f(10) = 2(10) - 8$$
$$f(10) = 12$$

b.

$$0 = 2x - 8$$
$$8 = 2x$$
$$\frac{8}{2} = \frac{2x}{2}$$
$$4 = x$$

c. Domain: All real numbers, range All real numbers

14. $h(x) = -\frac{2}{3}x + 14$

a.

$$h(12) = -\frac{2}{3}(12) + 14$$
$$h(12) = -8 + 14$$
$$h(12) = 6$$

b.

$$-8 = -\frac{2}{3}x + 14$$
$$-8 - 14 = -\frac{2}{3}x + 14 - 14$$
$$-22 = -\frac{2}{3}x$$
$$\left(-\frac{3}{2}\right)(-22) = \left(-\frac{3}{2}\right)\left(-\frac{2}{3}x\right)$$
$$33 = x$$

c. Domain: All real numbers, range All real numbers

15. $g(x) = 12$

a. $g(3) = 12$

b. $g(-20) = 12$

c. Domain: All real numbers, range $\{12\}$

16. $f(x) = 1.25x + 4.5$

a.

$$f(5) = 1.25(5) + 4.5$$
$$f(5) = 10.75$$

b.

$$-8.5 = 1.25x + 4.5$$
$$-8.5 - 4.5 = 1.25x + 4.5 - 4.5$$
$$-13 = 1.25x$$
$$\frac{-13}{1.25} = \frac{1.25x}{1.25}$$
$$-10.4 = x$$

c. Domain: All real numbers, range All real numbers

17. Let $P(v)$ be the monthly profit in dollars for Hope's Pottery, when v vases are produced and sold in the month.

$$P(v) = R(v) - C(v)$$
$$P(v) = 155v - (5000 + 65v)$$
$$P(v) = 155v - 5000 - 65v$$
$$P(v) = 90v - 5000$$

18. **a.** Let $P(t)$ be the total prescription drug sales in the United States in billions of dollars t years since 2000.

Using two points: $(2, 182.7)$ and $(6, 249.8)$:

find the slope: $m = \frac{249.8 - 182.7}{6 - 2} \approx 16.8$

$$P(t) = 16.8t + b$$
$$182.7 = 16.8(2) + b$$
$$182.7 = 33.6 + b$$
$$182.7 - 33.6 = 33.6 - 33.6 + b$$
$$149.1 = b$$
$$P(t) = 16.8t + 149.1$$

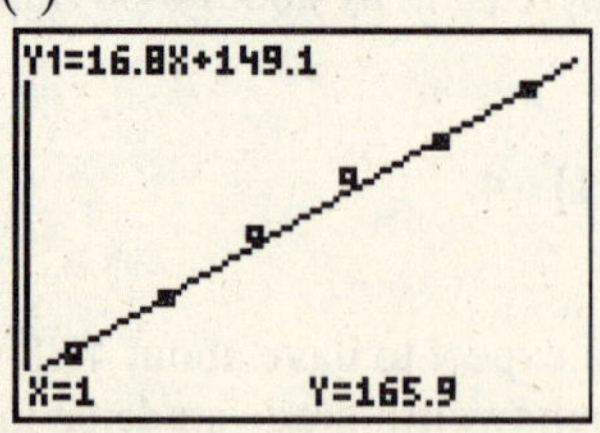

b. Let $M(t)$ be the amount of mail-order prescription drug sales in the United States in billions of dollars t years since 2000.

Using two points: $(5,45.5)$ and $(6,50.4)$:

find the slope: $m=\frac{50.4-45.5}{6-5}=4.9$

$$M(t)=4.9t+b$$

$45.5=4.9(5)+b$

$45.5=24.5+b$

$45.5-24.5=24.5-24.5+b$

$21=b$

$M(t)=4.9t+21$

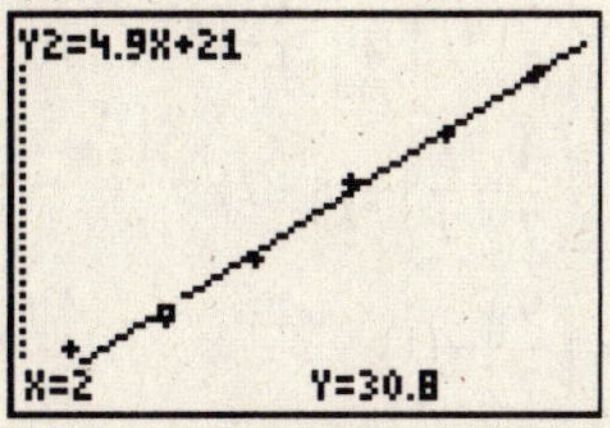

c. $P(10)=16.8(10)+149.1=317.1$. In 2010, the total prescription drug sales in the United States were about \$317.1 billion.

d. $M(10)=4.9(10)+21=70$. In 2010, the mail-order prescription drug sales in the United States were about \$70 billion.

e. Let $N(t)$ be the non-mail-order prescription drug sales in the United States, in billions of dollars, t years since 2000.

$N(t)=P(t)-M(t)$

$=(16.8t+149.1)-(4.9t+21)$

$=16.8t+149.1-4.9t-21$

$N(t)=11.9t+128.1$

f. $N(9)=11.9(9)+128.1=235.2$. In 2009, the non-mail-order prescription drug sales in the United States were about \$235.2 billion.

$N(15)=11.9(15)+128.1=306.6$. In 2015, the non-mail-order prescription drug sales in the United States will be about \$306.6 billion.

19. **a.**

$J(9)=16.27(9)^3-321.0(9)^2+2089.32(9)-3885.40$

$J(9)=778.31$

In 1999, the federal funding for research and development at Johns Hopkins University was about \$778.31 million.

$W(9)=2.89(9)^3-53.94(9)^2+339.66(9)-411.82=382.79$

$W(9)=382.79$

In 1999, the federal funding for research and development at the University of Washington was about \$382.79 million.

b. Let $D(t)$ be the difference between the federal funding at John Hopkins University and the University of Washington, in millions of dollars, t years since 1990.

$D(t)=J(t)-W(t)$

$D(t)=16.27t^3-321.0t^2+2089.32t-3885.40$

$\underline{-2.89t^3+53.94t^2-339.66t+411.82}$

$D(t)=13.38t^3-267.06t^2+1749.66t-3473.58$

c.

$D(10)=13.38(10)^3-267.06(10)^2$
$+1749.66(10)-3473.58$

$D(10)=697.02$

In 2000, John Hopkins University had about \$697.02 million more federal funding for research and development than the University of Washington.

20. $U(t)-C(t)=(U-C)(t)$

Input units: years since 1900
Output units: number of adults 20-years-old or older in the United States.

21. $R(l)-C(l)=(R-C)(l)$

Input units: number of limousines per year
Output units: profit in thousands of dollars

22. $E(t)S(t)=ES(t)$

Input units: the year
Output units: total number of sick days taken by Disneyland employees

23. $W(E(t))$

Input units: the year
Output units: annual worker's compensation insurance cost at Disneyland

24. $M(t)+R(t)=(M+R)(t)$

Input units: the year
Output units: total amount spent on cancer treatments and research in the U.S.

25. $f(x)=6x+3$ and $g(x)=-4x+8$

a.

$f(x)+g(x)=(6x+3)+(-4x+8)$
$=6x+3+(-4x)+8$
$=2x+11$

b.

$f(x)-g(x)=(6x+3)-(-4x+8)$
$=6x+3+4x-8$
$=10x-5$

c.

$f(x)g(x)=(6x+3)(-4x+8)$
$=-24x^2+48x-12x+24$
$=-24x^2+36x+24$

26. $f(x)=15x+34$ and $g(x)=-17x+34$

a.

$$\begin{aligned}f(x)+g(x)&=(15x+34)+(-17x+34)\\&=15x+34+(-17x)+34\\&=-2x+68\end{aligned}$$

Domain: all real numbers.

b.

$$\begin{aligned}g(x)-f(x)&=(-17x+34)-(15x+34)\\&=-17x+34-15x-34\\&=-32x\end{aligned}$$

Domain: all real numbers.

c.

$$\begin{aligned}f(x)g(x)&=(15x+34)(-17x+34)\\&=-255x^2+510x-578x+1156\\&=-255x^2-68x+1156\end{aligned}$$

Domain: all real numbers.

d.

$$\frac{f(x)}{g(x)}=\frac{15x+34}{-17x+34}$$

Domain: all real numbers such that $x\neq 2$.

27. $f(x)=\frac{2}{5}x+\frac{3}{5}$ and $g(x)=4x-7$

a.

$$\begin{aligned}f(x)+g(x)&=\left(\frac{2}{5}x+\frac{3}{5}\right)+(4x-7)\\&=\frac{2}{5}x+\frac{3}{5}+4x-7\\&=4.4x-6.4\end{aligned}$$

Domain: all real numbers.

b.

$$\begin{aligned}f(x)-g(x)&=\left(\frac{2}{5}x+\frac{3}{5}\right)-(4x-7)\\&=\frac{2}{5}x+\frac{3}{5}-4x+7\\&=-3.6x+7.6\end{aligned}$$

Domain: all real numbers.

c.

$$\begin{aligned}f(x)g(x)&=\left(\frac{2}{5}x+\frac{3}{5}\right)(4x-7)\\&=1.6x^2-2.8x+2.4x-4.2\\&=1.6x^2-0.4x-4.2\end{aligned}$$

Domain: all real numbers.

28. $f(x)=\frac{1}{7}x+\frac{5}{7}$ and $g(x)=2x-7$

a.

$$\begin{aligned}f(4)+g(4)&=\left[\frac{1}{7}(4)+\frac{5}{7}\right]+[2(4)-7]\\&=\left(\frac{4}{7}+\frac{5}{7}\right)+(8-7)\\&=\left(\frac{9}{7}\right)+(1)\\&=\frac{16}{7}\end{aligned}$$

b.

$$\begin{aligned}f(4)-g(4)&=\left[\frac{1}{7}(4)+\frac{5}{7}\right]-[2(4)-7]\\&=\left(\frac{4}{7}+\frac{5}{7}\right)-(8-7)\\&=\left(\frac{9}{7}\right)-(1)\\&=\frac{2}{7}\end{aligned}$$

c.

$$\begin{aligned}f(4)g(4)&=\left[\frac{1}{7}(4)+\frac{5}{7}\right][2(4)-7]\\&=\left(\frac{4}{7}+\frac{5}{7}\right)(8-7)\\&=\left(\frac{9}{7}\right)(1)\\&=\frac{9}{7}\end{aligned}$$

29. Adjust the data to fit the definitions in parts (a) and (b).

L1	L2	L3 1
6	15.4	1.17
7	15.75	1.06
8	16.22	.98
9	16.5	.93
10	16.86	.84
11	17.25	.81
	------	------

L1(7)=

a. Let $L(t)$ be the number of people in California's labor force, in millions, t years since 1990.

Using two points: $(6,15.4)$ and $(11,17.25)$

find the slope: $m=\frac{17.25-15.4}{11-6}=0.37$

$$L(t)=0.37t+b$$

$$\begin{aligned}15.4&=0.37(6)+b\\15.4&=2.22+b\\15.4-2.22&=2.22-2.22+b\\13.2&\approx b\end{aligned}$$

$$L(t)=0.37t+13.2$$

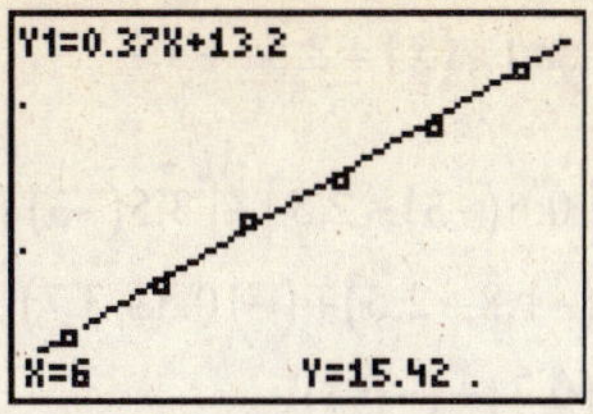

b. Let $U(t)$ be the number of people in California who are considered unemployed, in millions, t years since 1990.

Using two points: $(7,1.06)$ and $(9,0.93)$

Find the slope: $m=\dfrac{0.93-1.06}{9-7}\approx -0.07$

$$U(t)=-0.07t+b$$

$0.93=-0.07(9)+b$

$0.93=-0.63+b$

$0.93+0.63=-0.63+0.63+b$

$1.56=b$

$U(t)=-0.07t+1.56$

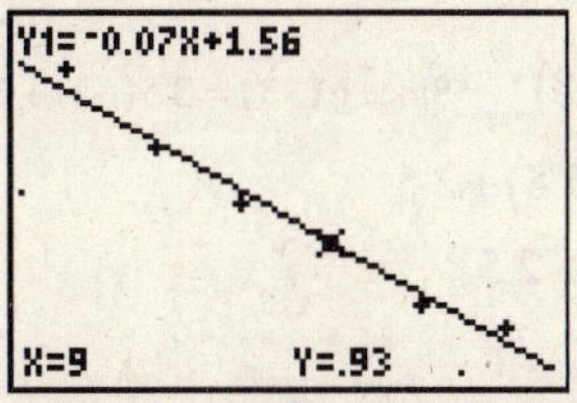

c. $L(15)=0.37(15)+13.2=18.75$. In 2005, there were about 18.75 million people in California's labor force.

d. $U(15)=-0.07(15)+1.56=0.51$. In 2005, there were about 0.51 million people (510,000 people) in California who are considered unemployed. This may be too low and therefore be considered model breakdown.

e. Let $P(t)$ be the percent of California's labor force that is considered unemployed, t years since 1990.

$$P(t)=\frac{U(t)}{L(t)}=\frac{-0.07t+1.56}{0.37t+13.2}$$

f.

$P(8)=\dfrac{-0.07(8)+1.56}{0.37(8)+13.2}\approx 0.062$. In 1998, 6.2% of California's labor force was considered unemployed.

$P(9)=\dfrac{-0.07(9)+1.56}{0.37(9)+13.2}\approx 0.056$. In 1999, 5.6% of California's labor force was considered unemployed.

$P(10)=\dfrac{-0.07(10)+1.56}{0.37(10)+13.2}\approx 0.051$. In 2000, 5.1% of California's labor force was considered unemployed.

$P(15)=\dfrac{-0.07(15)+1.56}{0.37(15)+13.2}\approx 0.027$. In 2005, 2.7% of California's labor force was considered unemployed.

This may be too low and therefore be considered model breakdown.

30. $f(x)=\dfrac{3x+2}{x-9}$

Domain: All real numbers such that $x\neq 9$.

31. $g(x)=\dfrac{x+5}{(x+5)(x-3)}$ There is a hole at $x=-5$.

Domain: All real numbers such that $x\neq 3$.

32.

$h(x)=\dfrac{5x+2}{2x^2-7x-15}$

$h(x)=\dfrac{5x+2}{(2x+3)(x-5)}$

$2x+3=0$ $\quad$ $x-5=0$

$2x=-3$ $\quad$ $x=5$

$\dfrac{2x}{2}=\dfrac{-3}{2}$

$x=\dfrac{-3}{2}$

Domain: All real numbers such that $x\neq -\dfrac{3}{2}$ and $x\neq 5$.

33.

$f(x)=\dfrac{x+3}{7x^2+5x+25}$

$7x^2+5x+25\neq 0$

Domain: All real numbers

34. Domain: All real numbers such that $x\neq -6$ and $x\neq 4$

35. Domain: All real numbers such that $x\neq -3$ and $x\neq 6$

36. $f(x)=6x+3$ and $g(x)=-4x+8$

a.

$$\begin{aligned} f(g(x)) &= f(-4x+8) \\ &= 6(-4x+8)+3 \\ &= -24x+48+3 \\ &= -24x+51 \end{aligned}$$

b.

$$\begin{aligned} g(f(x)) &= g(6x+3) \\ &= -4(6x+3)+8 \\ &= -24x-12+8 \\ &= -24x-4 \end{aligned}$$

37. $f(x)=15x+34$ and $g(x)=-17x+34$

a.

$$\begin{aligned} f(g(x)) &= f(-17x+34) \\ &= 15(-17x+34)+34 \\ &= -255x+510+34 \\ &= -255x+544 \end{aligned}$$

b.

$$\begin{aligned} g(f(x)) &= g(15x+34) \\ &= -17(15x+34)+34 \\ &= -255x-578+34 \\ &= -255x-544 \end{aligned}$$

38. $f(x)=\frac{2}{5}x+\frac{3}{5}$ and $g(x)=4x-7$

a.

$$\begin{aligned} (f\circ g)(x) &= f(g(x)) \\ &= \frac{2}{5}(4x-7)+\frac{3}{5} \\ &= \frac{8}{5}x-\frac{14}{5}+\frac{3}{5} \\ &= \frac{8}{5}x-\frac{11}{5} \end{aligned}$$

b.

$$\begin{aligned} (g\circ f)(x) &= g(f(x)) \\ &= 4\left(\frac{2}{5}x+\frac{3}{5}\right)-7 \\ &= \frac{8}{5}x+\frac{12}{5}-7 \\ &= \frac{8}{5}x-\frac{23}{5} \end{aligned}$$

39. $f(x)=\frac{1}{7}x+\frac{5}{7}$ and $g(x)=2x-7$

a.

$$\begin{aligned} (f\circ g)(x) &= f(g(4)) \\ &= f(1) \quad \text{*Since } g(4)=2(4)-7=1 \\ &= \frac{1}{7}(1)+\frac{5}{7} \\ &= \frac{1}{7}+\frac{5}{7} \\ &= \frac{6}{7} \end{aligned}$$

b.

$$\begin{aligned} (g\circ f)(4) &= g(f(4)) \\ &= g\left(\frac{9}{7}\right) \quad \text{*Since } f(4)=\frac{1}{7}(4)+\frac{5}{7}=\frac{9}{7} \\ &= 2\left(\frac{9}{7}\right)-7 \\ &= \frac{18}{7}-7 \\ &= -\frac{31}{7} \end{aligned}$$

40. $f(x)=0.6x+2.5$ and $g(x)=3.5x+3.7$

a.

$$\begin{aligned} f(-3)+g(-3) &= [0.6(-3)+2.5]+[3.5(-3)+3.7] \\ &= (-1.8+2.5)+(-10.5+3.7) \\ &= (0.7)+(-6.8) \\ &= -6.1 \end{aligned}$$

b.

$$\begin{aligned} g(-3)-f(-3) &= [3.5(-3)+3.7]-[0.6(-3)+2.5] \\ &= (-10.5+3.7)-(-1.8+2.5) \\ &= (-6.8)-(0.7) \\ &= -7.5 \end{aligned}$$

c.

$$\begin{aligned} f(-3)g(-3) &= [0.6(-3)+2.5][3.5(-3)+3.7] \\ &= (-1.8+2.5)(-10.5+3.7) \\ &= (0.7)(-6.8) \\ &= -4.76 \end{aligned}$$

d.

$$\begin{aligned} f(g(-3)) &= f(-6.8) \quad \text{*Since } g(-3)=3.5(-3)+3.7=-6.8 \\ &= 0.6(-6.8)+2.5 \\ &= -4.08+2.5 \\ &= -1.58 \end{aligned}$$

e.

$$\begin{aligned} g(f(-3)) &= g(0.7) \quad \text{*Since } f(-3)=0.6(-3)+2.5=0.7 \\ &= 3.5(0.7)+3.7 \\ &= 2.45+3.7 \\ &= 6.15 \end{aligned}$$

41. $f(x)=0.35x-2.78$ and $g(x)=2.4x-6.3$

a.

$$\begin{aligned} f(2)+g(2) &= [0.35(2)-2.78]+[2.4(2)-6.3] \\ &= (0.7-2.78)+(4.8-6.3) \\ &= (-2.08)+(-1.5) \\ &= -3.58 \end{aligned}$$

b.

$$\begin{aligned} g(2)-f(2) &= [2.4(2)-6.3]-[0.35(2)-2.78] \\ &= (4.8-6.3)-(0.7-2.78) \\ &= (-1.5)-(-2.08) \\ &= 0.58 \end{aligned}$$

c.

$$\begin{aligned} f(2)g(2) &= [0.35(2)-2.78][2.4(2)-6.3] \\ &= (0.7-2.78)(4.8-6.3) \\ &= (-2.08)(-1.5) \\ &= 3.12 \end{aligned}$$

d.

$f(g(2)) = f(-1.5)$ *Since $g(2) = 2.4(2) - 6.3 = -1.5$

$= 0.35(-1.5) - 2.78$

$= -0.525 - 2.78$

$= -3.305$

Chapter 8 Test

1. **a.** Let C be the number of work-related injury cases in thousands in U.S. private industry t years since 2000.

t	C
3	4095
4	4008
5	3972
6	3857

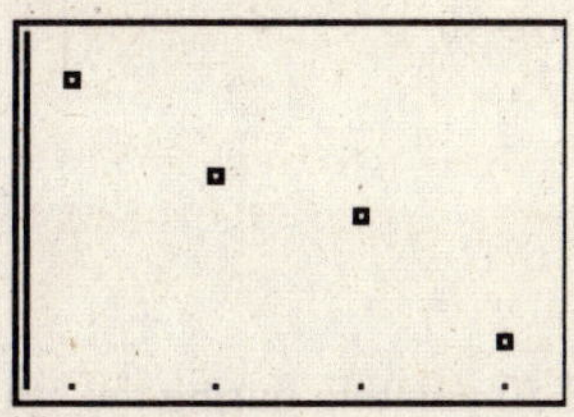

$C(t) = mt + b$

Using two points: (3,4095) and (6,3857)

Find the slope: $m = \dfrac{3857 - 4095}{6-3} \approx -79.3$

$C(t) = -79.3t + b$

$b = 4095 - (-79.3)(3)$

$b \approx 4333$

$C(t) = -79.3t + 4333$

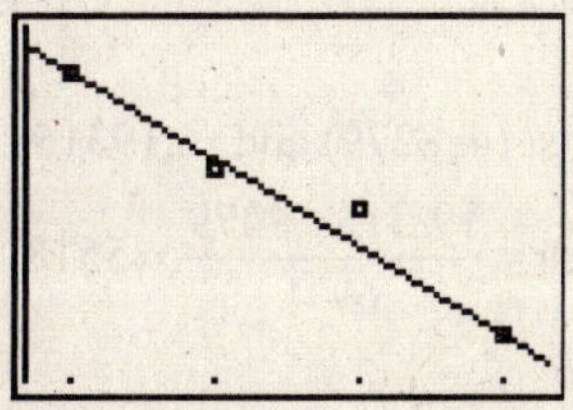

b.

$C(9) = -79.3(9) + 4333$

$C(9) = 3619.3$

In 2009, there were about 3619.3 thousand work-related injury cases in the United States private industry.

c. The slope is -79.3. The number of work-related injury cases in U.S. private industry is decreasing about 79.3 thousand cases per year.

d. Domain: $[0,10]$

$C(0) = -79.3(0) + 4333 = 4333$

$C(10) = -79.3(10) + 4333 = 3540$

Range: $[3540, 4333]$

e.

$5000 = -79.3t + 4333$

$5000 - 4333 = -79.3t + 4333 - 4333$

$667 = -79.3t$

$\dfrac{667}{-79.3} = \dfrac{-79.3t}{-79.3}$

$-8.4 \approx t$

In 1992, there were about 5 million work-related injury cases in U.S. private industry.

2. $P = 2m + 30$

a.

$$P = 2(6) + 30$$

$$P = 42$$

Six months after starting to sell paintings, John Clark will sell approximately 42 paintings.

b.

$$50 = 2m + 30$$

$$50 - 30 = 2m + 30 - 30$$

$$20 = 2m$$

$$\frac{20}{2} = \frac{2m}{2}$$

$$10 = m$$

John Clark will sell 50 painting during the 10th month after starting to sell painting.

3. Let R be the total revenue in millions of dollars for Apple Inc. t years since 2000.

a.

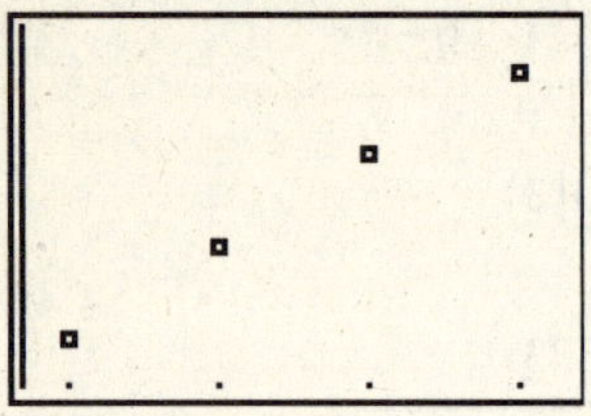

$$R(t) = mt + b$$

Using two points: (4,8279) and (6,19315)

Find the slope: $m = \dfrac{19{,}315 - 8279}{6 - 4} = 5518$

$$R(t) = 5518t + b$$

$$b = 8279 - 5518(4)$$

$$b = -13{,}793$$

$$R(t) = 5518t - 13{,}793$$

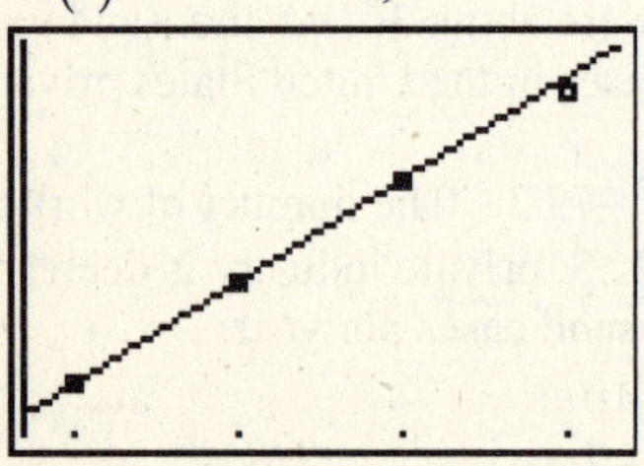

b. Vertical intercept is $(0, -13{,}793)$. In 2000, Apple, Inc. had total revenue of $-\$13{,}793$ million. This is model breakdown, since Apple did not have negative revenue in 2000.

c. Domain: $[3,9]$

$$R(3) = 5518(3) - 13{,}793 = 2761$$

$$R(9) = 5518(9) - 13{,}793 = 35{,}869$$

Range: $[2761, 35869]$

d.

$$R(10) = 5518(10) - 13{,}793$$

$$R(10) = 41{,}387$$

In 2010, Apple, Inc. had total revenue of about $41,387 million.

e.

$$30{,}000 = 5518t - 13{,}793$$

$$30{,}000 + 13{,}793 = 5518t - 13{,}793 + 13{,}793$$

$$43{,}793 = 5518t$$

$$\frac{43{,}793}{5518} = \frac{5518t}{5518}$$

$$8 \approx t$$

Apple, Inc.'s total revenue was about $30 billion in 2008.

f. The slope is 5518. Apple, Inc.'s total revenue increases by about $5518 million dollars per year.

4. Domain: All real numbers, range: All real numbers

5. In 2010, the population of New York will be about 19.4 million.

6. $H(p) = 0.20p + 5$

a.

$$H(70) = 0.20(70) + 5$$

$$H(70) = 19$$

To earn a 70% on the exam, you need to study about 19 hours.

b.

$$H(100) = 0.20(100) + 5$$

$$H(100) = 25$$

To earn a 100% on the exam, you need to study about 25 hours.

7. Domain: all real numbers.

8. **a.**

$$h(x) = \frac{4}{7}x + 6$$

$$h(\ ) = \frac{4}{7}(\ ) + 6$$

$$h(35) = \frac{4}{7}(35) + 6$$

$$h(35) = 20 + 6$$

$$h(35) = 26$$

b.

$$h(x) = \frac{4}{7}x + 6$$

$$4 = \frac{4}{7}x + 6$$

$$4 - 6 = \frac{4}{7}x + 6 - 6$$

$$-2 = \frac{4}{7}x$$

$$\frac{7}{4}\cdot(-2) = \frac{7}{4}\cdot\left(\frac{4}{7}x\right)$$

$$x = -\frac{7}{2}$$

c. Domain: All real numbers, range: All real numbers

9. **a.**

$$g(x) = -9$$
$$g(\ ) = -9$$
$$g(6) = -9$$

b. Domain: All real numbers, range: $\{-9\}$

10. A relation may not be a function if the input value was paired with more than one output value.

11. **a.** $T(9) = 1.1(9) + 6.84 = 16.74$. In 1999, New Zealand spent about 16.74 million New Zealand dollars on the treatment of diabetes.

b.

$$10 = 1.3t + 0.52$$
$$10 - 0.52 = 1.3t + 0.52 - 0.52$$
$$9.48 = 1.3t$$
$$\frac{9.48}{1.3} = \frac{1.3t}{1.3}$$
$$7.3 \approx t$$

New Zealand spent about 10 million New Zealand dollars on diabetes research in 1997.

c. Let $D(t)$ be the total New Zealand spent on diabetes research and treatment, in millions of New Zealand dollars, t years since 1990.

$$D(t) = T(t) + R(t)$$
$$D(t) = (1.1t + 6.84) + (1.3t + 0.52)$$
$$D(t) = 2.4t + 7.36$$

d. $D(10) = 2.4(10) + 7.36 = 31.36$. New Zealand spent about 31.36 million New Zealand dollars on research and treatment of diabetes in 2000.

12. $f(x) = 4x + 17 \quad g(x) = 2x - 7$

a.

$$\begin{aligned} f(x) - g(x) &= (4x + 17) - (2x - 7) \\ &= 4x + 17 - 2x + 7 \\ &= 2x + 24 \end{aligned}$$

b.

$$\begin{aligned} f(x)g(x) &= (4x + 17)(2x - 7) \\ &= 8x^2 - 28x + 34x - 119 \\ &= 8x^2 + 6x - 119 \end{aligned}$$

c.

$$\begin{aligned} f(g(x)) &= 4(2x - 7) + 17 \\ &= 8x - 28 + 17 \\ &= 8x - 11 \end{aligned}$$

13. $f(x) = 2.35x + 1.45 \quad g(x) = 2.4x - 6.3$

a.

$$\begin{aligned} (f + g)(4) &= (2.35(4) + 1.45) + (2.4(4) - 6.3) \\ &= 9.4 + 1.45 + 9.6 - 6.3 \\ &= 14.15 \end{aligned}$$

b.

$$\begin{aligned} fg(-2) &= (2.35(-2) + 1.45)(2.4(-2) - 6.3) \\ &= (-3.25)(-11.1) \\ &= 36.075 \end{aligned}$$

c.

$$\begin{aligned} (f \circ g)(6) &= 2.35(2.4(6) - 6.3) + 1.45 \\ &= 2.35(8.1) + 1.45 \\ &= 20.485 \end{aligned}$$

d.

$$\begin{aligned} g(f(0)) &= 2.4(2.35(0) + 1.45) - 6.3 \\ &= 2.4(1.45) - 6.3 \\ &= -2.82 \end{aligned}$$

14. **a.**

$$S(10) = 1.2(10)^2 + 4.8(10) + 376.6$$
$$S(10) = 544.6$$

In 2000, there were approximately 544.6 thousand science/engineering graduate students in doctoral programs.

b. Let $M(t)$ be the number of male science/engineering graduate students in doctoral programs, in thousands, t years since 2000.

$$\begin{aligned} M(t) &= S(t) - F(t) \\ &= 1.2t^2 + 4.8t + 376.6 \\ &\underline{-0.4t^2 - 4.1t - 150.9} \\ M(t) &= 0.8t^2 + 0.7t + 225.7 \end{aligned}$$

c.

$$M(10) = 0.8(10)^2 + 0.7(10) + 225.7$$
$$M(10) = 312.7$$

In 2010, there were approximately 312.7 thousand male science/engineering graduate students in doctoral programs.

15. **a.**

$$P(6) = -0.3(6)^3 + 3.55(6)^2 - 10.41(6) + 51.23$$
$$P(6) = 51.77$$

In 2006, about 51.8% of murders were committed by using a handgun.

b. Let $H(t)$ be the number of murders in thousands, that were committed using a handgun t years since 2000.

$$\begin{aligned} H(t) &= M(t)\frac{P(t)}{100} \\ &= (-0.02t^3 + 0.28t^2 - 11t + 13.23)\frac{-0.3t^3 + 3.55t^2 - 10.41t + 51.23}{100} \\ &= (-0.02t^3 + 0.28t^2 - 11t + 13.23)(-0.003t^3 + 0.0355t^2 - 0.1041t + 0.5123) \\ &= 0.00006t^6 - 0.0013t^5 + 0.0095t^4 - 0.075t^3 + 0.58t^2 - 1.43t + 6.78 \end{aligned}$$

c.

$$H(7)=0.00006(7)^6-0.0013(7)^5+0.0095(7)^4$$
$$-0.075(7)^3+0.58(7)^2-1.43(7)+6.78\approx 7.5$$

In 2007, about 7.5 thousand murders were committed using a handgun.

16. $(T-B)(t)$ is the number of girls attending the tennis camp in year t.

17. $M(B(t))$ is the number of boys' matches at the camp in year t.

18. $C(T(t))$ is the total cost for the camp in year t.

19. $A(t)\cdot M(T(t))$ is the time it takes for all the matches to be played in year t.

20. **a.**

$$h(x)=\frac{3x-7}{x^2+6x-27}=\frac{3x-7}{(x+9)(x-3)}$$

The denominator cannot equal zero.

Set the linear factors not equal to zero and solve:

$x+9\neq 0$ and $x-3\neq 0$

$x\neq -9$ and $x\neq 3$

Domain: x is all real numbers except $x\neq -9$ and $x\neq 3$.

b. Domain: x is all real numbers except $x\neq -2$ and $x\neq 4$.

CHAPTER 9 INEQUALITIES AND ABSOLUTE VALUES

Section 9.1 Absolute Value Equations

1. **a.** $|x+1|=2$

$x+1=2$ or $x+1=-2$

$x=1$ or $x=-3$

b.

x	$\lvert x+1\rvert$
-5	$\lvert -5+1\rvert=\lvert -4\rvert=4$
-3	$\lvert -3+1\rvert=\lvert -2\rvert=2$
-2	$\lvert -2+1\rvert=\lvert -1\rvert=1$
-1	$\lvert -1+1\rvert=\lvert 0\rvert=0$
0	$\lvert 0+1\rvert=\lvert 1\rvert=1$
1	$\lvert 1+1\rvert=\lvert 2\rvert=2$
2	$\lvert 2+1\rvert=\lvert 3\rvert=3$
3	$\lvert 3+1\rvert=\lvert 4\rvert=4$
5	$\lvert 5+1\rvert=\lvert 6\rvert=6$

c.

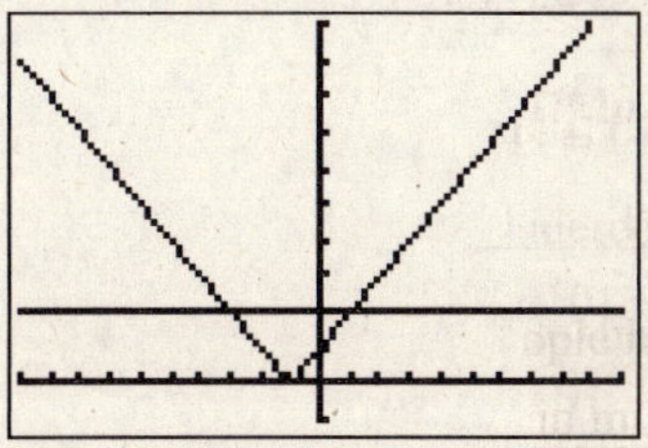

3. **a.** $|x-3|=1$

$x-3=1$ or $x-3=-1$

$x=4$ or $x=2$

b.

x	$\lvert x-3\rvert$
-2	$\lvert -2-3\rvert=\lvert -5\rvert=5$
-1	$\lvert -1-3\rvert=\lvert -4\rvert=4$
0	$\lvert 0-3\rvert=\lvert -3\rvert=3$
1	$\lvert 1-3\rvert=\lvert -2\rvert=2$
2	$\lvert 2-3\rvert=\lvert -1\rvert=1$
3	$\lvert 3-3\rvert=\lvert 0\rvert=0$
4	$\lvert 4-3\rvert=\lvert -1\rvert=1$
5	$\lvert 5-3\rvert=\lvert -2\rvert=2$
6	$\lvert 6-3\rvert=\lvert -3\rvert=3$

c.

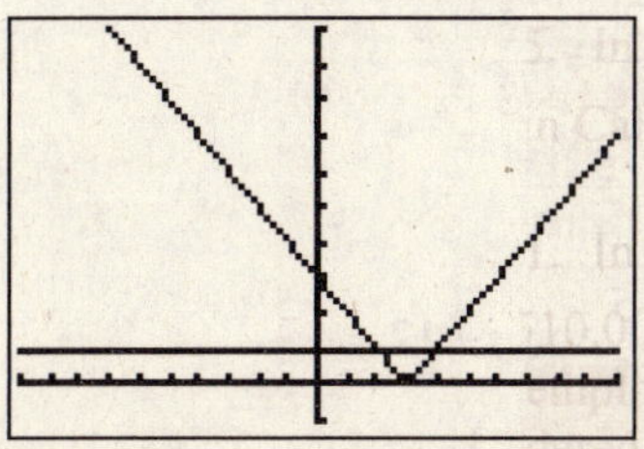

5. **a.** $|2x-4|=6$

$2x-4=6$ or $2x-4=-6$

$2x=10$ or $2x=-2$

$x=5$ or $x=-1$

b.

x	$\lvert 2x-4\rvert$
-2	$\lvert 2(-2)-4\rvert=\lvert -4-4\rvert=\lvert -8\rvert=8$
-1	$\lvert 2(-1)-4\rvert=\lvert -2-4\rvert=\lvert -6\rvert=6$
0	$\lvert 2(0)-4\rvert=\lvert 0-4\rvert=\lvert -4\rvert=4$
1	$\lvert 2(1)-4\rvert=\lvert 2-4\rvert=\lvert -2\rvert=2$
2	$\lvert 2(2)-4\rvert=\lvert 4-4\rvert=\lvert 0\rvert=0$
3	$\lvert 2(3)-4\rvert=\lvert 6-4\rvert=\lvert 2\rvert=2$
4	$\lvert 2(4)-4\rvert=\lvert 8-4\rvert=\lvert 4\rvert=4$
5	$\lvert 2(5)-4\rvert=\lvert 10-4\rvert=\lvert 6\rvert=6$
6	$\lvert 2(6)-4\rvert=\lvert 12-4\rvert=\lvert 8\rvert=8$

c.

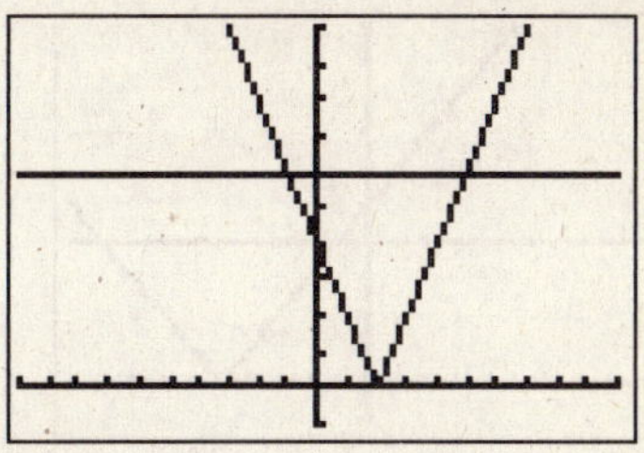

7. **a.** $|3x-1|=2$

$3x-1=2$ or $3x-1=-2$

$3x=3$ or $3x=-1$

$x=1$ or $x=-\frac{1}{3}$

b.

x	$\lvert 3x-1\rvert$
-3	$\lvert 3(-3)-1\rvert=\lvert -9-1\rvert=\lvert -10\rvert=10$
-2	$\lvert 3(-2)-1\rvert=\lvert -6-1\rvert=\lvert -7\rvert=7$
-1	$\lvert 3(-1)-1\rvert=\lvert -3-1\rvert=\lvert -4\rvert=4$
$-\frac{1}{3}$	$\left\lvert 3\left(-\frac{1}{3}\right)-1\right\rvert=\lvert -1-1\rvert=\lvert -2\rvert=2$
0	$\lvert 3(0)-1\rvert=\lvert 0-1\rvert=\lvert -1\rvert=1$
$\frac{1}{3}$	$\left\lvert 3\left(\frac{1}{3}\right)-1\right\rvert=\lvert 1-1\rvert=\lvert 0\rvert=0$
1	$\lvert 3(1)-1\rvert=\lvert 3-1\rvert=\lvert 2\rvert=2$
2	$\lvert 3(2)-1\rvert=\lvert 6-1\rvert=\lvert 5\rvert=5$
3	$\lvert 3(3)-1\rvert=\lvert 9-1\rvert=\lvert 8\rvert=8$

c.

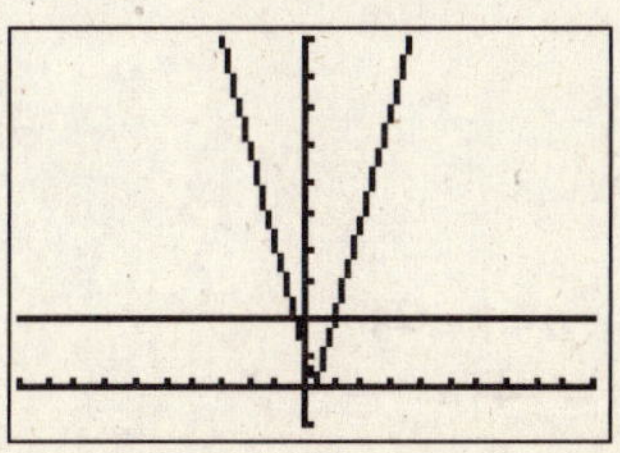

9. **a.** $|5-x|=4$

$5-x=4$ or $5-x=-4$

$-x=-1$ or $-x=-9$

$x=1$ or $x=9$

b.

x	$\lvert 5-x\rvert$
−1	$\lvert 5-(-1)\rvert=\lvert 5+1\rvert=\lvert 6\rvert=6$
0	$\lvert 5-(0)\rvert=\lvert 5\rvert=5$
1	$\lvert 5-(1)\rvert=\lvert 4\rvert=4$
3	$\lvert 5-(3)\rvert=\lvert 2\rvert=2$
5	$\lvert 5-(5)\rvert=\lvert 0\rvert=0$
7	$\lvert 5-(7)\rvert=\lvert -2\rvert=2$
8	$\lvert 5-(8)\rvert=\lvert -3\rvert=3$
9	$\lvert 5-(9)\rvert=\lvert -4\rvert=4$
10	$\lvert 5-(10)\rvert=\lvert -5\rvert=5$

c.

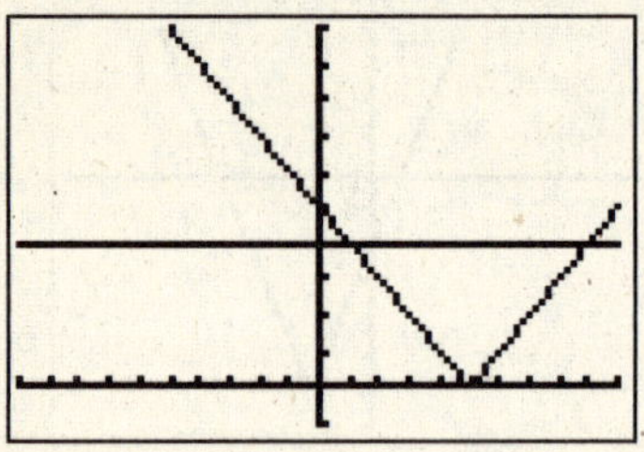

11.

$|x|=12$

$x=12$ or $x=-12$

$x=\pm 12$

13.

$|h+7|=15$

$h+7=15$ or $h+7=-15$

$h=8$ or $h=-22$

15.

$|b-12|=8$

$b-12=8$ or $b-12=-8$

$b=20$ or $b=4$

17.

$2|r+11|=36$

$\frac{2|r+11|}{2}=\frac{36}{2}$

$|r+11|=18$

$r+11=18$ or $r+11=-18$

$r=7$ or $r=-29$

19.

$3|a+1|=9$

$\frac{3|a+1|}{3}=\frac{9}{3}$

$|a+1|=3$

$a+1=3$ or $a+1=-3$

$a=2$ or $a=-4$

21.

$5|x-6|=25$

$\frac{5|x-6|}{5}=\frac{25}{5}$

$|x-6|=5$

$x-6=5$ or $x-6=-5$

$x=11$ or $x=1$

23.

$|m-3|+2=6$

$\quad -2 \; -2$

$|m-3|=4$

$m-3=4$ or $m-3=-4$

$m=7$ or $m=-1$

25.

$|y+4|-6=7$

$\quad +6 \; +6$

$|y+4|=13$

$y+4=13$ or $y+4=-13$

$y=9$ or $y=-17$

27.

$2|x-5|+1=11$

$\quad -1 \; -1$

$2|x-5|=10$

$\frac{2|x-5|}{2}=\frac{10}{2}$

$|x-5|=5$

$x-5=5$ or $x-5=-5$

$x=10$ or $x=0$

29.

$-2|x-5|+1=-5$

$\quad -1 \; -1$

$-2|x-5|=-6$

$\frac{-2|x-5|}{-2}=\frac{-6}{-2}$

$|x-5|=3$

$x-5=3$ or $x-5=-3$

$x=8$ or $x=2$

31.

$-3|x-12|=-5$

$\frac{-3|x-12|}{-3}=\frac{-5}{-3}$

$|x-12|=\frac{5}{3}$

$x-12=\frac{5}{3}$ or $x-12=-\frac{5}{3}$

$x=\frac{41}{3}$ or $x=\frac{31}{3}$

33. $|k+10|=-3$

There is no solution, because an absolute value cannot equal a negative number.

35.

$$|g-9|+12=8$$
$$|g-9|+12-12=8-12$$
$$|g-9|=-4$$

There is no solution, because an absolute value cannot equal a negative number.

37.

$$-2|d-8|+1=11$$
$$-2|d-8|+1-1=11-1$$
$$-2|d-8|=10$$
$$\frac{-2|d-8|}{-2}=\frac{10}{-2}$$
$$|d-8|=-5$$

There is no solution, because an absolute value cannot equal a negative number.

39.

$$-\frac{1}{3}|s-60|+25=20$$
$$-\frac{1}{3}|s-60|+25-25=20-25$$
$$-\frac{1}{3}|s-60|=-5$$
$$(-3)\left(-\frac{1}{3}\right)|s-60|=(-5)(-3)$$
$$|s-60|=15$$
$$s-60=15 \text{ or } s-60=-15$$
$$s=75 \text{ or } s=45$$

Ricardo should travel at an average speed of 45 mph or 75 mph to get gas mileage of 20 mpg during his trip.

41.

$$|700-60t|=50$$
$$700-60t=50 \text{ or } 700-60t=-50$$
$$-60t=-650 \quad \text{or } -60t=-750$$
$$\frac{-60t}{-60}=\frac{-650}{-60} \quad \text{or} \quad \frac{-60t}{-60}=\frac{-750}{-60}$$
$$t\approx 10.83 \text{ or } t=12.5$$

Ricardo will be fifty miles away from Fresno after about 10.83 hours and again after about 12.5 hours.

43.

$$|x-8|+3=3$$
$$\underline{\quad -3\ -3}$$
$$|x-8|=0$$
$$x-8=0$$
$$x=8$$

45. $|p+4|=-3$

There is no solution, because an absolute value cannot equal a negative number.

47.

$$|2x+1|-10=-4$$
$$\underline{\quad +10\ \ +10}$$
$$|2x+1|=6$$
$$2x+1=6 \text{ or } 2x+1=-6$$
$$2x=5 \text{ or } 2x=-7$$
$$x=\frac{5}{2} \text{ or } x=-\frac{7}{2}$$

49.

$$|x-7|+2=-5$$
$$\underline{\quad -2\ \ -2}$$
$$|x-7|=-7$$

There is no solution, because an absolute value cannot equal a negative number.

Section 9.2 Absolute Value Inequalities

1. $-6 < x < 6$

3. $x \le -3$ or $x \ge 3$

5. $x < -5$ or $x > -2$

7. $-3 \le x \le 8$

9. $x < -6$ or $x > 5$

11. **a.** For $f(x) < g(x): -4 < x < 4$

b. For $f(x) > g(x): x < -4$ or $x > 4$

13. **a.** For $f(x) \le g(x): 0 \le x \le 4$

b. For $f(x) \ge g(x): x \le 0$ or $x \ge 4$

15. **a.** For $f(x) < g(x): -5 < x < 5$

b. For $f(x) > g(x): x < -5$ or $x > 5$

17.

$|x| < 5$

$-5 < x < 5$

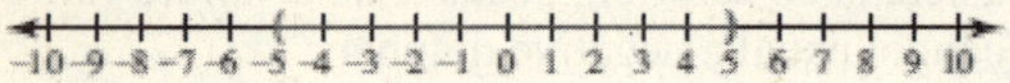

19.

$|h-3| \le 4$

$-4 \le h-3 \le 4$

$-4+3 \le h-3+3 \le 4+3$

$-1 \le h \le 7$

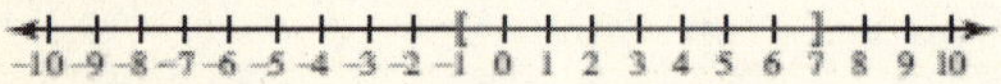

21.

$|b-1| < 6$

$-6 < b-1 < 6$

$-6+1 < b-1+1 < 6+1$

$-5 < b < 7$

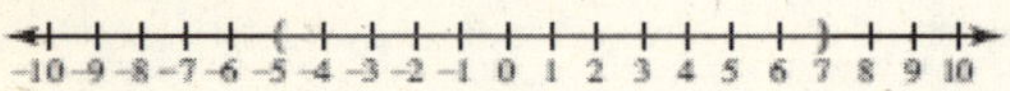

23.

$|L-3.00| \le 0.001$

$-0.001 \le L-3.00 \le 0.001$

$-0.001+3.00 \le L-3.00+3.00 \le 0.001+3.00$

$2.999 \le L \le 3.001$

The acceptable lengths of this coupler are from a low of 2.999 cm, to a high of 3.001 cm.

25.

$|P-20| \le 0.5$

$-0.5 \le P-20 \le 0.5$

$-0.5+20 \le P-20+20 \le 0.5+20$

$19.5 \le P \le 20.5$

The acceptable pressures for this experiment are from a low of 19.5 psi, to a high of 20.5 psi.

27.

$|x| > 3$

$x < -3$ or $x > 3$

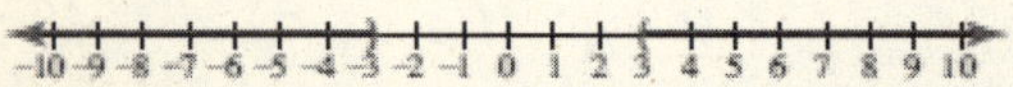

29.

$|p-4| \ge 2$

$p-4 \le -2$ or $p-4 \ge 2$

$p-4+4 \le -2+4$ or $p-4+4 \ge 2+4$

$p \le 2$ or $p \ge 6$

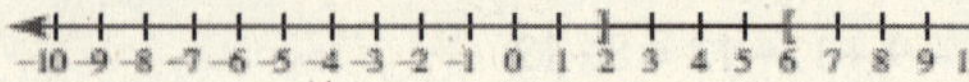

31.

$|y+6| > 3$

$y+6 < -3$ or $y+6 > 3$

$y+6-6 < -3-6$ or $y+6-6 > 3-6$

$y < -9$ or $y > -3$

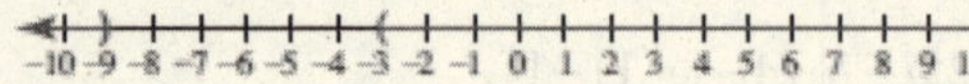

33.

$\left|\dfrac{h-69}{2.5}\right| > 2$

$\dfrac{h-69}{2.5} < -2$ or $\dfrac{h-69}{2.5} > 2$

$2.5\left(\dfrac{h-69}{2.5}\right) < (-2)(2.5)$ or $2.5\left(\dfrac{h-69}{2.5}\right) > (2)(2.5)$

$h-69 < -5$ or $h-69 > 5$

$h-69+69 < -5+69$ or $h-69+69 > 5+69$

$h < 64$ or $h > 74$

Men's heights that are less than 64 inches or greater than 74 inches are considered unusual.

35. **a.**

$\left|\dfrac{S-100}{16}\right| > 2$

$\dfrac{S-100}{16} < -2$ or $\dfrac{S-100}{16} > 2$

$16\left(\dfrac{S-100}{16}\right) < (-2)(16)$ or $16\left(\dfrac{S-100}{16}\right) > (2)(16)$

$S-100 < -32$ or $S-100 > 32$

$S-100+100 < -32+100$ or $S-100+100 > 32+100$

$S < 68$ or $S > 132$

IQ scores that are below 68 or above 132 would be considered unusual.

b. Yes, Albert Einstein's IQ was unusual. His IQ was greater than 132.

37.

$2|r+6| < 8$

$\dfrac{2|r+6|}{2} < \dfrac{8}{2}$

$|r+6| < 4$

$-4 < r+6 < 4$

$-4-6 < r+6-6 < 4-6$

$-10 < r < -2$

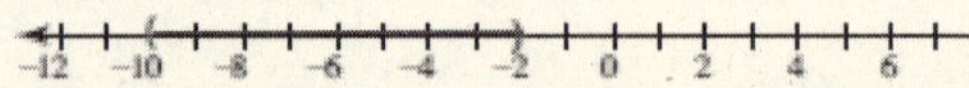

39.

$4x-20 > 12$

$4x-20+20 > 12+20$

$4x > 32$

$x > 8$

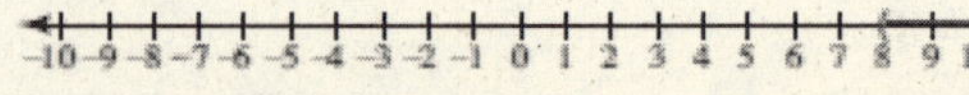

41.

$|d-5|+10 \le 7$

$|d-5|+10-10 \le 7-10$

$|d-5| \le -3$

There is no solution, because an absolute value cannot equal a negative number.

43.

$|m+1|-2>6$

$|m+1|-2+2>6+2$

$|m+1|>8$

$m+1<-8$ or $m+1>8$

$m+1-1<-8-1$ or $m+1-1>8-1$

$m<-9$ or $m>7$

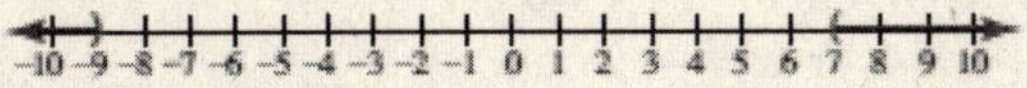

45.

$\frac{2}{3}x-8>\frac{5}{6}x-4$

$6\left(\frac{2}{3}x-8\right)>6\left(\frac{5}{6}x-4\right)$

$4x-48>5x-24$

$4x-48-5x>5x-24-5x$

$-x-48>-24$

$-x-48+48>-24+48$

$-x>24$

$\frac{-x}{-1}<\frac{24}{-1}$

$x<-24$

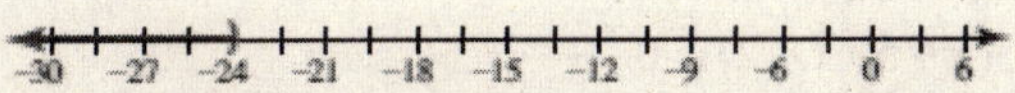

47.

$-6|p-4|+3 \le 16$

$-6|p-4|+3-3 \le 16-3$

$-6|p-4| \le 13$

$\frac{-6|p-4|}{-6} \ge \frac{13}{-6}$

$|p-4| \ge -\frac{13}{6}$

An absolute value will always be greater than a negative number. Therefore, the solution for p is all real numbers.

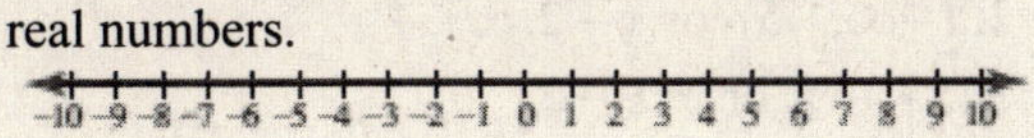

49.

$-|x-4| \ge -2$

$\frac{-|x-4|}{-1} \le \frac{-2}{-1}$

$|x-4| \le 2$

$-2 \le x-4 \le 2$

$-2+4 \le x-4+4 \le 2+4$

$2 \le x \le 6$

51.

$5+2|2x-4|>11$

$5+2|2x-4|-5>11-5$

$2|2x-4|>6$

$\frac{2|2x-4|}{2}>\frac{6}{2}$

$|2x-4|>3$

$2x-4<-3$ or $2x-4>3$

$2x-4+4<-3+4$ or $2x-4+4>3+4$

$2x<1$ or $2x>7$

$x<\frac{1}{2}$ or $x>\frac{7}{2}$

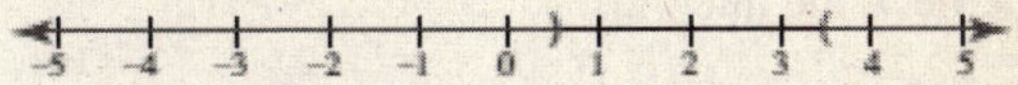

53.

$12+8|3d-8| \le 10$

$12+8|3d-8|-12 \le 10-12$

$8|3d-8| \le -2$

$\frac{8|3d-8|}{8} \le \frac{-2}{8}$

$|3d-8| \le -\frac{1}{4}$

There is no solution, because an absolute value cannot equal a negative number.

55.

$1.4x+3 \ge 4.4x-1$

$1.4x+3-4.4x \ge 4.4x-1-4.4x$

$-3x+3 \ge -1$

$-3x+3-3 \ge -1-3$

$-3x \ge -4$

$\frac{-3x}{-3} \le \frac{-4}{-3}$

$x \le \frac{4}{3}$

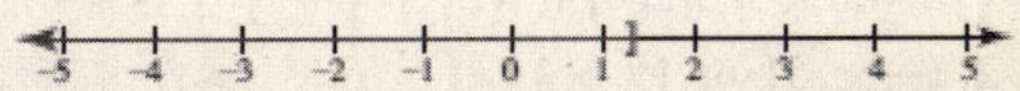

57.

$\left|\frac{1}{2}x+3\right| \le 5$

$-5 \le \frac{1}{2}x+3 \le 5$

$-5-3 \le \frac{1}{2}x+3-3 \le 5-3$

$-8 \le \frac{1}{2}x \le 2$

$2(-8) \le 2\left(\frac{1}{2}x\right) \le 2(2)$

$-16 \le x \le 4$

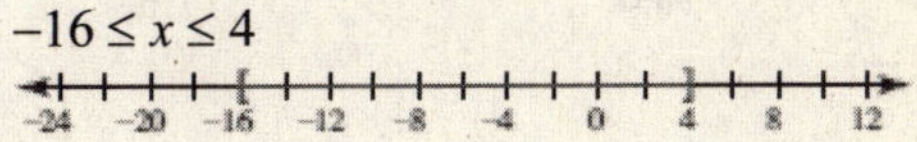

Section 9.3 Nonlinear Inequalities in One Variable

1. $2 < x < 7$

3. $x < 8$

5. $x \le 2$ or $8 \le x \le 11$

7. $(5, \infty)$

9. $[-5, 8]$

11. $(-\infty, 5)$ or $(9, \infty)$

13. $(x+3)(x-8) < 0$

$(x+3)(x-8) = 0$

$x+3=0$ or $x-8=0$

$x=-3$ or $x=8$

Check the intervals:

$x=-5$	$x=0$	$x=10$
$(-5+3)(-5-8)\overset{?}{<}0$	$(0+3)(0-8)\overset{?}{<}0$	$(10+3)(10-8)\overset{?}{<}0$
$(-2)(-13)\overset{?}{<}0$	$(3)(-8)\overset{?}{<}0$	$(13)(2)\overset{?}{<}0$
$26\overset{?}{<}0$	$-24\overset{?}{<}0$	$26\overset{?}{<}0$
False	True	False

1. 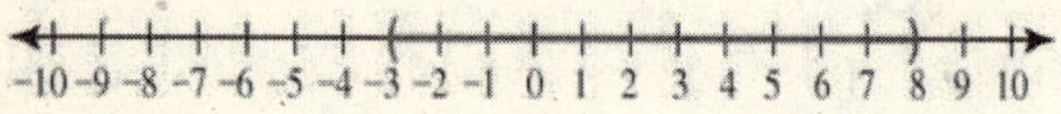

2. $(-3, 8)$

3. $-3 < x < 8$

15. $(x-5)(x-9) \ge 0$

$(x-5)(x-9) = 0$

$x-5=0$ or $x-9=0$

$x=5$ or $x=9$

Check the intervals:

$x=3$	$x=7$	$x=11$
$(3-5)(3-9)\overset{?}{\ge}0$	$(7-5)(7-9)\overset{?}{\ge}0$	$(11-5)(11-9)\overset{?}{\ge}0$
$(-2)(-6)\overset{?}{\ge}0$	$(2)(-2)\overset{?}{\ge}0$	$(6)(2)\overset{?}{\ge}0$
$12\overset{?}{\ge}0$	$-4\overset{?}{\ge}0$	$12\overset{?}{\ge}0$
True	False	True

1. 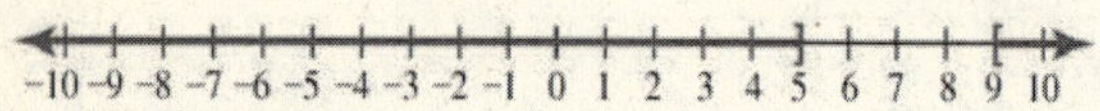

2. $(-\infty, 5]$ or $[9, \infty)$

3. $x \le 5$ or $x \ge 9$

17. $x(x+5)(x-6) > 0$

$x(x+5)(x-6) = 0$

$x=0$ or $x+5=0$ or $x-6=0$

$x=0$ or $x=-5$ or $x=6$

Check the intervals:

$x=-7$	$x=-2$	$x=3$	$x=8$
$-7(-7+5)(-7-6)\overset{?}{>}0$	$-2(-2+5)(-2-6)\overset{?}{>}0$	$3(3+5)(3-6)\overset{?}{>}0$	$8(8+5)(8-6)\overset{?}{>}0$
$-7(-2)(-13)\overset{?}{>}0$	$-2(3)(-8)\overset{?}{>}0$	$3(8)(-3)\overset{?}{>}0$	$8(13)(2)\overset{?}{>}0$
$-182\overset{?}{>}0$	$48\overset{?}{>}0$	$-72\overset{?}{>}0$	$208\overset{?}{>}0$
False	True	False	True

1. 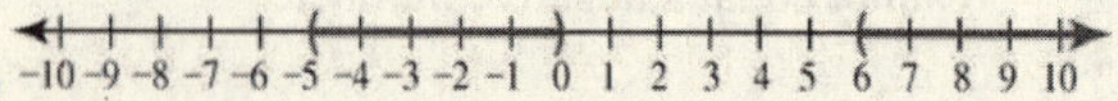

2. $(-5, 0)$ or $(6, \infty)$

3. $-5 < x < 0$ or $x > 6$

19. $x^2 - 16 \le 0$

$x^2 - 16 = 0$

$(x-4)(x+4) = 0$

$x-4=0$ or $x+4=0$

$x=4$ or $x=-4$

Check the intervals:

$x=-6$	$x=0$	$x=6$
$(-6-4)(-6+4)\overset{?}{\le}0$	$(0-4)(0+4)\overset{?}{\le}0$	$(6-4)(6+4)\overset{?}{\le}0$
$(-10)(-2)\overset{?}{\le}0$	$(-4)(4)\overset{?}{\le}0$	$(2)(10)\overset{?}{\le}0$
$20\overset{?}{\le}0$	$-16\overset{?}{\le}0$	$20\overset{?}{\le}0$
False	True	False

1. 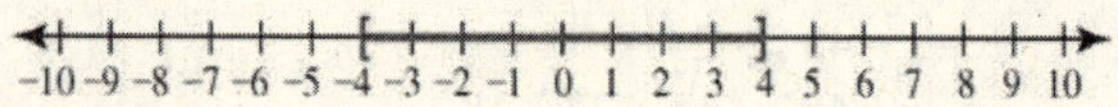

2. $[-4, 4]$

3. $-4 \le x \le 4$

21. $x^2 + 7x + 10 > 0$

$x^2 + 7x + 10 = 0$

$(x+5)(x+2) = 0$

$x+5=0$ or $x+2=0$

$x=-5$ or $x=-2$

Check the intervals:

$x=-6$	$x=-3$	$x=0$
$(-6+5)(-6+2)\overset{?}{>}0$	$(-3+5)(-3+2)\overset{?}{>}0$	$(0+5)(0+2)\overset{?}{>}0$
$(-1)(-4)\overset{?}{>}0$	$(2)(-1)\overset{?}{>}0$	$(5)(2)\overset{?}{>}0$
$4\overset{?}{>}0$	$-2\overset{?}{>}0$	$10\overset{?}{>}0$
True	False	True

1. 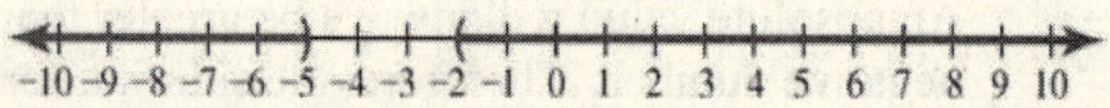

2. $(-\infty, -5)$ or $(-2, \infty)$

3. $x < -5$ or $x > -2$

23. $x^2 - 8x + 20 \ge 5$

$x^2 - 8x + 15 \ge 0$

$x^2 - 8x + 15 = 0$

$(x-5)(x-3) = 0$

$x-5=0$ or $x-3=0$

$x=5$ or $x=3$

Check the intervals:

$x=1$	$x=4$	$x=6$
$(1-5)(1-3)\overset{?}{\geq}0$	$(4-5)(4-3)\overset{?}{\geq}0$	$(6-5)(6-3)\overset{?}{\geq}0$
$(-4)(-2)\overset{?}{\geq}0$	$(-1)(1)\overset{?}{\geq}0$	$(1)(3)\overset{?}{\geq}0$
$8\overset{?}{\geq}0$	$-1\overset{?}{\geq}0$	$3\overset{?}{\geq}0$
True	False	True

1. 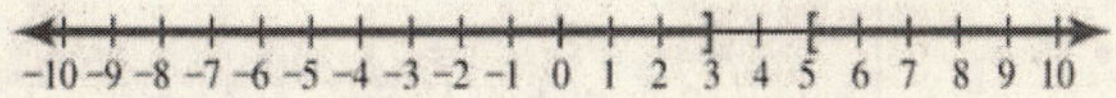

2. $(-\infty,3]$ or $[5,\infty)$

3. $x\leq 3$ or $x\geq 5$

25. $x^2-5x-10\leq 2x+8$

$x^2-7x-18\leq 0$

$x^2-7x-18=0$

$(x-9)(x+2)=0$

$x-9=0$ or $x+2=0$

$x=9$ or $x=-2$

Check the intervals:

$x=-3$	$x=0$	$x=10$
$(-3-9)(-3+2)\overset{?}{\leq}0$	$(0-9)(0+2)\overset{?}{\leq}0$	$(10-9)(10+2)\overset{?}{\leq}0$
$(-12)(-1)\overset{?}{\leq}0$	$(-9)(2)\overset{?}{\leq}0$	$(1)(12)\overset{?}{\leq}0$
$12\overset{?}{\leq}0$	$-18\overset{?}{\leq}0$	$12\overset{?}{\leq}0$
False	True	False

1. 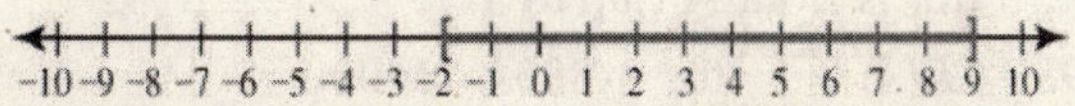

2. $[-2,9]$

3. $-2\leq x\leq 9$

27. $x^3+9x^2+18x>0$

$x(x^2+9x+18)>0$

$x(x^2+9x+18)=0$

$x(x+6)(x+3)=0$

$x=0$ or $x+6=0$ or $x+3=0$

$x=0$ or $x=-6$ or $x=-3$

Check the intervals:

$x=-7$	$x=-5$
$-7(-7+6)(-7+3)\overset{?}{>}0$	$-5(-5+6)(-5+3)\overset{?}{>}0$
$-7(-1)(-4)\overset{?}{>}0$	$-5(1)(-2)\overset{?}{>}0$
$-28\overset{?}{>}0$	$10\overset{?}{>}0$
False	True

$x=-1$	$x=1$
$-1(-1+6)(-1+3)\overset{?}{>}0$	$1(1+6)(1+3)\overset{?}{>}0$
$-1(5)(2)\overset{?}{>}0$	$1(7)(4)\overset{?}{>}0$
$-10\overset{?}{>}0$	$28\overset{?}{>}0$
False	True

1. 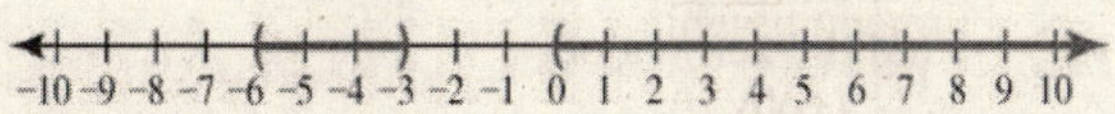

2. $(-6,-3)$ or $(0,\infty)$

3. $-6<x<-3$ or $x>0$

29. $2x^3+6x^2-20x<0$

$2x(x^2+3x-10)<0$

$2x(x^2+3x-10)=0$

$2x(x+5)(x-2)=0$

$2x=0$ or $x+5=0$ or $x-2=0$

$x=0$ or $x=-5$ or $x=2$

Check the intervals:

$x=-7$	$x=-1$
$2(-7)(-7+5)(-7-2)\overset{?}{<}0$	$2(-1)(-1+5)(-1-2)\overset{?}{<}0$
$-14(-2)(-9)\overset{?}{<}0$	$-1(4)(-3)\overset{?}{<}0$
$-252\overset{?}{<}0$	$12\overset{?}{<}0$
True	False

$x=1$	$x=3$
$2(1)(1+5)(1-2)\overset{?}{<}0$	$2(3)(3+5)(3-2)\overset{?}{<}0$
$2(6)(-1)\overset{?}{<}0$	$6(8)(1)\overset{?}{<}0$
$-12\overset{?}{<}0$	$48\overset{?}{<}0$
True	False

1. 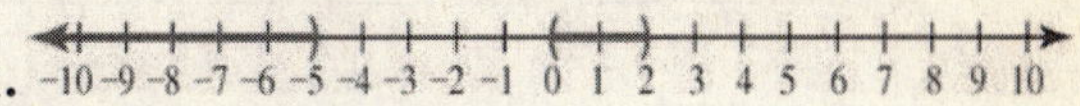

2. $(-\infty,-5)$ or $(0,2)$

3. $x<-5$ or $0<x<2$

31. $2x^2+11x+12\geq 0$

$2x^2+11x+12=0$

$a\cdot c=2\cdot 12=24$ Factors of 24 that add to 11 are 8 and 3. Factor by grouping.

$2x^2+8x+3x+12=0$

$(2x^2+8x)+(3x+12)=0$

$2x(x+4)+3(x+4)=0$

$(x+4)(2x+3)=0$

$x+4=0$ or $2x+3=0$

$x=-4$ or $2x=-3$

$x=-\frac{3}{2}$

Check the intervals:

$x=-5$	$x=-2$	$x=0$
$(-5+4)[2(-5)+3]\overset{?}{\geq}0$	$(-2+4)[2(-2)+3]\overset{?}{\geq}0$	$(0+4)[2(0)+3]\overset{?}{\geq}0$
$(-1)(-7)\overset{?}{\geq}0$	$(2)(-1)\overset{?}{\geq}0$	$(4)(3)\overset{?}{\geq}0$
$7\overset{?}{\geq}0$	$-2\overset{?}{\geq}0$	$12\overset{?}{\geq}0$
True	False	True

1. 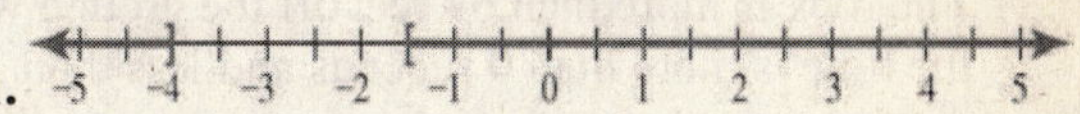

2. $(-\infty,-4]$ or $[-\frac{3}{2},\infty)$

3. $x\leq -4$ or $x\geq -\frac{3}{2}$

33. $6x^2+13x+5\le 0$

$6x^2+13x+5=0$

$a\cdot c=6\cdot 5=30$ Factors of 30 that add to 13 are 10 and 3. Factor by grouping.

$6x^2+10x+3x+5=0$

$(6x^2+10x)+(3x+5)=0$

$2x(3x+5)+(3x+5)=0$

$(3x+5)(2x+1)=0$

$3x+5=0$ or $2x+1=0$

$3x=-5$ or $2x=-1$

$x=-\dfrac{5}{3}$ or $x=-\dfrac{1}{2}$

Check the intervals:

$x=-2$: $[3(-2)+5][2(-2)+1]\overset{?}{\le}0$; $(-1)(-3)\overset{?}{\le}0$; $15\overset{?}{\le}0$; False

$x=-1$: $[3(-1)+5][2(-1)+1]\overset{?}{\le}0$; $(2)(-1)\overset{?}{\le}0$; $-15\overset{?}{\le}0$; True

$x=0$: $[3(0)+5][2(0)+1]\overset{?}{\le}0$; $(5)(1)\overset{?}{\le}0$; $6\overset{?}{\le}0$; False

1. 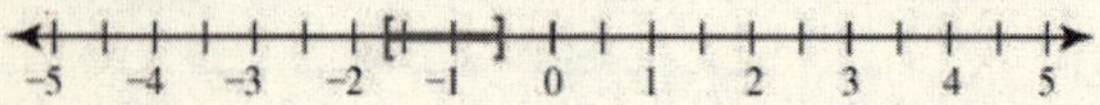

2. $\left[-\dfrac{5}{3},-\dfrac{1}{2}\right]$

3. $-\dfrac{5}{3}\le x\le -\dfrac{1}{2}$

35. $h(t)=-16t^2+48t+64$

$h(t)>64$

$-16t^2+48t+64>64$

$-16t^2+48t>0$

$-16t^2+48t=0$

$-16t(t-3)=0$

$-16t=0$ or $t-3=0$

$t=0$ or $t=3$

Check the intervals:

$t=-1$: $-16(-1)(-1-3)\overset{?}{>}0$; $16(-4)\overset{?}{>}0$; $-64\overset{?}{>}0$; False

$t=1$: $-16(1)(1-3)\overset{?}{>}0$; $-16(-2)\overset{?}{>}0$; $32\overset{?}{>}0$; True

$t=4$: $-16(4)(4-3)\overset{?}{>}0$; $-64(1)\overset{?}{>}0$; $-64\overset{?}{>}0$; False

$0<t<3$

The rock is more than 64 feet off the ground when the time is more than 0 seconds and less than 3 seconds after being thrown.

37. $P(n)=-400n^2+3200n+4800$

$P(n)\ge 8000$

$-400n^2+3200n+4800\ge 8000$

$-400n^2+3200n-3200\ge 0$

$-400n^2+3200n-3200=0$

$-400(n^2-8n+8)=0$

This cannot be factored further. Use the quadratic formula:

$n=\dfrac{8\pm\sqrt{(-8)^2-4(1)(8)}}{2(1)}$

$n=\dfrac{8\pm\sqrt{64-32}}{2}$

$n=\dfrac{8\pm\sqrt{32}}{2}=\dfrac{8\pm4\sqrt{2}}{2}=4\pm2\sqrt{2}$

Approximate the values of n to check the intervals.

$n=4\pm2\sqrt{2}$

$n\approx 6.83$ or $n\approx 1.17$

$n=1$: $-400[(1)^2-8(1)+8]\overset{?}{\ge}0$; $-400(1)\overset{?}{\ge}0$; $-400\overset{?}{\ge}0$; False

$n=3$: $-400[(3)^2-8(3)+8]\overset{?}{\ge}0$; $-400(-7)\overset{?}{\ge}0$; $2800\overset{?}{\ge}0$; True

$n=7$: $-400[(7)^2-8(7)+8]\overset{?}{\ge}0$; $-400(1)\overset{?}{\ge}0$; $-400\overset{?}{\ge}0$; False

$1.17<n<6.83$

Only whole numbers will work in the context of the problem. The price can be raised by \$1.50 above the original price at least 2 times, but no more than 6 times. The price range the club should sell the cards for is from \$11 up to \$17.

39. $P(s)=-40s^2+880s-1600$

$P(s)\ge 0$

$-40s^2+880s-1600\ge 0$

$-40s^2+880s-1600=0$

$-40(s^2-22+40)=0$

$-40(s-20)(s-2)=0$

$s-20=0$ or $s-2=0$

$s=20$ or $s=2$

Check the intervals:

$s=0$: $-40[(0-20)(0-2)]\overset{?}{\ge}0$; $-40(40)\overset{?}{\ge}0$; $-1600\overset{?}{\ge}0$; False

$s=4$: $-40[(4-20)(4-2)]\overset{?}{\ge}0$; $-40(-32)\overset{?}{\ge}0$; $1280\overset{?}{\ge}0$; True

$s=21$: $-40[(21-20)(21-2)]\overset{?}{\ge}0$; $-40(19)\overset{?}{\ge}0$; $-760\overset{?}{\ge}0$; False

$2\le s\le 20$

The company will break even or make a profit when they sell at least 2 surfboards and no more than 20 surfboards.

41. $\dfrac{x+5}{x+2}>0$

Find the zero value:

$x+5=0$

$x=-5$

Find the excluded value:

$x+2=0$

$x=-2$

Check the intervals:

$x=-7$	$x=-3$	$x=0$
$\dfrac{-7+5}{-7+2}\overset{?}{>}0$	$\dfrac{-3+5}{-3+2}\overset{?}{>}0$	$\dfrac{0+5}{0+2}\overset{?}{>}0$
$\dfrac{-2}{-5}\overset{?}{>}0$	$\dfrac{2}{-1}\overset{?}{>}0$	$\dfrac{5}{2}\overset{?}{>}0$
$\dfrac{2}{5}\overset{?}{>}0$	$-2\overset{?}{>}0$	True
True	False	

1. 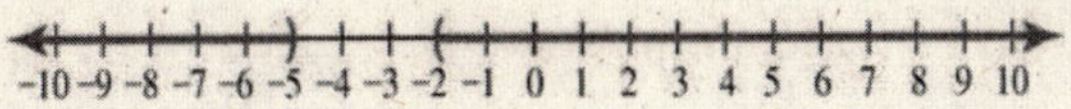

2. $(-\infty,-5)$ or $(-2,\infty)$

3. $x<-5$ or $x>-2$

43. $\dfrac{x-7}{x-3}<0$

Find the zero value:

$x-7=0$

$x=7$

Find the excluded value:

$x-3=0$

$x=3$

Check the intervals:

$x=0$	$x=5$	$x=9$
$\dfrac{0-7}{0-3}\overset{?}{<}0$	$\dfrac{5-7}{5-3}\overset{?}{<}0$	$\dfrac{9-7}{9-3}\overset{?}{<}0$
$\dfrac{-7}{-3}\overset{?}{<}0$	$\dfrac{-2}{2}\overset{?}{<}0$	$\dfrac{2}{6}\overset{?}{<}0$
$\dfrac{7}{3}\overset{?}{<}0$	$-1\overset{?}{<}0$	$\dfrac{1}{3}\overset{?}{<}0$
False	True	False

1. 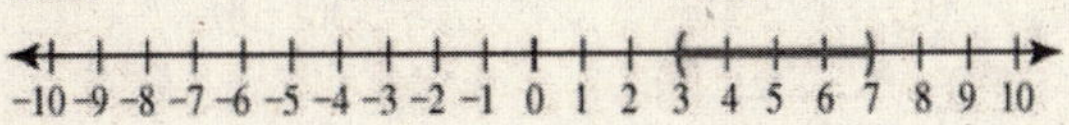

2. $(3,7)$

3. $3<x<7$

45. $\dfrac{x-1}{x+4}\leq 0$

Find the zero value:

$x-1=0$

$x=1$

Find the excluded value:

$x+4=0$

$x=-4$

Check the intervals:

$x=-5$	$x=0$	$x=2$
$\dfrac{-5-1}{-5+4}\overset{?}{\leq}0$	$\dfrac{0-1}{0+4}\overset{?}{\leq}0$	$\dfrac{2-1}{2+4}\overset{?}{\leq}0$
$\dfrac{-6}{-1}\overset{?}{\leq}0$	$\dfrac{-1}{4}\overset{?}{\leq}0$	$\dfrac{1}{6}\overset{?}{\leq}0$
$6\overset{?}{\leq}0$	True	False
False		

1. 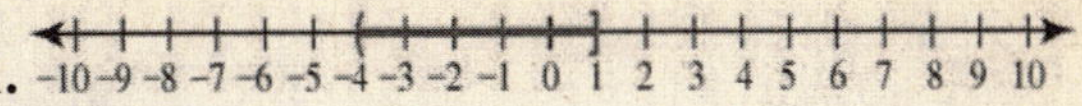

2. $(-4,1]$

3. $-4<x\leq 1$

47. $\dfrac{x-4}{x-9}\geq 0$

Find the zero value:

$x-4=0$

$x=4$

Find the excluded value:

$x-9=0$

$x=9$

Check the intervals:

$x=3$	$x=5$	$x=10$
$\dfrac{3-4}{3-9}\overset{?}{\geq}0$	$\dfrac{5-4}{5-9}\overset{?}{\geq}0$	$\dfrac{10-4}{10-9}\overset{?}{\geq}0$
$\dfrac{-1}{-6}\overset{?}{\geq}0$	$\dfrac{1}{-4}\overset{?}{\geq}0$	$\dfrac{6}{1}\overset{?}{\geq}0$
$\dfrac{1}{6}\overset{?}{\geq}0$	$-\dfrac{1}{4}\overset{?}{\geq}0$	$6\overset{?}{\geq}0$
True	False	True

1. 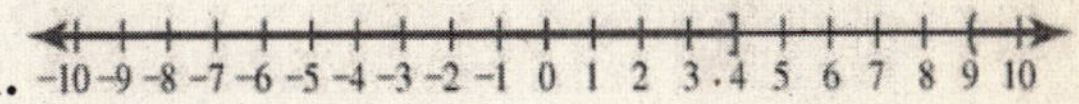

2. $(-\infty,4]$ or $(9,\infty)$

3. $x\leq 4$ or $x>9$

49. Henry chose the wrong interval and should have included 1 in the solution set.

$\dfrac{x-1}{x-7}\leq 0$

Find the zero value:

$x-1=0$

$x=1$

Find the excluded value:

$x-7=0$

$x=7$

Check the intervals:

$x=0$	$x=3$	$x=8$
$\dfrac{0-1}{0-7}\overset{?}{\leq}0$	$\dfrac{3-1}{3-7}\overset{?}{\leq}0$	$\dfrac{8-1}{8-7}\overset{?}{\leq}0$
$\dfrac{-1}{-7}\overset{?}{\leq}0$	$\dfrac{2}{-4}\overset{?}{\leq}0$	$\dfrac{7}{1}\overset{?}{\leq}0$
$\dfrac{1}{7}\overset{?}{\leq}0$	$-\dfrac{1}{2}\overset{?}{\leq}0$	$7\overset{?}{\leq}0$
False	True	False

$[1,7)$

51. Jennie should not have included -3 in the solution set.

$\dfrac{x-5}{x+3}\leq 0$

Find the zero value:

$x-5=0$

$x=5$

Find the excluded value:
$x+3=0$
$x=-3$

Check the intervals:

$x=-4$: $\frac{-4-5}{-4+3}\overset{?}{\le}0$, $\frac{-9}{-1}\overset{?}{\le}0$, $9\overset{?}{\le}0$ False

$x=0$: $\frac{0-5}{0+3}\overset{?}{\le}0$, $\frac{-5}{3}\overset{?}{\le}0$ True

$x=6$: $\frac{6-5}{6+3}\overset{?}{\le}0$, $\frac{1}{9}\overset{?}{\le}0$ False

$(-3,5]$

53. $$\frac{x-5}{(x+3)(x+7)}>0$$

Find the zero value:
$x-5=0$
$x=5$

Find the excluded values:
$x+3=0$ or $x+7=0$
$x=-3$ or $x=-7$

Check the intervals:

$x=-8$: $\frac{-8-5}{(-8+3)(-8+7)}\overset{?}{>}0$, $\frac{-13}{(-5)(-1)}\overset{?}{>}0$, $\frac{-13}{5}\overset{?}{>}0$ False

$x=-5$: $\frac{-5-5}{(-5+3)(-5+7)}\overset{?}{>}0$, $\frac{-10}{(-2)(2)}\overset{?}{>}0$, $\frac{5}{2}\overset{?}{>}0$ True

$x=0$: $\frac{0-5}{(0+3)(0+7)}\overset{?}{>}0$, $\frac{-5}{(3)(7)}\overset{?}{>}0$, $\frac{-5}{21}\overset{?}{>}0$ False

$x=6$: $\frac{6-5}{(6+3)(6+7)}\overset{?}{>}0$, $\frac{1}{(9)(13)}\overset{?}{>}0$, $\frac{1}{117}\overset{?}{>}0$ True

1. 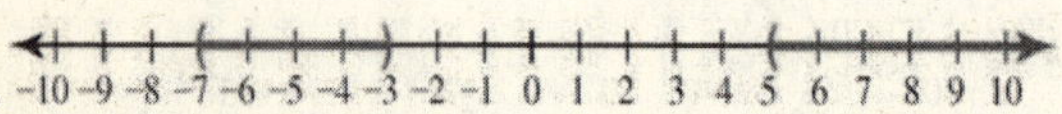

2. $(-7,-3)$ or $(5,\infty)$

3. $-7<x<-3$ or $x>5$

55. $$\frac{x+2}{(x-4)(x+5)}<0$$

Find the zero value:
$x+2=0$
$x=-2$

Find the excluded values:
$x-4=0$ or $x+5=0$
$x=4$ or $x=-5$

Check the intervals:

$x=-8$: $\frac{-8+2}{(-8-4)(-8+5)}\overset{?}{<}0$, $\frac{-6}{(-12)(-3)}\overset{?}{<}0$, $\frac{-1}{6}\overset{?}{<}0$ True

$x=-4$: $\frac{-4+2}{(-4-4)(-4+5)}\overset{?}{<}0$, $\frac{-2}{(-8)(1)}\overset{?}{<}0$, $\frac{1}{4}\overset{?}{<}0$ False

$x=0$: $\frac{0+2}{(0-4)(0+5)}\overset{?}{<}0$, $\frac{2}{(-4)(5)}\overset{?}{<}0$, $\frac{-1}{10}\overset{?}{<}0$ True

$x=6$: $\frac{6+2}{(6-4)(6+5)}\overset{?}{<}0$, $\frac{8}{(2)(11)}\overset{?}{<}0$, $\frac{4}{11}\overset{?}{<}0$ False

1. 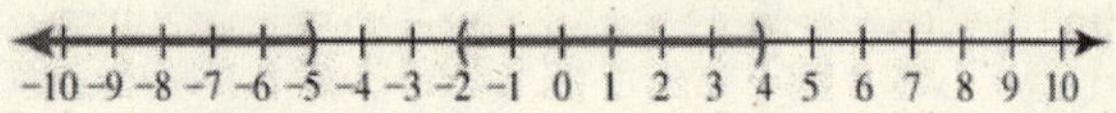

2. $(-\infty,-5)$ or $(-2,4)$

3. $x<-5$ or $-2<x<4$

57. $$\frac{x-4}{(x-2)(x-8)}\le 0$$

Find the zero value:
$x-4=0$
$x=4$

Find the excluded values:
$x-2=0$ or $x-8=0$
$x=2$ or $x=8$

Check the intervals:

$x=0$: $\frac{0-4}{(0-2)(0-8)}\overset{?}{\le}0$, $\frac{-4}{(-2)(-8)}\overset{?}{\le}0$, $\frac{-1}{4}\overset{?}{\le}0$ True

$x=3$: $\frac{3-4}{(3-2)(3-8)}\overset{?}{\le}0$, $\frac{-1}{(1)(-5)}\overset{?}{\le}0$, $\frac{1}{5}\overset{?}{\le}0$ False

$x=6$: $\frac{6-4}{(6-2)(6-8)}\overset{?}{\le}0$, $\frac{2}{(4)(-2)}\overset{?}{\le}0$, $\frac{-1}{4}\overset{?}{\le}0$ True

$x=9$: $\frac{9-4}{(9-2)(9-8)}\overset{?}{\le}0$, $\frac{5}{(7)(1)}\overset{?}{\le}0$, $\frac{5}{7}\overset{?}{\le}0$ False

1. 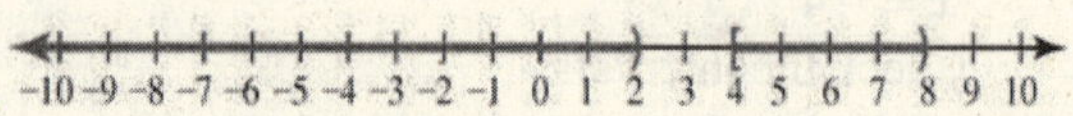

2. $(-\infty,2)$ or $[4,8)$

3. $x<2$ or $4\le x<8$

59. $$\frac{x}{x-4}\ge 2$$

$$\frac{x}{x-4}-2\ge 0$$

$$\frac{x}{x-4}-2\cdot\left(\frac{x-4}{x-4}\right)\geq 0$$

$$\frac{x}{x-4}+\frac{-2x+8}{x-4}\geq 0$$

$$\frac{-x+8}{x-4}\geq 0$$

Find the zero value:

$-x+8=0$

$-x=-8$

$x=8$

Find the excluded value:

$x-4=0$

$x=4$

Check the intervals:

$x=0$	$x=5$	$x=10$
$\frac{-(0)+8}{0-4}\overset{?}{\geq}0$	$\frac{-(5)+8}{5-4}\overset{?}{\geq}0$	$\frac{-(10)+8}{10-4}\overset{?}{\geq}0$
$\frac{8}{-4}\overset{?}{\geq}0$	$\frac{3}{1}\overset{?}{\geq}0$	$\frac{-2}{6}\overset{?}{\geq}0$
$-2\overset{?}{\geq}0$	$3\overset{?}{\geq}0$	$\frac{-1}{3}\overset{?}{\geq}0$
False	True	False

1. 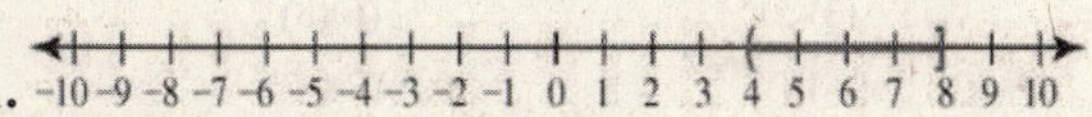

2. $(4,8]$

3. $4<x\leq 8$

61. $$\frac{x}{x+3}<1$$

$$\frac{x}{x+3}-1<0$$

$$\frac{x}{x+3}-1\cdot\left(\frac{x+3}{x+3}\right)<0$$

$$\frac{x}{x+3}+\frac{-x-3}{x+3}<0$$

$$\frac{-3}{x+3}<0$$

There is no variable in the numerator so there is no zero value.

Find the excluded value:

$x+3=0$

$x=-3$

Check the intervals:

$x=-5$	$x=0$
$\frac{-3}{-5+3}\overset{?}{<}0$	$\frac{-3}{0+3}\overset{?}{<}0$
$\frac{-3}{-2}\overset{?}{<}0$	$\frac{-3}{3}\overset{?}{<}0$
$\frac{3}{2}\overset{?}{<}0$	$-1\overset{?}{<}0$
False	True

1. 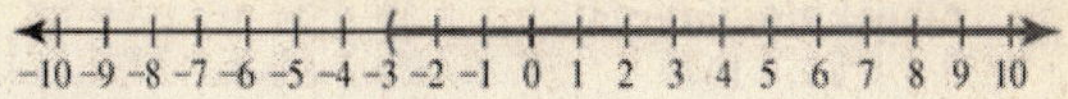

2. $(-3,\infty)$

3. $x>-3$

63. $$\frac{x}{x+3}>2$$

$$\frac{x}{x+3}-2>0$$

$$\frac{x}{x+3}-2\cdot\left(\frac{x+3}{x+3}\right)>0$$

$$\frac{x}{x+3}+\frac{-2x-6}{x+3}>0$$

$$\frac{-x-6}{x+3}>0$$

Find the zero value:

$-x-6=0$

$-x=6$

$x=-6$

Find the excluded value:

$x+3=0$

$x=-3$

Check the intervals:

$x=-7$	$x=-5$	$x=0$
$\frac{-(-7)-6}{-7+3}\overset{?}{>}0$	$\frac{-(-5)-6}{-5+3}\overset{?}{>}0$	$\frac{-(0)-6}{0+3}\overset{?}{>}0$
$\frac{1}{-4}\overset{?}{>}0$	$\frac{-1}{-2}\overset{?}{>}0$	$\frac{-6}{3}\overset{?}{>}0$
$\frac{-1}{4}\overset{?}{>}0$	$\frac{1}{2}\overset{?}{>}0$	$-2\overset{?}{>}0$
False	True	False

1. 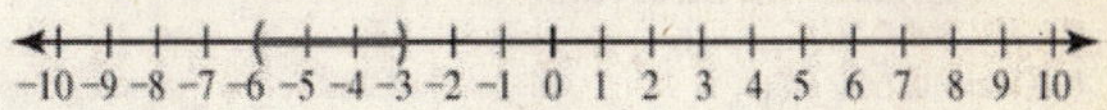

2. $(-6,-3)$

3. $-6<x<-3$

65. $$\frac{x+2}{x^2+4x-45}<0$$

$$\frac{x+2}{(x+9)(x-5)}<0$$

Find the zero value:

$x+2=0$

$x=-2$

Find the excluded values:

$x+9=0$ or $x-5=0$

$x=-9$ or $x=5$

Check the intervals:

$x=-10$

$\frac{-10+2}{(-10+9)(-10-5)}\overset{?}{<}0$

$\frac{-8}{(-1)(-15)}\overset{?}{<}0$

$\frac{-8}{15}\overset{?}{<}0$

True

$x=-4$

$\frac{-4+2}{(-4+9)(-4-5)}\overset{?}{<}0$

$\frac{-2}{(5)(-9)}\overset{?}{<}0$

$\frac{2}{45}\overset{?}{<}0$

False

$x=0$

$\frac{0+2}{(0+9)(0-5)}\overset{?}{<}0$

$\frac{2}{(9)(-5)}\overset{?}{<}0$

$\frac{-2}{45}\overset{?}{<}0$

True

$x=6$

$\frac{6+2}{(6+9)(6-5)}\overset{?}{<}0$

$\frac{8}{(15)(1)}\overset{?}{<}0$

$\frac{8}{15}\overset{?}{<}0$

False

1. 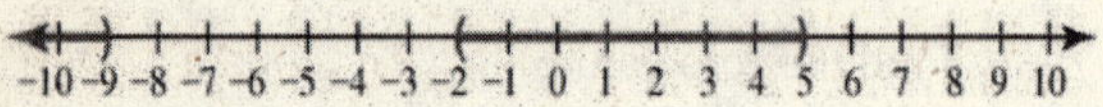

2. $(-\infty,-9)$ or $(-2,5)$

3. $x<-9$ or $-2<x<5$

67. $\frac{x+5}{x^2+12x+32}>0$

$\frac{x+5}{(x+8)(x+4)}>0$

Find the zero value:

$x+5=0$

$x=-5$

Find the excluded values:

$x+8=0$ or $x+4=0$

$x=-8$ or $x=-4$

Check the intervals:

$x=-10$

$\frac{-10+5}{(-10+8)(-10+4)}\overset{?}{>}0$

$\frac{-5}{(-2)(-6)}\overset{?}{>}0$

$\frac{-5}{12}\overset{?}{>}0$

False

$x=-7$

$\frac{-7+5}{(-7+8)(-7+4)}\overset{?}{>}0$

$\frac{-2}{(1)(-3)}\overset{?}{>}0$

$\frac{2}{3}\overset{?}{>}0$

True

$x=-4.5$

$\frac{-4.5+5}{(-4.5+8)(-4.5+4)}\overset{?}{>}0$

$\frac{0.5}{(3.5)(-0.5)}\overset{?}{>}0$

$\frac{-10}{35}\overset{?}{>}0$

False

$x=0$

$\frac{0+5}{(0+8)(0+4)}\overset{?}{>}0$

$\frac{5}{(8)(4)}\overset{?}{>}0$

$\frac{5}{32}\overset{?}{>}0$

True

1. 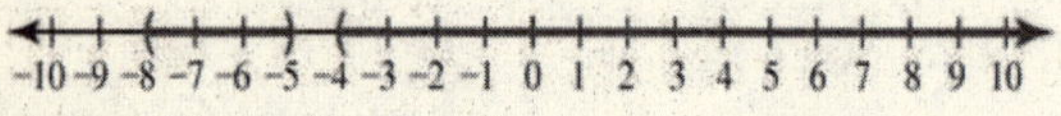

2. $(-8,-5)$ or $(-4,\infty)$

3. $-8<x<-5$ or $x>-4$

69. $\frac{x^2-1}{x^2-8x+15}\le 0$

$\frac{(x-1)(x+1)}{(x-5)(x-3)}\le 0$

Find the zero values:

$x-1=0$ or $x+1=0$

$x=1$ or $x=-1$

Find the excluded values:

$x-5=0$ or $x-3=0$

$x=5$ or $x=3$

Check the intervals:

$x=-3$

$\frac{(-3-1)(-3+1)}{(-3-5)(-3-3)}\overset{?}{\le}0$

$\frac{(-4)(-2)}{(-8)(-6)}\overset{?}{\le}0$

$\frac{1}{6}\overset{?}{\le}0$

False

$x=0$

$\frac{(0-1)(0+1)}{(0-5)(0-3)}\overset{?}{\le}0$

$\frac{(-1)(1)}{(-5)(-3)}\overset{?}{\le}0$

$\frac{-1}{15}\overset{?}{\le}0$

True

$x=2$

$\frac{(2-1)(2+1)}{(2-5)(2-3)}\overset{?}{\le}0$

$\frac{(1)(3)}{(-3)(-1)}\overset{?}{\le}0$

$1\overset{?}{\le}0$

False

$x=4$

$\frac{(4-1)(4+1)}{(4-5)(4-3)}\overset{?}{\le}0$

$\frac{(3)(5)}{(-1)(1)}\overset{?}{\le}0$

$-15\overset{?}{\le}0$

True

$x=7$

$\frac{(7-1)(7+1)}{(7-5)(7-3)}\overset{?}{\le}0$

$\frac{(6)(8)}{(2)(4)}\overset{?}{\le}0$

$6\overset{?}{\le}0$

False

1. 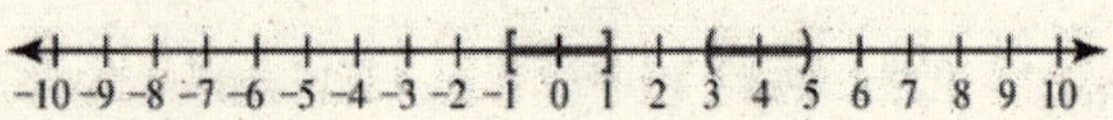

2. $[-1,1]$ or $(3,5)$

3. $-1\le x\le 1$ or $3<x<5$

71. $\frac{x^2-36}{x^2+3x+2}\ge 0$

$\frac{(x-6)(x+6)}{(x+2)(x+1)}\ge 0$

Find the zero values:

$x-6=0$ or $x+6=0$

$x=6$ or $x=-6$

Find the excluded values:

$x+2=0$ or $x+1=0$

$x=-2$ or $x=-1$

Check the intervals:

$x=-7$

$\frac{(-7-6)(-7+6)}{(-7+2)(-7+1)}\overset{?}{\ge}0$

$\frac{(-13)(-1)}{(-5)(-6)}\overset{?}{\ge}0$

$\frac{13}{30}\overset{?}{\ge}0$

True

$x=-3$

$\frac{(-3-6)(-3+6)}{(-3+2)(-3+1)}\overset{?}{\ge}0$

$\frac{(-9)(3)}{(-1)(-2)}\overset{?}{\ge}0$

$\frac{-27}{2}\overset{?}{\ge}0$

False

$x=-1.5$	$x=2$	$x=7$
$\frac{(-1.5-6)(-1.5+6)}{(-1.5+2)(-1.5+1)}\overset{?}{\geq}0$	$\frac{(2-6)(2+6)}{(2+2)(2+1)}\overset{?}{\geq}0$	$\frac{(9-6)(9+6)}{(9+2)(9+1)}\overset{?}{\geq}0$
$\frac{(-7.5)(4.5)}{(0.5)(-0.5)}\overset{?}{\geq}0$	$\frac{(-4)(8)}{(4)(3)}\overset{?}{\geq}0$	$\frac{(3)(15)}{(11)(10)}\overset{?}{\geq}0$
$135\overset{?}{\geq}0$	$\frac{-8}{3}\overset{?}{\geq}0$	$\frac{9}{22}\overset{?}{\geq}0$
True	False	True

1. 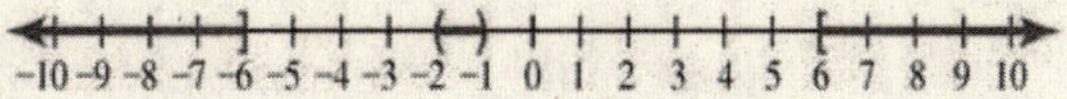

2. $(-\infty,-6]$ or $(-2,-1)$ or $[6,\infty)$

3. $x\leq -6$ or $-2<x<-1$ or $x\geq 6$

73. $$\frac{x^2-49}{x^2-14x+45}<0$$

$$\frac{(x-7)(x+7)}{(x-9)(x-5)}<0$$

Find the zero values:

$x-7=0$ or $x+7=0$

$x=7$ or $x=-7$

Find the excluded values:

$x-9=0$ or $x-5=0$

$x=9$ or $x=5$

Check the intervals:

$x=-8$	$x=0$	$x=6$
$\frac{(-8-7)(-8+7)}{(-8-9)(-8-5)}\overset{?}{<}0$	$\frac{(0-7)(0+7)}{(0-9)(0-5)}\overset{?}{<}0$	$\frac{(6-7)(6+7)}{(6-9)(6-5)}\overset{?}{<}0$
$\frac{(-15)(-1)}{(-17)(-13)}\overset{?}{<}0$	$\frac{(-7)(7)}{(-9)(-5)}\overset{?}{<}0$	$\frac{(-1)(13)}{(-3)(1)}\overset{?}{<}0$
$\frac{15}{221}\overset{?}{<}0$	$\frac{-49}{45}\overset{?}{<}0$	$\frac{13}{3}\overset{?}{<}0$
False	True	False

$x=8$	$x=11$
$\frac{(8-7)(8+7)}{(8-9)(8-5)}\overset{?}{<}0$	$\frac{(11-7)(11+7)}{(11-9)(11-5)}\overset{?}{<}0$
$\frac{(1)(15)}{(-1)(3)}\overset{?}{<}0$	$\frac{(4)(18)}{(2)(6)}\overset{?}{<}0$
$-5\overset{?}{<}0$	$6\overset{?}{<}0$
True	False

1. 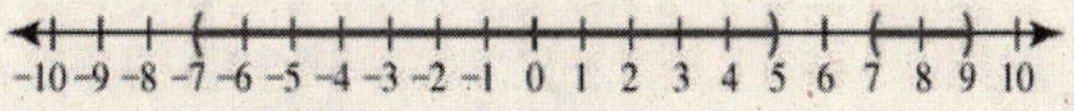

2. $(-7,5)$ or $(7,9)$

3. $-7<x<5$ or $7<x<9$

Section 9.4 Solving Systems of Linear Inequalities

1. A 61" tall person weighing 105 pounds falls into the normal range.

3. A 5'9"(69") tall person is considered overweight if he weighs above 172 pounds.

5. A person who weighs 150 pounds must be between 5'5" and 6'3" to fall in the normal range.

7. A person who is 67" tall has a normal weight range of 120 pounds to 160 pounds.

9. They can build at most 370 mountain bikes per month.

11. Yes, they can build 400 cruisers and 100 mountain bikes per month. This is at the 500 bike per month limit and under the cost limitations for a month.

13. **a.** Let C be the total number of cruisers Bicycles Galore produces monthly and let M be the total number of mountain bikes Bicycles Galore produces monthly.

$C+M\leq 500$

$65C+120M\geq 40000$

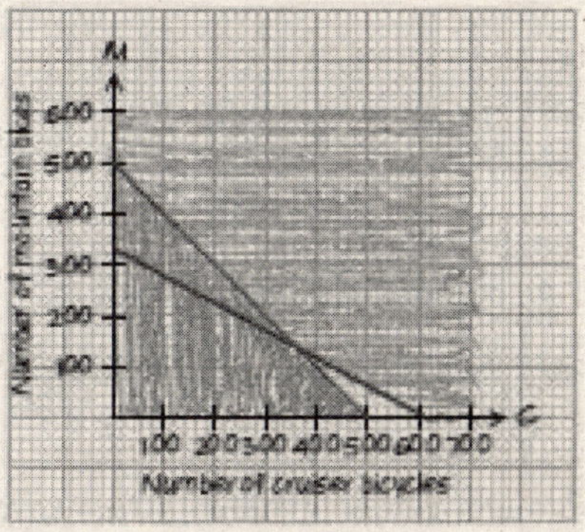

b. Bicycles Galore will be able to meet a demand by the board of directors for a profit of \$40,000 per month. Any combination of mountain bikes and cruisers in the overlapping shaded region will meet the board's demand.

c.

$$\begin{cases} C+M=500 \\ 65C+120M=40{,}000 \end{cases}$$

Multiply the first equation by -65 and add the result to the second equation.

$$(-65)(C+M)=(500)(-65)$$

$$-65C-65M=-32{,}500$$

$$\left.\begin{aligned} 65C+120M&=40{,}000 \\ -65C-65M&=-32{,}500 \end{aligned}\right\}\text{ add}$$

$$55M=7500$$

$$\frac{55M}{55}=\frac{7500}{55}$$

$$M\approx 136.36$$

Now solve for C:

$$C=500-136.36$$

$$C\approx 363.64$$

They must make 137 mountain bikes and 363 cruisers to make $40,000 profit. If they make more mountain bikes they will make more profit.

15. $y > 3x + 2$

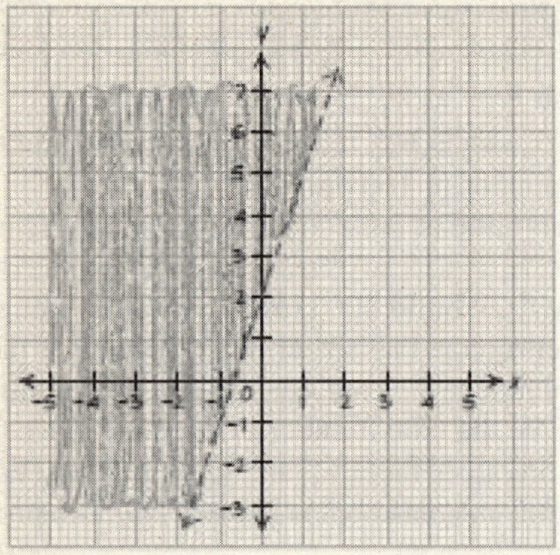

17. $y > \frac{5}{4}x - 2$

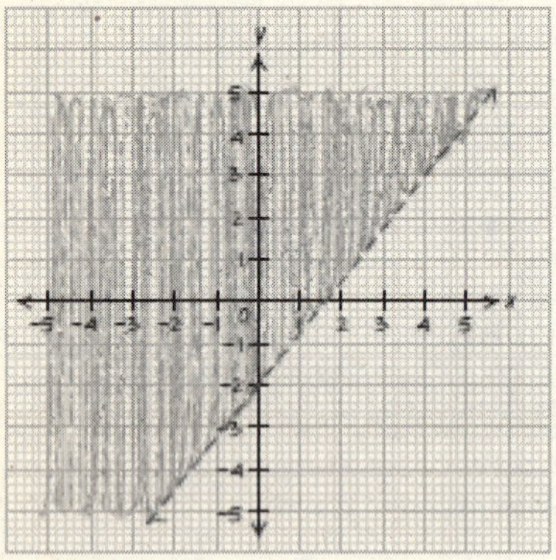

19. $y \geq \frac{2}{3}x + 6$

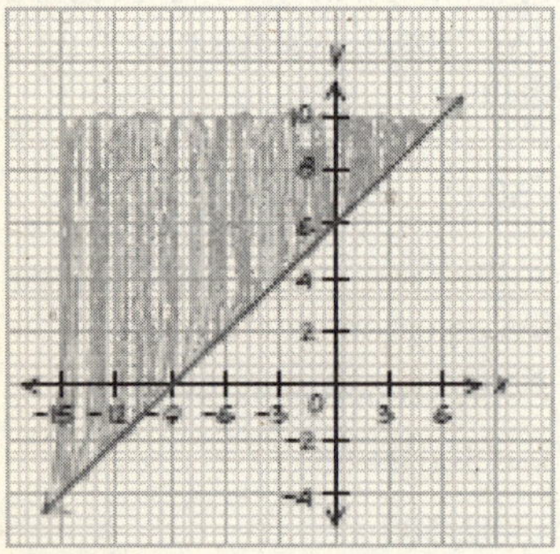

21. $y \leq -2x + 6$

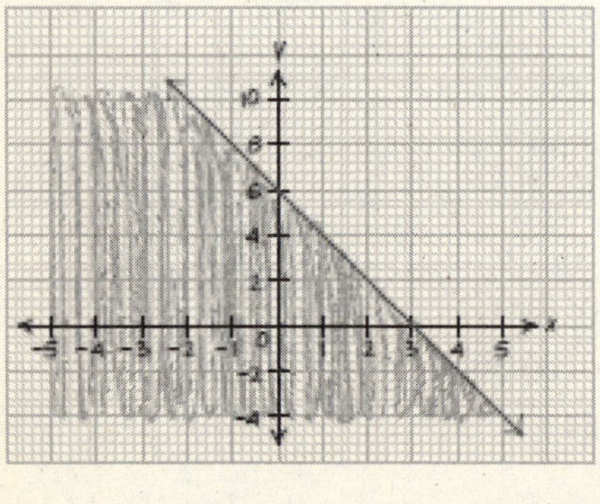

23.

$$2x + 5y > 10$$
$$2x + 5y - 2x > 10 - 2x$$
$$5y > -2x + 10$$
$$\frac{5y}{5} > \frac{-2x + 10}{5}$$
$$y > -\frac{2}{5}x + 2$$

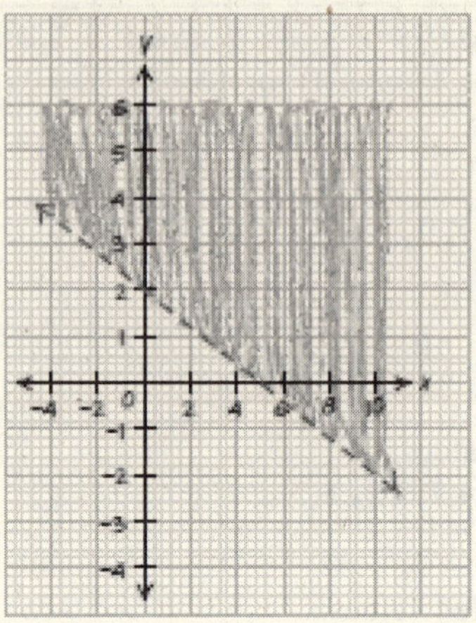

25.

$$12x - 4y > 8$$
$$12x - 4y - 12x > 8 - 12x$$
$$-4y > -12x + 8$$
$$\frac{-4y}{-4} < \frac{-12x + 8}{-4}$$
$$y < 3x - 2$$

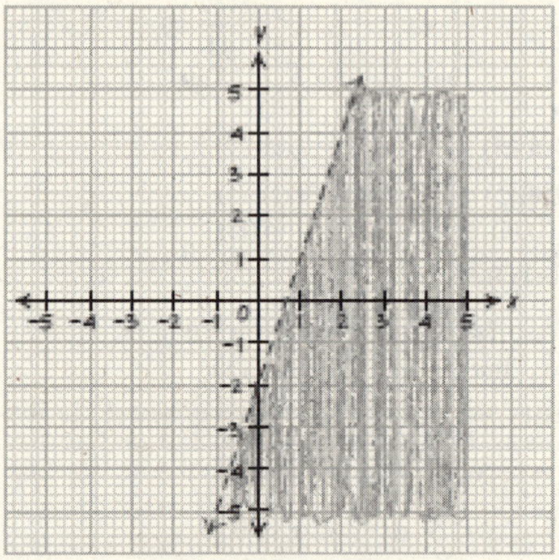

27. Using the points $(3, 3)$ and $(0, -3)$:

$$m = \frac{3 - (-3)}{3 - 0} = \frac{6}{3} = 2$$
$$b = -3$$

Since the solid line is shaded below use the $\leq$ symbol.

$$y \leq 2x - 3$$

29. Using the points $(4,0)$ and $(0,6)$:

$m=\dfrac{6-0}{0-4}=\dfrac{6}{-4}=-\dfrac{3}{2}$

$b=6$

Since the dashed line is shaded above use the $>$ symbol.

$y>-\dfrac{3}{2}x+6$

31. Using the points $(0,9)$ and $(6,6)$:

$m=\dfrac{9-6}{0-6}=\dfrac{3}{-6}=-\dfrac{1}{2}$

$b=9$

Since the dashed line is shaded above use the $>$ symbol.

$y>-\dfrac{1}{2}x+9$

33. Using the points $(1,0)$ and $(3,-2)$:

$m=\dfrac{-2-0}{3-1}=\dfrac{-2}{2}=-1$

Using $m=-1$ and $(1,0)$:

$0=-1(1)+b$

$0=-1+b$

$1=b$

Since the solid line is shaded below use the $\le$ symbol.

$y\le -x+1$

35. Let A be the amount (in dollars) Juanita invests in the account paying 4.5% interest. Let B be the amount (in dollars) invested in the account paying 3.75% interest.

$A+B\le 750{,}000$

$0.045A+0.0375B\ge 30{,}000$

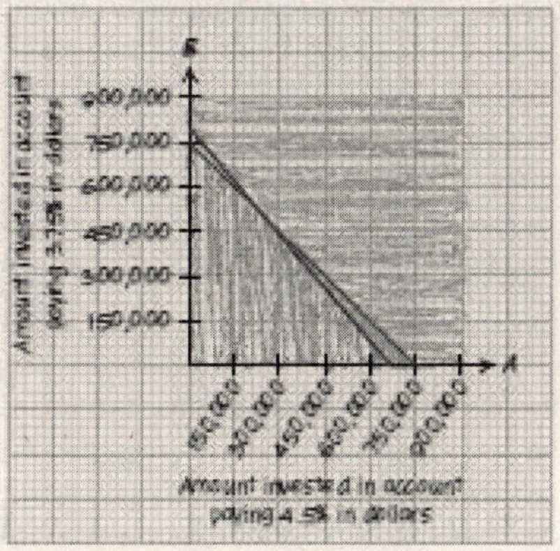

37. **a.** Let P be the number of power bars, and let D be the number of sports drinks that need to be consumed by the athletes.

$240P+300D\ge 2000$

$30P+70D\ge 350$

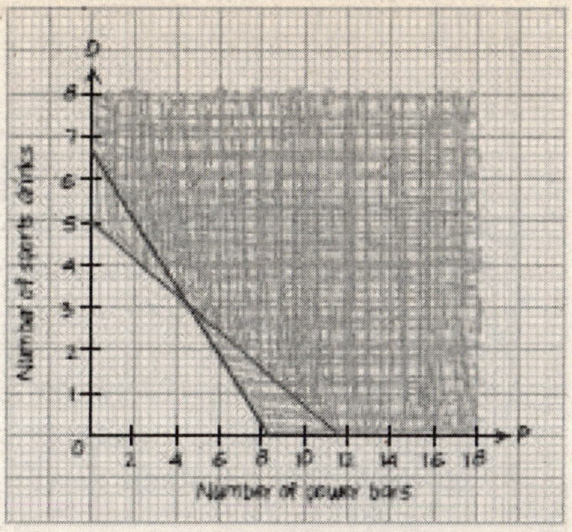

b. If a racer carries 4 drinks, he or she should eat 4 power bars.

b. According to the graph, the racers should eat 6 power bars if they can carry only 4 sports drinks.

39.

$\begin{cases} y>2x+4 \\ y<-3x+7 \end{cases}$

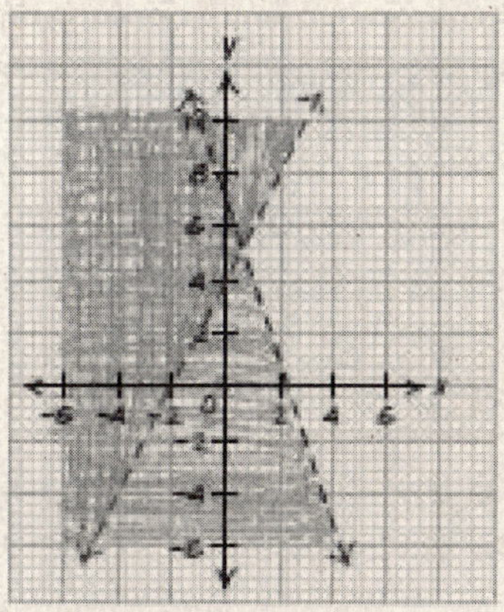

41.

$\begin{cases} y<\dfrac{2}{5}x-3 \\ y<-\dfrac{3}{4}x+6 \end{cases}$

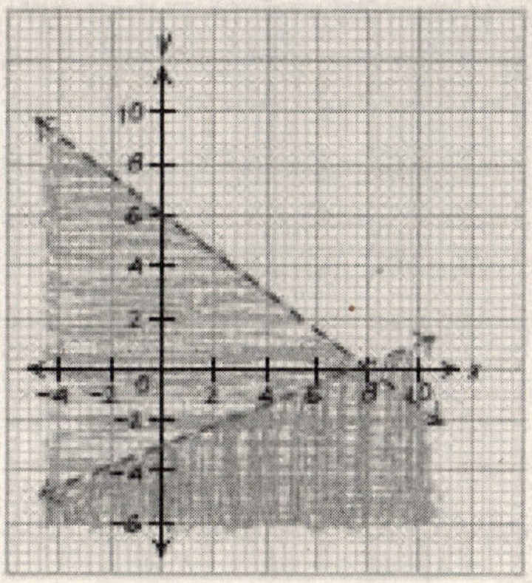

43.

$$\begin{cases} y \le 2x+5 \\ y \ge 2x+8 \end{cases}$$

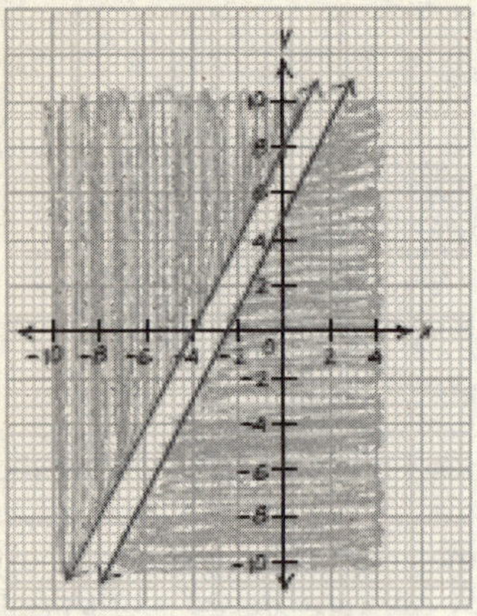

No solution because there is no overlap.

45.

$$\begin{cases} 2x+4y \le 5 \\ 2x-4y \le 5 \end{cases}$$

Solve for y in both equations to graph:

$2x+4y \le 5 \quad , \quad 2x-4y \le 5$

$4y \le -2x+5 \quad , \quad -4y \le -2x+5$

$\dfrac{4y}{4} \le \dfrac{-2x+5}{4} \quad , \quad \dfrac{-4y}{-4} \ge \dfrac{-2x+5}{-4}$

$y \le -\dfrac{1}{2}x+\dfrac{5}{4} \quad , \quad y \ge \dfrac{1}{2}x-\dfrac{5}{4}$

$$\begin{cases} y \le -\dfrac{1}{2}x+\dfrac{5}{4} \\ y \ge \dfrac{1}{2}x-\dfrac{5}{4} \end{cases}$$

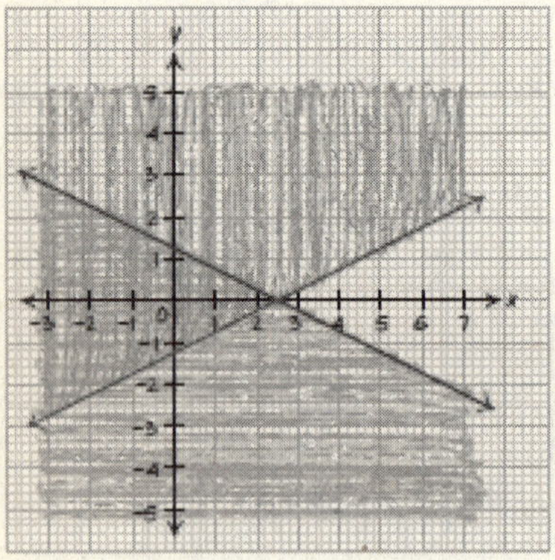

47.

$$\begin{cases} 5x+3y > 7 \\ 4x+2y < 10 \end{cases}$$

Solve for y in both equations to graph:

$5x+3y > 7 \quad , \quad 4x+2y < 10$

$3y > -5x+7 \quad , \quad 2y < -4x+10$

$\dfrac{3y}{3} > \dfrac{-5x+7}{3} \quad , \quad \dfrac{2y}{2} < \dfrac{-4x+10}{2}$

$y > -\dfrac{5}{3}x+\dfrac{7}{3} \quad , \quad y < -2x+5$

$$\begin{cases} y > -\dfrac{5}{3}x+\dfrac{7}{3} \\ y < -2x+5 \end{cases}$$

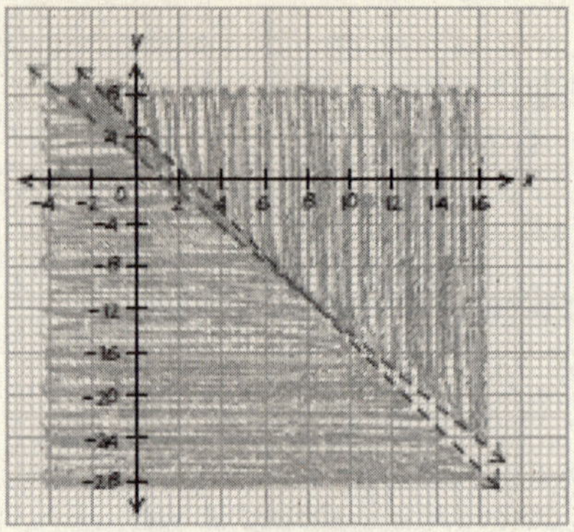

49.

$$\begin{cases} y > \dfrac{2}{3}x-12 \\ 6x-4y > 12 \end{cases}$$

Solve for y in the second equation to graph:

$6x-4y-6x > 12-6x$

$-4y > -6x+12$

$\dfrac{-4y}{-4} < \dfrac{-6x+12}{-4}$

$y < \dfrac{3}{2}x-3$

$$\begin{cases} y > \dfrac{2}{3}x-12 \\ y < \dfrac{3}{2}x-3 \end{cases}$$

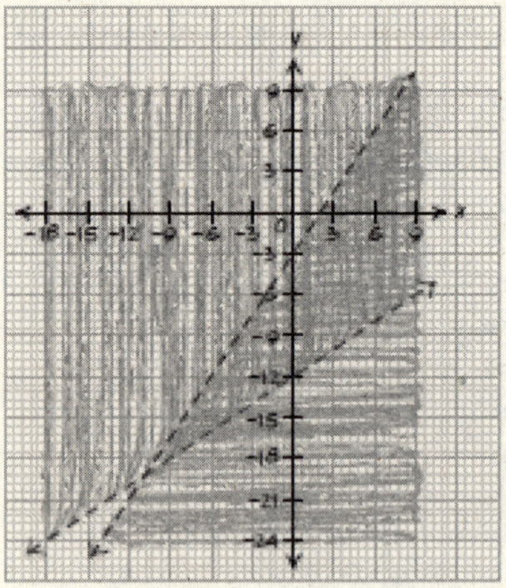

Chapter 9 Review

1.

$|x| = 5$

$x = 5$ or $x = -5$

$x = \pm 5$

2.

$|y| = 61$

$y = 61$ or $y = -61$

$y = \pm 61$

3.

$|a+6| = 7$

$a+6 = 7$ or $a+6 = -7$

$a = 1$ or $a = -13$

4.

$|x+19| = 30$

$x+19 = 30$ or $x+19 = -30$

$x = 11$ or $x = -49$

5.

$|x-7| = 20$

$x-7 = 20$ or $x-7 = -20$

$x = 27$ or $x = -13$

6.

$|m-7.1| = 3.4$

$m-7.1 = 3.4$ or $m-7.1 = -3.4$

$m = 10.5$ or $m = 3.7$

7. $|y-8| = -16$

There is no solution, because an absolute value cannot equal a negative number.

8. $|x-4| = -12$

There is no solution, because an absolute value cannot equal a negative number.

9.

$|2x-3| = 10$

$2x-3 = 10$ or $2x-3 = -10$

$2x = 13$ or $2x = -7$

$x = \frac{13}{2}$ or $x = -\frac{7}{2}$

10.

$|5x+1| = 7$

$5x+1 = 7$ or $5x+1 = -7$

$5x = 6$ or $5x = -8$

$x = \frac{6}{5}$ or $x = -\frac{8}{5}$

11.

$|2x+3| - 10 = 40$

$|2x+3| - 10 + 10 = 40 + 10$

$|2x+3| = 50$

$2x+3 = 50$ or $2x+3 = -50$

$2x = 47$ or $2x = -53$

$x = 23.5$ or $x = -26.5$

12.

$|3y-4| - 8 = 20$

$|3y-4| - 8 + 8 = 20 + 8$

$|3y-4| = 28$

$3y-4 = 28$ or $3y-4 = -28$

$3y = 32$ or $3y = -24$

$y = \frac{32}{3}$ or $y = -8$

13.

$5|x-7| + 10 = 95$

$5|x-7| + 10 - 10 = 95 - 10$

$5|x-7| = 85$

$\frac{5|x-7|}{5} = \frac{85}{5}$

$|x-7| = 17$

$x-7 = 17$ or $x-7 = -17$

$x = 24$ or $x = -10$

14.

$2|4t-3| - 7 = 15$

$2|4t-3| - 7 + 7 = 15 + 7$

$2|4t-3| = 22$

$\frac{2|4t-3|}{2} = \frac{22}{2}$

$|4t-3| = 11$

$4t-3 = 11$ or $4t-3 = -11$

$4t = 14$ or $4t = -8$

$t = \frac{7}{2}$ or $t = -2$

15.

$|374-65t| = 60$

$374-65t = 60$ or $374-65t = -60$

$-65t = -314$ or $-65t = -434$

$\frac{-65t}{-65} = \frac{-314}{-65}$ or $\frac{-65t}{-65} = \frac{-434}{-65}$

$t \approx 4.83$ or $t = 6.68$

Patsy will be 60 miles away from Davenport after about 4.83 hours and again after about 6.68 hours.

16.

$|1853-65t| = 100$

$1853-65t = 100$ or $1853-65t = -100$

$-65t = -1753$ or $-65t = -1953$

$\frac{-65t}{-65} = \frac{-1753}{-65}$ or $\frac{-65t}{-65} = \frac{-1953}{-65}$

$t \approx 26.97$ or $t = 30.05$

Tracy will be 100 miles away from Albuquerque after about 26.97 hours and again after about 30.05 hours.

17. $-5 < x < 4$

18. $-8 \le x \le -1$

19. $x < -8$ or $x > 1$

20. $x \le -2$ or $x \ge 1$

21.

$|x| < 8.5$

$-8.5 < x < 8.5$

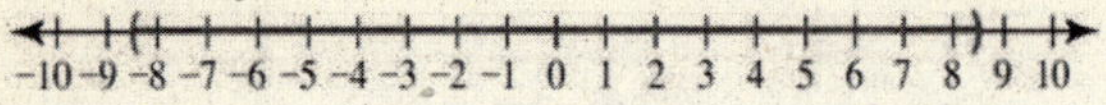

22.

$|w| \le 3$

$-3 \le w \le 3$

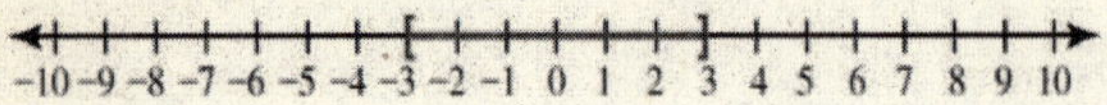

23.

$|x| > 6.5$

$x < -6.5$ or $x > 6.5$

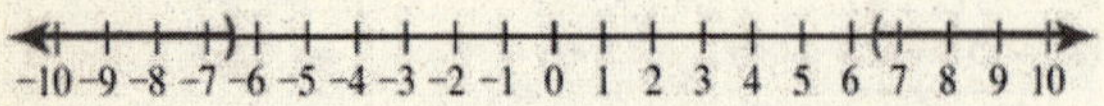

24.

$|y| \ge \frac{2}{3}$

$y \le \frac{-2}{3}$ or $y \ge \frac{2}{3}$

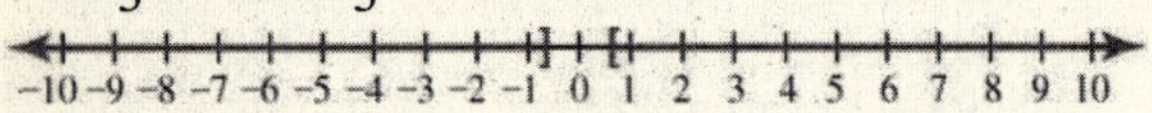

25.

$|x+3| \ge 12$

$x+3 \le -12$ or $x+3 \ge 12$

$x \le -15$ or $x \ge 9$

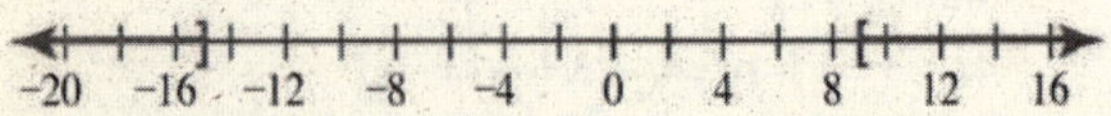

26.

$|x-5| \ge 17$

$x-5 \le -17$ or $x-5 \ge 17$

$x \le -12$ or $x \ge 22$

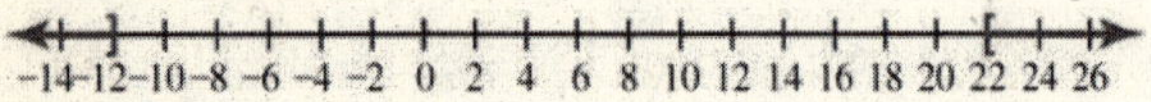

27.

$|x-7| \le 3$

$-3 \le x-7 \le 3$

$-3+7 \le x-7+7 \le 3+7$

$4 \le x \le 10$

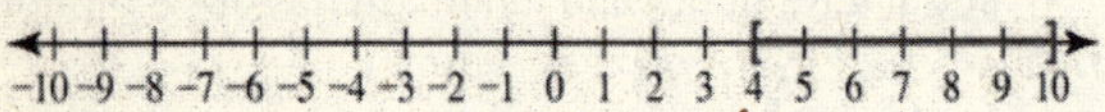

28.

$|w-5| < 4.9$

$w-5 > -4.9$ or $w-5 < 4.9$

$0.1 < w < 9.9$

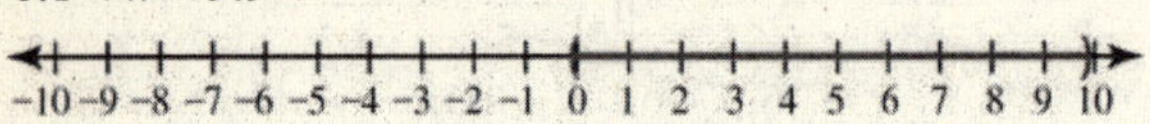

29.

$2|x+5| - 4 > 16$

$2|x+5| - 4 + 4 > 16 + 4$

$2|x+5| > 20$

$\frac{2|x+5|}{2} > \frac{20}{2}$

$|x+5| > 10$

$x+5 < -10$ or $x+5 > 10$

$x < -15$ or $x > 5$

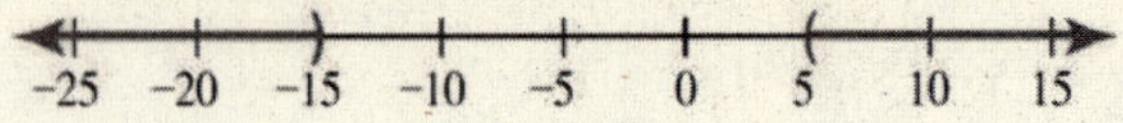

30.

$7|t-2| - 1 \ge 13$

$7|t-2| - 1 + 1 \ge 13 + 1$

$7|t-2| \ge 14$

$\frac{7|t-2|}{7} \ge \frac{14}{7}$

$|t-2| \ge 2$

$t-2 \le -2$ or $t-2 \ge 2$

$t \le 0$ or $t \ge 4$

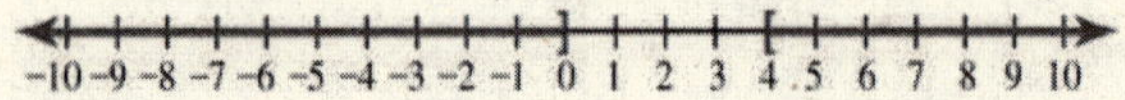

31.

$-3|x+5| + 7 \le 4$

$-3|x+5| + 7 - 7 \le 4 - 7$

$-3|x+5| \le -3$

$\frac{-3|x+5|}{-3} \ge \frac{-3}{-3}$

$|x+5| \ge 1$

$x+5 \le -1$ or $x+5 \ge 1$

$x \le -6$ or $x \ge -4$

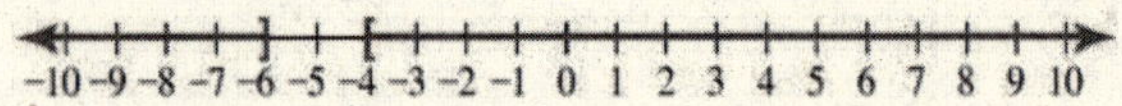

32.

$5|a-3| + 13 \le 8$

$5|a-3| + 13 - 13 \le 8 - 13$

$5|a-3| \le -5$

$\frac{5|a-3|}{5} \ge \frac{-5}{5}$

$|a-3| \ge -1$

There is no solution, because an absolute value cannot equal a negative number.

33. **a.** $-2 < x < 2$

b. $x < -2$ or $x > 2$

34. **a.** $-4 < x < -2$

b. $x < -4$ or $x > -2$

35. $\left|\frac{h-70}{3}\right| > 2$

$\frac{h-70}{3} < -2$ or $\frac{h-70}{3} > 2$

$h-70 < -6$ or $h-70 > 6$

$h < 64$ or $h > 76$

Men's heights below 64 inches or above 76 inches are considered unusual.

36. $\left|\frac{g-272}{9}\right|>2$

$\frac{g-272}{9}<-2$ or $\frac{g-272}{9}>2$

$g-272<-18$ or $g-272>18$

$g<254$ or $g>290$

Any gestational period for women of less than 254 days or more than 290 days is considered unusual.

37. $(x+2)(x+7)>0$

$(x+2)(x+7)=0$

$x+2=0$ or $x+7=0$

$x=-2$ or $x=-7$

Check the intervals:

$x=-8$	$x=-5$	$x=0$
$(-8+2)(-8+7)\overset{?}{>}0$	$(-5+2)(-5+7)\overset{?}{>}0$	$(0+2)(0+7)\overset{?}{>}0$
$(-6)(-1)\overset{?}{>}0$	$(-3)(2)\overset{?}{>}0$	$(2)(7)\overset{?}{>}0$
$6\overset{?}{>}0$	$-6\overset{?}{>}0$	$14\overset{?}{>}0$
True	False	True

$x<-7$ or $x>-2$

38. $x^2\le 100$

$x^2-100\le 0$

$x^2-100=0$

$(x-10)(x+10)=0$

$x-10=0$ or $x+10=0$

$x=10$ or $x=-10$

Check the intervals:

$x=-11$	$x=0$	$x=11$
$(-11-10)(-11+10)\overset{?}{\le}0$	$(0-10)(0+10)\overset{?}{\le}0$	$(11-10)(11+10)\overset{?}{\le}0$
$(-21)(-1)\overset{?}{\le}0$	$(-10)(10)\overset{?}{\le}0$	$(1)(21)\overset{?}{\le}0$
$21\overset{?}{\le}0$	$-100\overset{?}{\le}0$	$21\overset{?}{\le}0$
False	True	False

$-10\le x\le 10$

39. $x^2+12x+32<0$

$x^2+12x+32=0$

$(x+8)(x+4)=0$

$x+8=0$ or $x+4=0$

$x=-8$ or $x=-4$

Check the intervals:

$x=-9$	$x=-6$	$x=0$
$(-9+8)(-9+4)\overset{?}{<}0$	$(-6+8)(-6+4)\overset{?}{<}0$	$(0+8)(0+4)\overset{?}{<}0$
$(-1)(-5)\overset{?}{<}0$	$(2)(-2)\overset{?}{<}0$	$(8)(4)\overset{?}{<}0$
$5\overset{?}{<}0$	$-4\overset{?}{<}0$	$32\overset{?}{<}0$
False	True	False

$-8<x<-4$

40. $x^2+14x+48\ge 0$

$x^2+14x+48=0$

$(x+8)(x+6)=0$

$x+8=0$ or $x+6=0$

$x=-8$ or $x=-6$

Check the intervals:

$x=-9$	$x=-7$	$x=0$
$(-9+8)(-9+6)\overset{?}{\ge}0$	$(-7+8)(-7+6)\overset{?}{\ge}0$	$(0+8)(0+6)\overset{?}{\ge}0$
$(-1)(-3)\overset{?}{\ge}0$	$(1)(-1)\overset{?}{\ge}0$	$(8)(6)\overset{?}{\ge}0$
$3\overset{?}{\ge}0$	$-1\overset{?}{\ge}0$	$48\overset{?}{\ge}0$
True	False	True

$(-\infty,-8]$ or $[-6,\infty)$

41. $-4x^3+12x^2+40x<0$

$-4x^3+12x^2+40x=0$

$-4x(x^2-3x-10)=0$

$-4x(x-5)(x+2)=0$

$-4x=0$ or $x-5=0$ or $x+2=0$

$x=0$ or $x=5$ or $x=-2$

Check the intervals:

$x=-3$	$x=-1$
$-4\cdot-3(-3-5)(-3+2)\overset{?}{<}0$	$-4\cdot-1(-1-5)(-1+2)\overset{?}{<}0$
$12(-8)(-1)\overset{?}{<}0$	$4(-6)(1)\overset{?}{<}0$
$96\overset{?}{<}0$	$-24\overset{?}{<}0$
False	True

$x=1$	$x=7$
$-4\cdot1(1-5)(1+2)\overset{?}{<}0$	$-4\cdot7(7-5)(7+2)\overset{?}{<}0$
$-4(-4)(3)\overset{?}{<}0$	$-28(2)(9)\overset{?}{<}0$
$48\overset{?}{<}0$	$-504\overset{?}{<}0$
False	True

$(-2,0)$ or $(5,\infty)$

42. $6x^2+7x\le 5$

$6x^2+7x-5\le 0$

$6x^2+7x-5=0$

$a\cdot c=6\cdot-5=-30$ Factors of -30 that add to 7 are 10 and -3. Factor by grouping.

$6x^2+10x-3x-5=0$

$(6x^2+10x)-(3x+5)=0$

$2x(3x+5)-(3x+5)=0$

$(3x+5)(2x-1)=0$

$3x+5=0$ or $2x-1=0$

$3x=-5$ or $2x=1$

$x=-\frac{5}{3}$ or $x=\frac{1}{2}$

Check the intervals:

$x=-2$	$x=0$	$x=2$
$[3(-2)+5][2(-2)-1]\overset{?}{\le}0$	$[3(0)+5][2(0)-1]\overset{?}{\le}0$	$[3(2)+5][2(2)-1]\overset{?}{\le}0$
$(-1)(-5)\overset{?}{\le}0$	$(5)(-1)\overset{?}{\le}0$	$(11)(3)\overset{?}{\le}0$
$5\overset{?}{\le}0$	$-5\overset{?}{\le}0$	$33\overset{?}{\le}0$
False	True	False

$\left[-\frac{5}{3},\frac{1}{2}\right]$

43. $x^2+13x+40<0$

$x^2+13x+40=0$

$(x+8)(x+5)=0$

$x+8=0$ or $x+5=0$

$x=-8$ or $x=-5$

Check the intervals:

$x=-9$	$x=-7$	$x=0$
$(-9+8)(-9+5)\stackrel{?}{<}0$	$(-7+8)(-7+5)\stackrel{?}{<}0$	$(0+8)(0+5)\stackrel{?}{<}0$
$(-1)(-4)\stackrel{?}{<}0$	$(1)(-2)\stackrel{?}{<}0$	$(8)(5)\stackrel{?}{<}0$
$4\stackrel{?}{<}0$	$-2\stackrel{?}{<}0$	$40\stackrel{?}{<}0$
False	True	False

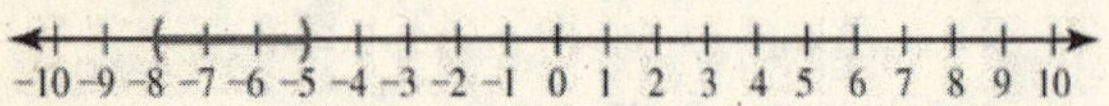

44. $x^3+2x^2-24x<0$

$x^3+2x^2-24x=0$

$x(x^2+2x-24)=0$

$x(x+6)(x-4)=0$

$x=0$ or $x+6=0$ or $x-4=0$

$x=-6$ or $x=4$

Check the intervals:

$x=-7$	$x=-1$
$-7(-7+6)(-7-4)\stackrel{?}{<}0$	$-1(-1+6)(-1-4)\stackrel{?}{<}0$
$-7(-1)(-11)\stackrel{?}{<}0$	$-1(5)(-5)\stackrel{?}{<}0$
$-77\stackrel{?}{<}0$	$10\stackrel{?}{<}0$
True	False

$x=1$	$x=7$
$1(1+6)(1-4)\stackrel{?}{<}0$	$7(7+6)(7-4)\stackrel{?}{<}0$
$1(7)(-3)\stackrel{?}{<}0$	$7(13)(3)\stackrel{?}{<}0$
$-21\stackrel{?}{<}0$	$273\stackrel{?}{<}0$
True	False

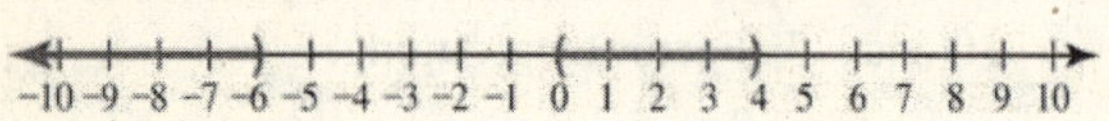

45. $\dfrac{x+3}{x-6}>0$

Find the zero value:

$x+3=0$

$x=-3$

Find the excluded value:

$x-6=0$

$x=6$

Check the intervals:

$x=-4$	$x=0$	$x=7$
$\dfrac{-4+3}{-4-6}\stackrel{?}{>}0$	$\dfrac{0+3}{0-6}\stackrel{?}{>}0$	$\dfrac{7+3}{7-6}\stackrel{?}{>}0$
$\dfrac{-1}{-10}\stackrel{?}{>}0$	$\dfrac{3}{-6}\stackrel{?}{>}0$	$\dfrac{10}{1}\stackrel{?}{>}0$
$\dfrac{1}{10}\stackrel{?}{>}0$	$\dfrac{-1}{2}\stackrel{?}{>}0$	$10\stackrel{?}{>}0$
True	False	True

$x<-3$ or $x>6$

46. $\dfrac{x-7}{x+2}\le 0$

Find the zero value:

$x-7=0$

$x=7$

Find the excluded value:

$x+2=0$

$x=-2$

Check the intervals:

$x=-4$	$x=0$	$x=8$
$\dfrac{-4-7}{-4+2}\stackrel{?}{\le}0$	$\dfrac{0-7}{0+2}\stackrel{?}{\le}0$	$\dfrac{8-7}{8+2}\stackrel{?}{\le}0$
$\dfrac{-11}{-2}\stackrel{?}{\le}0$	$\dfrac{-7}{2}\stackrel{?}{\le}0$	$\dfrac{1}{10}\stackrel{?}{\le}0$
$\dfrac{11}{2}\stackrel{?}{\le}0$	True	False
False		

$-2<x\le 7$

47. $\dfrac{x+5}{x^2+4x-12}\le 0$

$\dfrac{x+5}{(x+6)(x-2)}\le 0$

Find the zero value:

$x+5=0$

$x=-5$

Find the excluded values:

$x+6=0$ or $x-2=0$

$x=-6$ or $x=2$

Check the intervals:

$x=-7$	$x=-5.5$
$\dfrac{-7+5}{(-7+6)(-7-2)}\stackrel{?}{\le}0$	$\dfrac{-5.5+5}{(-5.5+6)(-5.5-2)}\stackrel{?}{\le}0$
$\dfrac{-2}{(-1)(-9)}\stackrel{?}{\le}0$	$\dfrac{-0.5}{(0.5)(-7.5)}\stackrel{?}{\le}0$
$\dfrac{-2}{9}\stackrel{?}{\le}0$	$\dfrac{1}{75}\stackrel{?}{\le}0$
True	False

$x=0$	$x=3$
$\dfrac{0+5}{(0+6)(0-2)}\stackrel{?}{\le}0$	$\dfrac{3+5}{(3+6)(3-2)}\stackrel{?}{\le}0$
$\dfrac{5}{(6)(-2)}\stackrel{?}{\le}0$	$\dfrac{8}{(9)(1)}\stackrel{?}{\le}0$
$\dfrac{-5}{12}\stackrel{?}{\le}0$	$\dfrac{8}{9}\stackrel{?}{\le}0$
True	False

$x<-6$ or $-5\le x<2$

48. $\dfrac{x-4}{x+8}\ge 0$

Find the zero value:

$x-4=0$

$x=4$

Find the excluded value:
$x+8=0$
$x=-8$
Check the intervals:

$x=-9$	$x=0$	$x=6$
$\frac{-9-4}{-9+8}\overset{?}{\geq}0$	$\frac{0-4}{0+8}\overset{?}{\geq}0$	$\frac{6-4}{6+8}\overset{?}{\geq}0$
$\frac{-13}{-1}\overset{?}{\geq}0$	$\frac{-4}{8}\overset{?}{\geq}0$	$\frac{2}{14}\overset{?}{\geq}0$
$13\overset{?}{\geq}0$	$\frac{-1}{2}\overset{?}{\geq}0$	$\frac{1}{7}\overset{?}{\geq}0$
True	False	True

$(-\infty,-8)$ or $[4,\infty)$

49. $\frac{x-6}{2x^2+9x+4}<0$

Factor the denominator.
$a\cdot c=2\cdot 4=8$ Factors of 8 that add to 9 are 8 and 1.
Factor by grouping.

$$\frac{x-6}{2x^2+8x+x+4}<0$$
$$\frac{x-6}{(2x^2+8x)+(x+4)}<0$$
$$\frac{x-6}{2x(x+4)+(x+4)}<0$$
$$\frac{x-6}{(x+4)(2x+1)}<0$$

Find the zero value:
$x-6=0$
$x=6$

Find the excluded values:
$x+4=0$ or $2x+1=0$
$x=-4$ or $2x=-1$
$x=-\frac{1}{2}$

Check the intervals:

$x=-5$	$x=-1$
$\frac{-5-6}{(-5+4)[2(-5)+1]}\overset{?}{<}0$	$\frac{-1-6}{(-1+4)[2(-1)+1]}\overset{?}{<}0$
$\frac{-11}{(-1)(-9)}\overset{?}{<}0$	$\frac{-7}{(3)(-1)}\overset{?}{<}0$
$-\frac{11}{9}\overset{?}{<}0$	$\frac{7}{3}\overset{?}{<}0$
True	False

$x=1$	$x=7$
$\frac{1-6}{(1+4)[2(1)+1]}\overset{?}{<}0$	$\frac{7-6}{(7+4)[2(7)+1]}\overset{?}{<}0$
$\frac{-5}{(5)(3)}\overset{?}{<}0$	$\frac{1}{(11)(15)}\overset{?}{<}0$
$-\frac{1}{3}\overset{?}{<}0$	$\frac{1}{165}\overset{?}{<}0$
True	False

$(-\infty,-4)$ or $\left(-\frac{1}{2},6\right)$

50. $\frac{x-9}{x+1}<0$

Find the zero value:
$x-9=0$
$x=9$

Find the excluded value:
$x+1=0$
$x=-1$

Check the intervals:

$x=-3$	$x=0$	$x=10$
$\frac{-3-9}{-3+1}\overset{?}{<}0$	$\frac{0-9}{0+1}\overset{?}{<}0$	$\frac{10-9}{10+1}\overset{?}{<}0$
$\frac{-12}{-2}\overset{?}{<}0$	$\frac{-9}{1}\overset{?}{<}0$	$\frac{1}{11}\overset{?}{<}0$
$6\overset{?}{<}0$	$-9\overset{?}{<}0$	False
False	True	

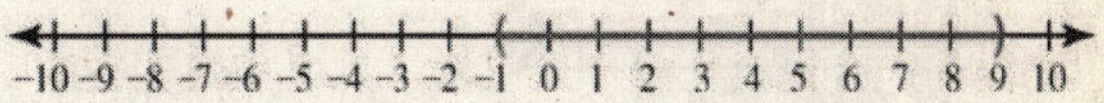

51. $\frac{x+4}{x^2-7x+6}<0$

$$\frac{x+4}{(x-6)(x-1)}<0$$

Find the zero value:
$x+4=0$
$x=-4$

Find the excluded values:
$x-6=0$ or $x-1=0$
$x=6$ or $x=1$

Check the intervals:

$x=-5$	$x=-1$
$\frac{-5+4}{(-5-6)(-5-1)}\overset{?}{<}0$	$\frac{-1+4}{(-1-6)(-1-1)}\overset{?}{<}0$
$\frac{-1}{(-11)(-6)}\overset{?}{<}0$	$\frac{3}{(-7)(-2)}\overset{?}{<}0$
$-\frac{1}{66}\overset{?}{<}0$	$\frac{3}{14}\overset{?}{<}0$
True	False

$x=3$	$x=7$
$\frac{3+4}{(3-6)(3-1)}\overset{?}{<}0$	$\frac{7+4}{(7-6)(7-1)}\overset{?}{<}0$
$\frac{7}{(-3)(2)}\overset{?}{<}0$	$\frac{11}{(1)(6)}\overset{?}{<}0$
$-\frac{7}{6}\overset{?}{<}0$	$\frac{11}{6}\overset{?}{<}0$
True	False

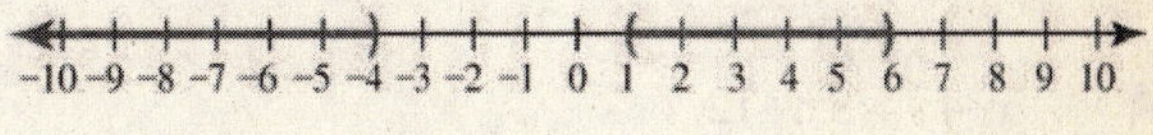

52. $y \le 4x - 10$

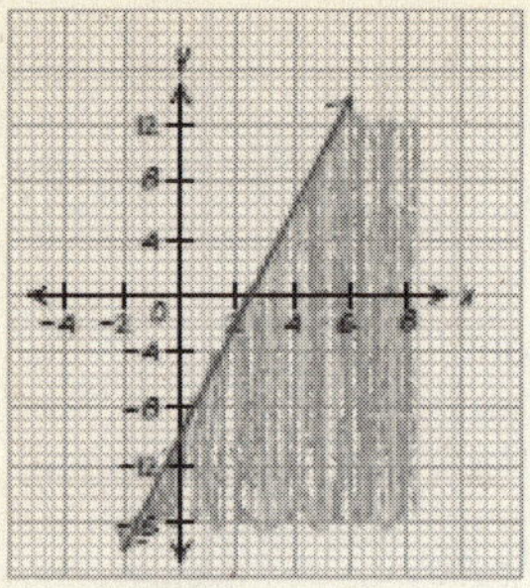

53. $y \ge 1.5x - 5$

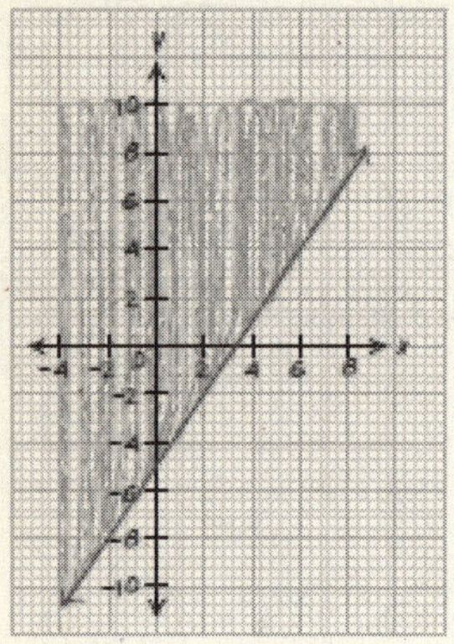

54.

$2x + 5y > 15$

$2x + 5y - 2x > 15 - 2x$

$5y > -2x + 15$

$\dfrac{5y}{5} > \dfrac{-2x+15}{5}$

$y > -\dfrac{2}{5}x + 3$

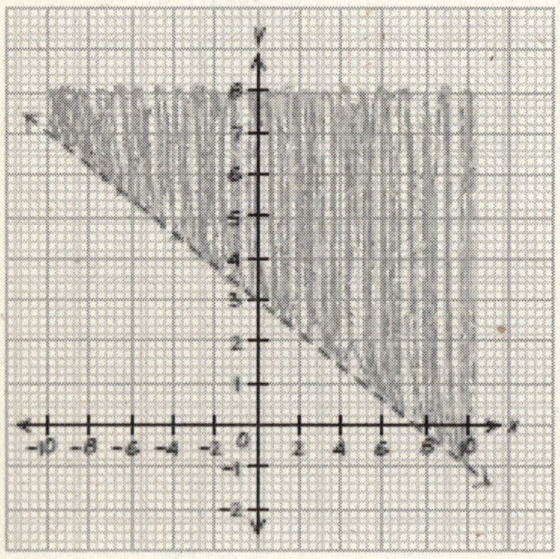

55.

$-3x - 4y > 12$

$-3x - 4y + 3x > 12 + 3x$

$-4y > 3x + 12$

$\dfrac{-4y}{-4} < \dfrac{3x+12}{-4}$

$y < -\dfrac{3}{4}x - 3$

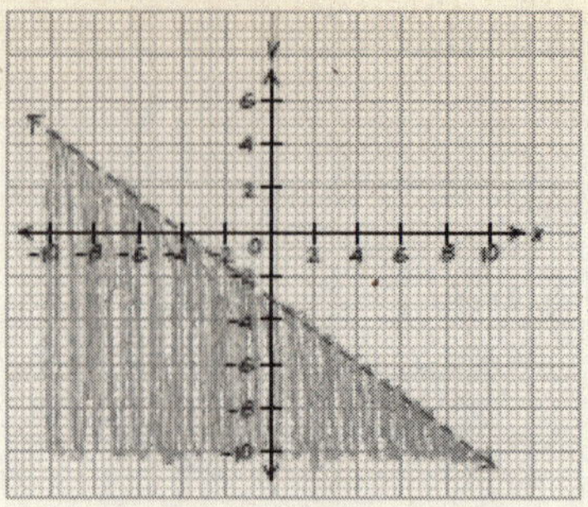

56.

$\begin{cases} y > 2x + 5 \\ y < -x + 7 \end{cases}$

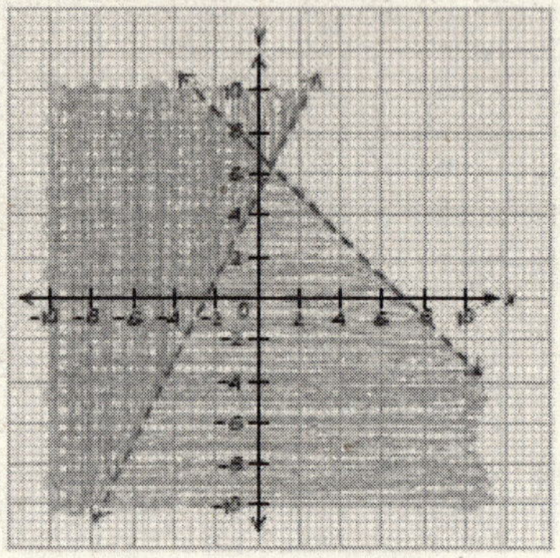

57.

$\begin{cases} y > 1.5x + 2 \\ 4.5x - 3y > 6 \end{cases}$

Solve for y in the second equation to graph:

$4.5x - 3y - 4.5x > 6 - 4.5x$

$-3y > -4.5x + 6$

$\dfrac{-3y}{-3} < \dfrac{-4.5x+6}{-3}$

$y < 1.5x - 2$

$\begin{cases} y > 1.5x + 2 \\ y < 1.5x - 2 \end{cases}$

The lines are parallel. The system has no solution.

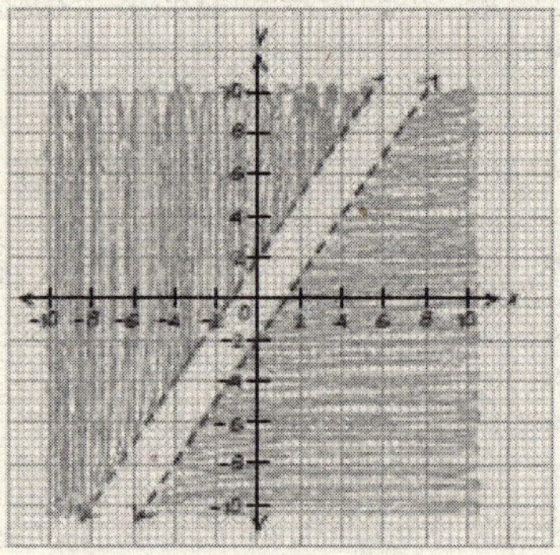

58.

$\begin{cases} 4x + 5y \le 12 \\ 2x + 8y \ge 5 \end{cases}$

Solve for y in both equations to graph:

$4x + 5y \le 12 \quad , \quad 2x + 8y \ge 5$

$5y \le -4x + 12 \quad , \quad 8y \ge -2x + 5$

$$\frac{5y}{5} \le \frac{-4x+12}{5}, \quad \frac{8y}{8} \ge \frac{-2x+5}{8}$$

$$y \le -\frac{4}{5}x + \frac{12}{5}, \quad y \ge -\frac{1}{4}x + \frac{5}{8}$$

$$\begin{cases} y \le -\frac{4}{5}x + \frac{12}{5} \\ y \ge -\frac{1}{4}x + \frac{5}{8} \end{cases}$$

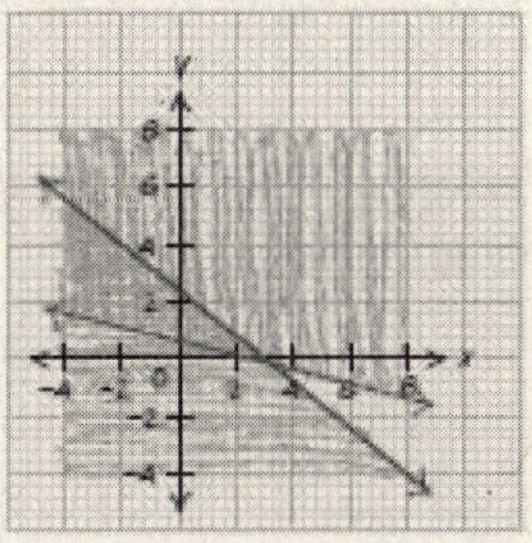

59.

$$\begin{cases} y > \frac{2}{3}x - 7 \\ y < \frac{2}{3}x + 5 \end{cases}$$

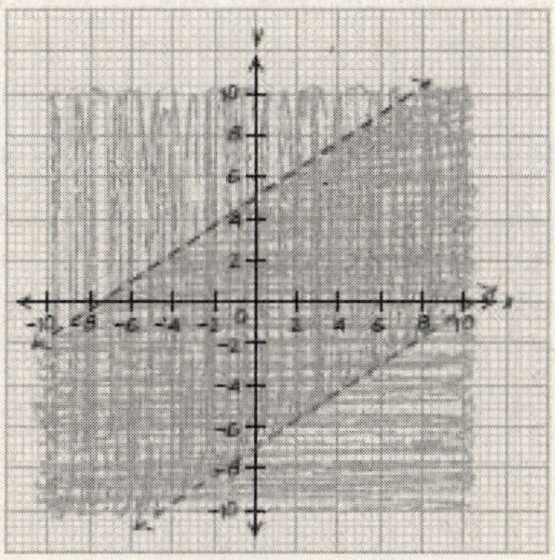

60. Let R be the number of regular tickets and let S be the number of student tickets.

$S + R \le 12000$

$9S + 15R \ge 140000$

61.

$(0,-6)$ and $(6,-4)$

$$m = \frac{-4-(-6)}{6-0} = \frac{2}{6} = \frac{1}{3}$$

This is a solid line, shaded below:

$$y \le \frac{1}{3}x - 6$$

62.

$(0,6)$ and $(4,4)$

$$m = \frac{4-6}{4-0} = \frac{-2}{4} = -\frac{1}{2}$$

This is a dashed line, shaded above:

$$y > -\frac{1}{2}x + 6$$

Chapter 9 Test

1.

$|w| = 5$

$w = 5$ or $w = -5$

$w = \pm 5$

2.

$|x+7| = 16$

$x + 7 = 16$ or $x + 7 = -16$

$x = 9$ or $x = -23$

3.

$-2\,|y-4| = 8$

$$\frac{-2\,|y-4|}{-2} = \frac{8}{-2}$$

$|y-4| = -4$

There is no solution, because an absolute value cannot equal a negative number.

4.

$|x+5| + 3 = 17$

$|x+5| + 3 - 3 = 17 - 3$

$|x+5| = 14$

$x + 5 = 14$ or $x + 5 = -14$

$x = 9$ or $x = -19$

5.

$-3|2x-1| + 19 = 10$

$-3|2x-1| + 19 - 19 = 10 - 19$

$-3|2x-1| = -9$

$$\frac{-3|2x-1|}{-3} = \frac{-9}{-3}$$

$|2x-1| = 3$

$2x - 1 = 3$ or $2x - 1 = -3$

$2x = 4$ or $2x = -2$

$x = 2$ or $x = -1$

6. $-3 < x < 9$

7.

$|t-8| \ge 7$

$t - 8 \le -7$ or $t - 8 \ge 7$

$t \le 1$ or $t \ge 15$

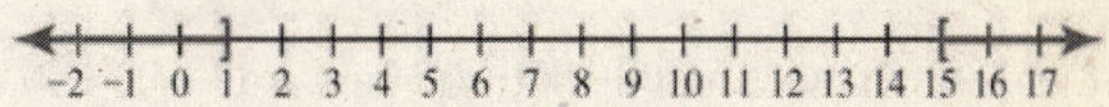

8.

$|3n+4| < 12$

$3n + 4 > -12$ or $3n + 4 < 12$

$3n > -16$ or $3n < 8$

$$n > -\frac{16}{3} \text{ or } n < \frac{8}{3}$$

$$\frac{-16}{3} < n < \frac{8}{3}$$

9.

$$2|3n-7|-12>4$$
$$2|3n-7|-12+12>4+12$$
$$2|3n-7|>16$$
$$\frac{2|3n-7|}{2}>\frac{16}{2}$$
$$|3n-7|>8$$
$$3n-7<-8 \text{ or } 3n-7>8$$
$$3n<-1 \text{ or } 3n>15$$
$$n<-\frac{1}{3} \text{ or } n>5$$

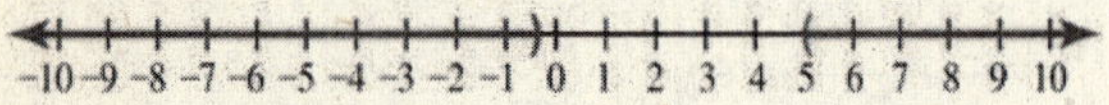

10.

$$|x-4|-8<5$$
$$|x-4|-8+8<5+8$$
$$|x-4|<13$$
$$x-4>-13 \text{ or } x-4<13$$
$$x>-9 \text{ or } x<17$$
$$-9<x<17$$

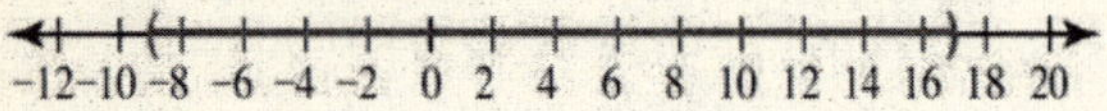

11.

$$\left|\frac{T-36.4}{0.114}\right|>2$$
$$\frac{T-36.4}{0.114}<-2 \text{ or } \frac{T-36.4}{0.114}>2$$
$$T-36.4<-0.228 \text{ or } T-36.4>0.228$$
$$T<36.172 \text{ or } T-36.4>36.628$$

Black-tailed prairie dogs have a body temperature that is considered unusual if it is less than 36.172°C or more than 36.628°C.

12. $x^2+3x-10>0$

$$x^2+3x-10=0$$
$$(x+5)(x-2)=0$$
$$x+5=0 \text{ or } x-2=0$$
$$x=-5 \text{ or } x=2$$

Check the intervals:

$x=-8$	$x=0$	$x=4$
$(-8+5)(-8-2)\overset{?}{>}0$	$(0+5)(0-2)\overset{?}{>}0$	$(4+5)(4-2)\overset{?}{>}0$
$(-3)(-10)\overset{?}{>}0$	$(5)(-2)\overset{?}{>}0$	$(9)(2)\overset{?}{>}0$
$30\overset{?}{>}0$	$-10\overset{?}{>}0$	$18\overset{?}{>}0$
True	False	True

$$x<-5 \text{ or } x>2$$

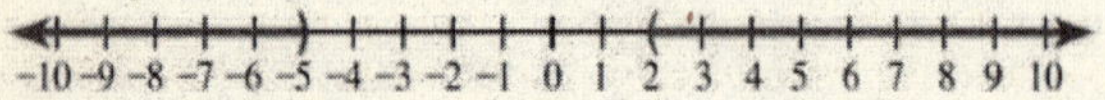

13. $x^2-3x-28<0$

$$x^2-3x-28=0$$
$$(x-7)(x+4)=0$$
$$x-7=0 \text{ or } x+4=0$$
$$x=7 \text{ or } x=-4$$

Check the intervals:

$x=-5$	$x=0$	$x=8$
$(-5-7)(-5+4)\overset{?}{<}0$	$(0-7)(0+4)\overset{?}{<}0$	$(8-7)(8+4)\overset{?}{<}0$
$(-12)(-1)\overset{?}{<}0$	$(-7)(4)\overset{?}{<}0$	$(1)(12)\overset{?}{<}0$
$12\overset{?}{<}0$	$-28\overset{?}{<}0$	$12\overset{?}{<}0$
False	True	False

$$-4<x<7$$

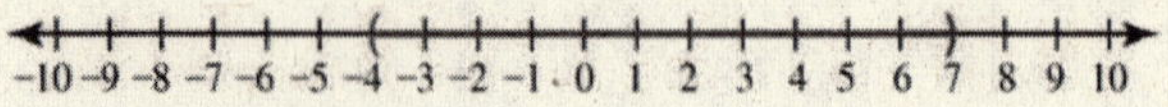

14. $2x^2-5x-12\ge 0$

$$2x^2-5x-12=0$$

$a\cdot c=2\cdot -12=-24$ Factors of -24 that add to -5 are -8 and 3. Factor by grouping.

$$2x^2-8x+3x-12=0$$
$$(2x^2-8x)+(3x-12)=0$$
$$2x(x-4)+3(x-4)=0$$
$$(x-4)(2x+3)=0$$
$$x-4=0 \text{ or } 2x+3=0$$
$$x=4 \text{ or } 2x=-3$$
$$x=-\frac{3}{2}$$

Check the intervals:

$x=-2$	$x=0$	$x=5$
$(-2-4)[2(-2)+3]\overset{?}{\ge}0$	$(0-4)[2(0)+3]\overset{?}{\ge}0$	$(5-4)[2(5)+3]\overset{?}{\ge}0$
$(-6)(-1)\overset{?}{\ge}0$	$(-4)(3)\overset{?}{\ge}0$	$(1)(13)\overset{?}{\ge}0$
$6\overset{?}{\ge}0$	$-12\overset{?}{\ge}0$	$13\overset{?}{\ge}0$
True	False	True

$$\left(-\infty,-\frac{3}{2}\right] \text{ or } [4,\infty)$$

15. $\frac{x+4}{x-3}\le 0$

Find the zero value:

$$x+4=0$$
$$x=-4$$

Find the excluded value:

$$x-3=0$$
$$x=3$$

Check the intervals:

$x=-5$	$x=0$	$x=5$
$\frac{-5+4}{-5-3}\overset{?}{\le}0$	$\frac{0+4}{0-3}\overset{?}{\le}0$	$\frac{5+4}{5-3}\overset{?}{\le}0$
$\frac{-1}{-8}\overset{?}{\le}0$	$\frac{4}{-3}\overset{?}{\le}0$	$\frac{9}{2}\overset{?}{\le}0$
$\frac{1}{8}\overset{?}{\le}0$	$\frac{-4}{3}\overset{?}{\le}0$	False
False	True	

$$[-4,3)$$

16. $\frac{x-1}{x^2-x-20}\le 0$

$$\frac{x-1}{(x-5)(x+4)}\le 0$$

Find the zero value:

$x-1=0$

$x=1$

Find the excluded values:

$x-5=0$ or $x+4=0$

$x=5$ or $x=-4$

Check the intervals:

$x=-7$	$x=-1$
$\frac{-7-1}{(-7-5)(-7+4)} \overset{?}{\le} 0$	$\frac{-1-1}{(-1-5)(-1+4)} \overset{?}{\le} 0$
$\frac{-8}{(-12)(-3)} \overset{?}{\le} 0$	$\frac{-2}{(-6)(3)} \overset{?}{\le} 0$
$\frac{-2}{9} \overset{?}{\le} 0$	$\frac{1}{9} \overset{?}{\le} 0$
True	False

$x=3$	$x=7$
$\frac{3-1}{(3-5)(3+4)} \overset{?}{\le} 0$	$\frac{7-1}{(7-5)(7+4)} \overset{?}{\le} 0$
$\frac{2}{(-2)(7)} \overset{?}{\le} 0$	$\frac{6}{(2)(11)} \overset{?}{\le} 0$
$\frac{-1}{7} \overset{?}{\le} 0$	$\frac{3}{11} \overset{?}{\le} 0$
True	False

$(-\infty,-4)$ or $[1,5)$

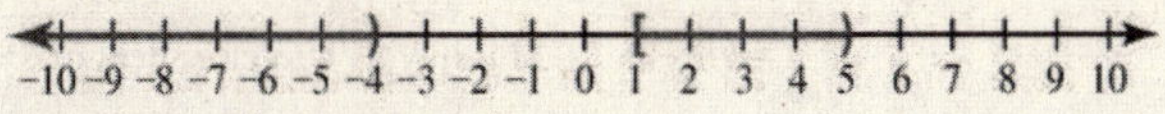

17. $-2x^3+6x^2+36x>0$

$-2x^3+6x^2+36x=0$

$-2x(x^2-3x-18)=0$

$-2x(x-6)(x+3)=0$

$-2x=0$ or $x-6=0$ or $x+3=0$

$x=0$ or $x=6$ or $x=-3$

Check the intervals:

$x=-5$	$x=-1$
$-2\cdot-5(-5-6)(-5+3)\overset{?}{>}0$	$-2\cdot-1(-1-6)(-1+3)\overset{?}{>}0$
$10(-11)(-2)\overset{?}{>}0$	$2(-7)(2)\overset{?}{>}0$
$220\overset{?}{>}0$	$-28\overset{?}{>}0$
True	False

$x=3$	$x=7$
$-2\cdot3(3-6)(3+3)\overset{?}{>}0$	$-2\cdot7(7-6)(7+3)\overset{?}{>}0$
$-6(-3)(6)\overset{?}{>}0$	$-14(1)(10)\overset{?}{>}0$
$108\overset{?}{>}0$	$-140\overset{?}{>}0$
True	False

$(-\infty,-3)$ or $(0,6)$

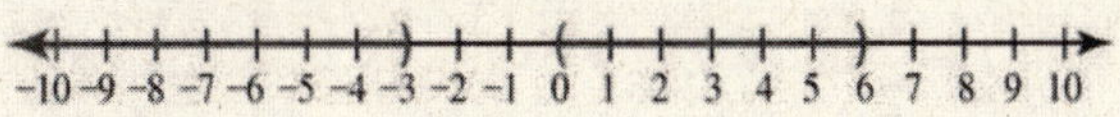

18. $y>-\frac{1}{2}x+6$

19.

$$\begin{cases} x\ge 2 \\ y<-\frac{3}{2}x+7 \end{cases}$$

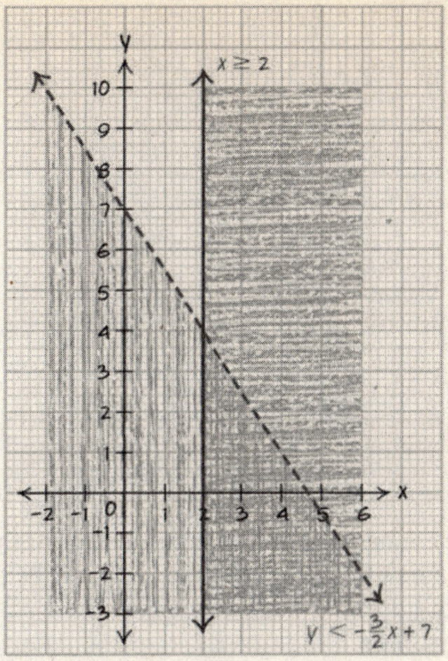

20.

$$\begin{cases} y\ge 2x-5 \\ y\le \frac{1}{3}x+4 \end{cases}$$

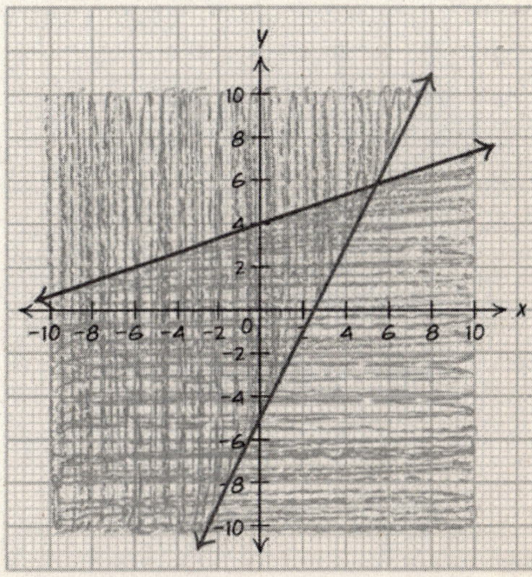

21.

$$\begin{cases} 4x-3y<6 \\ 3x+y>12 \end{cases}$$

Solve for y in both equations to graph:

$4x-3y<6$	$3x+y>12$
$-3y<-4x+6$	$y>-3x+12$
$\frac{-3y}{-3}>\frac{-4x+6}{-3}$	
$y>\frac{4}{3}x-2$	

$$\begin{cases} y>\frac{4}{3}x-2 \\ y>-3x+12 \end{cases}$$

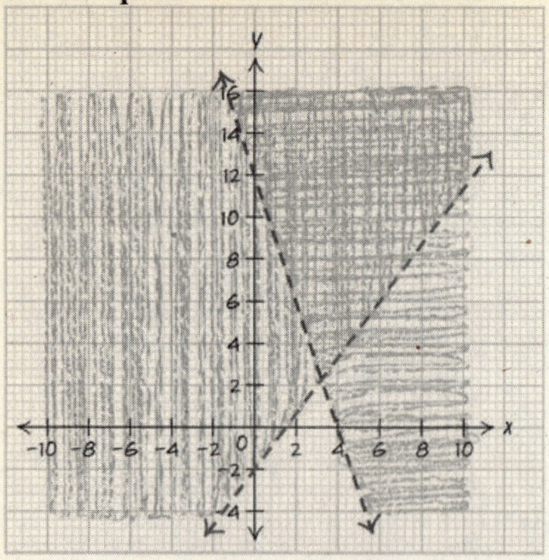

22. Let M be the number of hours per week Georgia needs to study for her math class, and let H be the number of hours per week Georgia needs to study for her history class.

$M + H \le 20$

$M \ge 12$

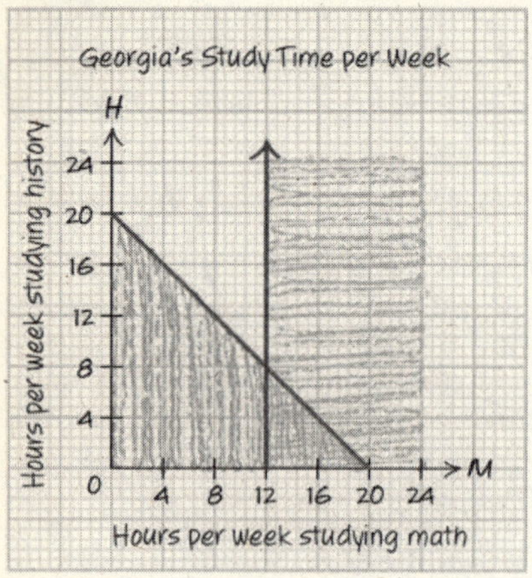

Cumulative Review Chapters 1-9

1. $(-3a)\cdot(-3a)\cdot(-3a)\cdot(-3a)=(-3a)^4$

2. $-\dfrac{2}{3}\cdot-\dfrac{2}{3}\cdot-\dfrac{2}{3}\cdot-\dfrac{2}{3}=\dfrac{16}{81}$

3.

$$\boxed{-6\cdot 4\div 12}+\boxed{(-4)^2}\boxed{-(-2-5)}$$
$$=-6\cdot 4\div 12+(-4)^2-(-7)$$
$$=-6\cdot 4\div 12+16+7$$
$$=-24\div 12+16+7$$
$$=-2+16+7$$
$$=21$$

4. Solve for y.

$-7x\boxed{+3y}=6$ Identify variable term to isolate.

$\underline{+7x \qquad +7x}$ Add $7x$ to both sides.

$3\boxed{y}=7x+6$ Identify variable to isolate.

$\dfrac{3y}{3}=\dfrac{7x+6}{3}$ Divide both sides by 3.

$y=\dfrac{7}{3}x+2$

5. Solve:

$\dfrac{7n}{2}-\dfrac{7}{4}=n+2$ Multiply by 4 to clear denom.

$4\left(\dfrac{7n}{2}-\dfrac{7}{4}\right)=4(n+2)$

$14n-7=4n+8$

$\underline{-4n \qquad -4n}$ Subtract $4n$ from both sides.

$10n-7=8$

$\underline{+7 \quad +7}$ Add 7 to both sides.

$10n=15$

$\dfrac{10n}{10}=\dfrac{15}{10}$ Divide both sides by 10.

$n=\dfrac{3}{2}$

6.

$$abc+b=a$$
$$abc=a-b$$
$$\frac{abc}{ab}=\frac{a-b}{ab}$$
$$c=\frac{a-b}{ab}$$

7.

$$P=\frac{1}{2}mn^2+3$$
$$P-3=\frac{1}{2}mn^2=\frac{mn^2}{2}$$
$$\left(\frac{2}{n^2}\right)(P-3)=\left(\frac{2}{n^2}\right)\left(\frac{mn^2}{2}\right)$$
$$m=\frac{2(P-3)}{n^2}$$

8. **a.** Substitute $h = 48$ into the equation $S = 11h$ and solve for *S*.

$S = 11h$ Given equation.

$S = 11(48)$ Substitute $h = 48$.

$S = 528$ Multiply & solve for *S*.

Gustavo's monthly salary is $528 if he works 48 hours.

b. Substitute $h = 32$ into the equation $S = 11h$ and solve for *S*.

$S = 11h$ Given equation.

$S = 11(32)$ Substitute $h = 32$.

$S = 352$ Multiply & solve for *S*.

Gustavo's monthly salary is $352 if he works 32 hours.

c. Substitute $S = 800$ into the equation $S = 11h$ and solve for *h*.

$S = 11h$ Given equation.

$800 = 11h$ Substitute $S = 800$.

$\frac{800}{11} = \frac{11h}{11}$ Divide & solve for *h*.

$72.73 \approx h$

Gustavo must work approximately 72.73 hours if he wants to earn $800.

9.

$-3(x-5) > 4$ Distribute on the left.

$-3x + 15 > 4$

$\underline{\quad -15 \quad -15}$ Subtract 15 from both sides.

$-3x > -11$ Divide both sides by − 3.

$\frac{-3x}{-3} < \frac{-11}{-3}$ Multiply by neg., reverse ineq.

$x < \frac{11}{3}$

10.

$2.5x - 4 < 12.8$

$\underline{\quad +4 \quad +4}$ Subtract 15 from both sides.

$2.5x < 16.8$

$\frac{2.5x}{2.5} < \frac{16.8}{2.5}$ Divide both sides by − 3.

$x < 6.72$

11. Let $c =$ the number of children.

$17.5c + 3c$ Substitute $c = 20$.

$17.5(20) + 3(20)$

$350 + 60 = 410$

It will cost $410 for Donna to take 20 children to see the 3-D movie.

12.

x	$y = \lvert x-3 \rvert$
−5	8
−3	6
−1	4
1	2
3	0
5	2
6	3
8	5
9	6

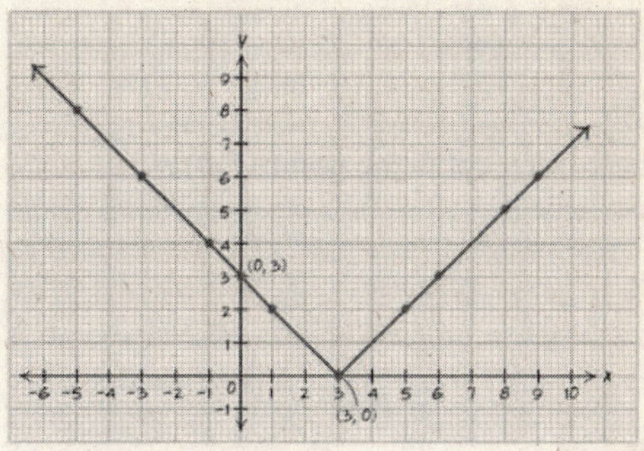

The graph is non-linear.

13. Find the slope of the line using the given points.

$$\text{slope} = \frac{6-(0)}{-3-(6)} = \frac{6}{-9} = -\frac{2}{3}$$

Find the *y*-intercept using the value for slope and one of the given points $(6,0)$.

$y = mx + b$ Substitute $m = -\frac{2}{3}$, $x = 6, y = 0$

$(0) = \left(-\frac{2}{3}\right)(6) + b$

$0 = -4 + b$ Solve for *b*.

$\underline{+4 \quad +4}$

$4 = b$

Using the values we found for the slope and *b* yields the equation of the line

$$y = -\frac{2}{3}x + 4$$

Graph the line.

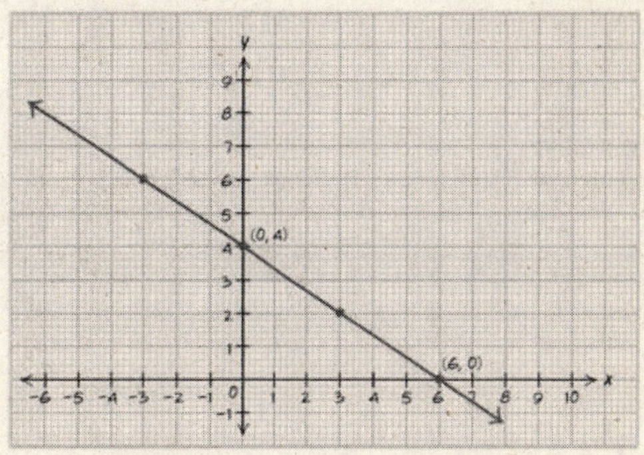

14. Find the y-intercept:

$$4(0)-5y=20 \quad \text{Substitute in } x=0.$$
$$-5y=20$$
$$\frac{-5y}{-5}=\frac{20}{-5}$$
$$y=-4$$

The y-intercept is the point $(0,-4)$.

Find the x-intercept:

$$4x-5(0)=20 \quad \text{Substitute in } y=0.$$
$$4x=20$$
$$\frac{4x}{4}=\frac{20}{4}$$
$$x=5$$

The x-intercept is the point $(5,0)$.

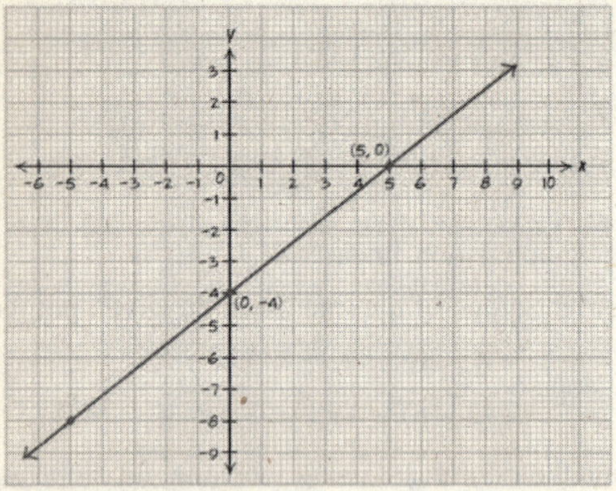

15. $y=3x-4$

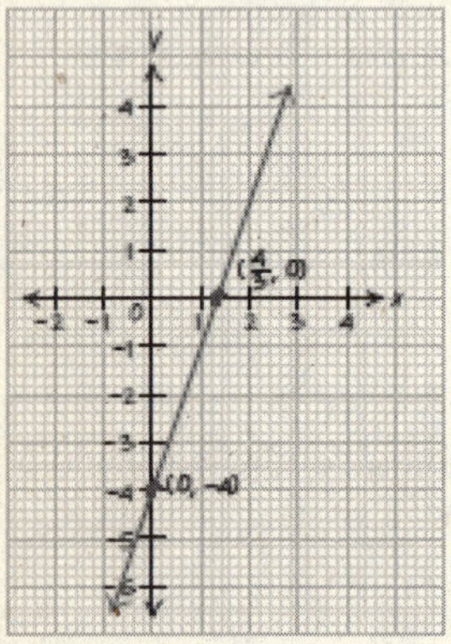

16. $y=-\frac{1}{3}x+2$

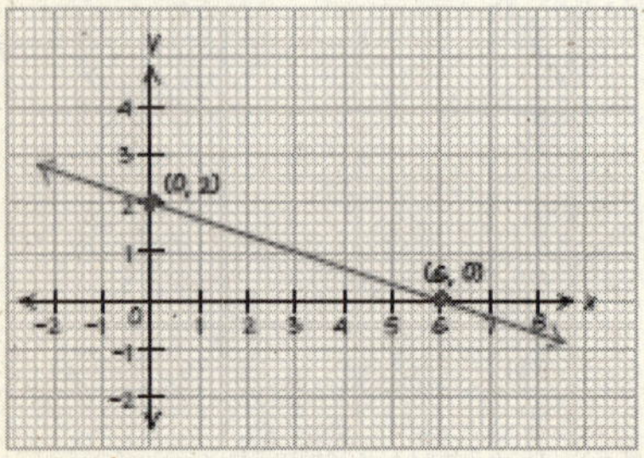

17. Find the slope (change to slope-intercept form):

$$3x-5y=10$$
$$\underline{-3x \qquad -3x} \quad \text{Subtract } 3x \text{ from both sides.}$$
$$-5y=-3x+10$$
$$\frac{-5y}{-5}=\frac{-3x+10}{-5} \quad \text{Divide both sides by } -5.$$
$$y=\frac{3}{5}x-2$$

The slope is $\frac{3}{5}$.

Find the y-intercept:

$$y=\frac{3}{5}(0)-2 \quad \text{Substitute in } x=0.$$
$$y=0-2$$
$$y=-2$$

The y-intercept is the point $(0,-2)$.

18. Using point-slope form:

$$y-y_1=m(x-x_1) \quad \text{Substitute in } m=\frac{-1}{5} \text{ and}$$
$$y-(6)=-\frac{1}{5}\left[x-(-5)\right] \quad (x_1,y_1)=(-5,6)$$
$$y-6=-\frac{1}{5}(x+5)$$
$$y-6=-\frac{1}{5}x-1$$
$$\underline{+6 \qquad +6} \quad \text{Add 6 to both sides.}$$
$$y=-\frac{1}{5}x+5$$

19.

$$3x+4y=7$$
$$4y=-3x+7$$
$$\frac{4y}{4}=\frac{-3x+7}{4}$$
$$y=-\frac{3}{4}x+\frac{7}{4}$$

Use the slope $m=-\frac{3}{4}$, and the point $(6,2)$.

$$b=y-mx$$
$$b=2-\left(-\frac{3}{4}\right)(6)$$
$$b=2+\frac{9}{2}=\frac{13}{2}$$
$$y=-\frac{3}{4}x+\frac{13}{2}$$

21. **a.** Use two points to find the slope: $(-6,0)$ and $(0,4)$.

$$m=\frac{4-0}{0-(-6)}=\frac{4}{6}=\frac{2}{3}$$

b. The y-intercept is $(0,4)$.

c. The x-intercept is $(-6,0)$.

d. The y-intercept is $(0,4)$, therefore $b=4$.

Since $m=\frac{2}{3}$ and $b=4$, the line is $y=\frac{2}{3}x+4$.

22. **a.** Use two points to find the slope: $(0,-5)$ and $(4,10)$.

$$m=\frac{10-(-5)}{4-0}=\frac{15}{4}$$

b. The vertical intercept is $(0,-5)$.

c. The horizontal intercept is $\left(\frac{4}{3},0\right)$.

d. The vertical intercept is $(0,-5)$, therefore $b=-5$.

Since $m=\frac{15}{4}$ and $b=-5$, the line is $y=\frac{15}{4}x-5$

23. **a.**

$$R=1.5(4)+5$$
$$R=11$$

Steve plans to run eleven miles in week four.

b.

$$20=1.5w+5$$
$$15=1.5w$$
$$\frac{15}{1.5}=\frac{1.5w}{1.5}$$
$$w=10$$

Steve plans to run twenty miles in week ten.

24. Substitute $y=\frac{3}{4}x+5$ into the second equation.

$$5x+4\left(\frac{3}{4}x+5\right)=-12$$
$$5x+3x+20=-12$$
$$8x+20=-12$$
$$\underline{-20\quad -20}\quad \text{Subtract 20 from both sides.}$$
$$8x=-32$$
$$\frac{8x}{8}=\frac{-32}{8}\quad \text{Divide both sides by 8.}$$
$$x=-4$$

Substitute $x=-4$ into one equation to find y.

$$y=\frac{3}{4}x+5$$
$$y=\frac{3}{4}(-4)+5$$
$$y=-3+5$$
$$y=2$$

The solution to the system is $(-4,2)$.

25. Multiply the second equation by -3:

$$\begin{aligned}5x-3y&=9\\-3(x-y)&=-3(5)\end{aligned}\quad \text{is now}\quad \begin{aligned}5x-3y&=9\\-3x+3y&=-15\end{aligned}$$

Add the equations vertically to eliminate y.

$$5x-3y=9\quad \text{Add the equations vertically.}$$
$$\underline{-3x+3y=-15}$$
$$2x=-6\quad \text{Solve for } x.$$
$$\frac{2x}{2}=\frac{-6}{2}$$
$$x=-3$$

Substitute $x=-3$ into either original equation to find the value of y.

$$x-y=5\quad \text{Substitute in } x=-3$$
$$(-3)-y=5$$
$$\underline{+3\qquad +3}$$
$$-y=8\quad \text{Solve for } y.$$
$$y=-8$$

The solution to the system is $(-3,-8)$.

26.

$$\begin{cases}m=3n-12\\4m-12n=10\end{cases}$$
$$4(3n-12)-12n=10$$
$$12n-48-12n=10$$
$$-48\neq 10$$

There is no solution. This is an inconsistent system.

27. Let $p=$ the rate of the plane (in mph) in still air and $w=$ the rate of the wind in mph.

The system of equations is:

$$3(p+w)=360$$
$$4(p-w)=360$$

Solve the second equation for p.

$$4(p-w)=360\quad \text{Distribute.}$$
$$4p-4w=360$$
$$\underline{+4w\quad +4w}$$
$$4p=4w+360$$
$$\frac{4p}{4}=\frac{4w+360}{4}$$
$$p=w+90$$

Now substitute $p=w+90$ into the first equation.

$$3(p+w)=360\quad \text{Distribute.}$$
$$3p+3w=360\quad \text{Substitute in } p=w+90.$$
$$3(w+90)+3w=360\quad \text{Distribute.}$$
$$3w+270+3w=360\quad \text{Simplify.}$$
$$6w+270=360$$
$$\underline{-270\quad -270}$$
$$6w=90\quad \text{Solve for } w.$$
$$\frac{6w}{6}=\frac{90}{6}$$
$$w=15$$

$p = w + 90$ Substitute in $w = 15$ to find b.

$p = (15) + 90$

$p = 105$

The rate of the plane is 105 mph and the rate of the wind is 15 mph.

28. Let x be the number of liters of 6% HCl solution and y be the number of liters of 24% HCl solution used to get the 4 liters of a 12% HCl solution that Hamid needs.

$$\begin{cases} x + y = 4 \\ 0.06x + 0.24y = (0.12)(4) = 0.48 \end{cases}$$

Solve for x in the first equation and use the result for x in the second equation.

$x = -y + 4$

$0.06(-y+4) + 0.24y = 0.48$

Solve for y.

$0.06(-y+4) + 0.24y = 0.48$

$-0.06y + 0.24 + 0.24y = 0.48$

$0.18y + 0.24 = 0.48$

$\underline{\quad -0.24 \quad -0.24}$

$0.18y = 0.24$

$\frac{0.18y}{0.18} = \frac{0.24}{0.18}$

$y = \frac{4}{3}$

Now solve for x:

$x = -\frac{4}{3} + 4 = 2\frac{2}{3}$

Hamid should mix $2\frac{2}{3}$ liters of 6% HCL solution with $1\frac{1}{3}$ liters of 24% HCL solution to get 4 liters of 12% HCL solution.

29. Change the first equation and substitute $y = 2x - 7$ into the second equation.

$-6x + 3(2x - 7) = -21$

$-6x + 6x - 21 = -21$

$-21 = -21$ This statement is true.

Because the statement is true, there are infinitely many solutions. The two equations graph as the same line.

30. Let $w =$ the width of the rectangle in cm, therefore, the length would be $x + 3$. Substitute these values into the formula for perimeter.

$P = 2l + 2w$

$28 = 2(w+3) + 2w$

$28 = 2w + 6 + 2w$

$28 = 2w + 6 + 2w$

$28 = 4w + 6$

$\underline{-6 \qquad -6}$

$22 = 4w$

$\frac{22}{4} = \frac{4w}{4}$

$5.5 = w$

$l = w + 3$ Substitute in $w = 5.5$ to find l.

$l = (5,5) + 3$

$l = 8.5$

The width of the rectangle is 5.5 cm and the length is 8.5 cm.

31. Let $x =$ the amount (in dollars) invested in the CD earning 3.5% interest and let $y =$ the amount (in dollars) invested in the bond fund earning 5% interest.

The system of equations is:

$x + y = 8000$

$0.035x + 0.05y = 325$

Solve for y in the first equation.

$x + y = 8000$

$\underline{-x \qquad -x}$

$y = -x + 8000$

Now substitute $y = -x + 8000$ into the second equation.

$0.035x + 0.05y = 325$ Substitute in $y = -x + 8000$.

$0.035x + 0.05(-x + 8000) = 325$ Distribute.

$0.035x - 0.05x + 400 = 325$ Simplify.

$-0.015x + 400 = 325$

$\underline{\qquad -400 \quad -400}$ Subtract 400 from both sides.

$-0.015x = -75$ Solve for x.

$\frac{-0.015x}{-0.015} = \frac{-75}{-0.015}$

$x = 5000$

$y = -x + 8000$ Substitute in $x = 5000$ to find y.

$y = -(5000) + 8000$

$y = 3000$

Aivy should invest \$5,000 in the CD and \$3,000 in the bond fund in order to earn \$325 a year in interest.

32. Change the equations to slope-intercept form $y < \frac{4}{5}x + 4$ then graph using the slope and y-intercept. This is a strict inequality (<) therefore the line is dashed and we shade below the line.

$4x-5y>-20$

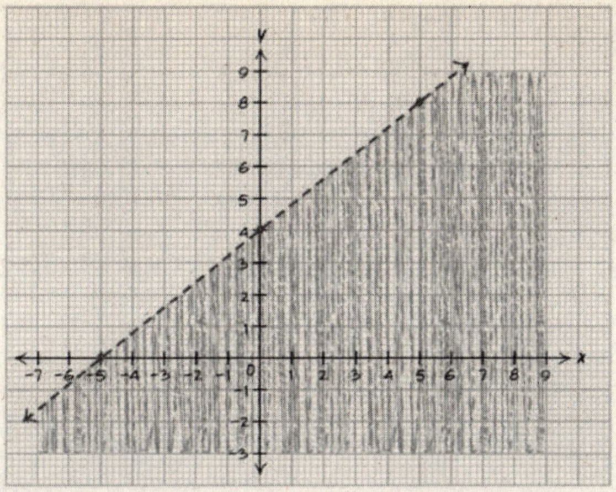

33. Graph using the slope and y-intercept. This is a strict inequality (>) therefore the line is dashed and we shade below the line.

$$y>-\frac{2}{3}x+9$$

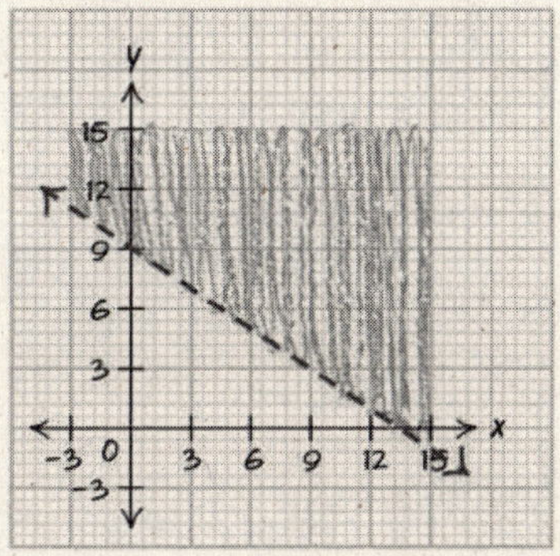

34. **a.**

$$\begin{aligned}(-3x^4y^3)^3&=(-3)^3(x^4)^3(y^3)^3\\&=-27x^{4\cdot 3}y^{3\cdot 3}\\&=-27x^{12}y^9\end{aligned}$$

b. $-6a^4\cdot 3ab^5=-18a^{4+1}b^5=-18a^5b^5$

c. $(-3xy^2)^3(2x^3y)^2$

$$\begin{aligned}&=(-3)^3(x)^3(y^2)^3\cdot(2)^2(x^3)^2(y)^2\\&=-27x^3y^6\cdot 4x^6y^2\\&=-108x^{3+6}y^{6+2}\\&=-108x^9y^8\end{aligned}$$

35. **a.** $\frac{t^6}{t^2}=t^{6-2}=t^4$

b. $\frac{3b^3}{-12b^2}=-\frac{b^{3-2}}{4}=-\frac{b}{4}$

c. $\frac{-15x^3y^2}{-3x^3y}=5x^{3-3}y^{2-1}=5y$

36. Simplify:

$$\left(\frac{-15x^4y^2}{50x^3y}\right)^2=\left(\frac{-3x^{4-3}y^{2-1}}{10}\right)^2$$

$$=\left(\frac{-3xy}{10}\right)^2$$

$$=\left(\frac{-3}{10}\right)^2(x)^2(y)^2$$

$$=\frac{9x^2y^2}{100}$$

37. **a.** $\frac{-4a^2b}{2a^3b^3}=-2a^{2-3}b^{1-3}=-2a^{-1}b^{-2}=\frac{-2}{ab^2}$

b.

$$\begin{aligned}(-2x^2y^{-3})^2&=(-2)^2(x^2)^2(y^{-3})^2\\&=4x^4y^{-6}\\&=\frac{4x^4}{y^6}\end{aligned}$$

c. $\left(\frac{5x}{y}\right)^{-3}=\left(\frac{y}{5x}\right)^3=\frac{(y)^3}{(5x)^3}=\frac{y^3}{125x^3}$

38. **a.** $0.000195=1.95\times10^{-4}$

b. Move the decimal 10 places to the left; the exponent of 10 will be positive.

$52{,}890{,}000{,}000=5.289\times10^{10}$

39. **a.** The exponent on 10 is -5. Move the decimal point 5 places to the *left* since the exponent is negative.

$3.004\times10^{-5}=0.00003004$

b. The exponent on 10 is 8. Move the decimal point 8 places to the right.

$9.27\times10^8=9.27\cdot100{,}000{,}000=927{,}000{,}000$

40. The polynomial has 3 terms. The highest exponent is 3 so the degree is 3.

41. The degree of the term $5a^3b$ is 4.

The degree of the term $3a^5$ is 5

The degree of the term $-6b$ is 1

The degree of the term ab^2 is 3

The highest degree term has a degree of 5, so the degree of the polynomial is 5.

42. $(-3b^2+2b+7)+(-b^2+5b-1)$

$$=-3b^2-b^2+2b+5b+7-1$$

$$=-4b^2+7b+6$$

43. $(3x-6)-(-x+5)$

$=3x-6+x-5$

$=3x+x-6-5$

$=4x-11$

44. **a.** $(2x+5)(x-7)$

$=2x^2-14x+5x-35$

$=2x^2-9x-35$

b. $(y-1)(3y^2+2y-1)$

$=3y^3+2y^2-y-3y^2-2y+1$

$=3y^3-y^2-3y+1$

c. $(3x-4)^2=9x^2-24x+16$

d. $3(7a-2)-4(-3a+5)$

$=21a-6+12a-20$

$=33a-26$

45.

$$\frac{3x^2-6x+9}{-3x}$$

$$=\frac{3x^{\not{2}1}}{-3\not{x}}-\frac{6\not{x}}{-3\not{x}}+\frac{9}{-3x}$$

$$=-x+2-\frac{3}{x}$$

46.

$$\begin{array}{r|l} & 3x-5 \\ x+1 & 3x^2-2x-5 \\ & \underline{-3x^2-3x} \\ & -5x-5 \\ & \underline{+5x+5} \\ & 0 \end{array}$$

47.

$$\begin{array}{r|rrr} 4 & 5 & -22 & 8 \\ & & 20 & 14 \\ \hline & 5 & -2 & 0 \end{array}$$

$(5x^2-22x+8)\div(x-4)=5x-2$

48. **a.** GCF $=6x$

$6x^3y-18xy+24x=6x(x^2y-3y+4)$

b. Factor by grouping.

$5xy+15x+2y+6=(5xy+15x)+(2y+6)$

$=5x(y+3)+2(y+3)$

$=(y+3)(5x+2)$

49.

$6ab+36a-4b-24$

$=2(3ab+18a-2b-12)$

$=2[(3ab+18a)-(2b+12)]$

$=2[3a(b+6)-2(b+6)]$

$=2(b+6)(3a-2)$

50. **a.** Factors of -14 that add to -5 are -7 and 2.

$x^2-5x-14=(x-7)(x+2)$

b. $-n^2-3n+18=-(n^2+3n-18)$

Factors of -18 that add to 3 are 6 and -3.

$-(n^2+3n-18)=-(n+6)(n-3)$

c. $a\cdot c=20\cdot -3=-60$ Factors of -60 that add to 11 are 15 and -4. Factor by grouping.

$20y^2+11y-3=20y^2+15y-4y-3$

$=(20y^2+15y)-(4y+3)$

$=5y(4y+3)-1(4y+3)$

$=(4y+3)(5y-1)$

51. **a.** Using $a^2-b^2=(a-b)(a+b)$.

$25-x^2=(5-x)(5+x)$

b. Using $a^2-b^2=(a-b)(a+b)$.

$9y^2-1=(3y-1)(3y+1)$

c. Factor using $a^2-2ab+b^2=(a-b)^2$.

$9t^2-12t+4=(3t-2)^2$

d. x^2+16 is prime.

e. $3y^2-12=3(y^2-4)$

Using $a^2-b^2=(a-b)(a+b)$.

$3(y^2-4)=3(y-2)(y+2)$

52. **a.** This is an equation of degree 1, therefore it is linear.

b. This is an equation of degree 5 (it has a variable with an exponent of 5), therefore it is neither linear nor quadratic.

c. This is an equation of degree 2 (the variable has an exponent of 2), therefore it is quadratic.

53. **a.** $a \cdot c = 3 \cdot -8 = -24$ Factors of -24 that add to -10 are -12 and 2. Factor by grouping.

$$3x^2 - 10x - 8 = 0$$
$$3x^2 - 12x + 2x - 8 = 0$$
$$(3x^2 - 12x) + (2x - 8) = 0$$
$$3x(x-4) + 2(x-4) = 0$$
$$(x-4)(3x+2) = 0$$

$$x - 4 = 0 \quad +4\ +4 \quad x = 4$$

$$3x + 2 = 0 \quad -2\ -2 \quad 3x = -2 \quad \frac{3x}{3} = \frac{-2}{3} \quad x = -\frac{2}{3}$$

b. $3x(x+5) = 0$

$$3x = 0 \quad x = 0$$

$$x + 5 = 0 \quad -5\ -5 \quad x = -5$$

c. Using $a^2 - b^2 = (a-b)(a+b)$.

$$n^2 - 81 = 0$$
$$(n-9)(n+9) = 0$$

$$n - 9 = 0 \quad +9\ +9 \quad n = 9$$

$$n + 9 = 0 \quad -9\ -9 \quad n = -9$$

54.

$$\frac{2(2)-4}{(2)+2}$$ Substitute in $x = 2$.

$$= \frac{4-4}{4}$$

$$= \frac{0}{4} = 0$$ Simplify.

55. **a.**

$$c = \frac{60}{(10)}$$ Substitute in $n = 10$.

$$c = 6$$

The cost is \$6 per person if 10 parents agree to help pay for the plaque.

b.

$$c = \frac{60}{(15)}$$ Substitute in $n = 15$.

$$c = 4$$

The lowest possible cost per parent is \$4.

56. **a.** Set denominator equal to 0 and solve:

$$\frac{3x+1}{2x-7}$$
$$2x - 7 = 0$$
$$2x = 7$$
$$\frac{2x}{2} = \frac{7}{2}$$
$$x = \frac{7}{2}$$ This is the excluded value.

b. Set denominator equal to 0 and solve:

$$\frac{5t}{(t+3)(3t-4)}$$
$$(t+3)(3t-4) = 0$$

$$t + 3 = 0 \quad t = -3$$

$$3t - 4 = 0 \quad 3t = 4 \quad \frac{3t}{3} = \frac{4}{3} \quad t = \frac{4}{3}$$

$$t = -3, \frac{4}{3}$$ These are the excluded values.

c. Set denominator equal to 0 and solve:

$$\frac{3y}{6y^2 + 13y - 5}$$
$$6y^2 + 13y - 5 = 0$$
$$6y^2 + 15y - 2y - 5 = 0$$
$$(6y^2 + 15y) - (2y + 5) = 0$$
$$3y(2y+5) - 1(2y+5) = 0$$
$$(2y+5)(3y-1) = 0$$

$$2y + 5 = 0 \quad 2y = -5 \quad \frac{2y}{2} = \frac{-5}{2} \quad y = -\frac{5}{2}$$

$$3y - 1 = 0 \quad 3y = 1 \quad \frac{3y}{3} = \frac{1}{3} \quad y = \frac{1}{3}$$

$$y = -\frac{5}{2}, \frac{1}{3}$$ These are the excluded values.

57. **a.** $\frac{-6\cancel{(2x+5)}}{x\cancel{(2x+5)}} = -\frac{6}{x}$

b. $\frac{n-4}{n^2 - 3n - 4} = \frac{\cancel{n-4}}{\cancel{(n-4)}(n+1)} = \frac{1}{n+1}$

c. $\dfrac{t-2}{t^2-4}=\dfrac{\cancel{t-2}}{(\cancel{t-2})(t+2)}=\dfrac{1}{t+2}$

58. **a.**

$$T=\frac{6(10)}{6+(10)} \qquad \text{Substitute in } x=10.$$

$$T=\frac{60}{16}=3.75$$

The two office specialists working together can complete the job in 3.75 hours, or 3 hours 45 minutes.

b.

$$T=\frac{6(8)}{6+(8)} \qquad \text{Substitute in } x=8.$$

$$T=\frac{48}{14}\approx 3.43$$

The two office specialists working together can complete the job in about 3.43 hours, or 3 hours 26 minutes.

59. **a.** $\dfrac{5n+7}{2n^2+17n+35}$ is simplified.

b.

$$\frac{4x^2-19x-5}{(x+7)(x-5)}=\frac{4x^2-20x+x-5}{(x+7)(x-5)}$$

$$=\frac{(4x^2-20x)+(x-5)}{(x+7)(x-5)}=\frac{4x(x-5)+1(x-5)}{(x+7)(x-5)}$$

$$=\frac{(\cancel{x-5})(4x+1)}{(x+7)(\cancel{x-5})}=\frac{4x+1}{x+7}$$

c. $\dfrac{12-4x}{x-3}=\dfrac{-4x+12}{x-3}=\dfrac{-4(\cancel{x-3})}{\cancel{x-3}}=-4$

60. **a.** $\dfrac{-4}{9}\cdot\dfrac{27}{30}=-\dfrac{\cancel{2}\cdot 2}{\cancel{3}\cdot\cancel{3}}\cdot\dfrac{\cancel{3}\cdot\cancel{3}\cdot\cancel{3}}{\cancel{2}\cdot\cancel{3}\cdot 5}=-\dfrac{2}{5}$

b. $\dfrac{-5xy}{15x^2}\cdot\dfrac{45x}{12y^2}$

$$=-\frac{\cancel{5}\cancel{x}\cancel{y}}{\cancel{3}\cdot\cancel{5}\cdot\cancel{x^2}}\cdot\frac{\cancel{3}\cdot\cancel{3}\cdot 5\cdot\cancel{x}}{2\cdot 2\cdot\cancel{3}\cdot y^{\cancel{2}1}}$$

$$=-\frac{5}{4y}$$

c. $\dfrac{x^2+x-12}{x^2+3x-10}\cdot\dfrac{(x-2)}{(x+4)}$

$$=\frac{(\cancel{x+4})(x-3)}{(x+5)(\cancel{x-2})}\cdot\frac{(\cancel{x-2})}{(\cancel{x+4})}=\frac{x-3}{x+5}$$

d. $\dfrac{y^2+5y+6}{y^2+y-20}\cdot\dfrac{y^2+4y-5}{y^2+8y+12}$

$$=\frac{(y+3)(\cancel{y+2})}{(\cancel{y+5})(y-4)}\cdot\frac{(\cancel{y+5})(y-1)}{(y+6)(\cancel{y+2})}$$

$$=\frac{(y+3)(y-1)}{(y-4)(y+6)}$$

61. $\dfrac{\$6.29}{1\ \cancel{\text{lb}}}\cdot\dfrac{1\ \cancel{\text{lb}}}{16\text{ oz}}=\dfrac{\$6.29}{16\text{ oz}}\approx \$0.39$ per oz.

The price for a 1-pound jar of barbeque sauce is \$0.39 per ounce.

62.

$$\frac{35\ \cancel{\text{mile}}}{1\text{ hr}}\cdot\frac{5280\text{ ft}}{1\ \cancel{\text{mile}}}=\frac{184{,}800\text{ ft}}{1\text{ hr}}$$

$$\frac{184{,}800\text{ ft}}{1\ \cancel{\text{hr}}}\cdot\frac{1\ \cancel{\text{hr}}}{60\text{ min}}=\frac{184{,}800\text{ ft}}{60\text{ min}}=3080\text{ ft/min}$$

The car's speed is 3080 feet per minute.

63. **a.** $\dfrac{-3}{7}\div\dfrac{9}{28}=\dfrac{-\cancel{3}}{\cancel{7}}\cdot\dfrac{2\cdot 2\cdot\cancel{7}}{\cancel{3}\cdot 3}=-\dfrac{4}{3}$

b. $\dfrac{-16x}{5y^2}\div\dfrac{24x^2}{35y}$

$$=-\frac{\cancel{2}\cdot\cancel{2}\cdot\cancel{2}\cdot 2\cdot\cancel{x}}{\cancel{5}y^{\cancel{2}1}}\cdot\frac{\cancel{5}\cdot 7\cdot\cancel{y}}{\cancel{2}\cdot\cancel{2}\cdot\cancel{2}\cdot 3\cdot x^{\cancel{2}1}}=-\frac{14}{3xy}$$

c. $\dfrac{(t+2)(t-3)}{(t+1)(t-4)}\div\dfrac{(t^2-9)}{(t^2-1)}$

$$=\frac{(t+2)(t-3)}{(t+1)(t-4)}\cdot\frac{(t^2-1)}{(t^2-9)}$$

$$=\frac{(t+2)(\cancel{t-3})}{(\cancel{t+1})(t-4)}\cdot\frac{(t-1)(\cancel{t+1})}{(\cancel{t-3})(t+3)}$$

$$=\frac{(t+2)(t-1)}{(t-4)(t+3)}=\frac{t^2+t-2}{(t-4)(t+3)}$$

d. $\dfrac{a^2+4a+3}{a^2+2a-8}\div\dfrac{a^2-a-2}{a^2-16}$

$$=\frac{a^2+4a+3}{a^2+2a-8}\cdot\frac{a^2-16}{a^2-a-2}$$

$$=\frac{(a+3)(\cancel{a+1})}{(\cancel{a+4})(a-2)}\cdot\frac{(a-4)(\cancel{a+4})}{(a-2)(\cancel{a+1})}$$

$$=\frac{(a+3)(a-4)}{(a-2)^2}=\frac{a^2-a-12}{(a-2)^2}$$

64. **a.**

$$\frac{\frac{-2x^2}{15y}}{\frac{12x^3}{5y^2}} = \frac{-2x^2}{15y} \div \frac{12x^3}{5y^2} = \frac{-2x^2}{15y} \cdot \frac{5y^2}{12x^3}$$

$$= \frac{-\cancel{2}\cancel{x^2}}{3 \cdot \cancel{5} \cdot \cancel{y}} \cdot \frac{\cancel{5}y^{\cancel{2}1}}{\cancel{2} \cdot 2 \cdot 3 \cdot x^{\cancel{3}1}} = -\frac{y}{18x}$$

b.

$$\frac{\frac{b+2}{2b^2-8}}{\frac{b+5}{b-2}} = \frac{b+2}{2b^2-8} \div \frac{b+5}{b-2} = \frac{b+2}{2b^2-8} \cdot \frac{b-2}{b+5}$$

$$= \frac{\cancel{b+2}}{2\cancel{(b-2)}\cancel{(b+2)}} \cdot \frac{\cancel{b-2}}{b+5} = \frac{1}{2(b+5)}$$

65. **a.** $\frac{6x}{x+1} + \frac{6}{x+1} = \frac{6x+6}{x+1} = \frac{6\cancel{(x+1)}}{\cancel{x+1}} = 6$

b. $\frac{6t^2}{t-5} - \frac{30t}{t-5} = \frac{6t^2-30t}{t-5} = \frac{6t\cancel{(t-5)}}{\cancel{t-5}} = 6t$

c. $\frac{y+6}{(y-2)(y+3)} - \frac{(y-2)}{(y-2)(y+3)}$

$$= \frac{y+6-y+2}{(y-2)(y+3)} = \frac{8}{(y-2)(y+3)}$$

d. $\frac{x-7}{(x-1)(x+5)} + \frac{3x}{(x-1)(x+5)}$

$$= \frac{x-7+3x}{(x-1)(x+5)} = \frac{4x-7}{(x-1)(x+5)}$$

66. **a.** $\frac{-4}{5x^2y} \qquad \frac{1}{2 \cdot 5 \cdot xy}$

$\text{LCD} = 2 \cdot 5 \cdot x^2y = 10x^2y$

b. $\frac{-7b}{b-1} \qquad \frac{6}{b-2}$

$\text{LCD} = (b-1)(b-2)$

c. $\frac{15}{4h(h-3)} \qquad \frac{-2}{h-3}$

$\text{LCD} = 4h(h-3)$

d. $\frac{-2}{(a-5)(a-3)} \qquad \frac{5}{(a+4)(a-3)}$

$\text{LCD} = (a-5)(a-3)(a+4)$

67. **a.** $\frac{3}{x-1} \qquad \frac{-2x}{x+4}$

$\text{LCD} = (x-1)(x+4)$

$$\frac{3}{x-1} \cdot \frac{x+4}{x+4} \qquad \frac{-2x}{x+4} \cdot \frac{x-1}{x-1}$$

$$\frac{3x+12}{(x-1)(x+4)} \qquad \frac{-2x^2+2x}{(x-1)(x+4)}$$

b. $\frac{5k}{(k-6)(k+6)} \qquad \frac{8}{(k-6)(k-5)}$

$\text{LCD} = (k-6)(k+6)(k-5)$

$$\frac{5k}{(k-6)(k+6)} \cdot \frac{k-5}{k-5} \qquad \frac{8}{(k-6)(k-5)} \cdot \frac{k+6}{k+6}$$

$$\frac{5k^2-25k}{(k-6)(k+6)(k-5)} \qquad \frac{8k+48}{(k-6)(k+6)(k-5)}$$

68. **a.**

$$\frac{3\cancel{x}}{2\cancel{x}y} + \frac{7\cancel{x}}{\cancel{x}y} \qquad \text{LCD} = 2y$$

$$= \frac{3}{2y} + \frac{7}{y} \cdot \frac{2}{2} = \frac{3}{2y} + \frac{14}{2y}$$

$$= \frac{3+14}{2y} = \frac{17}{2y}$$

b.

$$\frac{y}{y-1} - \frac{5y-2}{(y-4)(y-1)}$$

$\text{LCD} = (y-4)(y-1)$

$$= \frac{y}{y-1} \cdot \frac{y-4}{y-4} - \frac{5y-2}{(y-4)(y-1)}$$

$$= \frac{y^2-4y}{(y-4)(y-1)} - \frac{5y-2}{(y-4)(y-1)}$$

$$= \frac{y^2-4y-5y+2}{(y-4)(y-1)} = \frac{y^2-9y+2}{(y-4)(y-1)}$$

69. **a.**

$$\frac{\frac{10}{x}-5}{\frac{6}{x}-3} \cdot \frac{\frac{x}{1}}{\frac{x}{1}} \qquad \text{Multiply by LCD} = x.$$

$$= \frac{\frac{10}{x} \cdot \frac{x}{1} - 5 \cdot \frac{x}{1}}{\frac{6}{x} \cdot \frac{x}{1} - 3 \cdot \frac{x}{1}} = \frac{10-5x}{6-3x}$$

$$= \frac{-5x+10}{-3x+6} = \frac{-5\cancel{(x-2)}}{-3\cancel{(x-2)}} = \frac{5}{3}$$

b.

$$\frac{\frac{3}{x}-\frac{2}{y}}{\frac{1}{x}+\frac{4}{y}}\cdot\frac{\frac{xy}{1}}{\frac{xy}{1}} \quad \text{Multiply by LCD} = xy.$$

$$=\frac{\frac{3}{x}\cdot\frac{xy}{1}-\frac{2}{y}\cdot\frac{xy}{1}}{\frac{1}{x}\cdot\frac{xy}{1}+\frac{4}{y}\cdot\frac{xy}{1}}=\frac{3y-2x}{y+4x}=\frac{-2x+3y}{4x+y}$$

c.

$$\frac{k-2}{1}\left(\frac{2}{k-2}+4\right)=(k+3)\frac{k-2}{1}$$

$$\frac{k-2}{1}\left(\frac{2}{k-2}+4\right)=(k+3)\frac{k-2}{1}$$

$$2+4(k-2)=k^2-2k+3k-6$$

$$2+4k-8=k^2-2k+3k-6$$

$$4k-6=k^2+k-6$$

$$\underline{-4k+6 \qquad -4k+6}$$

$$0=k^2-3k$$

$$0=k(k-3)$$

$$k=0 \qquad k-3=0$$

$$k=3$$

70. a.

$$\frac{3x}{1}\left(6+\frac{3}{x}\right)=\left(\frac{23}{3}-\frac{2}{x}\right)\frac{3x}{1}$$

$$\frac{3x}{1}\cdot 6+\frac{3x}{1}\cdot\frac{3}{x}=\frac{3x}{1}\cdot\frac{23}{3}-\frac{3x}{1}\cdot\frac{2}{x}$$

$$18x+9=23x-6$$

$$\underline{-9 \qquad -9}$$

$$18x=23x-15$$

$$18x=23x-15$$

$$\underline{-23x \quad -23x}$$

$$-5x=-15$$

$$\frac{-5x}{-5}=\frac{-15}{-5}$$

$$x=3$$

b.

$$\frac{y-3}{1}\left(\frac{4}{y-3}+2\right)=\left(\frac{4}{y-3}\right)\frac{y-3}{1}$$

$$\frac{y-3}{1}\cdot\frac{4}{y-3}+\frac{y-3}{1}\cdot 2=\frac{y-3}{1}\cdot\frac{4}{y-3}$$

$$4+2y-6=4$$

$$2y-2=4$$

$$\underline{+2 \quad +2}$$

$$2y=6$$

$$\frac{2y}{2}=\frac{6}{2}$$

$$y=3$$

This is an excluded value

No solution

71. a.

$$C=\frac{100+10(25)}{(25)} \quad \text{Substitute } n = 25.$$

$$C=\frac{100+250}{25}=\frac{350}{25}=14$$

The average cost per person is \$14 if 25 people attend the retirement lunch.

b.

$$(12)=\frac{100+10n}{n} \quad \text{Substitute } C = 12.$$

$$12n=100+10n \quad \text{Cross multiply.}$$

$$\underline{-10n \qquad -10n}$$

$$2n=100$$

$$\frac{2n}{2}=\frac{100}{2}$$

$$n=50$$

There must be 50 people who attend the lunch for the average cost per person to be \$12.

72. a.

$$\frac{6}{a+1}=\frac{-3}{a-2} \quad \text{Cross multiply.}$$

$$-3(a+1)=6(a-2)$$

$$-3a-3=6a-12$$

$$\underline{-6a \qquad -6a}$$

$$-9a-3=-12$$

$$\underline{+3 \qquad +3}$$

$$-9a=-9$$

$$\frac{-9a}{-9}=\frac{-9}{-9}$$

$$a=1$$

b.

$$\frac{5b}{1}\left(\frac{3}{5b}-\frac{1}{5}\right)=\left(\frac{-1}{b}\right)\frac{5b}{1}$$

$$\frac{5b}{1}\cdot\frac{3}{5b}-\frac{5b}{1}\cdot\frac{1}{5}=\frac{5b}{1}\cdot\frac{-1}{b}$$

$$3-b=-5$$

$$\underline{-3\quad\ -3}$$

$$-b=-8$$

$$\frac{-b}{-1}=\frac{-8}{-1}$$

$$b=8$$

c. $\frac{(x+4)(x-2)}{1}\cdot\frac{3}{x+4}+\frac{4}{x-2}=\frac{x^2-8}{(x+4)(x-2)}\cdot\frac{(x+4)(x-2)}{1}$

$$3(x-2)+4(x+4)=x^2-8$$

$$3x-6+4x+16=x^2-8$$

$$7x+10=x^2-8$$

$$\underline{-x^2+8\ -x^2+8}$$

$$-x^2+7x+18=0$$

$$-(x^2-7x-18)=0$$

$$-(x-9)(x+2)=0$$

$$x-9=0 \qquad x+2=0$$

$$x=9 \qquad x=-2$$

73. Time $=3$ hr

Rate $=\frac{1}{3}$

Kenzo's work rate is $\frac{1}{3}$ per hour.

74. New copier's work time $=4$ hr. Rate $=\frac{1}{4}$

Old copier's work time $=6$ hr. Rate $=\frac{1}{6}$

$$\frac{1}{4}+\frac{1}{6}=\frac{1}{t}$$

$$\frac{12t}{1}\left(\frac{1}{4}+\frac{1}{6}\right)=\left(\frac{1}{t}\right)\frac{12t}{1}$$

$$\frac{12t}{1}\cdot\frac{1}{4}+\frac{12t}{1}\cdot\frac{1}{6}=\frac{12t}{1}\cdot\frac{1}{t}$$

$$3t+2t=12$$

$$5t=12$$

$$\frac{5t}{5}=\frac{12}{5}$$

$$t=\frac{12}{5}=2.4$$

It will take 2.4 hours, or 2 hours 24 minutes, to copy the announcements using both machines.

75. The ratio of 9 to 45 $=\frac{9}{45}=\frac{1}{5}$

76. 400 miles for 16 gal. $=\frac{400\text{mi}}{16\text{gal}}=25$ mi/gal

77. 36 oz of coffee beans for \$31.00 $=\frac{\$31}{36\text{ oz}}\approx\$0.86$ per ounce.

48 oz of coffee beans for \$25.00 $=\frac{\$25}{48\text{ oz}}\approx\$0.52$ per ounce.

The better buy is 48 oz of coffee beans for \$25.00.

78.

$$\frac{1}{16}=\frac{x}{40}$$

$$16x=40$$

$$\frac{16x}{16}=\frac{40}{16}$$

$$x=2.5$$

The length of the wall on the blueprint will be 2.5 in.

79. **a.**

$$\frac{3}{n}=\frac{-15}{2} \quad \text{Cross multiply.}$$

$$-15n=2\cdot3$$

$$-15n=6$$

$$\frac{-15n}{-15}=\frac{6}{-15}$$

$$n=-\frac{2}{5}$$

b.

$$\frac{x-2}{2}=\frac{13x+10}{14} \quad \text{Cross multiply.}$$

$$2(13x+10)=14(x-2)$$

$$26x+20=14x-28$$

$$\underline{-20\qquad -20}$$

$$26x=14x-48$$

$$\underline{-14x\ -14x}$$

$$12x=-48$$

$$\frac{12x}{12}=\frac{-48}{12}$$

$$x=-4$$

c.

$$\frac{3t-2}{t+5}=\frac{25}{14} \quad \text{Cross multiply.}$$
$$25(t+5)=14(3t-2)$$
$$25t+125=42t-28$$
$$\underline{-125 \qquad -125}$$
$$25t=42t-153$$
$$\underline{-42t \;\; -42t}$$
$$-17t=-153$$
$$\frac{-17t}{-17}=\frac{-153}{-17}$$
$$t=9$$

80.

$$d=kt \quad \text{Substitute given values for } d \text{ and } t.$$
$$(150)=k(3)$$
$$\frac{150}{3}=\frac{3k}{3}$$
$$50=k$$

$$d=50t \quad \text{Substitute values for } k \text{ and } t.$$
$$d=50(7.5)$$
$$d=375$$

81. **a.**

$$W=kt \quad \text{Substitute given values for } W \text{ and } t.$$
$$(231)=k(22)$$
$$\frac{231}{22}=\frac{22k}{22}$$
$$10.5=k$$
$$W=10.5t$$

The equation that expresses the relationship is $W=10.5t$

b.

$$W=10.5t \quad \text{Substitute values for } k \text{ and } t.$$
$$W=10.5(32)$$
$$W=336$$

The amount of weekly paycheck is \$336 for 32 hours of work.

82.

$$y=\frac{k}{x} \quad \text{Substitute given values for } x \text{ and } y.$$
$$\left(-\frac{1}{4}\right)=\frac{k}{(12)} \quad \text{Cross multiply.}$$
$$4k=-1\cdot 12$$
$$4k=-12$$
$$\frac{4k}{4}=\frac{-12}{4}$$
$$k=-3$$

$$y=\frac{-3}{x} \quad \text{Substitute values for } k \text{ and } x.$$
$$y=\frac{-3}{(2)}$$
$$y=-\frac{3}{2}$$

83. **a.**

$$s=\frac{k}{t} \quad \text{Substitute given values for } t \text{ and } s.$$
$$(50)=\frac{k}{(4)} \quad \text{Cross multiply.}$$
$$k=50\cdot 4$$
$$k=200$$
$$s=\frac{200}{t}$$

b.

$$s=\frac{200}{t} \quad \text{Substitute value for } t.$$
$$s=\frac{200}{(6)}$$
$$s=33\frac{1}{3}$$

The speed is $33\frac{1}{3}$ mph for 6 hours of driving.

84. **a.** Evaluate $f(x)=-5x+13$ for $x=-4$:

$$f(x)=-5x+13 \quad \text{Substitute in } x=-4.$$
$$f(-4)=-5(-4)+13 \quad \text{Simplify the right side.}$$
$$f(-4)=20+13$$
$$f(-4)=33$$

b. Evaluate $f(x)=-2x^2+5$ for $x=-1$:

$$f(x)=-2x^2+5 \quad \text{Substitute in } x=-1.$$
$$f(-1)=-2(-1)^2+5 \quad \text{Simplify the right side.}$$
$$f(-1)=-2+5$$
$$f(-1)=3$$

85. **a.** Let P the population of the United States in millions, and let t be the years since 2000.

$$P=mt+b$$

Using two points: $(4,293)$ and $(7,301)$

find the slope: $m=\dfrac{301-293}{7-4}\approx 2.7$

$$P=2.7t+b$$
$$b=301-2.7(7)$$
$$b=282.1$$
$$P=2.7t+282.1$$

b.

$P = 2.7(12) + 282.1$

$P = 314.5$

The population of the United States will be 314.5 million in 2012.

c. Domain $[0,13]$

$P = 2.7(13) + 282.1$

$P = 317.2$

Range $[282.1, 317.2]$

d. Vertical intercept: $(0, 282.1)$. The population of the United States was about 282.1 million in 2000.

e.

$325 = 2.7t + 282.1$

$-282.1 \quad -282.1$

$42.9 = 2.7t$

$\frac{42.9}{2.7} = \frac{2.7t}{2.7}$

$15.9 \approx t$

The U.S. population will reach 325 million in about 2016. This might be model breakdown.

86. Let $R(t)$ be the total revenue in billions of dollars for Cisco Systems, Inc t years since 2000.

a..

$R(t) = mt + b$

Using two points: $(5, 24.8)$ and $(8, 39.5)$

Find the slope: $m = \frac{39.5 - 24.8}{8 - 5} = 4.9$

$b = 39.5 - 4.9(8)$

$b = 0.3$

$R(t) = 4.9t + 0.3$

b. Vertical intercept: $(0, 0.3)$. The total revenue for Cisco Systems, Inc. in 2000, was about $300 million.

c. Domain $[3,10]$

$R(3) = 4.9(3) + 0.3 = 15$

$R(8) = 4.9(10) + 0.3 = 49.3$

Range $[15, 49.3]$

d.

$R(10) = 4.9(10) + 0.3 = 49.3$

The model predicts Cisco Systems, Inc.'s revenue at $49.3 billion in 2010.

e.

$4.9t + 0.3 = 64$

$4.9t + 0.3 - 0.3 = 64 - 0.3$

$4.9t = 63.7$

$\frac{4.9t}{4.9} = \frac{63.7}{4.9}$

$t = 13$

The model predicts Cisco Systems, Inc.'s revenue at $64 billion in 2013. This might be model breakdown.

f. The slope is 4.9, which means Cisco Systems, Inc.'s revenue is increasing at a rate of $4.9 billion per year.

87. Domain: All real numbers.

88. $g(x) = \frac{3x - 5}{2x^2 - x - 3} = \frac{3x - 5}{(2x - 3)(x + 1)}$

The denominator cannot equal zero. Set the linear factors not equal to zero and solve:

$2x - 3 \neq 0,\ x + 1 \neq 0$

$x \neq \frac{3}{2},\ x \neq -1$

Domain: x is all real numbers except $x \neq \frac{3}{2},\ x \neq -1$.

89. $H(6) = 7$

On the 6th day of the month, a person sleeps 7 hours.

90. $P(d) = -3d + 100$

a.

$P(14) = -3(14) + 100$

$P(14) = 58$

58 people out of 100 remain on the plan 14 days after starting the diet.

b. The slope is -3. The number of people remaining on the diet plan decreases by three people per day.

91. $f(x) = 9x - 45$

a.

$f(3) = 9(3) - 45$

$f(3) = 27 - 45$

$f(3) = -18$

b.

$9x - 45 = 54$

$9x = 99$

$\frac{9x}{9} = \frac{99}{9}$

$x = 11$

c. Domain: All real numbers, range: All real numbers

92. $h(x) = -\frac{5}{8}x + 11$

a.

$h(32) = -\frac{5}{8}(32) + 11$

$h(32) = -20 + 11$

$h(32) = -9$

b.

$$-\frac{5}{8}x+11=\frac{13}{8}$$
$$8\left(-\frac{5}{8}x+11\right)=8\left(\frac{13}{8}\right)$$
$$-5x+88=13$$
$$-5x+88-88=13-88$$
$$-5x=-75$$
$$\frac{-5x}{-5}=\frac{-75}{-5}$$
$$x=15$$

c. Domain: All real numbers, range: All real numbers

93. $g(x)=7$

a. $g(10)=7$

b. Domain: All real numbers, range: $\{7\}$

94. **a.** $f(6)=8$

b.

$$f(x)=0$$
$$f(-6)=0$$
$$x=-6$$

c. Domain: All real numbers, range: All real numbers

95. $f(x)=(2x-4)$ and $g(x)=(3x+7)$

a.

$$f(x)+g(x)=(2x-4)+(3x+7)$$
$$=5x+3$$

b.

$$f(x)-g(x)=(2x-4)-(3x+7)$$
$$=2x-4-3x-7$$
$$=-x-11$$

c.

$$f(x)g(x)=(2x-4)(3x+7)$$
$$=6x^2+14x-12x-28$$
$$=6x^2+2x-28$$

96. $f(x)=(x+3)$ and $g(x)=(x^2+5)$

a.

$$f(x)+g(x)=(x+3)+(x^2+5)$$
$$=x^2+x+8$$

b.

$$f(x)-g(x)=(x+3)-(x^2+5)$$
$$=x+3-x^2-5$$
$$=-x^2+x-2$$

c.

$$f(x)g(x)=(x+3)(x^2+5)$$
$$=x^3+5x+3x^2+15$$
$$=x^3+3x^2+5x+15$$

97.

$$(f\circ g)(x)=f(g(x))$$
$$=f(x^2+1)$$
$$=2(x^2+1)-4$$
$$=2x^2+2-4$$
$$=2x^2-2$$

98.

$$(f\circ g)(x)=f(g(x))$$
$$=f(3x+2)$$
$$=(3x+2)^2+(3x+2)$$
$$=9x^2+12x+4+3x+2$$
$$=9x^2+15x+6$$

99.

$$S(t)>C(t)$$
$$2t+17>-2t+82$$
$$\underline{+2t-17\quad +2t-17}$$
$$4t>65$$
$$\frac{4t}{4}>\frac{65}{4}$$
$$t>16.25$$

The percentage of Canadian TV subscribers that use satellite services will be greater than those who use cable services after 2016.

100. $Y_1<Y_2$ when $x<4$

101. $f(x)\ge g(x)$ when $x\le 3$.

102. Let $P(s)$ be Sandra's weekly salary for s dollars of sales.

$$P(s)=400+0.06s$$
$$400+0.06s\ge 565$$
$$400-400+0.06s\ge 565-400$$
$$0.06s\ge 165$$
$$\frac{0.06s}{0.06}\ge\frac{165}{0.06}$$
$$s\ge 2750$$

Sandra must have \$2750 or more per week in sales to earn at least \$565 per week.

103.

$$|x+2|+6=18$$
$$|x+2|+6-6=18-6$$
$$|x+2|=12$$
$$x+2=12 \text{ or } x+2=-12$$
$$x=10 \text{ or } x=-14$$

104.

$$3|a-8|-4=44$$
$$3|a-8|-4+4=44+4$$
$$3|a-8|=48$$
$$\frac{3|a-8|}{3}=\frac{48}{3}$$
$$|a-8|=16$$
$$a-8=16 \text{ or } a-8=-16$$
$$a=24 \text{ or } a=-8$$

105.

$$|4n+9|+20=7$$
$$|4n+9|+20-20=7-20$$
$$|4n+9|=-13$$

There is no solution, because an absolute value cannot equal a negative number.

106.

$$-2|3r+5|+15=7$$
$$-2|3r+5|+15-15=7-15$$
$$-2|3r+5|=-8$$
$$\frac{-2|3r+5|}{-2}=\frac{-8}{-2}$$
$$|3r+5|=4$$
$$3r+5=4 \text{ or } 3r+5=-4$$
$$3r=-1 \text{ or } 3r=-9$$
$$r=-\frac{1}{3} \text{ or } r=-3$$

107.

$$|x-2|<7$$
$$-7<x-2<7$$
$$-7+2<x-2+2<7+2$$
$$-5<x<9$$

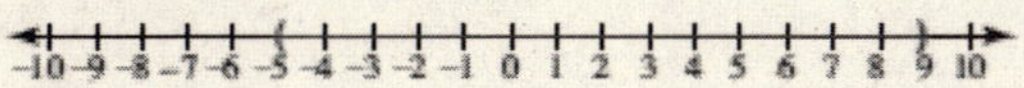

108.

$$|2x+3|\le 15$$
$$-15\le 2x+3\le 15$$
$$-15-3\le 2x+3-3\le 15-3$$
$$-18\le 2x\le 12$$
$$\frac{-18}{2}\le\frac{2x}{2}\le\frac{12}{2}$$
$$-9\le x\le 6$$

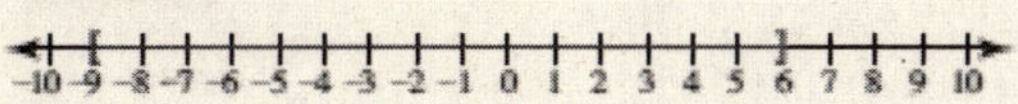

109.

$$|x+4|\ge 6$$
$$x+4\le -6 \text{ or } x+4\ge 6$$
$$x\le -10 \text{ or } x\ge 2$$

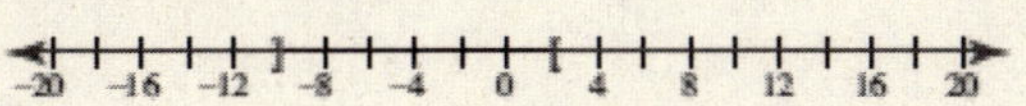

110.

$$|3x-7|+10>15$$
$$|3x-7|+10-10>15-10$$
$$|3x-7|>5$$
$$3x-7<-5 \text{ or } 3x-7>5$$
$$3x<2 \text{ or } 3x>12$$
$$x<\frac{2}{3} \text{ or } x>4$$

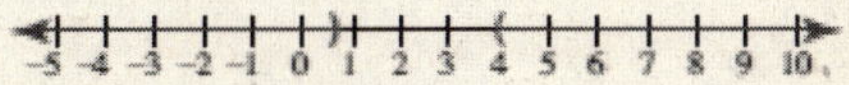

111. $x^2-7x+10>0$

$$x^2-7x+10=0$$
$$(x-5)(x-2)=0$$
$$x-5=0 \text{ or } x-2=0$$
$$x=5 \text{ or } x=2$$

Check the intervals:

$x=0$	$x=3$	$x=7$
$(0-5)(0-2)\overset{?}{>}0$	$(3-5)(3-2)\overset{?}{>}0$	$(7-5)(7-2)\overset{?}{>}0$
$(-5)(-2)\overset{?}{>}0$	$(-2)(1)\overset{?}{>}0$	$(2)(5)\overset{?}{>}0$
$10\overset{?}{>}0$	$-2\overset{?}{>}0$	$10\overset{?}{>}0$
True	False	True

$(-\infty,2)$ or $(5,\infty)$

112. $\dfrac{x+5}{x-8}\le 0$

Find the zero value:

$$x+5=0$$
$$x=-5$$

Find the excluded value:

$$x-8=0$$
$$x=8$$

Check the intervals:

$x=-7$	$x=0$	$x=10$
$\frac{-7+5}{-7-8}\overset{?}{\le}0$	$\frac{0+5}{0-8}\overset{?}{\le}0$	$\frac{10+5}{10-8}\overset{?}{\le}0$
$\frac{-2}{-15}\overset{?}{\le}0$	$\frac{5}{-8}\overset{?}{\le}0$	$\frac{15}{2}\overset{?}{\le}0$
$\frac{2}{15}\overset{?}{\le}0$	$\frac{-5}{8}\overset{?}{\le}0$	False
False	True	

$[-5,8)$

113. $x^2+8x-9\le 0$

$$x^2+8x-9=0$$
$$(x+9)(x-1)=0$$
$$x+9=0 \text{ or } x-1=0$$
$$x=-9 \text{ or } x=1$$

Check the intervals:

$x=-11$	$x=0$	$x=2$
$(-11+9)(-11-1)\overset{?}{\le}0$	$(0+9)(0-1)\overset{?}{\le}0$	$(2+9)(2-1)\overset{?}{\le}0$
$(-2)(-12)\overset{?}{\le}0$	$(9)(-1)\overset{?}{\le}0$	$(11)(1)\overset{?}{\le}0$
$24\overset{?}{\le}0$	$-9\overset{?}{\le}0$	$11\overset{?}{\le}0$
False	True	False

$-9 \le x \le 1$

114. $\dfrac{x-3}{x^2+9x+14} \ge 0$

$\dfrac{x-3}{(x+7)(x+2)} \ge 0$

Find the zero value:
$x-3=0$
$x=3$

Find the excluded values:
$x+7=0$ or $x+2=0$
$x=-7$ or $x=-2$

Check the intervals:

$x=-8$	$x=-4$
$\dfrac{-8-3}{(-8+7)(-8+2)}\overset{?}{\ge}0$	$\dfrac{-4-3}{(-4+7)(-4+2)}\overset{?}{\ge}0$
$\dfrac{-11}{(-1)(-6)}\overset{?}{\ge}0$	$\dfrac{-7}{(3)(-2)}\overset{?}{\ge}0$
$\dfrac{-11}{6}\overset{?}{\ge}0$	$\dfrac{7}{6}\overset{?}{\ge}0$
False	True

$x=0$	$x=4$
$\dfrac{0-3}{(0+7)(0+2)}\overset{?}{\ge}0$	$\dfrac{4-3}{(4+7)(4+2)}\overset{?}{\ge}0$
$\dfrac{-3}{(7)(2)}\overset{?}{\ge}0$	$\dfrac{1}{(11)(6)}\overset{?}{\ge}0$
$\dfrac{-3}{14}\overset{?}{\ge}0$	$\dfrac{1}{66}\overset{?}{\ge}0$
False	True

$-7 < x < -2$ or $x \ge 3$

115.

$$\begin{cases} y \ge x-5 \\ y \le \dfrac{1}{5}x+2 \end{cases}$$

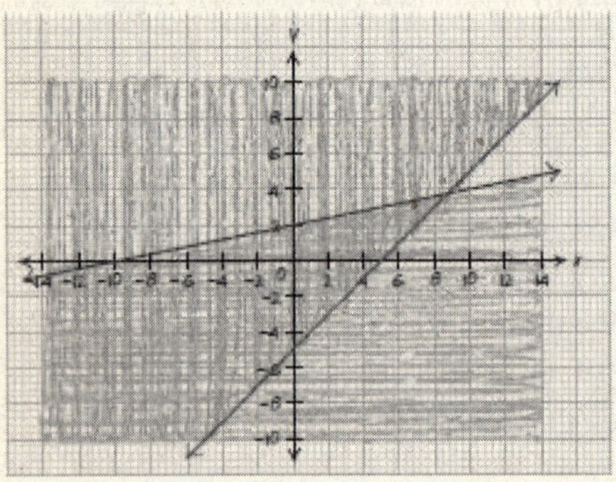

116.

$$\begin{cases} 3x-5y<10 \\ x+2y>12 \end{cases}$$

Solve for y in both equations to graph:

$3x-5y<10 \quad , \quad x+2y>12$

$-5y<-3x+10 \quad , \quad 2y>-x+12$

$\dfrac{-5y}{-5}>\dfrac{-3x+10}{-5} \quad , \quad \dfrac{2y}{2}>\dfrac{-x+12}{2}$

$y>\dfrac{3}{5}x-2 \quad , \quad y>-\dfrac{1}{2}x+6$

$$\begin{cases} y>\dfrac{3}{5}x-2 \\ y>-\dfrac{1}{2}x+6 \end{cases}$$

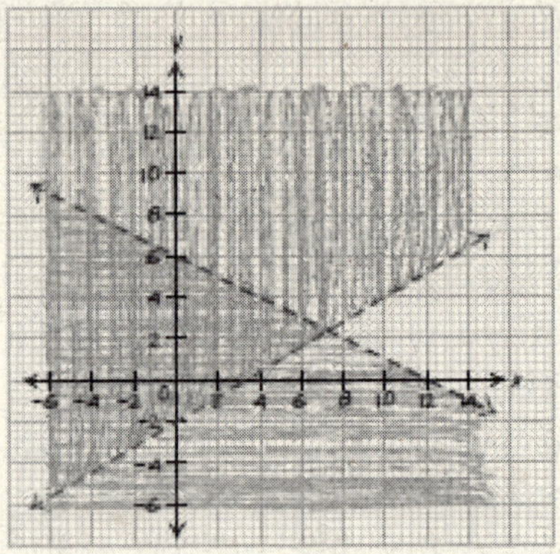

CHAPTER 10 RADICAL FUNCTIONS

Section 10.1 From Squaring a Number to Roots and Radicals

1. $\sqrt{25} = 5$ because $5^2 = 25$

3. $\sqrt{64} = 8$ because $8^2 = 64$

5. $\sqrt{\frac{16}{9}} = \frac{4}{3}$ because $\left(\frac{4}{3}\right)^2 = \frac{16}{9}$

7. $\sqrt{\frac{100}{49}} = \frac{10}{7}$ because $\left(\frac{10}{7}\right)^2 = \frac{100}{49}$

9. $\sqrt{\frac{1}{16}} = \frac{1}{4}$ because $\left(\frac{1}{4}\right)^2 = \frac{1}{16}$

11. $-\sqrt{49} = -1 \cdot \sqrt{49} = -1 \cdot 7 = -7$

13. $-\sqrt{\frac{25}{64}} = -1 \cdot \sqrt{\frac{25}{64}} = -1 \cdot \frac{5}{8} = -\frac{5}{8}$

15. $\sqrt{-9}$ not a real number

17. $\sqrt{\frac{-1}{4}}$ not a real number

19. $\sqrt{15} \approx 3.87\underline{2}98...$

$\sqrt{15} \approx 3.87$

21. $\sqrt{51} \approx 7.14142...$

$\sqrt{51} \approx 7.14$

23. $\sqrt{7} \approx 2.64\underline{5}75...$

$\sqrt{7} \approx 2.65$

25. $\sqrt{-15}$ not a real number

27. $\sqrt{4x}$ Substitute $x = 9$.

$= \sqrt{4(9)} = \sqrt{36} = 6$

29. $\sqrt{x+6}$ Substitute $x = -5$.

$= \sqrt{(-5)+6} = \sqrt{1} = 1$

31. $\sqrt{x+5}$ Substitute $x = -1$.

$= \sqrt{(-1)+5} = \sqrt{4} = 2$

33. $\sqrt{a^2 - 2a}$ Substitute $a = 1$.

$= \sqrt{(1)^2 - 2(1)} = \sqrt{1-2} = \sqrt{-1}$ not a real number

35. $\sqrt{h^2 - 36}$ Substitute $h = 2$.

$= \sqrt{(2)^2 - 36} = \sqrt{4-36} = \sqrt{-32}$ not a real number

37. $\sqrt{1-4a}$ Substitute $a = -3$.

$= \sqrt{1-4(-3)} = \sqrt{1+12} = \sqrt{13}$

39. **a.**

$t = \sqrt{\frac{h}{16}}$ Substitute $h = 50$.

$t = \sqrt{\frac{(50)}{16}} = \sqrt{3.125} \approx 1.77$

It will take the ball approximately 1.77 seconds to fall 50 feet.

b.

$t = \sqrt{\frac{h}{16}}$ Substitute $h = 80$.

$t = \sqrt{\frac{(80)}{16}} = \sqrt{5} \approx 2.24$

It will take the ball approximately 2.24 seconds to fall 80 feet.

41. **a.** The variable s represents the speed of the car in miles per hour and d represents the length in feet of the skid mark left by the car.

b.

$s = \sqrt{13.32d}$ Substitute $d = 200$.

$s = \sqrt{13.32(200)} = \sqrt{2664} \approx 51.61$

The speed of the car was approximately 51.61 mph when the length of the skid mark was 200 feet.

c.

$s = \sqrt{13.32d}$ Substitute $d = 145$.

$s = \sqrt{13.32(145)} = \sqrt{1931.4} \approx 43.95$

The speed of the car was approximately 43.95 mph when the length of the skid mark was 145 feet.

43.

$t = \frac{\sqrt{s}}{4}$ Substitute $s = 8$.

$t = \frac{\sqrt{(8)}}{4} \approx \frac{2.8284...}{4} \approx 0.71$

It takes the ball approximately 0.71 seconds to fall 8 feet.

45.

$P = \sqrt{4n+25}$ Substitute $n = 50$.

$P = \sqrt{4(50)+25} = \sqrt{225} = 15$

Eamond will earn $15 in profit from selling 50 wristbands.

47.

$x = \sqrt[3]{V}$ Substitute $V = 2.7$.

$x = \sqrt[3]{2.7} \approx 1.39$

The length of the side of the cube is approximately 1.39 inches.

49.

$d = 1.5\sqrt{h}$ Substitute $h = 36$.

$d = 1.5\sqrt{36} = 1.5 \cdot 6 = 9$

A person can see a distance of 9 miles if they are 36 feet above the surface of the earth.

51.

$d = \sqrt{(x_2 - x_1)^2 + (y_2 - y_1)^2}$

$d = \sqrt{(10-3)^2 + (8-4)^2}$

$d = \sqrt{(7)^2 + (4)^2} = \sqrt{49+16} = \sqrt{65} \approx 8.06$

53.

$d = \sqrt{(x_2 - x_1)^2 + (y_2 - y_1)^2}$

$d = \sqrt{(-6-[-4])^2 + (-3-7)^2}$

$d = \sqrt{(-2)^2 + (-10)^2} = \sqrt{4+100} = \sqrt{104} \approx 10.20$

55.

$d = \sqrt{(x_2 - x_1)^2 + (y_2 - y_1)^2}$

$d = \sqrt{(-5-0)^2 + (0-6)^2}$

$d = \sqrt{(-5)^2 + (-6)^2} = \sqrt{25+36} = \sqrt{61} \approx 7.81$

57.

$d = \sqrt{(x_2 - x_1)^2 + (y_2 - y_1)^2}$

$d = \sqrt{(-6-11)^2 + (-1-2)^2}$

$d = \sqrt{(-17)^2 + (-3)^2} = \sqrt{289+9} = \sqrt{298} \approx 17.26$

59. $\sqrt{x^2} = x$

61 $\sqrt{a^6} = \sqrt{(a^3)^2} = a^3$

63. $\sqrt{16x^2} = \sqrt{(4x)^2} = 4x$

65. $-\sqrt{x^{20}} = -\sqrt{(x^{10})^2} = -x^{10}$

67. $-\sqrt{a^2} = -a$

69. $-\sqrt{4y^6} = -\sqrt{(2y^3)^2} = -2y^3$

71. $\sqrt{25x^{22}} = \sqrt{(5x^{11})^2} = 5x^{11}$

73. $\sqrt{\frac{x^4}{49}} = \sqrt{\left(\frac{x^2}{7}\right)^2} = \frac{x^2}{7}$

75. $\sqrt{\frac{y^{10}}{36}} = \sqrt{\left(\frac{y^5}{6}\right)^2} = \frac{y^5}{6}$

77. $\sqrt[3]{8} = \sqrt[3]{2^3} = 2$

79. $\sqrt[3]{1} = \sqrt[3]{1^3} = 1$

81. $\sqrt[3]{-8} = \sqrt[3]{(-2)^3} = -2$

83. $\sqrt[3]{-125} = \sqrt[3]{(-5)^3} = -5$

85. $-\sqrt[3]{-8} = -\sqrt[3]{(-2)^3} = -(-2) = 2$

87. $-\sqrt[3]{-1} = -\sqrt[3]{(-1)^3} = -(-1) = 1$

89. $\sqrt[3]{64} = \sqrt[3]{(4)^3} = 4$

91. $\sqrt{100} = \sqrt{(10)^2} = 10$

93. $\sqrt{\frac{1}{49}} = \sqrt{\left(\frac{1}{7}\right)^2} = \frac{1}{7}$

95. $\sqrt{y^8} = \sqrt{(y^4)^2} = y^4$

97. $\sqrt{a^2} = a$

99. $-\sqrt[3]{64} = -\sqrt[3]{(4)^3} = -4$

101. $-\sqrt{25} = -\sqrt{5^2} = -5$

103. $\sqrt[3]{-8} = \sqrt[3]{(-2)^3} = -2$

105. $\sqrt{-4}$ is not a real number

107. $\sqrt{a^2b^8} = \sqrt{(ab^4)^2} = ab^4$

109. $\sqrt{49z^2} = \sqrt{(7z)^2} = 7z$

111. $\sqrt{\frac{x^4}{81}} = \sqrt{\left(\frac{x^2}{9}\right)^2} = \frac{x^2}{9}$

113. $\sqrt[4]{16} = \sqrt[4]{2^4} = 2$

115. $\sqrt[4]{-81}$ not a real number

117. $\sqrt[5]{100{,}000} = \sqrt[5]{10^5} = 10$

119. $\sqrt[5]{(-1)} = \sqrt[5]{(-1)^5} = -1$

Section 10.2 Basic Operations with Radical Expressions

1. $\sqrt{x} = x^{\frac{1}{2}}$

3. $\sqrt{c^3} = \left(c^3\right)^{\frac{1}{2}} = c^{\frac{3}{2}}$

5. $\left(\sqrt[3]{t}\right)^2 = \left(t^{\frac{1}{3}}\right)^2 = t^{\frac{2}{3}}$

7. $\sqrt{5xy} = (5xy)^{\frac{1}{2}} = 5^{\frac{1}{2}}x^{\frac{1}{2}}y^{\frac{1}{2}}$

9.

$$\begin{aligned}\sqrt[7]{4m^3n^6p^2} &= \left(4m^3n^6p^2\right)^{\frac{1}{7}}\\ &= 4^{\frac{1}{7}}\left(m^3\right)^{\frac{1}{7}}\left(n^6\right)^{\frac{1}{7}}\left(p^2\right)^{\frac{1}{7}}\\ &= 4^{\frac{1}{7}}m^{\frac{3}{7}}n^{\frac{6}{7}}p^{\frac{2}{7}}\end{aligned}$$

11. $r^{\frac{1}{3}} = \sqrt[3]{r}$

13. $n^{\frac{2}{3}} = \sqrt[3]{n^2}$

15. $x^{\frac{1}{3}}y^{\frac{1}{3}} = (xy)^{\frac{1}{3}} = \sqrt[3]{xy}$

17. $r^{\frac{1}{5}}s^{\frac{2}{5}} = \left(rs^2\right)^{\frac{1}{5}} = \sqrt[5]{rs^2}$

19. $\left(xy^3z\right)^{\frac{1}{2}} = \sqrt{xy^3z}$

21. $27^{\frac{1}{3}} = 27\text{^}(1/3) = 3$

23. $1{,}000{,}000^{\frac{1}{6}} = 1{,}000{,}000\text{^}(1/6) = 10$

25. $(-64)^{\frac{1}{3}} = -64\text{^}(1/3) = -4$

27. $25^{-\frac{1}{2}} = 25\text{^}(-1/2) = \frac{1}{5}$

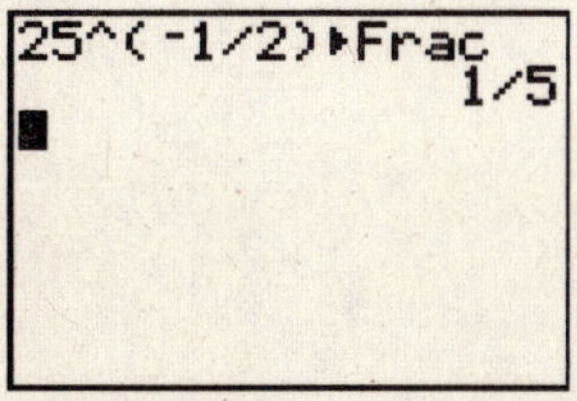

29. $\sqrt{18} = \sqrt{9\cdot 2} = \sqrt{9}\cdot\sqrt{2} = 3\sqrt{2}$

31. $\sqrt{125} = \sqrt{25\cdot 5} = \sqrt{25}\cdot\sqrt{5} = 5\sqrt{5}$

33. $\sqrt{80} = \sqrt{16\cdot 5} = \sqrt{16}\cdot\sqrt{5} = 4\sqrt{5}$

35. $\sqrt{75} = \sqrt{25\cdot 3} = \sqrt{25}\cdot\sqrt{3} = 5\sqrt{3}$

37. $\sqrt{34} = \sqrt{34}$ (no perfect-square factors)

39. $\sqrt{148} = \sqrt{4\cdot 37} = \sqrt{4}\cdot\sqrt{37} = 2\sqrt{37}$

41. **a.** $\sqrt{50} = \sqrt{25\cdot 2} = \sqrt{25}\cdot\sqrt{2} = 5\sqrt{2}$

b. $\sqrt{50} \approx 7.07106 \approx 7.07$

43. **a.** $\sqrt{12} = \sqrt{4\cdot 3} = \sqrt{4}\cdot\sqrt{3} = 2\sqrt{3}$

b. $\sqrt{12} \approx 3.46410 \approx 3.46$

45. **a.** $\sqrt{28} = \sqrt{4\cdot 7} = \sqrt{4}\cdot\sqrt{7} = 2\sqrt{7}$

b. $\sqrt{28} \approx 5.29150 \approx 5.29$

47. **a.** $\sqrt{54} = \sqrt{9\cdot 6} = \sqrt{9}\cdot\sqrt{6} = 3\sqrt{6}$

b. $\sqrt{54} \approx 7.34846 \approx 7.35$

49. $\sqrt{b^9} = \sqrt{b^8\cdot b} = \sqrt{b^8}\cdot\sqrt{b} = b^4\sqrt{b}$

51. $\sqrt{a^4} = \sqrt{\left(a^2\right)^2} = a^2$

53. $\sqrt{x^{15}} = \sqrt{x^{14}\cdot x} = \sqrt{x^{14}}\cdot\sqrt{x} = x^7\sqrt{x}$

55. $\sqrt{y^2} = y$

57. $\sqrt{16x^6} = \sqrt{\left(4x^3\right)^2} = 4x^3$

59. $\sqrt{64a^3} = \sqrt{64a^2\cdot a} = \sqrt{(8a)^2}\cdot\sqrt{a} = 8a\sqrt{a}$

61. $\sqrt{49a^4} = \sqrt{\left(7a^2\right)^2} = 7a^2$

63. $\sqrt{45a^4} = \sqrt{9a^4\cdot 5} = \sqrt{\left(3a^2\right)^2}\cdot\sqrt{5} = 3a^2\sqrt{5}$

65. $\sqrt{40x^3} = \sqrt{4x^2\cdot 10x} = \sqrt{(2x)^2}\cdot\sqrt{10x} = 2x\sqrt{10x}$

67. $\sqrt{75b^{11}} = \sqrt{25b^{10}\cdot 3b} = \sqrt{\left(5b^5\right)^2}\cdot\sqrt{3b} = 5b^5\sqrt{3b}$

69. $\sqrt[3]{8x^3y^6} = \left(2^3x^3y^6\right)^{\frac{1}{3}} = 2\left(x^3\right)^{\frac{1}{3}}\left(y^6\right)^{\frac{1}{3}} = 2xy^2$

71.

$$\begin{aligned}\sqrt[3]{3888x^3y^5z^8} &= \sqrt[3]{216x^3y^3z^6}\cdot\sqrt[3]{18y^2z^2}\\ &= \sqrt[3]{6^3x^3y^3z^6}\cdot\sqrt[3]{2\cdot 3^2y^2z^2}\\ &= 6xyz^2\sqrt[3]{18y^2z^2}\end{aligned}$$

73.

$$\begin{aligned}\sqrt[4]{960m^3n^4p^5} &= \sqrt[4]{16n^4p^4}\cdot\sqrt[4]{60m^3p}\\ &= \sqrt[4]{2^4n^4p^4}\cdot\sqrt[4]{2^2\cdot 3\cdot 5\cdot m^3p}\\ &= 2np\sqrt[4]{60m^3p}\end{aligned}$$

75. $\sqrt[5]{-32c^5d^{10}} = \left[(-2)^5c^5d^{10}\right]^{\frac{1}{5}} = -2cd^2$

77. $\sqrt[5]{32m^5n^{10}} = \left(2^5m^5n^{10}\right)^{\frac{1}{5}} = 2mn^2$

79. $-3\sqrt{5} + \sqrt{5} = -3\sqrt{5} + 1\sqrt{5} = -2\sqrt{5}$

81. $11\sqrt{3} + 7\sqrt{3} - 6\sqrt{3} = 12\sqrt{3}$

83. $2\sqrt{3} + 3\sqrt{2}$ (no like terms)

85. $2\sqrt{x} - 7\sqrt{2}$ (no like terms)

87. $8\sqrt{a} - 9\sqrt{a} = -\sqrt{a}$

89. $4a\sqrt{b} + 5b - a\sqrt{b} + b = 6b + 3a\sqrt{b}$

91. $-15 + 3x\sqrt{3} - \sqrt{3} - 17 - 2x\sqrt{3} = -32 + x\sqrt{3} - \sqrt{3}$

93. $P=\sqrt{2}+3\sqrt{2}+\sqrt{8}$

$P=\sqrt{2}+3\sqrt{2}+\sqrt{4\cdot 2}$

$P=1\sqrt{2}+3\sqrt{2}+2\sqrt{2}$

$P=6\sqrt{2}$

95. $P=2l+2w$

$P=2(5\sqrt{3})+2(\sqrt{12})$

$P=10\sqrt{3}+2\sqrt{4\cdot 3}$

$P=10\sqrt{3}+4\sqrt{3}$

$P=14\sqrt{3}$

97. $P=\sqrt{7}+3+\sqrt{7}$

$P=1\sqrt{7}+3+1\sqrt{7}$

$P=3+2\sqrt{7}$

99. $7\sqrt{2}+\sqrt{8}$

$=7\sqrt{2}+\sqrt{4\cdot 2}$

$=7\sqrt{2}+2\sqrt{2}$

$=9\sqrt{2}$

101. $4\sqrt{20}-3\sqrt{18}+5\sqrt{32}$

$=4\sqrt{4\cdot 5}-3\sqrt{9\cdot 2}+5\sqrt{16\cdot 2}$

$=8\sqrt{5}-9\sqrt{2}+20\sqrt{2}$

$=8\sqrt{5}+11\sqrt{2}$

103. a. $\sqrt{49x^2}-5x=\sqrt{(7x)^2}-5x=7x-5x$

b. $7x-5x=2x$

105. a. $-4a^2-\sqrt{16a^4}=-4a^2-\sqrt{\left(4a^2\right)^2}=-4a^2-4a^2$

b. $-4a^2-4a^2=-8a^2$

107. a. $-\sqrt{16}+\sqrt{4x^3}+\sqrt{25x^3}$

$=-\sqrt{16}+\sqrt{4x^2\cdot x}+\sqrt{25x^2\cdot x}$

$=-4+2x\sqrt{x}+5x\sqrt{x}$

b. $-4+2x\sqrt{x}+5x\sqrt{x}=-4+7x\sqrt{x}$

109.

$\sqrt[3]{8a}+4\sqrt[3]{a}=2\sqrt[3]{a}+4\sqrt[3]{a}$

$=6\sqrt[3]{a}$

111.

$5\sqrt{2x}+4\sqrt[3]{2x}+7\sqrt{2x}=5\sqrt{2x}+7\sqrt{2x}+4\sqrt[3]{2x}$

$=12\sqrt{2x}+4\sqrt[3]{2x}$

113.

$3\sqrt[5]{64a^5b^{10}c^3}-10ab\sqrt[5]{2b^5c^3}$

$=3\sqrt[5]{32\cdot 2a^5b^{10}c^3}-10ab\sqrt[5]{2b^5c^3}$

$=3\sqrt[5]{2^5a^5b^{10}\cdot 2c^3}-10ab\sqrt[5]{b^5\cdot 2c^3}$

$=3\cdot 2ab^2\sqrt[5]{2c^3}-10ab\cdot b\sqrt[5]{2c^3}$

$=6ab^2\sqrt[5]{2c^3}-10ab^2\sqrt[5]{2c^3}$

$=-4ab^2\sqrt[5]{2c^3}$

115.

$\sqrt[4]{16x^5y^8}+7xy^2\sqrt[4]{81x}=\sqrt[4]{16x^4xy^8}+7xy^2\sqrt[4]{81x}$

$=\sqrt[4]{2^4x^4y^8\cdot x}+7xy^2\sqrt[4]{3^4x}$

$=2xy^2\sqrt[4]{x}+7xy^2\cdot 3\sqrt[4]{x}$

$=2xy^2\sqrt[4]{x}+21xy^2\sqrt[4]{x}$

$=23xy^2\sqrt[4]{x}$

117.

$\sqrt[3]{7xy}+\sqrt[5]{7xy}-\sqrt[3]{448xy}+\sqrt[5]{224xy}$

$=\sqrt[3]{7xy}+\sqrt[5]{7xy}-\sqrt[3]{64\cdot 7xy}+\sqrt[5]{32\cdot 7xy}$

$=\sqrt[3]{7xy}+\sqrt[5]{7xy}-\sqrt[3]{4^3\cdot 7xy}+\sqrt[5]{2^5\cdot 7xy}$

$=\sqrt[3]{7xy}+\sqrt[5]{7xy}-4\sqrt[3]{7xy}+2\sqrt[5]{7xy}$

$=\sqrt[3]{7xy}-4\sqrt[3]{7xy}+\sqrt[5]{7xy}+2\sqrt[5]{7xy}$

$=-3\sqrt[3]{7xy}+3\sqrt[5]{7xy}$

Section 10.3 Multiplying and Dividing Radical Expressions

1. $\sqrt{3}\cdot\sqrt{12}=\sqrt{3\cdot 12}=\sqrt{36}=6$

3. $\sqrt{18}\cdot\sqrt{2}=\sqrt{18\cdot 2}=\sqrt{36}=6$

5. $\sqrt{2}\cdot\sqrt{22}=\sqrt{2\cdot 22}=\sqrt{44}=\sqrt{4\cdot 11}=2\sqrt{11}$

7. $\sqrt{8}\cdot\sqrt{10}=\sqrt{8\cdot 10}=\sqrt{80}=\sqrt{16\cdot 5}=4\sqrt{5}$

9. $\sqrt{18}\cdot\sqrt{8}=\sqrt{18\cdot 8}=\sqrt{144}=12$

11. $\sqrt{3}\cdot\sqrt{50}=\sqrt{3\cdot 50}=\sqrt{150}=\sqrt{25\cdot 6}=5\sqrt{6}$

13. $\sqrt{3x}\cdot\sqrt{6x}=\sqrt{3x\cdot 6x}=\sqrt{18x^2}$

$=\sqrt{9x^2\cdot 2}=3x\sqrt{2}$

15. $\sqrt{5a}\cdot\sqrt{8a^3}=\sqrt{5a\cdot 8a^3}=\sqrt{40a^4}$

$=\sqrt{4a^4\cdot 10}=2a^2\sqrt{10}$

17. $\sqrt{3x^2}\cdot\sqrt{12x^2}=\sqrt{3x^2\cdot 12x^2}=\sqrt{36x^4}=6x^2$

19. $\sqrt{50x^2}\cdot\sqrt{2x^3}=\sqrt{50x^2\cdot 2x^3}=\sqrt{100x^5}$

$=\sqrt{100x^4\cdot x}=10x^2\sqrt{x}$

21. $\sqrt{3a^2}\cdot\sqrt{20a^7}=\sqrt{3a^2\cdot 20a^7}=\sqrt{60a^9}$

$=\sqrt{4a^8\cdot 15a}=2a^4\sqrt{15a}$

23. $\sqrt{8x^{11}}\cdot\sqrt{10x^6}=\sqrt{8x^{11}\cdot 10x^6}=\sqrt{80x^{17}}$

$=\sqrt{16x^{16}\cdot 5x}=4x^8\sqrt{5x}$

25. $\sqrt{17}\cdot\sqrt{17}=17$

27. $\sqrt{6x}\cdot\sqrt{6x}=6x$

29. $\sqrt{xyz}\cdot\sqrt{xyz}=xyz$

31. $3\sqrt{y}\cdot 7\sqrt{y}=3\cdot 7\cdot\sqrt{y}\cdot\sqrt{y}=21y$

33. $-7\sqrt{a}\cdot(-2\sqrt{a^5})=14\sqrt{a\cdot a^5}=14\sqrt{a^6}=14a^3$

35. $\left(\sqrt{12a^2}\right)\left(7\sqrt{2a}\right)=7\sqrt{12a^2\cdot 2a}=7\sqrt{24a^3}$

$=7\sqrt{4a^2\cdot 6a}=7\cdot 2a\sqrt{6a}=14a\sqrt{6a}$

37. $\sqrt[3]{4x}\cdot\sqrt[3]{2x^2}=\sqrt[3]{4x\cdot 2x^2}=\sqrt[3]{8x^3}=2x$

39.

$$\begin{aligned}5n^3\sqrt[4]{m^2n^2}\cdot 7m^2n\sqrt[4]{m^3n^7}&=35m^2n^4\sqrt[4]{m^2n^2\cdot m^3n^7}\\&=35m^2n^4\sqrt[4]{m^5n^9}\\&=35m^2n^4\sqrt[4]{m^4n^8\cdot mn}\\&=35m^2n^4\cdot mn^2\sqrt[4]{mn}\\&=35m^3n^6\sqrt[4]{mn}\end{aligned}$$

41.

$$\begin{aligned}2x^2\sqrt[3]{5x^2y}\cdot 7y\sqrt[3]{4xy}&=14x^2y\sqrt[3]{5x^2y\cdot 4xy}\\&=14x^2y\sqrt[3]{20x^3y^2}\\&=14x^2y\cdot x\sqrt[3]{20y^2}\\&=14x^3y\sqrt[3]{20y^2}\end{aligned}$$

43.

$$\begin{aligned}12x^2y\sqrt[5]{7x^3y^4}\cdot 3xy^3\sqrt[5]{98x^4y^6}&=36x^3y^4\sqrt[5]{7x^3y^4\cdot 98x^4y^6}\\&=36x^3y^4\sqrt[5]{686x^7y^{10}}\\&=36x^3y^4\sqrt[5]{x^5y^{10}\cdot 686x^2}\\&=36x^3y^4\cdot xy^2\sqrt[5]{686x^2}\\&=36x^4y^6\sqrt[5]{686x^2}\end{aligned}$$

45. $A=\frac{1}{2}bh$

$A=\frac{1}{2}\left(4\sqrt{2}\right)\left(3\sqrt{2}\right)$

$A=\frac{1}{2}\cdot 4\cdot 3\cdot\sqrt{2}\cdot\sqrt{2}$

$A=6\cdot 2=12$

47. $A=\frac{1}{2}bh$

$A=\frac{1}{2}\left(\sqrt{3}\right)\left(\sqrt{3}\right)$

$A=\frac{1}{2}\cdot 3$

$A=\frac{3}{2}$

49. $\sqrt{2}(\sqrt{2}-5)=\sqrt{2}(\sqrt{2})-\sqrt{2}(5)=2-5\sqrt{2}$

51. $\sqrt{3}(\sqrt{6}-\sqrt{5})=\sqrt{3}(\sqrt{6})-\sqrt{3}(\sqrt{5})$

$=\sqrt{3\cdot 6}-\sqrt{3\cdot 5}=\sqrt{18}-\sqrt{15}=\sqrt{9\cdot 2}-\sqrt{15}$

$=3\sqrt{2}-\sqrt{15}$

53. $\sqrt{2}(\sqrt{x}-2)=\sqrt{2}(\sqrt{x})-\sqrt{2}(2)$

$=\sqrt{2\cdot x}-2\sqrt{2}=\sqrt{2x}-2\sqrt{2}$

55. $-2(3\sqrt{8}-\sqrt{5})=-2(3\sqrt{8})-2(-\sqrt{5})$

$=-6\sqrt{8}+2\sqrt{5}=-6\sqrt{4\cdot 2}+2\sqrt{5}$

$=-6\cdot 2\sqrt{2}+2\sqrt{5}=-12\sqrt{2}+2\sqrt{5}$

57. $\sqrt{x}(\sqrt{x}+\sqrt{11})=\sqrt{x}(\sqrt{x})+\sqrt{x}(\sqrt{11})$

$=x+\sqrt{x\cdot 11}=x+\sqrt{11x}$

59. $(2+\sqrt{6})(3-\sqrt{3})=2\cdot 3-2\sqrt{3}+3\sqrt{6}+\sqrt{6}(-\sqrt{3})$

$=6-2\sqrt{3}+3\sqrt{6}-\sqrt{6\cdot 3}=6-2\sqrt{3}+3\sqrt{6}-\sqrt{18}$

$=6-2\sqrt{3}+3\sqrt{6}-\sqrt{9\cdot 2}=6-2\sqrt{3}+3\sqrt{6}-3\sqrt{2}$

61. $(9-\sqrt{7})(3+\sqrt{7}) = 9\cdot 3 + 9\sqrt{7} - 3\sqrt{7} - \sqrt{7}(\sqrt{7})$

$= 27 + 9\sqrt{7} - 3\sqrt{7} - 7 = 20 + 6\sqrt{7}$

63. $(\sqrt{x}+\sqrt{3})(\sqrt{x}-\sqrt{5})$

$= \sqrt{x}(\sqrt{x}) - \sqrt{x}(\sqrt{5}) + \sqrt{3}(\sqrt{x}) - \sqrt{3}(\sqrt{5})$

$= x - \sqrt{5x} + \sqrt{3x} - \sqrt{15}$

65. $(1-\sqrt{2})(1+\sqrt{2}) = 1\cdot 1 + 1\sqrt{2} - 1\sqrt{2} - \sqrt{2}(\sqrt{2})$

$= 1 + \boxed{\sqrt{2}-\sqrt{2}} - 2 = 1 - 2 = -1$

67. $(3-\sqrt{x})(3+\sqrt{x}) = 3\cdot 3 + 3\sqrt{x} - 3\sqrt{x} - \sqrt{x}(\sqrt{x})$

$= 9 + \boxed{3\sqrt{x} - 3\sqrt{x}} - x = 9 - x$

69. $(4+\sqrt{3})(4+\sqrt{3}) = 4\cdot 4 + 4\sqrt{3} + 4\sqrt{3} + \sqrt{3}(\sqrt{3})$

$= 16 + \boxed{4\sqrt{3} + 4\sqrt{3}} + 3 = 19 + 8\sqrt{3}$

71. $(a+\sqrt{11})(a+\sqrt{11})$

$= a\cdot a + a\sqrt{11} + a\sqrt{11} + \sqrt{11}(\sqrt{11})$

$= a^2 + \boxed{a\sqrt{11} + a\sqrt{11}} + 11 = a^2 + 2a\sqrt{11} + 11$

73. $(2-\sqrt{y})(2-\sqrt{y})$

$= 2\cdot 2 - 2\sqrt{y} - 2\sqrt{y} - \sqrt{y}(-\sqrt{y})$

$= 4 \boxed{-2\sqrt{y} - 2\sqrt{y}} + y = 4 - 4\sqrt{y} + y$

75. $(x-\sqrt{5})(x-\sqrt{5})$

$= x\cdot x - x\sqrt{5} - x\sqrt{5} - \sqrt{5}(-\sqrt{5})$

$= x^2 \boxed{-x\sqrt{5} - x\sqrt{5}} + 5 = x^2 - 2x\sqrt{5} + 5$

77. $\frac{\sqrt{125}}{\sqrt{5}} = \sqrt{\frac{125}{5}} = \sqrt{25} = 5$

79. $\frac{\sqrt{72}}{\sqrt{2}} = \sqrt{\frac{72}{2}} = \sqrt{36} = 6$

81. $\frac{\sqrt{45x^3}}{\sqrt{5x}} = \sqrt{\frac{45x^3}{5x}} = \sqrt{9x^2} = 3x$

83. $\frac{\sqrt{3}}{\sqrt{27}} = \sqrt{\frac{3}{27}} = \sqrt{\frac{1}{9}} = \frac{\sqrt{1}}{\sqrt{9}} = \frac{1}{3}$

85. $\frac{\sqrt{3x^3}}{\sqrt{48x}} = \sqrt{\frac{3x^3}{48x}} = \sqrt{\frac{x^2}{16}} = \frac{\sqrt{x^2}}{\sqrt{16}} = \frac{x}{4}$

87. $\frac{4}{\sqrt{3}} = \frac{4}{\sqrt{3}}\cdot\frac{\sqrt{3}}{\sqrt{3}} = \frac{4\sqrt{3}}{3}$

89. $\frac{-15}{\sqrt{3}} = \frac{-15}{\sqrt{3}}\cdot\frac{\sqrt{3}}{\sqrt{3}} = \frac{-15\sqrt{3}}{3} = -5\sqrt{3}$

91. $\frac{\sqrt{5}}{3\sqrt{x}} = \frac{\sqrt{5}}{3\sqrt{x}}\cdot\frac{\sqrt{x}}{\sqrt{x}} = \frac{\sqrt{5x}}{3x}$

93. $\frac{\sqrt{2}}{5\sqrt{a}} = \frac{\sqrt{2}}{5\sqrt{a}}\cdot\frac{\sqrt{a}}{\sqrt{a}} = \frac{\sqrt{2a}}{5a}$

95. $\frac{2}{2+\sqrt{3}} = \frac{2}{(2+\sqrt{3})}\cdot\frac{(2-\sqrt{3})}{(2-\sqrt{3})}$

$= \frac{4-2\sqrt{3}}{4+2\sqrt{3}-2\sqrt{3}-\sqrt{3}\cdot\sqrt{3}}$

$= \frac{4-2\sqrt{3}}{4-3} = \frac{4-2\sqrt{3}}{1} = 4-2\sqrt{3}$

97. $\frac{7}{\sqrt{7}+\sqrt{3}} = \frac{7}{(\sqrt{7}+\sqrt{3})}\cdot\frac{(\sqrt{7}-\sqrt{3})}{(\sqrt{7}-\sqrt{3})}$

$= \frac{7\sqrt{7}-7\sqrt{3}}{\sqrt{7}\cdot\sqrt{7}-\sqrt{7}\cdot\sqrt{3}+\sqrt{7}\cdot\sqrt{3}-\sqrt{3}\cdot\sqrt{3}}$

$= \frac{7\sqrt{7}-7\sqrt{3}}{7-\sqrt{21}+\sqrt{21}-3} = \frac{7\sqrt{7}-7\sqrt{3}}{7-3} = \frac{7\sqrt{7}-7\sqrt{3}}{4}$

99. $\frac{4}{\sqrt{x}+3} = \frac{4}{(\sqrt{x}+3)}\cdot\frac{(\sqrt{x}-3)}{(\sqrt{x}-3)}$

$= \frac{4\sqrt{x}-12}{\sqrt{x}\cdot\sqrt{x}-3\sqrt{x}+3\sqrt{x}-9} = \frac{4\sqrt{x}-12}{x-9}$

101. $\frac{2}{\sqrt{7}+10} = \frac{2}{(\sqrt{7}+10)}\cdot\frac{(\sqrt{7}-10)}{(\sqrt{7}-10)}$

$= \frac{2\sqrt{7}-20}{\sqrt{7}\cdot\sqrt{7}-10\sqrt{7}+10\sqrt{7}-100}$

$= \frac{2\sqrt{7}-20}{7-100} = \frac{2\sqrt{7}-20}{-93} = \frac{-2\sqrt{7}+20}{93}$

103. $A = lw$ Substitute $A = 6\sqrt{3}$ and $w = \sqrt{2}$.

$6\sqrt{3} = \sqrt{2}\cdot l$

$\frac{6\sqrt{3}}{\sqrt{2}} = \frac{\sqrt{2}\cdot l}{\sqrt{2}}$

$l = \frac{6\sqrt{3}}{\sqrt{2}} = \frac{6\sqrt{3}}{\sqrt{2}}\cdot\frac{\sqrt{2}}{\sqrt{2}} = \frac{6\sqrt{6}}{2} = 3\sqrt{6}$

The length of the rectangle is $3\sqrt{6}$.

105. $A = lw$ Substitute $A = 12\sqrt{7}$ and $l = 3\sqrt{7}$.

$12\sqrt{7} = 3\sqrt{7}\cdot l$

$\frac{12\sqrt{7}}{3\sqrt{7}} = \frac{3\sqrt{7}\cdot l}{3\sqrt{7}}$

$l = \frac{12\sqrt{7}}{3\sqrt{7}} = 4\cdot\sqrt{\frac{7}{7}} = 4\sqrt{1} = 4$

The width of the rectangle is 4.

107.

$$\sqrt[3]{\frac{5x}{7x^2y^2}}=\sqrt[3]{\frac{5}{7xy^2}}$$
$$=\frac{\sqrt[3]{5}\cdot\sqrt[3]{7^2x^2y}}{\sqrt[3]{7xy^2}\cdot\sqrt[3]{7^2x^2y}}$$
$$=\frac{\sqrt[3]{5\cdot 49\cdot x^2\cdot y}}{\sqrt[3]{7^3x^3y^3}}$$
$$=\frac{\sqrt[3]{245x^2y}}{7xy}$$

109.

$$\frac{\sqrt[3]{9x^2y^2}}{7x\sqrt[3]{2xy^3}}=\frac{\sqrt[3]{9xy^2}}{7x\sqrt[3]{2y^3}}$$
$$=\frac{\sqrt[3]{9xy^2}\cdot\sqrt[3]{2^2}}{7x\sqrt[3]{2y^3}\cdot\sqrt[3]{2^2}}$$
$$=\frac{\sqrt[3]{4\cdot 9\cdot xy^2}}{7x\sqrt[3]{2^3y^3}}$$
$$=\frac{\sqrt[3]{36xy^2}}{7x\cdot 2y}$$
$$=\frac{\sqrt[3]{36xy^2}}{14xy}$$

111.

$$\sqrt[4]{\frac{5ab}{8a^3b^6}}=\sqrt[4]{\frac{5}{8a^2b^5}}$$
$$=\frac{\sqrt[4]{5}}{\sqrt[4]{2^3a^2b^5}}$$
$$=\frac{\sqrt[4]{5}}{\sqrt[4]{b^4}\sqrt[4]{2^3a^2b}}$$
$$=\frac{\sqrt[4]{5}}{b\sqrt[4]{2^3a^2b}}$$
$$=\frac{\sqrt[4]{5}\cdot\sqrt[4]{2a^2b^3}}{b\sqrt[4]{2^3a^2b}\cdot\sqrt[4]{2a^2b^3}}$$
$$=\frac{\sqrt[4]{2\cdot 5\cdot a^2b^3}}{b\sqrt[4]{2^4a^4b^4}}$$
$$=\frac{\sqrt[4]{10a^2b^3}}{b\cdot 2ab}$$
$$=\frac{\sqrt[4]{10a^2b^3}}{2ab^2}$$

113.

$$\frac{8\sqrt[4]{4xy}}{\sqrt[4]{144xy^3z^2}}=\frac{8}{\sqrt[4]{36y^2z^2}}$$
$$=\frac{8\cdot\sqrt[4]{2^2\cdot 3^2\cdot y^2\cdot z^2}}{\sqrt[4]{2^2\cdot 3^2\cdot y^2\cdot z^2}\cdot\sqrt[4]{2^2\cdot 3^2\cdot y^2\cdot z^2}}$$
$$=\frac{8\sqrt[4]{4\cdot 9\cdot y^2z^2}}{\sqrt[4]{2^43^4y^4z^4}}$$
$$=\frac{8\sqrt[4]{36y^2z^2}}{2\cdot 3\cdot y\cdot z}$$
$$=\frac{8\sqrt[4]{36y^2z^2}}{6yz}$$
$$=\frac{4\sqrt[4]{36y^2z^2}}{3yz}$$

Section 10.4 Radical Functions

1. **a.**

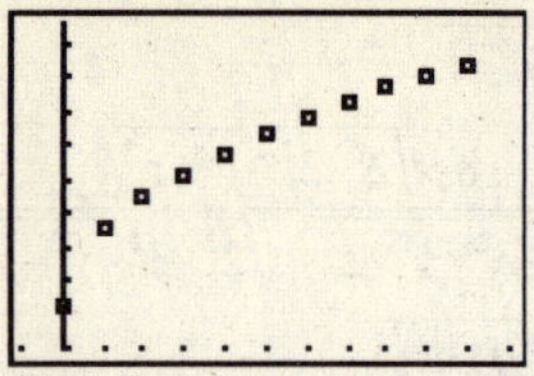

A radical function seems appropriate here because the scatterplot increases rapidly and curves slightly.

b.

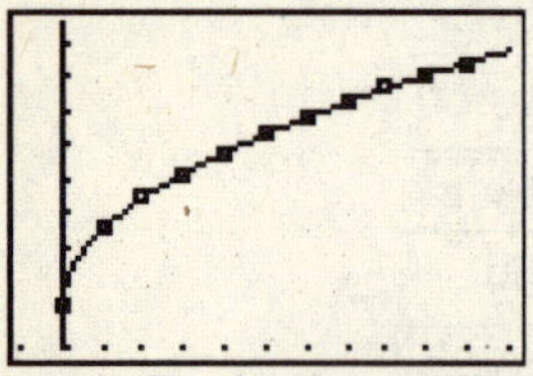

The graph fits the data well and seems to follow the pattern.

c. The cost to produce 550,000 heaters is approximately \$8.57 million.

d. The cost to produce 1,500,000 heaters is approximately \$12.07 million.

e. Domain: $[0,15]$ Range: $[3.2,12.07]$

3. **a.** The graph fits the data well and seems to follow the pattern.

b. A basilisk lizard with body mass of 50 grams will have a leg length of 0.061 meters.

c. A basilisk lizard with body mass of 100 grams will have a leg length of 0.077 meters.

d.

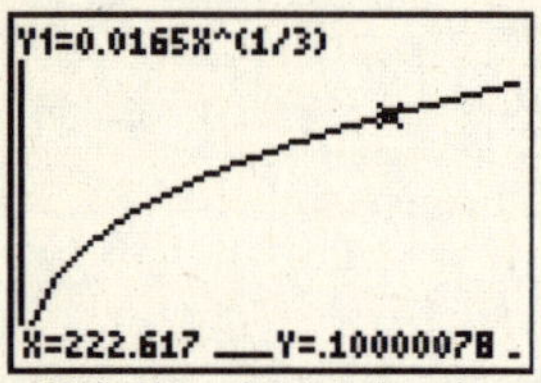

A basilisk lizard with leg length of 0.1 meter will have a body mass of 222.6 grams.

5. **a.** The airspeed of a butterfly with a thoracic mass of 0.05 gram is 2.92 meters per second.

b. The airspeed of a butterfly with a thoracic mass of 0.12 gram is 4.93 meters per second.

c. Range: $[0.28,12.95]$

7. **a.** $f(x)=2.5\sqrt{x}$

$f(81)=2.5\sqrt{81}$

$f(81)=2.5(9)$

$f(81)=22.5$

b. $f(25)=12.5$

X	Y1
21	11.456
22	11.726
23	11.99
24	12.247
25	12.5
26	12.748
27	12.99

X=25

9. **a.** $f(x)=6\sqrt[5]{x}$

$f(45)=6\sqrt[5]{45}$

$f(45)\approx 12.847$

b. $f(32)=12$

X	Y1
30	11.846
31	11.924
32	12
33	12.074
34	12.146
35	12.217
36	12.286

X=32

11. **a.** $f(x)=\sqrt{x+8}$

$f(28)=\sqrt{28+8}$

$f(28)=\sqrt{36}$

$f(28)=6$

b. $f(113)=11$

X	Y1
111	10.909
112	10.954
113	11
114	11.045
115	11.091
116	11.136
117	11.18

X=113

13. **a.** $f(x)=\sqrt{x-10}$

$f(50)=\sqrt{50-10}$

$f(50)=\sqrt{40}$

$f(50)\approx 6.325$

b. $f(59)=7$

X	Y1
56	6.7823
57	6.8557
58	6.9282
59	7
60	7.0711
61	7.1414
62	7.2111

X=59

15. **a.** $f(x)=\sqrt[7]{x+20}$

$f(-60)=\sqrt[7]{-60+20}$

$f(-60)=\sqrt[7]{-40}$

$f(-60)\approx -1.694$

b. $f(-2207)=-3$

X	Y1
-2237	-3.006
-2227	-3.004
-2217	-3.002
-2207	-3
-2197	-2.998
-2187	-2.996
-2177	-2.994

X=-2207

17. **a.** $f(x)=\sqrt[4]{x^3}$

$f(5)=\sqrt[4]{5^3}$

$f(5)=\sqrt[4]{125}$

$f(5)\approx 3.344$

b. $f(27.473) \approx 12$

X	Y_1
27.453	11.993
27.463	11.997
[illegible]	12
27.483	12.003
27.493	12.007
27.503	12.01
27.513	12.013
X=27.473	

19. **a.** $f(5) = 3$

b. $f(x) = 1$ when $x = -3$

21. **a.** $f(4) = 2$

b. $f(x) = -1.5$ when $x = -7.5$

23. **a.** This is an odd root.

b. Domain: $(-\infty, \infty)$

c. Range: $(-\infty, \infty)$

25. **a.** This is an even root.

b. Domain: $[-4, \infty)$

c. Range: $[0, \infty)$

27. **a.** This is an even root.

b. Domain: $(-\infty, -5]$

c. Range: $[0, \infty)$

29. **a.** This is an odd root.

b. Domain: $(-\infty, \infty)$

c. Range: $(-\infty, \infty)$

31. **a.** This is an even root.

b. Domain: $(-\infty, -2]$

c. Range: $(-\infty, 0]$

33. Since it is an odd root:
Domain: All real numbers
Range: All real numbers

35. Since it is an even root there is a restricted domain and range:
Domain: $x \geq 0$
Range: $[0, \infty)$

37. Since it is an odd root:
Domain: All real numbers
Range: All real numbers

39. Since it is an even root there is a restricted domain and range:
Domain: $x \geq 0$
Range: $[0, \infty)$

41. Since it is an even root there is a restricted domain and range:

$x + 12 \geq 0$

$x + 12 - 12 \geq 0 - 12$

$x \geq -12$

Domain: $x \geq -12$

Range: $[0, \infty)$

43. Since it is an even root there is a restricted domain and range:

$x + 7 \geq 0$

$x + 7 - 7 \geq 0 - 7$

$x \geq -7$

Domain: $x \geq -7$

Range: $[0, \infty)$

45. Since it is an odd root:
Domain: All real numbers
Range: All real numbers

47. Since it is an even root there is a restricted domain and range:

$x + 2 \geq 0$

$x + 2 - 2 \geq 0 - 2$

$x \geq -2$

Domain: $x \geq -2$

Range: $[0, \infty)$

49. Since it is an even root there is a restricted domain and range:

$x - 23 \geq 0$

$x - 23 + 23 \geq 0 + 23$

$x \geq 23$

Domain: $x \geq 23$

Range: $[0, \infty)$

51. Since it is an even root there is a restricted domain and range:

$-x + 3 \geq 0$

$-x + 3 - 3 \geq 0 - 3$

$-x \geq -3$

$\frac{-x}{-1} \leq \frac{-3}{-1}$

$x \leq 3$

Domain: $x \leq 3$

Range: $[0, \infty)$

53. Since it is an even root there is a restricted domain and range:

$-x - 10 \geq 0$

$-x - 10 + 10 \geq 0 + 10$

$-x \geq 10$

$\frac{-x}{-1} \leq \frac{10}{-1}$

$x \leq -10$

Domain: $x \leq -10$

Range: $[0, \infty)$

55. Since it is an odd root:
Domain: All real numbers
Range: All real numbers

57. Since it is an even root there is a restricted domain and range:

$3x+5\geq 0$

$3x+5-5\geq 0-5$

$3x\geq -5$

$\frac{3x}{3}\geq\frac{-5}{3}$

$x\geq -\frac{5}{3}$

Domain: $x\geq -\frac{5}{3}$

Range: $[0,\infty)$

59. Since it is an even root there is a restricted domain and range:

$-4x+9\geq 0$

$-4x+9-9\geq 0-9$

$-4x\geq -9$

$\frac{-4x}{-4}\leq\frac{-9}{-4}$

$x\leq\frac{9}{4}$

Domain: $x\leq\frac{9}{4}$

Range: $[0,\infty)$

61. $f(x)=\sqrt{x+7}$

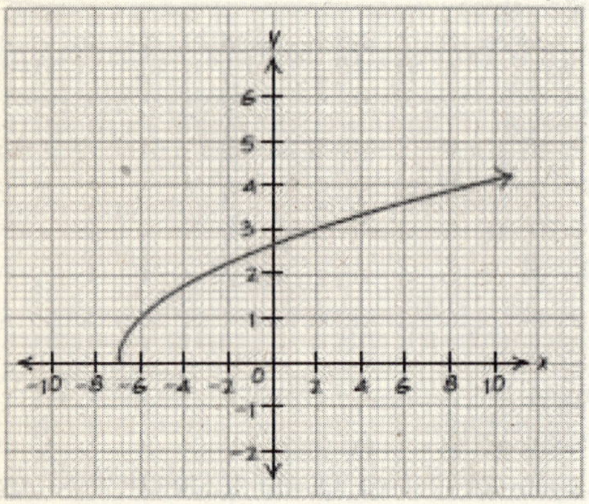

63. $g(x)=\sqrt{x-8}$

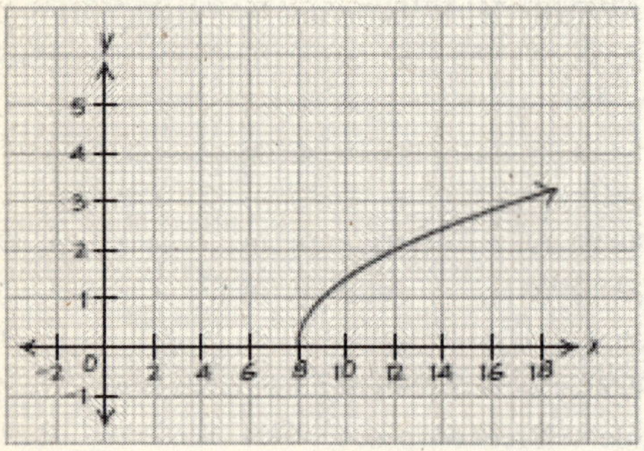

65. $f(x)=-\sqrt{x+3}$

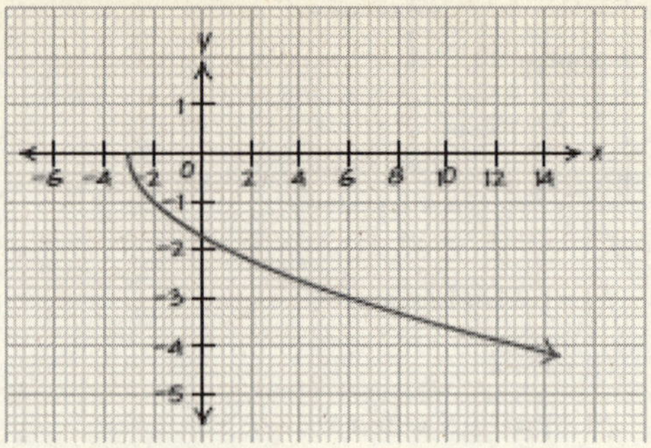

67. $f(x)=-\sqrt{x-6}$

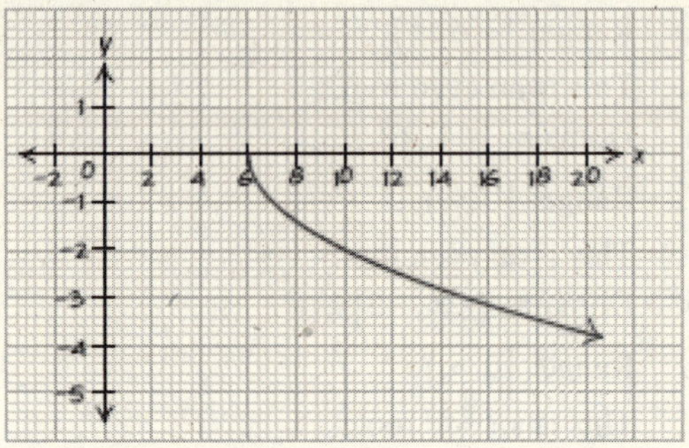

69. $f(x)=\sqrt[3]{x+2}$

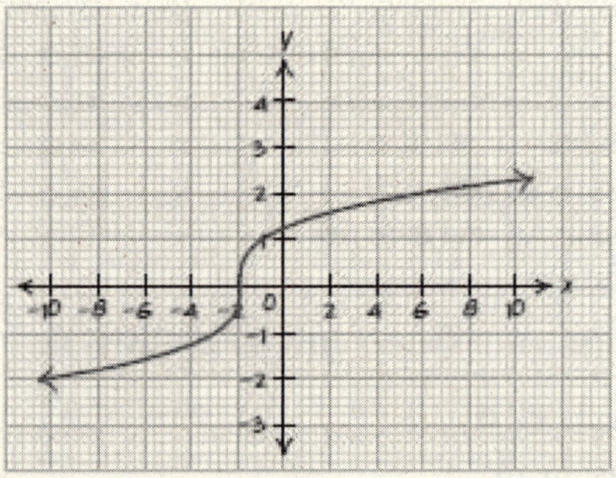

71. $h(x)=\sqrt[3]{x-11}$

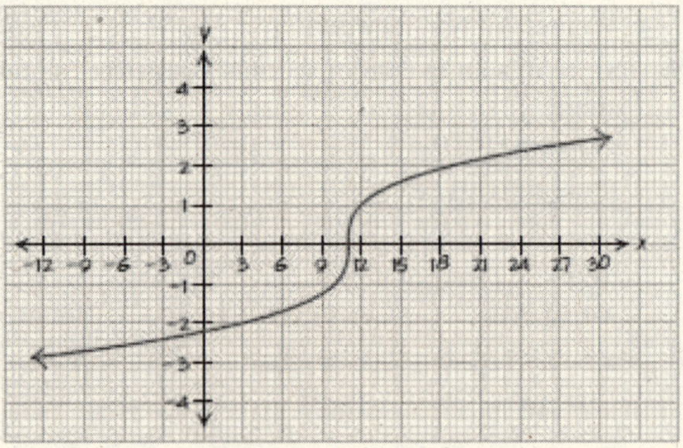

73. $f(x)=\sqrt{9-x}$

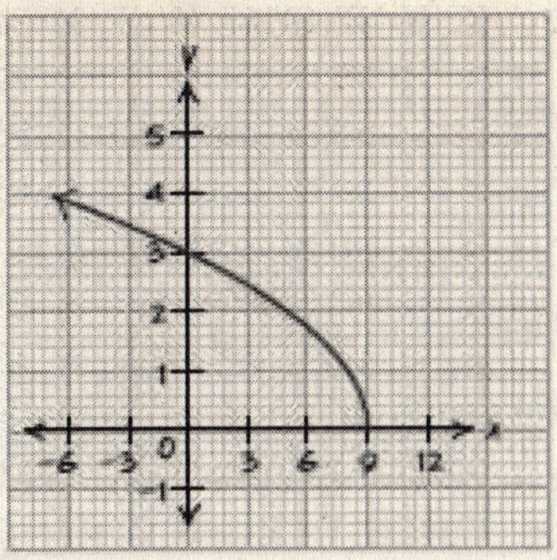

75. $g(x)=-\sqrt{x-3}$

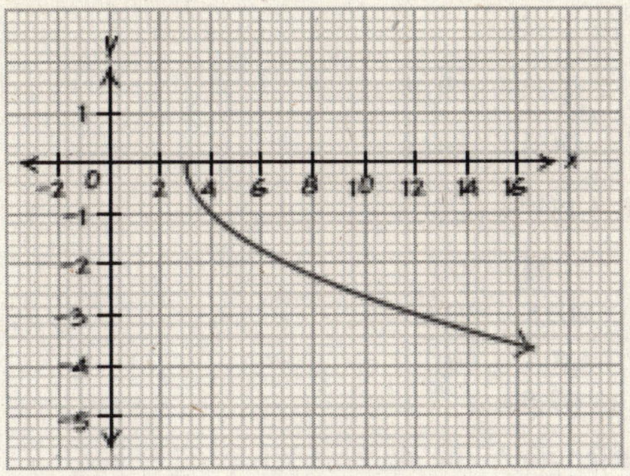

77. $h(x)=\sqrt[5]{2x}$

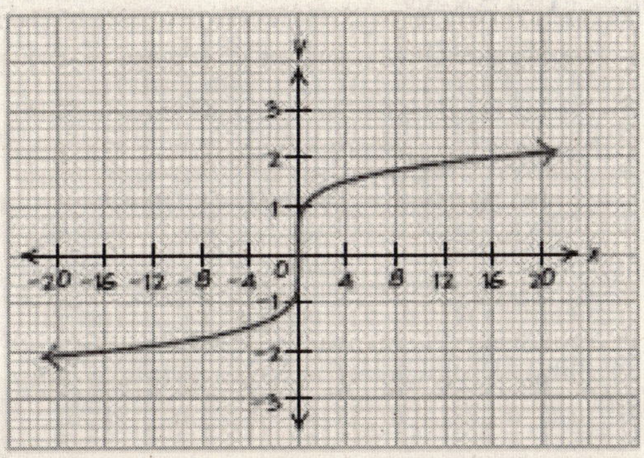

79. $g(x)=2\sqrt{x+3}$

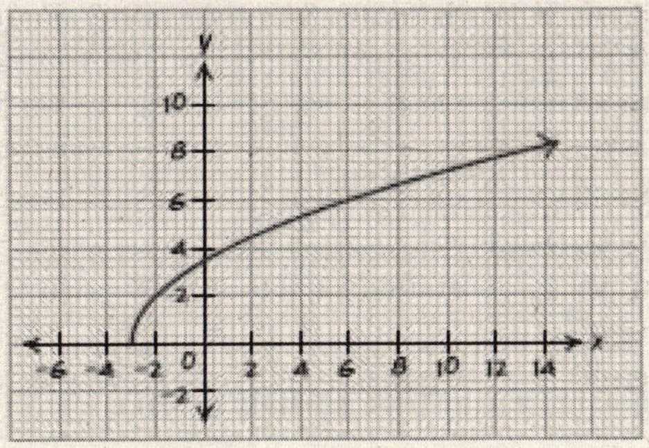

Section 10.5 Solving Radical Equations

1.

$\sqrt{x}-5=-1$ Substitute $x=16$

$\sqrt{16}-5=-1$

$4-5=-1$

$-1=-1$ This is a true statement.

Yes, $x=16$ is a solution.

3.

$\sqrt{x}=-9$ Substitute $x=81$

$\sqrt{81}=-9$

$9\stackrel{?}{=}-9$ This is a false statement.

No, $x=81$ is not a solution.

5.

$\sqrt{x-1}+9=4$ Substitute $x=26$

$\sqrt{(26)-1}+9=4$

$\sqrt{25}+9=4$

$5+9=4$

$14\stackrel{?}{=}4$ This is a false statement.

No, $x=26$ is not a solution.

7.

$\sqrt{x^2-7}=3$ Substitute $x=4$

$\sqrt{(4)^2-7}=3$

$\sqrt{16-7}=3$

$\sqrt{9}=3$

$3=3$ This is a true statement.

Yes, $x=4$ is a solution.

9. The student's work is not correct. The student concludes that $\sqrt{-9}=-3$, but this is not correct because a negative number under the radical sign ($\sqrt{-9}$) is not a real number and cannot be equal to a real number (-3).

11. The student's work is correct.

13. The student's work is not correct. The student concludes that $-\sqrt{-1}=1$, but this is not correct because a negative number under the radical sign ($\sqrt{-1}$) is not a real number and cannot be equal to a real number (1).

15.

$\sqrt{a}=4$

$\left(\sqrt{a}\right)^2=(4)^2$

$a=16$

17. $\sqrt{y}=-12$ no solution.

A square root cannot equal a negative number.

19.

$$\sqrt{x-1}=6$$
$$\left(\sqrt{x-1}\right)^2=(6)^2$$
$$x-1=36$$
$$+1\ \ +1$$
$$x=37$$

21.

$$\sqrt{-y+4}=2$$
$$\left(\sqrt{-y+4}\right)^2=(2)^2$$
$$-y+4=4$$
$$-4\ \ -4$$
$$-y=0$$
$$y=0$$

23.

$$\sqrt{a}-6=9$$
$$+6\ \ +6$$
$$\sqrt{a}=15$$
$$\left(\sqrt{a}\right)^2=(15)^2$$
$$a=225$$

25.

$$2\sqrt{a}+19=15$$
$$-19\ \ -19$$
$$2\sqrt{a}=-4$$
$$\frac{2\sqrt{a}}{2}=\frac{-4}{2}$$
$$\sqrt{a}=-2 \quad \text{no solution}$$

A square root cannot equal a negative number.

27.

$$\sqrt{3y+1}=5$$
$$\left(\sqrt{3y+1}\right)^2=(5)^2$$
$$3y+1=25$$
$$-1\ \ -1$$
$$3y=24$$
$$\frac{3y}{3}=\frac{24}{3}$$
$$y=8$$

29. a.

$$T(h)=\sqrt{\frac{h}{4.5}}$$
$$(3.4)=\sqrt{\frac{h}{4.5}}$$
$$(3.4)^2=\left(\sqrt{\frac{h}{4.5}}\right)^2$$
$$4.5\cdot 11.56=\frac{h}{4.5}\cdot 4.5$$
$$52.02=h$$

The ball was dropped from a height of 52.02 meters.

b.

$$T(h)=\sqrt{\frac{h}{4.5}}$$
$$(9)=\sqrt{\frac{h}{4.5}}$$
$$(9)^2=\left(\sqrt{\frac{h}{4.5}}\right)^2$$
$$4.5\cdot 81=\frac{h}{4.5}\cdot 4.5$$
$$364.5=h$$

The ball was dropped from a height of 364.5 meters.

31. a.

$$s(d)=\sqrt{30d}$$
$$(45)=\sqrt{30d}$$
$$(45)^2=\left(\sqrt{30d}\right)^2$$
$$2025=30d$$
$$\frac{2025}{30}=\frac{30d}{30}$$
$$67.5=d$$

The length of the skid mark was 67.5 feet.

b.

$$s(d)=\sqrt{30d}$$
$$(37)=\sqrt{30d}$$
$$(37)^2=\left(\sqrt{30d}\right)^2$$
$$1369=30d$$
$$\frac{1369}{30}=\frac{30d}{30}$$
$$45.63\approx d$$

The length of the skid mark was approximately 45.63 feet.

33.

$$P(x)=4\sqrt{x-2}$$
$$(20)=4\sqrt{x-2}$$
$$\frac{20}{4}=\frac{4\sqrt{x-2}}{4}$$
$$(5)^2=\left(\sqrt{x-2}\right)^2$$
$$25=x-2$$
$$+2\ \ \ +2$$
$$27=x$$

Andrew must sell 27 wristbands to make a profit of \$20.

35.

$$r(A) = 0.56\sqrt{A}$$
$$(200) = 0.56\sqrt{A}$$
$$\frac{200}{0.56} = \frac{0.56\sqrt{A}}{0.56}$$
$$(357.14)^2 \approx \left(\sqrt{A}\right)^2$$
$$127{,}551.02 \approx A$$

The irrigation system covers an area of approximately 127,551.02 square feet.

37.

$$\sqrt{x+2} = 4$$
$$\left(\sqrt{x+2}\right)^2 = (4)^2$$
$$x + 2 = 16$$
$$\underline{-2 \quad -2}$$
$$x = 14$$

39. $\sqrt{x+5} = -1$ no solution.

A square root cannot equal a negative number.

41.

$$\sqrt{z-2} = \sqrt{2z-5}$$
$$\left(\sqrt{z-2}\right)^2 = \left(\sqrt{2z-5}\right)^2$$
$$z - 2 = 2z - 5$$
$$\underline{+2 \qquad +2}$$
$$z = 2z - 3$$
$$\underline{-2z \quad -2z}$$
$$-z = -3$$
$$\frac{-z}{-1} = \frac{-3}{-1}$$
$$z = 3$$

43.

$$2 - \sqrt{a+14} = 1$$
$$\underline{-2 \qquad -2}$$
$$-\sqrt{a+14} = -1$$
$$\left(-\sqrt{a+14}\right)^2 = (-1)^2$$
$$a + 14 = 1$$
$$\underline{-14 \quad -14}$$
$$a = -13$$

45.

$$\sqrt{x-2} + 4 = 3$$
$$\underline{-4 \quad -4}$$
$$\sqrt{x-2} = -1 \quad \text{no solution}$$

A square root cannot equal a negative number.

47. The student's work is not correct. The student did not isolate the radical before squaring. The student then simplified $(2\sqrt{x})^2$ incorrectly. Correctly simplified:

$$2\sqrt{x} = 4$$
$$\frac{2\sqrt{x}}{2} = \frac{4}{2}$$
$$\sqrt{x} = 2$$
$$\left(\sqrt{x}\right)^2 = (2)^2$$
$$x = 4$$

49. The student's work is not correct. The student did not isolate the radical before squaring. The student then simplified $(\sqrt{x}+1)^2$ incorrectly. Correctly simplified:

$$\sqrt{x} + 1 = 5$$
$$\underline{-1 \quad -1}$$
$$\sqrt{x} = 4$$
$$\left(\sqrt{x}\right)^2 = (4)^2$$
$$x = 16$$

51. The student's work is not correct. A square root cannot equal a negative number. If the student had checked the solution, this would reveal that $x = 36$ is not a solution to the equation. Correctly simplified:

$$\sqrt{x} = -6$$
$$\left(\sqrt{x}\right)^2 = (-6)^2$$
$$x = 36$$

Check:

$$\sqrt{36} \neq -6$$

Therefore there is no solution

53. The student's work is incorrect. The term $\sqrt{-x}$ is not a real number, therefore there is no solution.

55. The student did not use the distance formula correctly. The differences between the x and y coordinates should be added under the radical sign. The distance formula does not use a quotient. Correctly simplified:

$$d = \sqrt{(x_2 - x_1)^2 + (y_2 - y_1)^2}$$
$$d = \sqrt{(-5-3)^2 + (2-6)^2}$$
$$d = \sqrt{(-8)^2 + (-4)^2}$$
$$d = \sqrt{64 + 16}$$
$$d = \sqrt{80} = \sqrt{16 \cdot 5} = 4\sqrt{5}$$

57. The student found the differences between the x and y of the ordered pairs ($y_1 - x_1$ and $y_2 - x_2$), rather than correctly finding the difference between the x coordinates and y coordinates ($x_2 - x_1$ and $y_2 - y_1$). Correctly simplified:

$$d = \sqrt{(x_2 - x_1)^2 + (y_2 - y_1)^2}$$
$$d = \sqrt{(9-3)^2 + (4-2)^2}$$
$$d = \sqrt{(6)^2 + (2)^2}$$
$$d = \sqrt{36 + 4}$$
$$d = \sqrt{40} = \sqrt{4 \cdot 10} = 2\sqrt{10}$$

59.

$$\sqrt{x^2-x+7}=3$$
$$\left(\sqrt{x^2-x+7}\right)^2=(3)^2$$
$$x^2-x+7=9$$
$$\underline{\quad -9\ -9}$$
$$x^2-x-2=0$$
$$(x-2)(x+1)=0$$
$$x-2=0 \qquad x+1=0$$
$$x=2 \qquad x=-1$$

61.

$$\sqrt{2x^2-x+1}=2$$
$$\left(\sqrt{2x^2-x+1}\right)^2=(2)^2$$
$$2x^2-x+1=4$$
$$\underline{\quad -4\ -4}$$
$$2x^2-x-3=0$$
$$(2x-3)(x+1)=0$$
$$2x-3=0$$
$$2x=3$$
$$\frac{2x}{2}=\frac{3}{2} \qquad x+1=0$$
$$x=\frac{3}{2} \qquad x=-1$$

63.

$$\sqrt{x+45}=\sqrt{x}+5$$
$$\left(\sqrt{x+45}\right)^2=\left(\sqrt{x}+5\right)^2$$
$$x+45=x+10\sqrt{x}+25$$
$$\underline{-x-25\ -x\qquad -25}$$
$$20=10\sqrt{x}$$
$$\frac{20}{10}=\frac{10\sqrt{x}}{10}$$
$$2=\sqrt{x}$$
$$(2)^2=\left(\sqrt{x}\right)^2$$
$$4=x$$

65.

$$\sqrt{y+3}=\sqrt{y}-3$$
$$\left(\sqrt{y+3}\right)^2=\left(\sqrt{y}-3\right)^2$$
$$y+3=y-6\sqrt{y}+9$$
$$\underline{-y-9\ -y\qquad -9}$$
$$-6=-6\sqrt{y}$$
$$\frac{-6}{-6}=\frac{-6\sqrt{y}}{-6}$$
$$1=\sqrt{y}$$
$$(1)^2=\left(\sqrt{y}\right)^2$$
$$1=y$$

Check:

$$\sqrt{(1)+3}=\sqrt{(1)}-3$$
$$2\neq-2$$

Therefore there is no solution

67. $-1=\sqrt{a^2-7a+12}$ no solution

A square root cannot equal a negative number.

69.

$$\sqrt{x+32}=4+\sqrt{x}$$
$$\left(\sqrt{x+32}\right)^2=\left(4+\sqrt{x}\right)^2$$
$$x+32=16+8\sqrt{x}+x$$
$$\underline{-x-16\ -16\qquad -x}$$
$$16=8\sqrt{x}$$
$$\frac{16}{8}=\frac{8\sqrt{x}}{8}$$
$$2=\sqrt{x}$$
$$(2)^2=\left(\sqrt{x}\right)^2$$
$$4=x$$

71. Use $T(L)=2\pi\sqrt{\frac{L}{32}}$ with $L=3$

$$T(3)=2\pi\sqrt{\frac{3}{32}}=1.92$$

The period of a pendulum of length 3 feet is 1.92 seconds.

73. Use $T(L)=2\pi\sqrt{\frac{L}{32}}$ with $\mathrm{T}(L)=1$

$$1=2\pi\sqrt{\frac{L}{32}}$$
$$\frac{1}{2\pi}=\frac{2\pi\sqrt{\frac{L}{32}}}{2\pi}$$
$$\frac{1}{2\pi}=\sqrt{\frac{L}{32}}$$
$$\left(\frac{1}{2\pi}\right)^2=\left(\sqrt{\frac{L}{32}}\right)^2$$
$$\left(\frac{1}{2\pi}\right)^2=\frac{L}{32}$$
$$32\left(\frac{1}{2\pi}\right)^2=L$$
$$L\approx0.81$$

For a period of 1 second, the pendulum needs to be about 0.81 feet (9.7 inches) in length.

75. $C(h)=2.29\sqrt{h}+3.2$

a.

$$\begin{aligned}15&=2.29\sqrt{h}+3.2\\15-3.2&=2.29\sqrt{h}+3.2-3.2\\11.8&=2.29\sqrt{h}\\\frac{11.8}{2.29}&=\frac{2.29\sqrt{h}}{2.29}\\\frac{11.8}{2.29}&=\sqrt{h}\\\left(\frac{11.8}{2.29}\right)^2&=\left(\sqrt{h}\right)^2\\h&\approx 26.552\end{aligned}$$

Jim Bob's can make 26.552 hundred thousand space heaters with a budget of \$15 million.

b.

$$\begin{aligned}12&=2.29\sqrt{h}+3.2\\12-3.2&=2.29\sqrt{h}+3.2-3.2\\8.8&=2.29\sqrt{h}\\\frac{8.8}{2.29}&=\frac{2.29\sqrt{h}}{2.29}\\\frac{8.8}{2.29}&=\sqrt{h}\\\left(\frac{8.8}{2.29}\right)^2&=\left(\sqrt{h}\right)^2\\h&\approx 14.767\end{aligned}$$

Jim Bob's can make 14.767 hundred thousand space heaters with a budget of \$12 million.

77. $L(M)=0.330\sqrt[3]{M}$

a.

$$\begin{aligned}1&=0.330\sqrt[3]{M}\\\frac{1}{0.330}&=\frac{0.330\sqrt[3]{M}}{0.330}\\\left(\frac{1}{0.330}\right)^3&=\left(\sqrt[3]{M}\right)^3\\M&\approx 27.8\end{aligned}$$

The body mass of a mammal with a body length of 1 meter is 27.8 kg.

b.

$$\begin{aligned}2.4&=0.330\sqrt[3]{M}\\\frac{2.4}{0.330}&=\frac{0.330\sqrt[3]{M}}{0.330}\\\left(\frac{2.4}{0.330}\right)^3&=\left(\sqrt[3]{M}\right)^3\\M&\approx 384.7\end{aligned}$$

The body mass of a mammal with a body length of 2.4 meters is 384.7 kg.

79.

$a=\sqrt[3]{P^2}$ with $P=84$

$a=\sqrt[3]{84^2}\approx 19.18$

The average distance from the Sun to Uranus is approximately 19.2 AU.

81.

$a=\sqrt[3]{P^2}$ with $a=39.5$

$$\begin{aligned}39.5&=\sqrt[3]{P^2}\\39.5&=P^{2/3}\\(39.5)^{3/2}&=\left(P^{2/3}\right)^{3/2}\\P&\approx 248.3\end{aligned}$$

It takes Pluto approximately 248.3 years to orbit the sun.

83.

$$\begin{aligned}\sqrt{x-7}-\sqrt{x}&=-1\\\sqrt{x-7}-\sqrt{x}&=-1\\\sqrt{x-7}&=\sqrt{x}-1\\\left(\sqrt{x-7}\right)^2&=\left(\sqrt{x}-1\right)^2\\x-7&=\left(\sqrt{x}-1\right)\left(\sqrt{x}-1\right)\\x-7&=x-2\sqrt{x}+1\\-7&=-2\sqrt{x}+1\\-8&=-2\sqrt{x}\\\frac{-8}{-2}&=\frac{-2\sqrt{x}}{-2}\\4&=\sqrt{x}\\4^2&=\left(\sqrt{x}\right)^2\\x&=16\end{aligned}$$

85.

$$\begin{aligned}\sqrt{w+2}&=4-w\\\left(\sqrt{w+2}\right)^2&=(4-w)^2\\w+2&=(4-w)(4-w)\\w+2&=16-8w+w^2\\0&=w^2-9w+14\\0&=(w-7)(w-2)\\w&=7,w=2\end{aligned}$$

The answer $w=7$ does not check out, therefore the solution is $w=2$.

87.

$\sqrt{6x+1}-\sqrt{9x}=-1$

$\sqrt{6x+1}=\sqrt{9x}-1$

$\left(\sqrt{6x+1}\right)^2=\left(\sqrt{9x}-1\right)^2$

$6x+1=\left(\sqrt{9x}-1\right)\left(\sqrt{9x}-1\right)$

$6x+1=9x-2\sqrt{9x}+1$

$6x=9x-2\sqrt{9x}$

$(-3x)^2=\left(-2\sqrt{9x}\right)^2$

$9x^2=4(9x)$

$9x^2=36x$

$9x^2-36x=0$

$9x(x-4)=0$

$9x=0, x-4=0$

$x=0, x=4$

The answer $x=0$ does not check out, therefore the solution is $x=4$.

89.

$\sqrt{6x+5}+\sqrt{3x+2}=5$

$\sqrt{6x+5}=5-\sqrt{3x+2}$

$\left(\sqrt{6x+5}\right)^2=\left(5-\sqrt{3x+2}\right)^2$

$6x+5=\left(5-\sqrt{3x+2}\right)\left(5-\sqrt{3x+2}\right)$

$6x+5=25-10\sqrt{3x+2}+3x+2$

$6x+5=27+3x-10\sqrt{3x+2}$

$3x+5=27-10\sqrt{3x+2}$

$3x-22=-10\sqrt{3x+2}$

$(3x-22)^2=\left(-10\sqrt{3x+2}\right)^2$

$(3x-22)(3x-22)=100(3x+2)$

$9x^2-132x+484=300x+200$

$9x^2-432x+284=0$

$(3x-142)(3x-2)=0$

$x=\frac{142}{3}, x=\frac{2}{3}$

The answer $x=\frac{142}{3}$ does not check out, therefore the solution is $x=\frac{2}{3}$.

91.

$\sqrt{8x+8}-\sqrt{8x-4}=-2$

$\sqrt{8x+8}=-2+\sqrt{8x-4}$

$\left(\sqrt{8x+8}\right)^2=\left(-2+\sqrt{8x-4}\right)^2$

$8x+8=\left(-2+\sqrt{8x-4}\right)\left(-2+\sqrt{8x-4}\right)$

$8x+8=4-4\sqrt{8x-4}+8x-4$

$8x+8=8x-4\sqrt{8x-4}$

$8=-4\sqrt{8x-4}$

$\frac{8}{-4}=\frac{-4\sqrt{8x-4}}{-4}$

$-2=\sqrt{8x-4}$

$(-2)^2=\left(\sqrt{8x-4}\right)^2$

$4=8x-4$

$8=8x$

$x=1$

The answer $x=1$ does not check out, therefore there is no solution.

93.

$\sqrt{2x+3}=1+\sqrt{x+1}$

$\left(\sqrt{2x+3}\right)^2=\left(1+\sqrt{x+1}\right)^2$

$2x+3=\left(1+\sqrt{x+1}\right)\left(1+\sqrt{x+1}\right)$

$2x+3=1+2\sqrt{x+1}+x+1$

$2x+3=2\sqrt{x+1}+x+2$

$x+1=2\sqrt{x+1}$

$(x+1)^2=\left(2\sqrt{x+1}\right)^2$

$(x+1)(x+1)=4(x+1)$

$x^2+2x+1=4x+4$

$x^2-2x-3=0$

$(x+1)(x-3)=0$

$x+1=0, x-3=0$

$x=-1, x=3$

95.

$\sqrt[3]{2x+5}=6$

$\left(\sqrt[3]{2x+5}\right)^3=(6)^3$

$2x+5=216$

$2x=211$

$x=105.5$

97.

$\sqrt[4]{5x-9}=2$

$\left(\sqrt[4]{5x-9}\right)^4=(2)^4$

$5x-9=16$

$5x=25$

$x=5$

99.

$$\sqrt[3]{5x+2}+7=12$$
$$\sqrt[3]{5x+2}=5$$
$$\left(\sqrt[3]{5x+2}\right)^3=(5)^3$$
$$5x+2=125$$
$$5x=123$$
$$x=24.6$$

101.

$$\sqrt[3]{-4x+3}=-5$$
$$\left(\sqrt[3]{-4x+3}\right)^3=(-5)^3$$
$$-4x+3=-125$$
$$-4x=-128$$
$$x=32$$

103. The variable is under the radical sign therefore the equation is radical.

105. The variable is in the denominator therefore the equation is rational.

107. The variable is degree 1 therefore the equation is linear.

109. The variable is degree 2 therefore the equation is quadratic.

111. a. There is an equal sign therefore it is an equation.

b.

$$\sqrt{x+6}=5$$
$$\left(\sqrt{x+6}\right)^2=(5)^2$$
$$x+6=25$$
$$-6\quad -6$$
$$x=19$$

113. a. There is an equal sign therefore it is an equation.

b.

$$x^2+8x+12=0$$
$$(x+6)(x+2)=0$$
$$x+6=0 \qquad x+2=0$$
$$x=-6 \qquad x=-2$$

115. a. There is no equal sign therefore it is an expression.

b.

$$(n+5)(3n-7)+8n^2-4n$$
$$=3n^2-7n+15n-35+8n^2-4n$$
$$=11n^2+4n-35$$

117. a. There is an equal sign therefore it is an equation.

b.

$$\sqrt{x-4}+1=\sqrt{x+3}$$
$$\left(\sqrt{x-4}+1\right)^2=\left(\sqrt{x+3}\right)^2$$
$$\left(\sqrt{x-4}\right)\left(\sqrt{x-4}\right)+2\sqrt{x-4}+1=x+3$$
$$x-4+2\sqrt{x-4}+1=x+3$$
$$x+2\sqrt{x-4}-3=x+3$$
$$-x+3\quad -x+3$$
$$2\sqrt{x-4}=6$$
$$\frac{2\sqrt{x-4}}{2}=\frac{6}{2}$$
$$\sqrt{x-4}=3$$
$$\left(\sqrt{x-4}\right)^2=(3)^2$$
$$x-4=9$$
$$x=13$$

Section 10.6 Complex Numbers

1. $\sqrt{-64}=\sqrt{-1}\cdot\sqrt{64}=i\cdot 8=8i$

3. $\sqrt{-81}=\sqrt{-1}\cdot\sqrt{81}=i\cdot 9=9i$

5. $-\sqrt{36}=-6$

7.
$$\begin{aligned}\sqrt{-20}&=\sqrt{-1}\cdot\sqrt{20}\\&=i\cdot\sqrt{4}\cdot\sqrt{5}\\&=2i\sqrt{5}\\&\approx 4.472i\end{aligned}$$

9.
$$\begin{aligned}\sqrt{-200}&=\sqrt{-1}\cdot\sqrt{200}\\&=i\cdot\sqrt{100}\cdot\sqrt{2}\\&=10i\sqrt{2}\\&\approx 14.142i\end{aligned}$$

11.
$$\begin{aligned}12-\sqrt{-100}&=12-\sqrt{-1}\cdot\sqrt{100}\\&=12-i\cdot 10\\&=12-10i\end{aligned}$$

13.
$$\begin{aligned}\sqrt{-4}+\sqrt{-36}&=\sqrt{-1}\cdot\sqrt{4}+\sqrt{-1}\cdot\sqrt{36}\\&=i\cdot 2+i\cdot 6\\&=2i+6i\\&=8i\end{aligned}$$

15.
$$\begin{aligned}\sqrt{9}+\sqrt{-16}&=\sqrt{9}+\sqrt{-1}\cdot\sqrt{16}\\&=3+i\cdot 4\\&=3+4i\end{aligned}$$

17.
$$\begin{aligned}\sqrt{-3}+\sqrt{-5.6}&=\sqrt{-1}\cdot\sqrt{3}+\sqrt{-1}\cdot\sqrt{5.6}\\&=i\sqrt{3}+i\sqrt{5.6}\\&\approx 1.732i+2.366i\\&\approx 4.098i\end{aligned}$$

19.
$$\begin{aligned}-4\left(3+\sqrt{-63}\right)&=-12-4\sqrt{-63}\\&=-12-4\sqrt{-1}\cdot\sqrt{63}\\&=-12-4i\cdot 3\sqrt{7}\\&=-12-4\cdot 3i\sqrt{7}\\&=-12-12i\sqrt{7}\\&\approx -12-31.749i\end{aligned}$$

21. Real part = 2 Imaginary part = 5

23. Real part = 2.3 Imaginary part = −4.9

25. Real part = 3 Imaginary part = 0

27. Real part = 0 Imaginary part = 7

29. $(2+5i)+(6+4i)=2+6+5i+4i=8+9i$

31.
$$\begin{aligned}(5.6+3.2i)+(2.3-4.9i)&=5.6+2.3+3.2i-4.9i\\&=7.9-1.7i\end{aligned}$$

33. $(5i)+(3-9i)=3+5i-9i=3-4i$

35. $(5+7i)-10=5-10+7i=-5+7i$

37. $(3+5i)-(7+8i)=3-7+5i-8i=-4-3i$

39.
$$\begin{aligned}(4.7-3.5i)+(1.8-5.7i)&=4.7+1.8-3.5i-5.7i\\&=6.5-9.2i\end{aligned}$$

41. $(2+8i)+(3-8i)=2+3+8i-8i=5$

43.
$$\begin{aligned}(2+3i)(4+7i)&=8+14i+12i+21i^2\\&=8+26i+21(-1)\\&=8+26i-21\\&=-13+26i\end{aligned}$$

45.
$$\begin{aligned}(2-4i)(3-5i)&=6-10i-12i+20i^2\\&=6-22i+20(-1)\\&=6-22i-20\\&=-14-22i\end{aligned}$$

47.
$$\begin{aligned}(7.5+3i)(4.5-9i)&=33.75-67.5i+13.5i-27i^2\\&=33.75-67.5i+13.5i-27(-1)\\&=33.75-54i+27\\&=60.75-54i\end{aligned}$$

49.
$$\begin{aligned}(2+8i)(2+8i)&=4+16i+16i+64i^2\\&=4+32i+64(-1)\\&=4+32i-64\\&=-60+32i\end{aligned}$$

51.
$$\begin{aligned}(3-7i)(3+7i)&=9+21i-21i-49i^2\\&=9-49(-1)\\&=9+49\\&=58\end{aligned}$$

53.
$$\begin{aligned}(7+6i)^2&=(7+6i)(7+6i)\\&=49+42i+42i+36i^2\\&=49+84i+36(-1)\\&=49+84i-36\\&=13+84i\end{aligned}$$

55.
$$\begin{aligned}\frac{12+9i}{3}&=\frac{12}{3}+\frac{9i}{3}\\&=4+3i\end{aligned}$$

57.

$$\frac{4+7i}{2i}=\left(\frac{4+7i}{2i}\right)\left(\frac{i}{i}\right)$$
$$=\frac{4i+7i^2}{2i^2}$$
$$=\frac{4i+7(-1)}{2(-1)}$$
$$=\frac{4i-7}{-2}$$
$$=\frac{-7}{-2}+\frac{4i}{-2}$$
$$=\frac{7}{2}-2i$$

59.

$$\frac{2.4+3.7i}{7.2i}=\left(\frac{2.4+3.7i}{7.2i}\right)\left(\frac{i}{i}\right)$$
$$=\frac{2.4i+3.7i^2}{7.2i^2}$$
$$=\frac{2.4i+3.7(-1)}{7.2(-1)}$$
$$=\frac{2.4i-3.7}{-7.2}$$
$$=\frac{-3.7}{-7.2}+\frac{2.4i}{-7.2}$$
$$=\frac{37}{72}-\frac{1}{3}i$$

61.

$$\frac{5+2i}{2+3i}=\left(\frac{5+2i}{2+3i}\right)\left(\frac{2-3i}{2-3i}\right)$$
$$=\frac{10-11i-6i^2}{4-9i^2}$$
$$=\frac{10-11i-6(-1)}{4-9(-1)}$$
$$=\frac{10-11i+6}{4+9}$$
$$=\frac{16-11i}{13}$$
$$=\frac{16}{13}-\frac{11}{13}i$$

63.

$$\frac{2-4i}{3-7i}=\left(\frac{2-4i}{3-7i}\right)\left(\frac{3+7i}{3+7i}\right)$$
$$=\frac{6+2i-28i^2}{9-49i^2}$$
$$=\frac{6+2i-28(-1)}{9-49(-1)}$$
$$=\frac{6+2i+28}{9+49}$$
$$=\frac{34+2i}{58}$$
$$=\frac{17}{29}+\frac{1}{29}i$$

65.

$$\frac{-8-5i}{-2+9i}=\left(\frac{-8-5i}{-2+9i}\right)\left(\frac{-2-9i}{-2-9i}\right)$$
$$=\frac{16+82i+45i^2}{4-81i^2}$$
$$=\frac{16+82i+45(-1)}{4-81(-1)}$$
$$=\frac{16+82i-45}{4+81}$$
$$=\frac{16+82i-45}{4+81}$$
$$=-\frac{29}{85}+\frac{82}{85}i$$

67.

$$\frac{-12-73i}{-5-14i}=\left(\frac{-12-73i}{-5-14i}\right)\left(\frac{-5+14i}{-5+14i}\right)$$
$$=\frac{60+197i-1022i^2}{25-196i^2}$$
$$=\frac{60+197i-1022(-1)}{25-196(-1)}$$
$$=\frac{60+197i+1022}{25+196}$$
$$=\frac{1082+197i}{221}$$
$$=\frac{1082}{221}+\frac{197}{221}i$$

69.

$$x^2+9=0$$
$$x^2=-9$$
$$x=\pm\sqrt{-9}$$
$$x=\pm 3i$$

71.

$$y^2+49=0$$
$$y^2=-49$$
$$y=\pm\sqrt{-49}$$
$$y=\pm 7i$$

73.

$$3(x-8)^2+12=4$$
$$3(x-8)^2=-8$$
$$\frac{3(x-8)^2}{3}=-\frac{8}{3}$$
$$(x-8)^2=-\frac{8}{3}$$
$$x-8=\pm\sqrt{-\frac{8}{3}}$$
$$x=8\pm i\sqrt{\frac{8}{3}}\cdot\sqrt{\frac{3}{3}}$$
$$x=8\pm i\sqrt{\frac{24}{9}}$$
$$x=8\pm\frac{2\sqrt{6}}{3}i$$

75.

$$2x^2+12=-8$$
$$2x^2=-20$$
$$\frac{2x^2}{2}=-\frac{20}{2}$$
$$x^2=-10$$
$$x=\pm\sqrt{-10}$$
$$x=\pm i\sqrt{10}$$

77.

$$5x^3+2x^2-7x=0$$
$$x\left(5x^2+2x-7\right)=0$$
$$x(5x+7)(x-1)=0$$
$$x=0,\ 5x+7=0,\ x-1=0$$
$$x=0,\ x=-\frac{7}{5},\ x=1$$

79.

$$3x^2=-48$$
$$\frac{3x^2}{3}=-\frac{48}{3}$$
$$x^2=-16$$
$$x=\pm\sqrt{-16}$$
$$x=\pm 4i$$

Chapter 10 Review

1. $-\sqrt{36}=-\sqrt{6^2}=-6$

2. $-\sqrt{100}=-\sqrt{10^2}=-10$

3. $\sqrt{-9}$ not a real number

4. $\sqrt{-25}$ not a real number

5. $\sqrt[3]{27}=\sqrt[3]{3^3}=3$

6. $\sqrt[3]{125}=\sqrt[3]{5^3}=5$

7. $\sqrt[3]{-64}=\sqrt[3]{(-4)^3}=-4$

8. $\sqrt[3]{-1}=\sqrt[3]{(-1)^3}=-1$

9. **a.** $t=\sqrt{\frac{h}{4.9}}$ Substitute $h=25$.

$$t=\sqrt{\frac{(25)}{4.9}}\approx\sqrt{5.1020}\approx 2.26$$

The ball will fall 25 meters in approximately 2.26 seconds.

b. $t=\sqrt{\frac{h}{4.9}}$ Substitute $h=12$.

$$t=\sqrt{\frac{(12)}{4.9}}\approx\sqrt{2.4490}\approx 1.56$$

The ball will fall 12 meters in approximately 1.56 seconds.

10. **a.** $t=\sqrt{\frac{h}{4.9}}$ Substitute $h=32$.

$$t=\sqrt{\frac{(32)}{4.9}}\approx\sqrt{6.5306}\approx 2.56$$

The ball will fall 32 meters in approximately 2.56 seconds.

b. $t=\sqrt{\frac{h}{4.9}}$ Substitute $h=9$.

$$t=\sqrt{\frac{(9)}{4.9}}\approx\sqrt{1.8367}\approx 1.36$$

The ball will fall 9 meters in approximately 1.36 seconds.

11. $d=\sqrt{(x_2-x_1)^2+(y_2-y_1)^2}$

$$d=\sqrt{(-2-[-3])^2+(-6-1)^2}$$
$$d=\sqrt{(1)^2+(-7)^2}=\sqrt{1+49}=\sqrt{50}\approx 7.07$$

12. $d=\sqrt{(x_2-x_1)^2+(y_2-y_1)^2}$

$d=\sqrt{(-3-[-5])^2+(1-[-2])^2}$

$d=\sqrt{(2)^2+(3)^2}=\sqrt{4+9}=\sqrt{13}\approx 3.61$

13. $\sqrt{4x-4}$ Substitute $x=5$.

$=\sqrt{4(5)-4}=\sqrt{20-4}=\sqrt{16}=4$

14. $\sqrt{-2x}$ Substitute $x=-2$.

$=\sqrt{-2(-2)}=\sqrt{4}=2$

15. $\sqrt{1-x^2}$ Substitute $x=3$.

$=\sqrt{1-(3)^2}=\sqrt{1-9}=\sqrt{-8}$ not a real number

16. $\sqrt{x^2-6}$ Substitute $x=4$.

$=\sqrt{(4)^2-6}=\sqrt{16-6}=\sqrt{10}$

17. $\sqrt{y^{10}}=\sqrt{\left(y^5\right)^2}=y^5$

18. $\sqrt{y^6}=\sqrt{\left(y^3\right)^2}=y^3$

19. $\sqrt{25x^4}=\sqrt{\left(5x^2\right)^2}=5x^2$

20. $\sqrt{36x^6}=\sqrt{\left(6x^3\right)^2}=6x^3$

21. $\sqrt{x^6y^2}=\sqrt{\left(x^3y\right)^2}=x^3y$

22. $\sqrt{x^2y^4}=\sqrt{\left(xy^2\right)^2}=xy^2$

23. $\sqrt[5]{32a^5b^{10}c^{30}}=\sqrt[5]{2^5a^5b^{10}c^{30}}=2ab^2c^6$

24. **a.** $x^{\frac{1}{4}}y^{\frac{3}{4}}=\left(xy^3\right)^{\frac{1}{4}}=\sqrt[4]{xy^3}$

b. $3^{\frac{1}{2}}a^{\frac{1}{2}}=(3a)^{\frac{1}{2}}=\sqrt{3a}$

25. **a.** $\sqrt[5]{4x^2}=\left(4x^2\right)^{\frac{1}{5}}=4^{\frac{1}{5}}\left(x^2\right)^{\frac{1}{5}}=4^{\frac{1}{5}}x^{\frac{2}{5}}$

b. $\sqrt{7ab}=(7ab)^{\frac{1}{2}}=7^{\frac{1}{2}}a^{\frac{1}{2}}b^{\frac{1}{2}}$

26. $\sqrt{75}=\sqrt{25\cdot 3}=5\sqrt{3}$

27. $\sqrt{54}=\sqrt{9\cdot 6}=3\sqrt{6}$

28. $\sqrt{x^7}=\sqrt{x^6\cdot x}=x^3\sqrt{x}$

29. $\sqrt{4y^3}=\sqrt{4y^2\cdot y}=2y\sqrt{y}$

30. $\sqrt{20x^3}=\sqrt{4x^2\cdot 5x}=2x\sqrt{5x}$

31. $\sqrt{24x^9}=\sqrt{4x^8\cdot 6x}=2x^4\sqrt{6x}$

32. $\sqrt{144x^2y^4}=\sqrt{12^2}\sqrt{x^2}\sqrt{y^4}=12xy^2$

33. $5mn^3\sqrt{180m^4n^5}=5mn^3\sqrt{36\cdot 5\cdot m^4n^4n}$

$=5mn^3\sqrt{6^2m^4n^4}\sqrt{5n}$

$=5mn^3\cdot 6m^2n^2\sqrt{5n}$

$=30m^3n^5\sqrt{5n}$

34. $2c^2d\sqrt{220c^7d^6}=2c^2d\sqrt{4\cdot 55c^6cd^6}$

$=2c^2d\sqrt{2^2c^6d^6}\sqrt{55c}$

$=2c^2d\cdot 2c^3d^3\sqrt{55c}$

$=4c^5d^4\sqrt{55c}$

35. $\sqrt[3]{-8x^3y^7}=\sqrt[3]{(-2)^3x^3y^6}\sqrt[3]{y}=-2xy^2\sqrt[3]{y}$

36. $\sqrt[3]{64x^3y^5}=\sqrt[3]{4^3x^3y^3}\sqrt[3]{y^2}=4xy\sqrt[3]{y^2}$

37. $\sqrt[4]{56a^2b^{10}}=\sqrt[4]{56a^2b^8b^2}$

$=\sqrt[4]{b^8}\sqrt[4]{56a^2b^2}$

$=b^2\sqrt[4]{56a^2b^2}$

38. $\sqrt[4]{25m^3n^{11}}=\sqrt[4]{25m^3n^8n^3}$

$=\sqrt[4]{n^8}\sqrt[4]{25m^3n^3}$

$=n^2\sqrt[4]{25m^3n^3}$

39. $\sqrt[5]{-40a^7b^3c^5}=\sqrt[5]{-40a^5a^2b^3c^5}$

$=\sqrt[5]{a^5c^5}\sqrt[5]{-40a^2b^3}$

$=ac\sqrt[5]{-40a^2b^3}$

40. $\sqrt{27}=\sqrt{9\cdot 3}=3\sqrt{3}\approx 5.20$

41. $\sqrt{45}=\sqrt{9\cdot 5}=3\sqrt{5}\approx 6.71$

42. $\sqrt{200}=\sqrt{100\cdot 2}=10\sqrt{2}\approx 14.14$

43. $\sqrt{250}=\sqrt{25\cdot 10}=5\sqrt{10}\approx 15.81$

44. $-\sqrt{3x}+7\sqrt{3x}=6\sqrt{3x}$

45. $2\sqrt{6y}+9\sqrt{6y}=11\sqrt{6y}$

46. $\sqrt{16x^3}+7-2x\sqrt{x}-3=\sqrt{16x^2\cdot x}-2x\sqrt{x}+4$

$=4x\sqrt{x}-2x\sqrt{x}+4=4+2x\sqrt{x}$

47. $11y-2\sqrt{y^3}+6y-y\sqrt{9y}$

$=11y-2\sqrt{y^2\cdot y}+6y-y\cdot 3\sqrt{y}$

$=11y-2y\sqrt{y}+6y-3y\sqrt{y}=17y-5y\sqrt{y}$

48. $4\sqrt{3a}+7\sqrt[3]{3a}+8\sqrt{3a}=12\sqrt{3a}+7\sqrt[3]{3a}$

49. $\sqrt[3]{27a^3b^2}-5a\sqrt[3]{b^2}=3a\sqrt[3]{b^2}-5a\sqrt[3]{b^2}=-2a\sqrt[3]{b^2}$

50. $\sqrt[5]{-64a^7b^3c^{10}}+4ac\sqrt[5]{2a^2b^3c^5}+19ac^2\sqrt[5]{2a^2b^3}$

$=\sqrt[5]{(-2)^5\cdot 2a^5a^2b^3c^{10}}+4ac\sqrt[5]{2a^2b^3c^5}+19ac^2\sqrt[5]{2a^2b^3}$

$=-2ac^2\sqrt[5]{2a^2b^3}+4ac^2\sqrt[5]{2a^2b^3}+19ac^2\sqrt[5]{2a^2b^3}$

$=21ac^2\sqrt[5]{2a^2b^3}$

51. $\sqrt{3}\cdot\sqrt{6}=\sqrt{3\cdot6}=\sqrt{18}=\sqrt{9\cdot2}=3\sqrt{2}$

52. $\sqrt{2y}\cdot\sqrt{8y}=\sqrt{2y\cdot8y}=\sqrt{16y^2}=4y$

53. $\sqrt{3x^3}\cdot\sqrt{8x^2}=\sqrt{3x^3\cdot8x^2}=\sqrt{24x^5}$
$=\sqrt{4x^4\cdot6x}=2x^2\sqrt{6x}$

54. $\sqrt{7a}\cdot\sqrt{7a}=7a$

55. $\sqrt{3b^3}\cdot\sqrt{3b^3}=3b^3$

56. $\sqrt[3]{3a^3b^2}\cdot\sqrt[3]{18b^2}=\sqrt[3]{54a^3b^4}=\sqrt[3]{3^3\cdot2a^3b^3b}=3ab\sqrt[3]{2b}$

57. $3x^2y\sqrt[3]{15xy^2z}\cdot2xy^3\sqrt[3]{18xyz^5}=6x^3y^4\sqrt[3]{270x^2y^3z^6}$
$=6x^3y^4\sqrt[3]{3^3y^3z^6}\sqrt[3]{10x^2}$
$=18x^3y^5z^2\sqrt[3]{10x^2}$

58. $\sqrt{5}(\sqrt{10}+\sqrt{5})=\sqrt{5}(\sqrt{10})+\sqrt{5}(\sqrt{5})$
$=\sqrt{50}+5=\sqrt{25\cdot2}+5=5+5\sqrt{2}$

59. $-7(4\sqrt{a}+\sqrt{7})=-7(4\sqrt{a})-7(\sqrt{7})=-28\sqrt{a}-7\sqrt{7}$

60. $(3-\sqrt{2})(1+\sqrt{6})$
$=3\cdot1+3(\sqrt{6})-1(\sqrt{2})-\sqrt{2}(\sqrt{6})$
$=3+3\sqrt{6}-\sqrt{2}-\sqrt{12}=3+3\sqrt{6}-\sqrt{2}-\sqrt{4\cdot3}$
$=3+3\sqrt{6}-\sqrt{2}-2\sqrt{3}$

61. $(4+\sqrt{6})(4-\sqrt{6})$
$=4\cdot4-4(\sqrt{6})+4(\sqrt{6})-\sqrt{6}(\sqrt{6})$
$=16-4\sqrt{6}+4\sqrt{6}-6=10$

62. The conjugate of $3+\sqrt{5}$ is $3-\sqrt{5}$.

63. The conjugate of $9-\sqrt{3}$ is $9+\sqrt{3}$.

64. The conjugate of $-5-\sqrt{x}$ is $-5+\sqrt{x}$.

65. $\dfrac{\sqrt{32}}{\sqrt{2}}=\sqrt{\dfrac{32}{2}}=\sqrt{16}=4$

66. $\dfrac{\sqrt{54x^3}}{\sqrt{6x}}=\sqrt{\dfrac{54x^3}{6x}}=\sqrt{9x^2}=3x$

67. $\dfrac{5}{\sqrt{7}}=\dfrac{5}{\sqrt{7}}\cdot\dfrac{\sqrt{7}}{\sqrt{7}}=\dfrac{5\sqrt{7}}{7}$

68. $\dfrac{-3}{\sqrt{5}}=\dfrac{-3}{\sqrt{5}}\cdot\dfrac{\sqrt{5}}{\sqrt{5}}=\dfrac{-3\sqrt{5}}{5}$

69. $\dfrac{5x}{\sqrt[3]{2x^2y}}=\dfrac{5x\cdot\sqrt[3]{2^2xy^2}}{\sqrt[3]{2x^2y}\cdot\sqrt[3]{2^2xy^2}}$
$=\dfrac{5x\sqrt[3]{4xy^2}}{\sqrt[3]{2^3x^3y^3}}$
$=\dfrac{5x\sqrt[3]{4xy^2}}{2xy}$
$=\dfrac{5\sqrt[3]{4xy^2}}{2y}$

70. $\dfrac{4}{3-\sqrt{2}}=\dfrac{4}{(3-\sqrt{2})}\cdot\dfrac{(3+\sqrt{2})}{(3+\sqrt{2})}$
$=\dfrac{12+4\sqrt{2}}{9+3\sqrt{2}-3\sqrt{2}-\sqrt{2}\cdot\sqrt{2}}$
$=\dfrac{12+4\sqrt{2}}{9-2}=\dfrac{12+4\sqrt{2}}{7}$

71. $\dfrac{-3}{1+\sqrt{5}}=\dfrac{-3}{(1+\sqrt{5})}\cdot\dfrac{(1-\sqrt{5})}{(1-\sqrt{5})}$
$=\dfrac{-3+3\sqrt{5}}{1\cdot1-1\sqrt{5}+1\sqrt{5}-\sqrt{5}\cdot\sqrt{5}}$
$=\dfrac{-3+3\sqrt{5}}{1-5}=\dfrac{-3+3\sqrt{5}}{-4}=\dfrac{3-3\sqrt{5}}{4}$

72. a. $P(3)=10.3+2\sqrt{3}$
$P(3)\approx13.76$
In the third month of the year Big Jim's Mart had a profit of about \$13.76 thousand dollars.
b. $P(8)=10.3+2\sqrt{8}$
$P(8)\approx15.96$
In August Big Jim's Mart had a profit of about \$15.96 thousand dollars.

73. $W(4)=22.37\sqrt[5]{4^3}$
$W(4)\approx51.39$
A four-year-old alpaca weighs about 51.39 kilograms.

74. $f(x)=-5\sqrt{x}$
Domain: $[0,\infty)$; Range: $(-\infty,0]$

75. $h(x)=5\sqrt{x}$
Domain: $[0,\infty)$; Range: $[0,\infty)$

76. $g(x)=\sqrt{2x-7}$
Domain: $\left[\dfrac{7}{2},\infty\right)$; Range: $[0,\infty)$

77. $h(x)=\sqrt{7-x}$
Domain: $(-\infty,7]$; Range: $[0,\infty)$

78. $f(x)=4\sqrt[3]{x+3}$
Domain: All real numbers; Range: All real numbers

79. $k(a)=3\sqrt[4]{a}$
Domain: $[0,\infty)$; Range: $[0,\infty)$

80. $M(p) = -2.3\sqrt[6]{p}$

Domain: $[0,\infty)$; Range: $(-\infty,0]$

81. $g(x) = 3.4\sqrt[5]{x}$

Domain: All real numbers; Range: All real numbers

82. Domain: $[5,\infty)$; Range: $(-\infty,0]$

83. Domain: $(-\infty,18]$; Range: $(-\infty,0]$

84. $f(x) = \sqrt{x+2}$

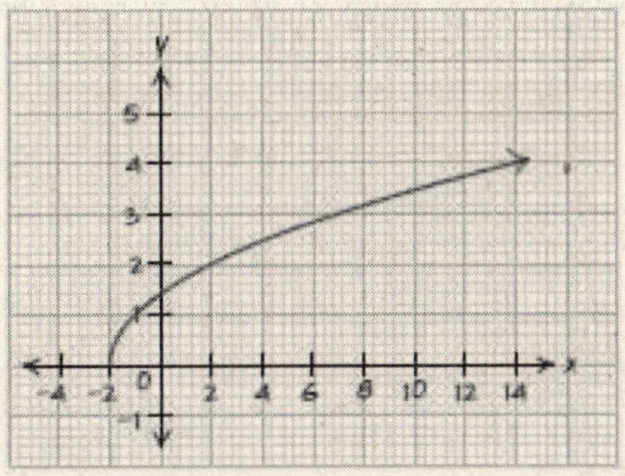

85. $f(x) = -\sqrt{x-4}$

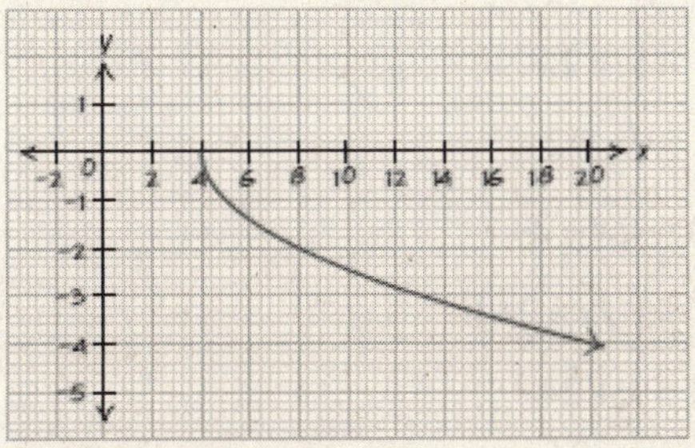

86. $f(x) = -\sqrt{x+12}$

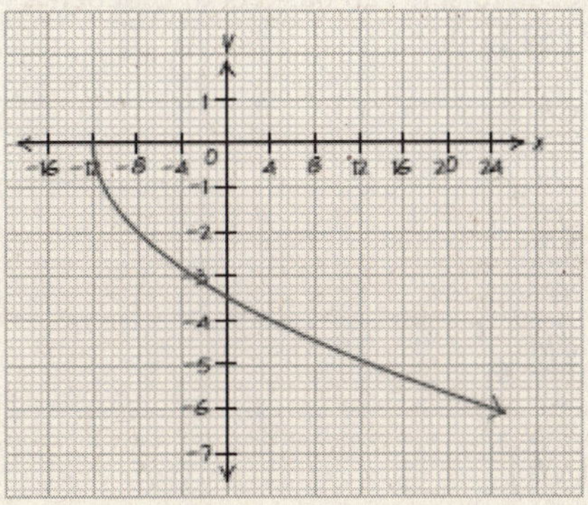

87. $f(x) = \sqrt[3]{x-10}$

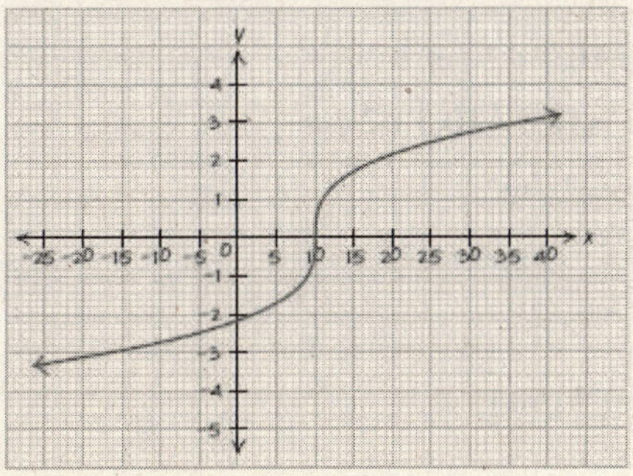

88. $f(x) = -\sqrt[3]{x+4}$

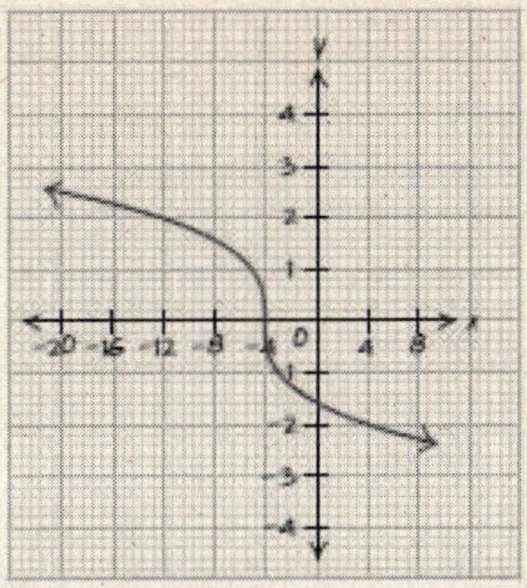

89. $f(x) = \sqrt[5]{6x}$

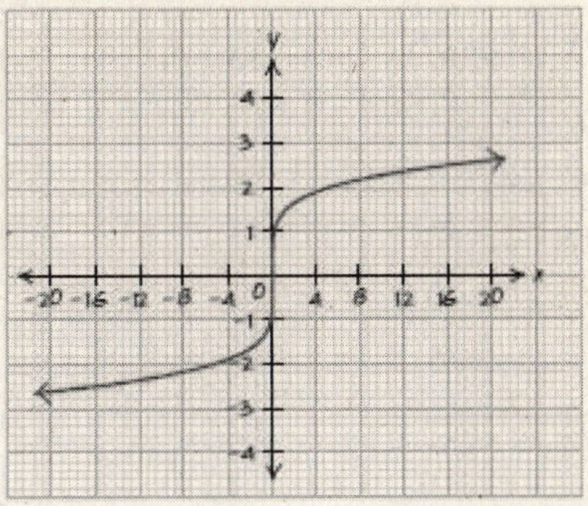

90. $\sqrt{3x-3}+5=8$ Substitute $x = 4$.

$$\sqrt{3(4)-3}+5=8$$
$$\sqrt{12-3}+5=8$$
$$\sqrt{9}+5=8$$
$$3+5=8$$
$$8=8$$

Yes, $x=4$ is a solution to the radical equation.

91. $\sqrt{5x-1}+1=2$ Substitute $x = 2$.

$$\sqrt{5(2)-1}+1=2$$
$$\sqrt{10-1}+1=2$$
$$\sqrt{9}+1=2$$
$$3+1=2$$
$$4 \neq 2$$

No, $x=2$ is not a solution to the radical equation.

92. $\sqrt{1-a}=3$

$$\left(\sqrt{1-a}\right)^2=(3)^2$$
$$1-a=9$$
$$\underline{-1 \quad -1}$$
$$-a=8$$
$$\frac{-a}{-1}=\frac{8}{-1}$$
$$a=-8$$

93. $\sqrt{5-2x}=7$

$\left(\sqrt{5-2x}\right)^2=(7)^2$

$5-2x=49$

$\underline{-5 \quad -5}$

$-2x=44$

$\frac{-2x}{-2}=\frac{44}{-2}$

$x=-22$

94. $\sqrt{x^2-10x+41}=4$

$\left(\sqrt{x^2-10x+41}\right)^2=(4)^2$

$x^2-10x+41=16$

$\underline{-16 \;\; -16}$

$x^2-10x+25=0$

$(x-5)(x-5)=0$

$x-5=0$

$x=5$

95. $\sqrt{x^2-14x+58}=3$

$\left(\sqrt{x^2-14x+58}\right)^2=(3)^2$

$x^2-14x+58=9$

$\underline{-9 \;\; -9}$

$x^2-14x+49=0$

$(x-7)(x-7)=0$

$x-7=0$

$x=7$

96. $\sqrt{3a-5}=-3$ no solution

A square root cannot equal a negative number.

97. $\sqrt{1-2a}=-1$

A square root cannot equal a negative number.

98. $\sqrt{x+5}+\sqrt{3x+4}=13$

$\sqrt{3x+4}=13-\sqrt{x+5}$

$\left(\sqrt{3x+4}\right)^2=\left(13-\sqrt{x+5}\right)^2$

$3x+4=\left(13-\sqrt{x+5}\right)\left(13-\sqrt{x+5}\right)$

$3x+4=169-13\sqrt{x+5}-13\sqrt{x+5}+x+5$

$3x+4=174-26\sqrt{x+5}+x$

$2x-170=-26\sqrt{x+5}$

$(2x-170)^2=\left(-26\sqrt{x+5}\right)^2$

$(2x-170)(2x-170)=676(x+5)$

$4x^2-680x+28900=676x+3380$

$4x^2-1356x+25520=0$

$x=\frac{1356\pm\sqrt{(-1356)^2-4(4)(25520)}}{2(4)}$

$x=\frac{1356\pm\sqrt{1430416}}{8}$

$x=\frac{1356\pm1196}{8}$

$x=319, x=20$

The answer $x=319$ does not check out, therefore the solution is $x=20$.

99. $\sqrt[3]{2x}=3$

$\left(\sqrt[3]{2x}\right)^3=3^3$

$2x=27$

$x=13.5$

100. $\sqrt[4]{5x+2}=2$

$\left(\sqrt[4]{5x+2}\right)^4=2^4$

$5x+2=16$

$5x=14$

$x=2.8$

101. $P=\sqrt{x+24}$

$(12)=\sqrt{x+24}$

$(12)^2=\left(\sqrt{x+24}\right)^2$

$144=x+24$

$\underline{-24 \quad\quad -24}$

$120=x$

Nelson needs to sell 120 baskets of blueberries to make a profit of \$12.00.

102. $t=\frac{\sqrt{h}}{4}$

$(6)=\frac{\sqrt{h}}{4}$

$4\cdot6=\frac{\sqrt{h}}{4}\cdot4$

$(24)^2=\left(\sqrt{h}\right)^2$

$576=h$

The rock falls 576 feet in 6 seconds.

103. $17=10.3+2\sqrt{m}$

$17-10.3=10.3-10.3+2\sqrt{m}$

$6.7=2\sqrt{m}$

$\frac{6.7}{2}=\frac{2\sqrt{m}}{2}$

$\left(\frac{6.7}{2}\right)^2=\left(\sqrt{m}\right)^2$

$m=11.2225$

Big Jim's Mart will have a profit of about \$17,000 in the 11th month of the year.

104. $$60 = 22.37\sqrt[5]{a^3}$$
$$\frac{60}{22.37} = \frac{22.37\sqrt[5]{a^3}}{22.37}$$
$$\frac{60}{22.37} = \sqrt[5]{a^3} = a^{3/5}$$
$$\left(\frac{60}{22.37}\right)^{5/3} = \left(a^{3/5}\right)^{5/3}$$
$$a \approx 5.18$$

A 5 year old alpaca will weigh about 60 kilograms.

105. $\sqrt{-25} = 5i$

106. $\sqrt{-32} = 4i\sqrt{2}$

107. $\sqrt{-4} + \sqrt{-25} = 2i + 5i = 7i$

108. $$\begin{aligned}\left(2+\sqrt{-5}\right)\left(3-\sqrt{-10}\right) &= \left(2+i\sqrt{5}\right)\left(3-i\sqrt{10}\right)\\ &= 6-2i\sqrt{10}+3i\sqrt{5}-i^2\sqrt{50}\\ &= 6-2i\sqrt{10}+3i\sqrt{5}-(-1)\cdot 5\sqrt{2}\\ &= 6-2i\sqrt{10}+3i\sqrt{5}+5\sqrt{2}\\ &= \left(6+5\sqrt{2}\right)+\left(-2\sqrt{10}+3\sqrt{5}\right)i\end{aligned}$$

109. $(2+3i)+(5+7i) = 7+10i$

110. $(3.5+1.2i)+(2.4-3.6i) = 5.9-2.4i$

111. $(4+15i)-(3+11i) = 4-3+15i-11i = 1+4i$

112. $$\begin{aligned}(4.5+2.9i)-(1.6-4.2i) &= 4.5-1.6+2.9i+4.2i\\ &= 2.9+7.1i\end{aligned}$$

113. $$\begin{aligned}(3+4i)(2-7i) &= 6-21i+8i-28i^2\\ &= 6-13i-28i^2\\ &= 6-13i-28\cdot(-1)\\ &= 6-13i+28\\ &= 34-13i\end{aligned}$$

114. $$\begin{aligned}(6+5i)(2+7i) &= 12+42i+10i+35i^2\\ &= 12+52i+35i^2\\ &= 12+52i+35\cdot(-1)\\ &= 12+52i-35\\ &= -23+52i\end{aligned}$$

115. $$\begin{aligned}(2-3i)(2+3i) &= 4+6i-6i-9i^2\\ &= 4-9i^2\\ &= 4-9\cdot(-1)\\ &= 4+9\\ &= 13\end{aligned}$$

116. $$\begin{aligned}\frac{12+7i}{3} &= \frac{12}{3}+\frac{7i}{3}\\ &= 4+\frac{7}{3}i\end{aligned}$$

117. $$\begin{aligned}\frac{4+7i}{5i} &= \frac{(4+7i)i}{(5i)i}\\ &= \frac{4i+7i^2}{5i^2}\\ &= \frac{4i+7(-1)}{5(-1)}\\ &= \frac{4i-7}{-5}\\ &= \frac{-7}{-5}+\frac{4i}{-5}\\ &= 1.4-0.8i\end{aligned}$$

118. $$\begin{aligned}\frac{7}{2+3i} &= \frac{7(2-3i)}{(2+3i)(2-3i)}\\ &= \frac{14-21i}{4-6i+6i-9i^2}\\ &= \frac{14-21i}{4-9i^2}\\ &= \frac{14-21i}{4-9(-1)}\\ &= \frac{14-21i}{4+9}\\ &= \frac{14-21i}{13}\\ &= \frac{14}{13}-\frac{21}{13}i\end{aligned}$$

119. $$\begin{aligned}\frac{5+7i}{3-4i} &= \frac{(5+7i)(3+4i)}{(3-4i)(3+4i)}\\ &= \frac{15+20i+21i+28i^2}{9+12i-12i-16i^2}\\ &= \frac{15+41i+28i^2}{9-16i^2}\\ &= \frac{15+41i+28(-1)}{9-16(-1)}\\ &= \frac{15+41i-28}{9+16}\\ &= \frac{-13+41i}{25}\\ &= -\frac{13}{25}+\frac{41}{25}i\end{aligned}$$

120. $\dfrac{2-9i}{4+3i}=\dfrac{(2-9i)(4-3i)}{(4+3i)(4-3i)}$

$=\dfrac{8-6i-36i+27i^2}{16-12i+12i-9i^2}$

$=\dfrac{8-42i+27i^2}{16-9i^2}$

$=\dfrac{8-42i+27(-1)}{16-9(-1)}$

$=\dfrac{8-42i-27}{16+9}$

$=\dfrac{-19-42i}{25}$

$=-\dfrac{19}{25}-\dfrac{42}{25}i$

121. $\dfrac{2.5+6.4i}{3.3+8.2i}=\dfrac{(2.5+6.4i)(3.3-8.2i)}{(3.3+8.2i)(3.3-8.2i)}$

$=\dfrac{8.25-20.5i+21.12i-52.48i^2}{10.89-27.06i+27.06i-67.24i^2}$

$=\dfrac{8.25+0.62i-52.48i^2}{10.89-67.24i^2}$

$=\dfrac{8.25+0.62i-52.48(-1)}{10.89-67.24(-1)}$

$=\dfrac{8.25+0.62i+52.48}{10.89+67.24}$

$=\dfrac{60.73+0.62i}{78.13}$

$=\dfrac{60.73}{78.13}+\dfrac{0.62}{78.13}i$

$\approx 0.777+0.008i$

Chapter 10 Test

1. $-\sqrt{81}=-\sqrt{(9)^2}=-9$

2. $\sqrt[3]{-27}=\sqrt[3]{(-3)^3}=-3$

3. $-\sqrt{36a^8}=-\sqrt{\left(6a^4\right)^2}=-6a^4$

4. $\sqrt{a^8b^2}=\sqrt{\left(a^4b\right)^2}=a^4b$

5. $d=\sqrt{(x_2-x_1)^2+(y_2-y_1)^2}$

$d=\sqrt{(4-[-3])^2+(-2-5)^2}$

$d=\sqrt{(7)^2+(-7)^2}=\sqrt{49+49}=\sqrt{98}\approx 9.90$

6. $\sqrt{6x+1}$ Substitute $x=4$.

$\sqrt{6(4)+1}=\sqrt{24+1}=\sqrt{25}=5$

7. $\sqrt{\dfrac{x^2}{9}}=\sqrt{\left(\dfrac{x}{3}\right)^2}=\dfrac{x}{3}$

8. $\sqrt{60}=\sqrt{4\cdot 15}=2\sqrt{15}$

9. $\sqrt{24}\approx 4.8\underline{9}897...\approx 4.90$

10. $\sqrt{28b^5}=\sqrt{4b^4\cdot 7b}=2b^2\sqrt{7b}$

11. $\sqrt{25a^3b^2}=\sqrt{25a^2b^2\cdot a}=5ab\sqrt{a}$

12. $\sqrt[5]{-32m^3n^7p^{10}}=\sqrt[5]{(-2)^5m^3n^7p^{10}}$

$=\sqrt[5]{(-2)^5n^5p^{10}}\cdot\sqrt[5]{m^3n^2}$

$=-2np^2\sqrt[5]{m^3n^2}$

13. $7\sqrt{3}-11-\sqrt{3}+8=-3+6\sqrt{3}$

14. $\sqrt{50}+9\sqrt{2}=\sqrt{25\cdot 2}+9\sqrt{2}$

$=5\sqrt{2}+9\sqrt{2}=14\sqrt{2}$

15. $\sqrt{24x^2}-x\sqrt{6}=\sqrt{4x^2\cdot 6}-x\sqrt{6}$

$=2x\sqrt{6}-x\sqrt{6}=x\sqrt{6}$

16. $\sqrt{6}\cdot\sqrt{8}=\sqrt{6\cdot 8}=\sqrt{48}=\sqrt{16\cdot 3}=4\sqrt{3}$

17. $\sqrt{5x^3}\cdot\sqrt{10x^2}=\sqrt{5x^3\cdot 10x^2}=\sqrt{50x^5}$

$=\sqrt{25x^4\cdot 2x}=5x^2\sqrt{2x}$

18. $\sqrt{8}(\sqrt{2}+\sqrt{5})=\sqrt{8}(\sqrt{2})+\sqrt{8}(\sqrt{5})$

$=\sqrt{16}+\sqrt{40}=4+2\sqrt{10}$

19. $(1+3\sqrt{5})(2-4\sqrt{5})$

$=1\cdot 2-4\sqrt{5}(1)+3\sqrt{5}(2)-3\sqrt{5}(4\sqrt{5})$

$=2-4\sqrt{5}+6\sqrt{5}-3\cdot 4\cdot\sqrt{5}\cdot\sqrt{5}$

$=2+2\sqrt{5}-12\cdot 5=2+2\sqrt{5}-60=-58+2\sqrt{5}$

20. $(5-\sqrt{2})(5+\sqrt{2})$
$= 5\cdot 5+5\sqrt{2}-5\sqrt{2}-\sqrt{2}\sqrt{2}$
$= 25-2=23$

21. $\dfrac{\sqrt{125}}{\sqrt{5}}=\sqrt{\dfrac{125}{5}}=\sqrt{25}=5$

22. $\dfrac{-3}{\sqrt{6}}=\dfrac{-3}{\sqrt{6}}\cdot\dfrac{\sqrt{6}}{\sqrt{6}}=\dfrac{-3\sqrt{6}}{6}=-\dfrac{\sqrt{6}}{2}$

23. $\dfrac{5m}{\sqrt[3]{3mn^2}}=\dfrac{5m}{\sqrt[3]{3mn^2}}\cdot\dfrac{\sqrt[3]{3^2m^2n}}{\sqrt[3]{3^2m^2n}}$
$=\dfrac{5m\sqrt[3]{3^2m^2n}}{\sqrt[3]{3^3m^3n^3}}$
$=\dfrac{5m\sqrt[3]{9m^2n}}{3mn}$
$=\dfrac{5\sqrt[3]{9m^2n}}{3n}$

24. $\dfrac{5}{2+\sqrt{3}}=\dfrac{5}{\left(2+\sqrt{3}\right)}\cdot\dfrac{\left(2-\sqrt{3}\right)}{\left(2-\sqrt{3}\right)}$
$=\dfrac{10-5\sqrt{3}}{4-2\sqrt{3}+2\sqrt{3}-\sqrt{3}\cdot\sqrt{3}}$
$=\dfrac{10-5\sqrt{3}}{4-3}=\dfrac{10-5\sqrt{3}}{1}=10-5\sqrt{3}$

25. $\sqrt{a}-7=15$
$\quad +7 \quad +7$
$\sqrt{a}=2$
$\left(\sqrt{a}\right)^2=(22)^2$
$a=484$

26. $2\sqrt{y}+6=3$
$\quad -6 \quad -6$
$2\sqrt{y}=-3$
$\dfrac{2\sqrt{y}}{2}=\dfrac{-3}{2}$
$\sqrt{y}=-\dfrac{3}{2}$ no solution

A square root cannot equal a negative number.

27. $\sqrt{x^2+2x-6}=3$
$\left(\sqrt{x^2+2x-6}\right)^2=(3)^2$
$x^2+2x-6=9$
$\quad -9 \quad -9$
$x^2+2x-15=0$
$(x+5)(x-3)=0$

$x+5=0 \qquad x-3=0$
$\quad -5 \quad -5 \qquad +3 \quad +3$
$x=-5 \qquad x=3$

28. $\sqrt{x-7}+\sqrt{4x-11}=13$
$\sqrt{4x-11}=13-\sqrt{x-7}$
$\left(\sqrt{4x-11}\right)^2=\left(13-\sqrt{x-7}\right)^2$
$4x-11=169-26\sqrt{x-7}+x-7$
$4x-11=162-26\sqrt{x-7}+x$
$3x-173=-26\sqrt{x-7}$
$(3x-173)^2=\left(-26\sqrt{x-7}\right)^2$
$9x^2-1038x+29{,}929=676(x-7)$
$9x^2-1038x+29{,}929=676x-4732$
$9x^2-1714x+34{,}661=0$
$x=\dfrac{1714\pm\sqrt{(-1714)^2-4(9)(34{,}661)}}{2(9)}$
$x=\dfrac{1714+\sqrt{1{,}690{,}000}}{18}, x=\dfrac{1714-\sqrt{1{,}690{,}000}}{18}$
$x=\dfrac{1507}{9}, x=23$

Since $x=\dfrac{1507}{9}$ does not check, $x=23$ is the only solution.

29. $\sqrt[3]{x+5}=4$
$\left(\sqrt[3]{x+5}\right)^3=4^3$
$x+5=64$
$x=59$

30. $v(20)=20.1\sqrt{273+20}\approx 344.1$
The speed of sound in air is about 344.1meters per second when the temperature is 20°C.

31. $90=0.243\sqrt{h}$
$\dfrac{90}{0.243}=\dfrac{0.243\sqrt{h}}{0.243}$
$\left(\dfrac{90}{0.243}\right)^2=\left(\sqrt{h}\right)^2$
$h\approx 137{,}174.2$

$T(h)=90$ when $h\approx 137174.2$, add 1500 to allow time to pull the cord. The plane would have to be at an altitude of about 138,674 feet. This is not reasonable so it is not possible for the daredevil to free-fall for 1.5 minutes.

32. Since it is an even root there is a restricted domain and range:

$x+7\geq 0$

$x+7-7\geq 0-7$

$x\geq -7$

Domain: $x\geq -7$; Range: $(-\infty,0]$

33. Since it is an odd root:

Domain: All real numbers; Range: All real numbers

34. $f(x)=-\sqrt{10-x}$

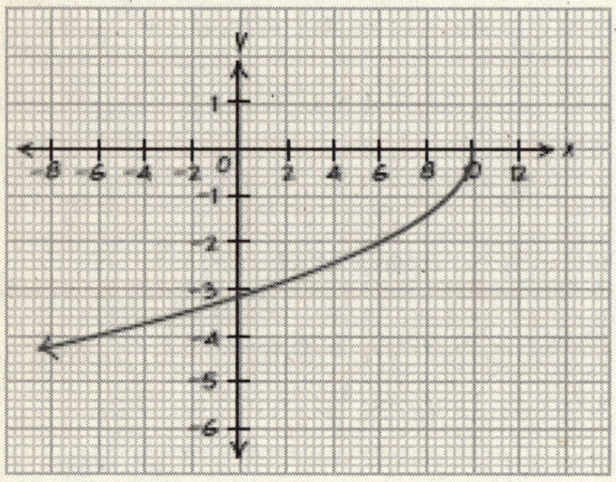

35. $g(x)=\sqrt[3]{x-7}$

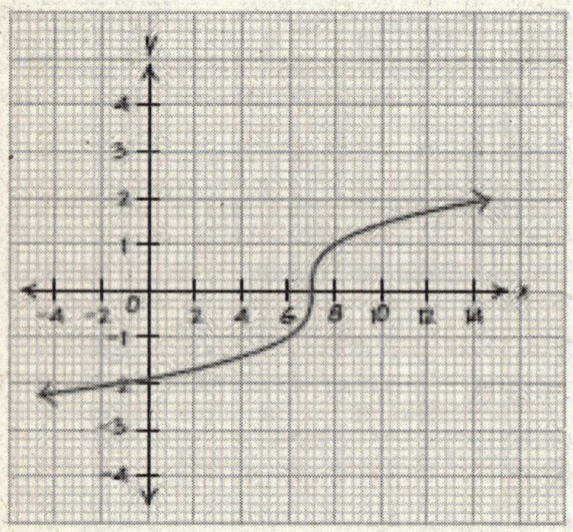

36. $(2.7+3.4i)+(1.4-4.8i)=4.1-1.4i$

37. $(5+11i)-(4-7i)=5+11i-4+7i=1+18i$

38. $(7+2i)(3-5i)=21-35i+6i-10i^2$

$=21-29i+10$

$=31-29i$

39. $(1.5-4.5i)(2.25-6.5i)$

$=3.375-9.75i-10.125i+29.25i^2$

$=3.375-19.875i-29.25$

$=-25.875-19.875i$

40. $\left(\frac{8}{4+5i}\right)\left(\frac{4-5i}{4-5i}\right)=\frac{32-40i}{16-20i+20i-25i^2}$

$=\frac{32-40i}{16+25}$

$=\frac{32}{41}-\frac{40}{41}i$

41. $\left(\frac{3+2i}{6-7i}\right)\left(\frac{6+7i}{6+7i}\right)=\frac{18+21i+12i+14i^2}{36+42i-42i-49i^2}$

$=\frac{18+33i-14}{36+49}$

$=\frac{4+33i}{85}$

$=\frac{4}{85}+\frac{33}{85}i$

CHAPTER 11 QUADRATIC FUNCTIONS

Section 11.1 Quadratic Functions and Parabolas

1. This function is a quadratic function, because the term with the highest degree is 2.

3. This function is a linear function, because the term with the highest degree is 1.

5. This function is a quadratic function, because the term with the highest degree is 2.

7. Other, because the variable is in the exponent.

9. This function is a quadratic function, because the term with the highest degree is 2 once the polynomial is simplified.

11. This function is a quadratic function, because the term with the highest degree is 2.

13. This function is a quadratic function, because the term with the highest degree is 2 once the polynomial is simplified.

15. **a.** $(-5,1)$

b. $x<-5$

c. $x>-5$

d. The graph crosses the horizontal x-axis at $(-7.5,0)$ and $(-2.5,0)$.

e. The graph crosses the vertical y-axis at $(0,-3)$.

17. **a.** (4, 25)

b. $x<4$

c. $x>4$

d. The graph crosses the x-axis at $(-1,0)$ and $(9,0)$.

e. The graph crosses the y-axis at $(0,9)$.

f. $f(2)=21$, because when $x=2, y=21$.

19. **a.** $(-1,4)$

b. $x>-1$

c. $x<-1$

d. None, because the horizontal intercept is the point at which the graph crosses the x-axis and the graph does not cross the x-axis.

e. The graph crosses the vertical axis at $(0,5)$.

f. $f(-4)=9.5$, because when $x=-4, y=9.5$.

g. $f(x)=14$ for $x=3$ and $x=-5$.

21. Horizontal intercepts: $(-4,0)$ and $(3,0)$, vertical intercept: $(0,6)$

23. Horizontal intercepts: $(-10,0)$ and $(15,0)$, vertical intercept: $(0,-15)$

25. **a.** Let $D(m)$ be the average number of days above 70°F in San Diego, California, during month m (i.e., $m=1$ represents January).

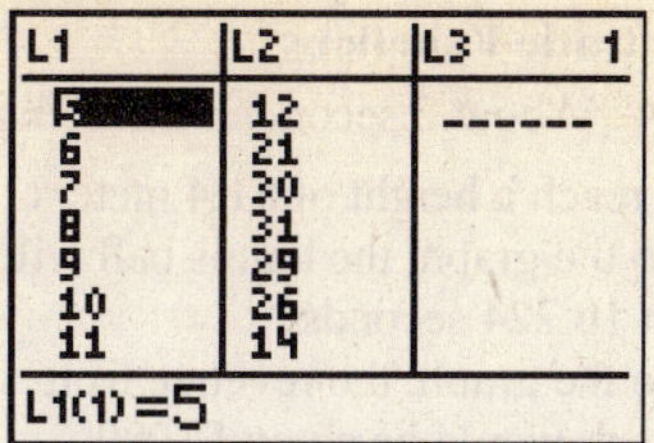

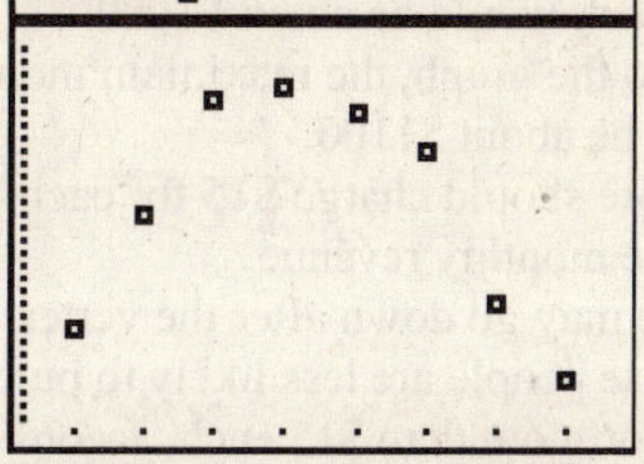

b. A quadratic model would fit best. The data looks like a downward facing parabola.

c. The vertex is $(8,31)$, which is a maximum point. Thus, August has the most number of days, on average, above 70°F in San Diego, California.

d. Using the shape of the distribution, there would be about seven days above 70°F in San Diego, California, during the month of April.

27. **a.**

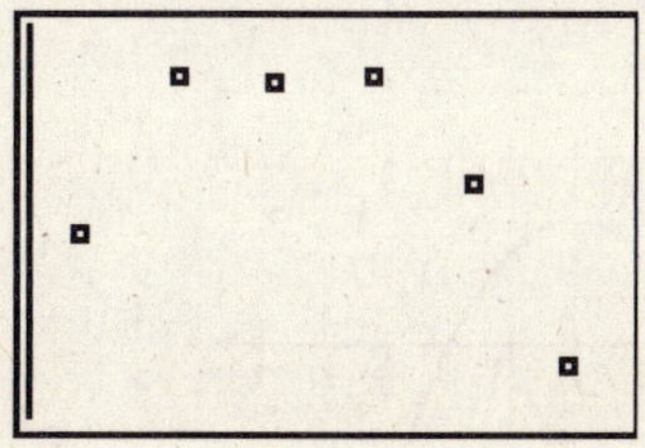

b.

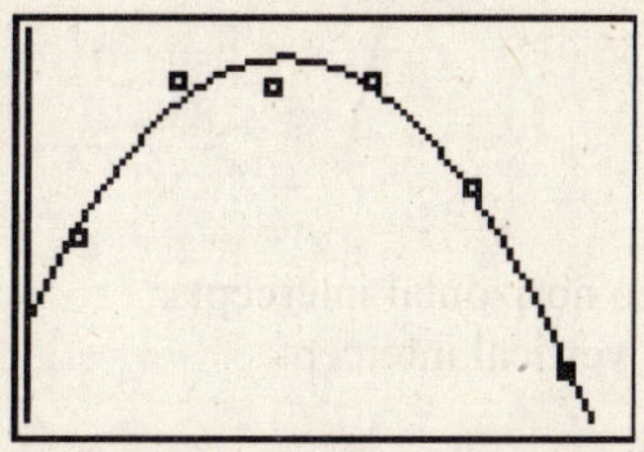

c. This model fits the data reasonably well; the point at the top is low for a vertex, but otherwise the data is shaped like a quadratic.

d. The vertex for the model is at about $(4, 11{,}238)$ and is a maximum point for this function.

e. This vertex represents that in 1994, the number of cable television systems in the United States reached a maximum of 11,238.

29. **a.**

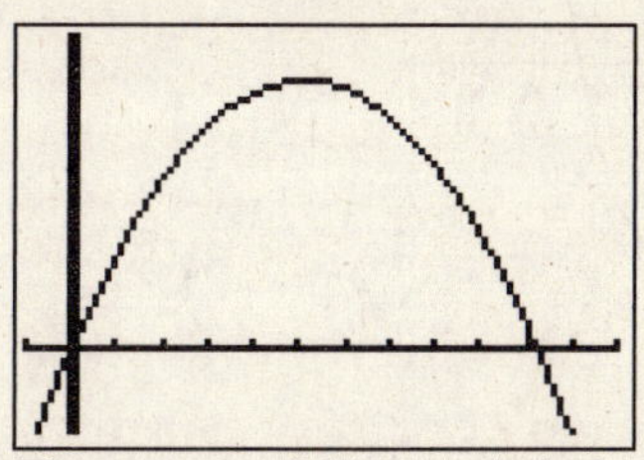

b. The vertex of this parabola is at about (5, 128.5), which means that the tennis ball will reach its highest point in the air at 128.5 meters after about 5 seconds.

c. $H(2)=81.4$. About 2 seconds after being hit, the tennis ball will reach a height of 81.4 meters.
d. According to the graph, the tennis ball will hit the ground at about 10.224 seconds.

31. **a.** According to the graph, the revenue from selling T-shirts for \$10 each would be about \$1000.
b. According to the graph, the maximum monthly revenue would be about \$1100.
c. The bookstore should charge \$15 for each T-shirt to maximize the monthly revenue.
d. The revenue may go down after the vertex in this situation because people are less likely to purchase these T-shirts for more than \$15 each, according to this graph.

33. A quadratic model would best fit the data. The vertex is $(-5,8)$, which is a maximum point.

35. A quadratic model would best fit the data. The vertex is $(5,-5)$, which is a minimum point.

37. A linear model would best fit the data.

39. A linear model would best fit the data.

41.

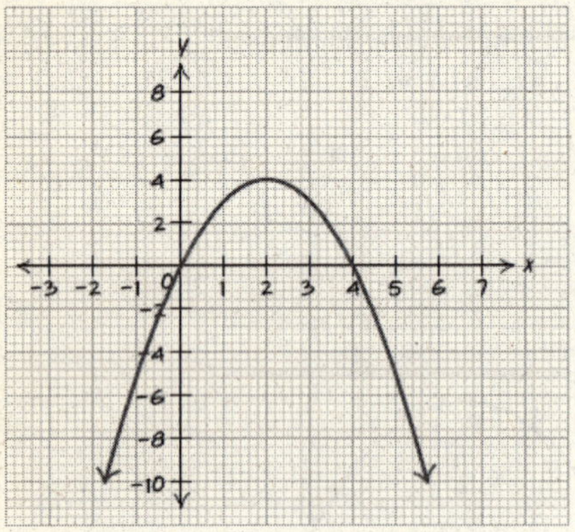

a. There are two horizontal intercepts.
b. There is one vertical intercept.

43.

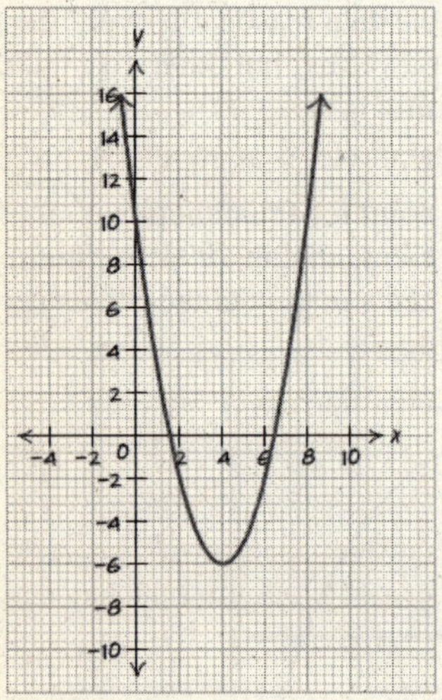

a. There are two horizontal intercepts.
b. There is one vertical intercept.

45.

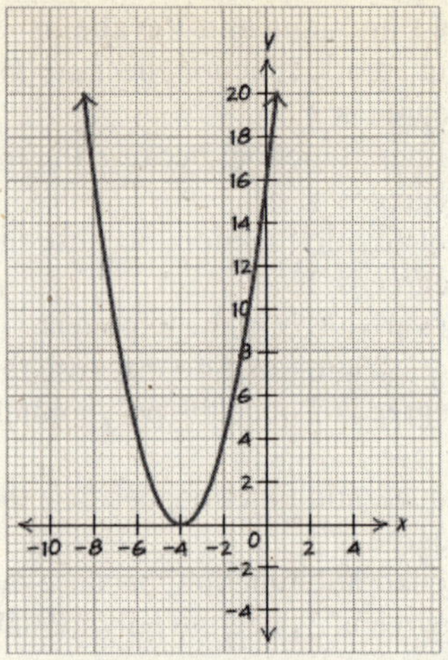

The vertex must be on the horizontal axis for the parabola to face upward and have one horizontal intercept.

47.

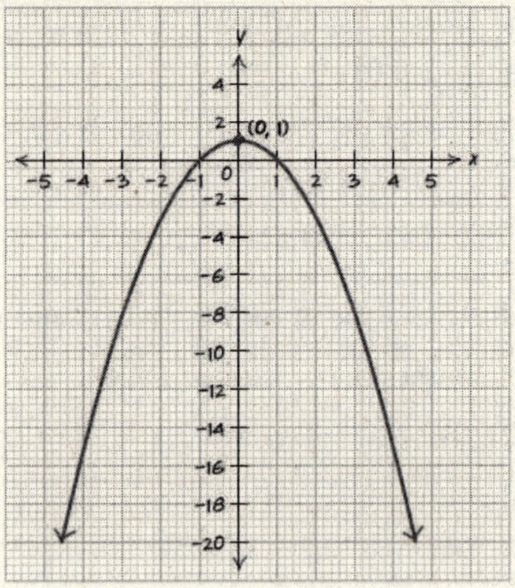

a. The vertex is above the horizontal axis.
b. The vertex must be below the horizontal axis for the parabola to face down and have no horizontal intercepts.

Section 11.2 Graphing Quadratics in Vertex Form

1. h; positive.
3. k; negative.
5. $|a|>1$
7. a; positive
9. $h=0,\ k=0$
11. **a.** $(-2.5,-3)$ **b.** $x=-2.5$ **c.** $(-5.5,4)$
13. **a.** $(5.5,-6)$ **b.** $x=5.5$ **c.** $(2,5)$
15. **a.** $(-4,-8)$ **b.** $x=-4$ **c.** positive
    d. $h=-4, k=-8$ **e.** $(2,15)$
17. **a.** $(2,10)$ **b.** $x=2$ **c.** negative
    d. $h=2, k=10$ **e.** $(5,-6)$
19. $g(x)=(x-7)^2+9$

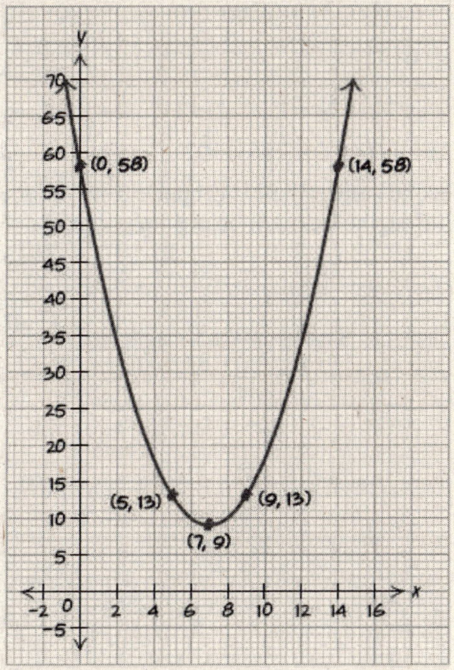

21. $f(x)=(x+4)^2+2$

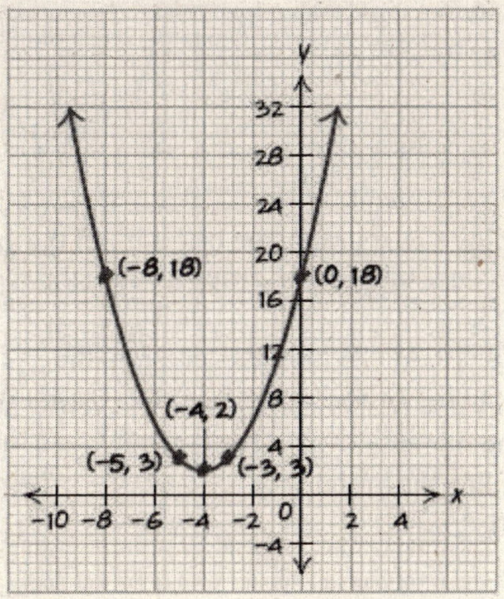

23. $f(x)=-1.5x^2$

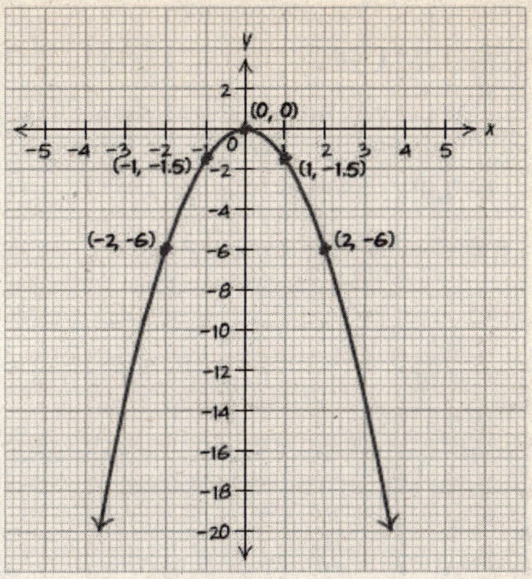

25. $h(x)=x^2+4$

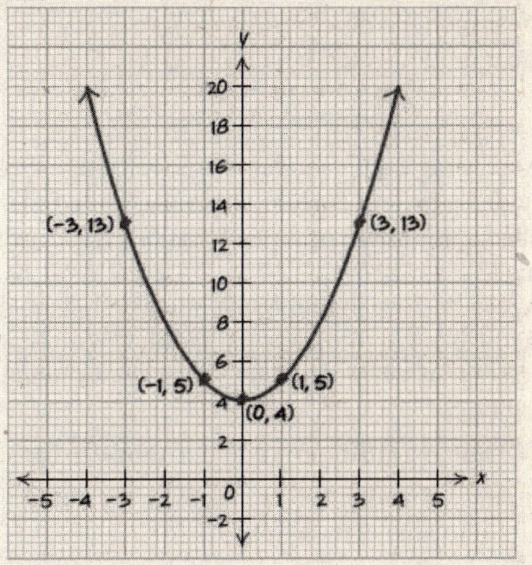

27. $g(x)=4(x-5)^2$

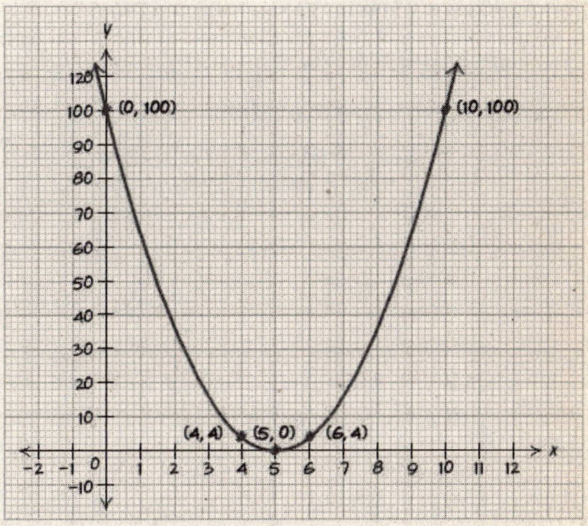

29. **a.**

$$P(90)=4.95(90-57)^2+1406$$
$$P(90)=6796.55$$

In 1990, the poverty threshold was about $6,796.55 for individuals under 65 years old in the United States.

b.

$$P(80)=4.95(80-57)^2+1406$$

$$P(80)=4024.55$$

In 1980, the poverty threshold was about $4,024.55 for individuals under 65 years old in the United States.

c. $P(t)=4.95(t-57)^2+1406$

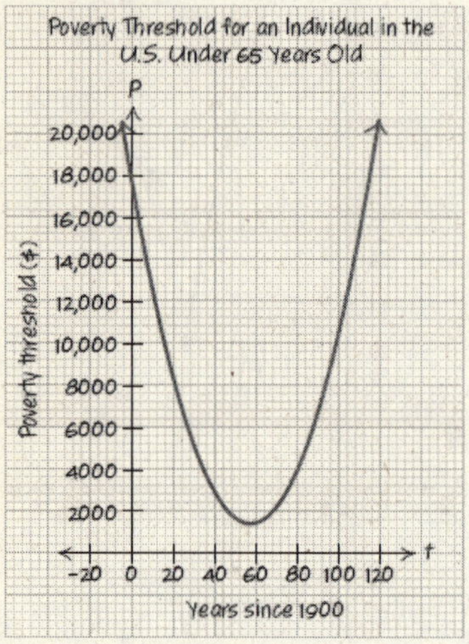

d. According to this model, the poverty threshold for individuals under 65 years old in the United States reached a minimum of about $1,406 in 1957, which seems to make sense.

e. According to the graph, the poverty threshold for individuals under 65 years old in the United States was $3000 in 1975 and 1939; 1939 shows model breakdown.

f. Domain: $[57,90]$. Within this domain, the minimum value of 1406 comes from the vertex. The maximum value comes from $t=90$.

$$P(57)=4.95(57-57)^2+1406=1406$$

$$P(90)=4.95(90-57)^2+1406=6796.55$$

Range: $[1406,6796.55]$

31. **a.** $D(t)=17(t-3)^2+2300$

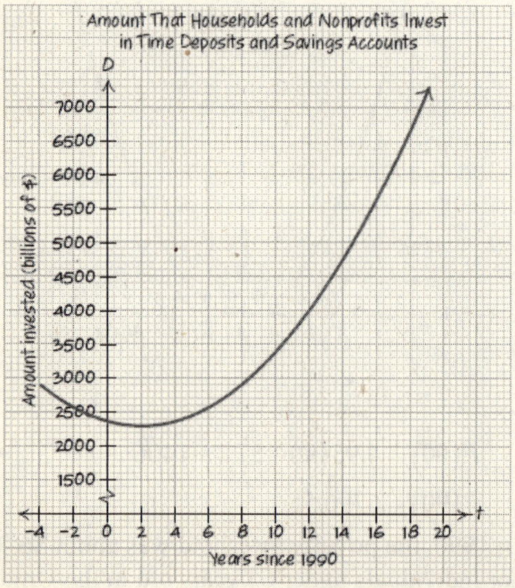

b.

$$D(6)=17(6-3)^2+2300$$

$$D(6)=2453$$

In 1996, households and nonprofit organizations invested about $2453 billion in time and savings accounts.

c. In about 1993, these time and savings accounts were at their lowest levels with about $2300 billion invested.

d. According to the graph, households and nonprofit organizations invested $3000 billion in about 1987 and again in 1999.

e. Domain: $[0,18]$. Within this domain, the minimum value of 2300 comes from the vertex. The maximum value comes from $t=18$.

$$D(18)=17(18-3)^2+2300$$

$$D(18)=6125$$

Range: $[2300\ ,6125]$

33. **a.** $h(2)=-16(2)^2+256=192$

Two seconds after a ball is dropped from the roof of the building, the ball will be at a height of about 192 feet.

b. $h(t)=-16t^2+256$

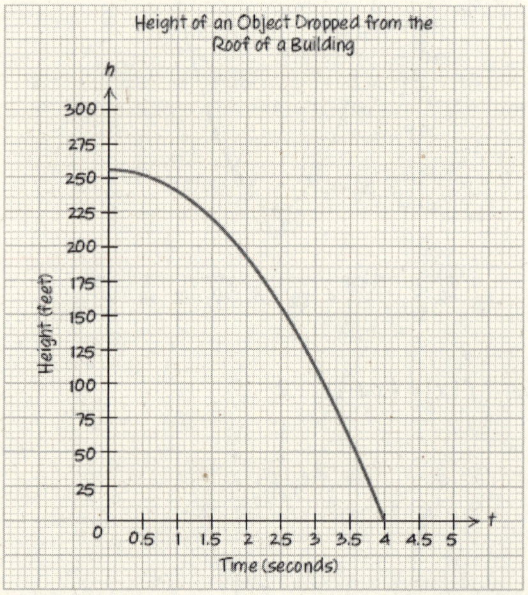

c. According to the graph, the ball will hit the ground about 4 seconds after being dropped from the roof of the building.

d. Domain: $[0,4]$. Within this domain, the maximum value of 256 comes from the vertex at time $t=0$. The minimum value comes from $t=4$ when the object hits the ground

Range: $[0,256]$

35. **a.**

$$P(100) = -0.000025(100-120)^2 + 2.75$$
$$P(100) = 2.74$$

The monthly profit made from selling round trip airline tickets from New York City to Orlando, Florida, when the tickets are sold for \$100 each is \$2.74 million.

b. From the vertex we see that selling round trip airline tickets from New York City to Orlando, Florida for \$120 each will produce a maximum profit of \$2.75 million.

c. $P(t) = -0.000025(t-120)^2 + 2.75$

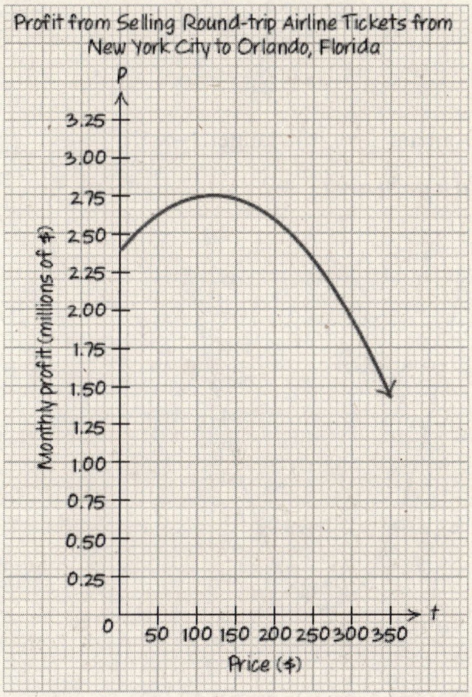

d. According to the graph, the company will have a monthly profit of about \$2 million when selling round trip airline tickets from New York City to Orlando, Florida, for \$295 each and $-\$50$; $-\$50$ is model breakdown.

e. Domain: $[60,300]$. Within this domain, the maximum value of 2.75 comes from the vertex. The minimum value comes from $t = 300$.

$$P(300) = -0.000025(300-120)^2 + 2.75$$
$$P(300) = 1.95$$

Range: $[1.95 , 2.75]$

37. **a.**

$$E(4) = -0.006(4-10)^2 + 8$$
$$E(4) = 7.784$$

The epidemic threshold for the fourth week of 2006 was 7.784%

b. $E(w) = -0.006(w-10)^2 + 8$

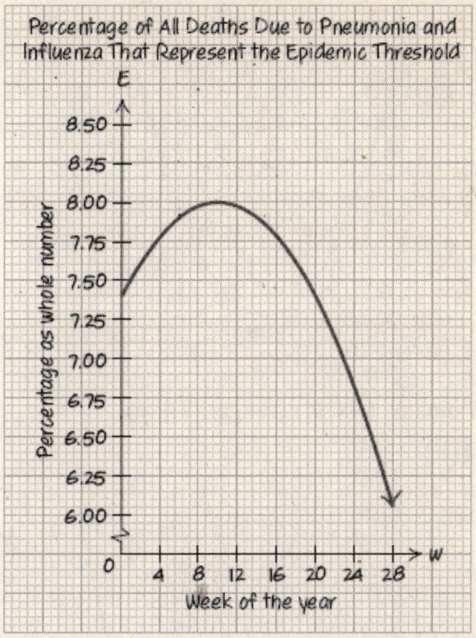

c. According to the graph, in the 23rd week of 2006, there was not a flu epidemic; the epidemic threshold was about 7%.

d. The vertex for this model is (10, 8), which means that in week 10 of 2006, the epidemic threshold reached a maximum of 8%.

e. Domain: $[1,26]$. Within this domain, the maximum value of 8% comes from the vertex. The minimum value comes from $n = 0$.

$$E(26) = -0.006(26-10)^2 + 8$$
$$E(26) = 6.464$$

Range: $[6.464,8]$

39. Domain: $(-\infty,\infty)$, range: $(-\infty,15]$

41. Domain: $(-\infty,\infty)$, range: $[-58,\infty)$

43. Domain: $(-\infty,\infty)$, range: $[0,\infty)$

45. Domain: $(-\infty,\infty)$, range: $(-\infty,4]$

47. Domain: $(-\infty,\infty)$, range: $y \ge -9$

The domain should be all real numbers. The range is using a greater than sign, but it should use a greater than or equal to sign.

49. Domain: $(-\infty,\infty)$, range: $(-\infty,7]$

The domain is correct, but the range is assuming that the parabola is facing upward and there is a parenthesis on the 7 instead of a bracket.

51. $h(t) = (t-3)^2 - 4$

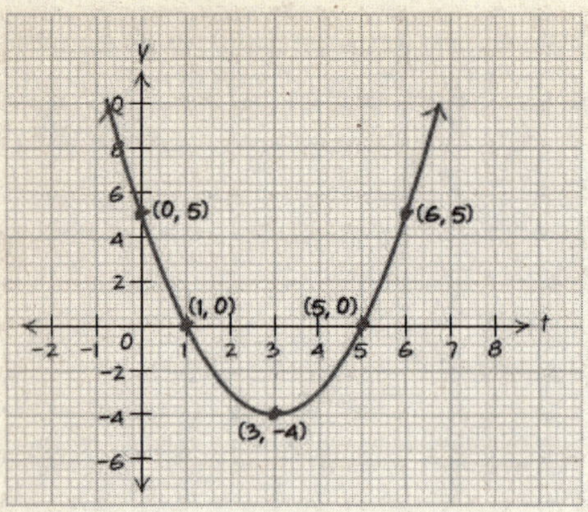

Domain: $(-\infty, \infty)$, range: $[-4, \infty)$

53. $k(n) = 2n^2$

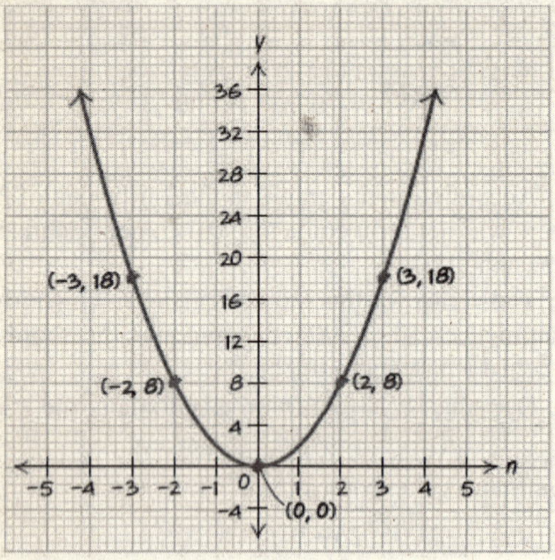

Domain: $(-\infty, \infty)$, range: $[0, \infty)$

55. $f(x) = 0.4x^2 + 2$

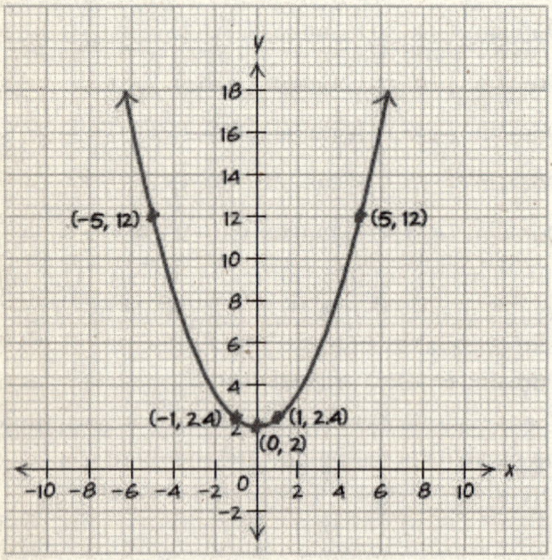

Domain: $(-\infty, \infty)$, range: $[2, \infty)$

57. $f(x) = -2(x+4)^2$

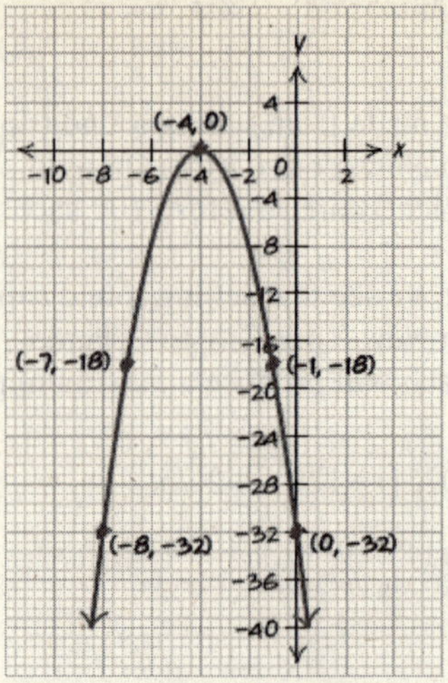

Domain: $(-\infty, \infty)$, range: $(-\infty, 0]$

59. $q(a) = 3(a+7)^2 + 2$

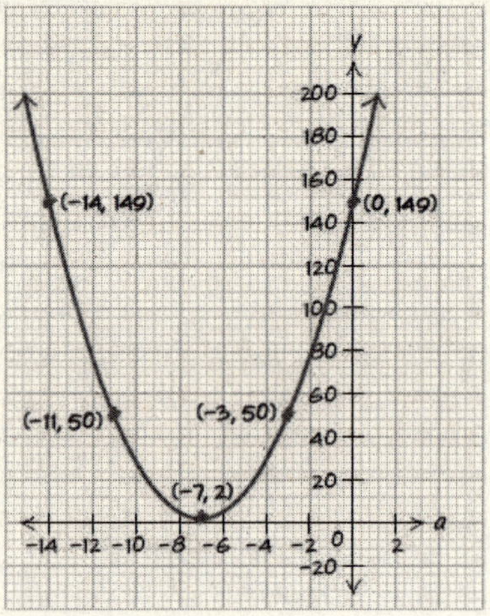

Domain: $(-\infty, \infty)$, range: $[2, \infty)$

61. $T(z) = 0.2(z+6)^2 - 8$

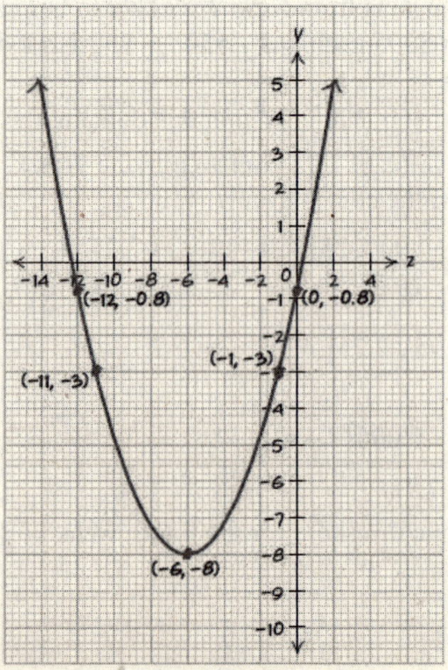

Domain: $(-\infty, \infty)$, range: $[-8, \infty)$

63. $h(x) = -5(x-17)^2 - 15$

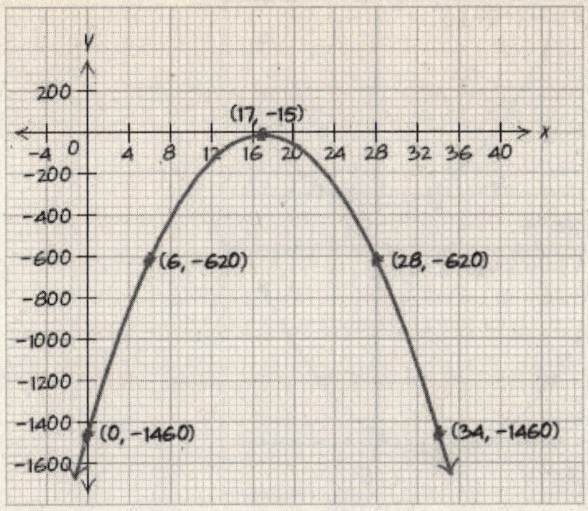

Domain: $(-\infty, \infty)$, range: $(-\infty, -15]$

65. $s(t) = -0.25(t+50)^2 + 25$

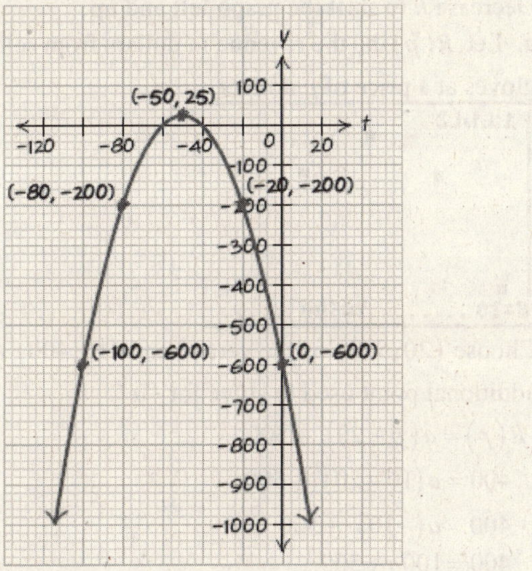

Domain: $(-\infty, \infty)$, range: $(-\infty, 25]$

67. $p(w) = 123(w-4)^2 - 2500$

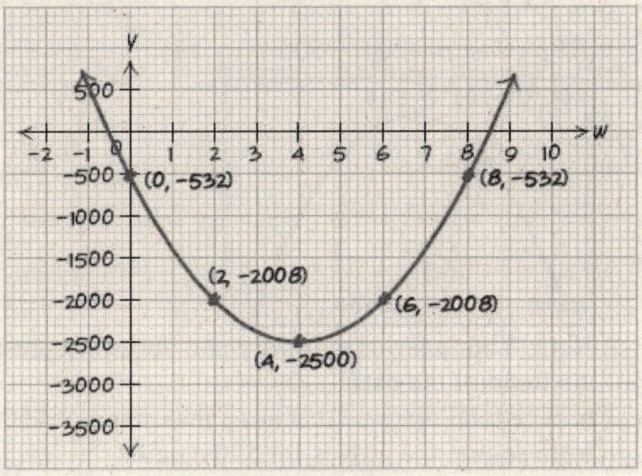

Domain: $(-\infty, \infty)$, range: $[-2500, \infty)$

69. $f(x) = 5x^2 + 100$

a. Vertex: $(0, 100)$ **b.** The parabola is narrow.

c. The parabola faces upward.

d. X Min: -10 X Max: 10
Y Min: 0 Y Max: 600

71. $f(x) = (x+30)^2 - 50$

a. Vertex: $(-30, -50)$

b. The parabola is neither wide nor narrow.

c. The parabola faces upward.

d. X Min: -45 X Max: -10
Y Min: -60 Y Max: 70

73. $f(x) = 0.002(x+20)^2 + 50$

a. Vertex: $(-20, 50)$ **b.** The parabola is wide.

c. The parabola faces upward.

d. X Min: -100 X Max: 60
Y Min: 45 Y Max: 60

75. $f(x) = 0.0005(x-1000)^2 + 1000$

a. Vertex: $(1000, 1000)$

b. The parabola is wide.

c. The parabola faces upward.

d. X Min:0 X Max: 2000
Y Min: 900 Y Max: 1500

77. $f(x) = -10(x+25{,}000)^2 - 10{,}000$

a. Vertex: $(-25{,}000, -10{,}000)$

b. The parabola is narrow.

c. The parabola faces downward.

d. X Min: $-25{,}010$ X Max: $-24{,}990$
Y Min: $-11{,}000$ Y Max: -9900

Section 11.3 Finding Quadratic Models

1. $f(x)=2.0(x+5)^2-15$

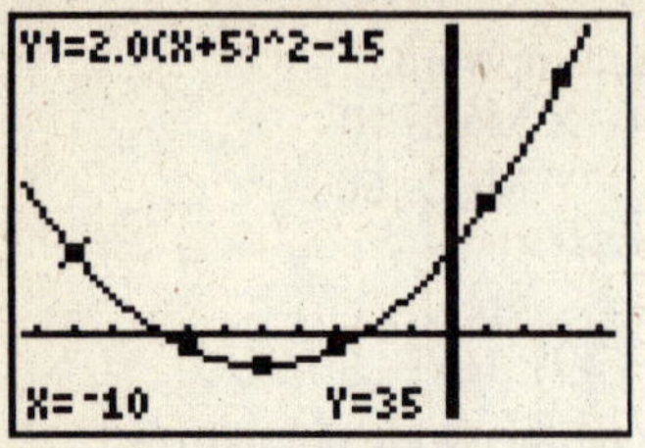

3. $f(x)=0.4(x-4)^2+3.2$

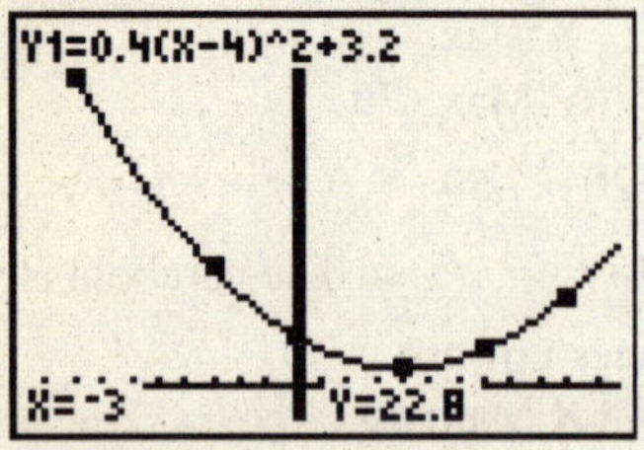

5. $f(x)=2(x+1.5)^2-20$

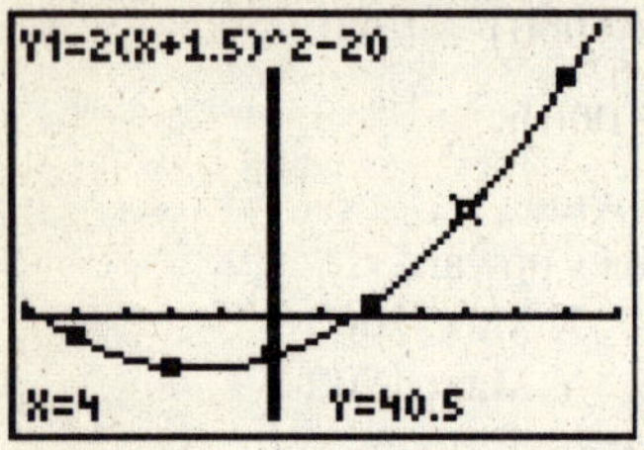

7. Adjusted to: $f(x)=2.7(x-5)^2-8$

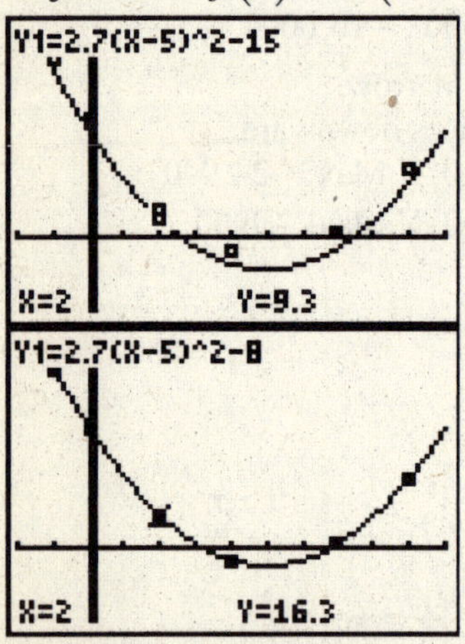

9. Adjusted to: $f(x)=-2(x-7)^2+15$

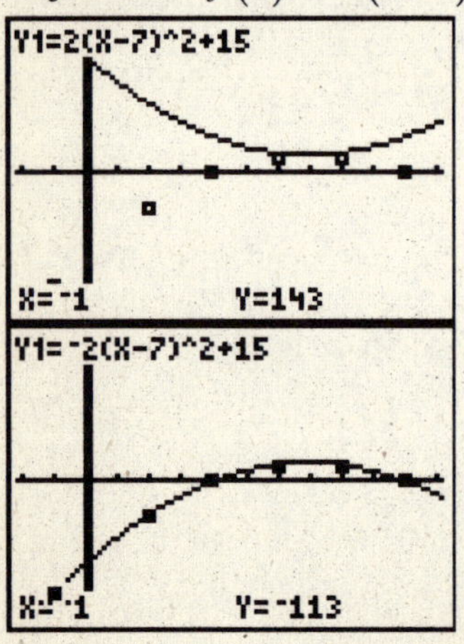

11. Adjusted to: $f(x)=4(x+2)^2-10$

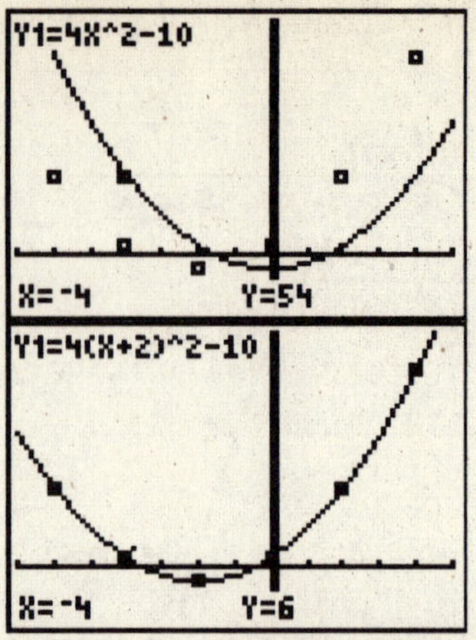

13. Increase k to shift the graph up.

15. Increase h to shift the graph right.

17. Decrease a to make the graph wider.

19. Decrease h to shift the graph left and up.

21. **a.** Let $R(p)$ be the revenue in dollars from selling gloves at a price of p dollars.

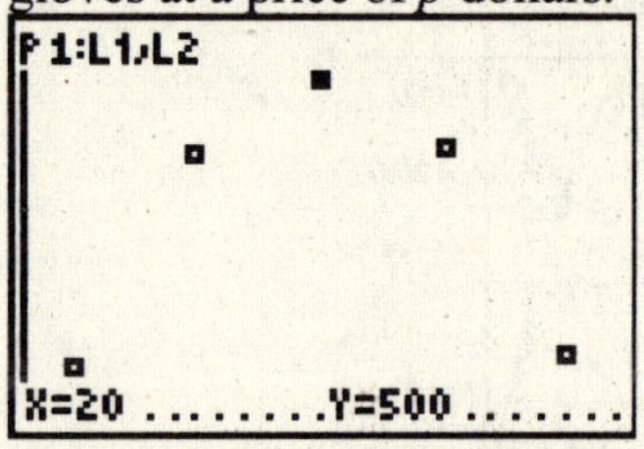

Choose $(20,500)$ as the vertex, and $(10,400)$ as the additional point used to solve for *"a"*.

$$R(p)=a(p-20)^2+500$$
$$400=a(10-20)^2+500$$
$$400=a(-10)^2+500$$
$$400=100a+500$$
$$-500 \qquad -500$$
$$-100=100a$$
$$a=-1$$
$$R(p)=-(p-20)^2+500$$

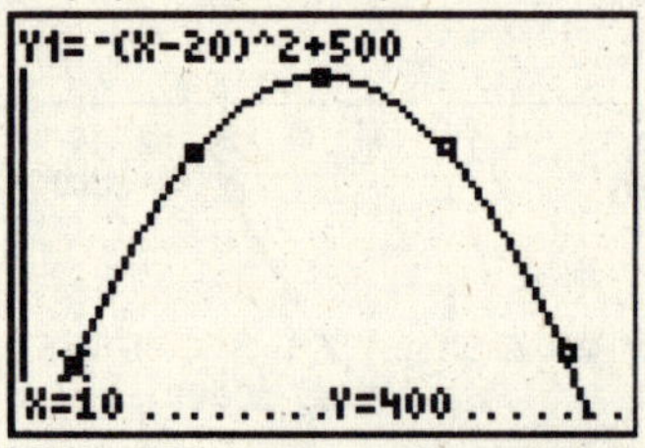

Note: Your answer may vary and yet still represent the trend of the data.

b. The total revenue reaches a maximum of \$500 from selling the gloves for \$20 a pair.

c.

$$R(22)=-(22-20)^2+500$$
$$R(22)=496$$

This model predicts the revenue will be about \$496 if the gloves sell for \$22 a pair.

d. Domain: $[6,35]$. Within this domain, the maximum value of \$500 comes from the vertex. The minimum value comes from $p=35$.

$$R(35)=-(35-20)^2+500=275$$

Range: $[275,500]$

23. **a.** Let $T(m)$ be the average monthly low temperatures (in degrees Fahrenheit) in Anchorage, Alaska, during month m (i.e., $m=1$ represents January).

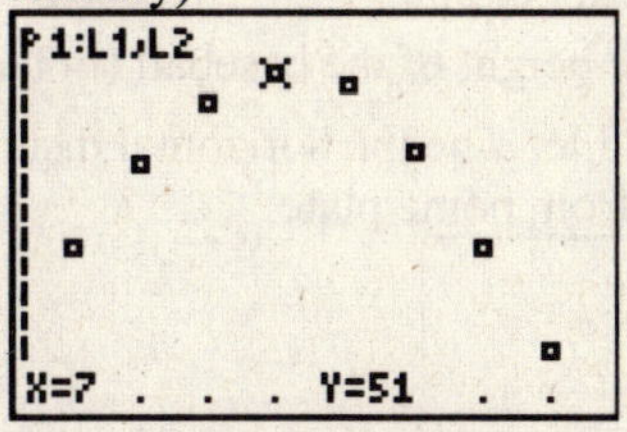

Choose $(7,51)$ as the vertex, and $(10,28)$ as the additional point used to solve for *"a"*.

$$T(m)=a(m-7)^2+51$$
$$28=a(10-7)^2+51$$
$$28=a(3)^2+51$$
$$28=9a+51$$
$$-51 \qquad -51$$
$$-23=9a$$
$$\left.\frac{-23}{9}=\frac{9a}{9}\right\} \quad a\approx -2.56$$
$$T(m)=-2.56(m-7)^2+51$$

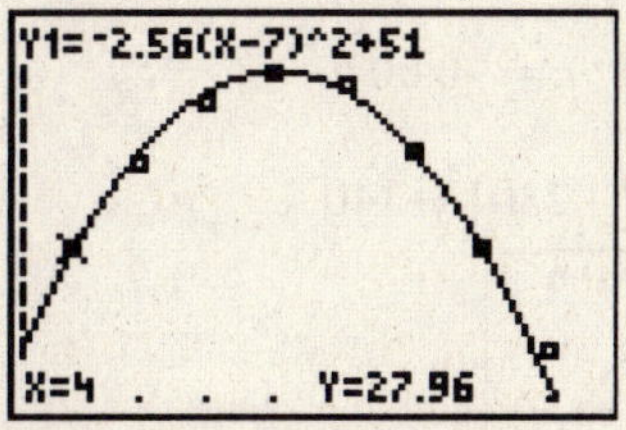

Note: Your answer may vary and yet still represent the trend of the data.

b. The highest average low temperature for Anchorage, Alaska, is about 51 °F in July.

c.

$$T(3)=-2.56(3-7)^2+51$$
$$T(3)=10.04$$

This model predicts the average low temperature in March to be 10 °F.

d. Domain: $[3,11]$. Within this domain, the maximum value of 51 °F comes from the vertex. The minimum value comes from $m=3$.

$$T(3)=-2.56(3-7)^2+51\approx 10$$

Range: $[10,51]$

25. **a.** Let $F(t)$ be the number of Hispanic families, in thousands, in the United States below the poverty level t years since 1990.

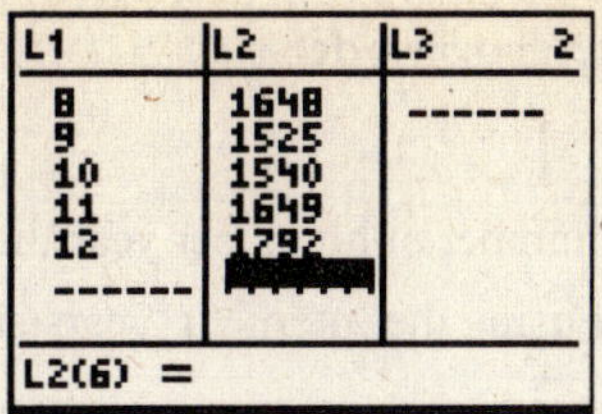

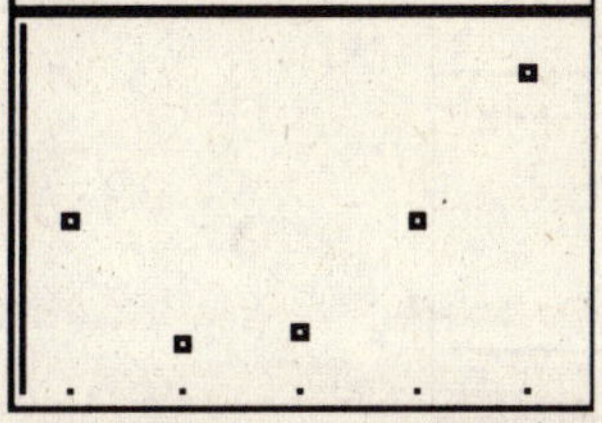

Choose $(9.5,1520)$ as the vertex, and $(12,1792)$ as the additional point used to solve for *"a"*.

$$F(t)=a(t-9.5)^2+1520$$
$$1792=a(12-9.5)^2+1520$$
$$1792=a(2.5)^2+1520$$
$$1792=6.25a+1520$$
$$-1520 \qquad -1520$$
$$272=6.25a$$
$$\left.\frac{272}{6.25}=\frac{6.25a}{6.25}\right\} \quad a=43.52$$
$$F(t)=43.52(t-9.5)^2+1520$$

For a better fit, allow $a=46$.

$$F(t)=46(t-9.5)^2+1520$$

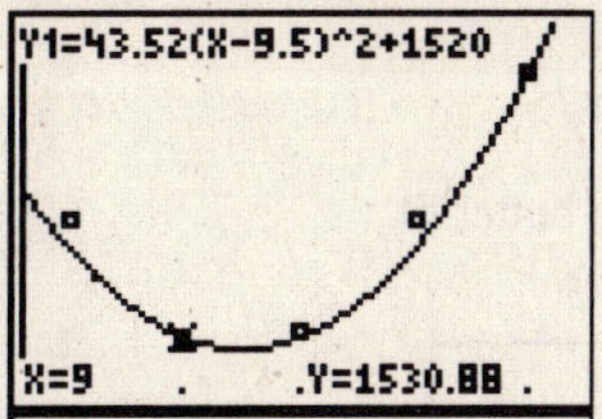

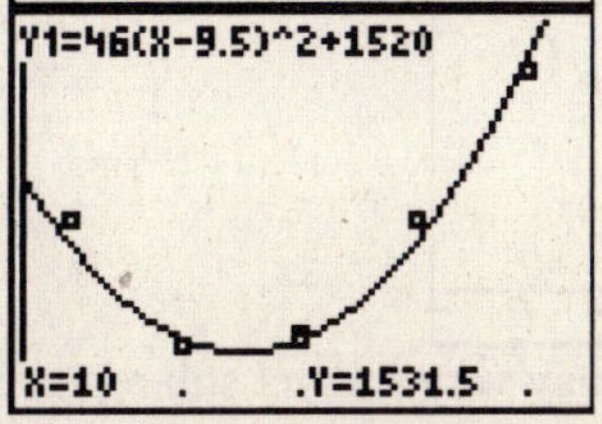

Note: Your answer may vary and yet still represent the trend of the data.

b.

$$F(15)=46(15-9.5)^2+1520$$
$$F(15)=2911.5$$

In 2005, there were approximately 2,911,500 Hispanic families in the United States under the poverty level.

c. In about 1999 or 2000, the number of Hispanic families in the United States under the poverty level reached a minimum of about 1,520,000. This is probably true only for this close time period.

d. Domain: $[6,15]$. Within this domain, the minimum value of 1520 comes from the vertex. The maximum value comes from $t=15$.

$F(15) = 46(15 - 9.5)^2 + 1520 = 2911.5$

Range: $[1520, 2911.5]$

27. **a.** Let $H(t)$ be the number of hours per year the average person spent using the Internet t years since 1990.

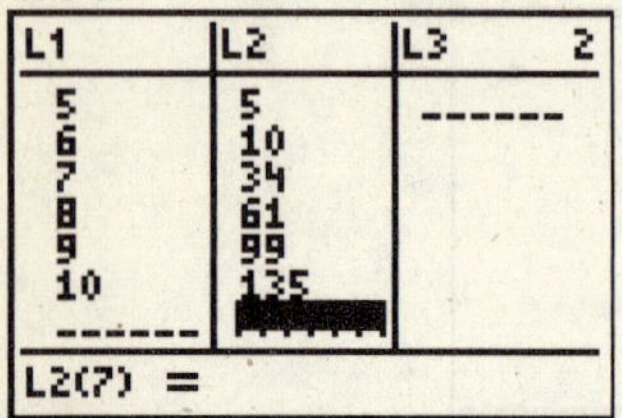

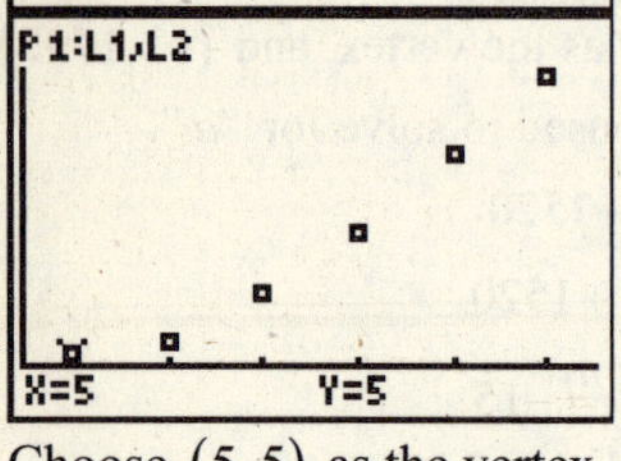

Choose $(5,5)$ as the vertex, and $(9,99)$ as the additional point used to solve for "a".

$$H(t) = a(t-5)^2 + 5$$

$$99 = a(9-5)^2 + 5$$

$$99 = a(4)^2 + 5$$

$$99 = 16a + 5$$
$$-5 \qquad -5$$

$$94 = 16a$$

$$\left.\frac{94}{16} = \frac{16a}{16}\right\} \quad a = 5.875$$

Adjusted to 5.7 for a better fit.

$$H(t) = 5.7(t-5)^2 + 5$$

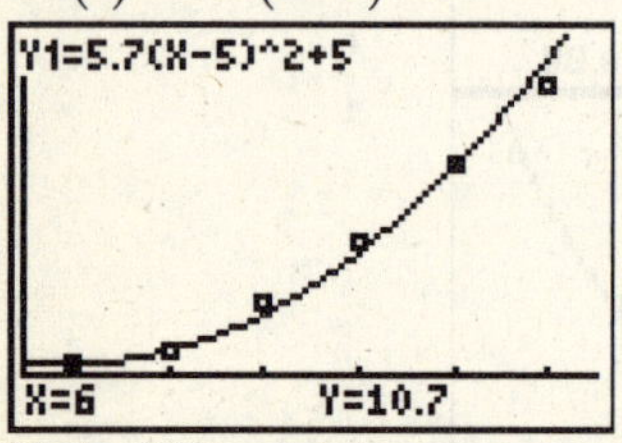

Note: Your answer may vary and yet still represent the trend of the data.

b. If you use this model for years before 1995, the model shows more and more hours spent on the Internet as you go back in time. This is model breakdown because the Internet was just beginning to take off in the early 1990s.

c. Domain: $[5,12]$. Within this domain, the minimum value of 5 comes from the vertex. The maximum value comes from $t = 12$.

$$H(12) = 5.7(12-5)^2 + 5 \approx 284$$

Range: $[5, 284]$

d.

$$H(11) = 5.7(11-5)^2 + 5$$

$$H(11) = 210.2$$

In 2001, the average person spent 210 hours using the Internet.

e.

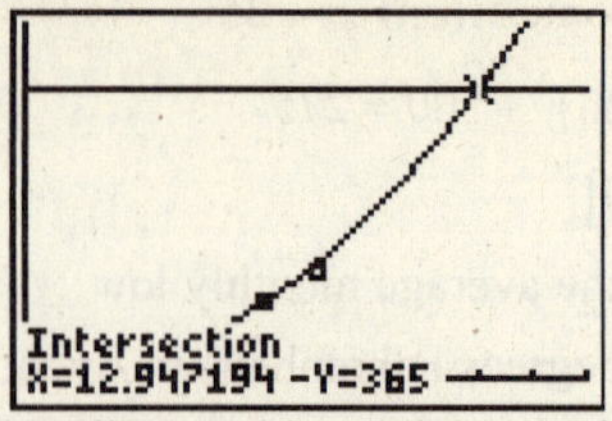

According to this model, people will spend 365 hours per year on the Internet in 2003.

29. **a.** Let $H(d)$ be the height of the baseball (in feet) from the ground, and let d be the horizontal distance of the ball (in feet) from home plate.

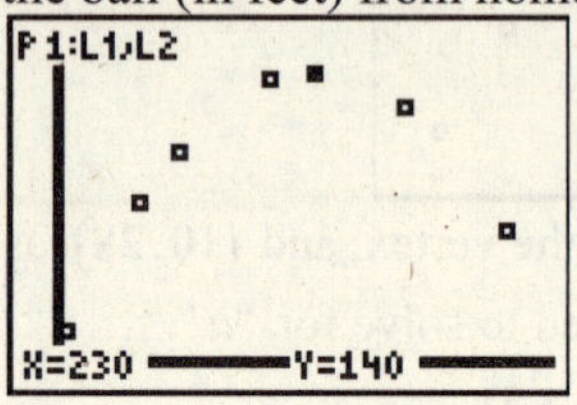

Choose $(230,140)$ as the vertex, and $(400,65)$ as the additional point used to solve for "a".

$$H(d) = a(d-230)^2 + 140$$

$$65 = a(400-230)^2 + 140$$

$$65 = a(170)^2 + 140$$

$$65 = 28{,}900a + 140$$
$$-140 \qquad -140$$

$$-75 = 28{,}900a$$

$$\left.\frac{-75}{28{,}900} = \frac{28{,}900a}{28{,}900}\right\} \quad a \approx -0.0026$$

$$H(d) = -0.0026(d-230)^2 + 140$$

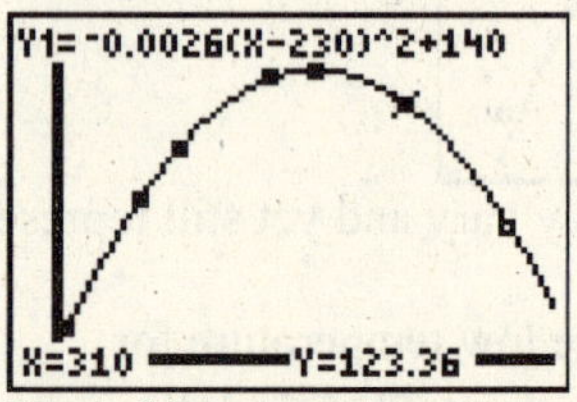

Note: Your answer may vary and yet still represent the trend of the data.

b. Domain: $[0,400]$. Within this domain, the maximum value of 140 comes from the vertex. The minimum value comes from $h = 0$.

$$H(0) = -0.0026(0-230)^2 + 140$$

$$H(0) = 2.46$$

Range: $[2.46, 140]$

c. This vertex of this model is (230, 140), which means that at a distance of 230 feet after being hit, the baseball will be at a maximum height of about 140 feet from the ground.

d.

$$H(450) = -0.0026(450-230)^2 + 140$$

$$H(450) = 14.16$$

If the center wall of the stadium is 450 feet from home plate and is 10 feet tall, the ball will make it over the wall.

31. Domain: $(-\infty,\infty)$, range: $(-\infty,8]$

Vertex $=(-5,8)$

Other point $=(-3,-8)$

$-8=a(-3+5)^2+8$

$-8=a(2)^2+8$

$-8=4a+8$

$-8-8=4a+8-8$

$-16=4a$

$\frac{-16}{4}=\frac{4a}{4}$

$a=-4$

Since $h=-5, k=8,$ and $a=-4$

$y=-4(x+5)^2+8$

33. Domain: $(-\infty,\infty)$, range: $[-4,\infty)$

Vertex $=(5,-4)$

Other point $=(2,5)$

$5=a(2-5)^2-4$

$5=a(-3)^2-4$

$5=9a-4$

$5+4=9a-4+4$

$9=9a$

$\frac{9}{9}=\frac{9a}{9}$

$a=1$

Since $h=5, k=-4,$ and $a=1$

$y=(x-5)^2-4$

35. Domain: $(-\infty,\infty)$, range: $[-9000,\infty)$

Vertex $=(-55,-9000)$

Other point $=(-30,-1500)$

$-1500=a(-30+55)^2-9000$

$-1500=a(25)^2-9000$

$-1500=625a-9000$

$-1500+9000=625a-9000+9000$

$7500=625a$

$\frac{7500}{625}=\frac{625a}{625}$

$a=12$

Since $h=-55, k=-9000,$ and $a=12$

$y=12(x+55)^2-9000$

37. Domain: $(-\infty,\infty)$, range: $(-\infty,14]$

Vertex $=(6,14)$

Other point $=(4,10)$

$10=a(4-6)^2+14$

$10=a(-2)^2+14$

$10=4a+14$

$10-14=4a+14-14$

$-4=4a$

$\frac{-4}{4}=\frac{4a}{4}$

$a=-1$

Since $h=6, k=14,$ and $a=-1$

$y=-(x-6)^2+14$

39. Domain: $(-\infty,\infty)$, range: $(-\infty,\infty)$

This is a linear function

First point $=(0,-15)$

Other point $=(6,4.2)$

$y-y_1=m(x-x_1)$

$4.2+15=m(6-0)$

$19.2=6m$

$\frac{19.2}{6}=\frac{6m}{6}$

$m=3.2$

Since $m=3.2$ and $b=-15$

$y=3.2x-15$

41. Domain: $(-\infty,\infty)$, range: $(-\infty,8]$

Vertex $=(0,8)$

Other point $=(2,2)$

$2=a(2-0)^2+8$

$2=a(2)^2+8$

$2=4a+8$

$2-8=4a+8-8$

$-6=4a$

$\frac{-6}{4}=\frac{4a}{4}$

$a=-1.5$

Since $h=0, k=8,$ and $a=-1.5$

$y=-1.5x^2+8$

43. Domain: $(-\infty,\infty)$, range: $(-\infty,0]$

Vertex $=(-5,0)$

Other point $=(0,-10)$

$-10=a(0+5)^2+0$

$-10=a(5)^2$

$-10=25a$

$\frac{-10}{25}=\frac{25a}{25}$

$a=-0.4$

Since $h=-5, k=0,$ and $a=-0.4$

$y=-0.4(x+5)^2$

45. Use: $x=-3$, $x=2$ with $f(0)=-6$.

$x+3=0$, $x-2=0$

$f(x)=a(x+3)(x-2)$

$f(x)=a(x^2+x-6)$

$f(0)=a(-6)=-6$

$-6a=-6 \rightarrow a=1$

$f(x)=x^2+x-6$

47. Use: $x=-3$, $x=3$ with $f(0)=9$.

$x+3=0$, $x-3=0$
$f(x)=a(x+3)(x-3)$
$f(x)=a(x^2-9)$
$f(0)=a(-9)=9$
$-9a=9 \rightarrow a=-1$
$f(x)=-(x^2-9)$
$f(x)=-x^2+9$

49. Use: $x=-2$, $x=2$ with $f(0)=8$.

$x+2=0$, $x-2=0$
$f(x)=a(x+2)(x-2)$
$f(x)=a(x^2-4)$
$f(0)=a(-4)=8$
$-4a=8 \rightarrow -2=a$
$f(x)=-2(x^2-4)$
$f(x)=-2x^2+8$

51. $f(x)=a(x-h)^2+k$
Vertex $=(-2,5)$
Second point $=(1,15)$
$f(x)=a(x+2)^2+5$
$15=a(1+2)^2+5$
$15=a(3)^2+5$
$-5 \qquad -5$
$10=9a$
$\frac{10}{9}=a$
$f(x)=\frac{10}{9}(x+2)^2+5$
$f(x)=\frac{10}{9}(x^2+4x+4)+5$
$f(x)=\frac{10}{9}x^2+\frac{40}{9}x+\frac{40}{9}+5$
$f(x)=\frac{10}{9}x^2+\frac{40}{9}x+\frac{85}{9}$

53. Given: $x=3$, $x=-7$.

$x-3=0$, $x+7=0$
$f(x)=(x-3)(x+7)$
$f(x)=x^2+4x-21$

55. Given: $x=2$, $x=4$.

$x-2=0$, $x-4=0$
$f(x)=(x-2)(x-4)$
$f(x)=x^2-6x+8$

57. Given: $x=\frac{2}{3}$, $x=4$.

$3x=2$
$3x-2=0$, $x-4=0$
$f(x)=(3x-2)(x-4)$
$f(x)=3x^2-14x+8$

59. Given: $x=4$, $x=2$ with $f(0)=40$.

$x-4=0$, $x-2=0$
$f(x)=a(x-4)(x-2)$
$f(x)=a(x^2-6x+8)$
$f(0)=a(8)=40$
$8a=40$
$a=5$
$f(x)=5(x^2-6x+8)$
$f(x)=5x^2-30x+40$

61. Given: $x=\frac{1}{4}$, $x=-2$ with $f(2)=-21$.

$4x=1$
$4x-1=0$, $x+2=0$
$f(x)=a(4x-1)(x+2)$
$f(x)=a(4x^2+7x-2)$
$f(2)=a(4(2)^2+7(2)-2)=-21$
$28a=-21$
$\frac{28a}{28}=\frac{-21}{28}$
$a=-\frac{3}{4}$
$f(x)=-\frac{3}{4}(4x^2+7x-2)$
$f(x)=-3x^2-5.25x+1.5$

63. Given: $x=-\frac{1}{3}$, $x=\frac{9}{2}$ with $f(3)=20$.

$3x=-1$, $2x=9$
$3x+1=0$, $2x-9=0$
$f(x)=a(3x+1)(2x-9)$
$f(x)=a(6x^2-25x-9)$
$f(3)=a(6(3)^2-25(3)-9)=20$
$-30a=20$
$\frac{-30a}{-30}=\frac{20}{-30}$
$a=-\frac{2}{3}$
$f(x)=-\frac{2}{3}(6x^2-25x-9)$
$f(x)=-4x^2+\frac{50}{3}x+6$

Section 11.4 Solving Quadratic Equations by the Square Root Property

1.

$x^2 = 100$

$x = \pm\sqrt{100}$

$x = 10\ ,\ x = -10$

3.

$2x^2 = 162$

$\frac{2x^2}{2} = \frac{162}{2} = 81$

$x^2 = 81$

$x = \pm\sqrt{81}$

$x = 9\ ,\ x = -9$

5.

$x^2 + 12 = 181$

$\underline{\ -12\ \ -12\ }$

$x^2 = 169$

$x = \pm\sqrt{169}$

$x = 13\ ,\ x = -13$

7. $x^2 = 64$

$x = \pm\sqrt{64}$

$x = \pm 8$

9. $a^2 = 169$

$a = \pm\sqrt{169}$

$a = \pm 13$

11. $a^2 - 25 = 0$

$\underline{\ +25\ \ +25\ }$

$a^2 = 25$

$a = \pm\sqrt{25}$

$a = \pm 5$

13. $x^2 - 49 = 0$

$\underline{\ +49\ \ +49\ }$

$x^2 = 49$

$x = \pm\sqrt{49}$

$x = \pm 7$

15. $4r^2 = 100$

$\frac{4r^2}{4} = \frac{100}{4}$

$r^2 = 25$

$r = \pm\sqrt{25}$

$r = \pm 5$

17. $3t^2 = 75$

$\frac{3t^2}{3} = \frac{75}{3}$

$t^2 = 25$

$t = \pm\sqrt{25}$

$t = \pm 5$

19. $2x^2 - 32 = 0$

$\underline{\ +32\ \ +32\ }$

$2x^2 = 32$

$\frac{2x^2}{2} = \frac{32}{2}$

$x^2 = 16$

$x = \pm\sqrt{16}$

$x = \pm 4$

21. $8b^2 - 128 = 0$

$\underline{\ +128\ \ +128\ }$

$8b^2 = 128$

$\frac{8b^2}{8} = \frac{128}{8}$

$b^2 = 16$

$b = \pm\sqrt{16}$

$b = \pm 4$

23.

$(x-5)^2 = 49$

$x - 5 = \pm\sqrt{49} = \pm 7$

$x = 5 + 7 = 12$

$x = 5 - 7 = -2$

$x = 12\ ,\ x = -2$

25.

$4(x+7)^2 = 400$

$\frac{4(x+7)^2}{4} = \frac{400}{4}$

$(x+7)^2 = 100$

$x + 7 = \pm\sqrt{100} = \pm 10$

$x = -7 + 10 = 3$

$x = -7 - 10 = -17$

$x = 3,\ \ x = -17$

27.

$3(x-7)^2 + 8 = 647.48$

$\underline{\ \ -8\ \ \ -8\ }$

$3(x-7)^2 = 639.48$

$\frac{33(x-7)^2}{3} = \frac{639.48}{3}$

$(x-7)^2 = 213.16$

$x - 7 = \pm\sqrt{213.16} = \pm 14.6$

$x = 7 + 14.6 = 21.6$

$x = 7 - 14.6 = -7.6$

$x = 21.6\ ,\ \ x = -7.6$

29.

$$-5(x-6)^2-15=-30$$
$$+15 \quad +15$$
$$-5(x-6)^2=-15$$
$$\frac{-5(x-6)^2}{-5}=\frac{-15}{-5}$$
$$(x-6)^2=3$$
$$x-6=\pm\sqrt{3}\approx\pm1.732$$
$$x\approx 6+1.732=7.732$$
$$x\approx 6-1.732=4.268$$
$$x\approx 4.268\ ,\ x\approx 7.732$$

31.

$$-0.5(x+4)^2-9=-20$$
$$+9 \quad +9$$
$$-0.5(x+4)^2=-11$$
$$\frac{-0.5(x+4)^2}{-0.5}=\frac{-11}{-0.5}$$
$$(x+4)^2=22$$
$$x+4=\pm\sqrt{22}\approx\pm4.690$$
$$x\approx -4+4.690=0.690$$
$$x\approx -4-4.690=-8.690$$
$$x\approx 0.690\ ,\ x\approx -8.690$$

33. The correct answers are $x=2,\ x=-10$. When applying the square root property, the student did not include the negative square root, so one answer was missed.

$$27=(x+4)^2-9$$
$$+9 \quad +9$$
$$36=(x+4)^2$$
$$\pm\sqrt{36}=x+4$$
$$\pm6=x+4$$

The two solutions are:

$$x=-4+6=2$$
$$x=-4-6=-10$$
$$x=2\ ,\ x=-10$$

35. **a.**

$$20=-16t^2+60$$
$$-60 \quad -60$$
$$-40=-16t^2$$
$$\frac{-40}{-16}=\frac{-16t^2}{-16}$$
$$2.5=t^2$$
$$t=\pm\sqrt{2.5}$$
$$t\approx\pm1.58$$

The negative value does not make sense. It will take approximately 1.58 seconds for the ball to fall to a height of 20 feet.

b.

$$0=-16t^2+60$$
$$-60 \quad -60$$
$$-60=-16t^2$$
$$\frac{-60}{-16}=\frac{-16t^2}{-16}$$
$$3.75=t^2$$
$$t=\pm\sqrt{3.75}$$
$$t\approx\pm1.94$$

The negative value does not make sense. It will take approximately 1.94 seconds for the ball to hit the ground.

37. **a.**

$$10=-4.9t^2+25$$
$$-25 \quad -25$$
$$-15=-4.9t^2$$
$$\frac{-15}{-4.9}=\frac{-4.9t^2}{-4.9}$$
$$3.0612\approx t^2$$
$$t\approx\pm\sqrt{3.0612}$$
$$t\approx\pm1.75$$

The negative value does not make sense. It will take approximately 1.75 seconds for the rock to fall to a height of 10 meters.

b.

$$0=-4.9t^2+25$$
$$-25 \quad -25$$
$$-25=-4.9t^2$$
$$\frac{-25}{-4.9}=\frac{-4.9t^2}{-4.9}$$
$$5.102\approx t^2$$
$$t\approx\pm\sqrt{5.102}$$
$$t\approx\pm2.26$$

The negative value does not make sense. It will take approximately 2.26 seconds for the rock to hit the ground.

39. $A=\pi r^2$ Substitute $A=3,000,000$.

$$3,000,000=\pi r^2$$
$$\frac{3,000,000}{\pi}=\frac{\pi r^2}{\pi}$$
$$\frac{3,000,000}{\pi}=r^2$$
$$r=\pm\sqrt{\frac{3,000,000}{\pi}}$$
$$r\approx\pm977.21$$

The negative value does not make sense. The radius of the circular region is approximately 977.21 feet.

41. $A = s^2$ Substitute $A = 225$.

$$225 = s^2$$
$$s = \pm\sqrt{225}$$
$$s = \pm 15$$

The negative value does not make sense. The length of each side of the room is 15 feet.

43. **a.**

$$C(18) = -2.336(18-13)^2 + 85.6$$
$$= -2.336(5)^2 + 85.6$$
$$= 27.2$$

In 1998, 27.2 million cassette singles were shipped by major recording media manufacturers.

b.

$$-2.336(t-13)^2 + 85.6 = 50$$
$$\underline{\quad -85.6 \; -85.6}$$
$$-2.336(t-13)^2 = -35.6$$
$$\frac{-2.336(t-13)^2}{-2.336} = \frac{-35.6}{-2.336}$$
$$(t-13)^2 \approx 15.240$$
$$t-13 \approx \pm\sqrt{15.240} \approx \pm 3.9$$
$$t \approx 13 - 3.9 = 9.1 \rightarrow 1989$$
$$t \approx 13 + 3.9 = 16.9 \rightarrow 1997$$

In about 1989 and 1997, about 50 million cassette singles were shipped by major recording media manufacturers.

c. Vertex $= (13, 85.6)$. In 1993, the number of cassette singles shipped by major recording media manufacturers reached a maximum of 85.6 million.

d. Anytime before about 1988 is probably model breakdown, and anytime after 1999 will also be model breakdown because the model gives a negative number of cassette singles being shipped.

45. **a.**

$$D(10) = 17(10-3)^2 + 2300$$
$$= 17(7)^2 + 2300$$
$$= 3133$$

In 2000, personal households and nonprofit organizations invested about \$3133 billion in time and savings accounts.

b.

$$17(t-3)^2 + 2300 = 7500$$
$$\underline{\quad -2300 \; -2300}$$
$$17(t-3)^2 = 5200$$
$$\frac{17(t-3)^2}{17} = \frac{5200}{17}$$
$$(t-3)^2 \approx 305.882$$
$$t-3 \approx \pm\sqrt{305.882} \approx 17.5$$
$$t \approx 3 - 17.5 = -14.5 \rightarrow 1976$$
$$t \approx 3 + 17.5 = 20.5 \rightarrow 2010$$

According to this model, personal households and nonprofit organizations invested about \$7500 billion in time and savings deposits in the years 1976 and 2010. 1976 might be model breakdown.

c. Vertex $= (3, 2300)$. In about 1993, personal households and nonprofit organizations invested about \$2300 billion in time and savings deposits. This was the minimum amount invested around this time.

47. **a.** Let $P(t)$ be the poverty threshold (in dollars) for a family of four t years since 1900.

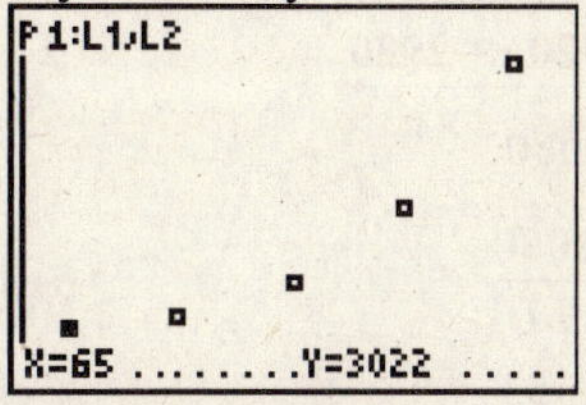

Choose $(65, 3022)$ as the vertex, and $(80, 5500)$ as the additional point used to solve for *"a"*.

$$P(t) = a(t-65)^2 + 3022$$
$$5500 = a(80-65)^2 + 3022$$
$$5500 = a(15)^2 + 3022$$
$$\underline{-3022 \qquad -3022}$$
$$2478 = 225a$$
$$\left.\frac{2478}{225} = \frac{225a}{225}\right\} \; a \approx 11$$

~~$P(t) = 11(t-65)^2 + 3022$~~

This model needs adjusting.

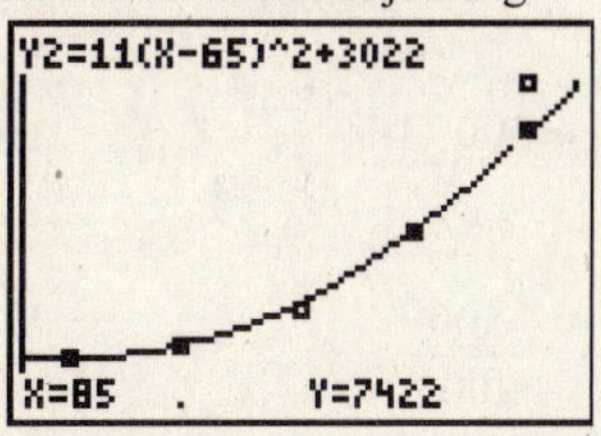

If we adjust the vertex to $(67, 2920)$, and make $a = 17$ slightly larger, the model has a much better fit.

$$P(t) = 17(t-67)^2 + 2920$$

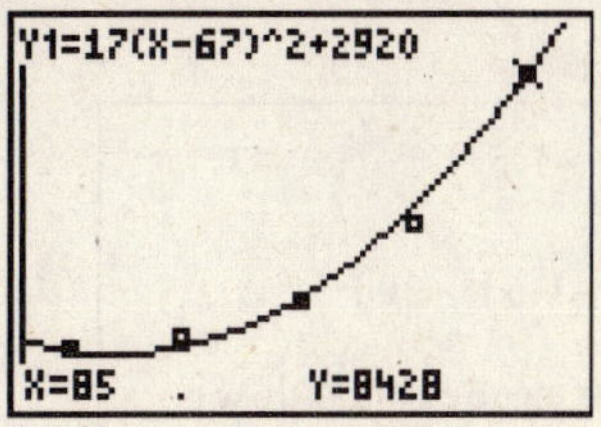

Note: Your answer may vary and yet still represent the trend of the data.

b.

$$P(90) = 17(90-67)^2 + 2920$$
$$= 17(23)^2 + 2920$$
$$= 11{,}913$$

The poverty threshold for a family of four in 1990 was about \$11,913.

c. Domain: $[67,95]$. Within this domain, the minimum value of 2920 comes from the vertex. The maximum value comes from $t=95$.

$$P(95)=17(95-67)^2+2920$$
$$P(95)=16,248$$

Range: $[2920, 16248]$

d.

$$17(t-67)^2+2920=5000$$
$$-2920 \quad -2920$$
$$17(t-67)^2=2080$$
$$\frac{17(t-67)^2}{17}=\frac{2080}{17}$$
$$(t-67)^2\approx 122.353$$
$$t-67\approx \pm\sqrt{122.353}\approx 11.1$$
$$t\approx 67-11.1=55.9 \rightarrow 1956$$

This is model breakdown.

$$t\approx 67+11.1=78.1 \rightarrow 1978$$

The poverty threshold for a family of four was $5000 in about 1978.

49. a.

$$V(5)=2000\left(1-\frac{5}{20}\right)^2$$
$$V(5)=1125$$

After 5 minutes, there would be about 1,125 gallons of water remaining in the tank.

b.

$$2000\left(1-\frac{t}{20}\right)^2=500$$
$$\frac{2000\left(1-\frac{t}{20}\right)^2}{2000}=\frac{500}{2000}$$
$$\left(1-\frac{t}{20}\right)^2=0.25$$
$$1-\frac{t}{20}=\pm\sqrt{0.25}=\pm 0.5$$

The first solution is:

$$-\frac{t}{20}=-1-0.5=-1.5$$
$$(-20)\left(-\frac{t}{20}\right)=(-1.5)(-20)$$

$t=30 \rightarrow$ This is model breakdown since the tank is empty after 20 minutes.

The second solution is:

$$-\frac{t}{20}=-1+0.5=-0.5$$
$$(-20)\left(-\frac{t}{20}\right)=(-0.5)(-20)$$

$t=10 \rightarrow$ This is within the domain of the function.

After about 10 minutes, there would be only 500 gallons of water remaining in the tank.

c.

$$2000\left(1-\frac{t}{20}\right)^2=20$$
$$\frac{2000\left(1-\frac{t}{20}\right)^2}{2000}=\frac{20}{2000}$$
$$\left(1-\frac{t}{20}\right)^2=0.01$$
$$1-\frac{t}{20}=\pm\sqrt{0.01}=\pm 0.1$$

The first solution is:

$$-\frac{t}{20}=-1-0.1=-1.1$$
$$(-20)\left(-\frac{t}{20}\right)=(-1.1)(-20)$$

$t=22 \rightarrow$ This is model breakdown since the tank is empty after 20 minutes.

The second solution is:

$$-\frac{t}{20}=-1+0.1=-0.9$$
$$(-20)\left(-\frac{t}{20}\right)=(-0.9)(-20)$$

$t=18 \rightarrow$ This is within the domain of the function.

After about 18 minutes, there would be only 20 gallons of water remaining in the tank.

51. $x^2+9=0$

$$x^2=-9$$
$$x=\pm\sqrt{-9}$$
$$x=\pm 3i$$

53. $3t^2+45=-3$

$$3t^2=-48$$
$$\frac{3t^2}{3}=\frac{-48}{3}$$
$$t^2=-16$$
$$t=\pm\sqrt{-16}$$
$$t=\pm 4i$$

55. $3(x-8)^2+12=4$

$$3(x-8)^2=-8$$
$$\frac{3(x-8)^2}{3}=-\frac{8}{3}$$
$$(x-8)^2=-\frac{8}{3}$$
$$x-8=\pm\sqrt{-\frac{8}{3}}$$
$$x=8\pm i\sqrt{\frac{8}{3}}\cdot\sqrt{\frac{3}{3}}$$
$$x=8\pm i\sqrt{\frac{24}{9}}$$
$$x=8\pm\frac{2\sqrt{6}}{3}i$$

57. Use the formula $A = P(1+r)^t$:

$$900 = 800(1+r)^2$$
$$\frac{900}{800} = \frac{800(1+r)^2}{800}$$
$$1.125 = (1+r)^2$$
$$1+r = \pm\sqrt{1.125}$$
$$r = -1+\sqrt{1.125} \approx 0.0607$$
$$r = -1-\sqrt{1.125} \approx \cancel{-2.0607}$$

A 6.07% interest rate would turn a deposit of \$800 into \$900 in 2 years.

59. Use the formula $A = P(1+r)^t$:

$$2150 = 2000(1+r)^2$$
$$\frac{2150}{2000} = \frac{2000(1+r)^2}{2000}$$
$$1.075 = (1+r)^2$$
$$1+r = \pm\sqrt{1.075}$$
$$r = -1+\sqrt{1.075} \approx 0.0368$$
$$r = -1-\sqrt{1.075} \approx \cancel{-2.0368}$$

A 3.68% interest rate would turn a deposit of \$2000 into \$2150 in 2 years.

61. $a^2 + b^2 = c^2$ Substitute given values.

$$(2)^2 + b^2 = (5)^2$$
$$4 + b^2 = 25$$
$$\underline{-4 \qquad -4}$$
$$b^2 = 21$$
$$b = \pm\sqrt{21}$$
$$b \approx 4.58 \quad \text{(Use positive value for length.)}$$

63. $a^2 + b^2 = c^2$ Substitute given values.

$$a^2 + (20)^2 = (25)^2$$
$$a^2 + 400 = 625$$
$$\underline{-400 \quad -400}$$
$$a^2 = 225$$
$$a = \pm\sqrt{225}$$
$$a = 15 \quad \text{(Use positive value for length.)}$$

65. $a^2 + b^2 = c^2$ Substitute given values.

$$(3)^2 + (3)^2 = c^2$$
$$9 + 9 = c^2$$
$$18 = c^2$$
$$c = \pm\sqrt{18}$$
$$c \approx 4.24 \quad \text{(Use positive value for length.)}$$

67. $a^2 + b^2 = c^2$ Substitute given values.

$$(6)^2 + (8)^2 = c^2$$
$$36 + 64 = c^2$$
$$100 = c^2$$
$$c = \pm\sqrt{100}$$
$$c = 10 \quad \text{(Use positive value for length.)}$$

69. **a.**

$$\frac{10}{12} = \frac{B}{15}$$
$$\frac{5}{6} = \frac{B}{15}$$
$$6B = 5 \cdot 15$$
$$6B = 75$$
$$\frac{6B}{6} = \frac{75}{6}$$
$$B = 12.5$$

The height of the roof is 12.5 feet.

b. $A^2 + B^2 = C^2$ Substitute $A = 15$ and $B = 12.5$.

$$(15)^2 + (12.5)^2 = C^2$$
$$225 + 156.25 = C^2$$
$$381.25 = C^2$$
$$C = \pm\sqrt{381.25}$$
$$C \approx 19.53 \quad \text{(Use positive value for length.)}$$

The common rafter is approximately 19.53 ft long.

71. **a.**

$$\frac{9}{12} = \frac{B}{17}$$
$$\frac{3}{4} = \frac{B}{17}$$
$$4B = 3 \cdot 17$$
$$4B = 51$$
$$\frac{4B}{4} = \frac{51}{4}$$
$$B = 12.75$$

The height of the roof is 12.75 feet.

b. $A^2 + B^2 = C^2$ Substitute $A = 17$ and $B = 12.75$.

$$(17)^2 + (12.75)^2 = C^2$$
$$289 + 162.5625 = C^2$$
$$451.5625 = C^2$$
$$C = \pm\sqrt{451.5625}$$
$$C = 21.25 \quad \text{(Use positive value for length.)}$$

The common rafter is 21.25 ft long.

73. $a^2 + b^2 = c^2$ Substitute $A = 16$ and $B = 12$.

$$(16)^2 + (12)^2 = c^2$$
$$256 + 144 = c^2$$
$$400 = c^2$$
$$c = \pm\sqrt{400}$$
$$c = 20 \quad \text{(Use positive value for length.)}$$

The diagonal of the room is 20 feet.

75. $a^2+b^2=c^2$ Substitute $A = 300$ and $B = 400$.

$$(300)^2+(400)^2=c^2$$
$$90,000+160,000=c^2$$
$$250,000=c^2$$
$$c=\pm\sqrt{250,000}$$
$$c=500 \text{ (Use positive value for length.)}$$

The diagonal of the warehouse is 500 feet.

$A=lw$ Substitute $l = 300$ and $w = 400$.
$$A=(300)(400)$$
$$A=120,000$$

The area of the warehouse is 120,000 square feet.

77. Use the Pythagorean Theorem.

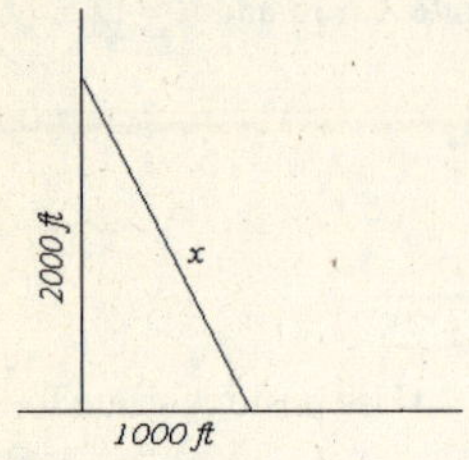

$$x^2=2000^2+1000^2$$
$$x^2=5,000,000$$
$$x=\sqrt{5,000,000}$$
$$x\approx 2236.07$$

The guy wire is approximately 2236.068 feet long.

79. Use the Pythagorean Theorem.

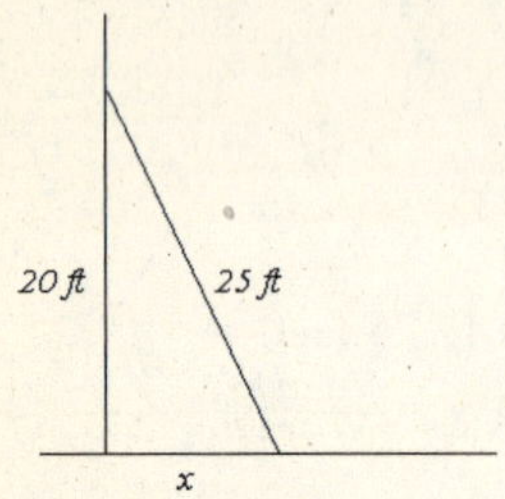

$$25^2=x^2+20^2$$
$$625=x^2+400$$
$$-400 \qquad -400$$
$$225=x^2$$
$$x=\sqrt{225}=15$$

Abida should put the base of the ladder 15 feet away from the house.

Section 11.5 Solving Quadratic Equations by Completing the Square

1. $x^2-5x+21=0$

$a=1 \quad b=-5 \quad c=21$

3. $-x^2+12x-16=0$

$a=-1 \quad b=12 \quad c=-16$

5. $t^2-6t=-12$

$a=1 \quad b=-6 \quad c=12$

7. $-4m^2+10=3m$

$a=-4 \quad b=-3 \quad c=10$

9. **a.** $\frac{1}{2}b=\frac{1}{2}(8)=4$

b. $\left(\frac{1}{2}b\right)^2=(4)^2=16$

11. **a.** $\frac{1}{2}b=\frac{1}{2}(-14)=-7$

b. $\left(\frac{1}{2}b\right)^2=(-7)^2=49$

13. **a.** $\frac{1}{2}b=\frac{1}{2}(5)=2.5$

b. $\left(\frac{1}{2}b\right)^2=(2.5)^2=6.25$

15.

$$x^2+10x+21=0$$
$$-21 \quad -21$$
$$x^2+10x=-21$$
$$+25 \quad +25$$
$$x^2+10x+25=4$$
$$(x+5)^2=4$$
$$x+5=\pm\sqrt{4}$$
$$x+5=\pm 2$$

$$x+5=2 \qquad\qquad x+5=-2$$
$$-5 \ -5 \qquad\qquad -5 \ -5$$
$$x=-3 \qquad\qquad x=-7$$

17.

$$x^2-8x+14=0$$
$$-14 \quad -14$$
$$x^2-8x=-14$$
$$+16 \quad +16$$
$$x^2-8x+16=2$$
$$(x-4)^2=2$$
$$x-4=\pm\sqrt{2}$$

$$x-4=\sqrt{2} \qquad x-4=-\sqrt{2}$$
$$\underline{\quad+4 \quad +4} \qquad \underline{\quad+4 \quad +4}$$
$$x=4+\sqrt{2} \qquad x=4-\sqrt{2}$$

$x \approx 5.41,\ x \approx 2.59$

19.

$$y^2-22y+111=0$$
$$\underline{\quad -111 \quad -111}$$
$$y^2-22y=-111$$
$$\underline{\quad +121 \quad +121}$$
$$y^2-22y+121=10$$
$$(y-11)^2=10$$
$$y-11=\pm\sqrt{10}$$

$$y-11=\sqrt{10} \qquad y-11=-\sqrt{10}$$
$$\underline{\quad+11 \quad +11} \qquad \underline{\quad+11 \quad +11}$$
$$y=11+\sqrt{10} \qquad y=11-\sqrt{10}$$

$y \approx 14.16,\ y \approx 7.84$

21.

$$a^2-16a=-61$$
$$\underline{\quad +64 \quad +64}$$
$$a^2-16a+64=3$$
$$(a-8)^2=3$$
$$a-8=\pm\sqrt{3}$$

$$a-8=\sqrt{3} \qquad a-8=-\sqrt{3}$$
$$\underline{\quad+8 \quad +8} \qquad \underline{\quad+8 \quad +8}$$
$$a=8+\sqrt{3} \qquad a=8-\sqrt{3}$$

$a \approx 9.73,\ a \approx 6.27$

23.

$$x^2-2x=63$$
$$\underline{\quad +1 \quad +1}$$
$$x^2-2x+1=64$$
$$(x-1)^2=64$$
$$x-1=\pm\sqrt{64}$$
$$x-1=\pm 8$$

$$x-1=8 \qquad x-1=-8$$
$$\underline{\quad+1 \ +1} \qquad \underline{\quad+1 \ +1}$$
$$x=9 \qquad x=-7$$

25. The student did not add $\left(\frac{1}{2}b\right)^2$ to the right side.

$$x^2+2x=3$$
$$\underline{\quad +1 \ +1}$$
$$x^2+2x+1=4$$
$$(x+1)^2=4$$
$$x+1=\pm\sqrt{4}$$
$$x+1=\pm 2$$

$$x+1=2 \qquad x+1=-2$$
$$\underline{\quad -1 \ -1} \qquad \underline{\quad -1 \ -1}$$
$$x=1 \qquad x=-3$$

27. The student did not include both positive and negative possibilities as solutions to $\sqrt{16}$.

$$x^2-4x=12$$
$$\underline{\quad +4 \quad +4}$$
$$x^2-4x+4=16$$
$$(x-2)^2=16$$
$$x-2=\pm\sqrt{16}$$
$$x-2=\pm 4$$

$$x-2=4 \qquad x-2=-4$$
$$\underline{\quad +2 \ +2} \qquad \underline{\quad +2 \ +2}$$
$$x=6 \qquad x=-2$$

29.

$x^2+6x=7$

$x^2+6x+\square=7+\square$

Completing the square: $\left(\frac{6}{2}\right)^2=9$

$x^2+6x+\boxed{9}=7+\boxed{9}$

$(x+3)^2=16$

$x+3=\pm\sqrt{16}=\pm 4$

The two solutions are:

$x+3=4\ ,\ x+3=-4$

$x=1\ ,\ x=-7$

31.

$k^2-16k=3$

$k^2-16k+\square=3+\square$

Completing the square: $\left(\frac{-16}{2}\right)^2=64$

$k^2-16k+\boxed{64}=3+\boxed{64}$

$(k-8)^2=67$

$k-8=\pm\sqrt{67}\approx\pm 8.185$

The two solutions are:

$k-8\approx 8.185\ ,\ k-8\approx -8.185$

$k\approx 16.185\ ,\ k\approx -0.185$

33.

$x^2+4x+7=0$

$x^2+4x=-7$

$x^2+4x+\square=-7+\square$

Completing the square: $\left(\frac{4}{2}\right)^2=4$

$x^2+4x+\boxed{4}=-7+\boxed{4}$

$(x+2)^2=-3$

$x+2=\pm\sqrt{-3}$

There are no real solutions.

35.

$t^2+11t=4$

$t^2+11t+\square=4+\square$

Completing the square: $\left(\frac{11}{2}\right)^2=30.25$

$t^2+11t+\boxed{30.25}=4+\boxed{30.25}$

$(t+5.5)^2=34.25$

$t+5.5=\pm\sqrt{34.25}\approx\pm5.852$

The two solutions are:

$t+5.5\approx5.852\ ,\ t+5.5\approx-5.852$

$t\approx0.352\ ,\ t\approx-11.352$

37.

$x^2-9x+14=0$

$x^2-9x=-14$

$x^2-9x+\square=-14+\square$

Completing the square: $\left(\frac{-9}{2}\right)^2=20.25$

$x^2-9x+\boxed{20.25}=-14+\boxed{20.25}$

$(x-4.5)^2=6.25$

$x-4.5=\pm\sqrt{6.25}=\pm2.5$

The two solutions are:

$x-4.5=2.5\ ,\ x-4.5=-2.5$

$x=7\ ,\ x=2$

39.

$m^2+8m+20=0$

$m^2+8m=-20$

$m^2+8m+\square=-20+\square$

Completing the square: $\left(\frac{8}{2}\right)^2=16$

$m^2+8m+\boxed{16}=-20+\boxed{16}$

$(m+4)^2=-4$

$m+4=\pm\sqrt{-4}$

There are no real solutions.

41.

$3x^2+12x-15=0$

$3x^2+12x=15$

$\frac{3x^2+12x}{3}=\frac{15}{3}$

$x^2+4x=5$

$x^2+4x+\square=5+\square$

Completing the square: $\left(\frac{4}{2}\right)^2=4$

$x^2+4x+\boxed{4}=5+\boxed{4}$

$(x+2)^2=9$

$x+2=\pm\sqrt{9}=\pm3$

The two solutions are:

$x+2=3\ ,\ x+2=-3$

$x=1\ ,\ x=-5$

43.

$5x^2+7x=0$

$\frac{5x^2+7x}{5}=\frac{0}{5}$

$x^2+\frac{7}{5}x=0$

$x^2+\frac{7}{5}x+\square=\square$

Completing the square: $\left(\frac{1}{2}\cdot\frac{7}{5}\right)^2=\frac{49}{100}$

$x^2+\frac{7}{5}x+\boxed{\frac{49}{100}}=\boxed{\frac{49}{100}}$

$\left(x+\frac{7}{10}\right)^2=\frac{49}{100}$

$x+\frac{7}{10}=\pm\sqrt{\frac{49}{100}}=\pm\frac{7}{10}$

The two solutions are:

$x+\frac{7}{10}=\frac{7}{10}\ ,\ x+\frac{7}{10}=-\frac{7}{10}$

$x=0\ ,\ x=-\frac{14}{10}=-\frac{7}{5}$

45.

$-7x^2+4x+20=0$

$-7x^2+4x=-20$

$\frac{-7x^2+4x}{-7}=\frac{-20}{-7}$

$x^2-\frac{4}{7}x=\frac{20}{7}$

$x^2-\frac{4}{7}x+\square=\frac{20}{7}+\square$

Completing the square: $\left(\frac{1}{2}\cdot\frac{-4}{7}\right)^2=\left(\frac{-2}{7}\right)^2=\frac{4}{49}$

$x^2-\frac{4}{7}x+\boxed{\frac{4}{49}}=\frac{20}{7}+\boxed{\frac{4}{49}}=\frac{140}{49}+\frac{4}{49}=\frac{144}{49}$

$\left(x-\frac{2}{7}\right)^2=\frac{144}{49}$

$x-\frac{2}{7}=\pm\sqrt{\frac{144}{49}}=\pm\frac{12}{7}$

The two solutions are:

$x-\frac{2}{7}=\frac{12}{7}\ ,\ x-\frac{2}{7}=-\frac{12}{7}$

$x=2\ ,\ x=-\frac{10}{7}$

47. $f(x)=x^2+6x+8$

$f(x)=(x^2+6x+\_\_)+8-\_\_$

Completing the square: $\left(\frac{6}{2}\right)^2=9$

$f(x)=(x^2+6x+9)+8-9$

$f(x)=(x+3)(x+3)-1$

$f(x)=(x+3)^2-1$

49. $g(t)=t^2-8t-20$

$g(t)=\left(t^2-8t+\_\_\right)-20-\_\_$

Completing the square: $\left(\frac{-8}{2}\right)^2=16$

$g(t)=\left(t^2-8t+16\right)-20-16$

$g(t)=(t-4)(t-4)-36$

$g(t)=(t-4)^2-36$

51. $f(x)=x^2-7x+10$

$f(x)=(x^2-7x+\_\_)+10-\_\_$

Completing the square: $\left(\frac{-7}{2}\right)^2=\frac{49}{4}$

$f(x)=\left(x^2-7x+\frac{49}{4}\right)+10-\frac{49}{4}$

$f(x)=\left(x-\frac{7}{2}\right)\left(x-\frac{7}{2}\right)-\frac{9}{4}$

$f(x)=\left(x-\frac{7}{2}\right)^2-\frac{9}{4}$

53. $h(x)=3x^2+12x+24$

$h(x)=3\left(x^2+4x\right)+24$

$h(x)=3\left(x^2+4x+\_\_-\_\_\right)+24$

Completing the square: $\left(\frac{4}{2}\right)^2=4$

$h(x)=3\left(x^2+4x+4-4\right)+24$

$h(x)=3\left(x^2+4x+4\right)-3(4)+24$

$h(x)=3(x+2)(x+2)-12+24$

$h(x)=3(x+2)^2+12$

55. $f(t)=2t^2-16t-12$

$f(t)=2\left(t^2-8t\right)-12$

$f(t)=2\left(t^2-8t+\_\_-\_\_\right)-12$

Completing the square: $\left(\frac{-8}{2}\right)^2=16$

$f(t)=2\left(t^2-8t+16-16\right)-12$

$f(t)=2\left(t^2-8t+16\right)-2(16)-12$

$f(t)=2(t-4)(t-4)-32-12$

$f(t)=2(t-4)^2-44$

57. $f(x)=4x^2+5x-20$

$f(x)=4\left(x^2+\frac{5}{4}x\right)-20$

$f(x)=4\left(x^2+\frac{5}{4}x+\_\_-\_\_\right)-20$

Completing the square: $\left(\frac{5}{8}\right)^2=\frac{25}{64}$

$f(x)=4\left(x^2+\frac{5}{4}x+\frac{25}{64}-\frac{25}{64}\right)-20$

$f(x)=4\left(x^2+\frac{5}{4}x+\frac{25}{64}\right)-4\left(\frac{25}{64}\right)-20$

$f(x)=4\left(x+\frac{5}{8}\right)\left(x+\frac{5}{8}\right)-\frac{25}{16}-20$

$f(x)=4\left(x+\frac{5}{8}\right)^2-\frac{345}{16}$

59. $g(x)=0.5x^2+7x-30$

$g(x)=0.5\left(x^2+14x\right)-30$

$g(x)=0.5\left(x^2+14x+\_\_-\_\_\right)-30$

Completing the square: $\left(\frac{14}{2}\right)^2=49$

$g(x)=0.5\left(x^2+14x+49-49\right)-30$

$g(x)=0.5\left(x^2+14x+49\right)-0.5(49)-30$

$g(x)=0.5(x+7)(x+7)-24.5-30$

$g(x)=0.5(x+7)^2-54.5$

61. $f(x)=0.2x^2-7x-10$

$f(x)=0.2\left(x^2-35x\right)-10$

$f(x)=0.2\left(x^2-35x+\_\_-\_\_\right)-10$

Completing the square: $\left(\frac{-35}{2}\right)^2=306.25$

$f(x)=0.2\left(x^2-35x+306.25-306.25\right)-10$

$f(x)=0.2\left(x^2-35x+306.25\right)-0.2(306.25)-10$

$f(x)=0.2(x-17.5)(x-17.5)-61.25-10$

$f(x)=0.2(x-17.5)^2-71.25$

63. $c(p)=\frac{2}{7}p^2-5p-\frac{3}{7}$

$c(p)=\frac{2}{7}\left(p^2-\frac{35}{2}p\right)-\frac{3}{7}$

$c(p)=\frac{2}{7}\left(p^2-\frac{35}{2}p+\_\_-\_\_\right)-\frac{3}{7}$

Completing the square: $\left(\frac{35}{4}\right)^2=\frac{1225}{16}$

$c(p)=\frac{2}{7}\left(p^2-\frac{35}{2}p+\frac{1225}{16}-\frac{1225}{16}\right)-\frac{3}{7}$

$c(p)=\frac{2}{7}\left(p^2-\frac{35}{2}p+\frac{1225}{16}\right)-\frac{2}{7}\left(\frac{1225}{16}\right)-\frac{3}{7}$

$c(p)=\frac{2}{7}\left(p-\frac{35}{4}\right)\left(p-\frac{35}{4}\right)-\frac{175}{8}-\frac{3}{7}$

$c(p)=\frac{2}{7}\left(p-\frac{35}{4}\right)^2-\frac{1249}{56}$

65. $f(x)=(x-5)^2-16$

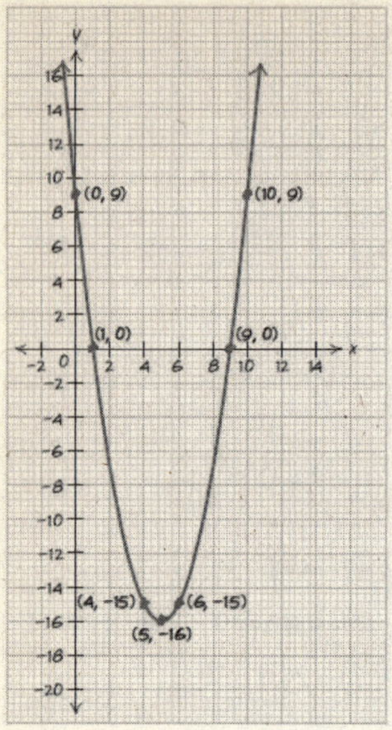

Domain: $(-\infty,\infty)$, range: $[-16,\infty)$

67. $h(x)=(x+2)^2-5$

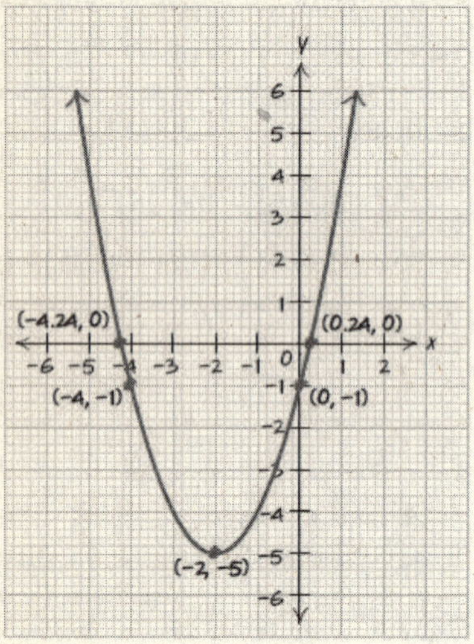

Domain: $(-\infty,\infty)$, range: $[-5,\infty)$

69. $g(x)=3(x-7)^2+10$

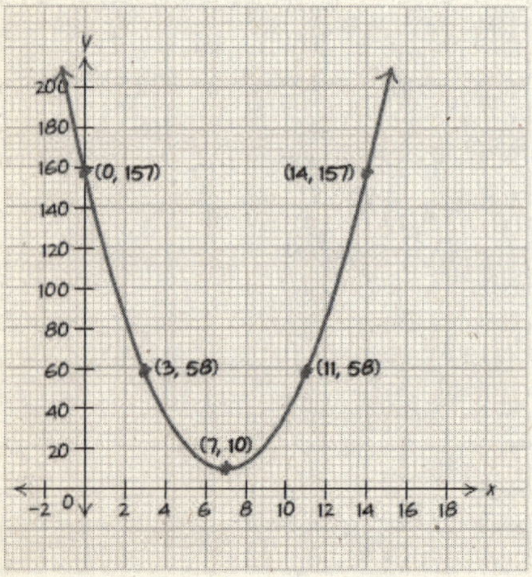

Domain: $(-\infty,\infty)$, range: $[10,\infty)$

71. $g(x)=-2(x-4)^2+18$

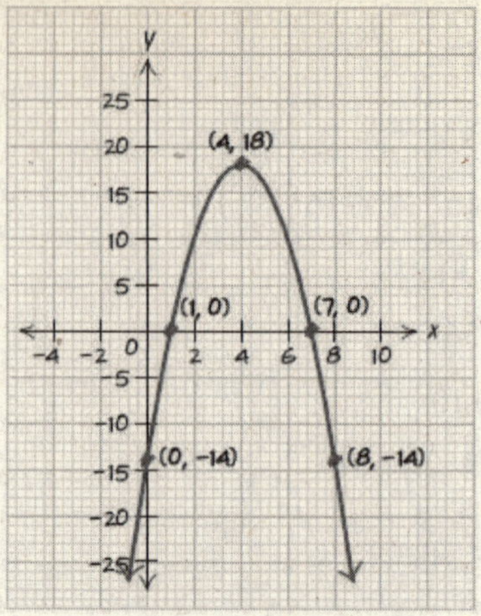

Domain: $(-\infty,\infty)$, range: $(-\infty,18]$

73. $h(x)=-0.5(x+3)^2+32$

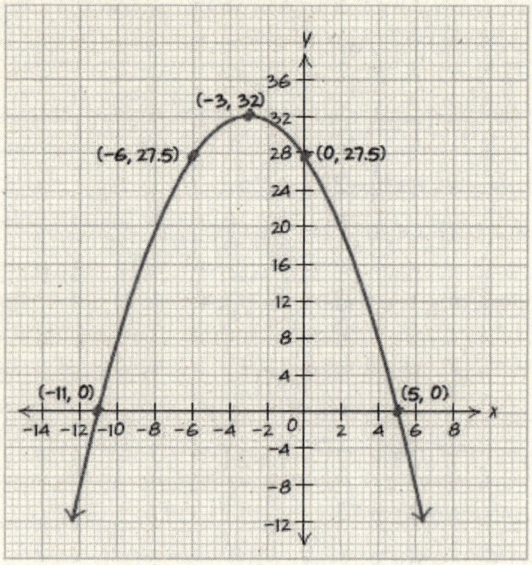

Domain: $(-\infty,\infty)$, range: $(-\infty,32]$

75. $f(x)=\frac{1}{3}(x-5)^2-12$

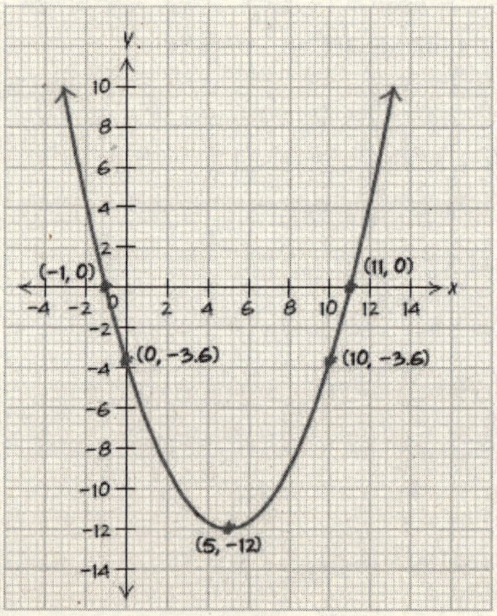

Domain: $(-\infty,\infty)$, range: $[-12,\infty)$

77. $f(x)=\frac{2}{5}(x-6)^2+\frac{1}{5}$

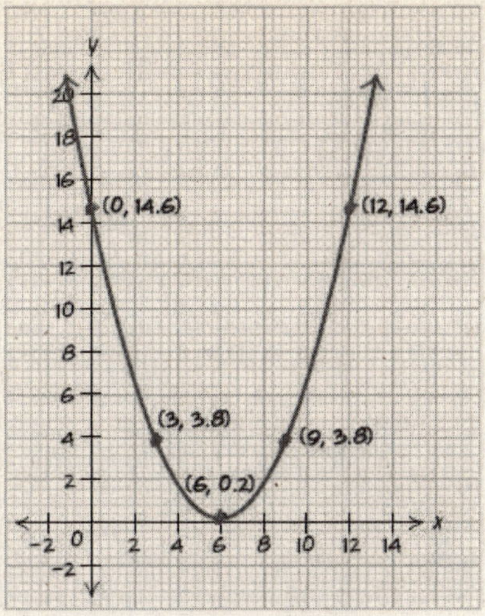

Domain: $(-\infty,\infty)$, range: $[0.2,\infty)$

79. $g(x)=4(x+3.5)^2-16$

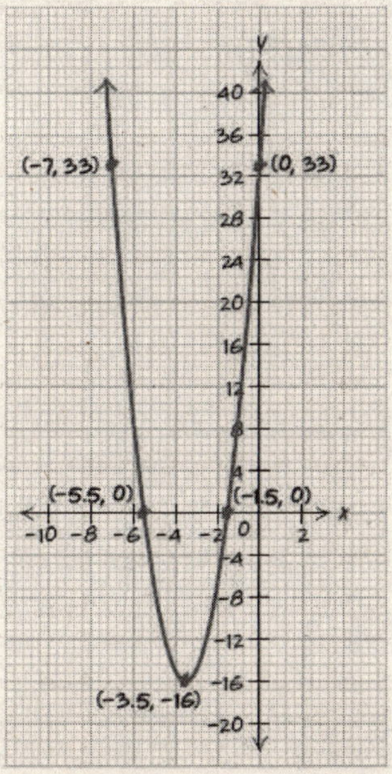

Domain: $(-\infty,\infty)$, range: $[-16,\infty)$

81. $f(x)=(x+3)^2-1$

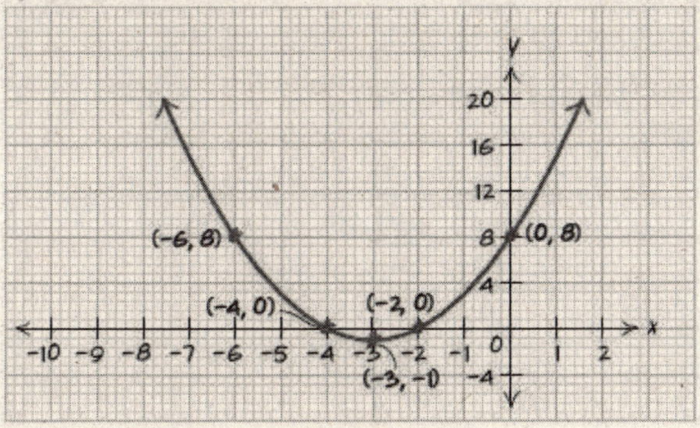

Domain: $(-\infty,\infty)$, range: $[-1,\infty)$

83. $g(t)=(t-4)^2-36$

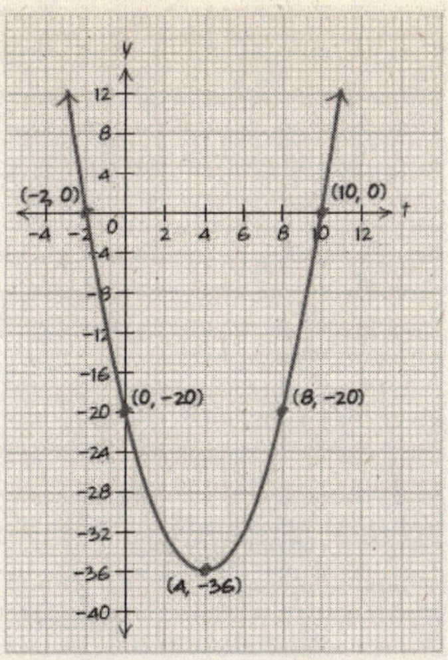

Domain: $(-\infty,\infty)$, range: $[-36,\infty)$

85. $f(x)=\left(x-\frac{7}{2}\right)^2-\frac{9}{4}$

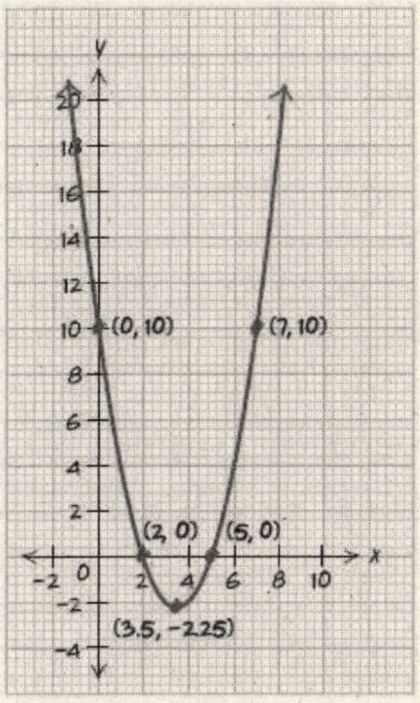

Domain: $(-\infty,\infty)$, range: $[-2.25,\infty)$

87. $h(x)=3(x+2)^2+12$

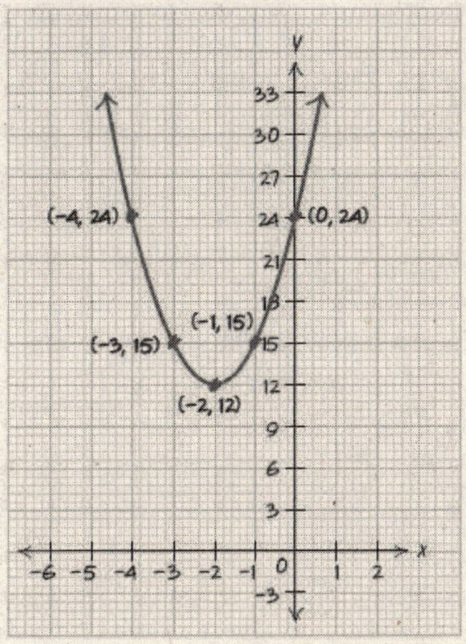

Domain: $(-\infty,\infty)$, range: $[12,\infty)$

89. $f(t)=2(t-4)^2-44$

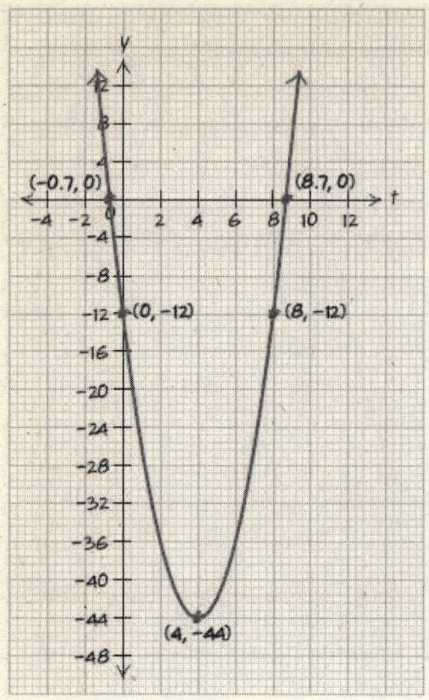

Domain: $(-\infty,\infty)$, range: $[-44,\infty)$

91. $f(x)=4\left(x+\frac{5}{8}\right)^2-\frac{345}{16}$

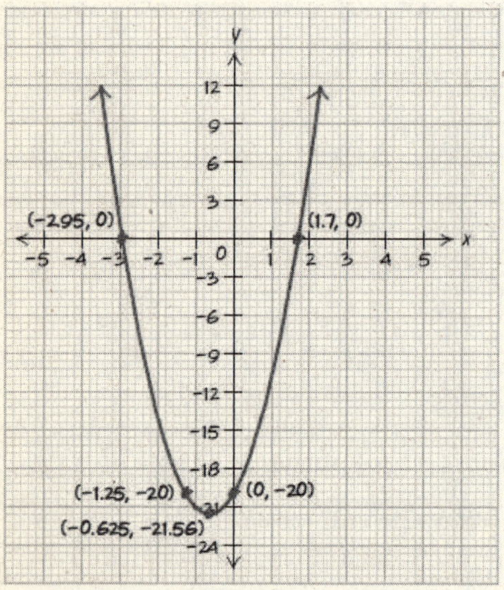

Domain: $(-\infty,\infty)$, range: $[-21.5625,\infty)$

93. $g(x)=0.5(x+7)^2-54.5$

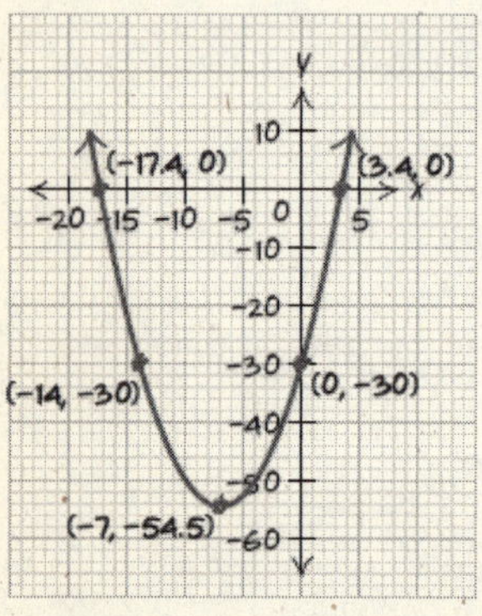

Domain: $(-\infty,\infty)$, range: $[-54.5,\infty)$

Section 11.6 Solving Quadratic Equations by Using the Quadratic Formula

1.

$x^2+8x+15=0$

$a=1, b=8, c=15$

$x=\dfrac{-8\pm\sqrt{8^2-4(1)(15)}}{2(1)}$

$x=\dfrac{-8\pm\sqrt{4}}{2}$

$x=\dfrac{-8\pm 2}{2}$

$x=\dfrac{-8+2}{2}, x=\dfrac{-8-2}{2}$

$x=\dfrac{-6}{2}, x=\dfrac{-10}{2}$

$x=-3, x=-5$

3.

$x^2-7x+10=0$

$a=1, b=-7, c=10$

$x=\dfrac{-(-7)\pm\sqrt{(-7)^2-4(1)(10)}}{2(1)}$

$x=\dfrac{7\pm\sqrt{9}}{2}$

$x=\dfrac{7\pm 3}{2}$

$x=\dfrac{7+3}{2}, x=\dfrac{7-3}{2}$

$x=\dfrac{10}{2}, x=\dfrac{4}{2}$

$x=5, x=2$

5. $a^2+2a-7=0$

$a=\dfrac{-(2)\pm\sqrt{(2)^2-4(1)(-7)}}{2(1)}$

$a=\dfrac{-2\pm\sqrt{4+28}}{2}=\dfrac{-2\pm\sqrt{32}}{2}$

$a=\dfrac{-2\pm 4\sqrt{2}}{2}=\dfrac{\cancel{2}(-1\pm 2\sqrt{2})}{\cancel{2}}=-1\pm 2\sqrt{2}$

$a\approx 1.83, a\approx -3.83$

7. $2x^2-14x+23=0$

$x=\dfrac{-(-14)\pm\sqrt{(-14)^2-4(2)(23)}}{2(2)}$

$x=\dfrac{14\pm\sqrt{196-184}}{4}=\dfrac{14\pm\sqrt{12}}{4}$

$x=\dfrac{14\pm 2\sqrt{3}}{4}=\dfrac{\cancel{2}(7\pm\sqrt{3})}{\cancel{4}2}=\dfrac{7\pm\sqrt{3}}{2}$

$x\approx 4.37, x\approx 2.63$

9. $9x^2-6x-4=0$

$$x=\frac{-(-6)\pm\sqrt{(-6)^2-4(9)(-4)}}{2(9)}$$
$$x=\frac{6\pm\sqrt{36+144}}{18}=\frac{6\pm\sqrt{180}}{18}$$
$$x=\frac{6\pm6\sqrt{5}}{18}=\frac{\not{6}(1\pm\sqrt{5})}{\not{18}3}=\frac{1\pm\sqrt{5}}{3}$$
$x\approx1.08,\ x\approx-0.41$

11. $3x^2+9x-12=0$

$a=3, b=9, c=-12$
$$x=\frac{-9\pm\sqrt{9^2-4(3)(-12)}}{2(3)}$$
$$x=\frac{-9\pm\sqrt{225}}{6}$$
$$x=\frac{-9\pm15}{6}$$
$$x=\frac{-9+15}{6}, x=\frac{-9-15}{6}$$
$$x=\frac{6}{6}, x=\frac{-24}{6}$$
$x=1, x=-4$

13. $2x^2-9x=5$

$2x^2-9x-5=0$
$a=2, b=-9, c=-5$
$$x=\frac{-(-9)\pm\sqrt{(-9)^2-4(2)(-5)}}{2(2)}$$
$$x=\frac{9\pm\sqrt{121}}{4}$$
$$x=\frac{9\pm11}{4}$$
$$x=\frac{9+11}{4}, x=\frac{9-11}{4}$$
$$x=\frac{20}{4}, x=\frac{-2}{4}$$
$x=5, x=-0.5$

15. $-4x^2+7x-8=-20$

$-4x^2+7x+12=0$
$a=-4, b=7, c=12$
$$x=\frac{-7\pm\sqrt{7^2-4(-4)(12)}}{2(-4)}$$
$$x=\frac{-7\pm\sqrt{241}}{-8}$$
$$x=\frac{-7+\sqrt{241}}{-8}, x=\frac{-7-\sqrt{241}}{-8}$$
$x\approx-1.07, x\approx2.82$

17. $15a^2-22a-5=0$

$$a=\frac{-(-22)\pm\sqrt{(-22)^2-4(15)(-5)}}{2(15)}$$
$$a=\frac{22\pm\sqrt{484+300}}{30}=\frac{22\pm\sqrt{784}}{30}$$
$$a=\frac{22\pm28}{30}=\frac{\not{2}(11\pm14)}{\not{30}15}=\frac{11\pm14}{15}$$
$$a=\frac{11+14}{15} \qquad a=\frac{11-14}{15}$$
$$a=\frac{25}{15}=\frac{5}{3} \qquad a=\frac{-3}{15}=-\frac{1}{5}$$

19. $25y^2+60y-29=0$

$$y=\frac{-(60)\pm\sqrt{(60)^2-4(25)(-29)}}{2(25)}$$
$$y=\frac{-60\pm\sqrt{3600+2900}}{50}=\frac{-60\pm\sqrt{6500}}{50}$$
$$y=\frac{-60\pm10\sqrt{65}}{50}=\frac{\not{10}(-6\pm\sqrt{65})}{\not{50}5}=\frac{-6\pm\sqrt{65}}{5}$$
$y\approx0.41, y\approx-2.81$

21. The student did not put the equation in the form $ax^2+bx+c=0$ and improperly identified the values for *a*, *b* and *c*.

$x^2+2x+1=0$
$a=1\ \ b=2\ \ c=1$
$$x=\frac{-(2)\pm\sqrt{(2)^2-4(1)(1)}}{2(1)}$$
$$x=\frac{-2\pm\sqrt{4-4}}{2}=\frac{-2\pm\sqrt{0}}{2}=\frac{-2}{2}=-1$$

23. The student did not take the opposite of *b* for the first step in using the formula. The opposite of -5 is 5.

$3x^2-5x-12=0$
$a=3\ \ b=-5\ \ c=-12$
$$x=\frac{-(-5)\pm\sqrt{(-5)^2-4(3)(-12)}}{2(3)}$$
$$x=\frac{5\pm\sqrt{25+144}}{6}=\frac{5\pm\sqrt{169}}{6}=\frac{5\pm13}{6}$$
$$x=\frac{5+13}{6}=\frac{18}{6}=3$$
$$x=\frac{5-13}{6}=\frac{-8}{6}=-\frac{4}{3}$$

25. **a.** $5000=-2x^2+250x$

b. $-2x^2+250x-5000=0$

$$x=\frac{-(250)\pm\sqrt{(250)^2-4(-2)(-5000)}}{2(-2)}$$

$$x=\frac{-250\pm\sqrt{62,500-40,000}}{-4}$$

$$x=\frac{-250\pm\sqrt{22,500}}{-4}=\frac{-250\pm150}{-4}$$

$$x=\frac{-250+150}{-4} \qquad x=\frac{-250-150}{-4}$$

$$x=\frac{-100}{-4}=25 \qquad x=\frac{-400}{-4}=100$$

c. The photographer must sell at least 25 copies of his photographs to make a profit of \$5,000.00.

27. $500=-2x^2+80x$

$$-2x^2+80x-500=0$$

$$x=\frac{-(80)\pm\sqrt{(80)^2-4(-2)(-500)}}{2(-2)}$$

$$x=\frac{-80\pm\sqrt{6400-4000}}{-4}=\frac{-80\pm\sqrt{2400}}{-4}$$

$$x=\frac{-80\pm20\sqrt{6}}{-4}=\frac{\cancel{-4}(20\pm5\sqrt{6})}{\cancel{-4}}=20\pm5\sqrt{6}$$

$$x=20+5\sqrt{6} \qquad x=20-5\sqrt{6}$$

$$x\approx32.25 \qquad x\approx7.75$$

The web designer must work at least 7.75 hours to make a revenue of \$500.00.

29. **a.** *When* is a reference to time t.

b. The value of the height h is 0 when the rocket hits the ground.

$$0=-16t^2+140t+5$$

c. The equation in part b is quadratic because the variable is degree 2.

d. $t=\dfrac{-(140)\pm\sqrt{(140)^2-4(-16)(5)}}{2(-16)}$

$$t=\frac{-140\pm\sqrt{19600+320}}{-32}$$

$$t=\frac{-140\pm\sqrt{19920}}{-32}$$

$$t=\frac{-140+\sqrt{19920}}{-32} \qquad t=\frac{-140-\sqrt{19920}}{-32}$$

$$t\approx-0.04 \qquad t\approx8.79$$

e. The negative solution $t\approx-0.04$ does not make sense for this problem, but $t\approx8.79$ does make sense. The rocket will hit the ground approximately 8.79 seconds after launch.

31. **a.**

$$52=-4.9t^2+60$$
$$\underline{-60 \qquad\quad -60}$$
$$-8=-4.9t^2$$

$$\frac{-8}{-4.9}=\frac{-4.9t^2}{-4.9}$$

$$\frac{8}{4.9}=t^2$$

$$t=\pm\sqrt{\frac{8}{4.9}}$$

$t\approx1.28$ (use only positive value for time)

The rock will fall to a height of 52 meters in approximately 1.28 seconds.

b.

$$0=-4.9t^2+60$$
$$\underline{-60 \qquad\quad -60}$$
$$-60=-4.9t^2$$

$$\frac{-60}{-4.9}=\frac{-4.9t^2}{-4.9}$$

$$\frac{60}{4.9}=t^2$$

$$t=\pm\sqrt{\frac{60}{4.9}}$$

$t\approx3.50$ (use only positive value for time)

The rock will hit the ground in approximately 3.50 seconds.

33. **a.** $L(t)=0.0572t^2-8.5587t+320.3243$

$$L(55)=0.0572(55)^2-8.5587(55)+320.3243$$
$$L(55)=22.63$$

In 1955, about 22.63 thousand tons of lead were used in paint in the United States

b.

$$5.5=0.0572t^2-8.5587t+320.3243$$
$$\underline{-5.5 \qquad\qquad\qquad\qquad -5.5}$$
$$0=0.0572t^2-8.5587t+314.8243$$

$$t=\frac{8.5587\pm\sqrt{(-8.5587)^2-4(0.0572)(314.8243)}}{2(0.0572)}$$

$$t=\frac{8.5587+\sqrt{1.21954585}}{0.1144}\approx\cancel{84.5}$$

$$t=\frac{8.5587-\sqrt{1.21954585}}{.01144}\approx65.2$$

In about 1965, 5500 tons of lead were used in paints in the United States. 1984 would be considered model breakdown.

c.

$$51=0.0572t^2-8.5587t+320.3243$$
$$\underline{-51 \qquad\qquad\qquad\qquad -51}$$
$$0=0.0572t^2-8.5587t+269.3243$$

$$t=\frac{8.5587\pm\sqrt{(-8.5587)^2-4(0.0572)(269.3243)}}{2(0.0572)}$$

$$t = \frac{8.5587 + \sqrt{11.62994585}}{0.1144} \approx 104.6$$

$$t = \frac{8.5587 - \sqrt{11.62994585}}{.01144} \approx 45.0$$

In 1945, about 51 thousand tons of lead was used in paints in the United States. 2004 would be model breakdown.

35. **a.** $I(t) = 3t^2 - 4t + 5$

$$I(0.5) = 3(0.5)^2 - 4(0.5) + 5$$
$$I(0.5) = 3.75$$

The current in the wire after 0.5 second would be about 3.75 amperes.

b.

$$\begin{aligned} 20 &= 3t^2 - 4t + 5 \\ -20 & \qquad\quad -20 \\ \hline 0 &= 3t^2 - 4t - 15 \end{aligned}$$

$$t = \frac{4 \pm \sqrt{(4)^2 - 4(3)(-15)}}{2(3)}$$

$$t = \frac{4 + \sqrt{196}}{6} = 3$$

$$t = \frac{4 - \sqrt{196}}{6} = \frac{5}{3}$$

After about 3 seconds, the current would reach 20 amperes.

c.

$$\begin{aligned} 60 &= 3t^2 - 4t + 5 \\ -60 & \qquad\quad -60 \\ \hline 0 &= 3t^2 - 4t - 55 \end{aligned}$$

$$t = \frac{4 \pm \sqrt{(4)^2 - 4(3)(-55)}}{2(3)}$$

$$t = \frac{4 + \sqrt{676}}{6} = 5$$

$$t = \frac{4 - \sqrt{676}}{6} = -\frac{11}{3}$$

After about 5 seconds, the current would reach 60 amperes.

37. **a.** $M(n) = 0.00045n^2 + 0.02n + 5$

$$M(200) = 0.00045(200)^2 + 0.02(200) + 5$$
$$M(200) = 27$$

The marginal cost of producing the 201st pair of shoes would be \$27.

b.

$$\begin{aligned} 20 &= 0.00045n^2 + 0.02n + 5 \\ -20 & \qquad\qquad\qquad\quad -20 \\ \hline 0 &= 0.00045n^2 + 0.02n - 15 \end{aligned}$$

$$n = \frac{-0.02 \pm \sqrt{0.02^2 - 4(0.00045)(-15)}}{2(0.02)}$$

$$n = \frac{-0.02 + \sqrt{0.0274}}{0.04} \approx 161.7 \approx 162$$

$$n = \frac{-0.02 - \sqrt{0.0274}}{0.04} \approx -206$$

For the marginal cost to be \$20 per pair of shoes you would be producing the 163rd pair of shoes.

39. **a.** $M(n) = 0.0005n^2 + 0.07n + 50$

$$M(400) = 0.0005(400)^2 + 0.07(400) + 50$$
$$M(400) = 158$$

The marginal cost of producing the 401st bike would be \$158.

b.

$$\begin{aligned} 700 &= 0.0005n^2 + 0.07n + 50 \\ -700 & \qquad\qquad\qquad\quad -700 \\ \hline 0 &= 0.0005n^2 + 0.07n - 650 \end{aligned}$$

$$n = \frac{-0.07 \pm \sqrt{0.07^2 - 4(0.0005)(-650)}}{2(0.0005)}$$

$$n = \frac{-0.07 + \sqrt{1.3049}}{0.001} \approx 1072$$

$$n = \frac{-0.07 - \sqrt{1.3049}}{0.001} \approx -1212$$

For the marginal cost to be \$700 per bike, you would be producing the 1073rd bike.

41. **a.** $h(t) = -16t^2 + 200t + 2$

$$h(1) = -16(1)^2 + 200(1) + 2$$
$$h(1) = 186$$

One second after launch, the rocket was 186 feet high.

b.

$$\begin{aligned} 450 &= -16t^2 + 200t + 2 \\ -450 & \qquad\qquad\quad -450 \\ \hline 0 &= -16t^2 + 200t - 448 \end{aligned}$$

$$t = \frac{-200 \pm \sqrt{200^2 - 4(-16)(-448)}}{2(-16)}$$

$$t = \frac{-200 + \sqrt{11{,}328}}{-32} \approx 2.9$$

$$t = \frac{-200 - \sqrt{11{,}328}}{-32} \approx 9.6$$

The rocket first reached 450 feet 2.9 seconds into the flight.

c.

$$\begin{aligned} 600 &= -16t^2 + 200t + 2 \\ -600 & \qquad\qquad\quad -600 \\ \hline 0 &= -16t^2 + 200t - 598 \end{aligned}$$

$$t = \frac{-200 \pm \sqrt{200^2 - 4(-16)(-598)}}{2(-16)}$$

$$t = \frac{-200 + \sqrt{1728}}{-32} \approx 4.95$$

$$t=\frac{-200-\sqrt{1728}}{-32}\approx 7.55$$

The rocket will first reach a height of 600 feet about 4.95 seconds into the flight.

d.

$$700=-16t^2+200t+2$$
$$\underline{-700 \qquad\qquad -700}$$
$$0=-16t^2+200t-698$$
$$t=\frac{-200\pm\sqrt{200^2-4(-16)(-698)}}{2(-16)}$$
$$t=\frac{-200\pm\sqrt{-4672}}{-32}$$

Nonreal answers.

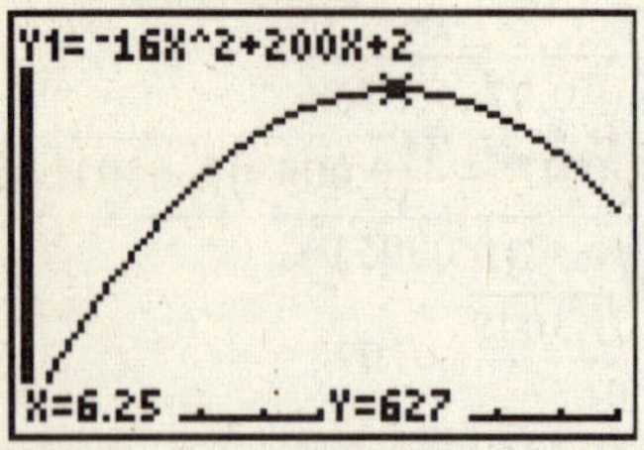

The rocket never reached a height of 700 feet. The rocket reached a maximum of 627 feet.

43. **a.** $I(t)=3.8t^2-29.5t+56.0$

$$I(10)=3.8(10)^2-29.5(10)+56.0$$
$$I(10)=141$$

In 2000, an average person spent 141 hours on the Internet.

b.

$$100=3.8t^2-29.5t+56.0$$
$$\underline{-100 \qquad\qquad -100}$$
$$0=3.8t^2-29.5t-44$$
$$t=\frac{29.5\pm\sqrt{(-29.5)^2-4(3.8)(-44)}}{2(3.8)}$$
$$t=\frac{29.5+\sqrt{1539.05}}{7.6}\approx 9.04$$
$$t=\frac{29.5-\sqrt{1539.05}}{7.6}\approx \text{~~-1.28~~}$$

In 1999, an average person spent 100 hours on the Internet. 1989 would be model breakdown.

c.

$$365=3.8t^2-29.5t+56.0$$
$$\underline{-365 \qquad\qquad -365}$$
$$0=3.8t^2-29.5t-309$$
$$t=\frac{29.5\pm\sqrt{(-29.5)^2-4(3.8)(-309)}}{2(3.8)}$$
$$t=\frac{29.5+\sqrt{5567.05}}{7.6}\approx 14$$
$$t=\frac{29.5-\sqrt{5567.05}}{7.6}\approx \text{~~-6~~}$$

In 2004, an average person will spend about 1 hour per day on the Internet. 1984 would be model breakdown.

45. $2x^2+5x+4=-4$

$$2x^2+5x+8=0$$
$$x=\frac{-5\pm\sqrt{5^2-4(2)(8)}}{2(2)}$$
$$x=\frac{-5\pm\sqrt{-39}}{4}$$
$$x=-\frac{5}{4}\pm\frac{\sqrt{39}}{4}i$$

47. $5x^3+2x^2-7x=0$

$$x(5x^2+2x-7)=0$$
$$x(5x+7)(x-1)=0$$
$$x=0,\ 5x+7=0,\ x-1=0$$
$$x=0,\ x=-\frac{7}{5},\ x=1$$

49. $0.25x^3-3.5x^2+16.25x=0$

$$x(0.25x^2-3.5x+16.25)=0$$
$$x=0,\ 0.25x^2-3.5x+16.25=0$$
$$x=0,\ x=\frac{3.5\pm\sqrt{(-3.5)^2-4(0.25)(16.25)}}{2(0.25)}$$
$$x=0,\ x=\frac{3.5\pm\sqrt{-4}}{0.5}$$
$$x=0,\ x=\frac{3.5\pm 2i}{0.5}$$
$$x=0, x=7\pm 4i$$

51. $2x^2-28x+45=-54$

$$2x^2-28x+99=0$$
$$x=\frac{28\pm\sqrt{(-28)^2-4(2)(99)}}{2(2)}$$
$$x=\frac{28\pm\sqrt{-8}}{4}$$
$$x=7\pm\frac{\sqrt{2}}{2}i$$

53. $a^2+2a=15$

$$a^2+2a-15=0$$
$$(a+5)(a-3)=0$$
$$a+5=0, a-3=0$$
$$a=-5, a=3$$

55. $5t^2-14=0$

$$5t^2=14$$
$$t^2=\frac{14}{5}$$
$$\sqrt{t^2}=\pm\sqrt{\frac{14}{5}}$$
$$t=\pm\sqrt{\frac{14}{5}}$$
$$t\approx\pm 1.67$$

57. $\frac{1}{7}x^2-\frac{5}{7}x=0$

$7\left(\frac{1}{7}x^2\right)-7\left(\frac{5}{7}x\right)=7(0)$

$x^2-5x=0$

$x(x-5)=0$

$x=0, x-5=0$

$x=0, x=5$

59. $4.7x^2-2.6x=0$

$x(4.7x-2.6)=0$

$x=0, 4.7x-2.6=0$

$x=0, 4.7x=2.6$

$x=0, x=\frac{2.6}{4.7}$

$x=0, x\approx 0.55$

61. $5x-12=80$

$5x=92$

$x=\frac{92}{5}$

$x=18.4$

63. $5x^2+3x=0$

$x(5x+3)=0$

$x=0, 5x+3=0$

$x=0, 5x=-3$

$x=0, x=-\frac{3}{5}$

65. $(x-9)^2+8=24$

$(x-9)^2=16$

$\sqrt{(x-9)^2}=\pm\sqrt{16}$

$x-9=\pm 4$

$x-9=4, x-9=-4$

$x=13, x=5$

67. $3(4x-5)^2+20=47$

$3(4x-5)^2=27$

$(4x-5)^2=9$

$\sqrt{(4x-5)^2}=\pm\sqrt{9}$

$4x-5=\pm 3$

$4x-5=3, 4x-5=-3$

$4x=8, 4x=2$

$x=2, x=0.5$

69. $33=-7(4-w)^2+59$

$-26=-7(4-w)^2$

$\frac{-26}{-7}=(4-w)^2$

$\pm\sqrt{\frac{26}{7}}=\sqrt{(4-w)^2}$

$\pm\sqrt{\frac{26}{7}}=4-w$

$\pm\sqrt{\frac{26}{7}}-4=4-w-4$

$\pm\sqrt{\frac{26}{7}}-4=-w$

$\frac{\pm\sqrt{\frac{26}{7}}-4}{-1}=w$

$w=4\pm\sqrt{\frac{26}{7}}$

$w=4+\sqrt{\frac{26}{7}}, w=4-\sqrt{\frac{26}{7}}$

$w\approx 5.93, w\approx 2.07$

71. $r^2+1.4r-14.9=0$

$a=1, b=1.4, c=-14.9$

$r=\frac{-1.4\pm\sqrt{1.4^2-4(1)(-14.9)}}{2(1)}$

$r=\frac{-1.4\pm\sqrt{61.56}}{2}$

$r=\frac{-1.4+\sqrt{61.56}}{2}, r=\frac{-1.4-\sqrt{61.56}}{2}$

$r\approx 3.22, r\approx -4.62$

73. $d^2+2d-35=0$

$(d+7)(d-5)=0$

$d+7=0, d-5=0$

$d=-7, d=5$

75. $b^2-3b=28$

$b^2-3b-28=0$

$(b+4)(b-7)=0$

$b+4=0, b-7=0$

$b=-4, b=7$

77. $23=-2(s+9)^2+5$

$18=-2(s+9)^2$

$-9=(s+9)^2$

$(s+9)^2=-9$

$\sqrt{(s+9)^2}=\pm\sqrt{-9}$

$s+9=\pm 3i$

$s=-9\pm 3i$

79. $105=5(17+d)^2-20$

$125=5(17+d)^2$

$25=(17+d)^2$

$\sqrt{(17+d)^2}=\pm\sqrt{25}$

$17+d=\pm 5$

$17+d=5,\ 17+d=-5$

$d=-12,\ d=-22$

81.

$$120 = -28f + 7f^2 - 939$$
$$0 = 7f^2 - 28f - 1059$$
$$a = 7, b = -28, c = -1059$$
$$f = \frac{-(-28) \pm \sqrt{(-28)^2 - 4(7)(-1059)}}{2(7)}$$
$$f = \frac{28 \pm \sqrt{30436}}{14}$$
$$f = \frac{28 + \sqrt{30436}}{14}, f = \frac{28 - \sqrt{30436}}{14}$$
$$f \approx 14.46, f \approx -10.46$$

83.

$$(3p-4)(p+3) = 15$$
$$3p^2 + 5p - 12 = 15$$
$$3p^2 + 5p - 27 = 0$$
$$a = 3, b = 5, c = -27$$
$$p = \frac{-5 \pm \sqrt{5^2 - 4(3)(-27)}}{2(3)}$$
$$p = \frac{-5 \pm \sqrt{349}}{6}$$
$$p = \frac{-5 + \sqrt{349}}{6}, p = \frac{-5 - \sqrt{349}}{6}$$
$$p \approx 2.28, p \approx -3.95$$

85.

$$3x^2 + 4x + 20 = 0$$
$$a = 3, b = 4, c = 20$$
$$x = \frac{-4 \pm \sqrt{4^2 - 4(3)(20)}}{2(3)}$$
$$x = \frac{-4 \pm \sqrt{-224}}{6}$$
$$x = \frac{-4 \pm \sqrt{-16 \cdot 14}}{6}$$
$$x = \frac{-4 \pm 4i\sqrt{14}}{6} = \frac{-2 \pm 2i\sqrt{14}}{3}$$

87.

$$\frac{3}{2}(x-8)^2 + 10 = 1$$
$$\frac{3}{2}(x-8)^2 = -9$$
$$\frac{2}{3}\left(\frac{3}{2}(x-8)^2\right) = \frac{2}{3}(-9)$$
$$(x-8)^2 = -6$$
$$\sqrt{(x-8)^2} = \pm\sqrt{-6}$$
$$x - 8 = \pm i\sqrt{6}$$
$$x = 8 \pm i\sqrt{6}$$

89.

$$3x^3 - 15x^2 = 252x$$
$$3x^3 - 15x^2 - 252x = 0$$
$$3x(x^2 - 5x - 84) = 0$$
$$3x(x-12)(x+7) = 0$$
$$3x = 0, x - 12 = 0, x + 7 = 0$$
$$x = 0, x = 12, x = -7$$

91.

$$7(x+3) - 8 = 20$$
$$7x + 21 - 8 = 20$$
$$7x + 13 = 20$$
$$7x = 7$$
$$x = 1$$

93.

$$-1.5(d-5) - 7 = 2d + 8$$
$$-1.5d - 7.5 - 7 = 2d + 8$$
$$-1.5d - 14.5 = 2d + 8$$
$$-1.5d = 2d + 22.5$$
$$-3.5d = 22.5$$
$$d \approx -6.43$$

95.

$$\frac{2}{3}(x-5)^2 - 7 = 10$$
$$\frac{2}{3}(x-5)^2 = 17$$
$$\frac{3}{2}\left(\frac{2}{3}(x-5)^2\right) = \frac{3}{2}(17)$$
$$(x-5)^2 = \frac{51}{2}$$
$$\sqrt{(x-5)^2} = \pm\sqrt{\frac{51}{2}}$$
$$x - 5 = \pm\sqrt{\frac{51}{2}}$$
$$x = 5 \pm \sqrt{\frac{51}{2}}$$
$$x = 5 + \sqrt{\frac{51}{2}}, x = 5 - \sqrt{\frac{51}{2}}$$
$$x \approx 10.05, x \approx -0.05$$

97.

$$\frac{1}{4}x^2 - \frac{3}{4}x + 7 = 13$$
$$4\left(\frac{1}{4}x^2 - \frac{3}{4}x + 7\right) = 4(13)$$
$$x^2 - 3x + 28 = 52$$
$$x^2 - 3x - 24 = 0$$
$$a = 1, b = -3, c = -24$$
$$x = \frac{-(-3) \pm \sqrt{(-3)^2 - 4(1)(-24)}}{2(1)}$$
$$x = \frac{3 \pm \sqrt{105}}{2}$$
$$x = \frac{3 + \sqrt{105}}{2}, x = \frac{3 - \sqrt{105}}{2}$$
$$x \approx 6.62, x \approx -3.62$$

99. $(3,10),(-4,10)$

101. No real solution.

103. $(-5,-1)$

105. $(-3,-25)$

107. $(-4,12),(3,12)$

109. $\approx(-3.71,0.295),(2.71,6.71)$

111.

$$5x-8=x^2+3x-9$$
$$\underline{-5x+8 \qquad -5x+8}$$
$$0=x^2-2x-1$$
$$x=\frac{2\pm\sqrt{(-2)^2-4(1)(-1)}}{2(1)}$$
$$x=\frac{2+\sqrt{8}}{2}\approx 2.41,\ x=\frac{2-\sqrt{8}}{2}\approx -0.41$$
$$y\approx 5(2.41)-8,\ y\approx 5(-0.41)-8$$
$$y\approx 4.05,\ y\approx -10.05$$
$$(2.41,4.05),(-0.41,-10.05)$$

113.

$$2x-10=x^2+5x-3$$
$$\underline{-2x+10 \qquad -2x+10}$$
$$0=x^2+3x+7$$
$$x=\frac{-3\pm\sqrt{3^2-4(1)(7)}}{2(1)}$$
$$x=\frac{-3\pm\sqrt{-19}}{2}$$

No real solution.

115.

$$-0.5x^2-3x+15=3x^2+5x-9$$
$$\underline{+0.5x^2+3x-15 \quad +0.5x^2+3x-15}$$
$$0=3.5x^2+8x-24$$
$$x=\frac{-8\pm\sqrt{(8)^2-4(3.5)(-24)}}{2(3.5)}$$
$$x=\frac{-8+\sqrt{400}}{7}\approx 1.71$$
$$x=\frac{-8-\sqrt{400}}{7}=-4$$
$$y=3(1.71)^2+5(1.71)-9\approx 8.32$$
$$y=3(-4)^2+5(-4)-9=19$$
$$(1.71,8.32),(-4,19)$$

117.

$$-x^2+7x-4=x^2-4x+11$$
$$\underline{+x^2-7x+4 \quad +x^2-7x+4}$$
$$0=2x^2-11x+15$$
$$0=(2x-5)(x-3)$$
$$2x-5=0\rightarrow x=\frac{5}{2}=2.5$$
$$y=(2.5)^2-4(2.5)+11=-3.75$$
$$x-3=0\rightarrow x=3$$
$$y=(3)^2-4(3)+11=8$$
$$(2.5,7.25),(3,8)$$

119.

$$2.5x^2+3.4x-8.5=-1.8x^2-2.3x+4.7$$
$$\underline{-2.5x^2-3.4x+8.5 \quad -2.5x^2-3.4x+8.5}$$
$$0=-4.3x^2-5.7x+13.2$$
$$x=\frac{5.7\pm\sqrt{(-5.7)^2-4(-4.3)(13.2)}}{2(-4.3)}$$
$$x=\frac{5.7+\sqrt{259.53}}{-8.6}\approx 1.2105\approx 1.21$$
$$x=\frac{5.7-\sqrt{259.53}}{-8.6}\approx -2.536\approx -2.54$$
$$y\approx 2.5(1.2105)^2+3.4(1.2105)-8.5\approx -0.72$$
$$y\approx 2.5(-2.536)^2+3.4(-2.536)-8.5\approx -1.04$$
$$(1.21,-0.72),(-2.54,-1.04)$$

121.

$$-x^2-6x-38=x^2+6x-20$$
$$\underline{+x^2+6x+38 \quad +x^2+6x+38}$$
$$0=2x^2+12x+18$$
$$\frac{0}{2}=\frac{2x^2+12x+18}{2}$$
$$0=x^2+6x+9$$
$$0=(x+3)^2$$
$$x+3=0\rightarrow x=-3$$
$$y=(-3)^2+6(-3)-20=-29$$
$$(-3,-29)$$

123.

$$-2x^2+2x-18=x^2+2x-8$$
$$\underline{+2x^2-2x+18 \quad +2x^2-2x+18}$$
$$0=3x^2+10$$
$$-3x^2=10$$
$$x^2=\frac{10}{-3}$$
$$x=\sqrt{\frac{10}{-3}}$$

No real solution.

125.

$$-x^2+6x-15=x^2-10x+30$$
$$\underline{+x^2-6x+15 \quad +x^2-6x+15}$$
$$0=2x^2-16x+45$$
$$x=\frac{16\pm\sqrt{(-16)^2-4(2)(45)}}{2(2)}$$
$$x=\frac{16\pm\sqrt{-104}}{4}$$

No real solution.

127.

$9x^2+2x-15=6x^2+2x-9$

$\underline{-9x^2-2x+15} \quad \underline{-9x^2-2x+15}$

$0=-3x^2+6$

$3x^2=6$

$x^2=2 \rightarrow x=\pm\sqrt{2}\approx\pm 1.41$

$y=6(-1.41)^2+2(-1.41)-9\approx 0.11$

$y=6(1.41)^2+2(1.41)-9\approx 5.75$

$(-1.41,0.11)$, $(1.41,5.75)$

Section 11.7 Graphing Quadratics from Standard Form

1. $y=-2x^2-6x-7$: C, downward, y-int: $(0,-7)$.

$y=-0.5x^2-x+7$: A, downward, y-int: $(0,7)$.

$y=x^2+4x+4$: B, upward, y-int: $(0,4)$.

$y=x^2-5x-6$: None, upward, y-int: $(0,-6)$.

3. $y=-2(x-3)^2-9$: None, downward.

$y=2(x-3)^2-9$: A, upward, vertex: $(3,-9)$.

$y=2(x+4)^2+1$: C, upward, vertex: $(-4,1)$.

$y=2(x+4)^2-9$: B, upward, vertex: $(-4,-9)$.

5. $f(x)=x^2+6x+8$

Vertex: (h,k)

$h=\dfrac{-6}{2}=-3$

$k=f(-3)=(-3)^2+6(-3)+8=-1$

$(-3,-1)$

Vertical intercept:

$f(0)=8$

$(0,8)$

Horizontal intercepts:

$x^2+6x+8=0$

$(x+4)(x+2)=0$

$x+4=0$, $x+2=0$

$x=-4$, $x=-2$

$(-4,0)$, $(-2,0)$

7. $h(x)=3x^2-18x+15$

Vertex: (h,k)

$h=\dfrac{18}{2(3)}=3$

$k=h(3)=3(3)^2-18(3)+15=-12$

$(3,-12)$

Vertical intercept:

$h(0)=15$

$(0,15)$

Horizontal intercepts:

$3x^2-18x+15=0$

$3(x^2-6x+5)=0$

$3(x-5)(x-1)=0$

$x-5=0$, $x-1=0$

$x=5$, $x=1$

$(5,0)$, $(1,0)$

9. $h(x)=4(x-8)^2-20$

Vertex: $(8,-20)$

Vertical intercept:

$h(0)=4(-8)^2-20$
$h(0)=236$
$(0,236)$
Horizontal intercepts:
$4(x-8)^2-20=0$
$4(x-8)^2=20$
$\frac{4(x-8)^2}{4}=\frac{20}{4}$
$(x-8)^2=5$
$x-8=\pm\sqrt{5}$
$x=8+\sqrt{5}\approx 10.236$
$x=8-\sqrt{5}\approx 5.764$
$(10.236,0)$, $(5.764,0)$

11. $g(x)=5x^2+12x+10$
Vertex: (h,k)
$h=\frac{-12}{2(5)}=-1.2$
$k=g(-1.2)=5(-1.2)^2+12(-1.2)+10=2.8$
$(-1.2,2.8)$
Vertical intercept:
$g(0)=10$
$(0,10)$
Horizontal intercepts:
$5x^2+12x+10=0$
$x=\frac{-12\pm\sqrt{(12)^2-4(5)(10)}}{2(5)}$
$x=\frac{-12+\sqrt{-56}}{10}$
There are no horizontal intercepts.

13. $f(x)=1.5x^2-6x+4$
Vertex: (h,k)
$h=\frac{6}{2(1.5)}=\frac{6}{3}=2$
$k=f(2)=1.5(2)^2-6(2)+4=-2$
$(2,-2)$
Vertical intercept:
$f(0)=4$
$(0,4)$
Horizontal intercepts:
$1.5x^2-6x+4=0$
$x=\frac{6\pm\sqrt{(-6)^2-4(1.5)(4)}}{2(1.5)}$
$x=\frac{6+\sqrt{12}}{3}\approx 3.155$
$x=\frac{6-\sqrt{12}}{3}\approx 0.845$
$(3.155,0)$, $(0.845,0)$

15. **a.** $C(u)=0.25u^2-25u+3500$
$C(30)=0.25(30)^2-25(30)+3500$
$C(30)=2975$
The cost to produce 30 uniforms would be about \$2975.
b. Vertex: (h,k)
$h=\frac{25}{2(0.25)}=\frac{25}{0.5}=50$
$k=C(50)=0.25(50)^2-25(50)+3500$
$C(50)=2875$
The vertex is $(50,2875)$, which means the minimum cost would be \$2875, producing 50 uniforms.
c.
$1600=0.25u^2-25u+3500$
$-1600 \qquad -1600$
$0=0.25u^2-25u+1900$
$u=\frac{25\pm\sqrt{(25)^2-4(0.25)(1900)}}{2(0.25)}$
$u=\frac{25\pm\sqrt{-2875}}{0.5}$
According to the model, the school can never produce uniforms for \$1600, which would be model breakdown.

17. **a.** $R(c)=-3c^2+90c$
$R(5)=-3(5)^2+90(5)$
$R(5)=375$
The revenue from selling 5 thousand digital cameras would be \$375 thousand.
b.
$600=-3c^2+90c$
$3c^2-90c+600=0$
$\frac{3c^2-90c+600}{3}=\frac{0}{3}$
$c^2-30c+200=0$
$(c-10)(c-20)=0$
$c=10$ or $c=20$
The company must sell 10 thousand cameras or 20 thousand cameras to generate revenue of \$600,000.
c. The maximum would occur at the input value h for the vertex of this model.
(h,k)
$h=\frac{-90}{2(-3)}=\frac{-90}{-6}=15$
The company must sell 15 thousand cameras to maximize its revenue.

19. **a.** $N(t)=376.5t^2+548.1t+2318.4$

$$N(5)=376.5(5)^2+548.1(5)+2318.4$$
$$N(5)=14{,}471.4$$

According to this model, the annual net sales for Home Depot in 1995, was about \$14,471 million.

b.

$$30{,}000=376.5t^2+548.1t+2318.4$$
$$-30{,}000 \qquad\qquad -30{,}000$$
$$0=376.5t^2+548.1t-27{,}681.6$$
$$t=\frac{-548.1\pm\sqrt{548.1^2-4(376.5)(-27{,}681.6)}}{2(376.5)}$$
$$t=\frac{-548.1+\sqrt{41{,}988{,}903.21}}{753}\approx 7.878$$
$$t=\frac{-548.1-\sqrt{41{,}988{,}903.21}}{753}\approx \cancel{-9.333}$$

Home Depot's net sales were \$30,000 million in about 1998.

c. Vertex: (h,k)

$$h=\frac{-548.1}{2(376.5)}=\frac{-10}{1.5}\approx -0.728$$
$$k\approx N(-0.728)=376.5(-0.728)^2+548.1(-0.728)+2318.4$$
$$N(-0.728)\approx 2118.9$$

The vertex is $(-0.728,2118.9)$, which means that in 1989, Home Depot's net sales were at a low point of \$2118.90 million.

21. **a.** $I(t)=-1.5t^2+32.3t-138.8$

$$I(9)=-1.5(9)^2+32.3(9)-138.8$$
$$I(9)=30.4$$

According to this model, the net income for Quicksilver in 1999 was about \$30.4 million.

b.

$$18=-1.5t^2+32.3t-138.8$$
$$-18 \qquad\qquad -18$$
$$0=-1.5t^2+32.3t-156.8$$
$$t=\frac{-32.3\pm\sqrt{32.3^2-4(-1.5)(-156.8)}}{2(-1.5)}$$
$$t=\frac{-32.3+\sqrt{102.49}}{-3}\approx 7.39$$
$$t=\frac{-32.3-\sqrt{102.49}}{-3}\approx 14.14$$

According to this model, Quicksilver's net income reached \$18 million in 1997 and again in 2004.

c. Vertex: (h,k)

$$h=\frac{-32.3}{2(-1.5)}\approx 10.77$$
$$k\approx I(10.77)=-1.5(10.77)^2+32.3(10.77)-138.8$$
$$I(10.77)\approx 35.08$$

The vertex is $(10.77,35.08)$, which means that in about 2001, Quicksilver reached a maximum net sales of \$35.08 million.

23. **a.** $h(t)=-16t^2+40t+4$

$$h(0)=-16(0)^2+40(0)+4$$
$$h(0)=4$$

The ball is at a height of 4 feet when it is hit.

b.

$$20=-16t^2+40t+4$$
$$-20 \qquad\qquad -20$$
$$0=-16t^2+40t-16$$
$$\frac{0}{-8}=\frac{-16t^2+40t-16}{-8}$$
$$0=2t^2-5t+2$$
$$0=(2t-1)(t-2)$$
$$2t-1=0\ ,\ t-2=0$$
$$2t=1$$
$$t=0.5\quad ,\ t=2$$

The ball reached a height of 20 feet about 0.5 second and again 2 seconds after being hit.

c. Find the input value for the vertex of this model.

$$\frac{-b}{2a}=\frac{-40}{2(-16)}=1.25$$

The ball reached its maximum height 1.25 seconds after being hit.

d. Find the output value for the vertex of this model when the input value is 1.25.

$$h(1.25)=-16(1.25)^2+40(1.25)+4$$
$$h(1.25)=29$$

The ball reached a maximum height of 29 feet.

e.

$$0=-16t^2+40t+4$$
$$t=\frac{-40\pm\sqrt{40^2-4(-16)(4)}}{2(-16)}$$
$$t=\frac{-40+\sqrt{1856}}{-32}\approx \cancel{-0.096}$$
$$t=\frac{-40-\sqrt{1856}}{-32}\approx 2.596$$

If the ball does not get caught, the ball will hit the ground 2.596 seconds after being hit.

25. **a.** $h(t)=-16t^2+60t+4.2$

$$h(0)=-16(0)^2+60(0)+4.2$$
$$h(0)=4.2$$

The ball is at a height of 4.2 feet when it is hit.

b.

$$40=-16t^2+60t+4.2$$
$$-40 \qquad\qquad -40$$
$$0=-16t^2+60t-35.8$$

$$t=\frac{-60\pm\sqrt{60^2-4(-16)(4.2)}}{2(-16)}$$
$$t=\frac{-60+\sqrt{1308.8}}{-32}\approx 0.744$$
$$t=\frac{-60-\sqrt{1308.8}}{-32}\approx 3.0055$$

The ball reached a height of 40 feet at 0.744 second and again 3.0055 seconds after being hit.

c. Find the vertex for this model.

$$\frac{-b}{2a}=\frac{-60}{2(-16)}=1.875$$
$$h(1.875)=-16(1.875)^2+60(1.875)+4.2$$
$$h(1.875)=60.45$$

The ball's maximum height is 60.45 feet.

d.

$$0=-16t^2+60t+4.2$$
$$t=\frac{-60\pm\sqrt{60^2-4(-16)(4.2)}}{2(-16)}$$
$$t=\frac{-60+\sqrt{3868.8}}{-32}\approx \cancel{-0.069}$$
$$t=\frac{-60-\sqrt{3868.8}}{-32}\approx 3.819$$

If the ball does not get caught, the ball will hit the ground 3.819 seconds after being hit.

27. **a.** $H(m)=0.9m^2-13m+104$

$$H(6)=0.9(6)^2-13(6)+104$$
$$H(6)=58.4$$

In June, the average high temperature in Melbourne, Australia is 58.4°F

b.

$$\frac{-b}{2a}=\frac{13}{2(0.9)}\approx 7.22$$
$$H(7.22)=0.9(7.22)^2-13(7.22)+104$$
$$H(7.22)\approx 57.056$$

The vertex is $(7.22, 57.056)$, which means that in July, the average high temperature in Melbourne, Australia, reaches its minimum of 57.056°F.

c.

$$60=0.9m^2-13m+104$$
$$\underline{-60\qquad\qquad -60}$$
$$0=0.9m^2-13m+44$$
$$m=\frac{13\pm\sqrt{(13)^2-4(0.9)(44)}}{2(.9)}$$
$$m=\frac{13+\sqrt{10.6}}{1.8}\approx 9.03$$
$$m=\frac{13-\sqrt{10.6}}{1.8}\approx 5.41$$

During September and May, the average temperature of Melbourne, Australia, is about 60°F.

29. $f(x)=x^2+2x-15$

To find the vertex use $\left(-\frac{b}{2a}, f\left(-\frac{b}{2a}\right)\right)$.

$$-\frac{b}{2a}=-\frac{2}{2(1)}=-1$$
$$f\left(-\frac{b}{2a}\right)=f(-1)=(-1)^2+2(-1)-15=-16$$

Vertex $=(-1,-16)$

To find the vertical intercept set $x=0$.

$$f(0)=(0)^2+2(0)-15=-15$$

Vertical intercept $=(0,-15)$

To find the horizontal intercepts set $f(x)=0$.

$$x^2+2x-15=0$$
$$(x+5)(x-3)=0$$
$$x+5=0; x-3=0$$
$$x=-5, x=3$$

Horizontal intercepts $=(-5,0)$ and $(3,0)$.

Domain: $(-\infty,\infty)$, range: $[-16,\infty)$

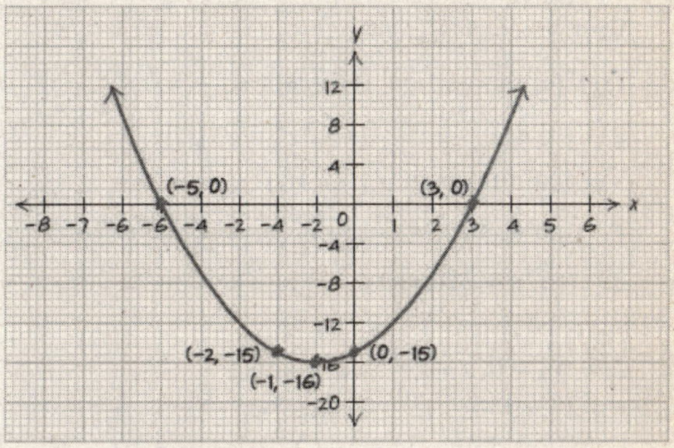

31. $m(b)=-b^2+11b-24$

To find the vertex use $\left(-\frac{b}{2a}, m\left(-\frac{b}{2a}\right)\right)$.

$$-\frac{b}{2a}=-\frac{11}{2(-1)}=5.5$$
$$m\left(-\frac{b}{2a}\right)=m(5.5)=-(5.5)^2+11(5.5)-24=6.25$$

Vertex $=(5.5, 6.25)$

To find the vertical intercept set $b=0$.

$$m(0)=-(0)^2+11(0)-24=-24$$

Vertical intercept $=(0,-24)$

To find the horizontal intercepts set $m(b)=0$.

$$-b^2+11b-24=0$$
$$(-b+8)(b-3)=0$$
$$-b+8=0, b-3=0$$
$$b=8, b=3$$

Horizontal intercepts $=(8,0)$ and $(3,0)$

Domain: $(-\infty,\infty)$, range: $(-\infty, 6.25]$

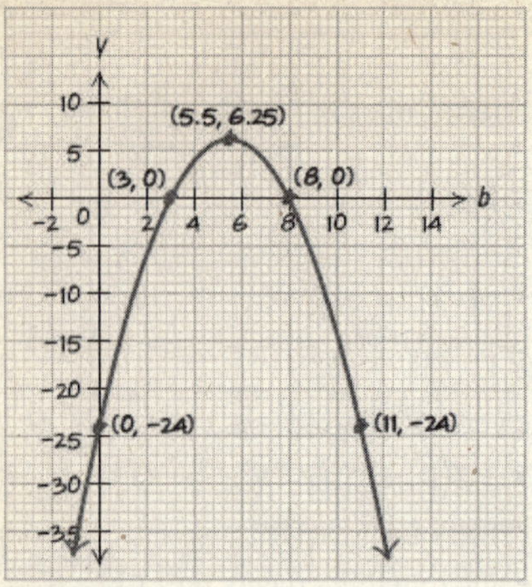

33. $g(s) = 2s^2 - 62s + 216$

To find the vertex use $\left(-\dfrac{b}{2a}, g\left(-\dfrac{b}{2a}\right)\right)$.

$$-\frac{b}{2a} = -\frac{-62}{2(2)} = 15.5$$

$$g\left(-\frac{b}{2a}\right) = g(15.5) = 2(15.5)^2 - 62(15.5) + 216 = -264.5$$

Vertex $= (15.5, -264.5)$

To find the vertical intercept set $s = 0$.

$g(0) = 2(0)^2 - 62(0) + 216 = 216$

Vertical intercept $= (0, 216)$

To find the horizontal intercepts set $g(s) = 0$

$2s^2 - 62s + 216 = 0$

$(2s - 54)(s - 4) = 0$

$2s - 54 = 0, s - 4 = 0$

$s = 27, s = 4$

Horizontal intercepts $= (27, 0)$ and $(4, 0)$

Domain: $(-\infty, \infty)$, range: $[-264.5, \infty)$

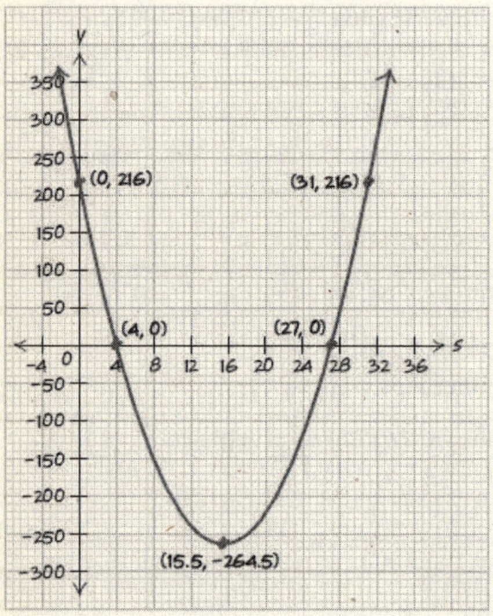

35. $f(x) = 2x^2 + 5$

To find the vertex use $\left(-\dfrac{b}{2a}, f\left(-\dfrac{b}{2a}\right)\right)$.

$$-\frac{b}{2a} = -\frac{0}{2(2)} = 0$$

$$f\left(-\frac{b}{2a}\right) = f(0) = 2(0)^2 + 5 = 5$$

Vertex $= (0, 5)$

To find the vertical intercept set $x = 0$.

$f(0) = 2(0)^2 + 5 = 5$

Vertical intercept $= (0, 5)$

To find the horizontal intercepts set $f(x) = 0$.

$2x^2 + 5 = 0$

$a = 2, b = 0, c = 5$

$$x = \frac{-0 \pm \sqrt{0^2 - 4(2)(5)}}{2(2)}$$

$x = \dfrac{\pm\sqrt{-40}}{4}$ This is not a real number.

There are no horizontal intercepts.

Domain: $(-\infty, \infty)$, range: $[5, \infty)$

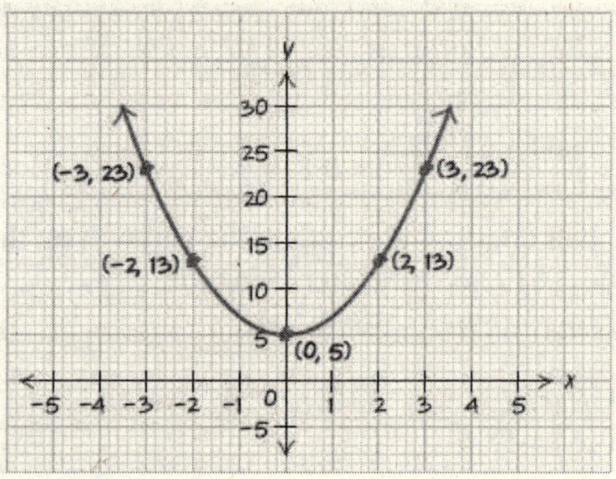

37. $d(p) = -1.5p^2 - 3$

To find the vertex use $\left(-\dfrac{b}{2a}, d\left(-\dfrac{b}{2a}\right)\right)$.

$$-\frac{b}{2a} = -\frac{0}{2(-1.5)} = 0$$

$$d\left(-\frac{b}{2a}\right) = d(0) = -1.5(0)^2 - 3 = -3$$

Vertex $= (0, -3)$

To find the vertical intercept set $p = 0$.

$d(0) = -1.5(0)^2 - 3 = -3$

Vertical intercept $= (0, -3)$

To find the horizontal intercepts set $d(p) = 0$.

$-1.5p^2 - 3 = 0$

$a = -1.5, b = 0, c = -3$

$$p = \frac{-0 \pm \sqrt{0^2 - 4(-1.5)(-3)}}{2(-1.5)}$$

$p = \dfrac{\pm\sqrt{-18}}{-3}$ This is not a real number.

There are no horizontal intercepts.

Domain: $(-\infty, \infty)$, range: $(-\infty, -3]$

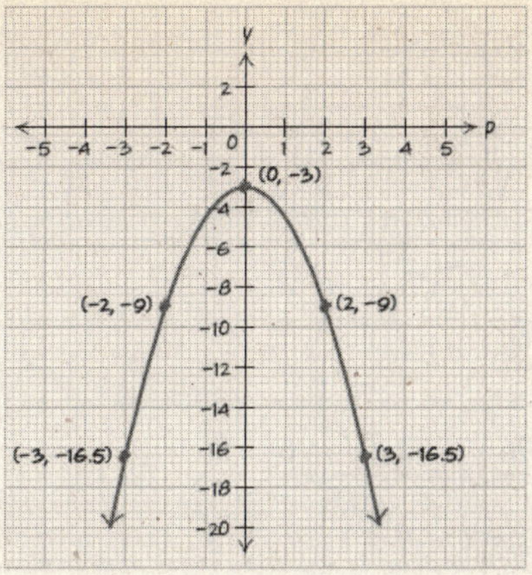

39. $h(x) = -\frac{1}{4}x^2 - 5$

To find the vertex use $\left(-\frac{b}{2a}, h\left(-\frac{b}{2a}\right)\right)$.

$-\frac{b}{2a} = -\frac{0}{2\left(-\frac{1}{4}\right)} = 0$

$h\left(-\frac{b}{2a}\right) = -\frac{1}{4}(0)^2 - 5 = -5$

Vertex = $(0, -5)$

To find the vertical intercept set $x = 0$.

$h(0) = -\frac{1}{4}(0)^2 - 5 = -5$

Vertical intercept = $(0, -5)$

To find the horizontal intercepts set $h(x) = 0$.

$-\frac{1}{4}x^2 - 5 = 0$

$a = -\frac{1}{4}, b = 0, c = -5$

$$x = \frac{-0 \pm \sqrt{0^2 - 4\left(-\frac{1}{4}\right)(-5)}}{2\left(-\frac{1}{4}\right)}$$

$x = \frac{\pm\sqrt{-5}}{-0.5}$ This is not a real number.

There are no horizontal intercepts.

Domain: $(-\infty, \infty)$, range: $(-\infty, -5]$

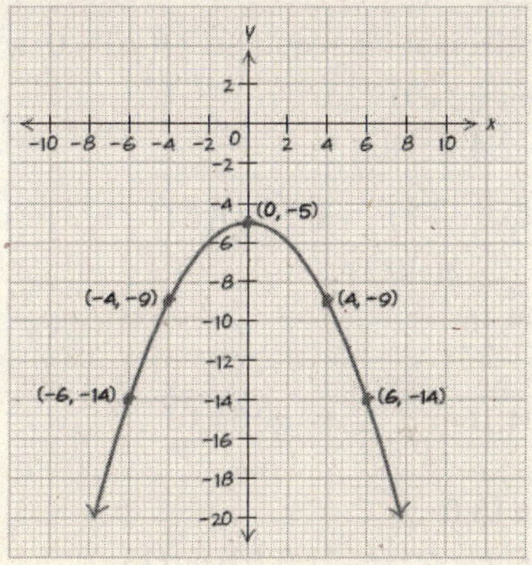

41. $p(k) = -5k^2 - 17.5k - 12.5$

To find the vertex use $\left(-\frac{b}{2a}, p\left(-\frac{b}{2a}\right)\right)$.

$-\frac{b}{2a} = -\frac{-17.5}{2(-5)} = -1.75$

$p\left(-\frac{b}{2a}\right) = p(-1.75)$

$= -5(-1.75)^2 - 17.5(-1.75) - 12.5 = 2.8125$

Vertex = $(-1.75, 2.8125)$

To find the vertical intercept set $k = 0$.

$p(0) = -5(0)^2 - 17.5(0) - 12.5$

Vertical intercept = $(0, -12.5)$

To find the horizontal intercepts set $p(k) = 0$.

$-5k^2 - 17.5k - 12.5 = 0$

$a = -5, b = -17.5, c = -12.5$

$$k = \frac{-(-17.5) \pm \sqrt{(-17.5)^2 - 4(-5)(-12.5)}}{2(-5)}$$

$k = \frac{17.5 \pm \sqrt{56.25}}{-10}$

$k = \frac{17.5 \pm 7.5}{-10}$

$k = \frac{17.5 + 7.5}{-10}, k = \frac{17.5 - 7.5}{-10}$

$k \approx -2.5, k \approx -1$

Horizontal intercepts = $(-2.5, 0)$ and $(-1, 0)$

Domain: $(-\infty, \infty)$, range: $(-\infty, 2.8125]$

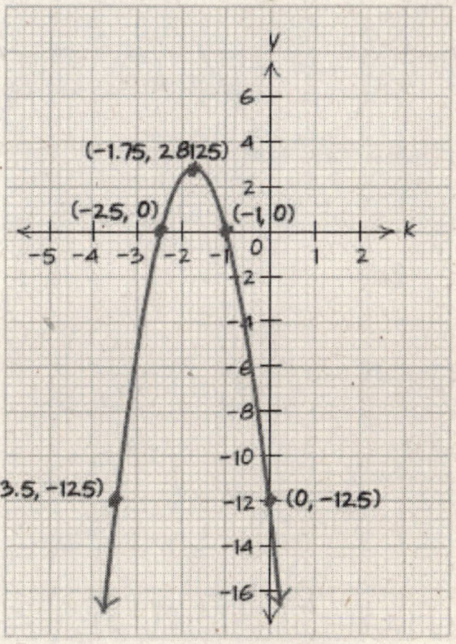

43. $h(w) = 0.4w^2 - 3.6w - 44.8$

To find the vertex use $\left(-\frac{b}{2a}, h\left(-\frac{b}{2a}\right)\right)$.

$-\frac{b}{2a} = -\frac{(-3.6)}{2(0.4)} = 4.5$

$h\left(-\frac{b}{2a}\right) = h(4.5) = 0.4(4.5)^2 - 3.6(4.5) - 44.8 = -52.9$

Vertex = $(4.5, -52.9)$

To find the vertical intercept set $w = 0$.

$h(0) = 0.4(0)^2 - 3.6(0) - 44.8 = -44.8$

Vertical intercept $= (0, -44.8)$

To find the horizontal intercepts set $h(w) = 0$.

$0.4w^2 - 3.6w - 44.8 = 0$

$a = 0.4, b = -3.6, c = -44.8$

$$w = \frac{-(-3.6) \pm \sqrt{(-3.6)^2 - 4(0.4)(-44.8)}}{2(0.4)}$$

$$w = \frac{3.6 \pm \sqrt{84.64}}{0.8}$$

$$w = \frac{3.6 \pm 9.2}{0.8}$$

$$w = \frac{3.6 - 9.2}{0.8}, w = \frac{3.6 + 9.2}{0.8}$$

$w = -7, w = 16$

Horizontal intercepts $= (-7, 0)$ and $(16, 0)$

Domain: $(-\infty, \infty)$, range: $[-52.9, \infty)$

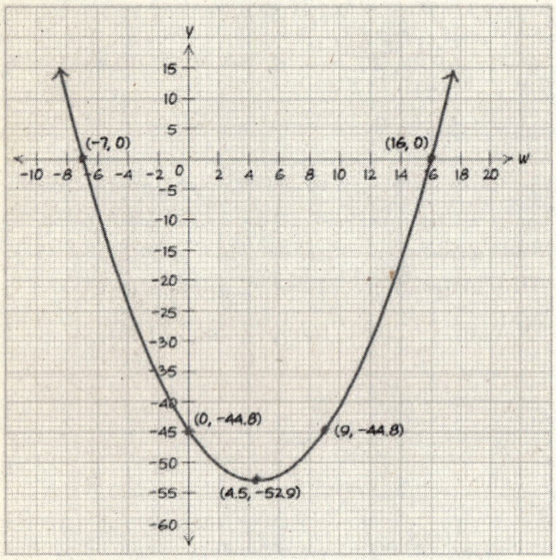

45. $p(x) = \frac{2}{5}x^2 - 2x - \frac{3}{5}$

To find the vertex use $\left(-\frac{b}{2a}, p\left(-\frac{b}{2a}\right)\right)$.

$$-\frac{b}{2a} = -\frac{(-2)}{2\left(\frac{2}{5}\right)} = 2.5$$

$$p\left(-\frac{b}{2a}\right) = p(2.5) = \frac{2}{5}(2.5)^2 - 2(2.5) - \frac{3}{5} = -3.1$$

Vertex $= (2.5, -3.1)$

To find the vertical intercept set $x = 0$.

$$p(0) = \frac{2}{5}(0)^2 - 2(0) - \frac{3}{5} = -0.6$$

Vertical intercept $= (0, -0.6)$

To find the horizontal intercepts set $p(x) = 0$.

$$\frac{2}{5}x^2 - 2x - \frac{3}{5} = 0$$

$$a = \frac{2}{5}, b = -2, c = -\frac{3}{5}$$

$$x = \frac{-(-2) \pm \sqrt{(-2)^2 - 4\left(\frac{2}{5}\right)\left(-\frac{3}{5}\right)}}{2\left(\frac{2}{5}\right)}$$

$$x = \frac{2 \pm \sqrt{4.96}}{0.8}$$

$$x = \frac{2 - \sqrt{4.96}}{0.8}, x = \frac{2 + \sqrt{4.96}}{0.8}$$

$x = -0.3, x = 5.3$

Horizontal intercepts $= (-0.3, 0)$ and $(5.3, 0)$

Domain: $(-\infty, \infty)$, range: $[-3.1, \infty)$

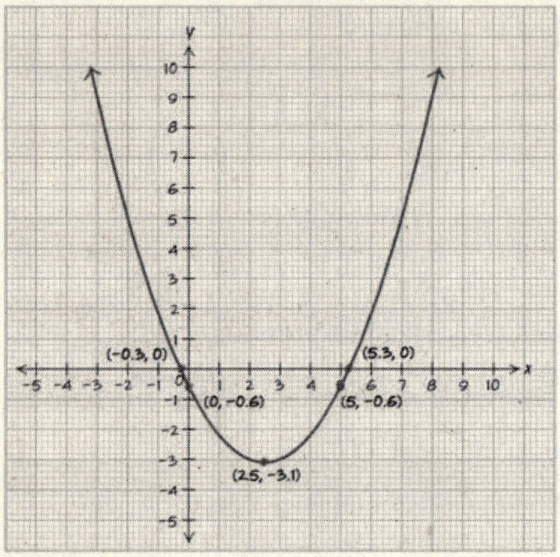

47. $Q(p) = -0.3p^2 - 2.4p + 82$

To find the vertex use $\left(-\frac{b}{2a}, Q\left(-\frac{b}{2a}\right)\right)$.

$$-\frac{b}{2a} = -\frac{(-2.4)}{2(-0.3)} = -4$$

$$Q\left(-\frac{b}{2a}\right) = Q(-4) = -0.3(-4)^2 - 2.4(-4) + 82 = 86.8$$

Vertex $= (-4, 86.8)$

To find the vertical intercept set $p = 0$.

$Q(0) = -0.3(0)^2 - 2.4(0) + 82 = 82$

Vertical intercept $= (0, 82)$

To find the horizontal intercepts set $Q(p) = 0$.

$-0.3p^2 - 2.4p + 82 = 0$

$a = -0.3, b = -2.4, c = 82$

$$p = \frac{-(-2.4) \pm \sqrt{(-2.4)^2 - 4(-0.3)(82)}}{2(-0.3)}$$

$$p = \frac{2.4 \pm \sqrt{104.16}}{-0.6}$$

$$p = \frac{2.4 + \sqrt{104.16}}{-0.6}, p = \frac{2.4 - \sqrt{104.16}}{-0.6}$$

$p \approx -21.0, p \approx 13.0$

Horizontal intercepts $= (-21.0, 0)$ and $(13, 0)$

Domain: $(-\infty,\infty)$, range: $(-\infty,86.8]$

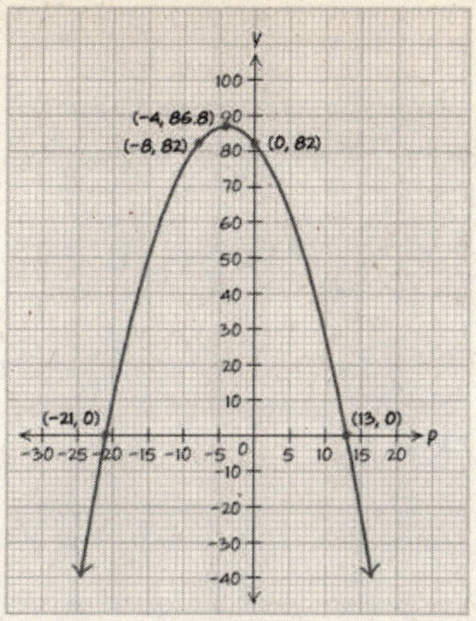

49. $W(g)=-0.3(g+2)^2+17$

Vertex $=(-2,17)$

To find the vertical intercept set $g=0$.

$W(0)=-0.3(0+2)^2+17=15.8$

Vertical intercept $=(0,15.8)$

To find the horizontal intercepts set $W(g)=0$.

$-0.3(g+2)^2+17=0$

$-0.3(g+2)^2=-17$

$(g+2)^2=\dfrac{17}{0.3}$

$\sqrt{(g+2)^2}=\pm\sqrt{\dfrac{17}{0.3}}$

$g+2=\pm\sqrt{\dfrac{17}{0.3}}$

$g=-2\pm\sqrt{\dfrac{17}{0.3}}$

$g=-2-\sqrt{\dfrac{17}{0.3}},g=-2+\sqrt{\dfrac{17}{0.3}}$

$g\approx-9.5,g\approx5.5$

Horizontal intercepts $=(-9.5,0)$ and $(5.5,0)$

Domain: $(-\infty,\infty)$, range: $(-\infty,17]$

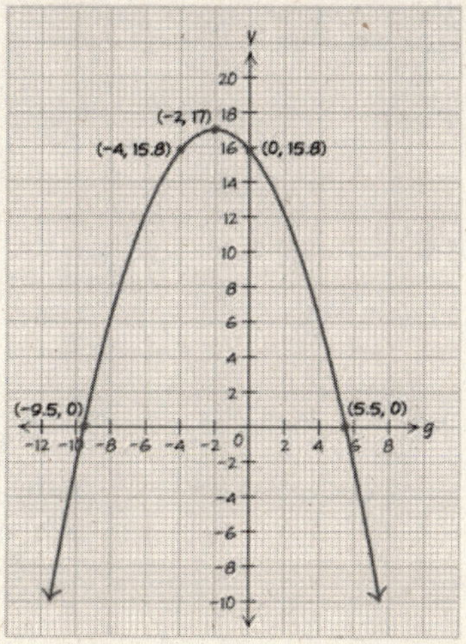

51. $y<x^2+8x+15$

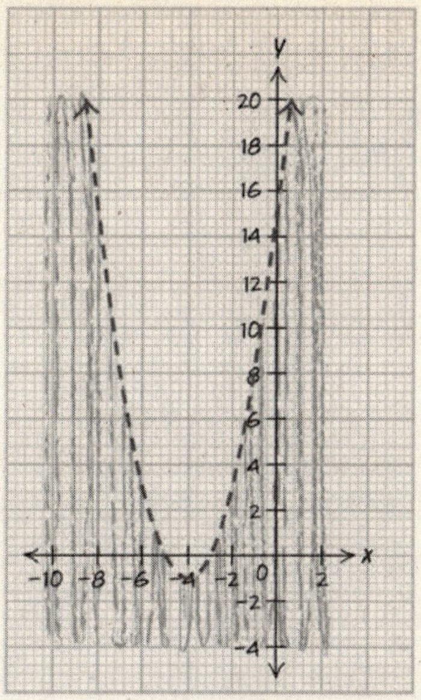

53. $y\le-2(x+3)^2+10$

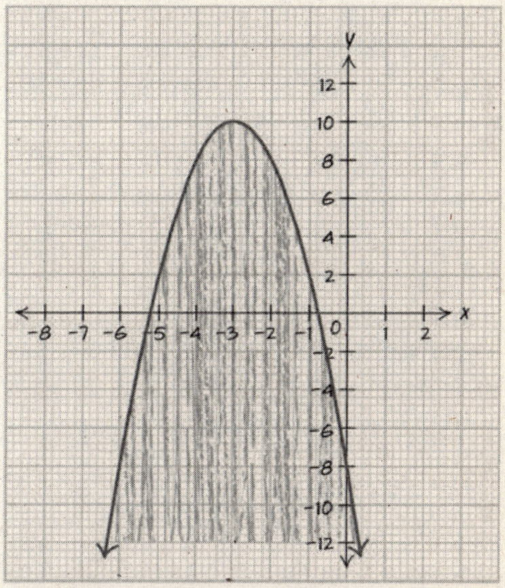

55. $y\ge1.5x^2+6x+10$

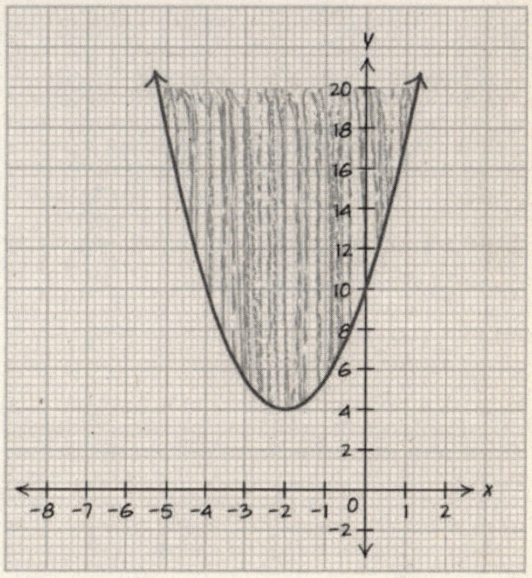

57. $y \geq -0.25x^2 + 3x - 2$

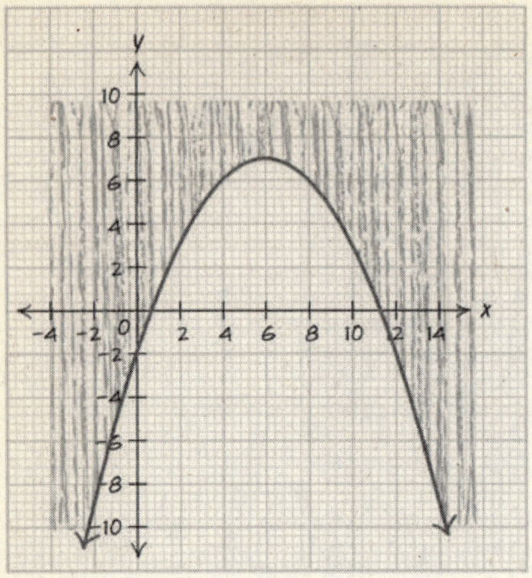

59. $y > 0.3(x-4)^2 - 6$

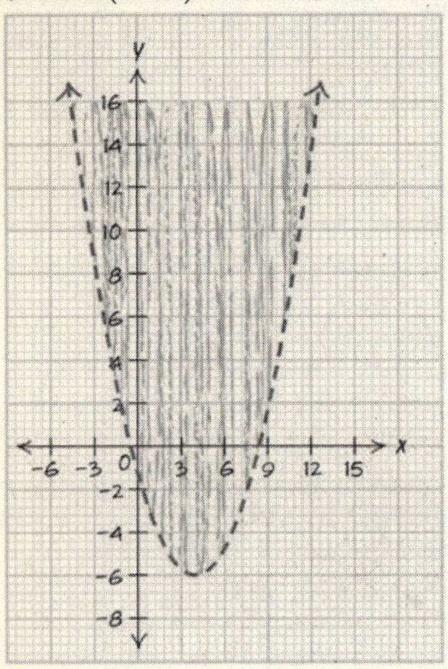

61. $y \leq 2.5x^2$

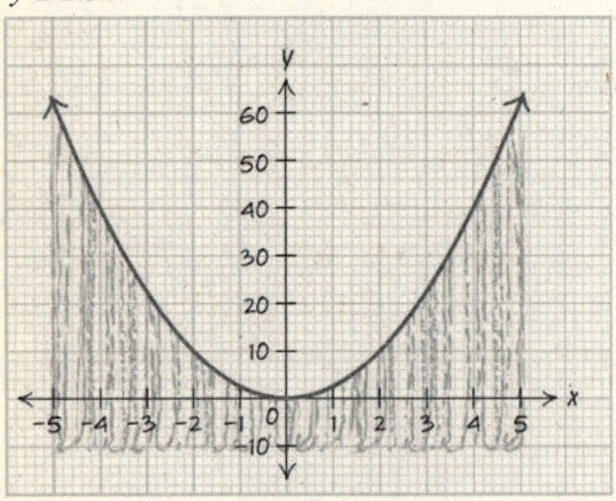

63. $y > 2(x+7)(x-4)$

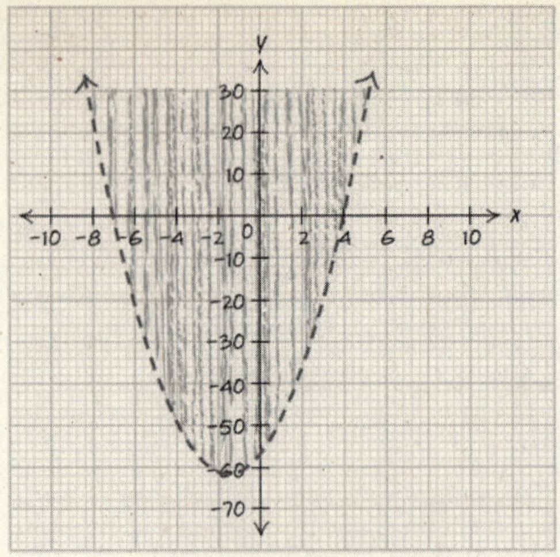

65. Note: Dashed boundary curve, shaded below, so use "$y <$" in the answer.

Use: $x = -4$, $x = 3$

with $f(0) = -12$.

$x + 4 = 0 \quad , \quad x - 3 = 0$

$f(x) = a(x+4)(x-3)$

$f(x) = a(x^2 + x - 12)$

$f(0) = a(-12) = -12$

$a = 1$

$f(x) = x^2 + x - 12$

$y < x^2 + x - 12$

67. Note: Solid boundary curve, shaded above so use "$y \geq$" in the answer.

Use: $x = -5$, $x = -1$

with $f(-3) = 2$.

$x + 5 = 0 \quad , \quad x + 1 = 0$

$f(x) = a(x+5)(x+1)$

$f(-3) = a(-3+5)(-3+1) = 2$

$a(2)(-2) = 2$

$-4a = 2 \rightarrow a = -0.5$

$f(x) = -0.5(x+5)(x+1)$

$y \geq -0.5(x+5)(x+1)$

or

$y \geq -0.5x^2 - 3x - 2.5$

Chapter 11 Review Exercises

1. **a.** Vertex: $(-1,9)$

b. $x<-1$

c. $x>-1$

d. $(-4,0)$, $(2,0)$

e. $(0,8)$

2. **a.** $x=-1$

b. a is negative, since the graph faces downward.

c. $h=-1$, $k=9$

d. $f(1)=5$

e. $x=-5.5$, $x=3.5$

3. **a.** $x=1$

b. a is positive, since the graph faces upward.

c. $h=1$, $k=-4$

d. $f(-1.5)=2.25$

e. $x=-1.6$, $x=3.6$

4. **a.** Vertex: $(1,-4)$

b. $x>1$

c. $x<1$

d. $(-1,0)$, $(3,0)$

e. $(0,-3)$

5. **a.** $P(t)=-0.89(t-5.6)^2+643$

$P(8)=-0.89(8-5.6)^2+643$

$P(8)=-0.89(2.4)^2+643$

$P(8)=637.87$

In 1998, the population of North Dakota was approximately 638 thousand.

b.

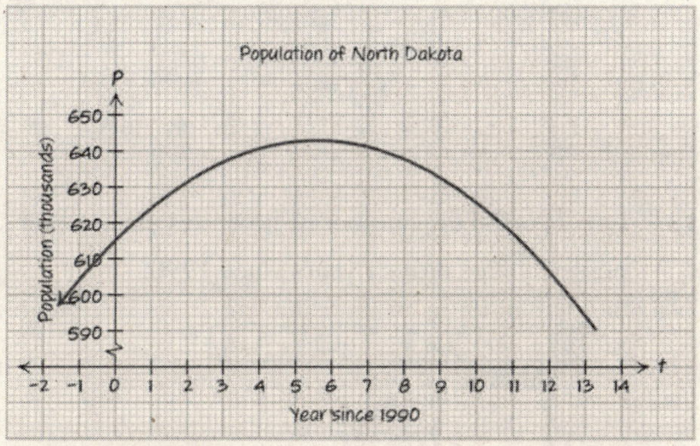

c. Vertex: $(5.6,643)$. In about 1996, the population of North Dakota reached a maximum of about 643 thousand people.

d. $-0.89(t-5.6)^2+643=640$

$\quad\quad\quad\quad\quad\; -643 \;\; -643$

$-0.89(t-5.6)^2=-3$

$\dfrac{-0.89(t-5.6)^2}{-0.89}=\dfrac{-3}{-0.89}$

$(t-5.6)^2\approx 3.3708$

$t-5.6\approx\pm\sqrt{3.3708}$

$t\approx 5.6+\sqrt{3.3708}\approx 7.44$

$t\approx 5.6-\sqrt{3.3708}\approx 3.76$

The population of North Dakota was about 640,000 in about 1993 and about 1997.

e. Domain: $0\le t\le 12$. Within this domain, the maximum value of 643 comes from the vertex. The minimum value comes from $t=12$.

$P(12)=-0.89(12-5.6)^2+643$

$P(12)\approx 606.55$

Range: $606.55\le P(t)\le 643$

6. **a.** $M(t)=-0.42(t-2.5)^2+23$

$M(6)=-0.42(6-2.5)^2+23$

$M(6)=-0.42(3.5)^2+23$

$M(6)=17.855$

There were approximately 17.9 thousand murders in the United States in 1996.

b. Vertex : $(2.5,23)$. In about 1993, the most murders occurred in the United States at 23 thousand.

c. $-0.42(t-2.5)^2+23=14.5$

$\quad\quad\quad\quad\quad\quad\; -23 \;\; -23$

$-0.42(t-2.5)^2=-8.5$

$\dfrac{-0.42(t-2.5)^2}{-0.42}=\dfrac{-8.5}{-0.42}$

$(t-2.5)^2\approx 20.2381$

$t-2.5\approx\pm\sqrt{20.2381}$

$t\approx 2.5+\sqrt{20.2381}\approx 7.00$

$t\approx 2.5-\sqrt{20.2381}\approx -2.00$

In 1997 and 1988, there were about 14,500 murders in the United States.

d. Domain: $[0,8]$. Within this domain, the maximum value of 23 comes from the vertex. The minimum value comes from $t=8$.

$M(8)=-0.42(8-2.5)^2+23$

$M(8)=10.295$

Range: $[10.295,23]$

7. $h(x)=(x+3)^2-16$

Domain: $(-\infty,\infty)$, range: $[-16,\infty)$

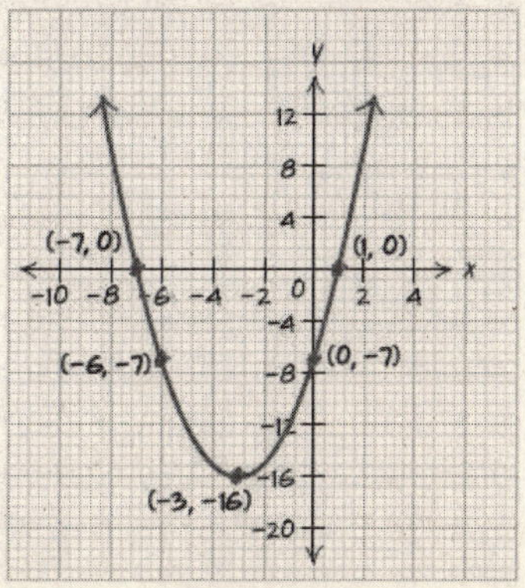

8. $f(x)=1.25(x-4)^2-20$

Domain: $(-\infty,\infty)$; range: $[-20,\infty)$

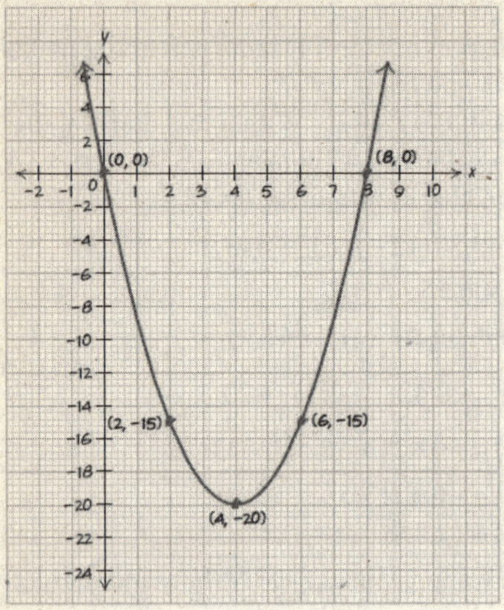

9. $g(x)=-\frac{1}{4}(x-8)^2+9$

Domain: $(-\infty,\infty)$; range: $(-\infty,9]$

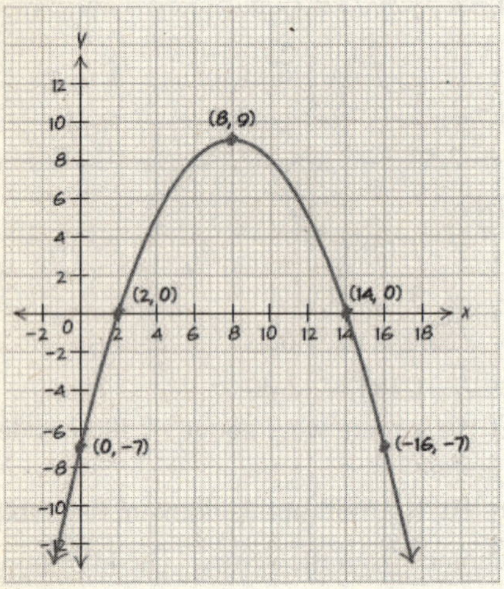

10. $h(x)=-2(x+4.5)^2+12.5$

Domain: $(-\infty,\infty)$; range: $(-\infty,12.5]$

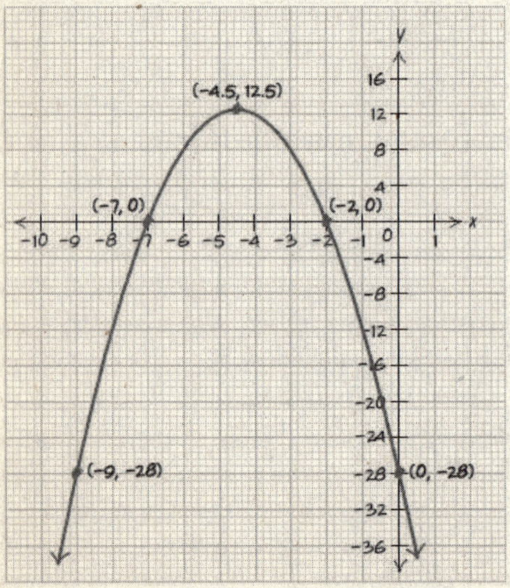

11. $g(x)=-0.5(x+7)^2-3$

Domain: $(-\infty,\infty)$; range: $(-\infty,-3]$

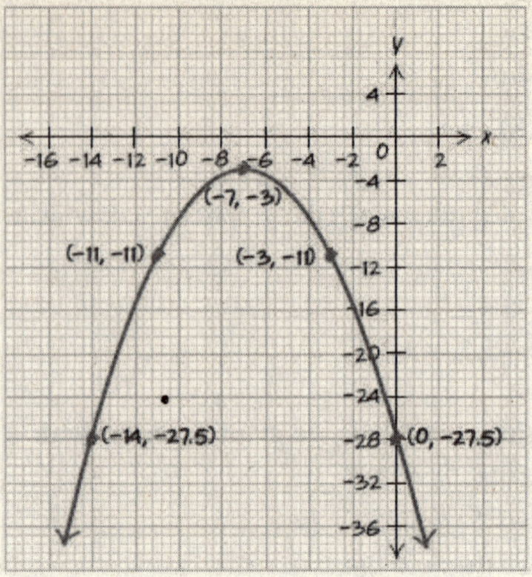

12. $f(x)=3(x-8)^2+4$

Domain: $(-\infty,\infty)$; range: $[4,\infty)$

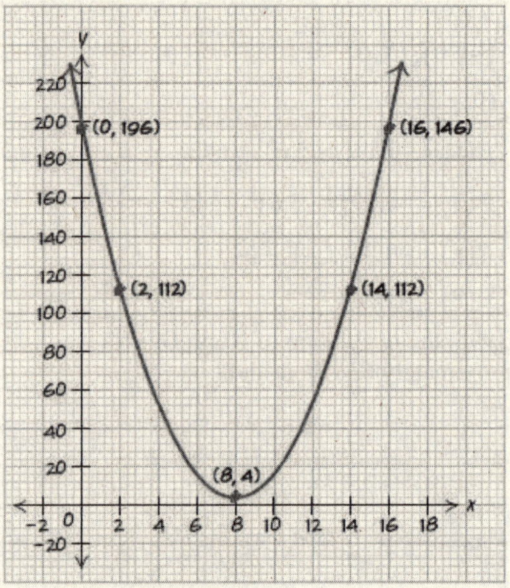

13. Let $J(t)$ be the number of juveniles arrested for possession of drugs t years since 1990.

a.

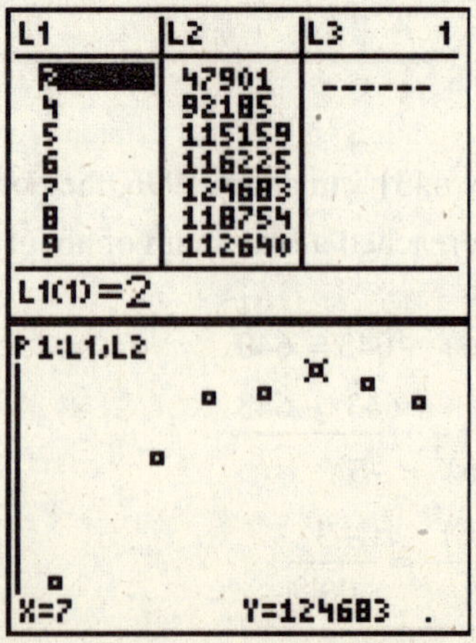

Choose $(7,124683)$ as the vertex, and $(2,47901)$ as the additional point used to solve for *"a"*.

$$J(t)=a(t-7)^2+124,683$$
$$47,901=a(2-7)^2+124,683$$
$$47,901=a(-5)^2+124,683$$
$$47,901=25a+124,683$$
$$\underline{-124,683 \qquad -124,683}$$
$$-76,782=25a$$
$$\left.\frac{-76,782}{25}=\frac{25a}{25}\right\}\ a=-3071.28$$
$$J(t)=-3071.28(t-7)^2+124,683$$

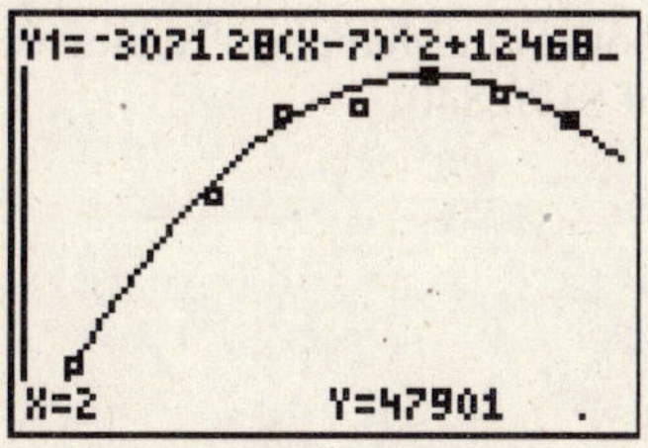

Note: Your answer may vary and yet still represent the trend of the data.

b. Domain: $[1,11]$. Within this domain, the maximum value of 124,683 comes from the vertex. The minimum value comes from $t=1$.

$$J(1)=-3071.28(1-7)^2+124,683$$
$$J(1)\approx 14,117$$
Range: $[14117,124683]$

c.

$$J(10)=-3071.28(10-7)^2+124,683$$
$$J(10)=-3071.28(3)^2+124,683$$
$$J(10)\approx 97,041$$

In 2000, there were approximately 97,041 juveniles arrested for drug possession.

d.

$$-3071.28(t-7)^2+124,683=50,000$$
$$\underline{\qquad -124,683 \ -124,683}$$
$$-3071.28(t-7)^2=-74,683$$
$$\frac{-3071.28(t-7)^2}{-3071.28}=\frac{-74,683}{-3071.28}$$
$$(t-7)^2\approx 24.3166$$
$$t-7\approx\pm\sqrt{24.3166}$$
$$t\approx 7+\sqrt{24.3166}\approx 11.9$$
$$t\approx 7-\sqrt{24.3166}\approx 2.1$$

The juvenile arrests were down to 50,000 again in about 2002.

14. Let $R(t)$ be the obligations of the U.S. Department of Commerce for research and development in millions of dollars t years since 1990.

a.

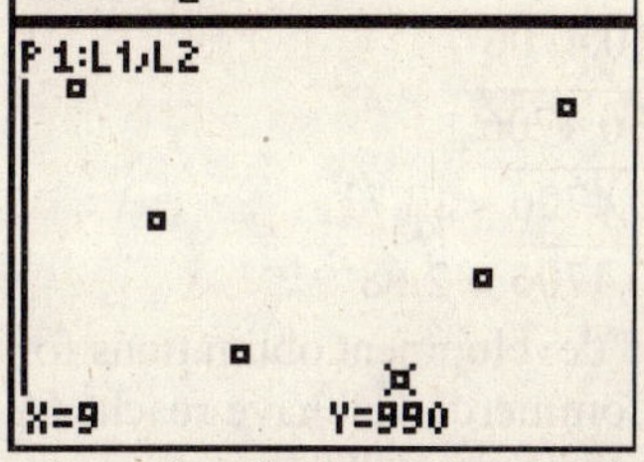

Note here that the lowest point does not look like a good choice for the vertex.

Choose $(8.2,982)$ (for example) as the vertex, and $(6,1068)$ as the additional point used to solve for *"a"*.

$$R(t)=a(t-8.2)^2+982$$
$$1068=a(6-8.2)^2+982$$
$$1068=a(-2.2)^2+982$$
$$1068=4.42a+982$$
$$\underline{-982 \qquad -982}$$
$$86=4.42a$$
$$\left.\frac{86}{4.42}=\frac{4.42a}{4.42}\right\}\ a\approx 17.8$$

Adjust "$a=17$" for a better fit.

$$R(t)=17(t-8.2)^2+982$$

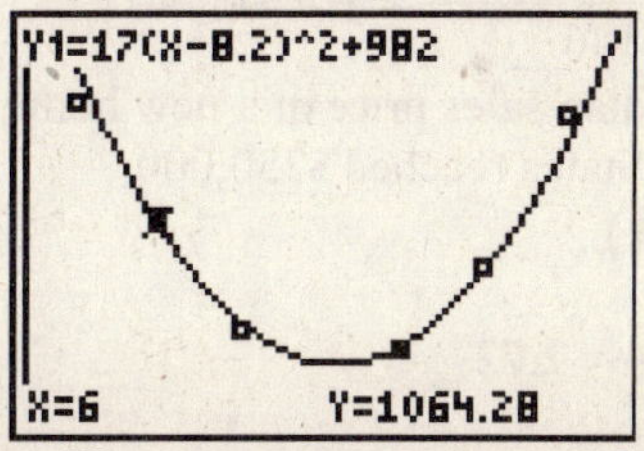

Note: Your answer may vary and yet still represent the trend of the data.

b. Domain: $[4,12]$. Within this domain, the minimum value of 982 comes from the vertex. The maximum value comes from $t=4$.

$$R(4)=17(4-8.2)^2+982$$
$$R(4)\approx 1282$$
Range: $[982,1282]$

c.

$$R(8)=17(8-8.2)^2+982$$
$$R(8)=17(-0.2)^2+982$$
$$R(8)=982.68$$

In 1998, the research and development obligations of the U.S. Department of Commerce, was approximately \$983 million.

d.

$$17(t-8.2)^2+982=1500$$
$$\underline{\quad -982 \quad -982}$$
$$17(t-8.2)^2=518$$
$$\frac{17(t-8.2)^2}{17}=\frac{518}{17}$$
$$(t-8.2)^2\approx 30.4706$$
$$t-8.2\approx \pm\sqrt{30.4706}$$
$$t\approx 8.2+\sqrt{30.4706}\approx 13.72$$
$$t\approx 8.2-\sqrt{30.4706}\approx 2.68$$

The research and development obligations for the U.S. Department of Commerce will have reached $1.5 billion in 1993 and again in 2004.

15. **a.** $P(t)=1.073t^2-5.84t+144.68$

$$P(5)=1.073(5)^2-5.84(5)+144.68$$
$$P(5)=142.31$$

In 1995, the median sales price of a new home in the western United States was $142,310.

b.

$$250=1.073t^2-5.84t+144.68$$
$$\underline{-250 \qquad\qquad\qquad -250}$$
$$0=1.073t^2-5.84t-105.32$$
$$t=\frac{5.84\pm\sqrt{(-5.84)^2-4(1.073)(-105.32)}}{2(1.073)}$$
$$t=\frac{5.84+\sqrt{486.13904}}{2.146}\approx 12.996$$
$$t=\frac{5.84-\sqrt{486.13904}}{2.146}\approx$$ ~~−7.553~~

In 2003, the median sales price of a new home in the western United States reached $250,000.

c. Vertex: (h,k).

$$h=\frac{5.84}{2(1.073)}\approx 2.72$$
$$k=P(2.72)=1.073(2.72)^2-5.84(2.72)+144.68$$
$$k\approx 136.73$$

Vertex: (2.72, 136.73). In about 2003, the median sales price of a new home in the western United States reached a low of $136,730.

16. **a.** $P(t)=-1.43t^2+19.89t+101.68$

$$P(4)=-1.43(4)^2+19.89(4)+101.68$$
$$P(4)=158.36$$

In 2004, the median asking price of houses in Memphis, Tennessee, was $158,360.

b.

$$165=-1.43t^2+19.89t+101.68$$
$$\underline{-165 \qquad\qquad\qquad -165}$$
$$0=-1.43t^2+19.89t-63.32$$
$$t=\frac{-19.89\pm\sqrt{19.89^2-4(-1.43)(-63.32)}}{2(-1.43)}$$
$$t=\frac{-19.89+\sqrt{33.4217}}{-2.86}\approx 4.933$$
$$t=\frac{-19.89-\sqrt{33.4217}}{-2.86}\approx 8.976$$

In 2005 and again in 2009, the median asking price of houses in Memphis, Tennessee, reached $165,000.

c. Vertex: (h,k).

$$h=\frac{-19.89}{2(-1.43)}\approx 6.95$$
$$k=P(6.95)=-1.43(6.95)^2+19.89(6.95)+101.68$$
$$k\approx 170.84$$

Vertex: (6.95, 170.84). In about 2007, the median asking price of houses in Memphis, Tennessee, reached a high of $170,840.

17.

$$t^2=169$$
$$\sqrt{t^2}=\pm\sqrt{169}$$
$$t=13,\ t=-13$$

18.

$$-6m^2+294=0$$
$$-6m^2=-294$$
$$m^2=49$$
$$\sqrt{m^2}=\pm\sqrt{49}$$
$$m=7,\ m=-7$$

19.

$$m^2-4=-20$$
$$m^2=-16$$
$$m=\pm\sqrt{-16}$$
$$m=\pm 4i$$

20.

$$4(x+7)^2-36=0$$
$$4(x+7)^2=36$$
$$(x+7)^2=9$$
$$\sqrt{(x+7)^2}=\pm\sqrt{9}$$
$$x+7=\pm 3$$
$$x+7=3,\ x+7=-3$$
$$x=-4,\ x=-10$$

21.

$$-0.25(x-6)^2+8=0$$
$$-0.25(x-6)^2=-8$$
$$(x-6)^2=\frac{-8}{-0.25}$$
$$(x-6)^2=32$$
$$\sqrt{(x-6)^2}=\pm\sqrt{32}$$
$$x-6=\pm\sqrt{32}$$
$$x-6=\sqrt{32},\ x-6=-\sqrt{32}$$
$$x=6+\sqrt{32},\ x=6-\sqrt{32}$$
$$x\approx 11.66,\ x\approx 0.34$$

22.

$$
\begin{aligned}
2(a-7)^2+11&=5\\
2(a-7)^2&=-6\\
(a-7)^2&=-3\\
\sqrt{(a-7)^2}&=\pm\sqrt{-3}\\
a-7&=\pm i\sqrt{3}\\
a&=7\pm i\sqrt{3}
\end{aligned}
$$

23.

$$
\begin{aligned}
5(g-9)^2+25&=5\\
5(g-9)^2&=-20\\
(g-9)^2&=-4\\
\sqrt{(g-9)^2}&=\pm\sqrt{-4}\\
g-9&=\pm 2i\\
g&=9\pm 2i
\end{aligned}
$$

24.

$$
\begin{aligned}
&\frac{1}{2}(c+6)^2-\frac{5}{2}=0\\
&\frac{1}{2}(c+6)^2=\frac{5}{2}\\
&2\left(\frac{1}{2}(c+6)^2\right)=2\left(\frac{5}{2}\right)\\
&(c+6)^2=5\\
&\sqrt{(c+6)^2}=\pm\sqrt{5}\\
&c+6=\pm\sqrt{5}\\
&c+6=\sqrt{5}, c+6=-\sqrt{5}\\
&c=-6+\sqrt{5}, c=-6-\sqrt{5}\\
&c\approx-3.76, c\approx-8.24
\end{aligned}
$$

25.

$$
\begin{aligned}
&\frac{2}{7}(p-4)^2-\frac{3}{14}=0\\
&\frac{2}{7}(p-4)^2=\frac{3}{14}\\
&\frac{7}{2}\left(\frac{2}{7}(p-4)^2\right)=\frac{7}{2}\left(\frac{3}{14}\right)\\
&(p-4)^2=\frac{3}{4}\\
&\sqrt{(p-4)^2}=\pm\sqrt{\frac{3}{4}}\\
&p-4=\pm\sqrt{\frac{3}{4}}\\
&p-4=\sqrt{\frac{3}{4}}, p-4=-\sqrt{\frac{3}{4}}\\
&p=4+\sqrt{\frac{3}{4}}, p=4-\sqrt{\frac{3}{4}}\\
&p\approx 4.87, p\approx 3.13
\end{aligned}
$$

26. $a^2+b^2=c^2$ Substitute given values.

$$
\begin{aligned}
(7)^2+b^2&=(15)^2\\
49+b^2&=225\\
-49\quad&\;\;-49\\
b^2&=176\\
b&=\pm\sqrt{176}\\
b&\approx 13.27 \text{ (Use positive value for length.)}
\end{aligned}
$$

27. $a^2+b^2=c^2$ Substitute given values.

$$
\begin{aligned}
a^2+(5)^2&=(11)^2\\
a^2+25&=121\\
-25\;&\;-25\\
a^2&=96\\
a&=\pm\sqrt{96}\\
a&\approx 9.80 \text{ (Use positive value for length.)}
\end{aligned}
$$

28.

$$
\begin{aligned}
&x^2+26x=30\\
&x^2+26x+\_\_=30+\_\_\\
&\text{Completing the square:}\left(\frac{26}{2}\right)^2=169\\
&x^2+26x+169=30+169\\
&(x+13)(x+13)=199\\
&(x+13)^2=199\\
&\sqrt{(x+13)^2}=\pm\sqrt{199}\\
&x+13=\pm\sqrt{199}\\
&x+13=\sqrt{199}, x+13=-\sqrt{199}\\
&x=-13+\sqrt{199}, x=-13-\sqrt{199}\\
&x\approx 1.11, x\approx -27.11
\end{aligned}
$$

29.

$$
\begin{aligned}
&x^2-12x-13=0\\
&x^2-12x=13\\
&x^2-12x+\_\_=13+\_\_\\
&\text{Completing the square:}\left(\frac{-12}{2}\right)^2=36\\
&x^2-12x+36=13+36\\
&(x-6)(x-6)=49\\
&(x-6)^2=49\\
&\sqrt{(x-6)^2}=\pm\sqrt{49}\\
&x-6=\pm 7\\
&x-6=7, x-6=-7\\
&x=13, x=-1
\end{aligned}
$$

30.

$$
\begin{aligned}
&3x^2+15x=198\\
&\frac{3x^2+15x}{3}=\frac{198}{3}\\
&x^2+5x=66\\
&x^2+5x+\_\_=66+\_\_\\
&\text{Completing the square:}\left(\frac{5}{2}\right)^2=\frac{25}{4}
\end{aligned}
$$

$$x^2+5x+\frac{25}{4}=66+\frac{25}{4}$$
$$\left(x+\frac{5}{2}\right)\left(x+\frac{5}{2}\right)=\frac{289}{4}$$
$$\left(x+\frac{5}{2}\right)^2=\frac{289}{4}$$
$$\sqrt{\left(x+\frac{5}{2}\right)^2}=\pm\sqrt{\frac{289}{4}}$$
$$x+\frac{5}{2}=\pm\frac{17}{2}$$
$$x+\frac{5}{2}=\frac{17}{2},x+\frac{5}{2}=-\frac{17}{2}$$
$$x=6,x=-11$$

31.

$$4x^2-6x+20=0$$
$$4x^2-6x=-20$$
$$\frac{4x^2-6x}{4}=\frac{-20}{4}$$
$$x^2-\frac{3}{2}x=-5$$
$$x^2-\frac{3}{2}x+\_\_=-5+\_\_$$

Completing the square: $\left(\frac{-\frac{3}{2}}{2}\right)^2=\frac{9}{16}$

$$x^2-\frac{3}{2}x+\frac{9}{16}=-5+\frac{9}{16}$$
$$\left(x-\frac{3}{4}\right)\left(x-\frac{3}{4}\right)=-\frac{71}{16}$$
$$\left(x-\frac{3}{4}\right)^2=-\frac{71}{16}$$
$$\sqrt{\left(x-\frac{3}{4}\right)^2}=\pm\sqrt{-\frac{71}{16}}$$
$$x-0.75=\pm\sqrt{-4.4375}$$
$$x\approx 0.75\pm 2.11i$$

32.

$$f(x)=x^2+8x+11$$
$$f(x)=\left(x^2+8x+\_\_\right)+11-\_\_$$

Completing the square: $\left(\frac{8}{2}\right)^2=16$

$$f(x)=\left(x^2+8x+16\right)+11-16$$
$$f(x)=(x+4)(x+4)-5$$
$$f(x)=(x+4)^2-5$$

33.

$$g(x)=-6x^2+20x-18$$
$$g(x)=-6\left(x^2-\frac{10}{3}x\right)-18$$
$$g(x)=-6\left(x^2-\frac{10}{3}x+\_\_-\_\_\right)-18$$

Completing the square: $\left(\frac{-\frac{10}{3}}{2}\right)^2=\frac{25}{9}$

$$g(x)=-6\left(x^2-\frac{10}{3}x+\frac{25}{9}-\frac{25}{9}\right)-18$$
$$g(x)=-6\left(x^2-\frac{10}{3}x+\frac{25}{9}\right)-(-6)\left(\frac{25}{9}\right)-18$$
$$g(x)=-6\left(x-\frac{5}{3}\right)\left(x-\frac{5}{3}\right)+\frac{50}{3}-18$$
$$g(x)=-6\left(x-\frac{5}{3}\right)^2-\frac{4}{3}$$

34. Vertex $=(-1,-9)$; Other point $=(2,0)$

$$y=a(x-h)^2+k$$
where $h=-1,k=-9,x=2,y=0$
$$0=a\left(2-(-1)\right)^2+(-9)$$
$$0=a(9)-9$$
$$9=9a$$
$$a=1$$
Since $a=1,h=-1,k=-9$
$$y=1\left(x-(-1)\right)^2+(-9)$$
$$y=(x+1)^2-9$$

35. Vertex $=(-1,12)$; Other point $=(1,0)$

$$y=a(x-h)^2+k$$
where $h=-1,k=12,x=1,y=0$
$$0=a\left(1-(-1)\right)^2+12$$
$$0=a(4)+12$$
$$-12=4a$$
$$a=-3$$
Since $a=-3,h=-1,k=12$
$$y=-3\left(x-(-1)\right)^2+12$$
$$y=-3(x+1)^2+12$$

36. $x^2-4x-12=0$

$$a=1,b=-4,c=-12$$
$$x=\frac{-(-4)\pm\sqrt{(-4)^2-4(1)(-12)}}{2(1)}$$
$$x=\frac{4\pm\sqrt{64}}{2}$$
$$x=\frac{4\pm 8}{2}$$
$$x=\frac{4-8}{2},x=\frac{4+8}{2}$$
$$x=\frac{-4}{2},x=\frac{12}{2}$$
$$x=-2,x=6$$

37. $2t^2+11t-63=0$

$$a=2,b=11,c=-63$$

$$t=\frac{-11\pm\sqrt{11^2-4(2)(-63)}}{2(2)}$$
$$t=\frac{-11\pm\sqrt{625}}{4}$$
$$t=\frac{-11\pm 25}{4}$$
$$t=\frac{-11-25}{4},t=\frac{-11+25}{4}$$
$$t=-9,t=\frac{7}{2}$$

38. $x^2+6x+18=0$
$a=1,b=6,c=18$
$$x=\frac{-6\pm\sqrt{6^2-4(1)(18)}}{2(1)}$$
$$x=\frac{-6\pm\sqrt{-36}}{2}$$
$$x=\frac{-6\pm 6i}{2}$$
$$x=-3\pm 3i$$

39. $3x^2+9x-20=14$
$3x^2+9x-34=0$
$a=3,b=9,c=-34$
$$x=\frac{-9\pm\sqrt{9^2-4(3)(-34)}}{2(3)}$$
$$x=\frac{-9\pm\sqrt{489}}{6}$$
$$x=\frac{-9-\sqrt{489}}{6},x=\frac{-9+\sqrt{489}}{6}$$
$$x\approx-5.19,x\approx 2.19$$

40. $-4.5x^2+3.5x+12.5=0$
$a=-4.5,b=3.5,c=12.5$
$$x=\frac{-3.5\pm\sqrt{3.5^2-4(-4.5)(12.5)}}{2(-4.5)}$$
$$x=\frac{-3.5\pm\sqrt{237.25}}{-9}$$
$$x=\frac{-3.5+\sqrt{237.25}}{-9},x=\frac{-3.5-\sqrt{237.25}}{-9}$$
$$x\approx-1.32,x\approx 2.10$$

41. $3.25n^2-4.5n-42.75=0$
$a=3.25,b=-4.5,c=-42.75$
$$n=\frac{-(-4.5)\pm\sqrt{(-4.5)^2-4(3.25)(-42.75)}}{2(3.25)}$$
$$n=\frac{4.5\pm\sqrt{576}}{6.5}$$
$$n=\frac{4.5\pm 24}{6.5}$$
$$n=\frac{4.5-24}{6.5},n=\frac{4.5+24}{6.5}$$
$$n=\frac{-19.5}{6.5},n=\frac{28.5}{6.5}=\frac{285}{65}$$
$$n=-3,n=\frac{57}{13}$$

42. $3x^2+8x=-2$
$3x^2+8x+2=0$
$a=3,b=8,c=2$
$$x=\frac{-8\pm\sqrt{8^2-4(3)(2)}}{2(3)}$$
$$x=\frac{-8\pm\sqrt{40}}{6}$$
$$x=\frac{-8-\sqrt{40}}{6},x=\frac{-8+\sqrt{40}}{6}$$
$$x\approx-2.39,x\approx-0.28$$

43. $2a^2+6a-10=6a+20$
$2a^2-30=0$
$a=2,b=0,c=-30$
$$a=\frac{0\pm\sqrt{0^2-4(2)(-30)}}{2(2)}$$
$$a=\frac{\pm\sqrt{240}}{4}$$
$$a\approx 3.87,\ a\approx-3.87$$

44. $3x^2+12x+15=2$
$3x^2+12x+13=0$
$$x=\frac{-12\pm\sqrt{12^2-4(3)(13)}}{2(3)}$$
$$x=\frac{-12\pm\sqrt{-12}}{6}$$
$$x=\frac{-12\pm 2i\sqrt{3}}{6}$$
$$x=-2\pm\frac{\sqrt{3}}{3}i$$

45. $-b^2-4b+6=15$
$0=b^2+4b+9$
$$b=\frac{-4\pm\sqrt{4^2-4(1)(9)}}{2(1)}$$
$$b=\frac{-4\pm\sqrt{-20}}{2}$$
$$b=\frac{-4\pm 2i\sqrt{5}}{2}$$
$$b=-2\pm i\sqrt{5}$$

46. $t^3-6t^2=-13t$
$t^3-6t^2+13t=0$
$t(t^2-6t+13)=0$
$t=0,t^2-6t+13=0$
$$t=0,t=\frac{6\pm\sqrt{(-6)^2-4(1)(13)}}{2(1)}$$
$$t=0,t=\frac{6\pm\sqrt{-16}}{2}$$
$$t=0,t=\frac{6\pm 4i}{2}$$
$$t=0,t=3\pm 2i$$

47. $h^3+8h^2+65h=0$

$h\left(h^2+8h+65\right)=0$

$h=0, h=\dfrac{-8\pm\sqrt{8^2-4(1)(65)}}{2(1)}$

$h=0, h=\dfrac{-8\pm\sqrt{-196}}{2}$

$h=0, h=\dfrac{-8\pm 14i}{2}$

$h=0, h=-4\pm 7i$

48. **a.** $4500=-2x^2+225x$

b. $-2x^2+225x-4500=0$

$x=\dfrac{-(225)\pm\sqrt{(225)^2-4(-2)(-4500)}}{2(-2)}$

$x=\dfrac{-225\pm\sqrt{50{,}625-36{,}000}}{-4}$

$x=\dfrac{-225\pm\sqrt{14{,}625}}{-4}$

$x=\dfrac{-225+\sqrt{14{,}625}}{-4}$ $\quad x=\dfrac{-225-\sqrt{14{,}625}}{-4}$

$x\approx 26.02$ $\quad x\approx 86.48$

c. The photographer must sell at least 27 prints to make a profit of $4,500.00.

49. **a.**

$$\begin{aligned} 13&=-4.9t^2+18\\ -18&\qquad\quad -18\\ \hline -5&=-4.9t^2 \end{aligned}$$

$\dfrac{-5}{-4.9}=\dfrac{-4.9t^2}{-4.9}$

$\dfrac{5}{4.9}=t^2$

$t=\pm\sqrt{\dfrac{5}{4.9}}$

$t\approx 1.01$ (use only positive value for time)

The rock will fall to a height of 13 meters in approximately 1.01 seconds.

b.

$$\begin{aligned} 0&=-4.9t^2+18\\ -18&\qquad\quad -18\\ \hline -18&=-4.9t^2 \end{aligned}$$

$\dfrac{-18}{-4.9}=\dfrac{-4.9t^2}{-4.9}$

$\dfrac{18}{4.9}=t^2$

$t=\pm\sqrt{\dfrac{18}{4.9}}$

$t\approx 1.92$ (use only positive value for time)

The rock will hit the ground in approximately 1.92 seconds.

50. Domain: $(-\infty,\infty)$, range: $(-\infty,2.5]$

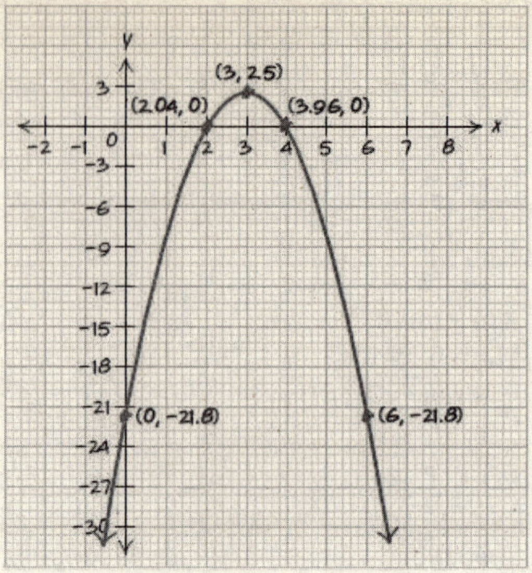

51. Domain: $(-\infty,\infty)$, range: $[7,\infty)$

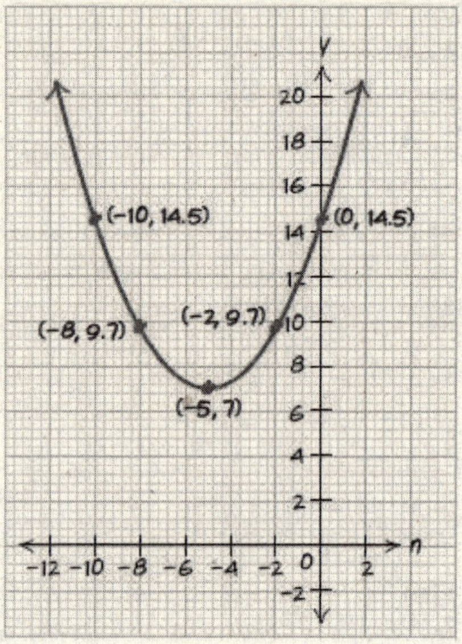

52. Domain: $(-\infty,\infty)$, range: $[-15,\infty)$

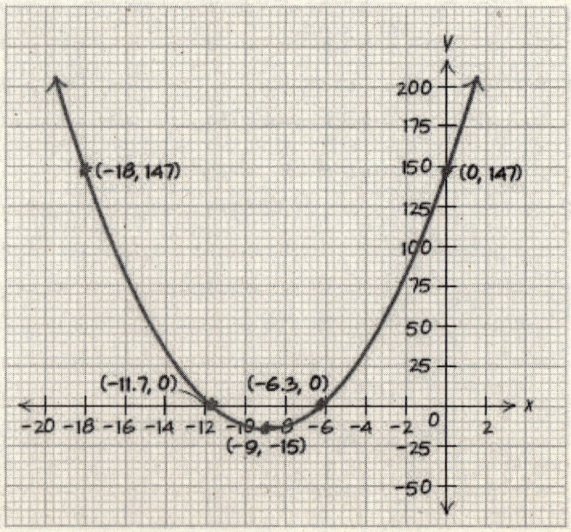

53. Domain: $(-\infty,\infty)$, range: $(-\infty,28]$

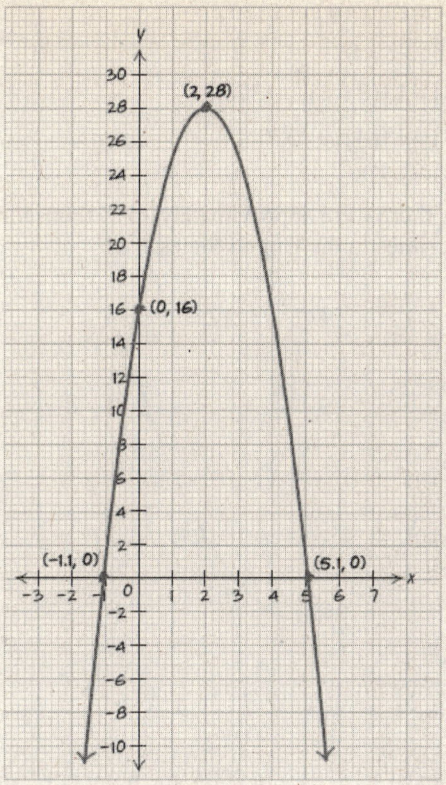

54.

$$5x-9=-2x^2+7x+15$$
$$\underline{-5x+9 \quad -5x+9}$$
$$0=-2x^2+2x+24$$
$$\frac{0}{-2}=\frac{-2x^2+2x+24}{-2}$$
$$0=x^2-x-12$$
$$0=(x+3)(x-4)$$
$$x+3=0\ ,\ x-4=0$$
$$x=-3\quad ,\ x=4$$
$$y=5(-3)-9=-24$$
$$y=5(4)-9=11$$
$$(-3,-24),\ (4,11)$$

55.

$$-4x+3=x^2-8x-9$$
$$\underline{+4x-3 \qquad +4x-3}$$
$$0=x^2-4x-12$$
$$0=(x+2)(x-6)$$
$$x+2=0\ ,\ x-6=0$$
$$x=-2\ ,\ x=6$$
$$y=-4(-2)+3=11$$
$$y=-4(6)+3=-21$$
$$(-2,11),\ (6,-21)$$

56.

$$6x^2-3x+10=4x^2+8x+1$$
$$\underline{-4x^2-8x-1 \quad -4x^2-8x-1}$$
$$2x^2-11x+9=0$$
$$(2x-9)(x-1)=0$$
$$2x-9=0\rightarrow x=\frac{9}{2}=4.5$$
$$y=4(4.5)^2+8(4.5)+1=118$$
$$x-1=0\rightarrow x=1$$
$$y=4(1)^2+8(1)+1=13$$
$$(1,13)\ ,\ (4.5,118)$$

57.

$$0.25x^2+4x-7=0.5x^2+3x-9$$
$$\underline{-0.25x^2-4x+7 \quad -0.25x^2-4x+7}$$
$$0=0.25x^2-x-2$$
$$x=\frac{1\pm\sqrt{(-1)^2-4(0.25)(-2)}}{2(0.25)}$$
$$x=\frac{1+\sqrt{3}}{0.5}\approx 5.46\ ,\ x=\frac{1-\sqrt{3}}{0.5}\approx -1.46$$
$$y\approx 0.5(5.46)^2+3(5.46)-9\approx 22.32$$
$$y\approx 0.5(-1.46)^2+3(-1.46)-9\approx -12.32$$
$$(-1.46,-12.32)\ ,\ (5.46,22.32)$$

58. $(3,9),\ (5,9)$

59. **a.** $2w+l=100\rightarrow l=100-2w$

$A=lw\rightarrow A=(100-2w)w$

Use function notation.

$A(w)=-2w^2+100w$

Find the vertex of this function.

$$\frac{-b}{2a}=\frac{-100}{4}=-25$$
$$A(-25)=(-25)^2+100(-25)$$
$$A(-25)=1250$$

The largest possible rectangular area is $1250\,\text{ft}^2$.

b. The largest possible area requires a length of 50 feet and a width of 25 feet.

60. $y<2x^2+24x-10$

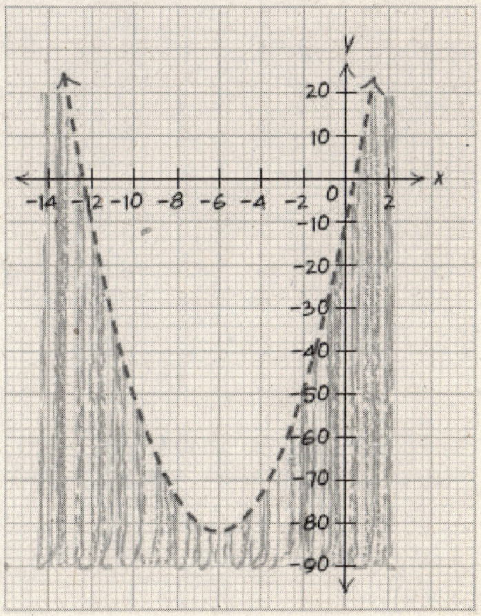

61. $y \ge 0.5x^2 - 4x - 6$

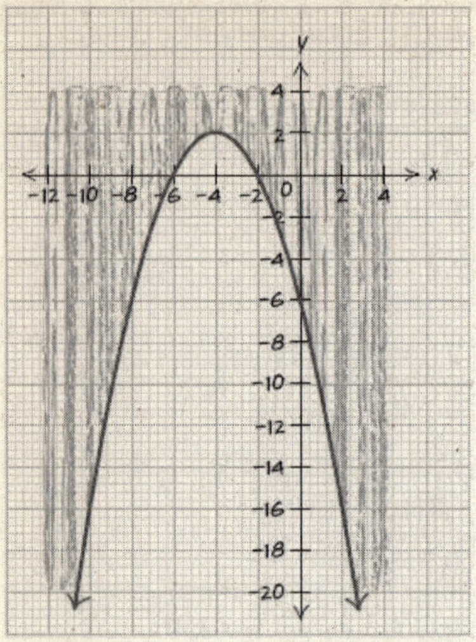

62. $y \le 2.5(x-4)^2 - 18$

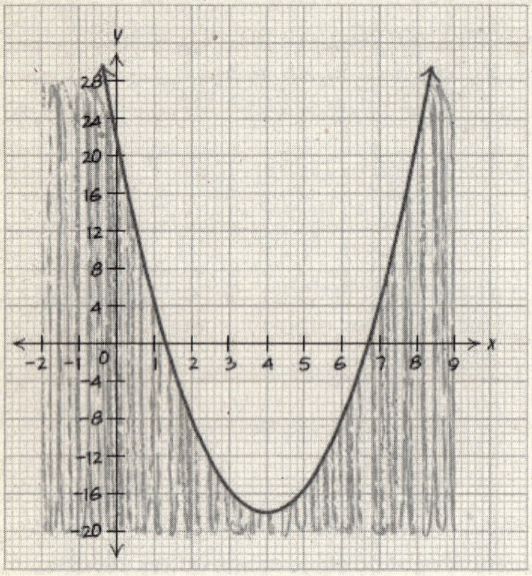

63. $y > -0.8(x+1)^2 - 3$

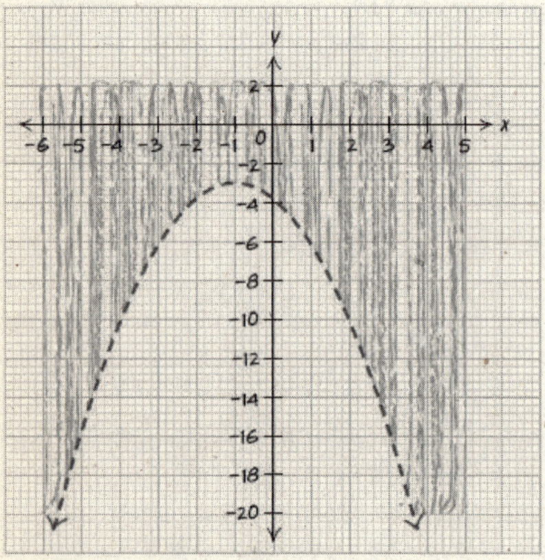

64. $(-4,3)$, $(2,6)$

65. $(-4,0)$, $(2,9)$

Chapter 11 Test

1. **a.** $V(t) = 13(t-4)^2 + 1370$

$2008 \to t = 8$

$V(8) = 13(8-4)^2 + 1370$

$V(8) = 13(4)^2 + 1370$

$V(8) = 1578$

In 2008, there were approximately 1578 thousand violent crimes in the United States.

b.

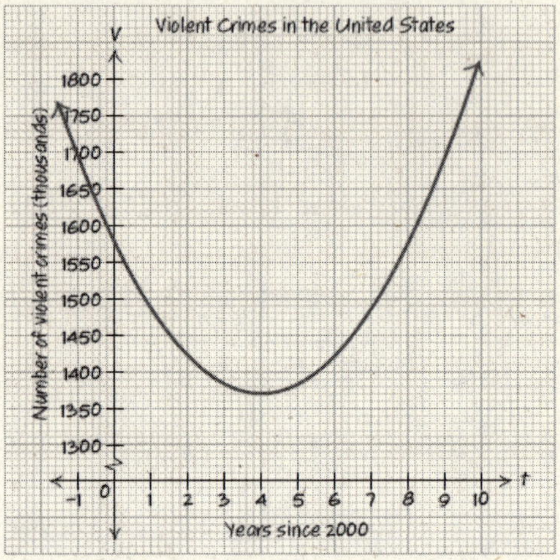

c. Vertex: $(4,1370)$. In 2004, the number of violent crimes in the United States was at a minimum of 1370 thousand.

d. According to this graph, there were 1.5 million violent crimes in 2001 and again in 2007.

2. $(-6,4)$ and $(4,8)$

3. $3x^2 - 5x + 5 = 33$

$3x^2 - 5x - 28 = 0$

$(3x+7)(x-4) = 0$

$3x + 7 = 0, x - 4 = 0$

$3x = -7, x = 4$

$x = -\frac{7}{3}, x = 4$

4. $8x^2 - 34x = -35$

$8x^2 - 34x + 35 = 0$

$(4x-7)(2x-5) = 0$

$4x - 7 = 0, 2x - 5 = 0$

$4x = 7, 2x = 5$

$x = \frac{7}{4}, x = \frac{5}{2}$

5. **a.** Let $O(t)$ be the total outlays in billions of dollars for national defense and veterans benefits by the United States t years since 1990.

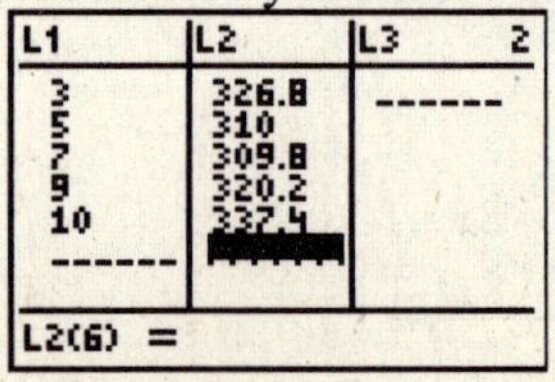

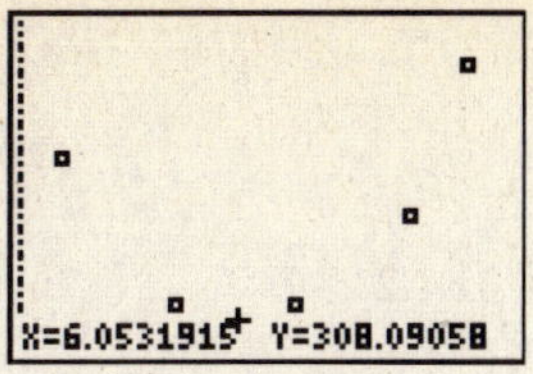

Choose $(6,308)$ as the vertex, and $(10,337.4)$ as the additional point used to solve for *"a"*.

$$O(t)=a(t-6)^2+308$$
$$337.4=a(10-6)^2+308$$
$$337.4=a(4)^2+308$$
$$337.4=16a+308$$
$$\underline{-308 \qquad -308}$$
$$29.4=16a$$
$$\frac{29.4}{16}=\frac{16a}{16}$$
$$a=1.8375$$

Adjusted to $a=1.9$.

$$O(t)=1.9(t-6)^2+308$$

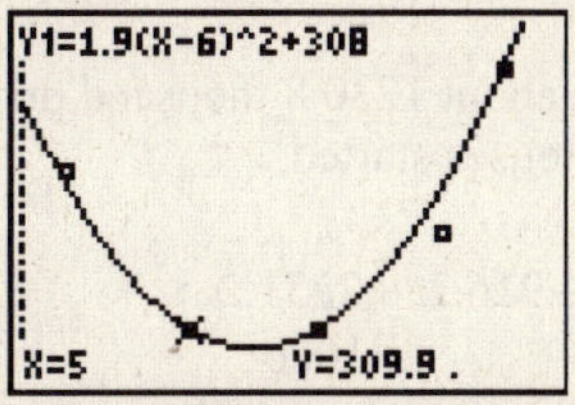

Note: Your answer may vary and yet still represent the trend of the data.

b. Domain: $[1,11]$. Within this domain, the minimum value of 308 comes from the vertex. The maximum value comes from $t=1$.

$$O(1)=1.9(1-6)^2+308$$
$$O(1)=355.5$$

Range: $[308, 355.5]$

c.

$$O(8)=1.9(8-6)^2+308$$
$$O(8)=315.6$$

In 1998, the total outlays for national defense and veterans benefits by the United States were about $315.6 billion.

d.

$$1.9(t-6)^2+308=500$$
$$\underline{-308 \quad -308}$$
$$1.9(t-6)^2=192$$
$$\frac{1.9(t-6)^2}{1.9}=\frac{192}{1.9}$$
$$(t-6)^2\approx 101.0526$$
$$t-6\approx \pm\sqrt{101.0526}$$
$$t\approx 6+\sqrt{101.0526}\approx 16.05$$

$t\approx 6-\sqrt{101.0526}\approx$ ~~−4.05~~

Total outlays for national defense and veterans benefits by the United States will reach half a trillion dollars in about 2006.

6. **a.** Let $R(t)$ be the revenue for the U.S. commercial space industry from satellite manufacturing in billions of dollars t years since 1990.

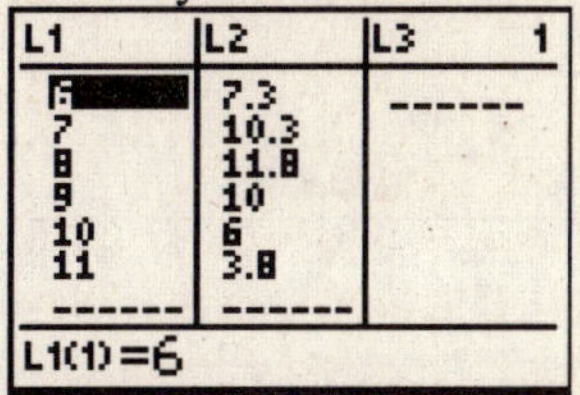

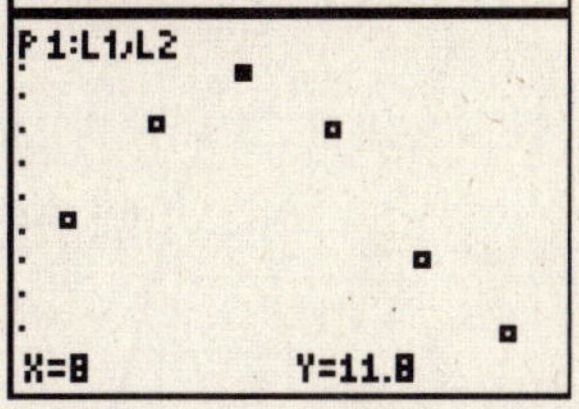

Choose $(8,11.8)$ as the vertex, and $(6,7.3)$ as the additional point used to solve for *"a"*.

$$R(t)=a(t-8)^2+11.8$$
$$7.3=a(6-8)^2+11.8$$
$$7.3=a(-2)^2+11.8$$
$$7.3=4a+11.8$$
$$\underline{-11.8 \qquad -11.8}$$
$$-4.5=4a$$
$$\left.\frac{-4.5}{4}=\frac{4a}{4}\right\} \quad a=-1.125$$
$$R(t)=-1.125(t-8)^2+11.8$$

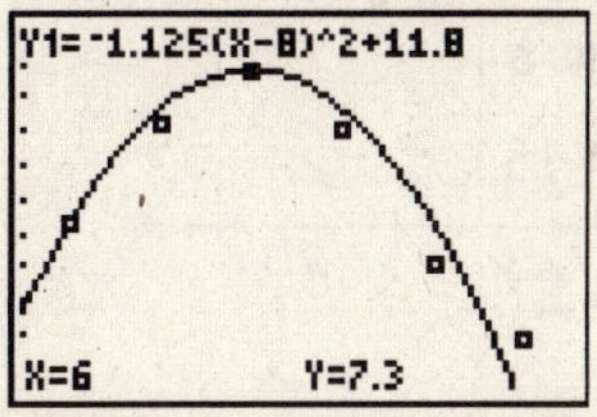

Note: Your answer may vary and yet still represent the trend of the data.

b. Domain: $[6,10]$. Within this domain, the maximum value of 11.8 comes from the vertex. The minimum value comes from $t=10$.

$$R(10)=-1.125(10-8)^2+11.8$$
$$R(10)=7.3$$

Range: $[7.3, 11.8]$

In this case, we did not go back from the data because it resulted in a very small revenue in 1995, and that seemed like model breakdown.

c.

$$R(10)=-1.125(10-8)^2+11.8$$
$$R(10)=-1.125(2)^2+11.8$$
$$R(10)=7.3$$

In 2000, the revenue from satellite manufacturing was about $7.3 billion.

d. Vertex: $(8,11.8)$. In 1998, the revenue from satellite manufacturing reached a maximum of $11.8 billion.

7. Domain: $(-\infty,\infty)$, range: $(-\infty,3.5]$

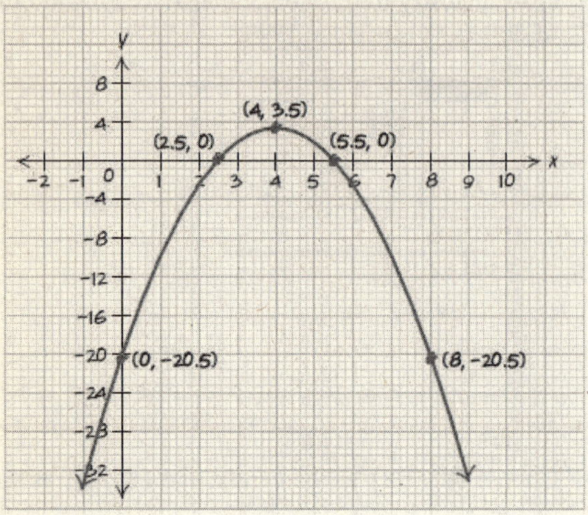

8. Domain: $(-\infty,\infty)$, range: $[10,\infty)$

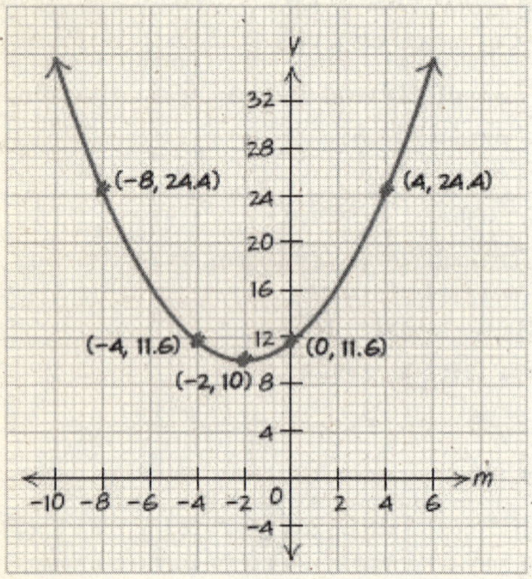

9. Domain: $(-\infty,\infty)$, range: $(-\infty,-17]$

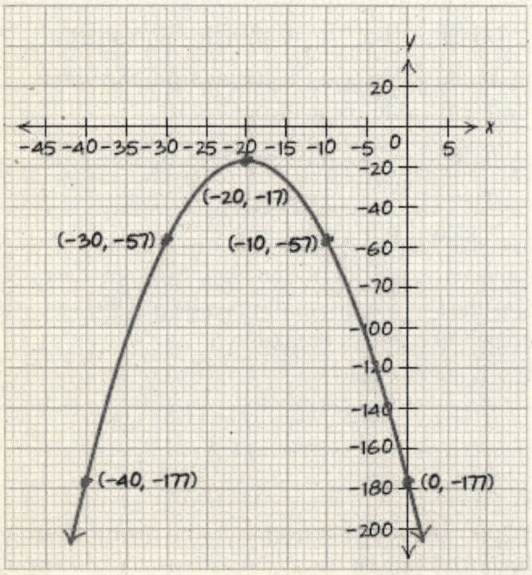

10.

$$\begin{cases} y = 2x^2+5x-9 \\ y = -7x^2+3x+2 \end{cases}$$

$$2x^2+5x-9=-7x^2+3x+2$$
$$9x^2+2x-11=0$$
$$(x-1)(9x+11)=0$$
$$x-1=0, 9x+11=0$$
$$x=1, x=-\frac{11}{9}$$

Substitute x back into either equation solve for y:

$$y=2(1)^2+5(1)-9=-2$$
$$y=2\left(-\frac{11}{9}\right)^2+5\left(-\frac{11}{9}\right)-9=-\frac{982}{81}$$

The solutions to the system are $(1,-2)$ and $\left(-\frac{11}{9},-\frac{982}{81}\right)$.

11. a. $H(t)=-52t^2+916.2t-2731.2$

$$H(10)=-52(10)^2+916.2(10)-2731.2$$
$$H(10)=1230.8$$

In 2000, there were about 1230.8 thousand privately owned single-unit houses started.

b.

$$1000=-52t^2+916.2t-2731.2$$
$$-1000 \qquad\qquad -1000$$
$$0=-52t^2+916.2t-3731.2$$
$$t=\frac{-916.2\pm\sqrt{916.2^2-4(-52)(-3731.2)}}{2(-52)}$$
$$t=\frac{-916.2+\sqrt{63{,}332.84}}{-104}\approx 6.390$$
$$t=\frac{-916.2-\sqrt{63{,}332.84}}{-104}\approx 11.229$$

There were 1000 thousand privately owned single-unit houses started in about 2001 and in about 1996.

12.

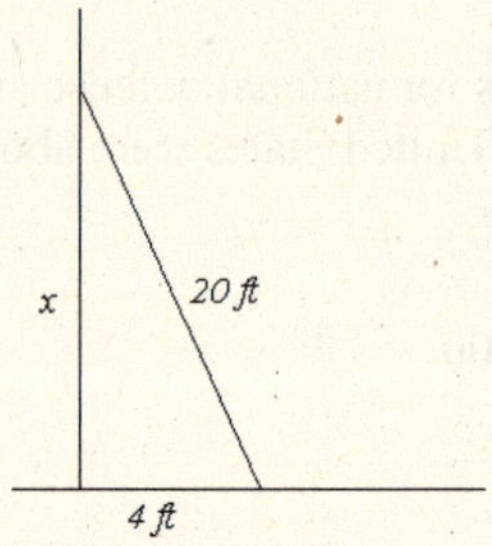

$$20^2=x^2+4^2$$
$$400=x^2+16$$
$$-16 \qquad -16$$
$$384=x^2$$
$$x=\sqrt{384}\approx 19.6$$

The ladder will reach about 19.6 feet up the house.

13. $(-7,4)$, $(-4,8)$

14. **a.** $(-3,-4)$ **b.** $x>-3$ **c.** $x<-3$

d. $x=-3$ **e.** $(0,5)$ **f.** $(-5,0),(-1,0)$

g. $f(x)=a(x+3)^2-4$

$f(0)=a(3)^2-4=5$

$9a-4=5$

$9a=9\rightarrow a=1$

$f(x)=(x+3)^2-4$

or

$f(x)=x^2+6x+5$

15. $x^2+12x+30=0$

$x^2+12x=-30$

$x^2+12x+\square=-30+\square$

Completing the square: $\left(\frac{12}{2}\right)^2=36$

$x^2+12x+\boxed{36}=-30+\boxed{36}$

$(x+6)^2=6$

$x+6=\pm\sqrt{6}\approx\pm2.449$

The two solutions are:

$x\approx-6+2.449$, $x\approx-6-2.449$

$x\approx-3.551$, $x\approx-8.449$

16. $4(x-9)^2-20=124$

$4(x-9)^2=144$

$(x-9)^2=36$

$\sqrt{(x-9)^2}=\pm\sqrt{36}$

$x-9=\pm6$

$x-9=-6, x-9=6$

$x=3, x=15$

17. $8x^2+15=65$

$8x^2=50$

$x^2=\frac{50}{8}$

$x^2=\frac{25}{4}$

$\sqrt{x^2}=\pm\sqrt{\frac{25}{4}}$

$x=\pm\frac{5}{2}$

$x=\pm2.5$

18. $a^2+b^2=c^2$ Substitute given values.

$(6)^2+b^2=(12)^2$

$36+b^2=144$

$-36 \quad -36$

$b^2=108$

$b=\pm\sqrt{108}$

$b\approx10.39$ (Use positive value for length.)

19.

$-4(x+5)^2+7=9$

$-4(x+5)^2=2$

$\frac{-4(x+5)^2}{-4}=\frac{2}{-4}$

$(x+5)^2=-0.5$

$x+5=\pm\sqrt{-0.5}$

$x=-5\pm i\sqrt{0.5}$

$x\approx-5\pm0.707i$

20.

$2.3x^2+4.6x+9=5$

$2.3x^2+4.6x+4=0$

$x=\frac{-4.6\pm\sqrt{4.6^2-4(2.3)(4)}}{2(2.3)}$

$x=\frac{-4.6\pm\sqrt{-15.64}}{4.6}$

$x\approx-1\pm0.86i$

21.

$f(x)=3x^2+24x-30$

$f(x)=3(x^2+8x)-30$

$f(x)=3(x^2+8x+\_\_-\_\_)-30$

Completing the square: $\left(\frac{8}{2}\right)^2=16$

$f(x)=3(x^2+8x+16-16)-30$

$f(x)=3(x^2+8x+16)-3(16)-30$

$f(x)=3(x+4)(x+4)-48-30$

$f(x)=3(x+4)^2-78$

22. $y<0.5x^2-3x+9$

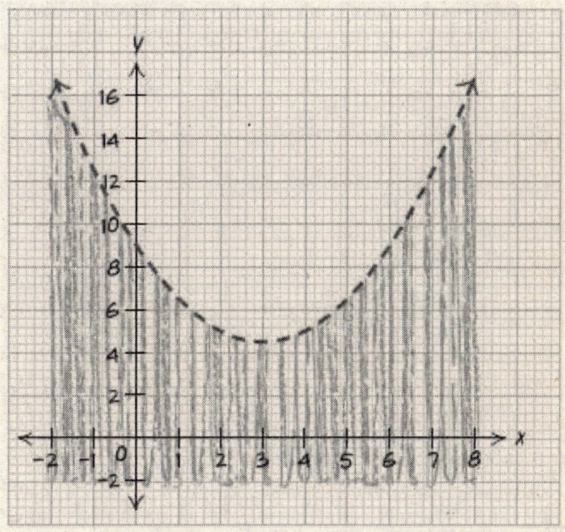

CHAPTER 11 Quadratic Functions

Cumulative Review for Chapters 1 through 11

1. $10m+4=3m+32$
$10m=3m+28$
$7m=28$
$m=4$

2. $t^2-11t+10=0$
$(t-1)(t-10)=0$
$t-1=0, t-10=0$
$t=1, t=10$

3. $4(x-7)^2+30=46$
$4(x-7)^2=16$
$\frac{4(x-7)^2}{4}=\frac{16}{4}$
$(x-7)^2=4$
$\sqrt{(x-7)^2}=\pm\sqrt{4}$
$x-7=\pm 2$
$x-7=2, x-7=-2$
$x=9, x=5$

4. $\frac{7}{15}a+\frac{3}{5}=\frac{2}{3}(a-4)-\frac{5}{3}$
$\frac{7}{15}a+\frac{3}{5}=\frac{2}{3}a-\frac{8}{3}-\frac{5}{3}$
$\frac{7}{15}a+\frac{3}{5}=\frac{2}{3}a-\frac{13}{3}$
$15\left(\frac{7}{15}a+\frac{3}{5}\right)=15\left(\frac{2}{3}a-\frac{13}{3}\right)$
$7a+9=10a-65$
$7a=10a-74$
$-3a=-74$
$\frac{-3a}{-3}=\frac{-74}{-3}$
$a=\frac{74}{3}$

5. $3x^2-8x=0$
$x(3x-8)=0$
$x=0, 3x-8=0$
$x=0, 3x=8$
$x=0, x=\frac{8}{3}$

6. $3.4m^2-4.6m=108$
$3.4m^2-4.6m-108=0$
$a=3.4, b=-4.6, c=-108$
$m=\frac{-(-4.6)\pm\sqrt{(-4.6)^2-4(3.4)(-108)}}{2(3.4)}$
$m=\frac{4.6\pm\sqrt{1489.96}}{6.8}$
$m=\frac{4.6\pm 38.6}{6.8}$
$m=\frac{4.6-38.6}{6.8}, m=\frac{4.6+38.6}{6.8}$
$m=\frac{-34}{6.8}, m=\frac{43.2}{6.8}$
$m=-5, m\approx 6.35$

7. $\frac{1}{3}(c-9)-6=\frac{5}{6}c-15$
$\frac{1}{3}c-3-6=\frac{5}{6}c-15$
$\frac{1}{3}c-9=\frac{5}{6}c-15$
$6\left(\frac{1}{3}c-9\right)=6\left(\frac{5}{6}c-15\right)$
$2c-54=5c-90$
$2c=5c-36$
$-3c=-36$
$c=12$

8. $2n^3+12n^2-432n=0$
$2n(n^2+6n-216)=0$
$2n(n+18)(n-12)=0$
$2n=0, n+18=0, n-12=0$
$n=0, n=-18, n=12$

9. $(x+3)(x-7)=-16$
$x^2-4x-21=-16$
$x^2-4x-5=0$
$(x+1)(x-5)=0$
$x+1=0, x-5=0$
$x=-1, x=5$

10. $3.8h-4.2=7.5h+27.9$
$3.8h=7.5h+32.1$
$-3.7h=32.1$
$\frac{-3.7h}{-3.7}=\frac{32.1}{-3.7}$
$h\approx -8.68$

11. $\frac{5}{c-3}=\frac{2}{c+7}$
$\left(\frac{5}{\cancel{c-3}}\right)(\cancel{c-3})(c+7)=\left(\frac{2}{\cancel{c+7}}\right)(c-3)(\cancel{c+7})$
$5(c+7)=2(c-3)$
$5c+35=2c-6$
$3c=-41$
$c=-\frac{41}{3}$

12. $\frac{w}{w-7}-\frac{3}{w-4}=\frac{9}{(w-7)(w-4)}$
$\left(\frac{w}{\cancel{w-7}}\right)(\cancel{w-7})(w-4)-\left(\frac{3}{\cancel{w-4}}\right)(w-7)(\cancel{w-4})=\frac{9}{(w-7)(w-4)}$
$w(w-4)-3(w-7)=\frac{9}{(\cancel{w-7})(\cancel{w-4})}(\cancel{w-7})(\cancel{w-4})$

$w(w-4)-3(w-7)=9$
$w^2-4w-3w+21=9$
$w^2-7w+21=9$
$w^2-7w+12=0$
$(w-4)(w-3)=0$
$w-4=0, w-3=0$
$w=4, w=3$

When w equals 4 the denominators are zero so the original fractions are undefined.
The final solution is $w = 3$.

13. $\sqrt{5-2x}=7$

$\left(\sqrt{5-2x}\right)^2=(7)^2$
$5-2x=49$
$-5 \quad -5$
$-2x=44$
$\frac{-2x}{-2}=\frac{44}{-2}$
$x=-22$

14. $\sqrt{x^2-10x+41}=4$

$\left(\sqrt{x^2-10x+41}\right)^2=(4)^2$
$x^2-10x+41=16$
$-16 \quad -16$
$x^2-10x+25=0$
$(x-5)(x-5)=0$
$x-5=0$
$x=5$

15. $4mn-5=n$
$4mn=n+5$
$\frac{4mn}{4n}=\frac{n+5}{4n}$
$m=\frac{n+5}{4n}$

16. $\frac{2}{5}a^2b+7a=c$

$\frac{2}{5}a^2b=c-7a$
$\frac{5}{2}\left(\frac{2}{5}a^2b\right)=\frac{5}{2}(c-7a)$
$a^2b=\frac{5}{2}c-\frac{35}{2}a$
$\frac{a^2b}{a^2}=\frac{\frac{5}{2}c-\frac{35}{2}a}{a^2}$
$b=\frac{\frac{5}{2}c-\frac{35}{2}a}{a^2}$
$b=\frac{\frac{5}{2}c}{a^2}-\frac{\frac{35}{2}a}{a^2}$
$b=\frac{5c}{2a^2}-\frac{35}{2a}$

17.

$\begin{cases}4x-7y=-8\\3x+10y=55\end{cases}$
$\begin{cases}10(4x-7y)=10(-8)\\7(3x+10y)=7(55)\end{cases}$
$\begin{cases}40x-\cancel{70y}=-80\\21x+\cancel{70y}=385\end{cases}$
$61x=305$
$x=5$

Substitute x back into either equation solve for y:
$3(5)+10y=55$
$15+10y=55$
$10y=40$
$y=4$

The solution to the system is $(5,4)$.

18.

$\begin{cases}y=3x-7\\y=x^2+6x-17\end{cases}$
$3x-7=x^2+6x-17$
$0=x^2+3x-10$
$0=(x+5)(x-2)$
$x+5=0, x-2=0$
$x=-5, x=2$

Substitute x back into either equation solve for y:
$y=3(-5)-7=-22$
$y=3(2)-7=-1$

The solutions to the system are $(-5,-22)$ and $(2,-1)$.

19.

$\begin{cases}y=2.5x-9\\-10x+4y=20\end{cases}$
$-10x+4(2.5x-9)=20$
$-10x+10x-36=20$
$-36=20$

Inconsistent system.
No solution.

20.

$\begin{cases}m=7n+8\\3m+2n=1\end{cases}$
$3(7n+8)+2n=1$
$21n+24+2n=1$
$23n+24=1$
$23n=-23$
$n=-1$

Substitute n back into either equation solve for m:
$m=7(-1)+8=1$

21.

$\begin{cases}b=3a^2+3a-18\\b=-8a^2+4a+30\end{cases}$
$3a^2+3a-18=-8a^2+4a+30$
$11a^2-a-48=0$

$a=11, b=-1, c=-48$

$a=\frac{-(-1)\pm\sqrt{(-1)^2-4(11)(-48)}}{2(11)}$

$a=\frac{1\pm\sqrt{2113}}{22}$

$a=\frac{1+\sqrt{2113}}{22}, a=\frac{1-\sqrt{2113}}{22}$

$a\approx 2.13, a\approx -2.04$

Substitute a back into either equation solve for b:

$b=3(2.13)^2+3(2.13)-18\approx 2.00$

$b=3(-2.04)^2+3(-2.04)-18\approx -11.64$

The solutions to the system are $(2.13, 2.00)$ and $(-2.04, -11.64)$.

22.

$\begin{cases} h=1.4g+2.6 \\ 7g-5h=-13 \end{cases}$

$7g-5(1.4g+2.6)=-13$

$7g-7g-13=-13$

$-13=-13$

Consistent system with dependent lines.
Infinite number of solutions.

23. Using two points: $(4,8)$ and $(10,35)$

find the slope: $m=\frac{35-8}{10-4}=\frac{27}{6}=\frac{9}{2}$.

Use $m=\frac{9}{2}$ and $(4,8)$.

$8=\frac{9}{2}(4)+b$

$8=18+b$

$b=-10$

Since $m=\frac{9}{2}$, and $b=-10$:

$y=\frac{9}{2}x-10.$

24. Since the two lines are perpendicular, $m_1=-\frac{1}{m_2}$.

Find the slope: $m_1=8$, therefore $m_2=-\frac{1}{8}$.

Use $m=-\frac{1}{8}$, and $(12, 4.5)$.

$4.5=12\left(-\frac{1}{8}\right)+b$

$4.5=-1.5+b$

$b=6$

Since $m=-\frac{1}{8}$, and $b=6$:

$y=-\frac{1}{8}x+6.$

25. Since the two lines are perpendicular, $m_1=m_2$.

$2x+5y=11\rightarrow$ Solve for y to find slope.

$2x+5y-2x=11-2x$

$5y=-2x+11$

$\frac{5y}{5}=\frac{-2x}{5}+\frac{11}{5}$

$y=-0.4x+2.2$

Find the slope: $m_1=-0.4$ therefore $m_2=-0.4$

Use $m=-0.4$, and $(-3,10)$.

$10=-0.4(-3)+b$

$10=1.2+b$

$b=8.8$

Since $m=-0.4$, and $b=8.8$:

$y=-0.4x+8.8.$

26. $\frac{4-m}{(m+5)(m-4)}=\frac{-1\cancel{(m-4)}}{(m+5)\cancel{(m-4)}}=\frac{-1}{m+5}$

27. $\frac{4a+8}{a^2+6a+8}=\frac{4\cancel{(a+2)}}{\cancel{(a+2)}(a+4)}=\frac{4}{a+4}$

28.

$$\frac{3+\frac{4}{a}}{4-\frac{7}{a}}=\frac{\left(3+\frac{4}{a}\right)(a)}{\left(4-\frac{7}{a}\right)(a)}=\frac{3a+\frac{4}{\cancel{a}}(\cancel{a})}{4a-\frac{7}{\cancel{a}}(\cancel{a})}=\frac{3a+4}{4a-7}$$

29. $\sqrt{x^6y^2}=\sqrt{(x^3y)^2}=x^3y$

30. $\sqrt{75}=\sqrt{25\cdot 3}=5\sqrt{3}$

31. $\sqrt{20x^3}=\sqrt{4x^2\cdot 5x}=2x\sqrt{5x}$

32. Let $C(n)$ be the cost to produce custom printed tote bag, and let n be the number of bags.

a. $(25, 138.75)$ & $(100, 405)$

Slope: $m=\frac{405-138.75}{100-25}=3.55$

Use the point slope form of the equation:

$C-405=3.55(n-100)$

$C-405=3.55n-355$

$\underline{+405 \qquad +405}$

$C=3.55n+50$

Using function notation:

$C(n)=3.55n+50$

b.

$C(200)=3.55(200)+50$

$C(200)=760$

200 bags will cost \$760.

c. Slope = 3.55. The cost to produce custom printed tote bags is increased by \$3.55 per bag.

33. a. Let $P(t)$ be the population of Pittsburgh's metro area in millions t years since 2000.

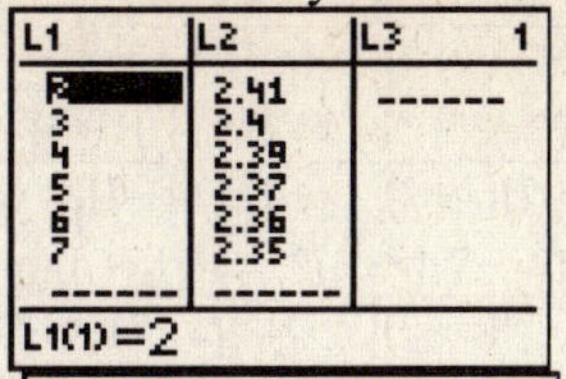

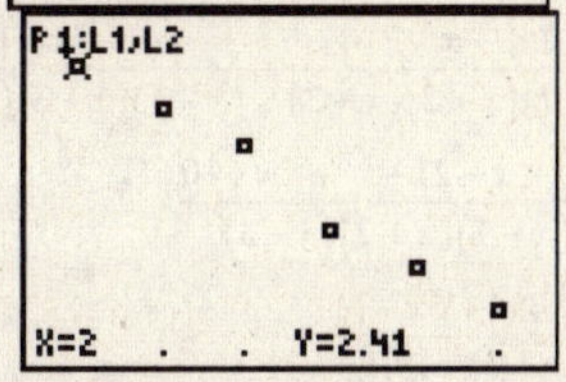

$P(t)=mt+b$

Using two points: $(2,2.41)$ and $(7,2.35)$,

find the slope: $m=\dfrac{2.35-2.41}{7-2}=-0.012.$

$P(t)=-0.012t+b$

$b=2.41-(-0.012)2$

$b=2.434$

$P(t)=-0.012t+2.434$

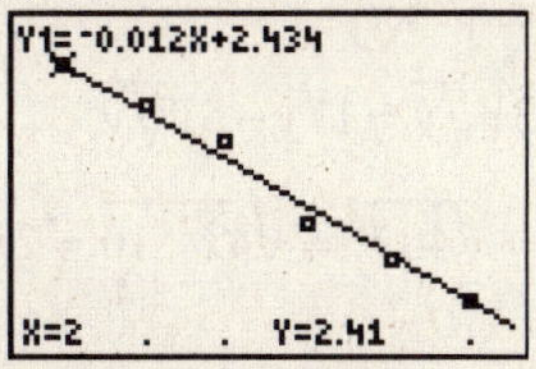

b. Domain: $[-2,12]$

$P(-2)=-0.012(-2)+2.434=2.458$

$P(12)=-0.012(12)+2.434=2.29$

Range: $[2.29,2.458]$

c. $P(12)=-0.012(12)+2.434=2.29$

The population of Pittsburgh's metro area will be about 2.29 million in 2012.

d. The slope is -0.012. The population of Pittsburgh's metro area is decreasing by 12,000 people per year.

e.
$$\begin{aligned}2.25&=-0.012t+2.434\\ -2.434&\qquad\quad -2.434\\ -0.184&=-0.012t\\ \frac{-0.184}{-0.012}&=\frac{-0.012t}{-0.012}\\ 15.33&\approx t\end{aligned}$$

In 2015, the population of Pittsburgh's metro area will be 2.25 million.

34. a. Let $R(p)$ be the revenue from selling four-person tents (in dollars) if they are priced at p dollars each.

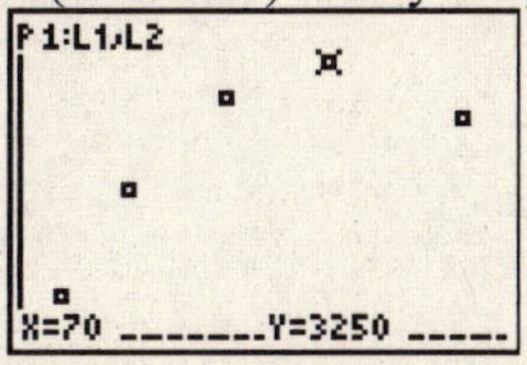

Choose $(70,3250)$ as the vertex, and $(40,2360)$ as the additional point used to solve for "a".

$$\begin{aligned}R(p)&=a(p-70)^2+3250\\ 2360&=a(40-70)^2+3250\\ 2360&=a(-30)^2+3250\\ 2360&=900a+3250\\ -3250&\qquad\quad -3250\\ -890&=900a\\ \frac{-890}{900}&=\frac{900a}{900}\quad a\approx-0.989\\ R(p)&=-0.989(p-70)^2+3250\end{aligned}$$

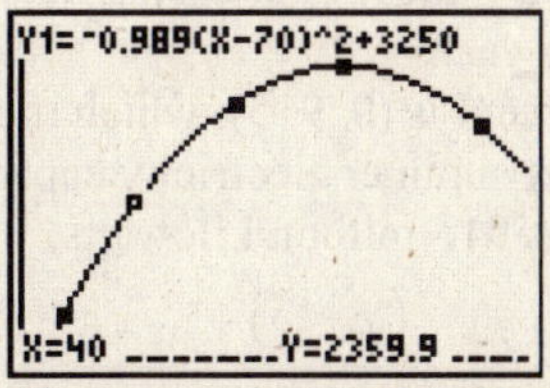

b. The vertex is (70, 3250). When the price is \$70 per four-person tent, the revenue generated is at a maximum of \$3250.

c.

$R(60)=-0.989(60-70)^2+3250$

$R(60)=3151.1$

When the price is \$60 per four-person tent, the revenue generated is \$3151.10.

d. Domain: $[20,100]$. Within this domain, the maximum value of \$3250 comes from the vertex. The minimum value comes from $p=20$.

$R(20)=-0.989(20-70)^2+3250=777.5$

Range: $[777.50,3250]$

35. a. Let $E(t)$ be the net summer electricity capacity in the United States (in millions of kilowatts) t years since 2000.

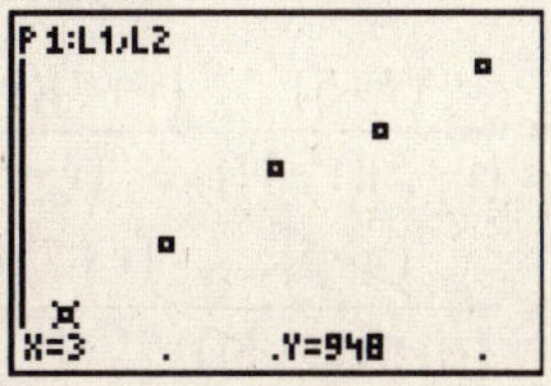

$E(t)=mt+b$

Using two points: $(4,963)$ and $(7,999)$,

find the slope: $m=\dfrac{999-963}{7-4}=12.$

$E(t)=12t+b$

$b=963-(12)4$

$b=915$

$E(t)=12t+915$

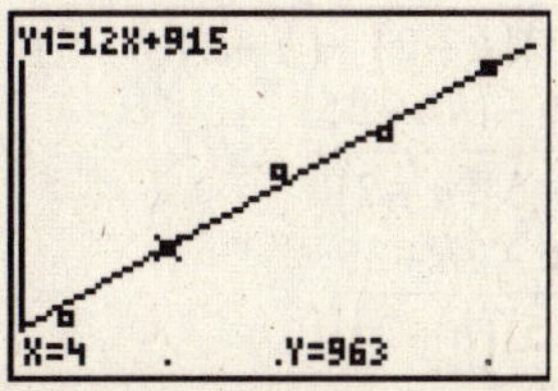

b. Domain: $[-2,12]$

$$E(-2)=12(-2)+915=891$$
$$E(12)=12(12)+915=1059$$

Range: $[891,1059]$

c.

$$E(15)=12(15)+915$$
$$E(15)=1095$$

In 2015, the net summer electricity capacity in the United States will be about 1095 million kilowatts.

d. The slope is 12. The net summer electricity capacity in the United States is increasing by 12 million kilowatts per year.

e. The vertical intercept is (0, 915), which means in the year 2000, the net summer electricity capacity in the United States was 915 million kilowatts.

36. $$\frac{x+3}{x-15}\cdot\frac{x-15}{x-2}=\frac{(x+3)}{\cancel{(x-15)}}\cdot\frac{\cancel{(x-15)}}{(x-2)}=\frac{x+3}{x-2}$$

37. $$\frac{h-5}{h+2}\div\frac{5-h}{h-4}=\frac{(h-5)}{(h+2)}\cdot\frac{(h-4)}{(5-h)}$$
$$=\frac{\cancel{(h-5)}}{(h+2)}\cdot\frac{(h-4)}{(-1)\cancel{(h-5)}}$$
$$=-\frac{h-4}{h+2}$$

38.

$$\frac{2n+6}{n^2+8n+15}\cdot\frac{3n+15}{n^2+11n+28}=\frac{2(n+3)}{(n+5)(n+3)}\cdot\frac{3(n+5)}{(n+4)(n+7)}$$
$$=\frac{2\cancel{(n+3)}}{\cancel{(n+5)}\cancel{(n+3)}}\cdot\frac{3\cancel{(n+5)}}{(n+4)(n+7)}$$
$$=\frac{6}{(n+4)(n+7)}$$

39.

$$\frac{v+5}{v^2-6v-55}\div\frac{v-8}{v^2-4v-77}=\frac{(v+5)}{(v+5)(v-11)}\cdot\frac{(v+7)(v-11)}{(v-8)}$$
$$=\frac{\cancel{(v+5)}}{\cancel{(v+5)}\cancel{(v-11)}}\cdot\frac{(v+7)\cancel{(v-11)}}{(v-8)}$$
$$=\frac{v+7}{v-8}$$

40. $$\frac{3}{h-5}-\frac{7}{h+2}=\frac{3}{(h-5)}\cdot\frac{(h+2)}{(h+2)}-\frac{7}{(h+2)}\cdot\frac{(h-5)}{(h-5)}$$
$$=\frac{3(h+2)}{(h-5)(h+2)}-\frac{7(h-5)}{(h+2)(h-5)}$$
$$=\frac{3h+6}{(h-5)(h+2)}-\frac{7h-35}{(h+2)(h-5)}$$
$$=\frac{3h+6-(7h-35)}{(h-5)(h+2)}$$
$$=\frac{3h+6-7h+35}{(h-5)(h+2)}$$
$$=\frac{-4h+41}{(h-5)(h+2)}$$

41.

$$\frac{2x-7}{x^2-5x-14}+\frac{x+5}{x^2-4x-21}=\frac{2x-7}{(x-7)(x+2)}+\frac{x+5}{(x-7)(x+3)}$$
$$=\frac{(2x-7)}{(x-7)(x+2)}\cdot\frac{(x+3)}{(x+3)}+\frac{(x+5)}{(x-7)(x+3)}\cdot\frac{(x+2)}{(x+2)}$$
$$=\frac{(2x-7)(x+3)}{(x-7)(x+2)(x+3)}+\frac{(x+5)(x+2)}{(x-7)(x+3)(x+2)}$$
$$\frac{2x^2-x-21}{(x-7)(x+2)(x+3)}+\frac{x^2+7x+10}{(x-7)(x+3)(x+2)}$$
$$=\frac{2x^2-x-21+x^2+7x+10}{(x-7)(x+2)(x+3)}$$
$$=\frac{3x^2+6x-11}{(x-7)(x+2)(x+3)}$$

42. $$\sqrt{16x^3}+7-2x\sqrt{x}-3$$
$$=\sqrt{16x^2\cdot x}-2x\sqrt{x}+4=4x\sqrt{x}-2x\sqrt{x}+4$$
$$=2x\sqrt{x}+4$$

43. $$11y-2\sqrt{y^3}+6y-y\sqrt{9y}$$
$$=11y-2\sqrt{y^2\cdot y}+6y-y\cdot 3\sqrt{y}$$
$$=11y-2y\sqrt{y}+6y-3y\sqrt{y}=17y-5y\sqrt{y}$$

44. $$\sqrt{5x}\cdot\sqrt{8x}=\sqrt{5x\cdot 8x}=\sqrt{40x^2}=\sqrt{4x^2\cdot 10}=2x\sqrt{10}$$

45. $$(5+\sqrt{3})(2-\sqrt{12})$$
$$=5\cdot 2-5(\sqrt{12})+2(\sqrt{3})-\sqrt{3}(\sqrt{12})$$
$$=10-5\sqrt{4\cdot 3}+2\sqrt{3}-\sqrt{36}$$
$$=10-10\sqrt{3}+2\sqrt{3}-6=4-8\sqrt{3}$$

46. $$\frac{\sqrt{54x^3}}{\sqrt{6x}}=\sqrt{\frac{54x^3}{6x}}=\sqrt{9x^2}=3x$$

47. $$y=\frac{2}{3}x-6$$

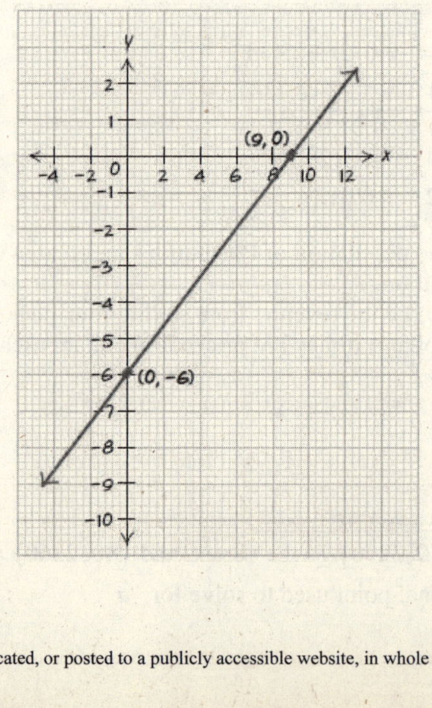

48. $y = -2x + 5$

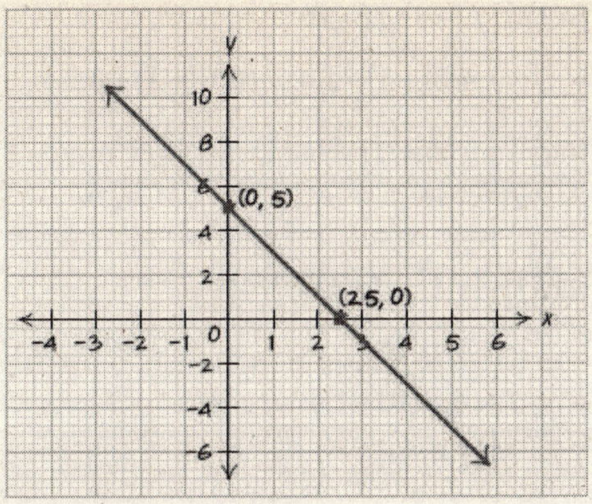

49. $y = 0.5(x-6)^2 - 8$

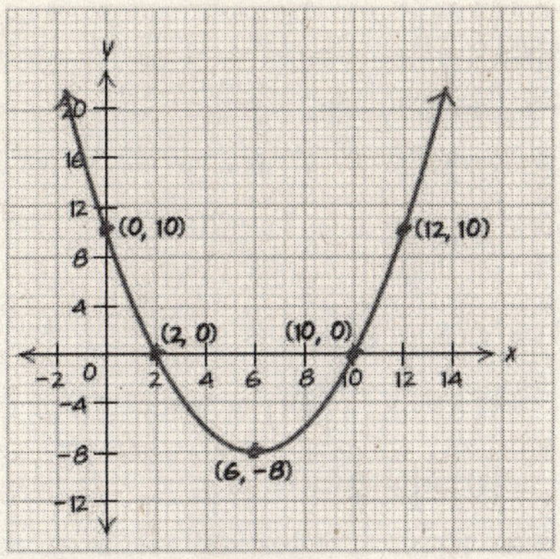

50. $y = -2x^2 + 10x + 12$

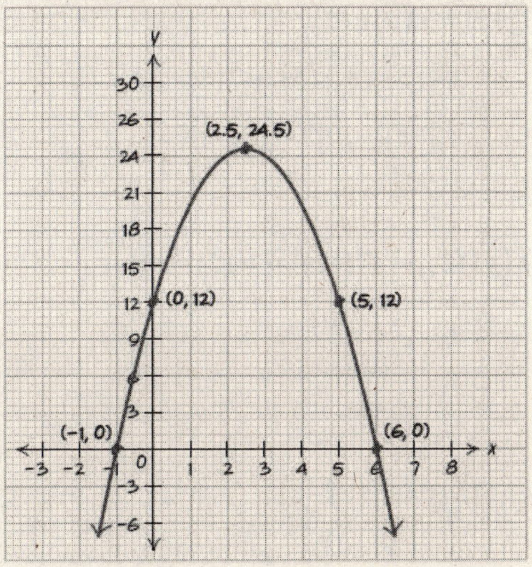

51. $4x + 7y = 12$

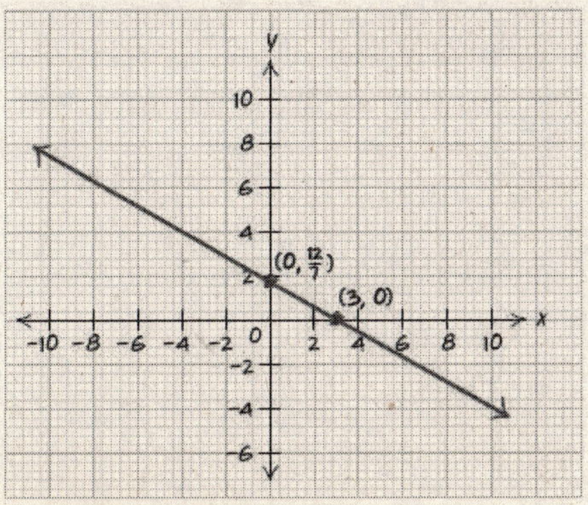

52. $x - 5y < 20$

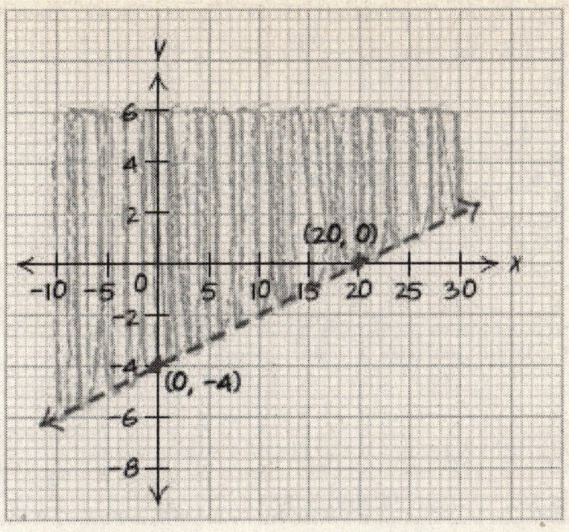

53. $y \geq \frac{5}{8}x + 7$

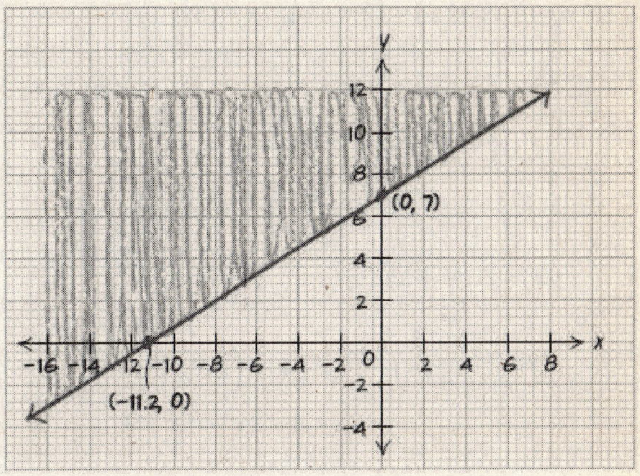

54. $y < x^2 - 5x - 6$

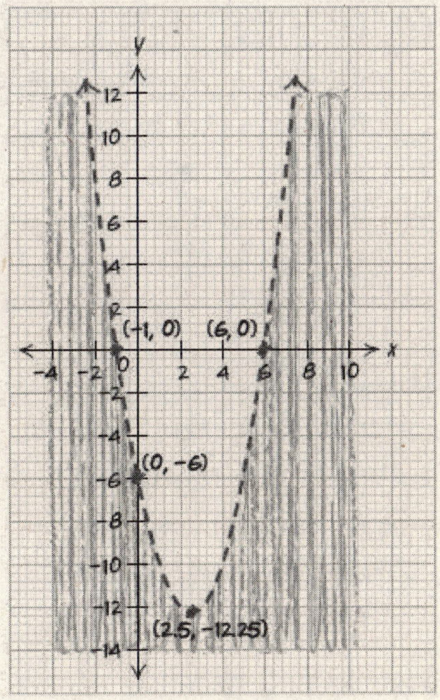

55. $\frac{5}{\sqrt{7}} = \frac{5}{\sqrt{7}} \cdot \frac{\sqrt{7}}{\sqrt{7}} = \frac{5\sqrt{7}}{7}$

56. $\frac{4}{3-\sqrt{2}}=\frac{4}{\left(3-\sqrt{2}\right)}\cdot\frac{\left(3+\sqrt{2}\right)}{\left(3+\sqrt{2}\right)}$

$=\frac{12+4\sqrt{2}}{9+3\sqrt{2}-3\sqrt{2}-\sqrt{2}\cdot\sqrt{2}}$

$=\frac{12+4\sqrt{2}}{9-2}=\frac{12+4\sqrt{2}}{7}$

57. $f(x)=2x+8$ and $g(x)=7x-18$

a.

$$\begin{aligned}f(x)-g(x)&=(2x+8)-(7x-18)\\&=2x+8-7x+18\\&=-5x+26\end{aligned}$$

b.

$$\begin{aligned}f(x)g(x)&=(2x+8)(7x-18)\\&=14x^2+20x-144\end{aligned}$$

c.

$$\begin{aligned}f(g(x))&=2(7x-18)+8\\&=14x-36+8\\&=14x-28\end{aligned}$$

58. $f(x)=3x-4$ and $g(x)=2x^2+5x-9$

a.

$$\begin{aligned}f(x)-g(x)&=(3x-4)-(2x^2+5x-9)\\&=3x-4-2x^2-5x+9\\&=-2x^2-2x+5\end{aligned}$$

b.

$$\begin{aligned}f(x)g(x)&=(3x-4)(2x^2+5x-9)\\&=6x^3+15x^2-27x-8x^2-20x+36\\&=6x^3+7x^2-47x+36\end{aligned}$$

c.

$$\begin{aligned}f(g(x))&=3(2x^2+5x-9)-4\\&=6x^2+15x-27-4\\&=6x^2+15x-31\end{aligned}$$

d.

$$\begin{aligned}g(f(x))&=2(3x-4)^2+5(3x-4)-9\\&=2(3x-4)(3x-4)+15x-20-9\\&=2(9x^2-24x+16)+15x-29\\&=18x^2-48x+32+15x-29\\&=18x^2-33x+3\end{aligned}$$

59. $f(x)=4.8x+1.4$ and $g(x)=-1.6x+3.2$

a.

$$\begin{aligned}(f+g)(5)&=f(5)+g(5)\\&=[4.8(5)+1.4]+[-1.6(5)+3.2]\\&=[25.4]+[-4.8]\\&=20.6\end{aligned}$$

b.

$$\begin{aligned}fg(-3)&=f(-3)g(-3)\\&=[4.8(-3)+1.4][-1.6(-3)+3.2]\\&=[-13][8]\\&=-104\end{aligned}$$

c.

$$\begin{aligned}f(g(2))&=f(-1.6(2)+3.2)\\&=f(0)\\&=4.8(0)+1.4\\&=1.4\end{aligned}$$

d.

$$\begin{aligned}g(f(1))&=g(4.8(1)+1.4)\\&=g(6.2)\\&=-1.6(6.2)+3.2\\&=-6.72\end{aligned}$$

60. $f(x)=x-8$ and $g(x)=x^2-7x-20$

a.

$$\begin{aligned}(f+g)(8)&=f(8)+g(8)\\&=[(8)-8]+[(8)^2-7(8)-20]\\&=[0]+[-12]\\&=-12\end{aligned}$$

b.

$$\begin{aligned}fg(5)&=f(5)g(5)\\&=[(5)-8][(5)^2-7(5)-20]\\&=[-3][-30]\\&=90\end{aligned}$$

c.

$$\begin{aligned}f(g(12))&=f((12)^2-7(12)-20)\\&=f(40)\\&=(40)-8\\&=32\end{aligned}$$

d.

$$\begin{aligned}g(f(6))&=g((6)-8)\\&=g(-2)\\&=(-2)^2-7(-2)-20\\&=-2\end{aligned}$$

61. $f(x)=-\frac{4}{3}x+5$

a.

$$\begin{aligned}f(21)&=-\frac{4}{3}(21)+5\\&=-28+5\\&=-23\end{aligned}$$

b.

$$-\frac{4}{3}x+5=\frac{41}{3}$$
$$3\left(-\frac{4}{3}x+5\right)=3\left(\frac{41}{3}\right)$$
$$-4x+15=41$$
$$-4x=26$$
$$x=-6.5$$

c. Domain: $(-\infty,\infty)$, range: $(-\infty,\infty)$

62. $g(x)=20$

a. $g(9)=20$

b. Domain: $(-\infty,\infty)$, range: $\{20\}$

63. $h(x)=4x+20$

a.

$$h(-8)=4(-8)+20$$
$$=-32+20$$
$$=-12$$

b.

$$4x+20=75$$
$$4x=55$$
$$x=13.75$$

c. Domain: $(-\infty,\infty)$, range: $(-\infty,\infty)$

64. During the first 8 minutes of a jog-a-thon, an eighth grader has run 3.5 laps.

65. **a.** When $x=0$ and $y=9$: $(0,9)$

b. When $x=15$ and $y=0$: $(15,0)$

c. Using two points: $(0,9.0)$ and $(15,0)$, find slope:

$$m=\frac{9.0-0}{0-15}=-\frac{9}{15}=-\frac{3}{5}$$

d. $m=-\frac{3}{5}$

y-intercept $=(0,9)$, therefore $b=9$.

$$y=-\frac{3}{5}x+9$$

66. **a.** $f(8)=4$

b. $x=-4$

c. Domain: $(-\infty,\infty)$, range: $(-\infty,\infty)$

67. **a.** Using two points: $(0,2)$ and $(9,0)$, find the slope:

$$m=\frac{2-0}{0-9}=-\frac{2}{9}$$

b. $(0,2)$

c. $(9,0)$

d. $m=-\frac{2}{9}$

y-intercept $=(0,2)$, therefore $b=2$.

$$y=-\frac{2}{9}x+2$$

68. Domain: All real numbers such that $x\neq-6$ and $x\neq4$

69. $f(x)=\sqrt{21-7x}$

Domain: $x\leq3$

70. Let x be the amount in dollars Greg is investing at 3% and y be the amount in dollars Greg is investing at 2.5%.

a.

$x+y=900{,}000\rightarrow$ Total amount invested.

$0.03x+0.025y=25{,}800\rightarrow$ Total interest earned.

b. Substitute $x=900{,}000-y$ in the second equation and solve.

$$0.03(900{,}000-y)+0.025y=25{,}800$$
$$27{,}000-0.03y+0.025y=25{,}800$$
$$27{,}000-0.005y=25{,}800$$
$$\underline{-27{,}000 \qquad -27{,}000}$$
$$-0.005y=1200$$
$$\frac{-0.005y}{-0.005}=\frac{1200}{-0.005}$$
$$y=240{,}000$$
$$x=900{,}000-240{,}000$$
$$x=660{,}000$$

\$240,000 should be invested at 2.5% and \$660,000 should be invested at 3%.

71. Let x be the number of pounds of bold roast coffee, and let y be the number of pounds of mild roast coffee.

$$x+y=200$$
$$8.99x+11.99y=(10.19)(200)=2038$$

Substitute $x=200-y$ in the second equation and solve.

$$8.99(200-y)+11.99y=2038$$
$$1798-8.99y+11.99y=2038$$
$$1798+3y=2038$$
$$\underline{-1798 \qquad -1798}$$
$$3y=240$$
$$\frac{3y}{3}=\frac{240}{3}$$
$$y=80$$
$$x=200-80=120$$

They need to mix 80 pounds of the \$11.99 per pound mild roast coffee with 120 pounds of the \$8.99 per pound bold roast coffee to make 200 pounds of a \$10.19 per pound blend.

72. Let P be the salary Deeva earns if she has s dollars in sales.

$$P=300+0.04s$$
$$410=300+0.04s$$
$$110=0.04s$$
$$\frac{110}{0.04}=\frac{0.04s}{0.04}$$
$$2750=s$$

Deeva needs to make at least \$2750 per week in sales to earn at least \$410 per week.

73. $24x^3y - 7x^2y^2 + 12y^3 - 10$

There are four terms.

$24x^3y$: variable term, coefficient 24

$-7x^2y^2$: variable term, coefficient -7

$12y^3$: variable term, coefficient 12

-10 : constant term

74. $-6x^5 + 4x^2 - 2x + 4$

There are four terms

$-6x^5$: variable term, coefficient -6

$4x^2$: variable term, coefficient 4

$-2x$: variable term, coefficient -2

4 : constant term

75. $24x^3y - 7x^2y^2 + 12y^3 - 10$

$24x^3y$ has degree 4

$-7x^2y^2$ has degree 4

$12y^3$ has degree 3

-10 has degree 0

The degree of the polynomial is 4.

76. $-6x^5 + 4x^2 - 2x + 4$

$-6x^5$ has degree 5

$4x^2$ has degree 2

$-2x$ has degree 1

4 has degree 0

The degree of the polynomial is 5.

77. Polynomial

78. Not a polynomial. The variable y has a negative exponent.

79.

$$\begin{array}{r} b+7 \\ b-4\overline{)b^2+3b+5} \\ -(b^2-4b) \\ \hline 7b+5 \\ -(7b-28) \\ \hline 33 \end{array}$$

$$(b^2+3b+5) \div (b-4) = b+7+\frac{33}{b-4}$$

80.

$$\begin{array}{r|rrr} 3 & 2 & -1 & -15 \\ & & 6 & 15 \\ \hline & 2 & 5 & 0 \end{array}$$

$$(2t^2 - t - 15) \div (t-3) = 2t+5$$

81.

$$4x - 11 > 10x + 1$$
$$-10x + 11 \quad -10x + 11$$
$$-6x > 12$$
$$\frac{-6x}{-6} < \frac{12}{-6}$$
$$x < -2$$

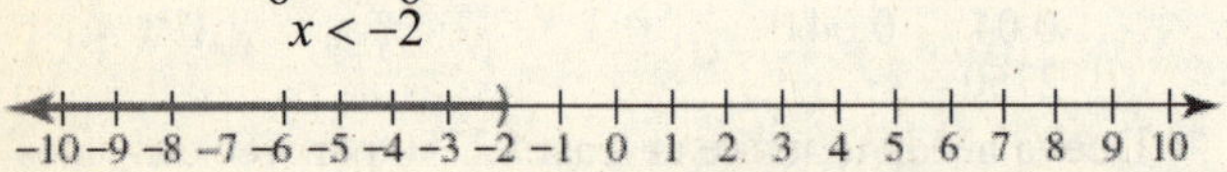

82. $\frac{1}{4}(x+3) - \frac{5}{12} \le \frac{1}{2}x + \frac{5}{6}$

$$\frac{x}{4} + \frac{3}{4} - \frac{5}{12} \le \frac{1}{2}x + \frac{5}{6}$$
$$(12)\left(\frac{x}{4} + \frac{3}{4} - \frac{5}{12}\right) \le \left(\frac{1}{2}x + \frac{5}{6}\right)(12)$$
$$3x + 9 - 5 \le 6x + 10$$
$$3x + 4 \le 6x + 10$$
$$-6x - 4 \quad -6x - 4$$
$$-3x \le 6$$
$$\frac{-3x}{-3} \ge \frac{6}{-3}$$
$$x \ge -2$$

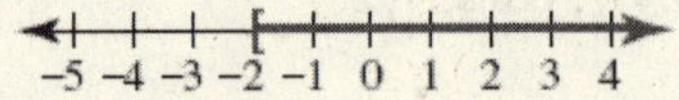

83. $|x-12| < 9$

$$-9 < x - 12 < 9$$
$$+12 \quad +12 \quad +12$$
$$3 < x < 21$$

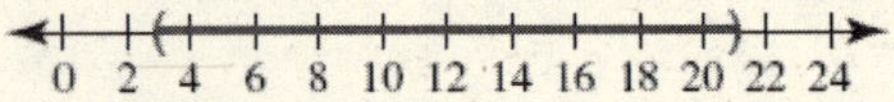

84. $|-3x+5| \le 11$

$$-11 \le -3x + 5 \le 11$$
$$-5 \quad -5 \quad -5$$
$$-16 \le -3x \le 6$$
$$\frac{-16}{-3} \ge \frac{-3x}{-3} \ge \frac{6}{-3}$$
$$\frac{16}{3} \ge x \ge -2$$
$$-2 \le x \le \frac{16}{3}$$

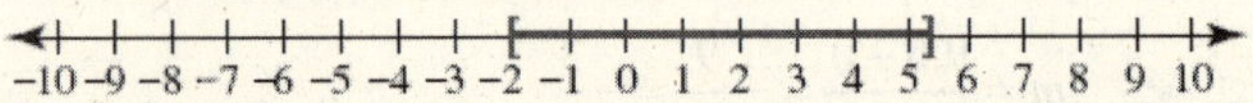

85. $|x-9| \ge 3$

$x - 9 \le -3$ or $x - 9 \ge 3$

$x - 9 + 9 \le -3 + 9$ or $x - 9 + 9 \ge 3 + 9$

$x \le 6$ or $x \ge 12$

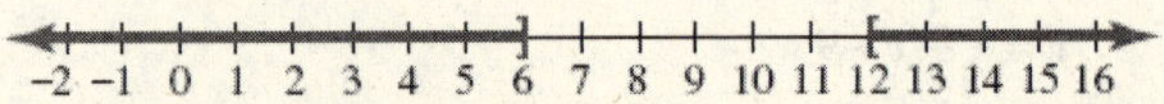

86. $-5|x-4| + 20 < -15$

$$-20 \quad -20$$
$$-5|x-4| < -35$$
$$\frac{-5|x-4|}{-5} > \frac{-35}{-5}$$
$$|x-4| > 7$$

$x - 4 < -7$ or $x - 4 > 7$

$x < -3$ or $x > 11$

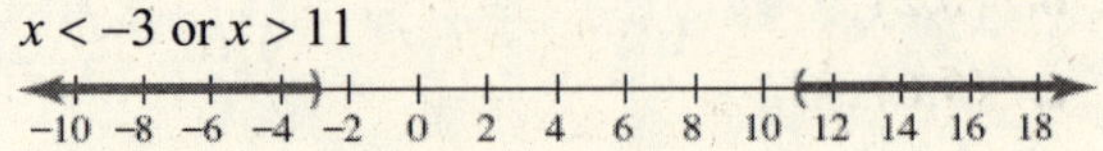

87. $y_1 < y_2$ for $x < 8$

88. $|2m-5| + 7 = 22$

$$-7 \quad -7$$
$$|2m-5| = 15$$

$$2m-5=-15 \text{ or } 2m-5=15$$
$$2m=-10 \text{ or } 2m=20$$
$$\frac{2m}{2}=\frac{-10}{2} \text{ or } \frac{2m}{2}=\frac{20}{2}$$
$$m=-5 \text{ or } m=10$$

89. $|a-9|-4=-2$
$$\underline{\quad +4 \quad +4}$$
$$|a-9|=2$$
$$a-9=-2 \text{ or } a-9=2$$
$$a=7 \text{ or } a=11$$

90. $y \ge 2x-8$

$y \le -\frac{1}{3}x+2$

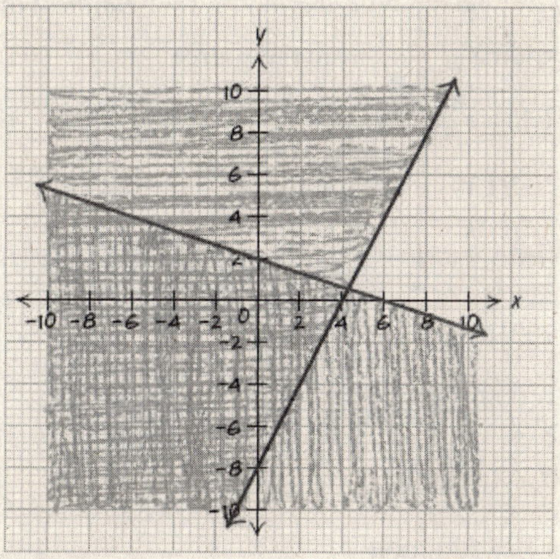

91. $2x-4y \le 16$

$-2x+y \ge 6$

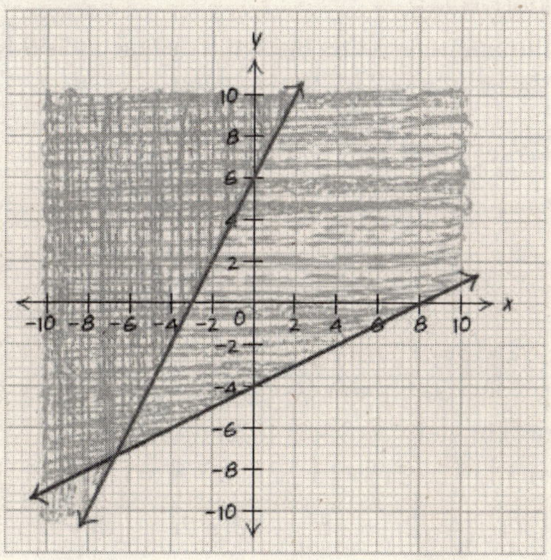

92. $a^2+b^2=c^2$ Substitute given values.
$$(7)^2+b^2=(15)^2$$
$$49+b^2=225$$
$$\underline{-49 \qquad -49}$$
$$b^2=176$$
$$b=\pm\sqrt{176}$$
$b \approx 13.27$ (Use positive value for length.)

93. $a^2+b^2=c^2$ Substitute given values.
$$a^2+(5)^2=(11)^2$$
$$a^2+25=121$$
$$\underline{-25 \quad -25}$$
$$a^2=96$$
$$a=\pm\sqrt{96}$$
$a \approx 9.80$ (Use positive value for length.)

94. $\left(\frac{3}{5}m^4n^5\right)^4=\frac{3^4}{5^4}m^{4\times4}n^{5\times4}=\frac{81m^{16}n^{20}}{625}$

95. $\left(10x^3y^7\right)\left(-4xy^2\right)=-40x^{3+1}y^{7+2}=-40x^4y^9$

96. $\frac{24g^3h^6}{16gh^2}=\frac{\cancel{8}\cdot 3g^{3-1}h^{6-2}}{\cancel{8}\cdot 2}=\frac{3g^2h^4}{2}$

97. $\frac{-14x^4yz^{-3}}{20xy^{-5}z^{-3}}=\frac{-7\cdot\cancel{2}x^{4-1}y^{1-(-5)}z^{-3-(-3)}}{\cancel{2}\cdot 10}=-\frac{7x^3y^6}{10}$

98. $\left(-27a^6b^{-3}\right)^{\frac{1}{3}}\left(16a^{12}b^{-20}\right)^{-\frac{1}{4}}$
$$=(-27)^{\frac{1}{3}}a^{\frac{6}{3}}b^{-\frac{3}{3}}\cdot 16^{-\frac{1}{4}}a^{\frac{-12}{4}}b^{\frac{-20}{-4}}$$
$$=-3a^2b^{-1}\cdot\frac{1}{2}a^{-3}b^5$$
$$=-\frac{3}{2}a^{2-3}b^{-1+5}$$
$$=-\frac{3a^{-1}b^4}{2}$$
$$=-\frac{3b^4}{2a}$$

99. $\left(\frac{72m^3n^{12}}{50m^{-1}n^8}\right)^{\frac{1}{2}}=\left(\frac{\cancel{2}\cdot 36m^{3-(-1)}n^{12-8}}{\cancel{2}\cdot 25}\right)^{\frac{1}{2}}$
$$=\frac{36^{\frac{1}{2}}m^{\frac{4}{2}}n^{\frac{4}{2}}}{25^{\frac{1}{2}}}$$
$$=\frac{6m^2n^2}{5}$$

100. a. $S(6)=1.88(6)+54.28$
$$S(6)=65.56$$
In 2006, Sweden's GDP per employed person was $65,560.

b. $S(t)>N(t)$
$$1.88t+\cancel{54.28}>\cancel{1.01t}+61.3$$
$$\underline{-1.01t \quad \cancel{-54.28} \quad \cancel{-1.01t} \quad -54.28}$$
$$0.87t>7.02$$
$$\frac{0.87t}{0.87}>\frac{7.02}{0.87}$$
$$t>8.07$$
Sweden's GDP per employed person will be greater than Netherlands in 2008.

101. a. $A(7)=1.26(7)-0.04$
$$A(7)=8.78$$
In 2007, there were about 8780 loans to African American owned small businesses.

b. Let $T(t)$ be the total number of loans (in thousands) given to African American and Hispanic American owned small businesses t years past 2000.

$$T(t) = A(t) + H(t)$$
$$T(t) = (1.26t - 0.04) + (1.33t + 2.31)$$
$$T(t) = 2.59t + 2.27$$

c. $T(10) = 2.59(10) + 2.27$

$$T(10) = 28.17$$

There will be about 28,170 loans given to African American and Hispanic American owned small businesses in 2010.

102. $a^2 + 11a + 24 = (a+3)(a+8)$

103.

$$\begin{aligned} 3n^3 - 27n^2 + 42n &= 3n(n^2 - 9n + 14) \\ &= 3n(n-7)(n-2) \end{aligned}$$

104.

$$\begin{aligned} 16x^2 - 40xy - 14x + 35y &= \underline{16x^2 - 40xy}\ \underline{-14x + 35y} \\ &= 8x(2x-5y) - 7(2x-5y) \\ &= (2x-5y)(8x-7) \end{aligned}$$

105. $25k^2 - 49 = (5k+7)(5k-7)$

106. $12t^2 - 19t - 21 = (4t+3)(3t-7)$

$ac = 12 \times -21 = -252$

Find the factors of -252 that sum to -19:

$-28 \times 9 = -252$

$-28 + 9 = -19$

We get:

$$\begin{aligned} 12t^2 - 19t - 21 &= \underline{12t^2 - 28t} + \underline{9t - 21} \\ &= 4t(3t-7) + 3(3t-7) \\ &= (4t+3)(3t-7) \end{aligned}$$

107.

$$\begin{aligned} 5z^3 - 40 &= 5(z^3 - 8) \\ &= 5(z^3 - 2^3) \\ &= 5(z-2)(z^2 + 2z + 2^2) \\ &= 5(z-2)(z^2 + 2z + 4) \end{aligned}$$

108. a. $P(15) = 2(15) + 48 = 78$

The population of this city will be about 78,000 in 2015.

b. $B(78) = 0.6(78) + 0.2 = 47$

There will be about 4700 burglaries in this city in 2015.

c. Let $K(t)$ be the number of burglaries in this city t years since 2000.

$$K(t) = B(P(t))$$
$$K(t) = 0.6(2t + 48) + 0.2$$
$$K(t) = 1.2t + 28.8 + 0.2$$
$$K(t) = 1.2t + 29$$

d. $K(20) = 1.2(20) + 29$

$$K(20) = 53$$

In 2020, there will be about 5300 burglaries in this city.

109. $f(x) = x^2 + 10x - 14$

$f(x) = (x^2 + 10x + \square - \square) - 14$

Completing the square: $\left(\frac{10}{2}\right)^2 = 25$

$f(x) = (x^2 + 10x + \boxed{25} - \boxed{25}) - 14$

$f(x) = (x^2 + 10x + 25) - 25 - 14$

$f(x) = (x+5)^2 - 39$

110. Let $I(d)$ be the illumination (in foot-candles) of a light source d feet from the source.

a. $I(d) = \dfrac{k}{d^2}$

Use $I(30) = 40$ to solve for k.

$$40 = \frac{k}{(30)^2}$$
$$(900)(40) = \left(\frac{k}{900}\right)(900)$$
$$36{,}000 = k$$

so

$$I(d) = \frac{36{,}000}{d^2}$$

b. $I(40) = \dfrac{36{,}000}{40^2}$

$$I(40) = 22.5$$

The illumination of this light at a distance of 40 feet from the source is 22.5 foot-candles.

c.

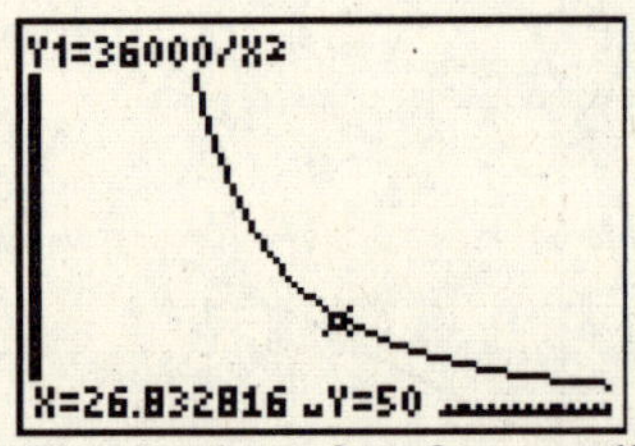

An illumination of 50 foot-candles occurs at a distance of approximately 26.8 feet from the source.

111. $(13 - 4i) + (-2 + 5i) = 13 - 2 - 4i + 5i = 11 + i$

112. $(-6 + 3i) - (4 - 2i) = -6 - 4 + 3i + 2i = -10 + 5i$

113.

$$\begin{aligned} (3+2i)(5-3i) &= 15 - 9i + 10i - 6i^2 \\ &= 15 + i - 6(-1) \\ &= 15 + i + 6 \\ &= 21 + i \end{aligned}$$

114. $\dfrac{-15 + 20i}{-5} = \dfrac{-15}{-5} + \dfrac{20i}{-5} = 3 - 4i$

115.

$$\frac{4+3i}{1+i}=\left(\frac{4+3i}{1+i}\right)\left(\frac{1-i}{1-i}\right)$$
$$=\frac{4-4i+3i-3i^2}{1-i^2}$$
$$=\frac{4-i-3(-1)}{1-1(-1)}$$
$$=\frac{4-i+3}{1+1}$$
$$=\frac{7-i}{2}=\frac{7}{2}-\frac{1}{2}i$$

116.

$$m^2=-49$$
$$m=\pm\sqrt{-49}$$
$$m=\pm 7i$$

117.

$$(y-1)^2-64=0$$
$$(y-1)^2=64$$
$$y-1=\pm\sqrt{64}$$
$$y=1\pm\sqrt{64}$$
$$y=1\pm 8$$
$$y=1+8,\ y=1-8$$
$$y=9,\ y=-7$$

118.

$$(x-3)^2+16=0$$
$$(x-3)^2=-16$$
$$x-3=\pm\sqrt{-16}$$
$$x=3\pm\sqrt{-16}$$
$$x=3\pm 4i$$

119.

$$x^2-2x-2=0$$
$$x=\frac{-(-2)\pm\sqrt{(-2)^2-4(1)(-2)}}{2(1)}$$
$$x=\frac{2\pm\sqrt{4+8}}{2}$$
$$x=\frac{2\pm\sqrt{12}}{2}$$
$$x=\frac{2\pm 2\sqrt{3}}{2}$$
$$x=1\pm\sqrt{3}$$

120. $\sqrt{5x^8}=\left(5x^8\right)^{\frac{1}{2}}$

121. $\left(16m^5\right)^{\frac{2}{3}}=\sqrt[3]{\left(16m^5\right)^2}=\sqrt[3]{256m^{10}}$

CHAPTER 12 EXPONENTIAL FUNCTIONS

Section 12.1 Exponential Functions: Patterns of Growth and Decay

1. This function is an exponential function because the variable is in the exponent.

3. This function is an exponential function because the variable is in the exponent.

5. This function is a quadratic function because the degree of the polynomial is 2.

7. This function is a quadratic function because the degree of the polynomial is 2.

9. Other, because there is a variable in the numerator and in the denominator.

11. Let $L(h)$ be the number of Lactobacillus acidophilus bacteria present after h hours have passed.

a. The bacteria double every hour:

h	L(h)
0	30
1	$30(2)^1$
2	$30(2)^2$
3	$30(2)^3$
h	$30(2)^h$

$L(h) = 30(2)^h$

b. $L(12) = 30(2)^{12} = 122{,}880$. After 12 hours, there are 122,880 Lactobacillus acidophilus bacteria present.

c. $L(24) = 30(2)^{24} = 503{,}316{,}480$. After 24 hours, there are 503,316,480 Lactobacillus acidophilus bacteria present.

d.

X	Y1
15.02	996763
15.021	997454
15.022	998145
15.023	998838
15.024	999530
15.025	1E6
15.026	1E6

Y1=1000223.23652

After just 15 hours, there are 1 million bacteria present.

13. Let $E(h)$ be the number of E coli bacteria present after h hours have passed.

a. The bacteria double 2 times each hour:

h	$E(h)$
0	3
1	$3(2)(2) = 3(2)^2$
2	$3(2^2)(2^2) = 3(2^2)^2 = 3(2)^{2\times 2} = 3(2)^4$
3	$3(2^2)(2^2)(2^2) = 3(2^2)^3 = 3(2)^{2\times 3} = 3(2)^6$
h	$3(2)^{2h}$

$E(h) = 3(2)^{2h}$

b.

$$B(5) = 3(2)^{2\times 5} = 3(2)^{10}$$
$$B(5) = 3072$$

After 5 hours, there are approximately 3072 bacteria present.

c.

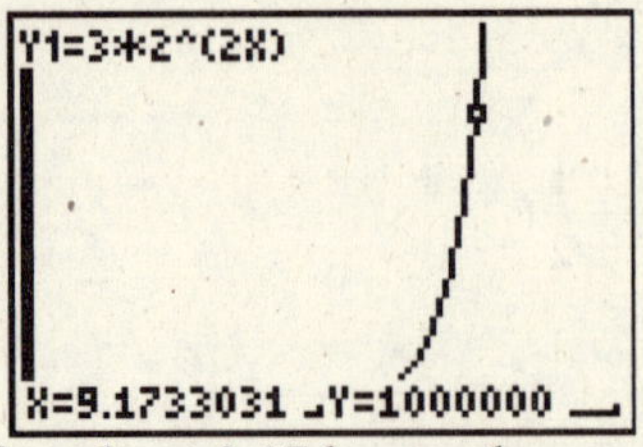

After about 9.17 hours, there are 1 million bacteria present.

15. Let $S(n)$ be the number of Streptococcus lactis bacteria present after n 26-minute time intervals have passed.

a. The bacteria double every 26-minute time period:

n	$S(n)$
0	8
1	$8(2)$
2	$8(2)(2) = 8(2)^2$
3	$8(2)(2)(2) = 8(2)^3$
n	$8(2)^n$

$S(n) = 8(2)^n$

b.

$$6.5\text{ hours} = (6.5\text{ hours})\left(\frac{60\text{ min}}{1\text{ hour}}\right) = 390\text{ min}$$
$$n = \frac{390\text{ min}}{26\text{ min}} = 15$$

There are 15 of the 26-minute time periods in 390 minutes.

$S(15) = 8(2)^{15} = 262{,}144$. After 6.5 hours, there are 262,144 bacteria present.

17. Let $B(h)$ be the number of bacteria present after h hours have passed.

a. The bacteria triple 2 times each hour:

h	$B(h)$
0	8
1	$8(3)(3) = 8(3)^2$
2	$8(3^2)(3^2) = 8(3^2)^2 = 8(3)^{2\times 2} = 8(3)^4$
3	$8(3^2)(3^2)(3^2) = 8(3^2)^3 = 8(3)^{2\times 3} = 8(3)^6$
h	$8(3)^{2h}$

$B(h) = 8(3)^{2h}$

b. $B(10)=8(3)^{2\times 10}=8(3)^{20}\approx 2.7894\times 10^{10}$.

After 10 hours, there are approximately 2.7894×10^{10} bacteria present.

19. Let $T(h)$ be the number of Treponema pallidum bacteria present after h hours have passed. Let n be the number of 33 hour time intervals.

a. The bacteria double every 33 hours:

h	$T(h)$
0	12
33	$12(2)$
$2(33)$	$12(2)^2$
$3(33)$	$12(2)^3$
$n(33)$	$12(2)^n$

$$n(33)=h$$
$$\frac{n(33)}{33}=\frac{h}{33}$$
$$n=\frac{h}{33}$$
$$T(h)=12(2)^{\frac{h}{33}}$$

b. $7\text{ days}=(7\text{ days})\left(\frac{24\text{ hours}}{1\text{ day}}\right)=168\text{ hours}$

$$T(168)=12(2)^{\frac{168}{33}}$$
$$T(168)\approx 408.98$$

$T(168)\approx 409$. After 1 week, there are approximately 409 bacteria present.

21. Let $F(n)$ be the number of square meters burnt after n 6-minute time intervals have passed.

a.

n	$F(n)$
0	2
1	$2(2)$
2	$2(2)^2=2^3$
3	$2(2)^3=2^4$
n	$2(2)^n=2^{n+1}$

$$F(n)=2^{n+1}$$

b. $\frac{1}{2}\text{hour}=30\text{ min}$

$$n=\frac{30\text{ min}}{6\text{ min}}=5$$

There are 5 of the 6-minute time intervals in $\frac{1}{2}$ hour.

$$F(5)=2^{5+1}=2^6$$
$$F(5)=64$$

After $\frac{1}{2}$ hour, there are 64 square meters burnt.

23. Let $H(t)$ be the number of Hispanic centenarians in the year t years since 1990. Let n be the number of 7.5 year time intervals.

a. The number of centenarians double every 7.5 years:

t	$H(t)$
0	2072
7.5	$2072(2)$
$2(7.5)$	$2072(2)^2$
$3(7.5)$	$2072(2)^3$
$n(7.5)$	$2072(2)^n$

$$n(7.5)=t$$
$$\frac{n(7.5)}{7.5}=\frac{t}{7.5}$$
$$n=\frac{t}{7.5}$$
$$H(t)=2072(2)^{\frac{t}{7.5}}$$

b. $H(60)=2072(2)^{\frac{60}{7.5}}$

$$H(60)=530,432.$$

In 2050, there will be approximately 530,432 Hispanic centenarians in the U.S.

25. Let $I(t)$ be the population of India (in millions) in the year t years since 1971. Let n be the number of 33 year time intervals.

a. The population of India doubles every 33 years:

t	$I(t)$
0	560
33	$560(2)$
$2(33)$	$560(2)^2$
$3(33)$	$560(2)^3$
$n(33)$	$560(2)^n$

$$n(33)=t$$
$$\frac{n(33)}{33}=\frac{t}{33}$$
$$n=\frac{t}{33}$$
$$I(t)=560(2)^{\frac{t}{33}}$$

b. $2015\rightarrow t=44$

$$I(44)=560(2)^{\frac{44}{33}}$$
$$I(44)\approx 1411$$

In 2015, there will be approximately 1411 million people in India.

c.

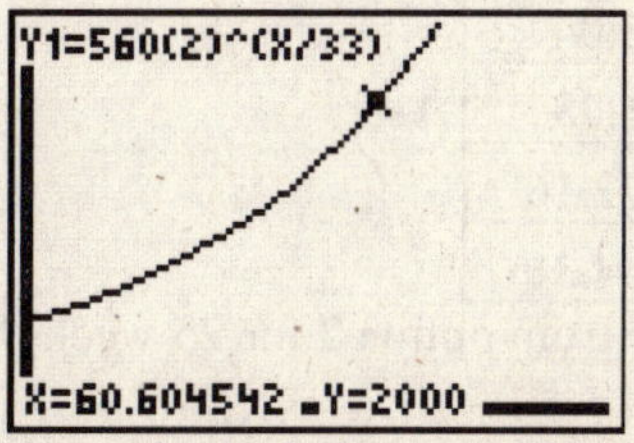

Midway through the year 2031, there will be approximately 2 billion people in India.

27. Internet search: approximately 65 years.

29. **a.** Option 1:

$$\begin{array}{r} \$2000 \\ \times\ \ 25 \\ \hline \$50{,}000 \end{array}$$

b. Your opinion.

c. Option 2: Let $S(w)$ be salary option 2 (in dollars) for week w of the job:

w	$S(w)$
1	0.01
2	$0.01(2)$
3	$0.01(2)^2$
4	$0.01(2)^3$
w	$0.01(2)^{w-1}$

$S(w)=0.01(2)^{w-1}$

d.

Week	Option 2
1	0.01
2	0.02
3	0.04
4	0.08
5	0.16
6	0.32
7	0.64
8	1.28
9	2.56
10	5.12
11	10.24
12	20.48
13	40.96
14	81.92
15	163.84
16	327.68
17	655.36
18	1310.72
19	2621.44
20	5242.88
21	10485.76
22	20971.52
23	41943.04
24	83886.08
25	167772.16
Total:	**335544.31**

The total salary from option 2 for 25 weeks of work is $335,544.31.

e. Option 2 is the best deal for 25 weeks of work.

f.

Week	Option1	Option2
1	2000	0.01
2	2000	0.02
3	2000	0.04
4	2000	0.08
5	2000	0.16
6	2000	0.32
7	2000	0.64
8	2000	1.28
9	2000	2.56
10	2000	5.12
11	2000	10.24
12	2000	20.48
13	2000	40.96
14	2000	81.92
15	2000	163.84
16	2000	327.68
17	2000	655.36
18	2000	1310.72
19	2000	2621.44
20	2000	5242.88
Totals:	**40000**	**10485.75**

Option 1 offers $40,000 for 20 weeks of work and option 2 offers $10,485.75 for 20 weeks of work therefore option 1 is the best deal for 20 weeks of work.

31. Let $P(t)$ be the percent of lead-210 left in a body t years after a person has died. Let n be the number of 22-year time intervals.

a. The percent of lead-210 is cut in half every 22 years:

t	$P(t)$
0	100
22	$100\left(\frac{1}{2}\right)$
2(22)	$100\left(\frac{1}{2}\right)^2$
3(22)	$100\left(\frac{1}{2}\right)^3$
$n(22)$	$100\left(\frac{1}{2}\right)^n$

$$n(22)=t$$
$$\frac{n(22)}{22}=\frac{t}{22}$$
$$n=\frac{t}{22}$$
$$P(t)=100\left(\tfrac{1}{2}\right)^{t/22}$$

b. $P(60)=100\left(\frac{1}{2}\right)^{60/22}\approx 15.10$. Approximately 60 years after death, there will be about 15.10% of lead-210 left in the body.

c.

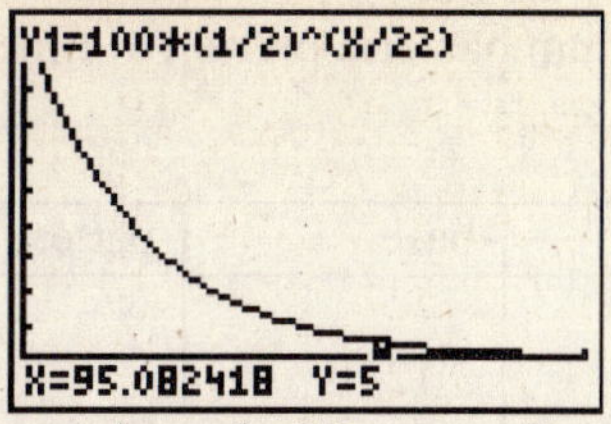

Approximately 95 years after death, there will be about 5% of lead-210 left in the body.

33. Let $P(d)$ be the percent of radon-222 left after d days. Let n be the number of 3.825-day time intervals.

a. The percent of radon-222 is cut in half every 3.825 days:

d	$P(d)$
0	100
3.825	$100\left(\frac{1}{2}\right)$
2(3.825)	$100\left(\frac{1}{2}\right)^2$
3(3.825)	$100\left(\frac{1}{2}\right)^3$
n(3.825)	$100\left(\frac{1}{2}\right)^n$

$$n(3.825) = d$$

$$\frac{n(3.825)}{3.825} = \frac{d}{3.825}$$

$$n = \frac{d}{3.825}$$

$$P(d) = 100\left(\tfrac{1}{2}\right)^{d/3.825}$$

b. $P(30) = 100\left(\tfrac{1}{2}\right)^{30/3.825} \approx 0.44$. After approximately 30 days, there will be about 0.44% of radon-222 left.

c.

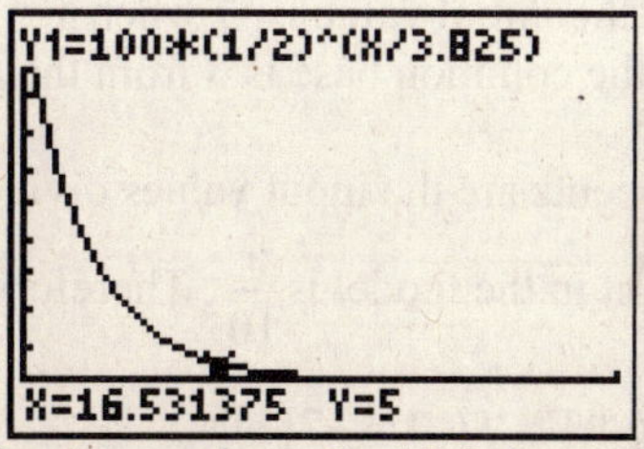

After approximately 16.5 days, there will be about 5% of radon-222 left.

35. **a.** The graph is decreasing.
b. This is an example of exponential decay.
c. $f(10) = 22$
d. $x = 6$ for $f(x) = 40$
e. $f(0) = 100$

37. **a.** The graph is increasing.
b. This is an example of exponential growth.
c. $h(5) = 20$
d. $x = 15$ for $h(x) = 100$
e. $h(0) = 10$

39. **a.** The graph is decreasing.
b. This is an example of exponential decay.
c. $g(7.5) = 20$
d. $x = 4$ for $g(x) = 30$
e. $g(0) = 60$

41.

x	$f(x)$	base
0	25	$\frac{100}{25} = 4$
1	100	$\frac{400}{100} = 4$
2	400	$\frac{1600}{400} = 4$
3	1600	$\frac{6400}{1600} = 4$
4	6400	

$a = 25$ because the initial value is 25 when $x = 0$.
$b = 4$ because the common base is 4 from the above table.

Therefore the exponential model is $f(x) = 25(4)^x$.

43.

x	$f(x)$	base
0	-35	$\frac{-245}{-35} = 7$
1	-245	$\frac{-1715}{-245} = 7$
2	-1715	$\frac{-12005}{-1715} = 7$
3	-12005	$\frac{-84035}{-12005} = 7$
4	-84035	

$a = -35$ because the initial value is -35 when $x = 0$.
$b = 7$ because the common base is 7 from the above table.

Therefore the exponential model is $f(x) = -35(7)^x$.

45.

x	$f(x)$	base
0	2000	$\frac{400}{2000} = 0.2$
1	400	$\frac{80}{400} = 0.2$
2	80	$\frac{16}{80} = 0.2$
3	16	$\frac{3.2}{16} = 0.2$
4	3.2	

$a = 2000$ because the initial value is 2000 when $x = 0$.
$b = 0.2$ because the common base is 0.2 from the above table.
Therefore the exponential model is $f(x) = 2000(0.2)^x$.

47.

x	$f(x)$	base
0	32	$\frac{48}{32}=1.5$
1	48	$\frac{72}{48}=1.5$
2	72	$\frac{108}{72}=1.5$
3	108	$\frac{162}{108}=1.5$
4	162	

$a = 32$ because the initial value is 32 when $x = 0$.
$b = 1.5$ because the common base is 1.5 from the above table.

Therefore the exponential model is $f(x)=32(1.5)^x$.

49.

x	$f(x)$	base
0	6400	$\frac{800}{6400}=0.125$
1	800	$\frac{100}{800}=0.125$
x	$f(x)$	base
2	100	$\frac{12.5}{100}=0.125$
3	12.5	$\frac{1.5625}{12.5}=0.125$
4	1.5625	

$a = 6400$ because the initial value is 6400 when $x = 0$.
$b = 0.125$ because the common base is 0.125 from the above table. Therefore the exponential model is $f(x)=6400(0.125)^x$.

51. For $f(x)=5(3^x)$: The base $b=3$ and the coefficient $a=5$.

53. For $f(x)=200\left(\frac{1}{5}\right)^x$: The base $b=\frac{1}{5}$ and the coefficient $a=200$.

55. For $f(x)=-4(2)^x$: The base $b=2$ and the coefficient $a=-4$.

57. Let $E(h)$ be the number of E coli bacteria present after h hours have passed: $E(h)=3(2)^{2h}$.

a. The base $b=2$.
b. The coefficient $a=3$.
c. The coefficient $a=3$ represents the number of E coli bacteria present at time $t=0$ (at the start).

59. Let $T(h)$ be the number of Treponema pallidum bacteria present after h hours have passed: $T(h)=12(2)^{\frac{h}{33}}$

a. The base $b=2$.
b. The coefficient $a=12$.
c. The coefficient $a=12$ represents the number of Treponema pallidum bacteria present at time $t=0$ (at the start).

61.

X	$f(x)$	base	pattern
0	3		3
5	12	$\frac{12}{3}=4$	$3(4)^1=12$
10	48	$\frac{48}{12}=4$	$3(4)^2=48$
15	192	$\frac{192}{48}=4$	$3(4)^3=192$
20	768	$\frac{768}{192}=4$	$3(4)^4=768$

$a = 3$ because the initial value is 3 when $x = 0$.
$b = 4$ because the common base is 4 from the above table.
Since the exponents are the input values divided by 5, the exponent in the model is $\frac{x}{5}$. Therefore the exponential model is $f(x)=3(4)^{\frac{x}{5}}$.

63.

x	$f(x)$	base	pattern
0	-7		-7
10	-21	$\frac{-21}{-7}=3$	$-7(3)^1=-21$
20	-63	$\frac{-63}{-21}=3$	$-7(3)^2=-63$
30	-189	$\frac{-189}{-63}=3$	$-7(3)^3=-189$
40	-567	$\frac{-567}{-189}=3$	$-7(3)^4=-567$

a = -7 because the initial value is -7 when $x = 0$.
$b = 3$ because the common base is 3 from the above table.
Since the exponents are the input values divided by 10, the exponent in the model is $\frac{x}{10}$. Therefore the exponential model is $f(x)=-7(3)^{\frac{x}{10}}$.

65.

x	$f(x)$	base	pattern
0	1701		1701
6	567	$\frac{567}{1701}=\frac{1}{3}$	$1701\left(\frac{1}{3}\right)^1=567$
12	189	$\frac{189}{567}=\frac{1}{3}$	$1701\left(\frac{1}{3}\right)^2=189$
18	63	$\frac{63}{189}=\frac{1}{3}$	$1701\left(\frac{1}{3}\right)^3=63$
24	21	$\frac{21}{63}=\frac{1}{3}$	$1701\left(\frac{1}{3}\right)^4=21$

$a = 1701$ because the initial value is 1701 when $x = 0$.

$b = \frac{1}{3}$ because the common base is $\frac{1}{3}$ from the above table.
Since the exponents are the input values divided by 6, the exponent in the model is $\frac{x}{6}$. Therefore the exponential model is $f(x) = 1701\left(\frac{1}{3}\right)^{\frac{x}{6}}$.

67.

x	$f(x)$	base	pattern
2	80		80
3	320	$\frac{320}{80} = 4$	$80(4)^1 = 320$
4	1280	$\frac{1280}{320} = 4$	$80(4)^2 = 1280$
5	5120	$\frac{5120}{1280} = 4$	$80(4)^3 = 5120$
6	20480	$\frac{20480}{5120} = 4$	$80(4)^4 = 20480$

$b = 4$ because the common base is 4 from the above table. The input values do not include $x = 0$ so we do not know the initial value of the function. The pattern in the above table does show that we have 80 times the common base 4 raised to an exponent. Since the exponents are the input values minus 2, the exponent in the model is $x - 2$. Therefore the exponential model is $f(x) = 80(4)^{x-2}$.

69.

x	$f(x)$	base	pattern
5	2		2
6	12	$\frac{12}{2} = 6$	$2(6)^1 = 12$
7	72	$\frac{72}{12} = 6$	$2(6)^2 = 72$
8	432	$\frac{432}{72} = 6$	$2(6)^3 = 432$
9	2592	$\frac{2592}{432} = 6$	$2(6)^4 = 2592$

$b = 6$ because the common base is 6 from the above table. The input values do not include $x = 0$ so we do not know the initial value of the function. The pattern in the above table does show that we have 2 times the common base 6 raised to an exponent. Since the exponents are the input values minus 5, the exponent in the model is $x - 5$. Therefore the exponential model is $f(x) = 2(6)^{x-5}$.

71.

x	$f(x)$	base	pattern
4	3584		3584
5	896	$\frac{896}{3584} = 0.25$	$3584(0.25)^1 = 896$
6	224	$\frac{224}{896} = 0.25$	$3584(0.25)^2 = 224$
7	56	$\frac{56}{224} = 0.25$	$3584(0.25)^3 = 56$
8	14	$\frac{14}{56} = 0.25$	$3584(0.25)^4 = 14$

$b = 0.25$ because the common base is 0.25 from the above table. The input values do not include $x = 0$ so we do not know the initial value of the function. The pattern in the above table does show that we have 3584 times the common base 0.25 raised to an exponent. Since the exponents are the input values minus 4, the exponent in the model is $x - 4$. Therefore the exponential model is $f(x) = 3584(0.25)^{x-4}$.

73. The function is increasing because the base is greater than 1.

75. The function is decreasing because the base is less than 1.

77. The function is increasing because the base is greater than 1.

79. The function is decreasing because the base is less than 1.

Section 12.2 Solving Equations Using Exponent Rules

1. This is an exponential equation because the variable is in the exponent.
3. This is a power equation because it has a constant (a number) as an exponent.
5. This is a power equation because it has a constant (a number) as an exponent.
7. This is an exponential equation because the variable is in the exponent.
9. This is a power equation because it has a constant (a number) as an exponent.

11.

$$2^x = 8$$
$$2^x = 2^3$$
$$x = 3$$

13.

$$5^c = 3125$$
$$5^c = 5^5$$
$$c = 5$$

15.

$$\frac{1}{9} = 3^t$$
$$\frac{1}{9} = \frac{1}{3^2} = 3^{-2}$$
$$3^t = 3^{-2}$$
$$t = -2$$

17.

$$3^x = \frac{1}{81}$$
$$\frac{1}{81} = \frac{1}{3^4} = 3^{-4}$$
$$3^x = 3^{-4}$$
$$x = -4$$

19.

$$(-2)^d = -32$$
$$(-2)^d = (-2)^5$$
$$d = 5$$

21.

$$(-6)^w = 36$$
$$(-6)^w = (-6)^2$$
$$w = 2$$

23.

$$\left(\frac{1}{2}\right)^x = \frac{1}{16}$$
$$\left(\frac{1}{2}\right)^x = \left(\frac{1}{2}\right)^4$$
$$x = 4$$

25.

$$\left(\frac{1}{2}\right)^x = 16$$
$$\left(2^{-1}\right)^x = 2^4$$
$$2^{-x} = 2^4$$
$$-x = 4$$
$$x = -4$$

27.

$$\left(\frac{1}{2}\right)^x = 32$$
$$\left(2^{-1}\right)^x = 2^5$$
$$2^{-x} = 2^5$$
$$-x = 5$$
$$x = -5$$

29.

$$10^x = 1000$$
$$10^x = 10^3$$
$$x = 3$$

31.

$$10^x = 1$$
$$x = 0$$

33.

$$(-5)^m = 1$$
$$m = 0$$

35.

$$\left(\frac{2}{3}\right)^t = 1$$
$$t = 0$$

37. Tom evaluated $\frac{1}{16} = 4^2$ as instead of $\frac{1}{16} = 4^{-2}$. So he missed the negative sign in the exponent.

$$4^x = \frac{1}{16}$$
$$4^x = \frac{1}{4^2} = 4^{-2}$$
$$x = -2$$

39. $B(h) = 100\left(2^h\right)$

a. $B(8) = 100\left(2^8\right) = 25,600$ There are 25,600 Lactobacillus acidophilus bacteria present after 8 hours have passed.

b.

$$800 = 100\left(2^h\right)$$
$$\frac{800}{100} = \frac{100\left(2^h\right)}{100}$$
$$8 = \left(2^h\right)$$

$$2^3 = 2^h$$
$$h = 3$$

There are 800 Lactobacillus acidophilus bacteria present after 3 hours have passed.

41. $R(h) = 5(3^h)$

a. $R(8) = 5(3^8) = 32{,}805$

After 8 hours have passed, 32,805 people have heard the rumor.

b.

$$405 = 5(3^h)$$
$$\frac{405}{5} = \frac{5(3^h)}{5}$$
$$81 = 3^h$$
$$3^4 = 3^h$$
$$h = 4$$

It takes 4 hours for the rumor to spread to 405 people.

43.

$$5(2^x) = 40$$
$$\frac{5(2^x)}{5} = \frac{40}{5}$$
$$2^x = 8$$
$$2^x = 2^3$$
$$x = 3$$

45.

$$-2(5^c) = -250$$
$$\frac{-2(5^c)}{-2} = \frac{-250}{-2}$$
$$5^c = 125$$
$$5^c = 5^3$$
$$c = 3$$

47.

$$3^c + 5 = 32$$
$$3^c + 5 - 5 = 32 - 5$$
$$3^c = 27$$
$$3^c = 3^3$$
$$c = 3$$

51.

$$-4(5^m) - 9 = -109$$
$$-4(5^m) - 9 + 9 = -109 + 9$$
$$-4(5^m) = -100$$
$$\frac{-4(5^m)}{-4} = \frac{-100}{-4}$$
$$5^m = 25 = 5^2$$
$$m = 2$$

53.

$$3(6^x) + 2(6^x) = 180$$
$$5(6^x) = 180$$
$$\frac{5(6^x)}{5} = \frac{180}{5}$$
$$6^x = 36$$
$$6^x = 6^2$$
$$x = 2$$

55.

$$10(2^x) - 7(2^x) = 48$$
$$3(2^x) = 48$$
$$\frac{3(2^x)}{3} = \frac{48}{3}$$
$$2^x = 16$$
$$2^x = 2^4$$
$$x = 4$$

57.

$$7(2^x) = 48 + 4(2^x)$$
$$7(2^x) - 4(2^x) = 48 + 4(2^x) - 4(2^x)$$
$$3(2^x) = 48$$
$$\frac{3(2^x)}{3} = \frac{48}{3}$$
$$2^x = 16$$
$$2^x = 2^4$$
$$x = 4$$

59.

$$13(7^w) + 20 = 5(7^w) + 2764$$
$$13(7^w) + 20 - 5(7^w) = 5(7^w) + 2764 - 5(7^w)$$
$$8(7^w) + 20 = 2764$$
$$8(7^w) + 20 - 20 = 2764 - 20$$
$$8(7^w) = 2744$$
$$\frac{8(7^w)}{8} = \frac{2744}{8}$$
$$7^w = 343$$
$$7^w = 7^3$$
$$w = 3$$

61.

$$x^5 = 32$$
$$(x^5)^{\frac{1}{5}} = (32)^{\frac{1}{5}}$$
$$x = 2$$

63.

$$x^6 = 15625$$
$$(x^6)^{\frac{1}{6}} = \pm(15625)^{\frac{1}{6}}$$
$$x = \pm 5$$

65.

$$12w^4 = 7500$$
$$\frac{12w^4}{12} = \frac{7500}{12}$$
$$w^4 = 625$$
$$\left(w^4\right)^{\frac{1}{4}} = \pm(625)^{\frac{1}{4}}$$
$$w = \pm 5$$

67.

$$-7x^3 = 1512$$
$$\frac{-7x^3}{-7} = \frac{1512}{-7}$$
$$x^3 = -216$$
$$\left(x^3\right)^{\frac{1}{3}} = (-216)^{\frac{1}{3}}$$
$$x = -6$$

69. Warrick forgot to put the $\pm$ symbol on the final answer. Both positive 6 and negative 6 are correct answers. The correct answer is $x = \pm 6$.

71. Frank divided 216 by 3 instead of raising 216 to the power $\frac{1}{3}$. The correct answer is $r = 6$.

73. **a.** $V = \frac{4}{3}\pi(12)^3 \approx 7238.2\,\text{cm}^3$

An NBA basketball with radius 12 cm has a volume of about $7238.2\,\text{cm}^3$.

b.

$$(8011) = \frac{4}{3}\pi r^3$$
$$\left(\frac{3}{4\pi}\right)(8011) = \left(\frac{4\pi}{3}r^3\right)\left(\frac{3}{4\pi}\right)$$
$$1912.49 \approx r^3$$
$$(1912.49)^{\frac{1}{3}} \approx \left(r^3\right)^{\frac{1}{3}}$$
$$r \approx 12.4$$

An International Basketball Federation basketball with a volume of $8011\,\text{cm}^3$ has a radius of about 12.4 cm.

75. The volume of the cube is given by the formula $V = s^3$.

Since the volume is given as 91.125 in^3 substituting this value into the formula we get:

$$91.125 = s^3$$
$$(91.125)^{\frac{1}{3}} = \left(s^3\right)^{\frac{1}{3}}$$
$$s = 4.5$$

The length of the side is 4.5 in.

77.

$$3x^5 + 94 = 190$$
$$3x^5 + 94 - 94 = 190 - 94$$
$$3x^5 = 96$$
$$x^5 = 32$$
$$\left(x^5\right)^{\frac{1}{5}} = (32)^{\frac{1}{5}}$$
$$x = 2$$

79.

$$5x^6 - 30 = 1475$$
$$5x^6 - 30 + 30 = 1475 + 30$$
$$5x^6 = 1505$$
$$x^6 = 301$$
$$\left(x^6\right)^{\frac{1}{6}} = \pm(301)^{\frac{1}{6}}$$
$$x \approx \pm 2.6$$

81.

$$96 - 24c^4 = 40$$
$$96 - 24c^4 - 96 = 40 - 96$$
$$-24c^4 = -56$$
$$\frac{-24c^4}{-24} = \frac{-56}{-24}$$
$$c^4 = \frac{7}{3}$$
$$\left(c^4\right)^{\frac{1}{4}} = \pm\left(\frac{7}{3}\right)^{\frac{1}{4}}$$
$$c \approx \pm 1.2$$

83.

$$\frac{4x^3 + 5}{8} = 63.125$$
$$8\left(\frac{4x^3 + 5}{8}\right) = 8(63.125)$$
$$4x^3 + 5 = 505$$
$$4x^3 + 5 - 5 = 505 - 5$$
$$4x^3 = 500$$
$$\frac{4x^3}{4} = \frac{500}{4}$$
$$x^3 = 125$$
$$\left(x^3\right)^{\frac{1}{3}} = (125)^{\frac{1}{3}}$$
$$x = 5$$

85.

$$\frac{7b^4 + 58}{8} = 1141.25$$
$$8\left(\frac{7b^4 + 58}{8}\right) = 8(1141.25)$$
$$7b^4 + 58 = 9130$$
$$7b^4 + 58 - 58 = 9130 - 58$$
$$7b^4 = 9072$$
$$\frac{7b^4}{7} = \frac{9072}{7}$$
$$b^4 = 1296$$
$$\left(b^4\right)^{\frac{1}{4}} = \pm(1296)^{\frac{1}{4}}$$
$$b = \pm 6$$

87. Exponential

$$3^k = 81$$
$$3^k = 3^4$$
$$k = 4$$

89. Power

$$150r^5 = 1725$$
$$\frac{150r^5}{150} = \frac{1725}{150}$$
$$r^5 = 11.5$$
$$\left(r^5\right)^{\frac{1}{5}} = (11.5)^{\frac{1}{5}}$$
$$r \approx 1.6$$

91. Exponential

$$3\left(4^x\right) + 20 = 212$$
$$3\left(4^x\right) + 20 - 20 = 212 - 20$$
$$3\left(4^x\right) = 192$$
$$\frac{3\left(4^x\right)}{3} = \frac{192}{3}$$
$$4^x = 64$$
$$4^x = 4^3$$
$$x = 3$$

93. Power

$$45 - 3.5g^5 = -67$$
$$45 - 3.5g^5 - 45 = -67 - 45$$
$$-3.5g^5 = -112$$
$$\frac{-3.5g^5}{-3.5} = \frac{-112}{-3.5}$$
$$g^5 = 32$$
$$\left(g^5\right)^{\frac{1}{5}} = (32)^{\frac{1}{5}}$$
$$g = 2$$

95. Power

$$\frac{3.6h^8 - 56}{33} = 58.168$$
$$33\left(\frac{3.6h^8 - 56}{33}\right) = 33(58.168)$$
$$3.6h^8 - 56 = 1919.544$$
$$3.6h^8 - 56 + 56 = 1919.544 + 56$$
$$3.6h^8 = 1975.544$$
$$\frac{3.6h^8}{3.6} = \frac{1975.544}{3.6}$$
$$h^8 \approx 548.76222$$
$$\left(h^8\right)^{\frac{1}{8}} \approx \pm(548.76222)^{\frac{1}{8}}$$
$$h \approx \pm 2.2$$

97. Exponential

$$\left(\frac{1}{5}\right)^x = \frac{1}{125}$$
$$\left(\frac{1}{5}\right)^x = \left(\frac{1}{5}\right)^3$$
$$x = 3$$

99. Power

$$5x^6 + 20 = 3x^6 + 1478$$
$$5x^6 + 20 - 3x^6 = 3x^6 + 1478 - 3x^6$$
$$2x^6 + 20 = 1478$$
$$2x^6 + 20 - 20 = 1478 - 20$$
$$2x^6 = 1458$$
$$\frac{2x^6}{2} = \frac{1458}{2}$$
$$x^6 = 729$$
$$\left(x^6\right)^{\frac{1}{6}} = \pm(729)^{\frac{1}{6}}$$
$$x = \pm 3$$

101. Exponential

$$5\left(\frac{1}{2}\right)^x - \frac{2}{32} = 3\left(\frac{1}{2}\right)^x - \frac{3}{64}$$
$$5\left(\frac{1}{2}\right)^x - \frac{2}{32} - 3\left(\frac{1}{2}\right)^x = 3\left(\frac{1}{2}\right)^x - \frac{3}{64} - 3\left(\frac{1}{2}\right)^x$$
$$2\left(\frac{1}{2}\right)^x - \frac{2}{32} = -\frac{3}{64}$$
$$2\left(\frac{1}{2}\right)^x - \frac{2}{32} + \frac{2}{32} = -\frac{3}{64} + \frac{2}{32}$$
$$2\left(\frac{1}{2}\right)^x = -\frac{3}{64} + \frac{2}{32}$$
$$2\left(\frac{1}{2}\right)^x = \frac{1}{64}$$
$$\frac{2\left(\frac{1}{2}\right)^x}{2} = \frac{\frac{1}{64}}{2}$$
$$\left(\frac{1}{2}\right)^x = \frac{1}{128}$$
$$\left(\frac{1}{2}\right)^x = \left(\frac{1}{2}\right)^7$$
$$x = 7$$

CHAPTER 12 Exponential Functions

Section 12.3 Graphing Exponential Functions

1. **a.** a is positive because the graph is above the x-axis.

b. $b > 1$ because we have exponential growth with $a > 0$.

c. The graph is increasing.

d. Domain: All real numbers $(-\infty, \infty)$

e. Range: $(0, \infty)$

3. **a.** a is positive because the graph is above the x-axis.

b. $b < 1$ because we have exponential decay with $a > 0$.

c. The graph is decreasing.

d. Domain: All real numbers $(-\infty, \infty)$

e. Range: $(0, \infty)$

5. **a.** a is negative because the graph is below the x-axis.

b. $b < 1$ because we have exponential growth with $a < 0$.

c. The graph is increasing.

d. Domain: All real numbers $(-\infty, \infty)$

e. Range: $(-\infty, 0)$

7. **a.** a is negative because the graph is below the x-axis.

b. $b > 1$ because we have exponential decay with $a < 0$.

c. The graph is decreasing.

d. Domain: All real numbers $(-\infty, \infty)$

e. Range: $(-\infty, 0)$

9. **a.** a is negative because the graph is below the x-axis.

b. $b > 1$ because we have exponential decay with $a < 0$.

c. The graph is decreasing.

d. Domain: All real numbers $(-\infty, \infty)$

e. Range: $(-\infty, 0)$

11.

$f(x) = 7(2)^x$

$a = 7$ so the y-intercept is $(0, 7)$.

$b = 2,\ b > 1$ and $a > 0$ so the graph is increasing.

Domain: $(-\infty, \infty)$

Range: $(0, \infty)$

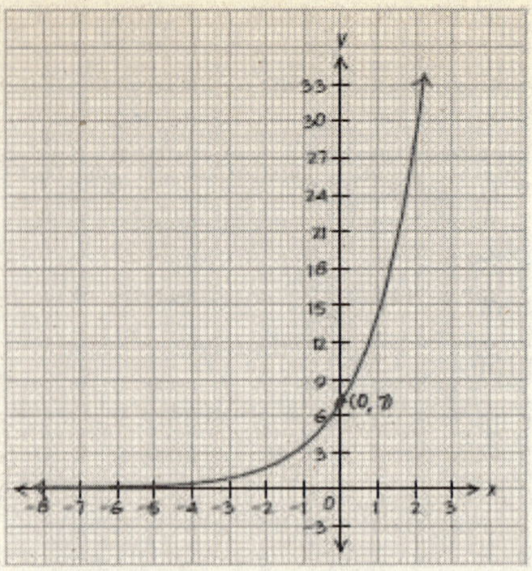

13.

$g(x) = 3(1.2)^x$

$a = 3$ so the y-intercept is $(0, 3)$.

$b = 1.2,\ b > 1$ and $a > 0$ so the graph is increasing.

Domain: $(-\infty, \infty)$

Range: $(0, \infty)$

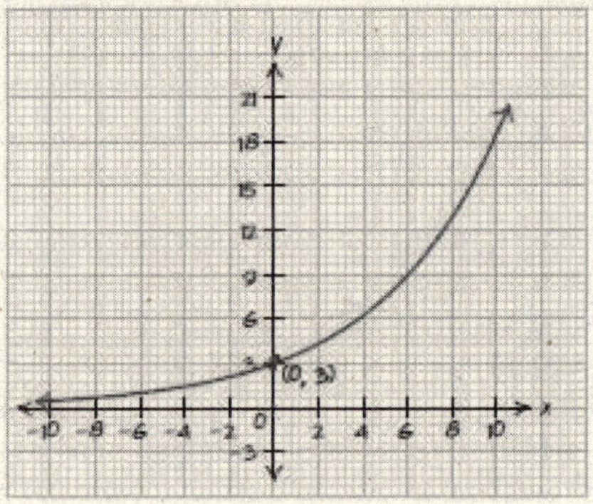

15.

$h(x) = 12(1.4)^x$

$a = 12$ so the y-intercept is $(0, 12)$.

$b = 1.4,\ b > 1$ and $a > 0$ so the graph is increasing.

Domain: $(-\infty, \infty)$

Range: $(0, \infty)$

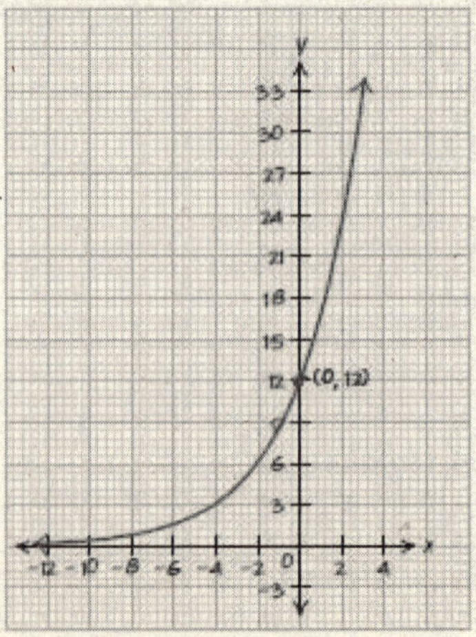

17.

$$p(t)=140\left(\frac{1}{2}\right)^t$$

$a=140$ so the t-intercept is $(0,140)$.

$b=\frac{1}{2}$, $b<1$ and $a>0$ so the graph is decreasing.

Domain: $(-\infty,\infty)$

Range: $(0,\infty)$

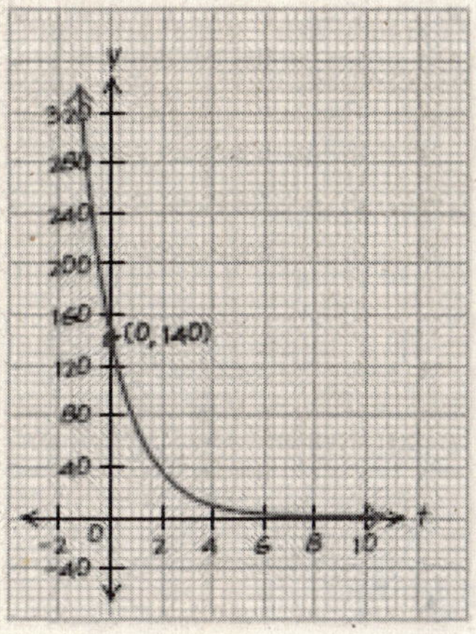

19.

$$f(x)=250\left(\frac{1}{4}\right)^x$$

$a=250$ so the x-intercept is $(0,250)$.

$b=\frac{1}{4}$, $b<1$ and $a>0$ so the graph is decreasing.

Domain: $(-\infty,\infty)$

Range: $(0,\infty)$

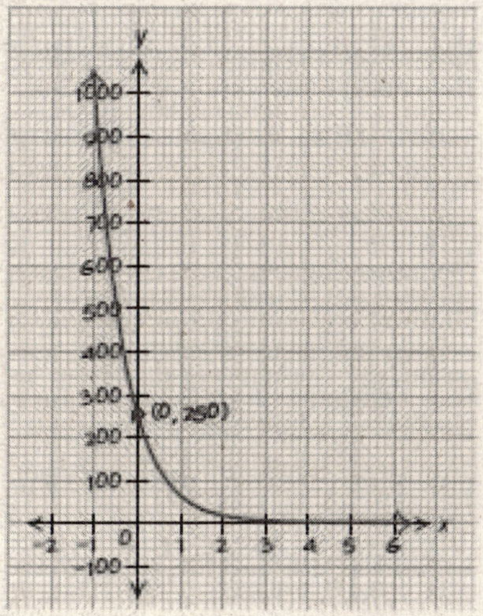

21. **a.** $B(n)=4(2)^n$

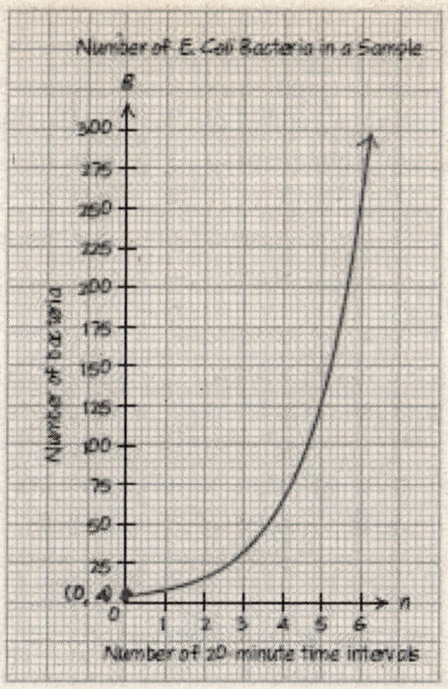

b.

$$B(0)=4(2)^0=4$$

The B-intercept is $(0,4)$.

At the start, there were 4 E coli bacteria present in the sample.

23. **a.** $N(t)=5000\left(\frac{1}{2}\right)^t$

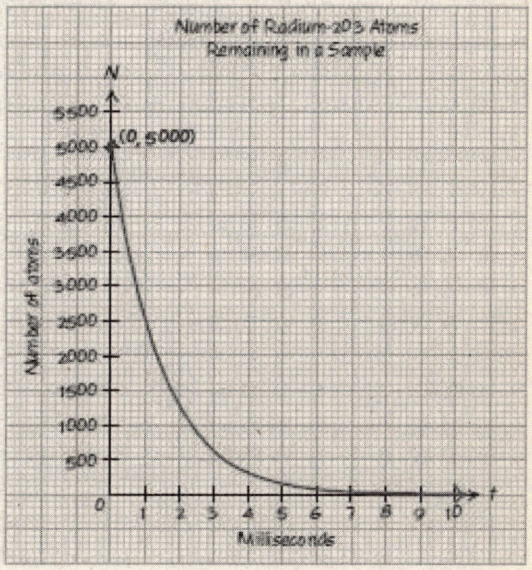

b.

$$N(0)=5000\left(\frac{1}{2}\right)^0=5000$$

The N-intercept is $(0,5000)$.

At the start, there were 5000 Radium-203 atoms present in the sample.

25.

$f(x) = -2(1.4)^x$

$a = -2$ so the y-intercept is $(0, -2)$.

$b = 1.4$, $b > 1$ and $a < 0$ so the graph is decreasing.

Domain: $(-\infty, \infty)$

Range: $(-\infty, 0)$

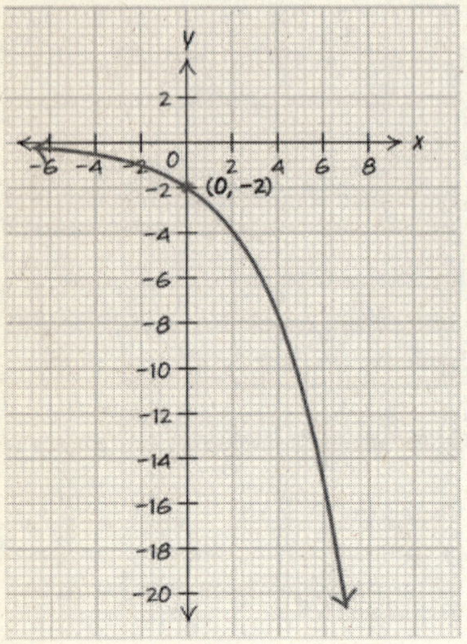

27.

$g(t) = -3(0.7)^t$

$a = -3$ so the vertical intercept is $(0, -3)$.

$b = 0.7$, $b < 1$ and $a < 0$ so the graph is increasing.

Domain: $(-\infty, \infty)$

Range: $(-\infty, 0)$

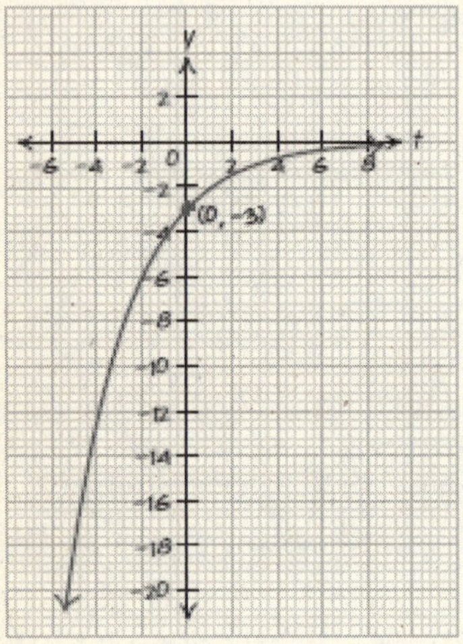

29.

$h(m) = 0.5(2.5)^m$

$a = 0.5$ so the vertical intercept is $(0, 0.5)$.

$b = 2.5$, $b > 1$ and $a > 0$ so the graph is increasing.

Domain: $(-\infty, \infty)$

Range: $(0, \infty)$

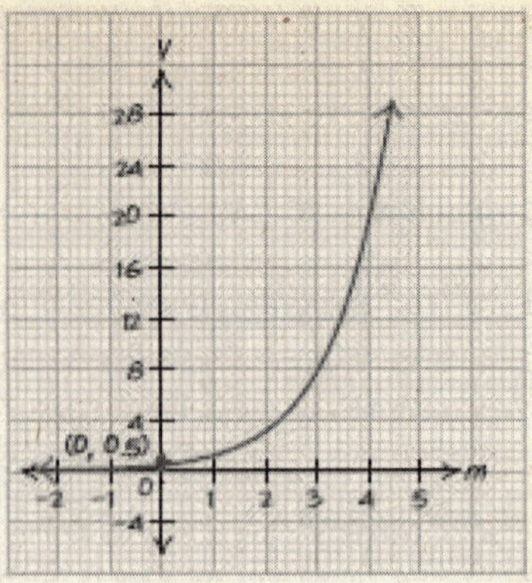

31.

$j(w) = -0.5(4)^w$

$a = -0.5$ so the vertical intercept is $(0, -0.5)$.

$b = 4$, $b > 1$ and $a < 0$ so the graph is decreasing.

Domain: $(-\infty, \infty)$

Range: $(-\infty, 0)$

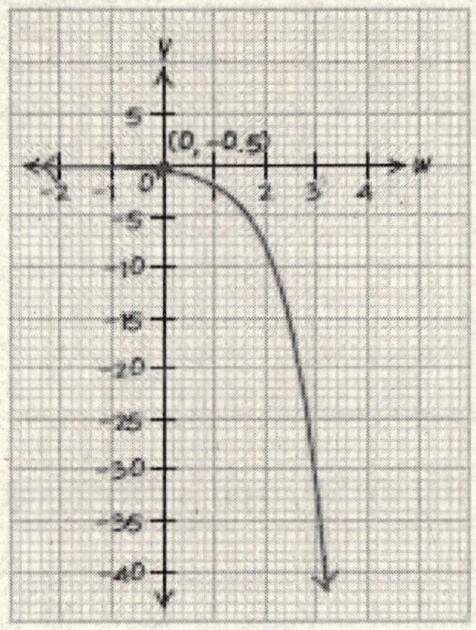

33.

$h(t) = -0.4(1.5)^t$

$a = -0.4$ so the vertical intercept is $(0, -0.4)$.

$b = 1.5$, $b > 1$ and $a < 0$ so the graph is decreasing.

Domain: $(-\infty, \infty)$

Range: $(-\infty, 0)$

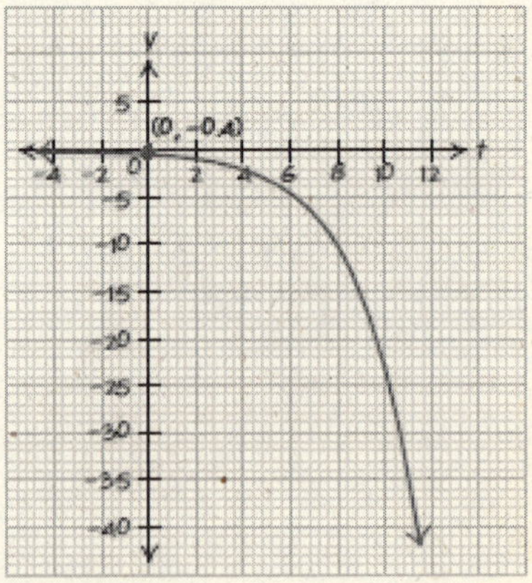

35.

$$c(n) = 550\left(\frac{3}{4}\right)^n$$

$a = 550$ so the vertical intercept is $(0, 550)$.

$b = \frac{3}{4}$, $b < 1$ and $a > 0$ so the graph is decreasing.

Domain: $(-\infty, \infty)$

Range: $(0, \infty)$

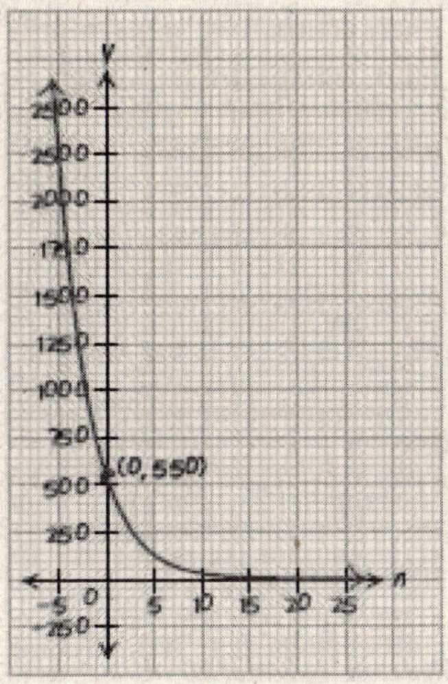

37.

$$f(x) = -500\left(\frac{1}{5}\right)^x$$

$a = -500$ so the vertical intercept is $(0, -500)$.

$b = \frac{1}{5}$, $b < 1$ and $a < 0$ so the graph is increasing.

Domain: $(-\infty, \infty)$

Range: $(-\infty, 0)$

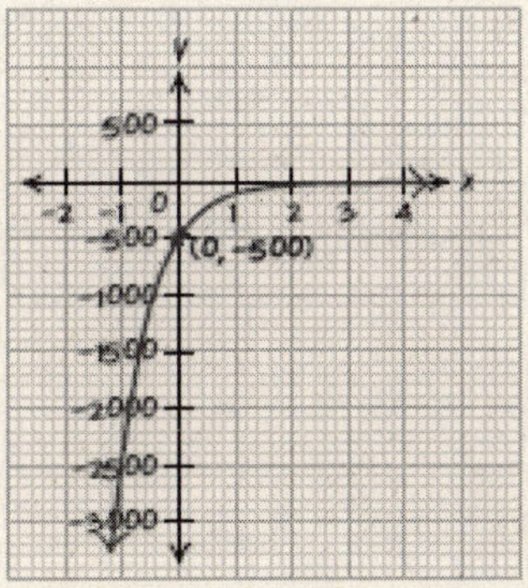

39.

$$f(x) = -700(0.95)^x$$

$a = -700$ so the vertical intercept is $(0, -700)$.

$b = 0.95$, $b < 1$ and $a < 0$ so the graph is increasing.

Domain: $(-\infty, \infty)$

Range: $(-\infty, 0)$

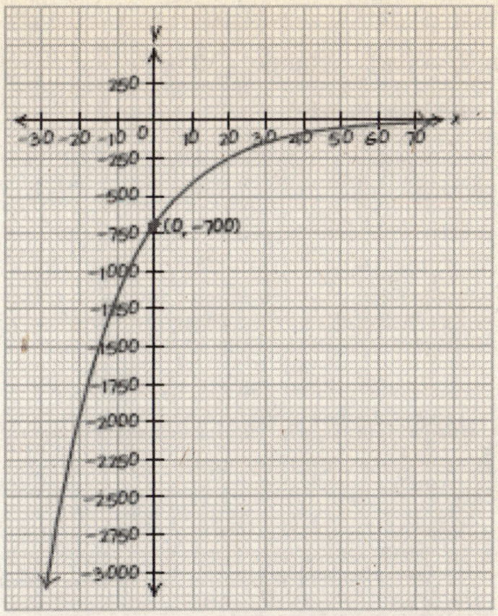

41. The student wrote the domain started at zero and the range started at 7. Domain: $(-\infty, \infty)$ Range: $(0, \infty)$

43. The student wrote the domain correctly but the range should be from zero to infinity.
Domain: $(-\infty, \infty)$ Range: $(0, \infty)$

45. **a.** $y = 10$
b. Domain: $(-\infty, \infty)$ **c.** Range: $(10, \infty)$

47. **a.** $y = -5$
b. Domain: $(-\infty, \infty)$ **c.** Range: $(-\infty, -5)$

49. $f(x) = 4(5)^x + 3$

a. The horizontal asymptote is $y = 3$. The graph is shifted up 3 units so the horizontal asymptote is also shifted up 3 units.

b. Domain: $(-\infty, \infty)$ The domain of an exponential function is all real numbers.

c. Range: $(3, \infty)$ The function is an exponential growth function because the base is greater than 1. It will also be above the x-axis since a is positive.

51. $f(x) = 7(4)^x$

a. The horizontal asymptote is $y = 0$.

b. Domain: $(-\infty, \infty)$ The domain of an exponential function is all real numbers.

c. Range: $(0, \infty)$ The function is an exponential growth function because the base is greater than 1. It will also be above the x-axis since a is positive.

53. $f(x) = -40(1.25)^x - 6$

a. The horizontal asymptote is $y = -6$. The graph is shifted down 6 units so the horizontal asymptote is also shifted down 6 units.

b. Domain: $(-\infty, \infty)$ The domain of an exponential function is all real numbers.

c. Range: $(-\infty, -6)$ The function is an exponential growth function because the base is greater than 1; a is negative so the function is flipped below the x-axis.

55. Horizontal asymptote: $y = 30$

Domain: $(-\infty, \infty)$ Range: $(30, \infty)$

$f(x) = 200(0.4)^x + 30$

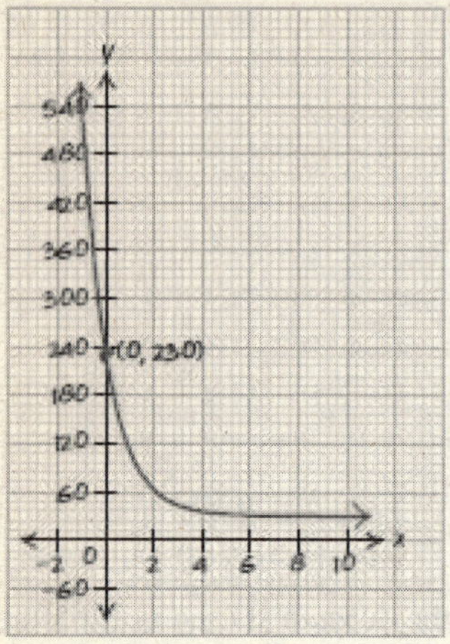

57. Horizontal asymptote: $y = 3$

Domain: $(-\infty, \infty)$ Range: $(3, \infty)$

$h(x) = 20(1.5)^x + 3$

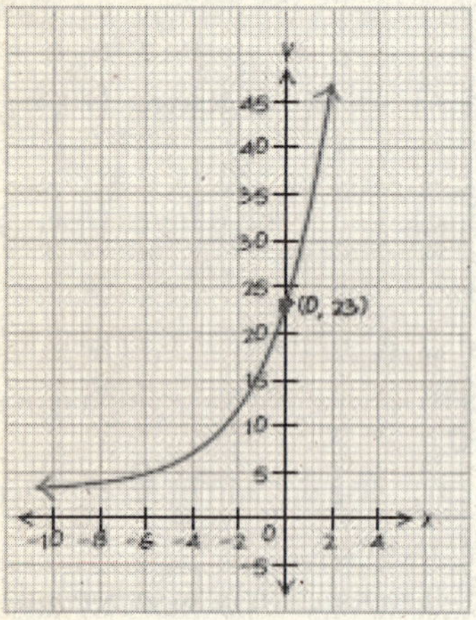

59. Horizontal asymptote: $y = -12$

Domain: $(-\infty, \infty)$ Range: $(-\infty, -12)$

$h(t) = -3(2)^t - 12$

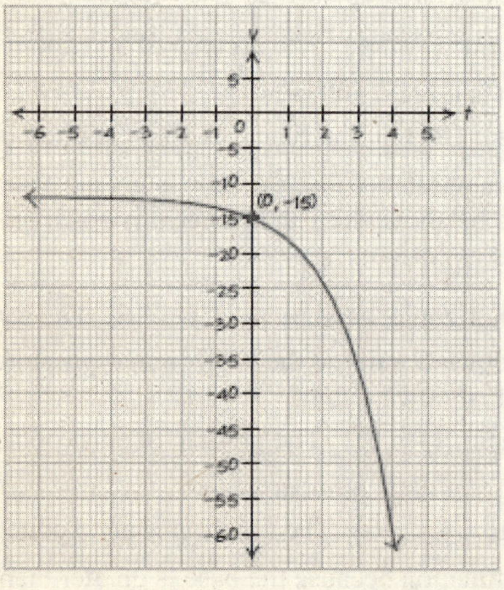

61. Horizontal asymptote: $y = -50$

Domain: $(-\infty, \infty)$ Range: $(-\infty, -50)$

$g(x) = -300(0.9)^x - 50$

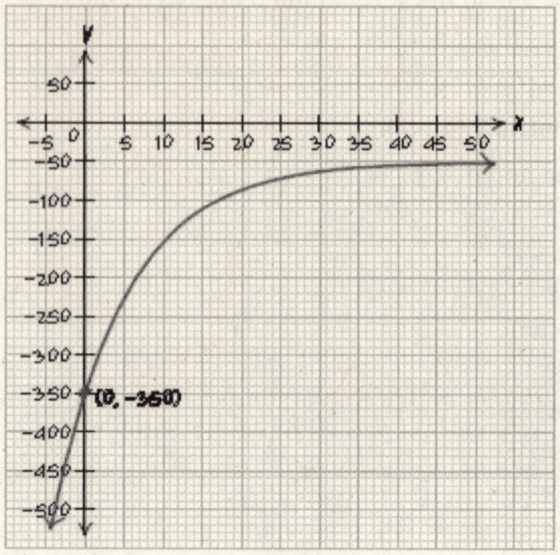

63. Horizontal asymptote: $y = 25$

Domain: $(-\infty, \infty)$ Range: $(25, \infty)$

$f(x) = 750\left(\frac{2}{5}\right)^x + 25$

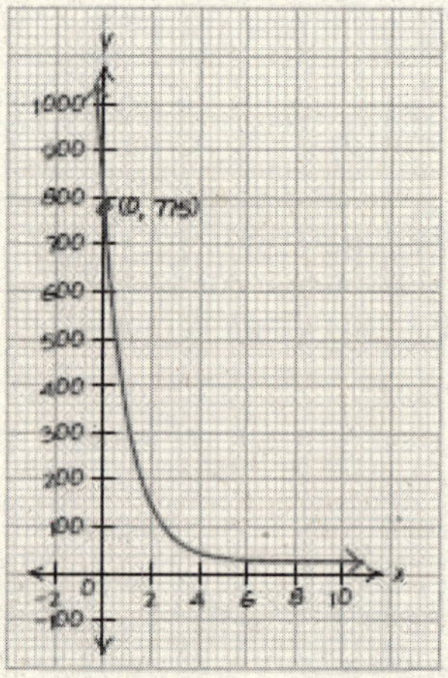

65. Domain: $(-\infty, \infty)$ Range: $(-\infty, \infty)$

$f(x) = 3x + 5$

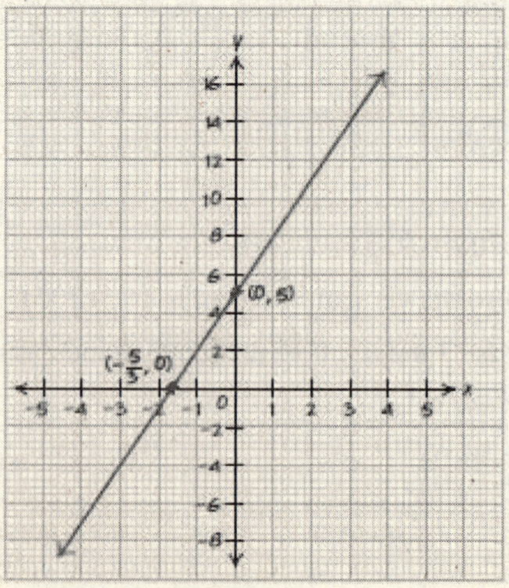

67. Domain: $(-\infty, \infty)$ Range: $[3, \infty)$

$f(x) = (x-4)^2 + 3$

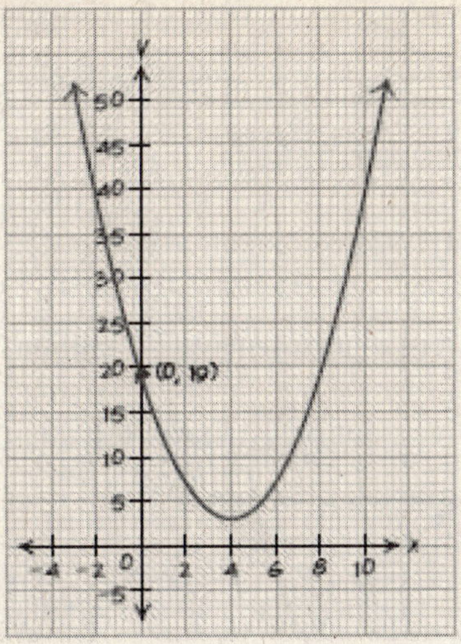

69. Domain: $(-\infty, \infty)$ Range: $(0, \infty)$

$f(x) = 6(3)^x$

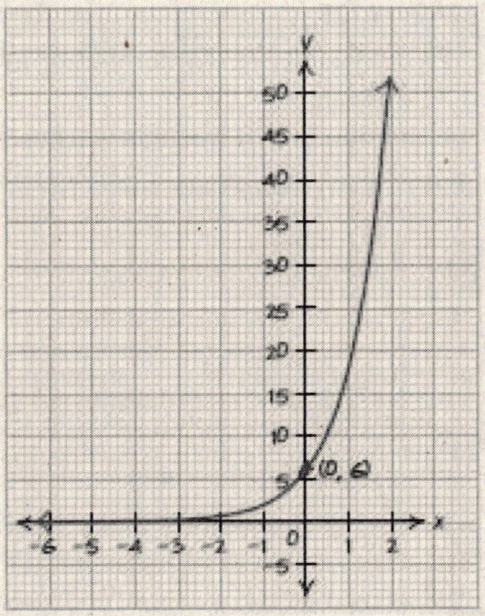

71. Domain: $(-\infty, \infty)$ Range: $(-\infty, \infty)$

$f(x) = 5 + 2(x+7)$

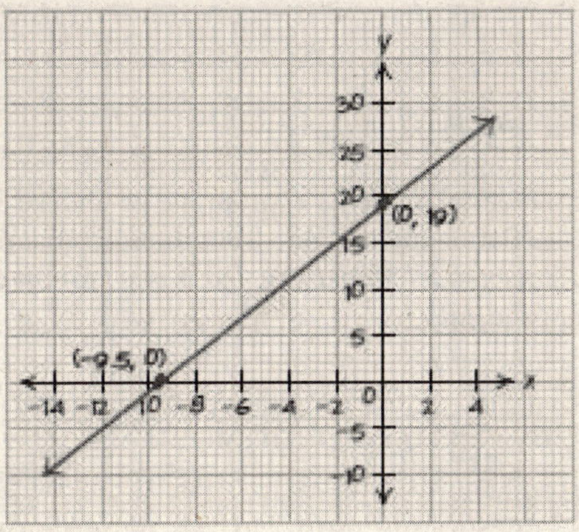

73. Domain: $(-\infty, \infty)$ Range: $[-75, \infty)$

$h(x) = \frac{1}{3}(x+6)^2 - 75$

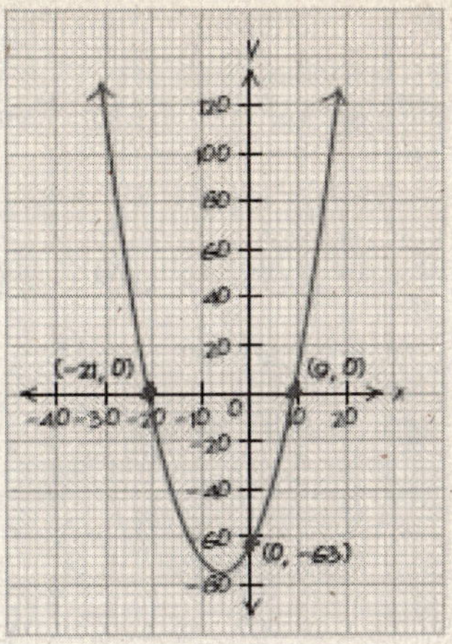

1. $f(x) = 25(1.5)^x$

3. $f(x) = 900(0.9)^x$

5. b, because the function is increasing too rapidly.

7. a, because the y-intercept is too low.

9. Adjust $a = 3.5$, because the value of the functions equals 3.5 when $x = 0$.
   The new equation is $f(x) = 3.5(1.2)^x$.

11. Adjust $a = 78$, because the value of the functions equals 78 when $x = 0$.
    The new equation is $f(x) = 78(0.45)^x$.

13. Adjust $b = 1.3$, because the original graph is increasing too fast.
    The new equation is $f(x) = 8(1.3)^x$.

15. Adjust $b = 0.5$, because the original graph is increasing too slowly.
    The new equation is $f(x) = 120(0.5)^x$.

17. Adjust $a = 5$, because the value of the functions equals 5 when $x = 0$.
    Adjust $b = 1.2$, because the original graph is increasing too fast.
    The new equation is $f(x) = 5(1.2)^x$.

19. **a.** Let $I(t)$ be the number of inter net hosts (in millions) in the year t years since 1990.

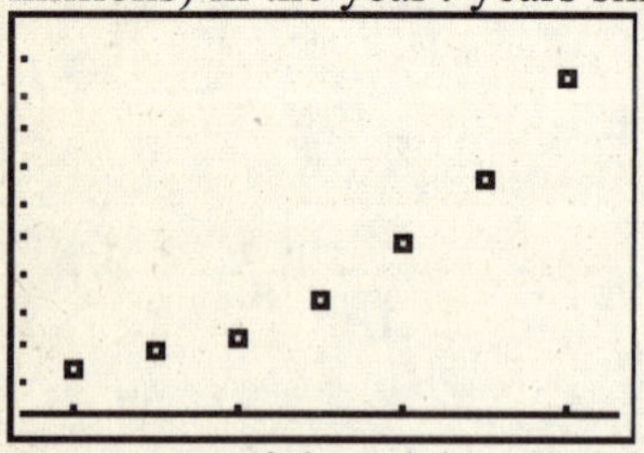

An exponential model is appropriate so choose a point from the flatter end and one from the steeper end.
Note: Your model may differ, but it must follow the trend of the trend of the data.

$I(t) = ab^t$

To find a and b choose two points $(3, 1.3)$ and $(5.5, 6.6)$.

$1.3 = ab^3$ and $6.6 = ab^{5.5}$

$$\frac{6.6}{1.3} = \frac{ab^{5.5}}{ab^3}$$

$$\frac{6.6}{1.3} = b^{2.5}$$

$$\left(\frac{6.6}{1.3}\right)^{\frac{1}{2.5}} = \left(b^{2.5}\right)^{\frac{1}{2.5}}$$

$b \approx 1.92$

Use b and one of the above equations to find a.

$1.3 = a(1.92)^3$

$$\frac{1.3}{(1.92)^3} = \frac{a(1.92)^3}{(1.92)^3}$$

$a \approx 0.184$

Then $I(t) = 0.184(1.92)^t$

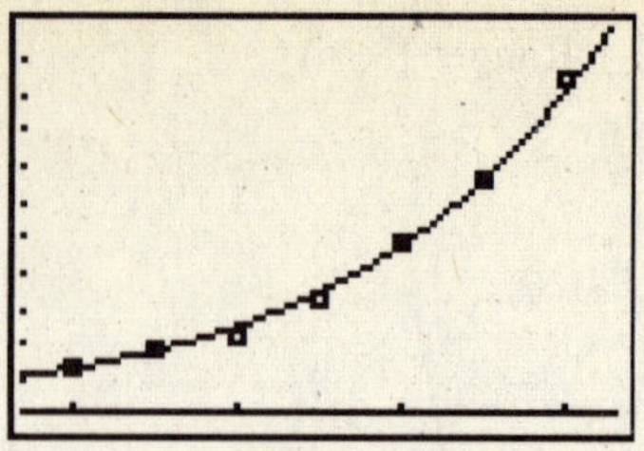

b.

$I(7) = 0.184(1.92)^7$

$I(7) \approx 17.7$

In 1997 there were approximately 17.7 million internet hosts.

c. Domain: $[0, 8]$ Range: $[0.18, 33.98]$

d.

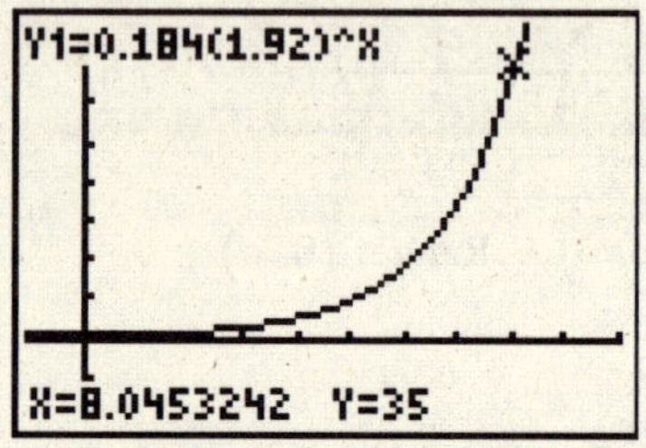

Partway through the year 1998 the number of internet hosts reached 35 million.

21. **a.** Let $N(t)$ be the number of nuclear warheads in the U.S. arsenal in the year t years since 1940.

t	$N(t)$
5	6
7	32
9	235
11	640
13	1436
15	3057
17	6444
19	15,468

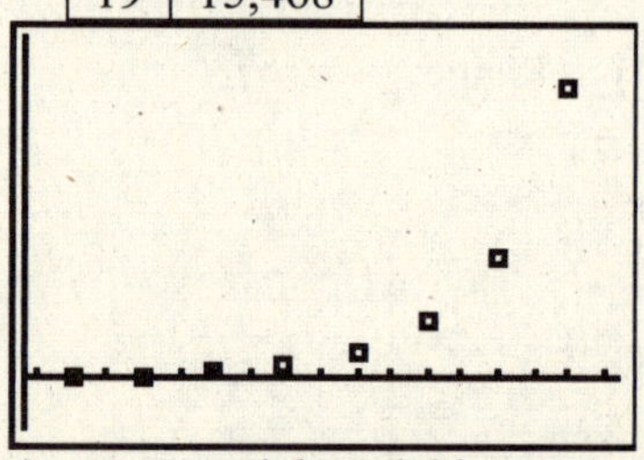

An exponential model is appropriate so choose a point from the flatter end and one from the steeper end.
Note: Your model may differ, but it must follow the trend of the data.

$N(t) = ab^t$

To find a and b choose two points $(7, 32)$ and $(17, 6444)$.

$32 = ab^7$ and $6444 = ab^{17}$

$$\frac{6444}{32} = \frac{ab^{17}}{ab^7}$$

$201.375 = b^{10}$

$(201.375)^{\frac{1}{10}} = \left(b^{10}\right)^{\frac{1}{10}}$

NOTE: ± not needed on answer since $b > 0$.

$b \approx 1.70$

Use b and one of the above equations to find a.

$32 = a(1.70)^7$

$$\frac{32}{(1.70)^7} = \frac{a(1.70)^7}{(1.70)^7}$$

$a \approx 0.78$

Then $N(t) = 0.78(1.70)^t$

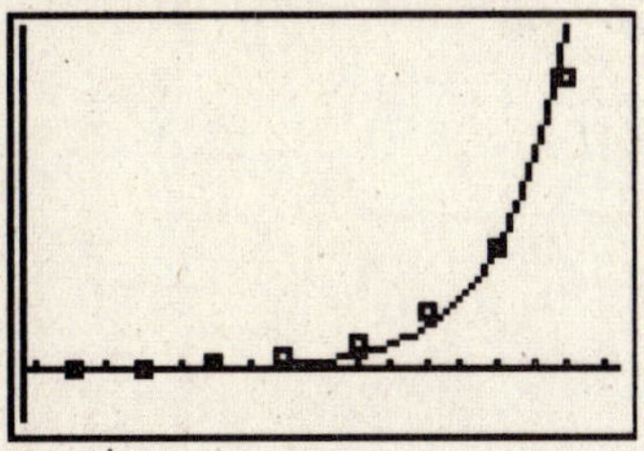

b.

$N(10) = 0.78(1.70)^{10}$

$N(10) \approx 157.25$

In 1970 there were approximately 157 nuclear warheads stockpiled in the U.S. arsenal.

c. Domain: $[5,20]$ Range: $[11,31701]$

d.

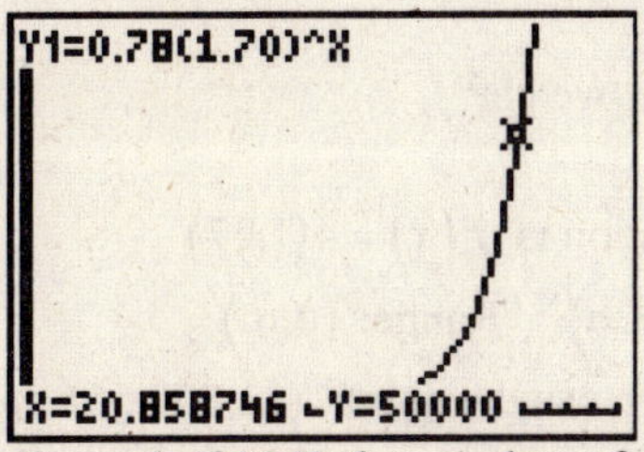

By the end of 1960 the number of stockpiled nuclear warheads in the U.S. arsenal surpassed 50,000.

23. **a.** Let $V(t)$ be the volume of water left in the cylinder t seconds after the start of the experiment.

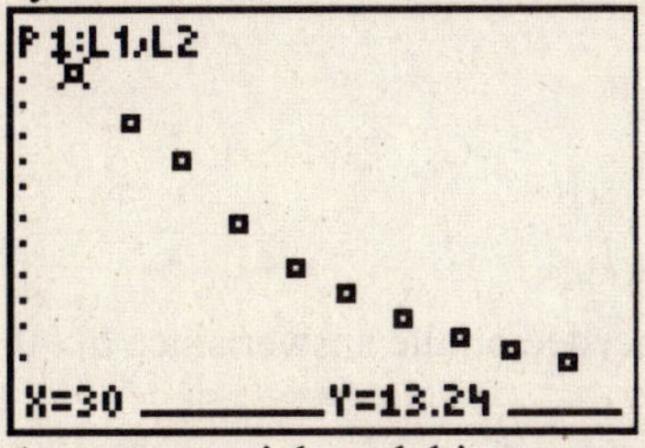

An exponential model is appropriate so choose a point from the flatter end and one from the steeper end.

Note: Your model may differ, but it must follow the trend of the data.

$V(t) = ab^t$

To find a and b choose two points $(30, 13.24)$ and $(300, 2.66)$.

$13.24 = ab^{30}$ and $2.66 = ab^{300}$

$$\frac{2.66}{13.24} = \frac{ab^{300}}{ab^{30}}$$

$$\frac{2.66}{13.24} = b^{270}$$

$$\left(\frac{2.66}{13.24}\right)^{\frac{1}{270}} = \left(b^{270}\right)^{\frac{1}{270}}$$

$b \approx 0.994$

Use b and one of the above equations to find a.

$13.24 = a(0.994)^{30}$

$$\frac{13.24}{(0.994)^{30}} = \frac{a(0.994)^{30}}{(0.994)^{30}}$$

$a \approx 15.86$

Then $V(t) = 15.86(0.994)^t$

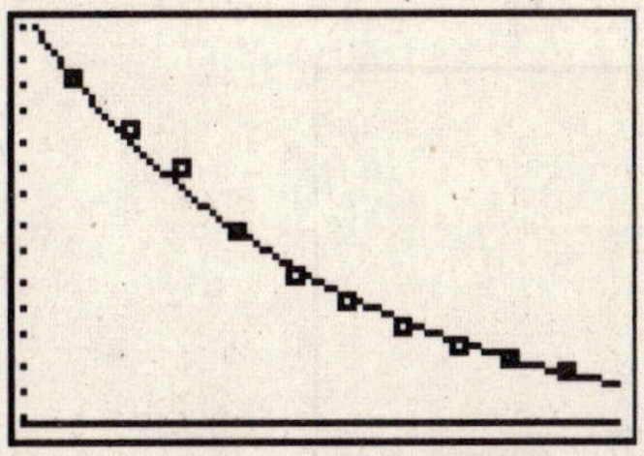

b.

6 minutes $\rightarrow t = 360$

$V(360) = 15.86(0.994)^{360}$

$V(360) \approx 1.82$

Six minutes after the start of the experiment, approximately 1.82 liters of water remain in the cylinder.

c. Domain: $[0,500]$ Range: $[0.78,15.86]$

d.

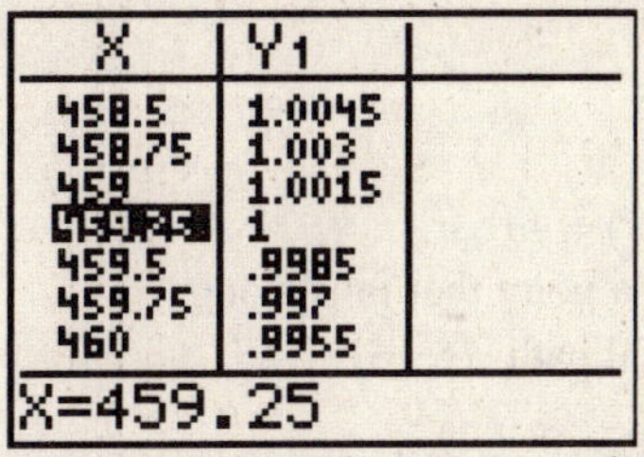

It took approximately 459 seconds for there to be only 1 liter of water remaining in the cylinder.

25. **a.** Let $R(d)$ be the river gage height, in feet above normal, d days after the rainfall event.

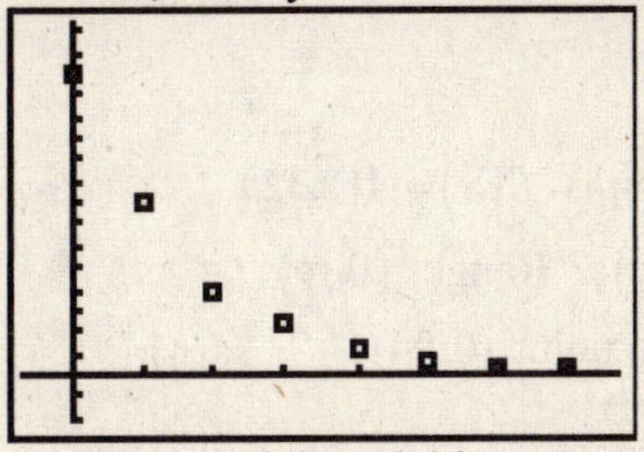

An exponential model is appropriate so choose the vertical intercept and a point from the flatter end.

Note: Your model may differ, but must it follow the trend of the data.

$R(d) = ab^d$

From $(0,14)$ we get $a = 14$.

Giving us: $R(d) = 14b^d$

To find b use a second point: (6, 0.4)

$0.4 = 14b^6$

$\frac{0.4}{14} = \frac{14b^6}{14}$

$\frac{0.4}{14} \approx b^6$

$\left(\frac{0.4}{14}\right)^{\frac{1}{6}} \approx \left(b^6\right)^{\frac{1}{6}}$

NOTE: ± not needed on the answer since $b > 0$.

$b \approx 0.553$

Then $R(d) = 14(0.553)^d$

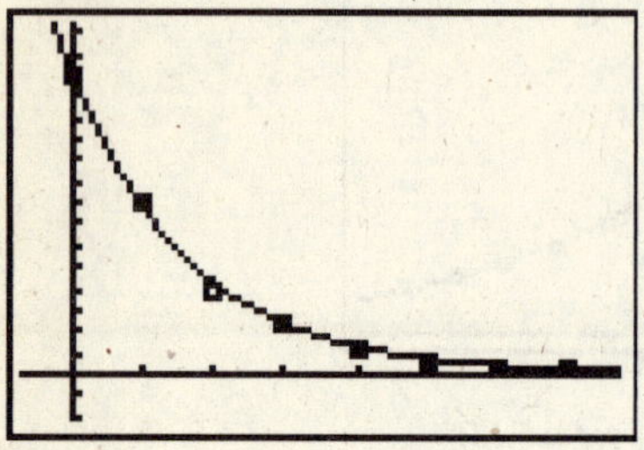

b.

$R(10) = 14(0.553)^{10}$

$R(10) \approx 0.037$

Ten days after the rainfall event the gage height will read 0.037 feet above normal.

c. Domain: $[0,12]$ Range: $[0.01,14]$

27. To find a use the point $(0,4)$.

$4 = ab^0$

$4 = a(1)$

$a = 4$

Giving us: $f(x) = 4b^x$

To find b use a point that falls along the exponential path: $(5,16)$

$16 = 4b^5$

$\frac{16}{4} = \frac{4b^5}{4}$

$4 = b^5$

$(4)^{\frac{1}{5}} = \left(b^5\right)^{\frac{1}{5}}$

$b \approx 1.32$

The new equation is $f(x) = 4(1.32)^x$

Domain: $(-\infty,\infty)$ Range: $(0,\infty)$

29. To find a use the point $(0,94)$.

$94 = ab^0$

$94 = a(1)$

$a = 94$

Giving us: $f(x) = 94b^x$

To find b use a point that falls along the exponential path: $(4,16)$

$16 = 94b^4$

$\frac{16}{94} = \frac{94b^4}{94}$

$\frac{8}{47} = b^4$

$\left(\frac{8}{47}\right)^{\frac{1}{4}} = \left(b^4\right)^{\frac{1}{4}}$

NOTE: ± not needed on the answer since $b > 0$.

$b \approx 0.63$

The new equation is $f(x) = 94(0.63)^x$

Domain: $(-\infty,\infty)$ Range: $(0,\infty)$

31. To find a and b use the points $(5,14)$ and $(12,76)$.

$14 = ab^5$ and $76 = ab^{12}$

$\frac{76}{14} = \frac{ab^{12}}{ab^5}$

$\frac{38}{7} = b^7$

$\left(\frac{38}{7}\right)^{\frac{1}{7}} = \left(b^7\right)^{\frac{1}{7}}$

$b \approx 1.27$

Use b and one of the above equations to find a.

$14 = a(1.27)^5$

$\frac{14}{(1.27)^5} = \frac{a(1.27)^5}{(1.27)^5}$

$\frac{14}{(1.27)^5} = a$

By rounding a we get:

$a \approx 4$

The new equation is $f(x) = 4(1.27)^x$

Domain: $(-\infty,\infty)$ Range: $(0,\infty)$

33. To find a and b use the points $(-1,855.56)$ and $(3,73.75)$.

$855.56 = ab^{-1}$ and $73.75 = ab^3$

$\frac{73.75}{855.56} = \frac{ab^3}{ab^{-1}}$

$\frac{73.75}{855.56} = b^4$

$\left(\frac{73.75}{855.56}\right)^{\frac{1}{4}} = \left(b^4\right)^{\frac{1}{4}}$

NOTE: ± not needed on the answer since $b > 0$.

$b \approx 0.54$

Use b and one of the above equations to find a.

$855.56 = a(0.54)^{-1}$

$\frac{855.56}{(0.54)^{-1}} = \frac{a(0.54)^{-1}}{(0.54)^{-1}}$

$\frac{855.56}{(0.54)^{-1}} = a$

By rounding a we get:

$a \approx 462$

The new equation is $f(x)=462(0.54)^x$

Domain: $(-\infty,\infty)$ Range: $(0,\infty)$

35. The data plot shows an increasing curve with positive values so we know that $a>0$ and $b>1$. Therefore, the student's model is incorrect. The student miscalculated the value of b by multiplying both sides of $3.58\approx b^7$ by $\frac{1}{7}$ instead of raising each side to the $\frac{1}{7}^{\text{th}}$ power. As a result the value of a is incorrect as well. The correct steps are:

$$3.58\approx b^7$$
$$(3.58)^{\frac{1}{7}}\approx\left(b^7\right)^{\frac{1}{7}}$$
$$1.2\approx b$$
$$1.44=a(1.2)^1$$
$$\frac{1.44}{1.2}=\frac{a(1.2)^1}{1.2}$$
$$a=1.2$$
$$R(t)=1.2(1.2)^t$$

37. To find a use the point $(0,-17)$.

$$-17=ab^0$$
$$-17=a(1)$$
$$a=-17$$

Giving us: $f(x)=-17b^x$

To find b use a point that falls along the exponential path: $(1,-21)$

$$-21=-17b^1$$
$$\frac{-21}{-17}=\frac{-17b}{-17}$$
$$\frac{21}{17}=b$$
$$b\approx1.22$$

The new equation is $f(x)=-17(1.22)^x$

Domain: $(-\infty,\infty)$ Range: $(-\infty,0)$

39. To find a use the point $(0,-56)$.

$$-56=ab^0$$
$$-56=a(1)$$
$$a=-56$$

Giving us: $f(x)=-56b^x$

To find b use a point that falls along the exponential path: $(3,-5.103)$

$$-5.103=-56b^3$$
$$\frac{-5.103}{-56}=\frac{-56b^3}{-56}$$
$$\frac{5.103}{56}=b^3$$
$$\left(\frac{5.103}{56}\right)^{\frac{1}{3}}=\left(b^3\right)^{\frac{1}{3}}$$
$$b\approx0.45$$

The new equation is $f(x)=-56(0.45)^x$

Domain: $(-\infty,\infty)$ Range: $(-\infty,0)$

41. To find a and b use the points $(1,-306.3)$ and $(8,-1380.61)$.

$$-306.3=ab^1 \quad\text{and}\quad -1380.61=ab^8$$
$$\frac{-1380.61}{-306.3}=\frac{ab^8}{ab^1}$$
$$\frac{1380.61}{306.3}=b^7$$
$$\left(\frac{1380.61}{306.3}\right)^{\frac{1}{7}}=\left(b^7\right)^{\frac{1}{7}}$$
$$b\approx1.24$$

Use b and one of the above equations to find a.

$$-306.3=a(1.24)^1$$
$$-306.3=1.24a$$
$$\frac{-306.3}{1.24}=\frac{1.24a}{1.24}$$

By rounding a we get:

$$a\approx-247$$

The new equation is $f(x)=-247(1.24)^x$

Domain: $(-\infty,\infty)$ Range: $(-\infty,0)$

43. To find a and b use the points $(-1,-282.54)$ and $(9,-2.78)$.

$$-282.54=ab^{-1} \quad\text{and}\quad -2.78=ab^9$$
$$\frac{-2.78}{-282.54}=\frac{ab^9}{ab^{-1}}$$
$$\frac{2.78}{282.54}=b^{10}$$
$$\left(\frac{2.78}{282.54}\right)^{\frac{1}{10}}=\left(b^{10}\right)^{\frac{1}{10}}$$
$$b\approx0.63$$

Use b and one of the above equations to find a.

$$-282.54=a(0.63)^{-1}$$
$$\frac{-282.54}{(0.63)^{-1}}=\frac{a(0.63)^{-1}}{(0.63)^{-1}}$$

By rounding a we get:

$$a\approx-178$$

The new equation is $f(x)=-178(0.63)^x$

Domain: $(-\infty,\infty)$ Range: $(-\infty,0)$

45. Exponential: Text messaging continues to grow rapidly.

47. Exponential: Music downloads continue to grow rapidly.

49. Linear: Steady growth of New Hampshire's population.

51. **a.** Let $V(m)$ be the number of visitors to MySpace.com (in millions) m months after the site started.

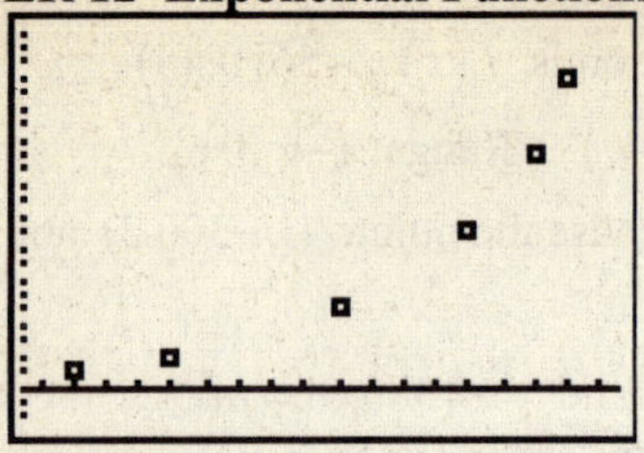

An exponential model is appropriate.

$V(m) = ab^m$

To find a and b choose two points (10,11) and (24,15).

$1 = ab^{10}$ and $15 = ab^{24}$

$\frac{15}{1} = \frac{ab^{24}}{ab^{10}}$

$15 = b^{14}$

$(15)^{\frac{1}{14}} = (b^{14})^{\frac{1}{14}}$

NOTE: $\pm$ not needed on the answer since $b > 0$.

$b \approx 1.21$

Use b and one of the above equations to find a.

$1 = a(1.21)^{10}$

$\frac{1}{(1.21)^{10}} = \frac{a(1.21)^{10}}{(1.21)^{10}}$

$a \approx 0.15$

Then $V(m) = 0.15(1.21)^m$

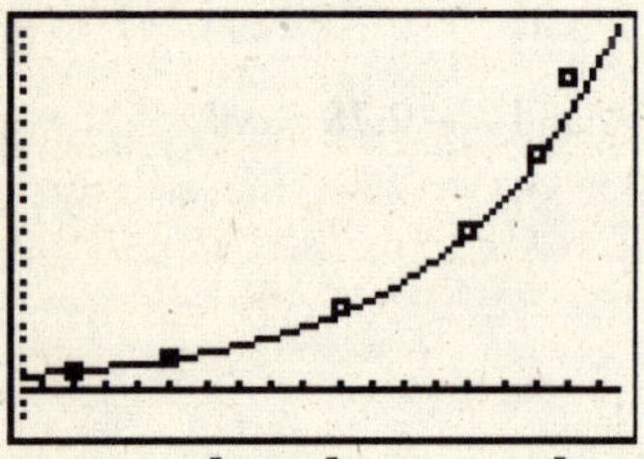

b. Domain: $[9,36]$ Range: $[0.8,143.3]$

c.

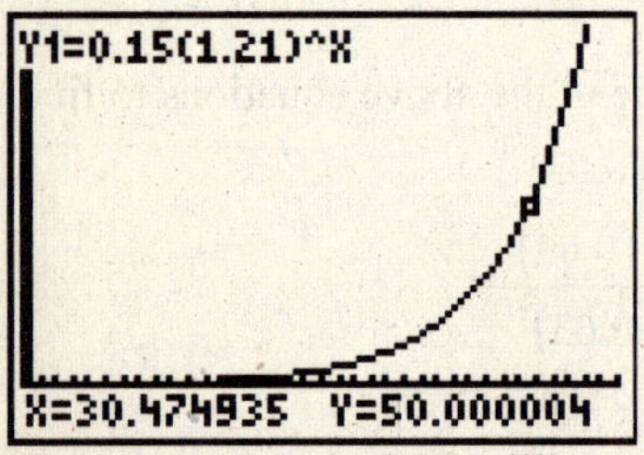

MySpace.com reached 50 million visitors after about 30 months.

53. **a.** Let $P(t)$ be the population of Georgia (in millions) in the year t years since 2000.

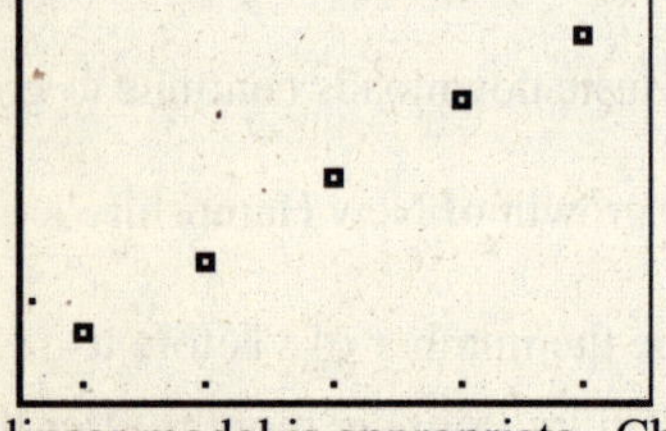

A linear model is appropriate. Choose two points: $(4,8.91)$ and $(8,9.69)$.

$\text{Slope} = \frac{9.69 - 8.91}{8 - 4} = 0.195$

So $P(t) = 0.195t + b$

$8.91 = (0.195)(4) + b$

$8.91 = 0.78 + b$

$b = 8.13$

Then $P(t) = 0.195t + 8.13$

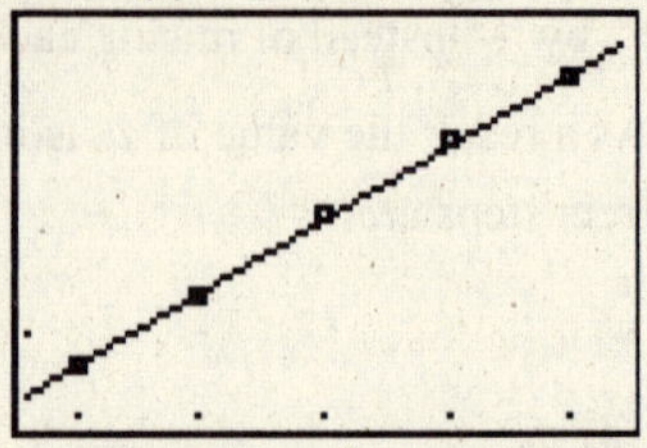

b. Domain: $[0,15]$ Range: $[8.13,11.06]$

c.

$11 = 0.195t + 8.13$

$11 - 8.13 = 0.195t + 8.13 - 8.13$

$2.87 = 0.195t$

$\frac{2.87}{0.195} = \frac{0.195t}{0.195}$

$14.7 \approx t$

The population of George will reach 11 million in the year 2014.

55.

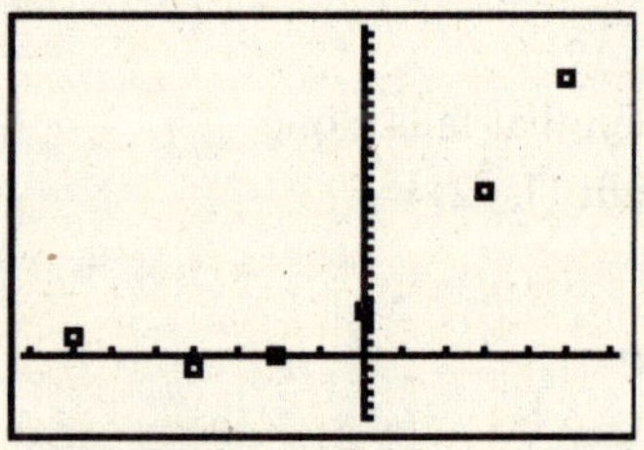

A quadratic model is appropriate. Choose the vertex and a second point.

Vertex: $(-4,-10)$

$f(x) = a(x+4)^2 - 10$

Use a second point to solve for a: $(5,233)$

$233 = a(5+4)^2 - 10$

$233 = 81a - 10$

$243 = 81a$

$\frac{243}{81} = \frac{81a}{81}$

$a = 3$

Then $f(x) = 3(x+4)^2 - 10$

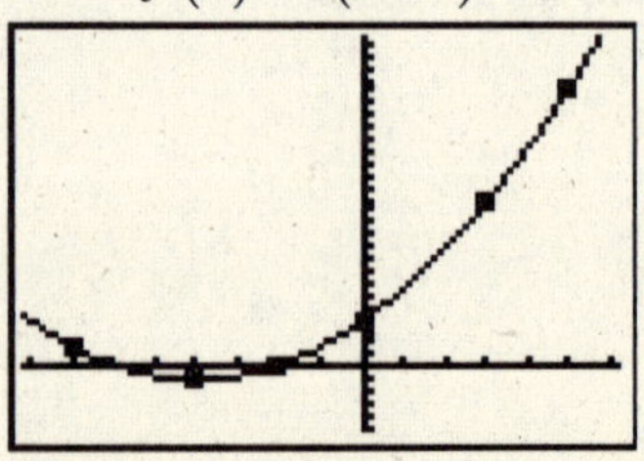

57.

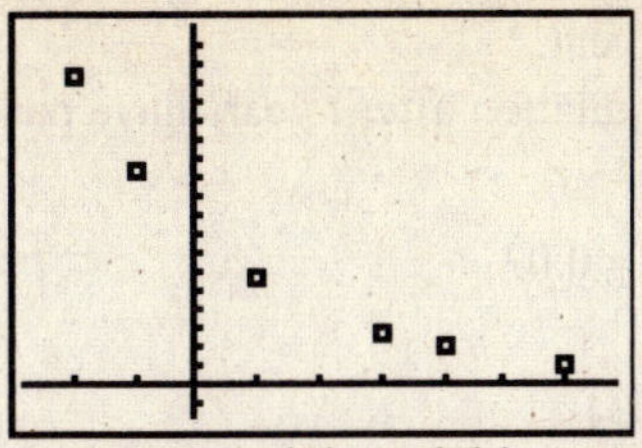

An exponential model is appropriate.

$f(x) = ab^x$

To find a and b choose two points $(-2,163.3)$ and $(6,9.4)$.

$163.3 = ab^{-2}$ and $9.4 = ab^6$

$$\frac{9.4}{163.3} = \frac{ab^6}{ab^{-2}}$$

$$\frac{9.4}{163.3} = b^{6-(-2)} = b^8$$

$$\left(\frac{9.4}{163.3}\right)^{\frac{1}{8}} = \left(b^8\right)^{\frac{1}{8}}$$

NOTE: $\pm$ not needed on the answer since $b > 0$.

$b \approx 0.70$

Use b and one of the above equations to find a.

$$9.4 = a(0.70)^6$$

$$\frac{9.4}{(0.70)^6} = \frac{a(0.70)^6}{(0.70)^6}$$

$a \approx 79.90$

Then $f(x) = 79.9(0.70)^x$

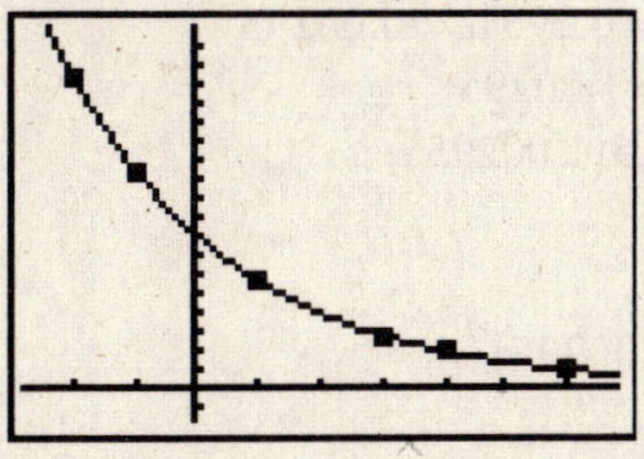

59.

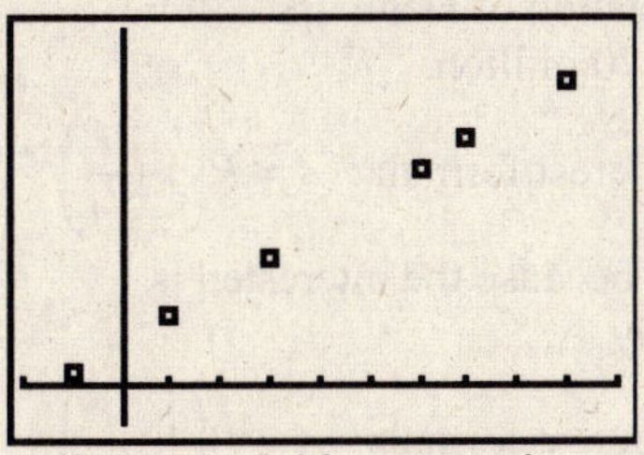

A linear model is appropriate. Choose two points $(-1, 0.1)$ and $(9, 2.1)$

$f(x) = mx + b$

$$m = \frac{2.1 - 0.1}{9-(-1)} = \frac{2}{10} = 0.2$$

$b = y - mx$

$b = 0.1 - (0.2)(-1) = 0.3$

Then $f(x) = 0.2x + 0.3$

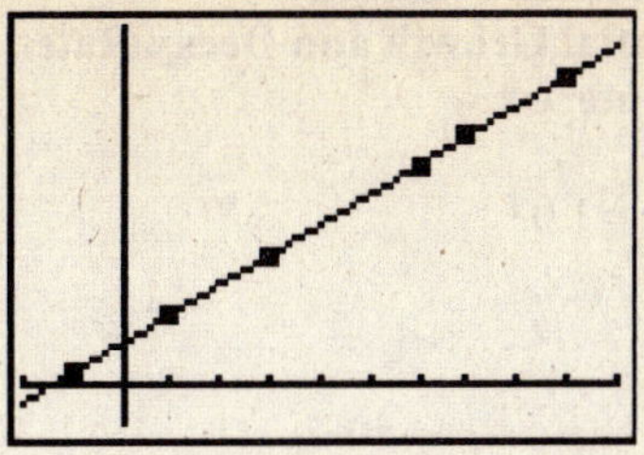

Section 12.5 Exponential Growth and Decay Rates and Compounding Interest

1.

$r = b - 1$ where $b = 1.03$
$r = 1.03 - 1$
$r = .03$
$r = 3\%$
3% growth

3.

$r = b - 1$ where $b = 1.25$
$r = 1.25 - 1$
$r = .25$
$r = 25\%$
25% growth

5.

$r = b - 1$ where $b = 3.5$
$r = 3.5 - 1$
$r = 2.5$
$r = 250\%$
250% growth

7.

$r = b - 1$ where $b = 0.95$
$r = 0.95 - 1$
$r = -0.05$
$r = -5\%$
5% decay

9.

$r = b - 1$ where $b = 0.36$
$r = 0.36 - 1$
$r = -0.64$
$r = -64\%$
64% decay

11. **a.** $W(t) = 1.19(1.08)^t$ where $1980 \to t = 0$

$1990 \to t = 10$
$W(10) = 1.19(1.08)^{10}$
$W(10) \approx 2.57$

In 1990, the population of white-tailed deer was approximately 2.57 million.

b.

$$\left.\begin{array}{l} b = 1.08 \\ r = b - 1 \end{array}\right\} r = 1.08 - 1 = 0.08$$

The population of white-tailed deer grows at a rate of 8% per year.

c. This population might stop growing at the rate of 8% per year because of limited food or space.

13. **a.** $V(t) = 109.8(0.99939)^t$ where $2010 \to t = 0$

$2015 \to t = 5$
$V(5) = 109.8(0.99939)^5$
$V(5) \approx 109.47$

In 2015, the population of the Virgin Islands will be approximately 109.47 thousand.

b.

$$\left.\begin{array}{l} b = 0.99939 \\ r = b - 1 \end{array}\right\} r = 0.99939 - 1 = -0.00061$$

The population of the Virgin Islands is decreasing at a rate of 0.061% per year.

15. Let $P(t)$ be the population after t years have passed.

$P(t) = ab^t$
$a = 40$ and $r = 3\% = 0.03$
$b = 1 + r$ so $b = 1.03$
Then $P(t) = 40(1.03)^t$

17. Let $P(t)$ be the population after t years have passed.

$P(t) = ab^t$
$a = 200$ and $r = 2\%$ decrease per year
$b = 1 + r$ with $r = -0.02$
$b = 1 - 0.02 = 0.98$
Then $P(t) = 200(0.98)^t$

19. **a.** Let $H(t)$ be the population of humpback whales t years after 1981.

$H(t) = ab^t$
$a = 350$ and $r = 14\% = 0.14$
$b = 1 + r$ so $b = 1.14$
Then $H(t) = 350(1.14)^t$

b. Domain: $[-2, 15]$ Range: $[269, 2498]$

c. $H(9) = 350(1.14)^9 \approx 1138$ In 1990 the humpback whale population was approximately 1138.

21. **a.** Let $D(t)$ be the population of Denmark (in millions) t years after 2008.

$D(t) = ab^t$
$a = 5.5$ and $r = 0.295\% = 0.00295$
$b = 1 + r$ so $b = 1.00295$
Then $D(t) = 5.5(1.00295)^t$

b.

$2020 \to t = 12$
$D(12) = 5.5(1.00295)^{12}$
$D(12) \approx 5.70$

In 2020 the population of Denmark will be approximately 5.70 million.

23. Compounding interest formula: $A = P\left(1 + \frac{r}{n}\right)^{nt}$

Use this formula because the interested is compounded daily.

25. Compounding interest formula: $A = P\left(1 + \frac{r}{n}\right)^{nt}$

Use this formula because the interested is compounded weekly.

27. Compounding interest formula: $A = P\left(1 + \frac{r}{n}\right)^{nt}$

Use this formula because the interested is compounded monthly.

29. Continuously compounding interest formula: $A = Pe^{rt}$

Use this formula because the interested is compounded continuously.

31. The missing variable is A.
$A = ?$
$P = 20{,}000$
$r = 0.03$
$n = 365$
$t = 5$

33. The missing variable is t.
$A = 8{,}000$
$P = 4{,}000$
$r = 0.0275$
$n = 12$
$t = ?$

35. The missing variable is r.
$A = 1{,}000$
$P = 500$
$r = ?$
$n = 365$
$t = 10$

37. Using the formula $A = P\left(1+\frac{r}{n}\right)^{nt}$ because the interest is compounded monthly where:
$P = \$10{,}000$
$r = 5\% = 0.05$
$t = 10$ years
$n = 12$
$A = ?$
Solve for A:
$$A = 10000\left(1+\frac{0.05}{12}\right)^{12\cdot 10}$$
$A \approx 16{,}470.09$
The amount of money in the account after 10 years is $16,470.09.

39. Using the formula $A = P\left(1+\frac{r}{n}\right)^{nt}$ because the interest is compounded daily where:
$P = \$10{,}000$
$r = 5\% = 0.05$
$t = 10$ years
$n = 365$
$A = ?$
Solve for A:
$$A = 10000\left(1+\frac{0.05}{365}\right)^{365\cdot 10}$$
$A \approx 16{,}486.65$
The amount of money in the account after 10 years is $16,486.65.

41. Using the formula $A = \mathrm{P}e^{rt}$ because the interest is compounded continuously where:
$P = \$500{,}000$
$r = 9\% = 0.09$
$t = 15$ years
$A = ?$
Solve for A:
$A = 500000e^{0.09\cdot 15}$
$A \approx 1{,}928{,}712.77$
The amount of money in the account after 15 years is $1,928,712.77.

43. Using the formula $A = P\left(1+\frac{r}{n}\right)^{nt}$ because the interest is compounded hourly where:
$P = \$500{,}000$
$r = 9\% = 0.09$
$t = 15$ years
$n = 8760$
$A = ?$
Solve for A:
$$A = 500000\left(1+\frac{0.09}{8760}\right)^{8760\cdot 15}$$
$A \approx 1{,}928{,}699.39$
The amount of money in the account after 15 years is $1,928,699.39.

45. Using the formula $A = P\left(1+\frac{r}{n}\right)^{nt}$ because the interest is compounded quarterly where:
$P = \$100{,}000$
$r = 7\% = 0.07$
$t = 20$ years
$n = 4$
$A = ?$
Solve for A:
$$A = 100000\left(1+\frac{0.07}{4}\right)^{4\cdot 20}$$
$A \approx 400{,}639.19$
The amount of money in the account after 20 years is $400,639.19.

47. Using the formula $A = \mathrm{P}e^{rt}$ because the interest is compounded continuously where:
$P = \$100{,}000$
$r = 7\% = 0.07$
$t = 20$ years
$A = ?$
Solve for A:
$A = 100000e^{0.07\cdot 20}$
$A \approx 405{,}520.00$
The amount of money in the account after 20 years is $405,520.00.

49. For Option #1 use the formula $A = P\left(1+\frac{r}{n}\right)^{nt}$
$P = \$400{,}000$
$r = 4\% = 0.04$
$t = 5$ years
$n = 1$

Solve for A:

$$A = 400{,}000\left(1+\frac{0.04}{1}\right)^{(1\times5)}$$

$$A \approx \$486{,}661.16$$

For Option #2 use the formula $A = Pe^{rt}$

$P = \$400{,}000$

$r = 3.95\% = 0.0395$

$t = 5$ years

Solve for A:

$$A = 400{,}000e^{(0.0395\times5)}$$

$$A \approx \$487{,}341.23$$

The second option, with an interest rate of 3.95% compounded continuously, will have the larger balance after 5 years.

Chapter 12 Review

1. **a.** Let $P(t)$ be the population of Africa (in millions) in the year t years since 1996. Let n be the number of 28 year time intervals.

t	$P(t)$
0	731.5
28	$731.5(2)$
$2(28)$	$731.5(2)^2$
$3(28)$	$731.5(2)^3$
$n(28)$	$731.5(2)^n$

$$n(28) = t$$

$$\frac{n(28)}{28} = \frac{t}{28}$$

$$n = \frac{t}{28}$$

$$P(t) = 731.5(2)^{t/28}$$

b. $2005 \to t = 9$

$$P(9) = 731.5(2)^{9/28}$$

$$P(9) = 914.06$$

In 2005, the population of Africa was approximately 914.06 million.

2. **a.** Let $F(d)$ be the number of people with flu symptoms after d days.

d	$F(d)$
0	1
1	$(3)^1$
2	$(3)(3) = (3)^2$
3	$(3)(3)(3) = (3)^3$
d	$(3)^d$

$$F(d) = (3)^d$$

b. Two weeks $\to d = 14$

$$F(14) = (3)^{14} = 4{,}782{,}969$$

After 2 weeks approximately 4,782,969 people have flu symptoms. This is possibly model breakdown.

3. **a.** Let $P(d)$ be the percentage of polonium-210 left in a sample after d days. Let n be the number of 138-day time intervals. The percent of polonium-210 is cut in half every 138 days:

d	$P(d)$
0	100
138	$100(\frac{1}{2})$
2(138)	$100(\frac{1}{2})^2$
3(138)	$100(\frac{1}{2})^3$
n(138)	$100(\frac{1}{2})^n$

$n(138) = d$

$\frac{n(138)}{138} = \frac{d}{138}$

$n = \frac{d}{138}$

$P(d) = 100\left(\frac{1}{2}\right)^{\frac{d}{138}}$

b.

$P(300) = 100\left(\frac{1}{2}\right)^{\frac{300}{138}}$

$P(300) = 22.16$

After 300 days, there will be about 22.16% of the polonium-210 sample left.

4. **a.** Let $P(t)$ be the percentage of thorium-228 left in a sample after t years. Let n be the number of 1.9-year time intervals. The percent of thorium-228 is cut in half every 1.9 years:

t	$P(t)$
0	100
1.9	$100\left(\frac{1}{2}\right)$
2(1.9)	$100\left(\frac{1}{2}\right)^2$
3(1.9)	$100\left(\frac{1}{2}\right)^3$
$n(1.9)$	$100\left(\frac{1}{2}\right)^n$

$n(1.9) = t$

$\frac{n(1.9)}{1.9} = \frac{t}{1.9}$

$n = \frac{t}{1.9}$

$P(t) = 100\left(\frac{1}{2}\right)^{\frac{t}{1.9}}$

b.

$P(50) = 100\left(\frac{1}{2}\right)^{\frac{50}{1.9}}$

$P(50) = 0.000001197$

After 50 years, there will be about 0.000001197% of the thorium-228 sample left. This amount approaches zero percent.

5.

x	$f(x)$	base
0	5000	$\frac{4000}{5000} = 0.8$
1	4000	$\frac{3200}{4000} = 0.8$
2	3200	$\frac{2560}{3200} = 0.8$
3	2560	$\frac{2048}{2560} = 0.8$
4	2048	

$a = 5000$ because the initial value is 5000 when $x = 0$.
$b = 0.8$ because the common base is 0.8 from the above table.
Therefore the exponential model is $f(x) = 5000(0.8)^x$

6.

x	$f(x)$	base
0	0.2	$\frac{1.52}{0.2} = 7.6$
1	1.52	$\frac{11.53}{1.52} = 7.6$
2	11.53	$\frac{87.58}{11.53} = 7.6$
3	87.58	$\frac{665.05}{87.05} = 7.6$
4	665.05	

$a = 0.2$ because the initial value is 0.2 when $x = 0$.
$b = 7.6$ because the common base is 7.6 from the above table.
Therefore the exponential model is $f(x) = 0.2(7.6)^x$

7.

$6^x = 216$

$6^x = 6^3$

$x = 3$

8.

$4^x = 1024$

$4^x = 4^5$

$x = 5$

9.

$6(7^x) = 14406$

$\frac{6(7^x)}{6} = \frac{14406}{6}$

$7^x = 2401$

$7^x = 7^4$

$x = 4$

10.

$5(3^x) = 10935$

$\frac{5(3^x)}{5} = \frac{10935}{5}$

$3^x = 2187$

$3^x = 3^7$

$x = 7$

11.

$2x^5 + 7 = 1057.4375$

$2x^5 + 7 - 7 = 1057.4375 - 7$

$2x^5 = 1050.4375$

$\frac{2x^5}{2} = \frac{1050.4375}{2}$

$x^5 = 525.21875$

$(x^5)^{\frac{1}{5}} = (525.21875)^{\frac{1}{5}}$

$x = 3.5$

12.

$3x^4 - 600 = 1275$

$3x^4 - 600 + 600 = 1275 + 600$

$3x^4 = 1875$

$$\frac{3x^4}{3} = \frac{1875}{3}$$
$$x^4 = 625$$
$$\left(x^4\right)^{\frac{1}{4}} = \pm(625)^{\frac{1}{4}}$$
$$x = \pm 5$$

13.

$$-\frac{1}{3}x^7 + 345 = -92967$$
$$-\frac{1}{3}x^7 + 345 - 345 = -92967 - 345$$
$$-\frac{1}{3}x^7 = -93312$$
$$-3\left(-\frac{1}{3}x^7\right) = -3(-93312)$$
$$x^7 = 279936$$
$$\left(x^7\right)^{\frac{1}{7}} = (279936)^{\frac{1}{7}}$$
$$x = 6$$

14.

$$-4x^8 + 2768 = -6715696$$
$$-4x^8 + 2768 - 2768 = -6715696 - 2768$$
$$-4x^8 = -6718464$$
$$\frac{-4x^8}{-4} = \frac{-6718464}{-4}$$
$$x^8 = 1679616$$
$$\left(x^8\right)^{\frac{1}{8}} = \pm(1679616)^{\frac{1}{8}}$$
$$x = \pm 6$$

15.

$$2^x = \frac{1}{32}$$
$$2^x = \frac{1}{2^5}$$
$$2^x = 2^{-5}$$
$$x = -5$$

16.

$$(-3)^x = \frac{1}{81}$$
$$(-3)^x = \frac{1}{(-3)^4}$$
$$(-3)^x = (-3)^{-4}$$
$$x = -4$$

17.

$$\left(\frac{2}{3}\right)^x = \frac{27}{8}$$
$$\left(\frac{2}{3}\right)^x = \left(\frac{3}{2}\right)^3$$
$$\left(\frac{2}{3}\right)^x = \left(\frac{2}{3}\right)^{-3}$$
$$x = -3$$

18.

$$\left(\frac{1}{4}\right)^x = 1024$$
$$\left(\frac{1}{4}\right)^x = 4^5$$
$$\left(\frac{1}{4}\right)^x = \left(\frac{1}{4}\right)^{-5}$$
$$x = -5$$

19. **a.** The value of a is positive because the y-intercept is above the origin.
b. b is greater than 1, because $a > 0$ and the function is increasing.
c. Increasing.
d. Domain: $(-\infty, \infty)$ Range: $(0, \infty)$

20. **a.** The value of a is negative because the y-intercept is below the origin.
b. b is greater than 1, because $a < 0$ and the function is decreasing.
c. Decreasing.
d. Domain: $(-\infty, \infty)$ Range: $(-\infty, 0)$

21. **a.** $a > c$, because the y-intercept for *f(x)* is greater than the y-intercept of *g(x)*.
b. $b < d$, because *f(x)* is increasing at a slower rate than *g(x)*.

22. **a.** Decreasing.
b. This is an example is exponential decay.
c. $f(20) = 5$
d. $x = 10$
e. $(0, 20)$, because when $x = 0$ $y = 20$.

23. **a.** Increasing, because $b > 1$ and $a > 0$.
b. This is an example is exponential growth, because $b > 0$.
c. $(0, 2)$, because when $x = 0$ $y = 2$.

24. Since $a > 0$ and $b > 1$ the function is increasing.

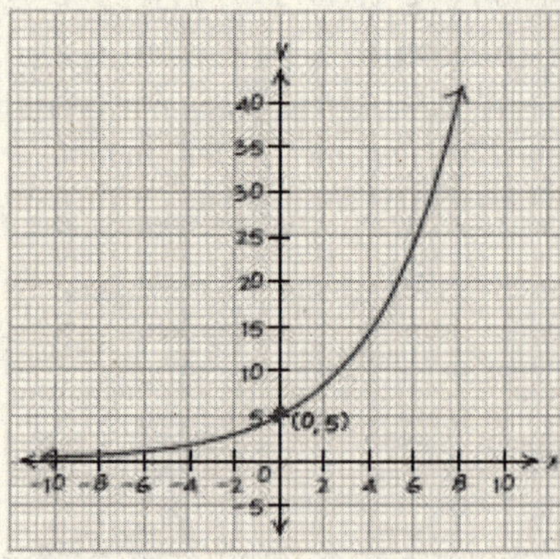

25. Since $a < 0$ and $b < 1$ the function is increasing.

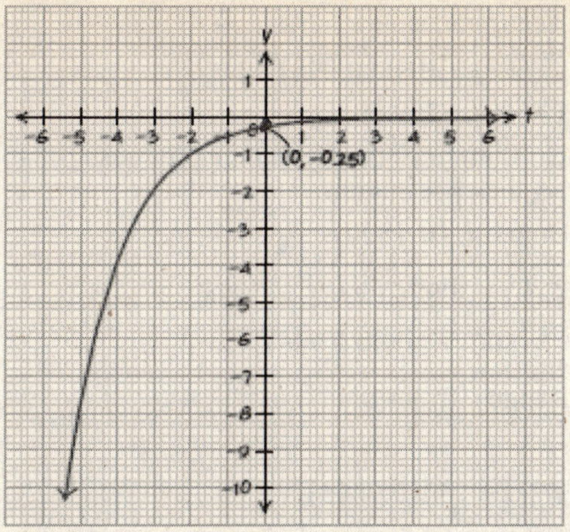

26. Since $a > 0$ and $b < 1$ the function is decreasing.

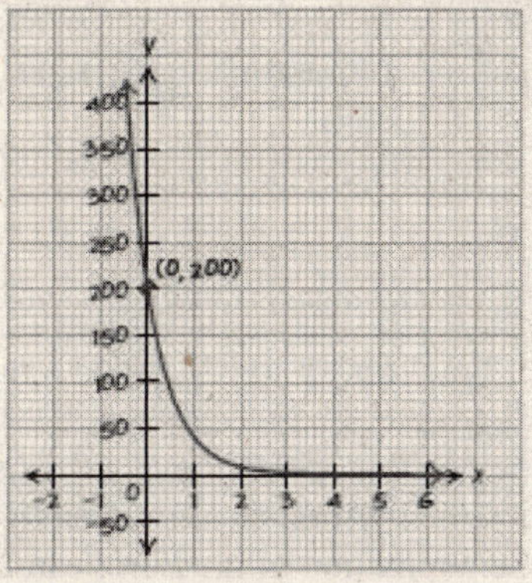

27. Since $a < 0$ and $b < 1$ the function is increasing.

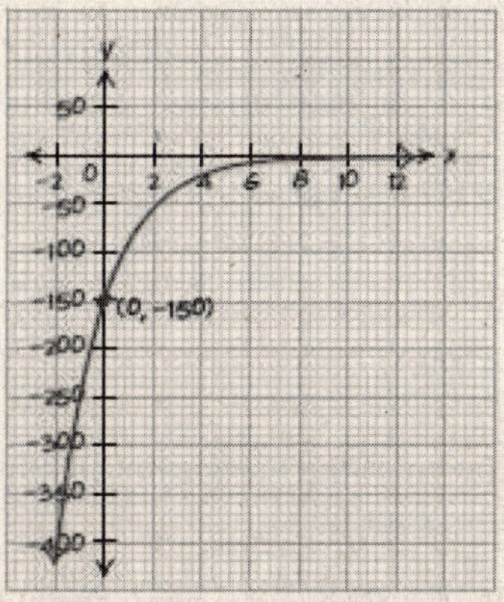

28. Since $a > 0$ and $b < 1$ the function is decreasing.

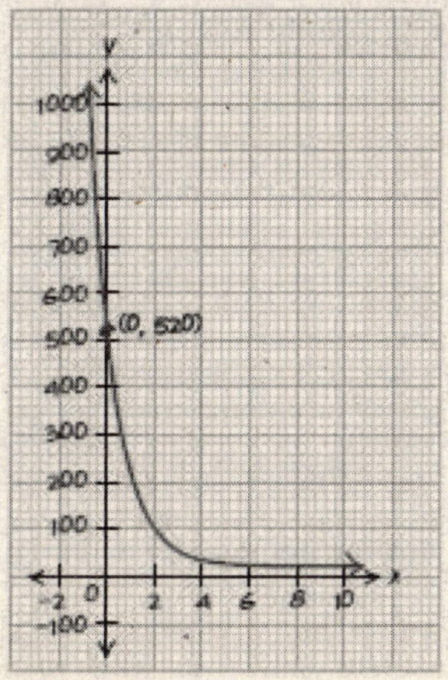

29. Since $a < 0$ and $b > 1$ the function is decreasing.

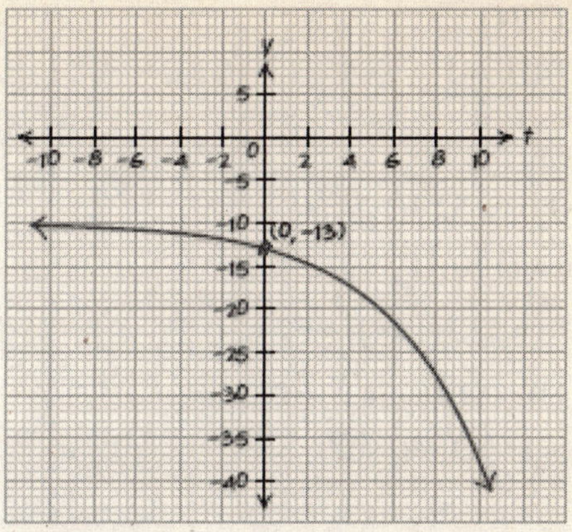

30. Domain: $(-\infty,\infty)$ Range: $(0,\infty)$

31. Domain: $(-\infty,\infty)$ Range: $(-\infty,0)$

32. Domain: $(-\infty,\infty)$ Range: $(20,\infty)$

33. Domain: $(-\infty,\infty)$ Range: $(-\infty,-10)$

34. **a.** Let $W(t)$ be the number of non-strategic warheads in the U.S. arsenal in the year t years since 1950.

t	$W(t)$
1	91
2	205
3	436
4	563
5	857
6	1,618
7	2,244
8	4,122
9	8,462
10	13,433

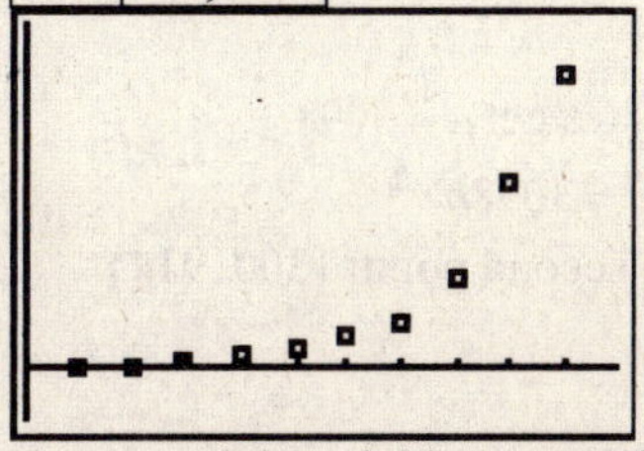

An exponential model is appropriate so choose a point from the flatter end and one from the steeper end.
Note: Your model may differ, but it must follow the trend of the data.

$W(t) = ab^t$

To find a and b choose two points $(1,91)$ and $(10,13433)$.

$91 = ab$ and $13{,}433 = ab^{10}$

$$\frac{13{,}433}{91} = \frac{ab^{10}}{ab}$$

$$\frac{13{,}433}{91} = b^9$$

$$\left(\frac{13{,}433}{91}\right)^{\frac{1}{9}} = \left(b^9\right)^{\frac{1}{9}}$$

$b \approx 1.74$

Use b and one of the above equations to find a.

$91 = a(1.74)$

$\frac{91}{1.74} = \frac{a(1.74)}{1.74}$

$a \approx 52.3$

Then $W(t) = 52.3(1.74)^t$

b. Domain: $[0,11]$

$W(0) = 52.3(1.74)^0 \approx 52$

$W(11) = 52.3(1.74)^{11} \approx 23{,}150$

Range: $[52,23150]$

c.

$1965 \rightarrow t = 15$

$W(15) = 52.3(1.74)^{15}$

$W(15) \approx 212{,}197.6$

The model predicts that there were approximately 212,198 non-strategic warheads in the U.S. arsenal in the year 1965. This could be model breakdown.

35. **a.** Let $C(s)$ be the number of counts from a Geiger counter from the decay of barium-137 after s seconds.

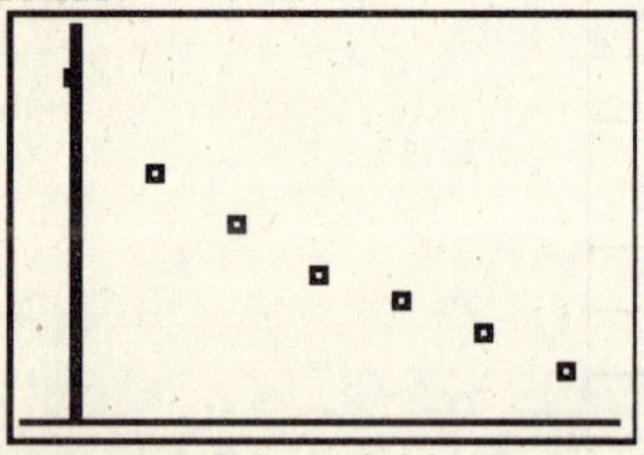

An exponential model is appropriate so choose the vertical intercept and a point from the flatter end. *Note: Your model may differ, but must it follow the trend of the data.*

$C(s) = ab^s$

From $(0,3098)$ we get $a = 3098$.

Giving us: $C(s) = 3098b^s$

To find b use a second point: $(300,914)$

$914 = 3098b^{300}$

$\frac{914}{3098} = \frac{3098b^{300}}{3098}$

$\frac{914}{3098} = b^{300}$

$\left(\frac{914}{3098}\right)^{\frac{1}{300}} = \left(b^{300}\right)^{\frac{1}{300}}$

$b \approx 0.996$

Then $C(s) = 3098(0.996)^s$

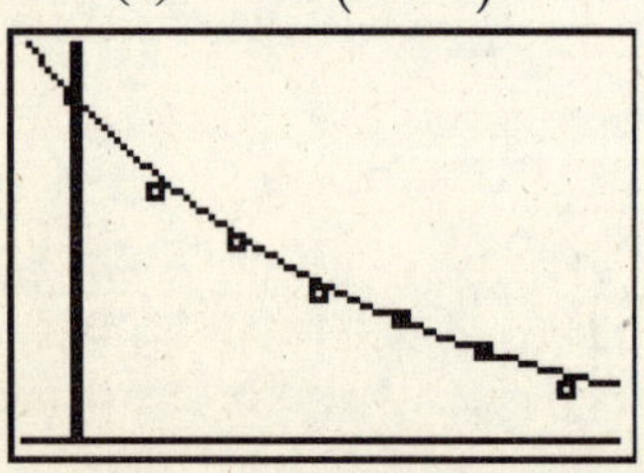

b.

Ten minutes $\rightarrow s = 600$

$C(600) = 3098(0.996)^{600}$

$C(600) \approx 279.69$

After 10 minutes the Geiger counter ticks off about 280 counts per minute.

c. Domain: $[0,1800]$

$C(0) = 3098(0.996)^0 = 3098$

$C(1800) = 3098(0.996)^{1800} \approx 2$

Range: $[2,3098]$

36. **a.**

$F(t) = 750(0.9405)^t$

$F(5) = 750(0.9405)^5$

$F(5) \approx 551.89$

In 2005 the population of Franklin's gulls was approximately 551.89 thousand.

b.

$r = b - 1$

$r = 0.9405 - 1$

$r = -0.0595$

The population of Franklin's gulls decays at a rate of 5.95% per year.

c. This population might stop declining at the rate of 5.95% per year due to environmental intervention.

37. **a.** Let $E(t)$ be the population of the European Union (in millions) t years after 2008.

$E(t) = ab^t$

$a = 491$ and $r = 0.12\% = 0.0012$

$b = 1 + r$ so $b = 1.0012$

Then $E(t) = 491(1.0012)^t$

b.

$2020 \rightarrow t = 12$

$E(12) = 491(1.0012)^{12}$

$E(12) \approx 498.12$

In 2020 the population of the European Union will be approximately 498.12 million.

38. Using the formula $A = P\left(1 + \frac{r}{n}\right)^{nt}$ because the interest is compounded monthly where:

$P = \$7{,}000$

$r = 6\% = 0.06$

$t = 10$ years

$n = 12$

$A = ?$

Solve for A:

$A = 7000\left(1 + \frac{0.06}{12}\right)^{(12\times10)}$

$A \approx 12{,}735.78$

The amount of money in the account after 10 years is \$12,735.78.

39. Using the formula $A = Pe^{rt}$ because the interest is compounded continuously where:
$P = \$100,000$
$r = 8.5\% = 0.085$
$t = 25$ years
$A = ?$
Solve for *A*:
$A = 100000e^{(0.085\times 25)}$
$A \approx 837,289.75$
The amount of money in the account after 25 years is $837,289.75.

40. Using the formula $A = P\left(1+\frac{r}{n}\right)^{nt}$ because the interest is compounded daily where:
$P = \$50,000$
$r = 4\% = 0.04$
$t = 15$ years
$n = 365$
$A = ?$
Solve for *A*:
$A = 50000\left(1+\frac{0.04}{365}\right)^{(365\times 15)}$
$A \approx 91,102.95$
The amount of money in the account after 15 years is $91,102.95.

Chapter 12 Test

1. Let *B*(*h*) be the number of bacteria present (in millions) *h* hours after 12 noon have passed.

a. The number of bacteria doubles every hour:
$B(h) = 5(2)^h$

h	$B(h)$
0	5
1	$5(2)^1$
2	$5(2)^2$
3	$5(2)^3$
h	$5(2)^h$

b. $B(6) = 5(2)^6 = 320$ By 6:00pm there are 320 million bacteria on the bathroom door handle.

c. The bacteria double 4 times each hour:
$B(h) = 5(2)^{4h}$

h	$B(h)$
0	5
1	$5(2)(2)(2)(2) = 5(2)^4$
2	$5(2)^4(2)^4 = 5(2)^{4\times 2} = 5(2)^8$
3	$5(2)^4(2)^4(2)^4 = 5(2)^{4\times 3} = 5(2)^{12}$
h	$5(2)^{4h}$

2. Let *P*(*m*) be the percent of thallium-210 left in a sample after *m* minutes. Let *n* be the number of 1.32-minute time intervals.

a. The percent of thallium-210 is cut in half every 1.32 minutes: $P(m) = 100\left(\frac{1}{2}\right)^{m/1.32}$

m	$P(m)$
0	100
1.32	$100\left(\frac{1}{2}\right)$
2(1.32)	$100\left(\frac{1}{2}\right)^2$
3(1.32)	$100\left(\frac{1}{2}\right)^3$
$n(1.32)$	$100\left(\frac{1}{2}\right)^n$

$n(1.32) = m$
$\frac{n(1.32)}{1.32} = \frac{m}{1.32}$
$n = \frac{m}{1.32}$
$P(m) = 100\left(\frac{1}{2}\right)^{\frac{m}{1.32}}$

b.
$P(15) = 100\left(\frac{1}{2}\right)^{\frac{15}{1.32}}$
$P(15) \approx 0.038$

After 15 minutes, there is 0.038% of the thallium-210 sample left, which means there is approximately 0% left.

3. Using the formula $A = P\left(1+\frac{r}{n}\right)^{nt}$ because the interest is compounded monthly where:

$P = \$1,000$
$r = 3.5\% = 0.035$
$t = 12$ years
$n = 12$
$A = ?$
Solve for A:

$$A = 1000\left(1+\frac{0.035}{12}\right)^{(12\times12)}$$

$A \approx 1521.03$
The amount of money in the account after 12 years is \$1521.03.

4. Using the formula $A = Pe^{rt}$ because the interest is compounded continuously where:

$P = \$40,000$
$r = 6\% = 0.06$
$t = 35$ years
$A = ?$
Solve for A:
$A = 40000e^{(0.06\times35)}$
$A \approx 326,646.80$
The amount of money in the account after 35 years is \$326,646.80.

5. **a.** a is negative because the graph is below the x-axis.
b. $b < 1$ because we have exponential growth with $a < 0$.
c. Domain: $(-\infty,\infty)$ Range: $(-\infty,0)$

6. **a.** The graph is increasing.
b. $f(1) = -4$
c. $f(x) = -36$ and $x = -2$
d. The vertical intercept is $(0,-8)$.

7. **a.** Let $C(t)$ be the number of CD singles shipped (in millions) in the mid-1990's in the year t years since 1990.

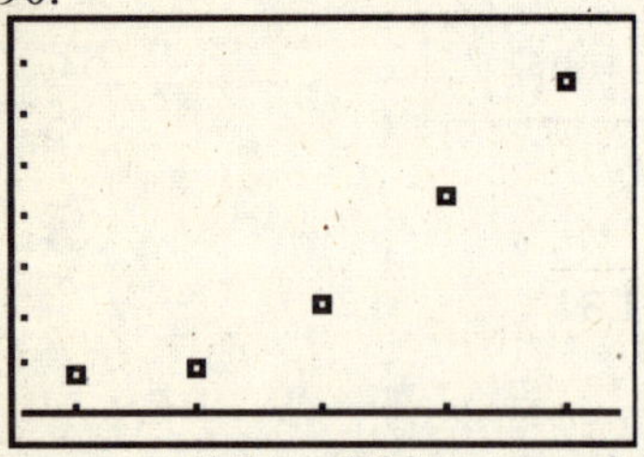

An exponential model is appropriate so choose a point from the flatter end and one from the steeper end.
Note: Your model may differ, but it must follow the trend of the data.
$C(t) = ab^t$
To find a and b choose two points $(3, 7.8)$ and $(7, 66.7)$.
$7.8 = ab^3$ and $66.7 = ab^7$

$$\frac{66.7}{7.8} = \frac{ab^7}{ab^3}$$

$$\frac{66.7}{7.8} = b^4$$

$$\left(\frac{66.7}{7.8}\right)^{\frac{1}{4}} = \left(b^4\right)^{\frac{1}{4}}$$

NOTE: $\pm$ not needed on the answer since $b > 0$.
$b \approx 1.71$
Use b and one of the above equations to find a.
$7.8 = ab^3$

$$\frac{7.8}{(1.71)^3} = \frac{a(1.71)^3}{(1.71)^3}$$

$a \approx 1.56$
Then $C(t) = 1.56(1.71)^t$

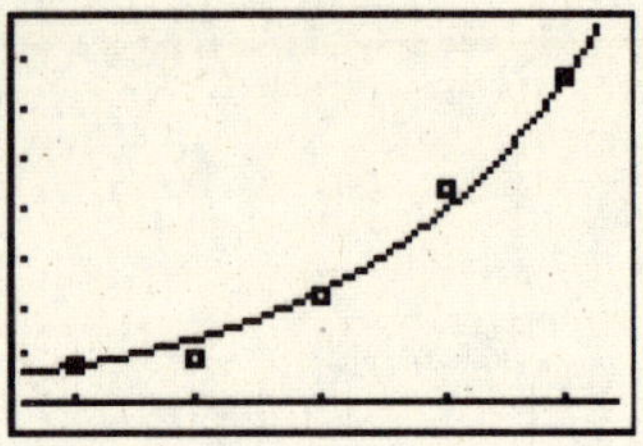

b. Domain: $[0,10]$ Range: $[1.56, 333.49]$
c. In 2000 there were approximately 333.49 million CD singles shipped.

8. **a.** Let $P(t)$ be the number of professionals in developing countries (in millions) in the year t years since 2009.

$P(t) = ab^t$
$a = 4$ and $r = 6.5\% = 0.065$
$b = 1 + r$ so $b = 1.065$
Then $P(t) = 4(1.065)^t$

b.
$2013 \rightarrow t = 4$
$P(4) = 4(1.065)^4$
$P(4) \approx 5.15$
By the year 2013 there will be approximately 5.15 million professionals in developing countries.

9. Since $a < 0$ and $b > 1$ the function is decreasing.
Domain: $(-\infty,\infty)$ Range: $(-\infty,0)$

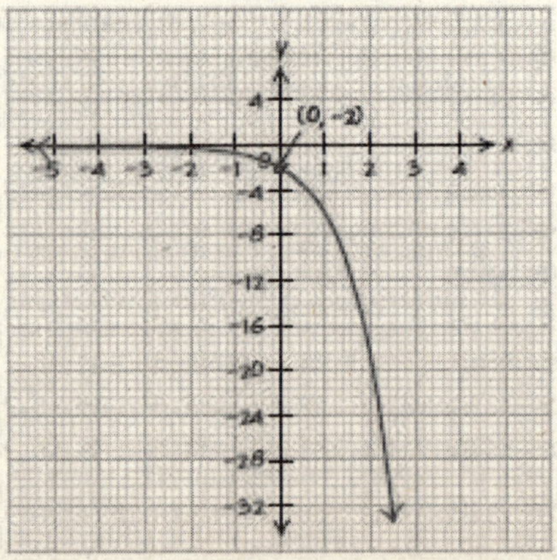

10. Since $a > 0$ and $b < 1$ the function is decreasing.
Domain: $(-\infty, \infty)$ Range: $(0, \infty)$

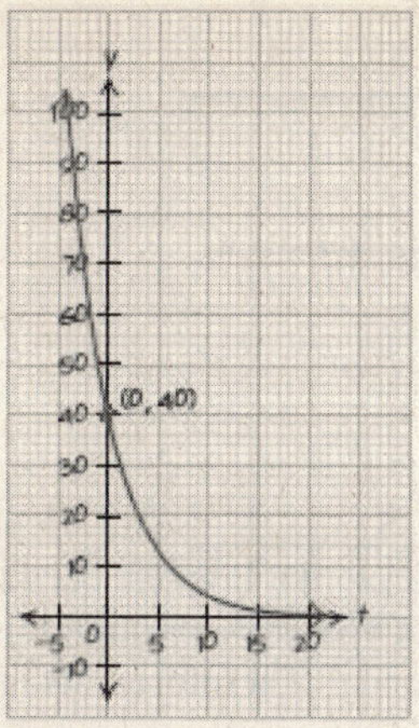

11. Since $a > 0$ and $b < 1$ the function is decreasing.
Domain: $(-\infty, \infty)$ Range: $(50, \infty)$

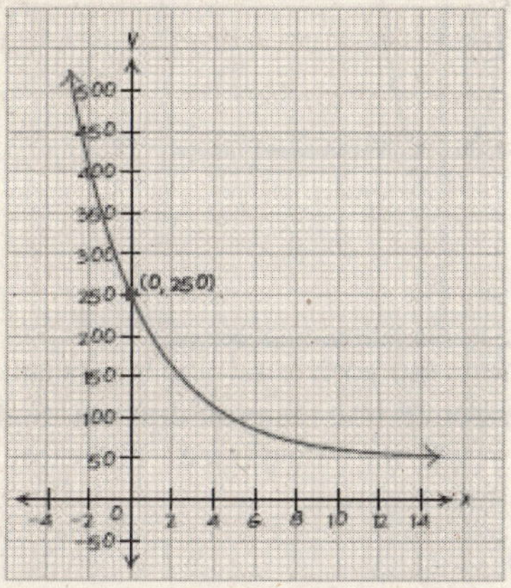

12. Since $a < 0$ and $b < 1$ the function is increasing.
Domain: $(-\infty, \infty)$ Range: $(-\infty, -20)$

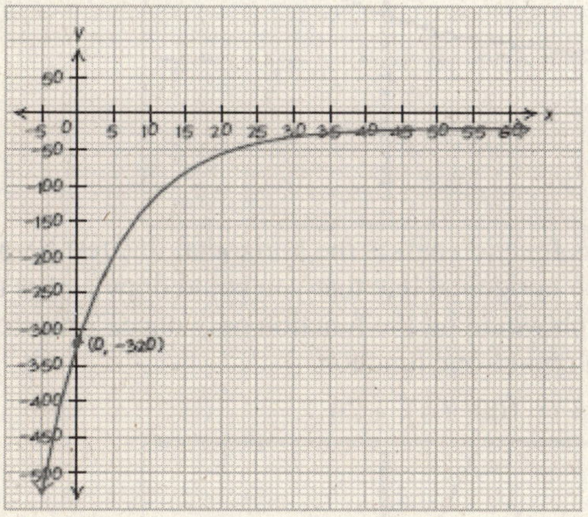

13.

$$12\left(4^x\right) = 12288$$
$$4^x = 1024$$
$$4^x = 4^5$$
$$x = 5$$

14.

$$2\left(3^x\right) - 557 = -503$$
$$2\left(3^x\right) - 557 + 557 = -503 + 557$$
$$2\left(3^x\right) = 54$$
$$3^x = 27$$
$$3^x = 3^3$$
$$x = 3$$

15.

$$-6\left(5^x\right) - 510 = -4260$$
$$-6\left(5^x\right) - 510 + 510 = -4260 + 510$$
$$-6\left(5^x\right) = -3750$$
$$5^x = 625$$
$$5^x = 5^4$$
$$x = 4$$

16.

$$x^3 = 216$$
$$\left(x^3\right)^{\frac{1}{3}} = (216)^{\frac{1}{3}}$$
$$x = 6$$

17.

$$-3x^5 + 650 = 3722$$
$$-3x^5 + 650 - 650 = 3722 - 650$$
$$-3x^5 = 3072$$
$$x^5 = -1024$$
$$\left(x^5\right)^{\frac{1}{5}} = (-1024)^{\frac{1}{5}}$$
$$x = -4$$

18.

$$4.5x^6 - 2865 = 415.5$$
$$4.5x^6 - 2865 + 2865 = 415.5 + 2865$$
$$4.5x^6 = 3280.5$$
$$x^6 = 729$$
$$\left(x^6\right)^{\frac{1}{6}} = \pm(729)^{\frac{1}{6}}$$
$$x = \pm 3$$

19.

$$300x^4 - 580 = 218768.48$$
$$300x^4 - 580 + 580 = 218768.48 + 580$$
$$300x^4 = 219348.48$$
$$\frac{300x^4}{300} = \frac{219348.48}{300}$$
$$x^4 = 731.1616$$
$$\left(x^4\right)^{\frac{1}{4}} = \pm(731.1616)^{\frac{1}{4}}$$
$$x = \pm 5.2$$

CHAPTER 13 LOGARITHMIC FUNCTIONS

Section 13.1 Functions and Their Inverses

1. This function passes the horizontal line test because each horizontal line that you could draw across the graph would hit the function only once. Therefore, it is a one-to-one function.

3. This function passes the horizontal line test because each horizontal line that you could draw across the graph would hit the function only once. Therefore, it is a one-to-one function.

5. This function fails the horizontal line test because almost any horizontal line that you draw through this graph hits the graph more than once. Therefore, this is not a one-to-one function.

7. This function passes the horizontal line test because each horizontal line that you could draw across the graph would hit the function only once. Therefore, it is a one-to-one function.

9. This line is horizontal, so when it is tested by using the horizontal line test, it fails to be one-to-one.

11. $f(x)=\frac{1}{2}x+3$

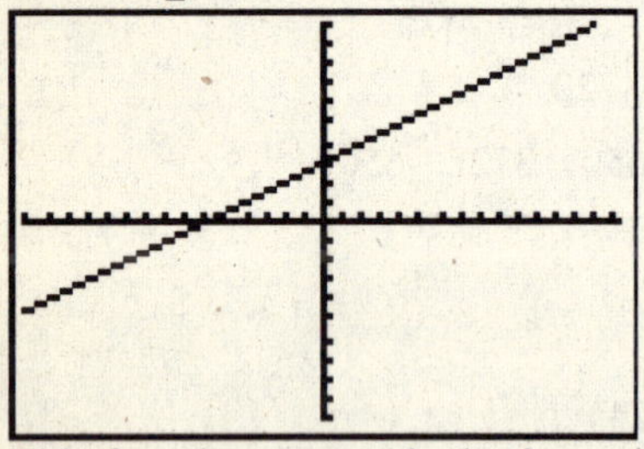

This function passes the horizontal line test because each horizontal line that you could draw across the graph would hit the function only once. Therefore, it is a one-to-one function.

13. $h(x)=2.5x^2+3x-9$

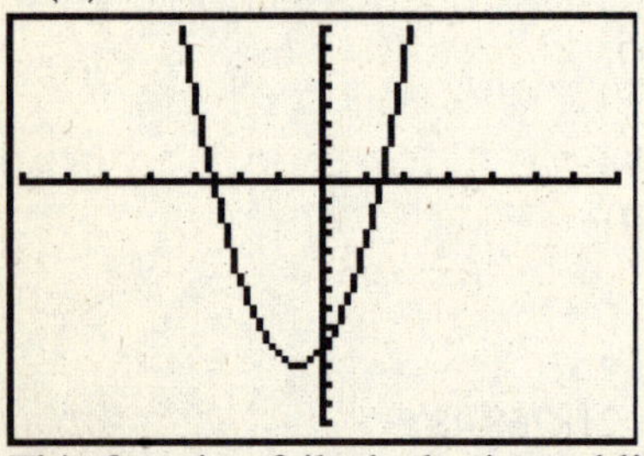

This function fails the horizontal line test because almost any horizontal line that you draw through this graph hits the graph more than once. Therefore, this is not a one-to-one function.

15. $f(x)=2x^3+4$

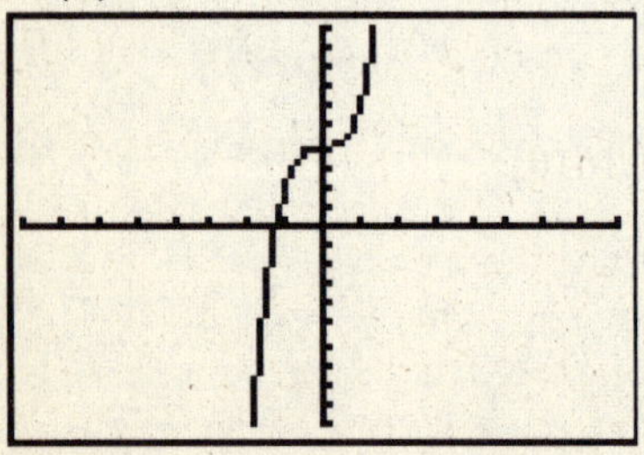

This function passes the horizontal line test because each horizontal line that you could draw across the graph would hit the function only once. Therefore, it is a one-to-one function.

17. $f(x)=4x^3+2x^2-5x-4$

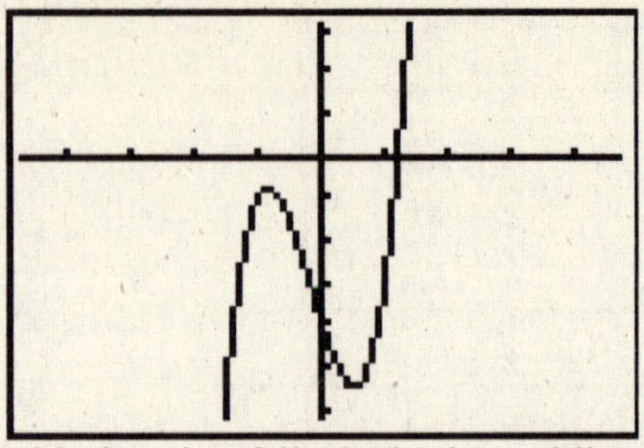

This function fails the horizontal line test because some horizontal lines that you draw through this graph hit the graph more than once. Therefore, this is not a one-to-one function.

19. $f(x)=2x^4$

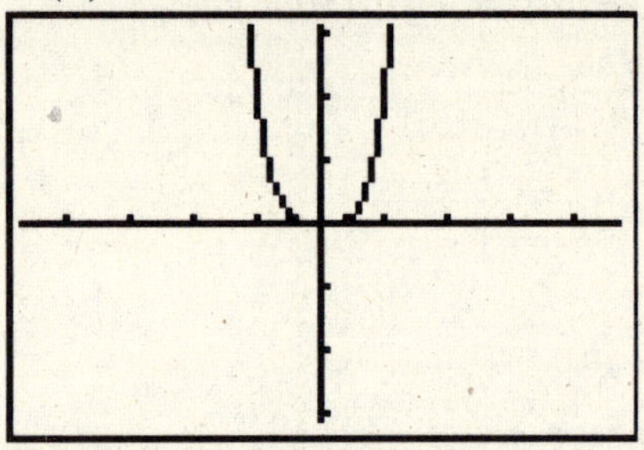

This function fails the horizontal line test because almost any horizontal line that you draw through this graph hits the graph more than once. Therefore, this is not a one-to-one function.

21. $g(x)=3(1.2)^x$

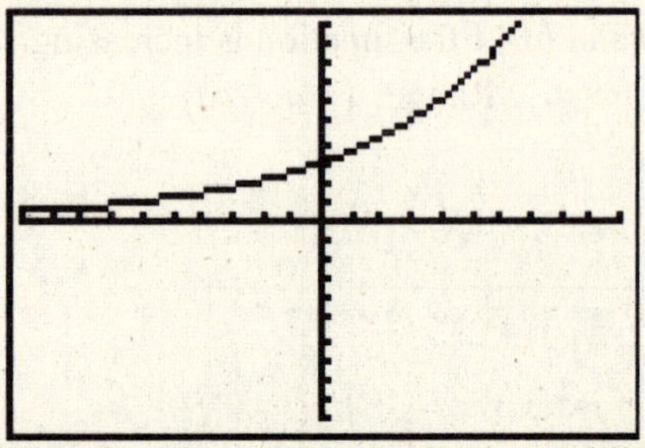

This function passes the horizontal line test because each horizontal line that you could draw across the graph would hit the function only once. Therefore, it is a one-to-one function.

23. $f(x)=100(0.4)^x+20$

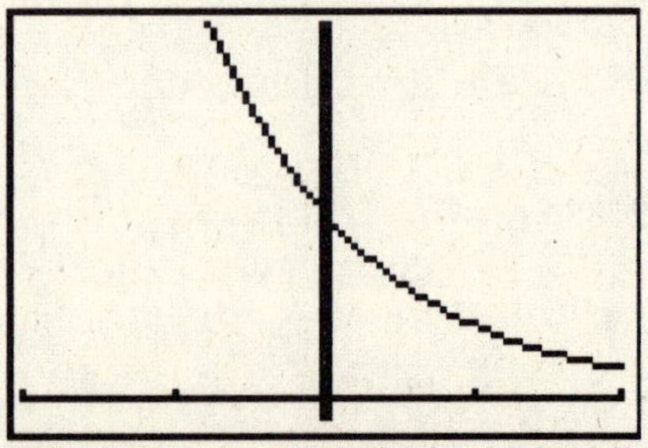

This function passes the horizontal line test because each horizontal line that you could draw across the graph would hit the function only once. Therefore, it is a one-to-one function.

25. $f(x) = 20$

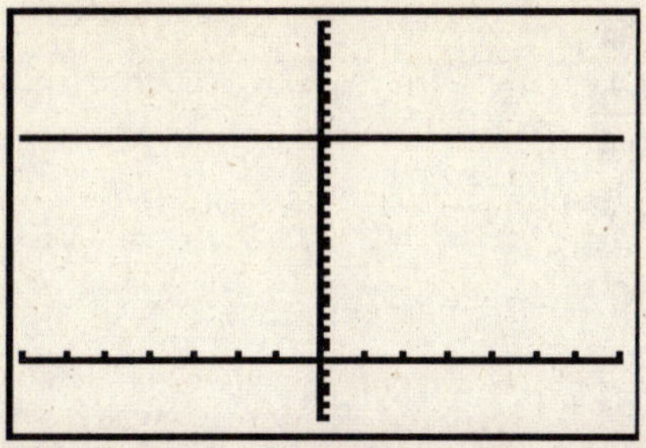

This line is horizontal, so when it is tested by using the horizontal line test, it fails to be one-to-one.

27. **a.**

$N(t) = -315.9t + 4809.8$
$N = -315.9t + 4809.8$
$N - 4809.8 = -315.9t$
$\frac{N - 4809.8}{-315.9} = \frac{-315.9t}{-315.9}$
$\frac{N}{-315.9} - \frac{4809.8}{-315.9} = t$
$-0.003N + 15.23 \approx t$
$t(N) = -0.003N + 15.23$

b. If the original model has a domain of $[-1,12]$ and a range of $[1019, 5125.7]$, then the inverse function has a domain of $[1019, 5125.7]$ and a range of $[-1,12]$.

c.

$t(2000) = -0.003(2000) + 15.23$
$t(2000) = 9.23$

In 1999, there were approximately 2000 homicides of 15-19 year olds in the United States.

29. **a.** If $P(b)$ has domain $0 \le b \le 5000$, then

$P(0) = 5.5(0) - 2500 = -2500$
$P(5000) = 5.5(5000) - 2500 = 25{,}000$

Therefore, the range is $-2500 \le P \le 25{,}000$.

b.

$P(b) = 5.5b - 2500$
$P = 5.5b - 2500$
$P + 2500 = 5.5b$
$\frac{P + 2500}{5.5} = \frac{5.5b}{5.5}$
$\frac{P}{5.5} + \frac{2500}{5.5} = b$
$0.18P + 454.55 \approx b$
$b(P) = 0.18P + 454.55$

c.

$P(1000) = 5.5(1000) - 2500$
$P(1000) = 3000$

There is a $3000 profit from selling 1000 books.

d.

$b(5000) = 0.18(5000) + 454.55$
$b(5000) = 1354.55$

You would need to sell 1355 books in order to make a $5000 profit.

e. Referring to part 'a' above, switch the roles of domain and range.

Domain: $-2500 \le P \le 25{,}000$

Range: $0 \le b \le 5000$

31. **a.**

$P(t) = 2.57t + 249.78$
$P = 2.57t + 249.78$
$P - 249.78 = 2.57t$
$\frac{P - 249.78}{2.57} = \frac{2.57t}{2.57}$
$\frac{P}{2.57} - \frac{249.78}{2.57} = t$
$0.389P - 97.191 \approx t$
$t(P) = 0.389P - 97.191$

b.

$t(260) = 0.389(260) - 97.191$
$t(260) = 3.95$

The U.S. population reached 260 million in about 1994.

c. For the inverse function $t(P) = 0.389P - 97.191$; the input variable P represents the U.S. population in millions, the output variable t represents the number of years since 1990.

33. **a.**

$M(c) = 10{,}000 + 1.5c$
$M = 10{,}000 + 1.5c$
$M - 10{,}000 = 1.5c$
$\frac{M - 10{,}000}{1.5} = \frac{1.5c}{1.5}$
$\frac{M}{1.5} - \frac{10{,}000}{1.5} = c$
$0.667M - 6666.667 \approx c$
$c(M) = 0.667M - 6666.667$

b.

$c(40{,}000) = 0.667(40{,}000) - 6666.667$
$c(40{,}000) \approx 20013.333$

They would have to sell about 20,014 CD's in order for "Math Dude" to earn $40,000.

c. If the domain of $M(c)$ is $10{,}000 \le c \le 50{,}000$, then

$M(10{,}000) = 10{,}000 + 1.5(10{,}000) = 25{,}000$
$M(50{,}000) = 10{,}000 + 1.5(50{,}000) = 85{,}000$

Therefore, the range of this original model is [25000, 85000]. Switching these roles gives the inverse domain and range.

Domain: $25{,}000 \le M \le 85{,}000$

Range: $10{,}000 \le c \le 50{,}000$

35.

$$f(x)=3x+5$$
$$y=3x+5$$
$$y-5=3x$$
$$\frac{1}{3}(y-5)=\frac{1}{3}(3x)$$
$$\frac{1}{3}y-\frac{5}{3}=x$$
$$\frac{1}{3}x-\frac{5}{3}=y$$
$$f^{-1}(x)=\frac{1}{3}x-\frac{5}{3}$$

37.

$$g(t)=-4t+8$$
$$g=-4t+8$$
$$g-8=-4t$$
$$\left(-\frac{1}{4}\right)(g-8)=\left(-\frac{1}{4}\right)(-4t)$$
$$-\frac{1}{4}g+2=t$$
$$-\frac{1}{4}t+2=g$$
$$g^{-1}(t)=-\frac{1}{4}t+2$$

39.

$$h(x)=\frac{2}{3}x-9$$
$$y=\frac{2}{3}x-9$$
$$y+9=\frac{2}{3}x$$
$$\left(\frac{3}{2}\right)(y+9)=\left(\frac{3}{2}\right)\left(\frac{2}{3}x\right)$$
$$\frac{3}{2}y+\frac{27}{2}=x$$
$$\frac{3}{2}x+\frac{27}{2}=y$$
$$h^{-1}(x)=\frac{3}{2}x+\frac{27}{2}$$

41.

$$f(x)=\frac{1}{5}x+\frac{3}{5}$$
$$y=\frac{1}{5}x+\frac{3}{5}$$
$$y-\frac{3}{5}=\frac{1}{5}x$$
$$5\left(y-\frac{3}{5}\right)=5\left(\frac{1}{5}x\right)$$
$$5y-3=x$$
$$5x-3=y$$
$$f^{-1}(x)=5x-3$$

43.

$$h(x)=0.4x-1.6$$
$$y=0.4x-1.6$$
$$y+1.6=0.4x$$
$$\frac{y+1.6}{0.4}=\frac{0.4x}{0.4}$$
$$\frac{y}{0.4}+\frac{1.6}{0.4}=x$$
$$2.5y+4=x$$
$$2.5x+4=y$$
$$h^{-1}(x)=2.5x+4$$

45.

$$P(t)=-2.5t-7.5$$
$$P=-2.5t-7.5$$
$$P+7.5=-2.5t$$
$$\frac{P+7.5}{-2.5}=\frac{-2.5t}{-2.5}$$
$$\frac{P}{-2.5}+\frac{7.5}{-2.5}=t$$
$$-0.4P-3=t$$
$$-0.4t-3=P$$
$$P^{-1}(t)=-0.4t-3$$

47. Yes, they are inverses.

$$f(g(x))=f\left(\frac{1}{3}x+3\right)$$
$$f(g(x))=3\left(\frac{1}{3}x+3\right)-9$$
$$f(g(x))=x+9-9$$
$$f(g(x))=x$$

$$g(f(x))=g(3x-9)$$
$$g(f(x))=\frac{1}{3}(3x-9)+3$$
$$g(f(x))=x-3+3$$
$$g(f(x))=x$$

49. Yes, they are inverses.

$$f(g(x))=f(0.25x-3)$$
$$f(g(x))=4(0.25x-3)+12$$
$$f(g(x))=x-12+12$$
$$f(g(x))=x$$

$$g(f(x))=g(4x+12)$$
$$g(f(x))=0.25(4x+12)-3$$
$$g(f(x))=x+3-3$$
$$g(f(x))=x$$

51. No, they are not inverses.

$$f(h(x))=f(7x+3)$$
$$f(h(x))=\frac{1}{7}(7x+3)+21$$
$$f(h(x))=x+\frac{3}{7}+21$$
$$f(h(x))=x+\frac{150}{7}$$
$$f(h(x))\neq x$$

53.

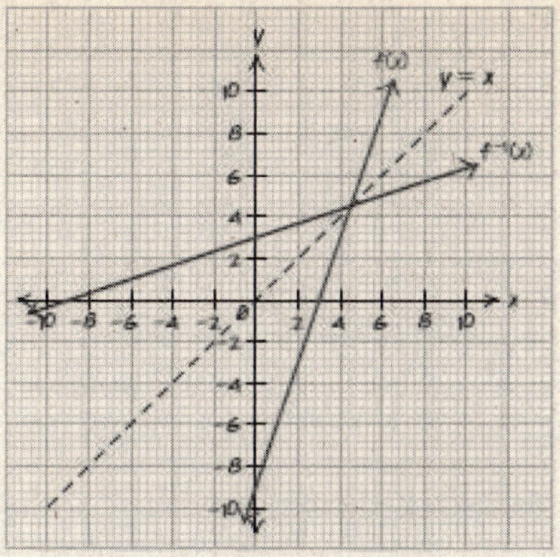

55.

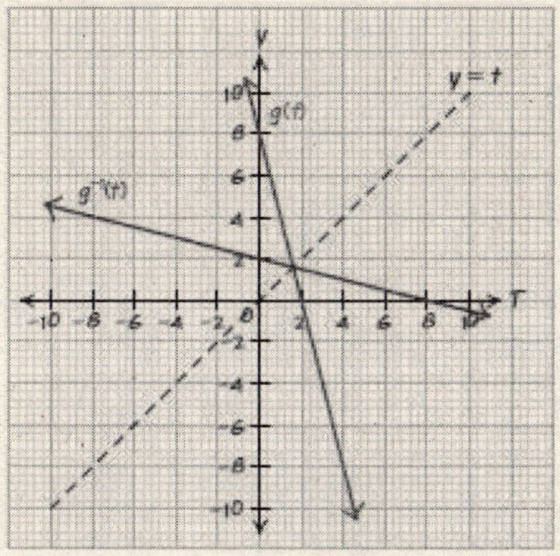

57.

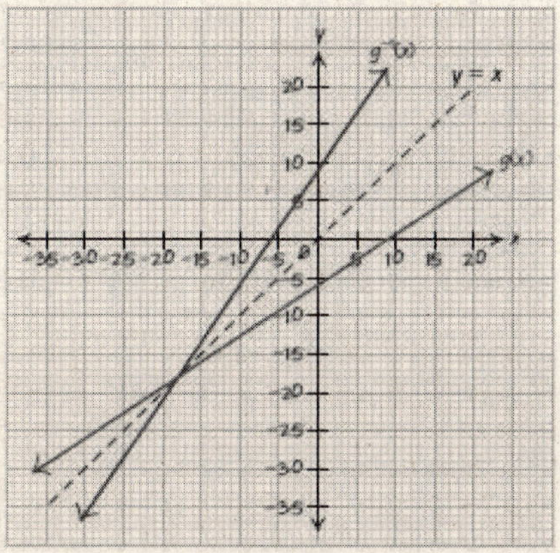

59.

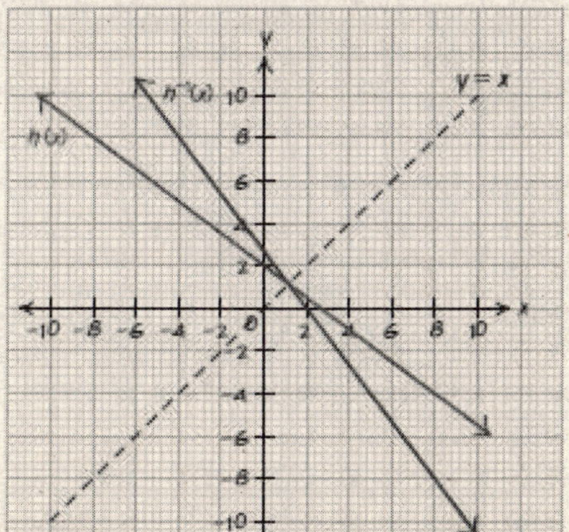

61.

x	$y=2^x$	$f(x)=(x,y)$	$f^{-1}(x)=(y,x)$
-1	$2^{-1}=\frac{1}{2}$	$\left(-1,\frac{1}{2}\right)$	$\left(\frac{1}{2},-1\right)$
0	$2^0=1$	$(0,1)$	$(1,0)$
1	$2^1=2$	$(1,2)$	$(2,1)$

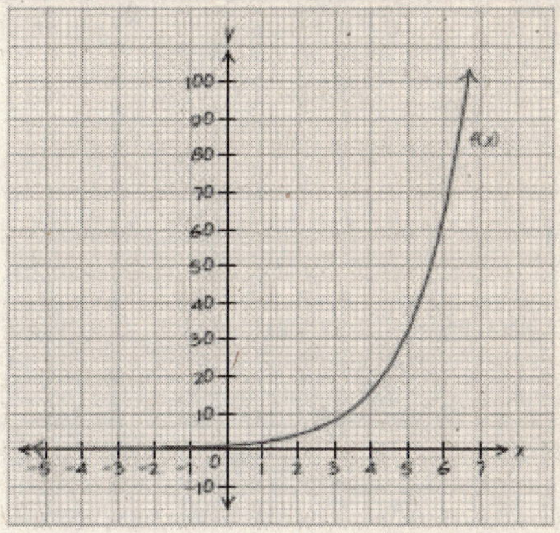

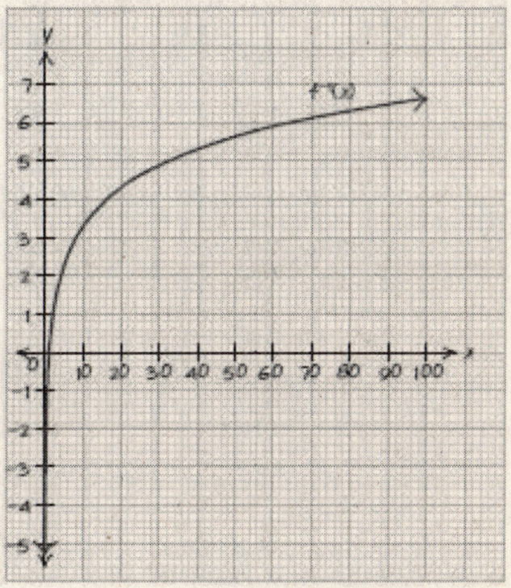

63.

x	$y=150(0.5)^x$	$h(x)=(x,y)$	$h^{-1}(x)=(y,x)$
0	$150(0.5)^0=150$	$(0,150)$	$(150,0)$
2	$150(0.5)^2=37.5$	$(2,37.5)$	$(37.5,2)$
4	$150(0.5)^4=9.375$	$(4,9.375)$	$(9.375,4)$

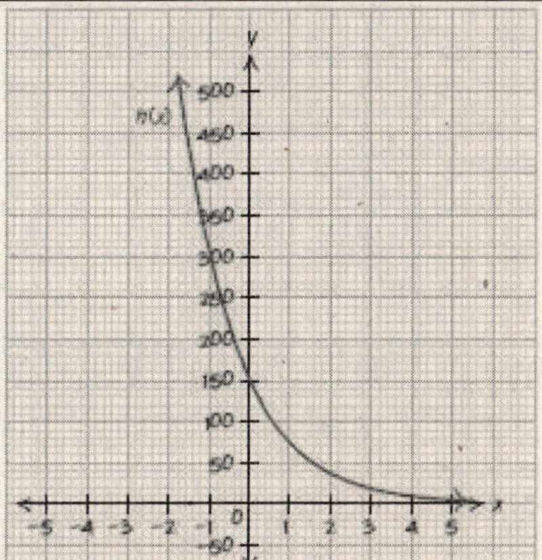

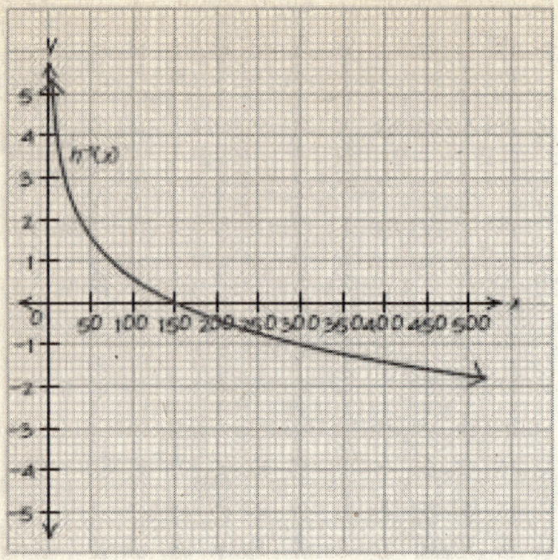

65.

x	$y=-20(1.25)^x$	(x,y)	(y,x)
-6	$-20(1.25)^{-6}\approx -5.2$	$(-6,-5.2)$	$(-5.2,-6)$
-2	$-20(1.25)^{-2}=-12.8$	$(-2,-12.8)$	$(-12.8,-2)$
0	$-20(1.25)^{0}=-20$	$(0,-20)$	$(-20,0)$
1	$-20(1.25)^{1}=-25$	$(1,-25)$	$(-25,1)$

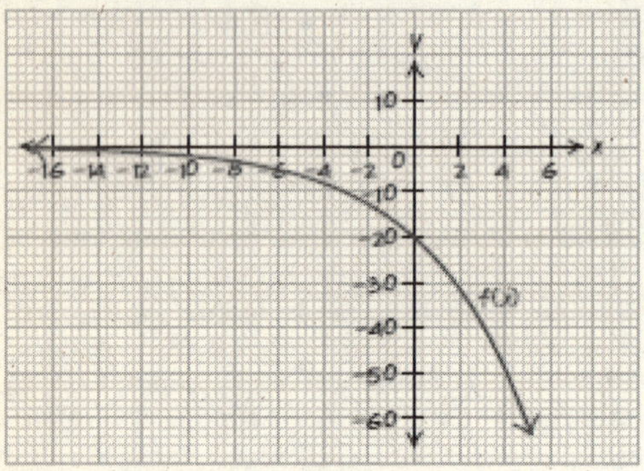

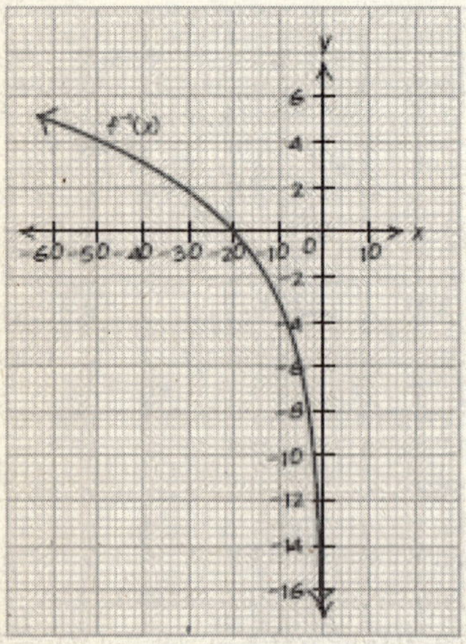

67.

x	$y=2(3)^x+5$	$h(x)=(x,y)$	$h^{-1}(x)=(y,x)$
0	$2(3)^0+5=7$	$(0,7)$	$(7,0)$
1	$2(3)^1+5=11$	$(1,11)$	$(11,1)$
2	$2(3)^2+5=23$	$(2,23)$	$(23,2)$

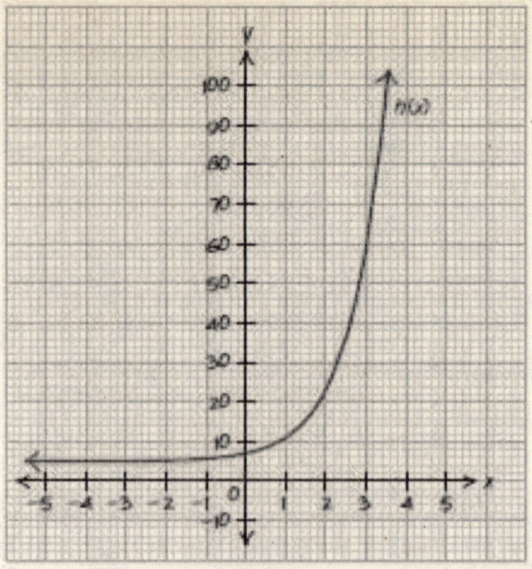

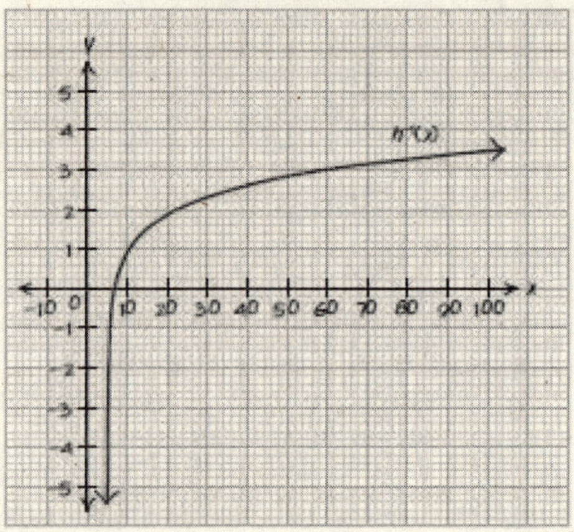

69.

x	$y=-2(2.5)^x-8$	$f(x)=(x,y)$	$f^{-1}(x)=(y,x)$
0	$-2(2.5)^0-8=-10$	$(0,-10)$	$(-10,0)$
1	$-2(2.5)^1-8=-13$	$(1,-13)$	$(-13,1)$
2	$-2(2.5)^2-8=-20.5$	$(2,-20.5)$	$(-20.5,2)$

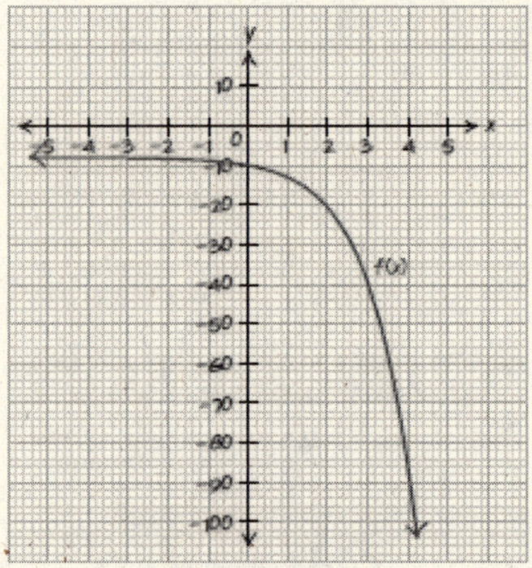

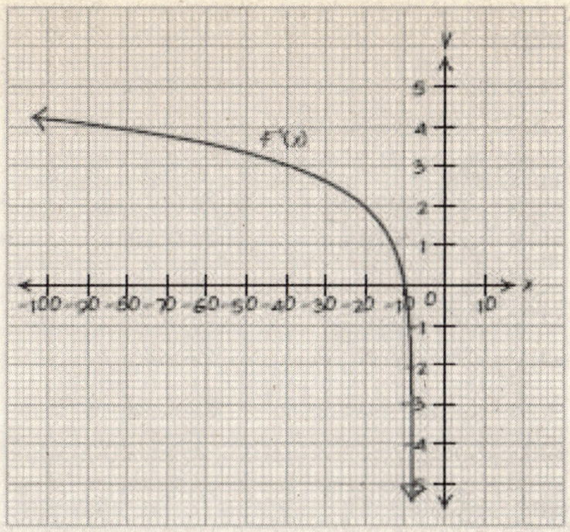

71.

x	$y=40\left(\frac{1}{10}\right)^x+15$	(x,y)	(y,x)
0	$40\left(\frac{1}{10}\right)^0+15=55$	$(0,55)$	$(55,0)$
1	$40\left(\frac{1}{10}\right)^1+15=19$	$(1,19)$	$(19,1)$
2	$40\left(\frac{1}{10}\right)^2+15=15.4$	$(2,15.4)$	$(15.4,2)$

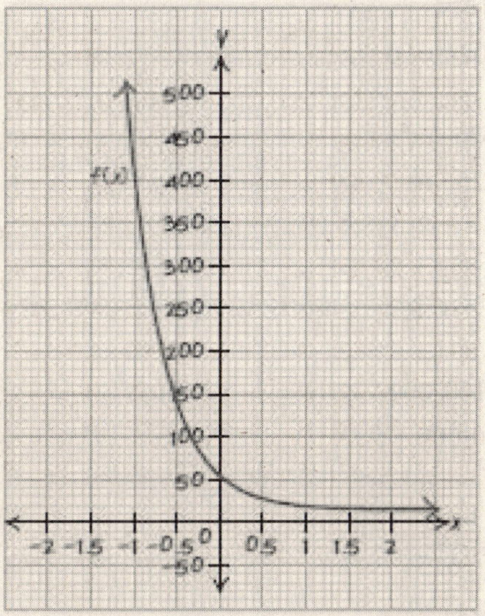

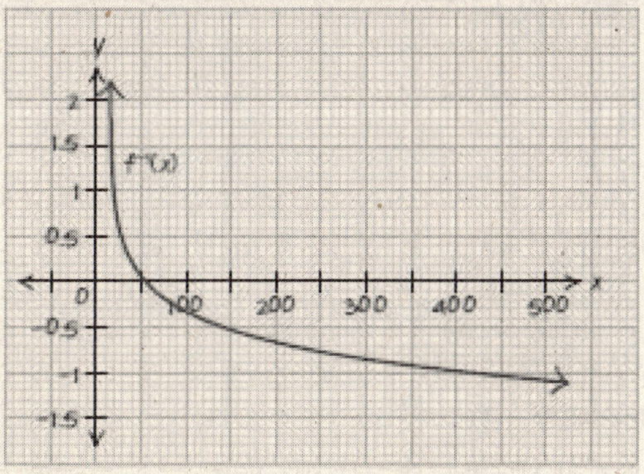

73.

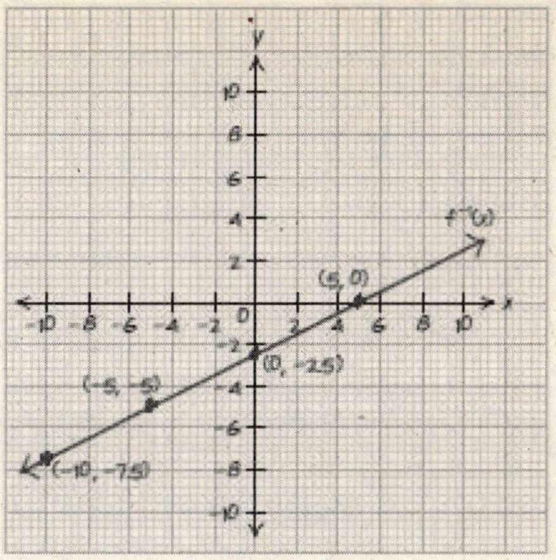

75.

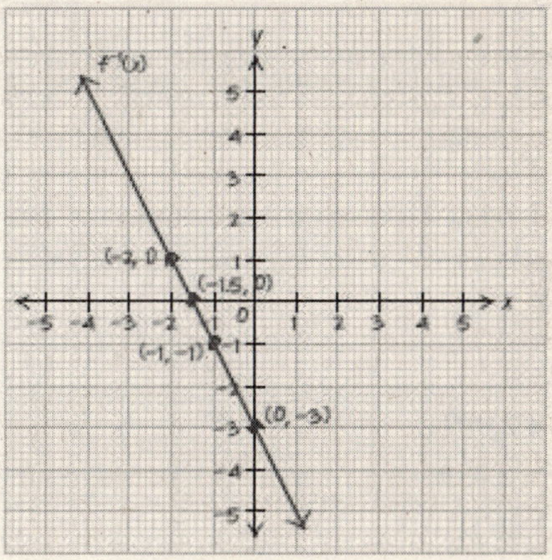

77.

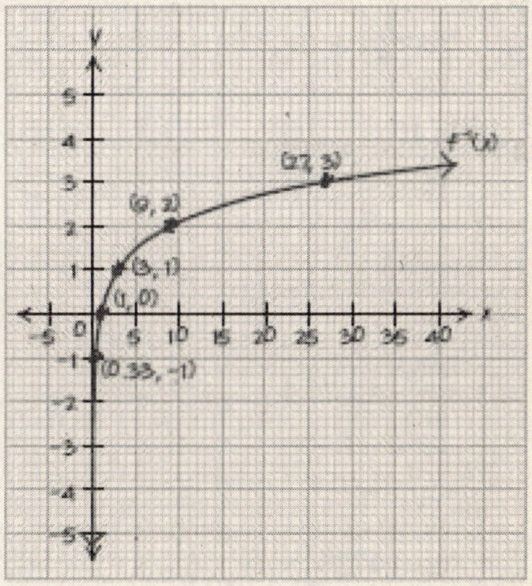

79. Yes, it is a function. Yes, it has an inverse. It passes the vertical line test so it is a function, and it passes the horizontal line test so it must have an inverse.

81. No, it is not a function. This is a vertical line and it will not pass the vertical line test.

CHAPTER 13 Logarithmic Functions

Section 13.2 Logarithmic Functions

1. The base is 5.
3. The base is 10.
5. The base is e.
7. $\log_3 27 = \log_3(3^3) = 3$
9. $\log_5 125 = \log_5(5^3) = 3$
11. $\log_2 256 = \log_2(2^8) = 8$
13. $\log 1000 = \log(10^3) = 3$
15. $\log_5\left(\frac{1}{5}\right) = \log_5(5^{-1}) = -1$
17. $\log 0.1 = \log\left(\frac{1}{10}\right) = \log(10^{-1}) = -1$
19. $\log_7\left(\frac{1}{49}\right) = \log_7\left(\frac{1}{7^2}\right) = \log_7(7^{-2}) = -2$
21. $\log_5\left(\frac{1}{25}\right) = \log_5\left(\frac{1}{5^2}\right) = \log_5(5^{-2}) = -2$
23. $\log_7 1 = 0$
25. $\log_8(8^4) = 4$
27. $\ln(e^3) = 3$
29. $\ln e = 1$
31. $\log_{19} 19 = 1$
33. $\log 125 \approx 2.097$
35. $\log 275 \approx 2.439$
37. $\ln 45 \approx 3.807$
39. $\ln 120 \approx 4.787$
41.
    a) $\log_3 40 = \frac{\log 40}{\log 3} \approx 3.358$
    b) $\log_3 40 = \frac{\ln 40}{\ln 3} \approx 3.358$
43.
    a) $\log_5 63 = \frac{\log 63}{\log 5} \approx 2.574$
    b) $\log_5 63 = \frac{\ln 63}{\ln 5} \approx 2.574$
45. $\log_7 25 = \frac{\log 25}{\log 7} \approx 1.654$
47. $\log_2 0.473 = \frac{\log 0.473}{\log 2} \approx -1.080$
49. $\log_{14} 478 = \frac{\log 478}{\log 14} \approx 2.338$
51. $\log_{11} 0.254 = \frac{\log 0.254}{\log 11} \approx -0.572$
53. $\log_{12} 36478 = \frac{\log 36478}{\log 12} \approx 4.227$
55. $\log 1000 = 3$
    $10^3 = 1000$
57.
    $\log 0.01 = -2$
    $10^{-2} = 0.01$
59.
    $\log_2(8) = 3$
    $2^3 = 8$
61.
    $\log_3(81) = 4$
    $3^4 = 81$
63.
    $\ln(e^5) = 5$
    $e^5 = e^5$
65.
    $\log_3\left(\frac{1}{9}\right) = -2$
    $3^{-2} = \frac{1}{9}$
67.
    $\log_5\left(\frac{1}{25}\right) = -2$
    $5^{-2} = \frac{1}{25}$
69.
    $2^{10} = 1024$
    $\log_2 1024 = 10$
71.
    $5^4 = 625$
    $\log_5 625 = 4$
73.
    $25^{0.5} = 5$
    $\log_{25} 5 = 0.5$
75.
    $10^5 = 100,000$
    $\log 100,000 = 5$
77.
    $\left(\frac{1}{5}\right)^4 = \frac{1}{625}$
    $\log_{\frac{1}{5}}\left(\frac{1}{625}\right) = 4$
79.
    $\left(\frac{1}{2}\right)^3 = \frac{1}{8}$
    $\log_{\frac{1}{2}}\left(\frac{1}{8}\right) = 3$
81.
    $3^{2x} = 729$
    $\log_3 729 = 2x$

83.

$$f(x)=7^x$$
$$y=7^x$$
$$\log_7 y=x$$
$$\log_7 x=y$$
$$\frac{\log x}{\log 7}=y$$
$$f^{-1}(x)=\log_7 x$$
$$f^{-1}(x)=\frac{\log x}{\log 7}$$

85.

$$h(c)=10^c$$
$$h=10^c$$
$$\log h=c$$
$$\log c=h$$
$$h^{-1}(c)=\log c$$

87.

$$m(r)=5^r$$
$$m=5^r$$
$$\log_5 m=r$$
$$\log_5 r=m$$
$$\frac{\log r}{\log 5}=m$$
$$m^{-1}(r)=\log_5 r$$
$$m^{-1}(r)=\frac{\log r}{\log 5}$$

89.

$$f(x)=\log_3 x$$
$$y=\log_3 x$$
$$3^y=x$$
$$3^x=y$$
$$f^{-1}(x)=3^x$$

91.

$$g(x)=\ln x$$
$$y=\ln x$$
$$e^y=x$$
$$e^x=y$$
$$g^{-1}(x)=e^x$$

93.

$$f(x)=3(4)^x$$
$$y=3(4)^x$$
$$\frac{y}{3}=(4)^x$$
$$\log_4\left(\frac{y}{3}\right)=x$$
$$\log_4\left(\frac{x}{3}\right)=y$$
$$\frac{\log\left(\frac{x}{3}\right)}{\log 4}=y$$
$$f^{-1}(x)=\log_4\left(\frac{x}{3}\right)$$
$$f^{-1}(x)=\frac{\log\left(\frac{x}{3}\right)}{\log 4}$$

95.

$$h(x)=\frac{1}{2}(9)^x$$
$$y=\frac{1}{2}(9)^x$$
$$2y=(9)^x$$
$$\log_9(2y)=x$$
$$\log_9(2x)=y$$
$$\frac{\log(2x)}{\log 9}=y$$
$$h^{-1}(x)=\log_9(2x)$$
$$h^{-1}(x)=\frac{\log(2x)}{\log 9}$$

97.

$$\log x=3$$
$$10^3=x$$
$$x=1000$$

99.

$$\ln t=2$$
$$e^2=t$$
$$t\approx 7.389$$

101.

$$\log_3 w=2.4$$
$$3^{2.4}=w$$
$$w\approx 13.967$$

103.

$$\log_{16} m = \frac{1}{2}$$

$$16^{\frac{1}{2}} = m$$

$$m = 4$$

105.

$$\log_{12} x = \frac{1}{5}$$

$$12^{\frac{1}{5}} = x$$

$$x \approx 1.644$$

107.

$$\log(8t) = 2$$
$$10^2 = 8t$$
$$\frac{10^2}{8} = t$$
$$\frac{100}{8} = t$$
$$t = \frac{25}{2}$$
$$t = 12.5$$

109.

$$\ln(3x) = 4$$
$$e^4 = 3x$$
$$\frac{e^4}{3} = x$$
$$x \approx 18.199$$

Section 13.3 Graphing Logarithmic Functions

1. **a)** $f(50) = 2$ **b)** $x = 125$

3. **a)** $h(250) = 4$ **b)** $x = 20$

5. **a)** $g(40) = -3$ **b)** $x = 13$

7. **a)** $h(3) = -11$ **b)** $x = 5$

9. The base is greater than 1, because the graph is increasing.

11. The base is less than 1, because the graph is decreasing.

13.

x	$f(x) = \log x$
1	0
2	0.301
5	0.699
10	1

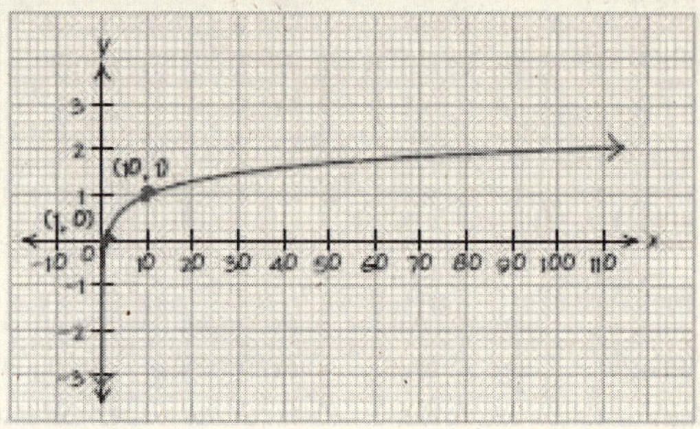

15.

x	$f(x) = \log_4 x$
1	0
2	.5
3	0.792
4	1

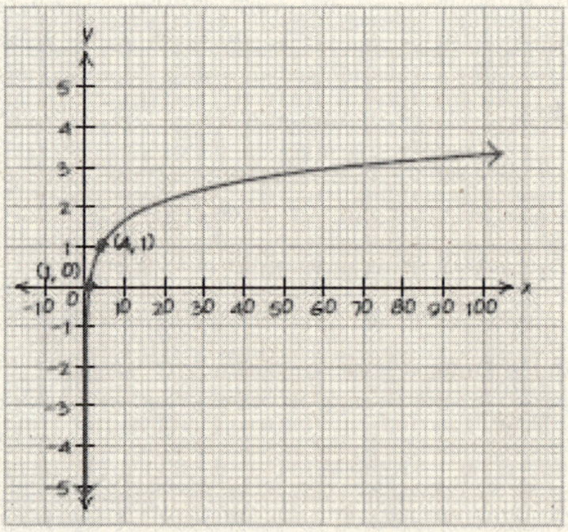

17.

x	$f(x) = \log_9 x$
1	0
2	0.315
3	0.5
9	1

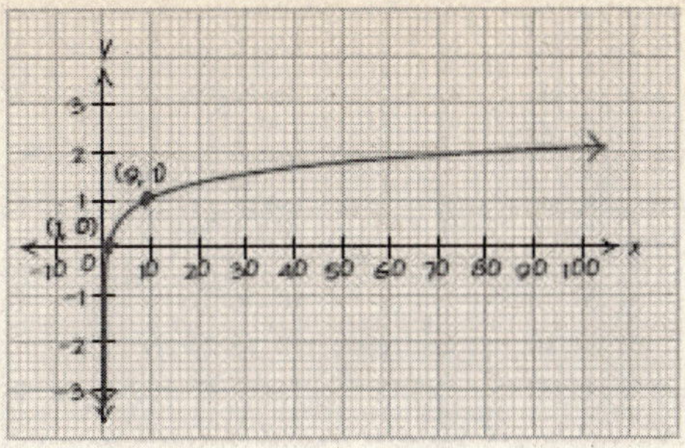

19.

x	$f(x) = \log_{20} x$
1	0
5	0.537
10	0.769
20	1

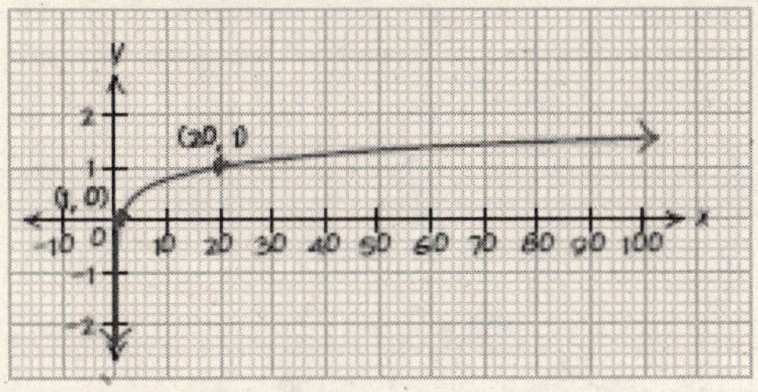

21.

x	$f(x) = \log_{0.2} x$
0.2	1
0.4	0.569
1	0
2	-0.431

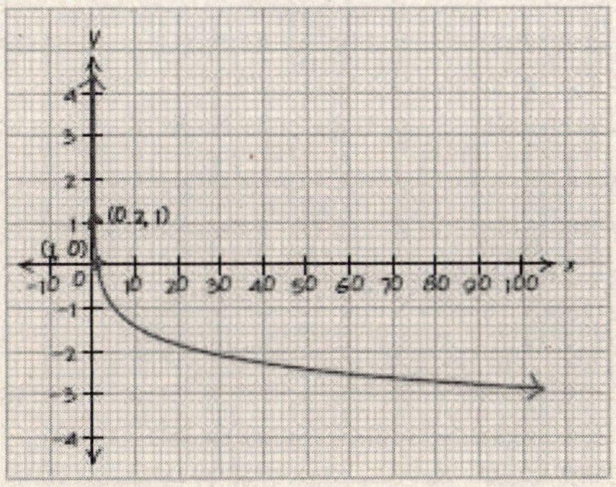

23.

x	$f(x) = \log_{0.9} x$
0.3	11.427
0.9	1
1	0
3	-10.427

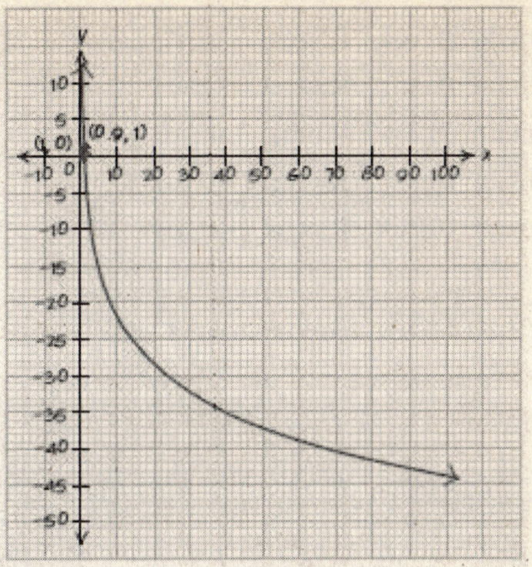

25.

x	$f(x) = \log_{0.6} x$
0.3	2.357
0.6	1
1	0
3	-2.151

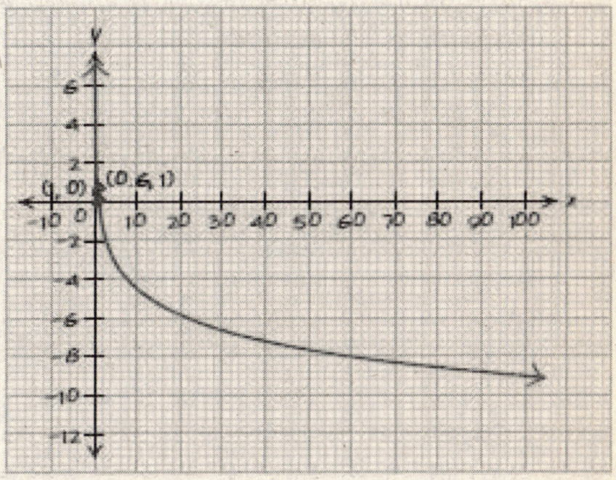

26.

x	$f(x) = \log_{0.1} x$
0.05	1.301
0.1	1
1	0
2	-0.301

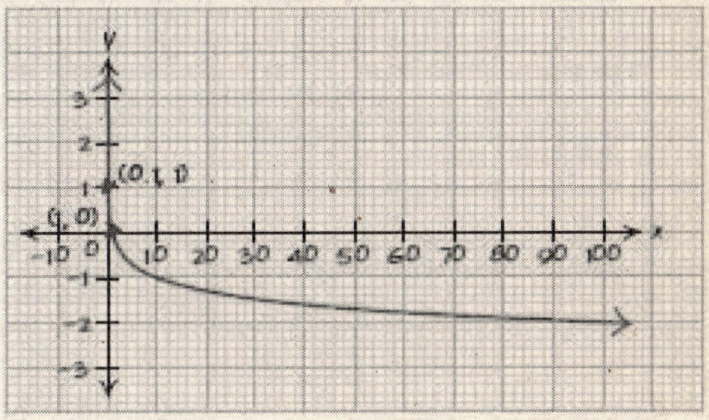

27. Domain: $(0,\infty)$ Range: $(-\infty,\infty)$

29. Domain: $(0,\infty)$ Range: $(-\infty,\infty)$

31. Domain: $(0,\infty)$ Range: $(-\infty,\infty)$

33. Domain: $(0,\infty)$ Range: $(-\infty,\infty)$

35. Domain: $(0,\infty)$ Range: $(-\infty,\infty)$

37. Domain: $(0,\infty)$ Range: $(-\infty,\infty)$

39.

$$f(x)=2x+12$$
$$y=2x+12$$
$$y-12=2x$$
$$\frac{1}{2}(y-12)=\frac{1}{2}(2x)$$
$$\frac{1}{2}y-6=x$$
$$\frac{1}{2}x-6=y$$
$$f^{-1}(x)=\frac{1}{2}x-6$$

41.

$$f(x)=4^x$$
$$y=4^x$$
$$\log_4 y=x$$
$$\log_4 x=y$$
$$\frac{\log x}{\log 4}=y$$
$$f^{-1}(x)=\log_4 x$$

or

$$f^{-1}(x)=\frac{\log x}{\log 4}$$

43.

$$g(x)=\log_9 x$$
$$y=\log_9 x$$
$$9^y=x$$
$$9^x=y$$
$$g^{-1}(x)=9^x$$

45. $f(x)=\frac{2}{3}x-9$

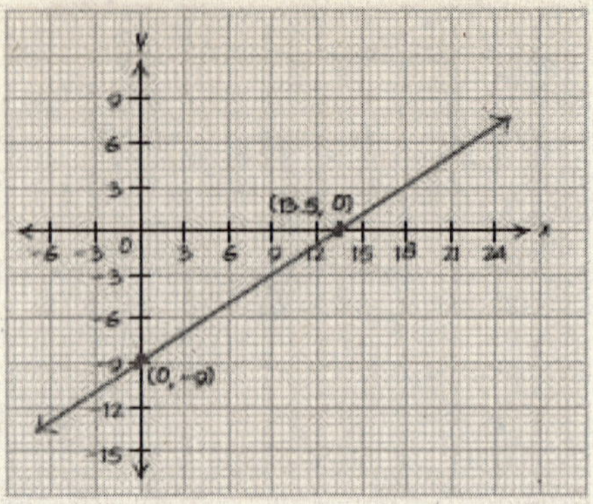

47. $f(x)=5(1.2)^x$

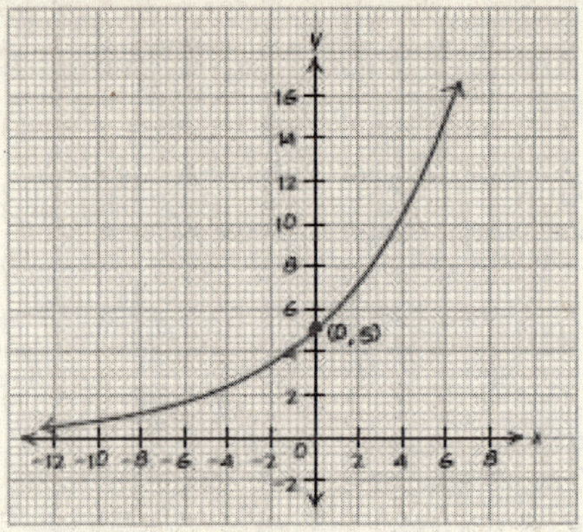

49. $g(x)=2(x+3)^2-8$

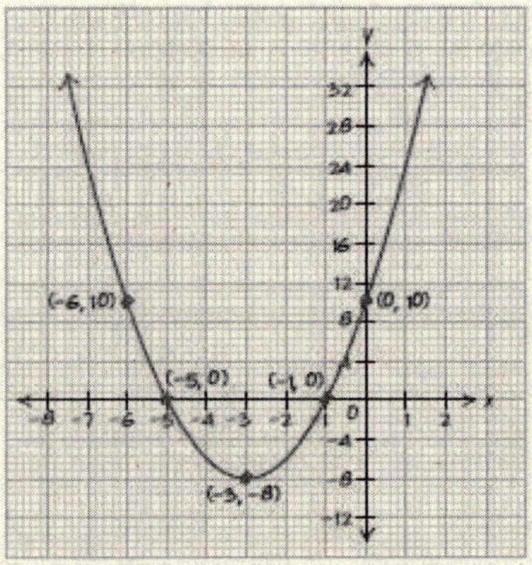

51. $f(x)=\log_5 x$

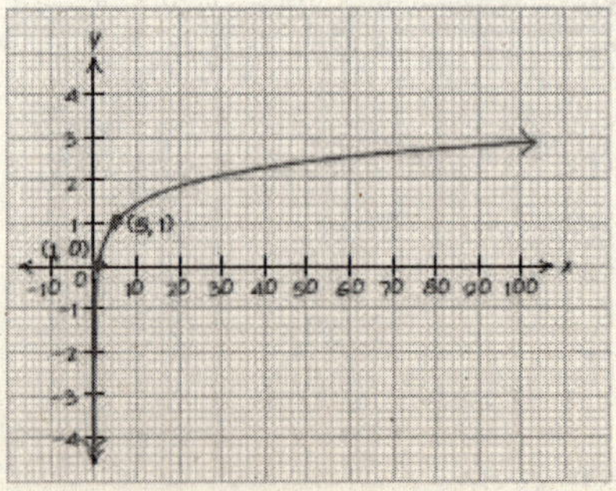

53. $\log 300 \approx 2.477$

55. $\ln 22 \approx 3.091$

57. $\log_5 630 = \dfrac{\log 630}{\log 5} \approx 4.005$

59. $\log_{4.5} 1256 = \dfrac{\log 1256}{\log 4.5} \approx 4.744$

61. $\log_6 0.45 = \dfrac{\log 0.45}{\log 6} \approx -0.446$

63.

$$\log_8 x = 3$$
$$8^3 = x$$
$$x = 512$$

65.

$$\log x = 7$$
$$10^7 = x$$
$$x = 10{,}000{,}000$$

67.

$$\log_4 x = -2$$
$$4^{-2} = x$$
$$x = \frac{1}{4^2}$$
$$x = \frac{1}{16}$$

Section 13.4 Properties of Logarithms

1. $\log(5x) = \log 5 + \log x$

3. $\ln(xy) = \ln x + \ln y$

5. $\ln(4ab^3) = \ln 4 + \ln a + \ln b^3 = \ln 4 + \ln a + 3\ln b$

7.

$$\begin{aligned}\ln(2h^2k^3) &= \ln 2 + \ln h^2 + \ln k^3\\ &= \ln 2 + 2\ln h + 3\ln k\end{aligned}$$

9.

$$\begin{aligned}\log_3\left(\frac{1}{2}ab^2c^3\right) &= \log_3\left(\frac{1}{2}\right) + \log_3 a + \log_3 b^2 + \log_3 c^3\\ &= \log_3 1 - \log_3 2 + \log_3 a + \log_3 b^2 + \log_3 c^3\\ &= 0 - \log_3 2 + \log_3 a + 2\log_3 b + 3\log_3 c\\ &= -\log_3 2 + \log_3 a + 2\log_3 b + 3\log_3 c\end{aligned}$$

11. $\log\left(\dfrac{x}{y}\right) = \log x - \log y$

13. $\log\left(\dfrac{12}{m}\right) = \log 12 - \log m$

15.

$$\begin{aligned}\log\left(\frac{2x^2}{y}\right) &= \log 2 + \log x^2 - \log y\\ &= \log 2 + 2\log x - \log y\end{aligned}$$

17.

$$\begin{aligned}\ln\left(\frac{3x^4y^3}{z}\right) &= \ln 3 + \ln x^4 + \ln y^3 - \ln z\\ &= \ln 3 + 4\ln x + 3\ln y - \ln z\end{aligned}$$

19.

$$\begin{aligned}\log\left(\sqrt{5x}\right) &= \log(5x)^{\frac{1}{2}}\\ &= \frac{1}{2}(\log 5x)\\ &= \frac{1}{2}(\log 5 + \log x)\\ &= \frac{1}{2}\log 5 + \frac{1}{2}\log x\end{aligned}$$

21.

$$\begin{aligned}\ln\left(\sqrt{7ab}\right) &= \ln(7ab)^{\frac{1}{2}}\\ &= \frac{1}{2}\ln(7ab)\\ &= \frac{1}{2}(\ln 7 + \ln a + \ln b)\\ &= \frac{1}{2}\ln 7 + \frac{1}{2}\ln a + \frac{1}{2}\ln b\end{aligned}$$

23.

$$\begin{aligned}\log_9\left(\sqrt{2w^2z^3}\right) &= \log_9\left(2w^2z^3\right)^{\frac{1}{2}}\\ &= \frac{1}{2}\log_9\left(2w^2z^3\right)\\ &= \frac{1}{2}\left(\log_9 2 + \log_9 w^2 + \log_9 z^3\right)\\ &= \frac{1}{2}\left(\log_9 2 + 2\log_9 w + 3\log_9 z\right)\\ &= \frac{1}{2}\log_9 2 + \log_9 w + \frac{3}{2}\log_9 z\end{aligned}$$

25.

$$\ln\left(\sqrt[5]{m^2p^3}\right)=\ln\left(m^2p^3\right)^{\frac{1}{5}}$$
$$=\frac{1}{5}\ln\left(m^2p^3\right)$$
$$=\frac{1}{5}\left(\ln m^2+\ln p^3\right)$$
$$=\frac{1}{5}(2\ln m+3\ln p)$$
$$=\frac{2}{5}\ln m+\frac{3}{5}\ln p$$

27.

$$\log_{15}\left(\frac{\sqrt{3x^4y^3}}{z^5}\right)=\log_{15}\left(\frac{\left(3x^4y^3\right)^{\frac{1}{2}}}{z^5}\right)$$
$$=\log_{15}\left(\frac{3^{\frac{1}{2}}x^2y^{\frac{3}{2}}}{z^5}\right)$$
$$=\log_{15}3^{\frac{1}{2}}+\log_{15}x^2+\log_{15}y^{\frac{3}{2}}-\log_{15}z^5$$
$$=\frac{1}{2}\log_{15}3+2\log_{15}x+\frac{3}{2}\log_{15}y-5\log_{15}z$$

29. $\ln x+\ln y=\ln(xy)$

31. $\log a^2+\log b^3=\log\left(a^2b^3\right)$

33. $\ln x-\ln y=\ln\left(\frac{x}{y}\right)$

35.

$$\log 5+2\log x+\log y+3\log z$$
$$=\log 5+\log x^2+\log y+\log z^3$$
$$=\log\left(5x^2yz^3\right)$$

37.

$$\log_3 a+2\log_3 b-5\log_3 c=\log_3 a+\log_3 b^2-\log_3 c^5$$
$$=\log_3\left(\frac{ab^2}{c^5}\right)$$

39.

$$\log_5 7+\frac{1}{2}\log_5 x+\frac{1}{2}\log_5 y=\log_5 7+\log_5 x^{1/2}+\log_5 y^{1/2}$$
$$=\log_5\left(7\sqrt{xy}\right)$$

41.

$$\frac{1}{2}\ln 7+\frac{1}{2}\ln a+\frac{3}{2}\ln b-4\ln c$$
$$=\ln 7^{1/2}+\ln a^{1/2}+\ln b^{3/2}-\ln c^4$$
$$=\ln\left(\frac{\sqrt{7ab^3}}{c^4}\right)$$

43.

$$5\ln 7+2\ln a+4\ln b-3\ln c-2\ln d$$
$$=\ln 7^5+\ln a^2+\ln b^4-\ln c^3-\ln d^2$$
$$=\ln\left(\frac{7^5a^2b^4}{c^3d^2}\right)$$
$$=\ln\left(\frac{16{,}807a^2b^4}{c^3d^2}\right)$$

45.

$$\log_3 a^2+2\log_3 bc-5\log_3 3c$$
$$=\log_3 a^2+\log_3(bc)^2-\log_3(3c)^5$$
$$=\log_3\left(\frac{a^2b^2c^2}{3^5c^5}\right)$$
$$=\log_3\left(\frac{a^2b^2c^{2-5}}{243}\right)$$
$$=\log_3\left(\frac{a^2b^2c^{-3}}{243}\right)$$
$$=\log_3\left(\frac{a^2b^2}{243c^3}\right)$$

47.

$$\log_5 7+2\log_5 xy+\log_5 xy=\log_5 7+3\log_5 xy$$
$$=\log_5 7+\log_5(xy)^3$$
$$=\log_5\left(7x^3y^3\right)$$

49.

$$\frac{1}{2}\log 5+\frac{1}{2}\log 3x+\frac{5}{2}\log y-4\log z$$
$$=\log 5^{1/2}+\log(3x)^{1/2}+\log y^{5/2}-\log z^4$$
$$=\log\left(\frac{\sqrt{5\cdot 3xy^5}}{z^4}\right)$$
$$=\log\left(\frac{\sqrt{15xy^5}}{z^4}\right)$$

51.

$$\log(x+6)+\log(x-2)=\log((x+6)(x-2))$$
$$=\log\left(x^2+4x-12\right)$$

53.

$$\ln(x-3)+\ln(x-7)=\ln((x-3)(x-7))$$
$$=\ln\left(x^2-10x+21\right)$$

55. There is no expansion for the expression $\log(x+5)$. We can expand when there is multiplication or division inside the logarithm, not for addition or subtraction inside the logarithm.

57. The student incorrectly multiplied two separate log functions instead of multiplying just inside one log function. The expression $\log x+\log y$ comes together as $\log(xy)$.

59.

Given $\log_b 2=3$ and $\log_b 16=4$

$$\log_b 32=\log_b(2\cdot 16)$$
$$=\log_b 2+\log_b 16$$
$$=3+4$$
$$=7$$

61.

Given $\log_b 2 = 3$ and $\log_b 16 = 4$

$$\begin{aligned}\log_b 8 &= \log_b\left(\frac{16}{2}\right)\\ &= \log_b 16 - \log_b 2\\ &= 4-3\\ &= 1\end{aligned}$$

63.

Given $\log_b 30 = 8$ and $\log_b 2 = 14$

$$\begin{aligned}\log_b 15 &= \log_b\left(\frac{30}{2}\right)\\ &= \log_b 30 - \log_b 2\\ &= 8-14\\ &= -6\end{aligned}$$

65.

Given $\log_b 100 = 4$ and $\log_b 500 = 6$

$$\begin{aligned}\log_b 50,000 &= \log_b(100\cdot 500)\\ &= \log_b 100 + \log_b 500\\ &= 4+6\\ &= 10\end{aligned}$$

67.

Given $\log_b 5 = 10$ and $\log_b 20 = 4$

$$\begin{aligned}\log_b 0.25 &= \log_b\left(\frac{5}{20}\right)\\ &= \log_b 5 - \log_b 20\\ &= 10-4\\ &= 6\end{aligned}$$

Section 13.5 Solving Exponential Equations

1.

$$\begin{aligned}3^w &= 125\\ \log_3 125 &= w\\ \frac{\log 125}{\log 3} &= w\\ w &\approx 4.395\end{aligned}$$

3.

$$\begin{aligned}5^x &= 373\\ \log_5 373 &= x\\ \frac{\log 373}{\log 5} &= x\\ x &\approx 3.679\end{aligned}$$

5.

$$\begin{aligned}6^x &= 0.48\\ \log_6 0.48 &= x\\ \frac{\log 0.48}{\log 6} &= x\\ x &\approx -0.410\end{aligned}$$

7.

$$\begin{aligned}0.4^x &= 0.13\\ \log_{0.4} 0.13 &= x\\ \frac{\log 0.13}{\log 0.4} &= x\\ x &\approx 2.227\end{aligned}$$

9.

$$\begin{aligned}0.6^x &= 24\\ \log_{0.6} 24 &= x\\ \frac{\log 24}{\log 0.6} &= x\\ x &\approx -6.221\end{aligned}$$

11. **a.** Let $C(t)$ be the number of hours the Cray C90 has been used per academic year t years since 1980.

t	$C(t)$
7	32
8	100
9	329
10	831
11	1685
12	2233
13	3084
14	8517
15	15,584
16	27,399

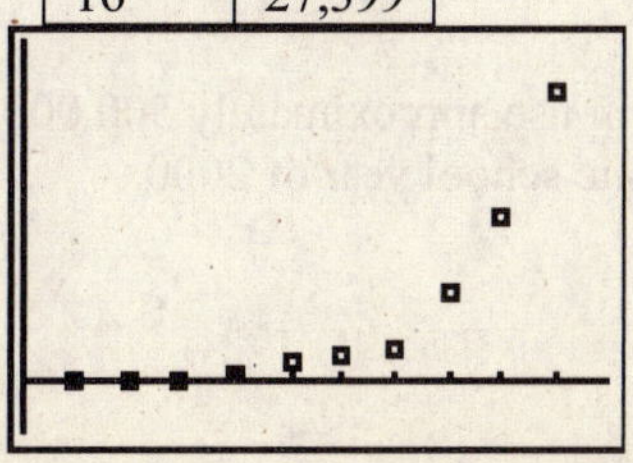

An exponential model is appropriate so choose a point from the flatter end and one from the steeper end.
Note: Your model may differ, but it must follow the trend of the data.

$C(t) = ab^t$

To find a and b choose two points (8,100) and (15,15584).

$100 = ab^8$ and $15584 = ab^{15}$

$$\frac{15584}{100} = \frac{ab^{15}}{ab^8}$$

$$155.84 = b^7$$

$$(155.84)^{\frac{1}{7}} = \left(b^7\right)^{\frac{1}{7}}$$

$$b \approx 2.057$$

Use b and one of the above equations to find a.

$$100 = a(2.057)^8$$

$$\frac{100}{(2.057)^8} = \frac{a(2.057)^8}{(2.057)^8}$$

$$a \approx 0.311$$

Then $C(t) = 0.311(2.057)^t$

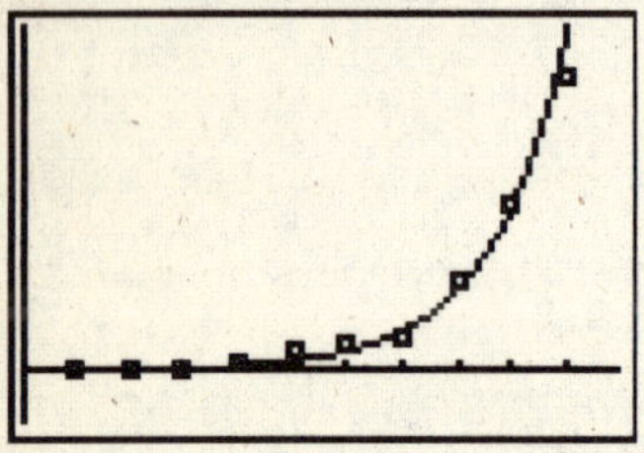

b. Domain: $[7,20]$ Range: $[48.5,572066]$

c.

$$C(17) = 0.311(2.057)^{17}$$

$$C(17) = 65,726.9$$

The Cray C90 was in use approximately 65,727 hours in the 1997 academic school year.

d.

$$500,000 = 0.311(2.057)^t$$

$$\frac{500,000}{0.311} = \frac{0.311(2.057)^t}{0.311}$$

$$\left(\frac{500,000}{0.311}\right) = (2.057)^t$$

$$\log_{2.057}\left(\frac{500,000}{0.311}\right) = t$$

$$\frac{\log\left(\frac{500,000}{0.311}\right)}{\log 2.057} = t$$

$$t \approx 19.813$$

The Cray C90 was in use approximately 500,000 hours in the academic school year of 2000.

13.

$$7^{x-3} = 16807$$

$$7^{x-3} = 7^5$$

$$x - 3 = 5$$

$$x = 8$$

15.

$$4^{x+7} = 3$$

$$\ln\left(4^{x+7}\right) = \ln 3$$

$$(x+7)\ln 4 = \ln 3$$

$$\frac{(x+7)\ln 4}{\ln 4} = \frac{\ln 3}{\ln 4}$$

$$x + 7 = \frac{\ln 3}{\ln 4}$$

$$x = \frac{\ln 3}{\ln 4} - 7$$

$$x \approx -6.208$$

17.

$$604(0.4)^x = 0.158$$

$$\frac{604(0.4)^x}{604} = \frac{0.158}{604}$$

$$(0.4)^x = \frac{0.158}{604}$$

$$\ln(0.4)^x = \ln\left(\frac{0.158}{604}\right)$$

$$x\ln(0.4) = \ln\left(\frac{0.158}{604}\right)$$

$$\frac{x\ln(0.4)}{\ln(0.4)} = \frac{\ln\left(\frac{0.158}{604}\right)}{\ln(0.4)}$$

$$x = \frac{\ln\left(\frac{0.158}{604}\right)}{\ln(0.4)}$$

$$x \approx 9.002$$

19.

$$5(3.2)^x = 74.2$$

$$\frac{5(3.2)^x}{5} = \frac{74.2}{5}$$

$$(3.2)^x = 14.84$$

$$\ln(3.2)^x = \ln 14.84$$

$$x\ln 3.2 = \ln 14.84$$

$$\frac{x\ln 3.2}{\ln 3.2} = \frac{\ln 14.84}{\ln 3.2}$$

$$x = \frac{\ln 14.84}{\ln 3.2}$$

$$x \approx 2.319$$

21.

$$2500\left(\frac{1}{5}\right)^x = 0.032$$

$$\frac{2500\left(\frac{1}{5}\right)^x}{2500} = \frac{0.032}{2500}$$

$$\left(\frac{1}{5}\right)^x = 0.0000128$$

$$\ln\left(\frac{1}{5}\right)^x = \ln 0.0000128$$

$$x\ln\left(\frac{1}{5}\right) = \ln 0.0000128$$

$$\frac{x\ln\left(\frac{1}{5}\right)}{\ln\left(\frac{1}{5}\right)} = \frac{\ln 0.0000128}{\ln\left(\frac{1}{5}\right)}$$

$$x = \frac{\ln 0.0000128}{\ln\left(\frac{1}{5}\right)}$$

$$x = 7$$

23.

$$2^t - 58 = 6$$

$$2^t = 64$$

$$2^t = 2^6$$

$$t = 6$$

25.

$$1.5^h + 20 = 106.5$$

$$1.5^h = 86.5$$

$$\ln 1.5^h = \ln 86.5$$

$$h\ln 1.5 = \ln 86.5$$

$$\frac{h\ln 1.5}{\ln 1.5} = \frac{\ln 86.5}{\ln 1.5}$$

$$h = \frac{\ln 86.5}{\ln 1.5}$$

$$h \approx 11.000$$

27.

$$-3(2.5)^m - 89 = -3262$$

$$-3(2.5)^m = -3173$$

$$\frac{-3(2.5)^m}{-3} = \frac{-3173}{-3}$$

$$(2.5)^m = \frac{3173}{3}$$

$$\ln(2.5)^m = \ln\left(\frac{3173}{3}\right)$$

$$m\ln(2.5) = \ln\left(\frac{3173}{3}\right)$$

$$\frac{m\ln(2.5)}{\ln(2.5)} = \frac{\ln\left(\frac{3173}{3}\right)}{\ln(2.5)}$$

$$m = \frac{\ln\left(\frac{3173}{3}\right)}{\ln(2.5)}$$

$$m \approx 7.600$$

29.

$$3(2)^{n+2} = 96$$

$$\frac{3(2)^{n+2}}{3} = \frac{96}{3}$$

$$2^{n+2} = 32$$

$$2^{n+2} = 2^5$$

$$n + 2 = 5$$

$$n = 3$$

31.

$$10 = 4(1.019)^t$$

$$\frac{10}{4} = \frac{4(1.019)^t}{4}$$

$$2.5 = (1.019)^t$$

$$\log 2.5 = \log(1.019)^t$$

$$\log 2.5 = t\log(1.019)$$

$$\frac{\log 2.5}{\log(1.019)} = t$$

$$t \approx 48.68$$

The world population will reach 10 billion, midway through the year 2023 (48.7 years after 1975).

33.

$$5 = 1.19(1.08)^t$$
$$\frac{5}{1.19} = \frac{1.19(1.08)^t}{1.19}$$
$$\frac{5}{1.19} = (1.08)^t$$
$$\log\left(\frac{5}{1.19}\right) = \log(1.08)^t$$
$$\log\left(\frac{5}{1.19}\right) = t\log(1.08)$$
$$\frac{\log\left(\frac{5}{1.19}\right)}{\log(1.08)} = t$$
$$t \approx 18.7$$

The white-tailed deer population will reach 5 million late in the year 1998 (18.7 years past 1980).

35. **a.**

$$40 = 43.8(0.995)^t$$
$$\frac{40}{43.8} = \frac{43.8(0.995)^t}{43.8}$$
$$\frac{40}{43.8} = (0.995)^t$$
$$\log\left(\frac{40}{43.8}\right) = \log(0.995)^t$$
$$\log\left(\frac{40}{43.8}\right) = t\log(0.995)$$
$$\frac{\log\left(\frac{40}{43.8}\right)}{\log(0.995)} = t$$
$$t \approx 18.1$$

South Africa's population will be down to 40 million by the year 2026 (18.1 years past 2008).

b. South Africa's population in 2008 was 43.8 million. Half of that is 21.9 million.

$$21.9 = 43.8(0.995)^t$$
$$\frac{21.9}{43.8} = \frac{43.8(0.995)^t}{43.8}$$
$$0.5 = (0.995)^t$$
$$\log(0.5) = \log(0.995)^t$$
$$\log(0.5) = t\log(0.995)$$
$$\frac{\log(0.5)}{\log(0.995)} = t$$
$$t \approx 138.3$$

South Africa's population will be down to 21.9 million by the year 2146 (138.3 years past 2008) if the trend continues.

37.

$$3^{5x-4} = 729$$
$$3^{5x-4} = 3^6$$
$$5x - 4 = 6$$
$$5x - 4 + 4 = 6 + 4$$
$$5x = 10$$
$$x = 2$$

39.

$$3(4.6)^{2m+7} = 17.3$$
$$\frac{3(4.6)^{2m+7}}{3} = \frac{17.3}{3}$$
$$(4.6)^{2m+7} = \frac{17.3}{3}$$
$$2m + 7 = \log_{4.6}\left(\frac{17.3}{3}\right)$$
$$2m + 7 = \frac{\log\left(\frac{17.3}{3}\right)}{\log(4.6)}$$
$$2m = \frac{\log\left(\frac{17.3}{3}\right)}{\log(4.6)} - 7$$
$$2m \approx -5.85188$$
$$m \approx \frac{-5.85188}{2}$$
$$m \approx -2.926$$

41.

$$2^{x^2+6} = 1024$$
$$2^{x^2+6} = 2^{10}$$
$$x^2 + 6 = 10$$
$$x^2 = 10 - 6$$
$$x^2 = 4$$
$$x^2 = \pm\sqrt{4}$$
$$x = \pm 2$$

43.

$$\left(\frac{1}{2}\right)^{-x^2+4} = 4$$
$$\left(2^{-1}\right)^{-x^2+4} = 2^2$$
$$2^{(-1)(-x^2+4)} = 2^2$$
$$2^{x^2-4} = 2^2$$
$$x^2 - 4 = 2$$
$$x^2 - 4 + 4 = 2 + 4$$
$$x^2 = 6$$
$$x = \pm\sqrt{6}$$
$$x = \pm 2.449$$

45.

$$10(2)^t - 450 = 6(2)^t + 15{,}934$$
$$10(2)^t - 6(2)^t = 15{,}934 + 450$$
$$4(2)^t = 16{,}384$$
$$\frac{4(2)^t}{4} = \frac{16{,}384}{4}$$
$$(2)^t = 4096$$
$$(2)^t = 2^{12}$$
$$t = 12$$

47.

$$5(4.3)^x + 7 = 2(4.3)^x + 89$$
$$5(4.3)^x - 2(4.3)^x = 89 - 7$$
$$3(4.3)^x = 82$$
$$\frac{3(4.3)^x}{3} = \frac{82}{3}$$
$$(4.3)^x = \frac{82}{3}$$
$$x = \log_{4.3}\left(\frac{82}{3}\right)$$
$$x = \frac{\log\left(\frac{82}{3}\right)}{\log 4.3}$$
$$x \approx 2.268$$

49.

Use $A = P\left(1+\frac{r}{n}\right)^{nt}$ with $n = 365$

$$80{,}000 = 40{,}000\left(1+\frac{r}{365}\right)^{365\times 8}$$
$$\frac{80{,}000}{40{,}000} = \frac{40{,}000\left(1+\frac{r}{365}\right)^{2920}}{40{,}000}$$
$$2 = \left(1+\frac{r}{365}\right)^{2920}$$
$$2^{\frac{1}{2920}} = \left(\left(1+\frac{r}{365}\right)^{2920}\right)^{\frac{1}{2920}}$$
$$2^{\frac{1}{2920}} = 1+\frac{r}{365}$$
$$2^{\frac{1}{2920}} - 1 = 1+\frac{r}{365} - 1$$
$$365\left(2^{\frac{1}{2920}} - 1\right) = 365\left(\frac{r}{365}\right)$$
$$365\left(2^{\frac{1}{2920}} - 1\right) = r$$
$$r \approx 0.08665$$
$$r \approx 8.67\%$$

To double their money in 8 years, they would need to invest at 8.67% compounded daily.

51.

Use $A = Pe^{rt}$

$$10{,}000 = 5000e^{12r}$$
$$\frac{10{,}000}{5000} = \frac{5000e^{12r}}{5000}$$
$$2 = e^{12r}$$
$$\ln 2 = 12r$$
$$\frac{\ln 2}{12} = \frac{12r}{12}$$
$$0.0578 \approx r$$
$$r = 5.78\%$$

To double your money in 12 years, you would need to invest at 5.78% compounded continuously.

53.

Use $A = P\left(1+\frac{r}{n}\right)^{nt}$ with $n = 12$

$$100{,}000 = 50{,}000\left(1+\frac{r}{12}\right)^{12\times 7}$$
$$\frac{100{,}000}{50{,}000} = \frac{50{,}000\left(1+\frac{r}{12}\right)^{84}}{50{,}000}$$
$$2 = \left(1+\frac{r}{12}\right)^{84}$$
$$(2)^{\frac{1}{84}} = \left(\left(1+\frac{r}{12}\right)^{84}\right)^{\frac{1}{84}}$$
$$1.008286 \approx 1+\frac{r}{12}$$
$$1.008286 - 1 \approx \frac{r}{12}$$
$$0.008286 \approx \frac{r}{12}$$
$$12(0.008286) \approx 12\left(\frac{r}{12}\right)$$
$$0.0994 \approx r$$
$$r = 9.94\%$$

To double your money in 7 years, you would need to invest at 9.94% compounded monthly.

55.

Use $A = P\left(1+\frac{r}{n}\right)^{nt}$ with $n = 12$

$$10{,}000 = 5000\left(1+\frac{0.07}{12}\right)^{12t}$$
$$\frac{10{,}000}{5000} = \frac{5000(1.005833)^{12t}}{5000}$$
$$2 = (1.005833)^{12t}$$
$$\log 2 = \log(1.005833)^{12t}$$
$$\log 2 = 12t\log(1.005833)$$
$$\frac{\log 2}{12\log(1.005833)} = \frac{12t\log(1.005833)}{12\log(1.005833)}$$
$$t \approx 9.9$$

It takes approximately 9.9 years to double your money at 7% compounded monthly.

57.

Use $A = Pe^{rt}$

$20{,}000 = 10{,}000e^{0.09t}$

$\frac{20{,}000}{10{,}000} = \frac{10{,}000e^{0.09t}}{10{,}000}$

$2 = e^{0.09t}$

$\ln 2 = 0.09t$

$\frac{\ln 2}{0.09} = \frac{0.09t}{0.09}$

$t \approx 7.7$

It takes approximately 7.7 years to double your money at 9% compounded continuously.

59.

Use $A = P\left(1+\frac{r}{n}\right)^{nt}$ with $n = 365$

$24{,}000 = 8000\left(1+\frac{0.02}{365}\right)^{365t}$

$\frac{24{,}000}{8000} = \frac{8000(1.0000548)^{365t}}{8000}$

$3 = (1.0000548)^{365t}$

$\log 3 = \log(1.0000548)^{365t}$

$\log 3 = 365t\log(1.0000548)$

$\frac{\log 3}{365\log(1.0000548)} = \frac{365t\log(1.0000548)}{365\log(1.0000548)}$

$t \approx 54.9$

It takes approximately 54.9 years to triple your money at 2% compounded daily.

61.

Use $A = Pe^{rt}$

$21{,}000 = 7000e^{0.025t}$

$\frac{21{,}000}{7000} = \frac{7000e^{0.025t}}{7000}$

$3 = e^{0.025t}$

$\ln 3 = 0.025t$

$\frac{\ln 3}{0.025} = \frac{0.025t}{0.025}$

$t \approx 43.9$

It takes approximately 43.9 years to triple your money at 2.5% compounded continuously.

63.

$f(x) = 5(3)^x$ Replace $f(x)$ with y.

$y = 5(3)^x$ Solve for x.

$\frac{y}{5} = \frac{5(3)^x}{3}$

$\frac{y}{5} = 3^x$

$\log_3\left(\frac{y}{5}\right) = \log_3 3^x$

$\log_3\left(\frac{y}{5}\right) = x\log_3 3$

$\log_3\left(\frac{y}{5}\right) = x$

$x = \log_3\left(\frac{y}{5}\right)$ Interchange variables x and y.

$y = \log_3\left(\frac{x}{5}\right)$

$y = \frac{\log\left(\frac{x}{5}\right)}{\log(3)}$ Replace y with $f^{-1}(x)$.

$f^{-1}(x) = \frac{\log\left(\frac{x}{5}\right)}{\log(3)}$

65.

$h(x) = -2.4(4.7)^x$ Replace $h(x)$ with y.

$y = -2.4(4.7)^x$ Solve for x.

$\frac{y}{-2.4} = \frac{-2.4(4.7)^x}{-2.4}$

$\frac{y}{-2.4} = 4.7^x$

$\ln\left(\frac{y}{-2.4}\right) = \ln 4.7^x$

$\ln\left(\frac{y}{-2.4}\right) = x\ln 4.7$

$\frac{\ln\left(\frac{y}{-2.4}\right)}{\ln 4.7} = \frac{x\ln 4.7}{\ln 4.7}$

$x = \frac{\ln\left(\frac{y}{-2.4}\right)}{\ln 4.7}$ Interchange the variables x and y.

$y = \frac{\ln\left(\frac{x}{-2.4}\right)}{\ln 4.7}$ Replace y with $h^{-1}(x)$.

$h^{-1}(x) = \frac{\ln\left(\frac{x}{-2.4}\right)}{\ln 4.7}$

67.

$g(x) = -3.4e^x$ Replace $g(x)$ with y.

$y = -3.4e^x$ Solve for x.

$\frac{y}{-3.4} = \frac{-3.4e^x}{-3.4}$

$\frac{y}{-3.4} = e^x$

$\ln\left(\frac{y}{-3.4}\right) = \ln e^x$

$\ln\left(\frac{y}{-3.4}\right) = x\ln e$

$\ln\left(\frac{y}{-3.4}\right) = x$

$x = \ln\left(\frac{y}{-3.4}\right)$ Interchange the variables x and y.

$$y = \ln\left(\frac{x}{-3.4}\right) \quad \text{Replace } y \text{ with } g^{-1}(x).$$
$$g^{-1}(x) = \ln\left(\frac{x}{-3.4}\right)$$

69.

$$\begin{aligned}
&4n^2 + 3n - 8 = 0 \\
&a = 4, b = 3, c = -8 \\
&n = \frac{-3 \pm \sqrt{(-3)^2 - 4(4)(-8)}}{2(4)} \\
&n = \frac{-3 \pm \sqrt{137}}{8} \\
&n = \frac{-3 + \sqrt{137}}{8}, n = \frac{-3 - \sqrt{137}}{8} \\
&n \approx 1.088, n \approx -1.838
\end{aligned}$$

71.

$$\begin{aligned}
&7d^5 + 20 = 300 \\
&7d^5 = 280 \\
&d^5 = 40 \\
&\left(d^5\right)^{\frac{1}{5}} = 40^{\frac{1}{5}} \\
&d = 40^{\frac{1}{5}} \\
&d \approx 2.091
\end{aligned}$$

73.

$$\begin{aligned}
&3(1.6)^x - 20 = 400 \\
&3(1.6)^x = 420 \\
&\frac{3(1.6)^x}{3} = \frac{420}{3} \\
&(1.6)^x = 140 \\
&\ln(1.6)^x = \ln(140) \\
&x\ln(1.6) = \ln(140) \\
&\frac{x\ln(1.6)}{\ln(1.6)} = \frac{\ln(140)}{\ln(1.6)} \\
&x = \frac{\ln(140)}{\ln(1.6)} \\
&x \approx 10.514
\end{aligned}$$

75.

$$\begin{aligned}
&4(n-3)^2 + 10 = 30 \\
&4(n-3)(n-3) + 10 = 30 \\
&4\left(n^2 - 6n + 9\right) + 10 = 30 \\
&4n^2 - 24n + 36 + 10 = 30 \\
&4n^2 - 24n + 46 = 30 \\
&4n^2 - 24n + 16 = 0 \\
&\frac{4n^2}{4} - \frac{24n}{4} + \frac{16}{4} = \frac{0}{4} \\
&n^2 - 6n + 4 = 0 \\
&a = 1, b = -6, c = 4 \\
&n = \frac{6 \pm \sqrt{(-6)^2 - 4(1)(4)}}{2(1)} \\
&n = \frac{6 \pm \sqrt{20}}{2} \\
&n = \frac{6 \pm 2\sqrt{5}}{2} \\
&n = 3 \pm \sqrt{5} \\
&n = 3 + \sqrt{5}, n = 3 - \sqrt{5} \\
&n \approx 5.236, n \approx 0.764
\end{aligned}$$

77. $(3.098, 75.145)$

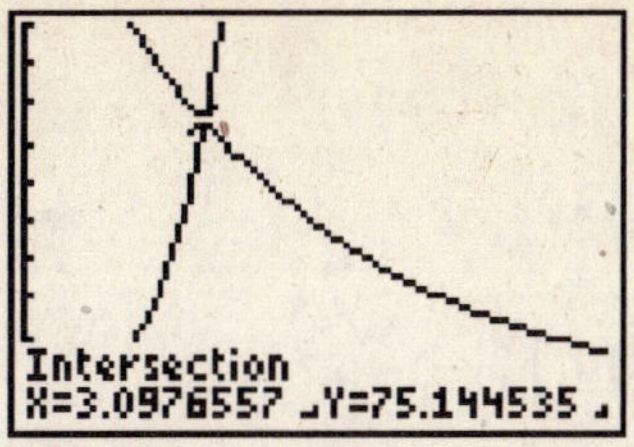

79. $(-4.130, -12.566)$

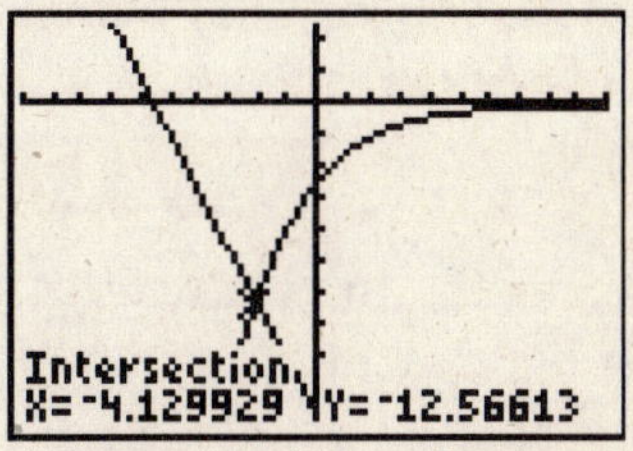

81. $(7.214, 8.013)$ and $(1.859, 54.111)$

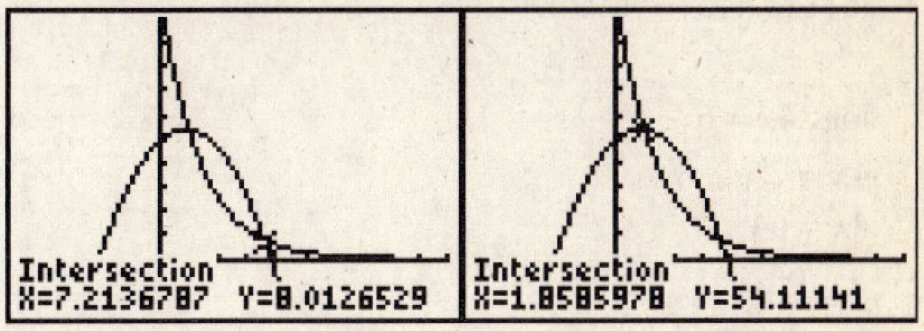

CHAPTER 13 Logarithmic Functions

Section 13.6 Solving Logarithmic Equations

1.

$$\log x = 4$$
$$x = 10^4$$
$$x = 10{,}000$$

Check your answer:

$$\log(10{,}000) = \log(10^4)$$
$$= 4$$

3.

$$\ln x = 5$$
$$x = e^5$$
$$x \approx 148.413$$

Check your answer:

$$\ln(e^5) = 5$$

or

$$\ln(148.413) \approx 4.99999$$

5.

$$\log_2 4x = 6$$
$$4x = 2^6$$
$$4x = 64$$
$$x = \frac{64}{4}$$
$$x = 16$$

Check your answer:

$$\log_2 4(16) = \log_2 64$$
$$= \log_2 2^6$$
$$= 6$$

7.

$$\log(5x+2) = 2$$
$$5x+2 = 10^2$$
$$5x+2 = 100$$
$$5x = 100-2$$
$$x = \frac{98}{5}$$
$$x = 19.6$$

Check your answer:

$$\log(5(19.6)+2) = \log(100)$$
$$= 2$$

9.

$$\ln(3x-5) = 2$$
$$3x-5 = e^2$$
$$3x = e^2+5$$
$$x = \frac{e^2+5}{3}$$
$$x \approx 4.1297$$

Check your answer:

$$\ln\left(3\left(\frac{e^2+5}{3}\right)-5\right) = \ln\left(\left(e^2+5\right)-5\right)$$
$$= \ln(e^2)$$
$$= 2$$

or

$$\ln(3(4.1297)-5) \approx 2.0000059$$

11.

$$M = \log\left(\frac{2000}{10^{-4}}\right)$$
$$M \approx 7.3$$

An earthquake that has an intensity of 2000 cm has a magnitude of about 7.3.

13.

$$M = \log\left(\frac{500{,}000}{10^{-4}}\right)$$
$$M \approx 9.7$$

An earthquake that has an intensity of 500,000 cm has a magnitude of about 9.7.

15.

$$8.0 = \log\left(\frac{I}{10^{-4}}\right)$$
$$\frac{I}{10^{-4}} = 10^8$$
$$10^{-4}\left(\frac{I}{10^{-4}}\right) = 10^{-4}(10^8)$$
$$I = 10^{-4+8} = 10^4$$
$$I = 10{,}000\,cm$$

The intensity of this 8.0 magnitude earthquake was 10,000 cm.

17.

$$6.6 = \log\left(\frac{I}{10^{-4}}\right)$$
$$\frac{I}{10^{-4}} = 10^{6.6}$$
$$10^{-4}\left(\frac{I}{10^{-4}}\right) = 10^{-4}(10^{6.6})$$
$$I = 10^{-4+6.6} = 10^{2.6}$$
$$I = 398.11\,cm$$

The intensity of this 6.6 magnitude earthquake was 398.11 cm.

19.

$$pH = -\log(1.0\times10^{-7})$$
$$pH = -\log(10^{-7}) = 7$$

Scope mouthwash, with a hydrogen ion concentration of 1.0×10^{-7} M, has a pH of 7.

21.

$$1 = -\log\left(H^{+}\right)$$
$$-1 = \log\left(H^{+}\right)$$
$$10^{-1} = H^{+}$$
$$H^{+} = 1.0\times10^{-1}$$

Car battery acid, with a pH of 1, has a hydrogen ion concentration of 1.0×10^{-1} M.

23.

$$14 = -\log\left(H^{+}\right)$$
$$-14 = \log\left(H^{+}\right)$$
$$10^{-14} = H^{+}$$
$$H^{+} = 1.0\times10^{-14}$$

Oven cleaner, with a pH of 14, has a hydrogen ion concentration of 1.0×10^{-14} M.

25.

$7.35 = -\log\left(H^{+}\right)$	$7.45 = -\log\left(H^{+}\right)$
$-7.35 = \log\left(H^{+}\right)$	$-7.45 = \log\left(H^{+}\right)$
$10^{-7.35} = H^{+}$	$10^{-7.45} = H^{+}$
$H^{+} = 4.47\times10^{-8}$	$H^{+} = 3.55\times10^{-8}$

Blood plasma, with a pH level between 7.35 and 7.45, has a hydrogen ion concentration between 3.55×10^{-8} M and 4.47×10^{-8} M.

27.

$$\begin{aligned} L\left(10^{-1}\right) &= 10\log\left(\frac{10^{-1}}{10^{-12}}\right) \\ &= 10\log\left(10^{-1}\cdot10^{12}\right) \\ &= 10\log\left(10^{11}\right) \\ &= 10(11) \\ &= 110 \text{ dB} \end{aligned}$$

The chain saw, with an intensity of 10^{-1} watts per m^2, has a decibel level of 110 dB.

29.

$$\begin{aligned} L\left(10^{-8}\right) &= 10\log\left(\frac{10^{-8}}{10^{-12}}\right) \\ &= 10\log\left(10^{-8}\cdot10^{12}\right) \\ &= 10\log\left(10^{4}\right) \\ &= 10(4) \\ &= 40 \text{ dB} \end{aligned}$$

Raindrops, with an intensity of 10^{-8} watts per m^2, has a decibel level of 40 dB.

31.

$$120 = 10\log\left(\frac{I}{10^{-12}}\right)$$
$$\frac{120}{10} = \log\left(\frac{I}{10^{-12}}\right)$$
$$12 = \log\left(\frac{I}{10^{-12}}\right)$$
$$10^{12} = \frac{I}{10^{-12}}$$
$$\left(10^{12}\right)10^{-12} = \left(\frac{I}{10^{-12}}\right)10^{-12}$$
$$I = 1$$

A jack hammer, with a decibel level of 120 dB, has an intensity of 1 watt/m^2.

33.

$$\log(2x) + 5 = 7$$
$$\log(2x) = 2$$
$$10^2 = 2x$$
$$100 = 2x$$
$$x = 50$$

35.

$$\log_4(-2h) + 40 = 47$$
$$\log_4(-2h) = 7$$
$$4^7 = -2h$$
$$16384 = -2h$$
$$h = -8192$$

37.

$$\ln(3t) + 5 = 3$$
$$\ln(3t) = -2$$
$$e^{-2} = 3t$$
$$\frac{e^{-2}}{3} = t$$
$$t \approx 0.045$$

39.

$$\ln(r+4) = 2$$
$$e^2 = r + 4$$
$$e^2 - 4 = r$$
$$r \approx 3.389$$

41.

$$\log(2p+1) = -0.5$$
$$10^{-0.5} = 2p + 1$$
$$10^{-0.5} - 1 = 2p$$
$$\frac{10^{-0.5} - 1}{2} = \frac{2p}{2}$$
$$p = \frac{10^{-0.5} - 1}{2}$$
$$p \approx -0.342$$

43.

$\log_5(3x)+\log_5 x=4$

$\log_5(3x^2)=4$

$5^4=3x^2$

$625=3x^2$

$x^2=\dfrac{625}{3}$

$x=\pm\sqrt{\dfrac{625}{3}}$

$x=\sqrt{\dfrac{625}{3}}, x=-\sqrt{\dfrac{625}{3}}$

$x\approx 14.434, x\approx -14.434$

The negative answer is not part of the domain.

The final answer is:

$x\approx 14.434$

45.

$\log_3(x+2)+\log_3(x-3)=2$

$\log_3((x+2)(x-3))=2$

$3^2=(x+2)(x-3)$

$9=x^2-x-6$

$0=x^2-x-15$

$x=\dfrac{1\pm\sqrt{(-1)^2-4(1)(-15)}}{2(1)}$

$x=\dfrac{1\pm\sqrt{61}}{2}$

$x=\dfrac{1+\sqrt{61}}{2}, x=\dfrac{1-\sqrt{61}}{2}$

$x\approx 4.405, x\approx -3.405$

The negative answer is not part of the domain.

The final answer is:

$x\approx 4.405$

47. Matt made the base 10 on the second step when it should have been *e*.

49. Frank added 3 and 5 on the second step when the student should have subtracted 5 from both sides of the equation.

51.

$\log(5t^3)-\log(2t)=2$

$\log\left(\dfrac{5t^3}{2t}\right)=2$

$\log(2.5t^2)=2$

$10^2=2.5t^2$

$100=2.5t^2$

$40=t^2$

$t=\pm\sqrt{40}$

$t=\pm 2\sqrt{10}$

$t=2\sqrt{10}, t=-2\sqrt{10}$

$t\approx 6.325, t=-6.325$

The negative answer is not part of the domain.

The final answer is:

$t\approx 6.325$

53.

$\log_5(x)+\log_5(2x+7)-3=0$

$\log_5(x)+\log_5(2x+7)=3$

$\log_5(x(2x+7))=3$

$5^3=x(2x+7)$

$125=2x^2+7x$

$0=2x^2+7x-125$

$x=\dfrac{-7\pm\sqrt{7^2-4(2)(-125)}}{2(2)}$

$x=\dfrac{-7\pm\sqrt{1049}}{4}$

$x=\dfrac{-7+\sqrt{1049}}{4}, x=\dfrac{-7-\sqrt{1049}}{4}$

$x\approx 6.347, x\approx -9.847$

The negative answer is not part of the domain.

The final answer is:

$x\approx 6.347$

55.

$$\log_4(-2x+5)+\log_4(x+21.5)=4$$
$$\log_4((-2x+5)(x+21.5))=4$$
$$4^4=(-2x+5)(x+21.5)$$
$$256=-2x^2-38x+107.5$$
$$0=-2x^2-38x-148.5$$
$$x=\frac{38\pm\sqrt{(-38)^2-4(-2)(-148.5)}}{2(-2)}$$
$$x=\frac{38\pm\sqrt{256}}{-4}$$
$$x=\frac{38+16}{-4},x=\frac{38-16}{-4}$$
$$x=-13.5,x=-5.5$$

57.

$$3\log_2(2x)+\log_2(16)=7$$
$$3\log_2(2x)+\log_2(2^4)=7$$
$$3\log_2(2x)+4=7$$
$$3\log_2(2x)=3$$
$$\frac{3\log_2(2x)}{3}=\frac{3}{3}$$
$$\log_2(2x)=1$$
$$2^1=2x$$
$$2=2x$$
$$x=1$$

59.

$$\ln(3x^2+5x)-3=2$$
$$\ln(3x^2+5x)=5$$
$$e^5=3x^2+5x$$
$$0=3x^2+5x-e^5$$
$$x=\frac{-5\pm\sqrt{5^2-4(3)(-e^5)}}{2(3)}$$
$$x=\frac{-5\pm\sqrt{25+12e^5}}{6}$$
$$x=\frac{-5+\sqrt{25+12e^5}}{6},x=\frac{-5-\sqrt{25+12e^5}}{6}$$
$$x\approx 6.249,x\approx -7.916$$

61.

$$2\log(3x)+8=2$$
$$2\log(3x)=-6$$
$$\frac{2\log(3x)}{2}=\frac{-6}{2}$$
$$\log(3x)=-3$$
$$10^{-3}=3x$$
$$\frac{1}{1000}=3x$$
$$\frac{1}{1000}\left(\frac{1}{3}\right)=3x\left(\frac{1}{3}\right)$$
$$x=\frac{1}{3000}$$
$$x\approx 0.0003$$

63.

$$3^x=27$$
$$3^x=3^3$$
$$x=3$$

65.

$$4x^2+6x-10=0$$
$$\frac{4x^2}{2}+\frac{6x}{2}-\frac{10}{2}=\frac{0}{2}$$
$$2x^2+3x-5=0$$
$$(2x+5)(x-1)=0$$
$$2x+5=0,x-1=0$$
$$x=-2.5,x=1$$

67.

$$t^3+6t^2=-5t$$
$$t^3+6t^2+5t=0$$
$$t(t^2+6t+5)=0$$
$$t(t+5)(t+1)=0$$
$$t=0,t+5=0,t+1=0$$
$$t=0,t=-5,t=-1$$

69.

$$3(d+4)-14=0$$
$$3d+12-14=0$$
$$3d-2=0$$
$$d=\frac{2}{3}$$
$$d\approx 0.667$$

CHAPTER 13 Logarithmic Functions

Chapter 13 Review Exercises

1.

$$f(x)=1.4x-7$$
$$y=1.4x-7$$
$$y+7=1.4x$$
$$y+7=\frac{14}{10}x$$
$$\left(\frac{10}{14}\right)(y+7)=\left(\frac{10}{14}\right)\left(\frac{14}{10}x\right)$$
$$\frac{5}{7}y+5=x$$
$$\frac{5}{7}x+5=y$$
$$f^{-1}(x)=\frac{5}{7}x+5$$

2.

$$g(t)=-2t+6$$
$$g=-2t+6$$
$$g-6=-2t$$
$$\left(-\frac{1}{2}\right)(g-6)=\left(-\frac{1}{2}\right)(-2t)$$
$$-\frac{1}{2}g+3=t$$
$$-\frac{1}{2}t+3=g$$
$$g^{-1}(x)=-\frac{1}{2}t+3$$

3.

$$h(x)=5^x$$
$$y=5^x$$
$$\log_5 y=x$$
$$\log_5 x=y$$
$$\frac{\log x}{\log 5}=y$$
$$h^{-1}(x)=\log_5 x$$
$$h^{-1}(x)=\frac{\log x}{\log 5}$$

4.

$$f(x)=3.5(6)^x$$
$$y=3.5(6)^x$$
$$\frac{y}{3.5}=\frac{3.5(6)^x}{3.5}$$
$$\frac{y}{3.5}=(6)^x$$
$$\log_6\left(\frac{y}{3.5}\right)=x$$
$$\log_6\left(\frac{x}{3.5}\right)=y$$
$$\frac{\log\left(\frac{x}{3.5}\right)}{\log 6}=y$$
$$f^{-1}(x)=\log_6\left(\frac{x}{3.5}\right)$$
$$f^{-1}(x)=\frac{\log\left(\frac{x}{3.5}\right)}{\log 6}$$

5.

$$g(x)=2e^x$$
$$y=2e^x$$
$$\frac{y}{2}=\frac{2e^x}{2}$$
$$\frac{y}{2}=e^x$$
$$\ln\left(\frac{y}{2}\right)=x$$
$$\ln\left(\frac{x}{2}\right)=y$$
$$g^{-1}(x)=\ln\left(\frac{x}{2}\right)$$

6.

$$f(x)=\log_5 x$$
$$y=\log_5 x$$
$$5^y=x$$
$$5^x=y$$
$$f^{-1}(x)=5^x$$

7.

$$h(t)=\log_{0.2} t$$
$$h=\log_{0.2} t$$
$$0.2^h=t$$
$$0.2^t=h$$
$$h^{-1}(t)=0.2^t$$

8. **a.**

$$D(e)=400e+50$$
$$D=400e+50$$
$$D-50=400e$$
$$\frac{D-50}{400}=\frac{400e}{400}$$
$$\frac{D}{400}-\frac{50}{400}=e$$
$$0.0025D-0.125=e$$
$$e(D)=0.0025D-0.125$$

b. $D(10)=400(10)+50=4050$

The AEM Toy company can manufacture 4050 dolls when they have 10 employees working.

c. $e(1000)=0.0025(1000)-0.125=2.375$

The AEM Toy company must have 3 employees working on a particular day to manufacture 1000 dolls.

d. Let the range of the inverse function be $5\le e\le 30$, then

$$D(5)=400(5)+50=2050$$
$$D(30)=400(30)+50=12{,}050$$

Therefore, the domain of the inverse function is $2050 \le D \le 12,050$.

9. **a.**

$$P(d) = 4.5d - 300$$
$$P = 4.5d - 300$$
$$P + 300 = 4.5d$$
$$\frac{P+300}{4.5} = \frac{4.5d}{4.5}$$
$$\frac{P}{4.5} + \frac{300}{4.5} = d$$
$$0.222P + 66.667 = d$$
$$d(P) = 0.222P + 66.667$$

b.

$$d(12,000) = 0.222(12,000) + 66.667$$
$$d(12,000) = 2730.667$$

The AEM Toy Company must sell 2731 dolls to make a $12,000 profit.

c. Let the range of the inverse function be $500 \le d \le 4000$, then

$$P(500) = 4.5(500) - 300 = 1950$$
$$P(4000) = 4.5(4000) - 300 = 17,700$$

Therefore, the domain of the inverse function is $1950 \le P \le 17,700$.

10. This function passes the horizontal line test because each horizontal line that you could draw across the graph would hit the function only once. Therefore, it is a one-to-one function.

11. This function fails the horizontal line test because many of the horizontal lines that you could draw through this graph hit the graph more than once. Therefore, this is not a one-to-one function.

12. $\log_4 64 = \log_4 4^3 = 3$

13. $\log_3 1 = 0$

14. $\log_8 8^5 = 5$

15. $\log_2\left(\frac{1}{8}\right) = \log_2\left(\frac{1}{2^3}\right)$

$$= \log_2 2^{-3}$$
$$= -3$$

16. $f(x) = \log_{2.5}(x)$. Find points using $2.5^y = x$.

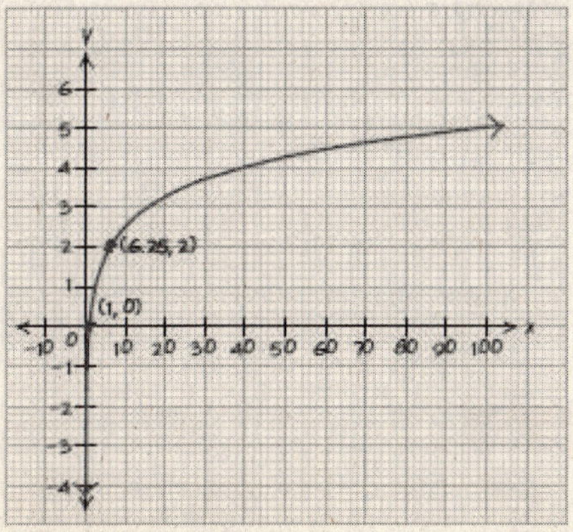

17. $f(x) = \log_{0.4}(x)$. Find points using $0.4^y = x$.

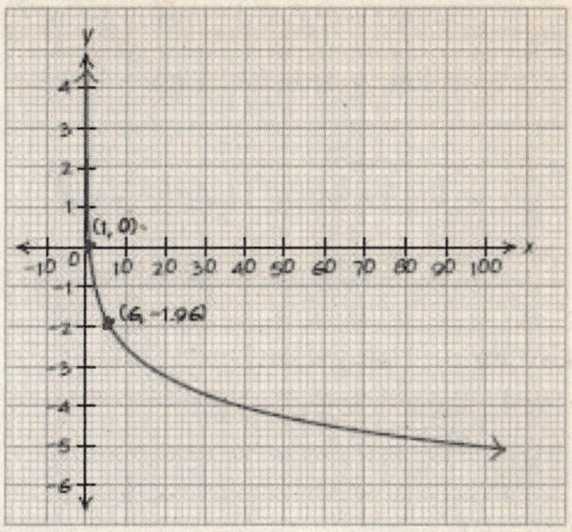

18. $f(x) = \log_{2.5}(x)$

Domain: $(0, \infty)$ Range: $(-\infty, \infty)$

19. $f(x) = \log_{0.4}(x)$

Domain: $(0, \infty)$ Range: $(-\infty, \infty)$

20. $\log(7xy) = \log 7 + \log x + \log y$

21.

$$\log_5(3a^3b^4) = \log_5 3 + \log_5(a^3) + \log_5(b^4)$$
$$= \log_5 3 + 3\log_5 a + 4\log_5 b$$

22.

$$\ln(2x^3y^4) = \ln 2 + \ln(x^3) + \ln(y^4)$$
$$= \ln 2 + 3\ln x + 4\ln y$$

23.

$$\log_7\left(\sqrt[5]{3x^3y}\right) = \log_7(3x^3y)^{\frac{1}{5}}$$
$$= \frac{1}{5}\log_7(3x^3y)$$
$$= \frac{1}{5}\left(\log_7 3 + \log_7(x^3) + \log_7 y\right)$$
$$= \frac{1}{5}\left(\log_7 3 + 3\log_7 x + \log_7 y\right)$$
$$= \frac{1}{5}\log_7 3 + \frac{3}{5}\log_7 x + \frac{1}{5}\log_7 y$$

24.

$$\log_3\left(\sqrt{5ab^2c^3}\right)$$
$$= \log_3(5ab^2c^3)^{\frac{1}{2}}$$
$$= \frac{1}{2}\log_3(5ab^2c^3)$$
$$= \frac{1}{2}\left(\log_3 5 + \log_3 a + \log_3(b^2) + \log_3(c^3)\right)$$
$$= \frac{1}{2}\left(\log_3 5 + \log_3 a + 2\log_3 b + 3\log_3 c\right)$$
$$= \frac{1}{2}\log_3 5 + \frac{1}{2}\log_3 a + \log_3 b + \frac{3}{2}\log_3 c$$

25.

$$\log\left(\frac{3x^3y^5}{z^4}\right) = \log 3 + \log\left(x^3\right) + \log\left(y^5\right) - \log\left(z^4\right)$$
$$= \log 3 + 3\log x + 5\log y - 4\log z$$

26.

$$\log\left(\frac{4a^5b}{c^3}\right) = \log 4 + \log\left(a^5\right) + \log b - \log\left(c^3\right)$$
$$= \log 4 + 5\log a + \log b - 3\log c$$

27.

$$\log 4 + 3\log x + 2\log y + \log z$$
$$= \log 4 + \log x^3 + \log y^2 + \log z$$
$$= \log\left(4x^3y^2z\right)$$

28.

$$\log 2 + 4\log x + 5\log y - 2\log z$$
$$= \log 2 + \log x^4 + \log y^5 - \log z^2$$
$$= \log\left(\frac{2x^4y^5}{z^2}\right)$$

29.

$$\log_3 a + 3\log_3 b - 2\log_3 c$$
$$= \log_3 a + \log_3 b^3 - \log_3 c^2$$
$$= \log_3\left(\frac{ab^3}{c^2}\right)$$

30.

$$\ln 7 + 2\ln x + \ln z - 5\ln y - 3\ln z$$
$$= \ln 7 + 2\ln x - 5\ln y - 2\ln z$$
$$= \ln 7 + \ln x^2 - \ln y^5 - \ln z^2$$
$$= \ln\left(\frac{7x^2}{y^5z^2}\right)$$

31.

$$\log(x+2) + \log(x-7) = \log\left((x+2)(x-7)\right)$$
$$= \log\left(x^2 - 5x - 14\right)$$

32. **a.**

$$2005 \rightarrow t = 5$$
$$F(5) = 750(0.9405)^5$$
$$F(5) \approx 551.89$$

In 2005 the population of Franklin's gulls was approximately 551.89 thousand.

b.

$$500 = 750(0.9405)^t$$
$$\frac{500}{750} = \frac{750(0.9405)^t}{750}$$
$$\frac{2}{3} = (0.9405)^t$$
$$\log\left(\frac{2}{3}\right) = \log(0.9405)^t$$
$$\log\left(\frac{2}{3}\right) = t\log(0.9405)$$
$$\frac{\log\left(\frac{2}{3}\right)}{\log(0.9405)} = \frac{t\log(0.9405)}{\log(0.9405)}$$
$$t \approx 6.6$$

According to the model, the population of Franklin's gulls reached 500,000 midway through the year 2006.

c. This population might stop declining at this rate due to environmental intervention.

33.

Use $A = P\left(1+\frac{r}{n}\right)^{nt}$ with $n = 12$

$$14{,}000 = 7000\left(1+\frac{0.06}{12}\right)^{12t}$$
$$\frac{14{,}000}{7000} = \frac{7000(1.005)^{12t}}{7000}$$
$$2 = (1.005)^{12t}$$
$$\log 2 = \log(1.005)^{12t}$$
$$\log 2 = 12t\log(1.005)$$
$$\frac{\log 2}{12\log(1.005)} = \frac{12t\log(1.005)}{12\log(1.005)}$$
$$t \approx 11.6$$

It takes approximately 11.6 years to double your money at 6% compounded monthly.

34.

Use $A = Pe^{rt}$

$$200{,}000 = 100{,}000e^{0.085t}$$
$$\frac{200{,}000}{100{,}000} = \frac{100{,}000e^{0.085t}}{100{,}000}$$
$$2 = e^{0.085t}$$
$$\ln 2 = 0.085t$$
$$\frac{\ln 2}{0.085} = \frac{0.085t}{0.085}$$
$$t \approx 8.2$$

It takes approximately 8.2 years to double your money at 8.5% compounded continuously.

35.

$$4^w = 12.5$$
$$\log 4^w = \log 12.5$$
$$w\log 4 = \log 12.5$$
$$\frac{w\log 4}{\log 4} = \frac{\log 12.5}{\log 4}$$
$$w = \frac{\log 12.5}{\log 4}$$
$$w \approx 1.822$$

36.

$$3^t = \frac{1}{243}$$
$$3^t = \frac{1}{3^5}$$
$$3^t = 3^{-5}$$
$$t = -5$$

37.

$$6^{3x-5} = 204$$
$$\ln 6^{3x-5} = \ln 204$$
$$(3x-5)\ln 6 = \ln 204$$
$$\frac{(3x-5)\ln 6}{\ln 6} = \frac{\ln 204}{\ln 6}$$
$$3x-5 = \frac{\ln 204}{\ln 6}$$
$$3x = \frac{\ln 204}{\ln 6}+5$$
$$\frac{3x}{3} = \frac{\frac{\ln 204}{\ln 6}+5}{3}$$
$$x = \frac{\frac{\ln 204}{\ln 6}+5}{3}$$
$$x \approx 2.656$$

38.

$$7^{x+2} = 47$$
$$\log 7^{x+2} = \log 47$$
$$(x+2)\log 7 = \log 47$$
$$\frac{(x+2)\log 7}{\log 7} = \frac{\log 47}{\log 7}$$
$$x+2 = \frac{\log 47}{\log 7}$$
$$x = \frac{\log 47}{\log 7}-2$$
$$x \approx -0.021$$

39.

$$17(1.9)^x = 55.4$$
$$\frac{17(1.9)^x}{17} = \frac{55.4}{17}$$
$$(1.9)^x = \frac{55.4}{17}$$
$$\ln(1.9)^x = \ln\left(\frac{55.4}{17}\right)$$
$$x\ln(1.9) = \ln\left(\frac{55.4}{17}\right)$$
$$\frac{x\ln(1.9)}{\ln(1.9)} = \frac{\ln\left(\frac{55.4}{17}\right)}{\ln(1.9)}$$
$$x = \frac{\ln\left(\frac{55.4}{17}\right)}{\ln(1.9)}$$
$$x \approx 1.841$$

40.

$$4\left(\frac{1}{3}\right)^x - 8 = 28$$
$$4\left(\frac{1}{3}\right)^x = 36$$
$$\frac{4\left(\frac{1}{3}\right)^x}{4} = \frac{36}{4}$$
$$\left(\frac{1}{3}\right)^x = 9$$
$$\left(3^{-1}\right)^x = 3^2$$
$$3^{-x} = 3^2$$
$$-x = 2$$
$$x = -2$$

41.

$$3(5)^{3x+2} - 700 = 11$$
$$3(5)^{3x+2} = 711$$
$$\frac{3(5)^{3x+2}}{3} = \frac{711}{3}$$
$$5^{3x+2} = 237$$
$$\log 5^{3x+2} = \log 237$$
$$(3x+2)\log 5 = \log 237$$
$$\frac{(3x+2)\log 5}{\log 5} = \frac{\log 237}{\log 5}$$
$$3x+2 = \frac{\log 237}{\log 5}$$
$$3x = \frac{\log 237}{\log 5}-2$$
$$\frac{3x}{3} = \frac{\frac{\log 237}{\log 5}-2}{3}$$
$$x = \frac{\frac{\log 237}{\log 5}-2}{3}$$
$$x \approx 0.466$$

42.

$$5^{x^2+4} = 15625$$
$$5^{x^2+4} = 5^6$$
$$x^2+4 = 6$$
$$x^2 = 2$$
$$x = \pm\sqrt{2}$$
$$x \approx \pm 1.414$$

43.

$$5.7 = \log\left(\frac{I}{10^{-4}}\right)$$
$$\frac{I}{10^{-4}} = 10^{5.7}$$
$$10^{-4}\left(\frac{I}{10^{-4}}\right) = 10^{-4}\left(10^{5.7}\right)$$
$$I = 10^{-4+5.7} = 10^{1.7}$$
$$I \approx 50.12\,cm$$

The intensity of this 5.7 magnitude earthquake was 50.12 cm.

44.

$$pH = -\log\left(3.7\times10^{-3}\right)$$
$$pH \approx 2.4$$

Cranberry juice, with a hydrogen ion concentration of 3.7×10^{-3} M, has a pH of 2.4.

45.

$$5.6 = -\log\left(H^+\right)$$
$$-5.6 = \log\left(H^+\right)$$
$$10^{-5.6} = H^+$$
$$H^+ = 2.5\times10^{-6}$$

Shampoo, with a pH of 5.6, has a hydrogen ion concentration of 2.5×10^{-6} M.

46.

$$\log x = -4$$
$$10^{-4} = x$$
$$x = 0.0001$$

47.

$$\log_8 t = 5$$
$$8^5 = t$$
$$t = 32768$$

48.

$$\log(10x+20) = 5$$
$$10^5 = 10x+20$$
$$100000 = 10x+20$$
$$99980 = 10x$$
$$\frac{99980}{10} = \frac{10x}{10}$$
$$x = 9998$$

49.

$$\ln(4.3t)+7.5 = 3$$
$$\ln(4.3t) = -4.5$$
$$e^{-4.5} = 4.3t$$
$$\frac{e^{-4.5}}{4.3} = \frac{4.3t}{4.3}$$
$$t = \frac{e^{-4.5}}{4.3}$$
$$t \approx 0.003$$

50.

$$\log_5(2x)+\log_5 x = 4$$
$$\log_5\left(2x^2\right) = 4$$
$$5^4 = 2x^2$$
$$625 = 2x^2$$
$$312.5 = x^2$$
$$x = \pm\sqrt{312.5}$$
$$x \approx \pm17.678$$

The negative answer is not part of the domain.

The final answer is:

$$x \approx 17.678$$

51.

$$\log_3(x-4)+\log_3(x+1) = 3$$
$$\log_3\left((x-4)(x+1)\right) = 3$$
$$3^3 = (x-4)(x+1)$$
$$27 = x^2-3x-4$$
$$0 = x^2-3x-31$$
$$x = \frac{3\pm\sqrt{(-3)^2-4(1)(-31)}}{2(1)}$$
$$x = \frac{3\pm\sqrt{133}}{2}$$
$$x = \frac{3+\sqrt{133}}{2}, x = \frac{3-\sqrt{133}}{2}$$
$$x \approx 7.266, x \approx -4.266$$

The negative answer is not part of the domain.

The final answer is:

$$x \approx 7.266$$

52.

$$\log_4(x+4)+\log_4(x+7) = 1$$
$$\log_4\left((x+4)(x+7)\right) = 1$$
$$4^1 = (x+4)(x+7)$$
$$4 = x^2+11x+28$$
$$0 = x^2+11x+24$$
$$x = \frac{-11\pm\sqrt{11^2-4(1)(24)}}{2(1)}$$
$$x = \frac{-11\pm\sqrt{25}}{2}$$
$$x = \frac{-11\pm5}{2}$$
$$x = \frac{-11+5}{2}, x = \frac{-11-5}{2}$$
$$x = -3, x = -8$$

The answer -8 is not part of the domain.

The final answer is:

$$x = -3$$

53.

$\log(8x^5)-\log(2x)=3$

$\log\left(\frac{8x^5}{2x}\right)=3$

$\log(4x^4)=3$

$10^3=4x^4$

$1000=4x^4$

$250=x^4$

$(x^4)^{\frac{1}{4}}=\pm 250^{\frac{1}{4}}$

$x\approx \pm 3.976$

The negative answer is not part of the domain.

The final answer is:

$x\approx 3.976$

Chapter 13 Test

1.

Use $A=P\left(1+\frac{r}{n}\right)^{nt}$ with $n=12$

$2000=1000\left(1+\frac{0.035}{12}\right)^{12t}$

$\frac{2000}{1000}=\frac{1000(1.002917)^{12t}}{1000}$

$2=(1.002917)^{12t}$

$\log 2=\log(1.002917)^{12t}$

$\log 2=12t\log(1.002917)$

$\frac{\log 2}{12\log(1.002917)}=\frac{12t\log(1.002917)}{12\log(1.002917)}$

$19.8\approx t$

It takes approximately 19.8 years to double your money at 3.5% compounded monthly.

2.

$7.0=\log\left(\frac{I}{10^{-4}}\right)$

$\frac{I}{10^{-4}}=10^{7.0}$

$10^{-4}\left(\frac{I}{10^{-4}}\right)=10^{-4}(10^{7.0})$

$I=10^{-4+7}=10^3$

$I=1000\,\text{cm}$

The intensity of this 7.0 magnitude earthquake was 1000 cm.

3.

$\log_{16}(2x+1)=-0.5$

$16^{-0.5}=2x+1$

$0.25=2x+1$

$-0.75=2x$

$x=-0.375$

4.

$\log_7 x+\log_7(2x+7)-3=0$

$\log_7 x+\log_7(2x+7)=3$

$\log_7(x(2x+7))=3$

$7^3=x(2x+7)$

$343=2x^2+7x$

$0=2x^2+7x-343$

$x=\frac{-7\pm\sqrt{7^2-4(2)(-343)}}{2(2)}$

$x=\frac{-7+\sqrt{2793}}{4},x=\frac{-7-\sqrt{2793}}{4}$

$x\approx 11.462, x\approx -14.962$

The negative answer is not part of the domain.

The final answer is:

$x\approx 11.462$

5.

$$20 = -3 + 4\left(2^x\right)$$
$$23 = 4\left(2^x\right)$$
$$\frac{23}{4} = \frac{4\left(2^x\right)}{4}$$
$$2^x = \frac{23}{4}$$
$$\ln 2^x = \ln\left(\frac{23}{4}\right)$$
$$x\ln 2 = \ln\left(\frac{23}{4}\right)$$
$$\frac{x\ln 2}{\ln 2} = \frac{\ln\left(\frac{23}{4}\right)}{\ln 2}$$
$$x = \frac{\ln\left(\frac{23}{4}\right)}{\ln 2}$$
$$x \approx 2.524$$

6.

$$8 = 3^{7x-1}$$
$$\ln 8 = \ln 3^{7x-1}$$
$$\ln 8 = (7x-1)\ln 3$$
$$\frac{\ln 8}{\ln 3} = \frac{(7x-1)\ln 3}{\ln 3}$$
$$\frac{\ln 8}{\ln 3} = 7x - 1$$
$$\frac{\ln 8}{\ln 3} + 1 = 7x$$
$$\frac{\frac{\ln 8}{\ln 3}+1}{7} = \frac{7x}{7}$$
$$x = \frac{\frac{\ln 8}{\ln 3}+1}{7}$$
$$x \approx 0.413$$

7. No, the graph is not one-to-one because it does not pass the horizontal line test.

8.

$$4\log\left(2a^3\right) + 5\log\left(ab^3\right) = \log\left(2a^3\right)^4 + \log\left(ab^3\right)^5$$
$$= \log\left(2^4a^{12}\right) + \log\left(a^5b^{15}\right)$$
$$= \log\left(2^4a^{12}\cdot a^5b^{15}\right)$$
$$= \log\left(16a^{17}b^{15}\right)$$

9.

$$\log 5 + \log x + 3\log y - 4\log z$$
$$= \log 5 + \log x + \log y^3 - \log z^4$$
$$= \log\left(\frac{5xy^3}{z^4}\right)$$

10.

$$\ln\left(4a^3bc^4\right) = \ln 4 + \ln\left(a^3\right) + \ln b + \ln\left(c^4\right)$$
$$= \ln 4 + 3\ln a + \ln b + 4\ln c$$

11.

$$\log\left(\frac{3xy^5}{\sqrt{z}}\right) = \log 3 + \log x + \log y^5 - \log\sqrt{z}$$
$$= \log 3 + \log x + \log y^5 - \log z^{\frac{1}{2}}$$
$$= \log 3 + \log x + 5\log y - \frac{1}{2}\log z$$

12.

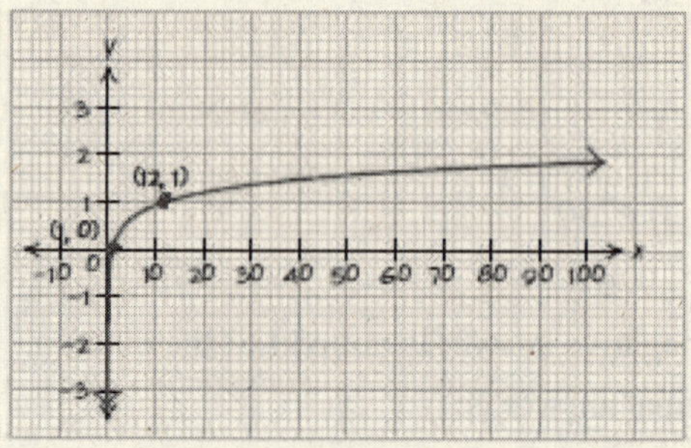

13. $f(x) = \log_{12}(x)$ Domain: $(0,\infty)$ Range: $(-\infty,\infty)$

14. **a.** Let $C(t)$ be the number of CD singles shipped (in millions) in the year t years since 1990.

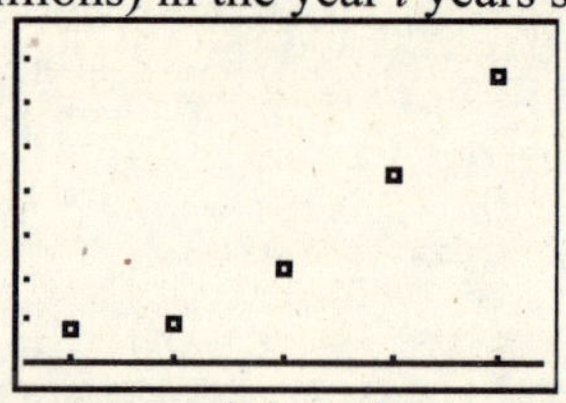

An exponential model is appropriate so choose a point from the flatter end and one from the steeper end.
Note: Your model may differ, but it must follow the trend of the data.

$C(t) = ab^t$

To find a & b choose two points $(3, 7.8)$ & $(7, 66.7)$.

$$7.8 = ab^3 \quad \text{and} \quad 66.7 = ab^7$$
$$\frac{66.7}{7.8} = \frac{ab^7}{ab^3}$$
$$\frac{66.7}{7.8} = b^4$$
$$\left(\frac{66.7}{7.8}\right)^{\frac{1}{4}} = \left(b^4\right)^{\frac{1}{4}}$$

NOTE: $\pm$ not needed on the answer since $b > 0$.

$b \approx 1.71$

Use b and one of the above equations to find a.

$$7.8 = ab^3$$
$$\frac{7.8}{(1.71)^3} = \frac{a(1.71)^3}{(1.71)^3}$$
$$a \approx 1.56$$

Then $C(t) = 1.56(1.71)^t$

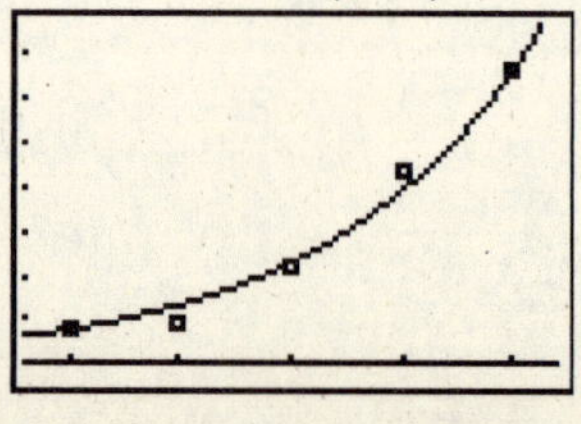

b. Domain: $[0,10]$ Range: $[1.56,333.49]$

c. $C(10)=1.56(1.71)^{10}\approx 333.49$ In 2000 there were approximately 333.49 million CD singles shipped.

d.

$$5=1.56(1.71)^t$$
$$\frac{5}{1.56}=\frac{1.56(1.71)^t}{1.56}$$
$$\frac{5}{1.56}=(1.71)^t$$
$$\log\left(\frac{5}{1.56}\right)=\log(1.71)^t$$
$$\log\left(\frac{5}{1.56}\right)=t\log(1.71)$$
$$\frac{\log\left(\frac{5}{1.56}\right)}{\log(1.71)}=\frac{t\log(1.71)}{\log(1.71)}$$
$$t\approx 2.17$$

In 1992 there were approximately 5 million CD singles shipped.

15.

$$f(x)=5x+2$$
$$y=5x+2$$
$$y-2=5x$$
$$\frac{1}{5}(y-2)=\frac{1}{5}(5x)$$
$$\frac{1}{5}y-\frac{2}{5}=x$$
$$\frac{1}{5}x-\frac{2}{5}=y$$
$$f^{-1}(x)=\frac{1}{5}x-\frac{2}{5}$$

16.

$$g(x)=\log_{15}x$$
$$y=\log_{15}x$$
$$15^y=x$$
$$15^x=y$$
$$g^{-1}(x)=15^x$$

17.

$$f(x)=5^x$$
$$y=5^x$$
$$\log_5 y=x$$
$$\log_5 x=y$$
$$\frac{\log x}{\log 5}=y$$
$$f^{-1}(x)=\log_5 x$$
$$f^{-1}(x)=\frac{\log x}{\log 5}$$

18.

$$h(x)=2(7)^x$$
$$y=2(7)^x$$
$$\frac{y}{2}=(7)^x$$
$$\log_7\left(\frac{y}{2}\right)=x$$
$$\log_7\left(\frac{x}{2}\right)=y$$
$$\frac{\log\left(\frac{x}{2}\right)}{\log 7}=y$$
$$h^{-1}(x)=\log_7\left(\frac{x}{2}\right)$$
$$h^{-1}(x)=\frac{\log\left(\frac{x}{2}\right)}{\log 7}$$

19.

$$2.58=-\log\left(H^+\right)$$
$$-2.58=\log\left(H^+\right)$$
$$10^{-2.58}=H^+$$
$$H^+=2.63\times10^{-3}$$

A solution with a pH of 2.58 has a hydrogen ion concentration of 2.63×10^{-3} M.

20.

$$\log_3 200=\frac{\log 200}{\log 3}$$
$$\approx 4.8227$$

Chapters 1–13 Cumulative Review

1. $3(4.2)^x + 7 = 23$

$3(4.2)^x = 16$

$(4.2)^x = \frac{16}{3}$

$\ln(4.2)^x = \ln\left(\frac{16}{3}\right)$

$x\ln(4.2) = \ln\left(\frac{16}{3}\right)$

$\frac{x\ln(4.2)}{\ln(4.2)} = \frac{\ln\left(\frac{16}{3}\right)}{\ln(4.2)}$

$x = \frac{\ln\left(\frac{16}{3}\right)}{\ln(4.2)}$

$x \approx 1.166$

2. $3x + 18 = 7x - 6$

$3x = 7x - 24$

$-4x = -24$

$x = 6$

3. $3(t-5)^2 - 18 = 30$

$3(t-5)^2 = 48$

$(t-5)^2 = 16$

$\sqrt{(t-5)^2} = \pm\sqrt{16}$

$t - 5 = \pm 4$

$t - 5 = -4, t - 5 = 4$

$t = 1, t = 9$

4. $\ln(2c-7) = 5$

$e^5 = 2c - 7$

$e^5 + 7 = 2c$

$c = \frac{e^5 + 7}{2}$

5. $\frac{2}{5}b + \frac{4}{5} = \frac{1}{5}(b+3)$

$\frac{2}{5}b + \frac{4}{5} = \frac{1}{5}b + \frac{3}{5}$

$5\left(\frac{2}{5}b + \frac{4}{5}\right) = 5\left(\frac{1}{5}b + \frac{3}{5}\right)$

$2b + 4 = b + 3$

$b + 4 = 3$

$b = -1$

6. $2.5n^2 + 1.4n - 3.8 = 0$

$n = \frac{-1.4 \pm \sqrt{(1.4)^2 - 4(2.5)(-3.8)}}{2(2.5)}$

$n = \frac{-1.4 \pm \sqrt{39.96}}{5}$

$n \approx 0.984,\ n \approx -1.544$

7. $4h^3 + 16h^2 - 32h = 0$

$4h(h^2 + 4h - 8) = 0$

$4h = 0, h^2 + 4h - 8 = 0$

$h = 0, h = \frac{-4 \pm \sqrt{4^2 - 4(1)(-8)}}{2(1)}$

$h = 0, h = \frac{-4 \pm \sqrt{48}}{2}$

$h = 0, h = \frac{-4 \pm 4\sqrt{3}}{2}$

$h = 0, h = -2 \pm 2\sqrt{3}$

8. $2^{3x-7} = 45$

$\ln(2^{3x-7}) = \ln(45)$

$(3x-7)\ln(2) = \ln(45)$

$\frac{(3x-7)\ln(2)}{\ln(2)} = \frac{\ln(45)}{\ln(2)}$

$3x - 7 = \frac{\ln(45)}{\ln(2)}$

$3x = \frac{\ln(45)}{\ln(2)} + 7$

$x = \frac{\frac{\ln 45}{\ln 2} + 7}{3}$

$x \approx 4.164$

9. $8.4g + 7.6 = 3.2(g - 4.2)$

$8.4g + 7.6 = 3.2g - 13.44$

$8.4g = 3.2g - 21.04$

$5.2g = -21.04$

$g \approx -4.046$

10. $(x+5)(x-4) = -6$

$x^2 + x - 20 = -6$

$x^2 + x - 14 = 0$

$x = \frac{-1 \pm \sqrt{1^2 - 4(1)(-14)}}{2(1)}$

$x = \frac{-1 \pm \sqrt{57}}{2}$

$x \approx 3.275,\ x \approx -4.275$

11. $\log(x+2) + \log(x-7) = 2$

$\log[(x+2)(x-7)] = 2$

$10^2 = (x+2)(x-7)$

$100 = x^2 - 5x - 14$

$0 = x^2 - 5x - 114$

$x = \frac{5 \pm \sqrt{(-5)^2 - 4(1)(-114)}}{2(1)}$

$x = \frac{5 \pm \sqrt{481}}{2}$

12. $e^{x+5} = 89$

$\ln(e^{x+5}) = \ln(89)$

$(x+5)\ln(e) = \ln(89)$

$(x+5) \cdot 1 = \ln(89)$

$x+5=\ln 89$
$x=-5+\ln 89$

13. $|b+7|=20$
$b+7=20, b+7=-20$
$b=13, b=-27$

14. $|2t-12|=6$
$2t-12=6,\ 2t-12=-6$
$2t=18,\ 2t=6$
$t=9,\ t=3$

15. $4(3^x)+25=8(3^x)-299$
$4(3^x)+324=8(3^x)$
$324=4(3^x)$
$\frac{324}{4}=\frac{4(3^x)}{4}$
$81=3^x$
$3^4=3^x$
$x=4$

16. $2.5x^8-3000=-2360$
$2.5x^8=640$
$x^8=256$
$(x^8)^{\frac{1}{8}}=\pm(256)^{\frac{1}{8}}$
$x=\pm 2$

17. $\frac{23}{4x}-3=-1+\frac{7}{4x}$
$4x\left(\frac{23}{4x}-3\right)=4x\left(-1+\frac{7}{4x}\right)$
$23-12x=-4x+7$
$-12x=-4x-16$
$-8x=-16$
$x=2$

18. $\frac{3}{b-2}=\frac{b-3}{(b-2)(b+5)}$
$\cancel{(b-2)}(b+5)\left(\frac{3}{\cancel{b-2}}\right)=\cancel{(b-2)}\cancel{(b+5)}\left(\frac{b-3}{\cancel{(b-2)}\cancel{(b+5)}}\right)$
$3(b+5)=b-3$
$3b+15=b-3$
$3b=b-18$
$2b=-18$
$b=-9$

19. $6+\sqrt{x-7}=11$
$\sqrt{x-7}=5$
$\left(\sqrt{x-7}\right)^2=(5)^2$
$x-7=25$
$x=32$

20. $\sqrt{z-16}-\sqrt{z}=-2$
$\sqrt{z-16}=\sqrt{z}-2$
$\left(\sqrt{z-16}\right)^2=\left(\sqrt{z}-2\right)^2$
$z-16=z-4\sqrt{z}+4$
$z-20=z-4\sqrt{z}$
$-20=-4\sqrt{z}$
$\frac{-20}{-4}=\frac{-4\sqrt{z}}{-4}$
$5=\sqrt{z}$
$(5)^2=\left(\sqrt{z}\right)^2$
$25=z$

21. $x^2-4x+5=0$
$x=\frac{4\pm\sqrt{(-4)^2-4(1)(5)}}{2(1)}$
$x=\frac{4\pm\sqrt{16-20}}{2}$
$x=\frac{4\pm\sqrt{-4}}{2}$
$x=\frac{4\pm 2i}{2}$
$x=2\pm i$
$x=2+i,\quad x=2-i$

22. $25x^2-20x+13=0$
$x=\frac{20\pm\sqrt{(-20)^2-4(25)(13)}}{2(25)}$
$x=\frac{20\pm\sqrt{400-1300}}{50}$
$x=\frac{20\pm\sqrt{-900}}{50}$
$x=\frac{20\pm 30i}{50}$
$x=\frac{2\pm 3i}{5}$
$x=\frac{2}{5}+\frac{3}{5}i,\quad x=\frac{2}{5}-\frac{3}{5}i$

23.
$\begin{cases} y=5x-20 \\ 2x-3y=-31 \end{cases}$
$2x-3(5x-20)=-31$
$2x-15x+60=-31$
$-13x+60=-31$
$-13x=-91$
$x=7$
Substituting x back into either equation solve for y:
$y=5(7)-20$
$y=35-20$
$y=15$
The solution to the system is $(7,15)$.

24.
$\begin{cases} 5x+12y=66.4 \\ -4x+10y=29.2 \end{cases}$
$\begin{cases} 4(5x+12y)=4(66.4) \\ 5(-4x+10y)=5(29.2) \end{cases}$

$$\begin{cases} 20x+48y=265.6 \\ -20x+50y=146 \end{cases}$$
$$98y=411.6$$
$y=4.2$

Substituting y back into either equation solve for x:

$5x+12(4.2)=66.4$
$5x+50.4=66.4$
$5x=16$
$x=3.2$

The solution to the system is $(3.2,4.2)$.

25.

$$\begin{cases} y=0.6x+3 \\ 3x-5y=-15 \end{cases}$$
$3x-5(0.6x+3)=-15$
$3x-3x-15=-15$
$-15=-15$

Infinite number of solutions.

26.

$$\begin{cases} h=3g+36 \\ h=2g^2+7g-12 \end{cases}$$
$2g^2+7g-12=3g+36$
$2g^2+4g-12=36$
$2g^2+4g-48=0$
$2(g^2+2g-24)=0$
$2(g+6)(g-4)=0$
$g=-6, g=4$

Substituting g back into either equation solve for h:

$h=3(-6)+36$
$h=-18+36$
$h=18$
$h=3(4)+36$
$h=12+36$
$h=48$

The solutions to the system are $(-6,18)$ are $(4,48)$.

27.

$$\begin{cases} y=x^2+3x-12 \\ y=4x^2-7x-9 \end{cases}$$
$4x^2-7x-9=x^2+3x-12$
$3x^2-7x-9=3x-12$
$3x^2-10x-9=-12$
$3x^2-10x+3=0$
$(3x-1)(x-3)=0$
$x=\frac{1}{3}, x=3$

Substituting x back into either equation solve for y:

$y=\left(\frac{1}{3}\right)^2+3\left(\frac{1}{3}\right)-12$
$y=\frac{1}{9}+1-12$
$y=\frac{1}{9}-11$
$y=-\frac{98}{9}$
$y=(3)^2+3(3)-12$
$y=9+9-12$
$y=6$

The solutions to the system are $\left(\frac{1}{3},-\frac{98}{9}\right)$ and $(3,6)$.

28.

$$\begin{cases} y=3x^2-4x-10 \\ y=-2x^2+6x-18 \end{cases}$$
$3x^2-4x-10=-2x^2+6x-18$
$5x^2-4x-10=6x-18$
$5x^2-10x-10=-18$
$5x^2-10x+8=0$
$x=\frac{10\pm\sqrt{(-10)^2-4(5)(8)}}{2(5)}$
$x=\frac{10\pm\sqrt{100-160}}{10}$
$x=\frac{10\pm\sqrt{-60}}{10}$

This system has no real solution.

29. **a.** Let $P(t)$ be the gross profit for UTstarcom, Inc.(in millions of dollars) t years since 1990.

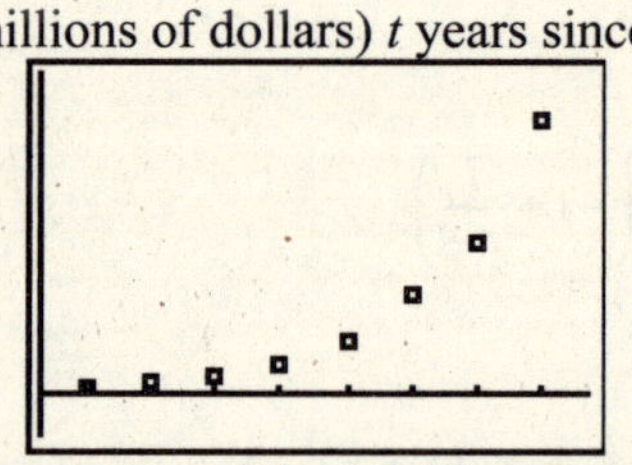

An exponential model is appropriate so choose a point from the flatter end and one from the steeper end. *Note: Your model may differ, but it must follow the trend of the data.*

$P(t)=ab^t$

To find a and b choose two points (6,13.2) and (13,636.2).

$13.2=ab^6$ and $636.2=ab^{13}$

$\frac{636.2}{13.2}=\frac{ab^{13}}{ab^6}$

$\frac{636.2}{13.2}=b^7$

$\left(\frac{636.2}{13.2}\right)^{\frac{1}{7}}=(b^7)^{\frac{1}{7}}$

$b\approx 1.74$

Use b and one of the above equations to find a.

$13.2=a(1.74)^6$

$\frac{1.3}{(1.74)^6}=\frac{a(1.74)^6}{(1.74)^6}$

$a\approx 0.48$

Then $P(t)=0.48(1.74)^t$

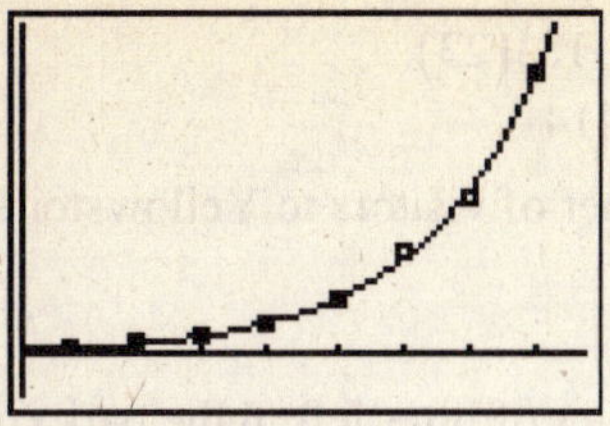

b. Domain: $[4,15]$

$P(4)=0.48(1.74)^4=4.4$

$P(15)=0.48(1.74)^{15}=1947.5$

Range: $[4.4,1947.5]$

c. \$2 billion $\rightarrow P(t)=2000$

$2000=0.48(1.74)^t$

$\frac{2000}{0.48}=\frac{0.48(1.74)^t}{0.48}$

$\left(\frac{2000}{0.48}\right)=1.74^t$

$t=\log_{1.74}\left(\frac{2000}{0.48}\right)$

$t=\frac{\log\left(\frac{2000}{0.48}\right)}{\log 1.74}$

$t\approx 15$

In 2005, the gross profit for UTstarcom, Inc reached \$2 billion.

30. **a.** Let $C(t)$ be the number of cocaine-related emergency department episodes (for people over the age of 35) t years since 1990.

t	$C(t)$
0	23,054
1	30,582
5	57,341
6	68,717
7	74,600
8	83,730
9	85,869
10	93,357
11	106,810

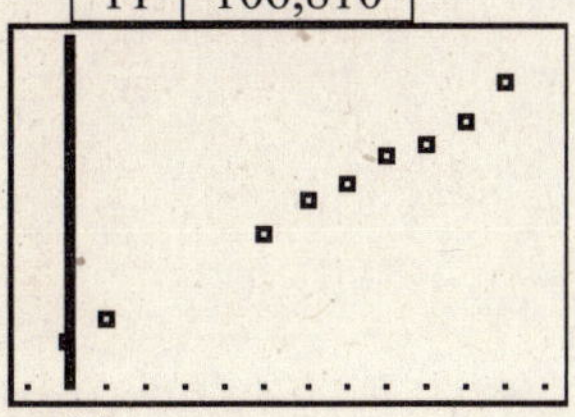

$C(t)=mt+b$

Using two points: (1,30582) and (7,74600)

Find the slope: $m=\frac{74,600-30,582}{7-1}\approx 7336.3$

$C(t)=mt+b$

$b=30,582-7336.3(1)$

$b=23,245.7$

$C(t)=7336.3t+23,245.7$

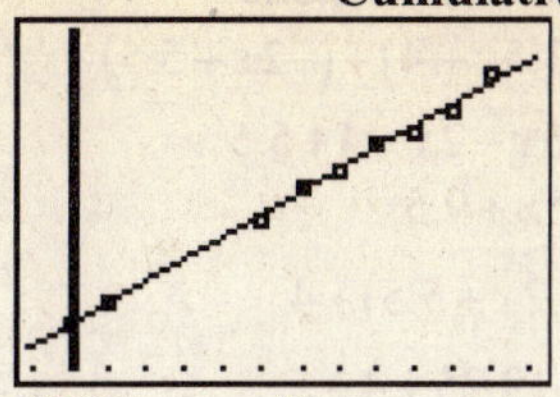

b. $2003\rightarrow t=13$

$C(13)=7336.3(13)+23,245.7$

$C(13)=118,617.6$

In the year 2003, the number of cocaine-related emergency department episodes (for people over the age of 35) was 118,618.

c. $140000=7336.3t+23245.7$

$116754.3=7336.3t$

$\frac{116754.3}{7336.3}=\frac{7336.3t}{7336.3}$

$t\approx 16$

In 2006, the number of cocaine-related emergency department episodes (for people over the age of 35) reached 140,000.

d. The slope is 7336.3. The number of cocaine-related emergency department episodes (for people over the age of 35) increases by 7336.3 cases each year.

e. Domain: $[-1,15]$

$C(-1)=7336.3(-1)+23,245.7=15,909.4$

$C(15)=7336.3(15)+23,245.7=133,290.2$

Range: $[15909,133290]$

31. **a.** $f(x)=4x-7$

$f(9)=4(9)-7$

$f(9)=29$

b. $-2=4x-7$

$5=4x$

$x=\frac{5}{4}$

Therefore, $f\left(\frac{5}{4}\right)=-2$.

c. Domain: $(-\infty,\infty)$ Range: $(-\infty,\infty)$

32. $f(x)=5\cdot 2^x$

The function is exponential. Domain: $(-\infty,\infty)$

33. $g(x)=\sqrt{5x-4}$

This is a radical function. Domain: $\left[\frac{4}{5},\infty\right)$

34. $h(x)=-6x^3+5x-10$

The function is a polynomial. Domain: $(-\infty,\infty)$

35. $g(x)=\ln(x+7)$

This is a logarithmic function. Domain: $(-7,\infty)$

36. $f(x)=\frac{3x+1}{x^2+3x-10}$

This is a rational function. Domain: $(-\infty,\infty)$ $x\neq 2,-5$

37. **a.** $f(x)+g(x)=(3.5x+4)+(-2x+5.5)$
$=3.5x-2x+4+5.5$
$=1.5x+9.5$

b. $f(g(x))=3.5(-2x+5.5)+4$
$=-7x+19.25+4$
$=-7x+23.25$

c. $f(x)g(x)=(3.5x+4)(-2x-5.5)$
$=-7x^2+19.25x-8x+22$
$=-7x^2+11.25x+22$

38. **a.** $f(3)-g(3)=[4(3)-15]-[-6(3)+3]$
$=(12-15)+(18-3)$
$=(-3)+(15)$
$=12$

b. $g(5)=-6(5)+3=-27$
$f(g(5))=f(-27)$
$=4(-27)-15$
$=-123$

c. $f(4)=4(4)-15$ and $g(4)=-6(4)+3$
$=1$ $\quad =-21$
$f(4)g(4)=(1)(-21)=-21$

39. **a.** $f(x)+g(x)=(3x^2+5x-2)+(4x-7)$
$=3x^2+9x-9$

b. $f(g(x))=3(4x-7)^2+5(4x-7)-2$
$=3(16x^2-56x+49)+20x-35-2$
$=48x^2-168x+147+20x-37$
$=48x^2-148x+110$

c. $f(x)g(x)=(3x^2+5x-2)(4x-7)$
$=12x^3-21x^2+20x^2-35x-8x+14$
$=12x^3-x^2-43x+14$

40. **a.** $f(5)=-4(5)^2+7(5)-3$ and $g(5)=5(5)+8$
$=-4(25)+35-3$ $\quad =25+8$
$=-100+32$ $\quad =33$
$=-68$

$f(5)-g(5)=(-68)-(33)$
$=-101$

b. $g(2)=5(2)+8=18$
$f(g(2))=f(18)$
$=-4(18)^2+7(18)-3$
$=-4(324)+126-3$
$=-1173$

c. $f(3)=-4(3)^2+7(3)-3$ and $g(3)=5(3)+8$
$=-4(9)+21-3$ $\quad =15+8$
$=-36+18$ $\quad =23$
$=-18$

$f(3)g(3)=(-18)(23)$
$=-414$

41. $V(t)$ is the number of visitors to Yellowstone National Park
t is the year
$L(p)$ is the amount of litter left in the park (in tons)
p is the number of visitors to Yellowstone National Park
Therefore $p=V(t)$ and $L(p)=L(V(t))$
$L(V(t))$ is the amount of litter (in tons) left in Yellowstone National Park in year t.

42. **a.** $W(5)=1.31(5)+13.94$
$=6.55+13.94$
$=20.49$
In 1995, about 20.5% of the USA triathlon members were women.

b.
$$W(t)>M(t)$$
$$1.31y+13.94>-1.31t+86.06$$
$$\underline{+1.31t-13.94}\quad \underline{+1.31t-13.94}$$
$$2.62t>72.12$$
$$\frac{2.62t}{2.62}>\frac{72.12}{2.62}$$
$$t>27.5$$

In the year 2017 the percent of USA Triathlon members that are women will be greater that that of men.

43. $2x-7<5x-15$
$\underline{-5x+7}\quad \underline{-5x+7}$
$-3x<-8$
$\frac{-3x}{-3}>\frac{-8}{-3}$
$x>\frac{8}{3}$

44. $3(4x-8)\geq 2x+8$
$12x-24\geq 2x+8$
$\underline{-2x+24}\quad \underline{-2x+24}$
$10x\geq 32$
$x\geq 3.2$

45. $|x-7|>5$
$x-7<-5$ or $x-7>5$
$\underline{+7\ \ +7}\quad \underline{+7\ +7}$
$x<2$ or $x>12$

46. $|3x+5|\leq 8$
$-8\leq 3x+5\leq 8$
$\underline{-5}\quad \underline{-5}\ \underline{-5}$
$-13\leq 3x\leq 3$
$\frac{-13}{3}\leq\frac{3x}{3}\leq\frac{3}{3}$
$-\frac{13}{3}\leq x\leq 1$

47. $\dfrac{x-1}{x+3} \le 0$

Find the zero value:
$x-1=0$
$x=1$

Find the excluded value:
$x+3=0$
$x=-3$

Check the intervals:

$x=-5$	$x=0$	$x=3$
$\dfrac{-5-1}{-5+3} \overset{?}{\le} 0$	$\dfrac{0-1}{0+3} \overset{?}{\le} 0$	$\dfrac{3-1}{3+3} \overset{?}{\le} 0$
$\dfrac{-6}{-2} \overset{?}{\le} 0$	$\dfrac{-1}{3} \overset{?}{\le} 0$	$\dfrac{2}{6} \overset{?}{\le} 0$
$3 \overset{?}{\le} 0$	True	$\dfrac{1}{3} \overset{?}{\le} 0$
False		False

$(-3,1]$

48. $x^2+x-56 \ge 0$

$x^2+x-56=0$
$(x+8)(x-7)=0$
$x=-8$ or $x=7$

Check the intervals:

$x=-9$	$x=0$	$x=8$
$(-9+8)(-9-7) \overset{?}{\ge} 0$	$(0+8)(0-7) \overset{?}{\ge} 0$	$(8+8)(8-7) \overset{?}{\ge} 0$
$(-1)(-16) \overset{?}{\ge} 0$	$(8)(-7) \overset{?}{\ge} 0$	$(16)(1) \overset{?}{\ge} 0$
$16 \overset{?}{\ge} 0$	$-56 \overset{?}{\ge} 0$	$16 \overset{?}{\ge} 0$
True	False	True

$(-\infty,-8]$ or $[7,\infty)$

49. **a.**

x	y	Slope
-2	10	$\dfrac{5-10}{0-(-2)}=-\dfrac{5}{2}$
0	5	$\dfrac{0-5}{2-0}=-\dfrac{5}{2}$
2	0	$\dfrac{-5-0}{4-2}=-\dfrac{5}{2}$
4	-5	$\dfrac{-10-(-5)}{6-4}=-\dfrac{5}{2}$
6	-10	

b. x-intercept: $(2,0)$

c. y-intercept: $(0,5)$

d. $y=mx+b$

$y=-\dfrac{5}{2}x+5$

50. Use the slope of the line $2x-3y=30$ and the point (24,4).

$$-3y=-2x+30$$
$$\frac{-3y}{-3}=\frac{-2x}{-3}+\frac{30}{-3}$$
$$y=\frac{2}{3}x-10 \rightarrow m=\frac{2}{3}$$
$$y-4=\frac{2}{3}(x-24)$$
$$y-4=\frac{2}{3}x-16$$
$$+4 \qquad +4$$
$$y=\frac{2}{3}x-12$$

51. Choosing points $(0,-9)$ and $(9,-3)$.

Dashed line, shaded below, means use the < symbol.

$$m=\frac{-3-(-9)}{9-0}=\frac{6}{9}=\frac{2}{3}$$
$$y<\frac{2}{3}x-9$$

52. **a.** Let $P(t)$ be the population of Detroit Michigan in thousands t years since 2000.

$(6,4487)$ and $(8,4425)$
$$P(t)=mt+b$$
$$m=\frac{4425-4487}{8-6}=\frac{-62}{2}=-31$$
$$4487=(-31)(6)+b$$
$$4487=-186+b$$
$$b=4673$$
$$P(t)=-31t+4673$$

b. $P(15)=-31(15)+4673$
$=4208$

In 2015, the population of Detroit Michigan will be 4208 thousand.

c. The slope is -31. The population of Detroit Michigan is decreasing by 31 thousand each year.

53. Let x be the number of ml of 15% HCl solution and y be the number of ml of 60% HCl solution used to get the 30 ml of a 40% HCl solution. Then:

$$\begin{cases} x+y=30 \\ 0.15x+0.60y=(0.40)(30)=12 \end{cases}$$
$$\begin{cases} x+y=30 \\ \left(\dfrac{-1}{0.15}\right)(0.15x+0.60y)=\left(\dfrac{-1}{0.15}\right)(12) \end{cases}$$
$$\begin{cases} x+y=30 \\ -x-4y=-80 \end{cases}$$
$$-3y=-50$$
$$\frac{-3y}{-3}=\frac{-50}{-3}$$
$$y=16\tfrac{2}{3}$$

Now solve for x:

$x = 30 - 16\frac{2}{3}$

$x = 13\frac{1}{3}$

Sarah needs $13\frac{1}{3}$ ml of 15% HCl solution and $16\frac{2}{3}$ ml of 60% HCl solution to make 30 ml of 60% HCl solution.

54. **a.** Let $W(t)$ be the number of women triathletes in the U.S. t years since 1990.

t	$W(t)$
0	1,100
5	4,600
10	13,000
13	19,100

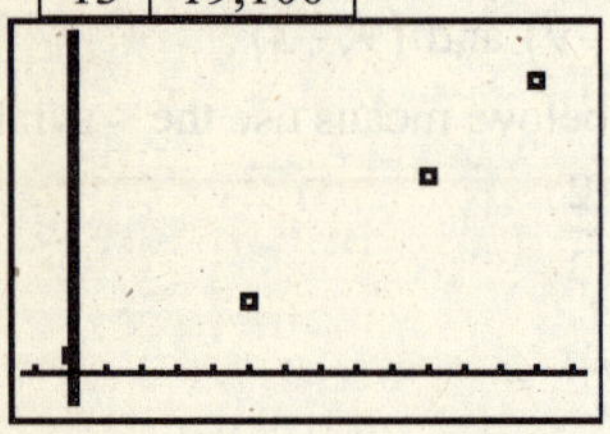

Choose $(0,1100)$ as the vertex and $(13,19100)$ as the additional point used to solve for "a". Then:

$$W(t) = a(t-0)^2 + 1{,}100$$

$$19{,}100 = a(13)^2 + 1{,}100$$

$$\underline{-1{,}100} \qquad \underline{-1{,}100}$$

$$18{,}000 = 169a$$

$$\left.\frac{18{,}000}{169} = \frac{169a}{169}\right\} \quad a \approx 106.5$$

$$W(t) = 106.5t^2 + 1{,}100$$

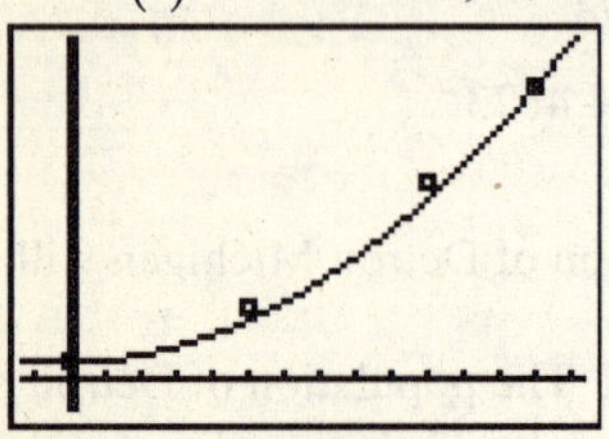

b. Domain: $[0,20]$

$W(0) = 106.5(0)^2 + 1{,}100 = 1{,}100$

$W(20) = 106.5(20)^2 + 1{,}100 = 43{,}700$

Range: $[1100, 43700]$

c. $W(15) = 106.5(15)^2 + 1{,}100$

$\approx 25{,}062.5$

In 2005 there were approximately 25,063 women triathletes in the United States.

d. The vertex is $(0,1100)$. In 1990, there were 1100 women triathletes in the United States. This was the lowest number during this time.

e. $40{,}000 = 106.5t^2 + 1{,}100$

$$\underline{-1{,}100} \qquad \underline{-1{,}100}$$

$$38{,}900 = 106.5t^2$$

$$\frac{38{,}900}{106.5} = \frac{106.5t^2}{106.5}$$

$$\frac{38{,}900}{106.5} = t^2$$

$$\sqrt{\frac{38{,}900}{106.5}} = t^2$$

$$19.1 \approx t$$

In 2009, there were about 40,000 women triathletes in the United States.

55.

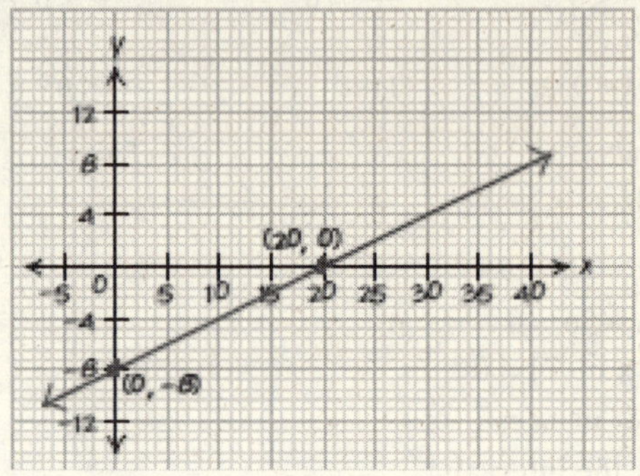

56.

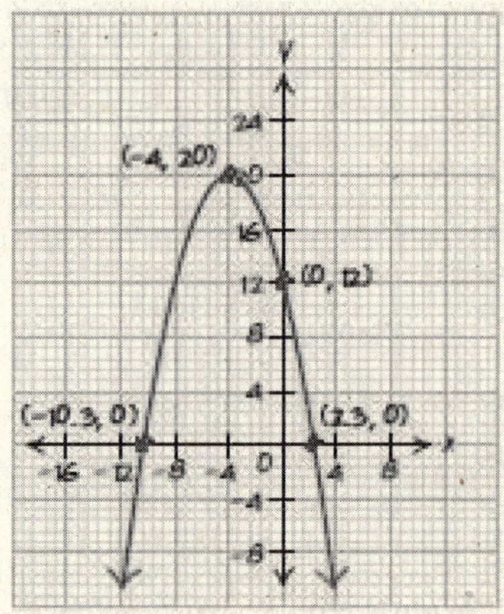

57.

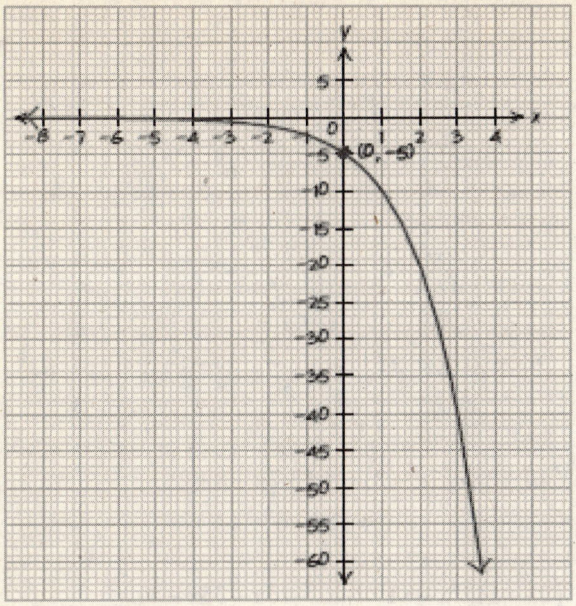

58.

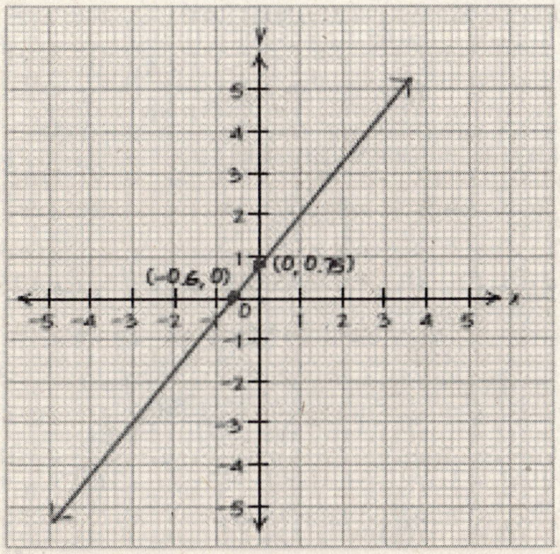

59.

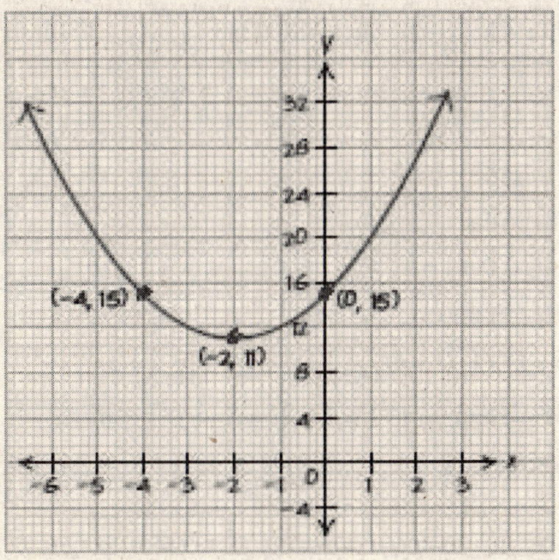

60.

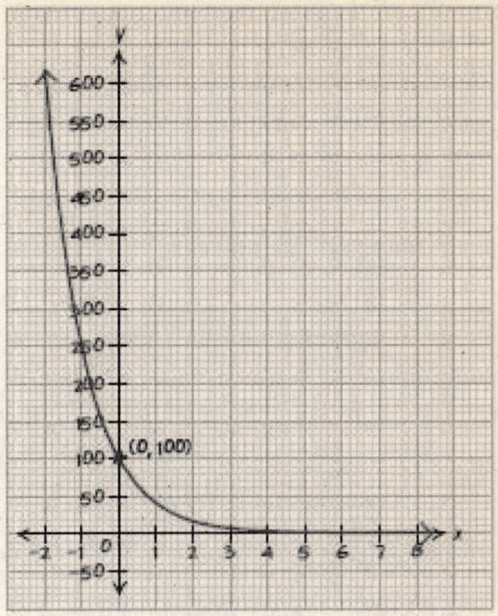

61.

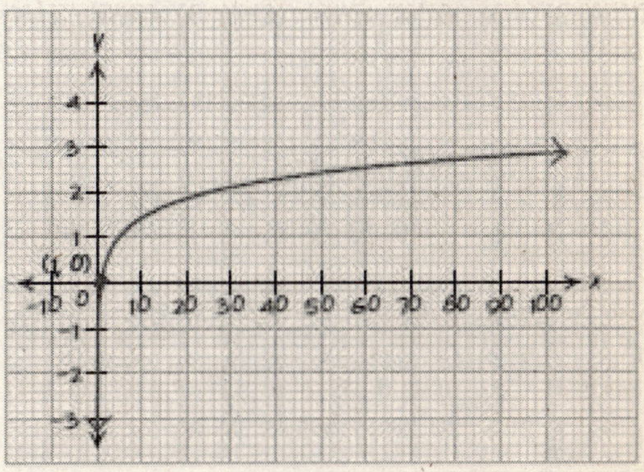

62.

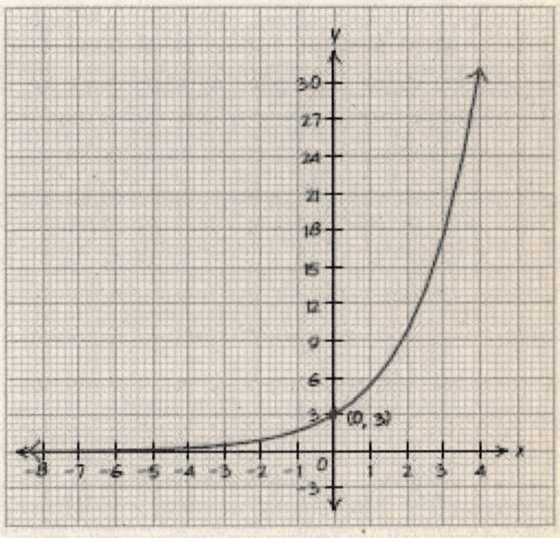

63.

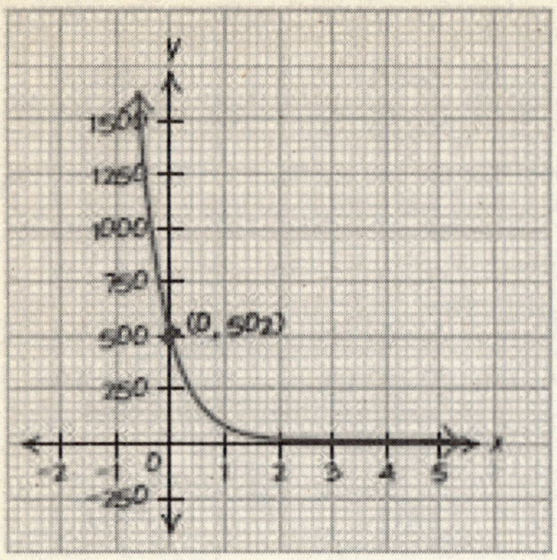

64.

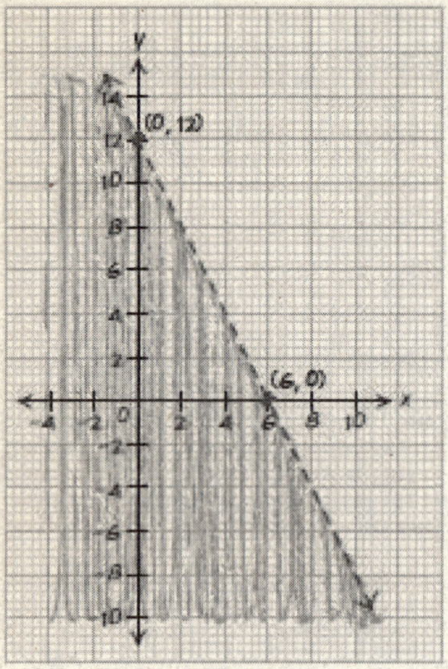

65.

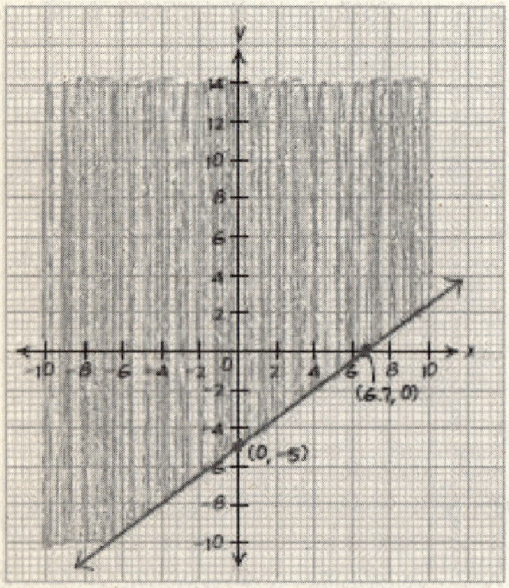

66.

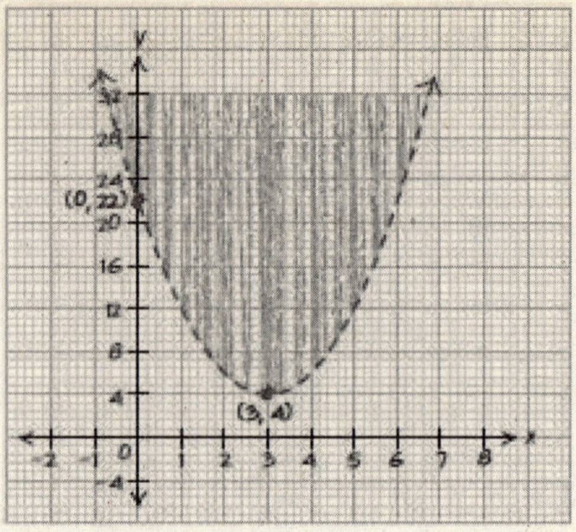

67. $(4d-8)+(7d+2)=4d+7d-8+2$
$=11d-6$

68. $(2x+5)(7x-8)=14x^2-16x+35x-40$
$=14x^2+19x-40$

69. $(3x^2+4x-12)-(x^2+9x-20)$
$=3x^2-x^2+4x-9x-12+20$
$=2x^2-5x+8$

70. $(3t-4)^2=(3t-4)(3t-4)$
$=9t^2-12t-12t+16$
$=9t^2-24t+16$

71. $4p^2+(2p+3)(5p-8)=4p^2+10p^2-16p+15p-24$
$=14p^2-p-24$

72. $\dfrac{2x-5}{2x^2+3x-20}=\dfrac{2x-5}{(2x-5)(x+4)}=\dfrac{1}{x+4}$

73. $\sqrt{9x^3}+2x\sqrt{x}-31=3x\sqrt{x}+2x\sqrt{x}-31=5x\sqrt{x}-31$

74. $-6y(3y^2-8y+2)+4y^3-9y$
$=-18y^3+48y^2-12y+4y^3-9y$
$=-14y^3+48y^2-21y$

75. $2\ln x+\ln(x-1)-\ln(x+3)=\ln x^2+\ln(x-1)-\ln(x+3)$
$=\ln\left[\dfrac{x^2(x-1)}{x+3}\right]$

76. $2^8=256$
$\log_2(256)=8$

77. $\log(10{,}000)=4$
$10^4=10{,}000$

78. $(5-i)+(-3+4i)=5-i-3+4i=2+3i$

79. $(4-2i)-(-3+i)=4-2i+3-i=7-3i$

80. $(3+i)(1-2i)=3-6i+i-2i^2=3-5i+2=5-5i$

81. $(5-i)\div(2+i)=\dfrac{5-i}{2+i}\cdot\dfrac{2-i}{2-i}$

$=\dfrac{10-5i-2i+i^2}{4-i^2}$

$=\dfrac{10-7i-1}{4+1}$

$=\dfrac{9-7i}{5}$

$=\dfrac{9}{5}-\dfrac{7}{5}i$

82. $\dfrac{2}{a-3}+\dfrac{1}{a+1}=\dfrac{2}{a-3}\cdot\dfrac{a+1}{a+1}+\dfrac{1}{a+1}\cdot\dfrac{a-3}{a-3}$

$=\dfrac{2a+2+a-3}{(a-3)(a+1)}$

$=\dfrac{3a-1}{(a-3)(a+1)}$

83. $\left(2\sqrt{3x}\right)\left(-4\sqrt{12x^3}\right)=-8\sqrt{36x^4}=-8\cdot 6x^2=-48x^2$

84. $\left(\dfrac{x+1}{x-2}\right)\left(\dfrac{x^2-4}{5x+5}\right)=\dfrac{\cancel{x+1}}{\cancel{x-2}}\cdot\dfrac{(x+2)\cancel{(x-2)}}{5\cancel{(x+1)}}=\dfrac{x+2}{5}$

85. $\dfrac{\sqrt{72x^7}}{\sqrt{2x}}=\sqrt{\dfrac{72x^7}{2x}}=\sqrt{36x^6}=6x^3$

86. $(5t-3)^2=(5t-3)(5t-3)=25t^2-30t+9$

87. $a^2+b^2=c^2$

$x^2+20^2=27^2$

$x^2+400=729$

$x^2=329$

$\sqrt{x^2}=\pm\sqrt{329}$

$x\approx 18.14$ cm use positive value for length

88. $y\le\dfrac{1}{4}x+2$

$y\ge-\dfrac{2}{5}x+4$

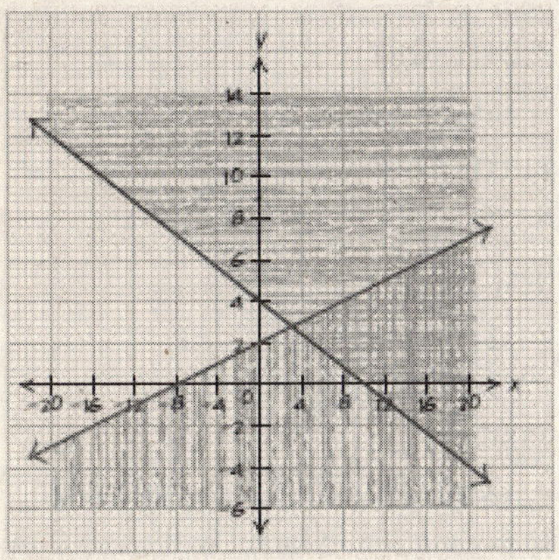

89. $2x^2-11x-21=(2x+3)(x-7)$

90. $25r^2-64t^2=(5r+8t)(5r-8t)$

91. $42g^3-7g^2-280g=7g\left(6g^2-g-40\right)$

$=7g(3g-8)(2g+5)$

92. $t^3+343=t^3+7^3$

$=(t+7)\left(t^2-7t+7^2\right)$

$=(t+7)\left(t^2-7t+49\right)$

93. $4x^2-12x+9=(2x-3)(2x-3)$

$=(2x-3)^2$

94. 2nd degree polynomial

95. $f(x)=3x^2+24x-10$

$f(x)=3\left(x^2+8x\right)-10$

$f(x)=3\left(x^2+8x+\square-\square\right)-10$

Completing the square: $\left(\dfrac{8}{2}\right)^2=16$

$f(x)=3\left(x^2+8x+\boxed{16}-\boxed{16}\right)-10$

$f(x)=3\left(x^2+8x+\boxed{16}\right)-(3)\boxed{16}-10$

$f(x)=3\left(x^2+8x+16\right)-48-10$

$f(x)=3(x+4)^2-58$

96. **a.** $P(t)=0.0237t^2+0.507t+8.384$

$P(20)=0.0237(20)^2+0.507(20)+8.384$

$P(20)=28.004$

The average price for baseball tickets in 2010 is \$28.00.

b. Vertex (h,k) with $h=\dfrac{-b}{2a}$ and $k=P(h)$.

$h=\dfrac{-0.507}{2(0.0237)}\approx-10.69$

$k=P(-10.69)$

$k=0.0237(-10.69)^2+0.507(10.69)+8.384$

$k\approx 5.67$

The vertex is $(-10.7, 5.67)$ In 1979, the average price for baseball tickets was \$5.67. This was a minimum for this model.

c.

$$\begin{aligned}40.77&=0.0237t^2+0.507t+8.384\\-40.77&\qquad\qquad\qquad\qquad-40.77\\\hline 0&=0.0237t^2+0.507t-32.386\end{aligned}$$

$t=\dfrac{-0.507\pm\sqrt{(0.507)^2-4(0.0237)(-32.386)}}{2(0.0237)}$

$t=\dfrac{-0.507-\sqrt{3.3272418}}{0.0474}\approx\cancel{-49.18}$

$t=\dfrac{-0.507+\sqrt{3.3272418}}{0.0474}\approx 27.79$

In 2017, the average price for baseball tickets will be \$40.77.

97. **a.** Let $T(m)$ be the number of thorium-233 atoms after *m* minutes. Let n be the number of 22-minute time intervals.

m	$T(m)$
0	5000
22	$5000\left(\frac{1}{2}\right)$
$2(22)$	$5000\left(\frac{1}{2}\right)^2$
$3(22)$	$5000\left(\frac{1}{2}\right)^3$
$n(22)$	$5000\left(\frac{1}{2}\right)^n$

$$n(22)=m$$
$$\frac{n(22)}{22}=\frac{m}{22}$$
$$n=\frac{m}{22}$$
$$T(m)=5000\left(\frac{1}{2}\right)^{m/22}$$

b. $1000=5000\left(\frac{1}{2}\right)^{m/22}$

$$\frac{1000}{5000}=\frac{5000\left(\frac{1}{2}\right)^{m/22}}{5000}$$
$$0.2=(0.5)^{m/22}$$
$$\frac{m}{22}=\log_{0.5}(0.2)$$
$$m=22\log_{0.5}(0.2)$$
$$m=\frac{22\log(0.2)}{\log(0.5)}\approx 51.08$$

After about 51 minutes there will be only 1000 of the thorium-233 atoms left.

c. 3 hours $\to$ 180 minutes

$$T(180)=5000\left(\frac{1}{2}\right)^{180/22}$$
$$T(180)\approx 17.22$$

By the end of the experiment, there will be only 17 of the thorium-233 atoms left.

98. $f(x)=3x+12$

$$y=3x+12$$
$$y-12=3x$$
$$\frac{1}{3}(y-12)=\frac{1}{3}(3x)$$
$$\frac{1}{3}y-4=x$$
$$\frac{1}{3}x-4=y$$
$$f^{-1}(x)=\frac{1}{3}x-4$$

99. $g(x)=4(3)^x$

$$y=4(3)^x$$
$$\frac{y}{4}=\frac{4(3)^x}{4}$$
$$\frac{y}{4}=(3)^x$$
$$\log_3\left(\frac{y}{4}\right)=x$$
$$\log_3\left(\frac{x}{4}\right)=y$$
$$g^{-1}(x)=\log_3\left(\frac{x}{4}\right)$$
$$g^{-1}(x)=\frac{\log\left(\frac{x}{4}\right)}{\log 3}$$

100. $6=-\log\left(H^+\right)$

$$-6=\log\left(H^+\right)$$
$$10^{-6}=H^+$$
$$H^+=1\times10^{-6}\text{ M}$$

Milk with a pH of 6, has a hydrogen ion concentration of 1×10^{-6} M.

101. **a.** Let $P(t)$ be the Muslim population in Israel (in millions) *t* years since 2008.

$P(t)=ab^t$

$a=1.24$ and $r=2.8\%=0.028$

$b=1+r$ so $b=1.028$

Then $P(t)=1.24(1.028)^t$

b. $2012\to t=4$

$$P(4)=1.24(1.028)^4$$
$$P(4)\approx 1.38$$

The Muslim population in Israel in 2012 will be 1.38 million.

c. $2=1.24(1.028)^t$

$$\frac{2}{1.24}=\frac{1.24(1.028)^t}{1.24}$$
$$\frac{2}{1.24}=(1.028)^t$$
$$t=\log_{1.028}\left(\frac{2}{1.24}\right)$$
$$t=\frac{\log\left(\frac{2}{1.24}\right)}{\log(1.028)}\approx 17.3$$

Seventeen years past 2008 is the year 2025.

The Muslim population in Israel will reach 2 million by the year 2025.

102. $\left(4m^5n^3\right)\left(15mn^2\right)=4\cdot15\cdot m^{5+1}\cdot n^{3+2}=60m^6n^5$

103. $\left(-3g^3h^{-4}\right)^4=(-3)^4\left(g^3\right)^4\left(h^{-4}\right)^4$
$$=81g^{12}h^{-16}$$
$$=\frac{81g^{12}}{h^{16}}$$

104. $3\left(17x^3y^7\right)^0+5\left(12x^4y^6\right)^0=3(1)+5(1)=8$

105. $\dfrac{300x^4y^{-3}z^2}{125x^2y^4z^{-5}}=\dfrac{12\cdot\cancel{25}x^{4-2}y^{-3-4}z^{2-(-5)}}{5\cdot\cancel{25}}$
$$=\frac{12x^2y^{-7}z^7}{5}$$
$$=\frac{12x^2z^7}{5y^7}$$

106. $\left(125a^3b^9\right)^{-\frac{1}{3}}\left(4ab^{-5}\right)^2=(125)^{-\frac{1}{3}}\left(a^3\right)^{-\frac{1}{3}}\left(b^9\right)^{-\frac{1}{3}}(4)^2(a)^2\left(b^{-5}\right)^2$
$$=\frac{1}{5}a^{-1}b^{-3}\cdot16a^2b^{-10}$$
$$=\frac{16ab^{-13}}{5}$$
$$=\frac{16a}{5b^{13}}$$

107. $\left(\dfrac{16m^{-5}n^3}{81m^{-1}n^{-5}}\right)^{-\frac{1}{4}}=\left(\dfrac{16m^{-5-(-1)}n^{3-(-5)}}{81}\right)^{-\frac{1}{4}}$
$$=\left(\frac{16m^{-4}n^8}{81}\right)^{-\frac{1}{4}}$$
$$=\frac{16^{-\frac{1}{4}}\left(m^{-4}\right)^{-\frac{1}{4}}\left(n^8\right)^{-\frac{1}{4}}}{81^{-\frac{1}{4}}}$$
$$=\frac{3mn^{-2}}{2}$$
$$=\frac{3m}{2n^2}$$

108. $a>0$. The graph is above the x-axis.

109. $b>1$. The graph is increasing and $a>0$.

110. Domain: $(-\infty,\infty)$ Range: $(0,\infty)$

111. $a>0$. The graph is facing upward.

112. $(h,k)=(-1,-2)$, where (h,k) is the vertex of the parabola.

113. $f(2.5)=18$

114. $f(2)=12$
$f(-4)=12$
$x=2$ and $x=-4$

115. Domain: $(-\infty,\infty)$ Range: $[-2,\infty)$

116. $m>0$. The slope is increasing.

117. $b=-1$. The y-intercept is $(0,-1)$.

118. $m=\dfrac{2}{3}$

119. $f(3)=1$
$x=3$

120. Domain: $(-\infty,\infty)$ Range: $(-\infty,\infty)$

CHAPTER 14 CONIC SECTIONS, SEQUENCES, AND SERIES

Section 14.1 Parabolas and Circles

1. $y = 3x^2 + 24x - 9$

$a = 3 > 0$ opens up

$h = \dfrac{-24}{2(3)} = -4$

$k = 3(-4)^2 + 24(-4) - 9$

$k = 48 - 96 - 9 = -57$

vertex: $(-4, -57)$

axis of symmetry: $x = -4$

3. $x = 2y^2 + 12y - 7$

$a = 3 > 0$ opens right

$k = \dfrac{-12}{2(3)} = -2$

$h = 3(-2)^2 + 12(-2) - 7$

$h = 12 - 24 - 7 = -19$

vertex: $(-19, -2)$

axis of symmetry: $y = -2$

5. $x = -\dfrac{1}{2}y^2 + 4$

$a = -\dfrac{1}{2} < 0$ opens left

vertex: $(4, 0)$

axis of symmetry: $y = 0$

7. $y = -0.2x^2 + 5x - 3$

$a = -0.2 < 0$ opens down

$h = \dfrac{-5}{2(-0.2)} = 12.5$

$k = -0.2(12.5)^2 + 5(12.5) - 3$

$k = -31.25 + 62.5 - 3 = 28.25$

vertex: $(12.5, 28.25)$

axis of symmetry: $x = 12.5$

9. $x = \dfrac{y^2}{32}$

$a = \dfrac{1}{32} > 0$ opens right

vertex: $(0, 0)$

axis of symmetry: $y = 0$

11.

$x = y^2 + 10y + 21$

$a = 1 > 0$ opens right

$k = \dfrac{-10}{2(1)} = -5$

$h = (-5)^2 + 10(-5) + 21 = -4$

vertex: $(-4, -5)$

$0 = y^2 + 10y + 21$

$y = \dfrac{-10 \pm \sqrt{10^2 - 4(1)(21)}}{2(1)}$

$y = \dfrac{-10 - \sqrt{16}}{2}$ and $y = \dfrac{-10 + \sqrt{16}}{2}$

$y = -7$ and $y = -3$

y-intercepts: $(0, -7)$ and $(0, -3)$

$x = (0)^2 + 10(0) + 21 = 21$

x-intercept: $(21, 0)$

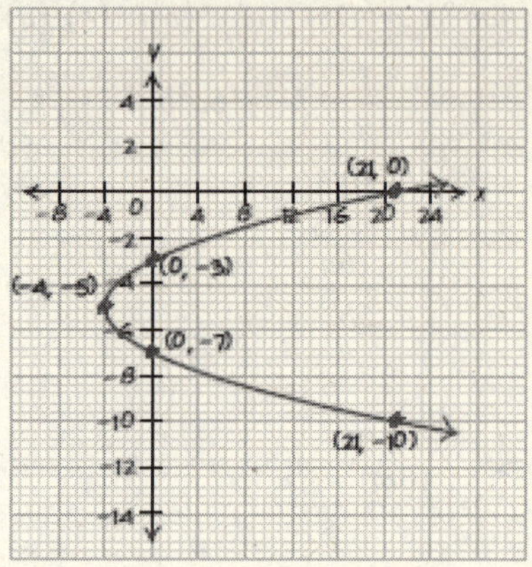

13.

$x = -2y^2 + 15y + 8$

$a = -2 < 0$ opens left

$k = \dfrac{-15}{2(-2)} = 3.75$

$h = -2(3.75)^2 + 15(3.75) + 8 = 36.125$

vertex: $(36.125, 3.75)$

$0 = -2y^2 + 15y + 8$

$y = \dfrac{-15 \pm \sqrt{15^2 - 4(-2)(8)}}{2(-2)}$

$y = \dfrac{-15 - \sqrt{289}}{-4}, y = \dfrac{-15 + \sqrt{289}}{-4}$

$y = 8$ and $y = -0.5$

y-intercepts: $(0, -0.5)$ and $(0, 8)$

$x = -2(0)^2 + 15(0) + 8 = 8$

x-intercept: $(8, 0)$

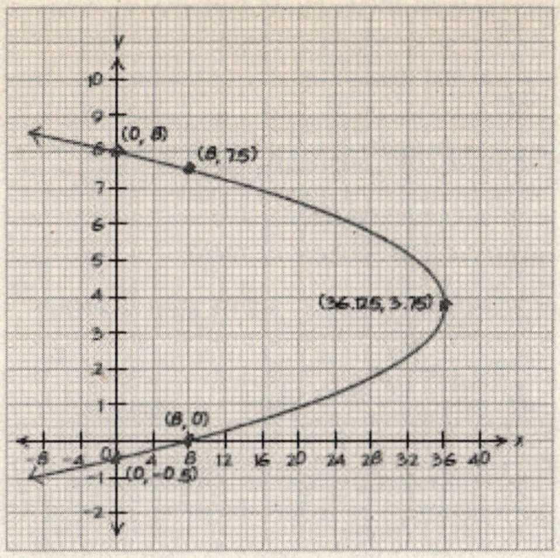

15.

$x=\frac{1}{2}(y-4)^2+5$

$a=\frac{1}{2}>0$ opens right

vertex: $(5,4)$

no y-intercepts

$x=\frac{1}{2}(0-4)^2+5=13$

x-intercept:$(13,0)$

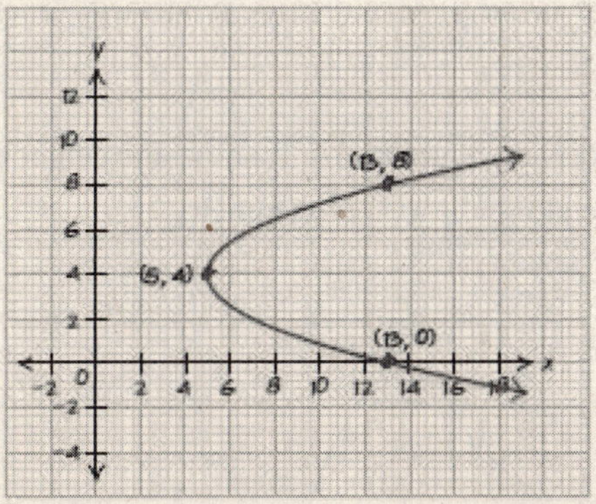

17.

$x=(y+2)^2-6$

$a=1>0$ opens right

vertex: $(-6,-2)$

$0=(y+2)^2-6$

$6=(y+2)^2$

$y+2=\pm\sqrt{6}$

$y=-2-\sqrt{6}, y=-2+\sqrt{6}$

$y\approx-4.45$ and $y\approx 0.45$

y-intercepts:$(0,-4.45)$ and $(0,0.45)$

$x=(0+2)^2-6=-2$

x-intercept:$(-2,0)$

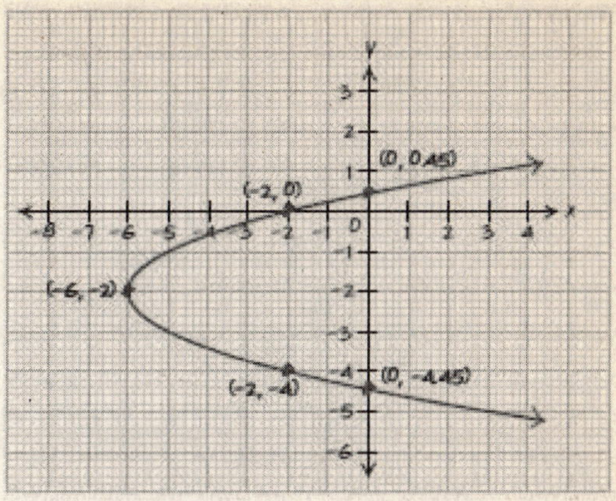

19.

$x=-2(y+5)^2+4$

$a=-2<0$ opens left

vertex: $(4,-5)$

$0=-2(y+5)^2+4$

$2(y+5)^2=4$

$\frac{2(y+5)^2}{2}=\frac{4}{2}$

$(y+5)^2=2$

$y+5=\pm\sqrt{2}$

$y=-5-\sqrt{2}, y=-5+\sqrt{2}$

$y\approx-6.4$ and $y\approx-3.6$

y-intercepts:$(0,-6.4)$ and $(0,-3.6)$

$x=-2(0+5)^2+4=-46$

x-intercept:$(-46,0)$

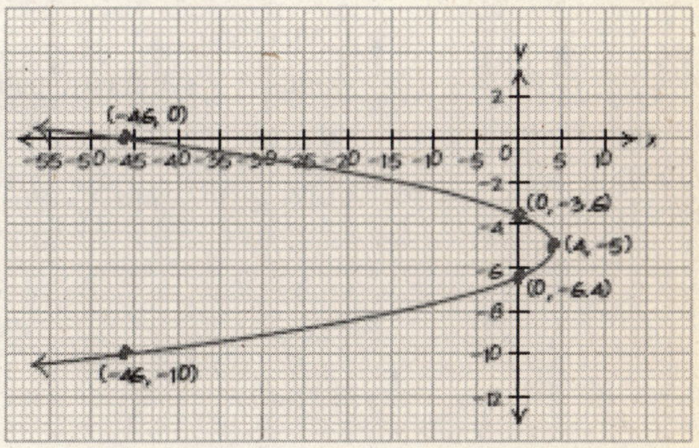

21.

$x=\frac{y^2}{36}=\frac{y^2}{4p}$

$4p=36$

$p=9>0$

focus: $(9,0)$

directrix: $x=-9$

vertex: $(0,0)$	
opens right	
points:	symmetric pairs:
$(1,6)$	$(1,-6)$
$(4,12)$	$(4,-12)$

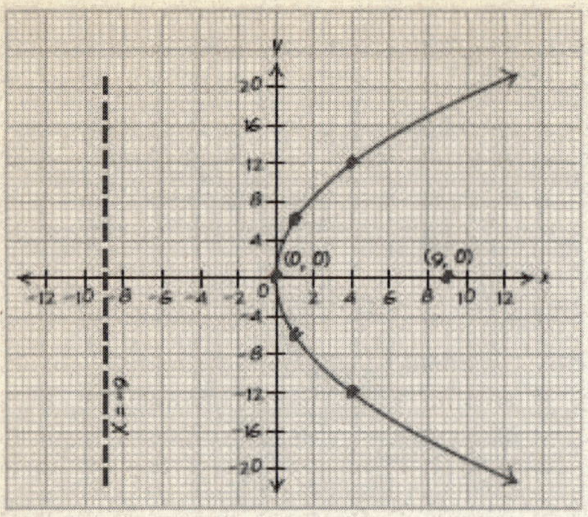

23.

$$y=\frac{x^2}{15}=\frac{x^2}{4p}$$
$$4p=15$$
$$p=\frac{15}{4}=3.75>0$$
focus: $(0,3.75)$
directrix: $y=-3.75$

vertex: $(0,0)$	
opens up	
points:	symmetric pairs:
$(6,2.4)$	$(-6,2.4)$
$(15,15)$	$(-15,15)$

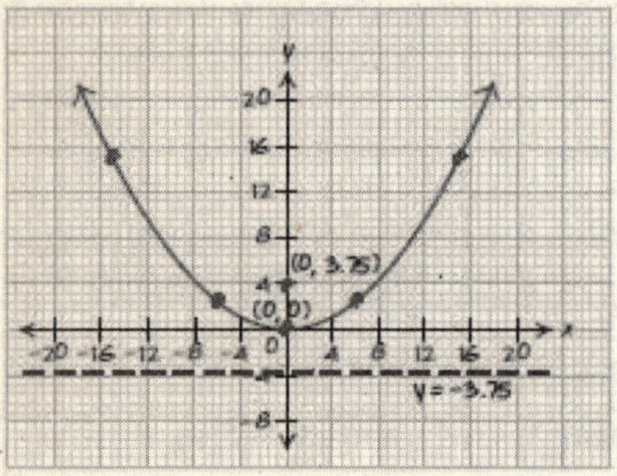

25.

$$28x=y^2$$
$$x=\frac{y^2}{28}=\frac{y^2}{4p}$$
$$4p=28$$
$$p=7>0$$
focus: $(7,0)$
directrix: $x=-7$

vertex: $(0,0)$	
opens right	
points:	symmetric pairs:
$(2.9,9)$	$(2.9,-9)$
$(14.3,20)$	$(14.3,-20)$

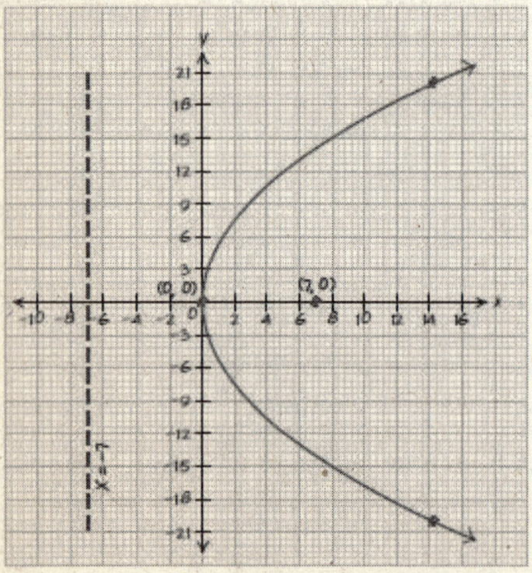

27. opens up with focus $(0,1)$ so $p=1$

$$y=\frac{x^2}{4(1)} \rightarrow y=\frac{x^2}{4}$$

29. opens right with focus $(6,0)$ so $p=6$

$$x=\frac{y^2}{4(6)} \rightarrow x=\frac{y^2}{24}$$

31. opens left with focus $(-10,0)$ so $p=-10$

$$x=\frac{y^2}{4(-10)} \rightarrow x=-\frac{y^2}{40}$$

33.

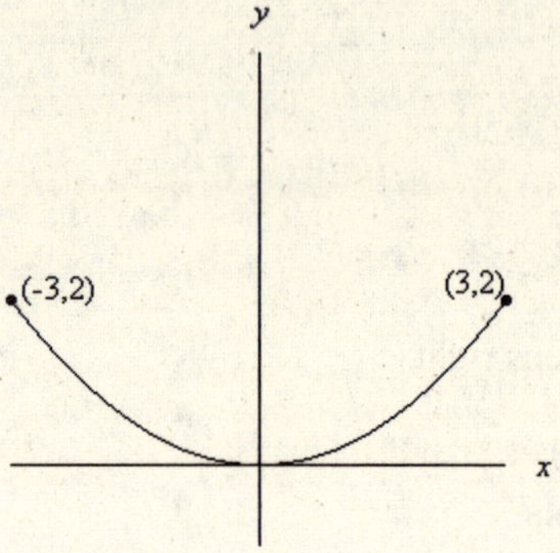

$$x^2=4py$$
$$3^2=4p(2)$$
$$9=8p$$
$$\frac{9}{8}=p=1\tfrac{1}{8}$$

The receiver should be $1\frac{1}{8}$ ft above the center of the satellite along the axis of symmetry.

35.

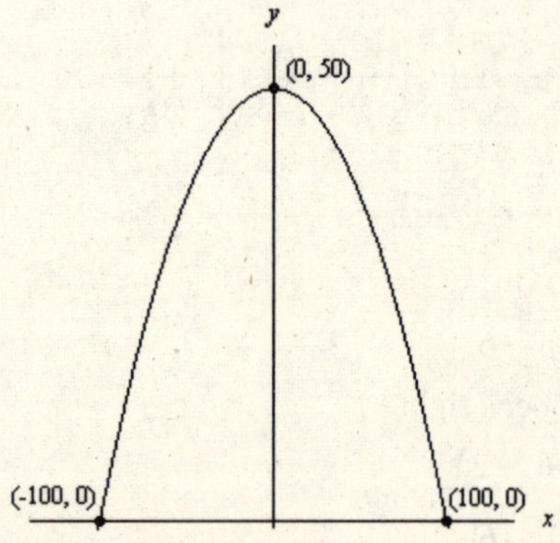

vertex: $(0,50)$
point on graph: $(100,0)$
Using: $y=a(x-h)^2+k$
Solve for a:
$$0=a(100-0)^2+50$$
$$0=10000a+50$$

$-50 = 10000a$
$a = -0.005$
The equation for the graph is:
$y = -0.005x^2 + 50$
Plug in $x = 40$
$y = -0.005(40)^2 + 50 = 42$
The arch, 40 feet from the center of the bridge, is 42 feet high.

37.

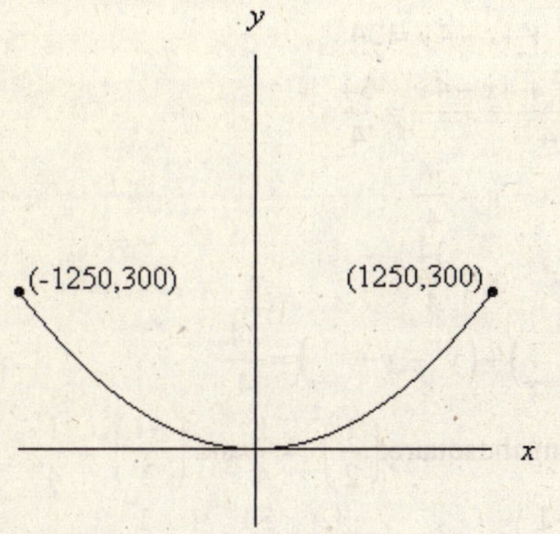

$x^2 = 4py$
$1250^2 = 4p(300)$
$1250^2 = 1200p$
$\frac{1250^2}{1200} = p$
so $x^2 = 4 \cdot \frac{1250^2}{1200} y$
$x^2 = \frac{1250^2}{300} y$
$x^2 = \frac{15,625}{3} y$ or $y = \frac{3x^2}{15,625}$
$y = \frac{3(1000)^2}{15,625} = 192$
The cables, 1000 feet from the center of the bridge, are 192 feet high.

39. Center $= (0,0)$ and Radius $= 6$
$h = 0, k = 0, r = 6$
$(x-0)^2 + (y-0)^2 = (6)^2$
$x^2 + y^2 = 36$

41. Center $= (4,9)$ and Radius $= 4$
$h = 4, k = 9, r = 4$
$(x-4)^2 + (y-9)^2 = (4)^2$
$(x-4)^2 + (y-9)^2 = 16$

43. Center $= (-2,3)$ and Radius $= 12$
$h = -2, k = 3, r = 12$
$(x-(-2))^2 + (y-3)^2 = (12)^2$
$(x+2)^2 + (y-3)^2 = 144$

45. Center $= (-1,-4)$ and Radius $= 0.5$
$h = -1, k = -4, r = 0.5$
$(x-(-1))^2 + (y-(-4))^2 = (0.5)^2$
$(x+1)^2 + (y+4)^2 = 0.25$

47. $x^2 + y^2 = 9$
$(x-0)^2 + (y-0)^2 = (3)^2$
$h = 0, k = 0, r = 3$
Center $= (0,0)$ and Radius $= 3$

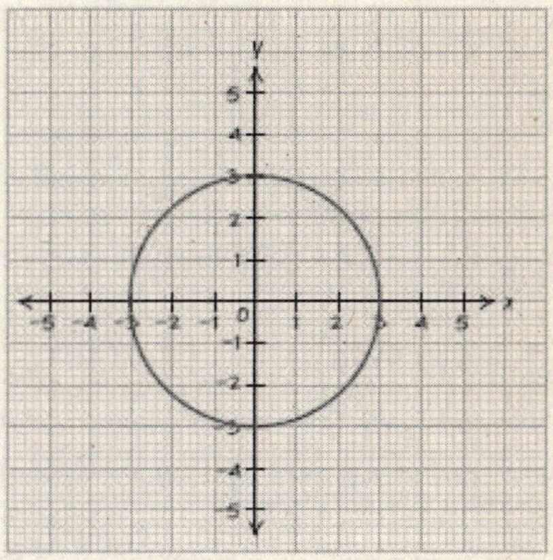

49. $(x-5)^2 + (y-4)^2 = 25$
$(x-5)^2 + (y-4)^2 = (5)^2$
$h = 5, k = 4, r = 5$
Center $= (5,4)$ and Radius $= 5$

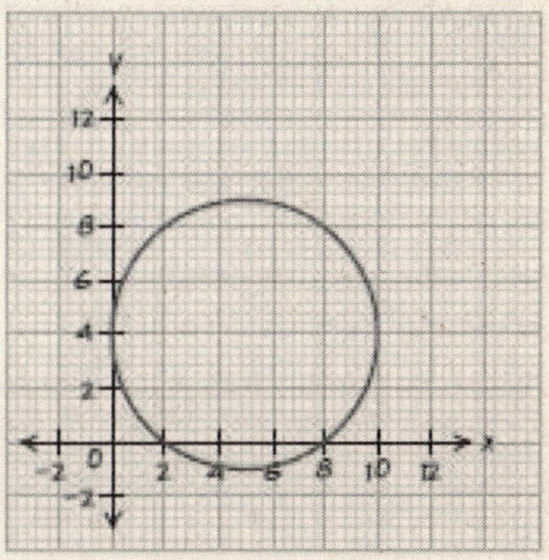

51. $(x+4)^2 + (y+1)^2 = 1$
$(x-(-4))^2 + (y-(-1))^2 = (1)^2$
$h = -4, k = -1, r = 1$
Center $= (-4,-1)$ and Radius $= 1$

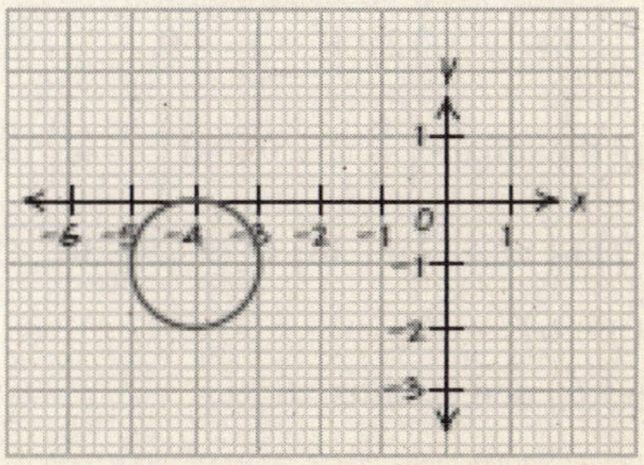

53. $(x-8)^2+(y+2)^2=36$

$(x-8)^2+(y-(-2))^2=(6)^2$

$h=8, k=-2, r=6$

Center $=(8,-2)$ and Radius $=6$

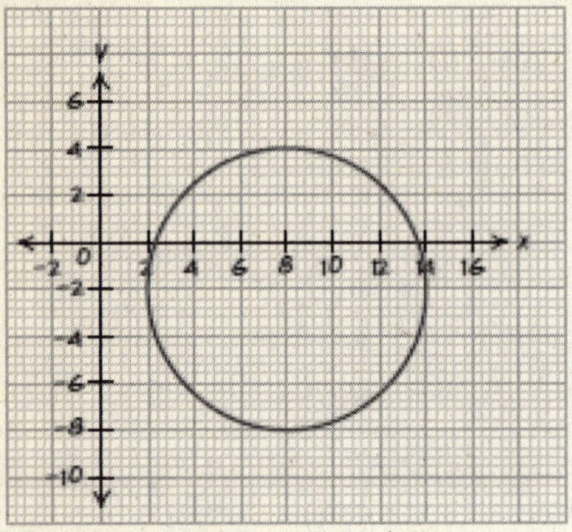

55. $x^2+y^2+2x-6y+9=0$

$x^2+2x+y^2-6y=-9$

$(x^2+2x+\_\_)+(y^2-6y+\_\_)=-9$

Completing the square: $\left(\frac{2}{2}\right)^2=1$ and $\left(\frac{-6}{2}\right)^2=9$

$(x^2+2x+1)+(y^2-6y+9)=-9+1+9$

$(x+1)^2+(y-3)^2=1$

Center $=(-1,3)$ and Radius $=1$

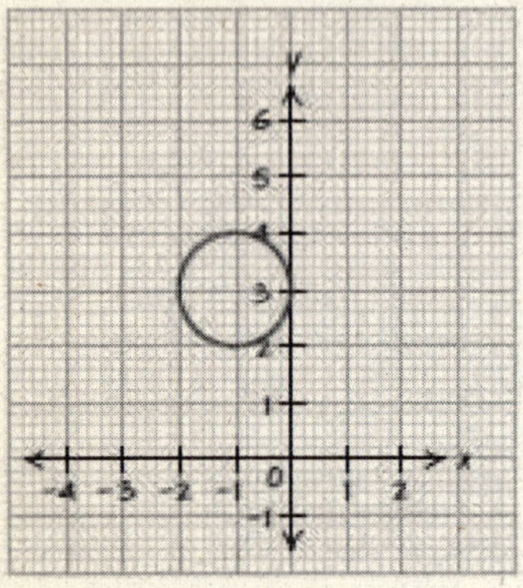

57. $3x^2+3y^2+18x-36y+27=0$

$\frac{3x^2+3y^2+18x-36y+27}{3}=\frac{0}{3}$

$x^2+y^2+6x-12y+9=0$

$x^2+6x+y^2-12y=-9$

$(x^2+6x+\_\_)+(y^2-12y+\_\_)=-9$

Completing the square: $\left(\frac{6}{2}\right)^2=9$ and $\left(\frac{-12}{2}\right)^2=36$

$(x^2+6x+9)+(y^2-12y+36)=-9+9+36$

$(x+3)^2+(y-6)^2=36$

Center $=(-3,6)$ and Radius $=6$

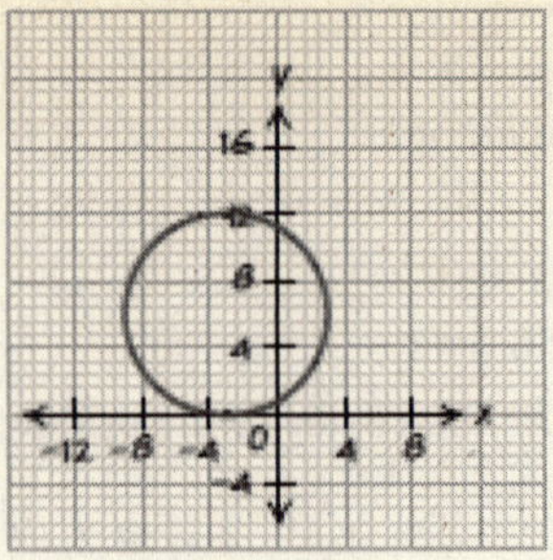

59. $4x^2+4y^2+4x-4y=34$

$\frac{4x^2+4y^2+4x-4y}{4}=\frac{34}{4}$

$x^2+y^2+x-y=\frac{34}{4}$

$x^2+x+y^2-y=\frac{34}{4}$

$(x^2+x+\_\_)+(y^2-y+\_\_)=\frac{34}{4}$

Completing the square: $\left(\frac{1}{2}\right)^2=\frac{1}{4}$ and $\left(\frac{-1}{2}\right)^2=\frac{1}{4}$

$\left(x^2+x+\frac{1}{4}\right)+\left(y^2-y+\frac{1}{4}\right)=\frac{34}{4}+\frac{1}{4}+\frac{1}{4}$

$\left(x+\frac{1}{2}\right)^2+\left(y-\frac{1}{2}\right)^2=9$

Center $=\left(-\frac{1}{2},\frac{1}{2}\right)$ and Radius $=3$

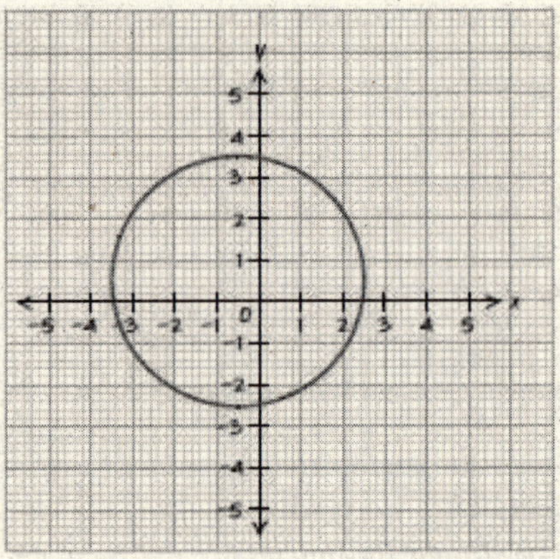

61.

Center $=(0,2)$ and Radius $=2$

$h=0, k=2, r=2$

$(x-0)^2+(y-2)^2=(2)^2$

$x^2+(y-2)^2=4$

63.

Center $=(-4,2)$ and Radius $=3$

$h=-4, k=2, r=3$

$(x-(-4))^2+(y-2)^2=(3)^2$

$(x+4)^2+(y-2)^2=9$

65.

Center $=(10,10)$ and Radius $=5$

$h=10, k=10, r=5$

$(x-10)^2+(y-10)^2=(5)^2$

$(x-10)^2+(y-10)^2=25$

67.

$2x^2+2y^2+4=22$

$2x^2+2y^2=18$

$\dfrac{2x^2+2y^2}{2}=\dfrac{18}{2}$

$x^2+y^2=9$

This is a circle with center $=(0,0)$ and radius $=3$.

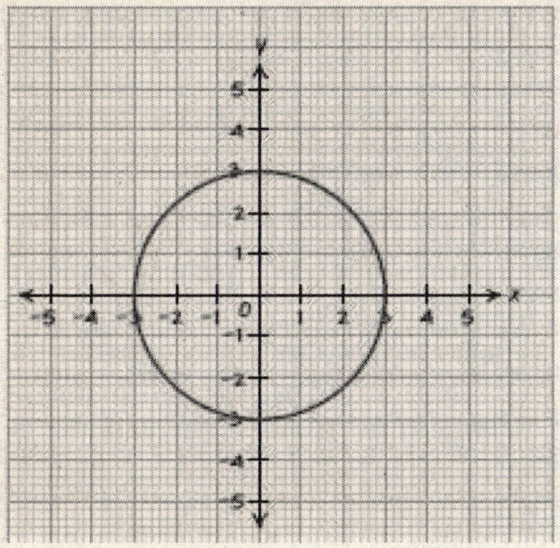

69.

$x^2-4x-y-12=0$

$y=x^2-4x-12$

For the vertex:

$x=\dfrac{-(-4)}{2(1)}=2$

Substitute the x-value into the equation to find y.

$y=(2)^2-4(2)-12$

$y=-16$

This is a parabola with vertex $=(2,-16)$.

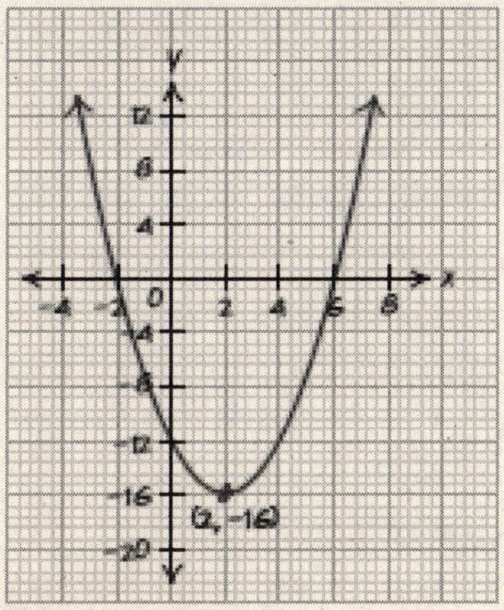

71.

$x-y^2-8y=15$

$x=y^2+8y+15$

For the vertex:

$y=\dfrac{-(8)}{2(1)}=-4$

Substitute the y-value into the equation to find x.

$x=(-4)^2+8(-4)+15$

$x=-1$

This is a parabola with vertex $=(-1,-4)$.

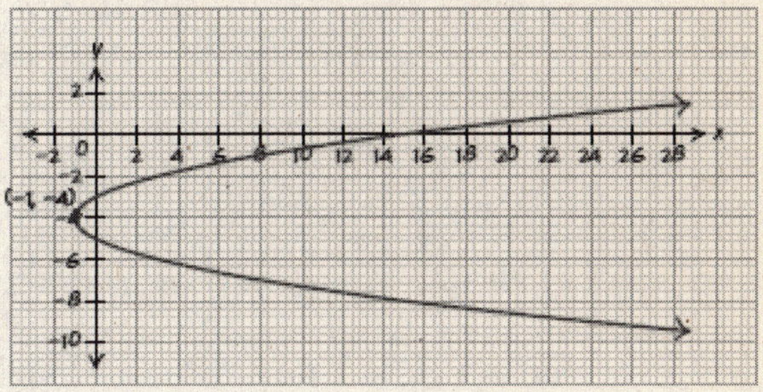

73.

$x^2=-(y-3)^2+49$

$x^2+(y-3)^2=49$

This is a circle with center $=(0,3)$ and radius $=7$.

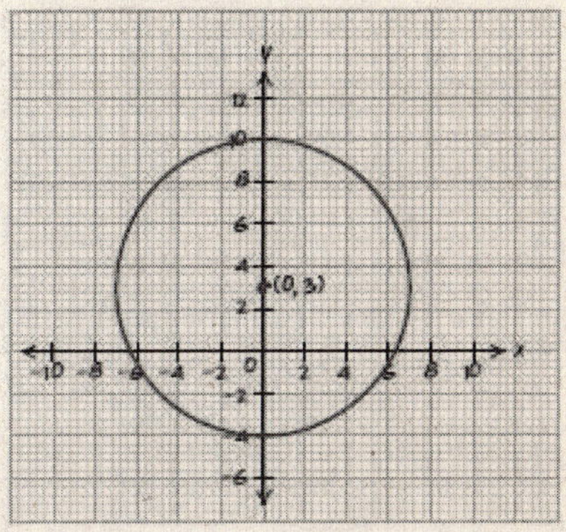

75.

$x=\dfrac{y^2}{32}$

$x=\dfrac{1}{32}y^2$

For the vertex:

$y=\dfrac{-(0)}{2\left(\dfrac{1}{32}\right)}=0$

Substitute the y-value into the equation to find x.

$x=\dfrac{1}{32}(0)^2$

$x=0$

This is a parabola with vertex $=(0,0)$.

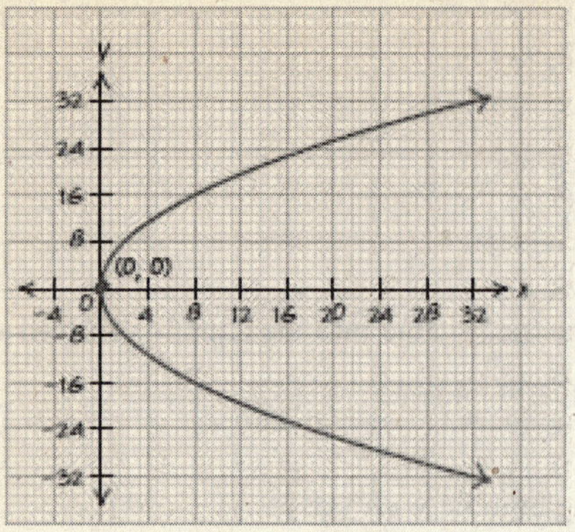

Section 14.2 Ellipses and Hyperbolas

1. $\frac{x^2}{4}+\frac{y^2}{16}=1$

$a^2=4\to a=2$

$b^2=16\to b=4$

major axis is vertical

$c^2=16-4$

$c^2=12$

$c=\pm\sqrt{12}\approx\pm3.46$

foci: $(0,-3.46)$ and $(0,3.46)$

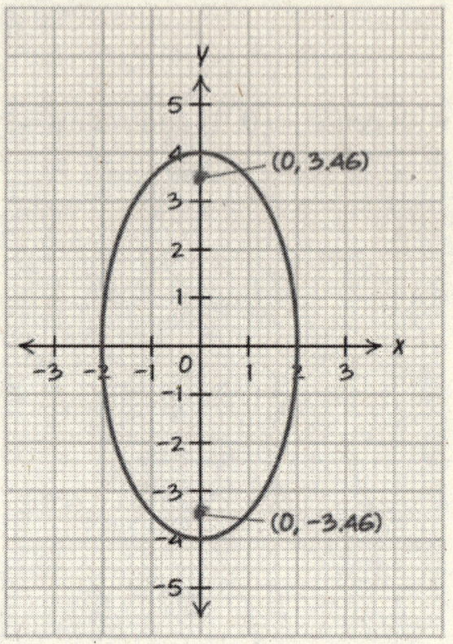

3. $\frac{x^2}{25}+\frac{y^2}{9}=1$

$a^2=25\to a=5$

$b^2=9\to b=3$

major axis is horizontal

$c^2=25-9$

$c^2=16$

$c=\pm\sqrt{16}=\pm4$

foci: $(-4,0)$ and $(4,0)$

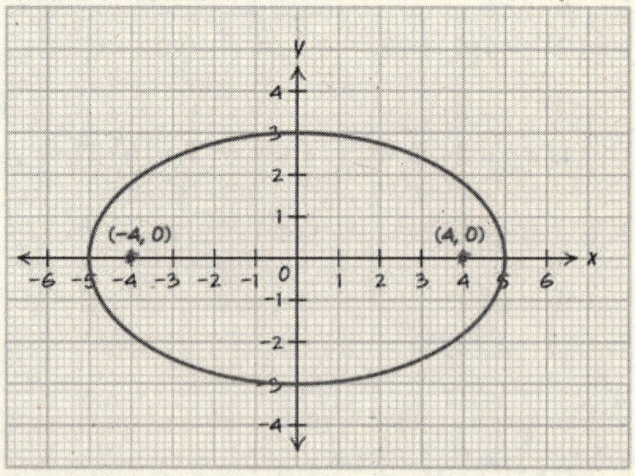

5. $x^2+\frac{y^2}{9}=1$

$a^2=1\to a=1$

$b^2=9\to b=3$

major axis is vertical

$c^2 = 9 - 1$

$c^2 = 8$

$c = \pm\sqrt{8} \approx \pm 2.83$

foci: $(0, -2.83)$ and $(0, 2.83)$

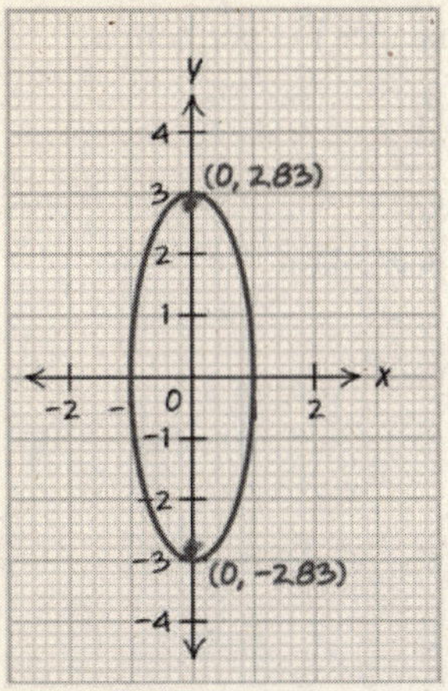

7. $\dfrac{x^2}{10} + \dfrac{y^2}{49} = 1$

$a^2 = 10 \to a \approx 3.16$

$b^2 = 49 \to b = 7$

major axis is vertical

$c^2 = 49 - 10$

$c^2 = 39$

$c = \pm\sqrt{39} \approx \pm 6.24$

foci: $(0, -6.24)$ and $(0, 6.24)$

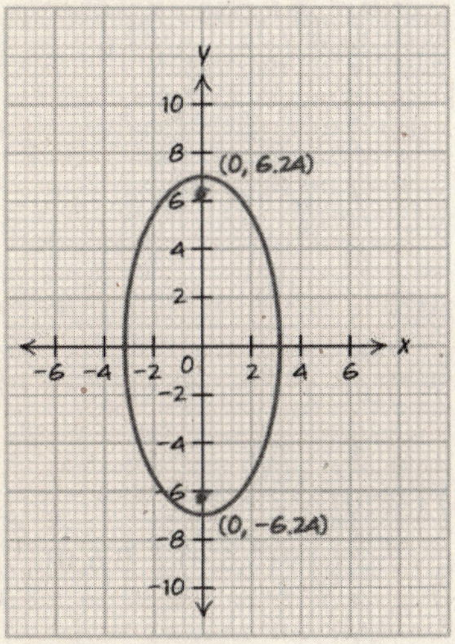

9. $4x^2 + 9y^2 = 36$

$\dfrac{4x^2}{36} + \dfrac{9y^2}{36} = \dfrac{36}{36}$

$\dfrac{x^2}{9} + \dfrac{y^2}{4} = 1$

$a^2 = 9 \to a = 3$

$b^2 = 4 \to b = 2$

major axis is horizontal

$c^2 = 9 - 4$

$c^2 = 5$

$c = \pm\sqrt{5} \approx \pm 2.24$

foci: $(-2.24, 0)$ and $(2.24, 0)$

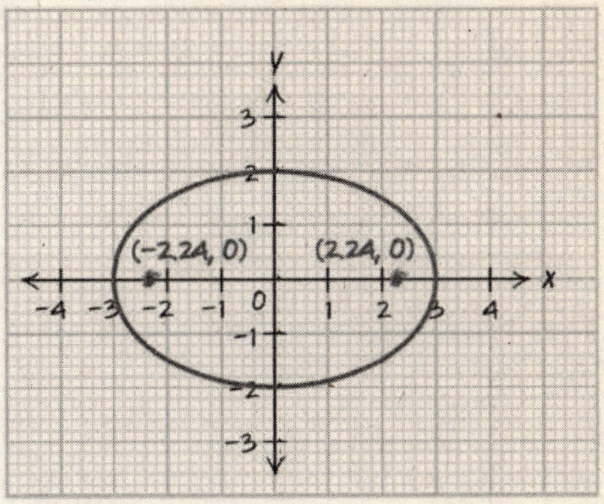

11. $x^2 + 36y^2 = 36$

$\dfrac{x^2}{36} + \dfrac{36y^2}{36} = \dfrac{36}{36}$

$\dfrac{x^2}{36} + y^2 = 1$

$a^2 = 36 \to a = 6$

$b^2 = 1 \to b = 1$

major axis is horizontal

$c^2 = 36 - 1$

$c^2 = 35$

$c = \pm\sqrt{35} \approx \pm 5.92$

foci: $(-5.92, 0)$ and $(5.92, 0)$

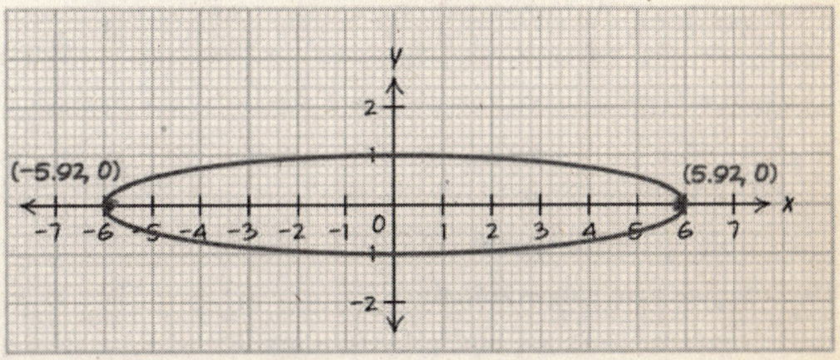

13. $20x^2 + 9y^2 = 180$

$\dfrac{20x^2}{180} + \dfrac{9y^2}{180} = \dfrac{180}{180}$

$\dfrac{x^2}{9} + \dfrac{y^2}{20} = 1$

$a^2 = 9 \to a = 3$

$b^2 = 20 \to b \approx 4.47$

major axis is vertical

$c^2 = 20 - 9$

$c^2 = 11$

$c = \pm\sqrt{11} \approx \pm 3.32$

foci: $(0, -3.32)$ and $(0, 3.32)$

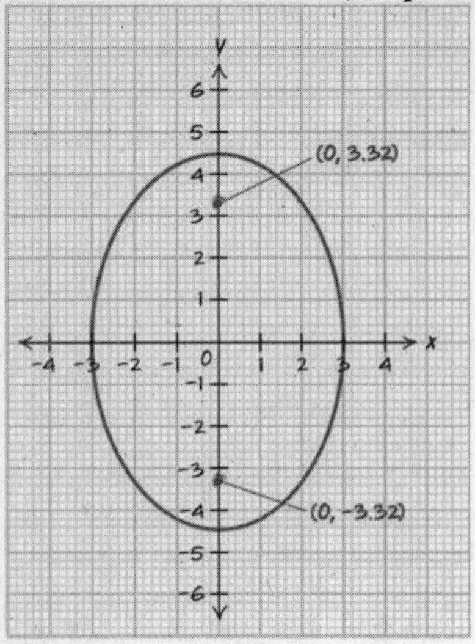

15. major axis is horizontal:
vertices $(-8,0)$ and $(8,0) \rightarrow a = 8$
minor axis end points:
$(0,-4)$ and $(0,4) \rightarrow b = 4$
equation: $\frac{x^2}{64}+\frac{y^2}{16}=1$
$c^2 = 64-16$
$c^2 = 48$
$c = \pm\sqrt{48} \approx \pm 6.93$
foci: $(-6.93,0)$ and $(6.93,0)$

17. major axis is vertical:
vertices $(0,-3)$ and $(0,3) \rightarrow b = 3$
minor axis end points:
$(-2,0)$ and $(2,0) \rightarrow a = 2$
equation: $\frac{x^2}{4}+\frac{y^2}{9}=1$
$c^2 = 9-4$
$c^2 = 5$
$c = \pm\sqrt{5} \approx \pm 2.24$
foci: $(0,-2.24)$ and $(0,2.24)$

19. major axis is horizontal:
vertices $(-15,0)$ and $(15,0) \rightarrow a = 15$
minor axis end points:
$(0,-10)$ and $(0,10) \rightarrow b = 10$
equation: $\frac{x^2}{225}+\frac{y^2}{100}=1$
$c^2 = 225-100$
$c^2 = 125$
$c = \pm\sqrt{125} \approx \pm 11.18$
foci: $(-11.18,0)$ and $(11.18,0)$

21. $a =$ mean distance $= 3671.5$
$c =$ mean distance $-$ perihelion
$c = 3671.5 - 2760 = 911.5$
$c^2 = a^2 - b^2$ so $b^2 = a^2 - c^2$
$b^2 = 3671.5^2 - 911.5^2$
$b^2 = 12{,}649{,}080$
$a^2 = 13{,}479{,}912.25$
$$\frac{x^2}{13{,}479{,}912.25}+\frac{y^2}{12{,}649{,}080}=1$$

23. aphelion $= 2(\text{mean distance}) -$ perihelion
$= 2(886) - 840$
$= 932$
$a =$ mean distance $= 886$
$c =$ mean distance $-$ perihelion
$c = 886 - 840 = 46$
$c^2 = a^2 - b^2$ so $b^2 = a^2 - c^2$
$b^2 = 886^2 - 46^2$
$b^2 = 782{,}880$
$a^2 = 784{,}996$
$$\frac{x^2}{784{,}996}+\frac{y^2}{782{,}880}=1$$

25. $\frac{w^2}{4556.25}+\frac{d^2}{945.5}=1$
$a^2 = 4556.25 \rightarrow a = 67.5$
$b^2 = 945.5 \rightarrow b \approx 30.75$
width $= 2(67.5) = 135$ cm
depth $= 30.75$ cm

27. Let w be half the width of the arch and h be the height of the arch.
$a = \frac{120}{2} = 60 \rightarrow a^2 = 3600$
$b = 35 \rightarrow b^2 = 1225$
$$\frac{w^2}{3600}+\frac{h^2}{1225}=1$$

29. $\frac{x^2}{4}-\frac{y^2}{9}=1$
$\frac{x^2}{2^2}-\frac{y^2}{3^2}=1$
$a = 2, b = 3$
$c^2 = 2^2 + 3^2$
$c = \pm\sqrt{13}$
Vertices $= (2,0),(-2,0)$
Foci $= (\sqrt{13},0),(-\sqrt{13},0) \approx (3.6,0),(-3.6,0)$

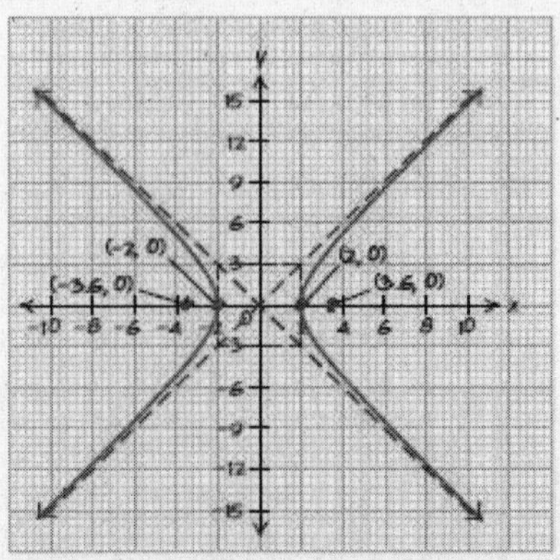

31. $\dfrac{y^2}{36}-\dfrac{x^2}{4}=1$

$\dfrac{y^2}{6^2}-\dfrac{x^2}{2^2}=1$

$a=2, b=6$

$c^2=2^2+6^2$

$c=\pm\sqrt{40}$

Vertices $=(0,6),(0,-6)$

Foci $=\left(0,\sqrt{40}\right),\left(0,-\sqrt{40}\right)\approx(0,6.3),(0,-6.3)$

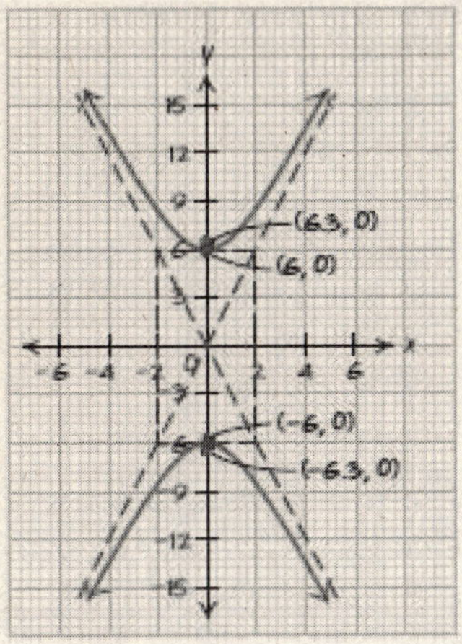

33. $x^2-25y^2=25$

$\dfrac{x^2-25y^2}{25}=\dfrac{25}{25}$

$\dfrac{x^2}{25}-\dfrac{y^2}{1}=1$

$\dfrac{x^2}{5^2}-\dfrac{y^2}{1^2}=1$

$a=5, b=1$

$c^2=5^2+1^2$

$c=\pm\sqrt{26}$

Vertices $=(5,0),(-5,0)$

Foci $=\left(\sqrt{26},0\right),\left(-\sqrt{26},0\right)\approx(5.1,0),(-5.1,0)$

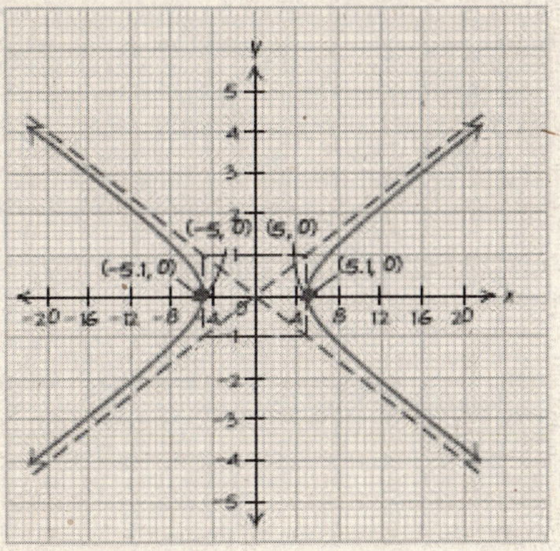

35. $9y^2-81x^2=81$

$\dfrac{9y^2-81x^2}{81}=\dfrac{81}{81}$

$\dfrac{y^2}{9}-\dfrac{x^2}{1}=1$

$\dfrac{y^2}{3^2}-\dfrac{x^2}{1^2}=1$

$a=1, b=3$

$c^2=1^2+3^2$

$c=\pm\sqrt{10}$

Vertices $=(0,3),(0,-3)$

Foci $=\left(0,\sqrt{10}\right),\left(0,-\sqrt{10}\right)\approx(0,3.2),(0,-3.2)$

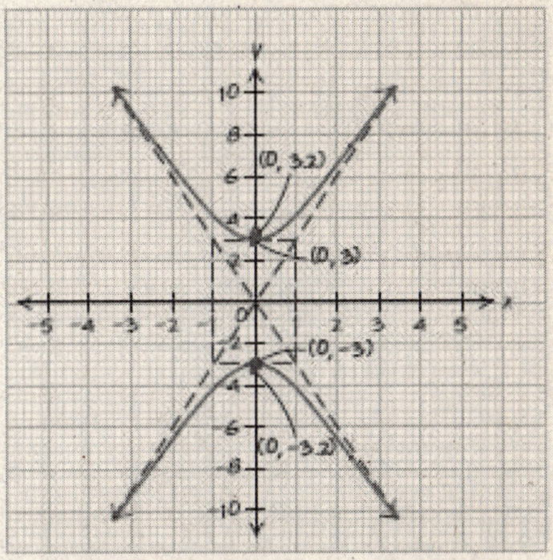

37. Since hyperbola opens left/right and $a=4$ and $b=3$ then

$c^2=4^2+3^2$

$c=\pm 5$

$\dfrac{x^2}{4^2}-\dfrac{y^2}{3^2}=1$

$\dfrac{x^2}{16}-\dfrac{y^2}{9}=1$

Foci $=(5,0),(-5,0)$

39. Since hyperbola opens up/down and $a=1$ and $b=2$ then

$c^2=1^2+2^2$

$c=\pm\sqrt{5}$

$\dfrac{y^2}{2^2}-\dfrac{x^2}{1^2}=1$

$\dfrac{y^2}{4}-\dfrac{x^2}{1}=1$

Foci $=\left(0,\sqrt{5}\right),\left(0,-\sqrt{5}\right)$

41. $3x^2+3y^2=27$ A = 3 and C = 3

Since A = C then this is a circle.

43. $4x^2-9y^2=36$ A = 4 and C = −9

Since A ≠ C, A < 0 or C < 0 then this is a hyperbola.

45. $6y^2 + 2y + x - 9 = 0 \quad A = 0 \text{ and } C = 6$

Since A = 0 or C = 0 then this is a parabola.

47. $64x^2 + 36y^2 = 576 \quad A = 64 \text{ and } C = 36$

Since $A \neq C$, $A > 0$ and $C > 0$ then this is an ellipse.

49. Ellipse

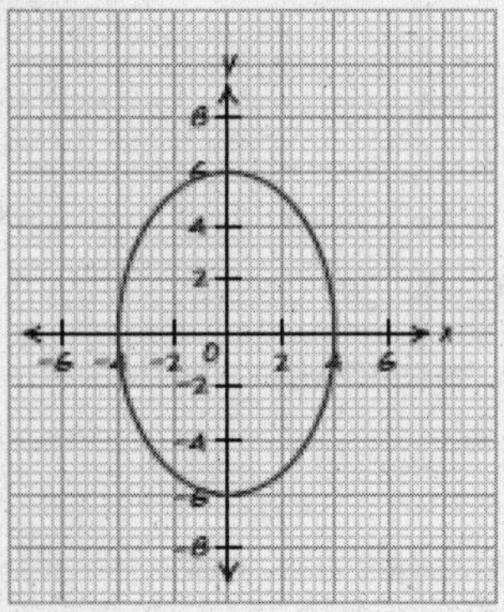

51. Hyperbola

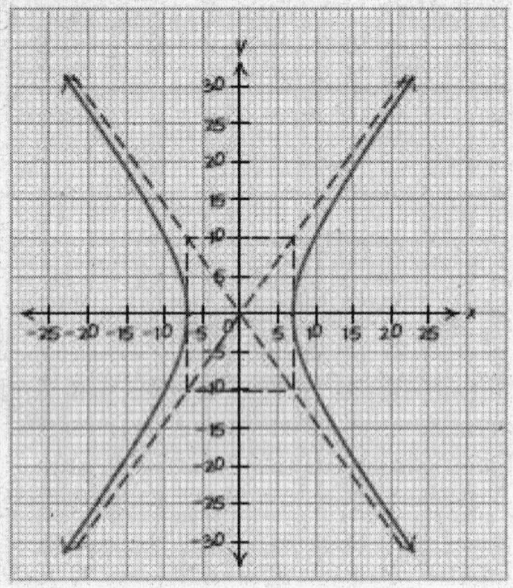

53. Ellipse

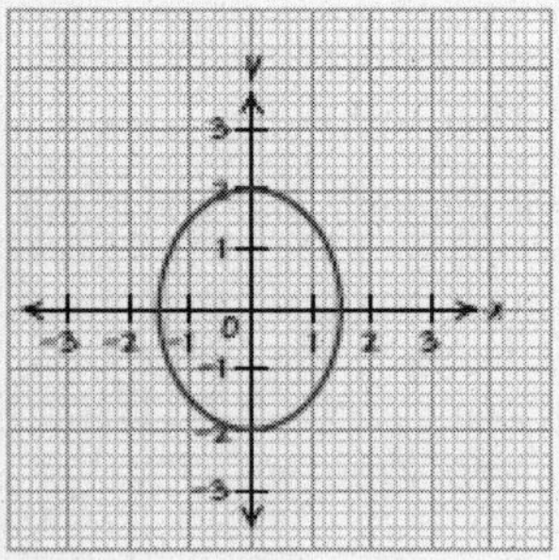

55. Circle

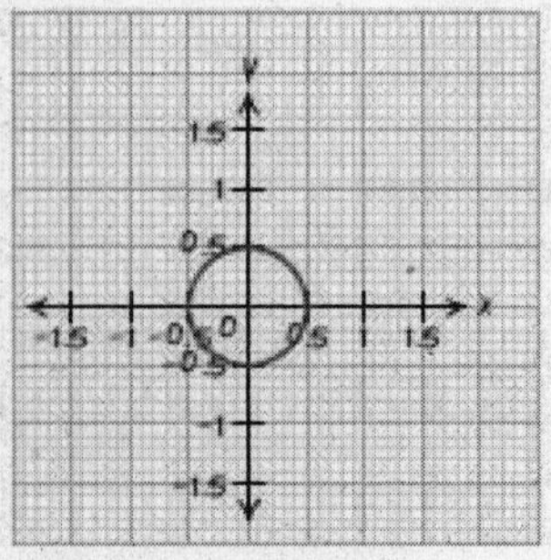

57. Circle

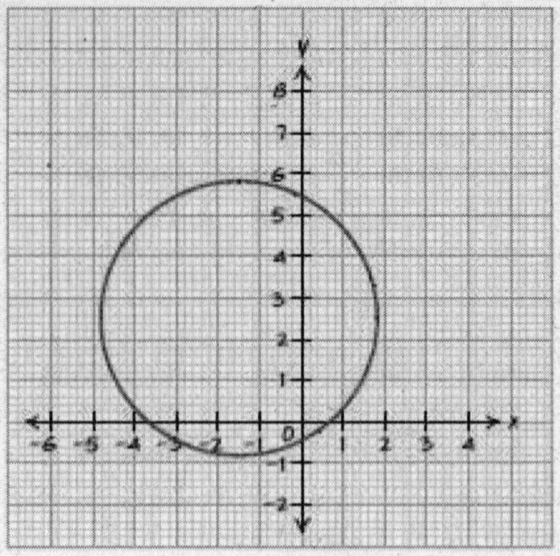

Section 14.3 Arithmetic Sequences

1.

n	□ EMBED Equation	a_n
1	$a_1 = 4+(1-1)6$	$a_1 = 4$
2	$a_2 = 4+(2-1)6$	$a_2 = 10$
3	$a_3 = 4+(3-1)6$	$a_3 = 16$
4	$a_4 = 4+(4-1)6$	$a_4 = 22$

3.

n	$f(n) = 2n^2$	$f(n)$
1	$f(1) = 2(1)^2$	$f(1) = 2$
2	$f(2) = 2(2)^2$	$f(2) = 8$
3	$f(3) = 2(3)^2$	$f(3) = 18$
4	$f(4) = 2(4)^2$	$f(4) = 32$

5.

n	$a_n = 4(-1)^n$	a_n
1	$a_1 = 4(-1)^1$	$a_1 = -4$
2	$a_2 = 4(-1)^2$	$a_2 = 4$
3	$a_3 = 4(-1)^3$	$a_3 = -4$
4	$a_4 = 4(-1)^4$	$a_4 = 4$
5	$a_5 = 4(-1)^5$	$a_4 = -4$

7.

n	$f(n) = 3^n - 20$
1	$f(1) = 3^1 - 20 = -17$
2	$f(2) = 3^2 - 20 = -11$
3	$f(3) = 3^3 - 20 = 7$
4	$f(4) = 3^4 - 20 = 61$
5	$f(5) = 3^5 - 20 = 223$

The first 5 terms of the sequence are -17, -11, 7, 61, 223.

9.

n	$a_n = \frac{4n}{n+7}$
1	$a_1 = \frac{4(1)}{(1)+7} = \frac{1}{2}$
2	$a_2 = \frac{4(2)}{(2)+7} = \frac{8}{9}$
3	$a_3 = \frac{4(3)}{(3)+7} = \frac{6}{5}$
4	$a_4 = \frac{4(4)}{(4)+7} = \frac{16}{11}$
5	$a_5 = \frac{4(5)}{(5)+7} = \frac{5}{3}$

The first 5 terms of the sequence are $\frac{1}{2}, \frac{8}{9}, \frac{6}{5}, \frac{16}{11}, \frac{5}{3}$.

11.

n	$a_n = \frac{1}{4^n}$
1	$a_1 = \frac{1}{4^1} = \frac{1}{4}$
2	$a_2 = \frac{1}{4^2} = \frac{1}{16}$
3	$a_3 = \frac{1}{4^3} = \frac{1}{64}$
4	$a_4 = \frac{1}{4^4} = \frac{1}{256}$
5	$a_5 = \frac{1}{4^5} = \frac{1}{1024}$

The first 5 terms of the sequence are $\frac{1}{4}, \frac{1}{16}, \frac{1}{64}, \frac{1}{256}, \frac{1}{1024}$.

13.

n	$a_n = \frac{(-1)^n}{n^2}$
1	$a_1 = \frac{(-1)^1}{(1)^2} = -1$
2	$a_2 = \frac{(-1)^2}{(2)^2} = \frac{1}{4}$
3	$a_3 = \frac{(-1)^3}{(3)^2} = -\frac{1}{9}$
4	$a_4 = \frac{(-1)^4}{(4)^2} = \frac{1}{16}$
5	$a_5 = \frac{(-1)^5}{(5)^2} = -\frac{1}{25}$

The first 5 terms of the sequence are $-1, \frac{1}{4}, -\frac{1}{9}, \frac{1}{16}, -\frac{1}{25}$.

15. $a_n = 4+(n-1)6$ where $a_n = 118$

$118 = 4+(n-1)6$
$118 = 4+6n-6$
$118 = 6n-2$
$120 = 6n$
$n = 20$

17. $a_n = 2n^2 + 5$ where $a_n = 397$

$397 = 2n^2 + 5$
$392 = 2n^2$
$196 = n^2$
$n = \pm\sqrt{196}$
$n = \pm 14$
Since n must be a natural number $n = 14$

$a_n = 2(3)^n - 24$ where $a_n = 354270$

$354270 = 2(3)^n - 24$

$354294 = 2(3)^n$

$177147 = (3)^n$

$\ln(177147) = \ln(3)^n$

$\ln(177147) = n\ln(3)$

$n = \dfrac{\ln(177147)}{\ln(3)}$

$n = 11$

21. $a_n = \dfrac{2n}{n+8}$ where $a_n = 1.2$

$1.2 = \dfrac{2n}{n+8}$

$1.2(n+8) = \dfrac{2n}{\cancel{n+8}}\cancel{(n+8)}$

$1.2(n+8) = 2n$

$1.2n + 9.6 = 2n$

$9.6 = 0.8n$

$n = 12$

23. $a_n = 50 + \sqrt{2n}$ where $a_n = 68$

$68 = 50 + \sqrt{2n}$

$18 = \sqrt{2n}$

$324 = 2n$

$n = 162$

25.

n	$a_n = 4n - 5$
1	$a_1 = 4(1) - 5 = -1$
2	$a_2 = 4(2) - 5 = 3$
3	$a_3 = 4(3) - 5 = 7$
4	$a_4 = 4(4) - 5 = 11$
5	$a_5 = 4(5) - 5 = 15$
6	$a_6 = 4(6) - 5 = 19$

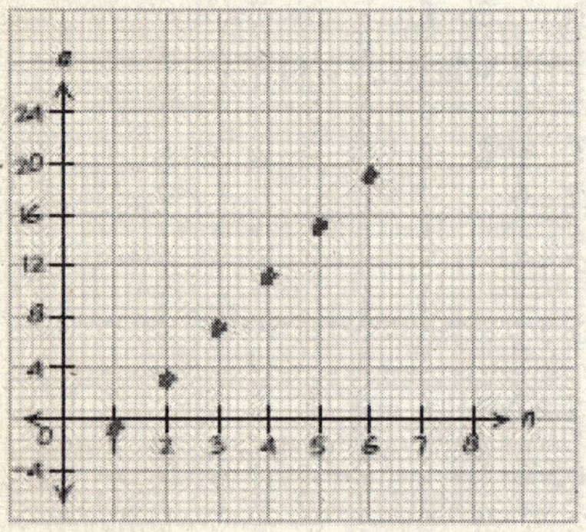

27.

n	$a_n = 4 + (n-1)6$
1	$a_1 = 4 + (1-1)6 = 4$
2	$a_2 = 4 + (2-1)6 = 10$
3	$a_3 = 4 + (3-1)6 = 16$
4	$a_4 = 4 + (4-1)6 = 22$
5	$a_5 = 4 + (5-1)6 = 28$
6	$a_6 = 4 + (6-1)6 = 34$

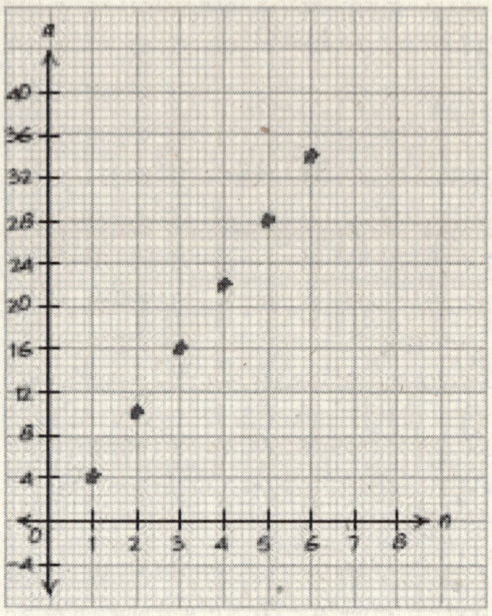

29.

n	$f(n) = 3^n - 20$
1	$f(1) = 3^1 - 20 = -17$
2	$f(2) = 3^2 - 20 = -11$
3	$f(3) = 3^3 - 20 = 7$
4	$f(4) = 3^4 - 20 = 61$
5	$f(5) = 3^5 - 20 = 223$
6	$f(6) = 3^6 - 20 = 709$

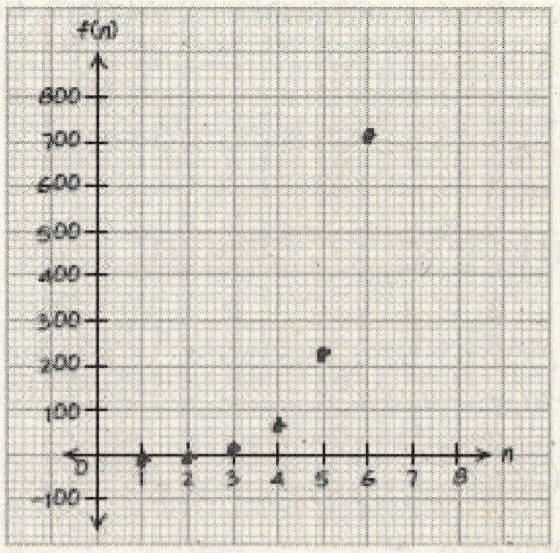

31.

n	$a_n = 4n(-1)^n$
1	$a_1 = 4(1)(-1)^1 = -4$
2	$a_2 = 4(2)(-1)^2 = 8$
3	$a_3 = 4(3)(-1)^3 = -12$
4	$a_4 = 4(4)(-1)^4 = 16$
5	$a_5 = 4(5)(-1)^5 = -20$
6	$a_6 = 4(6)(-1)^6 = 24$

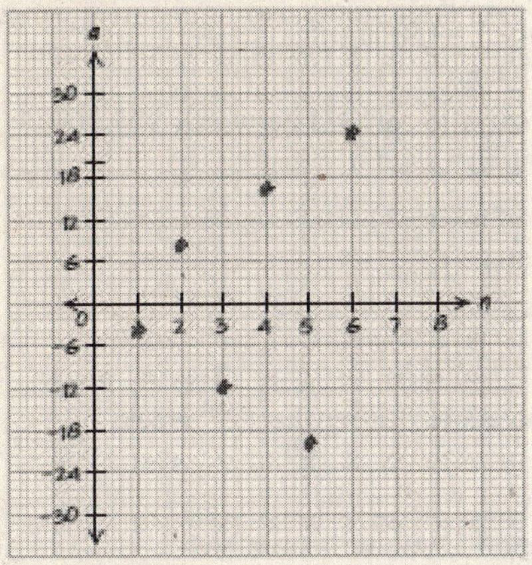

33.

n	$a_n = \dfrac{4n}{n+7}$
1	$a_1 = \dfrac{4(1)}{(1)+7} = \dfrac{1}{2}$
2	$a_2 = \dfrac{4(2)}{(2)+7} = \dfrac{8}{9}$
3	$a_3 = \dfrac{4(3)}{(3)+7} = \dfrac{6}{5}$
4	$a_4 = \dfrac{4(4)}{(4)+7} = \dfrac{16}{11}$
5	$a_5 = \dfrac{4(5)}{(5)+7} = \dfrac{5}{3}$
6	$a_5 = \dfrac{4(6)}{(6)+7} = \dfrac{24}{13}$

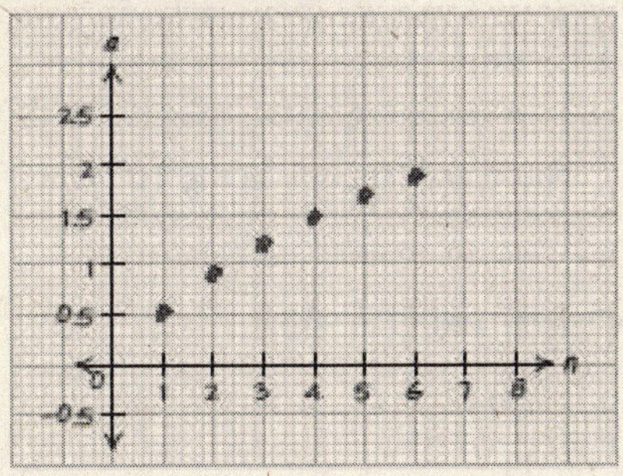

35. $a_2 - a_1 = 16 - 12 = 4$
$a_3 - a_2 = 20 - 16 = 4$
$a_4 - a_3 = 24 - 20 = 4$
$a_5 - a_4 = 28 - 24 = 4$
$a_6 - a_5 = 32 - 28 = 4$
The sequence is arithmetic. The common difference is 4.

37. $a_2 - a_1 = -4.3 - (-6.65) = -4.3 + 6.65 = 2.35$
$a_3 - a_2 = -1.95 - (-4.3) = -1.95 + 4.3 = 2.35$
$a_4 - a_3 = 0.4 - (-1.95) = 0.4 + 1.95 = 2.35$
$a_5 - a_4 = 2.75 - 0.4 = 2.35$
$a_6 - a_5 = 5.1 - 2.75 = 2.35$
The sequence is arithmetic. The common difference is 2.35.

39. $a_2 - a_1 = 80 - 100 = -20$
$a_3 - a_2 = 64 - 80 = -16$
$-20 \neq -16$
The differences between consecutive terms are not the same, so this is not an arithmetic sequence.

41. $a_2 - a_1 = 10.55 - 14 = -3.45$
$a_3 - a_2 = 7.1 - 10.55 = -3.45$
$a_4 - a_3 = 3.65 - 7.1 = -3.45$
$a_5 - a_4 = 0.2 - 3.65 = -3.45$
$a_6 - a_5 = -3.25 - 0.2 = -3.45$
The sequence is arithmetic. The common difference is −3.45.

43. $a_2 - a_1 = \frac{7}{2} - 4 = -\frac{1}{2}$
$a_3 - a_2 = 3 - \frac{7}{2} = -\frac{1}{2}$
$a_4 - a_3 = \frac{5}{2} - 3 = -\frac{1}{2}$
$a_5 - a_4 = 2 - \frac{5}{2} = -\frac{1}{2}$
$a_6 - a_5 = \frac{3}{2} - 2 = -\frac{1}{2}$
$a_7 - a_6 = 1 - \frac{3}{2} = -\frac{1}{2}$
The sequence is arithmetic. The common difference is $-\frac{1}{2}$.

45. $a_2 - a_1 = 10 - 4 = 6$
$a_3 - a_2 = 16 - 10 = 6$
$a_4 - a_3 = 22 - 16 = 6$
$a_5 - a_4 = 28 - 22 = 6$
$a_6 - a_5 = 34 - 28 = 6$
The sequence is arithmetic. The common difference is 6.
$a_n = 4 + (n-1)6$
$a_n = 4 + 6n - 6$
$a_n = -2 + 6n$

47. $a_2 - a_1 = 5.25 - 4 = 1.25$

$a_3 - a_2 = 6.5 - 5.25 = 1.25$

$a_4 - a_3 = 7.75 - 6.5 = 1.25$

$a_5 - a_4 = 9 - 7.75 = 1.25$

$a_6 - a_5 = 10.25 - 9 = 1.25$

The sequence is arithmetic. The common difference is 1.25.

$a_n = 4 + (n-1)1.25$

$a_n = 4 + 1.25n - 1.25$

$a_n = 2.75 + 1.25n$

49. $a_2 - a_1 = 9 - 3 = 6$

$a_3 - a_2 = -1 - 9 = -10$

$6 \neq -10$

The differences between consecutive terms are not the same, so this is not an arithmetic sequence.

51. $a_2 - a_1 = \frac{10}{3} - 4 = -\frac{2}{3}$

$a_3 - a_2 = \frac{8}{3} - \frac{10}{3} = -\frac{2}{3}$

$a_4 - a_3 = 2 - \frac{8}{3} = -\frac{2}{3}$

$a_5 - a_4 = \frac{4}{3} - 2 = -\frac{2}{3}$

$a_6 - a_5 = \frac{2}{3} - \frac{4}{3} = -\frac{2}{3}$

The sequence is arithmetic. The common difference is $-\frac{2}{3}$.

$a_n = 4 + (n-1)\left(-\frac{2}{3}\right)$

$a_n = 4 - \frac{2}{3}n + \frac{2}{3}$

$a_n = \frac{14}{3} - \frac{2}{3}n$

53. **a.** $a_1 = 36{,}000$ and $d = 900$

$a_n = 36{,}000 + (n-1)900$

$a_n = 36{,}000 + 900n - 900$

$a_n = 35{,}100 + 900n$

b. $a_{12} = 35{,}100 + 900(12)$

$a_{12} = 45{,}900$

Your salary will be \$45,900 in the 12th year of employment.

c. $60{,}300 = 35{,}100 + 900n$

$$\begin{aligned} -35{,}100 \quad & -35{,}100 \\ 25{,}200 &= 900n \\ \frac{25{,}200}{900} &= \frac{900n}{900} \\ 28 &= n \end{aligned}$$

In your 28th year of employment, your salary will be \$60,300.

55. **a.** $a_1 = 48{,}000$ and $d = -4500$

$a_n = 48{,}000 + (n-1)(-4500)$

$a_n = 48{,}000 - 4500n + 4500$

$a_n = 52{,}500 - 4500n$

b. $a_{11} = 52{,}500 - 4500(11)$

$a_{11} = 3000$

The car will be worth \$3000 in the 11th year.

c. $21{,}000 = 52{,}500 - 4500n$

$$\begin{aligned} -52{,}000 \quad & -52{,}000 \\ -31{,}500 &= -4500n \\ \frac{-31{,}500}{-4500} &= \frac{-4500n}{-4500} \\ 7 &= n \end{aligned}$$

In the 7th year, the car will be worth \$21,000.

57. **a.** $a_1 = 10$ and $d = 1$

$a_n = 10 + (n-1)(1)$

$a_n = 10 + n - 1$

$a_n = 9 + n$

b. $a_{32} = 9 + 32 = 41$

You will save \$41 during the 32nd week.

c. $61 = 9 + n$

$52 = n$

During week 52, you will save \$61.

59. **a.** $a_n = 400 + 0.15n$

$a_{100} = 400 + 0.15(100)$

$a_{100} = 415$

It costs \$415 to produce 100 candy bars.

b. $550 = 400 + 0.15n$

$$\begin{aligned} -400 \quad & -400 \\ 150 &= 0.15n \\ \frac{150}{0.15} &= \frac{0.15n}{0.15} \\ 1000 &= n \end{aligned}$$

One thousand candy bars can be produced for a cost of \$550.

c. The common difference is $d = 0.15$. The cost of production increases by \$0.15 per candy bar produced.

61. **a.** $a_1 = 1$ and $d = 1$

$a_n = 1 + (n-1)$

$a_n = n$

b. $a_{10} = 10$ There are 10 cups in the 10th row.

63.

$$\left.\begin{aligned} a_4 &= 9 \\ a_{15} &= -13 \end{aligned}\right\} (4,9) \text{ and } (15,-13)$$

$d = \frac{-13-9}{15-4} = \frac{-22}{11} = -2$

$$y-9=-2(x-4)$$
$$y-9=-2x+8$$
$$\underline{+9 \qquad +9}$$
$$y=-2x+17$$
so
$$a_n=-2n+17$$

65.

$$\left.\begin{array}{l}a_7=28\\a_{21}=77\end{array}\right\}(7,28)\text{ and }(21,77)$$
$$d=\frac{77-28}{21-7}=\frac{49}{14}=3.5$$
$$y-28=3.5(x-7)$$
$$y-28=3.5x-24.5$$
$$\underline{+28 \qquad +28}$$
$$y=3.5x+3.5$$
so
$$a_n=3.5n+3.5$$

67.

$$\left.\begin{array}{l}a_{10}=56.75\\a_{32}=161.25\end{array}\right\}(10,56.75)\text{ and }(32,161.25)$$
$$d=\frac{161.25-56.75}{32-10}=\frac{104.5}{22}=4.75$$
$$y-56.75=4.75(x-10)$$
$$y-56.75=4.75x-47.5$$
$$\underline{+56.75 \qquad +56.75}$$
$$y=4.75x+9.25$$
so
$$a_n=4.75n+9.25$$

69.

$$\left.\begin{array}{l}a_6=-36\\a_{25}=-32.2\end{array}\right\}(6,-36)\text{ and }(25,-32.2)$$
$$d=\frac{-32.2-(-36)}{25-6}=\frac{3.8}{19}=0.2$$
$$y-(-36)=0.2(x-6)$$
$$y+36=0.2x-1.2$$
$$\underline{-36 \qquad -36}$$
$$y=0.2x-37.2$$
so
$$a_n=0.2n-37.2$$

Section 14.4 Geometric Sequences

1. Yes, the sequence is geometric and the common ratio is 1.5.

$$\frac{a_2}{a_1}=\frac{3}{2}=1.5$$
$$\frac{a_3}{a_2}=\frac{4.5}{3}=1.5$$
$$\frac{a_4}{a_3}=\frac{6.75}{4.5}=1.5$$
$$\frac{a_5}{a_4}=\frac{10.125}{6.75}=1.5$$
$$\frac{a_6}{a_5}=\frac{15.1875}{10.125}=1.5$$

3. Yes, the sequence is geometric and the common ratio is $\frac{1}{3}$.

$$\frac{a_2}{a_1}=\frac{90}{270}=\frac{1}{3}$$
$$\frac{a_3}{a_2}=\frac{30}{90}=\frac{1}{3}$$
$$\frac{a_4}{a_3}=\frac{10}{30}=\frac{1}{3}$$
$$\frac{a_5}{a_4}=\frac{\frac{10}{3}}{10}=\frac{1}{3}$$
$$\frac{a_6}{a_5}=\frac{\frac{10}{9}}{\frac{10}{3}}=\frac{1}{3}$$

5. No, the sequence is not geometric.

$$\frac{a_2}{a_1}=\frac{-1}{-4}=0.25$$
$$\frac{a_3}{a_2}=\frac{4}{-1}=-4$$
$$\frac{a_4}{a_3}=\frac{11}{4}=2.75$$
$$\frac{a_5}{a_4}=\frac{20}{11}\approx1.818$$
$$\frac{a_6}{a_5}=\frac{31}{20}=1.55$$

7. Yes, the sequence is geometric and the common ratio is 2.

$$\frac{a_2}{a_1}=\frac{-12}{-6}=2$$
$$\frac{a_3}{a_2}=\frac{-24}{-12}=2$$
$$\frac{a_4}{a_3}=\frac{-48}{-24}=2$$
$$\frac{a_5}{a_4}=\frac{-96}{-48}=2$$
$$\frac{a_6}{a_5}=\frac{-192}{-96}=2$$

9. Yes, the sequence is geometric and the common ratio is -3.

$\frac{a_2}{a_1} = \frac{-12}{4} = -3$

$\frac{a_3}{a_2} = \frac{36}{-12} = -3$

$\frac{a_4}{a_3} = \frac{-108}{36} = -3$

$\frac{a_5}{a_4} = \frac{324}{-108} = -3$

$\frac{a_6}{a_5} = \frac{-972}{324} = -3$

11. Yes, the sequence is geometric.

$\frac{a_2}{a_1} = \frac{6}{3} = 2$

$\frac{a_3}{a_2} = \frac{12}{6} = 2$

$\frac{a_4}{a_3} = \frac{24}{12} = 2$

$\frac{a_5}{a_4} = \frac{48}{24} = 2$

$\frac{a_6}{a_5} = \frac{96}{48} = 2$

$r = 2, a_1 = 3$

$a_n = 3(2)^{n-1}$

13. Yes, the sequence is geometric.

$\frac{a_2}{a_1} = \frac{-8}{2} = -4$

$\frac{a_3}{a_2} = \frac{32}{-8} = -4$

$\frac{a_4}{a_3} = \frac{-128}{32} = -4$

$\frac{a_5}{a_4} = \frac{512}{-128} = -4$

$\frac{a_6}{a_5} = \frac{-2048}{512} = -4$

$r = -4, a_1 = 2$

$a_n = 2(-4)^{n-1}$

15. Yes, the sequence is geometric.

$\frac{a_2}{a_1} = \frac{125}{625} = 0.2$

$\frac{a_3}{a_2} = \frac{25}{125} = 0.2$

$\frac{a_4}{a_3} = \frac{5}{25} = 0.2$

$\frac{a_5}{a_4} = \frac{1}{5} = 0.2$

$\frac{a_6}{a_5} = \frac{0.2}{1} = 0.2$

$r = 0.2, a_1 = 625$

$a_n = 625(0.2)^{n-1}$

17. No, the sequence is not geometric.

$\frac{a_2}{a_1} = \frac{0}{-4.5} = 0$

$\frac{a_3}{a_2} = \frac{4.5}{0} = \textit{undefined}$

$\frac{a_4}{a_3} = \frac{9}{4.5} = 2$

$\frac{a_5}{a_4} = \frac{13.5}{9} = 1.5$

$\frac{a_6}{a_5} = \frac{18}{13.5} \approx 1.333$

19. Yes, the sequence is geometric.

$\frac{a_2}{a_1} = \frac{-2}{-4} = 0.5$

$\frac{a_3}{a_2} = \frac{-1}{-2} = 0.5$

$\frac{a_4}{a_3} = \frac{-\frac{1}{2}}{-1} = 0.5$

$\frac{a_5}{a_4} = \frac{-\frac{1}{4}}{-\frac{1}{2}} = 0.5$

$\frac{a_6}{a_5} = \frac{-\frac{1}{8}}{-\frac{1}{4}} = 0.5$

$r = 0.5, a_1 = -4$

$a_n = -4(0.5)^{n-1}$

21. **a.** $a_n = a_1 r^{n-1}$

$a_1 = 56{,}000$

$r = 1 + 0.015 = 1.015$

$a_n = 56{,}000(1.015)^{n-1}$

b. $a_7 = 56{,}000(1.015)^{7-1}$

$a_7 = 61{,}232.82$

Your salary will be \$61,232.82 in the 7th year of employment.

c. $80{,}052 = 56{,}000(1.015)^{n-1}$

$\frac{80{,}052}{56{,}000} = \frac{56{,}000(1.015)^{n-1}}{56{,}000}$

$1.4295 = 1.015^{n-1}$

$n - 1 = \log_{1.015}(1.4295)$

$n = \log_{1.015}(1.4295) + 1$

$n = \frac{\log 1.4295}{\log 1.015} + 1$

$n \approx 25$

In your 25th year of employment, your salary will be \$80,052.

23. **a.** $a_n = a_1 r^{n-1}$

$a_1 = 8$ and $r = \frac{1}{2}$

$a_n = 8\left(\frac{1}{2}\right)^{n-1}$

b. $a_3 = 8\left(\frac{1}{2}\right)^{3-1}$

$a_3 = 8\left(\frac{1}{2}\right)^2 = 8\left(\frac{1}{4}\right)$

$a_3 = 2$

The ball bounces to a height of 2 ft on the 3rd bounce.

25. **a.** $a_n = a_1 r^{n-1}$

$a_1 = 5$ and $r = 2$

$a_n = 5(2)^{n-1}$

b. $a_7 = 5(2)^{7-1}$

$a_7 = 5(2)^6 = 320$

On day 7, there were 320 people with the flu.

27. **a.** $a_n = a_1 r^{n-1}$

$a_1 = 21{,}000$

$r = 1 - 0.07 = 0.93$

$a_n = 21{,}000(0.93)^{n-1}$

b. $a_4 = 21{,}000(0.93)^{4-1}$

$a_4 = 21{,}000(0.93)^3$

$a_4 = 16{,}891.50$

A four year old Honda Civic hybrid is worth \$16,891.50.

29. **a.**

$$\left.\begin{aligned} a_0 &= 10 \\ a_2 &= 5 \end{aligned}\right\} a_n = a_1 r^n$$

$a_2 = 10r^2$

$5 = 10r^2$

$\frac{5}{10} = r^2$

$\left(\frac{1}{2}\right)^{1/2} = \left(r^2\right)^{1/2}$

$\left(\frac{1}{2}\right)^{1/2} = r$

$a_n = 10\left(\frac{1}{2}\right)^{\frac{n}{2}}$

b. $a_6 = 10\left(\frac{1}{2}\right)^{\frac{6}{2}}$

$a_6 = 10\left(\frac{1}{2}\right)^3 \approx 1.25$

After 6 hours, there is 1.25 mg of morphine left in the patient.

31. $a_4 = 250$ and $a_9 = 781{,}250$

$250 = a_1(r)^{4-1}$ and $781250 = a_1(r)^{9-1}$

$250 = a_1(r)^3$ and $781250 = a_1(r)^8$

$\frac{781250}{250} = \frac{a_1(r)^8}{a_1(r)^3}$

$3125 = r^5$

$(3125)^{\frac{1}{5}} = \left(r^5\right)^{\frac{1}{5}}$

$r = 5$

$250 = a_1(5)^{4-1}$

$250 = 125a_1$

$a_1 = 2$

$a_n = 2(5)^{n-1}$

33. $a_4 = 256$ and $a_{10} = \frac{1}{16}$

$256 = a_1(r)^{4-1}$ and $\frac{1}{16} = a_1(r)^{10-1}$

$256 = a_1(r)^3$ and $\frac{1}{16} = a_1(r)^9$

$\frac{\frac{1}{16}}{256} = \frac{a_1(r)^9}{a_1(r)^3}$

$\frac{1}{4096} = r^6$

$\left(\frac{1}{4096}\right)^{\frac{1}{6}} = \left(r^6\right)^{\frac{1}{6}}$

$r = \frac{1}{4}$

$256 = a_1\left(\frac{1}{4}\right)^{4-1}$

$256 = \frac{1}{64}a_1$

$a_1 = 16384$

$a_n = 16{,}384\left(\frac{1}{4}\right)^{n-1}$

35. $a_4 = -24$ and $a_{11} = 3072$

$-24 = a_1(r)^{4-1}$ and $3072 = a_1(r)^{11-1}$

$-24 = a_1(r)^3$ and $3072 = a_1(r)^{10}$

$\frac{3072}{-24} = \frac{a_1(r)^{10}}{a_1(r)^3}$

$-128 = r^7$

$(-128)^{\frac{1}{7}} = \left(r^7\right)^{\frac{1}{7}}$

$r = -2$

$-24 = a_1(-2)^{4-1}$

$-24 = -8a_1$

$a_1 = 3$

$a_n = 3(-2)^{n-1}$

37. $a_5 = 16$ and $a_{12} = -\frac{1}{8}$

$16 = a_1(r)^{5-1}$ and $-\frac{1}{8} = a_1(r)^{12-1}$

$16 = a_1(r)^4$ and $-\frac{1}{8} = a_1(r)^{11}$

$$\frac{-\frac{1}{8}}{16} = \frac{a_1(r)^{11}}{a_1(r)^4}$$

$$-\frac{1}{128} = r^7$$

$$\left(-\frac{1}{128}\right)^{\frac{1}{7}} = \left(r^7\right)^{\frac{1}{7}}$$

$$r = -\frac{1}{2}$$

$$16 = a_1\left(-\frac{1}{2}\right)^{5-1}$$

$$16 = \frac{1}{16}a_1$$

$$a_1 = 256$$

$$a_n = 256\left(-\frac{1}{2}\right)^{n-1}$$

39.

n	a_n	Ratio
1	11	
2	22	$\frac{a_2}{a_1} = \frac{22}{11} = 2$
3	44	$\frac{a_3}{a_2} = \frac{44}{22} = 2$
4	88	$\frac{a_4}{a_3} = \frac{88}{44} = 2$
5	176	$\frac{a_5}{a_4} = \frac{176}{88} = 2$
6	352	$\frac{a_6}{a_5} = \frac{352}{176} = 2$

This sequence has a common ratio of $r = 2$, so it is geometric.

$$a_n = a_1(r)^{n-1}$$
$$a_n = 11(2)^{n-1}$$

41.

n	a_n	Difference
1	14	
2	20	$a_2 - a_1 = 20 - 14 = 6$
3	26	$a_3 - a_2 = 26 - 20 = 6$
4	32	$a_4 - a_3 = 32 - 26 = 6$
5	38	$a_5 - a_4 = 38 - 32 = 6$
6	44	$a_6 - a_5 = 44 - 38 = 6$

This sequence has a common difference of $d = 6$, so it is arithmetic.

$$a_n = a_1 + (n-1)d$$
$$a_n = 14 + (n-1)6$$
$$a_n = 14 + 6n - 6$$
$$a_n = 8 + 6n$$

43.

n	a_n	Ratio
1	7776	
2	3888	$\frac{a_2}{a_1} = \frac{3888}{7776} = \frac{1}{2}$
3	1944	$\frac{a_3}{a_2} = \frac{1944}{3888} = \frac{1}{2}$
4	972	$\frac{a_4}{a_3} = \frac{972}{1944} = \frac{1}{2}$
5	486	$\frac{a_5}{a_4} = \frac{486}{972} = \frac{1}{2}$
6	243	$\frac{a_6}{a_5} = \frac{243}{486} = \frac{1}{2}$

This sequence has a common ratio of $r = \frac{1}{2}$, so it is geometric.

$$a_n = a_1(r)^{n-1}$$
$$a_n = 7776\left(\frac{1}{2}\right)^{n-1}$$

45.

n	a_n	Difference
1	−38	
2	−34.75	$a_2 - a_1 = -34.75 - (-38) = 3.25$
3	−31.5	$a_3 - a_2 = -31.5 - (-34.75) = 3.25$
4	−28.25	$a_4 - a_3 = -28.25 - (-31.5) = 3.25$
5	−25	$a_5 - a_4 = -25 - (-28.25) = 3.25$
6	−21.75	$a_6 - a_5 = -21.75 - (-25) = 3.25$

This sequence has a common difference of $d = 3.25$, so it is arithmetic.

$$a_n = a_1 + (n-1)d$$
$$a_n = -38 + (n-1)3.25$$
$$a_n = -38 + 3.25n - 3.25$$
$$a_n = -41.25 + 3.25n$$

47. **a.** $180, 360, 540, \ldots$

$a_2 - a_1 = 360 - 180 = 180$

$a_3 - a_2 = 540 - 360 = 180$

This is an arithmetic sequence with $d = 180$.

$$a_n = 180 + (n-1)180$$
$$a_n = 180 + 180n - 180$$
$$a_n = 180n$$

b. $a_{10} = 180(10) = 1800$ The sum of the interior angles of a 12 sided polygon is 1800°.

49. **a.** $a_n = 200{,}000 - 8000n$

b. $a_{10} = 200{,}000 - 8000(10)$

$a_{10} = 120{,}000$

When the machine is 10 years old it will be worth \$120,000.

51. **a.** $a_n = a_1 + (n-1)d$

$a_n = 15 + (n-1)5$

$a_n = 15 - 5n - 5$

$a_n = 10 + 5n$

b. $a_5 = 10 + 5(5) = 35$

You should be jogging 35 minutes during the 5th week after surgery.

Section 14.5 Series

1. **a.**

$$20.5+156+198+162+125+248+244+209+165+262+214+60.6+176+172+161+237+83+182+164+203+14.4$$

b.

$$20.5+156+198+162+125+248+244+209+165+262+214+60.6+176+172+161+237+83+182+164+203+14.4=3456.5$$

The 2009 Giro d' Italia professional bike race is 3456.5 km in length.

3.

$$\sum_{n=1}^{5}(2n+7)=(2(1)+7)+(2(2)+7)+(2(3)+7)+(2(4)+7)+(2(5)+7)=65$$

5. $$\sum_{n=1}^{4}(4^n)=4+4^2+4^3+4^4=340$$

7.

$$\sum_{n=1}^{6}\left(5n(-1)^n\right)=(5(-1))+\left(10(-1)^2\right)+\left(15(-1)^3\right)+\left(20(-1)^4\right)+\left(25(-1)^5\right)+\left(30(-1)^6\right)=15$$

9. $$\sum_{n=1}^{5}\left(\frac{1}{n}\right)=1+\frac{1}{2}+\frac{1}{3}+\frac{1}{4}+\frac{1}{5}=\frac{137}{60}$$

11.

$$\sum_{n=1}^{8}(n^2+5n)=(1+5)+(2^2+10)+(3^2+15)+(4^2+20)+(5^2+25)+(6^2+30)+(7^2+35)+(8^2+40)=384$$

13.

$$\sum_{n=1}^{8}\left(3(-2)^n\right)=(3(-2))+\left(3(-2)^2\right)+\left(3(-2)^3\right)+\left(3(-2)^4\right)+\left(3(-2)^5\right)+\left(3(-2)^6\right)+\left(3(-2)^7\right)+\left(3(-2)^8\right)=510$$

15. **a.**

$a_n = a_1 + (n-1)d$

$a_1 = 42{,}000$ and $d = 800$

$a_n = 42{,}000 + (n-1)800$

$a_n = 42{,}000 + 800n - 800$

$a_n = 41{,}200 + 800n$

b. $a_{15} = 41{,}200 + 800(15)$

$a_{15} = 53{,}200$

Your salary will be \$53,200 in the 15th year of employment.

c. Use $S_n = \dfrac{n(a_1 + a_n)}{2}$

$$S_{15} = \frac{15(42{,}000 + 53{,}200)}{2}$$
$$S_{15} = 714{,}000$$

You will have earned a total of $714,000 over a 15 year period of employment.

17. **a.** $a_n = a_1 + (n-1)d$

$$a_1 = 8 \text{ and } d = 2$$
$$a_n = 8 + (n-1)2$$
$$a_n = 8 + 2n - 2$$
$$a_n = 6 + 2n$$

b. $a_8 = 6 + 2(8) = 22$

You will save $22 during the 8th week.

c. Use $S_n = \dfrac{n(a_1 + a_n)}{2}$

$$a_{26} = 6 + 2(26) = 58$$
$$S_{26} = \frac{26(8 + 58)}{2}$$
$$S_{25} = 858$$

You will have saved a total of $858 during the first 26 weeks.

19. **a.** $a_n = a_1 + (n-1)d$

$$a_1 = 15 \text{ and } d = 1$$
$$a_n = 15 + (n-1)1$$
$$a_n = 15 + n - 1$$
$$a_n = 14 + n$$

b. Use $S_n = \dfrac{n(a_1 + a_n)}{2}$

$$a_{12} = 14 + 12 = 26$$
$$S_{12} = \frac{12(15 + 26)}{2}$$
$$S_{12} = 246$$

A total of 246 blocks are needed to build a wall with 12 rows.

21. $a_n = 4n + 1$

$$a_1 = 4(1) + 1$$
$$a_1 = 5$$
$$a_{30} = 4(30) + 1$$
$$a_{30} = 121$$
$$S_{30} = \frac{30(5 + 121)}{2}$$
$$S_{30} = 1{,}890$$

23. $a_n = 20 + (n-1)5$

$$a_1 = 20 + (1-1)5$$
$$a_1 = 20$$
$$a_{15} = 20 + (15-1)5$$
$$a_{15} = 90$$
$$S_{15} = \frac{15(20 + 90)}{2}$$
$$S_{15} = 825$$

25. $a_n = -9n - 2$

$$a_1 = -9(1) - 2$$
$$a_1 = -11$$
$$a_{40} = -9(40) - 2$$
$$a_{40} = -362$$
$$S_{40} = \frac{40(-11 + (-362))}{2}$$
$$S_{40} = -7{,}460$$

27. $a_n = 2.4n + 6.2$

$$a_1 = 2.4(1) + 6.2$$
$$a_1 = 8.6$$
$$a_{52} = 2.4(52) + 6.2$$
$$a_{52} = 131$$
$$S_{52} = \frac{52(8.6 + 131)}{2}$$
$$S_{52} = 3{,}629.6$$

29. $a_n = \dfrac{1}{3}n + 4$

$$a_1 = \frac{1}{3}(1) + 4$$
$$a_1 = \frac{13}{3}$$
$$a_{27} = \frac{1}{3}(27) + 4$$
$$a_{27} = 13$$
$$S_{27} = \frac{27\left(\frac{13}{3} + 13\right)}{2}$$
$$S_{27} = 234$$

31. **a.** $a_n = a_1 r^{n-1}$

$$a_1 = 42{,}000$$
$$r = 1 + 0.02 = 1.02$$
$$a_n = 42{,}000(1.02)^{n-1}$$

b. $a_{15} = 42{,}000(1.02)^{15-1}$

$$a_{15} = 42{,}000(1.02)^{14}$$
$$a_{15} = 55{,}418.11$$

Your salary will be $55,418.11 during the 15th year of employment.

c. Use $S_n = \dfrac{a_1(1 - r^n)}{1 - r}$

$$S_{15} = \frac{42{,}000(1 - 1.02^{15})}{1 - 1.02}$$
$$S_{15} = 726{,}323.51$$

You will have earned a total of $726,323.51 over a 15 year career.

33. **a.** $a_n = a_1r^{n-1}$

$a_1 = 10$

$r = 1 + 0.10 = 1.10$

$a_n = 10(1.10)^{n-1}$

b. $a_8 = 10(1.10)^{8-1}$

$a_8 = 10(1.10)^7 \approx 19.49$

You will save \$19.49 during the 8th week.

c. Use $S_n = \dfrac{a_1(1-r^n)}{1-r}$

$S_{26} = \dfrac{10(1-1.10^{26})}{1-1.10}$

$S_{26} \approx 1091.82$

You will have saved a total of \$1091.82 during the first 26 weeks

35. **a.** $a_n = a_1r^{n-1}$

$a_1 = 3$ and $r = 2$

$a_n = 3(2)^{n-1}$

b. $a_5 = 3(2)^{5-1}$

$a_5 = 3(2)^4 = 48$

There are 48 people diagnosed with the flu on the 5th day.

c. Use $S_n = \dfrac{a_1(1-r^n)}{1-r}$

$S_{14} = \dfrac{3(1-2^{14})}{1-2}$

$S_{14} = 49{,}149$

There were 49,149 people diagnosed with the flu during the first two weeks.

37. $a_n = 3^n$

$a_1 = 3^1 = 3$

$a_2 = 3^2 = 9$

$r = \dfrac{9}{3} = 3$

$S_9 = \dfrac{3(1-3^9)}{1-3}$

$S_9 = 29{,}523$

39. $a_n = 2(7)^n$

$a_1 = 2(7)^1 = 14$

$a_2 = 2(7)^2 = 98$

$r = \dfrac{98}{14} = 7$

$S_6 = \dfrac{14(1-7^6)}{1-7}$

$S_6 = 274{,}512$

41. $a_n = 2(-6)^n$

$a_1 = 2(-6)^1 = -12$

$a_2 = 2(-6)^2 = 72$

$r = \dfrac{72}{-12} = -6$

$S_{11} = \dfrac{-12\left(1-(-6)^{11}\right)}{1-(-6)}$

$S_{11} = -621{,}937{,}812$

43. $a_n = 900\left(\dfrac{1}{3}\right)^n$

$a_1 = 900\left(\dfrac{1}{3}\right)^1 = 300$

$a_2 = 900\left(\dfrac{1}{3}\right)^2 = 100$

$r = \dfrac{100}{300} = \dfrac{1}{3}$

$S_{12} = \dfrac{300\left(1-\left(\dfrac{1}{3}\right)^{12}\right)}{1-\left(\dfrac{1}{3}\right)}$

$S_{12} \approx 449.999$

45. $a_n = -1000\left(\dfrac{2}{5}\right)^n$

$a_1 = -1000\left(\dfrac{2}{5}\right)^1 = -400$

$a_2 = -1000\left(\dfrac{2}{5}\right)^2 = -160$

$r = \dfrac{-160}{-400} = \dfrac{2}{5}$

$S_{10} = \dfrac{-400\left(1-\left(\dfrac{2}{5}\right)^{10}\right)}{1-\left(\dfrac{2}{5}\right)}$

$S_{10} \approx -666.597$

47. Geometric

$a_n = 5(2)^n$

$a_1 = 5(2)^1 = 10$

$a_2 = 5(2)^2 = 20$

$r = \dfrac{20}{10} = 2$

$S_{10} = \dfrac{10(1-2^{10})}{1-2}$

$S_{10} = 10{,}230$

49. Geometric

$a_n = 2(-3)^n$

$a_1 = 2(-3)^1 = -6$

$a_2 = 2(-3)^2 = 18$

$r = \frac{18}{-6} = -3$

$S_6 = \frac{-6\left(1-(-3)^6\right)}{1-(-3)}$

$S_6 = 1,092$

51. Arithmetic

$a_n = 8n + 25$

$a_1 = 8(1) + 25 = 33$

$a_{23} = 8(23) + 25 = 209$

$S_{23} = \frac{23(33+209)}{2}$

$S_{23} = 2,783$

53. Geometric

$a_n = 900(0.94)^n$

$a_1 = 900(0.94)^1 = 846$

$a_2 = 900(0.94)^2 = 795.24$

$r = \frac{795.24}{846} = 0.94$

$S_{12} = \frac{846\left(1-(0.94)^{12}\right)}{1-(0.94)}$

$S_{12} \approx 7,389.524$

55. Arithmetic

$a_n = -\frac{1}{2}n + 30$

$a_1 = -\frac{1}{2}(1) + 30 = 29.5$

$a_{10} = -\frac{1}{2}(10) + 30 = 25$

$S_{10} = \frac{10(29.5+25)}{2}$

$S_{10} = 272.5$

Chapter 14 Review Exercises

1.

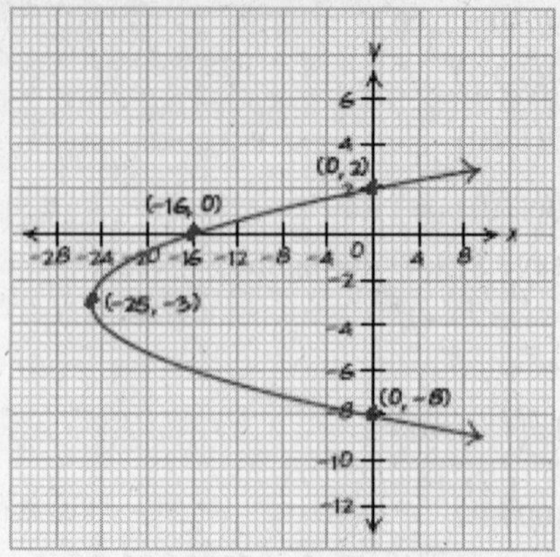

2.

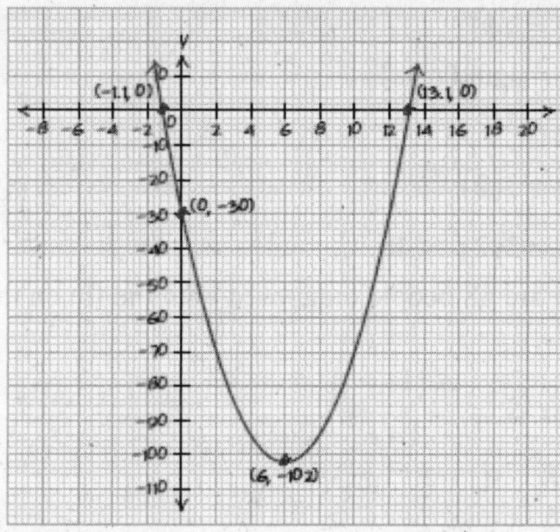

3.

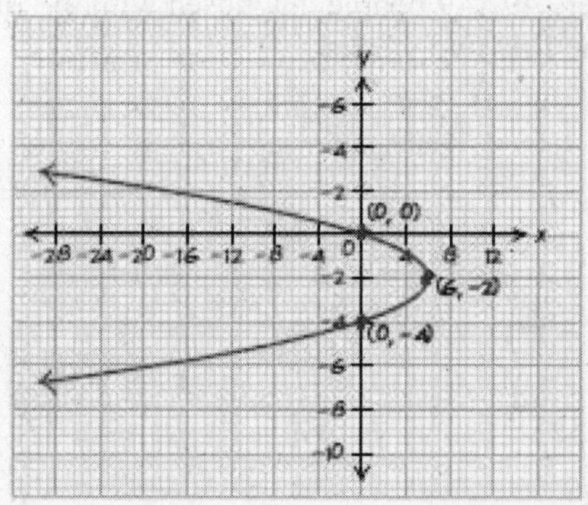

4.

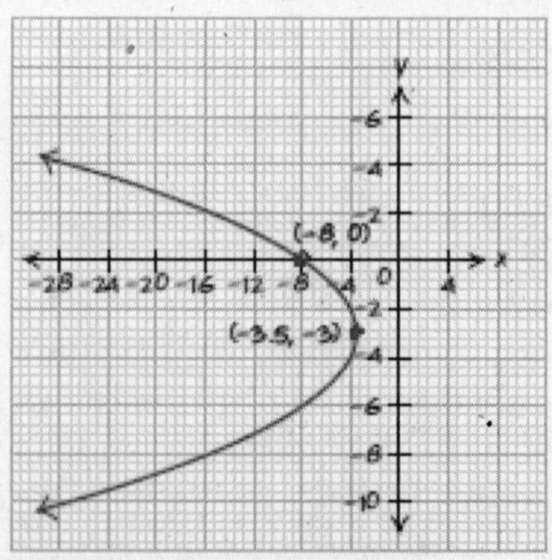

5.

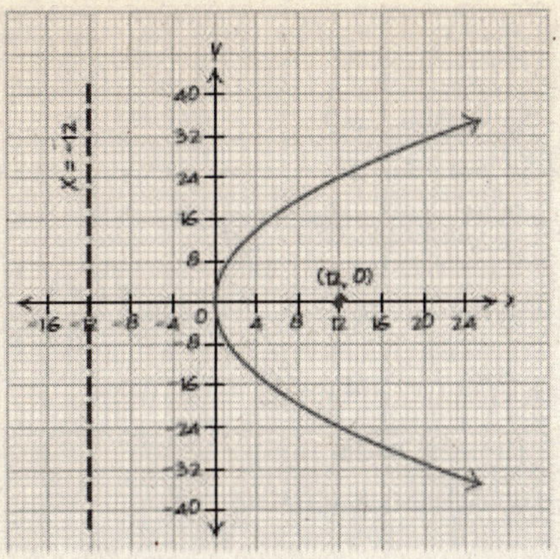

6.

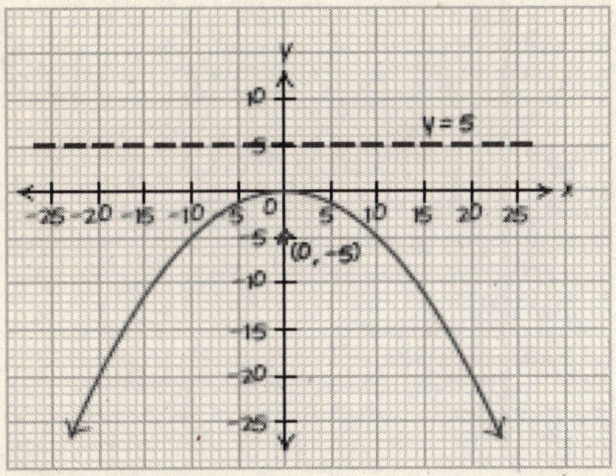

7.

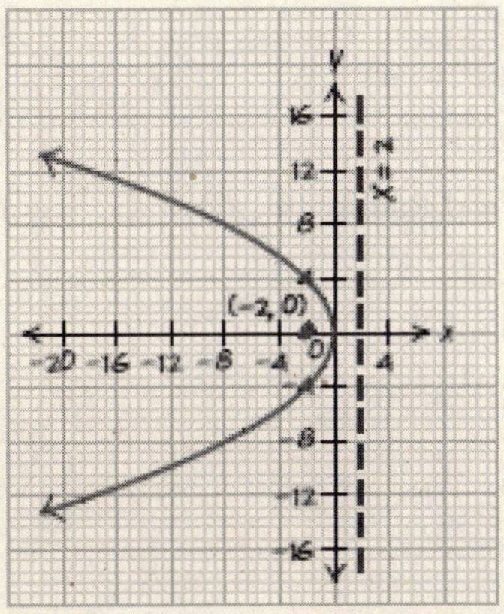

8.

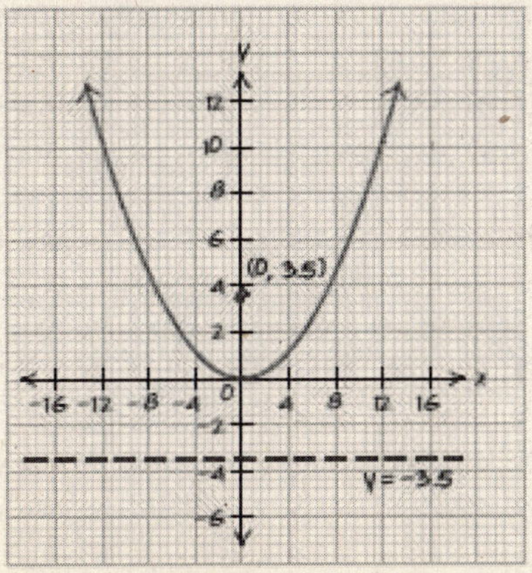

9. Center $=(2,8)$ and Radius $=3$

$h=2, k=8, r=3$

$(x-2)^2+(y-8)^2=(3)^2$

$(x-2)^2+(y-8)^2=9$

10. Center $=(5,-2)$ and Radius $=2.5$

$h=5, k=-2, r=2.5$

$(x-5)^2+(y-(-2))^2=(2.5)^2$

$(x-5)^2+(y+2)^2=6.25$

11. $x^2+y^2=16$

$(x-0)^2+(y-0)^2=(4)^2$

$h=0, k=0, r=4$

Center $=(0,0)$ and Radius $=4$

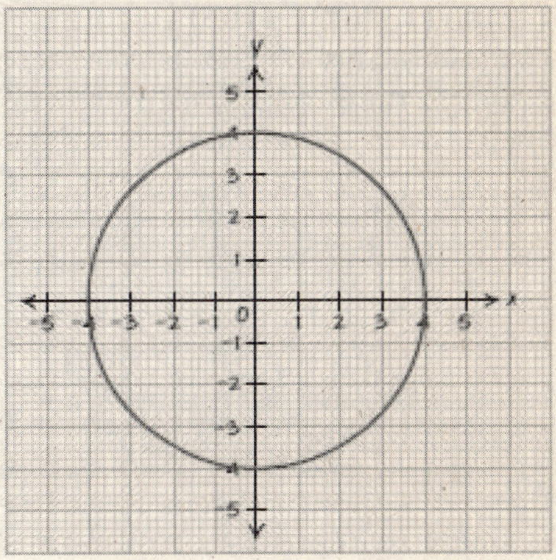

12. $(x-5)^2+(y-2)^2=4$

$(x-5)^2+(y-2)^2=(2)^2$

$h=5, k=2, r=2$

Center $=(5,2)$ and Radius $=2$

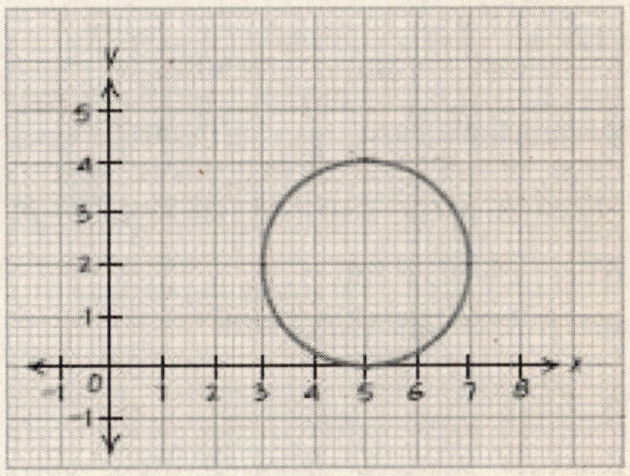

13. $(x+4)^2+(y-6)^2=1$

$(x-(-4))^2+(y-6)^2=(1)^2$

$h=-4, k=6, r=1$

Center $=(-4,6)$ and Radius $=1$

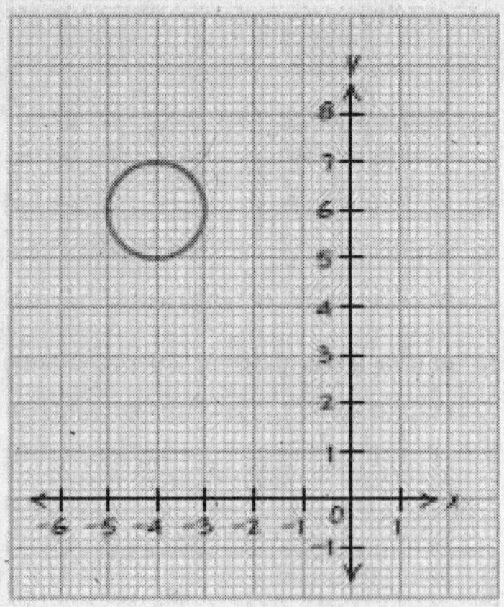

14. $(x+3)^2+(y+7)^2=0.25$

$(x-(-3))^2+(y-(-7))^2=(0.5)^2$

$h=-3, k=-7, r=0.5$

$\text{Center}=(-3,-7)$ and $\text{Radius}=0.5$

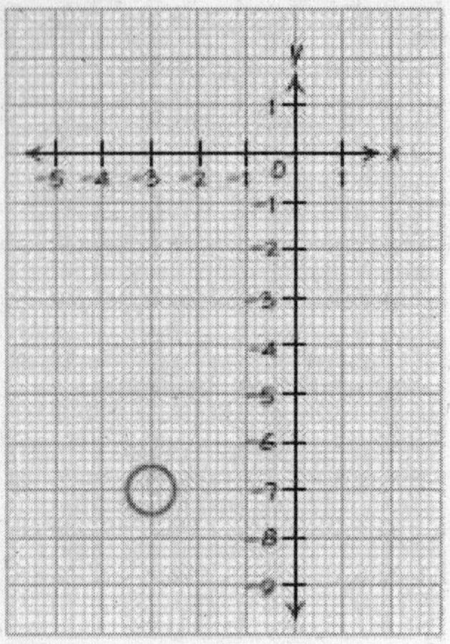

15. $x^2+y^2+12x-10y-39=0$

$x^2+12x+y^2-10y=39$

$(x^2+12x+\_\_)+(y^2-10y+\_\_)=39$

Completing the square: $\left(\frac{12}{2}\right)^2=36$ and $\left(\frac{-10}{2}\right)^2=25$

$(x^2+12x+36)+(y^2-10y+25)=39+36+25$

$(x+6)^2+(y-5)^2=100$

$\text{Center}=(-6,5)$ and $\text{Radius}=10$

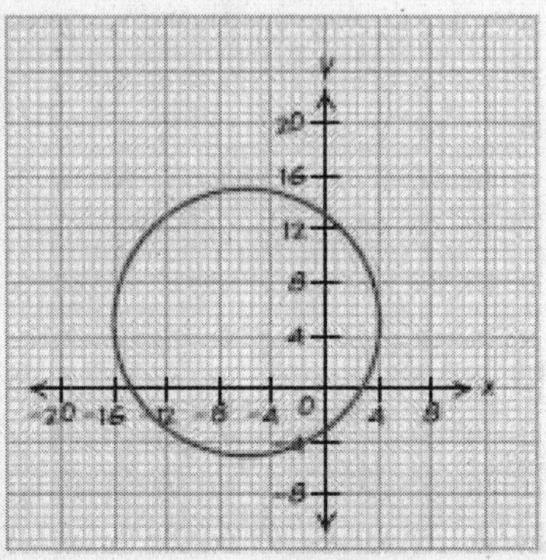

16. $3x^2+3y^2-24x+12y-15=0$

$\frac{3x^2+3y^2-24x+12y-15}{3}=\frac{0}{3}$

$x^2+y^2-8x+4y-5=0$

$x^2-8x+y^2+4y=5$

$(x^2-8x+\_\_)+(y^2+4y+\_\_)=5$

Completing the square: $\left(\frac{-8}{2}\right)^2=16$ and $\left(\frac{4}{2}\right)^2=4$

$(x^2-8x+16)+(y^2+4y+4)=5+16+4$

$(x-4)^2+(y+2)^2=25$

$\text{Center}=(4,-2)$ and $\text{Radius}=5$

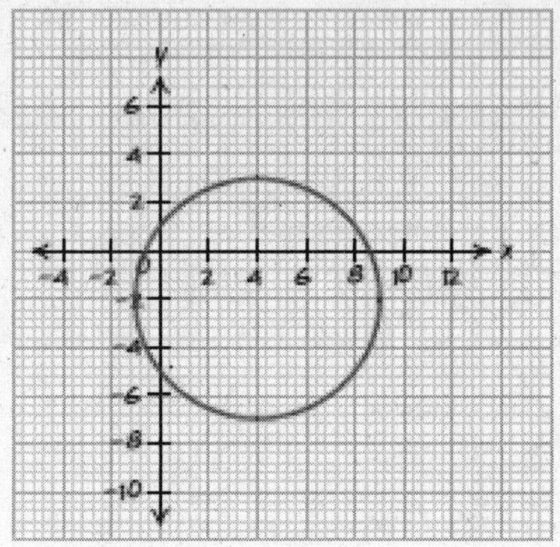

17. $\text{Center}=(2,6)$ and $\text{Radius}=4$

$h=2, k=6, r=4$

$(x-2)^2+(y-6)^2=(4)^2$

$(x-2)^2+(y-6)^2=16$

18. $\text{Center}=(-4,2)$ and $\text{Radius}=6$

$h=-4, k=2, r=6$

$(x-(-4))^2+(y-2)^2=(6)^2$

$(x+4)^2+(y-2)^2=36$

19. $\text{Focus}=(5,0)$

$p=5$

$x=\frac{y^2}{4(5)}$

$x=\frac{y^2}{20}$

20. $\text{Focus}=(0,-3)$

$p=-3$

$y=\frac{x^2}{4(-3)}$

$y=\frac{x^2}{-12}$

21. $\dfrac{x^2}{16}+\dfrac{y^2}{25}=1$

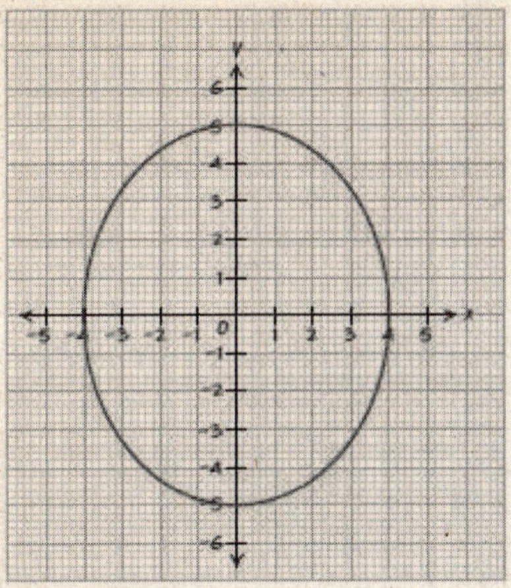

22. $\dfrac{x^2}{0.25}+y^2=1$

$\dfrac{x^2}{0.25}+\dfrac{y^2}{1}=1$

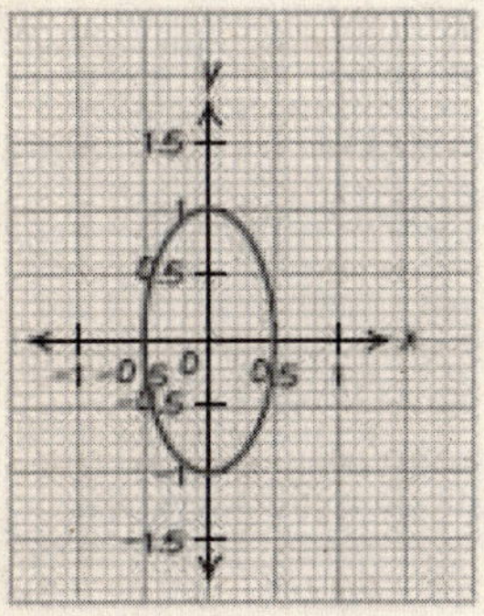

23. $100x^2+4y^2=400$

$\dfrac{100x^2}{400}+\dfrac{4y^2}{400}=\dfrac{400}{400}$

$\dfrac{x^2}{4}+\dfrac{y^2}{100}=1$

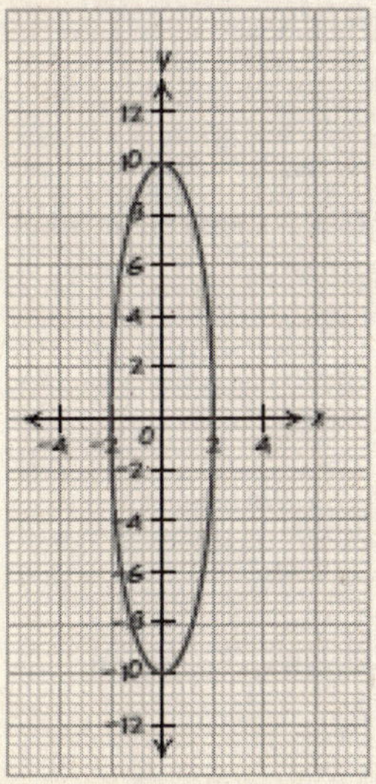

24. $10x^2+2y^2=40$

$\dfrac{10x^2}{40}+\dfrac{2y^2}{40}=\dfrac{40}{40}$

$\dfrac{x^2}{4}+\dfrac{y^2}{20}=1$

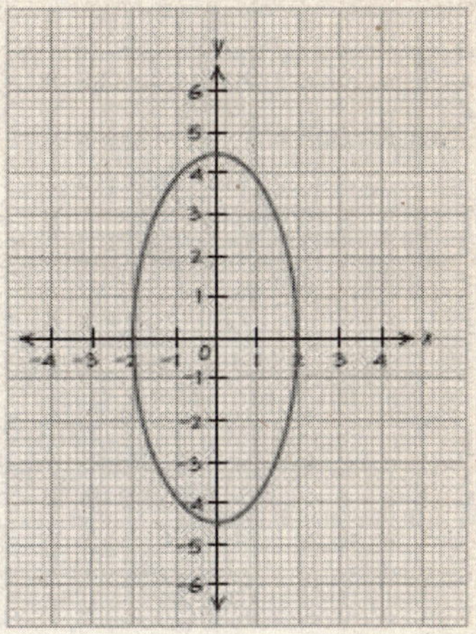

25. $\dfrac{x^2}{16}-\dfrac{y^2}{25}=1$

$\dfrac{x^2}{4^2}-\dfrac{y^2}{5^2}=1$

$a=4, b=5$

$c^2=4^2+5^2$

$c=\pm\sqrt{41}$

Vertices $=(4,0),(-4,0)$

Foci $=\left(\sqrt{41},0\right),\left(-\sqrt{41},0\right)\approx(6.4,0),(-6.4,0)$

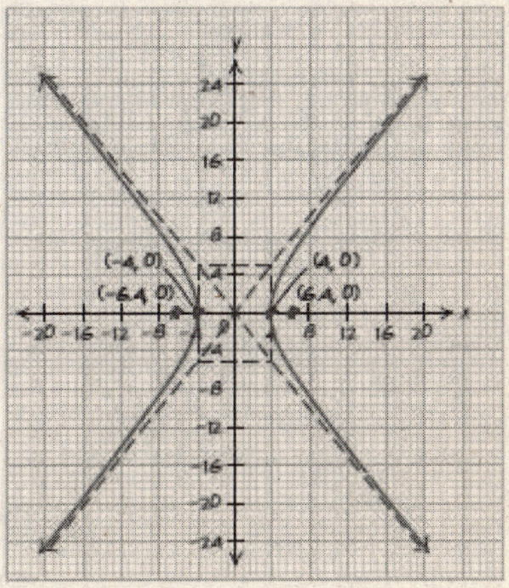

26. $\dfrac{y^2}{100}-\dfrac{x^2}{4}=1$

$\dfrac{y^2}{10^2}-\dfrac{x^2}{2^2}=1$

$a=2, b=10$

$c^2=2^2+10^2$

$c=\pm\sqrt{104}$

Vertices $=(0,10),(0,-10)$

Foci $=\left(0,\sqrt{104}\right),\left(0,-\sqrt{104}\right)\approx(0,10.2),(0,-10.2)$

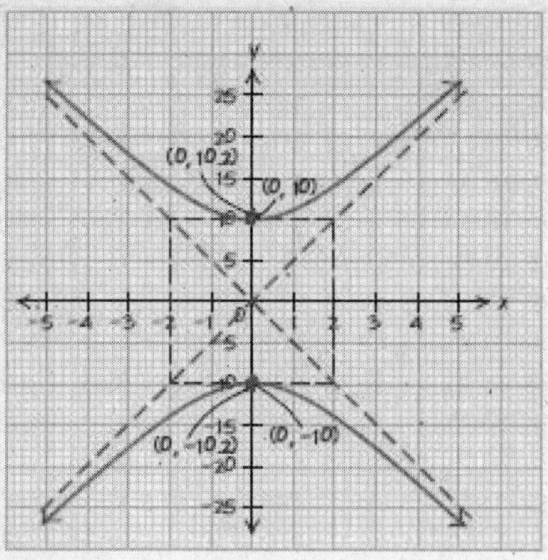

27. $x^2 - 16y^2 = 16$

$\dfrac{x^2}{16} - \dfrac{16y^2}{16} = \dfrac{16}{16}$

$\dfrac{x^2}{16} - \dfrac{y^2}{1} = 1$

$\dfrac{x^2}{4^2} - \dfrac{y^2}{1^2} = 1$

$a = 4, b = 1$

$c^2 = 4^2 + 1^2$

$c = \pm\sqrt{17}$

Vertices $= (4,0), (-4,0)$

Foci $= (\sqrt{17}, 0), (-\sqrt{17}, 0) \approx (4.1, 0), (-4.1, 0)$

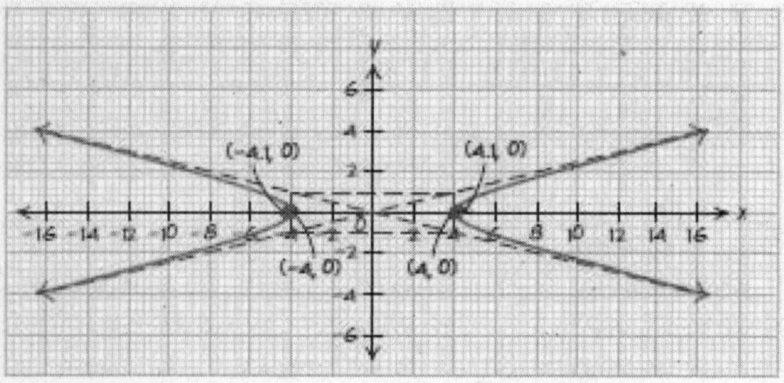

28.

$9y^2 - 25x^2 = 225$

$\dfrac{9y^2}{225} - \dfrac{25x^2}{225} = \dfrac{225}{225}$

$\dfrac{y^2}{25} - \dfrac{x^2}{9} = 1$

$\dfrac{y^2}{5^2} - \dfrac{x^2}{3^2} = 1$

$a = 3, b = 5$

$c^2 = 3^2 + 5^2$

$c = \pm\sqrt{34}$

Vertices $= (0,5), (0,-5)$

Foci $= (0, \sqrt{34}), (0, -\sqrt{34}) \approx (0, 5.8), (0, -5.8)$

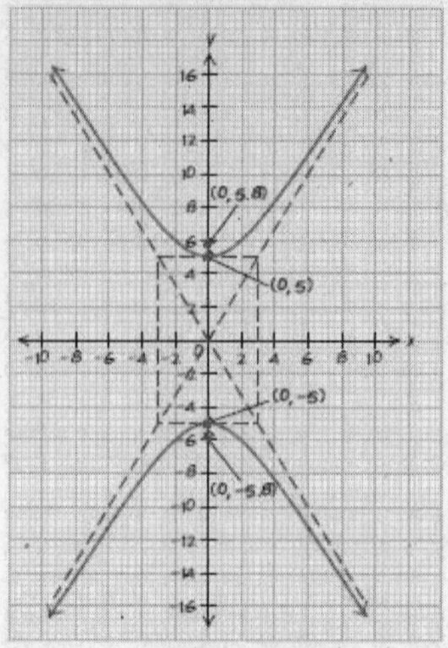

29. This is an ellipse.

Vertices $= (10,0), (-10,0), (0,15), (0,-15)$

$a = 10, b = 15$

$\dfrac{x^2}{10^2} + \dfrac{y^2}{15^2} = 1$

$\dfrac{x^2}{100} + \dfrac{y^2}{225} = 1$

30. This is an ellipse.

Vertices $= (20,0), (-20,0), (0,5), (0,-5)$

$a = 20, b = 5$

$\dfrac{x^2}{20^2} + \dfrac{y^2}{5^2} = 1$

$\dfrac{x^2}{400} + \dfrac{y^2}{25} = 1$

31. This is a hyperbola that opens up and down.

$a = 15, b = 10$

$\dfrac{y^2}{10^2} - \dfrac{x^2}{15^2} = 1$

$\dfrac{y^2}{100} - \dfrac{x^2}{225} = 1$

32. This is a hyperbola that opens left and right.

$a = 10, b = 8$

$\dfrac{x^2}{10^2} - \dfrac{y^2}{8^2} = 1$

$\dfrac{x^2}{100} - \dfrac{y^2}{64} = 1$

33. $4x^2 + 36y^2 = 36$ A = 4 and C = 36

Since A ≠ C, A > 0 and C > 0 then this is an ellipse.

$\dfrac{4x^2}{36} + \dfrac{36y^2}{36} = \dfrac{36}{36}$

$\dfrac{x^2}{9} + \dfrac{y^2}{1} = 1$

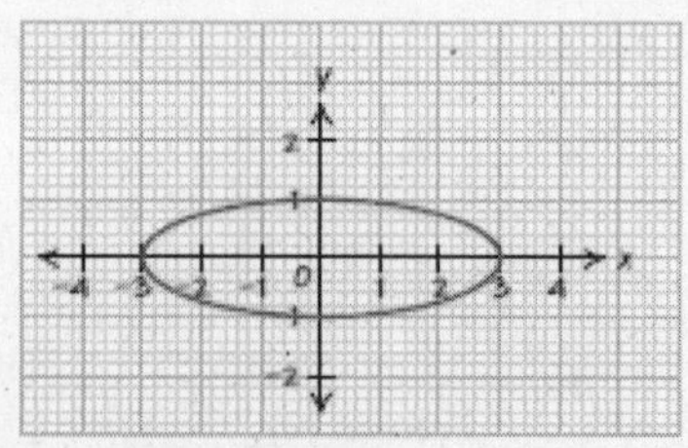

34. $(x-9)^2+(y-6)^2=9$

This is a circle in standard form.

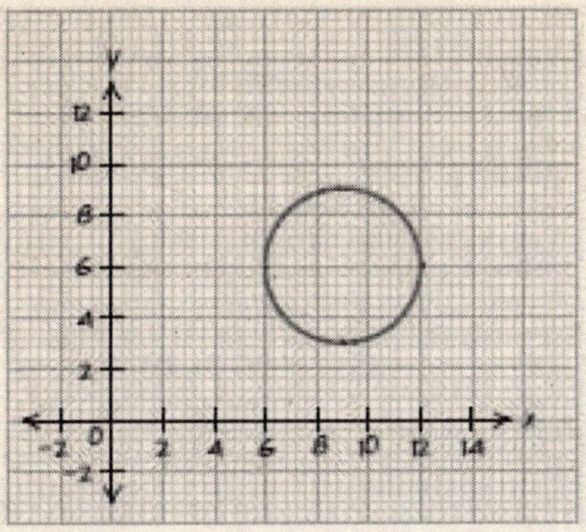

35. $2y^2-50x^2=50 \quad A=-50$ and $C=2$

Since $A \neq C$, $A<0$ or $C<0$ then this is a hyperbola.

$$\frac{2y^2}{50}-\frac{50x^2}{50}=\frac{50}{50}$$

$$\frac{y^2}{25}-\frac{x^2}{1}=1$$

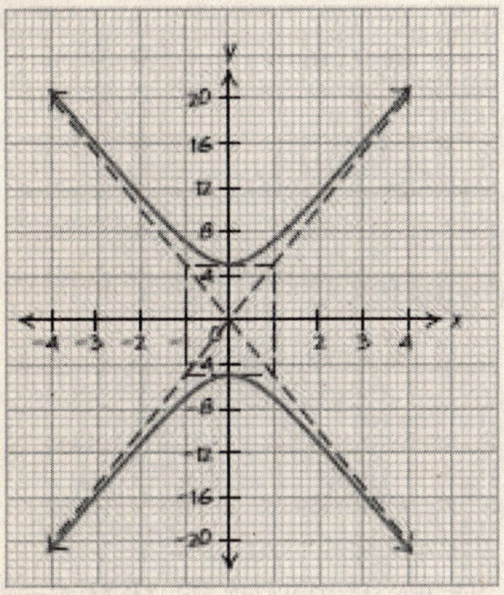

36. $-2y^2+3x+6y-21=0 \quad A=0$ and $C=-2$

Since $A=0$ or $C=0$ then this is a parabola.

$$-2y^2+3x+6y-21=0$$

$$3x=2y^2-6y+21$$

$$\frac{3x}{3}=\frac{2y^2-6y+21}{3}$$

$$x=\frac{2}{3}y^2-2y+7$$

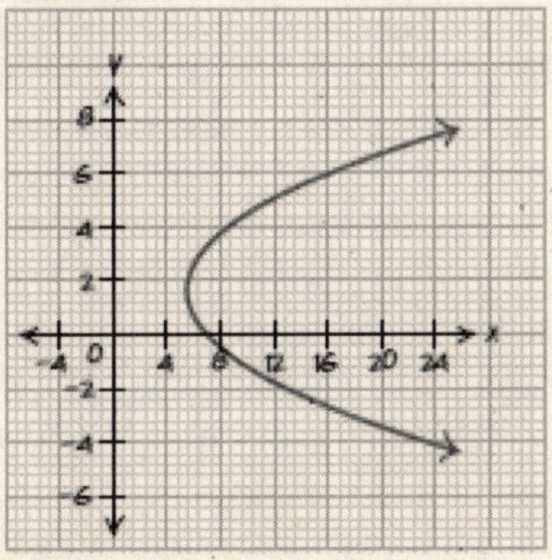

37. $4x^2+4y^2+24x=0 \quad A=4$ and $C=4$

Since $A=C$ then this is a circle.

$$\frac{4x^2+4y^2+24x}{4}=\frac{0}{4}$$

$$x^2+y^2+6x=0$$

$$\left(x^2+6x+\_\_\right)+y^2=0$$

Completing the square: $\left(\frac{6}{2}\right)^2=9$

$$\left(x^2+6x+9\right)+y^2=0+9$$

$$(x+3)^2+(y-0)^2=9$$

Center $=(-3,0)$ and Radius $=3$

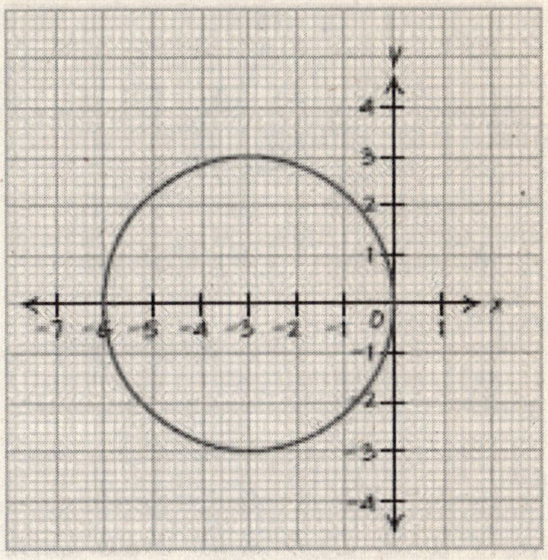

38. Hyperbola

$$9x^2-36y^2=324$$

$$\frac{9x^2}{324}-\frac{36y^2}{324}=\frac{324}{324}$$

$$\frac{x^2}{36}-\frac{y^2}{9}=1$$

$a^2=36 \rightarrow a=6$

$b^2=9 \rightarrow b=3$

$c=\pm\sqrt{36+9}=\pm\sqrt{45}\approx\pm 6.7$

Foci: $(-6.7,0)$ and $(6.7,0)$

Vertices: $(-6,0)$ and $(6,0)$

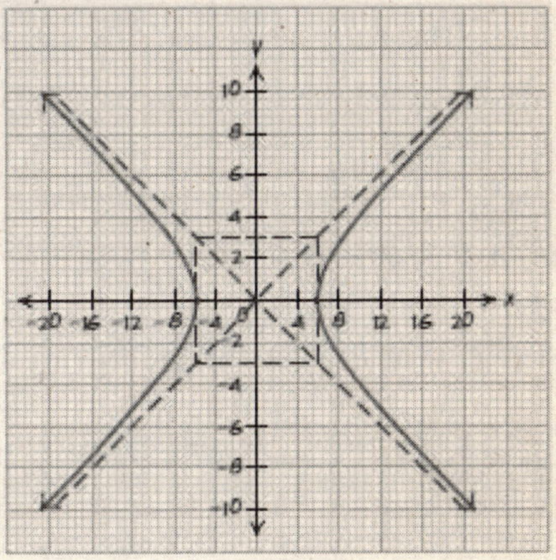

39. Parabola

$$0.5x^2+4x-y=6$$

$$0.5x^2+4x-6=y$$

$$\left.\begin{aligned} h &= \frac{-4}{2(0.5)} = -4 \\ k &= 0.5(4)^2 + 4(-4) - 6 = -14 \end{aligned}\right\}$$

vertex $(-4, -14)$

$$0.5(x+4)^2 = y+14$$
$$\frac{0.5(x+4)^2}{0.5} = \frac{y+14}{0.5}$$
$$(x+4)^2 = 2(y+14)$$
$$4p = 2$$
$$p = 0.5$$

Focus: $(-4, -14+0.5) \to (-4, -13.5)$

Directrix: $y = -14 - 0.5 = -14.5$

x – intercepts: $x = \dfrac{-4 \pm \sqrt{(4)^2 - 4(0.5)(-6)}}{2(0.5)}$

$x \approx -9.3$ and $x \approx 1.3$

$(-9.3, 0)$ and $(1.3, 0)$

y – intercept: $(0, -6)$

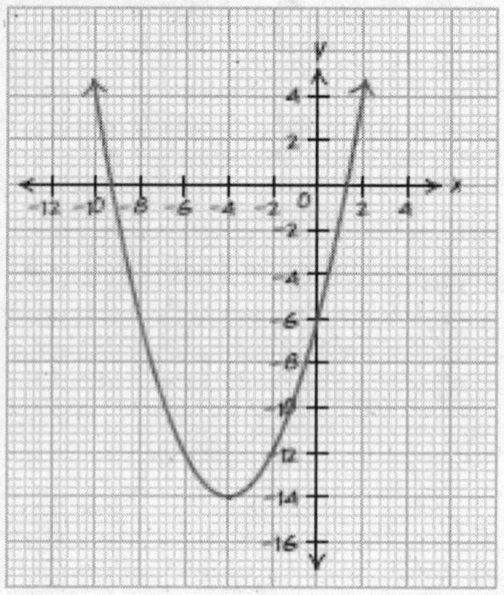

40. Ellipse

$$x^2 + 49y^2 = 49$$
$$\frac{x^2}{49} + \frac{49y^2}{49} = \frac{49}{49}$$
$$\frac{x^2}{49} + y^2 = 1$$

$a^2 = 49 \to a = 7$

$b^2 = 1 \to b = 1$

$c = \pm\sqrt{49-1} = \pm\sqrt{48} \approx \pm 6.9$

Foci: $(-6.9, 0)$ and $(6.9, 0)$

Vertices: $(-7, 0)$ and $(7, 0)$

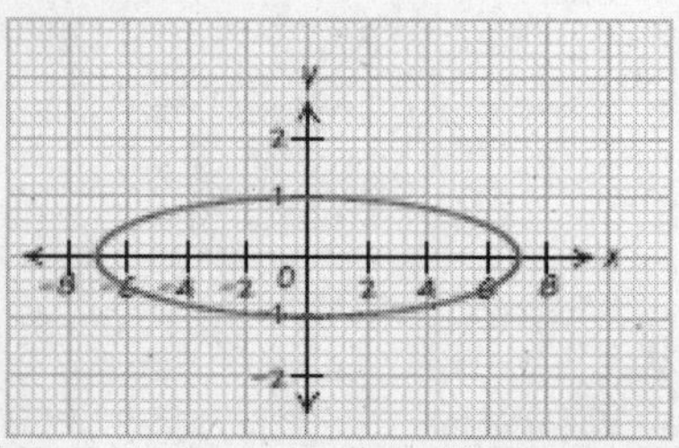

41. $a_n = 7n - 20$

$a_1 = 7(1) - 20 = -13$

$a_2 = 7(2) - 20 = -6$

$a_3 = 7(3) - 20 = 1$

$a_4 = 7(4) - 20 = 8$

$a_5 = 7(5) - 20 = 15$

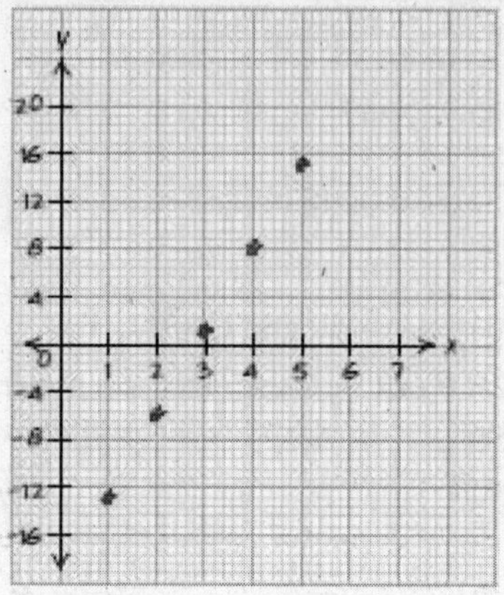

42. $a_n = 4(1.5)^n$

$a_1 = 4(1.5)^1 = 6$

$a_2 = 4(1.5)^2 = 9$

$a_3 = 4(1.5)^3 = 13.5$

$a_4 = 4(1.5)^4 = 20.25$

$a_5 = 4(1.5)^5 = 30.375$

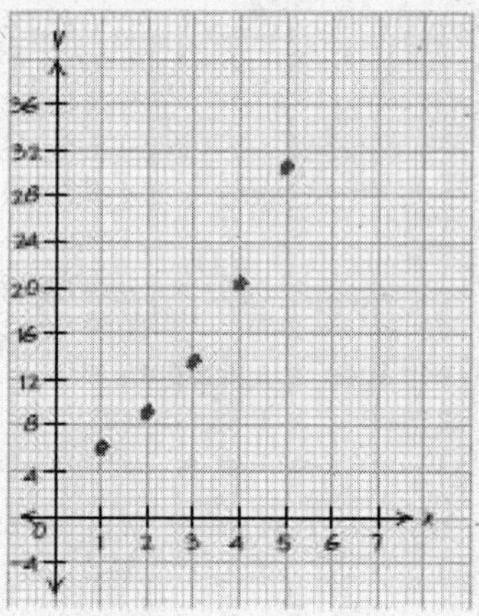

43. $a_n = \dfrac{(-1)^n}{8n}$

$a_1 = \dfrac{(-1)^1}{8(1)} = -\dfrac{1}{8}$

$a_2 = \dfrac{(-1)^2}{8(2)} = \dfrac{1}{16}$

$a_3 = \dfrac{(-1)^3}{8(3)} = -\dfrac{1}{24}$

$a_4 = \dfrac{(-1)^4}{8(4)} = \dfrac{1}{32}$

$$a_5 = \frac{(-1)^5}{8(5)} = -\frac{1}{40}$$

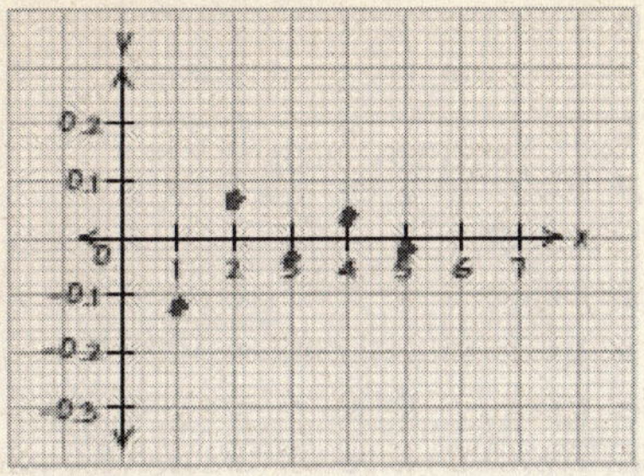

44. $a_n = \dfrac{2n}{n+8}$

$$a_1 = \frac{2(1)}{1+8} = \frac{2}{9} \approx 0.22$$
$$a_2 = \frac{2(2)}{2+8} = \frac{4}{10} = 0.4$$
$$a_3 = \frac{2(3)}{3+8} = \frac{6}{11} \approx 0.55$$
$$a_4 = \frac{2(4)}{4+8} = \frac{8}{12} \approx 0.67$$
$$a_5 = \frac{2(5)}{5+8} = \frac{10}{13} \approx 0.77$$

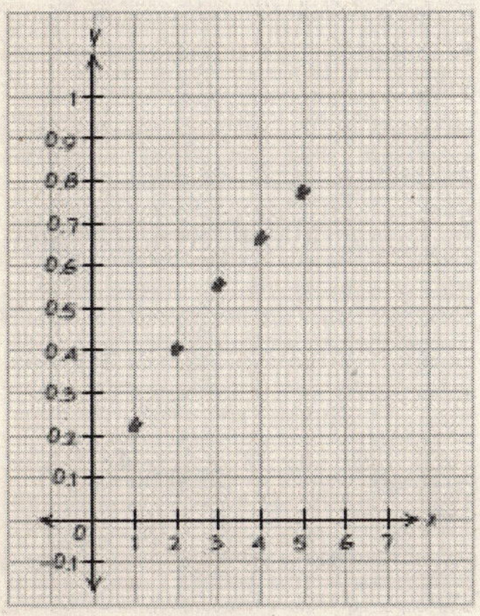

45. $a_n = 50 + 2(n-1)$

$$78 = 50 + 2(n-1)$$
$$28 = 2(n-1)$$
$$\frac{28}{2} = \frac{2(n-1)}{2}$$
$$14 = n - 1$$
$$n = 15$$

46. $a_n = 4n^2 + 3n$

$$351 = 4n^2 + 3n$$
$$0 = 4n^2 + 3n - 351$$
$$n = \frac{-3 \pm \sqrt{3^2 - 4(4)(-351)}}{2(4)}$$

$n = 9$ and ~~$n = -9.75$~~

47.

$$a_n = 0.25(2)^n$$
$$1{,}048{,}576 = 0.25(2)^n$$
$$\frac{1{,}048{,}576}{0.25} = \frac{0.25(2)^n}{0.25}$$
$$4{,}194{,}304 = 2^n$$
$$n = \log_2(4{,}194{,}304)$$
$$n = \frac{\log(4{,}194{,}304)}{\log 2}$$
$$n \approx 22$$

48. $a_n = \sqrt{6n+4} + 48$

$$68 = \sqrt{6n+4} + 48$$
$$20 = \sqrt{6n+4}$$
$$(20)^2 = \left(\sqrt{6n+4}\right)^2$$
$$400 = 6n + 4$$
$$396 = 6n$$
$$\frac{396}{6} = \frac{6n}{6}$$
$$n = 66$$

49. $a_2 - a_1 = 79 - 82 = -3$

$$a_3 - a_2 = 76 - 79 = -3$$
$$a_4 - a_3 = 73 - 76 = -3$$
$$a_5 - a_4 = 70 - 73 = -3$$
$$a_6 - a_5 = 67 - 70 = -3$$

The sequence is arithmetic. The common difference is -3.

$$a_n = 82 + (n-1)(-3)$$
$$= 82 - 3n + 3$$
$$a_n = 85 - 3n$$

50. $a_2 - a_1 = -1.5 - (-4) = 2.5$

$$a_3 - a_2 = 1 - (-1.5) = 2.5$$
$$a_4 - a_3 = 3.5 - 1 = 2.5$$
$$a_5 - a_4 = 6 - 3.5 = 2.5$$
$$a_6 - a_5 = 8.5 - 6 = 2.5$$

The sequence is arithmetic. The common difference is 2.5.

$$a_n = -4 + (n-1)2.5$$
$$= -4 + 2.5n - 2.5$$
$$a_n = -6.5 + 2.5n$$

51. $a_2 - a_1 = 72 - 48 = 24$

$$a_3 - a_2 = 108 - 72 = 36$$
$$24 \neq 36$$

The differences between consecutive terms are not the same, so this is not an arithmetic sequence.

52. $a_2 - a_1 = \dfrac{7}{2} - 6 = -\dfrac{5}{2}$

$$a_3 - a_2 = 1 - \frac{7}{2} = -\frac{5}{2}$$

$a_4 - a_3 = -\frac{3}{2} - 1 = -\frac{5}{2}$

$a_5 - a_4 = -4 + \frac{3}{2} = -\frac{5}{2}$

$a_6 - a_5 = -\frac{13}{2} + 4 = -\frac{5}{2}$

The sequence is arithmetic. The common difference is $-\frac{5}{2}$.

$$a_n = 6 + (n-1)\left(-\frac{5}{2}\right)$$
$$= 6 - \frac{5}{2}n + \frac{5}{2}$$
$$a_n = \frac{17}{2} - \frac{5}{2}n$$

53. **a.** $a_1 = 4000, a_2 = 4200, a_3 = 4400, \ldots$

$$a_n = 4000 + (n-1)200$$
$$a_n = 4000 + 200n - 200$$
$$a_n = 3800 + 200n$$

b. $a_{15} = 3800 + 200(15)$

$$a_{15} = 6800$$

In 2015, the attendance on Easter Sunday at North Coast Church will be 6800.

c. $3800 + 200n = 6000$

$$\underline{-3800 \qquad -3800}$$
$$200n = 2200$$
$$\frac{200n}{200} = \frac{2200}{200}$$
$$n = 11$$

The attendance on Easter Sunday at North Coast Church will be 6000 in the year 2011.

54. **a.** $a_1 = 27{,}000$

$$d = -2500$$
$$a_n = 27{,}000 + (n-1)(-2500)$$
$$a_n = 27{,}000 - 2500n + 2500$$
$$a_n = 29{,}500 - 2500n$$

b. $a_6 = 29{,}500 - 2500(6)$

$$a_6 = 14{,}500$$

The car will be worth $14,500 in the sixth year.

c. $29{,}500 - 2500n = 2000$

$$\underline{-29{,}500 \qquad -29{,}500}$$
$$-2500n = -27{,}500$$
$$\frac{-2500n}{-2500} = \frac{-27{,}500}{-2500}$$
$$n = 11$$

The car will be 11 years old when it is worth $2000.

55.

$$\left.\begin{aligned} a_5 &= 66 \\ a_{17} &= 144 \end{aligned}\right\} d = \frac{144 - 66}{17 - 5} = \frac{78}{12} = 6.5$$
$$y - 66 = 6.5(x - 5)$$
$$y - 66 = 6.5x - 32.5$$
$$\underline{+66 \qquad +66}$$
$$y = 6.5x + 33.5$$
$$a_n = 6.5n + 33.5$$

56.

$$\left.\begin{aligned} a_4 &= -12 \\ a_{16} &= -108 \end{aligned}\right\} d = \frac{-108 + 12}{16 - 4} = \frac{-96}{12} = -8$$
$$y + 12 = -8(x - 4)$$
$$y + 12 = -8x + 32$$
$$\underline{-12 \qquad -12}$$
$$y = -8x + 20$$
$$a_n = -8n + 20$$

57.

$$\frac{a_2}{a_1} = \frac{12}{3} = 4, \frac{a_3}{a_2} = \frac{48}{12} = 4, \frac{a_4}{a_3} = \frac{192}{48} = 4,$$
$$\frac{a_5}{a_4} = \frac{768}{192} = 4, \frac{a_6}{a_5} = \frac{3072}{768} = 4$$

This sequence is geometric with common ratio 4.

$$a_n = 3(4)^{n-1}$$

58.

$$\frac{a_2}{a_1} = \frac{12}{36} = \frac{1}{3}, \frac{a_3}{a_2} = \frac{4}{12} = \frac{1}{3},$$
$$\frac{a_4}{a_3} = \frac{4/3}{4} = \frac{1}{3}, \frac{a_5}{a_4} = \frac{4/9}{4/3} = \frac{1}{3},$$
$$\frac{a_6}{a_5} = \frac{4/27}{4/9} = \frac{1}{3}$$

This sequence is geometric with common ratio $\frac{1}{3}$.

$$a_n = 36\left(\frac{1}{3}\right)^{n-1}$$

59.

$$\frac{a_2}{a_1} = \frac{-1}{0.5} = -2, \frac{a_3}{a_2} = \frac{2}{-1} = -2, \frac{a_4}{a_3} = \frac{-4}{2} = -2,$$
$$\frac{a_5}{a_4} = \frac{8}{-4} = -2, \frac{a_6}{a_5} = \frac{-16}{8} = -2$$

This sequence is geometric with common ratio -2.

$$a_n = 0.5(-2)^{n-1}$$

60.

$$\frac{a_2}{a_1} = \frac{34}{6} = \frac{17}{3}, \frac{a_3}{a_2} = \frac{64}{34} = \frac{32}{17},$$

This sequence is not geometric.

61. **a.** $a_1 = 4000$

$$r = 1 + 0.02 = 1.02$$
$$a_n = 4000(1.02)^n$$

b. $a_8 = 4000(1.02)^8$

$$a_8 = 4686.64$$

After eight years, the product will cost $4686.64.

62. **a.** $a_1 = 42{,}000$

$r = 1 + 0.0125 = 1.0125$

$a_n = 42{,}000(1.0125)^{n-1}$

b. $a_5 = 42{,}000(1.0125)^4$

$a_5 = 44{,}139.70$

In the fifth year of employment, your salary will be \$44,139.70.

c. $50{,}603 = 42{,}000(1.0125)^{n-1}$

$$\frac{50{,}603}{42{,}000} = \frac{42{,}000(1.0125)^{n-1}}{42{,}000}$$

$$\frac{50{,}603}{42{,}000} = (1.0125)^{n-1}$$

$$\log_{1.0125}\left(\frac{50{,}603}{42{,}000}\right) = n-1$$

$$\frac{\log\left(\frac{50{,}603}{42{,}000}\right)}{\log(1.0125)} = n-1$$

$$n-1 \approx 15$$

$$n \approx 16$$

In the 16th year of employment, your salary will be approximately \$50,603.

63.

$$\left.\begin{array}{l} a_3 = 72 \\ a_6 = 243 \end{array}\right\} \rightarrow \begin{array}{l} 72 = a_1 r^{3-1} \\ 243 = a_1 r^{6-1} \end{array}$$

Solve for r:

$$\frac{243}{72} = \frac{\not{a_1} r^5}{\not{a_1} r^2} \rightarrow 3.375 = r^3$$

$$(3.375)^{1/3} = (r^3)^{1/3}$$

$$1.5 = r$$

Solve for a_1:

$$72 = a_1(1.5)^2$$

$$\frac{72}{(1.5)^2} = \frac{a_1(1.5)^2}{(1.5)^2}$$

$$a_1 = 32$$

$$a_n = 32(1.5)^{n-1}$$

64.

$$\left.\begin{array}{l} a_2 = 27 \\ a_5 = 1 \end{array}\right\} \rightarrow \begin{array}{l} 27 = a_1 r^{2-1} \\ 1 = a_1 r^{5-1} \end{array}$$

Solve for r:

$$\frac{1}{27} = \frac{\not{a_1} r^4}{\not{a_1} r^1} \rightarrow \frac{1}{27} = r^3$$

$$\left(\frac{1}{27}\right)^{1/3} = (r^3)^{1/3}$$

$$\frac{1}{3} = r$$

Solve for a_1:

$$27 = a_1\left(\frac{1}{3}\right)$$

$$(3)(27) = a_1\left(\frac{1}{3}\right)(3)$$

$$a_1 = 81$$

$$a_n = 81\left(\frac{1}{3}\right)^{n-1}$$

65.

$$\sum_{n=1}^{7}(4n-20) = (4(1)-20)+(4(2)-20)+(4(3)-20)+(4(4)-20)+(4(5)-20)+(4(6)-20)+(4(7)-20) = -28$$

66.

$$\sum_{n=1}^{6}\left(5(-2)^n\right) = \left(5(-2)\right)+\left(5(-2)^2\right)+\left(5(-2)^3\right)+\left(5(-2)^4\right)+\left(5(-2)^5\right)+\left(5(-2)^6\right) = 210$$

67. **a.** $a_1 = 38{,}000$, $d = 900$

$a_n = 38{,}000 + (n-1)900$

$a_n = 38{,}000 + 900n - 900$

$a_n = 37{,}100 + 900n$

b. $a_{15} = 37{,}100 + 900(15)$

$a_{15} = 50{,}600$

Your salary will be \$50,600 in the 15th year of employment.

c. $\sum_{n=1}^{15}(37{,}100 + 900n)$

$$S_{15} = \frac{15(38{,}000 + 50{,}600)}{2}$$

$$S_{15} = \frac{15(88{,}600)}{2}$$

$$S_{15} = 664{,}500$$

You will have earned a total of \$664,500 over a 15 year period of employment.

68. **a.** $a_1 = 5$, $d = 3$

$a_n = 5 + (n-1)3$

$a_n = 5 + 3n - 3$

$a_n = 2 + 3n$

b. $a_5 = 2 + 3(5) = 17$

On the 5th day you will pick 17 apricots.

c. $a_{14} = 2 + 3(14) = 44$

$$\sum_{n=1}^{14}(2+3n)$$

$$S_{14} = \frac{14(5+44)}{2}$$

$$S_{14} = 343$$

You will pick a total of 343 ripe apricots during the first two weeks.

69. $S_{15}=\dfrac{42,000\left(1-1.025^{15}\right)}{1-1.025}$

$S_{15}=688,226.05$

During a 15 year career, you will have earned a total of \$688,226.05.

70. **a.** $a_1=10,000$

$r=1+0.2=1.2$

$a_n=10,000(1.2)^{n-1}$

b. $S_7=\dfrac{10,000\left(1-1.2^7\right)}{1-1.2}$

$S_7=129,159$

During the seven day event, 129,159 people attended the car show.

71. $a_n=4n+7$

$a_1=11$, $a_{14}=4(14)+7=63$

$S_{14}=\dfrac{14(11+63)}{2}$

$S_{14}=518$

72. $a_n=300-2.5(n-1)$

$a_1=300$

$a_{23}=300-2.5(22)=245$

$S_{23}=\dfrac{23(300+245)}{2}$

$S_{23}=6267.5$

73. $a_n=7(1.8)^n$

$a_1=7(1.8)^1=12.6$

$S_{11}=\dfrac{12.6\left(1-1.8^{11}\right)}{1-1.8}$

$S_{11}=10,106.52459$

74. $a_n=48\left(-\dfrac{1}{2}\right)^n$

$a_1=48(-0.5)^1=-24$

$S_{20}=\dfrac{-24\left(1+0.5^{20}\right)}{1+0.5}$

$S_{20}=-15.99998$

Chapter 14 Test

1. Center$=(-7,4)$ and Radius$=8$

$h=-7, k=4, r=8$

$(x-(-7))^2+(y-4)^2=(8)^2$

$(x+7)^2+(y-4)^2=64$

2. $x=\dfrac{y^2}{36}$

$x=\dfrac{y^2}{4(9)}$

$p=9$

Focus$=(9,0)$, Directrix : $x=-9$

3. $\dfrac{x^2}{100}-\dfrac{y^2}{36}=1$

This is a hyperbola in standard form.

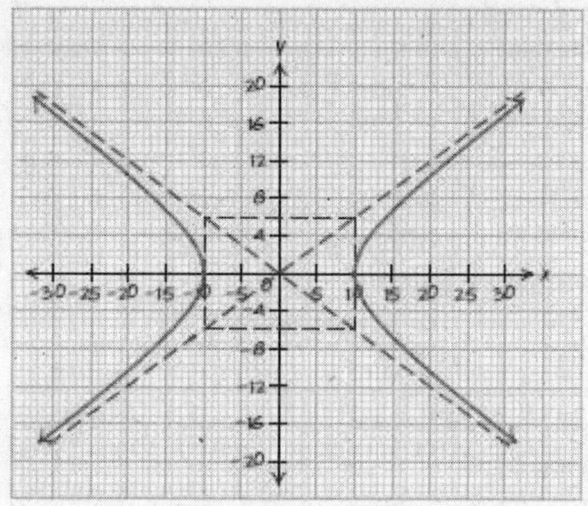

4. $y^2+8+x-20=0$ A = 0 and C = 1

Since A = 0 or C = 0 then this is a parabola.

$y^2+8+x-20=0$

$x=-y^2+12$

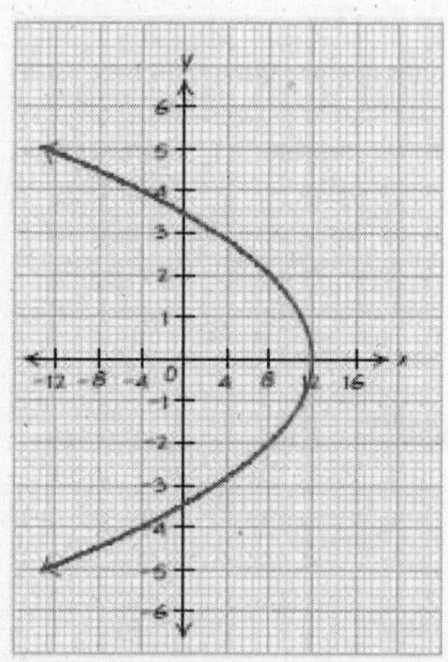

5. $49x^2+4y^2=196$ A = 49 and C = 4

Since A ≠ C, A > 0 and C > 0 then this is an ellipse.

$\dfrac{49x^2}{196}+\dfrac{4y^2}{196}=\dfrac{196}{196}$

$\dfrac{x^2}{4}+\dfrac{y^2}{49}=1$

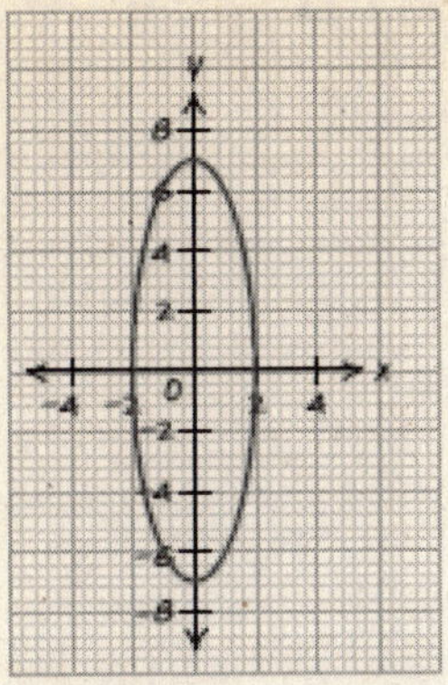

6. $(x-3)^2+(y-7)^2=81$

This is a circle in standard form.

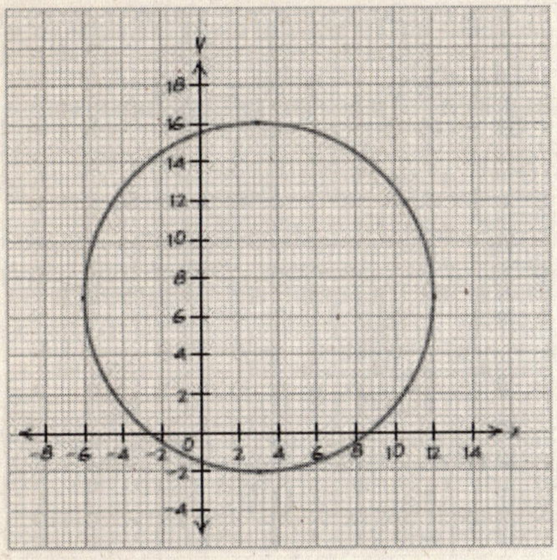

7. $\text{Focus}=(-2,0)$

$p=-2$

$x=\dfrac{y^2}{4(-2)}$

$x=\dfrac{y^2}{-8}$

8. $\text{Vertices}=(2,0),(-2,0),(0,6),(0,-6)$

$a=2, b=6$

$\dfrac{x^2}{2^2}+\dfrac{y^2}{6^2}=1$

$\dfrac{x^2}{4}+\dfrac{y^2}{36}=1$

9. An arithmetic sequence has a common difference therefore the difference between successive terms is constant. A geometric sequence has a common ratio therefore the ratio between successive terms is constant.

10. $a_n=4(-3)^n$

$a_1=4(-3)^1=-12$

$a_2=4(-3)^2=36$

$a_3=4(-3)^3=-108$

$a_4=4(-3)^4=324$

$a_5=4(-3)^5=-972$

The first five terms are $-12,36,-108,324,-972$.

11. $a_n=\dfrac{6n}{n+2}$

$a_1=\dfrac{6(1)}{1+2}=\dfrac{6}{3}=2$

$a_2=\dfrac{6(2)}{2+2}=\dfrac{12}{4}=3$

$a_3=\dfrac{6(3)}{3+2}=\dfrac{18}{5}$

$a_4=\dfrac{6(4)}{4+2}=\dfrac{24}{6}=4$

$a_5=\dfrac{6(5)}{5+2}=\dfrac{30}{7}$

The first five terms are $2,3,\dfrac{18}{5},4,\dfrac{30}{7}$

12. $a_n=400-11(n-1)=158$

$400-11n+11=158$

$411-11n=158$

$\underline{-411 \qquad -411}$

$-11n=-253$

$\dfrac{-11n}{-11}=\dfrac{-253}{-11}$

$n=23$

13. $a_n=20+\sqrt{n+8}=38$

$20+\sqrt{n+8}=38$

$\underline{-20 \qquad -20}$

$\sqrt{n+8}=18$

$\left(\sqrt{n+8}\right)^2=18^2$

$n+8=324$

$n=316$

14.

$\dfrac{a_2}{a_1}=\dfrac{50}{250}=\dfrac{1}{5}, \dfrac{a_3}{a_2}=\dfrac{10}{50}=\dfrac{1}{5}, \dfrac{a_4}{a_3}=\dfrac{2}{10}=\dfrac{1}{5},$

$\dfrac{a_5}{a_4}=\dfrac{0.4}{2}=\dfrac{1}{5}, \dfrac{a_6}{a_5}=\dfrac{0.08}{0.4}=\dfrac{1}{5}$

This sequence is geometric with common ratio $\dfrac{1}{5}$.

$a_n=250\left(\dfrac{1}{5}\right)^{n-1}$

15. $a_2-a_1=16.5-24=-7.5$

$a_3-a_2=9-16.5=-7.5$

$a_4-a_3=1.5-9=-7.5$

$a_5-a_4=-6-1.5=-7.5$

$a_6-a_5=-13.5-(-6)=-7.5$

This sequence is arithmetic with common difference $d=-7.5$.

$a_1=24$

$d=-7.5$

$a_n=24+(n-1)(-7.5)$

$a_n=24-7.5n+7.5$

$a_n=31.5-7.5n$

16. **a.** $a_1 = 5$

$d = 3$

$a_n = 5 + (n-1)3$

$a_n = 5 + 3n - 3$

$a_n = 2 + 3n$

b. $a_3 = 2 + 3(3)$

$a_3 = 11$

By week three, you should be stretching your injured knee for 11 minutes at one time.

c. $a_n = 2 + 3n$

$$\begin{aligned} 2 + 3n &= 30 \\ -2 \quad & \;\; -2 \\ \hline 3n &= 28 \\ n = \frac{28}{3} &= 9\frac{1}{3} \end{aligned}$$

In the tenth week, you will be stretching your injured knee for half an hour at one time.

17. **a.** $a_1 = 60{,}000$

$r = 1 + 0.025 = 1.025$

$a_n = 60{,}000(1.025)^{n-1}$

b. $a_{20} = 60{,}000(1.025)^{20-1} = 95{,}919.01$

Your salary will be \$95,919.01 in the 20th year.

c. $S_{20} = \dfrac{60{,}000(1 - 1.025^{20})}{1 - 1.025} = 1{,}532{,}679.46$

You will have earned a total of \$1,532,679.46 in salary over a 20 year career.

18.

$$\sum_{n=1}^{6}(2n^2) = 2 + 2(2)^2 + 2(3)^2 + 2(4)^2 + 2(5)^2 + 2(6)^2 = 182$$

19. $a_n = 600 - 18(n-1)$

$a_1 = 600$ and $d = -18$

$a_{15} = 600 - 18(14) = 348$

$S_{15} = \dfrac{15(600 + 348)}{2}$

$S_{15} = 7110$

20. $a_n = 20(1.75)^n$

$a_1 = 20(1.75)^1 = 35$

$r = 1.75$

$S_{12} = \dfrac{35(1 - 1.75^{12})}{1 - 1.75}$

$S_{12} = 38{,}453.56699$

Cumulative Review Chapters 1 – 14

1.
$$\begin{aligned} 7a^3 + 18 &= 1530 \\ 7a^3 &= 1512 \\ a^3 &= 216 \\ (a^3)^{\frac{1}{3}} &= 216^{\frac{1}{3}} \\ a &= 6 \end{aligned}$$

2.
$$\begin{aligned} -2(t+3)^2 - 8 &= -80 \\ -2(t+3)^2 &= -72 \\ (t+3)^2 &= 36 \\ \sqrt{(t+3)^2} &= \pm\sqrt{36} \\ t + 3 &= \pm 6 \\ t + 3 = 6, t + 3 &= -6 \\ t = 3, t &= -9 \end{aligned}$$

3.
$$\begin{aligned} 6x + 5 &= 2(4x - 9) \\ 6x + 5 &= 8x - 18 \\ -2x + 5 &= -18 \\ -2x &= -23 \\ x &= \frac{23}{2} \end{aligned}$$

4.
$$\begin{aligned} \frac{2}{m-3} &= \frac{5}{m+8} \\ \frac{2}{\cancel{m-3}}\cancel{(m-3)}(m+8) &= \frac{5}{\cancel{m+8}}(m-3)\cancel{(m+8)} \\ 2(m+8) &= 5(m-3) \\ 2m + 16 &= 5m - 15 \\ -3m + 16 &= -15 \\ -3m &= -31 \\ m &= \frac{31}{3} \end{aligned}$$

5.
$$\begin{aligned} -2\sqrt{3h+7} + 10 &= -4 \\ -2\sqrt{3h+7} &= -14 \\ \sqrt{3h+7} &= 7 \\ \left(\sqrt{3h+7}\right)^2 &= 7^2 \\ 3h + 7 &= 49 \\ 3h &= 42 \\ h &= 14 \end{aligned}$$

6.
$$\begin{aligned} 40\left(\frac{4}{5}\right)^n + 300 &= 320.48 \\ 40\left(\frac{4}{5}\right)^n &= 20.48 \\ \left(\frac{4}{5}\right)^n &= 0.512 \\ \ln\left(\frac{4}{5}\right)^n &= \ln(0.512) \\ n\ln\left(\frac{4}{5}\right) &= \ln(0.512) \end{aligned}$$

$$n = \frac{\ln(0.512)}{\ln\left(\frac{4}{5}\right)}$$
$$n = 3$$

7. $\ln(2x-5) = 4$

$$e^{\ln(2x-5)} = e^4$$
$$2x-5 = e^4$$
$$2x = e^4 + 5$$
$$x = \frac{e^4+5}{2}$$
$$x \approx 29.799$$

8. $$\frac{4}{b+3} + 2 = \frac{-5}{b-7}$$

$$\frac{4}{\cancel{b+3}}\cancel{(b+3)}(b-7) + 2(b+3)(b-7) = \frac{-5}{\cancel{b-7}}(b+3)\cancel{(b-7)}$$
$$4(b-7) + 2(b+3)(b-7) = -5(b+3)$$
$$4b - 28 + 2\left(b^2 - 4b - 21\right) = -5b - 15$$
$$4b - 28 + 2b^2 - 8b - 42 = -5b - 15$$
$$2b^2 + b - 55 = 0$$
$$(2b+11)(b-5) = 0$$
$$2b + 11 = 0, b - 5 = 0$$
$$b = -\frac{11}{2}, b = 5$$

9. $|5h-8| = 32$

$$5h - 8 = 32, 5h - 8 = -32$$
$$5h = 40, 5h = -24$$
$$h = 8, h = -\frac{24}{5}$$

10. $\log_2(x+3) + \log_2(x+5) = 3$

$$\log_2(x+3)(x+5) = 3$$
$$2^3 = (x+3)(x+5)$$
$$8 = x^2 + 8x + 15$$
$$0 = x^2 + 8x + 7$$
$$0 = (x+1)(x+7)$$
$$x + 1 = 0, x + 7 = 0$$
$$x = -1, x = -7$$

The answer $x = -7$ is not part of the domain.
The final answer is: $x = -1$

11. $\sqrt{t+12} + \sqrt{2t+1} = 7$

$$\sqrt{2t+1} = 7 - \sqrt{t+12}$$
$$\left(\sqrt{2t+1}\right)^2 = \left(7 - \sqrt{t+12}\right)^2$$
$$2t + 1 = \left(7 - \sqrt{t+12}\right)\left(7 - \sqrt{t+12}\right)$$
$$2t + 1 = 49 - 14\sqrt{t+12} + t + 12$$
$$t - 60 = -14\sqrt{t+12}$$
$$(t-60)^2 = \left(-14\sqrt{t+12}\right)^2$$
$$(t-60)(t-60) = 196(t+12)$$
$$t^2 - 120t + 3600 = 196t + 2352$$
$$t^2 - 316 + 1248 = 0$$
$$(t-312)(t-4) = 0$$
$$t - 312 = 0, t - 4 = 0$$
$$\cancel{t = 312}, t = 4$$

12. $30(2^n) - 8400 = 6960$

$$30(2^n) = 15360$$
$$2^n = 512$$
$$2^n = 2^9$$
$$n = 9$$

13. $3.2g^2 + 2.4g - 16.2 = 59.4$

$$3.2g^2 + 2.4g - 75.6 = 0$$
$$g = \frac{-2.4 \pm \sqrt{2.4^2 - 4(3.2)(-75.6)}}{2(3.2)}$$
$$g = \frac{-2.4 \pm \sqrt{973.44}}{6.4}$$
$$g = \frac{-2.4 \pm 31.2}{6.4}$$
$$g = \frac{-2.4 + 31.2}{6.4}, g = \frac{-2.4 - 31.2}{6.4}$$
$$g = 4.5, g = -5.25$$

14. $$\frac{44}{a^2 + 3a - 28} = \frac{7a}{a^2 - a - 12}$$

$$\frac{44}{(a+7)(a-4)} = \frac{7a}{(a+3)(a-4)}$$
$$\frac{44}{\cancel{(a+7)}\cancel{(a-4)}}\cancel{(a+7)}\cancel{(a-4)}(a+3)$$
$$= \frac{7a}{\cancel{(a+3)}\cancel{(a-4)}}(a+7)\cancel{(a-4)}\cancel{(a+3)}$$
$$44(a+3) = 7a(a+7)$$
$$44a + 132 = 7a^2 + 49a$$
$$0 = 7a^2 + 5a - 132$$
$$0 = (7a+33)(a-4)$$
$$7a + 33 = 0, a - 4 = 0$$
$$a = -\frac{33}{7}, \cancel{a = 4}$$

15. $e^{x+2} = 20$

$$\ln\left(e^{x+2}\right) = \ln(20)$$
$$(x+2)\ln(e) = \ln(20)$$
$$(x+2)\cdot 1 = \ln(20)$$
$$x + 2 = \ln 20$$
$$x = -2 + \ln 20$$
$$x \approx -2 + 2.996$$
$$x \approx 0.996$$

16. $$\frac{1}{5}(x+4) = \frac{3}{10}x - \frac{5}{4}$$

$$\frac{1}{5}x + \frac{4}{5} = \frac{3}{10}x - \frac{5}{4}$$
$$20\left(\frac{1}{5}x + \frac{4}{5}\right) = 20\left(\frac{3}{10}x - \frac{5}{4}\right)$$

$$4x+16=6x-25$$
$$-2x+16=-25$$
$$-2x=-41$$
$$x=\frac{41}{2}$$

17. $\sqrt[5]{7-3m}=3$

$$\left(\sqrt[5]{7-3m}\right)^5=3^5$$
$$7-3m=243$$
$$-3m=236$$
$$m=-\frac{236}{3}$$

18. $3r^3+9r^2-120r=0$

$$3r\left(r^2+3r-40\right)=0$$
$$3r(r+8)(r-5)=0$$
$$3r=0, r+8=0, r-5=0$$
$$r=0, r=-8, r=5$$

19. $-5|6h+4|=-30$

$$|6h+4|=6$$
$$6h+4=6, 6h+4=-6$$
$$6h=2, 6h=-10$$
$$h=\frac{1}{3}, h=-\frac{5}{3}$$

20. $2^{4x+5}=0.125$

$$\ln\left(2^{4x+5}\right)=\ln(0.125)$$
$$(4x+5)\ln(2)=\ln(0.125)$$
$$4x+5=\frac{\ln(0.125)}{\ln(2)}$$
$$4x+5=-3$$
$$4x=-8$$
$$x=-2$$

21. $5t^4+16=1269$

$$5t^4=1253$$
$$t^4=250.6$$
$$\left(t^4\right)^{\frac{1}{4}}=\pm 250.6^{\frac{1}{4}}$$
$$t\approx\pm 3.979$$

22.

$$\begin{cases}6x+8y=15\\4y=-3x+15\end{cases}$$

Solve for y in the second equation:

$$\frac{4y}{4}=\frac{-3x}{4}+\frac{15}{4}$$
$$y=-0.75x+3.75$$

Plug this into the first equation:

$$6x+8(-0.75x+3.75)=15$$
$$6x-6x+30=15$$
$$30=15$$

No Solution.

23.

$$\begin{cases}2x+7y=29\\10x-3y=-7\end{cases}$$

Solve for x in the first equation:

$$\frac{2x}{2}+\frac{7y}{2}=\frac{29}{2}$$
$$x+3.5y=14.5$$
$$x=14.5-3.5y$$

Plug this into the second equation:

$$10(14.5-3.5y)-3y=-7$$
$$145-35y-3y=-7$$
$$145-38y=-7$$
$$-38y=-152$$
$$y=4$$

Substituting y back into either equation solve for x:

$$x=14.5-3.5(4)=0.5$$

The solution to the system is $(0.5,4)$.

24.

$$\begin{cases}y=4x^2-6x+5\\y=\frac{1}{2}x+8\end{cases}$$

$$4x^2-6x+5=\frac{1}{2}x+8$$
$$2\left(4x^2-6x+5\right)=2\left(\frac{1}{2}x+8\right)$$
$$8x^2-12x+10=x+16$$
$$8x^2-13x-6=0$$
$$(8x+3)(x-2)=0$$
$$8x+3=0, x-2=0$$
$$x=-\frac{3}{8}, x=2$$

Substituting x back into either equation solve for y:

$$y=\frac{1}{2}\left(-\frac{3}{8}\right)+8=\frac{125}{16}$$
$$y=\frac{1}{2}(2)+8=9$$

The solutions to the system are $\left(-\frac{3}{8},\frac{125}{16}\right)$ and $(2,9)$.

25.

$$\begin{cases}y=-2x^2+3x+10\\y=x^2-5x-25\end{cases}$$

$$-2x^2+3x+10=x^2-5x-25$$
$$0=3x^2-8x-35$$
$$0=(3x+7)(x-5)$$
$$3x+7=0, x-5=0$$
$$x=-\frac{7}{3}, x=5$$

Substituting x back into either equation solve for y:

$$y=\left(-\frac{7}{3}\right)^2-5\left(-\frac{7}{3}\right)-25=-\frac{71}{9}$$
$$y=(5)^2-5(5)-25=-25$$

The solutions to the system are $\left(-\frac{7}{3},-\frac{71}{9}\right)$ and $(5,-25)$

26. Let x be the number of milliliters of 5% saline solution and y be the number of milliliters of 10% saline solution used to get the 20 ml of 8% saline solution that Rosemary needs.

$$\begin{cases} x+y=20 \\ 0.05x+0.10y=(0.08)(20)=1.6 \end{cases}$$

Solve for x in the first equation and use the result for x in the second equation.

$$x=-y+20$$
$$0.05(-y+20)+0.1y=1.6$$
$$-0.05y+1+0.1y=1.6$$
$$0.05y+1=1.6$$
$$\underline{-1\quad -1}$$
$$0.05y=0.6$$
$$\frac{0.05y}{0.05}=\frac{0.6}{0.05}$$
$$y=12$$

Now solve for x:

$$x=-(12)+20=8$$

Rosemary should mix 8 ml of 5% saline solution with 12 ml of 10% saline solution to get 20 ml of 8% saline solution.

27. **a.** Let $P(n)$ be the profit in dollars from manufacturing and selling n customs mp3 players.

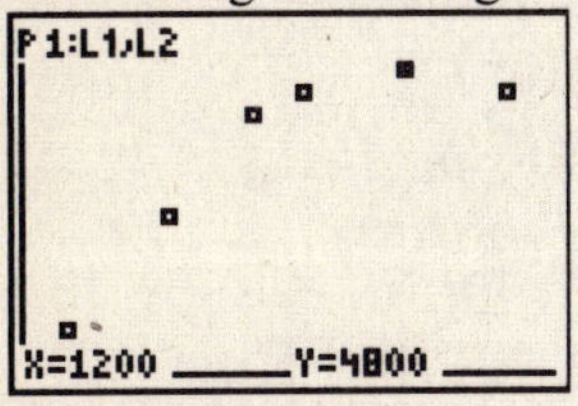

Choose $(1200,4800)$ as the vertex, and $(200,800)$ as the additional point used to solve for "a".

$$P(n)=a(n-1200)^2+4800$$
$$800=a(200-1200)^2+4800$$
$$800=a(1000)^2+4800$$
$$800=1{,}000{,}000a+4800$$
$$\underline{-4800\qquad\qquad -4800}$$
$$-4000=1{,}000{,}000a$$
$$\left.\frac{-4000}{1{,}000{,}000}=\frac{1{,}000{,}000a}{1{,}000{,}000}\right\}\ a\approx-\frac{1}{250}$$
$$P(n)=-\frac{1}{250}(n-1200)^2+4800$$

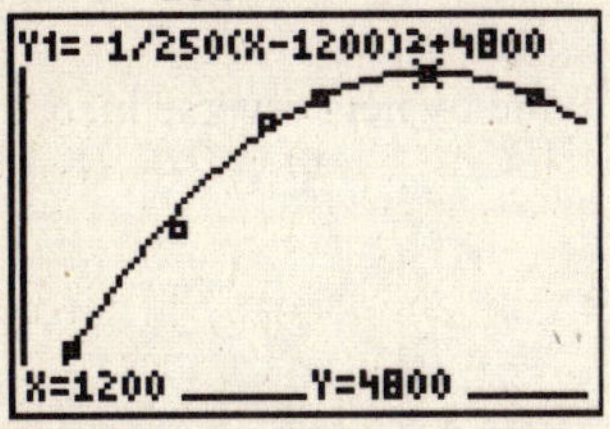

Note: Your answer may vary and yet still represent the trend of the data.

b. Domain: $[150,2000]$. Within this domain, the maximum value of 4800 comes from the vertex. The minimum value comes from $n=150$.

$$P(150)=-\frac{1}{250}(150-1200)^2+4800$$

Range: $[390,4800]$

c. $P(1000)=-\frac{1}{250}(1000-1200)^2+4800$

$$P(1000)=4640$$

The profit from manufacturing and selling 1000 custom mp3 players is \$4640.

d. Vertex: $(1200,4800)$ The profit from manufacturing and selling 1200 custom mp3 players is \$4800.

28. **a.** Let $R(t)$ be the primary revenue (in millions of dollars) t years since 2000.

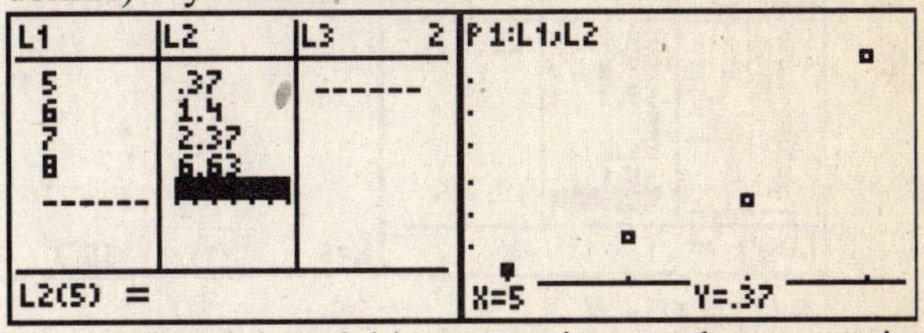

An exponential model is appropriate so choose a point from the flatter end and one from the steeper end. Note: Your model may differ, but it must follow the trend of the data.

$$R(t)=ab^t$$

To find a and b choose two points $(5,0.37)$ and $(8,6.63)$.

$$6.63=ab^8 \quad\text{and}\quad 0.37=ab^5$$
$$\frac{6.63}{0.37}=\frac{ab^8}{ab^5}$$
$$\frac{6.63}{0.37}=b^3$$
$$\left(\frac{6.63}{0.37}\right)^{\frac{1}{3}}=\left(b^3\right)^{\frac{1}{3}}$$
$$b\approx2.62$$

Use b and one of the above equations to find a.

$$0.37=a(2.62)^5$$
$$\frac{0.37}{(2.62)^5}=\frac{a(2.62)^5}{(2.62)^5}$$
$$a\approx0.003$$

Then $R(t)=0.003(2.62)^t$

b. Domain: $[2,10]$

$$R(2)=0.003(2.62)^2\approx0.021$$
$$R(10)=0.003(2.62)^{10}\approx45.723$$

Range: $[0.021,45.723]$

c. $R(10)=0.003(2.62)^{10}\approx45.723$

The primary revenue for the Wikimedia Foundation was approximately \$45.7 million in 2010.

d. $50 = 0.003(2.62)^t$

$$\frac{50}{0.003} = \frac{0.003(2.62)^t}{0.003}$$

$$\frac{50}{0.003} = (2.62)^t$$

$$\log_{2.62}\left(\frac{50}{0.003}\right) = t$$

$$t = \frac{\ln\left(\frac{50}{0.003}\right)}{\ln(2.62)} \approx 10.09$$

According to this model, the Wikimedia Foundation will reach \$50 million in annual primary revenue in the year 2010.

29. **a.** Let $N(t)$ be the total circulation of U.S. Sunday newspapers (in millions) t years since 2000.

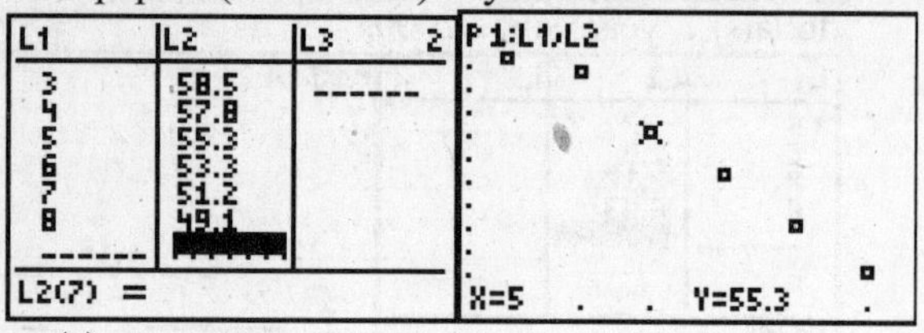

$N(t) = mt + b$

Using two points: $(5, 55.3)$ and $(7, 51.2)$

Find the slope: $m = \frac{51.2 - 55.3}{7 - 5} = -2.05$

$N(t) = -2.05t + b$

$b = 55.3 - (-2.05)(5)$

$b = 65.55$

$N(t) = -2.05t + 65.55$

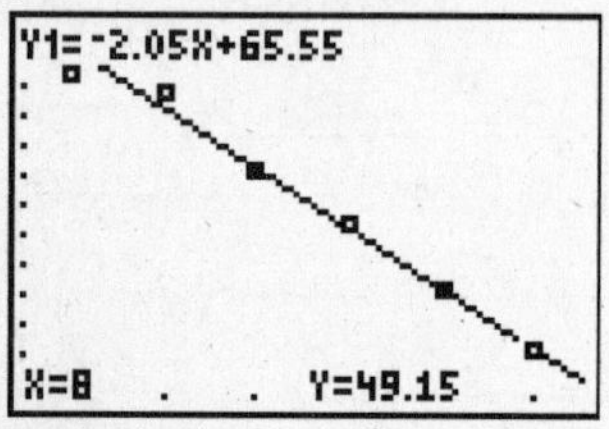

b. $N(10) = -2.05(10) + 65.55$

$N(10) = 45.05$

In 2010, the total circulation of U.S. Sunday newspapers was 45.05 million.

c. $25 = -2.05t + 65.55$

$$\underline{-65.55} \qquad \underline{-65.55}$$

$-40.55 = -2.05t$

$$\frac{-40.55}{-2.05} = \frac{-2.05t}{-2.05}$$

$19.78 \approx t$

The model predicts total circulation of U.S. Sunday newspapers will be 25 million by the year 2020.

d. Slope $= -2.05$ Each year the total U.S. Sunday newspaper circulation decreases by 2.05 million.

e. Domain: $[-5, 20]$

$N(-5) = -2.05(-5) + 65.55 = 75.8$

$N(20) = -2.05(20) + 65.55 = 24.55$

Range: $[24.55, 75.8]$

30. $f(x) = \frac{x-2}{x^2 + 7x - 18}$

$f(x) = \frac{x-2}{(x+9)(x-2)}$

Domain: All real numbers except for -9 and 2.

31. $f(x) = -\frac{1}{2}x + 5$ Domain: All real numbers.

32. $g(x) = 7x^2 - 10$ Domain: All real numbers.

33. $2x - 3y > 6$

$y < -x + 4$

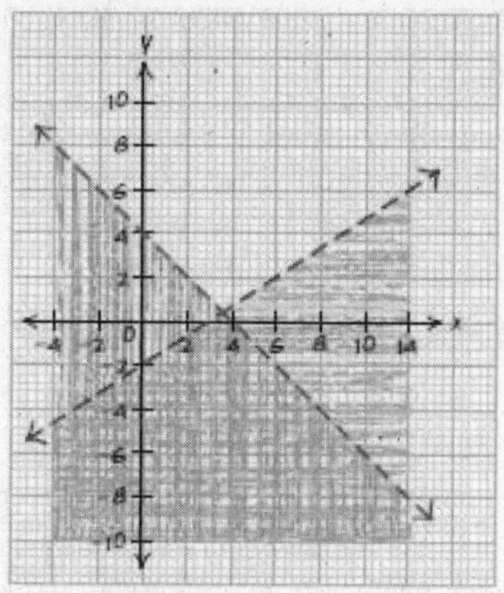

34. $6a - 15 < 11a + 20$

$$\underline{-11a + 15} \quad \underline{-11a + 15}$$

$-5a < 35$

$\frac{-5a}{-5} > \frac{35}{-5}$ Divide by neg. reverse ineq. symbol

$a > -7$

35. $\frac{1}{2}n + 7 \le \frac{3}{4}n + \frac{5}{12}$

Multiply both sides by LCD $= 12$.

$$(12)\left(\frac{1}{2}n + 7\right) \le (12)\left(\frac{3}{4}n + \frac{5}{12}\right)$$

$6n + 84 \le 9n + 5$

$$\underline{-9n - 84} \quad \underline{-9n - 84}$$

$-3n \le -79$

$\frac{-3n}{-3} \ge \frac{-79}{-3}$ Divide by neg. reverse ineq. symbol

$n \ge \frac{79}{3}$

36. $|4h-9|\ge 7$

$4h-9\le -7$ or $4h-9\ge 7$
$+9\ \ +9$ $\quad$ $+9\ +9$

$4h\le 2$ or $4h\ge 16$

$\frac{4h}{4}\le\frac{2}{4}$ or $\frac{4h}{4}\ge\frac{16}{4}$

$h\le\frac{1}{2}$ or $h\ge 4$

37. $-6|g-4|<30$

$\frac{-6|g-4|}{-6}>\frac{30}{-6}$ Divide by neg. reverse ineq. symbol

$|g-4|>-5$ which is always true.

Therefore, the solution is all real numbers.

38. $-4|-5x+7|\le -10$

$\frac{-4|-5x+7|}{-4}\ge\frac{-10}{-4}$ Divide by neg. reverse ineq. symbol

$|-5x+7|\ge 2.5$

$-5x+7\le -2.5$ or $-5x+7\ge 2.5$
$-7\ \ -7$ $\quad$ $-7\ -7$

$-5x\le -9.5$ or $-5x\ge -4.5$

$\frac{-5x}{-5}\ge\frac{-9.5}{-5}$ or $\frac{-5x}{-5}\le\frac{-4.5}{-5}$

$x\le 0.9$ or $x\ge 1.9$

39. $\frac{x+3}{x-7}\ge 0$

Find the zero value:

$x+3=0$

$x=-3$

Find the excluded value:

$x-7=0$

$x=7$

Check the intervals:

$x=-5$	$x=0$	$x=8$
$\frac{-5+3}{-5-7}\overset{?}{\ge}0$	$\frac{0+3}{0-7}\overset{?}{\ge}0$	$\frac{8+3}{8-7}\overset{?}{\ge}0$
$\frac{-2}{-13}\overset{?}{\ge}0$	$\frac{3}{-7}\overset{?}{\ge}0$	$\frac{11}{1}\overset{?}{\ge}0$
$\frac{2}{13}\overset{?}{\ge}0$	$-\frac{3}{7}\overset{?}{\ge}0$	$11\overset{?}{\ge}0$
True	False	True

$x\le -3$ or $x>7$

40. $2x^2-3x-20>0\rightarrow 2x^2-3x-20=0$

$(2x+5)(x-4)=0$

$2x+5=0$ or $x-4=0$

$2x=-5$ or $x=4$

$x=-\frac{5}{2}$

Check the intervals:

$x=-3$	$x=0$	$x=5$
$[2(-3)+5](-3-4)\overset{?}{>}0$	$[2(0)+5](0-4)\overset{?}{>}0$	$[2(5)+5](5-4)\overset{?}{>}0$
$(-1)(-7)\overset{?}{>}0$	$(5)(-4)\overset{?}{>}0$	$(15)(1)\overset{?}{>}0$
$7>0$	$-20>0$	$15>0$
True	False	True

$x<-2.5$ or $x>4$

41. **a.** $A(t)=1.4t^2-14.3t+202.5$

$A(18)=1.4(18)^2-14.3(18)+202.5$

$A(18)=398.7$

In 2008, there were 398.7 thousand apprenticeship training registrations for all major trade groups in Canada.

b. Vertex: (h,k)

$h=\frac{14.3}{2(1.4)}=\frac{143}{28}\approx 5.1$

$k=A\left(\frac{143}{28}\right)=1.4\left(\frac{143}{28}\right)^2-14.3\left(\frac{143}{28}\right)+202.5$

$k\approx 165.98$

Vertex: $(5.1,165.98)$

In 1995, there were 165.98 thousand apprenticeship training registrations for all major trade groups in Canada.

c. $300=1.4t^2-14.3t+202.5$
$-300 \qquad -300$

$0=1.4t^2-14.3t-97.5$

$t=\frac{14.3\pm\sqrt{(-14.3)^2-4(1.4)(-97.5)}}{2(1.4)}$

$t=\frac{14.3+\sqrt{750.49}}{2.8}\approx 14.89\rightarrow$ In the year 2005.

$t=\frac{14.3-\sqrt{750.49}}{2.8}\approx -4.68\rightarrow$ In the year 1985.

In 1985 and again in 2005, there were 300,000 apprenticeship training registrations for all major trade groups in Canada.

42. **a.** Let $B(h)$ be the number of bacteria present after h hours.

h	$B(h)$
0	2000
1	$2000(2)(2)(2)=2000(2)^3$
2	$2000(2)^3(2)^3=2000(2)^{3\cdot 2}$
3	$2000(2)^3(2)^3(2)^3=2000(2)^{3\cdot 3}$
h	$2000(2)^{3\cdot h}$

$B(h)=2000(2)^{3h}$

b. $1{,}000{,}000=2000(2)^{3h}$

$\frac{1{,}000{,}000}{2000}=\frac{2000(2)^{3h}}{2000}$

$500=(2)^{3h}=8^h$

$\log_8(500)=h$

$h=\frac{\ln(500)}{\ln(8)}\approx 2.99$

After approximately 3 hours, there will be 1 million bacteria.

c. $B(5)=2000(2)^{3(5)}$

$B(5)=65{,}536{,}000$

There will 65,536,000 bacteria present after 5 hours.

43. $f(x)=\frac{2}{3}x-8$ Rewrite without function notation.

$y=\frac{2}{3}x-8$ Switch x and y.

$x=\frac{2}{3}y-8$ Solve for y.

$x+8=\frac{2}{3}y$

$\left(\frac{3}{2}\right)(x+8)=\left(\frac{3}{2}\right)\left(\frac{2}{3}y\right)$

$\frac{3}{2}x+\frac{24}{2}=y$

$f^{-1}(x)=\frac{3}{2}x+12$

44. $g(x)=4(3)^x$ Rewrite without function notation.

$y=4(3)^x$ Switch x and y.

$x=4(3)^y$ Solve for y.

$\frac{x}{4}=\frac{4(3)^y}{4}$

$\frac{x}{4}=(3)^y$ Write in logaritmic form.

$\log_3\left(\frac{x}{4}\right)=y$ Use change of base formula.

$g^{-1}(x)=\frac{\log\left(\frac{x}{4}\right)}{\log(3)}$

45. **a.** $f(3)=2.5$

b. Domain: $(-\infty,\infty)$ Range: 2.5

46. **a.** $f(x)+g(x)=(2x-7)+(3.4x+2.7)$

$=2x-7+3.4x+2.7$

$=5.4x-4.3$

b. $f(g(x))=2(3.4x+2.7)-7$

$=6.8x+5.4-7$

$=6.8x-1.6$

c. $f(x)g(x)=(2x-7)(3.4x+2.7)$

$=6.8x^2+5.4x-23.8x-18.9$

$=6.8x^2-18.4x-18.9$

47. **a.** $f(2)-g(2)=\left(\frac{3}{4}(2)-10\right)-(-12(2)+5)$

$=1.5-10+24-5$

$=10.5$

b. $(f\circ g)(7)=f(g(7))$

$g(7)=-12(7)+5=-79$

$f(g(7))=f(-79)$

$f(-79)=\frac{3}{4}(-79)-10=-69.25$

$(f\circ g)(7)=f(g(7))=f(-79)=-69.25$

c. $f(2)g(2)=161.5$

48. **a.** $f(x)+g(x)=(3x^2+4x-6)+(3x-4)$

$=3x^2+4x-6+3x-4$

$=3x^2+7x-10$

b. $f(g(x))=3(3x-4)^2+4(3x-4)-6$

$=3(9x^2-24x+16)+12x-16-6$

$=27x^2-72x+48+12x-22$

$=27x^2-60x+26$

c. $f(x)g(x)=(3x^2+4x-6)(3x-4)$

$=9x^3-12x^2+12x^2-16x-18x+24$

$=9x^3-34x+24$

49. $6.5=\log\left(\frac{I}{10^{-4}}\right)$

$10^{6.5}=\frac{I}{10^{-4}}$

$(10^{-4})(10^{6.5})=(10^{-4})\left(\frac{I}{10^{-4}}\right)$

$10^{2.5}=I$

$I\approx 316.228$

The intensity of this earthquake is 316.228 cm.

50. Since the lines are perpendicular, $m_1=-\frac{1}{m_2}$.

$-3x+7y=4$

$7y=3x+4$

$y=\frac{3}{7}x+\frac{4}{7}\rightarrow$ slope $=\frac{3}{7}$

Using $m=-\frac{7}{3}$ and the point $(-5,2)$

$y-2=-\frac{7}{3}(x+5)$

$y=-\frac{7}{3}x-\frac{35}{3}+2$

$y=-\frac{7}{3}x-\frac{29}{3}$

51. $m=\frac{\text{rise}}{\text{run}}=\frac{6}{10}=\frac{3}{5}$

y-intercept $=(0,-6)$

Dashed line, shaded above so

$y>\frac{3}{5}x-6$

52. $y=\frac{1}{4}x-6$

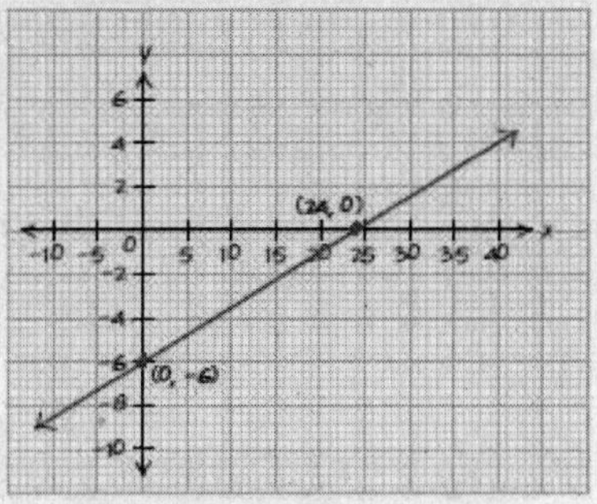

53. $y=-3x^2-24x+2$

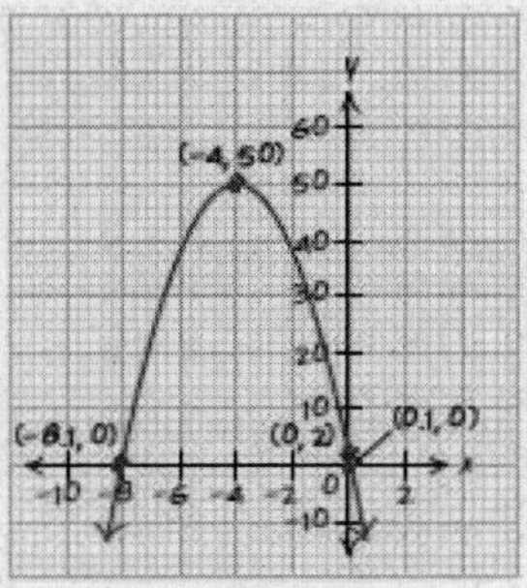

54. $y=-1.5(2.5)^x$

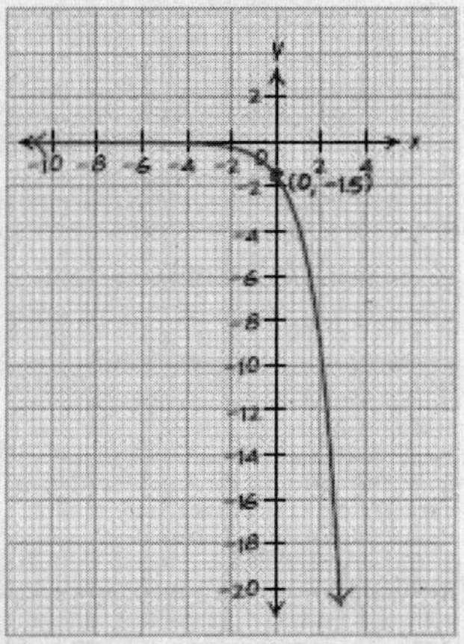

55. $y=\sqrt{x+17}$

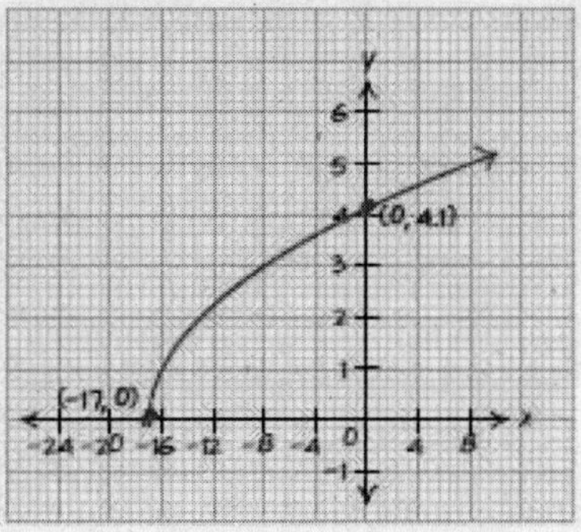

56. $y=\frac{1}{3}(x+4)^2-10$

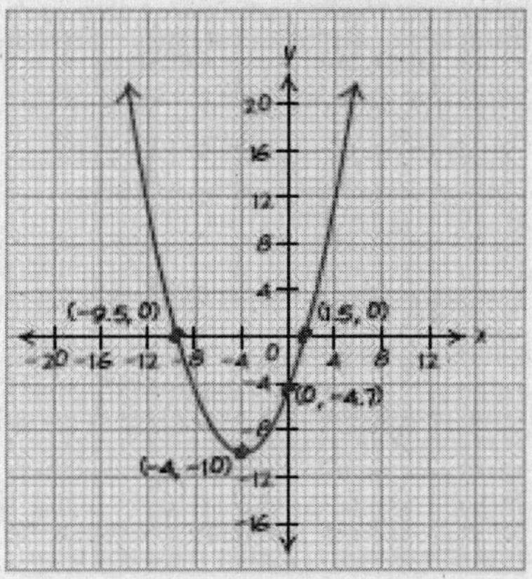

57. $5x-30y=180$

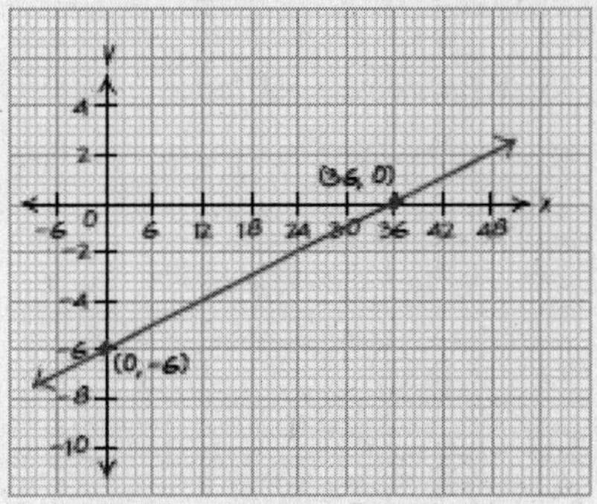

58. $y=800(0.2)^x$

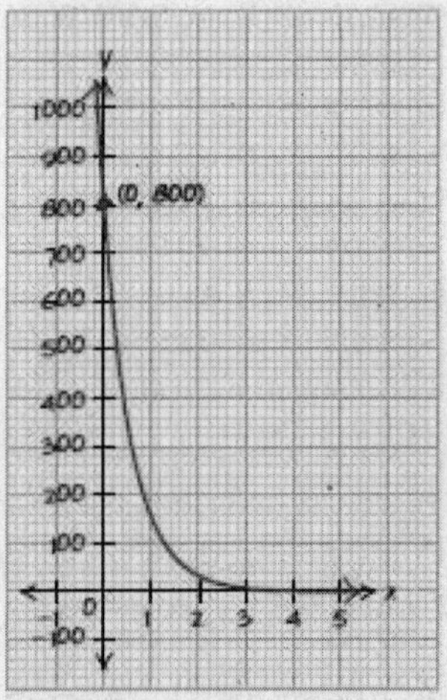

59. $y=\log_9 x$

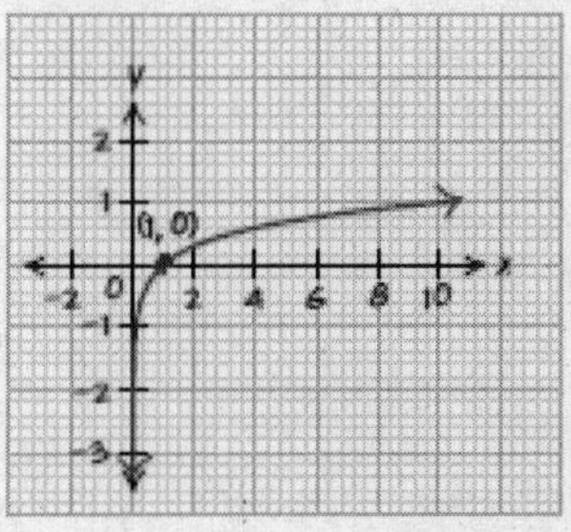

60. $y=\sqrt[3]{x+9}$

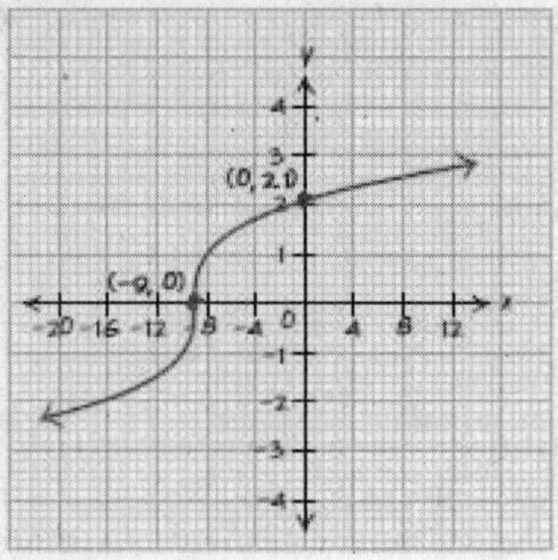

61. $y=3(1.6)^x+20$

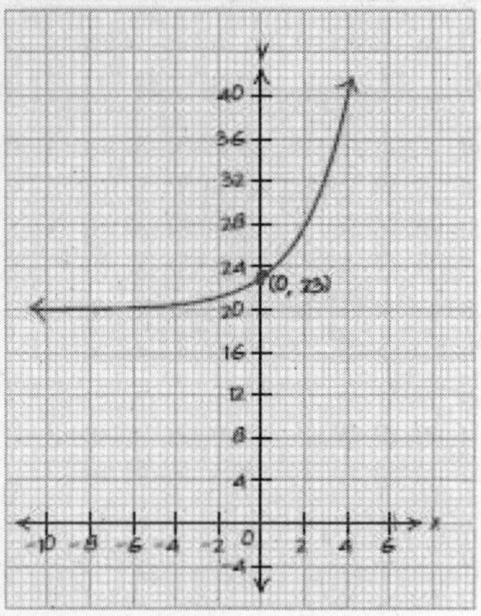

62. $y<x^2-5x-14$

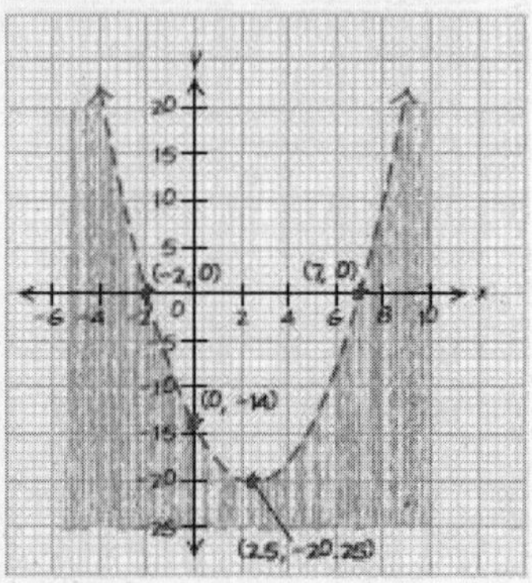

63. $y\ge 2x+3$

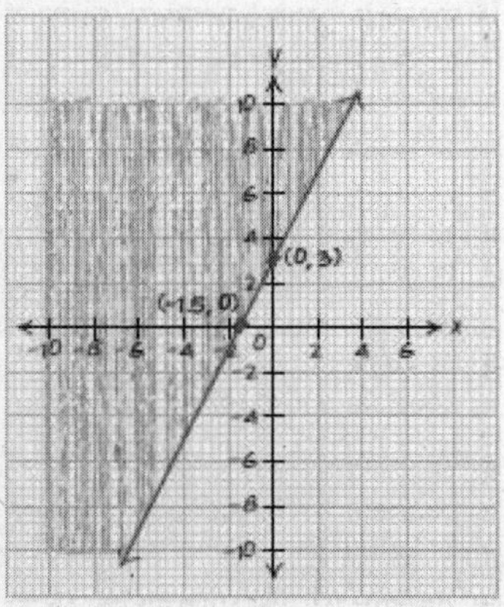

64. $5x-6y<6$

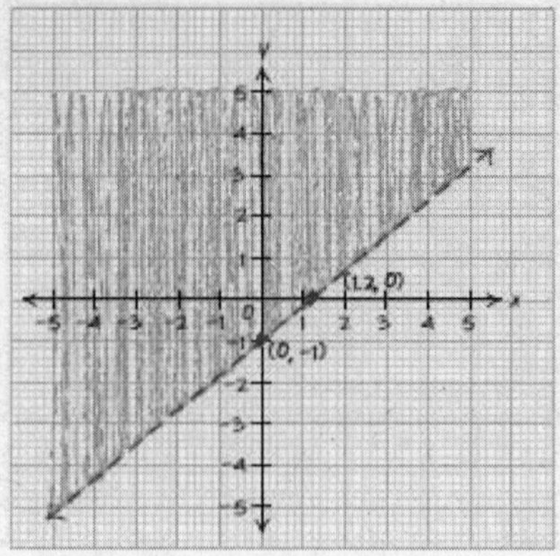

65. $y\ge -3(x-1)^2+8$

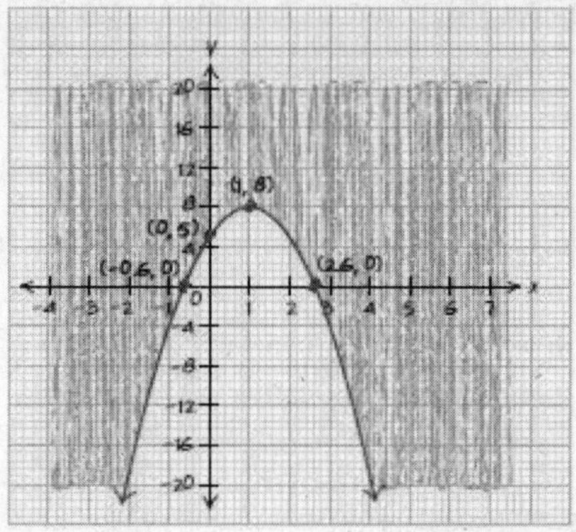

66. **a.** Let $P(t)$ be the population of Syria (in millions) t years since 2010.

$P(t)=ab^t$

$a=22.5$ and $r=0.913\%=0.00913$

$b=1+r$ so $b=1.00913$

Then $P(t)=22.5(1.00913)^t$

b. $2015\to t=5$

$P(5)=22.5(1.00913)^5$

$P(5)\approx 23.546$

The population in Syria in 2015 will be 23.5 million.

67. $\ln 5x+3\ln 2y-\ln 6z=\ln 5x+\ln(2y)^3-\ln 6z$

$$=\ln 5x+\ln 8y^3-\ln 6z$$

$$=\ln\left(\frac{5x\cdot 8y^3}{6z}\right)$$

$$=\ln\left(\frac{20xy^3}{3z}\right)$$

68. $7^5=16{,}807$

$\log_7(16{,}807)=5$

69. **a.** Let P be the profit in millions of dollars and t be the time in seconds that it takes to produce a product.

$$P=\frac{k}{t}$$
$$1.2=\frac{k}{12}\to k=12(1.2)$$
$$k=14.4$$
$$P=\frac{14.4}{t}$$

b. $P=\frac{14.4}{10}=1.44$

The profit is \$1.44 million if it takes 10 seconds each to produce the product.

70. $7\sqrt{12a}+8\sqrt{3a}-10\sqrt{a}$
$=7\sqrt{4\cdot 3a}+8\sqrt{3a}-10\sqrt{a}$
$=14\sqrt{3a}+8\sqrt{3a}-10\sqrt{a}$
$=22\sqrt{3a}-10\sqrt{a}$

71. $\left(14m^3+3m^2-2m+15\right)-\left(5m^3-8m^2+6\right)$
$=14m^3+3m^2-2m+15-5m^3+8m^2-6$
$=9m^3+11m^2-2m+9$

72. $\frac{x^2-16}{x-3}\cdot\frac{x+2}{x+4}$
$=\frac{(x+4)(x-4)(x+2)}{(x-3)(x+4)}$
$=\frac{(x-4)(x+2)}{(x-3)}$

73. $\left(3x^2-8x+7\right)+\left(5x-3\right)$
$=3x^2-8x+7+5x-3$
$=3x^2-3x+4$

74. $(6x+1)(5x-8)$
$=30x^2-48x+5x-8$
$=30x^2-43x-8$

75. $9x^2+2x-(x+7)(5x-6)$
$=9x^2+2x-\left(5x^2-6x+35x-42\right)$
$=9x^2+2x-5x^2+6x-35x+42$
$=4x^2-27x+42$

76. $\left(7+\sqrt{3w}\right)\left(-2+\sqrt{12w}\right)$
$=\left(7+\sqrt{3w}\right)\left(-2+2\sqrt{3w}\right)$
$=-14+14\sqrt{3w}-2\sqrt{3w}+2(3w)$
$=-14+12\sqrt{3w}+6w$

77.

$$\begin{array}{r} 4a^2+27a+35 \\ 3a-4\overline{\smash{)}12a^3+65a^2-3a-140} \\ \underline{-\left(12a^3-16a^2\right)} \\ 81a^2-3a \\ \underline{-\left(81a^2-108a\right)} \\ 105a-140 \\ \underline{-\left(105a-140\right)} \\ 0 \end{array}$$

$\left(12a^3+65a^2-3a-140\right)\div(3a-4)=4a^2+27a+35$

78. $(5x-6)^2$
$=(5x-6)(5x-6)$
$=25x^2-60x+36$

79. $\frac{3h-4}{h^2-5h+6}\div\frac{h+2}{h^2+4h-21}$
$=\frac{(3h-4)}{(h-3)(h-2)}\cdot\frac{(h+7)(h-3)}{h+2}$
$=\frac{(3h-4)(h+7)}{(h-2)(h+2)}$

80. $\frac{5m}{m-3}+\frac{4}{m+7}$
Use LCD $=(m+7)(m-3)$
$\left(\frac{5m}{m-3}\right)\left(\frac{m+7}{m+7}\right)+\left(\frac{4}{m+7}\right)\left(\frac{m-3}{m-3}\right)$
$=\frac{5m^2+35m}{(m+7)(m-3)}+\frac{4m-12}{(m+7)(m-3)}$
$=\frac{5m^2+39m-12}{(m+7)(m-3)}$

81. **a.** $G(12)=3400+155(12)$
$G(12)=5260$

There will be 5260 graduating students at this college in 2012.

b. $P(12)=0.86-0.03(12)$
$P(12)=0.5$

In 2012, 50% of the graduating students at this college will attend graduating ceremonies.

c. $G(12)P(12)=(5260)(0.5)$
$=2630$

There will be 2630 graduating students attending graduating ceremonies at this college in 2012.

d. Let $A(t)$ be the number of graduating students attending graduating ceremonies at this college t years since 2000.

$A(t)=G(t)P(t)$
$A(t)=(3400+155t)(0.86-0.03t)$
$A(t)=2924-102t+133.3t-4.65t^2$
$A(t)=2924+31.3t-4.65t^2$

e. $A(15) = 2924 + 31.3(15) - 4.65(15)^2$

$A(15) = 2347.25$

There will be approximately 2347 graduating students attending graduating ceremonies at this college in 2015.

82. $2x^2 - x - 28$

$ac = 2 \times -28 = -56$

Factors of -56 that add to -1 are -8 and 7

$= 2x^2 + 7x - 8x - 28$

$= (2x^2 + 7x) - (8x + 28)$

$= x(2x + 7) - 4(2x + 7)$

$= (2x + 7)(x - 4)$

83. $20a^2 - 45b^2$

$= 5(4a^2 - 9b^2)$

$= 5\left((2a)^2 - (3b)^2\right)$

$= 5(2a + 3b)(2a - 3b)$

84. $h^3 + 125$

$= h^3 + 5^3$

$= (h + 5)(h^2 - 5h + 25)$

85. $21m^3 - 163m^2 - 48m$

$3m(7m^2 - 54m - 16)$

$ac = 7 \times -16 = -112$

Factors of -112 that add to -54 are -56 and 2

$= 3m(7m^2 + 2m - 56m - 16)$

$= 3m\left((7m^2 + 2m) - (56m + 16)\right)$

$= 3m\left(m(7m + 2) - 8(7m + 2)\right)$

$= 3m(7m + 2)(m - 8)$

86. $60x^2 - 265x + 280$

$= 5(12x^2 - 53x + 56)$

$ac = 12 \times 56 = 672$

Factors of 672 that add to -53 are -32 and -21

$= 5(12x^2 - 32x - 21x + 56)$

$= 5\left((12x^2 - 32x) - (21x - 56)\right)$

$= 5\left(4x(3x - 8) - 7(3x - 8)\right)$

$= 5(3x - 8)(4x - 7)$

87. $21m^3 - 163m^2 - 48m$ has degree 3.

88. $f(x) = 4x^2 - 48x + 126$

$f(x) = 4(x^2 - 12x) + 126$

$f(x) = 4(x^2 - 12x + \square - \square) + 126$

Completing the square: $\left(\frac{-12}{2}\right)^2 = (-6)^2 = 36$

$f(x) = 4(x^2 - 12x + \boxed{36} - \boxed{36}) + 126$

$f(x) = 4(x^2 - 12x + 36) + 4(-36) + 126$

$f(x) = 4(x - 6)(x - 6) - 144 + 126$

$f(x) = 4(x - 6)^2 - 18$

89. $P(m) = 6.5 + 3\sqrt{m}$

a. $P(12) = 6.5 + 3\sqrt{12}$

$P(12) \approx 16.89$

The profit for Beach Shack Rentals one year after it opens was \$16.89 thousand.

b. $25 = 6.5 + 3\sqrt{m}$

$25 - 6.5 = 3\sqrt{m}$

$18.5 = 3\sqrt{m}$

$(18.5)^2 = (3\sqrt{m})^2$

$342.25 = 9m$

$\frac{342.25}{9} = \frac{9m}{9}$

$m \approx 38.028$

After approximately 38 months, the profit for Beach Shack Rentals will be \$25,000.

90. $\frac{5x^3y^2}{10xy^5} = \frac{x^{3-1}}{2y^{5-2}} = \frac{x^2}{2y^3}$

91. $(4a^2b^{-5}c)^2(-3ab^3c^{-7}) = \left(\frac{4a^2c}{b^5}\right)^2\left(-\frac{3ab^3}{c^7}\right)$

$= \left(\frac{16a^4c^2}{b^{10}}\right)\left(-\frac{3ab^3}{c^7}\right)$

$= \left(-\frac{48a^5}{b^{10-3}c^{7-2}}\right)$

$= -\frac{48a^5}{b^7c^5}$

92. $\left(\frac{-5m^2n^{-5}}{20m^4n^3}\right)^3 = \left(\frac{-1}{4m^{4-2}n^{3+5}}\right)^3 = \left(\frac{-1}{4m^2n^8}\right)^3 = \frac{-1}{64m^6n^{24}}$

93. $f(x) = \log(x + 3)$ Domain: $x > -3$

94. $g(x) = 6x + \sqrt{5 - x}$ Domain: $x \le 5$

95. $h(x) = 4e^x$ Domain: all real numbers.

96. $a^2 + b^2 = c^2$

$x^2 + 4^2 = 11^2$

$x^2 + 16 = 121$

$x^2 = 105$

$\sqrt{x^2} = \pm\sqrt{105}$

$x \approx 10.25$ use positive value for length

97. $x = -6$

98. Domain: $(-\infty, 10]$ Range: $[0, \infty)$

99. The value of a is negative. The y-intercept is below the x axis.

100. The value of b is less than 1. This is an increasing function with $a < 0$.

101. Domain: $(-\infty, \infty)$ Range: $(-\infty, 0)$

102. The value of a is positive because the parabola opens upward.

103. The vertex of the parabola is $(-2, -16)$ therefore $h = -2$ and $k = -16$.

104. $f(-4) = -12$

105. $x=-6$ and $x=2$ when $f(x)=0$.

106. Domain: $(-\infty,\infty)$ Range: $[-16,\infty)$

107. The value of m is positive because the slope of the line is positive.

108. $b=-3$

109. $m=\dfrac{2}{5}$

110. $x=-5$ when $f(x)=-5$.

111. Domain: $(-\infty,\infty)$ Range: $(-\infty,\infty)$

112. $\sqrt{36a^4b^6c^2}=\sqrt{6^2a^4b^6c^2}=6a^2b^3c$

113.
$$\begin{aligned}\sqrt[3]{54x^3y^5}&=\sqrt[3]{2\cdot 3^3x^3y^5}\\&=\sqrt[3]{3^3x^3y^3}\cdot\sqrt[3]{2y^2}\\&=3xy\sqrt[3]{2y^2}\end{aligned}$$

114. $\dfrac{\sqrt{7a}}{\sqrt{2b}}=\dfrac{\sqrt{7a}}{\sqrt{2b}}\cdot\dfrac{\sqrt{2b}}{\sqrt{2b}}=\dfrac{\sqrt{14ab}}{2b}$

115.
$$\begin{aligned}\frac{6+\sqrt{5}}{2-\sqrt{3}}&=\frac{(6+\sqrt{5})}{(2-\sqrt{3})}\cdot\frac{(2+\sqrt{3})}{(2+\sqrt{3})}\\&=\frac{12+6\sqrt{3}+2\sqrt{5}+\sqrt{5}\cdot\sqrt{3}}{4+2\sqrt{3}-2\sqrt{3}-\sqrt{3}\cdot\sqrt{3}}\\&=\frac{12+6\sqrt{3}+2\sqrt{5}+\sqrt{15}}{4-3}\\&=\frac{12+6\sqrt{3}+2\sqrt{5}+\sqrt{15}}{1}\\&=12+6\sqrt{3}+2\sqrt{5}+\sqrt{15}\end{aligned}$$

116. $(7+4i)+(2-7i)=7+2+4i-7i=9-3i$

117.
$$\begin{aligned}(6+2i)-(10-6i)&=6+2i-10+6i\\&=6-10+2i+6i\\&=-4+8i\end{aligned}$$

118.
$$\begin{aligned}(4+9i)(4-9i)&=16-36i+36i-81i^2\\&=16-81(-1)\\&=16+81\\&=97\end{aligned}$$

119.
$$\begin{aligned}(11-3i)(8+2i)&=88+22i-24i-6i^2\\&=88-2i-6(-1)\\&=88-2i+6\\&=94-2i\end{aligned}$$

120.
$$\begin{aligned}\frac{6+5i}{4i}&=\frac{(6+5i)}{4i}\cdot\frac{i}{i}\\&=\frac{6i+5i^2}{4i^2}\\&=\frac{6i+5(-1)}{4(-1)}\\&=\frac{6i-5}{-4}\\&=-\frac{6}{4}i+\frac{5}{4}\\&=\frac{5}{4}-\frac{3}{2}i\end{aligned}$$

121.
$$\begin{aligned}\frac{8-6i}{2-5i}&=\frac{(8-6i)}{(2-5i)}\cdot\frac{(2+5i)}{(2+5i)}\\&=\frac{16+40i-12i-30i^2}{4+10i-10i-25i^2}\\&=\frac{16+40i-12i-30(-1)}{4-25(-1)}\\&=\frac{16+28i+30}{4+25}\\&=\frac{46+28i}{29}\\&=\frac{46}{29}+\frac{28}{29}i\end{aligned}$$

122.
$$\begin{aligned}\frac{10}{3+4i}&=\frac{10}{(3+4i)}\cdot\frac{(3-4i)}{(3-4i)}\\&=\frac{30-40i}{9-12i+12i-16i^2}\\&=\frac{30-40i}{9-16(-1)}\\&=\frac{30-40i}{9+16}\\&=\frac{30-40i}{25}\\&=\frac{30}{25}-\frac{40}{25}i\\&=\frac{6}{5}-\frac{8}{5}i\end{aligned}$$

123. $x^2+4x+29=0$
$$\begin{aligned}x&=\frac{-4\pm\sqrt{4^2-4(1)(29)}}{2(1)}\\x&=\frac{-4\pm\sqrt{-100}}{2}\\x&=\frac{-4\pm 10i}{2}\\x&=-2\pm 5i\end{aligned}$$

124. $x^3+40x=-12x^2$

$x^3+12x^2+40x=0$

$x\left(x^2+12x+40\right)=0$

$x=0, x^2+12x+40=0$

$x=\dfrac{-12\pm\sqrt{12^2-4(1)(40)}}{2(1)}$

$x=\dfrac{-12\pm\sqrt{-16}}{2}$

$x=\dfrac{-12\pm 4i}{2}$

$x=-6\pm 2i, x=0$

125. $x^2+70=0$

$x^2=-70$

$x=\pm\sqrt{-70}$

$x=\pm i\sqrt{70}$

126. $-3(x-6)^2+20=32$

$-3(x-6)^2=12$

$(x-6)^2=-4$

$x-6=\pm\sqrt{-4}$

$x-6=\pm 2i$

$x=6\pm 2i$

127. Center$=(3,-9)$ and Radius$=12$

$h=3, k=-9, r=12$

$(x-3)^2+(y-(-9))^2=(12)^2$

$(x-3)^2+(y+9)^2=144$

128. $x=\dfrac{y^2}{-44}$

$x=\dfrac{y^2}{4(-11)}$

$p=-11$

Focus$=(-11,0)$, Directrix : $x=11$

129. $\dfrac{x^2}{144}-\dfrac{y^2}{64}=1$ hyperbola

The $\dfrac{x^2}{a^2}$ term is first so opens left and right.

$a^2=144\to a=12$

$b^2=64\to b=8$

Vertices at $(-12,0)$ and $(12,0)$.

The fundamental rectangle will have vertical sides at $x=\pm 12$ and horizontal sides at $y=\pm 8$.

$c^2=a^2+b^2$

$c^2=144+64=208$

$c=\pm\sqrt{208}\approx\pm 14.4$

Foci at $(\pm 14.4, 0)$

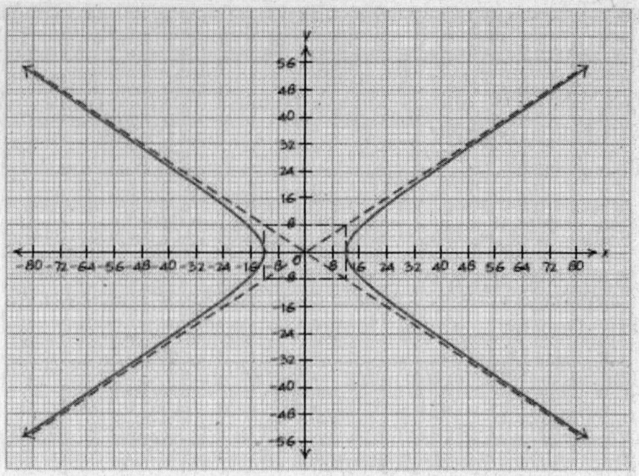

130. $121x^2+4y^2=484$ ellipse

$\dfrac{121x^2}{484}+\dfrac{4y^2}{484}=\dfrac{484}{484}$

$\dfrac{x^2}{4}+\dfrac{y^2}{121}=1$

The bigger number is under the y so the major axis is vertical.

$a^2=4\to a=2$

$b^2=121\to b=11$

Vertices at $(0,-11)$ and $(0,11)$.

$c^2=b^2-a^2$

$c^2=121+4=117$

$c=\pm\sqrt{117}\approx\pm 10.8$

Foci at $(0,\pm 10.8)$

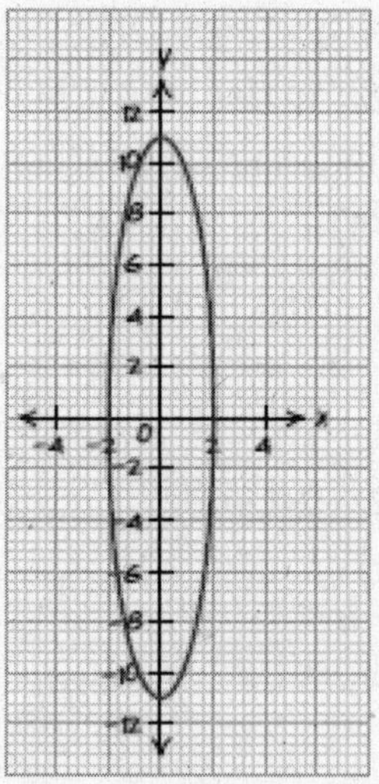

131. $y^2+2y+x-10=0$ parabola

$x=-y^2-2y+10$

Opens to the left $a=-1<0$.

Vertex (h,k).

$k=\dfrac{2}{2(-1)}=-1$

$h=-(-1)^2-2(-1)+10=11$

$(h,k)=(11,-1)$

x-intercept: $(10,0)$

y-intercept:

$$0=-y^2-2y+10$$

$$y=\frac{2\pm\sqrt{(-2)^2-4(-1)(10)}}{2(-1)}$$

$$y=\frac{2+\sqrt{44}}{-2}\approx -4.3\rightarrow(0,-4.3)$$

$$y=\frac{2-\sqrt{44}}{-2}\approx 2.3\rightarrow(0,2.3)$$

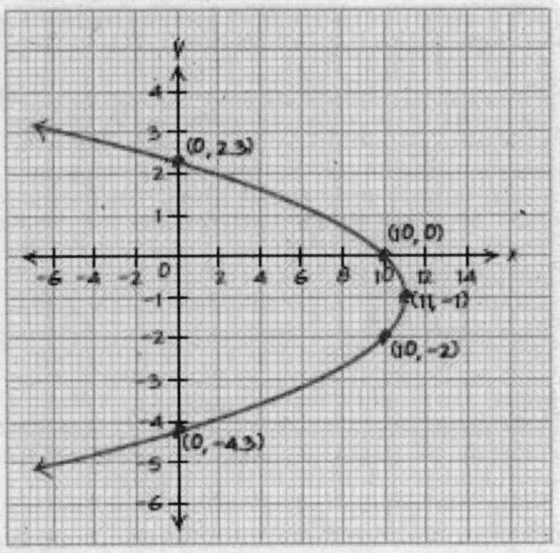

132. $(x+8)^2+(y-3)^2=36$ circle

Center at $(-8,3)$, radius $r=6$

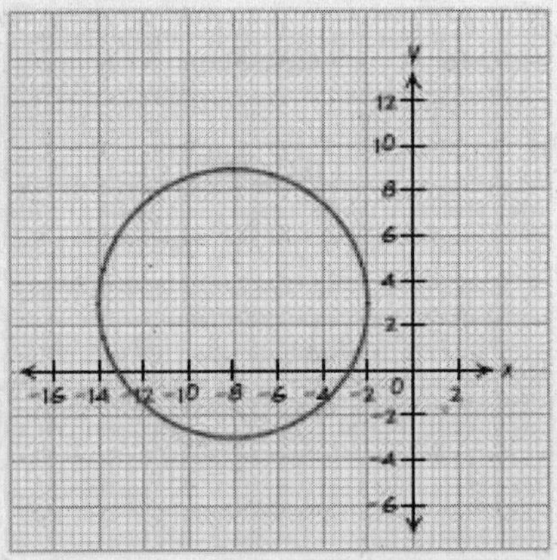

133. The parabola opens to the left so it is of the form $y^2=4px$. Use point $(-6,-12)$ to determine p.

$$(-12)^2=4p(-6)$$
$$144=-24p$$
$$\frac{144}{-24}=\frac{-24p}{-24}$$
$$-6=p$$

Therefore the equation is $y^2=-24x$.

134. The major axis is horizontal and it is centered at the origin so it is of the form $\frac{x^2}{a^2}+\frac{y^2}{b^2}=1$.

$$a=12\rightarrow a^2=144$$
$$b=5\rightarrow b^2=25$$
$$\frac{x^2}{144}+\frac{y^2}{25}=1$$

135. $a_n=34+(n-1)11$

$$a_n=34+11n-11$$
$$a_n=23+11n$$

136. $a_1=5(-2)^1=-10$

$$a_2=5(-2)^2=20$$
$$a_3=5(-2)^3=-40$$
$$a_4=5(-2)^4=80$$
$$a_5=5(-2)^5=-160$$
$$\{-10, 20, -40, 80, -160, ...\}$$

137. $a_1=\frac{-12(1)}{3(1)+5}=\frac{-12}{8}=-\frac{3}{2}$

$$a_2=\frac{-12(2)}{3(2)+5}=-\frac{24}{11}$$
$$a_3=\frac{-12(3)}{3(3)+5}=\frac{-36}{14}=-\frac{18}{7}$$
$$a_4=\frac{-12(4)}{3(4)+5}=-\frac{48}{17}$$
$$a_5=\frac{-12(5)}{3(5)+5}=\frac{-60}{20}=-3$$
$$\left\{-\frac{3}{2},-\frac{24}{11},-\frac{18}{7},-\frac{48}{17},-3,...\right\}$$

138. $26+7(n-1)=607$

$$26+7n-7=607$$
$$7n=588$$
$$\frac{7n}{7}=\frac{588}{7}$$
$$n=84$$

139. $-45+\sqrt{n+20}=-37$

$$\sqrt{n+20}=8$$
$$\left(\sqrt{n+20}\right)^2=(8)^2$$
$$n+20=64$$
$$n=44$$

140.

n	a_n	Common Difference
1	48	
2	45	$a_2-a_1=45-48=-3$
3	42	$a_3-a_2=42-45=-3$
4	39	$a_4-a_3=39-42=-3$
5	36	$a_5-a_4=36-39=-3$
6	33	$a_6-a_5=33-36=-3$

This is an arithmetic sequence with common difference $d=-3$.

$$a_n = 48+(n-1)(-3)$$
$$a_n = 48-3n+3$$
$$a_n = 51-3n$$

141.

n	a_n	Common Ratio
1	96	
2	384	$\frac{a_2}{a_1}=\frac{384}{96}=4$
3	1536	$\frac{a_3}{a_2}=\frac{1536}{384}=4$
4	6144	$\frac{a_4}{a_3}=\frac{6144}{1536}=4$
5	24,576	$\frac{a_5}{a_4}=\frac{24,576}{6144}=4$
6	98,304	$\frac{a_6}{a_5}=\frac{98,304}{24,576}=4$

This is a geometric sequence with common ratio $r=4$.

$$a_n = 96(4)^{n-1}$$

142. **a.** $a_1 = 55,000$ and $d = 825$

$$a_n = 55,000+(n-1)(825)$$
$$a_n = 55,000+825n-825$$
$$a_n = 54,175+825n$$

b. $a_{10} = 54,175+825(10)$

$$a_{10} = 62,425$$

Your salary will be \$62,425 in the 10th year of employment.

c. $S_{10} = \dfrac{10(55,000+62,425)}{2}$

$$S_{10} = 587,125$$

You will have earned a total of \$587,125 in the first ten year period of employment.

143. $a_1 = 55,000$ and $r = 1+0.015 = 1.015$

a. $a_n = 55,000(1.015)^{n-1}$

b. $a_{10} = 55,000(1.015)^{10-1}$

$$a_{10} = 55,000(1.015)^9$$
$$a_{10} \approx 62,886.45$$

Your salary will be \$62,886.45 in the 10th year of employment.

c. $S_{10} = \dfrac{55,000(1-1.015^{10})}{1-1.015}$

$$S_{10} \approx 588,649.69$$

You will have earned a total of \$588,649.69 in the first ten year period of employment.

144.

$$\sum_{n=1}^{6}(2n^3) = 2+2(2)^3+2(3)^3+2(4)^3+2(5)^3+2(6)^3 = 882$$

145. $a_1 = 75$

$$a_{15} = 75-6(15-1)$$
$$a_{15} = -9$$
$$S_{15} = \frac{15(75-9)}{2}$$
$$S_{15} = 495$$

146. $a_1 = 300(1.3)$

$$a_1 = 390$$
$$S_{12} = \frac{390(1-1.3^{12})}{1-1.3}$$
$$S_{12} \approx 28,987.51066$$